Cambridge - IISc Series

CONTINUUM MECHANICS

Foundations and Applications of Mechanics

Volume I, Third Edition

C. S. Jog

CAMBRIDGE UNIVERSITY PRESS

CAMBRIDGE
UNIVERSITY PRESS

4843/24, 2nd Floor, Ansari Road, Daryaganj, Delhi - 110002, India

Cambridge University Press is part of the University of Cambridge.

It furthers the University's mission by disseminating knowledge in the pursuit of education, learning and research at the highest international levels of excellence.

www.cambridge.org
Information on this title: www.cambridge.org/9781107091351

Third edition first published 2015

Printed in India by Shree Maitrey Printech Pvt. Ltd., Noida

A catalogue record for this publication is available from the British Library

Library of Congress Cataloging-in-Publication Data
Jog, C. S.
Continuum mechanics / C.S. Jog. – Third edition.
pages cm
– (Foundations and applications of mechanics; volume 1)
Summary: "Presents several advanced topics including fourth-order tensors,
differentiation of tensors, exponential and logarithmic tensors,
and their application to nonlinear elasticity"– Provided by publisher..
Includes bibliographical references and index.
ISBN 978-1-107-09135-1 (hardback)
1. Continuum mechanics. 2. Tensor algebra. I. Title.
QA808.2.J64 2015
531–dc23
2015001499

ISBN 978-1-107-09135-1 Hardback

To

My Parents and IISc

Contents

List of Figures

List of Tables

Preface

Apart from the usual goal of presenting a unified treatment of seemingly diverse branches such as solid or fluid mechanics, it is also our goal in this book to discuss some important topics, which have come into prominence in the latter half of the twentieth century, such as material symmetry, frame-indifference and thermomechanics. The use of these principles in tandem (of course, in addition to the usual axioms) in many cases delivers the most powerful and beautiful results.

Chapter 1 presents the necessary mathematical background for the following chapters in the form of a brief introduction to tensor analysis. Almost throughout this chapter, the assumed three-dimensional nature of the vector space is exploited to yield shorter proofs of results which in most (but not all) cases also hold for dimensions other than three. In keeping with modern practice, we emphasize the use of direct as opposed to indicial notation, although in many cases we present the proofs using indicial notation also. Presenting the equations in direct notation reinforces the idea that the governing equations are valid with respect to any set of coordinate axes in a given reference frame. It is beneficial, however, to be adept at the use of both, direct and indicial notation. When presenting equations in indicial form, we use only Cartesian components of tensors since we feel that they are sufficient to present the basic principles of the subject. Although other types of components are useful, especially when dealing with curvilinear coordinate systems, we believe that a detailed discussion of say covariant and contravariant components of tensors right at the outset can digress from the purpose of presenting the main physical concepts, and hence, we present them in an Appendix.

Chapter 2 discusses the branch of continuum mechanics known as kinematics, which deals with purely geometrical notions such as strain, rate of deformation, etc.

Chapter 3 discusses the basic axioms of continuum mechanics, namely the principles of conservation of mass, the principles of balance of linear and angular momentum, and the laws of thermodynamics. The principles of balance of linear and angular momentum in the form that are used today are due to Euler. The fascinating history of his struggle to derive the governing equations for a continuum using the linear balance momentum principle alone, before finally realizing the need for accepting the angular momentum balance as a separate axiom, and how using these two axioms simultaneously yielded a rich cornucopia of results, is outlined in [336].

Euler's axioms are formulated directly for a 'large' or 'massy' body. By large or massy, we mean bodies that are large in comparison to atomic dimensions; thus, a speck of dust, for example, would also constitute a large body. In the classical mechanics literature, one often finds 'derivations' of Euler's axioms by considering a continuum as a collection of particles, and by assuming Newton's laws to hold good for each particle. This approach might seem logical, but is, in fact, completely fallacious [336]. To begin with, Newton's laws become poorer and poorer approximations as the size of the body is reduced so that one needs to use quantum mechanics to predict the behavior of a particle. Thus, it does not make sense to deduce laws for a large body by using Newton's laws for a particle, which in fact do not hold. Worse still, Newton himself never meant a particle when he used the Latin word *particula*–he used it to mean a small body or an element of integration. On the other hand, Euler's axioms which are formulated directly for large bodies are known to be very accurate. *Thus, it seems that the particle-based approach can be avoided altogether*. As demonstrated in Chapter 8, Euler's axioms yield all the classical equations for rigid body motion (and much more). Of course, in cases where the motion of a body with respect to its center of mass can be neglected, it might be convenient to *approximate* the body as a particle with its entire mass lumped at a point–this approach, for example, is usually used in deducing the motion of planets around the sun.

Another anachronism in the view of this author, particularly in this post-Einstein era, is the use of inertial and non-inertial frames of reference. In this book we treat all reference frames on an equal footing in the sense that we *require* that the governing equations have the same form in all reference frames. This requirement yields a relation for the way body forces should transform under a change of frame (see Eqn. (4.18)), which, quite happily, also agrees with physical experience.

A reader may wonder as to what exact role Newton's laws, which one generally learns during highschool, and which are proclaimed to be basic, play in continuum mechanics. The answer is that although one does not use Newton's laws directly, one does obtain statements close in spirit to Newton's laws from Euler's axioms. To see this, let us consider Newton's laws. The first law states that 'A body continues in a state of rest or of uniform motion in a straight line unless acted upon by a net external force'. The part about the 'straight line motion' is not strictly true for bodies, since a body which is rotating in the absence of any forces would continue to rotate. Hence, modern treatments modify this statement and replace 'body' by 'particle'. Since our purpose in continuum mechanics is to study large bodies and not particles, and since we have already indicated that classical laws may no longer hold when one considers particles, this law is not directly used in our formulation. The second law which is again claimed to hold good for a particle, and which states that 'force is equal to mass times acceleration of a particle' is now obtained in a form which holds directly for the body; the acceleration is that of the center of mass (see Eqn. (3.16)). The third law which states that 'action and reaction are equal and opposite', is obtained in the form given by Eqn. (3.25).

It should be recognized that Euler's axioms are nontrivial generalizations of Newton's laws. In particular, the axioms are required to hold for any arbitrary sub-part of a body. It is precisely this feature (along with the Cauchy principle) that yields the governing differential equations from the integral form of the equations.

A new era in continuum mechanics began around 1955 with the works of Walter Noll. He proposed an axiomatization for forces, the principle of material-frame indifference, and group-theoretic definitions for simple solids and fluids, among numerous other things. Along with Bernard Coleman, he was also instrumental in using the second law of thermodynamics to deduce restrictions on constitutive relations. Constitutive relations had been treated in a very empirical way until their seminal work in which they showed how the use of material frame-indifference, material symmetry and the second law of thermodynamics can help restrict (and in some cases, even yield the final form of) constitutive relations. Chapter 4 discusses the concepts of material frame-indifference and material symmetry, and shows how these concepts can be applied to elastic materials when thermal effects are ignored. In Chapters 5, the nonlinear theory of elasticity is presented, while in Chapter 6, the classical theory of linear elasticity is obtained by linearizing the governing equations of nonlinear elasticity.

In Chapter 7, we discuss the very important topic of thermomechanics. To those seeking a rigorous treatment of thermodynamics, the treatment in standard works, with their liberal use of d's, Δ's and δ's, has always seemed obscure [335]. In particular, the usual treatment of the second law of thermodynamics has come in for severe criticism, with the definition of the entropy function itself in terms of reversible processes, while the processes that one is trying to model are usually highly irreversible. A lot of work has been done in the past few decades to remedy this unsatisfactory state of affairs–the modern theory may be found, for example, in [295]. Although the mathematics used in proving the various theorems is nontrivial, those concerned with applications may effectively assume (as we do) the Clausius–Duhem inequality as a statement of the second law of thermodynamics. We discuss the restrictions imposed by this inequality on solids and fluids. Quite amazingly, all the classical results, which had historically been obtained either empirically or through ad hoc means, are now recovered using rigorous arguments. In particular, in the case of fluids, we recover the entire classical theory as a special case, including the Gibbs relation, the fact that pressure is a state function of density and temperature, the constitutive relation for the viscous stress, the relation between the entropy and the free energy, and so on.

A point of elementary logic that we wish to stress is that the statement "A implies B" is logically equivalent to the statement "not B implies not A". Thus, we shall often prove that A implies B by assuming 'not B' and deducing 'not A'.

Continuum mechanics and fluid mechanics are classical subjects with a huge amount of literature, and this author has naturally benefitted by reading a part of this literature. The main works that have been consulted in writing this book are stated in the bibliographies in the two volumes; the author is greatly indebted to their authors. The author is also indebted to his teachers at the Theoretical and Applied Mechanics Department (now merged with Mechanical Engineering), University of Illinois at Urbana-Champaign, in particular, Professor D. E. Carlson, R. B. Haber, R. E. Johnson and R. T. Shield. Professor D. E. Carlson was kind enough to give the author a copy of his unpublished notes; Sections 1.12, 4.6 and 4.7 are based on these notes, and the author thanks him for granting permission to use them[1]. The author is thankful to the Mechanical Engineering Department at the Indian Institute of Science for having provided a stimulating environment during the development of this book. The author gratefully acknowledges the tremendous effort put in by Professor Annem Narayana Reddy of IIT(G) in designing and developing a web-based NPTEL course based primarily on this book. Students who have taken the continuum mechanics course offered by this author or worked on projects offered by him have made many valuable suggestions and comments which have resulted in a vast improvement in the book, and the author wishes to thank all of them. The author expresses his thanks to Cambridge University Press for agreeing to publish this book, and for their kind cooperation. Special thanks go to the Commissioning Editor Gauravjeet Singh Reen, Assistant Editors (Academic) Sana Bhanot and Shikha Vats, and their team for their tireless efforts in bringing this project to fruition. The entire book was typeset in the Linux environment using LaTeX–the author expresses his gratitude to the creators of these fine pieces of software. He also expresses his

[1]Professor Carlson passed away in August 2010. His notes are now available online at http://imechanica.org/node/15845, courtesy of Professor Amit Acharya.

gratitude to the creators of that jewel in the crown, Mathematica [353], which he used extensively in writing Chapter 6 and parts of Volume II. Finally, the author wishes to thank his family for having supported him through the writing of this book.

Suggestions and comments for improving this book are welcome.

Notation

$\boldsymbol{a}$	acceleration vector
$\boldsymbol{b}$	body force vector per unit mass
$\boldsymbol{B}$	left Cauchy–Green strain tensor
c_p	specific heat at constant pressure
c_v	specific heat at constant volume
$\mathbf{C}$	first elasticity tensor
$\mathbb{C}$	second elasticity tensor
$\boldsymbol{C}$	right Cauchy–Green strain tensor
χ	mapping characterizing the deformation
D/Dt	material derivative
$\boldsymbol{D}$	rate-of-deformation tensor
e	specific internal energy
$\boldsymbol{e}_i$	Cartesian basis vectors
$\boldsymbol{E}$	Green strain tensor
$\bar{\boldsymbol{E}}$	Almansi strain tensor
$\boldsymbol{F}$	deformation gradient
$\boldsymbol{g}$	temperature gradient
γ	ratio of specific heats (c_p/c_v)
Γ	circulation
h	specific enthalpy
$\boldsymbol{h}$	angular momentum
η	specific entropy
J	determinant of $\boldsymbol{F}$
$\boldsymbol{J}$	inertia tensor relative to the center of mass
K	kinetic energy
κ	bulk viscosity
$\boldsymbol{l}$	linear momentum
$\boldsymbol{L}$	velocity gradient tensor
λ	dilatational viscosity; also Lame constant

λ	unit tangent to a contour
μ	shear or dynamic viscosity; also Lame constant
$\boldsymbol{n}$	unit normal to a surface
ν	kinematic viscosity
p	thermodynamic pressure
$\hat{p}$	mean pressure
$\boldsymbol{q}$	heat flux vector
Q_h	heat generated per unit mass per unit time
$\boldsymbol{Q}$	orthogonal tensor
θ	absolute temperature
R	universal gas constant
ρ	density
S	control surface
$S(t)$	surface of material volume
$\boldsymbol{S}$	second Piola–Kirchhoff stress tensor
$\boldsymbol{\sigma}$	viscous stress tensor
t	time
$\boldsymbol{t}$	traction vector
$\boldsymbol{T}$	first Piola–Kirchhoff stress tensor
$\boldsymbol{\tau}$	Cauchy stress tensor
$\boldsymbol{u}$	displacement vector
$\boldsymbol{v}$	velocity vector
v	specific volume
V	vector space
$V(t)$	material volume
V_t	control volume
$\hat{W}$	stored energy function
$\hat{\boldsymbol{W}}$	vorticity tensor
$\boldsymbol{W}$	the skew tensor $\dot{\boldsymbol{Q}}\boldsymbol{Q}^T$
$\boldsymbol{\omega}$	vorticity vector (axial vector of $2\hat{\boldsymbol{W}}$)
$\boldsymbol{w}$	angular velocity (axial vector of $\dot{\boldsymbol{Q}}\boldsymbol{Q}^T$)
$\boldsymbol{\Omega}$	axial vector of $\boldsymbol{Q}^T\dot{\boldsymbol{Q}}$ (equal to $\boldsymbol{Q}^T\boldsymbol{w}$)
$\boldsymbol{x}$	spatial coordinates
$\bar{\boldsymbol{x}}$	center of mass
$\boldsymbol{X}$	material coordinates
ψ	free energy

Sets

Lin $=$ Set of all tensors

Lin^+ $=$ Set of all tensors $\boldsymbol{T}$ with $\det \boldsymbol{T} > 0$

Sym $=$ Set of all symmetric tensors

Psym $=$ Set of all symmetric, positive definite tensors

Orth $=$ Set of all orthogonal tensors

Orth^+ $=$ Set of all rotations ($\boldsymbol{Q}\boldsymbol{Q}^T = \boldsymbol{I}$ and $\det \boldsymbol{Q} = +1$)

Skw $=$ Set of all skew-symmetric tensors

Unim $=$ Set of all unimodular tensors ($|\det \boldsymbol{H}| = +1$)

Unim^+ $=$ Set of all proper unimodular tensors ($\det \boldsymbol{H} = +1$)

1 Introduction to Tensors

In elementary physics, we often come across two classes of quantities, namely scalars and vectors. Mass, density and temperature are examples of scalar quantities, while velocity and acceleration are examples of vector quantities. A scalar quantity's value does not depend on the choice of the coordinate system. Similarly, although the *components* of a vector depend on a particular choice of coordinate system, the vector itself is invariant, and has an existence independent of the choice of coordinate system. In this chapter, we generalize the concept of a scalar and vector, to that of a tensor. In this general framework, scalars are considered as zeroth-order tensors and vectors as first-order tensors. Tensor quantities of order two and greater, similar to scalars and vectors, have an existence independent of the coordinate system. Their components, however, just as in the case of vectors, depend on the choice of coordinate system. We will see that the governing field equations of continuum mechanics can be written as tensorial equations. The advantage of writing the field equations in such 'coordinate-free' notation is that it is immediately obvious that these equations are valid no matter what the choice of coordinate system is. A particular coordinate system is invoked only while solving a particular problem, whence the appropriate form of the differential operators and the components of the tensors with respect to the chosen coordinate system are used. It must be borne in mind, however, that although using tensorial notation shows the 'coordinate-free' nature of the governing equations in a *given frame of reference*, it does not address the issue of how the equations transform under a change of frame of reference. This aspect will be discussed in greater detail later in this book.

We now present a review of tensors. Throughout the text, scalars are denoted by lightface letters, vectors are denoted by boldface lower-case letters, while second and higher-order tensors are denoted by boldface capital letters. As a notational issue, summation over repeated indices is assumed, with the indices ranging from 1 to 3. Thus, for example, u_iv_i represents $u_1v_1 + u_2v_2 + u_3v_3$, and $T_{ij}n_j$ represents $T_{i1}n_1 + T_{i2}n_2 + T_{i3}n_3$. The quantity on the right-hand side of a ':=' symbol defines the quantity on its left-hand side. A function on '$V \times V$ to V' means that the function is defined in terms of two elements that belong to V, and the result is also in V.

1.1 Vector Spaces

In what follows, we consider only real vector spaces. We denote the set of real numbers by $\Re$. A *vector space* (or linear space) is a set, say V, equipped with an addition function on $V \times V$ to V (denoted by $+$), and a scalar multiplication function on $\Re \times V$ to V, which satisfy the following conditions:

1. Commutativity: For all $\boldsymbol{u}, \boldsymbol{v} \in V$,

$$\boldsymbol{u} + \boldsymbol{v} = \boldsymbol{v} + \boldsymbol{u}.$$

2. Associativity: For all $\boldsymbol{u}, \boldsymbol{v}, \boldsymbol{w} \in V$,

$$(\boldsymbol{u} + \boldsymbol{v}) + \boldsymbol{w} = \boldsymbol{u} + (\boldsymbol{v} + \boldsymbol{w}).$$

3. Existence of a zero element: There exists $\boldsymbol{0} \in V$ such that

$$\boldsymbol{u} + \boldsymbol{0} = \boldsymbol{u}.$$

4. Existence of negative elements: For each $\boldsymbol{u} \in V$, there exists a negative element denoted $-\boldsymbol{u}$ in V such that

$$\boldsymbol{u} - \boldsymbol{u} = \boldsymbol{0}.$$

5. Distributivity with respect to addition of vectors: For all $\alpha \in \Re$, and $\boldsymbol{u}, \boldsymbol{v} \in V$,

$$\alpha(\boldsymbol{u} + \boldsymbol{v}) = \alpha\boldsymbol{u} + \alpha\boldsymbol{v}.$$

6. Distributivity with respect to scalar addition: For all $\alpha, \beta \in \Re$, and for all $\boldsymbol{u} \in V$,

$$(\alpha + \beta)\boldsymbol{u} = \alpha\boldsymbol{u} + \beta\boldsymbol{u}.$$

7. Associativity: For all $\alpha, \beta \in \Re$, and for all $\boldsymbol{u} \in V$,

$$\alpha(\beta\boldsymbol{u}) = (\alpha\beta)\boldsymbol{u}.$$

8. Identity in scalar multiplication: For all $\boldsymbol{u} \in V$,

$$1\boldsymbol{u} = \boldsymbol{u}.$$

Since a vector space is a group with respect to addition (see Section 1.12) with $\boldsymbol{0}$ and $-\boldsymbol{u}$ playing the roles of the neutral and reverse elements, respectively, all the results derived for groups are applicable for vector spaces. In particular, the zero element of V, and the negative element $-\boldsymbol{u}$ corresponding to a given $\boldsymbol{u} \in V$ are unique. Note also that $\alpha\boldsymbol{u} + \beta\boldsymbol{v} \in V$ for all $\alpha, \beta \in \Re$, and all $\boldsymbol{u}, \boldsymbol{v} \in V$.

Perhaps, the most famous example of a vector space is the n-dimensional coordinate space $\Re^n$, which is defined by

$$\Re^n := \{(u_1, u_2, \ldots, u_n) : u_i \in \Re\}.$$

For $\Re^n$, addition and scalar multiplication are defined by the relations

$$\begin{aligned} \boldsymbol{u} + \boldsymbol{v} &:= (u_1 + v_1, u_2 + v_2, \ldots, u_n + v_n), \\ \alpha\boldsymbol{u} &:= (\alpha u_1, \alpha u_2, \ldots, \alpha u_n). \end{aligned}$$

If p is a positive real number, then another example of a vector space is the space

$$L^p[0,1] := \left\{ f : \int_0^1 |f|^p \, dx < \infty \right\},$$

with addition and scalar multiplication defined by

$$(f+g)(x) := f(x) + g(x), \; x \in [0,1],$$
$$(\alpha f)(x) := \alpha[f(x)], \; x \in [0,1].$$

To show that $L^p[0,1]$ is a vector space, we need to show that $f+g \in L^p[0,1]$ if $f,g \in L^p[0,1]$. This follows from the inequality

$$|f+g|^p \leq [2\max(|f|,|g|)]^p \leq 2^p \left[|f|^p + |g|^p\right].$$

A subset $\{\boldsymbol{u}_1, \boldsymbol{u}_2, \ldots, \boldsymbol{u}_m\}$ of V is said to be *linearly dependent* if and only if there exist scalars $\alpha_1, \alpha_2, \ldots, \alpha_m$, not all zero, such that

$$\alpha_1 \boldsymbol{u}_1 + \alpha_2 \boldsymbol{u}_2 + \cdots + \alpha_m \boldsymbol{u}_m = \boldsymbol{0}.$$

Thus, a subset $\{\boldsymbol{u}_1, \boldsymbol{u}_2, \ldots, \boldsymbol{u}_m\}$ of V is linearly independent if and only if the equation

$$\alpha_1 \boldsymbol{u}_1 + \alpha_2 \boldsymbol{u}_2 + \cdots + \alpha_m \boldsymbol{u}_m = \boldsymbol{0},$$

implies that $\alpha_1 = \alpha_2 = \cdots = \alpha_m = 0$. A subset, say $\{\boldsymbol{u}_1, \boldsymbol{u}_2, \ldots, \boldsymbol{u}_m, \boldsymbol{0}\}$, which includes the zero element is always linearly dependent even when the subset $\{\boldsymbol{u}_1, \boldsymbol{u}_2, \ldots, \boldsymbol{u}_m\}$ is linearly independent, since the coefficient of the zero element can be taken to be nonzero.

A subset $\{\boldsymbol{e}_1, \boldsymbol{e}_2, \ldots, \boldsymbol{e}_n\}$ of V is said to be a basis for V if

1. $\{\boldsymbol{e}_1, \boldsymbol{e}_2, \ldots, \boldsymbol{e}_n\}$ is linearly independent, and
2. Any element of V can be expressed as a linear combination of $\{\boldsymbol{e}_1, \boldsymbol{e}_2, \ldots, \boldsymbol{e}_n\}$, i.e., if $\boldsymbol{u} \in V$, then

   $$\boldsymbol{u} = u_1 \boldsymbol{e}_1 + u_2 \boldsymbol{e}_2 + \cdots + u_n \boldsymbol{e}_n,$$

   where the scalars $u_1, u_2, \ldots, u_n$ are known as the *components* of $\boldsymbol{u}$ with respect to the basis $\{\boldsymbol{e}_1, \boldsymbol{e}_2, \ldots, \boldsymbol{e}_n\}$.

If the bases have a finite number of elements, we have the following theorem:

Theorem 1.1.1. *All bases for a given vector space contain the same number of elements.*

Proof. Suppose that $\{\boldsymbol{e}_1, \boldsymbol{e}_2, \ldots, \boldsymbol{e}_n\}$ and $\{\boldsymbol{e}_1^*, \boldsymbol{e}_2^*, \ldots, \boldsymbol{e}_m^*\}$ are bases for a vector space. Every $\boldsymbol{e}_i^*$, since it is an element of the vector space, can be expressed in terms of the basis $\{\boldsymbol{e}_1, \boldsymbol{e}_2, \ldots, \boldsymbol{e}_n\}$ as $\beta_{ij}\boldsymbol{e}_j$, $\beta_{ij} \in \Re$. Thus,

$$\alpha_i \boldsymbol{e}_i^* = \beta_{ij} \alpha_i \boldsymbol{e}_j = \boldsymbol{0}$$

implies, by the linear independence of $\{e_1, e_2, \dots, e_n\}$, that

$$\beta_{ij}\alpha_i = 0 \quad i = 1, 2, \dots, m, \quad j = 1, 2, \dots, n.$$

Let $m > n$. Then, the number of unknowns α_i is more than the number of equations, so that it is possible to find a nontrivial solution. Thus, there exist α_1, α_2, ..., α_m, not all zero such that

$$\alpha_i e_i^* = \mathbf{0},$$

i.e., $\{e_1^*, e_2^*, \dots, e_m^*\}$ is linearly dependent, contradicting the fact that it is a basis. Hence, $m \leq n$. Next, suppose that $m < n$. Now reverse the roles of $\{e_i^*\}$ and $\{e_i\}$ in the above argument to conclude that $m \geq n$. Hence, $m = n$. □

In view of the above result, a vector space V is said to be n-dimensional if it contains a basis with n elements. If no such finite integer n exists, then the vector space is said to be infinite-dimensional. For example, $\Re^n$ is finite-dimensional with

$$\begin{aligned} e_1 &= (1, 0, 0, \dots, 0), \\ e_2 &= (0, 1, 0, \dots, 0), \\ &\dots \\ e_n &= (0, 0, 0, \dots, 1), \end{aligned}$$

as the 'canonical' or natural basis. On the other hand, $L^p[0, 1]$ is an infinite-dimensional vector space. Note that even in the finite-dimensional case, the basis need not be unique. When V is finite-dimensional, using the linear independence of $\{e_1, e_2, \dots, e_n\}$, it is easy to show that the components of any element $\boldsymbol{u}$ are unique.

Let V be an n-dimensional vector space. A subset $\{f_1, f_2, \dots, f_m\}$ of V, where

1. $m > n$ cannot be a basis for V, since, as can be shown by following the same methodology as used in the proof of Theorem 1.1.1, it is linearly dependent.

2. $m < n$ cannot be a basis for V, since by Theorem 1.1.1, any basis has n elements. Although, the subset $\{f_1, f_2, \dots, f_m\}$ can be linearly independent in this case, an arbitrary element of V cannot be expressed in terms of the elements of this subset.

Thus, $m = n$ is a necessary condition for the subset $\{f_1, f_2, \dots, f_m\}$ to be a basis. The following theorem is useful in finding if a given subset with n elements is a basis for an n-dimensional vector space:

Theorem 1.1.2. *Let $\{f_1, f_2, \dots, f_n\}$ be a subset of an n-dimensional vector space V, one of whose bases is given by $\{e_1, e_2, \dots, e_n\}$. Since $\{f_1, f_2, \dots, f_n\}$ is a subset of V, each f_i can be expressed as a linear combination of the basis vectors $\{e_1, e_2, \dots, e_n\}$, i.e., $f_i = \beta_{ij}e_j$, $i, j = 1, 2, \dots, n$, $\beta_{ij} \in \Re$. The following statements are equivalent:*

1. *$\{f_1, f_2, \dots, f_n\}$ is a basis.*

2. $\{f_1, f_2, \ldots, f_n\}$ *is linearly independent.*

3. $\det[\beta_{ij}] \neq 0$.

Proof. We prove that (i) $\Longrightarrow$ (ii) $\Longrightarrow$ (iii) $\Longrightarrow$ (i).

If $\{f_1, f_2, \ldots, f_n\}$ is a basis, then it is linearly independent by definition.

To prove that (ii) $\Longrightarrow$ (iii), we show that "not (iii)" $\Longrightarrow$ "not (ii)". Thus, let $\det[\beta_{ij}] = 0$. From matrix algebra, it follows that there exist α_i, $i = 1, 2, \ldots, n$, not all zero, such that $\beta_{ij}\alpha_i = 0$. Thus, there exist α_i, not all zero, such that

$$\alpha_i f_i = \alpha_i \beta_{ij} e_j = \mathbf{0},$$

which implies that the set $\{f_1, f_2, \ldots, f_n\}$ is linearly dependent.

To prove that (iii) $\Longrightarrow$ (i), note that $\alpha_i f_i = \beta_{ij}\alpha_i e_j = \mathbf{0}$, implies, since $\{e_1, e_2, \ldots, e_n\}$ is a basis (and hence a linearly independent set), that $\beta_{ij}\alpha_i = 0$, $i, j = 1, 2, \ldots, n$, and since $\det[\beta_{ij}] \neq 0$, it follows that all the α_i are zero, thus showing that $\{f_1, f_2, \ldots, f_n\}$ is linearly independent. In addition, since $\det[\beta_{ij}] \neq 0$, the matrix $[\beta_{ij}]$ is invertible. Let $[\gamma_{ij}]$ denote the components of this inverse, i.e., $\gamma_{ij}\beta_{jk} = \delta_{ik}$. Then

$$\gamma_{ij} f_j = \gamma_{ij}\beta_{jk} e_k = \delta_{ik} e_k = e_i.$$

Since $\{e_1, e_2, \ldots, e_n\}$ is a basis of V, every element u of V can be expressed as $u_i e_i$, so that on using the above relation, we have

$$u = u_i e_i = \gamma_{ij} u_i f_j.$$

Thus, the set $\{f_1, f_2, \ldots, f_n\}$ satisfies the two conditions for qualifying as a basis. □

A nonempty subset V_s of a vector space V is said to be a *linear subspace* if a linear combination of any two of its elements also lies in V_s, i.e., if $(\alpha u + \beta v) \in V_s$ for any arbitrary $u, v \in V_s$, and $\alpha, \beta \in \Re$. For example, $\Re^2$ is a linear subspace of $\Re^3$. By assuming the operations of addition and scalar multiplication to be the same as that for the 'parent space' V, it can be shown by verifying the defining axioms that a linear subspace is itself a vector space.

The set of all linear combinations of a subset $\{u_1, u_2, \ldots, u_m\}$ of V is called the *linear span* of $\{u_1, u_2, \ldots, u_m\}$, i.e.,

$$\mathrm{Lsp}\{u_1, u_2, \ldots, u_m\} := \{u : u = \alpha_1 u_1 + \alpha_2 u_2 + \cdots + \alpha_m u_m, \alpha_i \in \Re\}.$$

From this definition, it immediately follows that $\mathrm{Lsp}\{u_1, u_2, \ldots, u_m\}$ is a linear subspace of V.

An *inner product space* (or Euclidean space) is a vector space V equipped with a function on $V \times V$ to $\Re$, denoted by (u, v), and called the *inner product* (also called scalar product or dot product) of u and v, that satisfies the following conditions:

$$(u, v) = (v, u) \quad \forall u, v \in V, \tag{1.1a}$$

$$(\alpha u + \beta v, w) = \alpha(u, w) + \beta(v, w) \quad \forall \alpha, \beta \in \Re \text{ and } u, v, w \in V, \tag{1.1b}$$

$$(u,u) \geq 0 \quad \forall u \in V \text{ with } (u,u) = 0 \text{ if and only if } u = \mathbf{0}, \tag{1.1c}$$

The magnitude $|u|$ is defined by

$$|u| := (u,u)^{1/2}.$$

From the above definition, and the properties of the inner product it is obvious that

$$\begin{aligned} |\alpha u| &= |\alpha|\,|u| \quad \forall \alpha \in \Re \text{ and } u \in V, \\ |u| &\geq 0 \text{ with } |u| = 0 \text{ if and only if } u = \mathbf{0}. \end{aligned} \tag{1.2}$$

Both $\Re^n$ and $L^2[0,1]$ are inner product spaces. In the case of $\Re^n$, the inner product can be defined by

$$(u,v) := u_1 v_1 + u_2 v_2 + \cdots + u_n v_n.$$

The choice of inner product is not unique. If S is a symmetric, positive definite $n \times n$ matrix, then another choice of inner product for $\Re^n$ is

$$(u,v) := \sum_{i=1}^{n} \sum_{j=1}^{n} S_{ij} u_i v_j.$$

The canonical inner product for $L^2[0,1]$ is defined by

$$(f,g) := \int_0^1 f(x) g(x)\, dx.$$

We now prove the following important inequality:

Theorem 1.1.3 (Cauchy–Schwartz Inequality). *Let V be an inner product space. Then*

$$(u,v)^2 \leq (u,u)(v,v) \quad \forall u, v \in V, \tag{1.3}$$

with equality if and only if u and v are linearly dependent.

Proof. If $(u,v) = 0$, then the above inequality is obvious. If it is not zero, then u and v are nonzero vectors, so that (u,u) and (v,v) are nonzero. By Eqn. (1.1c), we have

$$(u - \alpha v, u - \alpha v) \geq 0 \quad \forall \alpha \in \Re.$$

Expanding this equation using the properties of the inner product, we get

$$(u,u) - 2\alpha(u,v) + \alpha^2 (v,v) \geq 0 \quad \forall \alpha \in \Re. \tag{1.4}$$

The left-hand side of the above inequality is a quadratic function in α, with a minimum at $\alpha = (u,v)/(v,v)$. Substituting this value of α into Eqn. (1.4), we get Eqn. (1.3). Alternatively, one can use a modified form of Eqn. (5.128a) that includes equality to directly obtain Eqn. (1.3).

If $\boldsymbol{u}$ and/or $\boldsymbol{v}$ is the zero element, then by the remark following the definition of linear independence, $\boldsymbol{u}$ and $\boldsymbol{v}$ are linearly dependent, and we also have equality in Eqn. (1.3). If they are linearly dependent and nonzero, then there exists $\alpha \in \Re$ such that $\boldsymbol{u} = \alpha\boldsymbol{v}$, and equality in Eqn. (1.3) follows immediately. Conversely, if there is equality in Eqn. (1.3), and if $\boldsymbol{u}$ and/or $\boldsymbol{v}$ is the zero element, then they are linearly dependent. If there is equality, and both $\boldsymbol{u}$ and $\boldsymbol{v}$ are nonzero, then by letting $\alpha = (\boldsymbol{u},\boldsymbol{v})/(\boldsymbol{v},\boldsymbol{v})$, we see that

$$(\boldsymbol{u} - \alpha\boldsymbol{v}, \boldsymbol{u} - \alpha\boldsymbol{v}) = 0,$$

which in turn implies that $\boldsymbol{u} = \alpha\boldsymbol{v}$. □

Applying the Cauchy–Schwartz inequality to $\Re^n$ and $L^2[0,1]$, we get

$$\left(\sum_{i=1}^{n} u_i v_i\right)^2 \le \left(\sum_{i=1}^{n} u_i^2\right)\left(\sum_{i=1}^{n} v_i^2\right),$$

$$\left(\int_0^1 f(x)g(x)\,dx\right)^2 \le \left(\int_0^1 f(x)^2\,dx\right)\left(\int_0^1 g(x)^2\,dx\right).$$

Using the Cauchy–Schwartz inequality, we now prove the following:

Theorem 1.1.4 (Triangle inequality). *Let V be an inner product space. Then*

$$|\boldsymbol{u}+\boldsymbol{v}| \le |\boldsymbol{u}| + |\boldsymbol{v}| \quad \forall \boldsymbol{u},\boldsymbol{v} \in V.$$

Proof.

$$\begin{aligned} |\boldsymbol{u}+\boldsymbol{v}|^2 &= (\boldsymbol{u}+\boldsymbol{v},\boldsymbol{u}+\boldsymbol{v}) \\ &= |\boldsymbol{u}|^2 + 2(\boldsymbol{u},\boldsymbol{v}) + |\boldsymbol{v}|^2 \\ &\le |\boldsymbol{u}|^2 + 2\,|(\boldsymbol{u},\boldsymbol{v})| + |\boldsymbol{v}|^2 \\ &\le |\boldsymbol{u}|^2 + 2\,|\boldsymbol{u}|\,|\boldsymbol{v}| + |\boldsymbol{v}|^2 \quad \text{(by the Cauchy–Schwartz inequality)} \\ &= (|\boldsymbol{u}| + |\boldsymbol{v}|)^2. \end{aligned}$$

Taking the square-root of both sides, we get the desired result. □

A vector space V is a *normed vector space*, if we assign a nonnegative real number, $\|\boldsymbol{u}\|$, called the *norm* of $\boldsymbol{u}$, to each $\boldsymbol{u} \in V$ such that

$$\begin{aligned} &\|\alpha\boldsymbol{u}\| = |\alpha|\,\|\boldsymbol{u}\| \quad \forall\alpha \in \Re \text{ and } \forall\boldsymbol{u} \in V, \\ &\|\boldsymbol{u}\| \ge 0 \quad \forall\boldsymbol{u} \in V \text{ with } \|\boldsymbol{u}\| = 0 \text{ if and only if } \boldsymbol{u} = \boldsymbol{0}, \\ &\|\boldsymbol{u}+\boldsymbol{v}\| \le \|\boldsymbol{u}\| + \|\boldsymbol{v}\|. \end{aligned}$$

One can show that $L^p[0,1]$, $p \geq 1$ is a normed vector space. From Eqns. (1.2) and the triangle inequality, it is clear that an inner product space is a normed vector space with the norm defined by

$$\|\boldsymbol{u}\| := |\boldsymbol{u}| = (\boldsymbol{u}, \boldsymbol{u})^{1/2}.$$

Since continuum mechanics is primarily the study of deformable bodies in three-dimensional space, we now specialize the results of this section to the case $n = 3$, and henceforth present results only for this case.

1.2 Vectors in $\Re^3$

From now on, V denotes the three-dimensional Euclidean space $\Re^3$. Let $\{\boldsymbol{e}_1, \boldsymbol{e}_2, \boldsymbol{e}_3\}$ be a fixed set of orthonormal vectors that constitute the Cartesian basis. We have

$$\boldsymbol{e}_i \cdot \boldsymbol{e}_j = \delta_{ij},$$

where δ_{ij}, known as the Kronecker delta, is defined by

$$\delta_{ij} := \begin{cases} 0 & \text{when } i \neq j, \\ 1 & \text{when } i = j. \end{cases} \tag{1.5}$$

The Kronecker delta is also known as the substitution operator, since, from the definition, we can see that $x_i = \delta_{ij} x_j$, $\tau_{ij} = \tau_{ik}\delta_{kj}$, and so on. Note that $\delta_{ij} = \delta_{ji}$, and $\delta_{ii} = \delta_{11} + \delta_{22} + \delta_{33} = 3$.

Any vector $\boldsymbol{u}$ can be written as

$$\boldsymbol{u} = u_1 \boldsymbol{e}_1 + u_2 \boldsymbol{e}_2 + u_3 \boldsymbol{e}_3, \tag{1.6}$$

or, using the summation convention, as

$$\boldsymbol{u} = u_i \boldsymbol{e}_i.$$

The inner product of two vectors is given by

$$(\boldsymbol{u}, \boldsymbol{v}) = \boldsymbol{u} \cdot \boldsymbol{v} := u_i v_i = u_1 v_1 + u_2 v_2 + u_3 v_3. \tag{1.7}$$

Using Eqn. (1.6), the components of the vector $\boldsymbol{u}$ can be written as

$$u_i = \boldsymbol{u} \cdot \boldsymbol{e}_i. \tag{1.8}$$

Substituting Eqn. (1.8) into Eqn. (1.6), we have

$$\boldsymbol{u} = (\boldsymbol{u} \cdot \boldsymbol{e}_i)\boldsymbol{e}_i. \tag{1.9}$$

We define the cross product of two base vectors $\boldsymbol{e}_j$ and $\boldsymbol{e}_k$ by

$$\boldsymbol{e}_j \times \boldsymbol{e}_k := \epsilon_{ijk} \boldsymbol{e}_i, \tag{1.10}$$

where ϵ_{ijk} are the components of a third-order tensor $\mathcal{E}$ known as the alternate tensor (which we will discuss in greater detail in Section 1.7), and are given by

$$\begin{aligned}\epsilon_{123} &= \epsilon_{231} = \epsilon_{312} = 1\\ \epsilon_{132} &= \epsilon_{213} = \epsilon_{321} = -1\\ \epsilon_{ijk} &= 0 \quad \text{otherwise.}\end{aligned}$$

Taking the dot product of both sides of Eqn. (1.10) with $\boldsymbol{e}_m$, we get

$$\boldsymbol{e}_m \cdot (\boldsymbol{e}_j \times \boldsymbol{e}_k) = \epsilon_{ijk}\delta_{im} = \epsilon_{mjk}.$$

Using the index i in place of m, we have

$$\epsilon_{ijk} = \boldsymbol{e}_i \cdot (\boldsymbol{e}_j \times \boldsymbol{e}_k). \tag{1.11}$$

The cross product of two vectors is assumed to be distributive, i.e.,

$$(\alpha \boldsymbol{u}) \times (\beta \boldsymbol{v}) = \alpha\beta(\boldsymbol{u} \times \boldsymbol{v}) \quad \forall \alpha, \beta \in \Re \text{ and } \boldsymbol{u}, \boldsymbol{v} \in V.$$

If $\boldsymbol{w}$ denotes the cross product of $\boldsymbol{u}$ and $\boldsymbol{v}$, then by using this property and Eqn. (1.10), we have

$$\begin{aligned}\boldsymbol{w} &= \boldsymbol{u} \times \boldsymbol{v}\\ &= (u_j \boldsymbol{e}_j) \times (v_k \boldsymbol{e}_k)\\ &= \epsilon_{ijk} u_j v_k \boldsymbol{e}_i.\end{aligned} \tag{1.12}$$

It is clear from Eqn. (1.12) that

$$\boldsymbol{u} \times \boldsymbol{v} = -\boldsymbol{v} \times \boldsymbol{u}.$$

Taking $\boldsymbol{v} = \boldsymbol{u}$, we get $\boldsymbol{u} \times \boldsymbol{u} = \boldsymbol{0}$.

The scalar triple product of three vectors $\boldsymbol{u}$, $\boldsymbol{v}$, $\boldsymbol{w}$, denoted by $[\boldsymbol{u}, \boldsymbol{v}, \boldsymbol{w}]$, is defined by

$$[\boldsymbol{u}, \boldsymbol{v}, \boldsymbol{w}] := \boldsymbol{u} \cdot (\boldsymbol{v} \times \boldsymbol{w}).$$

In indicial notation, we have

$$[\boldsymbol{u}, \boldsymbol{v}, \boldsymbol{w}] = \epsilon_{ijk} u_i v_j w_k. \tag{1.13}$$

From Eqn. 1.13, it is clear that

$$[\boldsymbol{u}, \boldsymbol{v}, \boldsymbol{w}] = [\boldsymbol{v}, \boldsymbol{w}, \boldsymbol{u}] = [\boldsymbol{w}, \boldsymbol{u}, \boldsymbol{v}] = -[\boldsymbol{v}, \boldsymbol{u}, \boldsymbol{w}] = -[\boldsymbol{u}, \boldsymbol{w}, \boldsymbol{v}] = -[\boldsymbol{w}, \boldsymbol{v}, \boldsymbol{u}] \quad \forall \boldsymbol{u}, \boldsymbol{v}, \boldsymbol{w} \in V. \tag{1.14}$$

If any two elements in the scalar triple product are the same, then its value is zero, as can be seen by interchanging the identical elements, and using the above formula. From Eqn. (1.13), it is also clear that the scalar triple product is linear in each of its argument variables, so that, for example,

$$[\alpha \boldsymbol{u} + \beta \boldsymbol{v}, \boldsymbol{x}, \boldsymbol{y}] = \alpha [\boldsymbol{u}, \boldsymbol{x}, \boldsymbol{y}] + \beta [\boldsymbol{v}, \boldsymbol{x}, \boldsymbol{y}] \quad \forall \boldsymbol{u}, \boldsymbol{v}, \boldsymbol{x}, \boldsymbol{y} \in V. \tag{1.15}$$

As can be easily verified, Eqn. (1.13) can be written in determinant form as

$$[\boldsymbol{u}, \boldsymbol{v}, \boldsymbol{w}] = \det \begin{bmatrix} u_1 & u_2 & u_3 \\ v_1 & v_2 & v_3 \\ w_1 & w_2 & w_3 \end{bmatrix}. \tag{1.16}$$

Using Eqns. (1.11) and (1.16), the components of the alternate tensor can be written in determinant form as follows:

$$\epsilon_{ijk} = [\boldsymbol{e}_i, \boldsymbol{e}_j, \boldsymbol{e}_k] = \det \begin{bmatrix} \boldsymbol{e}_i \cdot \boldsymbol{e}_1 & \boldsymbol{e}_i \cdot \boldsymbol{e}_2 & \boldsymbol{e}_i \cdot \boldsymbol{e}_3 \\ \boldsymbol{e}_j \cdot \boldsymbol{e}_1 & \boldsymbol{e}_j \cdot \boldsymbol{e}_2 & \boldsymbol{e}_j \cdot \boldsymbol{e}_3 \\ \boldsymbol{e}_k \cdot \boldsymbol{e}_1 & \boldsymbol{e}_k \cdot \boldsymbol{e}_2 & \boldsymbol{e}_k \cdot \boldsymbol{e}_3 \end{bmatrix} = \det \begin{bmatrix} \delta_{i1} & \delta_{i2} & \delta_{i3} \\ \delta_{j1} & \delta_{j2} & \delta_{j3} \\ \delta_{k1} & \delta_{k2} & \delta_{k3} \end{bmatrix}. \tag{1.17}$$

Thus, we have

$$\begin{aligned}
\epsilon_{ijk}\epsilon_{pqr} &= \det \begin{bmatrix} \delta_{i1} & \delta_{i2} & \delta_{i3} \\ \delta_{j1} & \delta_{j2} & \delta_{j3} \\ \delta_{k1} & \delta_{k2} & \delta_{k3} \end{bmatrix} \det \begin{bmatrix} \delta_{p1} & \delta_{p2} & \delta_{p3} \\ \delta_{q1} & \delta_{q2} & \delta_{q3} \\ \delta_{r1} & \delta_{r2} & \delta_{r3} \end{bmatrix} \\
&= \det \begin{bmatrix} \delta_{i1} & \delta_{i2} & \delta_{i3} \\ \delta_{j1} & \delta_{j2} & \delta_{j3} \\ \delta_{k1} & \delta_{k2} & \delta_{k3} \end{bmatrix} \det \begin{bmatrix} \delta_{p1} & \delta_{q1} & \delta_{r1} \\ \delta_{p2} & \delta_{q2} & \delta_{r2} \\ \delta_{p3} & \delta_{q3} & \delta_{r3} \end{bmatrix} \qquad (\text{since } \det \boldsymbol{T} = \det(\boldsymbol{T}^T)) \\
&= \det \left\{ \begin{bmatrix} \delta_{i1} & \delta_{i2} & \delta_{i3} \\ \delta_{j1} & \delta_{j2} & \delta_{j3} \\ \delta_{k1} & \delta_{k2} & \delta_{k3} \end{bmatrix} \begin{bmatrix} \delta_{p1} & \delta_{q1} & \delta_{r1} \\ \delta_{p2} & \delta_{q2} & \delta_{r2} \\ \delta_{p3} & \delta_{q3} & \delta_{r3} \end{bmatrix} \right\} \qquad (\text{since } (\det \boldsymbol{R})(\det \boldsymbol{S}) = \det(\boldsymbol{RS})) \\
&= \det \begin{bmatrix} \delta_{im}\delta_{mp} & \delta_{im}\delta_{mq} & \delta_{im}\delta_{mr} \\ \delta_{jm}\delta_{mp} & \delta_{jm}\delta_{mq} & \delta_{jm}\delta_{mr} \\ \delta_{km}\delta_{mp} & \delta_{km}\delta_{mq} & \delta_{km}\delta_{mr} \end{bmatrix} \\
&= \det \begin{bmatrix} \delta_{ip} & \delta_{iq} & \delta_{ir} \\ \delta_{jp} & \delta_{jq} & \delta_{jr} \\ \delta_{kp} & \delta_{kq} & \delta_{kr} \end{bmatrix}.
\end{aligned} \tag{1.18}$$

From Eqn. (1.18) and the relation $\delta_{ii} = 3$, we obtain the following identities (the first of which is known as the ϵ–δ identity):

$$\epsilon_{ijk}\epsilon_{iqr} = \delta_{jq}\delta_{kr} - \delta_{jr}\delta_{kq}, \tag{1.19a}$$

$$\epsilon_{ijk}\epsilon_{ijm} = 2\delta_{km}, \tag{1.19b}$$

$$\epsilon_{ijk}\epsilon_{ijk} = 6. \tag{1.19c}$$

Using Eqn. (1.12) and the ϵ–δ identity, we get

$$\begin{aligned}
(\boldsymbol{u} \times \boldsymbol{v}) \cdot (\boldsymbol{u} \times \boldsymbol{v}) &= \epsilon_{ijk}\epsilon_{imn} u_j v_k u_m v_n \\
&= (\delta_{jm}\delta_{kn} - \delta_{jn}\delta_{km}) u_j v_k u_m v_n
\end{aligned}$$

$$= (u_m v_n u_m v_n - u_j v_m u_m v_j)$$
$$= (\boldsymbol{u} \cdot \boldsymbol{u})(\boldsymbol{v} \cdot \boldsymbol{v}) - (\boldsymbol{u} \cdot \boldsymbol{v})^2. \tag{1.20}$$

We now have the following result:

Theorem 1.2.1. *For any two vectors $\boldsymbol{u}, \boldsymbol{v} \in V$, $\boldsymbol{u} \times \boldsymbol{v} = \boldsymbol{0}$ if and only if $\boldsymbol{u}$ and $\boldsymbol{v}$ are linearly dependent.*

Proof. If $\boldsymbol{u}$ and $\boldsymbol{v}$ are linearly dependent, and $\boldsymbol{u}$ and/or $\boldsymbol{v}$ is zero, then it is obvious that $\boldsymbol{u} \times \boldsymbol{v} = \boldsymbol{0}$. If they are linearly dependent and nonzero, then there exists a scalar α such that $\boldsymbol{v} = \alpha \boldsymbol{u}$. Hence

$$\boldsymbol{u} \times \boldsymbol{v} = \alpha \boldsymbol{u} \times \boldsymbol{u} = \boldsymbol{0}.$$

Conversely, if $\boldsymbol{u} \times \boldsymbol{v} = \boldsymbol{0}$, then

$$0 = (\boldsymbol{u} \times \boldsymbol{v}) \cdot (\boldsymbol{u} \times \boldsymbol{v})$$
$$= (\boldsymbol{u} \cdot \boldsymbol{u})(\boldsymbol{v} \cdot \boldsymbol{v}) - (\boldsymbol{u} \cdot \boldsymbol{v})^2, \quad \text{(by Eqn. (1.20))}$$

which implies that

$$(\boldsymbol{u}, \boldsymbol{v})^2 = (\boldsymbol{u}, \boldsymbol{u})(\boldsymbol{v}, \boldsymbol{v}).$$

But by Theorem 1.1.3, this implies that $\boldsymbol{u}$ and $\boldsymbol{v}$ are linearly dependent. □

The vector triple products $\boldsymbol{u} \times (\boldsymbol{v} \times \boldsymbol{w})$ and $(\boldsymbol{u} \times \boldsymbol{v}) \times \boldsymbol{w}$, defined as the cross product of $\boldsymbol{u}$ with $\boldsymbol{v} \times \boldsymbol{w}$, and the cross product of $\boldsymbol{u} \times \boldsymbol{v}$ with $\boldsymbol{w}$, respectively, are different in general, and are given by

$$\boldsymbol{u} \times (\boldsymbol{v} \times \boldsymbol{w}) = (\boldsymbol{u} \cdot \boldsymbol{w})\boldsymbol{v} - (\boldsymbol{u} \cdot \boldsymbol{v})\boldsymbol{w}, \tag{1.21a}$$
$$(\boldsymbol{u} \times \boldsymbol{v}) \times \boldsymbol{w} = (\boldsymbol{u} \cdot \boldsymbol{w})\boldsymbol{v} - (\boldsymbol{v} \cdot \boldsymbol{w})\boldsymbol{u}. \tag{1.21b}$$

The first relation is proved by noting that

$$\begin{aligned}
\boldsymbol{u} \times (\boldsymbol{v} \times \boldsymbol{w}) &= \epsilon_{ijk} u_j (\boldsymbol{v} \times \boldsymbol{w})_k \boldsymbol{e}_i \\
&= \epsilon_{ijk} \epsilon_{kmn} u_j v_m w_n \boldsymbol{e}_i \\
&= \epsilon_{kij} \epsilon_{kmn} u_j v_m w_n \boldsymbol{e}_i \\
&= (\delta_{im}\delta_{jn} - \delta_{in}\delta_{jm}) u_j v_m w_n \boldsymbol{e}_i \\
&= (u_n w_n v_i - u_m v_m w_i) \boldsymbol{e}_i \\
&= (\boldsymbol{u} \cdot \boldsymbol{w})\boldsymbol{v} - (\boldsymbol{u} \cdot \boldsymbol{v})\boldsymbol{w}.
\end{aligned}$$

The second relation is proved in an analogous manner.

For scalar triple products, we have the following useful result:

Theorem 1.2.2. *The scalar triple product* $[u, v, w]$ *is zero if and only if* u, v *and* w *are linearly dependent.*

Proof. Since

$$u = u_1 e_1 + u_2 e_2 + u_3 e_3,$$
$$v = v_1 e_1 + v_2 e_2 + v_3 e_3,$$
$$w = w_1 e_1 + w_2 e_2 + w_3 e_3,$$

by Theorem 1.1.2, $\{u, v, w\}$ is linearly independent if and only if (see Eqn. (1.16))

$$[u, v, w] = \det \begin{bmatrix} u_1 & u_2 & u_3 \\ v_1 & v_2 & v_3 \\ w_1 & w_2 & w_3 \end{bmatrix} \neq 0,$$

which is equivalent to the statement of the theorem. □

Another proof of the above theorem may be found in [43].

1.3 Second-Order Tensors

A second-order tensor is a linear transformation that maps vectors to vectors. We shall denote the set of second-order tensors by Lin. If T is a second-order tensor that maps a vector u to a vector v, then we write it as

$$v = Tu. \tag{1.22}$$

T satisfies the property

$$T(ax + by) = aTx + bTy, \quad \forall x, y \in V \text{ and } a, b \in \Re.$$

By choosing $a = 1$, $b = -1$, and $x = y$, we get

$$T(0) = 0.$$

From the definition of a second-order tensor, it follows that the sum of two second-order tensors defined by

$$(R + S)u := Ru + Su \quad \forall u \in V,$$

and the scalar multiple of T by $\alpha \in \Re$, defined by

$$(\alpha T)u := \alpha(Tu) \quad \forall u \in V,$$

are both second-order tensors. The two second-order tensors R and S are said to be equal if

$$Ru = Su \quad \forall u \in V. \tag{1.23}$$

The above condition is equivalent to the condition

$$(v, Ru) = (v, Su) \quad \forall u, v \in V. \tag{1.24}$$

To see this, note that if Eqn. (1.23) holds, then clearly Eqn. (1.24) holds. On the other hand, if Eqn. (1.24) holds, then using the bilinearity property of the inner product, we have

$$(v, (Ru - Su)) = 0 \quad \forall u, v \in V.$$

Choosing $v = Ru - Su$, and using Eqn. (1.1c), we get Eqn. (1.23).

If we define the element Z as

$$Zu = 0 \quad \forall u \in V,$$

then we see that $Z \in$ Lin, and is the zero element of Lin since

$$(T + Z)u = Tu + Zu = Tu \quad \forall u \in V,$$

which, by the definition of the equality of tensors, implies that

$$T + Z = T.$$

Henceforth, we simply use the symbol $\mathbf{0}$ to denote the zero elements of both Lin and V. With the above definitions of addition, scalar multiplication, and zero, Lin is a vector space, and hence, all the results that we have derived for vector spaces in Section 1.1 are valid for Lin.

If we define the function $I : V \to V$ by

$$Iu := u \quad \forall u \in V, \tag{1.25}$$

then it is clear that $I \in$ Lin. I is called as the *identity tensor*.

Choosing $u = e_1, e_2$ and e_3 in Eqn. (1.22), we get three vectors that can be expressed as a linear combination of the base vectors e_i as

$$\begin{aligned} Te_1 &= \alpha_1 e_1 + \alpha_2 e_2 + \alpha_3 e_3 \\ Te_2 &= \alpha_4 e_1 + \alpha_5 e_2 + \alpha_6 e_3 \\ Te_3 &= \alpha_7 e_1 + \alpha_8 e_2 + \alpha_9 e_3, \end{aligned} \tag{1.26}$$

where α_i, $i = 1$ to 9, are scalar constants. Renaming the α_i as T_{ij}, $i = 1,3$, $j = 1,3$, we get

$$Te_j = T_{ij} e_i. \tag{1.27}$$

The elements T_{ij} are called the components of the tensor T with respect to the base vectors e_j; as seen from Eqn. (1.27), T_{ij} is the component of Te_j in the e_i direction. Taking the dot product of both sides of Eqn. (1.27) with e_k for some particular k, we get

$$e_k \cdot Te_j = T_{ij} \delta_{ik} = T_{kj},$$

or, replacing k by i,

$$T_{ij} = e_i \cdot Te_j. \tag{1.28}$$

By choosing $\boldsymbol{v} = \boldsymbol{e}_i$ and $\boldsymbol{u} = \boldsymbol{e}_j$ in Eqn. (1.24), it is clear that the components of two equal tensors are equal. From Eqn. (1.28), the components of the identity tensor in any orthonormal coordinate system $\boldsymbol{e}_i$ are

$$I_{ij} = \boldsymbol{e}_i \cdot \boldsymbol{I}\boldsymbol{e}_j = \boldsymbol{e}_i \cdot \boldsymbol{e}_j = \delta_{ij}. \tag{1.29}$$

Thus, the components of the identity tensor are scalars that are independent of the Cartesian basis. Using Eqn. (1.27), we write Eqn. (1.22) in component form (where the components are with respect to a particular orthonormal basis $\{\boldsymbol{e}_i\}$) as

$$v_i \boldsymbol{e}_i = \boldsymbol{T}(u_j \boldsymbol{e}_j) = u_j \boldsymbol{T}\boldsymbol{e}_j = u_j T_{ij} \boldsymbol{e}_i,$$

which, by virtue of the uniqueness of the components of any element of a vector space, yields

$$v_i = T_{ij} u_j. \tag{1.30}$$

Thus, the components of the vector $\boldsymbol{v}$ are obtained by a matrix multiplication of the components of $\boldsymbol{T}$, and the components of $\boldsymbol{u}$.

The transpose of $\boldsymbol{T}$, denoted by $\boldsymbol{T}^T$, is defined using the inner product as

$$(\boldsymbol{T}^T \boldsymbol{u}, \boldsymbol{v}) := (\boldsymbol{u}, \boldsymbol{T}\boldsymbol{v}) \quad \forall \boldsymbol{u}, \boldsymbol{v} \in V. \tag{1.31}$$

Once again, it follows from the definition that $\boldsymbol{T}^T$ is a second-order tensor. The transpose has the following properties:

$$\begin{aligned} (\boldsymbol{T}^T)^T &= \boldsymbol{T}, \\ (\alpha \boldsymbol{T})^T &= \alpha \boldsymbol{T}^T, \\ (\boldsymbol{R} + \boldsymbol{S})^T &= \boldsymbol{R}^T + \boldsymbol{S}^T. \end{aligned}$$

If (T_{ij}) represent the components of the tensor $\boldsymbol{T}$, then the components of $\boldsymbol{T}^T$ are

$$\begin{aligned} (\boldsymbol{T}^T)_{ij} &= \boldsymbol{e}_i \cdot \boldsymbol{T}^T \boldsymbol{e}_j \\ &= \boldsymbol{T}\boldsymbol{e}_i \cdot \boldsymbol{e}_j \\ &= T_{ji}. \end{aligned} \tag{1.32}$$

The tensor $\boldsymbol{T}$ is said to be symmetric if

$$\boldsymbol{T}^T = \boldsymbol{T},$$

and skew-symmetric (or anti-symmetric) if

$$\boldsymbol{T}^T = -\boldsymbol{T}.$$

Any tensor $\boldsymbol{T}$ can be decomposed uniquely into a symmetric and an skew-symmetric part as (see Problem 5)

$$\boldsymbol{T} = \boldsymbol{T}_s + \boldsymbol{T}_{ss}, \tag{1.33}$$

where

$$T_s = \frac{1}{2}(T + T^T),$$
$$T_{ss} = \frac{1}{2}(T - T^T).$$

The product of two second-order tensors $\boldsymbol{RS}$ is the composition of the two operations $\boldsymbol{R}$ and $\boldsymbol{S}$, with $\boldsymbol{S}$ operating first, and defined by the relation

$$(\boldsymbol{RS})\boldsymbol{u} := \boldsymbol{R}(\boldsymbol{Su}) \quad \forall \boldsymbol{u} \in V. \tag{1.34}$$

Since $\boldsymbol{RS}$ is a linear transformation that maps vectors to vectors, we conclude that the product of two second-order tensors is also a second-order tensor. From the definition of the identity tensor given by (1.25) it follows that $\boldsymbol{RI} = \boldsymbol{IR} = \boldsymbol{R}$. If $\boldsymbol{T}$ represents the product $\boldsymbol{RS}$, then its components are given by

$$\begin{aligned} T_{ij} &= \boldsymbol{e}_i \cdot (\boldsymbol{RS})\boldsymbol{e}_j \\ &= \boldsymbol{e}_i \cdot \boldsymbol{R}(\boldsymbol{Se}_j) \\ &= \boldsymbol{e}_i \cdot \boldsymbol{R}(S_{kj}\boldsymbol{e}_k) \\ &= \boldsymbol{e}_i \cdot S_{kj}\boldsymbol{Re}_k \\ &= S_{kj}(\boldsymbol{e}_i \cdot \boldsymbol{Re}_k) \\ &= S_{kj}R_{ik} \\ &= R_{ik}S_{kj}, \end{aligned} \tag{1.35}$$

which is consistent with matrix multiplication. Also consistent with the results from matrix theory, we have $(\boldsymbol{RS})^T = \boldsymbol{S}^T\boldsymbol{R}^T$, which follows from Eqns. (1.24), (1.31) and (1.34).

1.3.1 The tensor product

We now introduce the concept of a tensor product, which is convenient for working with tensors of rank higher than two. We first define the dyadic or tensor product of two vectors $\boldsymbol{a}$ and $\boldsymbol{b}$ by

$$(\boldsymbol{a} \otimes \boldsymbol{b})\boldsymbol{c} := (\boldsymbol{b} \cdot \boldsymbol{c})\boldsymbol{a} \quad \forall \boldsymbol{c} \in V. \tag{1.36}$$

Note that the tensor product $\boldsymbol{a} \otimes \boldsymbol{b}$ cannot be defined except in terms of its operation on a vector $\boldsymbol{c}$. We now prove that $\boldsymbol{a} \otimes \boldsymbol{b}$ defines a second-order tensor. The above rule obviously maps a vector into another vector. All that we need to do is to prove that it is a linear map. For arbitrary scalars c and d, and arbitrary vectors $\boldsymbol{x}$ and $\boldsymbol{y}$, we have

$$\begin{aligned} (\boldsymbol{a} \otimes \boldsymbol{b})(c\boldsymbol{x} + d\boldsymbol{y}) &= [\boldsymbol{b} \cdot (c\boldsymbol{x} + d\boldsymbol{y})]\,\boldsymbol{a} \\ &= [c\boldsymbol{b} \cdot \boldsymbol{x} + d\boldsymbol{b} \cdot \boldsymbol{y}]\,\boldsymbol{a} \\ &= c(\boldsymbol{b} \cdot \boldsymbol{x})\boldsymbol{a} + d(\boldsymbol{b} \cdot \boldsymbol{y})\boldsymbol{a} \\ &= c[(\boldsymbol{a} \otimes \boldsymbol{b})\boldsymbol{x}] + d\,[(\boldsymbol{a} \otimes \boldsymbol{b})\boldsymbol{y}], \end{aligned}$$

which proves that $\boldsymbol{a} \otimes \boldsymbol{b}$ is a linear function. Hence, $\boldsymbol{a} \otimes \boldsymbol{b}$ is a second-order tensor. Any second-order tensor $\boldsymbol{T}$ can be written as

$$\boldsymbol{T} = T_{ij}\boldsymbol{e}_i \otimes \boldsymbol{e}_j, \tag{1.37}$$

where the components of the tensor, T_{ij} are given by Eqn. (1.28). To see this, we consider the action of $\boldsymbol{T}$ on an arbitrary vector $\boldsymbol{u}$:

$$\begin{aligned}
\boldsymbol{Tu} = (\boldsymbol{Tu})_i\boldsymbol{e}_i &= [\boldsymbol{e}_i \cdot (\boldsymbol{Tu})]\, \boldsymbol{e}_i \\
&= \{\boldsymbol{e}_i \cdot [\boldsymbol{T}(u_j\boldsymbol{e}_j)]\}\, \boldsymbol{e}_i \\
&= \{u_j\, [\boldsymbol{e}_i \cdot (\boldsymbol{Te}_j)]\}\, \boldsymbol{e}_i \\
&= \{(\boldsymbol{u} \cdot \boldsymbol{e}_j)\, [\boldsymbol{e}_i \cdot (\boldsymbol{Te}_j)]\}\, \boldsymbol{e}_i \\
&= [\boldsymbol{e}_i \cdot (\boldsymbol{Te}_j)]\, [(\boldsymbol{u} \cdot \boldsymbol{e}_j)\boldsymbol{e}_i] \\
&= [\boldsymbol{e}_i \cdot (\boldsymbol{Te}_j)]\, [(\boldsymbol{e}_i \otimes \boldsymbol{e}_j)\boldsymbol{u}] \\
&= \{[\boldsymbol{e}_i \cdot (\boldsymbol{Te}_j)]\, \boldsymbol{e}_i \otimes \boldsymbol{e}_j\}\, \boldsymbol{u}.
\end{aligned}$$

Hence, we conclude that any second-order tensor admits the representation given by Eqn. (1.37), with the nine components T_{ij}, $i = 1,2,3$, $j = 1,2,3$, given by Eqn. (1.28). The dyadic products $\{\boldsymbol{e}_i \otimes \boldsymbol{e}_j\}$, $i = 1,2,3$, $j = 1,2,3$ constitute a basis of Lin since, as just mentioned, any element of Lin can be expressed in terms of them, and since they are linearly independent ($T_{ij}\boldsymbol{e}_i \otimes \boldsymbol{e}_j = \boldsymbol{0}$ implies that $T_{ij} = \boldsymbol{e}_i \cdot \boldsymbol{0}\boldsymbol{e}_j = 0$).

From Eqns. (1.29) and (1.37), it follows that

$$\boldsymbol{I} = \bar{\boldsymbol{e}}_i \otimes \bar{\boldsymbol{e}}_i, \tag{1.38}$$

where $\{\bar{\boldsymbol{e}}_1, \bar{\boldsymbol{e}}_2, \bar{\boldsymbol{e}}_3\}$ is *any* orthonormal coordinate frame. If $\boldsymbol{T}$ is represented as given by Eqn. (1.37), it follows from Eqn. (1.32) that the transpose of $\boldsymbol{T}$ can be represented as

$$\boldsymbol{T}^T = T_{ji}\boldsymbol{e}_i \otimes \boldsymbol{e}_j. \tag{1.39}$$

From Eqns. (1.37) and (1.39), we deduce that a tensor is symmetric ($\boldsymbol{T} = \boldsymbol{T}^T$) if and only if $T_{ij} = T_{ji}$ for all possible i and j. We now show how all the properties of a second-order tensor derived so far can be derived using the dyadic product.

Using Eqn. (1.27), we see that the components of a dyad $\boldsymbol{a} \otimes \boldsymbol{b}$ are given by

$$\begin{aligned}
(\boldsymbol{a} \otimes \boldsymbol{b})_{ij} &= \boldsymbol{e}_i \cdot (\boldsymbol{a} \otimes \boldsymbol{b})\boldsymbol{e}_j \\
&= \boldsymbol{e}_i \cdot (\boldsymbol{b} \cdot \boldsymbol{e}_j)\boldsymbol{a} \\
&= a_i b_j.
\end{aligned} \tag{1.40}$$

The components of a vector $\boldsymbol{v}$ obtained by a second-order tensor $\boldsymbol{T}$ operating on a vector $\boldsymbol{u}$ are obtained by noting that

$$v_i\boldsymbol{e}_i = T_{ij}(\boldsymbol{e}_i \otimes \boldsymbol{e}_j)\boldsymbol{u} = T_{ij}(\boldsymbol{u} \cdot \boldsymbol{e}_j)\boldsymbol{e}_i = T_{ij}u_j\boldsymbol{e}_i, \tag{1.41}$$

which in equivalent to Eqn. (1.30).

Convention regarding complex vectors: In what follows, we will often encounter the dyadic product of two complex-valued vectors. We note here that we use the same definition for the dyadic product of complex-valued vectors as that for real-valued vectors, namely, the definition given by Eqn. (1.36). Using Eqn. (1.40), we get the components of $\boldsymbol{a} \otimes \boldsymbol{b}$ as $a_i b_j$. However, it is possible to follow another convention in defining the dyadic product, namely $(\boldsymbol{a} \otimes \boldsymbol{b})\boldsymbol{c} = (\hat{\boldsymbol{b}} \cdot \boldsymbol{c})\boldsymbol{a}$ (with the hat denoting complex conjugation), whereby one obtains the components as $a_i \hat{b}_j$. We will follow this rule of using the same definitions for the real and complex case even when we define the dyadic products $\boldsymbol{A} \otimes \boldsymbol{B}$ and $\boldsymbol{A} \boxtimes \boldsymbol{B}$ of two complex-valued second-order tensors $\boldsymbol{A}$ and $\boldsymbol{B}$. The advantage of our convention is that all relations that we derive for the real case are also valid for the complex case, since the definitions are the same. If we used the alternate convention, one would need to re-derive all the relations for the complex case, and this can be quite cumbersome, since we will be deriving dozens of relations (including rules for differentiation) involving tensor products. We emphasize that both conventions are correct, and, as long as one sticks to one convention consistently in all the derivations, the particular choice made is just a matter of convenience.

1.3.2 Principal invariants of a second-order tensor

Theorem 1.3.1. *If $(\boldsymbol{u}, \boldsymbol{v}, \boldsymbol{w})$ and $(\boldsymbol{a}, \boldsymbol{b}, \boldsymbol{c})$ are two pairs of linearly independent vectors, and $\boldsymbol{T}$ is an arbitrary tensor, then*

$$\frac{[\boldsymbol{Tu}, \boldsymbol{v}, \boldsymbol{w}] + [\boldsymbol{u}, \boldsymbol{Tv}, \boldsymbol{w}] + [\boldsymbol{u}, \boldsymbol{v}, \boldsymbol{Tw}]}{[\boldsymbol{u}, \boldsymbol{v}, \boldsymbol{w}]} = \frac{[\boldsymbol{Ta}, \boldsymbol{b}, \boldsymbol{c}] + [\boldsymbol{a}, \boldsymbol{Tb}, \boldsymbol{c}] + [\boldsymbol{a}, \boldsymbol{b}, \boldsymbol{Tc}]}{[\boldsymbol{a}, \boldsymbol{b}, \boldsymbol{c}]}, \tag{1.42a}$$

$$\frac{[\boldsymbol{Tu}, \boldsymbol{Tv}, \boldsymbol{w}] + [\boldsymbol{u}, \boldsymbol{Tv}, \boldsymbol{Tw}] + [\boldsymbol{Tu}, \boldsymbol{v}, \boldsymbol{Tw}]}{[\boldsymbol{u}, \boldsymbol{v}, \boldsymbol{w}]} = \frac{[\boldsymbol{Ta}, \boldsymbol{Tb}, \boldsymbol{c}] + [\boldsymbol{a}, \boldsymbol{Tb}, \boldsymbol{Tc}] + [\boldsymbol{Ta}, \boldsymbol{b}, \boldsymbol{Tc}]}{[\boldsymbol{a}, \boldsymbol{b}, \boldsymbol{c}]}, \tag{1.42b}$$

$$\frac{[\boldsymbol{Tu}, \boldsymbol{Tv}, \boldsymbol{Tw}]}{[\boldsymbol{u}, \boldsymbol{v}, \boldsymbol{w}]} = \frac{[\boldsymbol{Ta}, \boldsymbol{Tb}, \boldsymbol{Tc}]}{[\boldsymbol{a}, \boldsymbol{b}, \boldsymbol{c}]}. \tag{1.42c}$$

Proof. Since $\boldsymbol{u} = u_i \boldsymbol{e}_i$, $\boldsymbol{v} = v_j \boldsymbol{e}_j$ and $\boldsymbol{w} = w_k \boldsymbol{e}_k$, the left-hand side of Eqn. (1.42a) can be written as

$$\begin{aligned}
\text{LHS} &= \frac{u_i v_j w_k \left\{ [\boldsymbol{Te}_i, \boldsymbol{e}_j, \boldsymbol{e}_k] + [\boldsymbol{e}_i, \boldsymbol{Te}_j, \boldsymbol{e}_k] + [\boldsymbol{e}_i, \boldsymbol{e}_j, \boldsymbol{Te}_k] \right\}}{[\boldsymbol{u}, \boldsymbol{v}, \boldsymbol{w}]} \\
&= \frac{\epsilon_{ijk} u_i v_j w_k \left\{ [\boldsymbol{Te}_1, \boldsymbol{e}_2, \boldsymbol{e}_3] + [\boldsymbol{e}_1, \boldsymbol{Te}_2, \boldsymbol{e}_3] + [\boldsymbol{e}_1, \boldsymbol{e}_2, \boldsymbol{Te}_3] \right\}}{[\boldsymbol{u}, \boldsymbol{v}, \boldsymbol{w}]} \\
&= [\boldsymbol{Te}_1, \boldsymbol{e}_2, \boldsymbol{e}_3] + [\boldsymbol{e}_1, \boldsymbol{Te}_2, \boldsymbol{e}_3] + [\boldsymbol{e}_1, \boldsymbol{e}_2, \boldsymbol{Te}_3].
\end{aligned}$$

Since the choice of vectors $\boldsymbol{u}$, $\boldsymbol{v}$ and $\boldsymbol{w}$ is arbitrary, Eqn. (1.42a) follows. Equations (1.42b) and (1.42c) are proved in a similar manner. □

As a consequence of Theorem 1.3.1, corresponding to an arbitrary tensor $\boldsymbol{T}$, there exist three scalars, I_1, I_2 and I_3 such that

$$[\boldsymbol{Tu}, \boldsymbol{v}, \boldsymbol{w}] + [\boldsymbol{u}, \boldsymbol{Tv}, \boldsymbol{w}] + [\boldsymbol{u}, \boldsymbol{v}, \boldsymbol{Tw}] = I_1 [\boldsymbol{u}, \boldsymbol{v}, \boldsymbol{w}], \tag{1.43a}$$

$$[\boldsymbol{Tu}, \boldsymbol{Tv}, \boldsymbol{w}] + [\boldsymbol{u}, \boldsymbol{Tv}, \boldsymbol{Tw}] + [\boldsymbol{Tu}, \boldsymbol{v}, \boldsymbol{Tw}] = I_2 [\boldsymbol{u}, \boldsymbol{v}, \boldsymbol{w}], \tag{1.43b}$$

$$[\boldsymbol{Tu}, \boldsymbol{Tv}, \boldsymbol{Tw}] = I_3 [\boldsymbol{u}, \boldsymbol{v}, \boldsymbol{w}]. \tag{1.43c}$$

The above equations are valid even when $[\boldsymbol{u}, \boldsymbol{v}, \boldsymbol{w}] = 0$. To see this, consider Eqn. (1.43a). Following the proof of Theorem 1.3.1, we see that if $[\boldsymbol{u}, \boldsymbol{v}, \boldsymbol{w}] = 0$, then

$$[\boldsymbol{Tu}, \boldsymbol{v}, \boldsymbol{w}] + [\boldsymbol{u}, \boldsymbol{Tv}, \boldsymbol{w}] + [\boldsymbol{u}, \boldsymbol{v}, \boldsymbol{Tw}] = [\boldsymbol{u}, \boldsymbol{v}, \boldsymbol{w}] \left\{ [\boldsymbol{Te}_1, \boldsymbol{e}_2, \boldsymbol{e}_3] + [\boldsymbol{e}_1, \boldsymbol{Te}_2, \boldsymbol{e}_3] + [\boldsymbol{e}_1, \boldsymbol{e}_2, \boldsymbol{Te}_3] \right\} = 0.$$

Equations (1.43b) and (1.43c) can be proved analogously.

The scalars I_1, I_2 and I_3 are called the *principal* invariants of $\boldsymbol{T}$. The reason for calling them as the principal invariants is that any other scalar invariant of $\boldsymbol{T}$ can be expressed in terms of them. The first and third invariants are referred to as the trace and determinant of $\boldsymbol{T}$, respectively, and written as

$$I_1 = \operatorname{tr}\boldsymbol{T}, \quad I_3 = \det\boldsymbol{T}.$$

Using the properties of the scalar triple product, it is clear that the trace is a linear operation, i.e.,

$$\operatorname{tr}(\alpha\boldsymbol{R} + \beta\boldsymbol{S}) = \alpha\operatorname{tr}\boldsymbol{R} + \beta\operatorname{tr}\boldsymbol{S} \quad \forall\alpha,\beta \in \Re \text{ and } \boldsymbol{R},\boldsymbol{S} \in \text{Lin}.$$

To get the component form for $\operatorname{tr}\boldsymbol{T}$, choose $\boldsymbol{u} = \boldsymbol{e}_1$, $\boldsymbol{v} = \boldsymbol{e}_2$ and $\boldsymbol{w} = \boldsymbol{e}_3$ in Eqn. (1.43a):

$$\operatorname{tr}\boldsymbol{T} = \boldsymbol{e}_1 \cdot \boldsymbol{T}\boldsymbol{e}_1 + \boldsymbol{e}_2 \cdot \boldsymbol{T}\boldsymbol{e}_2 + \boldsymbol{e}_3 \cdot \boldsymbol{T}\boldsymbol{e}_3 = T_{11} + T_{22} + T_{33} = T_{ii}. \tag{1.44}$$

Similarly, we have

$$\begin{aligned}\operatorname{tr}\boldsymbol{T}^T &= \boldsymbol{e}_1 \cdot \boldsymbol{T}^T\boldsymbol{e}_1 + \boldsymbol{e}_2 \cdot \boldsymbol{T}^T\boldsymbol{e}_2 + \boldsymbol{e}_3 \cdot \boldsymbol{T}^T\boldsymbol{e}_3 \\ &= \boldsymbol{e}_1 \cdot \boldsymbol{T}\boldsymbol{e}_1 + \boldsymbol{e}_2 \cdot \boldsymbol{T}\boldsymbol{e}_2 + \boldsymbol{e}_3 \cdot \boldsymbol{T}\boldsymbol{e}_3 \\ &= \operatorname{tr}\boldsymbol{T}.\end{aligned} \tag{1.45}$$

By letting $\boldsymbol{T} = \boldsymbol{a} \otimes \boldsymbol{b}$ in Eqn. (1.44), we obtain

$$\operatorname{tr}(\boldsymbol{a} \otimes \boldsymbol{b}) = a_1b_1 + a_2b_2 + a_3b_3 = a_ib_i = \boldsymbol{a} \cdot \boldsymbol{b}. \tag{1.46}$$

Using the linearity of the trace operator, and Eqn. (1.37), we get

$$\operatorname{tr}\boldsymbol{T} = \operatorname{tr}\left(T_{ij}\boldsymbol{e}_i \otimes \boldsymbol{e}_j\right) = T_{ij}\operatorname{tr}(\boldsymbol{e}_i \otimes \boldsymbol{e}_j) = T_{ij}\boldsymbol{e}_i \cdot \boldsymbol{e}_j = T_{ii},$$

which agrees with Eqn. (1.44).

To prove

$$\operatorname{tr}(\boldsymbol{R}\boldsymbol{S}) = \operatorname{tr}(\boldsymbol{S}\boldsymbol{R}),$$

we choose $\boldsymbol{u}$, $\boldsymbol{v}$ and $\boldsymbol{w}$ to be an orthonormal basis, so that

$$\begin{aligned}\operatorname{tr}(\boldsymbol{R}\boldsymbol{S}) &= (\boldsymbol{R}\boldsymbol{S}\boldsymbol{u},\boldsymbol{u}) + (\boldsymbol{R}\boldsymbol{S}\boldsymbol{v},\boldsymbol{v}) + (\boldsymbol{R}\boldsymbol{S}\boldsymbol{w},\boldsymbol{w}) \\ &= (\boldsymbol{S}\boldsymbol{u},\boldsymbol{R}^T\boldsymbol{u}) + (\boldsymbol{S}\boldsymbol{v},\boldsymbol{R}^T\boldsymbol{v}) + (\boldsymbol{S}\boldsymbol{w},\boldsymbol{R}^T\boldsymbol{w}).\end{aligned}$$

Substituting

$$\begin{aligned}\boldsymbol{R}^T\boldsymbol{u} &= (\boldsymbol{R}^T\boldsymbol{u},\boldsymbol{u})\boldsymbol{u} + (\boldsymbol{R}^T\boldsymbol{u},\boldsymbol{v})\boldsymbol{v} + (\boldsymbol{R}^T\boldsymbol{u},\boldsymbol{w})\boldsymbol{w} \\ &= (\boldsymbol{u},\boldsymbol{R}\boldsymbol{u})\boldsymbol{u} + (\boldsymbol{u},\boldsymbol{R}\boldsymbol{v})\boldsymbol{v} + (\boldsymbol{u},\boldsymbol{R}\boldsymbol{w})\boldsymbol{w},\end{aligned}$$

and similar expressions for $\boldsymbol{R}^T\boldsymbol{v}$ and $\boldsymbol{R}^T\boldsymbol{w}$, we get

$$\begin{aligned}\operatorname{tr}(\boldsymbol{R}\boldsymbol{S}) =&(\boldsymbol{u},\boldsymbol{R}\boldsymbol{u})(\boldsymbol{u},\boldsymbol{S}\boldsymbol{u}) + (\boldsymbol{u},\boldsymbol{R}\boldsymbol{v})(\boldsymbol{v},\boldsymbol{S}\boldsymbol{u}) + (\boldsymbol{u},\boldsymbol{R}\boldsymbol{w})(\boldsymbol{w},\boldsymbol{S}\boldsymbol{u}) + \\ &(\boldsymbol{v},\boldsymbol{R}\boldsymbol{u})(\boldsymbol{u},\boldsymbol{S}\boldsymbol{v}) + (\boldsymbol{v},\boldsymbol{R}\boldsymbol{v})(\boldsymbol{v},\boldsymbol{S}\boldsymbol{v}) + (\boldsymbol{v},\boldsymbol{R}\boldsymbol{w})(\boldsymbol{w},\boldsymbol{S}\boldsymbol{v}) +\end{aligned}$$

$$(w, Ru)(u, Sw) + (w, Rv)(v, Sw) + (w, Rw)(w, Sw).$$

Interchanging R and S in the above equation gives an expression for tr (SR), which is the same as that for tr (RS), thus yielding the desired result. The proof using indicial notation is left as an exercise (Problem 6).

Similar to the vector inner product given by Eqn. (1.7), we can define a tensor inner product of two second-order tensors R and S, denoted by $R : S$, by

$$(R, S) = R : S := \operatorname{tr}(R^T S) = \operatorname{tr}(RS^T) = \operatorname{tr}(SR^T) = \operatorname{tr}(S^T R) = R_{ij}S_{ij}. \tag{1.47}$$

The fact that $R : S$ satisfies the inner product conditions in Eqn. (1.1) can be easily verified. Thus, Lin equipped with the above inner product is an inner product space, and enjoys all the properties derived in Section 1.1. In particular, if the magnitude associated with the inner product is given by

$$|T| = (T : T)^{1/2},$$

then by the Cauchy–Schwartz inequality, we get

$$|(R, S)| \leq |R|\,|S|\,.$$

We have the following useful property:

$$R : (ST) = (S^T R) : T = (RT^T) : S = (TR^T) : S^T, \tag{1.48}$$

since

$$\begin{aligned} R : (ST) &= \operatorname{tr}(ST)^T R = \operatorname{tr} T^T (S^T R) = (S^T R) : T = (R^T S) : T^T \\ &= \operatorname{tr} S^T (RT^T) = (RT^T) : S = (TR^T) : S^T. \end{aligned}$$

From Eqn. (1.43c), it can be seen that

$$\begin{aligned} \det I &= 1, \\ \det(\alpha T) &= \alpha^3 \det T. \end{aligned}$$

It also follows from Eqn. (1.43c) that

$$\begin{aligned} \det(RS)\,[u, v, w] &= [RSu, RSv, RSw] \\ &= \det R\,[Su, Sv, Sw] \\ &= (\det R)(\det S)\,[u, v, w]\,. \end{aligned}$$

Choosing u, v and w to be linearly independent, the above equation leads us to the relation

$$\det(RS) = (\det R)(\det S), \tag{1.49}$$

from which we also conclude that $\det(RS) = \det(SR)$.

By choosing $u = e_j \times e_k = \epsilon_{ijk} e_i$, $v = e_j$ and $w = e_k$ in Eqn. (1.43c), so that $[u, v, w] = \epsilon_{ijk}\epsilon_{ijk} = 6$ (by Eqn. (1.19c)), and using the fact that $Te_i = T_{pi} e_p$, we get

$$\det T = \frac{1}{6}\epsilon_{ijk}\epsilon_{pqr} T_{pi} T_{qj} T_{rk} = \frac{1}{6}\epsilon_{ijk}\epsilon_{pqr} T_{ip} T_{jq} T_{kr}, \tag{1.50}$$

where the second relation is obtained by interchanging the dummy indices. From Eqn. (1.50), it is immediately obvious that (see Section 1.3.4 for a proof without using indicial notation)

$$\det \boldsymbol{T}^T = \det \boldsymbol{T}. \tag{1.51}$$

By choosing $\boldsymbol{u} = \boldsymbol{e}_p$, $\boldsymbol{v} = \boldsymbol{e}_q$ and $\boldsymbol{w} = \boldsymbol{e}_r$ in Eqn. (1.43c), and using Eqn. (1.51), we also have

$$\epsilon_{pqr}(\det \boldsymbol{T}) = \epsilon_{ijk} T_{ip} T_{jq} T_{kr} = \epsilon_{ijk} T_{pi} T_{qj} T_{rk}. \tag{1.52}$$

By choosing $(p,q,r) = (1,2,3)$ in the above equation, we get

$$\det \boldsymbol{T} = \epsilon_{ijk} T_{i1} T_{j2} T_{k3} = \epsilon_{ijk} T_{1i} T_{2j} T_{3k}.$$

Using Eqns. (1.16), (1.49), (1.51), (1.340) and (1.341), we have

$$\begin{aligned}
[\boldsymbol{u},\boldsymbol{v},\boldsymbol{w}]\,[\boldsymbol{p},\boldsymbol{q},\boldsymbol{r}] &= \{\det[\boldsymbol{e}_1 \otimes \boldsymbol{u} + \boldsymbol{e}_2 \otimes \boldsymbol{v} + \boldsymbol{e}_3 \otimes \boldsymbol{w}]\}\,\{\det[\boldsymbol{e}_1 \otimes \boldsymbol{p} + \boldsymbol{e}_2 \otimes \boldsymbol{q} + \boldsymbol{e}_3 \otimes \boldsymbol{r}]\} \\
&= \det\{[\boldsymbol{u} \otimes \boldsymbol{e}_1 + \boldsymbol{v} \otimes \boldsymbol{e}_2 + \boldsymbol{w} \otimes \boldsymbol{e}_3]\,[\boldsymbol{e}_1 \otimes \boldsymbol{p} + \boldsymbol{e}_2 \otimes \boldsymbol{q} + \boldsymbol{e}_3 \otimes \boldsymbol{r}]\} \\
&= \det[\boldsymbol{u} \otimes \boldsymbol{p} + \boldsymbol{v} \otimes \boldsymbol{q} + \boldsymbol{w} \otimes \boldsymbol{r}].
\end{aligned} \tag{1.53}$$

We now prove the following important theorem:

Theorem 1.3.2. *Given a tensor $\boldsymbol{T}$, there exists a nonzero vector $\boldsymbol{n}$ such that $\boldsymbol{T}\boldsymbol{n} = \boldsymbol{0}$ if and only if* $\det \boldsymbol{T} = 0$.

Proof. If $\det \boldsymbol{T} = 0$, then by Eqn. (1.43c),

$$[\boldsymbol{T}\boldsymbol{u}, \boldsymbol{T}\boldsymbol{v}, \boldsymbol{T}\boldsymbol{w}] = 0 \quad \forall \boldsymbol{u},\boldsymbol{v},\boldsymbol{w} \in V,$$

which, by Theorem 1.2.2, implies that $\boldsymbol{T}\boldsymbol{u}$, $\boldsymbol{T}\boldsymbol{v}$, $\boldsymbol{T}\boldsymbol{w}$ are linearly dependent. This means that there exist scalars α, β and γ, not all zero, such that

$$\alpha \boldsymbol{T}\boldsymbol{u} + \beta \boldsymbol{T}\boldsymbol{v} + \gamma \boldsymbol{T}\boldsymbol{w} = \boldsymbol{T}(\alpha \boldsymbol{u} + \beta \boldsymbol{v} + \gamma \boldsymbol{w}) = \boldsymbol{0}.$$

Thus, $\boldsymbol{T}\boldsymbol{n} = \boldsymbol{0}$, where $\boldsymbol{n} = \alpha \boldsymbol{u} + \beta \boldsymbol{v} + \gamma \boldsymbol{w}$.

Conversely, if there exists a nonzero vector $\boldsymbol{n}$ such that $\boldsymbol{T}\boldsymbol{n} = \boldsymbol{0}$, choose $\boldsymbol{v}$ and $\boldsymbol{w}$ such that $\boldsymbol{n}$,$\boldsymbol{v}$ and $\boldsymbol{w}$ are linearly independent, i.e., $[\boldsymbol{n},\boldsymbol{v},\boldsymbol{w}] \neq 0$. Then, by Eqn. (1.43c), we have

$$\det \boldsymbol{T}\,[\boldsymbol{n},\boldsymbol{v},\boldsymbol{w}] = [\boldsymbol{T}\boldsymbol{n}, \boldsymbol{T}\boldsymbol{v}, \boldsymbol{T}\boldsymbol{w}] = [\boldsymbol{0}, \boldsymbol{T}\boldsymbol{v}, \boldsymbol{T}\boldsymbol{w}] = 0,$$

which implies that $\det \boldsymbol{T} = 0$. □

1.3.3 Inverse of a tensor

In order to define the concept of a inverse of a tensor, it is convenient to first introduce the cofactor tensor, denoted by **cof** $\boldsymbol{T}$, and defined by the relation

$$\mathbf{cof}\,\boldsymbol{T}(\boldsymbol{u} \times \boldsymbol{v}) := \boldsymbol{T}\boldsymbol{u} \times \boldsymbol{T}\boldsymbol{v} \quad \forall \boldsymbol{u},\boldsymbol{v} \in V. \tag{1.54}$$

$\mathbf{cof}\,\boldsymbol{T}$ obviously maps a vector to a vector. To prove linearity, note that every nonzero vector $\boldsymbol{w}$ can be expressed as $\boldsymbol{u} \times \boldsymbol{v}$ for some nonzero $\boldsymbol{u}, \boldsymbol{v} \in V$ (e.g., take $\boldsymbol{v}$ as a unit vector perpendicular to $\boldsymbol{w}$, and $\boldsymbol{u} = \boldsymbol{v} \times \boldsymbol{w}$). We now show that if $\boldsymbol{w}_1$ and $\boldsymbol{w}_2$ are two arbitrary nonzero vectors, they can be expressed as $\boldsymbol{u}_1 \times \boldsymbol{v}$ and $\boldsymbol{u}_2 \times \boldsymbol{v}$. If $\boldsymbol{w}_1$ and $\boldsymbol{w}_2$ are linearly dependent, then $\boldsymbol{w}_2 = \alpha \boldsymbol{w}_1$, $\alpha \in \Re$, and then $\boldsymbol{u}_2 = \alpha \boldsymbol{u}_1$. If $\boldsymbol{w}_1$ and $\boldsymbol{w}_2$ are linearly independent, then let $\boldsymbol{v} = (\boldsymbol{w}_1 \times \boldsymbol{w}_2)/\left|\boldsymbol{w}_1 \times \boldsymbol{w}_2\right|^2$. Using Eqn. (1.21a), we can show that $\boldsymbol{u}_1 = -(\boldsymbol{w}_1 \cdot \boldsymbol{w}_2)\boldsymbol{w}_1 + \left|\boldsymbol{w}_1\right|^2 \boldsymbol{w}_2$ and $\boldsymbol{u}_2 = -\left|\boldsymbol{w}_2\right|^2 \boldsymbol{w}_1 + (\boldsymbol{w}_1 \cdot \boldsymbol{w}_2)\boldsymbol{w}_2$ yield $\boldsymbol{u}_1 \times \boldsymbol{v} = \boldsymbol{w}_1$ and $\boldsymbol{u}_2 \times \boldsymbol{v} = \boldsymbol{w}_2$. Noting that $\boldsymbol{T} \in \text{Lin}$, we now have

$$\begin{aligned}
\mathbf{cof}\,\boldsymbol{T}(\alpha \boldsymbol{w}_1 + \beta \boldsymbol{w}_2) &= \mathbf{cof}\,\boldsymbol{T}[(\alpha \boldsymbol{u}_1 + \beta \boldsymbol{u}_2) \times \boldsymbol{v}] \\
&= [\boldsymbol{T}(\alpha \boldsymbol{u}_1 + \beta \boldsymbol{u}_2)] \times \boldsymbol{T}\boldsymbol{v} \\
&= [\alpha \boldsymbol{T}\boldsymbol{u}_1 + \beta \boldsymbol{T}\boldsymbol{u}_2] \times \boldsymbol{T}\boldsymbol{v} \\
&= \alpha(\boldsymbol{T}\boldsymbol{u}_1 \times \boldsymbol{T}\boldsymbol{v}) + \beta(\boldsymbol{T}\boldsymbol{u}_2 \times \boldsymbol{T}\boldsymbol{v}) \\
&= \alpha \mathbf{cof}\,\boldsymbol{T}(\boldsymbol{u}_1 \times \boldsymbol{v}) + \beta \mathbf{cof}\,\boldsymbol{T}(\boldsymbol{u}_2 \times \boldsymbol{v}) \\
&= \alpha \mathbf{cof}\,\boldsymbol{T}\,\boldsymbol{w}_1 + \beta \mathbf{cof}\,\boldsymbol{T}\,\boldsymbol{w}_2,
\end{aligned}$$

thus proving that $\mathbf{cof}\,\boldsymbol{T} \in \text{Lin}$.

Using Eqn. (1.54), we now prove the following explicit formula for the cofactor:

$$(\mathbf{cof}\,\boldsymbol{T})^T = I_2 \boldsymbol{I} - (\text{tr}\,\boldsymbol{T})\boldsymbol{T} + \boldsymbol{T}^2. \tag{1.55}$$

Replacing $\boldsymbol{u}$ in Eqn. (1.43a) by $\boldsymbol{T}\boldsymbol{u}$, we get

$$\text{tr}\,\boldsymbol{T}\,[\boldsymbol{T}\boldsymbol{u}, \boldsymbol{v}, \boldsymbol{w}] = \left[\boldsymbol{T}^2\boldsymbol{u}, \boldsymbol{v}, \boldsymbol{w}\right] + [\boldsymbol{T}\boldsymbol{u}, \boldsymbol{T}\boldsymbol{v}, \boldsymbol{w}] + [\boldsymbol{T}\boldsymbol{u}, \boldsymbol{v}, \boldsymbol{T}\boldsymbol{w}],$$

which on rearranging yields

$$[\boldsymbol{T}\boldsymbol{u}, \boldsymbol{T}\boldsymbol{v}, \boldsymbol{w}] + [\boldsymbol{T}\boldsymbol{u}, \boldsymbol{v}, \boldsymbol{T}\boldsymbol{w}] = \text{tr}\,\boldsymbol{T}\,[\boldsymbol{T}\boldsymbol{u}, \boldsymbol{v}, \boldsymbol{w}] - \left[\boldsymbol{T}^2\boldsymbol{u}, \boldsymbol{v}, \boldsymbol{w}\right]. \tag{1.56}$$

Using Eqn. (1.43b) and (1.54), we have

$$I_2\,[\boldsymbol{u}, \boldsymbol{v}, \boldsymbol{w}] = [\boldsymbol{T}\boldsymbol{u}, \boldsymbol{T}\boldsymbol{v}, \boldsymbol{w}] + [\boldsymbol{T}\boldsymbol{u}, \boldsymbol{v}, \boldsymbol{T}\boldsymbol{w}] + \left[(\mathbf{cof}\,\boldsymbol{T})^T\boldsymbol{u}, \boldsymbol{v}, \boldsymbol{w}\right]. \tag{1.57}$$

Substituting Eqn. (1.56) into (1.57), using the fact that $\boldsymbol{u}$, $\boldsymbol{v}$ and $\boldsymbol{w}$ are arbitrary, and invoking equality of tensors, we get Eqn. (1.55).

It immediately follows from Eqn. (1.55) that $\mathbf{cof}\,\boldsymbol{T}$ corresponding to a given $\boldsymbol{T}$ is unique. We also observe that

$$\mathbf{cof}\,\boldsymbol{T}^T = (\mathbf{cof}\,\boldsymbol{T})^T, \tag{1.58}$$

and that

$$(\mathbf{cof}\,\boldsymbol{T})^T\boldsymbol{T} = \boldsymbol{T}(\mathbf{cof}\,\boldsymbol{T})^T. \tag{1.59}$$

The components of the cofactor tensor are given by

$$(\mathbf{cof}\,\boldsymbol{T})_{ij} = \frac{1}{2}\epsilon_{imn}\epsilon_{jpq}T_{mp}T_{nq}. \tag{1.60}$$

To prove this, let $\boldsymbol{u} = \boldsymbol{e}_q \times \boldsymbol{e}_j$ and $\boldsymbol{v} = \boldsymbol{e}_q$ in Eqn. (1.54), and use Eqn. (1.21b) to get $\boldsymbol{u} \times \boldsymbol{v} = 2\boldsymbol{e}_j$. On the right-hand side of Eqn. (1.54), use the relations $\boldsymbol{e}_q \times \boldsymbol{e}_j = \epsilon_{pqj}\boldsymbol{e}_p$, $\boldsymbol{T}\boldsymbol{e}_p = T_{mp}\boldsymbol{e}_m$ and $\boldsymbol{T}\boldsymbol{e}_q = T_{nq}\boldsymbol{e}_n$. Taking the dot product with $\boldsymbol{e}_i$ of both sides of the relation so obtained, we get the desired result. Equation (1.60) when written out explicitly reads

$$[\mathbf{cof}\,\boldsymbol{T}] = \begin{bmatrix} T_{22}T_{33} - T_{23}T_{32} & T_{23}T_{31} - T_{21}T_{33} & T_{21}T_{32} - T_{22}T_{31} \\ T_{32}T_{13} - T_{33}T_{12} & T_{33}T_{11} - T_{31}T_{13} & T_{31}T_{12} - T_{32}T_{11} \\ T_{12}T_{23} - T_{13}T_{22} & T_{13}T_{21} - T_{11}T_{23} & T_{11}T_{22} - T_{12}T_{21} \end{bmatrix}. \tag{1.61}$$

Equation (1.54) can be used to get a simple expression for I_2, the second invariant of a tensor. We first write Eqn. (1.43b) as

$$I_2\,[\boldsymbol{u}, \boldsymbol{v}, \boldsymbol{w}] = [\boldsymbol{u}, \boldsymbol{T}\boldsymbol{v}, \boldsymbol{T}\boldsymbol{w}] + [\boldsymbol{v}, \boldsymbol{T}\boldsymbol{w}, \boldsymbol{T}\boldsymbol{u}] + [\boldsymbol{w}, \boldsymbol{T}\boldsymbol{u}, \boldsymbol{T}\boldsymbol{v}],$$

and then use Eqn. (1.54) to write the above equation as

$$\begin{aligned} I_2\,[\boldsymbol{u}, \boldsymbol{v}, \boldsymbol{w}] &= \left[(\mathbf{cof}\,\boldsymbol{T})^T\boldsymbol{u}, \boldsymbol{v}, \boldsymbol{w}\right] + \left[(\mathbf{cof}\,\boldsymbol{T})^T\boldsymbol{v}, \boldsymbol{w}, \boldsymbol{u}\right] + \left[(\mathbf{cof}\,\boldsymbol{T})^T\boldsymbol{w}, \boldsymbol{u}, \boldsymbol{v}\right] \\ &= \left[(\mathbf{cof}\,\boldsymbol{T})^T\boldsymbol{u}, \boldsymbol{v}, \boldsymbol{w}\right] + \left[\boldsymbol{u}, (\mathbf{cof}\,\boldsymbol{T})^T\boldsymbol{v}, \boldsymbol{w}\right] + \left[\boldsymbol{u}, \boldsymbol{v}, (\mathbf{cof}\,\boldsymbol{T})^T\boldsymbol{w}\right] \\ &= \operatorname{tr}(\mathbf{cof}\,\boldsymbol{T})^T\,[\boldsymbol{u}, \boldsymbol{v}, \boldsymbol{w}], \end{aligned}$$

where the last step follows from Eqn. (1.43a). Using Eqn. (1.45), and choosing $\boldsymbol{u}$, $\boldsymbol{v}$ and $\boldsymbol{w}$ to be linearly independent (so that $[\boldsymbol{u}, \boldsymbol{v}, \boldsymbol{w}] \neq 0$), we get

$$I_2 = \operatorname{tr}(\mathbf{cof}\,\boldsymbol{T}). \tag{1.62}$$

An alternative expression for the I_2 can be found by directly taking the trace of both sides of Eqn. (1.55). Using Eqns. (1.45) and (1.62), and the fact that the trace is a linear operator, we get

$$I_2 = 3I_2 - (\operatorname{tr}\boldsymbol{T})^2 + \operatorname{tr}\boldsymbol{T}^2,$$

which implies that

$$I_2 = \frac{1}{2}\left[(\operatorname{tr}\boldsymbol{T})^2 - \operatorname{tr}\boldsymbol{T}^2\right]. \tag{1.63}$$

The equivalence of the two formulae for I_2 given by Eqns. (1.62) and (1.63) can also be shown using indicial notation (see Problem 10).

Substituting Eqn. (1.54) into Eqn. (1.43c), we get

$$(\boldsymbol{T}\boldsymbol{u}, \mathbf{cof}\,\boldsymbol{T}(\boldsymbol{v} \times \boldsymbol{w})) = \det\boldsymbol{T}(\boldsymbol{I}\boldsymbol{u}, \boldsymbol{v} \times \boldsymbol{w}) \quad \forall \boldsymbol{u}, \boldsymbol{v}, \boldsymbol{w} \in V,$$

which implies that

$$((\mathbf{cof}\,\boldsymbol{T})^T\boldsymbol{T}\boldsymbol{u}, \boldsymbol{v} \times \boldsymbol{w}) = ((\det\boldsymbol{T})\boldsymbol{I}\boldsymbol{u}, \boldsymbol{v} \times \boldsymbol{w}) \quad \forall \boldsymbol{u}, \boldsymbol{v}, \boldsymbol{w} \in V.$$

Any arbitrary vector $\boldsymbol{z}$ can be expressed as $\boldsymbol{v} \times \boldsymbol{w}$ for some $\boldsymbol{v}, \boldsymbol{w} \in V$ (take $\boldsymbol{w}$ as a unit vector perpendicular to $\boldsymbol{z}$, and $\boldsymbol{v} = \boldsymbol{w} \times \boldsymbol{z}$), so that from Eqn. (1.24), it follows that $(\mathbf{cof}\,\boldsymbol{T})^T\boldsymbol{T} = (\det\boldsymbol{T})\boldsymbol{I}$, which when combined with Eqn. (1.59) yields

$$\boldsymbol{T}(\mathbf{cof}\,\boldsymbol{T})^T = (\mathbf{cof}\,\boldsymbol{T})^T\boldsymbol{T} = (\det\boldsymbol{T})\boldsymbol{I}. \tag{1.64}$$

Similar to the result for determinants, the cofactor of the product of two tensors is the product of the cofactors of the tensors (see also Problem 9), i.e.,

$$\mathbf{cof}\,(\boldsymbol{RS}) = (\mathbf{cof}\,\boldsymbol{R})(\mathbf{cof}\,\boldsymbol{S}).$$

This is because

$$\begin{aligned}
[\mathbf{cof}\,\boldsymbol{R}\,\mathbf{cof}\,\boldsymbol{S}]^T\boldsymbol{RS} &= (\mathbf{cof}\,\boldsymbol{S})^T(\mathbf{cof}\,\boldsymbol{R})^T\boldsymbol{RS} \\
&= (\det\boldsymbol{R})(\mathbf{cof}\,\boldsymbol{S})^T\boldsymbol{IS} && \text{(by Eqn. (1.64))} \\
&= (\det\boldsymbol{R})(\det\boldsymbol{S})\boldsymbol{I} && \text{(by Eqn. (1.64))} \\
&= \det(\boldsymbol{RS})\boldsymbol{I}, && \text{(by Eqn. (1.49))}
\end{aligned}$$

and since the cofactor matrix is unique, the result follows.

The inverse of a second-order tensor $\boldsymbol{T}$, denoted by $\boldsymbol{T}^{-1}$, is defined by

$$\boldsymbol{T}^{-1}\boldsymbol{T} = \boldsymbol{I}, \tag{1.65}$$

where $\boldsymbol{I}$ is the identity tensor. A characterization of an invertible tensor is the following:

Theorem 1.3.3. *A tensor $\boldsymbol{T}$ is invertible if and only if* $\det\boldsymbol{T} \neq 0$. *The inverse, if it exists, is unique.*

Proof. Assuming $\boldsymbol{T}^{-1}$ exists, from Eqns. (1.49) and (1.65), we have $(\det\boldsymbol{T})(\det\boldsymbol{T}^{-1}) = 1$, and hence $\det\boldsymbol{T} \neq 0$.

Conversely, if $\det\boldsymbol{T} \neq 0$, then from Eqn. (1.64), we see that at least one inverse exists, and is given by

$$\boldsymbol{T}^{-1} = \frac{1}{\det\boldsymbol{T}}(\mathbf{cof}\,\boldsymbol{T})^T. \tag{1.66}$$

Let $\boldsymbol{T}_1^{-1}$ and $\boldsymbol{T}_2^{-1}$ be two inverses that satisfy $\boldsymbol{T}_1^{-1}\boldsymbol{T} = \boldsymbol{T}_2^{-1}\boldsymbol{T} = \boldsymbol{I}$, from which it follows that $(\boldsymbol{T}_1^{-1} - \boldsymbol{T}_2^{-1})\boldsymbol{T} = \boldsymbol{0}$. Choose $\boldsymbol{T}_2^{-1}$ to be given by the expression in Eqn. (1.66) so that, by virtue of Eqn. (1.64), we also have $\boldsymbol{T}\boldsymbol{T}_2^{-1} = \boldsymbol{I}$. Multiplying both sides of $(\boldsymbol{T}_1^{-1} - \boldsymbol{T}_2^{-1})\boldsymbol{T} = \boldsymbol{0}$ by $\boldsymbol{T}_2^{-1}$, we get $\boldsymbol{T}_1^{-1} = \boldsymbol{T}_2^{-1}$, which establishes the uniqueness of $\boldsymbol{T}^{-1}$. □

From Eqns. (1.64) and (1.66), we have

$$\boldsymbol{T}^{-1}\boldsymbol{T} = \boldsymbol{T}\boldsymbol{T}^{-1} = \boldsymbol{I}. \tag{1.67}$$

Another characterization of an invertible tensor is the following:

Theorem 1.3.4. *A tensor $\boldsymbol{T}$ is invertible if and only if the equation*

$$\boldsymbol{Tu} = \boldsymbol{v},$$

has a unique solution $\boldsymbol{u} \in V$ for each $\boldsymbol{v} \in V$.

Proof. If T is invertible, then multiplying $v = Tu$ by T^{-1}, we get

$$\begin{aligned} T^{-1}v &= T^{-1}Tu \\ &= Iu \\ &= u \end{aligned}$$

Since T^{-1} is unique, u is unique.

To prove the converse, note that if u_1 and u_2 are two solutions of the equation $Tu = v$, then, by the assumed uniqueness, $u_1 = u_2$. Also, by assumption, for every $v \in V$, there exists a u such that $Tu = v$. Thus, the mapping $T : V \to V$ is one-to-one and onto, and hence, $T^{-1} : V \to V$ exists. □

Summarizing, if T is invertible, we have

$$Tu = v \iff u = T^{-1}v, \quad u, v \in V.$$

By the above property, T^{-1} clearly maps vectors to vectors. Hence, to prove that T^{-1} is a second-order tensor, we just need to prove linearity. Let $a, b \in V$ be two arbitrary vectors, and let $u = T^{-1}a$ and $v = T^{-1}b$. Since $I = T^{-1}T$, we have

$$\begin{aligned} I(\alpha u + \beta v) &= T^{-1}T(\alpha u + \beta v) \\ &= T^{-1}[T(\alpha u + \beta v)] \\ &= T^{-1}[\alpha Tu + \beta Tv] \\ &= T^{-1}(\alpha a + \beta b), \end{aligned}$$

which implies that

$$T^{-1}(\alpha a + \beta b) = \alpha T^{-1}a + \beta T^{-1}b \quad \forall a, b \in V \text{ and } \alpha, \beta \in \Re.$$

The inverse of the product of two invertible tensors R and S is

$$(RS)^{-1} = S^{-1}R^{-1}, \tag{1.68}$$

since the inverse is unique, and

$$S^{-1}R^{-1}RS = S^{-1}IS = S^{-1}S = I.$$

Similarly, if T is invertible, then T^T is invertible since $\det T^T = \det T \neq 0$. The inverse of the transpose is given by

$$(T^T)^{-1} = (T^{-1})^T, \tag{1.69}$$

since

$$(T^{-1})^T T^T = (TT^{-1})^T = I^T = I.$$

Hence, without fear of ambiguity, we can write

$$T^{-T} := (T^T)^{-1} = (T^{-1})^T.$$

From Eqn. (1.69), it follows that if $T \in \text{Sym}$, then $T^{-1} \in \text{Sym}$.

Although, no easy expression for $(T_1 + T_2)^{-1}$ is known, the following Woodbury formula gives an expression for a particular kind of perturbation on an invertible tensor T, and which holds for any underlying space dimension n:

$$(T + UCV)^{-1} = T^{-1} - T^{-1}U(C^{-1} + VT^{-1}U)^{-1}VT^{-1}. \tag{1.70}$$

To prove this formula, consider the inversion of the matrix $\begin{bmatrix} T & -U \\ V & C^{-1} \end{bmatrix}$. If $\begin{bmatrix} A_1 & B_1 \\ C_1 & D_1 \end{bmatrix}$ represents the inverse of this matrix, then the condition

$$\begin{bmatrix} A_1 & B_1 \\ C_1 & D_1 \end{bmatrix} \begin{bmatrix} T & -U \\ V & C^{-1} \end{bmatrix} = \begin{bmatrix} I & 0 \\ 0 & I \end{bmatrix},$$

leads to the equations

$$A_1 T + B_1 V = I, \tag{1.71a}$$

$$-A_1 U + B_1 C^{-1} = 0. \tag{1.71b}$$

From Eqn. (1.71b), we get $B_1 = A_1 UC$, which on substituting into Eqn. (1.71a) yields

$$A_1 = (T + UCV)^{-1}. \tag{1.72}$$

On the other hand, from Eqn. (1.71a), we have

$$A_1 = T^{-1} - B_1 V T^{-1}, \tag{1.73}$$

which on substituting into Eqn. (1.71b) yields

$$B_1 = T^{-1}U(C^{-1} + VT^{-1}U)^{-1}.$$

Substituting this expression into Eqn. (1.73) yields

$$A_1 = T^{-1} - T^{-1}U(C^{-1} + VT^{-1}U)^{-1}VT^{-1}. \tag{1.74}$$

Comparing Eqns. (1.72) and (1.74), we get Eqn. (1.70). If $C = 1$, then we get the following Sherman–Morrison formula:

$$(T + u \otimes v)^{-1} = T^{-1} - \frac{T^{-1}(u \otimes v)T^{-1}}{1 + v \cdot T^{-1}u}. \tag{1.75}$$

$1 + v \cdot T^{-1}u$ should be nonzero for $T + u \otimes v$ to be invertible.

Let $S \in \text{Sym}$ be an invertible tensor, and let $\tilde{u} = S^{-1}u$ and $\tilde{v} = S^{-1}v$. Note that because of the symmetry of S, we have $u \cdot \tilde{v} = \tilde{u} \cdot v$. By applying Eqn. (1.75) twice, we get

$$(S + u \otimes v + v \otimes u)^{-1} = S^{-1} + \frac{1}{D}\left[(u \cdot \tilde{u})\tilde{v} \otimes \tilde{v} + (v \cdot \tilde{v})\tilde{u} \otimes \tilde{u} - (1 + u \cdot \tilde{v})(\tilde{u} \otimes \tilde{v} + \tilde{v} \otimes \tilde{u})\right],$$

where

$$D = (1 + u \cdot \tilde{v})^2 - (u \cdot \tilde{u})(v \cdot \tilde{v}).$$

D should be nonzero for $\boldsymbol{S}+\boldsymbol{u}\otimes\boldsymbol{v}+\boldsymbol{v}\otimes\boldsymbol{u}$ to be invertible.

As a corollary, for $\boldsymbol{S}=\boldsymbol{I}$, we get

$$(\boldsymbol{I}+\boldsymbol{u}\otimes\boldsymbol{v}+\boldsymbol{v}\otimes\boldsymbol{u})^{-1}=\boldsymbol{I}+\frac{1}{D}\left[|\boldsymbol{u}|^2\,\boldsymbol{v}\otimes\boldsymbol{v}+|\boldsymbol{v}|^2\,\boldsymbol{u}\otimes\boldsymbol{u}-(1+\boldsymbol{u}\cdot\boldsymbol{v})(\boldsymbol{u}\otimes\boldsymbol{v}+\boldsymbol{v}\otimes\boldsymbol{u})\right],$$

where

$$D=(1+\boldsymbol{u}\cdot\boldsymbol{v})^2-|\boldsymbol{u}|^2\,|\boldsymbol{v}|^2.$$

Consider the tensor

$$\boldsymbol{H}=\boldsymbol{I}+\boldsymbol{u}\otimes\boldsymbol{v}, \tag{1.76}$$

and let the underlying space dimension be n. Let $\{\boldsymbol{e}_1^*,\boldsymbol{e}_2^*,\ldots,\boldsymbol{e}_{n-1}^*\}$ be $n-1$ unit vectors that are perpendicular to $\boldsymbol{v}$, and mutually perpendicular to each other. Then from Eqn. (1.76), it is immediately evident that $(1,\boldsymbol{e}_1^*)$, $(1,\boldsymbol{e}_2^*),\ldots,(1,\boldsymbol{e}_{n-1}^*)$, $(1+\boldsymbol{u}\cdot\boldsymbol{v},\boldsymbol{u}/\,|\boldsymbol{u}|)$ are eigenvalues/eigenvectors of $\boldsymbol{H}$. If $\boldsymbol{u}\cdot\boldsymbol{v}\neq 0$, then all the eigenvectors are linearly independent, while if $\boldsymbol{u}\cdot\boldsymbol{v}=0$, then the first $n-1$ eigenvectors are linearly independent, while the last can be expressed in terms of them. The characteristic equation is

$$(\lambda-1)^{n-1}(\lambda-1-\boldsymbol{u}\cdot\boldsymbol{v})=0.$$

On applying the Cayley–Hamilton theorem (see Theorem 1.3.5), we get

$$[\boldsymbol{H}-\boldsymbol{I}]^{n-1}[\boldsymbol{H}-(1+\boldsymbol{u}\cdot\boldsymbol{v})\boldsymbol{I}]=\boldsymbol{0}.$$

The minimal polynomial is

$$[\boldsymbol{H}-\boldsymbol{I}][\boldsymbol{H}-(1+\boldsymbol{u}\cdot\boldsymbol{v})\boldsymbol{I}]=\boldsymbol{0},$$

which can be immediately verified by using Eqn. (1.76). Multiplying the eigenvalues of $\boldsymbol{H}$, we get

$$\det\boldsymbol{H}=1+\boldsymbol{u}\cdot\boldsymbol{v}.$$

Now consider the tensor that occurs in Eqn. (1.75), which is a generalization of the tensor $\boldsymbol{H}$, namely,

$$\boldsymbol{M}=\boldsymbol{T}+\boldsymbol{u}\otimes\boldsymbol{v},$$

which can be written as $\boldsymbol{T}[\boldsymbol{I}+(\boldsymbol{T}^{-1}\boldsymbol{u})\otimes\boldsymbol{v}]$, assuming $\boldsymbol{T}$ to be invertible. Thus,

$$\det\boldsymbol{M}=\det\boldsymbol{T}[1+(\boldsymbol{T}^{-1}\boldsymbol{u})\cdot\boldsymbol{v}].$$

The result for arbitrary $\boldsymbol{T}$ is obtained by taking the limit in the above equation. We get

$$\det\boldsymbol{M}=\det\boldsymbol{T}+\boldsymbol{u}\cdot[(\mathbf{cof}\,\boldsymbol{T})\boldsymbol{v}],$$

with $\mathbf{cof}\,\boldsymbol{T}$ given by Eqn. (J.12).

Consider a symmetric perturbation to $\boldsymbol{I}$:

$$\boldsymbol{K}=\boldsymbol{I}+\boldsymbol{u}\otimes\boldsymbol{v}+\boldsymbol{v}\otimes\boldsymbol{u}.$$

Let $\{e_1^*, e_2^*, \ldots, e_{n-2}^*\}$ be $n-2$ unit vectors that are perpendicular to $\boldsymbol{u}$ and $\boldsymbol{v}$. Then $(1, e_1^*)$, $(1, e_2^*), \ldots, (1, e_{n-2}^*)$, $(1 + \boldsymbol{u}\cdot\boldsymbol{v} + |\boldsymbol{u}|\,|\boldsymbol{v}|, \boldsymbol{u}/\,|\boldsymbol{u}| + \boldsymbol{v}/\,|\boldsymbol{v}|)$, $(1 + \boldsymbol{u}\cdot\boldsymbol{v} - |\boldsymbol{u}|\,|\boldsymbol{v}|, \boldsymbol{u}/\,|\boldsymbol{u}| - \boldsymbol{v}/\,|\boldsymbol{v}|)$ are the eigenvalues/eigenvectors of $\boldsymbol{K}$, so that

$$\det \boldsymbol{K} = (1 + \boldsymbol{u}\cdot\boldsymbol{v} + |\boldsymbol{u}|\,|\boldsymbol{v}|)(1 + \boldsymbol{u}\cdot\boldsymbol{v} - |\boldsymbol{u}|\,|\boldsymbol{v}|) = (1 + \boldsymbol{u}\cdot\boldsymbol{v})^2 - |\boldsymbol{u}|^2\,|\boldsymbol{v}|^2. \tag{1.77}$$

By virtue of the Cauchy–Schwartz inequality,

$$\det \boldsymbol{K} \le 1 + 2\boldsymbol{u}\cdot\boldsymbol{v}.$$

Let $\boldsymbol{S}$ be a symmetric positive definite tensor, and let $\boldsymbol{U} = \sqrt{\boldsymbol{S}}$ be its symmetric positive definite square-root. If $\boldsymbol{Z}$ is defined by

$$\boldsymbol{Z} := \boldsymbol{S} + \boldsymbol{u}\otimes\boldsymbol{v} + \boldsymbol{v}\otimes\boldsymbol{u},$$

then it can be written as $\boldsymbol{U}[\boldsymbol{I} + (\boldsymbol{U}^{-1}\boldsymbol{u})\otimes(\boldsymbol{U}^{-1}\boldsymbol{v}) + (\boldsymbol{U}^{-1}\boldsymbol{v})\otimes(\boldsymbol{U}^{-1}\boldsymbol{u})]\boldsymbol{U}$, so that on using Eqn. (1.77), we get

$$\det \boldsymbol{Z} = \det \boldsymbol{S}[\left(1 + (\boldsymbol{S}^{-1}\boldsymbol{u})\cdot\boldsymbol{v}\right)^2 - (\boldsymbol{u}\cdot\boldsymbol{S}^{-1}\boldsymbol{u})(\boldsymbol{v}\cdot\boldsymbol{S}^{-1}\boldsymbol{v})].$$

Although our proof assumed $\boldsymbol{S}$ to be positive definite, actually, the above result holds for any invertible $\boldsymbol{S} \in \mathrm{Sym}$.

1.3.4 Eigenvalues and eigenvectors of tensors

If $\boldsymbol{T}$ is an arbitrary tensor, a vector $\boldsymbol{n}$ is said to be an eigenvector of $\boldsymbol{T}$ if there exists λ such that

$$\boldsymbol{T}\boldsymbol{n} = \lambda\boldsymbol{n}. \tag{1.78}$$

Writing the above equation as $(\boldsymbol{T} - \lambda\boldsymbol{I})\boldsymbol{n} = \boldsymbol{0}$, we see from Theorem 1.3.2 that a nontrivial eigenvector $\boldsymbol{n}$ exists if and only if

$$\det(\boldsymbol{T} - \lambda\boldsymbol{I}) = 0.$$

This is known as the *characteristic* equation of $\boldsymbol{T}$. Using Eqn. (1.43c), the characteristic equation can be written as

$$[\boldsymbol{T}\boldsymbol{u} - \lambda\boldsymbol{u}, \boldsymbol{T}\boldsymbol{v} - \lambda\boldsymbol{v}, \boldsymbol{T}\boldsymbol{w} - \lambda\boldsymbol{w}] = 0,$$

for linearly independent vectors $\boldsymbol{u}$, $\boldsymbol{v}$ and $\boldsymbol{w}$. Using Eqns. (1.14) and (1.15), we can write the characteristic equation as

$$\lambda^3 - I_1\lambda^2 + I_2\lambda - I_3 = 0, \tag{1.79}$$

where I_1, I_2 and I_3 are the principal invariants given by Eqns. (1.43). Since the principal invariants are real, Eqn. (1.79) has either one or three real roots. If one of the eigenvalues is complex, then it follows from Eqn. (1.78) that the corresponding eigenvector is also complex. By taking the complex conjugate of both sides of Eqn. (1.78), we see that the complex conjugate of the complex eigenvalue, and the corresponding complex eigenvector are also eigenvalues and eigenvectors, respectively. Thus, eigenvalues and eigenvectors, if complex, occur in complex conjugate pairs. If $\lambda_1, \lambda_2, \lambda_3$ are the roots of the characteristic equation, then from Eqn. (1.79), it follows that

$$\begin{aligned} I_1 &= \operatorname{tr} \boldsymbol{T} = T_{11} + T_{22} + T_{33}, \\ &= \lambda_1 + \lambda_2 + \lambda_3, \end{aligned} \tag{1.80a}$$

$$\begin{aligned} I_2 &= \operatorname{tr} \mathbf{cof}\, \boldsymbol{T} = \frac{1}{2}\left[(\operatorname{tr} \boldsymbol{T})^2 - \operatorname{tr}(\boldsymbol{T}^2)\right] = \begin{vmatrix} T_{11} & T_{12} \\ T_{21} & T_{22} \end{vmatrix} + \begin{vmatrix} T_{22} & T_{23} \\ T_{32} & T_{33} \end{vmatrix} + \begin{vmatrix} T_{11} & T_{13} \\ T_{31} & T_{33} \end{vmatrix} \\ &= \lambda_1\lambda_2 + \lambda_2\lambda_3 + \lambda_1\lambda_3, \end{aligned} \tag{1.80b}$$

$$\begin{aligned} I_3 &= \det(\boldsymbol{T}) = \frac{1}{6}\left[(\operatorname{tr} \boldsymbol{T})^3 - 3(\operatorname{tr} \boldsymbol{T})(\operatorname{tr} \boldsymbol{T}^2) + 2\operatorname{tr} \boldsymbol{T}^3\right] = \epsilon_{ijk} T_{i1} T_{j2} T_{k3} \\ &= \lambda_1\lambda_2\lambda_3, \end{aligned} \tag{1.80c}$$

where $|.|$ denotes the determinant. The set of eigenvalues $\{\lambda_1, \lambda_2, \lambda_3\}$ is known as the spectrum of $\boldsymbol{T}$.

If λ is an eigenvalue, and $\boldsymbol{n}$ is the associated eigenvector of $\boldsymbol{T}$, then λ^2 is the eigenvalue of $\boldsymbol{T}^2$, and $\boldsymbol{n}$ is the associated eigenvector, since

$$\boldsymbol{T}^2\boldsymbol{n} = \boldsymbol{T}(\boldsymbol{T}\boldsymbol{n}) = \boldsymbol{T}(\lambda\boldsymbol{n}) = \lambda\boldsymbol{T}\boldsymbol{n} = \lambda^2\boldsymbol{n}.$$

In general, λ^n is an eigenvalue of $\boldsymbol{T}^n$ with associated eigenvector $\boldsymbol{n}$. The eigenvalues of $\boldsymbol{T}^T$ and $\boldsymbol{T}$ are the same since their characteristic equations are the same.

An extremely important result is the following:

Theorem 1.3.5 (Cayley–Hamilton Theorem). *A tensor $\boldsymbol{T}$ satisfies an equation having the same form as its characteristic equation, i.e.,*

$$\boldsymbol{T}^3 - I_1\boldsymbol{T}^2 + I_2\boldsymbol{T} - I_3\boldsymbol{I} = \boldsymbol{0} \quad \forall \boldsymbol{T}. \tag{1.81}$$

Proof. Multiplying Eqn. (1.55) by $\boldsymbol{T}$, we get

$$(\mathbf{cof}\, \boldsymbol{T})^T\boldsymbol{T} = I_2\boldsymbol{T} - I_1\boldsymbol{T}^2 + \boldsymbol{T}^3.$$

Since by Eqn. (1.64), $(\mathbf{cof}\, \boldsymbol{T})^T\boldsymbol{T} = (\det \boldsymbol{T})\boldsymbol{I} = I_3\boldsymbol{I}$, the result follows. □

By taking the trace of both sides of Eqn. (1.81), and using Eqn. (1.63), we get

$$\det \boldsymbol{T} = \frac{1}{6}\left[(\operatorname{tr} \boldsymbol{T})^3 - 3(\operatorname{tr} \boldsymbol{T})(\operatorname{tr} \boldsymbol{T}^2) + 2\operatorname{tr} \boldsymbol{T}^3\right]. \tag{1.82}$$

From the above expression and the properties of the trace operator, Eqn. (1.51) follows.

We have

$$\lambda_i = 0,\ i = 1,2,3 \iff \mathcal{I}_T = \boldsymbol{0} \iff \operatorname{tr}(\boldsymbol{T}) = \operatorname{tr}(\boldsymbol{T}^2) = \operatorname{tr}(\boldsymbol{T}^3) = 0. \tag{1.83}$$

The proof is as follows. If all the invariants are zero, then from the characteristic equation given by Eqn. (1.79), it follows that all the eigenvalues are zero. If all the eigenvalues are zero, then from Eqns. (1.80a)–(1.80c), it follows that all the principal invariants $\mathcal{I}_T$ are zero.

If $\mathrm{tr}\,(\boldsymbol{T}) = \mathrm{tr}\,(\boldsymbol{T}^2) = \mathrm{tr}\,(\boldsymbol{T}^3) = 0$, then again from Eqns. (1.80a)–(1.80c) it follows that the principal invariants are zero. Conversely, if all the principal invariants are zero, then all the eigenvalues are zero from which it follows that $\mathrm{tr}\,\boldsymbol{T}^j = \sum_{i=1}^{3} \lambda_i^j$, $j = 1,2,3$ are zero.

Consider the second-order tensor $\boldsymbol{u} \otimes \boldsymbol{v}$. By Eqn. (1.60) or by writing $\boldsymbol{u} \otimes \boldsymbol{v}$ as $\boldsymbol{u} \otimes \boldsymbol{v} + \boldsymbol{0} \otimes \boldsymbol{0} + \boldsymbol{0} \otimes \boldsymbol{0}$ and using Eqn. (1.333), it follows that

$$\mathbf{cof}\,(\boldsymbol{u} \otimes \boldsymbol{v}) = \boldsymbol{0}, \tag{1.84}$$

so that the second invariant, which is the trace of the above tensor, is zero. Similarly, on using Eqn. (1.53), we get the third invariant as zero. The first invariant is given by $\boldsymbol{u} \cdot \boldsymbol{v}$. Thus, from the characteristic equation, it follows that the eigenvalues of $\boldsymbol{u} \otimes \boldsymbol{v}$ are $(0, 0, \boldsymbol{u} \cdot \boldsymbol{v})$. If $\boldsymbol{u}$ and $\boldsymbol{v}$ are perpendicular, $\boldsymbol{u} \otimes \boldsymbol{v}$ is an example of a nonzero tensor all of whose eigenvalues are zero.

1.4 Skew-Symmetric Tensors

Let $\boldsymbol{W} \in \mathrm{Skw}$ and let $\boldsymbol{u}, \boldsymbol{v} \in V$. Then

$$(\boldsymbol{u}, \boldsymbol{W}\boldsymbol{v}) = (\boldsymbol{W}^T\boldsymbol{u}, \boldsymbol{v}) = -(\boldsymbol{W}\boldsymbol{u}, \boldsymbol{v}) = -(\boldsymbol{v}, \boldsymbol{W}\boldsymbol{u}). \tag{1.85}$$

On setting $\boldsymbol{v} = \boldsymbol{u}$, we get

$$(\boldsymbol{u}, \boldsymbol{W}\boldsymbol{u}) = -(\boldsymbol{u}, \boldsymbol{W}\boldsymbol{u}),$$

which implies that

$$(\boldsymbol{u}, \boldsymbol{W}\boldsymbol{u}) = 0. \tag{1.86}$$

Thus, $\boldsymbol{W}\boldsymbol{u}$ is always orthogonal to $\boldsymbol{u}$ for any arbitrary vector $\boldsymbol{u}$. By choosing $\boldsymbol{u} = \boldsymbol{e}_i$ and $\boldsymbol{v} = \boldsymbol{e}_j$, we see from the above results that any skew-symmetric tensor $\boldsymbol{W}$ has only three independent components (in each coordinate frame), which suggests that it might be replaced by a vector. This observation leads us to the following result:

Theorem 1.4.1. *Given any skew-symmetric tensor* $\boldsymbol{W}$, *there exists a unique vector* $\boldsymbol{w}$, *known as the* axial vector *or* dual vector, *corresponding to* $\boldsymbol{W}$ *such that*

$$\boldsymbol{W}\boldsymbol{u} = \boldsymbol{w} \times \boldsymbol{u} \quad \forall \boldsymbol{u} \in V. \tag{1.87}$$

Conversely, given any vector $\boldsymbol{w}$, *there exists a unique skew-symmetric second-order tensor* $\boldsymbol{W}$ *such that Eqn. (1.87) holds.*

Proof. A skew-symmetric tensor $\boldsymbol{W}$, like any other tensor, has at least one real eigenvalue, say λ. Let $\boldsymbol{p}$ be the associated eigenvector. Then $\boldsymbol{W}\boldsymbol{p} = \lambda\boldsymbol{p}$, and on taking the dot product of both sides with $\boldsymbol{p}$, we get $\lambda = \boldsymbol{p} \cdot (\boldsymbol{W}\boldsymbol{p})$. By virtue of Eqn. (1.86), $\lambda = 0$ (we show later in this section that this is the only real eigenvalue). Let $\boldsymbol{q}, \boldsymbol{r}$ be unit vectors that form an orthonormal basis with $\boldsymbol{p}$. We have

$$\boldsymbol{p} = \boldsymbol{q} \times \boldsymbol{r}, \quad \boldsymbol{q} = \boldsymbol{r} \times \boldsymbol{p}, \quad \boldsymbol{r} = \boldsymbol{p} \times \boldsymbol{q}. \tag{1.88}$$

Since $\boldsymbol{W}\boldsymbol{p} = \lambda\boldsymbol{p} = \boldsymbol{0}$, we have $(\boldsymbol{q}, \boldsymbol{W}\boldsymbol{p}) = (\boldsymbol{r}, \boldsymbol{W}\boldsymbol{p}) = 0$, and by Eqn. (1.85), we also have $(\boldsymbol{p}, \boldsymbol{W}\boldsymbol{q}) = (\boldsymbol{p}, \boldsymbol{W}\boldsymbol{r}) = 0$. Thus, using $\boldsymbol{W} = W_{ij}\boldsymbol{e}_i \otimes \boldsymbol{e}_j$ referred to the basis $\{\boldsymbol{p}, \boldsymbol{q}, \boldsymbol{r}\}$, we get

$$\boldsymbol{W} = \gamma(\boldsymbol{r} \otimes \boldsymbol{q} - \boldsymbol{q} \otimes \boldsymbol{r}),$$

where $\gamma = \boldsymbol{r} \cdot (\boldsymbol{W}\boldsymbol{q})$ must be nonzero if $\boldsymbol{W} \neq \boldsymbol{0}$. Let $\boldsymbol{w} = \gamma\boldsymbol{p}$, and let $\boldsymbol{u}$ be an arbitrary vector. Then

$$\begin{aligned}\boldsymbol{W}\boldsymbol{u} - \boldsymbol{w} \times \boldsymbol{u} &= \gamma\{(\boldsymbol{r} \otimes \boldsymbol{q})\boldsymbol{u} - (\boldsymbol{q} \otimes \boldsymbol{r})\boldsymbol{u} - \boldsymbol{p} \times [(\boldsymbol{u} \cdot \boldsymbol{p})\boldsymbol{p} + (\boldsymbol{u} \cdot \boldsymbol{q})\boldsymbol{q} + (\boldsymbol{u} \cdot \boldsymbol{r})\boldsymbol{r}]\} \\ &= \gamma\left[(\boldsymbol{u} \cdot \boldsymbol{q})(\boldsymbol{r} - \boldsymbol{p} \times \boldsymbol{q}) - (\boldsymbol{u} \cdot \boldsymbol{r})(\boldsymbol{q} + \boldsymbol{p} \times \boldsymbol{r})\right] \\ &= \boldsymbol{0},\end{aligned}$$

where the last step follows from Eqn. (1.88). Thus, associated with every skew-symmetric tensor $\boldsymbol{W}$, there is a vector $\boldsymbol{w}$ such that Eqn. (1.87) holds.

Conversely, given $\boldsymbol{w}$, let $\boldsymbol{q}$ and $\boldsymbol{r}$ be unit vectors such that $\boldsymbol{w}/\left|\boldsymbol{w}\right|$, $\boldsymbol{q}$ and $\boldsymbol{r}$ form an orthonormal basis. Then

$$\boldsymbol{W} = \left|\boldsymbol{w}\right|(\boldsymbol{r} \otimes \boldsymbol{q} - \boldsymbol{q} \otimes \boldsymbol{r}), \tag{1.89}$$

is a skew-symmetric tensor with $\boldsymbol{w}$ as the axial vector of $\boldsymbol{W}$, a fact that can be checked easily by verifying that $\boldsymbol{W}\boldsymbol{u} = \boldsymbol{w} \times \boldsymbol{u}$ for any arbitrary vector $\boldsymbol{u}$.

Given $\boldsymbol{W}$, uniqueness of the axial vector $\boldsymbol{w}$ can be shown by assuming the existence of two vectors $\boldsymbol{w}_1$ and $\boldsymbol{w}_2$. Now we have $\boldsymbol{W}\boldsymbol{u} = \boldsymbol{w}_1 \times \boldsymbol{u}$ and $\boldsymbol{W}\boldsymbol{u} = \boldsymbol{w}_2 \times \boldsymbol{u}$ for all vectors $\boldsymbol{u}$. Hence,

$$(\boldsymbol{w}_1 - \boldsymbol{w}_2) \times \boldsymbol{u} = \boldsymbol{0} \quad \forall \boldsymbol{u} \in V.$$

If $\boldsymbol{w}_1 - \boldsymbol{w}_2$ is a nonzero vector, then we can choose $\boldsymbol{u}$ to be perpendicular to $\boldsymbol{w}_1 - \boldsymbol{w}_2$, so that $(\boldsymbol{w}_1 - \boldsymbol{w}_2) \times \boldsymbol{u}$ is nonzero. But this leads to a contradiction with the above property. Hence $\boldsymbol{w}_1 = \boldsymbol{w}_2$, and the uniqueness of the axial vector is established.

Similarly, given $\boldsymbol{w}$, uniqueness of the corresponding $\boldsymbol{W}$ can be established by assuming the existence of two tensors $\boldsymbol{W}_1$ and $\boldsymbol{W}_2$, such that $\boldsymbol{W}_1\boldsymbol{u} = \boldsymbol{w} \times \boldsymbol{u}$ and $\boldsymbol{W}_2\boldsymbol{u} = \boldsymbol{w} \times \boldsymbol{u}$ for all vectors $\boldsymbol{u}$. Now we have

$$(\boldsymbol{W}_1 - \boldsymbol{W}_2)\boldsymbol{u} = \boldsymbol{0} \quad \forall \boldsymbol{u} \in V,$$

which implies that $\boldsymbol{W}_1 = \boldsymbol{W}_2$. □

Note that $\boldsymbol{W}\boldsymbol{u} = \boldsymbol{0}$ if and only if $\boldsymbol{u} = \alpha\boldsymbol{w}$, $\alpha \in \Re$, since if $\boldsymbol{u} = \alpha\boldsymbol{w}$ then $\boldsymbol{W}\boldsymbol{u} = \alpha\boldsymbol{w} \times \boldsymbol{w} = \boldsymbol{0}$, while conversely, if $\boldsymbol{W}\boldsymbol{u} = \boldsymbol{w} \times \boldsymbol{u} = \boldsymbol{0}$, then $\boldsymbol{u}$ is a scalar multiple of $\boldsymbol{w}$ by Theorem 1.2.1. This result justifies the use of the terminology 'axial vector' used for $\boldsymbol{w}$. Also note that by virtue of the uniqueness of $\boldsymbol{w}$, the vector $\alpha\boldsymbol{w}$, $\alpha \in \Re$, is a one-dimensional subspace of V. Similarly, $\boldsymbol{W}^2\boldsymbol{u} = \boldsymbol{0}$ if and only if $\boldsymbol{u} = \alpha\boldsymbol{w}$, $\alpha \in \Re$, since if $\boldsymbol{W}^2\boldsymbol{u} = \boldsymbol{0}$, by the above result, $\boldsymbol{W}\boldsymbol{u} = \boldsymbol{w} \times \boldsymbol{u} = \beta\boldsymbol{w}$, which is only possible when $\boldsymbol{u} = \alpha\boldsymbol{w}$ and $\beta = 0$.

By choosing $u = e_j$ and taking the dot product of both sides with e_i, Eqn. (1.87) can be expressed in component form as

$$\begin{aligned} W_{ij} &= -\epsilon_{ijk} w_k, \\ w_i &= -\frac{1}{2}\epsilon_{ijk} W_{jk}. \end{aligned} \tag{1.90}$$

More explicitly, if $w = (w_1, w_2, w_3)$, then

$$W = \begin{bmatrix} 0 & -w_3 & w_2 \\ w_3 & 0 & -w_1 \\ -w_2 & w_1 & 0 \end{bmatrix}.$$

From Eqns. (1.89) or (1.90), it follows that $W : W = -\mathrm{tr}\,(W^2) = 2w \cdot w$. Since $\mathrm{tr}\, W = \mathrm{tr}\, W^T = -\mathrm{tr}\, W$ and $\det W = \det W^T = -\det W$, we have $\mathrm{tr}\, W = \det W = 0$. The second invariant is given by $I_2 = [(\mathrm{tr}\, W)^2 - \mathrm{tr}\,(W^2)]/2 = (W : W)/2 = w \cdot w$. Thus, from the characteristic equation, we get the eigenvalues of W as $(0, i\,|w|\,, -i\,|w|)$, and the 'spectral resolution' as

$$W = i\,|w|\,(n \otimes \hat{n} - \hat{n} \otimes n),$$

where n is the eigenvector corresponding to $i\,|w|$, and $\hat{n}$ is its complex conjugate. Note that we have used the convention outlined at the end of Section 1.3.1 in writing the above result.

1.5 Orthogonal Tensors

A second-order tensor Q is said to be orthogonal if $Q^T = Q^{-1}$, or, alternatively by Eqn. (1.67), if

$$Q^T Q = Q Q^T = I, \tag{1.91}$$

where I is the identity tensor.

Theorem 1.5.1. *A tensor Q is orthogonal if and only if it has any of the following properties of preserving inner products, lengths and distances:*

$$(Qu, Qv) = (u, v) \quad \forall u, v \in V, \tag{1.92a}$$
$$|Qu| = |u| \quad \forall u \in V, \tag{1.92b}$$
$$|Qu - Qv| = |u - v| \quad \forall u, v \in V. \tag{1.92c}$$

Proof. Assuming that Q is orthogonal, Eqn. (1.92a) follows since

$$(Qu, Qv) = (Q^T Qu, v) = (Iu, v) = (u, v) \quad \forall u, v \in V.$$

Conversely, if Eqn. (1.92a) holds, then

$$0 = (Qu, Qv) - (u, v) = (u, Q^T Qv) - (u, v) = (u, (Q^T Q - I)v) \quad \forall u, v \in V,$$

which implies that $Q^T Q = I$ (by Eqn. (1.24)), and hence Q is orthogonal.

By choosing $v = u$ in Eqn. (1.92a), we get Eqn. (1.92b). Conversely, if Eqn. (1.92b) holds, i.e., if $(Qu, Qu) = (u, u)$ for all $u \in V$, then

$$((Q^T Q - I)u, u) = 0 \quad \forall u \in V,$$

which, by virtue of Problem 29, leads us to the conclusion that $Q \in$ Orth.

By replacing u by $(u - v)$ in Eqn. (1.92b), we obtain Eqn. (1.92c), and, conversely, by setting v to zero in Eqn. (1.92c), we get Eqn. (1.92b). □

As a corollary of the above results, it follows that the 'angle' between two vectors u and v, defined by $\theta := \cos^{-1}(u \cdot v)/(|u|\,|v|)$, is also preserved. Thus, physically speaking, multiplying the position vectors of all points in a domain by Q corresponds to rigid body rotation of the domain about the origin.

From Eqns. (1.49), (1.51) and (1.91), we have $\det Q = \pm 1$. Orthogonal tensors with determinant $+1$ are said to be proper orthogonal or rotations (henceforth, this set is denoted by Orth$^+$). Since $\det(-R) = (-1)^3 \det R = -1$ for $R \in$ Orth$^+$, $-R$ is an orthogonal tensor that is not a rotation. On the other hand, if Q is an orthogonal tensor that is not a rotation, $-Q$ is a rotation. Thus, when the underlying vector space dimension is odd (which includes $n = 3$), every orthogonal tensor is either a rotation, or the product of a rotation with $-I$:

$$\text{Orth} = \{\pm R;\ R \in \text{Orth}^+\}.$$

This result is not true when the dimension n is even, as the example $Q = \left[\begin{smallmatrix} 0 & 1 \\ 1 & 0 \end{smallmatrix}\right]$, which cannot be written as $-R$, where $R \in$ Orth$^+$, shows. For $Q \in$ Orth$^+$, we have

$$\mathbf{cof}\, Q = (\det Q) Q^{-T} = Q, \tag{1.93}$$

so that by Eqn. (1.54),

$$Q(u \times v) = (Qu) \times (Qv) \quad \forall u, v \in V. \tag{1.94}$$

A characterization of a rotation is as follows:

Theorem 1.5.2. *Let $\{\bar{e}_1, \bar{e}_2, \bar{e}_3\}$ and $\{e_1^*, e_2^*, e_3^*\}$ be two orthonormal bases. Then*

$$Q = \bar{e}_1 \otimes e_1^* + \bar{e}_2 \otimes e_2^* + \bar{e}_3 \otimes e_3^*,$$

is a proper orthogonal tensor. Conversely, every $Q \in$ Orth$^+$ can be represented in the above manner.

Proof. If $\{\bar{e}_1, \bar{e}_2, \bar{e}_3\}$ and $\{e_1^*, e_2^*, e_3^*\}$ are two orthonormal bases, then

$$QQ^T = [\bar{e}_1 \otimes e_1^* + \bar{e}_2 \otimes e_2^* + \bar{e}_3 \otimes e_3^*]\,[e_1^* \otimes \bar{e}_1 + e_2^* \otimes \bar{e}_2 + e_3^* \otimes \bar{e}_3]$$

$$= [\bar{e}_1 \otimes \bar{e}_1 + \bar{e}_2 \otimes \bar{e}_2 + \bar{e}_3 \otimes \bar{e}_3] \qquad \text{(by Eqn. (1.341))}$$

$$= I, \qquad \text{(by Eqn. (1.38))}$$

and, by Eqn. (1.53),

$$\det Q = [\bar{e}_1, \bar{e}_2, \bar{e}_3]\, [e_1^*, e_2^*, e_3^*] = 1.$$

Conversely, if $Q \in \text{Orth}^+$, and $\{e_1^*, e_2^*, e_3^*\}$ is an orthonormal basis, then

$$\begin{aligned} Q &= QI \\ &= Q\,[e_1^* \otimes e_1^* + e_2^* \otimes e_2^* + e_3^* \otimes e_3^*] && \text{(by Eqn. (1.38))} \\ &= (Qe_1^*) \otimes e_1^* + (Qe_2^*) \otimes e_2^* + (Qe_3^*) \otimes e_3^* && \text{(by Eqn. (1.343))} \\ &= \bar{e}_1 \otimes e_1^* + \bar{e}_2 \otimes e_2^* + \bar{e}_3 \otimes e_3^*, \end{aligned}$$

where $\{\bar{e}_1, \bar{e}_2, \bar{e}_3\} = \{Qe_1^*, Qe_2^*, Qe_3^*\}$ is an orthonormal basis since $|\bar{e}_i| = |Qe_i^*| = |e_i^*| = 1$ for each i, and since, by virtue of Eqn. (1.94),

$$\bar{e}_j \times \bar{e}_k = (Qe_j^*) \times (Qe_k^*) = Q(e_j^* \times e_k^*) = \epsilon_{ijk} Qe_i^* = \epsilon_{ijk} \bar{e}_i. \qquad \square$$

As a result of the above characterization, we have the following theorem.

Theorem 1.5.3. *Every $T \in Lin$ can be expressed as a linear combination of proper orthogonal tensors, i.e.,*

$$Lin = span\{Orth^+\}.$$

Proof. Let $\{e_1, e_2, e_3\}$ denote the canonical Cartesian basis vectors. Then

$$e_1 \otimes e_1 = \frac{1}{2}\left[I + (e_1 \otimes e_1 - e_2 \otimes e_2 - e_3 \otimes e_3)\right],$$

$$e_1 \otimes e_2 = \frac{1}{2}\left[(e_1 \otimes e_2 + e_2 \otimes e_3 + e_3 \otimes e_1) + (e_1 \otimes e_2 - e_2 \otimes e_3 - e_3 \otimes e_1)\right],$$

with similar expressions for the remaining combinations of dyadic products. Thus, the tensor $T = T_{ij} e_i \otimes e_j$ can be expressed as $\sum_i \phi_i Q_i$, where $Q_i \in \text{Orth}^+$. $\square$

If $\{e_i\}$ and $\{\bar{e}_i\}$ are two sets of orthonormal basis vectors, then they are related as

$$\bar{e}_i = Q^T e_i, \quad i = 1, 2, 3, \tag{1.95}$$

where $Q = e_k \otimes \bar{e}_k$ is a proper orthogonal tensor by virtue of Theorem 1.5.2. The components of Q with respect to the $\{e_i\}$ basis are given by $Q_{ij} = e_i \cdot (e_k \otimes \bar{e}_k) e_j = \delta_{ik} \bar{e}_k \cdot e_j = \bar{e}_i \cdot e_j$. Thus, if $\bar{e}$ and e are two unit vectors, we can always find $Q \in \text{Orth}^+$ (not necessarily unique), which rotates $\bar{e}$ to e, i.e., $e = Q\bar{e}$. Let u and v be two vectors. Since $u/|u|$ and $v/|v|$ are unit vectors, there exists $Q \in \text{Orth}^+$ such that

$$\frac{v}{|v|} = Q\left(\frac{u}{|u|}\right).$$

Thus, if u and v have the same magnitude, i.e., if $|u| = |v|$, then there exists $Q \in \text{Orth}^+$ such that $u = Qv$.

We now study the transformation laws for the components of tensors under an orthogonal transformation of the basis vectors. Let e_i and $\bar{e}_i$ represent the original and new orthonormal basis vectors, and let Q be the proper orthogonal tensor in Eqn. (1.95). From Eqn. (1.9), we have

$$\bar{e}_i = (\bar{e}_i \cdot e_j)e_j = Q_{ij}e_j, \tag{1.96a}$$

$$e_i = (e_i \cdot \bar{e}_j)\bar{e}_j = Q_{ji}\bar{e}_j. \tag{1.96b}$$

Using Eqn. (1.8) and Eqn. (1.96a), we get the transformation law for the components of a vector as

$$\bar{v}_i = v \cdot \bar{e}_i = v \cdot (Q_{ij}e_j) = Q_{ij}v \cdot e_j = Q_{ij}v_j. \tag{1.97}$$

In a similar fashion, using Eqn. (1.28), Eqn. (1.96a), and the fact that a tensor is a linear transformation, we get the transformation law for the components of a second-order tensor as

$$\bar{T}_{ij} = \bar{e}_i \cdot T\bar{e}_j = Q_{im}Q_{jn}T_{mn}. \tag{1.98}$$

Conversely, if the components of a matrix transform according to Eqn. (1.98), then they all generate the same tensor. To see this, let $\bar{T} = \bar{T}_{ij}\bar{e}_i \otimes \bar{e}_j$ and $T = T_{mn}e_m \otimes e_n$. Then

$$\begin{aligned}\bar{T} &= \bar{T}_{ij}\bar{e}_i \otimes \bar{e}_j \\ &= Q_{im}Q_{jn}T_{mn}\bar{e}_i \otimes \bar{e}_j \\ &= T_{mn}(Q_{im}\bar{e}_i) \otimes (Q_{jn}\bar{e}_j) \\ &= T_{mn}e_m \otimes e_n \quad \text{(by Eqn. (1.96b))} \\ &= T.\end{aligned}$$

Due to this property, often, an alternate viewpoint is followed by many authors who take the transformation law given by Eqn. (1.98) as the *definition* of second-order tensors.

We can write Eqns. (1.97) and (1.98) as

$$[\bar{v}] = Q[v], \tag{1.99}$$

$$[\bar{T}] = Q[T]Q^T. \tag{1.100}$$

where $[\bar{v}]$ and $[\bar{T}]$ represent the components of the vector v and tensor T, respectively, with respect to the $\bar{e}_i$ coordinate system. Using the orthogonality property of Q, we can write the reverse transformations as

$$[v] = Q^T[\bar{v}], \tag{1.101}$$

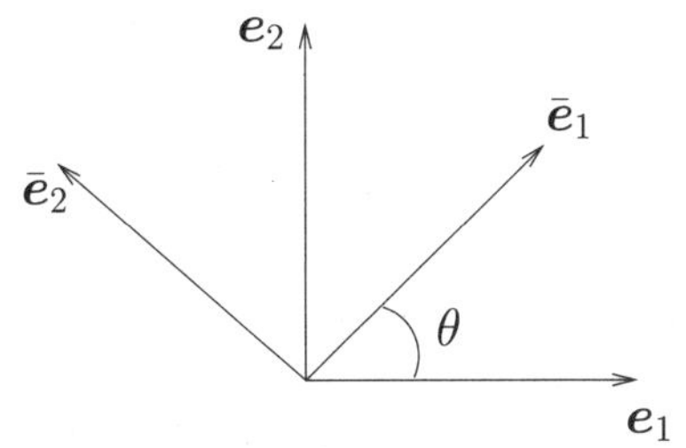

Fig. 1.1 *Example of a coordinate system obtained from an existing one by a rotation about the 3-axis.*

$$[T] = Q^T[\bar{T}]Q. \tag{1.102}$$

As an example, the Q matrix for the configuration shown in Fig. 1.1 is

$$Q = \begin{bmatrix} \bar{e}_1 \\ \bar{e}_2 \\ \bar{e}_3 \end{bmatrix} = \begin{bmatrix} \cos\theta & \sin\theta & 0 \\ -\sin\theta & \cos\theta & 0 \\ 0 & 0 & 1 \end{bmatrix}.$$

The only real eigenvalues of $Q \in$ Orth can be either $+1$ or -1, since if λ and n denote the eigenvalue and eigenvector of Q, i.e., $Qn = \lambda n$, then

$$(n, n) = (Qn, Qn) = \lambda^2(n, n),$$

which implies that $(n, n)(\lambda^2 - 1) = 0$. If λ and n are real, then $(n, n) \neq 0$ and $\lambda = \pm 1$, while if λ is complex, then $(n, n) = 0$. Let $\hat{\lambda}$ and $\hat{n}$ denote the complex conjugates of λ and n, respectively. To see that the complex eigenvalues have a magnitude of unity observe that $\lambda\hat{\lambda}(\hat{n}, n) = (Q\hat{n}, Qn) = (\hat{n}, n)$, which implies that $\lambda\hat{\lambda} = 1$ since $(\hat{n}, n) \neq 0$. Now, using a similar argument as used in showing $(n, n) = 0$, we also have $(n_1, n_2) = 0$ for n_1 and n_2 corresponding to distinct complex eigenvalues of Q (complex conjugates not being considered as 'distinct'–thus, this result is relevant only when the dimension $n > 3$), and $(n, e) = (\hat{n}, e) = 0$, where e is the eigenvector of Q corresponding to the eigenvalue of $+1$ or -1.

If $R \neq I$ is a rotation, then the set of all vectors e such that

$$Re = e \tag{1.103}$$

forms a one-dimensional subspace of V called the *axis* of R. To prove that such a vector always exists, we first show that $+1$ is always an eigenvalue of R, and -1 is always an eigenvalue of an improper rotation. Since $\det R = 1$,

$$\begin{aligned} \det(R - I) &= \det(R - RR^T) = (\det R)\det(I - R^T) = \det(I - R^T)^T \\ &= \det(I - R) = -\det(R - I), \end{aligned}$$

which implies that $\det(R - I) = 0$, or that $+1$ is an eigenvalue. If e is the eigenvector corresponding to the eigenvalue $+1$, then $Re = e$. Since every improper rotation is a product of $-I$ times a proper orthogonal tensor, -1 is an eigenvalue of an improper orthogonal tensor, with the same eigenvector as the proper orthogonal tensor from which it is generated.

To show that the set of vectors $\boldsymbol{e}$ satisfying Eqn. (1.103) is a one-dimensional subspace of V, premultiply both sides of this equation with $\boldsymbol{R}^T$ to get $\boldsymbol{R}^T\boldsymbol{e} = \boldsymbol{e}$. Combining this relation with Eqn. (1.103), we get $(\boldsymbol{R} - \boldsymbol{R}^T)\boldsymbol{e} = \boldsymbol{0}$. There are two possibilities: (i) $\boldsymbol{R}$ is unsymmetric, or (ii) $\boldsymbol{R}$ is symmetric. If $\boldsymbol{R}$ is unsymmetric, then $\boldsymbol{R} - \boldsymbol{R}^T$ is a nonzero skew-symmetric tensor with $\boldsymbol{e}$ a scalar multiple of the axial vector of $\boldsymbol{R} - \boldsymbol{R}^T$. Since we have shown that a scalar multiple of the axial vector of a skew-symmetric tensor constitutes a one-dimensional subspace, $\alpha\boldsymbol{e}$, $\alpha \in \Re$ is a one-dimensional subspace of V. Now consider the other possibility where $\boldsymbol{R}$ is symmetric. By Theorem 1.6.1 all the eigenvalues of $\boldsymbol{R}$ are real, and as shown above the only real eigenvalues can be ± 1. Taking into account that $\det \boldsymbol{R} = 1$, $\boldsymbol{R} \neq \boldsymbol{I}$ and that one of the real eigenvalues is $+1$, the set of eigenvalues in this case has to be $\{+1, -1, -1\}$. Since the eigenvalue $+1$ is distinct from the other two, again invoking Theorem 1.6.1, it follows that the corresponding eigenvector $\boldsymbol{e}$ is unique. Using Eqn.(1.123), it also follows that $\boldsymbol{R} \in \text{Orth}^+ \cap \text{Sym} - \{\boldsymbol{I}\}$ has to be of the form $2\boldsymbol{e} \otimes \boldsymbol{e} - \boldsymbol{I}$, where $\boldsymbol{e}$ is a unit vector.

Conversely, given a vector $\boldsymbol{w}$, there exists a proper orthogonal tensor $\boldsymbol{R}$, and an improper one $\boldsymbol{R}'$ such that $\boldsymbol{R}\boldsymbol{w} = \boldsymbol{w}$ and $\boldsymbol{R}'\boldsymbol{w} = -\boldsymbol{w}$. To see this, consider the family of tensors

$$\boldsymbol{R}(\boldsymbol{w}, \alpha) = \boldsymbol{I} + \frac{1}{|\boldsymbol{w}|} \sin\alpha\, \boldsymbol{W} + \frac{1}{|\boldsymbol{w}|^2}(1 - \cos\alpha)\boldsymbol{W}^2, \tag{1.104}$$

where $\boldsymbol{W}$ is the skew-symmetric tensor with $\boldsymbol{w}$ as its axial vector, i.e., $\boldsymbol{W}\boldsymbol{w} = \boldsymbol{0}$. Using the Cayley–Hamilton theorem, we have $\boldsymbol{W}^3 = -|\boldsymbol{w}|^2 \boldsymbol{W}$, from which it follows that $\boldsymbol{W}^4 = -|\boldsymbol{w}|^2 \boldsymbol{W}^2$. Using this result, we get

$$\boldsymbol{R}^T\boldsymbol{R} = \left[\boldsymbol{I} - \frac{\sin\alpha}{|\boldsymbol{w}|}\boldsymbol{W} + \frac{(1-\cos\alpha)}{|\boldsymbol{w}|^2}\boldsymbol{W}^2\right]\left[\boldsymbol{I} + \frac{\sin\alpha}{|\boldsymbol{w}|}\boldsymbol{W} + \frac{(1-\cos\alpha)}{|\boldsymbol{w}|^2}\boldsymbol{W}^2\right] = \boldsymbol{I}.$$

Since $\boldsymbol{R}$ has now been shown to be orthogonal, $\det \boldsymbol{R} = \pm 1$. However, since $\det[\boldsymbol{R}(\boldsymbol{w}, 0)] = \det \boldsymbol{I} = 1$, by continuity, we have $\det[\boldsymbol{R}(\boldsymbol{w}, \alpha)] = 1$ for any α. Thus, $\boldsymbol{R}$ is a proper orthogonal tensor that satisfies $\boldsymbol{R}\boldsymbol{w} = \boldsymbol{w}$. Taking $\boldsymbol{R}' = -\boldsymbol{R}$, we get the required improper orthogonal tensor. It is easily seen that $\boldsymbol{R}\boldsymbol{w} = \boldsymbol{w}$. Essentially, $\boldsymbol{R}$ rotates any vector in the plane perpendicular to $\boldsymbol{w}$ through an angle α. By Eqn. (1.89), $\boldsymbol{W}^2 = \boldsymbol{w} \otimes \boldsymbol{w} - |\boldsymbol{w}|^2 \boldsymbol{I}$. Thus, from Eqn. (1.104), we get $\boldsymbol{R}(\boldsymbol{w}, 0) = \boldsymbol{I}$ and $\boldsymbol{R}(\boldsymbol{w}, \pi) = 2\boldsymbol{e} \otimes \boldsymbol{e} - \boldsymbol{I}$, where $\boldsymbol{e} = \boldsymbol{w}/|\boldsymbol{w}|$.

Using this result, we have the following representation theorem for proper orthogonal tensors.

Theorem 1.5.4. *Let $\boldsymbol{W}$ be a skew-symmetric tensor with an axial vector of unit magnitude, and let $\{\boldsymbol{e}, \boldsymbol{q}, \boldsymbol{r}\}$ be an orthonormal set of vectors. Then tensors of the form*

$$\boldsymbol{R} = \boldsymbol{I} + \sin\alpha\, \boldsymbol{W} + (1 - \cos\alpha)\boldsymbol{W}^2, \tag{1.105a}$$

$$\boldsymbol{R} = \boldsymbol{e} \otimes \boldsymbol{e} + \cos\alpha(\boldsymbol{I} - \boldsymbol{e} \otimes \boldsymbol{e}) + \sin\alpha(\boldsymbol{r} \otimes \boldsymbol{q} - \boldsymbol{q} \otimes \boldsymbol{r}). \tag{1.105b}$$

are proper orthogonal. Conversely, there exist $\boldsymbol{W} \in$ Skw, and orthonormal vectors $\{\boldsymbol{e}, \boldsymbol{q}, \boldsymbol{r}\}$, such that every proper orthogonal tensor can be expressed in the above forms.

Proof. The tensor $\boldsymbol{R}$ in Eqn. (1.105a) is just a special case of the tensor $\boldsymbol{R}$ in Eqn. (1.104), which we have already shown to be proper orthogonal. Let $\{\boldsymbol{e},\boldsymbol{q},\boldsymbol{r}\}$ be an orthonormal set of vectors. Choose $\boldsymbol{W}=\boldsymbol{r}\otimes\boldsymbol{q}-\boldsymbol{q}\otimes\boldsymbol{r}$, so that $\boldsymbol{W}^2=\boldsymbol{e}\otimes\boldsymbol{e}-\boldsymbol{I}$. Substituting for $\boldsymbol{W}$ and $\boldsymbol{W}^2$ into Eqn. (1.105a), we get the representation in Eqn. (1.105b). Note that since $\boldsymbol{Re}=\boldsymbol{e}$, the vector $\boldsymbol{e}$ is the axis of $\boldsymbol{R}$.

Conversely, we now show that every proper orthogonal tensor can be represented as Eqns. (1.105a) and (1.105b). Let $\boldsymbol{e}$ be the axis of $\boldsymbol{R}$, and let $\boldsymbol{q}$, $\boldsymbol{r}$ be such that $\{\boldsymbol{e},\boldsymbol{q},\boldsymbol{r}\}$ form an orthonormal basis. Using the fact that $\boldsymbol{Re}=\boldsymbol{R}^T\boldsymbol{e}=\boldsymbol{e}$, we have $(\boldsymbol{Re},\boldsymbol{q})=(\boldsymbol{Re},\boldsymbol{r})=(\boldsymbol{e},\boldsymbol{Rq})=(\boldsymbol{e},\boldsymbol{Rr})=0$. Thus, $\boldsymbol{Rq}$ and $\boldsymbol{Rr}$ lie in the plane perpendicular to $\boldsymbol{e}$. If $(\boldsymbol{Rq},\boldsymbol{q})=(\boldsymbol{Rr},\boldsymbol{r})=\cos\alpha$, then

$$\begin{aligned}
\boldsymbol{R}&=\boldsymbol{RI}\\
&=\boldsymbol{R}(\boldsymbol{e}\otimes\boldsymbol{e}+\boldsymbol{q}\otimes\boldsymbol{q}+\boldsymbol{r}\otimes\boldsymbol{r})\\
&=(\boldsymbol{Re})\otimes\boldsymbol{e}+(\boldsymbol{Rq})\otimes\boldsymbol{q}+(\boldsymbol{Rr})\otimes\boldsymbol{r}\\
&=\boldsymbol{e}\otimes\boldsymbol{e}+[(\boldsymbol{Rq},\boldsymbol{e})\boldsymbol{e}+(\boldsymbol{Rq},\boldsymbol{q})\boldsymbol{q}+(\boldsymbol{Rq},\boldsymbol{r})\boldsymbol{r}]\otimes\boldsymbol{q}+\\
&\qquad[(\boldsymbol{Rr},\boldsymbol{e})\boldsymbol{e}+(\boldsymbol{Rr},\boldsymbol{q})\boldsymbol{q}+(\boldsymbol{Rr},\boldsymbol{r})\boldsymbol{r}]\otimes\boldsymbol{r}\\
&=\boldsymbol{e}\otimes\boldsymbol{e}+\cos\alpha(\boldsymbol{q}\otimes\boldsymbol{q}+\boldsymbol{r}\otimes\boldsymbol{r})+\sin\alpha(\boldsymbol{r}\otimes\boldsymbol{q}-\boldsymbol{q}\otimes\boldsymbol{r})\\
&=\boldsymbol{e}\otimes\boldsymbol{e}+\cos\alpha(\boldsymbol{I}-\boldsymbol{e}\otimes\boldsymbol{e})+\sin\alpha(\boldsymbol{r}\otimes\boldsymbol{q}-\boldsymbol{q}\otimes\boldsymbol{r}),
\end{aligned}$$

which is Eqn. (1.105b). By taking $\boldsymbol{W}=\boldsymbol{r}\otimes\boldsymbol{q}-\boldsymbol{q}\otimes\boldsymbol{r}$, we have $\boldsymbol{W}^2=\boldsymbol{e}\otimes\boldsymbol{e}-\boldsymbol{I}$, from which we get Eqn. (1.105a). □

Using the results for the eigenvalues/eigenvectors of $\boldsymbol{R}$, it is also possible to write the following "spectral resolutions" (using the convention outlined at the end of Section 1.3.1):

$$\begin{aligned}
\boldsymbol{I}&=\boldsymbol{e}\otimes\boldsymbol{e}+\boldsymbol{n}\otimes\hat{\boldsymbol{n}}+\hat{\boldsymbol{n}}\otimes\boldsymbol{n},\\
\boldsymbol{R}&=\boldsymbol{e}\otimes\boldsymbol{e}+\lambda\boldsymbol{n}\otimes\hat{\boldsymbol{n}}+\hat{\lambda}\hat{\boldsymbol{n}}\otimes\boldsymbol{n},\\
\boldsymbol{R}^2&=\boldsymbol{e}\otimes\boldsymbol{e}+\lambda^2\boldsymbol{n}\otimes\hat{\boldsymbol{n}}+\hat{\lambda}^2\hat{\boldsymbol{n}}\otimes\boldsymbol{n},
\end{aligned}\tag{1.106}$$

where $\lambda=\hat{\lambda}=-1$ in case all eigenvalues of $\boldsymbol{R}\neq\boldsymbol{I}$ are real (so that $\boldsymbol{R}=2\boldsymbol{e}\otimes\boldsymbol{e}-\boldsymbol{I}$), and $\boldsymbol{n}\cdot\boldsymbol{n}=\hat{\boldsymbol{n}}\cdot\hat{\boldsymbol{n}}=\boldsymbol{e}\cdot\boldsymbol{n}=\boldsymbol{e}\cdot\hat{\boldsymbol{n}}=0$ and $\boldsymbol{n}\cdot\hat{\boldsymbol{n}}=1$ in case λ is complex.

Note from Eqn. (1.105b) that $\operatorname{tr}\boldsymbol{R}=1+2\cos\alpha$. It is easily seen by multiplying the expression

$$\phi_1\boldsymbol{e}\otimes\boldsymbol{e}+\phi_2(\boldsymbol{q}\otimes\boldsymbol{q}+\boldsymbol{r}\otimes\boldsymbol{r})+\phi_3(\boldsymbol{r}\otimes\boldsymbol{q}-\boldsymbol{q}\otimes\boldsymbol{r})=\boldsymbol{0}$$

successively by $\boldsymbol{e}$, $\boldsymbol{q}$ and $\boldsymbol{r}$ that $\phi_1=\phi_2=\phi_3=0$, which implies that $\{\boldsymbol{e}\otimes\boldsymbol{e},\boldsymbol{q}\otimes\boldsymbol{q}+\boldsymbol{r}\otimes\boldsymbol{r},\boldsymbol{r}\otimes\boldsymbol{q}-\boldsymbol{q}\otimes\boldsymbol{r}\}$ are linearly independent. Using the representation given by Eqn. (1.105b), we have

$$\begin{bmatrix}\boldsymbol{I}\\ \boldsymbol{R}\\ \boldsymbol{R}^2\end{bmatrix}=\begin{bmatrix}1&1&0\\1&\cos\alpha&\sin\alpha\\1&\cos2\alpha&\sin2\alpha\end{bmatrix}\begin{bmatrix}\boldsymbol{e}\otimes\boldsymbol{e}\\ \boldsymbol{q}\otimes\boldsymbol{q}+\boldsymbol{r}\otimes\boldsymbol{r}\\ \boldsymbol{r}\otimes\boldsymbol{q}-\boldsymbol{q}\otimes\boldsymbol{r}\end{bmatrix}.$$

The determinant of the square matrix in the above equation is $2\sin\alpha(1-\cos\alpha)$. Thus, from Theorem 1.1.2 it follows that $\{\boldsymbol{I},\boldsymbol{R},\boldsymbol{R}^2\}$ are linearly independent for all α except when (i)

$\alpha = 0$, in which case $\boldsymbol{R} = \boldsymbol{R}^2 = \boldsymbol{I}$ and (ii) $\alpha = \pi$, in which case $\boldsymbol{R} = 2\boldsymbol{e} \otimes \boldsymbol{e} - \boldsymbol{I} \in \text{Sym}$, and $\boldsymbol{R}^2 = \boldsymbol{I}$ (in this case $\{\boldsymbol{I}, \boldsymbol{R}\}$ are linearly independent).

Using the representation in Eqn. (1.104), let

$$\boldsymbol{R}_1(\boldsymbol{w}_1, \alpha) = \boldsymbol{I} + \frac{1}{|\boldsymbol{w}_1|}\sin\alpha\, \boldsymbol{W}_1 + \frac{1}{|\boldsymbol{w}_1|^2}(1-\cos\alpha)\boldsymbol{W}_1^2,$$
$$\boldsymbol{R}_2(\boldsymbol{w}_2, \phi) = \boldsymbol{I} + \frac{1}{|\boldsymbol{w}_2|}\sin\phi\, \boldsymbol{W}_2 + \frac{1}{|\boldsymbol{w}_2|^2}(1-\cos\phi)\boldsymbol{W}_2^2,$$

be two proper orthogonal tensors. If the axes of these two tensors coincide, i.e., if $\boldsymbol{e}_1 = \boldsymbol{w}_1/|\boldsymbol{w}_1| = \boldsymbol{w}_2/|\boldsymbol{w}_2| = \boldsymbol{e}_2$, then we have $\boldsymbol{W}_1/|\boldsymbol{w}_1| = \boldsymbol{W}_2/|\boldsymbol{w}_2|$, and hence $\boldsymbol{R}_1$ and $\boldsymbol{R}_2$ commute, i.e., $\boldsymbol{R}_1\boldsymbol{R}_2 = \boldsymbol{R}_2\boldsymbol{R}_1$. However, the converse is not true since, for example, $\boldsymbol{R}_1 = \textbf{diag}[1,-1,-1]$ and $\boldsymbol{R}_2 = \textbf{diag}[-1,1,-1]$ commute, but do not have the same axes. If $\boldsymbol{e}_1$ and $\boldsymbol{e}_2$ are the axes of $\boldsymbol{R}_1$ and $\boldsymbol{R}_2$, respectively, then we assume the (unnormalized) axis of $\boldsymbol{R}_2\boldsymbol{R}_1$ to be of the form $\boldsymbol{w} = \hat{\alpha}\boldsymbol{e}_1 + \hat{\beta}\boldsymbol{e}_2 + \hat{\gamma}\boldsymbol{e}_1 \times \boldsymbol{e}_2$ (since $\{\boldsymbol{e}_1, \boldsymbol{e}_2, \boldsymbol{e}_1 \times \boldsymbol{e}_2\}$ constitute a basis when $\boldsymbol{e}_1$ and $\boldsymbol{e}_2$ are linearly independent) and find $(\hat{\alpha}, \hat{\beta}, \hat{\gamma})$ using the condition $\boldsymbol{R}_2\boldsymbol{R}_1\boldsymbol{w} = \boldsymbol{w}$. Using this procedure, we obtain the axis of $\boldsymbol{R}_2\boldsymbol{R}_1$ as $\boldsymbol{e} = \boldsymbol{w}/|\boldsymbol{w}|$, where

$$\boldsymbol{w} = (1-\cos\alpha)(1+\cos\phi)\boldsymbol{e}_1 + \sin\alpha\sin\phi\,\boldsymbol{e}_2 - (1-\cos\alpha)\sin\phi\,\boldsymbol{e}_1 \times \boldsymbol{e}_2. \tag{1.107}$$

In the representation given by Eqn. (1.104), $\boldsymbol{w}$ is a fixed vector; however, in problems of rigid-body dynamics, we shall need to consider the case when the angular velocity vector is a function of time, and which, in general, will not coincide with the axis of $\boldsymbol{Q}$. In such a situation, it is helpful to express $\boldsymbol{Q} \in \text{Orth}^+$ in terms of generalized coordinates. Problem 26 shows that 3 parameters are required to characterize an orthogonal tensor in three-dimensional space, and presents one such characterization. However, the most commonly used characterization is the one using the *Euler angles* θ, ϕ and ψ. The representation of $\boldsymbol{Q}$ in terms of these angles is

$$\begin{aligned}\boldsymbol{Q} &= \begin{bmatrix}\cos\psi & -\sin\psi & 0\\ \sin\psi & \cos\psi & 0\\ 0 & 0 & 1\end{bmatrix}\begin{bmatrix}1 & 0 & 0\\ 0 & \cos\theta & -\sin\theta\\ 0 & \sin\theta & \cos\theta\end{bmatrix}\begin{bmatrix}\cos\phi & -\sin\phi & 0\\ \sin\phi & \cos\phi & 0\\ 0 & 0 & 1\end{bmatrix}\\ &= \begin{bmatrix}\cos\phi\cos\psi - \cos\theta\sin\psi\sin\phi & -\sin\phi\cos\psi - \cos\theta\sin\psi\cos\phi & \sin\theta\sin\psi\\ \cos\phi\sin\psi + \cos\theta\cos\psi\sin\phi & -\sin\phi\sin\psi + \cos\theta\cos\psi\cos\phi & -\sin\theta\cos\psi\\ \sin\phi\sin\theta & \cos\phi\sin\theta & \cos\theta\end{bmatrix}.\end{aligned} \tag{1.108}$$

The axis of $\boldsymbol{Q}$ is $[0,0,1]$ when $\theta = 0$, and

$$\boldsymbol{e} = \begin{bmatrix}(1-\cos\theta)(\sin\phi + \sin\psi)\\ (1-\cos\theta)(\cos\phi - \cos\psi)\\ \sin\theta[1-\cos(\phi+\psi)]\end{bmatrix} \quad (\theta \neq 0). \tag{1.109}$$

We now present an expression for the eigenvalues of $\boldsymbol{Q}$. Since from Eqn. (1.105b), $\operatorname{tr}\boldsymbol{Q} = 1 + 2\cos\alpha$, and since we have shown that the magnitude of the eigenvalues is 1, the eigenvalues of $\boldsymbol{Q} \in \text{Orth}^+$ are $\{1, e^{i\alpha}, e^{-i\alpha}\}$. Since $-1 \le \cos\alpha \le 1$, we have $-1 \le \operatorname{tr}\boldsymbol{Q} \le 3$, with $\operatorname{tr}\boldsymbol{Q} = 3$ if and only if $\boldsymbol{Q} = \boldsymbol{I}$ ($\alpha = 0$ in Eqn. (1.105b)). The other bound $\operatorname{tr}\boldsymbol{Q} = -1$ is attained when $\alpha = \pi$ (or when θ or $\phi + \psi$ is equal to π in Eqn. (1.108)), and from Eqn. (1.105b), we see that $Q = 2e \otimes e - I \in \text{Sym}$ with eigenvalues $\{1, -1, -1\}$. When $-1 < \text{tr}\, Q < 3$, two eigenvalues and the corresponding eigenvectors are complex.

Since for $Q \in \text{Orth}^+$, $I_2 = \text{tr}\,(\textbf{cof}\, Q) = \text{tr}\,[(\det Q)Q^{-T}] = \text{tr}\, Q$, by the Cayley–Hamilton theorem, we get

$$Q^{-1} = Q^T = Q^2 - (\text{tr}\, Q)Q + (\text{tr}\, Q)I. \tag{1.110}$$

If $Q_1, Q_2 \in \text{Orth}^+$, which have the representations (see Theorem 1.5.2)

$$Q_1 = a_1 \otimes b_1 + a_2 \otimes b_2 + a_3 \otimes b_3,$$
$$Q_2 = c_1 \otimes d_1 + c_2 \otimes d_2 + c_3 \otimes d_3,$$

where $\{a_i\}$, $\{b_i\}$, $\{c_i\}$, $\{d_i\}$ are orthonormal sets of vectors, then there exist proper orthogonal tensors $R_1 = c_k \otimes a_k$ and $R_2 = d_k \otimes b_k$ such that $Q_2 = R_1 Q_1 R_2^T$. However, if the trace of Q_1 and Q_2 are the same (which implies that all their principal invariants are also the same), the following stronger result holds.

Theorem 1.5.5. *Two proper orthogonal tensors Q_1 and Q_2 have the same trace, i.e., $\text{tr}\, Q_1 = \text{tr}\, Q_2$, if and only if there exists an orthogonal tensor Q_0 (which means that either Q_0 or $-Q_0$ is a rotation) such that $Q_2 = Q_0 Q_1 Q_0^T$.*

Proof. If there exists $Q_0 \in \text{Orth}$ such that $Q_2 = Q_0 Q_1 Q_0^T$, then using the same method of proof as used to prove Eqn. (1.137), we see that all the principal invariants of Q_1 and Q_2 are the same. To prove the converse, note that if all the principal invariants of Q_1 and Q_2 are the same, then their characteristic equations, and hence their eigenvalues are the same. Thus, using Eqn. (1.106), we can write

$$Q_1 = e \otimes e + \lambda n \otimes \hat{n} + \hat{\lambda} \hat{n} \otimes n,$$
$$Q_2 = f \otimes f + \lambda g \otimes \hat{g} + \hat{\lambda} \hat{g} \otimes g.$$

The tensor

$$Q_0 = f \otimes e + g \otimes \hat{n} + \hat{g} \otimes n$$

is orthogonal since $Q_0^T Q_0 = e \otimes e + n \otimes \hat{n} + \hat{n} \otimes n = I$, and using the relations following Eqn. (1.106), we also see that $Q_2 = Q_0 Q_1 Q_0^T$. In the above treatment, we have assumed that two eigenvalues are complex. If $(1, -1, -1)$ are the eigenvalues of Q_1 and Q_2, then they have the representations $Q_1 = 2e \otimes e - I$ and $Q_2 = 2f \otimes f - I$, so that any tensor that rotates e into f can be taken to be Q_0.

An alternative proof can also be given using Eqn. (1.105b) (recall that $\text{tr}\, R = 1 + 2\cos\alpha$) by writing

$$Q_1 = e \otimes e + \cos\alpha (I - e \otimes e) + \sin\alpha (r \otimes q - q \otimes r),$$
$$Q_2 = f \otimes f + \cos\alpha (I - f \otimes f) \pm \sin\alpha (\tilde{r} \otimes \tilde{q} - \tilde{q} \otimes \tilde{r}),$$

where $\{e, q, r\}$ and $\{f, \tilde{q}, \tilde{r}\}$ are orthonormal sets of axes, and then taking Q_0 as $f \otimes e + \tilde{q} \otimes q + \tilde{r} \otimes r$. □

1.6 Symmetric Tensors

In this section, we examine some properties of symmetric second-order tensors. We first discuss the properties of the principal values (eigenvalues) and principal directions (eigenvectors) of a symmetric second-order tensor.

1.6.1 Principal values and principal directions

We have the following result:

Theorem 1.6.1. *Every symmetric tensor $\boldsymbol{S}$ has at least one principal frame, i.e., a right-handed triplet of orthogonal principal directions, and at most three distinct principal values. The principal values are always real. For the principal directions three possibilities exist:*

- *If all the three principal values are distinct, the principal axes are unique (modulo sign reversal).*
- *If two eigenvalues are equal, then there is one unique principal direction, and the remaining two principal directions can be chosen arbitrarily in the plane perpendicular to the first one, and mutually perpendicular to each other.*
- *If all three eigenvalues are the same, then every right-handed frame is a principal frame, and $\boldsymbol{S}$ is of the form $\boldsymbol{S} = \lambda \boldsymbol{I}$.*

The components of the tensor in the principal frame are

$$S^* = \begin{bmatrix} \lambda_1 & 0 & 0 \\ 0 & \lambda_2 & 0 \\ 0 & 0 & \lambda_3 \end{bmatrix}. \tag{1.111}$$

Proof. We seek λ and $\boldsymbol{n}$ such that

$$(\boldsymbol{S} - \lambda \boldsymbol{I})\boldsymbol{n} = \boldsymbol{0}. \tag{1.112}$$

But this is nothing but an eigenvalue problem. For a nontrivial solution, we need to satisfy the condition that

$$\det(\boldsymbol{S} - \lambda \boldsymbol{I}) = 0,$$

or, by Eqn. (1.79),

$$\lambda^3 - I_1\lambda^2 + I_2\lambda - I_3 = 0, \tag{1.113}$$

where I_1, I_2 and I_3 are the principal invariants of $\boldsymbol{S}$.

We now show that the principal values given by the three roots of the cubic equation Eqn. (1.113) are real. Suppose that two roots, and hence the eigenvectors associated with them, are complex. Denoting the complex conjugates of λ and $\boldsymbol{n}$ by $\hat{\lambda}$ and $\hat{\boldsymbol{n}}$, we have

$$\boldsymbol{S}\boldsymbol{n} = \lambda \boldsymbol{n}, \tag{1.114a}$$

$$S\hat{n} = \hat{\lambda}\hat{n}, \tag{1.114b}$$

where Eqn. (1.114b) is obtained by taking the complex conjugate of Eqn. (1.114a) (S being a real matrix is not affected). Taking the dot product of both sides of Eqn. (1.114a) with $\hat{n}$, and of both sides of Eqn. (1.114b) with n, we get

$$\hat{n} \cdot Sn = \lambda n \cdot \hat{n}, \tag{1.115}$$

$$n \cdot S\hat{n} = \hat{\lambda}\hat{n} \cdot n. \tag{1.116}$$

Using the definition of a transpose of a tensor, and subtracting the second relation from the first, we get

$$S^T\hat{n} \cdot n - n \cdot S\hat{n} = (\lambda - \hat{\lambda})n \cdot \hat{n}. \tag{1.117}$$

Since S is symmetric, $S^T = S$, and we have

$$(\lambda - \hat{\lambda})n \cdot \hat{n} = 0.$$

Since $n \cdot \hat{n} \neq 0$, $\lambda = \hat{\lambda}$, and hence the eigenvalues are real.

The principal directions n_1, n_2 and n_3, corresponding to distinct eigenvalues λ_1, λ_2 and λ_3, are mutually orthogonal and unique (modulo sign reversal). We now prove this. Taking the dot product of

$$Sn_1 = \lambda_1 n_1, \tag{1.118}$$

$$Sn_2 = \lambda_2 n_2, \tag{1.119}$$

with n_2 and n_1, respectively, and subtracting, we get

$$0 = (\lambda_1 - \lambda_2)n_1 \cdot n_2,$$

where we have used the fact that S being symmetric, $n_2 \cdot Sn_1 - n_1 \cdot Sn_2 = 0$. Thus, since we assumed that $\lambda_1 \neq \lambda_2$, we get $n_1 \perp n_2$. Similarly, we have $n_2 \perp n_3$ and $n_1 \perp n_3$. If n_1 satisfies $Sn_1 = \lambda_1 n_1$, then we see that $-n_1$ also satisfies the same equation. This is the only other choice possible that satisfies $Sn_1 = \lambda_1 n_1$. To see this, let r_1, r_2 and r_3 be another set of mutually perpendicular eigenvectors corresponding to the distinct eigenvalues λ_1, λ_2 and λ_3. Then r_1 has to be perpendicular to not only r_2 and r_3, but to n_2 and n_3 as well. Similar comments apply to r_2 and r_3. This is only possible when $r_1 = \pm n_1$, $r_2 = \pm n_2$ and $r_3 = \pm n_3$. Thus, the principal axes are unique modulo sign reversal.

To prove that the components of S in the principal frame are given by Eqn. (1.111), assume that n_1, n_2, n_3 have been normalized to unit length, and then let $e_1^* = n_1$, $e_2^* = n_2$ and $e_3^* = n_3$. Using Eqn. (1.28), and taking into account the orthonormality of e_1^* and e_2^*, the components S_{11}^* and S_{12}^* are given by

$$S_{11}^* = e_1^* \cdot Se_1^* = e_1^* \cdot (\lambda_1 e_1^*) = \lambda_1,$$

$$S_{12}^* = e_1^* \cdot Se_2^* = e_1^* \cdot (\lambda_2 e_2^*) = 0.$$

Similarly, on computing the other components, we see that the matrix representation of S with respect to e^* is given by Eqn. (1.111).

If there are two repeated roots, say, $\lambda_2 = \lambda_3$, and the third root $\lambda_1 \neq \lambda_2$, then let e_1^* coincide with n_1, so that $Se_1^* = \lambda_1 e_1^*$. Choose e_2^* and e_3^* such that e_1^*-e_3^* form a right-handed orthogonal coordinate system. The components of S with respect to this coordinate system are

$$S^* = \begin{bmatrix} \lambda_1 & 0 & 0 \\ 0 & S_{22}^* & S_{23}^* \\ 0 & S_{23}^* & S_{33}^* \end{bmatrix}. \tag{1.120}$$

By Eqn. (1.80), we have

$$\begin{aligned} S_{22}^* + S_{33}^* &= 2\lambda_2, \\ \lambda_1 S_{22}^* + (S_{22}^* S_{33}^* - (S_{23}^*)^2) + \lambda_1 S_{33}^* &= 2\lambda_1\lambda_2 + \lambda_2^2, \\ \lambda_1 \left[S_{22}^* S_{33}^* - (S_{23}^*)^2 \right] &= \lambda_1 \lambda_2^2. \end{aligned} \tag{1.121}$$

Substituting for λ_2 from the first equation into the second, we get

$$(S_{22}^* - S_{33}^*)^2 = -4(S_{23}^*)^2.$$

Since the components of S are real, the above equation implies that $S_{23}^* = 0$ and $\lambda_2 = S_{22}^* = S_{33}^*$. This shows that Eqn. (1.120) reduces to Eqn. (1.111), and that $Se_2^* = S_{12}^* e_1^* + S_{22}^* e_2^* + S_{32}^* e_3^* = \lambda_2 e_2^*$ and $Se_3^* = \lambda_2 e_3^*$ (thus, e_2^* and e_3^* are eigenvectors corresponding to the eigenvalue λ_2). However, in this case the choice of the principal frame e^* is not unique, since any vector lying in the plane of e_2^* and e_3^*, given by $n^* = c_1 e_2^* + c_2 e_3^*$ where c_1 and c_2 are arbitrary constants, is also an eigenvector. The choice of e_1^* is unique (modulo sign reversal), since it has to be perpendicular to e_2^* and e_3^*. Though the choice of e_2^* and e_3^* is not unique, we can choose e_2^* and e_3^* arbitrarily in the plane perpendicular to e_1^*, and such that $e_2^* \perp e_3^*$.

Finally, if $\lambda_1 = \lambda_2 = \lambda_3 = \lambda$, then the tensor S is of the form $S = \lambda I$. To show this choose e_1^*–e_3^*, and follow a procedure analogous to that in the previous case. We now get $S_{22}^* = S_{33}^* = \lambda$ and $S_{23}^* = 0$, so that $[S^*] = \lambda I$. Using the transformation law for second-order tensors, we have

$$[S] = Q^T [S^*] Q = \lambda Q^T Q = \lambda I.$$

Thus, any arbitrary vector n is a solution of $Sn = \lambda n$, and hence every right-handed frame is a principal frame. □

As a result of Eqn. (1.111), we can write

$$S = \lambda_1 e_1^* \otimes e_1^* + \lambda_2 e_2^* \otimes e_2^* + \lambda_3 e_3^* \otimes e_3^*, \tag{1.122}$$

which is called as the *spectral resolution* of S. The spectral resolution of S is unique since

- If all the eigenvalues are distinct, then the eigenvectors are unique, and consequently the representation given by Eqn. (1.122) is unique.
- If two eigenvalues are repeated, then, by virtue of Eqn. (1.38), Eqn. (1.122) reduces to

$$\begin{aligned} S &= \lambda_1 e_1^* \otimes e_1^* + \lambda_2 e_2^* \otimes e_2^* + \lambda_2 e_3^* \otimes e_3^* \\ &= \lambda_1 e_1^* \otimes e_1^* + \lambda_2 (I - e_1^* \otimes e_1^*), \end{aligned} \tag{1.123}$$

from which the asserted uniqueness follows, since e_1^* is unique.
- If all the eigenvalues are the same then $S = \lambda I$.

Although the eigenvalues of $S \in \text{Sym}$ are real, its eigenvectors can be complex (note that in this case, real eigenvectors, obtained by combining the complex eigenvectors, *also* exist). For example, if $S = \mathbf{diag}[1,2,3]$, then $n = (1+i)[1,0,0]/\sqrt{2}$ and $\hat{n} = (1-i)[1,0,0]/\sqrt{2}$ are eigenvectors corresponding to the eigenvalue 1. If $S = \mathbf{diag}[1,1,2]$, then $n = [1,i,0]/\sqrt{2}$ and $\hat{n} = [1,-i,0]/\sqrt{2}$ are eigenvectors corresponding to the repeated eigenvalue 1. Note that the eigenvectors in such a case occur in complex conjugate pairs (which is proved simply by taking the complex conjugate of the relation $Sn = \lambda n$, so that $(n, \hat{n})$ are the eigenvectors corresponding to the eigenvalue λ). Note, however, that although the eigenvectors can be complex, the eigenprojections P_i as given by Eqn. (J.4) are real.

Theorem 1.6.2. *Let S be given by Eqn. (1.122), and let $\{e_k\}$ denote the canonical basis of V. Then $Q = e_k \otimes e_k^*$ is a proper orthogonal tensor that diagonalizes S, i.e., the matrix representation of QSQ^T with respect to the canonical basis is diagonal.*

Conversely, if $Q \in \text{Orth}^+$ diagonalizes a tensor S, i.e., $QSQ^T = \Lambda$ where Λ is a diagonal matrix, then S is symmetric, the diagonal matrix Λ comprises the eigenvalues of S, and the rows of Q are the eigenvectors of S.

Proof. By Theorem 1.5.2, $Q = e_k \otimes e_k^* \in \text{Orth}^+$, and since $e_i = Qe_i^*$, we have

$$\begin{aligned} QSQ^T &= Q\left[\sum_i \lambda_i e_i^* \otimes e_i^*\right] Q^T \\ &= \sum_i \lambda_i (Qe_i^*) \otimes (Qe_i^*) \\ &= \sum_i \lambda_i e_i \otimes e_i, \end{aligned}$$

which is a diagonal matrix with λ_i along the diagonals.

Conversely, if $QSQ^T = \Lambda$ where Λ is a diagonal matrix, then $S = Q^T \Lambda Q$ is clearly a symmetric tensor. By Theorem 1.6.10, S and Λ have the same eigenvalues, and hence Λ has the eigenvalues of S along the diagonal. If follows that

$$S = Q^T \Lambda Q = Q^T \left[\sum_i \lambda_i e_i \otimes e_i\right] Q = \sum_i \lambda_i (Q^T e_i) \otimes (Q^T e_i),$$

which implies that $S(Q^T e_i) = \lambda_i (Q^T e_i)$ (no sum on i); thus, $Q^T e_i$ are the eigenvectors of S. Since $Q = IQ$, we have

$$Q = (e_k \otimes e_k) Q = e_k \otimes (Q^T e_k) = e_k \otimes e_k^*,$$

which shows that the rows of Q are the eigenvectors of S. □

The Cayley–Hamilton theorem (Theorem 1.3.5) applied to $S \in \text{Sym}$ yields

$$S^3 - I_1 S^2 + I_2 S - I_3 I = 0. \tag{1.124}$$

We have already proved this result for any arbitrary tensor. However, the following simpler proof can be given for symmetric tensors. Using Eqn. (1.341), the spectral resolutions of S, S^2 and S^3 are

$$\begin{aligned} S &= \lambda_1 e_1^* \otimes e_1^* + \lambda_2 e_2^* \otimes e_2^* + \lambda_3 e_3^* \otimes e_3^*, \\ S^2 &= \lambda_1^2 e_1^* \otimes e_1^* + \lambda_2^2 e_2^* \otimes e_2^* + \lambda_3^2 e_3^* \otimes e_3^*, \\ S^3 &= \lambda_1^3 e_1^* \otimes e_1^* + \lambda_2^3 e_2^* \otimes e_2^* + \lambda_3^3 e_3^* \otimes e_3^*. \end{aligned} \tag{1.125}$$

Substituting these expressions into the left-hand side of Eqn. (1.124), we get

$$\begin{aligned} \text{LHS} = (\lambda_1^3 - I_1 \lambda_1^2 + I_2 \lambda_1 - I_3)(e_1^* \otimes e_1^*) + (\lambda_2^3 - I_1 \lambda_2^2 + I_2 \lambda_2 - I_3)(e_2^* \otimes e_2^*) \\ + (\lambda_3^3 - I_1 \lambda_3^2 + I_2 \lambda_3 - I_3)(e_3^* \otimes e_3^*) = 0, \end{aligned}$$

since $\lambda_i^3 - I_1 \lambda_i^2 + I_2 \lambda_i - I_3 = 0$ for $i = 1, 2, 3$.

$\{S - I_1 I/3 : S \in \text{Sym}\}$ constitutes the set of traceless symmetric tensors, while tensors of the form $(a \otimes b + b \otimes a)/2$, $a, b \in V$ have determinant zero (a fact that can be verified by writing $(a \otimes b + b \otimes a)/2$ as $(a \otimes b + b \otimes a + 0 \otimes 0)/2$ and then using Eqn. (1.53)). However, not all symmetric tensors with determinant zero can be represented as $(a \otimes b + b \otimes a)/2$ since, from the Cauchy–Schwartz inequality it follows that the second invariant of $(a \otimes b + b \otimes a)/2$ computed using Eqn. (1.63) is always negative.

The following explicit expressions can be given for the eigenvalues of $S \in \text{Sym}$ when it is not of the form λI [298]:

$$\begin{aligned} \lambda_1 &= \frac{I_1}{3} + 2\sqrt{p} \cos\phi, \\ \lambda_2 &= \frac{I_1}{3} - \sqrt{p}(\cos\phi + \sqrt{3}\sin\phi), \\ \lambda_3 &= \frac{I_1}{3} - \sqrt{p}(\cos\phi - \sqrt{3}\sin\phi), \end{aligned}$$

where

$$\begin{aligned} p &= \frac{1}{6}\left(S - \frac{I_1}{3} I\right) : \left(S - \frac{I_1}{3} I\right), \\ q &= \frac{1}{2} \det\left(S - \frac{I_1}{3} I\right), \end{aligned}$$

$$\phi = \frac{1}{3}\tan^{-1}\frac{\sqrt{p^3 - q^2}}{q}, \quad 0 \le \phi \le \pi.$$

As can be seen from the above expressions, $\lambda_2 = \lambda_3$ when $p^3 = q^2$.

Alternative expressions for finding the eigenvalues are

$$\begin{aligned} \lambda_1 &= \frac{I_1}{3} - 2\sqrt{Q}\cos\alpha, \\ \lambda_2 &= \frac{I_1}{3} - 2\sqrt{Q}\cos\left(\alpha + \frac{2\pi}{3}\right), \\ \lambda_3 &= \frac{I_1}{3} - 2\sqrt{Q}\cos\left(\alpha - \frac{2\pi}{3}\right), \end{aligned} \tag{1.126}$$

where

$$\begin{aligned} Q &= \frac{I_1^2 - 3I_2}{9}, \\ U &= \frac{1}{54}\left[9I_1I_2 - 2I_1^3 - 27I_3\right], \\ \alpha &= \frac{1}{3}\cos^{-1}\left(\frac{U}{\sqrt{Q^3}}\right). \end{aligned}$$

If $Q = 0$ (i.e., if $I_1^2 = 3I_2$), then all the eigenvalues are repeated, $\lambda_1 = \lambda_2 = \lambda_3 = I_1/3$. If $Q \neq 0$ and $Q^3 = U^2$ (i.e., if $I_1^2I_2^2 - 4I_2^3 - 4I_1^3I_3 + 18I_1I_2I_3 - 27I_3^2 = 0$), then $\lambda_1 = I_1/3 - 2\sqrt{Q}$ and $\lambda_2 = \lambda_3 = I_1/3 + \sqrt{Q}$ if $U > 0$, and $\lambda_2 = I_1/3 + 2\sqrt{Q}$ and $\lambda_1 = \lambda_3 = I_1/3 - \sqrt{Q}$ if $U < 0$. If $Q \neq 0$ and $Q^3 \neq U^2$, then the eigenvalues are distinct.

We saw in Theorem 1.6.1 that the component form of a symmetric tensor is a diagonal matrix in a particular basis. The following theorem states the conditions under which the matrix representation has all the diagonal terms zero [18], [27], [117]:

Theorem 1.6.3. *There exists an orthonormal basis $\{\boldsymbol{n}_i\}$, $i = 1,2,3$, for which*

$$\boldsymbol{n}_1 \cdot \boldsymbol{S}\boldsymbol{n}_1 = \boldsymbol{n}_2 \cdot \boldsymbol{S}\boldsymbol{n}_2 = \boldsymbol{n}_3 \cdot \boldsymbol{S}\boldsymbol{n}_3 = 0, \tag{1.127}$$

if and only if $\operatorname{tr}\boldsymbol{S} = 0$.

Proof. If Eqn. (1.127) holds then clearly $\operatorname{tr}\boldsymbol{S} = \boldsymbol{n}_1 \cdot \boldsymbol{S}\boldsymbol{n}_1 + \boldsymbol{n}_2 \cdot \boldsymbol{S}\boldsymbol{n}_2 + \boldsymbol{n}_3 \cdot \boldsymbol{S}\boldsymbol{n}_3 = 0$. Thus, we just need to prove the converse. If $\operatorname{tr}\boldsymbol{S} = \lambda_1 + \lambda_2 + \lambda_3 = 0$, then the spectral resolution of $\boldsymbol{S}$ can be written as

$$\boldsymbol{S} = \lambda_1(\boldsymbol{e}_1^* \otimes \boldsymbol{e}_1^* - \boldsymbol{e}_3^* \otimes \boldsymbol{e}_3^*) + \lambda_2(\boldsymbol{e}_2^* \otimes \boldsymbol{e}_2^* - \boldsymbol{e}_3^* \otimes \boldsymbol{e}_3^*).$$

We have seen that $\{\boldsymbol{e}_i^*\}$, $i = 1,2,3$ can always be chosen to be an orthonormal basis, and we assume that such a choice has been made. Note that when $\operatorname{tr}\boldsymbol{S} = 0$, at most one eigenvalue can be zero, since if two or more are zero, the tensor $\boldsymbol{S}$

is the zero tensor. Thus, we can assume, without loss in generality, that both λ_1 and λ_2 are nonzero, and $\lambda_3 \equiv -(\lambda_1 + \lambda_2)$ can be either zero or nonzero. If we choose

$$n_1 = (e_1^* + e_2^* + e_3^*)/\sqrt{3}, \tag{1.128}$$

it is clear that $n_1 \cdot Sn_1 = 0$. We now try and find n_2 and n_3 in the plane perpendicular to n_1 such that $n_2 \cdot Sn_2 = n_3 \cdot Sn_3 = 0$. Since $n_1 \cdot n_2 = 0$, n_2 and n_3 are of the form

$$\begin{aligned} n_2 &= \frac{\gamma_1 e_1^* + \gamma_2 e_2^* - (\gamma_1 + \gamma_2) e_3^*}{\sqrt{\gamma_1^2 + \gamma_2^2 + (\gamma_1 + \gamma_2)^2}}, \\ n_3 = n_1 \times n_2 &= \frac{-(\gamma_1 + 2\gamma_2) e_1^* + (2\gamma_1 + \gamma_2) e_2^* + (\gamma_2 - \gamma_1) e_3^*}{\sqrt{3}\sqrt{\gamma_1^2 + \gamma_2^2 + (\gamma_1 + \gamma_2)^2}}, \end{aligned} \tag{1.129}$$

where γ_1, γ_2 are constants to be determined. Both the conditions $n_2 \cdot Sn_2 = 0$ and $n_3 \cdot Sn_3 = 0$ yield the equation

$$\lambda_1 \gamma_2^2 + 2(\lambda_1 + \lambda_2)\gamma_1\gamma_2 + \lambda_2\gamma_1^2 = 0,$$

which on solving yields (recall that both λ_1 and λ_2 are nonzero)

$$\gamma_2 = \left[\frac{-(\lambda_1 + \lambda_2) \pm \sqrt{\lambda_1^2 + \lambda_1\lambda_2 + \lambda_2^2}}{\lambda_1} \right] \gamma_1.$$

It suffices to consider only the solution with the positive sign, since it can be shown that the solution obtained with the negative sign is simply a rearrangement of the one obtained with the positive one, i.e., if $(\tilde{n}_1, \tilde{n}_2, \tilde{n}_3)$ is the solution obtained with the positive sign, then $(\tilde{n}_1, -\tilde{n}_3, \tilde{n}_2)$ is the solution obtained with the negative one. Substituting for γ_2 into Eqns. (1.129), we get

$$\begin{aligned} n_2 &= \frac{g_1 e_1^* + g_2 e_2^* + g_3 e_3^*}{\sqrt{g_1^2 + g_2^2 + g_3^2}}, \\ n_3 &= \frac{h_1 e_1^* + h_2 e_2^* + h_3 e_3^*}{\sqrt{h_1^2 + h_2^2 + h_3^2}}, \end{aligned} \tag{1.130}$$

where

$$\begin{aligned} g_1 &= \lambda_1, \\ g_2 &= \sqrt{\lambda_1^2 + \lambda_1\lambda_2 + \lambda_2^2} - (\lambda_1 + \lambda_2), \\ g_3 &= \lambda_2 - \sqrt{\lambda_1^2 + \lambda_1\lambda_2 + \lambda_2^2}, \end{aligned}$$

$$h_1 = \lambda_1 + 2\lambda_2 - 2\sqrt{\lambda_1^2 + \lambda_1\lambda_2 + \lambda_2^2},$$
$$h_2 = \lambda_1 - \lambda_2 + \sqrt{\lambda_1^2 + \lambda_1\lambda_2 + \lambda_2^2},$$
$$h_3 = -2\lambda_1 - \lambda_2 + \sqrt{\lambda_1^2 + \lambda_1\lambda_2 + \lambda_2^2}.$$

Equations (1.128) and (1.130) are the desired solutions.

The solution that has been presented above is only one choice in an infinite one-parameter family of solutions in terms of a parameter θ. To find this infinite one-parameter family, assume a unit vector in the principal basis $\{e_i^*\}$ to be given by

$$n_1 = (\sin\theta\cos\phi, \sin\theta\sin\phi, \cos\theta),$$

and also assume without loss in generality that λ_1 and λ_2 are chosen such that $\lambda_1 \neq \lambda_2$. Using the fact that S in this coordinate system is given by $\mathbf{diag}[\lambda_1, \lambda_2, -(\lambda_1 + \lambda_2)]$, and eliminating ϕ by using the condition $n_1 \cdot Sn_1 = 0$ yields

$$n_1(\theta) = \begin{bmatrix} \pm\sqrt{\frac{(\lambda_1+2\lambda_2)\cos^2\theta-\lambda_2}{\lambda_1-\lambda_2}} \\ \pm\sqrt{\frac{(\lambda_2+2\lambda_1)\cos^2\theta-\lambda_1}{\lambda_2-\lambda_1}} \\ \pm\cos\theta \end{bmatrix}.$$

The second vector n_2 also satisfies $n_2 \cdot Sn_2 = 0$, and hence has to be of the form $n_1(\alpha)$, where α is found in terms of θ by imposing the condition that $n_1 \cdot n_2 = 0$. Denoting $t \equiv \cos^2\theta$, we get

$$\cos^2\alpha = \frac{h_1 \pm \sqrt{h_2}}{D}, \tag{1.131}$$

where

$$h_1 = \lambda_1\lambda_2(-2 + 7t - 5t^2) + (\lambda_1^2 + \lambda_2^2)(-1 + 3t - 2t^2),$$
$$h_2 = (\lambda_1^4 + \lambda_2^4)(1 - 2t)^2 t^2 + (\lambda_1^3\lambda_2 + \lambda_1\lambda_2^3)(-1 + 8t - 22t^2 + 20t^3)t +$$
$$\lambda_1^2\lambda_2^2(-2 + 15t - 38t^2 + 33t^3)t,$$
$$D = 2(\lambda_1^2 + \lambda_2^2)(-1 + 2t) + 2\lambda_1\lambda_2(-2 + 5t).$$

Denoting the two roots for $\cos^2\alpha$ corresponding to the plus and minus signs in Eqn. (1.131) by $\cos^2\alpha_1$ and $\cos^2\alpha_2$, we get $n_2 = n_1(\alpha_1)$ and $n_3 = n_1(\alpha_2)$. The plus or minus signs for the components of n_1 can be chosen arbitrarily; the corresponding signs for n_2 are obtained by imposing the condition $n_1 \cdot n_2 = 0$, and then n_3 is found as $n_1 \times n_2$. □

The result in Theorem 1.6.3 actually holds for an arbitrary traceless tensor $\boldsymbol{T}$. This is seen by splitting $\boldsymbol{T}$ into its symmetric and antisymmetric parts, and since the diagonal elements of the antisymmetric part are zero in any basis, the basis found for the symmetric part works. As a simple example of the above theorem, let the matrix representation of $\boldsymbol{S}$ with respect to a basis be $\mathbf{diag}[1,-1,0]$. Then, by convention, we choose λ_1 and λ_2 to be the nonzero eigenvalues, i.e., $\lambda_1 = 1$, $\lambda_2 = -1$. The corresponding eigenvectors are $\boldsymbol{e}_1^* = (1,0,0)$, $\boldsymbol{e}_2^* = (0,1,0)$ and $\boldsymbol{e}_3^* = (0,0,1)$. From Eqns. (1.128) and (1.130), we get $\boldsymbol{n}_1 = (1,1,1)/\sqrt{3}$, $\boldsymbol{n}_2 = (1,1,-2)/\sqrt{6}$, and $\boldsymbol{n}_3 = (-1,1,0)/\sqrt{2}$. A more complicated example is given in Section 3.7.

The decomposition of a matrix into two symmetric components, i.e., $\boldsymbol{T} = \boldsymbol{S}_1\boldsymbol{S}_2$, where $\boldsymbol{S}_1, \boldsymbol{S}_2 \in \text{Sym}$, and where one of the factors can be taken to be invertible [26], can be found as follows. Assuming $\boldsymbol{S}_2$ to be invertible, we get $\boldsymbol{S}_1 = \boldsymbol{T}\boldsymbol{S}_2^{-1}$. Since $\boldsymbol{S}_1 \in \text{Sym}$, we get $\boldsymbol{S}_2^{-1}\boldsymbol{T}^T = \boldsymbol{T}\boldsymbol{S}_2^{-1}$, or, alternatively, $\boldsymbol{T}^T\boldsymbol{S}_2 = \boldsymbol{S}_2\boldsymbol{T}$. This results in at most $n(n-1)/2$ independent equations for the $n(n+1)/2$ components of $\boldsymbol{S}_2$. For example, for dimension $n = 2$, we get

$$\begin{bmatrix} -T_{12} & T_{21} & T_{11} - T_{22} \\ 0 & 0 & 0 \\ 0 & 0 & 0 \end{bmatrix} \begin{bmatrix} S_{11} \\ S_{22} \\ S_{12} \end{bmatrix} = \boldsymbol{0}, \tag{1.132}$$

while for dimension $n = 3$, we get

$$\begin{bmatrix} -T_{12} & T_{21} & 0 & T_{11} - T_{22} & T_{31} & -T_{32} \\ -T_{13} & 0 & T_{31} & -T_{23} & T_{21} & T_{11} - T_{33} \\ 0 & -T_{23} & T_{32} & -T_{13} & T_{22} - T_{33} & T_{12} \\ 0 & 0 & 0 & 0 & 0 & 0 \\ 0 & 0 & 0 & 0 & 0 & 0 \\ 0 & 0 & 0 & 0 & 0 & 0 \end{bmatrix} \begin{bmatrix} S_{11} \\ S_{22} \\ S_{33} \\ S_{12} \\ S_{23} \\ S_{13} \end{bmatrix} = \boldsymbol{0}.$$

The decomposition is obviously not unique since, e.g., $\boldsymbol{I}$ can be factorized as $(\boldsymbol{I})(\boldsymbol{I})$ or as $(-\boldsymbol{I})(-\boldsymbol{I})$.

In general, it is not possible to express the factors $\boldsymbol{S}_1$ and $\boldsymbol{S}_2$ as a function of $\boldsymbol{T}$. To see this, let $\boldsymbol{T} = \left[\begin{smallmatrix} 0 & w \\ -w & 0 \end{smallmatrix}\right]$, whose determinant is nonzero, so that both $\boldsymbol{S}_1$ and $\boldsymbol{S}_2$ are invertible. Using Eqn. (1.132), we get the most general solution as

$$\begin{bmatrix} 0 & w \\ -w & 0 \end{bmatrix} = \frac{1}{a^2 + b^2} \begin{bmatrix} bw & -aw \\ -aw & -bw \end{bmatrix} \begin{bmatrix} a & b \\ b & -a \end{bmatrix}, \quad a, b \in \Re.$$

One can see that it is not possible to express either $\boldsymbol{S}_1$ or $\boldsymbol{S}_2$ as a function of $\boldsymbol{T}$.

1.6.2 Positive definite tensors and the polar decomposition

A second-order symmetric tensor $\boldsymbol{S}$ is positive definite if

$$(\boldsymbol{u}, \boldsymbol{S}\boldsymbol{u}) \geq 0 \quad \forall \boldsymbol{u} \in V \text{ with } (\boldsymbol{u}, \boldsymbol{S}\boldsymbol{u}) = 0 \text{ if and only if } \boldsymbol{u} = \boldsymbol{0}.$$

We denote the set of symmetric, positive definite tensors by Psym. Since by virtue of Eqn. (1.33), all tensors T can be decomposed into a symmetric part T_s, and a skew-symmetric part T_{ss}, we have

$$\begin{aligned}(u, Tu) &= (u, T_s u) + (u, T_{ss} u),\\ &= (u, T_s u),\end{aligned}$$

because $(u, T_{ss} u) = 0$ by Eqn. (1.86). Thus, the positive definiteness of a tensor is decided by the positive definiteness of its symmetric part. In Theorem 1.6.4, we show that a symmetric tensor is positive definite if and only if its eigenvalues are positive. Although the eigenvalues of the *symmetric part* of T should be positive in order for T to be positive definite, positiveness of the eigenvalues of T itself does not ensure its positive definiteness as the following counterexample shows. If $T = \begin{bmatrix} 1 & -10 \\ 0 & 1 \end{bmatrix}$, then T is not positive definite since $u \cdot Tu < 0$ for $u = (1, 1)$, but the eigenvalues of T are $(1, 1)$. Conversely, if T is positive definite, then by choosing u to be the real eigenvectors n of T, it follows that its real eigenvalues $\lambda = (n \cdot Tn)$ are positive.

Theorem 1.6.4. *Let $S \in$ Sym. Then the following are equivalent:*

1. *S is positive definite.*
2. *The principal values of S are strictly positive.*
3. *The principal invariants of S are strictly positive.*

Proof. We first prove the equivalence of (1) and (2). Suppose S is positive definite. If λ and n denote the principal values and principal directions, respectively, of S, then $Sn = \lambda n$, which implies that $\lambda = (n, Sn) > 0$ since $n \neq 0$.

Conversely, suppose that the principal values of S are greater than 0. Assuming that e_1^*, e_2^* and e_3^* denote the principal axes, the representation of S in the principal coordinate frame is (see Eqn. (1.122))

$$S = \lambda_1 e_1^* \otimes e_1^* + \lambda_2 e_2^* \otimes e_2^* + \lambda_3 e_3^* \otimes e_3^*.$$

Then

$$\begin{aligned}Su &= (\lambda_1 e_1^* \otimes e_1^* + \lambda_2 e_2^* \otimes e_2^* + \lambda_3 e_3^* \otimes e_3^*)u\\ &= \lambda_1 (e_1^* \cdot u) e_1^* + \lambda_2 (e_2^* \cdot u) e_2^* + \lambda_3 (e_3^* \cdot u) e_3^*\\ &= \lambda_1 u_1^* e_1^* + \lambda_2 u_2^* e_2^* + \lambda_3 u_3^* e_3^*,\end{aligned}$$

and

$$(u, Su) = u \cdot Su = \lambda_1 (u_1^*)^2 + \lambda_2 (u_2^*)^2 + \lambda_3 (u_3^*)^2, \tag{1.133}$$

which is greater than or equal to zero since $\lambda_i > 0$. Suppose that $(u, Su) = 0$. Then by Eqn. (1.133), $u_i^* = 0$, which implies that $u = 0$. Thus, S is a positive definite tensor.

To prove the equivalence of (2) and (3), note that, by Eqn. (1.80), if all the principal values are strictly positive, then the principal invariants are also strictly positive. Conversely, if all the principal invariants are positive, then $I_3 = \lambda_1\lambda_2\lambda_3$, is positive, so that all the λ_i are nonzero in addition to being real. Each λ_i has to satisfy the characteristic equation

$$\lambda_i^3 - I_1\lambda_i^2 + I_2\lambda_i - I_3 = 0, \quad i = 1,2,3.$$

If λ_i is negative, then, since I_1, I_2, I_3 are positive, the left-hand side of the above equation is negative, and hence the above equation cannot be satisfied. We have already mentioned that λ_i cannot be zero. Hence, each λ_i has to be positive. □

Theorem 1.6.5. *If $\boldsymbol{S} \in Psym$, then there exists a unique $\boldsymbol{H} \in Psym$, such that $\boldsymbol{H}^2 := \boldsymbol{H}\boldsymbol{H} = \boldsymbol{S}$. The tensor $\boldsymbol{H}$ is called the positive definite square root of $\boldsymbol{S}$, and we write $\boldsymbol{H} = \sqrt{\boldsymbol{S}}$.*

Proof. Before we begin the proof, we note that a positive definite, symmetric tensor can have square roots that are not positive definite. For example, **diag**$[1, -1, 1]$ is a non-positive definite square root of $\boldsymbol{I}$. Here, we are interested only in those square roots that are positive definite. The short proof of uniqueness given below is due to Stephenson [305].

Since

$$\boldsymbol{S} = \sum_{i=1}^{3} \lambda_i \boldsymbol{e}_i^* \otimes \boldsymbol{e}_i^*,$$

is positive definite, by Theorem 1.6.4, all $\lambda_i > 0$. Define

$$\boldsymbol{H} := \sum_{i=1}^{3} \sqrt{\lambda_i} \boldsymbol{e}_i^* \otimes \boldsymbol{e}_i^*. \tag{1.134}$$

Since the $\{\boldsymbol{e}_i^*\}$ are orthonormal, it is easily seen that $\boldsymbol{H}\boldsymbol{H} = \boldsymbol{S}$. Since the eigenvalues of $\boldsymbol{H}$ given by $\sqrt{\lambda_i}$ are all positive, $\boldsymbol{H}$ is positive definite. Thus, we have shown that a positive definite square root tensor of $\boldsymbol{S}$ given by Eqn. (1.134) exists. We now prove uniqueness.

With λ and $\boldsymbol{n}$ denoting the principal value and principal direction of $\boldsymbol{S}$, we have

$$\begin{aligned} \boldsymbol{0} &= (\boldsymbol{S} - \lambda\boldsymbol{I})\boldsymbol{n} \\ &= (\boldsymbol{H}^2 - \lambda\boldsymbol{I})\boldsymbol{n} \\ &= (\boldsymbol{H} + \sqrt{\lambda}\boldsymbol{I})(\boldsymbol{H} - \sqrt{\lambda}\boldsymbol{I})\boldsymbol{n}. \end{aligned}$$

Calling $(\boldsymbol{H} - \sqrt{\lambda}\boldsymbol{I})\boldsymbol{n} = \bar{\boldsymbol{n}}$, we have

$$(\boldsymbol{H} + \sqrt{\lambda}\boldsymbol{I})\bar{\boldsymbol{n}} = \boldsymbol{0}.$$

This implies that $\bar{\boldsymbol{n}} = \boldsymbol{0}$. For, if not, $-\sqrt{\lambda}$ is a principal value of $\boldsymbol{H}$, which contradicts the fact that $\boldsymbol{H}$ is positive definite. Therefore,

$$(\boldsymbol{H} - \sqrt{\lambda}\boldsymbol{I})\boldsymbol{n} = \boldsymbol{0};$$

i.e., n is also a principal direction of H with associated principal values $\sqrt{\lambda}$. If $H = \sqrt{S}$, then it must have the form given by Eqn. (1.134) (since the spectral decomposition is unique), which establishes its uniqueness. □

Now we prove the polar decomposition theorem, which plays a crucial role in the characterization of frame-indifferent constitutive relations.

Theorem 1.6.6 (Polar Decomposition Theorem). *Let F be an invertible tensor. Then, it can be factored in a unique fashion as*

$$F = RU = VR,$$

where R is an orthogonal tensor, and U, V are symmetric and positive definite tensors. One has

$$U = \sqrt{F^T F}$$
$$V = \sqrt{FF^T}.$$

Proof. The tensor $F^T F$ is obviously symmetric. It is positive definite since

$$(u, F^T F u) = (Fu, Fu) \geq 0,$$

with equality if and only if $u = 0$ ($Fu = 0$ implies that $u = 0$, since F is invertible). Let $U = \sqrt{F^T F}$. U is unique, symmetric and positive definite by Theorem 1.6.5. Define $R = FU^{-1}$, so that $F = RU$. The tensor R is orthogonal, because

$$\begin{aligned} R^T R &= (FU^{-1})^T (FU^{-1}) \\ &= U^{-T} F^T F U^{-1} \\ &= U^{-1}(F^T F)U^{-1} \quad \text{(since } U \text{ is symmetric)} \\ &= U^{-1}(UU)U^{-1} \\ &= I. \end{aligned}$$

Since $\det U > 0$, we have $\det U^{-1} > 0$. Hence, $\det R$ and $\det F$ have the same sign. Usually, the polar decomposition theorem is applied to the deformation gradient F satisfying $\det F > 0$. In such a case $\det R = 1$, and R is a rotation.

Next, let $V = FUF^{-1} = FR^{-1} = RUR^{-1} = RUR^T$. Thus, V is symmetric since U is symmetric. V is positive definite since

$$\begin{aligned} (u, Vu) &= (u, RUR^T u) \\ &= (R^T u, UR^T u) \\ &\geq 0, \end{aligned}$$

with equality if and only if $u = 0$ (again since R^T is invertible). Note that $FF^T = VV = V^2$, so that $V = \sqrt{FF^T}$.

Finally, to prove the uniqueness of the polar decomposition, we note that since U is unique, $R = FU^{-1}$ is unique, and hence so is V. □

Let $(\lambda_i, \boldsymbol{e}_i^*)$ denote the eigenvalues/eigenvectors of $\boldsymbol{U}$. Then, since $\boldsymbol{V}\boldsymbol{R}\boldsymbol{e}_i^* = \boldsymbol{R}\boldsymbol{U}\boldsymbol{e}_i^* = \lambda_i(\boldsymbol{R}\boldsymbol{e}_i^*)$, the pairs $(\lambda_i, \boldsymbol{f}_i^*)$, where $\boldsymbol{f}_i \equiv \boldsymbol{R}\boldsymbol{e}_i^*$ are the eigenvalues/eigenvectors of $\boldsymbol{V}$. Thus, $\boldsymbol{F}$ and $\boldsymbol{R}$ can be represented as

$$\begin{aligned}
\boldsymbol{F} &= \boldsymbol{R}\boldsymbol{U} = \boldsymbol{R}\sum_{i=1}^{3}\lambda_i \boldsymbol{e}_i^* \otimes \boldsymbol{e}_i^* = \sum_{i=1}^{3}\lambda_i(\boldsymbol{R}\boldsymbol{e}_i^*) \otimes \boldsymbol{e}_i^* = \sum_{i=1}^{3}\lambda_i \boldsymbol{f}_i^* \otimes \boldsymbol{e}_i^*, \\
\boldsymbol{R} &= \boldsymbol{R}\boldsymbol{I} = \boldsymbol{R}\sum_{i=1}^{3}\boldsymbol{e}_i^* \otimes \boldsymbol{e}_i^* = \sum_{i}(\boldsymbol{R}\boldsymbol{e}_i^*) \otimes \boldsymbol{e}_i^* = \sum_{i=1}^{3}\boldsymbol{f}_i^* \otimes \boldsymbol{e}_i^*.
\end{aligned} \tag{1.135}$$

When $\boldsymbol{F}$ is invertible, the polar decomposition can be given in explicit form in terms of the eigenvalues $\{\lambda_1, \lambda_2, \lambda_3\}$ of $\boldsymbol{U}$, and powers of $\boldsymbol{C}$ or $\boldsymbol{B}$. First, one determines the eigenvalues $\{\lambda_1^2, \lambda_2^2, \lambda_3^2\}$ of $\boldsymbol{C} = \boldsymbol{F}^T\boldsymbol{F}$ (which are also the eigenvalues of $\boldsymbol{B} = \boldsymbol{F}\boldsymbol{F}^T = \boldsymbol{R}\boldsymbol{C}\boldsymbol{R}^T$), using Eqn. (1.126). Then, depending on whether the eigenvalues are repeated or distinct, the relevant expressions for the component matrices in the polar decomposition are ([126], [153], [326])

All eigenvalues distinct ($\lambda_1 \neq \lambda_2 \neq \lambda_3 \neq \lambda_1$):

$$\begin{aligned}
\boldsymbol{R} &= \boldsymbol{F}\left[\frac{a_1}{a}\boldsymbol{I} + \frac{a_2}{a}\boldsymbol{C} + \frac{a_3}{a}\boldsymbol{C}^2\right], \\
\boldsymbol{U}^{-1} &= \frac{a_1}{a}\boldsymbol{I} + \frac{a_2}{a}\boldsymbol{C} + \frac{a_3}{a}\boldsymbol{C}^2, \\
\boldsymbol{V}^{-1} &= \frac{a_1}{a}\boldsymbol{I} + \frac{a_2}{a}\boldsymbol{B} + \frac{a_3}{a}\boldsymbol{B}^2, \\
\boldsymbol{U} &= \frac{b_1}{b}\boldsymbol{I} + \frac{b_2}{b}\boldsymbol{C} + \frac{b_3}{b}\boldsymbol{C}^2, \\
\boldsymbol{V} &= \frac{b_1}{b}\boldsymbol{I} + \frac{b_2}{b}\boldsymbol{B} + \frac{b_3}{b}\boldsymbol{B}^2,
\end{aligned}$$

where

$$\begin{aligned}
a_1 &= (\lambda_1 + \lambda_2 + \lambda_3)(\lambda_1^2\lambda_2^2 + \lambda_2^2\lambda_3^2 + \lambda_1^2\lambda_3^2) + \lambda_1\lambda_2\lambda_3(\lambda_1^2 + \lambda_2^2 + \lambda_3^2 + \lambda_1\lambda_2 \\
&\quad + \lambda_2\lambda_3 + \lambda_1\lambda_3), \\
a_2 &= -\left[(\lambda_1 + \lambda_2 + \lambda_3)(\lambda_1^2 + \lambda_2^2 + \lambda_3^2) + \lambda_1\lambda_2\lambda_3\right], \\
a_3 &= \lambda_1 + \lambda_2 + \lambda_3, \\
a &= \lambda_1\lambda_2\lambda_3\left[(\lambda_1 + \lambda_2 + \lambda_3)(\lambda_1\lambda_2 + \lambda_2\lambda_3 + \lambda_1\lambda_3) - \lambda_1\lambda_2\lambda_3\right], \\
b_1 &= \lambda_1\lambda_2\lambda_3(\lambda_1 + \lambda_2 + \lambda_3), \\
b_2 &= \lambda_1^2 + \lambda_2^2 + \lambda_3^2 + \lambda_1\lambda_2 + \lambda_2\lambda_3 + \lambda_1\lambda_3, \\
b_3 &= -1, \\
b &= (\lambda_1 + \lambda_2 + \lambda_3)(\lambda_1\lambda_2 + \lambda_2\lambda_3 + \lambda_1\lambda_3) - \lambda_1\lambda_2\lambda_3.
\end{aligned}$$

Two distinct eigenvalues ($\lambda_1 \neq \lambda_2 = \lambda_3$):

$$\begin{aligned}
\boldsymbol{R} &= \boldsymbol{F}\left[\frac{a_1}{a}\boldsymbol{I} + \frac{a_2}{a}\boldsymbol{C}\right], \\
\boldsymbol{U}^{-1} &= \frac{a_1}{a}\boldsymbol{I} + \frac{a_2}{a}\boldsymbol{C},
\end{aligned}$$

$$V^{-1} = \frac{a_1}{a} I + \frac{a_2}{a} B,$$
$$U = \frac{b_1}{b} I + \frac{b_2}{b} C,$$
$$V = \frac{b_1}{b} I + \frac{b_2}{b} B,$$

where

$$a_1 = \lambda_1^2 + \lambda_1\lambda_2 + \lambda_2^2,$$
$$a_2 = -1,$$
$$a = \lambda_1\lambda_2(\lambda_1 + \lambda_2),$$
$$b_1 = \lambda_1\lambda_2,$$
$$b_2 = 1,$$
$$b = \lambda_1 + \lambda_2.$$

All eigenvalues repeated ($\lambda_1 = \lambda_2 = \lambda_3 \equiv \lambda$):

$$R = \frac{1}{\lambda} F, \quad U = V = \lambda I.$$

Rates of the stretch and rotation tensors are discussed in Section 1.9.4.

1.6.3 Isotropic functions

We shall frequently make use of the following theorem:

Theorem 1.6.7. *Let X, Y and Z be normed vector spaces, and let $f : X \to Z$ and $g : X \to Y$ be two mappings such that*

$$\{x, x' \in X \text{ and } g(x) = g(x')\} \implies f(x) = f(x').$$

Then, there exists a mapping $h : g(X) \subset Y \to Z$ such that

$$f(x) = h(g(x)) \quad \forall x \in X.$$

A scalar function $\phi : V \to \Re$ is isotropic if $\phi(v) = \phi(Qv)$ for all $v \in V$ and $Q \in \text{Orth}^+$. We have the following characterization:

Theorem 1.6.8. *ϕ is isotropic if and only if there exist a function $\hat{\phi}$ such that*

$$\phi(v) = \hat{\phi}(|v|) \quad \forall v \in V.$$

Proof. If $\hat{\phi}$ exists, then

$$\phi(Qv) = \hat{\phi}(|Qv|) = \hat{\phi}(|v|) = \phi(v) \quad \forall v \in V.$$

To prove the converse, we need to show that $|u| = |v|$ implies $\phi(u) = \phi(v)$, and then the result follows from Theorem 1.6.7. As noted after the proof of

Theorem 1.5.2, if $|u| = |v|$, there exists a proper orthogonal tensor that rotates u into v, i.e.,

$$v = Qu.$$

Now by the assumed isotropy, we have

$$\phi(v) = \phi(Qu) = \phi(u). \qquad \square$$

A vector function $g : V \to V$ is isotropic if $Qg(v) = g(Qv)$ for all $v \in V$ and $Q \in \text{Orth}^+$. We have the following characterization:

Theorem 1.6.9. *g is isotropic if and only if there exists a function ϕ such that*

$$g(v) = \phi(|v|)v \quad \forall v \in V.$$

Proof. If the function ϕ exists, then

$$g(Qv) = \phi(|Qv|)Qv = \phi(|v|)Qv = Qg(v).$$

To prove the converse, let v be an arbitrary vector, and let $Q \neq I$ be a proper orthogonal tensor with $v/|v|$ as its axis so that $Qv = v$. By the assumed isotropy, we have

$$Qg(v) = g(Qv) = g(v).$$

Since the axis of Q is a one-dimensional subspace, $g(v)$ is parallel to v, i.e., $g(v) = \alpha(v)v$. Again by the assumed isotropy, $\alpha(v) = \alpha(Qv)$, and the desired result follows by Theorem 1.6.8. $\square$

In what follows, we restrict the domain of the functions under consideration to a subset of Sym, and denote the list of principal invariants (I_1, I_2, I_3) by $\mathcal{I}$. A scalar function $\phi : \text{Psym} \to \Re$ is an isotropic function if it remains invariant under orthogonal transformations, i.e.,

$$\phi(S) = \phi(QSQ^T) \quad \forall S \in \text{Psym}, \ Q \in \text{Orth}^+. \tag{1.136}$$

An isotropic tensor function $G : \text{Psym} \to \text{Sym}$ is one that satisfies

$$QG(S)Q^T = G(QSQ^T) \quad \forall S \in \text{Psym}, \ Q \in \text{Orth}^+.$$

Although we have restricted S to be in Psym in the above definitions, all the following results will hold even if S lies in *any* subset $\mathcal{A}$ of Sym that is invariant under Orth^+, i.e., if

$$QAQ^T \in \mathcal{A} \quad \text{when} \quad A \in \mathcal{A}, \ Q \in \text{Orth}^+.$$

This condition is imposed in order that $\phi(QSQ^T)$ and $G(QSQ^T)$ make sense. As is obvious from the above definition, Sym itself is invariant under Orth^+. To show that Psym is invariant under Orth^+, let $S \in \text{Psym}$. By Theorem 1.6.4, all the eigenvalues of S are strictly positive. The following theorem shows that the eigenvalues of QSQ^T are the same as the eigenvalues of S. Then, again appealing to Theorem 1.6.4, we conclude that $QSQ^T \in$ Psym.

Theorem 1.6.10. *The list of principal invariants $\mathcal{I}$ is isotropic, i.e.,*

$$\mathcal{I}_S = \mathcal{I}_{QSQ^T} \quad \forall S \in Psym, Q \in Orth^+. \tag{1.137}$$

Proof. Note that

$$\begin{aligned}
\det(QSQ^T - \lambda I) &= \det(QSQ^T - \lambda QQ^T) \\
&= \det[Q(S - \lambda I)Q^T] \\
&= (\det Q)(\det Q^T)\det(S - \lambda I) \\
&= [\det(QQ^T)]\det(S - \lambda I) \\
&= \det(S - \lambda I).
\end{aligned}$$

Thus, the characteristic equation for QSQ^T and S is the same, leading to the same principal invariants (and hence, also to the same eigenvalues). □

Theorem 1.6.11. *The tensors $A, B \in Sym$ have the same principal invariants if and only if there exists a proper orthogonal tensor Q such that $A = QBQ^T$.*

Proof. If there exists $Q \in$ Orth$^+$ such that $A = QBQ^T$, then by Eqn. (1.137) (which also holds when $S \in$ Sym), A and B have the same principal invariants. Conversely, if A and B have the same principal invariants, then by Eqn. (1.113) they have the same eigenvalues. By the spectral resolution theorem, we can write

$$A = \sum_i \lambda_i e_i^* \otimes e_i^*,$$

$$B = \sum_i \lambda_i \bar{e}_i \otimes \bar{e}_i.$$

Since $\{e_i^*\}$ and $\{\bar{e}_i\}$ are orthonormal sets of vectors, by Eqn. (1.95) there exists $Q = e_k^* \otimes \bar{e}_k \in$ Orth$^+$ such that $e_i^* = Q\bar{e}_i$. Using Eqn. (1.344), we now have

$$QBQ^T = \sum_i \lambda_i Q(\bar{e}_i \otimes \bar{e}_i)Q^T = \sum_i \lambda_i (Q\bar{e}_i) \otimes (Q\bar{e}_i) = \sum_i \lambda_i e_i^* \otimes e_i^* = A. \; □$$

Theorem 1.6.12. *A function ϕ : $Psym \to \Re$ is isotropic if and only if there exists a function $\tilde{\phi} : \mathcal{I}(S) \to \Re$ such that*

$$\phi(S) = \tilde{\phi}(\mathcal{I}_S) \quad \forall S \in Psym. \tag{1.138}$$

Proof. Assuming Eqn. (1.138) holds, we conclude that ϕ is isotropic from Eqn. (1.137).

To prove the converse, we need to show that

$$\mathcal{I}_A = \mathcal{I}_B \implies \phi(A) = \phi(B),$$

for $A, B \in$ Psym, and then the desired result follows from Theorem 1.6.7. Since A and B have the same invariants, by Theorem 1.6.11 there exists $Q \in$ Orth$^+$ such that $A = QBQ^T$. Since ϕ is isotropic,

$$\phi(A) = \phi(QBQ^T) = \phi(B). \qquad \square$$

The generalization of the above theorem to arbitrary tensors is given by the following [246, 294]:

Theorem 1.6.13. *A function ϕ : Lin $\to \Re$ is isotropic if and only if there exists a function $\tilde{\phi}$ such that*

$$\phi(T) = \tilde{\phi}(\mathrm{tr}\, T, \mathrm{tr}\, T^2, \mathrm{tr}\, TT^T), \quad \forall T \in Lin \quad (n = 2),$$
$$\phi(T) = \tilde{\phi}(\mathrm{tr}\, T, \mathrm{tr}\, T^2, \mathrm{tr}\, T^3, \mathrm{tr}\, TT^T, \mathrm{tr}\, T^2T^T, \mathrm{tr}\, T^2(T^T)^2, \mathrm{tr}\, TT^TT^2(T^T)^2)$$
$$\forall T \in Lin \quad (n = 3).$$

Now we state and prove the Rivlin–Ericksen representation theorem for isotropic tensor functions (the generalization to an isotropic symmetric tensor-valued function of two symmetric tensors is stated in Eqn. (7.56)).

Theorem 1.6.14 (Rivlin–Ericksen Representation Theorem). *A function G : Psym $\to$ Sym is isotropic if and only if there exist scalar functions $\phi_0, \phi_1, \phi_2 : \mathcal{I}(S) \to \Re$, such that*

$$G(S) = \phi_0(\mathcal{I}_S)I + \phi_1(\mathcal{I}_S)S + \phi_2(\mathcal{I}_S)S^2 \quad \forall S \in Psym. \qquad (1.139)$$

Proof. Assume Eqn. (1.139) holds. Choose $S \in$ Psym and $Q \in$ Orth$^+$. By Eqn. (1.137), we have

$$\begin{aligned} G(QSQ^T) &= \phi_0(\mathcal{I}_{QSQ^T})I + \phi_1(\mathcal{I}_{QSQ^T})QSQ^T + \phi_2(\mathcal{I}_{QSQ^T})QSQ^TQSQ^T \\ &= \phi_0(\mathcal{I}_S)QQ^T + \phi_1(\mathcal{I}_S)QSQ^T + \phi_2(\mathcal{I}_S)QS^2Q^T \\ &= QG(S)Q^T, \end{aligned}$$

which proves that G is isotropic.

To prove the converse, assume that G is isotropic, i.e., $G(QSQ^T) = QG(S)Q^T$ for all $Q \in$ Orth$^+$. We first prove that every eigenvector of S is an eigenvector of $G(S)$, or, alternatively, any matrix that diagonalizes S also diagonalizes $G(S)$. Let the spectral resolution of S be given by $S = \sum_i \lambda_i e_i^* \otimes e_i^*$. If we form a matrix Q with the eigenvectors of S along its rows, i.e., if $Q = e_k \otimes e_k^*$, then by Theorem 1.6.2, Q diagonalizes S, and is proper orthogonal. Let

$$Q_1 = e_1 \otimes e_1 - e_2 \otimes e_2 - e_3 \otimes e_3,$$
$$Q_2 = -e_1 \otimes e_1 + e_2 \otimes e_2 - e_3 \otimes e_3.$$

Then, we see that Q_1, Q_2, Q_1Q and Q_2Q are all proper orthogonal tensors. In addition,

$$Q_\beta QSQ^TQ_\beta^T = \sum_i \lambda_i e_i \otimes e_i = QSQ^T, \quad \beta = 1,2 \text{ (no sum on } \beta).$$

By the assumed property of isotropy, we have

$$Q_\beta QG(S)Q^TQ_\beta^T = G(Q_\beta QSQ^TQ_\beta^T) = G(QSQ^T) = QG(S)Q^T, \quad \beta = 1,2. \tag{1.140}$$

A straightforward computation shows that the above equations imply that $QG(S)Q^T$ is diagonal, or, in other words, again appealing to Theorem 1.6.2, the eigenvectors of S and $G(S)$ are the same. Hence

$$G(S) = \mu_1 e_1^* \otimes e_1^* + \mu_2 e_2^* \otimes e_2^* + \mu_3 e_3^* \otimes e_3^*, \tag{1.141}$$

where μ_i are the eigenvalues of $G(S)$.

Let us now establish that $G(S)$ can be written as

$$G(S) = \alpha_0(S)I + \alpha_1(S)S + \alpha_2(S)S^2 \quad \forall S \in \text{Psym}, \tag{1.142}$$

where α_0, α_1, α_2 are real-valued functions of S. Assume first that S has three distinct eigenvalues λ_i, with associated eigenvectors e_i^*. Then the two sets $\{I, S, S^2\}$ and $\{e_1^* \otimes e_1^*, e_2^* \otimes e_2^*, e_3^* \otimes e_3^*\}$ span the same subspace of the vector space of symmetric tensors. To see this, we first observe that $e_1^* \otimes e_1^*$, $e_2^* \otimes e_2^*$, and $e_3^* \otimes e_3^*$ are symmetric tensors. They are also linearly independent, since if

$$\alpha e_1^* \otimes e_1^* + \beta e_2^* \otimes e_2^* + \gamma e_3^* \otimes e_3^* = 0,$$

then carrying out the operation $e_1^* \cdot ()e_1^*$, where $()$ denotes the left- and right-hand sides of the above equation, and using the orthogonality of e_1^*, e_2^* and e_3^*, we get $\alpha = 0$; similarly, the operations $e_2^* \cdot ()e_2^*$ and $e_3^* \cdot ()e_3^*$ yield $\beta = 0$ and $\gamma = 0$. Thus, $\text{Lsp}\{e_1^* \otimes e_1^*, e_2^* \otimes e_2^*, e_3^* \otimes e_3^*\}$ is a three-dimensional subspace of the vector space of symmetric tensors, and, hence, is a vector space itself. Since, by Eqns. (1.38) and (1.125),

$$I = e_1^* \otimes e_1^* + e_2^* \otimes e_2^* + e_3^* \otimes e_3^*,$$
$$S = \lambda_1 e_1^* \otimes e_1^* + \lambda_2 e_2^* \otimes e_2^* + \lambda_3 e_3^* \otimes e_3^*,$$
$$S^2 = \lambda_1^2 e_1^* \otimes e_1^* + \lambda_2^2 e_2^* \otimes e_2^* + \lambda_3^2 e_3^* \otimes e_3^*,$$

and since the van der Monde determinant

$$\det \begin{bmatrix} 1 & 1 & 1 \\ \lambda_1 & \lambda_2 & \lambda_3 \\ \lambda_1^2 & \lambda_2^2 & \lambda_3^2 \end{bmatrix} = (\lambda_1 - \lambda_2)(\lambda_2 - \lambda_3)(\lambda_3 - \lambda_1) \neq 0,$$

by virtue of the eigenvalues being assumed to be distinct, Theorem 1.1.2 yields the result

$$\mathrm{Lsp}\{e_1^* \otimes e_1^*, e_2^* \otimes e_2^*, e_3^* \otimes e_3^*\} = \mathrm{Lsp}\{I, S, S^2\}.$$

Hence, Eqn. (1.141) can be written in the form given by Eqn. (1.142).

Assume, next, that the tensor S has a double eigenvalue, say $\lambda_2 = \lambda_3 \neq \lambda_1$. Then, the two sets $\{I, S\}$ and $\{e_1^* \otimes e_1^*, e_2^* \otimes e_2^* + e_3^* \otimes e_3^*\}$ span the same subspace, since, in this case, we can write

$$I = e_1^* \otimes e_1^* + (e_2^* \otimes e_2^* + e_3^* \otimes e_3^*),$$
$$S = \lambda_1 e_1^* \otimes e_1^* + \lambda_2 (e_2^* \otimes e_2^* + e_3^* \otimes e_3^*),$$

and

$$\det \begin{bmatrix} 1 & 1 \\ \lambda_1 & \lambda_2 \end{bmatrix} = \lambda_2 - \lambda_1 \neq 0.$$

All the vectors in the subspace spanned by e_2^* and e_3^* are eigenvectors of S, and hence those of $G(S)$ as already proved. Thus,

$$G(S)e_2^* = \mu_2 e_2^*, \quad G(S)e_3^* = \mu_3 e_3^*,$$
$$G(S)(e_2^* + e_3^*) = \mu(e_2^* + e_3^*),$$

which implies that $\mu = \mu_2 = \mu_3$. Therefore, we can write

$$G(S) = \mu_1 e_1^* \otimes e_1^* + \mu(e_2^* \otimes e_2^* + e_3^* \otimes e_3^*),$$

as

$$G(S) = \alpha_0(S) I + \alpha_1(S) S.$$

Assume, finally, that S has three repeated eigenvalues. Now, the eigenvalues of $G(S)$ are also all repeated (proved in the same way as in the previous case), and all unit vectors are eigenvectors of S, and hence also of $G(S)$. Thus, by using the spectral resolution of $G(S)$, we conclude that it is a multiple I, i.e.,

$$G(S) = \alpha_0(S) I,$$

proving the assertion in all cases. Note from the above discussion that $\alpha_2(S) = 0$, when there are two repeated eigenvalues, and $\alpha_1(S) = \alpha_2(S) = 0$, when all eigenvalues are identical.

It remains to show that the functions α_0, α_1 and α_2 are only functions of $\mathcal{I}_S$. By virtue of Theorem 1.6.12, it suffices to prove that α_0, α_1 and α_2 are isotropic functions. This follows from the isotropy of $G(S)$:

$$\begin{aligned}G(QSQ^T) &= Q\left[\alpha_0(QSQ^T)I + \alpha_1(QSQ^T)S + \alpha_2(QSQ^T)S^2\right]Q^T \\ &= QG(S)Q^T = Q\left[\alpha_0(S)I + \alpha_1(S)S + \alpha_2(S)S^2\right] \\ &\qquad Q^T \quad \forall S \in \text{Psym}, Q \in \text{Orth}^+,\end{aligned}$$

which implies that

$$\alpha_i(QSQ^T) = \alpha_i(S) \quad \forall S \in \text{Psym}, Q \in \text{Orth}^+;\ \text{for } i = 0,1,2,$$

again by the uniqueness of the expansion of $G(S)$ in the spaces spanned by $\{I, S, S^2\}$, $\{I, S\}$ or $\{I\}$ according to which case is being considered. Thus, by Theorem 1.6.12, there exist functions $\phi(\mathcal{I}_S)$ such that

$$\alpha_i(S) = \phi_i(\mathcal{I}_S) \quad \forall S \in \text{Psym};\ i = 0,1,2. \qquad \square$$

Note that the above theorem *does not* say that an isotropic function is a quadratic function of the elements of S, since the functions $\phi_i(\mathcal{I}_S)$, $i = 0, 1, 2$, can be arbitrary.

For linear functions, we have the following simpler result:

Theorem 1.6.15. (Representation Theorem for Linear Isotropic Tensor Functions). *A linear function $G(S)$: Sym $\rightarrow$ Sym is isotropic if and only if there exist scalars μ and λ such that*

$$G(S) = \lambda(\text{tr}\, S)I + 2\mu S \quad \forall S \in \text{Sym}.$$

Proof. Although one could try and specialize the result from the previous theorem to the present case, it is easier to proceed directly; the following short proof is due to Gurtin [108]. Let V_1 be the set of all unit vectors. For $e \in V_1$, the tensor $e \otimes e$ has eigenvalues $(0, 0, 1)$ (since there exists $Q \in \text{Orth}^+$ such that $e = Qe_1$, which implies that $e \otimes e = Q(e_1 \otimes e_1)Q^T$, so that by Eqn. (1.137), the principal invariants, and hence the eigenvalues, of $e \otimes e$ and $e_1 \otimes e_1$ are the same). Since two eigenvalues of $e \otimes e$ are the same, analogous to the case of repeated eigenvalues in the previous theorem's proof, $G(e \otimes e)$ also has a repeated eigenvalue, and thus by Eqn. (1.123) we have

$$G(e \otimes e) = \lambda(e)I + 2\mu(e)e \otimes e \quad \forall e \in V_1. \tag{1.143}$$

Choose $e, f \in V_1$, and let Q be a proper orthogonal tensor such that $Qe = f$. Since

$$Q(e \otimes e)Q^T = f \otimes f,$$

and since G is isotropic,

$$0 = QG(e \otimes e)Q^T - G(f \otimes f) = [\lambda(e) - \lambda(f)]I + 2[\mu(e) - \mu(f)]f \otimes f.$$

By multiplying both sides of $\alpha \boldsymbol{I} + \beta \boldsymbol{f} \otimes \boldsymbol{f} = \boldsymbol{0}$ by a vector $\boldsymbol{g}$ perpendicular to $\boldsymbol{f}$, it follows that $\alpha = \beta = 0$, implying in turn that $\boldsymbol{I}$ and $\boldsymbol{f} \otimes \boldsymbol{f}$ are linearly independent; thus,

$$\lambda(\boldsymbol{e}) = \lambda(\boldsymbol{f}), \quad \mu(\boldsymbol{e}) = \mu(\boldsymbol{f}).$$

Therefore, λ and μ are scalar constants, and we conclude from Eqn. (1.143) that

$$\boldsymbol{G}(\boldsymbol{e} \otimes \boldsymbol{e}) = \lambda \boldsymbol{I} + 2\mu \boldsymbol{e} \otimes \boldsymbol{e}. \tag{1.144}$$

The spectral resolution of $\boldsymbol{S} \in \text{Sym}$ is

$$\boldsymbol{S} = \sum_i \alpha_i \boldsymbol{e}_i^* \otimes \boldsymbol{e}_i^*,$$

with $\boldsymbol{e}_i^*$ orthonormal; therefore, in view of Eqn. (1.144), and the linearity of $\boldsymbol{G}$,

$$\boldsymbol{G}(\boldsymbol{S}) = \sum_i \alpha_i \boldsymbol{G}(\boldsymbol{e}_i^* \otimes \boldsymbol{e}_i^*) = \lambda(\alpha_1 + \alpha_2 + \alpha_3)\boldsymbol{I} + 2\mu \boldsymbol{S}.$$

Since $\operatorname{tr} \boldsymbol{S} = \alpha_1 + \alpha_2 + \alpha_3$, the above equation implies that

$$\boldsymbol{G}(\boldsymbol{S}) = \lambda(\operatorname{tr} \boldsymbol{S})\boldsymbol{I} + 2\mu \boldsymbol{S}.$$

The converse assertion that the above representation is an isotropic function is left as an exercise (see Problem 37). □

1.7 Higher-Order Tensors

A detailed discussion about the decomposition of higher-order tensors into irreducible tensors, and their eigenvalues and eigentensors can be found in [8], [150], [151], [260], [304] and [360].

Similar to the definition of a second-order tensor, a third-order tensor, which we represent by $\boldsymbol{B}$, is defined as a linear transformation that transforms a vector $\boldsymbol{v}$ to a second-order tensor $\boldsymbol{T}$, i.e.,

$$\boldsymbol{B}\boldsymbol{v} = \boldsymbol{T}. \tag{1.145}$$

In addition, the map is linear, i.e.,

$$\boldsymbol{B}(a\boldsymbol{v}_1 + b\boldsymbol{v}_2) = a\boldsymbol{B}\boldsymbol{v}_1 + b\boldsymbol{B}\boldsymbol{v}_2 \quad \forall a, b \in \Re \text{ and } \boldsymbol{v}_1, \boldsymbol{v}_2 \in V.$$

In order to find the representation for a third-order tensor in terms of its components, we define the dyadic or tensor product of three vectors $\boldsymbol{a}$, $\boldsymbol{b}$ and $\boldsymbol{c}$ as

$$(\boldsymbol{a} \otimes \boldsymbol{b} \otimes \boldsymbol{c})\boldsymbol{d} := (\boldsymbol{a} \otimes \boldsymbol{b})(\boldsymbol{c} \cdot \boldsymbol{d}) \quad \forall \boldsymbol{d} \in V. \tag{1.146}$$

Since $\boldsymbol{a} \otimes \boldsymbol{b}$ is a second-order tensor, and since the above transformation is linear, the tensor product of three vectors is a linear transformation that transforms vectors into second-order tensors, and hence is a third-order tensor. Just as a second-order tensor can be expressed as the sum of 9 dyads (Eqn. (1.37)), a third-order tensor can be expressed as the sum of 27 dyads as follows:

$$B = B_{ijk} e_i \otimes e_j \otimes e_k, \tag{1.147}$$

where

$$B_{ijk} = e_i \cdot (Be_k) e_j. \tag{1.148}$$

In order to prove this, we consider the action of B on an arbitrary vector u:

$$\begin{aligned}
Bu &= (Bu)_{ij}(e_i \otimes e_j) \\
&= e_i \cdot \left[(Bu)e_j\right](e_i \otimes e_j) \\
&= e_i \cdot \left\{\left[B(u_k e_k)\right] e_j\right\}(e_i \otimes e_j) \\
&= e_i \cdot \left\{\left[B((e_k \cdot u)e_k)\right] e_j\right\}(e_i \otimes e_j) \\
&= e_i \cdot \left[(Be_k)e_j\right](e_k \cdot u)(e_i \otimes e_j) \\
&= e_i \cdot \left[(Be_k)e_j\right]\left[(e_i \otimes e_j \otimes e_k)u\right].
\end{aligned}$$

Since u is arbitrary, we have proved that the tensor B can be represented by Eqn. (1.147), with the components B_{ijk} of this tensor given by Eqn. (1.148). Note that

$$Be_m = B_{ijm} e_i \otimes e_j.$$

In [360], using the identity $\epsilon_{pjr}\epsilon_{pbr} = 2\delta_{jb}$, the authors write

$$B_{ijk} = \frac{1}{2}\epsilon_{pjr}\epsilon_{pbr}B_{ibk},$$

then, assuming all eigenvalues to be distinct, use the spectral resolution of a fourth-order tensor to write $B_{ibk}\epsilon_{pbr}$ in the form $\sum_{m=1}^{9} \lambda_m (P_m)_{ipkr}$, from which it follows that

$$B_{ijk} = \sum_{m=1}^{9} \lambda_m \left[\frac{1}{2}\epsilon_{pjr}(P_m)_{ipkr}\right]. \tag{1.149}$$

Based on this representation, the authors claim that the λ_i's (not all independent, since $B_{ibi}\epsilon_{pbp} = \sum_{m=1}^{9} \lambda_m = 0$) are eigenvalues of B. However, since the eigenvalue problem for a third-order tensor is never defined in this work, it is not clear how the eigenvalues of the fourth-order tensor are also the eigenvalues of the third-order tensor.

In general, an nth order tensor is defined as a linear transformation that transforms a vector to an $n-1$th order tensor. It can be proved by induction that an nth order tensor can be represented by the sum of 3^n dyads as

$$T = T_{i_1 i_2 i_3 \ldots i_n} e_{i_1} \otimes e_{i_2} \otimes e_{i_3} \cdots \otimes e_{i_n},$$

where

$$T_{i_1 i_2 i_3 \ldots i_n} = e_{i_1} \cdot \left[(((Te_{i_n})e_{i_{n-1}}) \cdots)e_{i_2}\right].$$

As in the case of second-order tensors, we define the sum of two nth order tensors $\boldsymbol{S}$ and $\boldsymbol{T}$ as

$$(\boldsymbol{S}+\boldsymbol{T})\boldsymbol{u} := \boldsymbol{S}\boldsymbol{u}+\boldsymbol{T}\boldsymbol{u} \quad \forall \boldsymbol{u} \in V,$$

and the scalar multiple of $\boldsymbol{T}$ by $\alpha \in \Re$ as

$$(\alpha\boldsymbol{T})\boldsymbol{u} := \alpha(\boldsymbol{T}\boldsymbol{u}) \quad \forall \boldsymbol{u} \in V.$$

It is easily seen from the definition of a tensor as a linear transformation that $\boldsymbol{S}+\boldsymbol{T}$ and $\alpha\boldsymbol{T}$ are nth order tensors.

The transformation law for third order tensors can be deduced from Eqn. (1.148) as follows:

$$\begin{aligned}\bar{B}_{ijk} &= Q_{il}\boldsymbol{e}_l \cdot [\boldsymbol{B}(Q_{kn}\boldsymbol{e}_n)]Q_{jm}\boldsymbol{e}_m \\ &= Q_{il}Q_{jm}Q_{kn}\boldsymbol{e}_l \cdot (\boldsymbol{B}\boldsymbol{e}_n)\boldsymbol{e}_m \\ &= Q_{il}Q_{jm}Q_{kn}B_{lmn}.\end{aligned}$$

The alternate tensor $\boldsymbol{\mathcal{E}}$ whose components are given by Eqn. (1.11) is an example of a third-order tensor. This can be proved by noting that

$$\begin{aligned}\bar{\epsilon}_{ijk} &= \bar{\boldsymbol{e}}_i \cdot (\bar{\boldsymbol{e}}_j \times \bar{\boldsymbol{e}}_k) \\ &= Q_{il}Q_{jm}Q_{kn}\boldsymbol{e}_l \cdot (\boldsymbol{e}_m \times \boldsymbol{e}_n) \\ &= Q_{il}Q_{jm}Q_{kn}\epsilon_{lmn},\end{aligned}$$

which is nothing but the transformation law for the components of a third-order tensor. In fact, it is an isotropic third-order tensor, as we show in Section 1.8. Note that $[[\boldsymbol{\mathcal{E}}\boldsymbol{w}]\boldsymbol{v}]\boldsymbol{u} = [\boldsymbol{u},\boldsymbol{v},\boldsymbol{w}]$.

Similarly, the transformation law for a fourth-order tensor is

$$\bar{\mathbb{C}}_{ijkl} = Q_{ip}Q_{jq}Q_{kr}Q_{ls}\mathbb{C}_{pqrs}. \tag{1.150}$$

Generalizing the above arguments, the transformation law for an nth order tensor is given by

$$\bar{T}_{i_1i_2i_3\ldots i_n} = Q_{i_1j_1}Q_{i_2j_2}Q_{i_3j_3}\cdots Q_{i_nj_n}T_{j_1j_2j_3\ldots j_n}. \tag{1.151}$$

We have seen how to construct fourth-order tensors as a dyadic product of vectors. However, another alternative way that is very widely used is to construct it as a dyadic product of two second-order tensors. This procedure is analogous to constructing second-order tensors using the dyadic product of vectors. We now discuss this methodology, which should be compared with the one for second-order tensors presented in Section 1.3.

A fourth-order tensor is defined as a linear transformation that maps second-order tensors to second-order tensors. Thus, if $\mathbb{C}$ is a fourth-order tensor that maps a second-order tensor $\boldsymbol{A}$ to a second-order tensor $\boldsymbol{B}$, we write it as

$$\boldsymbol{B} = \mathbb{C}\boldsymbol{A}. \tag{1.152}$$

$\mathbb{C}$ satisfies the property $\mathbb{C}(\alpha A_1 + \beta A_2) = \alpha\mathbb{C}A_1 + \beta\mathbb{C}A_2$, for all $\alpha, \beta \in \Re$ and $A_1, A_2 \in \text{Lin}$. The sum, scalar multiple and composition of fourth-order tensors is defined as

$$(\mathbb{C} + \mathbb{D})A = \mathbb{C}A + \mathbb{D}A \quad \forall A \in \text{Lin},$$
$$(\lambda\mathbb{C})A = \lambda(\mathbb{C}A) \quad \forall A \in \text{Lin},$$
$$(\mathbb{C}\mathbb{D})A = \mathbb{C}(\mathbb{D}A) \quad \forall A \in \text{Lin}.$$

For presenting the component form of fourth-order tensors, it is convenient to introduce the notation $E_{ij} \equiv e_i \otimes e_j$. We have

$$T : E_{ij} = \text{tr}\,(T(e_j \otimes e_i)) = \text{tr}\,((Te_j) \otimes e_i) = T_{kj}\text{tr}\,(e_k \otimes e_i) = T_{kj}\delta_{ki} = T_{ij}. \tag{1.153}$$

By letting $T \equiv E_{kl}$, we get

$$E_{ij} : E_{kl} = \delta_{ik}\delta_{jl}. \tag{1.154}$$

Analogous to the definition of the components of a second-order tensor (see Eqn. (1.27)), we now define the components via Eqn. (1.152) as

$$\mathbb{C}E_{kl} = \mathbb{C}_{ijkl}E_{ij}. \tag{1.155}$$

Taking the dot product on either side with E_{mn} and using Eqn. (1.154), we get $\mathbb{C}_{mnkl} = E_{mn} : \mathbb{C}E_{kl}$, which, on replacing the indices (m, n) with (i, j) can be written as

$$\mathbb{C}_{ijkl} = E_{ij} : \mathbb{C}E_{kl}. \tag{1.156}$$

Two tensors $\mathbb{C}$ and $\mathbb{D}$ are said to be equal if

$$\mathbb{C}B = \mathbb{D}B \quad \forall B \in \text{Lin}, \tag{1.157}$$

or, alternatively, if

$$(A, \mathbb{C}B) = (A, \mathbb{D}B) \quad \forall A, B \in \text{Lin}.$$

By choosing $A \equiv E_{ij}$ and $B \equiv E_{kl}$ in the above equation, we see that equal tensors have equal components.

Using Eqn. (1.155) and the linearity of fourth-order tensors, Eqn. (1.152) can be written as

$$B_{ij}E_{ij} = \mathbb{C}(A_{kl}E_{kl}) = A_{kl}\mathbb{C}E_{kl} = \mathbb{C}_{ijkl}A_{kl}E_{ij},$$

which by virtue of the linear independence of the $\{E_{ij}\}$ implies that

$$B_{ij} = \mathbb{C}_{ijkl}A_{kl}.$$

Similarly, we have

$$(\mathbb{C}\mathbb{D})_{ijkl} = (\mathbb{C})_{ijmn}(\mathbb{D})_{mnkl}.$$

We now give some examples of fourth-order tensors. The fourth-order identity tensor $\mathbb{I}$ is defined as one that maps a second-order tensor into itself. The transposer $\mathbb{T}$, symmetrizer

$\mathbb{S}$ and skew-symmetrizer $\mathbb{W}$ are defined as ones that map a second-order tensor into its transpose, symmetric and skew-symmetric part, respectively, i.e.,

$$\mathbb{I}A = A \quad \forall A \in \text{Lin},$$
$$\mathbb{T}A = A^T \quad \forall A \in \text{Lin},$$
$$\mathbb{S}A = \frac{1}{2}(A + A^T) \quad \forall A \in \text{Lin},$$
$$\mathbb{W}A = \frac{1}{2}(A - A^T) \quad \forall A \in \text{Lin}.$$

Note that

$$\begin{aligned} \mathbb{S} &= (\mathbb{I} + \mathbb{T})/2, & \mathbb{I} &= \mathbb{S} + \mathbb{W}, \\ \mathbb{W} &= (\mathbb{I} - \mathbb{T})/2, & \mathbb{T} &= \mathbb{S} - \mathbb{W}. \end{aligned} \tag{1.158}$$

We have

$$\mathbb{S}\mathbb{T} = \frac{1}{2}(\mathbb{I} + \mathbb{T})\mathbb{T} = \frac{1}{2}(\mathbb{T} + \mathbb{T}\mathbb{T}) = \frac{1}{2}(\mathbb{T} + \mathbb{I}) = \mathbb{S} = \mathbb{T}\mathbb{S}, \tag{1.159}$$
$$\mathbb{W}\mathbb{T} = \frac{1}{2}(\mathbb{I} - \mathbb{T})\mathbb{T} = \frac{1}{2}(\mathbb{T} - \mathbb{T}\mathbb{T}) = \frac{1}{2}(\mathbb{T} - \mathbb{I}) = -\mathbb{W} = \mathbb{T}\mathbb{W},$$
$$\mathbb{S}\mathbb{S} = \frac{1}{2}(\mathbb{I} + \mathbb{T})\mathbb{S} = \frac{1}{2}(\mathbb{S} + \mathbb{T}\mathbb{S}) = \frac{1}{2}(\mathbb{S} + \mathbb{S}) = \mathbb{S},$$
$$\mathbb{W}\mathbb{W} = \frac{1}{2}(\mathbb{I} - \mathbb{T})\mathbb{W} = \frac{1}{2}(\mathbb{W} - \mathbb{T}\mathbb{W}) = \frac{1}{2}(\mathbb{W} + \mathbb{W}) = \mathbb{W},$$
$$\mathbb{S}\mathbb{W} = \frac{1}{2}(\mathbb{I} + \mathbb{T})\mathbb{W} = \frac{1}{2}(\mathbb{W} + \mathbb{T}\mathbb{W}) = \frac{1}{2}(\mathbb{W} - \mathbb{W}) = \mathbf{0} = \mathbb{W}\mathbb{S}. \tag{1.160}$$

The components of $\mathbb{I}$, $\mathbb{T}$, $\mathbb{S}$ and $\mathbb{W}$ are obtained using Eqn. (1.156), and are given by

$$\begin{aligned} \mathbb{I}_{ijkl} &= \delta_{ik}\delta_{jl}, \\ \mathbb{T}_{ijkl} &= \delta_{il}\delta_{jk}, \\ \mathbb{S}_{ijkl} &= \frac{1}{2}(\delta_{ik}\delta_{jl} + \delta_{il}\delta_{jk}), \\ \mathbb{W}_{ijkl} &= \frac{1}{2}(\delta_{ik}\delta_{jl} - \delta_{il}\delta_{jk}). \end{aligned} \tag{1.161}$$

Using the above component form of $\mathbb{S}$, we have

$$[\mathbb{S}\mathbb{A}\mathbb{S}]_{ijkl} = \frac{1}{4}\left[\mathbb{A}_{ijkl} + \mathbb{A}_{ijlk} + \mathbb{A}_{jikl} + \mathbb{A}_{jilk}\right], \tag{1.162}$$

where $\mathbb{A}$ is any fourth-order tensor.

The transpose of $\mathbb{C}$, denoted by $\mathbb{C}^T$ is defined using the inner product as

$$(A, \mathbb{C}^T B) := (B, \mathbb{C}A) \quad \forall A, B \in \text{Lin}. \tag{1.163}$$

The components of $\mathbb{C}^T$ are obtained by taking $A \equiv E_{ij}$ and $B \equiv E_{kl}$ in the above equation, and are given by

$$(\mathbb{C}^T)_{ijkl} = \mathbb{C}_{klij}. \tag{1.164}$$

Similar to second-order tensors, one has

$$(\mathbb{C}^T)^T = \mathbb{C}, \tag{1.165a}$$
$$(\mathbb{C}\mathbb{D})^T = \mathbb{D}^T\mathbb{C}^T. \tag{1.165b}$$

Analogous to the dyadic product of two vectors, the dyadic product of two tensors $\boldsymbol{A}$ and $\boldsymbol{B}$ can be defined as

$$(\boldsymbol{A} \otimes \boldsymbol{B})\boldsymbol{C} := (\boldsymbol{B} : \boldsymbol{C})\boldsymbol{A} \quad \forall \boldsymbol{C} \in \text{Lin}. \tag{1.166}$$

The component form of this dyadic product is

$$(\boldsymbol{A} \otimes \boldsymbol{B})_{ijkl} = \boldsymbol{E}_{ij} : (\boldsymbol{A} \otimes \boldsymbol{B})\boldsymbol{E}_{kl} = (\boldsymbol{A} : \boldsymbol{E}_{ij})(\boldsymbol{B} : \boldsymbol{E}_{kl}) = A_{ij}B_{kl}.$$

The transpose of $\boldsymbol{A} \otimes \boldsymbol{B}$ is $\boldsymbol{B} \otimes \boldsymbol{A}$ since

$$(\boldsymbol{C}, (\boldsymbol{A} \otimes \boldsymbol{B})^T\boldsymbol{D}) = (\boldsymbol{D}, (\boldsymbol{A} \otimes \boldsymbol{B})\boldsymbol{C}) = (\boldsymbol{B} : \boldsymbol{C})(\boldsymbol{A} : \boldsymbol{D}) = (\boldsymbol{C}, (\boldsymbol{B} \otimes \boldsymbol{A})\boldsymbol{D}).$$

By taking $\boldsymbol{C} \equiv \boldsymbol{u} \otimes \boldsymbol{v}$ in Eqn. (1.166), we get

$$(\boldsymbol{A} \otimes \boldsymbol{B})(\boldsymbol{u} \otimes \boldsymbol{v}) = (\boldsymbol{u} \cdot \boldsymbol{B}\boldsymbol{v})\boldsymbol{A}.$$

A tensor $\mathbb{C}$ can be written as $C_{ijkl}\boldsymbol{E}_{ij} \otimes \boldsymbol{E}_{kl}$ since, for an arbitrary $\boldsymbol{A} \in \text{Lin}$, we have

$$\begin{aligned}
\mathbb{C}\boldsymbol{A} &= (\mathbb{C}\boldsymbol{A})_{ij}\boldsymbol{E}_{ij} \\
&= [(\mathbb{C}\boldsymbol{A}) : \boldsymbol{E}_{ij}]\boldsymbol{E}_{ij} \\
&= A_{kl}[\boldsymbol{E}_{ij} : (\mathbb{C}\boldsymbol{E}_{kl})]\boldsymbol{E}_{ij} \\
&= \mathbb{C}_{ijkl}(\boldsymbol{A} : \boldsymbol{E}_{kl})\boldsymbol{E}_{ij} \\
&= \mathbb{C}_{ijkl}[\boldsymbol{E}_{ij} \otimes \boldsymbol{E}_{kl}]\boldsymbol{A}.
\end{aligned}$$

Also analogous to the results for dyadic products of vectors, we have

$$\begin{aligned}
\mathbb{C}(\boldsymbol{A} \otimes \boldsymbol{B}) &= (\mathbb{C}\boldsymbol{A}) \otimes \boldsymbol{B}, \\
(\boldsymbol{A} \otimes \boldsymbol{B})\mathbb{C} &= \boldsymbol{A} \otimes (\mathbb{C}^T\boldsymbol{B}).
\end{aligned}$$

Another tensor product that is of great use is the "square tensor product" defined as

$$(\boldsymbol{A} \boxtimes \boldsymbol{B})\boldsymbol{C} := \boldsymbol{A}\boldsymbol{C}\boldsymbol{B}^T \quad \forall \boldsymbol{C} \in \text{Lin}.$$

The components of $\boldsymbol{A} \boxtimes \boldsymbol{B}$ are given by

$$\begin{aligned}
(\boldsymbol{A} \boxtimes \boldsymbol{B})_{ijkl} &= \boldsymbol{E}_{ij} : (\boldsymbol{A} \boxtimes \boldsymbol{B})\boldsymbol{E}_{kl} = \boldsymbol{E}_{ij} : (\boldsymbol{A}\boldsymbol{E}_{kl}\boldsymbol{B}^T) = (\boldsymbol{e}_i \otimes \boldsymbol{e}_j) : [(\boldsymbol{A}\boldsymbol{e}_k) \otimes (\boldsymbol{B}\boldsymbol{e}_l)] \\
&= (\boldsymbol{e}_i \cdot \boldsymbol{A}\boldsymbol{e}_k)(\boldsymbol{e}_j \cdot \boldsymbol{B}\boldsymbol{e}_l) = A_{ik}B_{jl}.
\end{aligned}$$

Since $(\boldsymbol{I} \boxtimes \boldsymbol{I})\boldsymbol{A} = \boldsymbol{I}\boldsymbol{A}\boldsymbol{I}^T = \boldsymbol{A}$, for all $\boldsymbol{A} \in \text{Lin}$, it is clear from Eqn. (1.157) that the fourth-order identity tensor can be written as

$$\mathbb{I} = \boldsymbol{I} \boxtimes \boldsymbol{I}, \tag{1.167}$$

with components $(\mathbb{I})_{ijkl} = \delta_{ik}\delta_{jl}$.

The transpose of $A \boxtimes B$ is given by $A^T \boxtimes B^T$ since

$$(C, (A \boxtimes B)^T D) = (D, (A \boxtimes B)C) = (D, ACB^T) = (C, A^T DB) = (C, (A^T \boxtimes B^T)D).$$

From the definition of the square tensor product, it follows that

$$\begin{aligned}(H \boxtimes I)A &= HA,\\ (I \boxtimes H^T)A &= AH.\end{aligned}$$

As an example of the use of these equations, note that the tensor equation $HA + AH = 0$ can be written as $(H \boxtimes I + I \boxtimes H^T)A = 0$.

The composition rule is given by

$$(A \boxtimes B)(C \boxtimes D) = (AC) \boxtimes (BD), \tag{1.168}$$

since

$$(A \boxtimes B)(C \boxtimes D)E = (A \boxtimes B)(CED^T) = ACED^T B^T = [(AC) \boxtimes (BD)]E \quad \forall E \in \text{Lin}.$$

Some additional properties are

$$(A \otimes B)(C \boxtimes D) = A \otimes (C^T BD), \tag{1.169}$$
$$(A \boxtimes B)(C \otimes D) = (ACB^T) \otimes D, \tag{1.170}$$
$$(A \boxtimes B)(u \otimes v) = (Au) \otimes (Bv), \tag{1.171}$$
$$(a \otimes b) \boxtimes (u \otimes v) = (a \otimes u) \otimes (b \otimes v). \tag{1.172}$$

As per our convention outlined at the end of Section 1.3.1, the above relations are valid even if the involved second-order tensors and vectors are complex-valued. The first relation is proved as follows:

$$\begin{aligned}(A \otimes B)(C \boxtimes D)E &= (A \otimes B)(CED^T) = [B : (CED^T)]A = [(C^T BD) : E]A\\ &= [A \otimes (C^T BD)]E \quad \forall E \in \text{Lin}.\end{aligned}$$

The remaining relations are proved in a similar manner.

The transposer $\mathbb{T}$ commutes with the tensor product $A \boxtimes B$ in the following way:

$$\mathbb{T}(A \boxtimes B) = (B \boxtimes A)\mathbb{T}, \tag{1.173}$$

since

$$\mathbb{T}(A \boxtimes B)E = \mathbb{T}(AEB^T) = BE^T A^T = (B \boxtimes A)E^T = (B \boxtimes A)\mathbb{T}E \quad \forall E \in \text{Lin}.$$

In an identical fashion, one can show that

$$\mathbb{S}(A \boxtimes B)\mathbb{S} = \mathbb{S}(B \boxtimes A)\mathbb{S}, \tag{1.174}$$
$$\mathbb{S}(A \boxtimes B + B \boxtimes A) = (A \boxtimes B + B \boxtimes A)\mathbb{S} = \mathbb{S}(A \boxtimes B + B \boxtimes A)\mathbb{S} = 2\mathbb{S}(A \boxtimes B)\mathbb{S}, \tag{1.175}$$
$$\mathbb{S}(A \boxtimes A) = (A \boxtimes A)\mathbb{S} = \mathbb{S}(A \boxtimes A)\mathbb{S}, \tag{1.176}$$

where the second relation in Eqn. (1.175) follows from the first by multiplying both sides by $\mathbb{S}$, and using the fact that $\mathbb{S}\mathbb{S} = \mathbb{S}$, while the third relation in Eqn. (1.175) follows from Eqn. (1.174). Equation (1.176) follows from Eqn. (1.175) by taking $\boldsymbol{B} = \boldsymbol{A}$.

The component form of $(\boldsymbol{A} \boxtimes \boldsymbol{B})\mathbb{T}$ is $A_{il}B_{jk}$ while that of $\mathbb{T}(\boldsymbol{A} \boxtimes \boldsymbol{B})$ is $A_{jk}B_{il}$. A summary of the component forms of the various dyadic products is

$$\begin{aligned} [\boldsymbol{A} \otimes \boldsymbol{B}]_{ijkl} &= A_{ij}B_{kl}, \\ [\boldsymbol{A} \boxtimes \boldsymbol{B}]_{ijkl} &= A_{ik}B_{jl}, \\ [(\boldsymbol{A} \boxtimes \boldsymbol{B})\mathbb{T}]_{ijkl} &= A_{il}B_{jk}, \\ [\mathbb{T}(\boldsymbol{A} \boxtimes \boldsymbol{B})]_{ijkl} &= A_{jk}B_{il}. \end{aligned} \tag{1.177}$$

The tensor $\mathbb{C}$ is said to have the first minor symmetry if $\mathbb{C} = \mathbb{T}\mathbb{C}$, or, in indicial notation, if

$$\mathbb{C}_{ijkl} = \mathbb{T}_{ijmn}\mathbb{C}_{mnkl} = \delta_{in}\delta_{jm}\mathbb{C}_{mnkl} = \mathbb{C}_{jikl},$$

and is said to have the second minor symmetry if $\mathbb{C} = \mathbb{C}\mathbb{T}$, which in indicial notation reads

$$\mathbb{C}_{ijkl} = \mathbb{C}_{ijlk}.$$

Note that the first minor symmetry condition can be written as $(\mathbb{I} - \mathbb{T})\mathbb{C} = 2\mathbb{W}\mathbb{C} = \boldsymbol{0}$, while the second minor symmetry condition can be written as $\mathbb{C}\mathbb{W} = \boldsymbol{0}$. Equivalently, using $\mathbb{S} = (\mathbb{I} + \mathbb{T})/2$, one can also write the two symmetry conditions as $\mathbb{C} = \mathbb{S}\mathbb{C}$ and $\mathbb{C} = \mathbb{C}\mathbb{S}$, respectively. One has

$$\mathbb{C} = \mathbb{S}\mathbb{C} \text{ and } \mathbb{C} = \mathbb{C}\mathbb{S} \iff \mathbb{C} = \mathbb{S}\mathbb{C}\mathbb{S},$$

since given the conditions on the left, we directly get $\mathbb{C} = \mathbb{S}\mathbb{C}\mathbb{S}$, while, conversely, if $\mathbb{C} = \mathbb{S}\mathbb{C}\mathbb{S}$, then $\mathbb{S}\mathbb{C} = \mathbb{S}\mathbb{S}\mathbb{C}\mathbb{S} = \mathbb{S}\mathbb{C}\mathbb{S} = \mathbb{C}$, and $\mathbb{C}\mathbb{S} = \mathbb{S}\mathbb{C}\mathbb{S}\mathbb{S} = \mathbb{S}\mathbb{C}\mathbb{S} = \mathbb{C}$. Thus a tensor $\mathbb{C}$ has both the minor symmetries if and only if $\mathbb{C} = \mathbb{S}\mathbb{C}\mathbb{S}$.

The tensor $\mathbb{C}$ is said to have major symmetry if $\mathbb{C} = \mathbb{C}^T$, or, in indicial notation (using Eqn. (1.164)), if $\mathbb{C}_{ijkl} = \mathbb{C}_{klij}$. The symmetric and skew-symmetric parts of $\mathbb{C}$ are given by $(\mathbb{C} + \mathbb{C}^T)/2$ and $(\mathbb{C} - \mathbb{C}^T)/2$. Using Eqn. (1.161) or directly, it can be shown that $\mathbb{I}$, $\mathbb{T}$, $\mathbb{S}$ and $\mathbb{W}$ are all symmetric.

Assume that $\mathbb{C}$ has the first minor symmetry, i.e., $\mathbb{C} = \mathbb{T}\mathbb{C}$. By transposing this equation, and using Eqn. (1.165b) and the symmetry of $\mathbb{T}$, we get

$$\mathbb{C}^T = \mathbb{C}^T\mathbb{T}.$$

Thus, if $\mathbb{C}$ has the first minor symmetry, then $\mathbb{C}^T$ has the second minor symmetry. It also follows that if $\mathbb{C}$ has major symmetry and one of the minor symmetries, it also has the other minor symmetry.

The eigenvalue problem for a fourth-order tensor $\mathbb{A}$ that does not possess minor symmetries, involves finding the eigenvalue and eigentensor pair $(\lambda, \boldsymbol{A})$, $\boldsymbol{A} \in \text{Lin}$, which satisfies

$$\mathbb{A}\boldsymbol{A} = \lambda\boldsymbol{A}. \tag{1.178}$$

By applying the Ψ-mapping as per Eqn. (I.1e) to the above equation, we get

$$[\Psi(\mathbb{A}) - \lambda \boldsymbol{I}]\Psi(\boldsymbol{A}) = \boldsymbol{0}.$$

This equation has a nontrivial solution $\Psi(\boldsymbol{A}) \neq \boldsymbol{0}$, if and only if the characteristic equation $\det[\Psi(\mathbb{A}) - \lambda \boldsymbol{I}] = 0$ is satisfied. The characteristic equation can also be written in expanded form as

$$\lambda^9 - I_1(\mathbb{A})\lambda^8 + I_2(\mathbb{A})\lambda^7 + \cdots - I_9(\mathbb{A}) = 0,$$

where (see Eqn. (J.17))

$$I_1(\mathbb{A}) = \operatorname{tr}\Psi(\mathbb{A}) = \lambda_1 + \lambda_2 + \cdots + \lambda_9,$$
$$I_2(\mathbb{A}) = \frac{1}{2}[(\operatorname{tr}\Psi(\mathbb{A}))^2 - \operatorname{tr}(\Psi(\mathbb{A}))^2] = \lambda_1\lambda_2 + \lambda_2\lambda_3 + \cdots + \lambda_1\lambda_9,$$
$$\ldots$$
$$I_9(\mathbb{A}) = \det\Psi(\mathbb{A}) = \lambda_1\lambda_2\ldots\lambda_9,$$

are a strict subset of the complete set of the principal invariants of $\mathbb{A}$ (see the references below for the case where $\mathbb{A}$ has minor symmetries, denoted by $\mathbb{C}$ there). Similar to the result for second-order tensors,

$$\det\mathbb{A}^T = \det\Psi(\mathbb{A}^T) = \det[\Psi(\mathbb{A})^T] = \det\Psi(\mathbb{A}) = \det\mathbb{A}. \tag{1.179}$$

If $\mathbb{A} = \mathbb{A}_1\mathbb{A}_2$, then, again, similar to the result for second-order tensors,

$$\det\mathbb{A} = \det\Psi(\mathbb{A}) = \det[\Psi(\mathbb{A}_1)\Psi(\mathbb{A}_2)] = [\det\Psi(\mathbb{A}_1)][\det\Psi(\mathbb{A}_2)] = (\det\mathbb{A}_1)(\det\mathbb{A}_2). \tag{1.180}$$

By virtue of Eqn. (1.171), we have

Theorem 1.7.1. *If $(\lambda_i, \boldsymbol{u}_i)$ and $(\gamma_i, \boldsymbol{v}_i)$, $i = 1,2,\ldots,n$, where n is the underlying space dimension, are the (possibly complex-valued) eigenvalues/eigenvectors of diagonalizable tensors $\boldsymbol{A}, \boldsymbol{B}$, then $(\lambda_i\gamma_j, \boldsymbol{u}_i \otimes \boldsymbol{v}_j)$, $i, j = 1,2,\ldots,n$, are the eigenvalues/eigentensors of $\boldsymbol{A} \boxtimes \boldsymbol{B}$.*

The theorem may not hold for non-diagonalizable tensors. For example, if $\boldsymbol{A}$ is non-diagonalizable but invertible, and $\boldsymbol{B} = \boldsymbol{A}^{-T}$, then $(1, \boldsymbol{I})$ is an eigenvalue/eigentensor of $\boldsymbol{A} \boxtimes \boldsymbol{A}^{-T}$, but $\boldsymbol{u}_i \otimes \boldsymbol{v}_j$ (or their linear combinations) corresponding to $\lambda_i\gamma_j = 1$ may not equal $\boldsymbol{I}$. An application of the above theorem is presented in Theorem 1.7.2.

An orthogonal fourth-order tensor $\mathbb{Q}$ is defined as one for which

$$\mathbb{Q}^T\mathbb{Q} = \mathbb{Q}\mathbb{Q}^T = \mathbb{I}. \tag{1.181}$$

A fourth-order tensor is orthogonal if and only if it preserves inner products, i.e., $\mathbb{Q}$ is orthogonal if and only if

$$(\mathbb{Q}\boldsymbol{A}, \mathbb{Q}\boldsymbol{B}) = (\boldsymbol{A}, \boldsymbol{B}) \quad \forall \boldsymbol{A}, \boldsymbol{B} \in \text{Lin},$$

the proof being identical to that for second-order orthogonal tensors. An example of an orthogonal tensor is $Q_1 \boxtimes Q_2$, where $Q_1, Q_2 \in$ Orth. This follows since

$$(Q_1 \boxtimes Q_2)^T(Q_1 \boxtimes Q_2) = (Q_1^T \boxtimes Q_2^T)(Q_1 \boxtimes Q_2) = (Q_1^T Q_1) \boxtimes (Q_2^T Q_2) = I \boxtimes I = \mathbb{I}.$$

A special class of fourth-order orthogonal tensors are fourth-order rotations obtained by choosing $Q_1 = Q_2 = Q$, where $Q \in \text{Orth}^+$. Equation (1.150) can be expressed as

$$[\bar{\mathbb{A}}] = [\mathbb{Q}][\mathbb{A}][\mathbb{Q}]^T, \tag{1.182}$$

where $\mathbb{Q} := Q \boxtimes Q$, $Q \in \text{Orth}^+$, is a fourth-order rotation.

Since $\Psi(\mathbb{I}) = I_{9\times 9}$, we have $\det \mathbb{I} = 1$. It follows from Eqns. (1.179)–(1.181) that

$$\det \mathbb{Q} = \pm 1.$$

From Eqns. (1.180)–(1.182), we get

$$\begin{aligned}\det(\bar{\mathbb{A}} - \lambda\mathbb{I}) = \det[\mathbb{Q}(\mathbb{A} - \lambda\mathbb{I})\mathbb{Q}^T] &= \det(\mathbb{Q}\mathbb{Q}^T)\det(\mathbb{A} - \lambda\mathbb{I}) = (\det\mathbb{I})\det(\mathbb{A} - \lambda\mathbb{I}) \\ &= \det(\mathbb{A} - \lambda\mathbb{I}),\end{aligned} \tag{1.183}$$

showing that the invariants that occur in the characteristic equation (and hence the eigenvalues) of $\mathbb{A}$ are independent (as they should be) of the choice of basis.

The Cayley–Hamilton theorem reads

$$\begin{aligned}\mathbf{0} &= \mathbb{A}^9 - I_1\mathbb{A}^8 + I_2\mathbb{A}^7 + \cdots - I_9\mathbb{I} \\ &= \prod_{i=1}^{9}(\mathbb{A} - \lambda_i\mathbb{I}).\end{aligned}$$

The minimal polynomial of $\mathbb{A}$ is given by

$$q(\mathbb{A}) = \prod_{i=1}^{k}(\mathbb{A} - \lambda_i\mathbb{I})^{m_i},$$

where k is the number of distinct eigenvalues, and $1 \le m_i \le n_i$, with n_i denoting the algebraic multiplicity of each eigenvalue λ_i. If $\mathbb{A}$ is "normal", i.e., if $\mathbb{A}^T\mathbb{A} = \mathbb{A}\mathbb{A}^T$, or, in particular, symmetric, then each $m_i = 1$.

The tensor $\mathbb{A}$ is said to be invertible if there exists another tensor, denoted by $\mathbb{A}^{-1}$, such that

$$\mathbb{A}^{-1}\mathbb{A} = \mathbb{A}\mathbb{A}^{-1} = \mathbb{I}.$$

Similar to Eqns. (1.68) and (1.69), we have

$$(\mathbb{U}\mathbb{V})^{-1} = \mathbb{V}^{-1}\mathbb{U}^{-1}, \tag{1.184a}$$

$$(\mathbb{C}^T)^{-1} = (\mathbb{C}^{-1})^T =: \mathbb{C}^{-T}, \tag{1.184b}$$

From Eqn. (1.184b), it follows that if $\mathbb{C}$ has major symmetric then so does its inverse. The inverse of $A \boxtimes B$ (assuming that A and B are invertible) is given by

$$(A \boxtimes B)^{-1} = A^{-1} \boxtimes B^{-1}, \tag{1.185}$$

since, by Eqns. (1.167) and (1.168), we have

$$(A^{-1} \boxtimes B^{-1})(A \boxtimes B) = (A^{-1}A) \boxtimes (B^{-1}B) = I \boxtimes I = \mathbb{I}.$$

A fourth-order tensor $\mathbb{A}$ is invertible if and only if $I_9(\mathbb{A}) = \prod_{i=1}^{9} \lambda_i$ is nonzero, in which case,

$$\mathbb{A}^{-1} = \frac{1}{I_9}\left[\mathbb{A}^8 - I_1\mathbb{A}^7 + \cdots + I_8\mathbb{I}\right].$$

By virtue of the relation $(A^T)^T = A$, we have $\mathbb{T}\mathbb{T}A = A$ for all $A \in \text{Lin}$, which implies that

$$\mathbb{T}\mathbb{T} = \mathbb{I}. \tag{1.186}$$

Thus, the inverse of the transposer tensor $\mathbb{T}$ is $\mathbb{T}$ itself.

If $\mathbb{A}$ has major symmetry (but not necessarily the minor symmetries), then $[\Upsilon(\mathbb{A})]^T = \Upsilon(\mathbb{A}^T) = \Upsilon(\mathbb{A})$, so that, similar to the case of second-order symmetric tensors, all the nine eigenvalues are real, and the corresponding eigentensors A_i, $i = 1, 2, \ldots, 9$ are orthogonal in the sense that

$$(A_i, A_j) = A_i : A_j = \delta_{ij}. \tag{1.187}$$

Thus, we can write the spectral representation of $\mathbb{A}$ as

$$\mathbb{A} = \sum_{i=1}^{9} \lambda_i A_i \otimes A_i, \quad \text{with } \sum_{i=1}^{9} A_i \otimes A_i = \mathbb{I}.$$

If k is the number of distinct eigenvalues, then, similar to Eqn. (J.4), we have the following explicit formula:

$$A_i \otimes A_i = \begin{cases} \prod_{\substack{j=1 \\ j\neq i}}^{k} \frac{\mathbb{A} - \lambda_j \mathbb{I}}{\lambda_i - \lambda_j}, & k > 1 \\ \mathbb{I}, & k = 1 \end{cases} \tag{1.188}$$

When $\mathbb{A}$ is invertible, the spectral decomposition of $\mathbb{A}^{-1}$ is given by

$$\mathbb{A}^{-1} = \sum_{i=1}^{9} \frac{1}{\lambda_i} A_i \otimes A_i.$$

$\mathbb{A}$ is said to be positive definite if $T : \mathbb{A}T > 0$ for all nonzero $T \in \text{Lin}$. Similar to symmetric second-order tensors (see Theorem 1.6.4), one can show that $\mathbb{A}$ is positive definite if and only if all the eigenvalues λ_i (or all the invariants that occur in the characteristic equation) are strictly positive. If $\mathbb{A}$ is positive semi-definite, the square root of $\mathbb{A}$ defined via the relation $\mathbb{H}\mathbb{H} = \mathbb{A}$ is given by

$$\mathbb{H} = \sum_{i=1}^{9} \sqrt{\lambda_i} A_i \otimes A_i.$$

Now consider the case when the fourth-order tensor $\mathbb{C}$ has both minor symmetries, i.e., $\mathbb{C} = \mathbb{S}\mathbb{C}\mathbb{S}$. The transpose is still defined by Eqn. (1.163), except that $A, B \in \mathrm{Sym}$. Using this definition, one can show that when $\mathbb{C}$ and $\mathbb{D}$ have both minor symmetries, then Eqns. (1.165) still hold. The eigenvalue problem given by Eqn. (1.178) is now written as $\mathbb{C}S = \lambda S = \lambda \mathbb{S}S$, $S \in \mathrm{Sym}$, or, alternatively, using Eqn. (I.14), as

$$\left[\chi(\mathbb{C})H^{-1} - \lambda I_{6\times 6}\right]\chi(S) = \mathbf{0}.$$

Thus, $\mathbb{C}$ has only six eigenvalues, with the remaining three eigenvalues zero, and the corresponding eigentensors being linearly independent skew-symmetric tensors. The above equation has a nontrivial solution $\chi(S) \neq \mathbf{0}$, if and only if $\det[\chi(\mathbb{C})H^{-1} - \lambda I] = 0$. This characteristic equation can be written in expanded form as

$$\lambda^6 - I_1(\mathbb{C})\lambda^5 + I_2(\mathbb{C})\lambda^4 + \cdots + I_6(\mathbb{C}) = 0,$$

where

$$I_1(\mathbb{C}) = \operatorname{tr}[\chi(\mathbb{C})H^{-1}] = \lambda_1 + \lambda_2 + \cdots + \lambda_6,$$

$$I_2(\mathbb{C}) = \frac{1}{2}[(\operatorname{tr}[\chi(\mathbb{C})H^{-1}])^2 - \operatorname{tr}(\chi(\mathbb{C})H^{-1})^2] = \lambda_1\lambda_2 + \lambda_2\lambda_3 + \cdots + \lambda_1\lambda_6,$$

$$\ldots$$

$$I_6(\mathbb{C}) = \det[\chi(\mathbb{C})H^{-1}] = 8\det\chi(\mathbb{C}) = \lambda_1\lambda_2\ldots\lambda_6,$$

are a strict subset of the set of principal invariants of $\mathbb{C}$ [4, 20, 141, 238] (the results of Ahmad and Norris on the one hand, and Betten on the other regarding the number of quadratic principal invariants seem to be contradictory). As an example,

$$\det\mathbb{S} = \det[\chi(\mathbb{S})H^{-1}] = \det[HH^{-1}] = 1. \tag{1.189}$$

Using Eqn. (I.11c), we now have, similar to Eqn. (1.179),

$$\det\mathbb{C}^T = \det\mathbb{C}. \tag{1.190}$$

If $\mathbb{C} = \mathbb{C}_1\mathbb{C}_2$, then using Eqn. (I.11f), we again have similar to Eqn. (1.180),

$$\det\mathbb{C} = (\det\mathbb{C}_1)(\det\mathbb{C}_2). \tag{1.191}$$

If $\mathbb{C}$ has both the minor symmetries, then $[\bar{\mathbb{C}}] = [\mathbb{S}][\bar{\mathbb{C}}][\mathbb{S}]$ and $[\mathbb{C}] = [\mathbb{S}][\mathbb{C}][\mathbb{S}]$, so that Eqn. (1.182) can be written as

$$[\bar{\mathbb{C}}] = [\mathbb{Q}][\mathbb{C}][\mathbb{Q}]^T, \tag{1.192}$$

where $\mathbb{Q} := [\mathbb{S}][Q \boxtimes Q][\mathbb{S}]$, $Q \in \mathrm{Orth}^+$. Now applying Eqns. (1.165b), (I.11c) and (I.11f), and using the symmetry of $\mathbb{S}$ and H, we get

$$\chi([\bar{\mathbb{C}}]) = \chi(\mathbb{Q})H^{-1}\chi([\mathbb{C}])H^{-1}\chi(\mathbb{Q})^T = A\chi([\mathbb{C}])A^T,$$

where $A := \chi(\mathbb{Q})H^{-1}$ is found by using Eqn. (I.5); the final expression for A is given by Eqn. (I.12) with $J \equiv Q^T$. Using Eqns. (1.165b) and (1.176), we get

$$\mathbb{Q}\mathbb{Q}^T = \mathbb{Q}^T\mathbb{Q} = \mathbb{S}(I \boxtimes I)\mathbb{S} = \mathbb{S}\mathbb{I}\mathbb{S} = \mathbb{S}. \tag{1.193}$$

It follows from Eqns. (1.189)–(1.191) that

$$\det \mathbb{Q} = \pm 1.$$

From Eqns. (1.189) and (1.191)–(1.193), we get

$$\begin{aligned}\det(\bar{\mathbb{C}} - \lambda\mathbb{S}) &= \det[\mathbb{Q}(\mathbb{C} - \lambda\mathbb{S})\mathbb{Q}^T] = \det(\mathbb{Q}\mathbb{Q}^T)\det(\mathbb{C} - \lambda\mathbb{S}) \\ &= (\det\mathbb{S})\det(\mathbb{C} - \lambda\mathbb{S}) = \det(\mathbb{C} - \lambda\mathbb{S}),\end{aligned} \tag{1.194}$$

showing that the invariants that occur in the characteristic equation (and hence the eigenvalues) of $\mathbb{C}$ are independent of the choice of basis.

The spectral decomposition of $\mathbb{C}$ reads

$$\mathbb{C} = \sum_{i=1}^{6} \lambda_i A_i \otimes A_i, \quad \text{with } A_i = A_i^T \text{ and } \sum_{i=1}^{6} A_i \otimes A_i = \mathbb{S}.$$

In place of Eqn. (1.188), we now have

$$A_i \otimes A_i = \begin{cases} \prod_{\substack{j=1 \\ j\neq i}}^{k} \frac{\mathbb{C}-\lambda_j\mathbb{S}}{\lambda_i - \lambda_j}, & k > 1 \\ \mathbb{S}. & k = 1 \end{cases}$$

$\mathbb{C}$ is invertible if there exists another tensor, denoted by $\mathbb{C}^{-1}$, such that

$$\mathbb{C}^{-1}\mathbb{C} = \mathbb{C}\mathbb{C}^{-1} = \mathbb{S}.$$

The necessary and sufficient condition for $\mathbb{C}$ to be invertible is that $I_6(\mathbb{C})$ be nonzero. If this condition is satisfied, then

$$\mathbb{C}^{-1} = \sum_{i=1}^{6} \frac{1}{\lambda_i} A_i \otimes A_i.$$

As an example, since $\mathbb{S}\mathbb{S} = \mathbb{S}$, we have $\mathbb{S}^{-1} = \mathbb{S}$. Let $\mathbb{U}$, $\mathbb{V}$ and $\mathbb{C}$ be tensors that have minor symmetries. Then, since $\mathbb{S}^T = \mathbb{S}$, both relations given by Eqns. (1.184) hold. From Eqn. (1.176), it follows (provided A is invertible) that

$$[\mathbb{S}(A \boxtimes A)\mathbb{S}]^{-1} = \mathbb{S}(A^{-1} \boxtimes A^{-1})\mathbb{S}.$$

Note that, in general, $[\mathbb{S}(A \boxtimes B)\mathbb{S}]^{-1} \neq \mathbb{S}(A^{-1} \boxtimes B^{-1})\mathbb{S}$.

$\mathbb{C}$ is said to be positive semi-definite if $S : \mathbb{C}S \geq 0$ for all $S \in \text{Sym}$. If $\mathbb{C}$ is positive semi-definite, its square root $\mathbb{H}$ is given by

$$\mathbb{H} = \sum_{i=1}^{6} \sqrt{\lambda_i} A_i \otimes A_i, \quad \text{with } A_i = A_i^T.$$

$\mathbb{C}$ is positive definite if and only if all the λ_i, $i = 1, 2, \ldots, 6$, are strictly positive.

Examples of some spectral resolutions are

$$\mathbb{I} = \sum_{i=1}^{9} \boldsymbol{A}_i \otimes \boldsymbol{A}_i,$$

$$\mathbb{T} = \sum_{i=1}^{6} \boldsymbol{A}_i \otimes \boldsymbol{A}_i - \sum_{i=7}^{9} \boldsymbol{A}_i \otimes \boldsymbol{A}_i,$$

$$\mathbb{S} = \sum_{i=1}^{6} \boldsymbol{A}_i \otimes \boldsymbol{A}_i, \quad \boldsymbol{A}_i^T = \boldsymbol{A}_i$$

$$\mathbb{W} = \sum_{i=1}^{3} \boldsymbol{A}_i \otimes \boldsymbol{A}_i, \quad \boldsymbol{A}_i^T = -\boldsymbol{A}_i$$

$$\boldsymbol{I} \otimes \boldsymbol{I} = 3\left(\frac{\boldsymbol{I}}{\sqrt{3}} \otimes \frac{\boldsymbol{I}}{\sqrt{3}}\right).$$

Let $\boldsymbol{e}$ be the axial vector of $\boldsymbol{Q} \in \text{Orth}^+ \backslash \text{Sym}$, and let $(\lambda, \boldsymbol{n})$ and $(\hat{\lambda}, \hat{\boldsymbol{n}})$ denote the remaining eigenvalue/eigenvector pairs, with the "hat" denoting a complex conjugate. The following result, which follows directly from Theorem 1.7.1, holds:

Theorem 1.7.2. *The eigenvalue/eigentensor pairs of* $\mathbb{Q} := \boldsymbol{Q} \boxtimes \boldsymbol{Q}$, *where* $\boldsymbol{Q} \in$ *Orth*$^+\backslash$*Sym, are* $(1, \boldsymbol{I})$, $(1, \boldsymbol{Q})$, $(1, \boldsymbol{Q}^2)$, $(\lambda, \boldsymbol{e} \otimes \boldsymbol{n})$, $(\lambda, \boldsymbol{n} \otimes \boldsymbol{e})$, $(\hat{\lambda}, \boldsymbol{e} \otimes \hat{\boldsymbol{n}})$, $(\hat{\lambda}, \hat{\boldsymbol{n}} \otimes \boldsymbol{e})$, $(\lambda^2, \boldsymbol{n} \otimes \boldsymbol{n})$, *and* $(\hat{\lambda}^2, \hat{\boldsymbol{n}} \otimes \hat{\boldsymbol{n}})$.

If the dimension of the underlying space is four, and $(\lambda_1, \boldsymbol{n}_1)$, $(\hat{\lambda}_1, \hat{\boldsymbol{n}}_1)$, $(\lambda_2, \boldsymbol{n}_2)$, $(\hat{\lambda}_2, \hat{\boldsymbol{n}}_2)$ denote the eigenvalue/eigenvector pairs of $\boldsymbol{Q}$, then the eigenvalue/eigentensor pairs of $\mathbb{Q}$ are $(1, \boldsymbol{I})$, $(1, \boldsymbol{Q})$, $(1, \boldsymbol{Q}^2)$, $(1, \boldsymbol{Q}^3)$, $(\lambda_1^2, \boldsymbol{n}_1 \otimes \boldsymbol{n}_1)$, $(\hat{\lambda}_1^2, \hat{\boldsymbol{n}}_1 \otimes \hat{\boldsymbol{n}}_1)$, $(\lambda_2^2, \boldsymbol{n}_2 \otimes \boldsymbol{n}_2)$, $(\hat{\lambda}_2^2, \hat{\boldsymbol{n}}_2 \otimes \hat{\boldsymbol{n}}_2)$, $(\lambda_1\lambda_2, \boldsymbol{n}_1 \otimes \boldsymbol{n}_2)$, $(\lambda_1\lambda_2, \boldsymbol{n}_2 \otimes \boldsymbol{n}_1)$, $(\lambda_1\hat{\lambda}_2, \boldsymbol{n}_1 \otimes \hat{\boldsymbol{n}}_2)$, $(\lambda_1\hat{\lambda}_2, \hat{\boldsymbol{n}}_2 \otimes \boldsymbol{n}_1)$, $(\hat{\lambda}_1\lambda_2, \hat{\boldsymbol{n}}_1 \otimes \boldsymbol{n}_2)$, $(\hat{\lambda}_1\lambda_2, \boldsymbol{n}_2 \otimes \hat{\boldsymbol{n}}_1)$, $(\hat{\lambda}_1\hat{\lambda}_2, \hat{\boldsymbol{n}}_1 \otimes \hat{\boldsymbol{n}}_2)$ and $(\hat{\lambda}_1\hat{\lambda}_2, \hat{\boldsymbol{n}}_2 \otimes \hat{\boldsymbol{n}}_1)$. In general, when the underlying space dimension is n, the eigenvalue one is repeated n times, with the corresponding eigentensors given by $\boldsymbol{I}$, $\boldsymbol{Q}$, $\ldots$, $\boldsymbol{Q}^{n-1}$, and the remaining $n(n-1)$ eigenvalue/eigentensor pairs are formed by combinations of each complex eigenvalue/eigenvector of $\boldsymbol{Q}$ with itself and with all the remaining eigenvalues/eigenvectors of $\boldsymbol{Q}$ barring the complex conjugate of that eigenvalue/eigenvector.

We now discuss the solution of the tensor equation $\boldsymbol{AXB} + \boldsymbol{CXD} = \boldsymbol{H}$, where $\boldsymbol{A}$, $\boldsymbol{B}$, $\boldsymbol{C}$, $\boldsymbol{D}$ and $\boldsymbol{H}$ are given second-order tensors. We write this equation as

$$[\boldsymbol{A} \boxtimes \boldsymbol{B}^T + \boldsymbol{C} \boxtimes \boldsymbol{D}^T]\boldsymbol{X} = \boldsymbol{H}.$$

In component form, the above equation reads

$$\mathbb{T}_{ijkl} X_{kl} = H_{ij},$$

where

$$\mathbb{T}_{ijkl} = A_{ik} B_{lj} + C_{ik} D_{lj}, \tag{1.195}$$

Applying Eqn. (I.1e), we get

$$\begin{bmatrix} X_{11} \\ X_{22} \\ X_{33} \\ X_{12} \\ X_{23} \\ X_{31} \\ X_{21} \\ X_{32} \\ X_{13} \end{bmatrix} = \begin{bmatrix} \mathbb{T}_{1111} & \mathbb{T}_{1122} & \mathbb{T}_{1133} & \mathbb{T}_{1112} & \mathbb{T}_{1123} & \mathbb{T}_{1131} & \mathbb{T}_{1121} & \mathbb{T}_{1132} & \mathbb{T}_{1113} \\ \mathbb{T}_{2211} & \mathbb{T}_{2222} & \mathbb{T}_{2233} & \mathbb{T}_{2212} & \mathbb{T}_{2223} & \mathbb{T}_{2231} & \mathbb{T}_{2221} & \mathbb{T}_{2232} & \mathbb{T}_{2213} \\ \mathbb{T}_{3311} & \mathbb{T}_{3322} & \mathbb{T}_{3333} & \mathbb{T}_{3312} & \mathbb{T}_{3323} & \mathbb{T}_{3331} & \mathbb{T}_{3321} & \mathbb{T}_{3332} & \mathbb{T}_{3313} \\ \mathbb{T}_{1211} & \mathbb{T}_{1222} & \mathbb{T}_{1233} & \mathbb{T}_{1212} & \mathbb{T}_{1223} & \mathbb{T}_{1231} & \mathbb{T}_{1221} & \mathbb{T}_{1232} & \mathbb{T}_{1213} \\ \mathbb{T}_{2311} & \mathbb{T}_{2322} & \mathbb{T}_{2333} & \mathbb{T}_{2312} & \mathbb{T}_{2323} & \mathbb{T}_{2331} & \mathbb{T}_{2321} & \mathbb{T}_{2332} & \mathbb{T}_{2313} \\ \mathbb{T}_{3111} & \mathbb{T}_{3122} & \mathbb{T}_{3133} & \mathbb{T}_{3112} & \mathbb{T}_{3123} & \mathbb{T}_{3131} & \mathbb{T}_{3121} & \mathbb{T}_{3132} & \mathbb{T}_{3113} \\ \mathbb{T}_{2111} & \mathbb{T}_{2122} & \mathbb{T}_{2133} & \mathbb{T}_{2112} & \mathbb{T}_{2123} & \mathbb{T}_{2131} & \mathbb{T}_{2121} & \mathbb{T}_{2132} & \mathbb{T}_{2113} \\ \mathbb{T}_{3211} & \mathbb{T}_{3222} & \mathbb{T}_{3233} & \mathbb{T}_{3212} & \mathbb{T}_{3223} & \mathbb{T}_{3231} & \mathbb{T}_{3221} & \mathbb{T}_{3232} & \mathbb{T}_{3213} \\ \mathbb{T}_{1311} & \mathbb{T}_{1322} & \mathbb{T}_{1333} & \mathbb{T}_{1312} & \mathbb{T}_{1323} & \mathbb{T}_{1331} & \mathbb{T}_{1321} & \mathbb{T}_{1332} & \mathbb{T}_{1313} \end{bmatrix}^{-1} \begin{bmatrix} H_{11} \\ H_{22} \\ H_{33} \\ H_{12} \\ H_{23} \\ H_{31} \\ H_{21} \\ H_{32} \\ H_{13} \end{bmatrix}, \tag{1.196}$$

with the components $\mathbb{T}_{ijkl}$ given by Eqn. (1.195). It is clear that this method can be extended to solve a matrix equation of the type $\sum_i \boldsymbol{A}_i \boldsymbol{X} \boldsymbol{B}_i = \boldsymbol{H}$ of any dimension n by inverting an $n^2 \times n^2$ matrix.

Of particular importance is the case when $\boldsymbol{B} = \boldsymbol{C} = \boldsymbol{I}$ and $\boldsymbol{D} = \boldsymbol{A} \equiv \boldsymbol{S} \in \text{Sym}$. In this case, $\mathbb{T}_{ijkl} = S_{ik}\delta_{lj} + \delta_{ik}S_{lj}$.

Although the above approach is suitable for numerical purposes, a tensorial solution is more suitable for proving properties of the solution $\boldsymbol{X}$. We now present a method, based on [155], for determining $(\boldsymbol{S} \boxtimes \boldsymbol{I} + \boldsymbol{I} \boxtimes \boldsymbol{S})^{-1}$, $\boldsymbol{S} \in \text{Sym}$, which is applicable for any dimension n (see Rosati [273] and references therein for other methods).

Consider the tensor $\boldsymbol{T} \boxtimes \boldsymbol{I} + \boldsymbol{I} \boxtimes \boldsymbol{T}^T$, $\boldsymbol{T} \in \text{Lin}$. If $(\lambda_i, \boldsymbol{u}_i)$ and $(\lambda_i, \boldsymbol{v}_i)$ are the eigenvalue/eigenvectors of $\boldsymbol{T}$ and $\boldsymbol{T}^T$, respectively, then

$$\begin{aligned} (\boldsymbol{T} \boxtimes \boldsymbol{I} + \boldsymbol{I} \boxtimes \boldsymbol{T}^T)(\boldsymbol{u}_i \otimes \boldsymbol{v}_j) &= \boldsymbol{T}(\boldsymbol{u}_i \otimes \boldsymbol{v}_j) + (\boldsymbol{u}_i \otimes \boldsymbol{v}_j)\boldsymbol{T} \\ &= (\boldsymbol{T}\boldsymbol{u}_i) \otimes \boldsymbol{v}_j + \boldsymbol{u}_i \otimes (\boldsymbol{T}^T \boldsymbol{v}_j) \\ &= (\lambda_i + \lambda_j)(\boldsymbol{u}_i \otimes \boldsymbol{v}_j), \end{aligned}$$

which shows that $(\lambda_i + \lambda_j, \boldsymbol{u}_i \otimes \boldsymbol{v}_j)$, $i, j = 1, 2, \ldots, n$, are eigenvalue/eigentensor pairs of $\boldsymbol{T} \boxtimes \boldsymbol{I} + \boldsymbol{I} \boxtimes \boldsymbol{T}^T$. However, unfortunately, these are *not the only* eigenvalue/eigentensor pairs, especially when the eigenvectors $\{\boldsymbol{u}_i\}$ are not linearly independent. For example, when $\boldsymbol{T} = \boldsymbol{e}_1 \otimes \boldsymbol{e}_2 + \boldsymbol{e}_3 \otimes \boldsymbol{e}_4 + \boldsymbol{e}_4 \otimes \boldsymbol{e}_5$, with $n = 5$, all the 25 eigenvalues of $\boldsymbol{T} \boxtimes \boldsymbol{I} + \boldsymbol{I} \boxtimes \boldsymbol{T}^T$ are zero, and there are nine linearly independent eigentensors given by $\boldsymbol{e}_1 \otimes \boldsymbol{e}_2$, $\boldsymbol{e}_1 \otimes \boldsymbol{e}_5$, $\boldsymbol{e}_3 \otimes \boldsymbol{e}_2$, $\boldsymbol{e}_3 \otimes \boldsymbol{e}_5$, $\boldsymbol{e}_3 \otimes \boldsymbol{e}_3 - \boldsymbol{e}_4 \otimes \boldsymbol{e}_4 + \boldsymbol{e}_5 \otimes \boldsymbol{e}_5$, $-\boldsymbol{e}_3 \otimes \boldsymbol{e}_4 + \boldsymbol{e}_4 \otimes \boldsymbol{e}_5$, $\boldsymbol{e}_3 \otimes \boldsymbol{e}_1 + \boldsymbol{e}_4 \otimes \boldsymbol{e}_2$, $-\boldsymbol{e}_1 \otimes \boldsymbol{e}_4 + \boldsymbol{e}_2 \otimes \boldsymbol{e}_5$ and $-\boldsymbol{e}_1 \otimes \boldsymbol{e}_1 + \boldsymbol{e}_2 \otimes \boldsymbol{e}_2$. Although the first four eigentensors are of the form $\boldsymbol{u}_i \otimes \boldsymbol{v}_j$, $i, j = 1, 2$, the remaining five are not. A similar situation arises even in the case of the tensor $\boldsymbol{T} \boxtimes \boldsymbol{I} + \boldsymbol{I} \boxtimes \boldsymbol{T}$, which again has nine linearly independent eigentensors. Nevertheless, as we now show, $\lambda_i + \lambda_j$, $i, j = 1, 2, \ldots, n$, are the eigenvalues of $\boldsymbol{T} \boxtimes \boldsymbol{I} + \boldsymbol{I} \boxtimes \boldsymbol{T}^T$ (or $\boldsymbol{T} \boxtimes \boldsymbol{I} + \boldsymbol{I} \boxtimes \boldsymbol{T}$). Thus, the necessary and sufficient condition for $\boldsymbol{T} \boxtimes \boldsymbol{I} + \boldsymbol{I} \boxtimes \boldsymbol{T}^T$ to be invertible is that $(\det \boldsymbol{T}) \prod^{n}_{\substack{i=1 \\ j=i+1}} (\lambda_i + \lambda_j) \neq 0$. For example, if $n = 3$, this condition is given by $\lambda_1 \lambda_2 \lambda_3 (\lambda_1 + \lambda_2)(\lambda_2 + \lambda_3)(\lambda_1 + \lambda_3) \neq 0$.

Let $T = H_1 J_1 H_1^{-1}$ and $T^T = H_2 J_2 H_2^{-1}$, where J_1 and J_2 are matrices in Jordan normal form. Then

$$\begin{aligned}
T \boxtimes I + I \boxtimes T^T &= (H_1 J_1 H_1^{-1}) \boxtimes I + I \boxtimes (H_2 J_2 H_2^{-1}) \\
&= (H_1 \boxtimes H_2)(J_1 \boxtimes I + I \boxtimes J_2)(H_1^{-1} \boxtimes H_2^{-1}) \\
&= (H_1 \boxtimes H_2)(J_1 \boxtimes I + I \boxtimes J_2)(H_1 \boxtimes H_2)^{-1}.
\end{aligned}$$

Thus, the eigenvalues of $T \boxtimes I + I \boxtimes T^T$ and $J_1 \boxtimes I + I \boxtimes J_2$ are the same. Since the matrix form of $J_1 \boxtimes I + I \boxtimes J_2$ is an upper triangular matrix with $\lambda_i + \lambda_j$, $i, j = 1, 2, \ldots, n$, on the diagonals, we get the desired result. A similar proof can be devised for $T \boxtimes I + I \boxtimes T$.

Let (λ_i, n_i) be the eigenvalues/eigenvectors of $S \in \text{Sym}$. Then, the eigenvalue/eigentensor pairs of $\mathbb{S}(S \boxtimes I + I \boxtimes S)\mathbb{S}$ when all the λ_i are distinct are $\lambda_i + \lambda_j, n_i \otimes n_j + n_j \otimes n_i$, $i, j = 1, 2, \ldots, n$ (the eigenvalue/eigentensors when the eigenvalues of S are repeated can be found similar to Eqns. (1.258) and (1.259)). Thus, the necessary and sufficient condition that $\mathbb{S}(S \boxtimes I + I \boxtimes S)\mathbb{S}$ be invertible is that $(\det S) \prod^{n}_{\substack{i=1 \\ j=i+1}} (\lambda_i + \lambda_j) \neq 0$.

Let λ_i, $i = 1, 2, \ldots, k$, be the *distinct* eigenvalues of $S \in \text{Sym}$, and let P_i be the associated projections, which are given by Eqn. (J.4). We have

$$\begin{aligned}
\mathbb{S}(S \boxtimes I + I \boxtimes S)\mathbb{S} &= \mathbb{S}\left[\left(\sum_{i=1}^{k} \lambda_i P_i\right) \boxtimes \left(\sum_{j=1}^{k} P_j\right) + \left(\sum_{i=1}^{k} P_i\right) \boxtimes \left(\sum_{j=1}^{k} \lambda_j P_j\right)\right]\mathbb{S} \\
&= \mathbb{S}\left[\sum_{i=1}^{k}\sum_{j=1}^{k} (\lambda_i + \lambda_j) P_i \boxtimes P_j\right]\mathbb{S}.
\end{aligned}$$

Assuming that the necessary and sufficient condition for invertibility is satisfied, the desired inverse is

$$[\mathbb{S}(S \boxtimes I + I \boxtimes S)\mathbb{S}]^{-1} = \mathbb{S}\left[\sum_{i=1}^{k}\sum_{j=1}^{k} \frac{P_i \boxtimes P_j}{\lambda_i + \lambda_j}\right]\mathbb{S}, \tag{1.197}$$

since, on using Eqn. (1.175),

$$\begin{aligned}
\mathbb{S}\left[\sum_{i=1}^{k}\sum_{j=1}^{k} \frac{P_i \boxtimes P_j}{\lambda_i + \lambda_j}\right]\mathbb{S}\,[S \boxtimes I + I \boxtimes S]\,\mathbb{S} &= \mathbb{S}\left[\sum_{i=1}^{k}\sum_{j=1}^{k} \frac{(P_i S) \boxtimes P_j + P_i \boxtimes (P_j S)}{\lambda_i + \lambda_j}\right]\mathbb{S} \\
&= \mathbb{S}\left[\sum_{i=1}^{k}\sum_{j=1}^{k} P_i \boxtimes P_j\right]\mathbb{S} \\
&= \mathbb{S}\left[\left(\sum_{i=1}^{k} P_i\right) \boxtimes \left(\sum_{j=1}^{k} P_j\right)\right]\mathbb{S} \\
&= \mathbb{S}\,[I \boxtimes I]\,\mathbb{S} \\
&= \mathbb{S},
\end{aligned}$$

where we have used the properties $P_m P_i = \delta_{im} P_i$ (no sum on i), $\sum_{i=1}^k P_i = I$ and $SP_i = (\sum_{m=1}^k \lambda_m P_m) P_i = \lambda_i P_i = P_i S$ (no sum on i). Thus, the solution of the tensor equation $\mathbb{S}(S \boxtimes I + I \boxtimes S)\mathbb{S}X = H$ is

$$X = \sum_{i=1}^k \sum_{j=1}^k \frac{P_i (H + H^T) P_j}{2(\lambda_i + \lambda_j)}. \tag{1.198}$$

The above method can obviously be generalized to the case when T is diagonalizable. Thus, if the tensor equation to be solved is $(T \boxtimes I + I \boxtimes T)X = TX + XT^T = H$, where $T = \sum_{i=1}^k \lambda_i P_i$, with P_i given by Eqn. (J.26), then assuming that T is such that the conditions for invertibility of $T \boxtimes I + I \boxtimes T$ are satisfied, we have

$$\begin{aligned} X &= (T \boxtimes I + I \boxtimes T)^{-1} H \\ &= \left[\sum_{i=1}^k \sum_{j=1}^k \frac{P_i \boxtimes P_j}{\lambda_i + \lambda_j} \right] H \\ &= \sum_{i=1}^k \sum_{j=1}^k \frac{P_i H P_j^T}{\lambda_i + \lambda_j}. \end{aligned}$$

We now show how the solutions given by Eqns. (1.198) reduce to some of the solutions presented in the literature for the case $H = W \in \text{Skw}$. When $n = 3$, and when the eigenvalues are distinct, for a given i, we have $P_i W P_i = (e_i^* \cdot W e_i^*) P_i = \mathbf{0}$. Multiplying both sides of Eqn. (1.198) by the determinant of

$$\tilde{S} := (\operatorname{tr} S) I - S = (\lambda_2 + \lambda_3) P_1 + (\lambda_1 + \lambda_3) P_2 + (\lambda_1 + \lambda_2) P_3,$$

which is given by $(\lambda_1 + \lambda_2)(\lambda_2 + \lambda_3)(\lambda_1 + \lambda_3)$, we get

$$(\det \tilde{S}) X = \tilde{S} W \tilde{S}, \tag{1.199}$$

which is the solution presented by Scheidler [282]. Since the solution has no explicit dependence on the eigenvalues λ_i, it holds even under the case when the eigenvalues are repeated. Alternatively, by writing W as $(P_1 + P_2 + P_3) W (P_1 + P_2 + P_3)$, $S^2 W$ as $(\lambda_1^2 P_1 + \lambda_2^2 P_2 + \lambda_3^2 P_3) W (P_1 + P_2 + P_3)$, etc. and noting that $(\lambda_1 + \lambda_2)(\lambda_2 + \lambda_3)(\lambda_1 + \lambda_3) = I_1(S) I_2(S) - I_3(S)$, one can also show that when $H = W$, Eqn. (1.198) reduces to

$$(I_1 I_2 - I_3) X = (I_1^2 - I_2) W - (S^2 W + W S^2),$$

which is the solution presented by Guo [104].

Finally, consider the case when the dimension of the underlying vector space n is two, and H is arbitrary. Using $I_1(S) = \lambda_1 + \lambda_2$, $I_2(S) = \lambda_1 \lambda_2$, and writing $H = (P_1 + P_2) H (P_1 + P_2)$, $SH = (\lambda_1 P_1 + \lambda_2 P_2) H (P_1 + P_2)$, etc., one can show that Eqn. (1.198) can be written as

$$(2 I_1 I_2) X = (I_1^2 + I_2) H - I_1 (SH + HS) + SHS,$$

which is the solution presented by Hoger and Carlson [127]. Alternatively, if S is invertible, writing $S^{-1} = P_1/\lambda_1 + P_2/\lambda_2$, we can also show that

$$2I_1 X = H + I_2(S^{-1}HS^{-1}),$$

which is the solution presented by Ting [327].

1.8 Isotropic Tensors

A second-order tensor is said to be isotropic if it's components are the same in all orthonormal bases, i.e., if $T = T_{ij}e_i \otimes e_j = T_{ij}\bar{e}_i \otimes \bar{e}_j$. Since $\bar{e}_i = Q^T e_i$, this condition can also be written as

$$QTQ^T = T \quad \forall Q \in \text{Orth}^+, \tag{1.200}$$

or, alternatively, as

$$QT = TQ \quad \forall Q \in \text{Orth}^+. \tag{1.201}$$

We have the following characterization for isotropic tensors.

Theorem 1.8.1. *A tensor T is isotropic if and only if it is of the form $T = \lambda I$, where $\lambda \in \Re$.*

Proof. If $T = \lambda I$, then obviously Eqn. (1.200) holds.

To prove the converse, let v be an arbitrary vector, and let R be the proper orthogonal tensor for which v is along its axis, i.e., $Rv = v$ (see Eqn. (1.104)). Since $RT = TR$, we have

$$RTv = TRv = Tv,$$

i.e., Tv also lies along the axis of R. Since the axis of R is a one-dimensional subspace, we have

$$Tv = \lambda v = \lambda I v.$$

Since the vector v is arbitrary, we have $T = \lambda I$ by the definition of equality of tensors. □

Another proof can be given as follows. Express Eqn. (1.200) as

$$\mathbb{Q}T = T,$$

where $\mathbb{Q} = Q \boxtimes Q$. By Theorem 1.7.2, I, Q and Q^2 are the eigentensors of $\mathbb{Q}$ corresponding to the eigenvalue one. Hence, the above equation implies that

$$T = \alpha_0(Q)I + \alpha_1(Q)Q + \alpha_2(Q)Q,$$

where α_0, α_1 and α_2 are, in general, constants that depend on Q. Since T is independent of Q, we have

$$T = \alpha_0 I + \alpha_1 Q_0 + \alpha_2 Q_0^2,$$

where, now, α_0, α_1 and α_2 are constants. Substituting this expression into Eqn. (1.201), we get $\alpha_1 = \alpha_2 = 0$, since, in general, rotations do not commute. This methodology can be extended to other dimensions as well. For example, if the dimension n is four, then we get $T = \alpha_0 I + \alpha_1 Q_0 + \alpha_2 Q_0^2 + \alpha_3 Q_0^3$, but again, since rotations do not commute, we get all the constants other than α_0 to be zero. Now consider the case when n is two. In this case, $T = \alpha_0 I + \alpha_1 Q_0$, but now since rotations do commute when $n = 2$, both α_0 and α_1 are nonzero. Using the representation of Q_0 in terms of $\cos\theta_0$ and $\sin\theta_0$, we can write the most general form of an isotropic second-order tensor when $n = 2$ as $T = \begin{bmatrix} \alpha & \beta \\ -\beta & \alpha \end{bmatrix}$.

A third-order tensor is isotropic if and only if it is of the form $\lambda\mathcal{E}$, where $\mathcal{E}$ is the alternate tensor with components ϵ_{ijk}. To see the 'if' part, note that

$$\bar{\mathcal{E}}_{ijk} = \bar{e}_i \cdot (\bar{e}_j \times \bar{e}_k) = Q^T e_i \cdot [\mathbf{cof}\, Q^T (e_j \times e_k)] = Q^T e_i \cdot [Q^T (e_j \times e_k)] = e_i \cdot (e_j \times e_k) = \mathcal{E}_{ijk}.$$

The reverse implication is proved by making special choices of Q (such as $e_1 \otimes e_1 + e_2 \otimes e_3 - e_3 \otimes e_2$, $e_2 \otimes e_2 + e_3 \otimes e_1 - e_1 \otimes e_3$, etc.) in the relation

$$B_{ijk} = Q_{ip} Q_{jq} Q_{kr} B_{pqr},$$

and showing that B_{ijk} has to be a scalar multiple of ϵ_{ijk}.

A fourth-order tensor is said to be isotropic if it has the same components in all orthonormal bases, i.e., if

$$\mathbb{C} = \mathbb{Q}\mathbb{C}\mathbb{Q}^T, \tag{1.202}$$

where $\mathbb{Q} := Q \boxtimes Q$, with $Q \in \text{Orth}^+$. Alternatively, the above condition can also be written as

$$\mathbb{Q}(\mathbb{C}A) = \mathbb{C}(QAQ^T) \quad \forall A \in \text{Lin}. \tag{1.203}$$

We have the following characterization for isotropic fourth-order tensors:

Theorem 1.8.2. *A fourth-order tensor $\mathbb{C}$: Lin $\to$ Lin is isotropic if and only if it is of the form*

$$\mathbb{C} = \lambda I \otimes I + (\mu + \gamma)\mathbb{I} + (\mu - \gamma)\mathbb{T} \tag{1.204}$$
$$= \lambda I \otimes I + 2\mu\mathbb{S} + 2\gamma\mathbb{W},$$

or, in other words, a fourth-order isotropic tensor is a linear combination of tensors with components $\delta_{ij}\delta_{kl}$, $\delta_{ik}\delta_{jl}$ and $\delta_{il}\delta_{jk}$.

If $\mathbb{C}$ has the first or second minor symmetry, i.e., if $\mathbb{C} = \mathbb{S}\mathbb{C}$ or $\mathbb{C} = \mathbb{C}\mathbb{S}$, then using Eqn. (1.160), it follows as a corollary that

$$\mathbb{C} = \lambda I \otimes I + 2\mu\mathbb{S}. \tag{1.205}$$

Proof. The proof is based on [157]. To see the "if" part, note that

$$\mathbb{Q}(\boldsymbol{I}\otimes\boldsymbol{I})\mathbb{Q}^T = (\mathbb{Q}\boldsymbol{I})\otimes(\mathbb{Q}\boldsymbol{I}) = [(\boldsymbol{Q}\boxtimes\boldsymbol{Q})\boldsymbol{I}]\otimes[(\boldsymbol{Q}\boxtimes\boldsymbol{Q})\boldsymbol{I}] = \boldsymbol{I}\otimes\boldsymbol{I},$$

$$\mathbb{Q}\mathbb{I}\mathbb{Q}^T = \mathbb{Q}\mathbb{Q}^T = \mathbb{I},$$

$$\mathbb{Q}\mathbb{T}\mathbb{Q}^T = (\boldsymbol{Q}\boxtimes\boldsymbol{Q})\mathbb{T}\mathbb{Q}^T = \mathbb{T}(\boldsymbol{Q}\boxtimes\boldsymbol{Q})\mathbb{Q}^T = \mathbb{T}\mathbb{Q}\mathbb{Q}^T = \mathbb{T}. \quad \text{(by Eqn. (1.173))}$$

To prove the converse, take $\boldsymbol{A}$ to be $\boldsymbol{Q}\in \text{Orth}^+\backslash\text{Sym}$ in Eqn. (1.203) to get

$$\mathbb{Q}(\mathbb{C}\boldsymbol{Q}) = \mathbb{C}\boldsymbol{Q}.$$

Thus, $\mathbb{C}\boldsymbol{Q}$ is a linear combination of the eigentensors of $\mathbb{Q}$ corresponding to the eigenvalue one, i.e., by virtue of Theorem 1.7.2,

$$\mathbb{C}\boldsymbol{Q} = \alpha_0(\boldsymbol{Q})\boldsymbol{I} + \alpha_1(\boldsymbol{Q})\boldsymbol{Q} + \alpha_2(\boldsymbol{Q})\boldsymbol{Q}^2 \quad \forall \boldsymbol{Q}\in \text{Orth}^+\backslash\text{Sym}.$$

In the limiting cases following Eqn. (1.105b), namely, (i) $\alpha\to 0$, we have $\boldsymbol{Q}=\boldsymbol{I}$, so that α_1 and α_2 can be taken to be zero, and (ii) $\alpha\to\pi$, we have $\boldsymbol{Q}^2=\boldsymbol{I}$, so that α_2 can be taken to be zero in the above expression. Thus, by virtue of the continuous dependence of $\boldsymbol{Q}$ on α, we can write

$$\mathbb{C}\boldsymbol{Q} = \alpha_0(\boldsymbol{Q})\boldsymbol{I} + \alpha_1(\boldsymbol{Q})\boldsymbol{Q} + \alpha_2(\boldsymbol{Q})\boldsymbol{Q}^2 \quad \forall \boldsymbol{Q}\in \text{Orth}^+, \tag{1.206}$$

with $\alpha_1=\alpha_2=0$ when $\boldsymbol{Q}=\boldsymbol{I}$, and $\alpha_2=0$, when $\boldsymbol{Q}=2\boldsymbol{e}\otimes\boldsymbol{e}-\boldsymbol{I}$.

We now show that the functions α_i, $i=0,1,2$, are in fact functions of the principal invariants of $\boldsymbol{Q}$ (which is just the trace, since, for proper orthogonal tensors, the determinant is equal to unity, and the second invariant is equal to the trace). By letting $\boldsymbol{A}\equiv\boldsymbol{R}_0\in\text{Orth}^+$ in Eqn. (1.203), we get

$$\mathbb{C}\boldsymbol{R}_0 = \boldsymbol{Q}^T\mathbb{C}(\boldsymbol{Q}\boldsymbol{R}_0\boldsymbol{Q}^T)\boldsymbol{Q}, \quad \forall\boldsymbol{Q}\in\text{Orth}^+.$$

Substituting the expressions for $\mathbb{C}\boldsymbol{R}_0$ and $\mathbb{C}(\boldsymbol{Q}\boldsymbol{R}_0\boldsymbol{Q}^T)$ obtained from Eqn. (1.206) into the above equation, we get

$$\left[\alpha_0(\boldsymbol{R}_0)-\alpha_0(\boldsymbol{Q}\boldsymbol{R}_0\boldsymbol{Q}^T)\right]\boldsymbol{I} + \left[\alpha_1(\boldsymbol{R}_0)-\alpha_1(\boldsymbol{Q}\boldsymbol{R}_0\boldsymbol{Q}^T)\right]$$
$$\boldsymbol{R}_0 + \left[\alpha_2(\boldsymbol{R}_0)-\alpha_2(\boldsymbol{Q}\boldsymbol{R}_0\boldsymbol{Q}^T)\right]\boldsymbol{R}_0^2 = \boldsymbol{0}.$$

Since, as shown in the discussion on page 38, $\{\boldsymbol{I},\boldsymbol{R}_0,\boldsymbol{R}_0^2\}$, $\{\boldsymbol{I},\boldsymbol{R}_0\}$ and $\{\boldsymbol{I}\}$ are linearly independent sets, depending on whether $\boldsymbol{R}_0\in\text{Orth}^+\backslash\text{Sym}$, $\boldsymbol{R}_0\in\text{Orth}^+\cap\text{Sym}-\{\boldsymbol{I}\}$ ($\alpha_2=0$), or $\boldsymbol{R}_0=\boldsymbol{I}$ ($\alpha_1=\alpha_2=0$), we have

$$\alpha_i(\boldsymbol{R}_0) = \alpha_i(\boldsymbol{Q}\boldsymbol{R}_0\boldsymbol{Q}^T),\ i=0,1,2,\ \forall\boldsymbol{Q}\in\text{Orth}^+, \tag{1.207}$$

i.e., the α_i's are isotropic scalar-valued functions of $\boldsymbol{R}_0$. To show that they are functions of $I_1(\boldsymbol{R}_0)$, we have to show that

$$I_1(\boldsymbol{Q}_1) = I_1(\boldsymbol{Q}_2) \implies \alpha_i(\boldsymbol{Q}_1) = \alpha_i(\boldsymbol{Q}_2),\ i=0,1,2,$$

and then the conclusion follows from Theorem 1.6.7. If $I_1(Q_1) = I_1(Q_2)$, then by Theorem 1.5.5, there exists $Q_0 \in \text{Orth}^+$ such that $Q_2 = Q_0 Q_1 Q_0^T$. Thus, using Eqn. (1.207), we get $\alpha_i(Q_2) = \alpha_i(Q_0 Q_1 Q_0^T) = \alpha_i(Q_1)$. It follows from Theorem 1.6.7 that there exist functions $\hat{\alpha}_i$ such that $\alpha_i(Q) = \hat{\alpha}_i(I_1(Q))$.

Equation (1.206) can now be written as

$$\mathbb{C}Q = \hat{\alpha}_0(I_1)I + \hat{\alpha}_1(I_1)Q + \hat{\alpha}_2(I_1)Q^2. \tag{1.208}$$

Eliminating Q^2 from Eqn. (1.208) using Eqn. (1.110), we get

$$\mathbb{C}Q = [\hat{\alpha}_0(I_1) - \hat{\alpha}_2(I_1)I_1]I + [\hat{\alpha}_1(I_1) + \hat{\alpha}_2(I_1)I_1]Q + \hat{\alpha}_2(I_1)Q^T.$$

Since $\mathbb{C}$ is independent of Q, and a linear transformation from Lin to Lin, the right-hand side of the above expression has to be linear in Q, i.e., the coefficients of Q and Q^T are constants, and the coefficient of I has to be a multiple of $\text{tr}\, Q$. Thus, set $\hat{\alpha}_0(I_1) \equiv (\lambda + \mu - \gamma)I_1$, $\hat{\alpha}_1(I_1) \equiv (\mu + \gamma) - (\mu - \gamma)I_1$, and $\hat{\alpha}_2(I_1) \equiv \mu - \gamma$, where λ, μ and γ are constants. This leads to

$$\begin{aligned}\mathbb{C}Q &= \lambda(\text{tr}\, Q)I + (\mu + \gamma)Q + (\mu - \gamma)Q^T \\ &= [\lambda(I \otimes I) + (\mu + \gamma)\mathbb{I} + (\mu - \gamma)\mathbb{T}]\, Q \quad \forall Q \in \text{Orth}^+. \end{aligned} \tag{1.209}$$

The last step is to show that the above equation implies the expression given in Eqn. (1.204). Thus, by virtue of Eqn. (1.157), we have to show that Eqn. (1.209) holds for an arbitrary tensor T in place of Q. Let $\mathbb{C}_{\text{iso}} := [\lambda(I \otimes I) + (\mu + \gamma)\mathbb{I} + (\mu - \gamma)\mathbb{T}]$. Since fourth-order tensors are a linear transformation from Lin to Lin, Eqn. (1.209) implies that

$$[\mathbb{C} - \mathbb{C}_{\text{iso}}]\,(\sum_i \phi_i Q_i) = \mathbf{0} \quad \forall \phi_i \in \Re,\ Q_i \in \text{Orth}^+.$$

Since by Theorem 1.5.3, any $T \in \text{Lin}$ can be expressed as a linear combination of $Q \in \text{Orth}^+$, the proof is complete. □

For various other proofs, see, e.g., [24], [102], [103], [105], [110], [142], [148], [175], [210], [250] and [251]. Note that this result does not hold when the space dimension $n = 2$, as the counterexample $\mathbb{C} = T \boxtimes T$, where $T = \begin{bmatrix} \alpha & \beta \\ -\beta & \alpha \end{bmatrix}$ shows (the above proof fails in the last step for this case, since an arbitrary tensor cannot be expressed as a linear combination of rotations). However, the result is applicable for any n if we assume $\mathbb{C}$ to be symmetric.

1.9 Differentiation of Tensors

In the subsequent chapters, we shall deal with tensor fields, where a tensor T is a function of a scalar quantity t, and a vector quantity x (e.g., t representing time, and x the position vector). In this section, we define the concept of the derivative of a tensor quantity. Let X

and Y denote normed vector spaces, and let $f : \Omega \subset X \rightarrow Y$. We say that $f(\boldsymbol{u})$ approaches zero faster than $\boldsymbol{u}$, and write $f(\boldsymbol{u}) = \mathrm{o}(\boldsymbol{u})$ if

$$\lim_{\boldsymbol{u}\rightarrow \boldsymbol{0}} \frac{\|f(\boldsymbol{u})\|}{\|\boldsymbol{u}\|} = 0.$$

1.9.1 The directional derivative

Let g be a function whose values are scalars, vectors or tensors, and whose domain Ω is an open interval of $\Re$. The derivative $Dg(t)$ is defined by

$$Dg(t) := \frac{d}{dt}g(t) = \lim_{\alpha\rightarrow 0} \frac{1}{\alpha}\left[g(t+\alpha) - g(t)\right]. \tag{1.210}$$

From Eqn. (1.210), it follows that

$$\lim_{\alpha\rightarrow 0} \frac{1}{\alpha}\left[g(t+\alpha) - g(t) - \alpha Dg(t)\right] = 0,$$

or, equivalently,

$$g(t+\alpha) = g(t) + \alpha Dg(t) + \mathrm{o}(\alpha). \tag{1.211}$$

Thus, we see that the derivative of a vector function is a vector, or the derivative of a tensor function is a tensor.

We now extend the above concept of a derivative to domains that lie in spaces of dimensions higher than one. Let $L(X,Y)$ denote the vector space of all continuous linear mappings from X to Y, and let $g : \Omega \subset X \rightarrow Y$. We say that g is differentiable at $\boldsymbol{x} \in \Omega$ if there exists a linear transformation $Dg(\boldsymbol{x}) \in L(X,Y)$ such that

$$g(\boldsymbol{x}+\boldsymbol{u}) = g(\boldsymbol{x}) + Dg(\boldsymbol{x})[\boldsymbol{u}] + \mathrm{o}(\boldsymbol{u}). \tag{1.212}$$

The quantity $Dg(\boldsymbol{x}) \in L(X,Y)$ is called the *Frechet derivative*, while the quantity $Dg(\boldsymbol{x})[\boldsymbol{u}] \in Y$ is called the *directional derivative* or *Gateaux derivative*. Note that the directional derivative is computed by the action of the Frechet derivative on elements of X. If the directional derivative exists, it is unique, and can be written as

$$Dg(\boldsymbol{x})[\boldsymbol{u}] = \lim_{\alpha\rightarrow 0} \frac{1}{\alpha}\left[g(\boldsymbol{x}+\alpha\boldsymbol{u}) - g(\boldsymbol{x})\right] = \left.\frac{d}{d\alpha}g(\boldsymbol{x}+\alpha\boldsymbol{u})\right|_{\alpha=0}. \tag{1.213}$$

When $\Omega \subset \Re$, by comparing Eqns. (1.211) and (1.212), we see that

$$Dg(t)[\alpha] = \alpha\frac{dg}{dt} \quad \forall\alpha \in \Re.$$

Often, the easiest way of computing the directional derivative is to appeal directly to its definition. We now consider some examples. Consider the function $\phi(\boldsymbol{v}) = \boldsymbol{v}\cdot\boldsymbol{v}$. Then

$$\begin{aligned}\phi(\boldsymbol{v}+\boldsymbol{u}) &= (\boldsymbol{v}+\boldsymbol{u})\cdot(\boldsymbol{v}+\boldsymbol{u})\\ &= \boldsymbol{v}\cdot\boldsymbol{v} + 2\boldsymbol{v}\cdot\boldsymbol{u} + \boldsymbol{u}\cdot\boldsymbol{u}\\ &= \phi(\boldsymbol{v}) + 2\boldsymbol{v}\cdot\boldsymbol{u} + \mathrm{o}(\boldsymbol{u}),\end{aligned}$$

so that

$$D\phi(v)[u] = 2v \cdot u.$$

Now consider the function $G : \text{Lin} \to \text{Lin}$ defined by $G(T) = T^2$, where T is a second-order tensor. Then we have

$$G(T+U) = (T+U)^2 = T^2 + TU + UT + U^2 = G(T) + TU + UT + o(U),$$

which implies that $DG(T)[U] = TU + UT$.

Similarly for $G(T) = T^3$, we have

$$G(T+U) = G(T) + T^2U + TUT + UT^2 + o(U),$$

which implies that $DG(T)[U] = T^2U + TUT + UT^2$.

For $\phi(T) = \text{tr}\,(T)$, we have $\phi(T+U) = \phi(T) + \text{tr}\,U$, which yields

$$D(\text{tr}\,T)[U] = \text{tr}\,U = I : U. \tag{1.214}$$

If $\phi(T) = \text{tr}\,T^k$, then (with $T^0 \equiv I$)

$$(T+U)^k = T^k + \sum_{i=1}^{k} T^{k-i}UT^{i-1} + o(U),$$

and on using the linearity of the trace operator, and the fact that $\text{tr}\,AB = \text{tr}\,BA$, we get

$$\begin{aligned}
\phi(T+U) &= \phi(T) + \sum_{i=1}^{k} \text{tr}\,[T^{k-i}UT^{i-1}] + o(U) \\
&= \phi(T) + \sum_{i=1}^{k} \text{tr}\,[T^{k-1}U] + o(U) \\
&= \phi(T) + k(T^{k-1})^T : U + o(U),
\end{aligned}$$

so that

$$D\phi(T)[U] = k(T^{k-1})^T : U. \tag{1.215}$$

If $\phi(T) = \det T$, then from Problem 14, we have

$$\begin{aligned}
\det(T+U) &= \det T + \mathbf{cof}\,T : U + T : \mathbf{cof}\,U + \det U \\
&= \det T + \mathbf{cof}\,T : U + o(U),
\end{aligned}$$

which yields

$$D\phi(T)[U] = \mathbf{cof}\,T : U. \tag{1.216}$$

Recall that when T is invertible, $\mathbf{cof}\,T = (\det T)T^{-T}$.

We now discuss the invariance properties of the derivative of a tensor function.

Theorem 1.9.1. *Let $G : Sym \to Sym$ be an isotopic tensor function. Then the derivative is invariant in the following sense:*

$$QDG(S)[U]Q^T = DG(QSQ^T)[QUQ^T] \quad \forall Q \in Orth^+,\ S, U \in Sym. \tag{1.217}$$

Proof. We have already remarked that Sym is invariant under Orth$^+$. Thus, we simply need to prove Eqn. (1.217). Since $G(S)$ is isotropic, we have

$$\begin{aligned} G\left(Q(S+U)Q^T\right) &= QG(S+U)Q^T \\ &= QG(S)Q^T + QDG(S)[U]Q^T + o(U) \\ &= G(QSQ^T) + QDG(S)[U]Q^T + o(U). \end{aligned}$$

On the other hand, we also have

$$\begin{aligned} G\left(Q(S+U)Q^T\right) &= G(QSQ^T + QUQ^T) \\ &= G(QSQ^T) + DG(QSQ^T)[QUQ^T] + o(U). \end{aligned}$$

From the above equations, and the uniqueness of the derivative, we get Eqn. (1.217). □

1.9.2 Product rule

Quite often, we will be required to compute the directional derivative of a bilinear map $\pi(f, g)$. Examples of bilinear maps are [108]

- the product of a scalar and a vector

  $$\pi(\phi, v) = \phi v,$$

- the inner product of two vectors

  $$\pi(u, v) = u \cdot v,$$

- the tensor product of two vectors

  $$\pi(u, v) = u \otimes v,$$

- the action of a tensor on a vector

  $$\pi(T, v) = Tv,$$

and so forth. Assuming that f and g are differentiable at a point x in the common domain of f and g, the product $\pi(f(x), g(x))$ is differentiable, and its directional derivative is given by (see [108] for the proof)

$$D\pi(f(x), g(x))[u] = \pi(f(x), Dg(x)[u]) + \pi(Df(x)[u], g(x)) \quad \forall u. \tag{1.218}$$

When $x \in \Re$, then we simply have (with x replaced by t)

$$\dot{\pi}(f(t), g(t)) = \pi\left(f(t), \dot{g}(t)\right) + \pi\left(\dot{f}(t), g(t)\right).$$

Another result that is used frequently is the chain rule, which we now state.

1.9.3 Chain rule

Let $f : \Omega_2 \to \Omega_3$ and $g : \Omega_1 \to \Omega_2$, and let $g(x)$ be differentiable at $x \in \Omega_1$. Further, let f be differentiable at $y = g(x)$. Then the composition $f \circ g$ is differentiable at x, and

$$D(f \circ g)(x)[u] = Df(g(x))[Dg(x)[u]]. \tag{1.219}$$

In case $x \in \Re$, then writing t in place of x, we have

$$\begin{aligned} D(f \circ g)(t)[\alpha] &= \alpha \frac{d}{dt}(f \circ g), \\ Dg(t)[\alpha] &= \alpha \frac{dg}{dt}, \end{aligned}$$

and hence,

$$\frac{d}{dt} f(g(t)) = Df(g(t)) \left[\frac{dg}{dt} \right]. \tag{1.220}$$

We have already shown how to find the directional derivatives of the first and third invariants. To illustrate the product and chain rules, we now compute the directional derivative of the second invariant of a second-order tensor $I_2(T) = (\det T)(\operatorname{tr} T^{-1})$ given by

$$DI_2(T)[U] = \det T \left[\operatorname{tr}(T^{-1}U) \operatorname{tr} T^{-1} - \operatorname{tr}(T^{-1}UT^{-1}) \right] = \left[(\operatorname{tr} T)I - T^T \right] : U. \tag{1.221}$$

First we calculate the derivative of T^{-1}. Since $TT^{-1} = I$, we have

$$DT[U]T^{-1} + TDT^{-1}[U] = 0,$$

and since $DT[U] = U$, we get $DT^{-1}[U] = -T^{-1}UT^{-1}$. Next, denoting $\operatorname{tr} T^{-1}$ by $\phi(G(T))$, where $G(T) = T^{-1}$, we have

$$D\phi(T)[U] = D\phi(G)[DG(T)[U]] = -D\phi(G)[T^{-1}UT^{-1}] = -\operatorname{tr}(T^{-1}UT^{-1}).$$

Finally, writing I_2 as $(\det T)\phi(T)$, and applying the product rule, we get

$$\begin{aligned} DI_2(T)[U] &= D(\det T)(T)[U]\phi(T) + \det T D\phi(T)[U] \\ &= \det T \operatorname{tr}(T^{-1}U) \operatorname{tr} T^{-1} - \det T \operatorname{tr}(T^{-1}UT^{-1}), \end{aligned}$$

which proves the first relation in Eqn. (1.221). To get the second relation, we note from Eqn. (1.80) that I_2 is also given by

$$\frac{1}{2}\left[(\operatorname{tr} T)^2 - \operatorname{tr}(T^2) \right].$$

Hence, we have

$$\begin{aligned} DI_2[U] &= \frac{1}{2}\left[2(\operatorname{tr} T)\operatorname{tr} U - 2\operatorname{tr}(TU)\right] \\ &= \left[(\operatorname{tr} T)I - T^T\right] : U. \end{aligned}$$

The second relation can also be obtained directly from the first using the Cayley–Hamilton theorem.

One other useful result is the following:

$$\frac{d}{dt}(\det T) = \mathbf{cof}\, T : \frac{dT}{dt}. \tag{1.222}$$

To see this note that

$$\begin{aligned} \frac{d}{dt}(\det T) &= D\det(T)\left[\frac{dT}{dt}\right] && \text{(by Eqn. (1.220))} \\ &= \mathbf{cof}\, T : \frac{dT}{dt} && \text{(by Eqn. (1.216)).} \end{aligned}$$

When T is invertible, then using $\mathbf{cof}\, T = (\det T)T^{-T}$, we can also write Eqn. (1.222) as

$$\frac{d}{dt}(\det T) = \det T \operatorname{tr}\left(\frac{dT}{dt}T^{-1}\right). \tag{1.223}$$

By taking the directional derivative of the Cayley–Hamilton equation, we get the following "Rivlin's identity" [68]:

$$\begin{aligned} T^2U + TUT + UT^2 - (\operatorname{tr} U)T^2 - (\operatorname{tr} T)(TU + UT) + \frac{1}{2}\left[(\operatorname{tr} T)^2 - \operatorname{tr} T^2\right] U \\ + \left[(\operatorname{tr} T)(\operatorname{tr} U) - \operatorname{tr}(TU)\right] T - (\mathbf{cof}\, T : U)I = \mathbf{0}. \end{aligned}$$

Similar identities can be derived in other space dimensions by taking the directional derivatives of the relevant Cayley–Hamilton equation. Identities involving three tensors can be derived by taking the directional derivative twice. Other identities can be derived by taking the directional derivative of the Cayley–Hamilton equation with a tensor function [68].

1.9.4 Gradient, divergence and curl

When X is a Hilbert space, the directional derivative of a real-valued function $\phi : \Omega \subset X \to \Re$ can be identified with an element of the space X. The derivative $D\phi(x)$ is an element of the dual space $X' = L(X, \Re)$ and thus, since the space X is a Hilbert space, there exists by the Riesz representation theorem [63] a unique element $\nabla\phi(x)$ in the space X that satisfies

$$(\nabla\phi(x), u) = D\phi(x)[u], \tag{1.224}$$

where $(.,.)$ denotes the inner product of the space X. The element $\nabla\phi(x)$ is called the *gradient* of ϕ. Note that while $D\phi(x)$ belongs to the dual space X', the gradient $\nabla\phi(x)$ belongs to the space X.

As an example, consider the case when $X = \Re^3$ equipped with the Euclidean inner product $(\boldsymbol{u}, \boldsymbol{v}) = \boldsymbol{u} \cdot \boldsymbol{v}$. Then we have

$$\phi(\boldsymbol{x} + \boldsymbol{u}) = \phi(\boldsymbol{x}) + \boldsymbol{\nabla}\phi(\boldsymbol{x}) \cdot \boldsymbol{u} + \mathrm{o}(\boldsymbol{u}).$$

To find the component form of the gradient, we take $\boldsymbol{u} = \boldsymbol{e}_i$ to get

$$(\boldsymbol{\nabla}\phi)_i = D\phi(\boldsymbol{x})[\boldsymbol{e}_i] = \lim_{\alpha \to 0} \frac{1}{\alpha} [\phi(\boldsymbol{x} + \alpha \boldsymbol{e}_i) - \phi(\boldsymbol{x})] = \frac{\partial \phi(\boldsymbol{x})}{\partial x_i}.$$

Thus, we have

$$D\phi(\boldsymbol{x})[\boldsymbol{u}] = \boldsymbol{\nabla}\phi \cdot \boldsymbol{u} = \frac{\partial \phi}{\partial x_i} u_i.$$

As another example, consider the case when $X = \text{Lin}$, with the matrix inner product $(\boldsymbol{A}, \boldsymbol{B}) = \boldsymbol{A} : \boldsymbol{B}$. Then the gradient of $\phi(\boldsymbol{T})$ is the matrix $(\partial\phi/\partial T_{ij})$. By definition, we have

$$D\phi(\boldsymbol{T})[\boldsymbol{U}] = \frac{\partial \phi}{\partial \boldsymbol{T}} : \boldsymbol{U} = \frac{\partial \phi}{\partial T_{ij}} U_{ij}.$$

Similarly, if $\boldsymbol{V}(\boldsymbol{T})$ is a tensor-valued function of $\boldsymbol{T}$, then the gradient and directional derivative of $\boldsymbol{V}$ are related as

$$D\boldsymbol{V}(\boldsymbol{T})[\boldsymbol{U}] = \frac{\partial \boldsymbol{V}}{\partial \boldsymbol{T}} \boldsymbol{U}. \tag{1.225}$$

The gradients of the invariants I_1, I_2 and I_3 are obtained from Eqns. (1.214), (1.216) and (1.221), and are given by

$$\frac{\partial I_1}{\partial \boldsymbol{T}} = \boldsymbol{I}, \tag{1.226a}$$

$$\frac{\partial I_2}{\partial \boldsymbol{T}} = (\operatorname{tr} \boldsymbol{T})\boldsymbol{I} - \boldsymbol{T}^T, \tag{1.226b}$$

$$\frac{\partial I_3}{\partial \boldsymbol{T}} = \mathbf{cof}\, \boldsymbol{T} = \left[I_2 \boldsymbol{I} - I_1 \boldsymbol{T} + \boldsymbol{T}^2 \right]^T. \tag{1.226c}$$

The gradient of $\phi(\boldsymbol{T}) = \operatorname{tr} \boldsymbol{T}^k$ is obtained using Eqn. (1.215), and is given by

$$\frac{\partial}{\partial \boldsymbol{T}} (\operatorname{tr} \boldsymbol{T}^k) = k(\boldsymbol{T}^{k-1})^T. \tag{1.227}$$

Since $\partial T_{ij}/\partial T_{kl} = \partial T_{ji}/\partial T_{lk} = \delta_{ik}\delta_{jl}$ and $\partial T_{ji}/\partial T_{kl} = \partial T_{ij}/\partial T_{lk} = \delta_{jk}\delta_{il}$, we have

$$\begin{aligned} \frac{\partial \boldsymbol{T}}{\partial \boldsymbol{T}} &= \frac{\partial \boldsymbol{T}^T}{\partial \boldsymbol{T}^T} = \mathbb{I}, \\ \frac{\partial \boldsymbol{T}^T}{\partial \boldsymbol{T}} &= \frac{\partial \boldsymbol{T}}{\partial \boldsymbol{T}^T} = \mathbb{T}. \end{aligned} \tag{1.228}$$

In general, if $\phi_1(\boldsymbol{T})$ and $\phi_2(\boldsymbol{T})$ are scalar-valued functions of $\boldsymbol{T}$, then

$$\frac{\partial(\phi_1 \phi_2)}{\partial \boldsymbol{T}} = \phi_1 \frac{\partial \phi_2}{\partial \boldsymbol{T}} + \phi_2 \frac{\partial \phi_1}{\partial \boldsymbol{T}}.$$

If $\phi(\boldsymbol{T})$ and $\boldsymbol{U}(\boldsymbol{T})$ are scalar and tensor-valued functions of $\boldsymbol{T}$, then

$$\frac{\partial \phi}{\partial \boldsymbol{T}^T} = \mathbb{T}\frac{\partial \phi}{\partial \boldsymbol{T}}, \tag{1.229}$$

$$\frac{\partial \boldsymbol{U}}{\partial \boldsymbol{T}^T} = \frac{\partial \boldsymbol{U}}{\partial \boldsymbol{T}}\mathbb{T}. \tag{1.230}$$

Since $(\phi \boldsymbol{U})_{ij} = \phi U_{ij}$, we have

$$\frac{\partial(\phi \boldsymbol{U})}{\partial \boldsymbol{T}} = \boldsymbol{U} \otimes \frac{\partial \phi}{\partial \boldsymbol{T}} + \phi \frac{\partial \boldsymbol{U}}{\partial \boldsymbol{T}}. \tag{1.231}$$

If $\boldsymbol{U}(\boldsymbol{T})$ and $\boldsymbol{V}(\boldsymbol{T})$ are tensor-valued functions of $\boldsymbol{T}$, then writing $\boldsymbol{U} : \boldsymbol{V}$ as $U_{mn}V_{mn}$, we get

$$\frac{\partial(\boldsymbol{U} : \boldsymbol{V})}{\partial \boldsymbol{T}} = \left(\frac{\partial \boldsymbol{U}}{\partial \boldsymbol{T}}\right)^T \boldsymbol{V} + \left(\frac{\partial \boldsymbol{V}}{\partial \boldsymbol{T}}\right)^T \boldsymbol{U}. \tag{1.232}$$

Similarly, using the relation $(\boldsymbol{U}\boldsymbol{V})_{ij} = U_{im}V_{mj}$, we get

$$\begin{aligned}
\left[\frac{\partial(\boldsymbol{U}\boldsymbol{V})}{\partial \boldsymbol{T}}\right]_{ijkl} &= U_{im}\frac{\partial V_{mj}}{\partial T_{kl}} + \frac{\partial U_{im}}{\partial T_{kl}} V_{mj} \\
&= U_{im}\delta_{nj}\frac{\partial V_{mn}}{\partial T_{kl}} + \frac{\partial U_{nm}}{\partial T_{kl}}\delta_{in}V_{mj} \\
&= U_{im}\delta_{nj}\frac{\partial V_{mn}}{\partial T_{kl}} + \frac{\partial U_{mn}}{\partial T_{kl}}\delta_{im}V_{nj},
\end{aligned}$$

which in tensorial form reads

$$\frac{\partial(\boldsymbol{U}\boldsymbol{V})}{\partial \boldsymbol{T}} = (\boldsymbol{U} \boxtimes \boldsymbol{I})\frac{\partial \boldsymbol{V}}{\partial \boldsymbol{T}} + (\boldsymbol{I} \boxtimes \boldsymbol{V}^T)\frac{\partial \boldsymbol{U}}{\partial \boldsymbol{T}}. \tag{1.233}$$

Using Eqns. (1.158), (1.173), (1.228) and (1.233), we have the following results:

$$\frac{\partial(\boldsymbol{T}^2)}{\partial \boldsymbol{T}} = \boldsymbol{T} \boxtimes \boldsymbol{I} + \boldsymbol{I} \boxtimes \boldsymbol{T}^T,$$

$$\frac{\partial(\boldsymbol{T}^T\boldsymbol{T})}{\partial \boldsymbol{T}} = (\boldsymbol{I} \boxtimes \boldsymbol{T}^T)\mathbb{T} + \boldsymbol{T}^T \boxtimes \boldsymbol{I} = 2\mathbb{S}(\boldsymbol{T}^T \boxtimes \boldsymbol{I}), \tag{1.234}$$

$$\frac{\partial(\boldsymbol{T}\boldsymbol{T}^T)}{\partial \boldsymbol{T}} = (\boldsymbol{I} \boxtimes \boldsymbol{T}) + (\boldsymbol{T} \boxtimes \boldsymbol{I})\mathbb{T} = 2\mathbb{S}(\boldsymbol{I} \boxtimes \boldsymbol{T}), \tag{1.235}$$

$$\frac{\partial(\boldsymbol{T}^{-1})}{\partial \boldsymbol{T}} = -\boldsymbol{T}^{-1} \boxtimes \boldsymbol{T}^{-T} = -(\boldsymbol{T} \boxtimes \boldsymbol{T}^T)^{-1}, \tag{1.236}$$

$$\frac{\partial(\boldsymbol{T}^{-T})}{\partial \boldsymbol{T}} = -\mathbb{T}(\boldsymbol{T}^{-1} \boxtimes \boldsymbol{T}^{-T}) = -(\boldsymbol{T}^{-T} \boxtimes \boldsymbol{T}^{-1})\mathbb{T} = -\mathbb{T}(\boldsymbol{T} \boxtimes \boldsymbol{T}^T)^{-1}. \tag{1.237}$$

The fourth relation is obtained by taking $\boldsymbol{U}$ and $\boldsymbol{V}$ to be $\boldsymbol{T}^{-1}$ and $\boldsymbol{T}$, respectively, and then differentiating $\boldsymbol{T}^{-1}\boldsymbol{T} = \boldsymbol{I}$. We get

$$\boldsymbol{0} = \boldsymbol{T}^{-1} \boxtimes \boldsymbol{I} + (\boldsymbol{I} \boxtimes \boldsymbol{T}^T)\frac{\partial \boldsymbol{T}^{-1}}{\partial \boldsymbol{T}}$$

$$\begin{aligned}
\implies \frac{\partial \boldsymbol{T}^{-1}}{\partial \boldsymbol{T}} &= -(\boldsymbol{I} \boxtimes \boldsymbol{T}^T)^{-1}(\boldsymbol{T}^{-1} \boxtimes \boldsymbol{I}) \\
&= -(\boldsymbol{I} \boxtimes \boldsymbol{T}^{-T})(\boldsymbol{T}^{-1} \boxtimes \boldsymbol{I}) && \text{(by Eqn. (1.185))} \\
&= -\boldsymbol{T}^{-1} \boxtimes \boldsymbol{T}^{-T}. && \text{(by Eqn. (1.168))}
\end{aligned}$$

The fifth relation is obtained by writing $\boldsymbol{T}^{-T}$ as $\mathbb{T}\boldsymbol{T}^{-1}$ and then using Eqn. (1.236).

If $\phi(\boldsymbol{G}(\boldsymbol{T}))$ is a scalar-valued function of a tensor function $\boldsymbol{G}(\boldsymbol{T})$, then by the chain rule, we have

$$\frac{\partial \phi}{\partial T_{ij}} = \frac{\partial \phi}{\partial G_{kl}} \frac{\partial G_{kl}}{\partial T_{ij}},$$

which implies that

$$\frac{\partial \phi}{\partial \boldsymbol{T}} = \left(\frac{\partial \boldsymbol{G}}{\partial \boldsymbol{T}} \right)^T \frac{\partial \phi}{\partial \boldsymbol{G}}. \tag{1.238}$$

Similarly, if $\boldsymbol{H}(\boldsymbol{G}(\boldsymbol{T}))$ is a tensor-valued function of a tensor function $\boldsymbol{G}(\boldsymbol{T})$, then

$$\frac{\partial \boldsymbol{H}}{\partial \boldsymbol{T}} = \frac{\partial \boldsymbol{H}}{\partial \boldsymbol{G}} \frac{\partial \boldsymbol{G}}{\partial \boldsymbol{T}}. \tag{1.239}$$

Special care has to be exercised in computing the gradients with respect to a symmetric tensor $\boldsymbol{S}$. As a specific example, let us consider the computation of the first and second-order derivatives of a scalar-valued function $\tilde{W}(\boldsymbol{C})$, where $\boldsymbol{C} \in \text{Sym}$. Let $\tilde{\boldsymbol{S}}$ and $\mathbb{C}$ denote these derivatives. We have

$$\begin{aligned}
\tilde{\boldsymbol{S}}(\boldsymbol{C}) &= \frac{1}{2}\left(\frac{\partial \tilde{W}}{\partial \boldsymbol{C}} + \frac{\partial \tilde{W}}{\partial \boldsymbol{C}^T} \right) \\
&= \frac{1}{2}\left(\frac{\partial \tilde{W}}{\partial \boldsymbol{C}} + \mathbb{T}\frac{\partial \tilde{W}}{\partial \boldsymbol{C}} \right) && \text{(by Eqn. (1.229))} \\
&= \frac{1}{2}(\mathbb{I} + \mathbb{T})\frac{\partial \tilde{W}}{\partial \boldsymbol{C}} \\
&= \mathbb{S}\frac{\partial \tilde{W}}{\partial \boldsymbol{C}}. && \text{(by Eqn. (1.158))}
\end{aligned}$$

In indicial notation, the above equation reads

$$\tilde{S}_{ij} = \frac{1}{2}\left(\frac{\partial \tilde{W}}{\partial C_{ij}} + \frac{\partial \tilde{W}}{\partial C_{ji}} \right).$$

The second-order gradient $\mathbb{C} = \partial \tilde{\boldsymbol{S}}/\partial \boldsymbol{C}$ is computed as follows:

$$\begin{aligned}
\mathbb{C} &= \frac{1}{2}\left(\frac{\partial(\mathbb{S}\tilde{\boldsymbol{S}})}{\partial \boldsymbol{C}} + \frac{\partial(\mathbb{S}\tilde{\boldsymbol{S}})}{\partial \boldsymbol{C}^T} \right) \\
&= \frac{1}{2}\left(\frac{\partial(\mathbb{S}\tilde{\boldsymbol{S}})}{\partial \boldsymbol{C}} + \frac{\partial(\mathbb{S}\tilde{\boldsymbol{S}})}{\partial \boldsymbol{C}}\mathbb{T} \right) && \text{(by Eqn. (1.230))}
\end{aligned}$$

$$
\begin{aligned}
&= \frac{1}{2}\mathbb{S}\frac{\partial \tilde{S}}{\partial C}(\mathbb{I}+\mathbb{T}) \\
&= \mathbb{S}\frac{\partial \tilde{S}}{\partial C}\mathbb{S} && (1.240)\\
&= \mathbb{S}\frac{\partial^2 \tilde{W}}{\partial C \partial C}\mathbb{S}. && (1.241)
\end{aligned}
$$

In other words, to find $\partial^2 \tilde{W}/\partial C \partial C$ when $C \in \text{Sym}$, we compute it as if C is not symmetric, and then pre- and post-multiply the result with the symmetrizer $\mathbb{S}$. In indicial notation, the above equations read

$$
\begin{aligned}
\mathbb{C}_{ijkl} &= \frac{1}{4}\left(\frac{\partial \tilde{S}_{ij}}{\partial C_{kl}} + \frac{\partial \tilde{S}_{ji}}{\partial C_{kl}} + \frac{\partial \tilde{S}_{ij}}{\partial C_{lk}} + \frac{\partial \tilde{S}_{ji}}{\partial C_{lk}}\right) \\
&= \frac{1}{4}\left(\frac{\partial^2 \tilde{W}}{\partial C_{ij}\partial C_{kl}} + \frac{\partial^2 \tilde{W}}{\partial C_{ji}\partial C_{kl}} + \frac{\partial^2 \tilde{W}}{\partial C_{ij}\partial C_{lk}} + \frac{\partial^2 \tilde{W}}{\partial C_{ji}\partial C_{lk}}\right).
\end{aligned}
$$

Now consider the evaluation of the stretch and rotation tensors of F, and their derivatives for arbitrary space dimension n. Let λ_i^2, $i = 1, 2, \ldots, k$ be the distinct eigenvalues, and P_i be the corresponding projections of $C \in \text{Psym}$, so that $C = \sum_{i=1}^k \lambda_i^2 P_i$ and $U = \sum_{i=1}^k \lambda_i P_i$, with P_i given via Eqn. (J.4) as

$$
P_i = \begin{cases} \prod_{\substack{j=1\\ j\neq i}}^k \frac{C-\lambda_j^2 I}{\lambda_i^2-\lambda_j^2} = \prod_{\substack{j=1\\ j\neq i}}^k \frac{U-\lambda_j I}{\lambda_i-\lambda_j}, & k>1 \\ I, & k=1. \end{cases} \tag{1.242}
$$

The rotation tensor R is now obtained as $FU^{-1} = F\sum_{i=1}^k P_i/\lambda_i$. Note that $|\det F| = \det U = \sqrt{\det C}$.

Now we evaluate the derivatives of these tensors. Differentiating the relation $C = UU$ using Eqn. (1.233), we get

$$
\mathbb{S} = \mathbb{S}\left[U \boxtimes I + I \boxtimes U\right]\mathbb{S}\frac{\partial U}{\partial C}\mathbb{S}.
$$

Since the eigenvalues λ_i, $i = 1, 2, \ldots, n$, of U are positive, the necessary and sufficient condition that $[U \boxtimes I + I \boxtimes U]$ be invertible is satisfied[1]. Thus, using (1.197), we get the unique solution $\partial U/\partial C = \mathbb{S}/(2\lambda)$ for $k = 1$, while for $k > 1$, we have

$$
\begin{aligned}
\frac{\partial U}{\partial C} &= \left\{\mathbb{S}\left[U \boxtimes I + I \boxtimes U\right]\mathbb{S}\right\}^{-1} \\
&= \mathbb{S}\left[\sum_{i=1}^k \sum_{j=1}^k \frac{P_i \boxtimes P_j}{\lambda_i + \lambda_j}\right]\mathbb{S}, && (1.243)
\end{aligned}
$$

Now using Eqns. (1.175), (1.234) and (1.239), we get

$$
\frac{\partial U}{\partial F} = \mathbb{S}\frac{\partial U}{\partial C}\mathbb{S}\frac{\partial C}{\partial F}
$$

[1] If C is positive semi-definite instead of positive definite, then $\partial U/\partial C$ does not exist when C is singular.

$$= 2\mathbb{S}\left[\sum_{i=1}^{k}\sum_{j=1}^{k}\frac{P_i \boxtimes P_j}{\lambda_i+\lambda_j}\right]\mathbb{S}\left[F^T \boxtimes I\right]$$

$$= 2\mathbb{S}\left[\sum_{i=1}^{k}\sum_{j=1}^{k}\frac{(P_iF^T)\boxtimes P_j}{\lambda_i+\lambda_j}\right]. \tag{1.244}$$

By differentiating $F = RU$, and using Eqns. (1.158), (1.168), (1.185), (1.173), we get

$$\frac{\partial R}{\partial F} = (I\boxtimes U)^{-1}\left[\mathbb{I} - (R\boxtimes I)\frac{\partial U}{\partial F}\right]$$

$$= I\boxtimes U^{-1} - 2[R\boxtimes U^{-1}]\mathbb{S}\left[\sum_{i=1}^{k}\sum_{j=1}^{k}\frac{(P_iF^T)\boxtimes P_j}{\lambda_i+\lambda_j}\right]$$

$$= I\boxtimes U^{-1} - \sum_{i=1}^{k}\sum_{j=1}^{k}\frac{(RP_iF^T)\boxtimes(U^{-1}P_j) + \mathbb{T}\left[(U^{-1}P_iF^T)\boxtimes(RP_j)\right]}{\lambda_i+\lambda_j}$$

$$= \sum_{i=1}^{k}\sum_{j=1}^{k}\frac{(\lambda_i+\lambda_j)(RP_iR^T)\boxtimes(U^{-1}P_j) - \lambda_i(RP_iR^T)\boxtimes(U^{-1}P_j) - [(RP_i)\boxtimes(P_jR^T)]\mathbb{T}}{\lambda_i+\lambda_j}$$

$$= \sum_{i=1}^{k}\sum_{j=1}^{k}\frac{(RP_iR^T)\boxtimes P_j - [(RP_i)\boxtimes(P_jR^T)]\mathbb{T}}{\lambda_i+\lambda_j}. \tag{1.245}$$

Note that we have interchanged i and j in the last term in the second-to-last step. Now using Eqn. (1.243), and the fact that $\dot{U} = (\partial U/\partial C)\dot{C}$ (or, alternatively, Eqn. (1.244) and the fact that $\dot{U} = (\partial U/\partial F)\dot{F}$), and in a similar manner using (1.245), we get $\dot{U} = \dot{C}/(2\lambda)$ and $R^T\dot{R} = (R^T\dot{F} - \dot{F}^T R)/(2\lambda)$ for $k = 1$, while for $k > 1$,

$$\dot{U} = \sum_{i=1}^{k}\sum_{j=1}^{k}\frac{P_i\dot{C}P_j}{\lambda_i+\lambda_j} = \sum_{i=1}^{k}\sum_{j=1}^{k}\frac{P_i(F^T\dot{F}+\dot{F}^TF)P_j}{\lambda_i+\lambda_j}. \tag{1.246}$$

$$\dot{R} = R\sum_{i=1}^{k}\sum_{j=1}^{k}\frac{P_i(R^T\dot{F}-\dot{F}^TR)P_j}{\lambda_i+\lambda_j}. \tag{1.247}$$

In a similar fashion, one can find $D_F U(F)[Z] = (\partial U/\partial F)Z$. If F is the deformation gradient, then see Eqns. (2.70) and (2.77).

The corresponding results for the left stretch tensor V are

$$\frac{\partial V}{\partial B} = \mathbb{S}\left[\sum_{i=1}^{k}\sum_{j=1}^{k}\frac{G_i\boxtimes G_j}{\lambda_i+\lambda_j}\right]\mathbb{S},$$

$$\frac{\partial V}{\partial F} = 2\mathbb{S}\left[\sum_{i=1}^{k}\sum_{j=1}^{k}\frac{G_i\boxtimes(G_jF)}{\lambda_i+\lambda_j}\right],$$

$$\dot{V} = \sum_{i=1}^{k}\sum_{j=1}^{k}\frac{G_i\dot{B}G_j}{\lambda_i+\lambda_j} = \sum_{i=1}^{k}\sum_{j=1}^{k}\frac{G_i(\dot{F}F^T+F\dot{F}^T)G_j}{\lambda_i+\lambda_j}, \tag{1.248}$$

$$\dot{R} = \left[\sum_{i=1}^{k}\sum_{j=1}^{k}\frac{G_i(\dot{F}R^T-R\dot{F}^T)G_j}{\lambda_i+\lambda_j}\right]R, \tag{1.249}$$

where $G_i = RP_iR^T$ are the projections of V, which are obtained using Eqn. (1.242) by replacing C and U by B and V, respectively. If F is the deformation gradient, then see Eqns. (2.71) and (2.76).

We now verify the following relations derived by Chen and Wheeler [47] and Wheeler [352] when $n = 3$:

$$\left(\frac{\partial U}{\partial F}\right)^T U = F,$$

$$\left(\frac{\partial R}{\partial F}\right)^T F = 0,$$

$$\frac{\partial U}{\partial F}L = R^T L - \frac{1}{\det \tilde{U}}\tilde{U}(R^T L - L^T R)\tilde{U}U,$$

$$\frac{\partial R}{\partial F}L = \frac{1}{\det \tilde{U}}R\tilde{U}(R^T L - L^T R)\tilde{U},$$

where $L \in \text{Lin}$, and

$$\tilde{U} := (\operatorname{tr} U)I - U = (\lambda_2 + \lambda_3)P_1 + (\lambda_1 + \lambda_3)P_2 + (\lambda_1 + \lambda_2)P_3.$$

Note that $\det \tilde{U} = (\lambda_1 + \lambda_2)(\lambda_2 + \lambda_3)(\lambda_1 + \lambda_3)$, and that

$$\tilde{U}U = \lambda_1(\lambda_2 + \lambda_3)P_1 + \lambda_2(\lambda_1 + \lambda_3)P_2 + \lambda_3(\lambda_1 + \lambda_2)P_3. \tag{1.250}$$

Using the facts $(A \boxtimes B)^T = A^T \boxtimes B^T$, $(\mathbb{C}_1\mathbb{C}_2)^T = \mathbb{C}_2^T\mathbb{C}_1^T$, $\mathbb{T}A = A^T$, $\mathbb{S}^T = \mathbb{S}$, $\mathbb{T}^T = \mathbb{T}$, $P_iP_j = \delta_{ij}P_j$ (no sum on j), $P_i^T = P_i$, $\sum_{i=1}^k P_i = I$, $UP_j = \left[\sum_{m=1}^k \lambda_m P_m\right] P_j = \lambda_j P_j = P_j U$ (no sum on j), we have

$$\begin{aligned}
\left(\frac{\partial U}{\partial F}\right)^T U &= \sum_{i=1}^k\sum_{j=1}^k \frac{FP_jUP_i + FP_iUP_j}{\lambda_i + \lambda_j} \\
&= F\left[\sum_{i=1}^k\sum_{j=1}^k \frac{P_jUP_i + P_iUP_j}{\lambda_i + \lambda_j}\right] \\
&= F\left[\sum_{i=1}^k\sum_{j=1}^k \frac{\lambda_i(P_jP_i + P_iP_j)}{\lambda_i + \lambda_j}\right] \\
&= F\sum_{i=1}^k \frac{\lambda_i P_i}{\lambda_i} \\
&= F,
\end{aligned}$$

$$\begin{aligned}
\left(\frac{\partial R}{\partial F}\right)^T F &= \sum_{i=1}^k\sum_{j=1}^k \frac{RP_iR^TFP_j - \mathbb{T}(P_iR^TFP_jR^T)}{\lambda_i + \lambda_j} \\
&= \sum_{i=1}^k\sum_{j=1}^k \frac{RP_iUP_j - RP_jUP_i}{\lambda_i + \lambda_j}
\end{aligned}$$

$$
\begin{aligned}
&= \sum_{i=1}^{k}\sum_{j=1}^{k} \frac{\lambda_j R(P_iP_j - P_jP_i)}{\lambda_i + \lambda_j} \\
&= \sum_{i=1}^{k} \frac{\lambda_i R(P_i - P_i)}{2\lambda_i} \\
&= \mathbf{0},
\end{aligned}
$$

$$
\begin{aligned}
\frac{\partial U}{\partial F} L &= \sum_{i=1}^{k}\sum_{j=1}^{k} \frac{P_iL^TFP_j + P_iF^TLP_j}{\lambda_i + \lambda_j} \\
&= \sum_{i=1}^{k}\sum_{j=1}^{k} \frac{P_iL^TRUP_j + P_iUR^TLP_j}{\lambda_i + \lambda_j} \\
&= \sum_{i=1}^{k}\sum_{j=1}^{k} \frac{\lambda_jP_iL^TRP_j + \lambda_iP_iR^TLP_j}{\lambda_i + \lambda_j} \\
&= \sum_{i=1}^{k}\sum_{j=1}^{k} P_iR^TLP_j - \sum_{i=1}^{k}\sum_{j=1}^{k} \frac{\lambda_jP_i(R^TL - L^TR)P_j}{\lambda_i + \lambda_j} \\
&= \left(\sum_{i=1}^{k} P_i\right) R^TL \left(\sum_{j=1}^{k} P_j\right) - \sum_{i=1}^{k}\sum_{j=1}^{k} \frac{\lambda_jP_i(R^TL - L^TR)P_j}{\lambda_i + \lambda_j} \\
&= R^TL - \sum_{i=1}^{k}\sum_{j=1}^{k} \frac{\lambda_jP_i(R^TL - L^TR)P_j}{\lambda_i + \lambda_j} \\
&= R^TL - \frac{1}{\det \tilde{U}}\tilde{U}(R^TL - L^TR)\tilde{U}U, \qquad \text{(by Eqn. (1.250))}
\end{aligned}
$$

$$
\begin{aligned}
\frac{\partial R}{\partial F} L &= (I \boxtimes U^{-1})L - (R \boxtimes U^{-1})\left[R^TL - \frac{1}{\det \tilde{U}}\tilde{U}(R^TL - L^TR)\tilde{U}U\right] \\
&= LU^{-1} - RR^TLU^{-1} + \frac{1}{\det \tilde{U}} R\tilde{U}(R^TL - L^TR)\tilde{U} \\
&= \frac{1}{\det \tilde{U}} R\tilde{U}(R^TL - L^TR)\tilde{U}.
\end{aligned}
$$

We now discuss the computation of the gradients of the eigenvalues λ_1, λ_2, λ_3 of a *diagonalizable* tensor T for the case $n = 3$, and then specialize the results to the case of a symmetric tensor–a more comprehensive discussion of the derivatives of the eigenvalues of a symmetric tensor for the case $n = 3$ may be found in [344]. By Eqn. (J.25), T can be expressed as $\sum_{i=1}^{k} \lambda_i P_i$, where k is the number of distinct eigenvalues, P_i are projections given by Eqn. (J.26), and $\sum_{i=1}^{k} P_i = I$. Consider the case when all eigenvalues are distinct, i.e., $\lambda_1 \neq \lambda_2 \neq \lambda_3 \neq \lambda_1$. It is known (see, e.g., the remark following (2.2.4) in [37]) that the eigenvalues and associated eigenprojections are differentiable in this case. Using Eqn. (1.60), we can show, for example, that $\mathbf{cof}\,(T - \lambda_1 I) = (T - \lambda_2 I)^T(T - \lambda_3 I)^T = P_1^T$. Since $\det(T - \gamma I - (\lambda_i - \gamma)I) = \det(T - \lambda_i I) = 0$, $i = 1,2,3$ and $\gamma \in \Re$, we see that $\lambda_i - \gamma$, $i = 1,2,3$ are the eigenvalues of $T - \gamma I$, $\gamma \in \Re$. Thus, $(\lambda_1 - \lambda_1)$, $(\lambda_2 - \lambda_1)$ and

$(\lambda_3 - \lambda_1)$ are the eigenvalues of $(T - \lambda_1 I)$, which implies that $(0, 0, (\lambda_2 - \lambda_1)(\lambda_3 - \lambda_1))$ are the eigenvalues of $\mathbf{cof}\,(T - \lambda_1 I)$, which in turn leads to $\operatorname{tr}\mathbf{cof}\,(T - \lambda_1 I) = (\lambda_1 - \lambda_2)(\lambda_1 - \lambda_3)$. By differentiating the relation $\det(T - \lambda_i I) = 0$, $i = 1, 2, 3$, and using the chain rule, we get

$$(\mathbb{I} - \frac{\partial \lambda_1}{\partial T} \otimes I)\mathbf{cof}\,(T - \lambda_1 I) = \mathbf{0},\ i = 1, 2, 3, \tag{1.251}$$

or, in other words,

$$\frac{\partial \lambda_1}{\partial T} = \frac{\mathbf{cof}\,(T - \lambda_1 I)}{\operatorname{tr}\mathbf{cof}\,(T - \lambda_1 I)} = \frac{(T - \lambda_2 I)^T (T - \lambda_3 I)^T}{(\lambda_1 - \lambda_2)(\lambda_1 - \lambda_3)} = P_1^T. \tag{1.252}$$

The corresponding results for $\partial \lambda_2 / \partial T$ and $\partial \lambda_3 / \partial T$ are obtained by permuting the indices 1, 2 and 3. Equation (1.252) can also be directly obtained by differentiating the relation $\lambda_1^3 - I_1 \lambda_1^2 + I_2 \lambda_1 - I_3 = 0$ with respect to T, and using Eqns. (1.226a)–(1.226c); this shows that the second expression for $\partial \lambda_1 / \partial T$ in Eqn. (1.252) is valid even for an *arbitrary* tensor, provided the denominator is nonzero.

By differentiating the expression $P_1 = (T - \lambda_2 I)(T - \lambda_3 I)/[(\lambda_1 - \lambda_2)(\lambda_1 - \lambda_3)]$ using Eqns. (1.170), (1.231) and (1.233), we get

$$\begin{aligned}\frac{\partial P_1}{\partial T} = \frac{1}{(\lambda_1 - \lambda_2)(\lambda_1 - \lambda_3)} \Big[& T \boxtimes I + I \boxtimes T^T - (\lambda_2 + \lambda_3)\mathbb{I} \\ & + (\lambda_2 + \lambda_3 - 2\lambda_1) P_1 \otimes P_1^T - (\lambda_2 - \lambda_3) \left[P_2 \otimes P_2^T - P_3 \otimes P_3^T \right] \Big].\end{aligned}$$

By using Eqn. (J.29) and the fact that $I = \sum_{i=1}^3 P_i = \sum_{i=1}^3 P_i^T$, the above relation simplifies to

$$\frac{\partial P_1}{\partial T} = \frac{P_1 \boxtimes P_2^T + P_2 \boxtimes P_1^T}{\lambda_1 - \lambda_2} + \frac{P_1 \boxtimes P_3^T + P_3 \boxtimes P_1^T}{\lambda_1 - \lambda_3}. \tag{1.253}$$

The gradients of P_2 and P_3 are obtained by cyclically permuting the indices 1, 2 and 3. The above expressions are valid for an arbitrary tensor with distinct eigenvalues (since such a tensor is diagonalizable).

Now consider the case when two of the eigenvalues are repeated, i.e., $\lambda_1 \neq \lambda_2 = \lambda_3$. Note that now, Eqn. (J.4) yields $P_1 = (T - \lambda_2 I)/(\lambda_1 - \lambda_2)$. Since $P_1 P_1 = P_1$, and since $\lambda_2 = \lambda_3$, (1.252) also holds. Using the fact that $\lambda_2 + \lambda_3 = I_1 - \lambda_1$, we get

$$\frac{\partial}{\partial T}(\lambda_2 + \lambda_3) = I - P_1^T = \frac{\lambda_1 I - T^T}{(\lambda_1 - \lambda_2)}. \tag{1.254}$$

The gradient of P_1 is obtained by setting $\lambda_2 = \lambda_3$ and using $P_2 + P_3 = I - P_1$ in Eqn. (1.253) as

$$\frac{\partial P_1}{\partial T} = \frac{P_1 \boxtimes (I - P_1)^T + (I - P_1) \boxtimes P_1^T}{(\lambda_1 - \lambda_2)}.$$

Since $P_2 + P_3 = I - P_1$, we get

$$\frac{\partial}{\partial T}(P_2 + P_3) = -\frac{\partial P_1}{\partial T}.$$

When all three eigenvalues are repeated, the gradient of the sum of the eigenvalues is $\boldsymbol{I}$, but the individual gradients are undefined. Since symmetric second-order tensors are diagonalizable, all the above results can be applied to them by setting $\boldsymbol{P}_i = \boldsymbol{P}_i^T$. For a generalization to any underlying space dimension n, see Eqn. (J.27).

The foregoing development can be used to find the gradients of scalar or tensor-valued functions of a diagonalizable tensor $\boldsymbol{T} = \sum_{i=1}^{k} \lambda_i \boldsymbol{P}_i$, where $(\lambda_i, \boldsymbol{P}_i)$, $i = 1, 2, \ldots, k$, are the distinct eigenvalues and corresponding eigenprojections of $\boldsymbol{T}$ (see, e.g., [37], [47], [104], [127], [244], [272], [352] and [355] for various other methods). For example, if $\boldsymbol{G}(\boldsymbol{T}) = \sum_{i=1}^{k} f(\lambda_i)\boldsymbol{P}_i$ is a tensor-valued function of $\boldsymbol{T}$, then using the above results, we get $\partial \boldsymbol{G}/\partial \boldsymbol{T} = f'(\lambda)\mathbb{I}$ for the case $k = 1$, while for $k > 1$,

$$\frac{\partial \boldsymbol{G}}{\partial \boldsymbol{T}} = \sum_{i=1}^{k} \frac{\partial f}{\partial \lambda_i} \boldsymbol{P}_i \boxtimes \boldsymbol{P}_i^T + \sum_{i=1}^{k} \sum_{\substack{j=1 \\ j \neq i}}^{k} \frac{f(\lambda_i) - f(\lambda_j)}{\lambda_i - \lambda_j} \boldsymbol{P}_i \boxtimes \boldsymbol{P}_j^T. \tag{1.255}$$

The above formula is derived by first considering the case of *distinct* eigenvalues (i.e., $k = n$), and then appropriate limits are taken (e.g., $\lambda_3 \to \lambda_2$ in case $\lambda_2 = \lambda_3$) to obtain the results for repeated eigenvalues. By virtue of Eqn. (J.27), the above result holds for any arbitrary underlying space dimension n. In case $\boldsymbol{G}(\boldsymbol{S}) = \sum_{i=1}^{k} f(\lambda_i)\boldsymbol{P}_i$ is a symmetric-valued function of $\boldsymbol{S} \in \mathrm{Sym}$, then the above result reduces to $\partial \boldsymbol{G}/\partial \boldsymbol{S} = f'(\lambda)\mathbb{S}$ for the case $k = 1$, while for $k > 1$,

$$\frac{\partial \boldsymbol{G}}{\partial \boldsymbol{S}} = \mathbb{S}\left[\sum_{i=1}^{k} \frac{\partial f}{\partial \lambda_i} \boldsymbol{P}_i \boxtimes \boldsymbol{P}_i + \sum_{i=1}^{k} \sum_{\substack{j=1 \\ j \neq i}}^{k} \frac{f(\lambda_i) - f(\lambda_j)}{\lambda_i - \lambda_j} \boldsymbol{P}_i \boxtimes \boldsymbol{P}_j\right]\mathbb{S}. \tag{1.256}$$

An application of Eqn. (1.256) can be found towards the end of Section 2.4. We now find the eigenvalue/eigentensors of $\partial \boldsymbol{G}/\partial \boldsymbol{S}$ in the above equation. First consider the case when all the eigenvalues λ_i of $\boldsymbol{S}$ are distinct. Then, if $\boldsymbol{n}_i$ represent the eigenvectors of $\boldsymbol{S}$, each $\boldsymbol{P}_i$ can be represented as $\boldsymbol{n}_i \otimes \boldsymbol{n}_i$ (no sum on i). Using Eqns. (1.172) and (1.342), the eigenvalues/eigentensors (the eigentensors have not been normalized) are found to be

$$\begin{gathered}
\left(\frac{\partial f}{\partial \lambda_1}, \boldsymbol{n}_1 \otimes \boldsymbol{n}_1\right), \quad \left(\frac{\partial f}{\partial \lambda_2}, \boldsymbol{n}_2 \otimes \boldsymbol{n}_2\right), \quad \left(\frac{\partial f}{\partial \lambda_3}, \boldsymbol{n}_3 \otimes \boldsymbol{n}_3\right), \\
\left(\frac{f(\lambda_1) - f(\lambda_2)}{\lambda_1 - \lambda_2}, \boldsymbol{n}_1 \otimes \boldsymbol{n}_2 + \boldsymbol{n}_2 \otimes \boldsymbol{n}_1\right), \left(\frac{f(\lambda_2) - f(\lambda_3)}{\lambda_2 - \lambda_3}, \boldsymbol{n}_2 \otimes \boldsymbol{n}_3 + \boldsymbol{n}_3 \otimes \boldsymbol{n}_2\right), \\
\left(\frac{f(\lambda_1) - f(\lambda_3)}{\lambda_1 - \lambda_3}, \boldsymbol{n}_1 \otimes \boldsymbol{n}_3 + \boldsymbol{n}_3 \otimes \boldsymbol{n}_1\right), \qquad \lambda_1 \neq \lambda_2 \neq \lambda_3 \neq \lambda_1.
\end{gathered} \tag{1.257}$$

Next consider the case when two eigenvalues of $\boldsymbol{S}$ are repeated, say, $\lambda_1 \neq \lambda_2 = \lambda_3$. We now have

$$\begin{aligned}
\frac{\partial \boldsymbol{G}}{\partial \boldsymbol{S}} = \mathbb{S}\Bigg[&\frac{\partial f}{\partial \lambda_1} \boldsymbol{P}_1 \boxtimes \boldsymbol{P}_1 + \frac{\partial f}{\partial \lambda_2} (\boldsymbol{I} - \boldsymbol{P}_1) \boxtimes (\boldsymbol{I} - \boldsymbol{P}_1) \\
&+ \frac{f(\lambda_1) - f(\lambda_2)}{\lambda_1 - \lambda_2} \left[\boldsymbol{P}_1 \boxtimes (\boldsymbol{I} - \boldsymbol{P}_1) + (\boldsymbol{I} - \boldsymbol{P}_1) \boxtimes \boldsymbol{P}_1\right]\Bigg]\mathbb{S}.
\end{aligned}$$

The distinct eigenvalue/eigentensor pairs of $\partial G/\partial S$ in this case are

$$\left(\frac{\partial f}{\partial \lambda_1}, \boldsymbol{n}_1 \otimes \boldsymbol{n}_1\right), \quad \left(\frac{\partial f}{\partial \lambda_2}, \boldsymbol{I} - \boldsymbol{n}_1 \otimes \boldsymbol{n}_1\right),$$
$$\left(\frac{f(\lambda_1) - f(\lambda_2)}{\lambda_1 - \lambda_2}, \boldsymbol{n}_1 \otimes [\boldsymbol{n} - (\boldsymbol{n} \cdot \boldsymbol{n}_1)\boldsymbol{n}_1] + [\boldsymbol{n} - (\boldsymbol{n} \cdot \boldsymbol{n}_1)\boldsymbol{n}_1] \otimes \boldsymbol{n}_1\right), \quad \lambda_1 \neq \lambda_2 = \lambda_3, \tag{1.258}$$

where $\boldsymbol{n}$ is such that $\boldsymbol{n} \times \boldsymbol{n}_1 \neq \boldsymbol{0}$, but is otherwise arbitrary. Finally, for the case when all eigenvalues are repeated, i.e., $\lambda_1 = \lambda_2 = \lambda_3 \equiv \lambda$, the distinct eigenvalue/eigentensor pair is

$$\left(\frac{\partial f}{\partial \lambda}, \boldsymbol{A}\right), \quad \lambda_1 = \lambda_2 = \lambda_3 \equiv \lambda, \tag{1.259}$$

where $\boldsymbol{A} \in \text{Sym}$.

The derivatives of the eigenvectors $\{\boldsymbol{e}_i^*\}$ of $\boldsymbol{S} \in \text{Sym}$ are given by [244]

$$\frac{\partial \boldsymbol{e}_1^*}{\partial \boldsymbol{S}} = \frac{\boldsymbol{e}_2^* \otimes \boldsymbol{e}_1^* \otimes \boldsymbol{e}_2^* + \boldsymbol{e}_2^* \otimes \boldsymbol{e}_2^* \otimes \boldsymbol{e}_1^*}{2(\lambda_1 - \lambda_2)} + \frac{\boldsymbol{e}_3^* \otimes \boldsymbol{e}_1^* \otimes \boldsymbol{e}_3^* + \boldsymbol{e}_3^* \otimes \boldsymbol{e}_3^* \otimes \boldsymbol{e}_1^*}{2(\lambda_1 - \lambda_3)},$$
$$\frac{\partial \boldsymbol{e}_2^*}{\partial \boldsymbol{S}} = \frac{\boldsymbol{e}_1^* \otimes \boldsymbol{e}_1^* \otimes \boldsymbol{e}_2^* + \boldsymbol{e}_1^* \otimes \boldsymbol{e}_2^* \otimes \boldsymbol{e}_1^*}{2(\lambda_2 - \lambda_1)} + \frac{\boldsymbol{e}_3^* \otimes \boldsymbol{e}_2^* \otimes \boldsymbol{e}_3^* + \boldsymbol{e}_3^* \otimes \boldsymbol{e}_3^* \otimes \boldsymbol{e}_2^*}{2(\lambda_2 - \lambda_3)},$$
$$\frac{\partial \boldsymbol{e}_3^*}{\partial \boldsymbol{S}} = \frac{\boldsymbol{e}_1^* \otimes \boldsymbol{e}_1^* \otimes \boldsymbol{e}_3^* + \boldsymbol{e}_1^* \otimes \boldsymbol{e}_3^* \otimes \boldsymbol{e}_1^*}{2(\lambda_3 - \lambda_1)} + \frac{\boldsymbol{e}_2^* \otimes \boldsymbol{e}_2^* \otimes \boldsymbol{e}_3^* + \boldsymbol{e}_2^* \otimes \boldsymbol{e}_3^* \otimes \boldsymbol{e}_2^*}{2(\lambda_3 - \lambda_2)}.$$

Similar to the gradient of a scalar field, we define the *gradient of a vector field* $\boldsymbol{v}$ as

$$\boldsymbol{\nabla} \boldsymbol{v}(\boldsymbol{x})\boldsymbol{u} := D\boldsymbol{v}(\boldsymbol{x})[\boldsymbol{u}]. \tag{1.260}$$

The terminology 'gradient' used for the above quantity is misleading since the gradient was defined in terms of the inner product, whereas the above quantity is not [58]. The quantity $\boldsymbol{\nabla} \boldsymbol{v}$ is in reality simply the matrix representation of the Frechet derivative of $\boldsymbol{v}$, and hence a second-order tensor. Taking $\boldsymbol{u} = \boldsymbol{e}_j$, and taking the dot product of the above equation with $\boldsymbol{e}_i$, we get the components of $\boldsymbol{\nabla} \boldsymbol{v}$ as

$$(\boldsymbol{\nabla} \boldsymbol{v})_{ij} = \frac{\partial v_i}{\partial x_j}.$$

Thus, we have

$$\boldsymbol{\nabla} \boldsymbol{v} = \frac{\partial v_i}{\partial x_j} \boldsymbol{e}_i \otimes \boldsymbol{e}_j.$$

The scalar field

$$\boldsymbol{\nabla} \cdot \boldsymbol{v} := \operatorname{tr} \boldsymbol{\nabla} \boldsymbol{v} = \frac{\partial v_i}{\partial x_j} \operatorname{tr}(\boldsymbol{e}_i \otimes \boldsymbol{e}_j) = \frac{\partial v_i}{\partial x_j} \boldsymbol{e}_i \cdot \boldsymbol{e}_j = \frac{\partial v_i}{\partial x_j} \delta_{ij} = \frac{\partial v_i}{\partial x_i}, \tag{1.261}$$

is called the divergence of $\boldsymbol{v}$.

The gradient of a second-order tensor $\boldsymbol{T}$ is a third-order tensor defined in a way similar to the gradient of a vector field as

$$\nabla \boldsymbol{T}(\boldsymbol{x})\boldsymbol{u} := D\boldsymbol{T}(\boldsymbol{x})[\boldsymbol{u}]. \tag{1.262}$$

$\nabla \boldsymbol{T}$ can be written in terms of its components as

$$\nabla \boldsymbol{T} = \frac{\partial T_{ij}}{\partial x_k} \boldsymbol{e}_i \otimes \boldsymbol{e}_j \otimes \boldsymbol{e}_k.$$

The *divergence of a second-order tensor* $\boldsymbol{T}$, denoted as $\nabla \cdot \boldsymbol{T}$, is defined as

$$(\nabla \cdot \boldsymbol{T}) \cdot \boldsymbol{u} := \nabla \cdot (\boldsymbol{T}^T \boldsymbol{u}) \quad \forall \text{ constant } \boldsymbol{u} \in V, \tag{1.263}$$

leading to the component form

$$\nabla \cdot \boldsymbol{T} = \frac{\partial T_{ij}}{\partial x_j} \boldsymbol{e}_i.$$

The *curl of a vector* $\boldsymbol{v}$, denoted as $\nabla \times \boldsymbol{v}$, is defined by

$$(\nabla \times \boldsymbol{v}) \times \boldsymbol{u} := [\nabla \boldsymbol{v} - (\nabla \boldsymbol{v})^T]\boldsymbol{u} \quad \forall \boldsymbol{u} \in V. \tag{1.264}$$

Thus, $\nabla \times \boldsymbol{v}$ is the axial vector corresponding to the skew tensor $[\nabla \boldsymbol{v} - (\nabla \boldsymbol{v})^T]$. In component form, we have

$$\nabla \times \boldsymbol{v} = \epsilon_{ijk} (\nabla \boldsymbol{v})_{kj} \boldsymbol{e}_i = \epsilon_{ijk} \frac{\partial v_k}{\partial x_j} \boldsymbol{e}_i = \frac{\partial v_k}{\partial x_j} \boldsymbol{e}_j \times \boldsymbol{e}_k.$$

The *curl of a tensor* $\boldsymbol{T}$, denoted by $\nabla \times \boldsymbol{T}$, is defined by

$$(\nabla \times \boldsymbol{T})\boldsymbol{u} := \nabla \times (\boldsymbol{T}^T \boldsymbol{u}) \quad \forall \text{ constant } \boldsymbol{u} \in V. \tag{1.265}$$

In component form,

$$\nabla \times \boldsymbol{T} = \epsilon_{irs} \frac{\partial T_{js}}{\partial x_r} \boldsymbol{e}_i \otimes \boldsymbol{e}_j = \frac{\partial T_{js}}{\partial x_r} (\boldsymbol{e}_r \times \boldsymbol{e}_s) \otimes \boldsymbol{e}_j. \tag{1.266}$$

The *Laplacian* of a scalar function $\phi(\boldsymbol{x})$ is defined by

$$\nabla^2 \phi := \nabla \cdot (\nabla \phi). \tag{1.267}$$

In component form, the Laplacian is given by

$$\nabla^2 \phi = \frac{\partial^2 \phi}{\partial x_i \partial x_i}.$$

If $\nabla^2 \phi = 0$, then ϕ is said to be harmonic.

The Laplacian of a tensor function $\boldsymbol{T}(\boldsymbol{x})$, denoted by $\nabla^2 \boldsymbol{T}$, is defined by

$$(\nabla^2 \boldsymbol{T}) : \boldsymbol{H} := \nabla^2 (\boldsymbol{T} : \boldsymbol{H}) \quad \forall \text{ constant } \boldsymbol{H} \in \text{Lin}. \tag{1.268}$$

In component form,

$$(\nabla^2 T)_{ij} = \frac{\partial^2 T_{ij}}{\partial x_k \partial x_k}.$$

In the remaining chapters, we shall assume that a function is sufficiently smooth (without explicitly stating what these smoothness properties are), so that all the equations presented make sense. For example, in the equation

$$\nabla \cdot \tau + \rho b = 0,$$

if we are given that ρ and b are continuous functions of the position vector x, then we implicitly assume τ to be continuously differentiable.

1.9.5 Examples

Although it is possible to derive tensor identities involving differentiation using the above definitions of the operators, the proofs can be quite cumbersome, and hence we prefer to use indicial notation instead. In what follows, u and v are vector fields, and $\nabla \equiv \frac{\partial}{\partial x_i} e_i$ (this is to be interpreted as the 'del' operator acting on a scalar, vector or tensor-valued field, e.g., $\nabla \phi = \frac{\partial \phi}{\partial x_i} e_i$):

1. Show that

$$\nabla \times \nabla \phi = 0. \tag{1.269}$$

2. Show that

$$\frac{1}{2} \nabla (u \cdot u) = (\nabla u)^T u. \tag{1.270}$$

3. Show that $\nabla \cdot [(\nabla u) v] = (\nabla u)^T : \nabla v + v \cdot [\nabla (\nabla \cdot u)]$.

4. Show that

$$\nabla \cdot (\nabla u)^T = \nabla (\nabla \cdot u), \tag{1.271a}$$

$$\nabla^2 u := \nabla \cdot (\nabla u) = \nabla (\nabla \cdot u) - \nabla \times (\nabla \times u). \tag{1.271b}$$

From Eqns. (1.271a) and (1.271b), it follows that

$$\nabla \cdot [(\nabla u) - (\nabla u)^T] = -\nabla \times (\nabla \times u).$$

From Eqn. (1.271b), it follows that if $\nabla \cdot u = 0$ and $\nabla \times u = 0$, then $\nabla^2 u = 0$, i.e., u is harmonic.

5. Show that $\nabla \cdot (u \times v) = v \cdot (\nabla \times u) - u \cdot (\nabla \times v)$.

6. Let $W \in$ Skw, and let w be its axial vector. Then show that

$$\nabla \cdot W = -\nabla \times w,$$

$$\nabla \times W = (\nabla \cdot w) I - \nabla w. \tag{1.272}$$

Solution:

1. Consider the ith component of the left-hand side:

$$(\nabla \times \nabla\phi)_i = \epsilon_{ijk}\frac{\partial^2\phi}{\partial x_j \partial x_k}$$
$$= \epsilon_{ikj}\frac{\partial^2\phi}{\partial x_k \partial x_j} \qquad \text{(interchanging } j \text{ and } k\text{)}$$
$$= -\epsilon_{ijk}\frac{\partial^2\phi}{\partial x_j \partial x_k},$$

which implies that $\nabla \times \nabla\phi = \mathbf{0}$.

2. $\frac{1}{2}\nabla(\boldsymbol{u}\cdot\boldsymbol{u}) = \frac{1}{2}\frac{\partial(u_i u_i)}{\partial x_j}\boldsymbol{e}_j = u_i\frac{\partial u_i}{\partial x_j}\boldsymbol{e}_j = (\nabla\boldsymbol{u})^T\boldsymbol{u}.$

3. We have

$$\nabla\cdot[(\nabla\boldsymbol{u})\boldsymbol{v}] = \nabla_j((\nabla\boldsymbol{u})\boldsymbol{v})_j$$
$$= \frac{\partial}{\partial x_j}\left(\frac{\partial u_j}{\partial x_i}v_i\right)$$
$$= \frac{\partial v_i}{\partial x_j}\frac{\partial u_j}{\partial x_i} + v_i\frac{\partial^2 u_j}{\partial x_i \partial x_j}$$
$$= (\nabla\boldsymbol{u})^T : \nabla\boldsymbol{v} + \boldsymbol{v}\cdot\nabla(\nabla\cdot\boldsymbol{u}).$$

4. The first identity is proved as follows:

$$[\nabla\cdot(\nabla\boldsymbol{u})^T]_i = \frac{\partial}{\partial x_j}\left(\frac{\partial u_j}{\partial x_i}\right)$$
$$= \frac{\partial}{\partial x_i}\left(\frac{\partial u_j}{\partial x_j}\right)$$
$$= [\nabla(\nabla\cdot\boldsymbol{u})]_i.$$

To prove the second identity, consider the last term

$$\nabla\times(\nabla\times\boldsymbol{u}) = \epsilon_{ijk}\nabla_j(\nabla\times\boldsymbol{u})_k\boldsymbol{e}_i$$
$$= \epsilon_{ijk}\nabla_j(\epsilon_{kmn}\nabla_m u_n)\boldsymbol{e}_i$$
$$= \epsilon_{ijk}\epsilon_{mnk}\frac{\partial^2 u_n}{\partial x_j \partial x_m}\boldsymbol{e}_i$$
$$= (\delta_{im}\delta_{jn} - \delta_{in}\delta_{jm})\frac{\partial^2 u_n}{\partial x_j \partial x_m}\boldsymbol{e}_i$$
$$= \left[\frac{\partial^2 u_j}{\partial x_i \partial x_j} - \frac{\partial^2 u_i}{\partial x_j \partial x_j}\right]\boldsymbol{e}_i$$
$$= \nabla(\nabla\cdot\boldsymbol{u}) - \nabla\cdot(\nabla\boldsymbol{u}).$$

5. We have

$$
\begin{aligned}
\nabla \cdot (\boldsymbol{u} \times \boldsymbol{v}) &= \frac{\partial (\boldsymbol{u} \times \boldsymbol{v})_i}{\partial x_i} \\
&= \epsilon_{ijk} \frac{\partial (u_j v_k)}{\partial x_i} \\
&= \epsilon_{ijk} v_k \frac{\partial u_j}{\partial x_i} + \epsilon_{ijk} u_j \frac{\partial v_k}{\partial x_i} \\
&= \epsilon_{kij} v_k \frac{\partial u_j}{\partial x_i} - \epsilon_{jik} u_j \frac{\partial v_k}{\partial x_i} \\
&= \boldsymbol{v} \cdot (\nabla \times \boldsymbol{u}) - \boldsymbol{u} \cdot (\nabla \times \boldsymbol{v}).
\end{aligned}
$$

6. Using the relation $W_{ij} = -\epsilon_{ijk} w_k$, we have

$$
\begin{aligned}
(\nabla \cdot \boldsymbol{W}) &= -\epsilon_{ijk} \frac{\partial w_k}{\partial x_j} \boldsymbol{e}_i \\
&= -\nabla \times \boldsymbol{w}. \\
(\nabla \times \boldsymbol{W})_{ij} &= \epsilon_{imn} \frac{\partial W_{jn}}{\partial x_m} \\
&= -\epsilon_{imn} \epsilon_{jnr} \frac{\partial w_r}{\partial x_m} \\
&= (\delta_{ij} \delta_{mr} - \delta_{ir} \delta_{mj}) \frac{\partial u_r}{\partial x_m} \\
&= \frac{\partial u_r}{\partial x_r} \delta_{ij} - \frac{\partial u_i}{\partial x_j},
\end{aligned}
$$

which is the indicial version of Eqn. (1.272).

1.10 The Exponential and Logarithmic Functions

The exponential of a tensor $\boldsymbol{T}(t)$ can be defined either in terms of its series representation as

$$
e^{\boldsymbol{T}(t)} := \boldsymbol{I} + \boldsymbol{T}(t) + \frac{1}{2!}[\boldsymbol{T}(t)]^2 + \cdots, \tag{1.273}
$$

or in terms of a solution of the initial value problem

$$
\dot{\boldsymbol{X}}(\xi) = \boldsymbol{T}(t)\boldsymbol{X}(\xi) = \boldsymbol{X}(\xi)\boldsymbol{T}(t), \quad \xi > 0, \tag{1.274}
$$

$$
\boldsymbol{X}(0) = \boldsymbol{I}, \tag{1.275}
$$

for the tensor function $\boldsymbol{X}(\xi)$, where the parameter ξ is independent of t. Note that the superposed dot in the above equation denotes differentiation with respect to ξ. The existence theorem for linear differential equations tells us that this problem has exactly one solution $\boldsymbol{X} : [0, \infty) \to \text{Lin}$, which we write in the form

$$
\boldsymbol{X}(\xi) = e^{\boldsymbol{T}(t)\xi}.
$$

From Eqn. (1.273), it is immediately evident that

$$e^{[T(t)]^T} = (e^{T(t)})^T, \tag{1.276}$$

and that if $A \in \text{Lin}$ is invertible, then $e^{(A^{-1}BA)} = A^{-1}e^{B}A$ for all $B \in \text{Lin}$.

Theorem 1.10.1. *For each $t \geq 0$, $e^{T(t)}$ belongs to Lin^+, and*

$$\det(e^{T(t)}) = e^{tr\, T(t)}. \tag{1.277}$$

Proof. If (λ_i, n) is an eigenvalue/eigenvector pair of $T(t)$, then from Eqn. (1.273), it follows that $(e^{\lambda_i(t)}, n)$ is an eigenvalue/eigenvector pair of $e^{T(t)}$. Hence, the determinant of $e^{T(t)}$, which is just the product of the eigenvalues, is given by

$$\det(e^{T(t)}) = \Pi_{i=1}^{n} e^{\lambda_i(t)} = e^{\sum_{i=1}^{n} \lambda_i(t)} = e^{\text{tr}\, T(t)}.$$

Since $e^{\text{tr}\, T(t)} > 0$ for all t, $e^{T(t)} \in \text{Lin}^+$. □

From Eqn. (1.277), it directly follows that

$$\det(e^{A}e^{B}) = \det(e^{A})\det(e^{B}) = e^{\text{tr}\, A}e^{\text{tr}\, B} = e^{\text{tr}\,(A+B)} = \det(e^{A+B}).$$

Theorem 1.10.2. *Let $A(t), B(t) \in Lin$. If*

$$e^{[A(t)+B(t)]\xi} = e^{A(t)\xi}e^{B(t)\xi} \quad \forall \xi \tag{1.278}$$

then $A(t)B(t) = B(t)A(t)$. Conversely, if $A(t)B(t) = B(t)A(t)$, then

$$e^{[A(t)+B(t)]\xi} = e^{A(t)\xi}e^{B(t)\xi} = e^{B(t)\xi}e^{A(t)\xi} \quad \forall \xi \tag{1.279}$$

Proof. We shall write $A(t)$ and $B(t)$ simply as A and B for notational convenience. Let $X_A(\xi) := e^{A\xi}$, $X_B(\xi) := e^{B\xi}$, and $X_{A+B} := e^{(A+B)\xi}$. Then, we have

$$\dot{X}_{A+B} = (A+B)X_{A+B}, \tag{1.280}$$

$$\begin{aligned} \overline{X_A X_B}^{\,\cdot} &= \dot{X}_A X_B + X_A \dot{X}_B \\ &= AX_A X_B + X_A B X_B, \end{aligned} \tag{1.281}$$

where, as before, the superposed dot denotes differentiation with respect to ξ. Let $Z := X_{A+B} - X_A X_B$. From Eqns. (1.280) and (1.281), we get

$$\dot{Z} = (A+B)X_{A+B} - AX_A X_B - X_A B X_B. \tag{1.282}$$

If $AB = BA$, then $X_A B = BX_A$, so that Eqn. (1.282) becomes

$$\dot{Z} = (A+B)\,[X_{A+B} - X_A X_B] = (A+B)Z. \tag{1.283}$$

The above differential equation is to be solved subject to the initial condition $Z(0) = 0$. The general solution of the differential equation is $Z = e^{(A+B)\xi}C$, where the tensor C is a constant tensor. On using the initial condition, we get $C = 0$, from which it follows that $Z = 0$, thus proving Eqn. (1.279). In an analogous fashion, by taking $Z := X_{A+B} - X_B X_A$, one can also show that if $AB = BA$, then $e^{(A+B)\xi} = e^{B\xi}e^{A\xi}$. Thus, Eqn. (1.279) follows.

Conversely, if Eqn. (1.278) holds, i.e., if $X_{A+B} = X_A X_B$, then Eqn. (1.282) reduces to

$$BX_A X_B = X_A B X_B \quad \forall \xi.$$

By Theorem 1.10.1, $X_B \in \mathrm{Lin}^+$. Hence multiplying the above equation by X_B^{-1}, we get $BX_A = X_A B$ for all ξ. Differentiating this relation with respect to ξ, and evaluating at $\xi = 0$, we get $BA = AB$. □

As a corollary of the above theorem, it follows, by taking $\xi = 1$ in Eqn. (1.279), that

$$A(t)B(t) = B(t)A(t) \implies e^{A(t)+B(t)} = e^{A(t)}e^{B(t)} = e^{B(t)}e^{A(t)}. \tag{1.284}$$

However, the converse of the above statement may not be true. Indeed, if $AB \neq BA$, one can have $e^{A+B} = e^A = e^B = e^A e^B$, or $e^A e^B = e^{A+B} \neq e^B e^A$ or even $e^A e^B = e^B e^A \neq e^{A+B}$ as the following examples show:

1. Wood [354] presents the following example:

$$A = 2\pi \begin{bmatrix} 0 & 0 & \frac{\sqrt{3}}{2} \\ 0 & 0 & -\frac{1}{2} \\ -\frac{\sqrt{3}}{2} & \frac{1}{2} & 0 \end{bmatrix}, \quad B = 2\pi \begin{bmatrix} 0 & 0 & -\frac{\sqrt{3}}{2} \\ 0 & 0 & -\frac{1}{2} \\ \frac{\sqrt{3}}{2} & \frac{1}{2} & 0 \end{bmatrix},$$

so that $e^{A+B} = e^A = e^B = e^A e^B = I$. The eigenvalues of A and B are $(0, 2\pi i, -2\pi i)$. Another example presented by Horn and Johnson [136] is

$$A = \begin{bmatrix} 0 & 0 & 0 & 0 \\ 0 & 0 & 0 & 0 \\ 0 & 0 & 0 & -2\pi \\ 0 & 0 & 2\pi & 0 \end{bmatrix}, \quad B = \begin{bmatrix} 0 & 0 & 1 & 0 \\ 0 & 0 & 0 & 1 \\ 0 & 0 & 0 & -2\pi \\ 0 & 0 & 2\pi & 0 \end{bmatrix},$$

for which, again, one has $e^{A+B} = e^A = e^B = e^A e^B = I$. The eigenvalues of A and B are $(0, 0, 2\pi i, -2\pi i)$.

2. Wermuth [350] presents the following example where $e^A e^B = e^{A+B} \neq e^B e^A$: if $z = a + ib$ is a solution of $e^z - z = 1$, e.g., $a = 2.088843\ldots$ and $b = 7.461489\ldots$, then for

$$A = \begin{bmatrix} 0 & 0 & 0 & 0 \\ 0 & 0 & 0 & 0 \\ 0 & 0 & a & -b \\ 0 & 0 & b & a \end{bmatrix}, \quad B = \begin{bmatrix} 0 & 0 & 1 & 0 \\ 0 & 0 & 0 & 1 \\ 0 & 0 & 0 & 0 \\ 0 & 0 & 0 & 0 \end{bmatrix},$$

we have

$$e^A = \begin{bmatrix} 1 & 0 & 0 & 0 \\ 0 & 1 & 0 & 0 \\ 0 & 0 & a+1 & -b \\ 0 & 0 & b & a+1 \end{bmatrix}, \quad e^B = \begin{bmatrix} 1 & 0 & 1 & 0 \\ 0 & 1 & 0 & 1 \\ 0 & 0 & 1 & 0 \\ 0 & 0 & 0 & 1 \end{bmatrix}, \tag{1.285}$$

so that

$$e^A e^B = e^{A+B} = \begin{bmatrix} 1 & 0 & 1 & 0 \\ 0 & 1 & 0 & 1 \\ 0 & 0 & a+1 & -b \\ 0 & 0 & b & a+1 \end{bmatrix}, \quad e^B e^A = \begin{bmatrix} 1 & 0 & a+1 & -b \\ 0 & 1 & b & a+1 \\ 0 & 0 & a+1 & -b \\ 0 & 0 & b & a+1 \end{bmatrix}.$$

3. An example for the case $e^A e^B = e^B e^A \neq e^{A+B}$, again provided in [136], is

$$A = \begin{bmatrix} 0 & -\pi & 0 & 0 \\ \pi & 0 & 0 & 0 \\ 0 & 0 & 0 & \pi \\ 0 & 0 & -\pi & 0 \end{bmatrix}, \quad B = \begin{bmatrix} 0 & 0 & 1 & 0 \\ 0 & 0 & 0 & 1 \\ 0 & 0 & 0 & 0 \\ 0 & 0 & 0 & 0 \end{bmatrix},$$

which yields

$$e^A = e^{A+B} = -I, \quad e^B = \begin{bmatrix} 1 & 0 & 1 & 0 \\ 0 & 1 & 0 & 1 \\ 0 & 0 & 1 & 0 \\ 0 & 0 & 0 & 1 \end{bmatrix},$$

so that $e^A e^B = e^B e^A \neq e^{A+B}$.

As an application of Eqn. (1.284), since $T(t)$ and $-T(t)$ commute, we have $e^{T(t)-T(t)} = I = e^{T(t)} e^{-T(t)}$. Thus,

$$(e^{T(t)})^{-1} = e^{-T(t)}. \tag{1.286}$$

In fact, one can extend this result to get

$$(e^{T(t)})^n = e^{nT(t)} \quad \forall \text{ integer } n.$$

As an application of Eqn. (1.286), consider the solution of the tensorial differential equation

$$\dot{X} + AX + XB^T = F(t), \tag{1.287}$$

subject to the initial condition $X(0) = C$. The differentiation is with respect to t, and A and B are assumed to be independent of t. Multiplying Eqn. (1.287) by $e^{At} \boxtimes e^{Bt}$, we get

$$(e^{At} \boxtimes e^{Bt})\dot{X} + A e^{At} X e^{B^T t} + e^{At} X B^T e^{B^T t} = e^{At} F(t) e^{B^T t}.$$

The above equation can be written as

$$\frac{d}{dt}\left[\left(e^{At}\boxtimes e^{Bt}\right)X\right]=e^{At}F(t)e^{B^Tt}.$$

Integrating the above equation between the limits $[0,t]$, we get

$$(e^{At}\boxtimes e^{Bt})X(t)=\int_0^t e^{A\xi}F(\xi)e^{B^T\xi}\,d\xi+C,$$

which, by virtue of Eqn. (1.286), leads to

$$\begin{aligned}X(t)&=\int_0^t e^{-A(t-\xi)}F(\xi)e^{-B^T(t-\xi)}\,d\xi+e^{-At}Ce^{-B^Tt}\\&=\int_0^t e^{-A\xi}F(t-\xi)e^{-B^T\xi}\,d\xi+e^{-At}Ce^{-B^Tt}.\end{aligned}\tag{1.288}$$

What we have effectively shown is that

$$e^{(A\boxtimes I+I\boxtimes B)t}=e^{At}\boxtimes e^{Bt}\quad\forall t.$$

The result given by Eqn. (1.288) also holds if $B=0$, and X, F and C are vectors. This observation allows us to solve the tensorial differential equation

$$\dot{X}+AXB=F(t),$$

which can be written as

$$\dot{X}+(A\boxtimes B^T)X=F(t),$$

Using the $\hat{\Psi}$ mapping of Appendix I, the above equation can be written as

$$\hat{\Psi}(\dot{X})+\hat{\Psi}(A\boxtimes B^T)\hat{\Psi}(X)=\hat{\Psi}(F),$$

whose solution is given by

$$\hat{\Psi}[X(t)]=\int_0^t e^{-\hat{\Psi}(A\boxtimes B^T)\xi}\hat{\Psi}[F(t-\xi)]\,d\xi+e^{-\hat{\Psi}(A\boxtimes B^T)t}\hat{\Psi}(C).$$

For the exponential of a skew-symmetric tensor, we have the following theorem:

Theorem 1.10.3. *Let $W(t)\in Skw$ for all t. Then $e^{W(t)}$ is a rotation for each $t\geq 0$. Conversely, for $R(t)\in Orth^+$, there exists a $W(t)\in Skw$ such that $R(t)=e^{W(t)}$.*

Proof. By Eqn. (1.286),

$$(e^{W(t)})^{-1}=e^{-W(t)}=e^{W^T(t)}=(e^{W(t)})^T,$$

where the last step follows from Eqn. (1.276). Thus, $e^{W(t)}$ is a orthogonal tensor. By Theorem 1.10.1, $\det(e^{W(t)})=e^{\operatorname{tr}W(t)}=e^0=1$, and hence $e^{W(t)}$ is a rotation.

The converse is proved simply by taking $W(t) = \log R(t)$, which we show to be a skew-symmetric tensor later in this section. □

In the three-dimensional case, by using the Cayley–Hamilton theorem, we get $W^3(t) = -|w(t)|^2 W(t)$, where $w(t)$ is the axial vector of $W(t)$. Thus, $W^4(t) = -|w(t)|^2 W^2(t)$, $W^5(t) = |w(t)|^4 W(t)$, and so on. Substituting these terms into the series expansion of the exponential function, and using the representations of sine and cosine functions, we get[2]

$$R(t) = e^{W(t)} = I + \frac{\sin(|w(t)|)}{|w(t)|} W(t) + \frac{[1-\cos(|w(t)|)]}{|w(t)|^2} W^2(t). \tag{1.289}$$

The directional derivative $DR(W)[U]$ can be obtained from the above expression using the fact that $|w|^2 = w \cdot w = W : W/2$, so that $2|w|\, D|w|\,(W)[U] = W : U$.

Another proof of Eqn. (1.289) is obtained by assuming $R(\xi) \equiv e^{W(t)\xi}$ to be of the form $h_1(\xi)I + h_2(\xi)W + h_3(\xi)W^2$, and then determining the unknown functions $h_1(\xi)$, $h_2(\xi)$, and $h_3(\xi)$ by using the governing equation $\dot{R} = RW$ (where the dot denotes differentiation with respect to ξ) subject to the initial conditions $R(0) = I$ and $\dot{R}(0) = W$. By using the fact that $W^3(t) = -|w(t)|^2 W(t)$, and that $\{I, W, W^2\}$ is a linearly independent set (see Problem 4), we get

$$\begin{aligned}
\dot{h}_1(\xi) &= 0,\\
\dot{h}_2(\xi) &= h_1 - |w(t)|^2 h_3(\xi),\\
\dot{h}_3(\xi) &= h_2(\xi),
\end{aligned}$$

which are to be solved subject to the initial conditions

$$\begin{aligned}
h_1(0) &= 1, & \dot{h}_1(0) &= 0,\\
h_2(0) &= 0, & \dot{h}_2(0) &= 1,\\
h_3(0) &= 0, & \dot{h}_3(0) &= 0.
\end{aligned}$$

We get the solution as

$$h_1(\xi) = 1, \qquad h_2(\xi) = \frac{\sin(|w(t)\xi|)}{|w(t)|}, \qquad h_3(\xi) = \frac{[1-\cos(|w(t)\xi|)]}{|w(t)|^2},$$

[2]Similarly, in the two-dimensional case, if

$$W(t) = \begin{bmatrix} 0 & \gamma(t) \\ -\gamma(t) & 0 \end{bmatrix},$$

where γ is a parameter which is a function of t, then

$$R(t) = e^{W(t)} = \cos\gamma(t) I + \frac{\sin\gamma(t)}{\gamma(t)} W(t) = \begin{bmatrix} \cos\gamma(t) & \sin\gamma(t) \\ -\sin\gamma(t) & \cos\gamma(t) \end{bmatrix}.$$

Let $\tilde{\gamma}(t) = \int_0^t \gamma(\xi)\, d\xi$. Then the solution of $dz/dt = W(t)z$ in this two-dimensional case is

$$z(t) = \begin{bmatrix} \cos\tilde{\gamma}(t) & \sin\tilde{\gamma}(t) \\ -\sin\tilde{\gamma}(t) & \cos\tilde{\gamma}(t) \end{bmatrix} z(0).$$

This result is a special case of the result in Theorem 1.10.4.

which, on setting $\xi = 1$, yields Eqn. (1.289).

Yet another proof is obtained by using Eqn. (1.292) with $\boldsymbol{T} \equiv \boldsymbol{W}(t)$, $\lambda = 0$, $\mu = i\,|\boldsymbol{w}(t)|$, $\nu = -i\,|\boldsymbol{w}(t)|$ and $\xi = 1$ (which implies that the eigenvalues of $e^{\boldsymbol{W}(t)}$ are $(1, e^{i|\boldsymbol{w}(t)|}, e^{-i|\boldsymbol{w}(t)|})$).

Not surprisingly, Eqn. (1.289) has the same form as Eqn. (1.104) with $\alpha = |\boldsymbol{w}(t)|$. Equation (1.289) is known as Rodrigues formula. If $\boldsymbol{W}(t)$ is a complex-valued skew-symmetric tensor, replace $|\boldsymbol{w}(t)|$ in Eqn. (1.289) by $\sqrt{\boldsymbol{w}(t)\cdot\boldsymbol{w}(t)}$ if it is nonzero; if it is zero, then $\boldsymbol{W}$ is nilpotent, and using the series expansion, we get $e^{\boldsymbol{W}(t)} = \boldsymbol{I} + \boldsymbol{W}(t) + \boldsymbol{W}^2(t)/2$. A generalization of Rodrigues formula to the case when $n \geq 4$ was developed by Gallier and Xu [86]; we shall present this formula later in this section (see Eqn. (1.295)), although we use a different method for deriving it than the one that they have used.

Let $a = \sin(|\boldsymbol{w}(t)|)/\,|\boldsymbol{w}(t)|$, $b = [1-\cos(|\boldsymbol{w}(t)|)]/\,|\boldsymbol{w}(t)|^2$ and $c = |\boldsymbol{w}(t)|^2$, and let $\boldsymbol{R}(t)$ be given by Eqn. (1.289). Using the facts $\dot{\boldsymbol{R}}\boldsymbol{R}^T + \boldsymbol{R}\dot{\boldsymbol{R}}^T = \boldsymbol{0}$, $2b - b^2c - a^2 = 0$, $\boldsymbol{W}\dot{\boldsymbol{W}}\boldsymbol{W} = -\dot{c}\boldsymbol{W}/2$, and $\boldsymbol{W}\dot{\boldsymbol{W}}\boldsymbol{W}^2 - \boldsymbol{W}^2\dot{\boldsymbol{W}}\boldsymbol{W} = \boldsymbol{0}$ (where now the superposed dot denotes differentiation with respect to t), we get

$$\dot{\boldsymbol{R}}\boldsymbol{R}^T = \frac{1}{c}(\boldsymbol{w}\cdot\dot{\boldsymbol{w}})(1-a)\boldsymbol{W} + a\dot{\boldsymbol{W}} + b(\boldsymbol{W}\dot{\boldsymbol{W}} - \dot{\boldsymbol{W}}\boldsymbol{W}), \tag{1.290a}$$

$$\boldsymbol{R}^T\dot{\boldsymbol{R}} = \frac{1}{c}(\boldsymbol{w}\cdot\dot{\boldsymbol{w}})(1-a)\boldsymbol{W} + a\dot{\boldsymbol{W}} - b(\boldsymbol{W}\dot{\boldsymbol{W}} - \dot{\boldsymbol{W}}\boldsymbol{W}). \tag{1.290b}$$

The axial vectors of $\dot{\boldsymbol{R}}\boldsymbol{R}^T$ and $\boldsymbol{R}^T\dot{\boldsymbol{R}}$ are, thus, $(\boldsymbol{w}\cdot\dot{\boldsymbol{w}})(1-a)\boldsymbol{w}/c + a\dot{\boldsymbol{w}} + b(\boldsymbol{w}\times\dot{\boldsymbol{w}})$ and $(\boldsymbol{w}\cdot\dot{\boldsymbol{w}})(1-a)\boldsymbol{w}/c + a\dot{\boldsymbol{w}} - b(\boldsymbol{w}\times\dot{\boldsymbol{w}})$, respectively. Of course, for the special case when $\boldsymbol{W}(t) = \boldsymbol{W}_0 t$, where $\boldsymbol{W}_0$ is independent of t, the above formulae reduce to $\dot{\boldsymbol{R}}\boldsymbol{R}^T = \boldsymbol{R}^T\dot{\boldsymbol{R}} = \boldsymbol{W}_0$. In this connection, one has the following theorem [354]:

Theorem 1.10.4. *For $\boldsymbol{T}(t) \in Lin$,*

$$\frac{d(e^{\boldsymbol{T}(t)})}{dt} = \dot{\boldsymbol{T}}e^{\boldsymbol{T}(t)} = e^{\boldsymbol{T}(t)}\dot{\boldsymbol{T}}, \tag{1.291}$$

if and only if $\dot{\boldsymbol{T}}\boldsymbol{T} = \boldsymbol{T}\dot{\boldsymbol{T}}$. Thus, the solution of $\dot{\boldsymbol{Z}} = \boldsymbol{T}(t)\boldsymbol{Z}$ is $\boldsymbol{Z}(t) = \left(e^{\int_0^t \boldsymbol{T}(\xi)\,d\xi}\right)\boldsymbol{Z}(0)$ if and only if $\dot{\boldsymbol{T}}\boldsymbol{T} = \boldsymbol{T}\dot{\boldsymbol{T}}$.

Proof. If $\dot{\boldsymbol{T}}\boldsymbol{T} = \boldsymbol{T}\dot{\boldsymbol{T}}$, then $\dot{\overline{\boldsymbol{T}^2}} = \dot{\boldsymbol{T}}\boldsymbol{T} + \boldsymbol{T}\dot{\boldsymbol{T}} = 2\dot{\boldsymbol{T}}\boldsymbol{T} = 2\boldsymbol{T}\dot{\boldsymbol{T}}$, and so on for derivatives of higher powers of $\boldsymbol{T}$. Thus, by simply differentiating Eqn. (1.273), we get Eqn. (1.291).

To prove the converse, note from Eqn. (1.273) that $e^{\boldsymbol{T}}\boldsymbol{T} = \boldsymbol{T}e^{\boldsymbol{T}}$. Differentiating both sides of this relation, and using the fact that Eqn. (1.291) holds, we get

$$\dot{\boldsymbol{T}}e^{\boldsymbol{T}}\boldsymbol{T} = \boldsymbol{T}\dot{\boldsymbol{T}}e^{\boldsymbol{T}}.$$

Noting again that $e^{\boldsymbol{T}}\boldsymbol{T} = \boldsymbol{T}e^{\boldsymbol{T}}$, and multiplying both sides by $(e^{\boldsymbol{T}})^{-1} = e^{-\boldsymbol{T}}$, we get the desired result.

By replacing $\boldsymbol{T}(t)$ by $\int_0^t \boldsymbol{T}(\xi)\,d\xi$ in the first relation in Eqn. (1.291), and post-multiplying by $\boldsymbol{Z}(0)$, we get the stated solution of the differential equation $\dot{\boldsymbol{Z}} = \boldsymbol{T}(t)\boldsymbol{Z}$. □

In what follows $\boldsymbol{T}(t)$ is written simply as $\boldsymbol{T}$.

Explicit formulae for $e^{\boldsymbol{T}\xi}$ (where $\boldsymbol{T}$ can be complex-valued) when $n = 3$ and $n = 4$ are given by Cheng and Yau [52] in terms of the eigenvalues and powers of $\boldsymbol{T}$.

Theorem 1.10.5. *For $n = 3$, we have the following cases:*

1. $\lambda_1 = \lambda_2 = \lambda_3 \equiv \lambda$

   (a) *If $q(\boldsymbol{T}) = (\boldsymbol{T} - \lambda \boldsymbol{I})$, then*

   $$e^{\boldsymbol{T}\xi} = e^{\lambda\xi}\boldsymbol{I}.$$

   (b) *If $q(\boldsymbol{T}) = (\boldsymbol{T} - \lambda \boldsymbol{I})^2$, then*

   $$e^{\boldsymbol{T}\xi} = e^{\lambda\xi}\left[\xi(\boldsymbol{T} - \lambda\boldsymbol{I}) + \boldsymbol{I}\right].$$

   (c) *If $q(\boldsymbol{T}) = (\boldsymbol{T} - \lambda \boldsymbol{I})^3$, then*

   $$e^{\boldsymbol{T}\xi} = e^{\lambda\xi}\left[\frac{\xi^2}{2}(\boldsymbol{T} - \lambda\boldsymbol{I})^2 + \xi(\boldsymbol{T} - \lambda\boldsymbol{I}) + \boldsymbol{I}\right].$$

2. $\lambda_1 \equiv \mu \neq \lambda_2 = \lambda_3 \equiv \lambda$

   (a) *If $q(\boldsymbol{T}) = (\boldsymbol{T} - \lambda \boldsymbol{I})(\boldsymbol{T} - \mu \boldsymbol{I})$, then*

   $$e^{\boldsymbol{T}\xi} = \frac{e^{\lambda\xi} - e^{\mu\xi}}{\lambda - \mu}\boldsymbol{T} + \frac{\lambda e^{\mu\xi} - \mu e^{\lambda\xi}}{\lambda - \mu}\boldsymbol{I}.$$

   (b) *If $q(\boldsymbol{T}) = (\boldsymbol{T} - \mu \boldsymbol{I})(\boldsymbol{T} - \lambda \boldsymbol{I})^2$, then*

   $$e^{\boldsymbol{T}\xi} = \left[\frac{\xi e^{\lambda\xi}}{\lambda - \mu} - \frac{e^{\lambda\xi} - e^{\mu\xi}}{(\lambda - \mu)^2}\right](\boldsymbol{T} - \lambda\boldsymbol{I})^2 + e^{\lambda\xi}\left[\xi(\boldsymbol{T} - \lambda\boldsymbol{I}) + \boldsymbol{I}\right].$$

3. $\lambda_1 \equiv \lambda \neq \lambda_2 \equiv \mu \neq \lambda_3 \equiv \nu \neq \lambda_1$. *We have $q(\boldsymbol{T}) = (\boldsymbol{T} - \lambda\boldsymbol{I})(\boldsymbol{T} - \mu\boldsymbol{I})(\boldsymbol{T} - \nu\boldsymbol{I})$, and*

$$e^{\boldsymbol{T}\xi} = e^{\lambda\xi}\frac{(\boldsymbol{T} - \mu\boldsymbol{I})(\boldsymbol{T} - \nu\boldsymbol{I})}{(\lambda - \mu)(\lambda - \nu)} + e^{\mu\xi}\frac{(\boldsymbol{T} - \lambda\boldsymbol{I})(\boldsymbol{T} - \nu\boldsymbol{I})}{(\mu - \lambda)(\mu - \nu)} + e^{\nu\xi}\frac{(\boldsymbol{T} - \lambda\boldsymbol{I})(\boldsymbol{T} - \mu\boldsymbol{I})}{(\nu - \lambda)(\nu - \mu)}. \tag{1.292}$$

We now present two methods for the explicit determination of the exponential function for arbitrary n (for a detailed survey, see [225]). The first method is based on the Jordan canonical form given by Eqn. (J.37), while the second method is based on the definition given by Eqn. (1.274). The first method yields the formula

$$e^{\boldsymbol{T}\xi} = \sum_{i=1}^{k}\left[e^{\lambda_i\xi}\left(\boldsymbol{P}_i + \sum_{j=1}^{m_i-1}\frac{\xi^j}{j!}\boldsymbol{N}_i^j\right)\right], \tag{1.293}$$

where m_i is as in Eqn. (J.24). To prove this result, we first note that $\boldsymbol{P}_i$ and $\boldsymbol{N}_i$ commute since they are polynomials in $\boldsymbol{T}$. Using Eqns. (J.37) and (1.279), and the properties $\boldsymbol{P}_i\boldsymbol{P}_j = \delta_{ij}\boldsymbol{P}_i$ (no sum on i), $\boldsymbol{N}_i\boldsymbol{N}_j = \boldsymbol{P}_i\boldsymbol{N}_j = \boldsymbol{0}$ for $i \neq j$, $\boldsymbol{P}_i\boldsymbol{N}_i = \boldsymbol{N}_i$ (no sum on i), and $\boldsymbol{N}_i^j = \boldsymbol{0}$, $j \geq m_i$, we have

$$\begin{aligned} e^{\boldsymbol{T}\xi} &= \left(e^{\sum_{i=1}^k \lambda_i \boldsymbol{P}_i \xi}\right)\left(e^{\sum_{i=1}^k \boldsymbol{N}_i \xi}\right) \\ &= \left[\sum_{i=1}^k e^{\lambda_i \xi}\boldsymbol{P}_i\right]\left[\boldsymbol{I} + \sum_{i=1}^k \sum_{j=1}^{m_i-1} \frac{\xi^j}{j!}\boldsymbol{N}_i^j\right] \\ &= \sum_{i=1}^k \left[e^{\lambda_i \xi}\left(\boldsymbol{P}_i + \sum_{j=1}^{m_i-1} \frac{\xi^j}{j!}\boldsymbol{N}_i^j\right)\right]. \end{aligned}$$

As an example, if $\boldsymbol{T}$ is given by Eqn. (J.39), we get

$$e^{\boldsymbol{T}} = e^{\alpha}\boldsymbol{P}_1 + e^{\beta}(\boldsymbol{P}_2 + \boldsymbol{N}_2) = \begin{bmatrix} e^{\alpha} & \frac{e^{\alpha}-e^{\beta}}{\alpha-\beta} & \frac{e^{\alpha}+(\beta-\alpha-1)e^{\beta}}{(\alpha-\beta)^2} \\ 0 & e^{\beta} & e^{\beta} \\ 0 & 0 & e^{\beta} \end{bmatrix}.$$

The exponential tensor $e^{\boldsymbol{T}}$ for a diagonalizable (in particular, symmetric) tensor $\boldsymbol{T}$ is obtained either as a special case of Eqn. (1.293), or by substituting Eqn. (J.25) into Eqn. (1.273), and is given by

$$e^{\boldsymbol{T}\xi} = \sum_{i=1}^k e^{\lambda_i \xi}\boldsymbol{P}_i, \tag{1.294}$$

with $\boldsymbol{P}_i$ given by Eqn. (J.26). The method proposed by Putzer [259] essentially yields the same result as that given by Eqn. (1.293), although more indirectly.

Writing λ_m as $(\lambda_r)_m + i(\lambda_s)_m$, we can write Eqn. (1.293) as

$$e^{\boldsymbol{T}} = \sum_{m=1}^k e^{(\lambda_r)_m}\left[\sin((\lambda_s)_m)\boldsymbol{B}_m - \cos((\lambda_s)_m)\boldsymbol{B}_m^2 + [\cos((\lambda_s)_m) + i\sin((\lambda_s)_m)]\,\boldsymbol{G}_m\right],$$

where $\boldsymbol{B}_m = i\boldsymbol{P}_m$ and $\boldsymbol{G}_m = \sum_{j=1}^{n_m-1}\frac{1}{j!}\boldsymbol{N}_m^j$. Note that $\boldsymbol{B}_m^3 = -i\boldsymbol{P}_m = -\boldsymbol{B}_m$. The above form allows us to write a Rodrigues-type formula for any dimension n when $\boldsymbol{T}$ is a skew-symmetric tensor [86]. As discussed earlier, the eigenvalues of a real-valued $\boldsymbol{W} \in$ Skw are either zero or of the form $\pm i\lambda_m$, $m = 1, 2, \ldots, k$, where k is the number of distinct nonzero eigenvalues. Also, since $\boldsymbol{W}$ is normal, it is diagonalizable. Let $\boldsymbol{B}_m := i[\boldsymbol{P}_m^{(1)} - \boldsymbol{P}_m^{(2)}]$, where $\boldsymbol{P}_m^{(1)}$ and $\boldsymbol{P}_m^{(2)}$ are the projections associated with the eigenvalues $i\lambda_m$ and $-i\lambda_m$, respectively. Since $\boldsymbol{B}_m^2 = -[\boldsymbol{P}_m^{(1)} + \boldsymbol{P}_m^{(2)}]$, it follows that $\boldsymbol{I} = -\sum_{m=1}^k \boldsymbol{B}_m^2$. From Eqn. (1.294), we get

$$e^{\boldsymbol{W}} = \boldsymbol{I} + \sum_{i=1}^k \left[\sin(\lambda_i)\boldsymbol{B}_i + [1 - \cos(\lambda_i)]\,\boldsymbol{B}_i^2\right], \tag{1.295}$$

where $W = \sum_{i=1}^{k} \lambda_i B_i$, with each B_i being a real-valued skew tensor. The λ_i are determined by taking the square roots of the eigenvalues of $-W^2$, while the matrices B_i are determined by solving the following system of k equations

$$\begin{aligned} W &= \sum_{i=1}^{k} \lambda_i B_i, \\ -W^3 &= \sum_{i=1}^{k} \lambda_i^3 B_i, \\ &\dots, \\ (-1)^{k+1} W^{2k-1} &= \sum_{i=1}^{k} \lambda_i^{2k-1} B_i, \end{aligned}$$

by inverting the van der Monde matrix [153]

$$\begin{bmatrix} \lambda_1 & \lambda_2 & \dots & \lambda_k \\ \lambda_1^3 & \lambda_2^3 & \dots & \lambda_k^3 \\ \cdots & \cdots & \cdots & \cdots \\ \cdots & \cdots & \cdots & \cdots \\ \lambda_1^{2k-1} & \lambda_2^{2k-1} & \dots & \lambda_k^{2k-1} \end{bmatrix},$$

to obtain

$$B_i(W) = \begin{cases} (-1)^{k+1} \frac{W}{\lambda_i} \prod_{\substack{j=1 \\ j \neq i}}^{k} \frac{W^2 + \lambda_j^2 I}{\lambda_i^2 - \lambda_j^2}, & k > 1 \\ \frac{W}{\lambda}, & k = 1. \end{cases}$$

An identical procedure can be used to find the exponential of a skew-Hermitian tensor (which is unitary) since, again, in this case, the real part of the eigenvalues and the G_i's are zero. We have already noted that some complex-valued skew-symmetric tensors can be nilpotent, and hence, no Rodrigues-type formula can be derived for its exponential (which is an orthogonal tensor), in general.

The second method that we discuss is the one devised by Fulmer [85]. Note that since $d^k/d\xi^k(e^{T\xi}) = T^k e^{T\xi}$, for $k = 1, 2, \dots, n$, we have

$$e^{T\xi}\Big|_{\xi=0} = I \text{ and } \left[\frac{d^k}{d\xi^k}(e^{T\xi})\right]_{\xi=0} = T^k. \tag{1.296}$$

Also, by virtue of the Cayley–Hamilton theorem,

$$(T^n - I_1 T^{n-1} + \dots + (-1)^n I_n I)e^{T\xi} = 0,$$

or, alternatively,

$$\left(\frac{d^n}{d\xi^n} - I_1 \frac{d^{n-1}}{d\xi^{n-1}} + \dots + (-1)^n I_n\right) e^{T\xi} = 0. \tag{1.297}$$

As a result of Eqns. (1.296) and (1.297), we see that $e^{T\xi}$ is the unique solution of the nth order initial value problem

$$DG(\xi) = 0, \quad G(0) = I, \quad \left.\frac{d^k G(\xi)}{d\xi^k}\right|_{\xi=0} = T^k, \; k = 1, 2, \ldots, n-1, \tag{1.298}$$

where D is the differential operator in parenthesis in Eqn. (1.297).

If T has the characteristic polynomial given by Eqn. (J.23), then the general solution of $DG(\xi) = \mathbf{0}$ is

$$G(\xi) = e^{\lambda_1 \xi}(C_{11} + \xi C_{12} + \cdots + \xi^{n_1 - 1} C_{1n_1}) + \cdots + e^{\lambda_k \xi}(C_{k1} + \xi C_{k2} + \cdots + \xi^{n_k - 1} C_{kn_k}), \tag{1.299}$$

where the C_{ij} are n matrices to be determined from the n initial conditions

$$\begin{aligned}
I &= C_{11} + C_{21} + \cdots + C_{k1}, \\
T &= (\lambda_1 C_{11} + C_{12}) + \cdots + (\lambda_k C_{k1} + C_{k2}), \\
T^2 &= (\lambda_1^2 C_{11} + 2\lambda_1 C_{12}) + \cdots + (\lambda_k^2 C_{k1} + 2\lambda_k C_{k2}), \\
&\ldots \\
T^{n-1} &= \left(\lambda_1^{n-1} C_{11} + (n-1)\lambda_1^{n-2} C_{12} + \cdots + \frac{(n-1)!}{(n-n_1)!}\lambda_1^{n-n_1} C_{1n_1}\right) + \cdots \\
&\quad + \left(\lambda_k^{n-1} C_{k1} + (n-1)\lambda_k^{n-2} C_{k2} + \cdots + \frac{(n-1)!}{(n-n_k)!}\lambda_k^{n-n_k} C_{kn_k}\right).
\end{aligned}$$

The C_{ij} are obtained as polynomials in T by inverting the $n \times n$ coefficient matrix in the above equations; substituting these expressions into Eqn. (1.299), we get the desired expression for $e^{T\xi}$. Note that the above method is based on the characteristic polynomial[3], whereas the first method is based on the minimal polynomial.

As an illustration, let us again find the expression for e^T, with T given by Eqn. (J.39). Since the characteristic polynomial is given by $(T - \alpha I)(T - \beta I)^2$, the general solution is $G(\xi) = e^{\alpha\xi} C_{11} + e^{\beta\xi}(C_{21} + \xi C_{22})$, where the C_{ij} are obtained from the initial conditions

$$\begin{aligned}
I &= C_{11} + C_{21}, \\
T &= \alpha C_{11} + \beta C_{21} + C_{22}, \\
T^2 &= \alpha^2 C_{11} + \beta^2 C_{21} + 2\beta C_{22}.
\end{aligned}$$

Solving for C_{ij} and substituting into $e^T = e^\alpha C_{11} + e^\beta (C_{21} + C_{22})$, we get the same expression as before.

Now we discuss the differentiation of e^T with respect to T, where $T \in \text{Lin}$, for any underlying space dimension n; the treatment is based on [155], alternative treatments may be found in [143] and [144]. We shall first give an alternate derivation of a formula that

[3] Actually, the method can be modified quite easily to work with the minimal polynomial as well–in this case the number of C_{ij} matrices to be determined is $m = \sum_{i=1}^{k} m_i$ instead of n. This is particularly advantageous, for example, if T is diagonalizable, since then $m_i = 1$ for $i = 1, 2, \ldots, k$, so that $C_{ij} = \mathbf{0}$ for $i = 1, 2, \ldots, k$ and $j \geq 2$.

appears in Mathias [211] and Ortiz et al. [241]. Let $H \equiv \partial X/\partial T$, where $X = e^{T\xi}$. Differentiating the expression $\dot{X} = TX$ with respect to T using Eqn. (1.233), we get

$$\dot{H} = (T \boxtimes I)H + I \boxtimes X^T.$$

Multiplying this equation by the "integrating factor" $e^{-T\xi} \boxtimes I$, we get

$$\frac{d}{d\xi}\left[(e^{-T\xi} \boxtimes I)H\right] = e^{-T\xi} \boxtimes (e^{T\xi})^T.$$

Integrating this equation subject to the condition $H|_{\xi=0} = \mathbf{0}$, and using Eqn. (1.286), we get

$$\begin{aligned} H(\xi) &= (e^{T\xi} \boxtimes I) \int_0^{\xi} \left[e^{-T\eta} \boxtimes (e^{T\eta})^T\right] d\eta \\ &= \int_0^{\xi} \left[e^{T(t-\eta)} \boxtimes (e^{T\eta})^T\right] d\eta. \end{aligned}$$

Similarly, by differentiating $\dot{X} = XT$, we also get

$$H(\xi) = \int_0^{xi} \left[e^{T\eta} \boxtimes e^{T^T(t-\eta)}\right] d\eta.$$

Since $\partial e^T/\partial T = H|_{\xi=1}$, we obtain

$$\begin{aligned} \frac{\partial(e^T)}{\partial T} &= \int_0^1 \left[e^{T(1-\eta)} \boxtimes (e^{T\eta})^T\right] d\eta \\ &= \int_0^1 \left[e^{T\eta} \boxtimes \left(e^{T(1-\eta)}\right)^T\right] d\eta. \end{aligned} \tag{1.300}$$

Explicit expressions for $e^{T\eta}$ such as those given by Eqn. (1.293) (or those presented in Theorem 1.10.5 for $n = 3$) should be substituted into the above expressions to get explicit formulae for $\partial e^T/\partial T$. Note that these formulae are valid for any $T \in \text{Lin}$ and any n.

Simpler expressions for a diagonalizable tensor T can be obtained using Eqns. (1.294) and (1.300), or directly by using Eqn. (1.255) with $f(\lambda_i) = e^{\lambda_i}$. The final result that we obtain for a diagonalizable tensor (for any n) is $e^{\lambda}\mathbb{I}$ for $k = 1$, and

$$\frac{\partial(e^T)}{\partial T} = \left[\sum_{i=1}^{k} e^{\lambda_i} P_i \boxtimes P_i^T + \sum_{i=1}^{k}\sum_{\substack{j=1\\ j\neq i}}^{k} \frac{e^{\lambda_i} - e^{\lambda_j}}{\lambda_i - \lambda_j} P_i \boxtimes P_j^T\right], \quad k > 1$$

where k is the number of distinct eigenvalues of T, and P_i is given by Eqn. (J.26). When $T \in \text{Sym}$, we obviously have $P_i^T = P_i$, and the above result should be pre- and post-multiplied by $\mathbb{S}$.

In what follows, we shall continue to denote $T(t)$ as T; the treatment is based on [158]. Recall that difficulties arise in defining a unique scalar-valued logarithmic function since $e^{\lambda} = e^{\lambda + 2\pi i n}$ where n is an integer. Similar difficulties arise in defining the logarithm of a tensor. If one defines the solution X of $e^X = T$ as $\log T$, then a non-singular T may have an infinite number of real and complex logarithms. In order to avoid this nonuniqueness and ensure that the logarithm of a real tensor is real, we define the principal logarithmic function, denoted by $\log \boldsymbol{T}$, of a matrix $\boldsymbol{T}$ with no eigenvalues that are negative or zero (i.e. $\lambda_i \notin (-\infty, 0]$, $i = 1, 2, \ldots, n$), as the unique function such that $e^{\log \boldsymbol{T}} = \boldsymbol{T}$. The eigenvalues of $\log \boldsymbol{T}$ have imaginary parts that lie strictly between $-\pi$ and π, since as we shall show below, if $re^{i\theta}$, $-\pi < \theta < \pi$, are the eigenvalues of $\boldsymbol{T}$, then $\log(re^{i\theta}) = \log r + i\theta$, $-\pi < \theta < \pi$, are the eigenvalues of $\log \boldsymbol{T}$. Thus, if λ_i denote the eigenvalues of $\boldsymbol{T}$, then

1. $e^{\log \boldsymbol{T}} = \boldsymbol{T}$ if and only if $\lambda_i \notin (-\infty, 0]$ $\forall i$.
2. $\log e^{\boldsymbol{T}} = \boldsymbol{T}$ if and only if $-\pi < \operatorname{Im}(\lambda_i) < \pi$ $\forall i$.

In series form the logarithmic function is given by

$$\log(\boldsymbol{I} + \boldsymbol{T}) = \sum_{i=1}^{\infty} \frac{(-1)^{i-1}}{i} (\boldsymbol{T})^i, \tag{1.301}$$

or, alternatively as

$$\log(\boldsymbol{T}) = \sum_{i=1}^{\infty} \frac{(-1)^{i-1}}{i} (\boldsymbol{T} - \boldsymbol{I})^i. \tag{1.302}$$

The series given by Eqn. (1.301) is absolutely convergent only if $\|\boldsymbol{T}\| < 1$. It is immediately evident from the above representation that $\log(\boldsymbol{I}) = \boldsymbol{0}$ and $\log(\boldsymbol{T}^T) = (\log \boldsymbol{T})^T$. Since, for all invertible $\boldsymbol{A} \in \mathrm{Lin}$, $e^{\boldsymbol{A}^{-1}\boldsymbol{B}\boldsymbol{A}} = \boldsymbol{A}^{-1} e^{\boldsymbol{B}} \boldsymbol{A}$ for all $\boldsymbol{B} \in \mathrm{Lin}$, we get $\log(\boldsymbol{A}^{-1}\boldsymbol{B}\boldsymbol{A}) = \boldsymbol{A}^{-1}(\log \boldsymbol{B})\boldsymbol{A}$.

If $\boldsymbol{H}(\xi) : [0, 1] \to \mathrm{Lin}$ is a tensor-valued function of ξ with no eigenvalues on the closed negative real axis for any $\xi \in [0, 1]$, an alternative definition can be given as the solution of the differential equation

$$\boldsymbol{H}(\xi) \frac{d\boldsymbol{X}}{d\xi} = \frac{d\boldsymbol{H}}{d\xi}, \quad \boldsymbol{X}(0) = \boldsymbol{0}, \tag{1.303}$$

with the solution denoted by $\boldsymbol{X}(\xi) = \log \boldsymbol{H}$. For example, if $\boldsymbol{H}(\xi) = \boldsymbol{I} + (\boldsymbol{T} - \boldsymbol{I})\xi$, with no eigenvalues of $\boldsymbol{T}$ on the closed negative real axis, then, from the above definition, we get

$$\log[\boldsymbol{I} + (\boldsymbol{T} - \boldsymbol{I})\xi] = \int_0^{\xi} [\boldsymbol{I} + (\boldsymbol{T} - \boldsymbol{I})\eta]^{-1} (\boldsymbol{T} - \boldsymbol{I}) \, d\eta. \tag{1.304}$$

If one violates the restriction on the eigenvalues of $\boldsymbol{T}$, then, from the above definition, one can get a complex-valued logarithm even for a real tensor as is seen, for example, by taking $\boldsymbol{T} = \operatorname{diag}[1, -1, 1]$. Let $(\lambda, \boldsymbol{n})$ denote an eigenvalues/eigenvector pair of $\boldsymbol{T}$. Then, by using the fact that

$$\boldsymbol{I}\boldsymbol{n} = [\boldsymbol{I} + (\boldsymbol{T} - \boldsymbol{I})\eta]^{-1} [\boldsymbol{I} + (\boldsymbol{T} - \boldsymbol{I})\eta] \boldsymbol{n} = (1 + (\lambda - 1)\eta) [\boldsymbol{I} + (\boldsymbol{T} - \boldsymbol{I})\eta]^{-1} \boldsymbol{n},$$

we get from Eqn. (1.304),

$$\log[\boldsymbol{I} + (\boldsymbol{T} - \boldsymbol{I})\xi]\boldsymbol{n} = \left[\int_0^{\xi} \frac{\lambda - 1}{1 + (\lambda - 1)\eta} \, d\eta \right] \boldsymbol{n} = \log[1 + (\lambda - 1)\xi] \boldsymbol{n},$$

so that $(\log[1+(\lambda-1)\xi], \boldsymbol{n})$ are the corresponding eigenvalue/eigenvector of $\log[\boldsymbol{I}+(\boldsymbol{T}-\boldsymbol{I})\xi]$. In particular, for $\xi=1$, $(\log\lambda, \boldsymbol{n})$ are the eigenvalue/eigenvector pair of $\log\boldsymbol{T}$. Similarly, one can show that $(-\log\lambda, \boldsymbol{n})$ are the eigenvalue/eigenvector pair of $\log(\boldsymbol{T}^{-1})$. Analogous to Eqn. (1.277), we get

$$\text{tr}\,\log\boldsymbol{T} = \sum_{i=1}^{n}\log\lambda_i = \log(\det\boldsymbol{T}). \tag{1.305}$$

We have the following analogue of Theorem 1.10.2:

Theorem 1.10.6. *Let* **A**, **B** *be such that no eigenvalues of* **A**, **B** *and* **AB** *are on the closed negative real axis. If*

$$\log\{[\boldsymbol{I}+(\boldsymbol{A}-\boldsymbol{I})\xi][\boldsymbol{I}+(\boldsymbol{B}-\boldsymbol{I})\xi]\} = \log[\boldsymbol{I}+(\boldsymbol{A}-\boldsymbol{I})\xi]+\log[\boldsymbol{I}+(\boldsymbol{B}-\boldsymbol{I})\xi] \quad \forall\xi\in[0,1] \tag{1.306}$$

then $\boldsymbol{AB}=\boldsymbol{BA}$.

Conversely, if $\boldsymbol{AB}=\boldsymbol{BA}$ *(so that* **A**, **B** *and* **AB** *have a common set of eigenvectors), and if* $-\pi < Im(\log\alpha_i+\log\beta_i) < \pi$ *for each* i,[a] *where* α_i *and* β_i *denote the eigenvalues of* **A** *and* **B** *corresponding to the same eigenvector, then*

$$\log\{[\boldsymbol{I}+(\boldsymbol{A}-\boldsymbol{I})\xi][\boldsymbol{I}+(\boldsymbol{B}-\boldsymbol{I})\xi]\} = \log\{[\boldsymbol{I}+(\boldsymbol{B}-\boldsymbol{I})\xi][\boldsymbol{I}+(\boldsymbol{A}-\boldsymbol{I})\xi]\} = \log[\boldsymbol{I}+(\boldsymbol{A}-\boldsymbol{I})\xi]+\log[\boldsymbol{I}+(\boldsymbol{B}-\boldsymbol{I})\xi] \quad \forall\xi\in[0,1] \tag{1.307}$$

Proof. If Eqn. (1.306) holds, then by differentiating it with respect to ξ using Eqn. (1.303), we get

$$[\boldsymbol{I}+(\boldsymbol{B}-\boldsymbol{I})\xi]^{-1}[\boldsymbol{I}+(\boldsymbol{A}-\boldsymbol{I})\xi]^{-1}[\boldsymbol{A}+\boldsymbol{B}-2\boldsymbol{I}+2\xi(\boldsymbol{A}-\boldsymbol{I})(\boldsymbol{B}-\boldsymbol{I})] = [\boldsymbol{I}+(\boldsymbol{A}-\boldsymbol{I})\xi]^{-1}(\boldsymbol{A}-\boldsymbol{I})+[\boldsymbol{I}+(\boldsymbol{B}-\boldsymbol{I})\xi]^{-1}(\boldsymbol{B}-\boldsymbol{I}).$$

Differentiating the above relation once again with respect to ξ (using the fact that $d(\boldsymbol{T}^{-1})/d\xi = -\boldsymbol{T}^{-1}[d\boldsymbol{T}/d\xi]\boldsymbol{T}^{-1}$), and evaluating the resulting expression at $\xi=0$, we get $\boldsymbol{AB}=\boldsymbol{BA}$.

Conversely, given that $\boldsymbol{AB}=\boldsymbol{BA}$, using Eqn. (1.303), we have

$$\begin{aligned}
&\log\{[\boldsymbol{I}+(\boldsymbol{A}-\boldsymbol{I})\xi][\boldsymbol{I}+(\boldsymbol{B}-\boldsymbol{I})\xi]\}\\
&= \int_0^\xi [\boldsymbol{I}+(\boldsymbol{B}-\boldsymbol{I})\eta]^{-1}[\boldsymbol{I}+(\boldsymbol{A}-\boldsymbol{I})\eta]^{-1}[\boldsymbol{A}+\boldsymbol{B}-2\boldsymbol{I}+2\eta(\boldsymbol{A}-\boldsymbol{I})(\boldsymbol{B}-\boldsymbol{I})]\,d\eta\\
&= \int_0^\xi [\boldsymbol{I}+(\boldsymbol{B}-\boldsymbol{I})\eta]^{-1}[\boldsymbol{I}+(\boldsymbol{A}-\boldsymbol{I})\eta]^{-1}\{[\boldsymbol{I}+(\boldsymbol{A}-\boldsymbol{I})\eta](\boldsymbol{B}-\boldsymbol{I})+[\boldsymbol{I}+(\boldsymbol{B}-\boldsymbol{I})\eta]\\
&\qquad (\boldsymbol{A}-\boldsymbol{I})\}\,d\eta\\
&= l\int_0^\xi [\boldsymbol{I}+(\boldsymbol{B}-\boldsymbol{I})\eta]^{-1}(\boldsymbol{B}-\boldsymbol{I})\,d\eta + \int_0^\xi [\boldsymbol{I}+(\boldsymbol{A}-\boldsymbol{I})\eta]^{-1}(\boldsymbol{A}-\boldsymbol{I})\,d\eta\\
&= \log[\boldsymbol{I}+(\boldsymbol{A}-\boldsymbol{I})\xi]+\log[\boldsymbol{I}+(\boldsymbol{B}-\boldsymbol{I})\xi].
\end{aligned}$$

In a similar manner, one gets the same expression for $\log\{[\boldsymbol{I}+(\boldsymbol{B}-\boldsymbol{I})\xi][\boldsymbol{I}+(\boldsymbol{A}-\boldsymbol{I})\xi]\}$, leading to Eqn. (1.307). □

[a]This condition is imposed since, as stated, the imaginary part of the eigenvalues of the log of any tensor has to lie between $-\pi$ and π.

As a corollary, it follows by taking $\xi = 1$ in Eqn. (1.307), that

$$AB = BA \text{ and } -\pi < \text{Im}(\log \alpha_i + \log \beta_i) < \pi \; \forall i \implies \log(AB) = \log(BA) = \log A + \log B. \quad (1.308)$$

The above implication need not be true if $AB = BA$, but the constraints on the eigenvalues of A and B are violated. Cheng et al. [53] have presented the following (complex-valued) scalar example: if $a = b = e^{(\pi-\epsilon)i}$, where ϵ is small and positive, then

$$\log(ab) = -2\epsilon i \neq 2(\pi - \epsilon)i = \log(a) + \log(b).$$

We present a (real-valued) matrix example that is a modification of the example of Wood [354]. If

$$A = (\pi - 0.01)\begin{bmatrix} 0 & 0 & \frac{\sqrt{3}}{2} \\ 0 & 0 & -\frac{1}{2} \\ -\frac{\sqrt{3}}{2} & \frac{1}{2} & 0 \end{bmatrix}, \quad H = e^A \approx \begin{bmatrix} -0.499963 & 0.866004 & 0.00866011 \\ 0.866004 & 0.500012 & -0.00499992 \\ -0.00866011 & 0.00499992 & -0.99995 \end{bmatrix},$$

then

$$\log H \approx \begin{bmatrix} 0 & 0 & 2.71204 \\ 0 & 0 & -1.5658 \\ -2.71204 & 1.5658 & 0 \end{bmatrix}, \quad \log H^2 \approx \begin{bmatrix} 0 & 0 & -0.0173205 \\ 0 & 0 & 0.01 \\ 0.0173205 & -0.01 & 0 \end{bmatrix},$$

so that $\log H^2 \neq 2\log H$. The eigenvalues of $\log H$ and $\log H^2$ are (approximately) $(0, (\pi - 0.01)i, -(\pi - 0.01)i)$ and $(0, 0.02i, -0.02i)$, respectively.

The converse assertion of Eqn. (1.308) is true for dimension $n = 2$, as seen in [226, 227], and also for symmetric tensors (for any n), as is evident by taking the transpose. For $n = 4$, it may appear that by taking $X = e^A$, $Y = e^B$, where e^A and e^B are given by Eqn. (1.285), we have $\log(XY) = \log X + \log Y$, but $XY \neq YX$. However, since both A and $A + B$ do not satisfy the constraint $-\pi < \text{Im}(\lambda_i) < -\pi$, we have $\log(e^A) \neq A$ and $\log(e^{A+B}) \neq A + B$. Thus, the question of whether the converse assertion of Eqn. (1.308) is true for $n > 2$ remains unresolved.

Let T be a tensor with no eigenvalues λ_i on the closed negative real axis. Since T and T^{-1} commute, and since $\log \lambda_i + \log(\lambda_i)^{-1} = 0$, by taking $A \equiv T$ and $B \equiv T^{-1}$, we get

$$\log(T^{-1}) = -\log T. \quad (1.309)$$

If $R \in \text{Orth}^+$ is such that it does not have any eigenvalues on the closed negative real axis, i.e., if $R \in \text{Orth}^+/\text{Sym} + \{I\}$, then $\log R \in \text{Skw}$. This is proved by noting that

$$[\log R]^T = \log(R^T) = \log(R^{-1}) = -\log(R),$$

where the last step follows from Eqn. (1.309).

We now discuss the explicit determination of the logarithm of a tensor; an alternative method may be found in [34]. Analogous to Eqn. (J.38), one can prove that

$$[\boldsymbol{I}+(\boldsymbol{T}-\boldsymbol{I})\eta]^{-1} = \sum_{i=1}^{k} \frac{1}{[1+(\lambda_i-1)\eta]}\left[\boldsymbol{P}_i + \sum_{j=1}^{m_i-1}\left(\frac{-\eta \boldsymbol{N}_i}{[1+(\lambda_i-1)\eta]}\right)^j\right].$$

Substituting this relation into Eqn. (1.304), we get

$$\begin{aligned}\log[\boldsymbol{I}+(\boldsymbol{T}-\boldsymbol{I})\xi] &= \int_0^{\xi}\sum_{i=1}^{k}\left\{\frac{\lambda_i-1}{[1+(\lambda_i-1)\eta]}\boldsymbol{P}_i + \sum_{j=1}^{m_i-1}\frac{(-1)^{j-1}\eta^{j-1}\boldsymbol{N}_i^j}{[1+(\lambda_i-1)\eta]^{j+1}}\right\}d\eta \\ &= \sum_{i=1}^{k}\left\{\log[1+(\lambda_i-1)\xi]\boldsymbol{P}_i + \sum_{j=1}^{m_i-1}\frac{(-1)^{j-1}\xi^j\boldsymbol{N}_i^j}{j[1+(\lambda_i-1)\xi]^j}\right\}. \end{aligned} \tag{1.310}$$

In particular, for $\xi = 1$, we get

$$\log \boldsymbol{T} = \sum_{i=1}^{k}\left\{(\log\lambda_i)\boldsymbol{P}_i + \sum_{j=1}^{m_i-1}\frac{(-1)^{j-1}\boldsymbol{N}_i^j}{j\lambda_i^j}\right\}. \tag{1.311}$$

An alternative derivation of Eqn. (1.310) can be given as follows. Since

$$\{\boldsymbol{P}_1, \boldsymbol{P}_2, \ldots, \boldsymbol{P}_k, \boldsymbol{N}_1, \ldots, \boldsymbol{N}_1^{m_1-1}, \ldots, \boldsymbol{N}_k, \ldots, \boldsymbol{N}_k^{m_k-1}\}$$

constitutes a basis for any tensor, we write

$$\log[\boldsymbol{I}+(\boldsymbol{T}-\boldsymbol{I})\xi] = \sum_{i=1}^{k}\left[f_i(\xi)\boldsymbol{P}_i + \sum_{j=1}^{m_i-1} f_{ij}(\xi)(\boldsymbol{N}_i)^j\right],$$

where the $f_i(\xi)$ and $f_{ij}(\xi)$ are functions to be determined. Substituting into the definition given by Eqn. (1.303), and using the linear independence of the basis, we get for each i the equations

$$\begin{aligned}&[1+(\lambda_i-1)\xi]\dot{f}_i = (\lambda_i-1),\\ &[1+(\lambda_i-1)\xi]\dot{f}_{i1} + \xi\dot{f}_i = 1,\\ &[1+(\lambda_i-1)\xi]\dot{f}_{i2} + \xi\dot{f}_{i1} = 0,\\ &\vdots\\ &[1+(\lambda_i-1)\xi]\dot{f}_{i(m_i-1)} + \xi\dot{f}_{i(m_i-2)} = 0,\end{aligned}$$

which are to be solved subject to the condition that $f_i(0) = f_{ij}(0) = 0$. We first solve for f_i using the first equation, then for f_{i1} using the second one, and so on. The solution is given by $f_i = \log[1+(\lambda_i-1)\xi]$ and $f_{ij} = (-1)^{j-1}\xi^j/[j(1+(\lambda_i-1)\xi)^j]$, which yields Eqn. (1.310).

For a diagonalizable tensor $\boldsymbol{T}$, Eqn. (1.311) simplifies to

$$\log \boldsymbol{T} = \sum_{i=1}^{k}(\log\lambda_i)\boldsymbol{P}_i, \tag{1.312}$$

with $\boldsymbol{P}_i$ given by Eqn. (J.26). By specializing Eqn. (1.311) to the case $n = 3$, we get the following analogue of Theorem 1.10.5:

Theorem 1.10.7. *For $n = 3$, we have the following cases:*

1. $\lambda_1 = \lambda_2 = \lambda_3 \equiv \lambda$
   (a) *If* $q(\boldsymbol{T}) = (\boldsymbol{T} - \lambda \boldsymbol{I})$, *then*
   $$\log \boldsymbol{T} = (\log \lambda)\boldsymbol{I}.$$
   (b) *If* $q(\boldsymbol{T}) = (\boldsymbol{T} - \lambda \boldsymbol{I})^2$, *then*
   $$\log \boldsymbol{T} = (\log \lambda)\boldsymbol{I} + \frac{\boldsymbol{T} - \lambda \boldsymbol{I}}{\lambda}.$$
   (c) *If* $q(\boldsymbol{T}) = (\boldsymbol{T} - \lambda \boldsymbol{I})^3$, *then*
   $$\log \boldsymbol{T} = (\log \lambda)\boldsymbol{I} + \frac{\boldsymbol{T} - \lambda \boldsymbol{I}}{\lambda} - \frac{(\boldsymbol{T} - \lambda \boldsymbol{I})^2}{2\lambda^2}.$$
2. $\lambda_1 \equiv \mu \neq \lambda_2 = \lambda_3 \equiv \lambda$
   (a) *If* $q(\boldsymbol{T}) = (\boldsymbol{T} - \lambda \boldsymbol{I})(\boldsymbol{T} - \mu \boldsymbol{I})$, *then*
   $$\log \boldsymbol{T} = \log \mu \left(\frac{\boldsymbol{T} - \lambda \boldsymbol{I}}{\mu - \lambda} \right) + \log \lambda \left(\frac{\boldsymbol{T} - \mu \boldsymbol{I}}{\lambda - \mu} \right).$$
   (b) *If* $q(\boldsymbol{T}) = (\boldsymbol{T} - \mu \boldsymbol{I})(\boldsymbol{T} - \lambda \boldsymbol{I})^2$, *then*
   $$\log \boldsymbol{T} = \log \mu \frac{(\boldsymbol{T} - \lambda \boldsymbol{I})^2}{(\lambda - \mu)^2} - \log \lambda \frac{(\boldsymbol{T} - \mu \boldsymbol{I})(\boldsymbol{T} - (2\lambda - \mu)\boldsymbol{I})}{(\lambda - \mu)^2} + \frac{(\boldsymbol{T} - \lambda \boldsymbol{I})(\boldsymbol{T} - \mu \boldsymbol{I})}{\lambda(\lambda - \mu)}.$$
3. $\lambda_1 \equiv \lambda \neq \lambda_2 \equiv \mu \neq \lambda_3 \equiv \nu \neq \lambda_1$. *We have* $q(\boldsymbol{T}) = (\boldsymbol{T} - \lambda \boldsymbol{I})(\boldsymbol{T} - \mu \boldsymbol{I})(\boldsymbol{T} - \nu \boldsymbol{I})$, *and*
   $$\log \boldsymbol{T} = \log \lambda \frac{(\boldsymbol{T} - \mu \boldsymbol{I})(\boldsymbol{T} - \nu \boldsymbol{I})}{(\lambda - \mu)(\lambda - \nu)} + \log \mu \frac{(\boldsymbol{T} - \lambda \boldsymbol{I})(\boldsymbol{T} - \nu \boldsymbol{I})}{(\mu - \lambda)(\mu - \nu)} + \log \nu \frac{(\boldsymbol{T} - \lambda \boldsymbol{I})(\boldsymbol{T} - \mu \boldsymbol{I})}{(\nu - \lambda)(\nu - \mu)}.$$

Some examples are

$$\boldsymbol{T} = \begin{bmatrix} e & 1 & 0 \\ 0 & e & 0 \\ 0 & 0 & e \end{bmatrix}, \qquad \log \boldsymbol{T} = \begin{bmatrix} 1 & \frac{1}{e} & 0 \\ 0 & 1 & 0 \\ 0 & 0 & 1 \end{bmatrix}, \qquad \text{(Case 1(b))},$$

$$\boldsymbol{T} = \begin{bmatrix} e & 1 & 1 \\ 0 & e & 1 \\ 0 & 0 & e \end{bmatrix}, \qquad \log \boldsymbol{T} = \begin{bmatrix} 1 & \frac{1}{e} & \frac{2e-1}{2e^2} \\ 0 & 1 & \frac{1}{e} \\ 0 & 0 & 1 \end{bmatrix}, \qquad \text{(Case 1(c))},$$

$$\boldsymbol{T} = \begin{bmatrix} e & 1 & 1 \\ 0 & e^2 & 1 \\ 0 & 0 & e^2 \end{bmatrix}, \qquad \log \boldsymbol{T} = \begin{bmatrix} 1 & \frac{1}{e(e-1)} & \frac{e^3-e^2-1}{e^3(e-1)^2} \\ 0 & 2 & \frac{1}{e^2} \\ 0 & 0 & 2 \end{bmatrix}, \qquad \text{(Case 2(b))}.$$

As an application, the logarithm of $\boldsymbol{R} \in \text{Orth}^+/\text{Sym}$, with eigenvalues $\{1, e^{i\alpha}, e^{-i\alpha}\}$ (and hence $\operatorname{tr} \boldsymbol{R} = 1 + 2\cos\alpha$) is obtained using Case 3 in the above theorem as

$$\begin{aligned}
\log \boldsymbol{R} &= i\alpha(\boldsymbol{P}_2 - \boldsymbol{P}_3) \\
&= \frac{i\alpha(\boldsymbol{R} - \boldsymbol{I})}{e^{i\alpha} - e^{-i\alpha}} \left[\frac{\boldsymbol{R} - e^{-i\alpha}\boldsymbol{I}}{e^{i\alpha} - 1} + \frac{\boldsymbol{R} - e^{i\alpha}\boldsymbol{I}}{e^{-i\alpha} - 1} \right] \\
&= \frac{\alpha}{2\sin\alpha}(\boldsymbol{R} - \boldsymbol{R}^T) \\
&= \sqrt{2}\alpha \frac{\boldsymbol{R} - \boldsymbol{R}^T}{\left\| \boldsymbol{R} - \boldsymbol{R}^T \right\|}.
\end{aligned}$$

By substituting $\boldsymbol{W} \equiv \log \boldsymbol{R}$ ($|\boldsymbol{w}| = \alpha$) into the right-hand side of Eqn. (1.289), we can easily verify that $e^{\log \boldsymbol{R}} = \boldsymbol{R}$.

By differentiating Eqn. (1.304) with respect to $\boldsymbol{T}$ and using Eqns. (1.233) and (1.236), we get (for any underlying space dimension n)

$$\frac{\partial}{\partial \boldsymbol{T}} \log(\boldsymbol{I} + (\boldsymbol{T} - \boldsymbol{I})\xi) = \int_0^{\xi} \left[(\boldsymbol{I} + (\boldsymbol{T} - \boldsymbol{I})\eta)^{-1} \boxtimes (\boldsymbol{I} + (\boldsymbol{T} - \boldsymbol{I})\eta)^{-T} \right] d\eta.$$

For a diagonalizable tensor, the above equation simplifies to

$$\begin{aligned}
&\frac{\partial}{\partial \boldsymbol{T}} \log(\boldsymbol{I} + (\boldsymbol{T} - \boldsymbol{I})\xi) = \\
&\sum_{i=1}^{k} \frac{\xi}{1 + (\lambda_i - 1)\xi} \boldsymbol{P}_i \boxtimes \boldsymbol{P}_i^T + \sum_{i=1}^{k} \sum_{\substack{j=1 \\ j \neq i}}^{k} \frac{\log[1 + (\lambda_i - 1)\xi] - \log[1 + (\lambda_j - 1)\xi]}{\lambda_i - \lambda_j} \boldsymbol{P}_i \boxtimes \boldsymbol{P}_j^T,
\end{aligned}$$

a result that we could also have obtained directly by differentiating Eqn. (1.312) using Eqn. (1.255). By setting $\xi = 1$ in the above equation, we get

$$\frac{\partial}{\partial \boldsymbol{T}} \log \boldsymbol{T} = \sum_{i=1}^{k} \frac{1}{\lambda_i} \boldsymbol{P}_i \boxtimes \boldsymbol{P}_i^T + \sum_{i=1}^{k} \sum_{\substack{j=1 \\ j \neq i}}^{k} \frac{\log \lambda_i - \log \lambda_j}{\lambda_i - \lambda_j} \boldsymbol{P}_i \boxtimes \boldsymbol{P}_j^T.$$

For $\boldsymbol{A} \in \text{Sym}$,

$$\begin{aligned}
\frac{\partial}{\partial \boldsymbol{A}} \log \boldsymbol{A} &= \mathbb{S} \left[\sum_{i=1}^{k} \frac{1}{\lambda_i} \boldsymbol{P}_i \boxtimes \boldsymbol{P}_i + \sum_{i=1}^{k} \sum_{\substack{j=1 \\ j \neq i}}^{k} \frac{\log \lambda_i - \log \lambda_j}{\lambda_i - \lambda_j} \boldsymbol{P}_i \boxtimes \boldsymbol{P}_j \right] \mathbb{S}, \quad k > 1, \\
&= \frac{1}{\lambda} \mathbb{S} \qquad \lambda_1 = \lambda_2 = \lambda_3 \equiv \lambda,
\end{aligned} \tag{1.313}$$

with $\boldsymbol{P}_i$ given by Eqn. (J.4).

An application is presented in Section 2.4.

1.11 Divergence and Stokes' Theorems

We state the divergence, Stokes', potential and localization theorems that are used quite frequently in the following development. The divergence theorem relates a volume integral to a surface integral, while the Stokes' theorem relates a contour integral to a surface integral. Let S represent the surface of a volume V, $\boldsymbol{n}$ represent the unit outward normal to the surface, ϕ a scalar field, $\boldsymbol{u}$ a vector field, and $\boldsymbol{T}$ a second-order tensor field. Then we have

Divergence theorem (also known as the Gauss' theorem)

$$\int_V \boldsymbol{\nabla}\phi \, dV = \int_S \phi \boldsymbol{n} \, dS. \tag{1.314}$$

Applying Eqn. (1.314) to the components u_i of a vector $\boldsymbol{u}$, we get

$$\begin{aligned} \int_V \boldsymbol{\nabla}\cdot\boldsymbol{u} \, dV &= \int_S \boldsymbol{u}\cdot\boldsymbol{n} \, dS, \\ \int_V \boldsymbol{\nabla}\times\boldsymbol{u} \, dV &= \int_S \boldsymbol{n}\times\boldsymbol{u} \, dS, \\ \int_V \boldsymbol{\nabla}\boldsymbol{u} \, dV &= \int_S \boldsymbol{u}\otimes\boldsymbol{n} \, dS. \end{aligned} \tag{1.315}$$

Similarly, on applying Eqn. (1.314) to $\boldsymbol{\nabla}\cdot\boldsymbol{T}$, we get the vector equation

$$\int_V \boldsymbol{\nabla}\cdot\boldsymbol{T} \, dV = \int_S \boldsymbol{T}\boldsymbol{n} \, dS. \tag{1.316}$$

Note that the divergence theorem is applicable even for multiply connected domains provided the surfaces are closed.

Stokes' theorem

Let C be a contour, and S be the area of any arbitrary surface enclosed by the contour C. Then

$$\oint_C \boldsymbol{u}\cdot d\boldsymbol{x} = \int_S (\boldsymbol{\nabla}\times\boldsymbol{u})\cdot\boldsymbol{n} \, dS, \tag{1.317}$$

$$\oint_C \boldsymbol{u}\times d\boldsymbol{x} = \int_S \left[(\boldsymbol{\nabla}\cdot\boldsymbol{u})\boldsymbol{n} - (\boldsymbol{\nabla}\boldsymbol{u})^T\boldsymbol{n}\right] dS. \tag{1.318}$$

In what follows, we assume $\boldsymbol{n}$ to be C^0 continuous over the entire surface S (so that $\boldsymbol{\nabla}\cdot\boldsymbol{n}$ makes sense). By taking $\boldsymbol{u}$ as $\boldsymbol{w}\times\boldsymbol{n}$ in Eqn. (1.317), using $(\boldsymbol{w}\times\boldsymbol{n})\cdot\boldsymbol{\lambda} = \boldsymbol{w}\cdot(\boldsymbol{n}\times\boldsymbol{\lambda})$, where $\boldsymbol{\lambda} = d\boldsymbol{x}/ds$ is the unit tangent to C, and then applying Eqn. (1.354), we get

$$\oint_C \boldsymbol{w}\cdot(\boldsymbol{n}\times\boldsymbol{\lambda}) \, ds = \int_S [(\boldsymbol{\nabla}\cdot\boldsymbol{n})(\boldsymbol{w}\cdot\boldsymbol{n}) - \boldsymbol{\nabla}\cdot\boldsymbol{w} + \boldsymbol{n}\cdot(\boldsymbol{\nabla}\boldsymbol{w})\boldsymbol{n}] \, dS. \tag{1.319}$$

Thus, for a closed surface S, we get

$$\int_S (\boldsymbol{\nabla}\cdot\boldsymbol{n})(\boldsymbol{w}\cdot\boldsymbol{n}) \, dS = \int_S [\boldsymbol{I} - \boldsymbol{n}\otimes\boldsymbol{n}] : \boldsymbol{\nabla}\boldsymbol{w} \, dS. \tag{1.320}$$

By choosing $\boldsymbol{w}$ to be a constant vector $\boldsymbol{c}$ in the above equation, we get $\boldsymbol{c} \cdot \int_S (\boldsymbol{\nabla} \cdot \boldsymbol{n})\boldsymbol{n}\, dS = 0$. Since the choice of $\boldsymbol{c}$ is arbitrary, we obtain

$$\int_S \kappa \boldsymbol{n}\, dS = \boldsymbol{0}, \tag{1.321}$$

where $\kappa = \boldsymbol{\nabla} \cdot \boldsymbol{n}$ is the curvature. For a more detailed discussion of this result, see [22]. By choosing $\boldsymbol{w} = \boldsymbol{x}$, $\boldsymbol{w} = \boldsymbol{c} \times \boldsymbol{x}$, $\boldsymbol{w} = (\boldsymbol{x} \cdot \boldsymbol{x})\boldsymbol{c}$ and $\boldsymbol{w} = (\boldsymbol{x} \otimes \boldsymbol{x})\boldsymbol{c}$ in Eqn. (1.320), we get

$$\int_S \kappa(\boldsymbol{x} \cdot \boldsymbol{n})\, dS = 2S, \tag{1.322a}$$

$$\int_S \kappa(\boldsymbol{x} \times \boldsymbol{n})\, dS = \boldsymbol{0}, \tag{1.322b}$$

$$\int_S \kappa(\boldsymbol{x} \cdot \boldsymbol{x})\boldsymbol{n}\, dS = 2\int_S [\boldsymbol{I} - \boldsymbol{n} \otimes \boldsymbol{n}]\, \boldsymbol{x}\, dS, \tag{1.322c}$$

$$\int_S \kappa(\boldsymbol{x} \cdot \boldsymbol{n})\boldsymbol{x}\, dS = \int_S [3\boldsymbol{I} - \boldsymbol{n} \otimes \boldsymbol{n}]\, \boldsymbol{x}\, dS. \tag{1.322d}$$

Note with relation to Eqns. (1.321) and (1.322b) that $\int_S \boldsymbol{n}\, dS$ and $\int_S (\boldsymbol{x} \times \boldsymbol{n})\, dS$ are also zero, as can be seen by using the divergence theorem.

Potential Theorems

1. Let V be a simply connected region, and let $\boldsymbol{u} = \boldsymbol{\nabla}\phi$ be a vector field on V. Then $\boldsymbol{\nabla} \times \boldsymbol{u} = \boldsymbol{0}$. Conversely, if $\boldsymbol{\nabla} \times \boldsymbol{u} = \boldsymbol{0}$, then there exists a scalar potential ϕ such that $\boldsymbol{u} = \boldsymbol{\nabla}\phi$.
   We have already proved the forward assertion (see Eqn. (1.269)). To prove the converse, we note from Eqn. (1.317) that $\oint_C \boldsymbol{u} \cdot d\boldsymbol{x} = 0$, and thus, the value of $\int_{\boldsymbol{x}_0}^{\boldsymbol{x}} \boldsymbol{u}(\boldsymbol{y}) \cdot d\boldsymbol{y}$ evaluated along any curve in V joining a fixed point $\boldsymbol{x}_0$ and $\boldsymbol{x}$, just depends on $\boldsymbol{x}$. Let

   $$\phi(\boldsymbol{x}) = \int_{\boldsymbol{x}_0}^{\boldsymbol{x}} \boldsymbol{u}(\boldsymbol{y}) \cdot d\boldsymbol{y}.$$

   Then $\boldsymbol{u} = \boldsymbol{\nabla}\phi$.
2. If $\boldsymbol{T}$ is a tensor field such that $\boldsymbol{\nabla} \times \boldsymbol{T} = \boldsymbol{0}$, then there exists a vector field $\boldsymbol{u}$ such that $\boldsymbol{T} = \boldsymbol{\nabla}\boldsymbol{u}$.
   To see this, using the definition of the curl of a tensor, we have $\boldsymbol{\nabla} \times (\boldsymbol{T}^T\boldsymbol{w}) = (\boldsymbol{\nabla} \times \boldsymbol{T})\boldsymbol{w} = \boldsymbol{0}$, where $\boldsymbol{w}$ is a constant unit vector. By the above result this implies that there exists a scalar field ϕ such that $\boldsymbol{T}^T\boldsymbol{w} = \boldsymbol{\nabla}\phi$, or, alternatively, $\boldsymbol{T} = \boldsymbol{w} \otimes \boldsymbol{\nabla}\phi = \boldsymbol{\nabla}(\phi\boldsymbol{w})$. Letting $\boldsymbol{u} = \phi\boldsymbol{w}$, we get the desired result.
3. If $\boldsymbol{T}$ is a tensor field such that $\boldsymbol{\nabla} \times \boldsymbol{T} = \boldsymbol{0}$, and in addition, if $\operatorname{tr}\boldsymbol{T} = 0$, then there exists $\boldsymbol{W} \in \mathrm{Skw}$ such that $\boldsymbol{T} = \boldsymbol{\nabla} \times \boldsymbol{W}$.
   To prove this, we take the trace of the relation $\boldsymbol{T} = \boldsymbol{\nabla}\boldsymbol{u}$ to get $\boldsymbol{\nabla} \cdot \boldsymbol{u} = 0$. Let $\boldsymbol{W}^T = -\boldsymbol{W}$ be the skew-symmetric tensor whose axial vector is $\boldsymbol{u}$. Then by Eqn. (1.272), we get $\boldsymbol{T} = \boldsymbol{\nabla}\boldsymbol{u} = \boldsymbol{\nabla} \times \boldsymbol{W}$.
4. If $\boldsymbol{v}$ and $\boldsymbol{T}$ are vector and tensor fields such that

   $$\int_S \boldsymbol{v} \cdot \boldsymbol{n}\, dS = 0,$$

   $$\int_S \boldsymbol{T}^T\boldsymbol{n}\, dS = \boldsymbol{0},$$

for every closed surface S in V, then there exist vector and tensor fields $\boldsymbol{u}$ and $\boldsymbol{U}$ such that $\boldsymbol{v} = \boldsymbol{\nabla} \times \boldsymbol{u}$ and $\boldsymbol{T} = \boldsymbol{\nabla} \times \boldsymbol{U}$, respectively.

To prove the second statement from the first, take the dot product of the equation $\int_S \boldsymbol{T}^T \boldsymbol{n}\, dS = 0$ with a constant unit vector $\boldsymbol{w}$. Then, by the definition of the transpose, we get $\int_S \boldsymbol{T}\boldsymbol{w} \cdot \boldsymbol{n}\, dS = 0$, which by virtue of the first result implies the existence of a vector field $\boldsymbol{u}$ such that $\boldsymbol{T}\boldsymbol{w} = \boldsymbol{\nabla} \times \boldsymbol{u}$, or, alternatively, $\boldsymbol{T} = (\boldsymbol{\nabla} \times \boldsymbol{u}) \otimes \boldsymbol{w} = \boldsymbol{\nabla} \times \boldsymbol{U}$, where $\boldsymbol{U} = \boldsymbol{w} \otimes \boldsymbol{u}$.

If the region V does not contain any 'holes' in the interior (i.e., the boundary of V is comprised of only the outer surface), then by means of the divergence theorem, we conclude that $\int_S \boldsymbol{v} \cdot \boldsymbol{n}\, dS = 0$ for every closed surface S is equivalent to $\boldsymbol{\nabla} \cdot \boldsymbol{v} = 0$. Thus, for a vector field $\boldsymbol{v}$ that satisfies $\boldsymbol{\nabla} \cdot \boldsymbol{v} = 0$ on such a body, there exists a vector field $\boldsymbol{u}$ such that $\boldsymbol{v} = \boldsymbol{\nabla} \times \boldsymbol{u}$.

Now we state the localization theorem, which is used in Chapter 3 to obtain the differential equations from the integral form of the governing equations.

Theorem 1.11.1 (Localization theorem). *Let ϕ be a continuous scalar, vector or tensor field on V. Then for any given $\boldsymbol{x}_0 \in V$,*

$$\phi(\boldsymbol{x}_0) = \lim_{r \to 0} \frac{1}{V(B_r)} \int_{B_r} \phi\, dV,$$

where B_r is the closed ball of radius $r > 0$ centered at $\boldsymbol{x}_0$, and $V(B_r)$ denotes its volume. It follows that if $\int_B \phi\, dV = 0$ for every closed ball B, then $\phi = 0$.

Proof. We have

$$\begin{aligned}\left|\phi(\boldsymbol{x}_0) - \lim_{r \to 0} \frac{1}{V(B_r)} \int_{B_r} \phi\, dV\right| &= \frac{1}{V(B_r)} \left|\int_{B_r} [\phi(\boldsymbol{x}_0) - \phi(\boldsymbol{x})]\, dV\right| \\ &\le \frac{1}{V(B_r)} \int_{B_r} |\phi(\boldsymbol{x}_0) - \phi(\boldsymbol{x})|\, dV \\ &\le \sup_{\boldsymbol{x} \in B_r} |\phi(\boldsymbol{x}_0) - \phi(\boldsymbol{x})|,\end{aligned}$$

which tends to zero as $r \to 0$ since ϕ is continuous. □

Note that there is no localization theorem for surfaces, i.e., $\int_S \phi\, dS = 0$ for every closed surface S *does not* imply that $\phi = 0$. To see this, take, for example, $\phi = \boldsymbol{a} \cdot \boldsymbol{n}$, where $\boldsymbol{a}$ is a constant vector.

1.12 Groups

The treatment in this section is based on [35]. A group is a set, say $\mathcal{G}$, equipped with a function from $\mathcal{G} \times \mathcal{G}$, called combination and denoted by

$$(a, b) \to a * b,$$

which satisfies the following axioms:

1. Associativity: For all $a, b, c \in \mathcal{G}$

$$(a * b) * c = a * (b * c).$$

2. Existence of a neutral element: There exists $n \in \mathcal{G}$ such that

$$n * a = a * n = a \quad \forall a \in \mathcal{G}.$$

3. Existence of reverse elements: For each $a \in \mathcal{G}$, there exists $\overset{r}{a} \in \mathcal{G}$ such that

$$\overset{r}{a} * a = a * \overset{r}{a} = n.$$

If, in addition, the commutative property

$$a * b = b * a \quad \forall a, b \in \mathcal{G}$$

is satisfied, then $\mathcal{G}$ is said to be a *commutative* or *Abelian* group.

Since the combination function in the definition of a group may be thought of as taking two elements from $\mathcal{G}$ and producing an element also in $\mathcal{G}$, it is often referred to as a *closed binary operation*. The *closure property*, i.e., the fact that the function value $a * b$ is also in $\mathcal{G}$, is often called the *fundamental closure property*.

As an example, the set of real numbers is a group with respect to addition. We write $a + b$ instead of $a * b$, i.e., the combination function is defined on $\Re \times \Re$ to $\Re$ by

$$(a, b) \rightarrow a + b.$$

The three defining axioms are satisfied since

$$(a + b) + c = a + (b + c) \quad \forall a, b, c \in \Re,$$

$$0 + a = a + 0 \quad \forall a \in \Re,$$

$$(-a) + a = a + (-a) = 0 \quad \forall a \in \Re.$$

Thus, 0 is the neutral element, and reverse elements are the negatives. In fact, $\Re$ is a commutative group since

$$a + b = b + a \quad \forall a, b \in \Re.$$

We have the following useful results:

Theorem 1.12.1. *Let $\mathcal{G}$ be a group. Then for any $a, b \in \mathcal{G}$, there exists a unique $x \in \mathcal{G}$, such that $a * x = b$. In fact, $x = \overset{r}{a} * b$. Similarly, there exists a unique $y \in \mathcal{G}$ such that $y * a = b$. In fact, $y = b * \overset{r}{a}$.*

Proof. Suppose, tentatively, that there exists a $x \in \mathcal{G}$ such that $a * x = b$. Then combining on the left with $\overset{r}{a}$, we have

$$\begin{aligned} \overset{r}{a} * b &= \overset{r}{a} * (a * x) \\ &= (\overset{r}{a} * a) * x && \text{(associativity)} \\ &= n * x && \text{(reverse)} \\ &= x. && \text{(neutral)} \end{aligned}$$

Thus, if x exists, it must be precisely $\overset{r}{a} * b$. This establishes the uniqueness. To prove existence, one must show that this x has the desired property. First of all, we note that by closure $\overset{r}{a} * b \in \mathcal{G}$. Next, consider

$$\begin{aligned} a * (\overset{r}{a} * b) &= (a * \overset{r}{a}) * b && \text{(associativity)} \\ &= n * b && \text{(reverse)} \\ &= b. && \text{(neutral)} \end{aligned}$$

Hence, $x = \overset{r}{a} * b$ does the job. The result for y can be proved similarly. □

Theorem 1.12.2 (Cancellation Property). *Let $\mathcal{G}$ be a group. Then for $a, b, c \in \mathcal{G}$*

$$\begin{aligned} b * a = c * a &\implies b = c. \\ a * b = a * c &\implies b = c. \end{aligned}$$

Proof. We have

$$(b * a) * \overset{r}{a} = (c * a) * \overset{r}{a}.$$

Using the associativity, reverse and neutral properties, we get

$$b = c.$$

The second statement can be proved in a similar manner. □

Theorem 1.12.3. *A given group has only one neutral element.*

Proof. Suppose n and n' are both neutral elements for a group $\mathcal{G}$, i.e., $n, n' \in \mathcal{G}$ and for all $a \in \mathcal{G}$

$$a * n = n * a = a \quad \text{and} \quad n' * a = a * n' = a.$$

Consider $a * n' = a$. We have

$$n' = \overset{r}{a} * a = n.$$

□

Theorem 1.12.4. *A given element of a group has only one reverse.*

Proof. Let $a \in \mathcal{G}$. Suppose $\overset{r}{a}$ and $\overset{r'}{a}$ are both reverses of a, i.e.,

$$\overset{r}{a} * a = a * \overset{r}{r} = n \quad \text{and} \quad \overset{r'}{a} * a = a * \overset{r'}{a} = n.$$

Consider $a * \overset{r'}{a} = n$. We have

$$\overset{r'}{a} = \overset{r}{a} * n = \overset{r}{a}.$$

□

Theorem 1.12.5. *Let n be a neutral element of a group. Then*

$$\overset{r}{n} = n.$$

Proof. For every element $a \in \mathcal{G}$, the reverse axiom requires $a * \overset{r}{a} = n$. Taking $a = n$, we get $n * \overset{r}{n} = n$. Hence,

$$\overset{r}{n} = \overset{r}{n} * n = n.$$

□

Theorem 1.12.6. *Let a be any element of a group. Then*

$$\overset{r}{\overset{r}{a}} = a.$$

Proof. We have $\overset{r}{a} * a = n$. Hence,

$$a = \overset{r}{\overset{r}{a}} * n = \overset{r}{\overset{r}{a}}.$$

□

Theorem 1.12.7. *Let a and b be any two elements of a group. Then*

$$\overset{r}{a * b} = \overset{r}{b} * \overset{r}{a}.$$

Proof. We have

$$\begin{aligned}
(a * b) * (\overset{r}{b} * \overset{r}{a}) &= a * (b * (\overset{r}{b} * \overset{r}{a})) && \text{(associativity)}\\
&= a * ((b * \overset{r}{b}) * \overset{r}{a}) && \text{(associativity)}\\
&= a * (n * \overset{r}{a}) && \text{(reverse)}\\
&= a * \overset{r}{a} && \text{(neutral)}\\
&= n. && \text{(reverse)}
\end{aligned}$$

In the same way

$$(\overset{r}{b} * \overset{r}{a}) * (a * b) = n.$$

Thus, $\overset{r}{b} * \overset{r}{a}$ is a reverse element of $a * b$. By the uniqueness of the inverse, it is *the* reverse. □

In general, subsets of groups will not be groups, e.g., the subset $\mathcal{G} - \{n\}$. We introduce the following terminology:

Definition: Let $\mathcal{H}$ be a nonempty [proper] subset of a group $\mathcal{G}$. Then, if $\mathcal{H}$ is a group with respect to the combination function of $\mathcal{G}$, it is called a [proper] subgroup of $\mathcal{G}$.[4]

Since $\mathcal{H} \times \mathcal{H} \subset \mathcal{G} \times \mathcal{G}$, the combination function for $\mathcal{H}$ is the restriction of the combination function for $\mathcal{G}$ to $\mathcal{H} \times \mathcal{H}$. However, the function values, which necessarily belong to $\mathcal{G}$, are not automatically in $\mathcal{H}$, i.e., for an arbitrary subset, closure could fail. An example of a subgroup of a group $\mathcal{G}$ with neutral element n is the singleton set $\{n\}$. The following theorem provides a sufficient condition for a subset to be a subgroup.

Theorem 1.12.8. *Let $\mathcal{H}$ be a nonempty subset of a group $\mathcal{G}$. If for all $a, b \in \mathcal{H}$, $a * \overset{r}{b} \in \mathcal{H}$, then $\mathcal{H}$ is a subgroup of $\mathcal{G}$.*

Proof. As noted, the combination function for $\mathcal{G}$ makes sense for $\mathcal{H}$. Hence, the associativity requirement is automatically satisfied. We have to show that the neutral and reverse elements of all $a \in \mathcal{H}$ lie in $\mathcal{H}$, and that $\mathcal{H}$ is closed under the combination operation.

Assume that for $a, b \in \mathcal{H}$

$$a * \overset{r}{b} \in \mathcal{H}. \tag{1.323}$$

Choosing $b = a$, we get $a * \overset{r}{a} \in \mathcal{H}$. But $a * \overset{r}{a} = n$. Thus, $n \in \mathcal{H}$.

Next, choose $a = n$. Then, we have $n * \overset{r}{b} \in \mathcal{H}$. But $n * \overset{r}{b} = \overset{r}{b}$. Thus, $b \in \mathcal{H}$ implies $\overset{r}{b} \in \mathcal{H}$. So $\mathcal{H}$ contains the reverses of all its elements. The reverses necessarily exist because of the properties of the 'parent' group $\mathcal{G}$.

Finally, let $a, b \in \mathcal{H}$. We have shown that $\overset{r}{b} \in \mathcal{H}$. Letting $\overset{r}{b}$ play the role of b in Eqn. (1.323), we have

$$a * \overset{r}{\overset{r}{b}} \in \mathcal{H}.$$

But $\overset{r}{\overset{r}{b}} = b$. Thus, $a * b \in \mathcal{H}$, and this proves that $\mathcal{H}$ is closed under the combination operation. □

[4] The bracket device above is used to make two statements simultaneously. To get the first statement, include the bracketed material. To get the second leave out the bracketed material.

We now come to a result that plays a key role in the classification of materials. First, we define the set of unimodular tensors Unim, and the set of proper unimodular tensors Unim^+ as

$$\text{Unim} := \{\boldsymbol{H} : |\det \boldsymbol{H}| = 1\},$$

$$\text{Unim}^+ := \{\boldsymbol{H} : \det \boldsymbol{H} = 1\}.$$

Theorem 1.12.9. *Unim and Unim^+ are groups with respect to tensor multiplication, and Unim^+ is a proper subgroup of Unim. Hence, Unim is called the unimodular group of tensors, while Unim^+ is called the proper unimodular group.*

Proof. Closure holds since $\boldsymbol{H}_1, \boldsymbol{H}_2 \in$ Unim implies that $|\det(\boldsymbol{H}_1\boldsymbol{H}_2)| = |\det \boldsymbol{H}_1|\,|\det \boldsymbol{H}_2| = 1$. Associativity follows since tensor multiplication is associative. The reverse element is simply the inverse, while the neutral element is the identity tensor. Similar arguments can be made for Unim^+. Unim^+ is a proper subgroup of Unim, since there are elements in Unim which are not in Unim^+, e.g., $\mathbf{diag}[-1, 1, 1]$. □

Theorem 1.12.10. *The orthogonal group Orth is a proper subgroup of the unimodular group Unim, and the proper orthogonal group Orth^+ is a proper subgroup of the proper unimodular group Unim^+.*

Proof. The closure property of Orth follows since $\boldsymbol{Q}_1, \boldsymbol{Q}_2 \in$ Orth implies that $\boldsymbol{Q}_1\boldsymbol{Q}_2 \in$ Orth. The other properties are proved as in the previous proof. Orth is a proper subgroup since there are unimodular tensors that are not orthogonal. e.g.,

$$\begin{bmatrix} 1 & 0 & 0 \\ 1 & 1 & 0 \\ 0 & 0 & 1 \end{bmatrix}.$$

A similar proof can be given for Orth^+. □

Before stating the next theorem, we prove the following lemma:

Lemma: Let $\boldsymbol{S}_0 \in \text{Psym} \cap \text{Unim}^+$, and let it have at least two distinct eigenvalues, say λ_u and λ_l, with $\lambda_u > \lambda_l$. Then, for every $\xi \in \left[\left(\frac{\lambda_l}{\lambda_u}\right)^2, \left(\frac{\lambda_u}{\lambda_l}\right)^2\right]$, there exists $\boldsymbol{R} \in \text{Orth}^+$ such that the symmetric and proper unimodular tensor

$$\boldsymbol{T} = \boldsymbol{S}_0^{-1}\boldsymbol{R}\boldsymbol{S}_0^2\boldsymbol{R}^{-1}\boldsymbol{S}_0^{-1}$$

has the spectrum $\{\xi, \xi^{-1}, 1\}$.

Proof. Since $\boldsymbol{S}_0 \in \text{Psym} \cap \text{Unim}^+$, all the eigenvalues of $\boldsymbol{S}_0$ are positive, and we can write the spectral resolution of $\boldsymbol{S}_0$ as

$$\boldsymbol{S}_0 = \lambda_u \boldsymbol{e}_1^* \otimes \boldsymbol{e}_1^* + \lambda_l \boldsymbol{e}_2^* \otimes \boldsymbol{e}_2^* + \frac{1}{\lambda_l \lambda_u} \boldsymbol{e}_3^* \otimes \boldsymbol{e}_3^*.$$

Let $\boldsymbol{R}(\theta)$ be the usual rotation matrix about the $\boldsymbol{e}_3^*$ axis, i.e.,

$$\boldsymbol{R}(\theta) = \begin{bmatrix} \cos\theta & \sin\theta & 0 \\ -\sin\theta & \cos\theta & 0 \\ 0 & 0 & 1 \end{bmatrix}.$$

In terms of its action on the basis $\{\boldsymbol{e}_1^*, \boldsymbol{e}_2^*, \boldsymbol{e}_3^*\}$, we have

$$\begin{aligned} \boldsymbol{R}(\theta)\boldsymbol{e}_1^* &= \cos\theta \boldsymbol{e}_1^* + \sin\theta \boldsymbol{e}_2^*, \\ \boldsymbol{R}(\theta)\boldsymbol{e}_2^* &= -\sin\theta \boldsymbol{e}_1^* + \cos\theta \boldsymbol{e}_2^*, \\ \boldsymbol{R}(\theta)\boldsymbol{e}_3^* &= \boldsymbol{e}_3^*. \end{aligned}$$

Note that $\boldsymbol{R}^{-1}(\theta) = \boldsymbol{R}^T(\theta) = \boldsymbol{R}(-\theta)$. Now, consider

$$\boldsymbol{T}(\theta) = \boldsymbol{S}_0^{-1}\boldsymbol{R}(\theta)\boldsymbol{S}_0^2\boldsymbol{R}^{-1}(\theta)\boldsymbol{S}_0^{-1}. \tag{1.324}$$

That $\boldsymbol{T}(\theta)$ is symmetric and proper unimodular follows from straightforward calculations. Hence, its eigenvalues are all real. Moreover, $+1$ is an eigenvalue of $\boldsymbol{T}(\theta)$ for all θ since

$$\begin{aligned} \boldsymbol{T}(\theta)\boldsymbol{e}_3^* &= \boldsymbol{S}_0^{-1}\boldsymbol{R}(\theta)\boldsymbol{S}_0^2\boldsymbol{R}(-\theta)\boldsymbol{S}_0^{-1}\boldsymbol{e}_3^* \\ &= \lambda_l\lambda_u\boldsymbol{S}_0^{-1}\boldsymbol{R}(\theta)\boldsymbol{S}_0^2\boldsymbol{R}(-\theta)\boldsymbol{e}_3^* \\ &= \lambda_l\lambda_u\boldsymbol{S}_0^{-1}\boldsymbol{R}(\theta)\boldsymbol{S}_0^2\boldsymbol{e}_3^* \\ &= \frac{1}{\lambda_l\lambda_u}\boldsymbol{S}_0^{-1}\boldsymbol{R}(\theta)\boldsymbol{e}_3^* \\ &= \frac{1}{\lambda_l\lambda_u}\boldsymbol{S}_0^{-1}\boldsymbol{e}_3^* \\ &= \boldsymbol{e}_3^*. \end{aligned}$$

For $\theta = 0$, we have $\boldsymbol{R} = \boldsymbol{I}$, which when substituted into Eqn. (1.324) yields $\boldsymbol{T} = \boldsymbol{I}$, i.e., the spectrum of $\boldsymbol{T}(0)$ is $\{1, 1, 1\}$. At $\theta = \pi/2$, we have $\boldsymbol{R}(\pi/2)\boldsymbol{e}_1^* = \boldsymbol{e}_2^*$, $\boldsymbol{R}(\pi/2)\boldsymbol{e}_2^* = -\boldsymbol{e}_1^*$, and $\boldsymbol{R}(\pi/2)\boldsymbol{e}_3^* = \boldsymbol{e}_3^*$. Hence,

$$\begin{aligned} \boldsymbol{T}\left(\frac{\pi}{2}\right)\boldsymbol{e}_1^* &= \boldsymbol{S}_0^{-1}\boldsymbol{R}\left(\frac{\pi}{2}\right)\boldsymbol{S}_0^2\boldsymbol{R}\left(-\frac{\pi}{2}\right)\boldsymbol{S}_0^{-1}\boldsymbol{e}_1^* \\ &= \frac{1}{\lambda_u}\boldsymbol{S}_0^{-1}\boldsymbol{R}\left(\frac{\pi}{2}\right)\boldsymbol{S}_0^2\boldsymbol{R}\left(-\frac{\pi}{2}\right)\boldsymbol{e}_1^* \\ &= -\frac{1}{\lambda_u}\boldsymbol{S}_0^{-1}\boldsymbol{R}\left(\frac{\pi}{2}\right)\boldsymbol{S}_0^2\boldsymbol{e}_2^* \\ &= -\frac{\lambda_l^2}{\lambda_u}\boldsymbol{S}_0^{-1}\boldsymbol{R}\left(\frac{\pi}{2}\right)\boldsymbol{e}_2^* \\ &= \frac{\lambda_l^2}{\lambda_u}\boldsymbol{S}_0^{-1}\boldsymbol{e}_1^* \\ &= \frac{\lambda_l^2}{\lambda_u^2}\boldsymbol{e}_1^*. \end{aligned}$$

Similarly, $T(\pi/2)e_2^* = \frac{\lambda_u^2}{\lambda_l^2}e_2^*$. Thus, the ordered spectrum of $T(\pi/2)$ is $\left\{\left(\frac{\lambda_l^2}{\lambda_u^2}\right), 1, \left(\frac{\lambda_u^2}{\lambda_l^2}\right)\right\}$. Since +1 is always an eigenvalue, and since $T(\theta)$ depends continuously on θ, we see that given any $\xi \in \left[\left(\frac{\lambda_l^2}{\lambda_u^2}\right), \left(\frac{\lambda_u^2}{\lambda_l^2}\right)\right]$, there exists $\theta \in [0, \pi/2]$ for which the lemma holds. □

Thus, if e_i' are the eigenvectors of T, then the spectral resolution of T is

$$T = \xi e_1' \otimes e_1' + \frac{1}{\xi} e_2' \otimes e_2' + e_3' \otimes e_3', \tag{1.325}$$

where $\xi \in \left[\left(\frac{\lambda_l^2}{\lambda_u^2}\right), \left(\frac{\lambda_u^2}{\lambda_l^2}\right)\right]$. Now we are in a position to prove the main result of this section. The proof that we present is due to Noll (see [237], page 200).

Theorem 1.12.11. (Maximality of the Orthogonal Group in the Unimodular Group). *If $\mathcal{G}$ is a group with respect to tensor multiplication such that*

$$Orth \subset \mathcal{G} \subset Unim,$$

then either $\mathcal{G}$ = Orth or $\mathcal{G}$ = Unim. In other words, there is no group between the orthogonal group and the unimodular group. The corresponding result for Orth$^+$ and Unim$^+$ is that Orth$^+$ is maximal in Unim$^+$.

Proof. There are two possibilities:

1. $\mathcal{G}$ contains a tensor $S_0 \in$ Psym ∩ Unim with at least two distinct eigenvalues, say $\lambda_u > \lambda_l$.
2. $\mathcal{G}$ does not contain such a tensor, i.e., the only tensor from Psym ∩ Unim that it contains is the identity tensor.

Consider the first possibility. Using the polar decomposition, any tensor $H \in$ Unim can be decomposed as $H = QU$, where Q is an orthogonal tensor and $U \in$ Psym ∩ Unim. Hence, if $\hat{e}_i$ are the eigenvectors of U, we can write the spectral resolution of U as

$$U = \gamma_1 \hat{e}_1 \otimes \hat{e}_1 + \gamma_2 \hat{e}_2 \otimes \hat{e}_2 + \frac{1}{\gamma_1 \gamma_2} \hat{e}_3 \otimes \hat{e}_3,$$

where $\gamma_1, \gamma_2 > 0$. We can decompose U as

$$U = U_1 U_2,$$

where

$$U_1 = \gamma_1 \hat{e}_1 \otimes \hat{e}_1 + \frac{1}{\gamma_1} \hat{e}_2 \otimes \hat{e}_2 + \hat{e}_3 \otimes \hat{e}_3,$$

$$U_2 = \hat{e}_1 \otimes \hat{e}_1 + \gamma_1 \gamma_2 \hat{e}_2 \otimes \hat{e}_2 + \frac{1}{\gamma_1 \gamma_2} \hat{e}_3 \otimes \hat{e}_3.$$

Now choose integers m_1, m_2 large enough so that $\zeta_1 := \gamma_1^{1/m_1}$ and $\zeta_2 := (\gamma_1\gamma_2)^{1/m_2}$ satisfy

$$\left(\frac{\lambda_l^2}{\lambda_u^2}\right) \leq \zeta_i \left(\frac{\lambda_u^2}{\lambda_l^2}\right).$$

(Choosing $m_1 \geq |\ln \gamma_1/[2\ln(\lambda_u/\lambda_l)]|$ and $m_2 \geq |\ln(\gamma_1\gamma_2)/[2\ln(\lambda_u/\lambda_l)]|$ does the job.) By the Lemma, there exist tensors $\boldsymbol{T}_1$ and $\boldsymbol{T}_2$ whose spectral resolution is given by

$$\begin{aligned}
\boldsymbol{T}_1 &= \zeta_1 \bar{\boldsymbol{e}}_1 \otimes \bar{\boldsymbol{e}}_1 + \frac{1}{\zeta_1}\bar{\boldsymbol{e}}_2 \otimes \bar{\boldsymbol{e}}_2 + \bar{\boldsymbol{e}}_3 \otimes \bar{\boldsymbol{e}}_3, \\
\boldsymbol{T}_2 &= \boldsymbol{e}'_1 \otimes \boldsymbol{e}'_1 + \zeta_2 \boldsymbol{e}'_2 \otimes \boldsymbol{e}'_2 + \frac{1}{\zeta_2}\boldsymbol{e}'_3 \otimes \boldsymbol{e}'_3.
\end{aligned}$$

Thus,

$$\begin{aligned}
\boldsymbol{T}_1^{m_1} &= \gamma_1 \bar{\boldsymbol{e}}_1 \otimes \bar{\boldsymbol{e}}_1 + \frac{1}{\gamma_1}\bar{\boldsymbol{e}}_2 \otimes \bar{\boldsymbol{e}}_2 + \bar{\boldsymbol{e}}_3 \otimes \bar{\boldsymbol{e}}_3, \\
\boldsymbol{T}_2^{m_2} &= \boldsymbol{e}'_1 \otimes \boldsymbol{e}'_1 + \gamma_1\gamma_2 \boldsymbol{e}'_2 \otimes \boldsymbol{e}'_2 + \frac{1}{\gamma_1\gamma_2}\boldsymbol{e}'_3 \otimes \boldsymbol{e}'_3.
\end{aligned}$$

Let $\boldsymbol{R}_1$ and $\boldsymbol{R}_2$ be the two proper orthogonal tensors that rotate $\bar{\boldsymbol{e}}_i$ and $\boldsymbol{e}'_i$ into $\hat{\boldsymbol{e}}_i$. Then, we have

$$\begin{aligned}
\boldsymbol{R}_1 \boldsymbol{T}_1^{m_1} \boldsymbol{R}_1^T &= \boldsymbol{R}_1(\gamma_1 \bar{\boldsymbol{e}}_1 \otimes \bar{\boldsymbol{e}}_1 + \frac{1}{\gamma_1}\bar{\boldsymbol{e}}_2 \otimes \bar{\boldsymbol{e}}_2 + \bar{\boldsymbol{e}}_3 \otimes \bar{\boldsymbol{e}}_3)\boldsymbol{R}_1^T \\
&= \gamma_1(\boldsymbol{R}_1\bar{\boldsymbol{e}}_1) \otimes (\boldsymbol{R}_1\bar{\boldsymbol{e}}_1) + \frac{1}{\gamma_1}(\boldsymbol{R}\bar{\boldsymbol{e}}_2) \otimes (\boldsymbol{R}_1\bar{\boldsymbol{e}}_2) + (\boldsymbol{R}_1\bar{\boldsymbol{e}}_3) \otimes (\boldsymbol{R}_1\bar{\boldsymbol{e}}_3) \\
&= \gamma_1 \hat{\boldsymbol{e}}_1 \otimes \hat{\boldsymbol{e}}_1 + \frac{1}{\gamma_1}\hat{\boldsymbol{e}}_2 \otimes \hat{\boldsymbol{e}}_2 + \hat{\boldsymbol{e}}_3 \otimes \hat{\boldsymbol{e}}_3 \\
&= \boldsymbol{U}_1.
\end{aligned}$$

Similarly,

$$\boldsymbol{R}_2 \boldsymbol{T}_2^{m_2} \boldsymbol{R}_2^T = \boldsymbol{U}_2.$$

Hence,

$$\begin{aligned}
\boldsymbol{H} &= \boldsymbol{Q}\boldsymbol{U} \\
&= \boldsymbol{Q}\boldsymbol{U}_1\boldsymbol{U}_2 \\
&= \boldsymbol{Q}\left(\boldsymbol{R}_1 \boldsymbol{T}_1^{m_1} \boldsymbol{R}_1^T\right)\left(\boldsymbol{R}_2 \boldsymbol{T}_2^{m_2} \boldsymbol{R}_2^T\right)
\end{aligned}$$

$$= QR_1 \left[S_0^{-1} R(\theta_1) S_0^2 R^{-1}(\theta_1) S_0^{-1} \right]^{m_1} R_1^T R_2 \left[S_0^{-1} R(\theta_2) S_0^2 R^{-1}(\theta_2) S_0^{-1} \right]^{m_2} R_2^T, \tag{1.326}$$

for some $\theta_1, \theta_2 \in [0, \pi/2]$. Thus, any $H \in$ Unim can be expressed in terms of powers of $S_0 \in$ Psym $\cap$ Unim, $Q \in$ Orth, and $R(\theta_1), R(\theta_2), R_1, R_2 \in$ Orth$^+$. But Orth $\subset \mathcal{G}$ and $S_0 \in \mathcal{G}$, and since $\mathcal{G}$ is a group, any element generated by powers of S_0 and elements of Orth has to lie in $\mathcal{G}$. Thus, we have Unim $\subset \mathcal{G}$. But, by hypothesis, $\mathcal{G} \subset$ Unim. Hence, for the case when S_0 has at least two distinct eigenvalues $\mathcal{G} =$ Unim.

Now consider the case when the only member from Psym $\cap$ Unim in $\mathcal{G}$ is the identity tensor. For any $H \in \mathcal{G}$, we have $H = QU$, or, alternatively, $U = Q^{-1}H$, where $Q \in$ Orth $\subset \mathcal{G}$, and $U \in$ Psym $\cap$ Unim. Since $\mathcal{G}$ is a group, and since $H, Q \in \mathcal{G}$, we have $U = Q^{-1}H \in \mathcal{G}$. However, since the only member from Psym $\cap$ Unim in $\mathcal{G}$ is I, we have $U = I$. Therefore, $H = Q \in$ Orth, i.e., $\mathcal{G} \subset$ Orth. But, by hypothesis, Orth $\subset \mathcal{G}$. Hence, in this case, we have $\mathcal{G} =$ Orth.

To prove that Orth$^+$ is maximal in Unim$^+$, note that if $H \in$ Unim$^+$, then Q in Eqn. (1.326) belongs to Orth$^+$. Thus, any $H \in$ Unim$^+$ can be expressed in terms of powers of $S_0 \in$ Psym $\cap$ Unim and tensors in Orth$^+$, and the result follows. □

EXERCISES

1. Show that

$$(w \times u) \times (w \times v) = [w \otimes w](u \times v). \tag{1.327}$$

$$[u \times v, v \times w, w \times u] = [u, v, w]^2. \tag{1.328}$$

From Eqn. (1.327), it follows that if n is a unit vector, then

$$[n, n \times u, n \times v] = [n, u, v].$$

2. Which of the spaces Sym, Psym, Skw, Orth$^+$ are linear subspaces of Lin? Justify. Evaluate if the matrices

$$\begin{bmatrix} 0 & 1 & -1 \\ -1 & 0 & 0 \\ 1 & 0 & 0 \end{bmatrix}, \quad \begin{bmatrix} 0 & 0 & 1 \\ 0 & 0 & -1 \\ -1 & 1 & 0 \end{bmatrix}, \quad \begin{bmatrix} 0 & -1 & 0 \\ 1 & 0 & 1 \\ 0 & -1 & 0 \end{bmatrix},$$

constitute a basis for Skw, and if the matrices

$$\begin{bmatrix} 1 & 1 & 0 \\ 1 & 1 & 0 \\ 0 & 0 & 0 \end{bmatrix}, \quad \begin{bmatrix} 1 & 0 & 1 \\ 0 & 0 & 0 \\ 1 & 0 & 1 \end{bmatrix}, \quad \begin{bmatrix} 0 & 0 & 0 \\ 0 & 1 & 1 \\ 0 & 1 & 1 \end{bmatrix},$$

constitute a basis for Sym. If not, then give a 'canonical' basis for Skw and Sym. Also give a basis for deviatoric symmetric tensors.

3. Let $\{p_i\}$ and $\{q_i\}$, $i = 1, 2, \ldots, n$, be sets of vectors in $\Re^n$. Determine if the set $\{p_i \otimes q_i\}$, $i = 1, 2, \ldots, n$, is linearly dependent or independent if

   (a) $\{p_i\}$ is linearly independent, and $\{q_i\}$ is linearly independent;

   (b) $\{p_i\}$ is linearly independent, but $\{q_i\}$ is not;

   (c) $\{q_i\}$ is linearly independent, but $\{p_i\}$ is not;

   (d) $\{p_i\}$ is linearly dependent, and $\{q_i\}$ is also linearly dependent.

4. Let $W \in \text{Skw}$. Determine if the set $\{I, W, W^2\}$ is linearly dependent or linearly independent. Deduce from this result and Eqn. (1.105a) if $\{I, Q, Q^T\}$, where $Q \in \text{Orth}^+$, is linearly dependent or independent.

5. Show that the decomposition of a tensor into a symmetric and skew-symmetric part as given by Eqn. (1.33) is unique.

6. Show that $\text{tr}\, RS = \text{tr}\, SR = \text{tr}\, R^T S^T = \text{tr}\, S^T R^T$.

7. If S and W are symmetric and skew-symmetric tensors, respectively, and T is an arbitrary second-order tensor prove that

$$S : T = S : T^T = S : \left[\frac{1}{2}(T^T + T)\right], \tag{1.329}$$

$$W : T = -W : T^T = W : \left[\frac{1}{2}(T - T^T)\right],$$

$$S : W = 0. \tag{1.330}$$

8. Prove the following:
   (i) If $A : B = 0$ for every symmetric tensor B, then $A \in \text{Skw}$.
   (ii) If $A : B = 0$ for every skew tensor B, then $A \in \text{Sym}$.

9. Using Eqns. (1.35) and (1.52) prove that

$$\textbf{cof}\,(RS) = (\textbf{cof}\, R)(\textbf{cof}\, S),$$

$$\det(RS) = (\det R)(\det S).$$

10. Using indicial notation prove that

$$\text{tr}\,(\textbf{cof}\, T) = \frac{1}{2}\left[(\text{tr}\, T)^2 - \text{tr}\,(T^2)\right], \tag{1.331}$$

$$[(\textbf{cof}\, T)u] \times v = T(u \times T^T v), \tag{1.332}$$

$$\textbf{cof}\,(\textbf{cof}\, T) = (\det T)T.$$

11. Show that an arbitrary tensor T cannot be represented as

$$\alpha a \otimes b + \beta b \otimes c + \gamma c \otimes a,$$

by showing, in particular, that the identity tensor I cannot be represented in this way.

12. Let $T = u \otimes p + v \otimes q + w \otimes r$. Show using Eqn. (1.60) that

$$\mathbf{cof}\, T = (u \times v) \otimes (p \times q) + (v \times w) \otimes (q \times r) + (w \times u) \otimes (r \times p). \quad (1.333)$$

Using Eqn. (1.64), deduce that

$$(u \times v) \otimes w + (v \times w) \otimes u + (w \times u) \otimes v = [u, v, w]\, I.$$

Using Eqn. (1.53), show that Eqn. (1.328) is recovered for the case $[u, v, w] \neq 0$ (Eqn. (1.328) holds even when $[u, v, w] = 0$). Note that the principal invariants of T are

$$I_1 = \operatorname{tr} T = u \cdot p + v \cdot q + w \cdot r,$$
$$I_2 = \operatorname{tr}(\mathbf{cof}\, T) = (u \times v) \cdot (p \times q) + (v \times w) \cdot (q \times r) + (w \times u) \cdot (r \times p),$$
$$I_3 = \det T = [u, v, w]\, [p, q, r],$$

and that T^{-1} (when it exists) is given by $(\mathbf{cof}\, T)^T / \det T$.

13. Let λ_i, $i = 1, 2, \cdots, n$, denote the eigenvalues of T for space-dimension n. Using Eqn. (J.11), show that the eigenvalues of $\mathbf{cof}\, T$ are $\prod_{\substack{j=1 \\ j\neq i}}^{n} \lambda_j$, $i = 1, 2, \ldots, n$. For the case $n = 3$, deduce Eqn. (1.331), $I_2(\mathbf{cof}\, T) = \operatorname{tr} T(\det T)$ and $I_3(\mathbf{cof}\, T) = (\det T)^2$. Now consider the case when T is diagonalizable with *distinct* eigenvalues, so that $T = \sum_{i=1}^{n} \lambda_i P_i$ (thus, $T - \gamma I = \sum_{i=1}^{n} (\lambda_i - \gamma) P_i$; $\operatorname{tr} P_i = e^* \cdot e^* = 1$). Using Eqn. (J.11), show that $\mathbf{cof}\, T = \sum_{i=1}^{n} (\prod_{\substack{j=1 \\ j\neq i}}^{n} \lambda_j) P_i^T$, and $\mathbf{cof}\,(T - \gamma I) = \sum_{i=1}^{n} \left[\prod_{\substack{j=1 \\ j\neq i}}^{n} (\lambda_j - \gamma)\right] P_i^T$. Thus, for $\gamma = \lambda_1$ say, $\mathbf{cof}\,(T - \lambda_1 I) = \prod_{j=2}^{n} (\lambda_j - \lambda_1) P_1^T$, and $\operatorname{tr} \mathbf{cof}\,(T - \lambda_1 I) = \prod_{j=2}^{n} (\lambda_j - \lambda_1)$.

14. If T, R and S are second-order tensors, then show that

$$\det T = \frac{1}{3} \operatorname{tr}\left[(\mathbf{cof}\, T)^T T\right] = \frac{1}{3} T : \mathbf{cof}\, T, \quad (1.334)$$

$$\det(R + S) = \det R + \mathbf{cof}\, R : S + R : \mathbf{cof}\, S + \det S, \quad (1.335)$$

$$\begin{aligned}\mathbf{cof}\,(R + S) = {} & \mathbf{cof}\, R + \mathbf{cof}\, S + \left[(\operatorname{tr} R)(\operatorname{tr} S) - \operatorname{tr}(RS)\right] I - (\operatorname{tr} R) S^T - (\operatorname{tr} S) R^T \\ & + (SR)^T + (RS)^T. \end{aligned} \quad (1.336)$$

By putting $R = I$ in Eqn. (1.336), deduce that

$$\mathbf{cof}\,(I + S) = (1 + \operatorname{tr} S) I - S^T + \mathbf{cof}\, S. \quad (1.337)$$

By putting $S = u \otimes v$ in the above equation and using Eqn. (1.84), deduce that

$$\mathbf{cof}\,(I + u \otimes v) = (1 + u \cdot v) I - v \otimes u.$$

Using Eqns. (1.335) and (1.337), show that for $W \in \text{Skw}$,

$$(I + W)^{-1} = \frac{I - W + w \otimes w}{1 + |w|^2}.$$

15. Apply the Cayley–Hamilton theorem to $W \in \text{Skw}$, and find the axial vector of W^3.

16. If w is the axial vector of $W \in \text{Skw}$, show that $w \cdot [u \times Wu] \geq 0 \ \forall u$, with equality if and only if (u, w) are linearly dependent.

17. Let $W \in \text{Skw}$, and let w be its axial vector. Using Eqn. (1.332), show that $(\textbf{cof}\, T)w$ is the axial vector of TWT^T. If $T \equiv Q \in \text{Orth}^+$, then deduce using Eqn. (1.93) that Qw is the axial vector of QWQ^T. Also, find the axial vector of QW^3Q^T.

18. Show that a (not necessarily symmetric) tensor T commutes with every skew tensor W if and only if $T = \lambda I$.

19. Either using the result of the previous problem, or independently, show that a (not necessarily symmetric) tensor T commutes with every orthogonal tensor Q if and only if $T = \lambda I$.

20. Show that a (not necessarily symmetric) tensor T commutes with every symmetric tensor S if and only if $T = \lambda I$.

21. Show that $(u, Tv) = 0 \ \forall \ u, v$ that are mutually orthogonal, if an only if $T = \lambda I$.

22. Let a, b be arbitrary vectors, and u, v be unit vectors such that $u \cdot v \neq -1$.

    (a) Show that $a \times b$ is the axial vector of the skew tensor $b \otimes a - a \otimes b$. It follows that if $b \otimes a = a \otimes b$, then $a \times b = 0$, so that by Theorem 1.2.1, a and b are linearly dependent.

    (b) Using (a) and Eqn. (1.104), show that the unique orthogonal tensor that rotates u into v about an axis perpendicular to u and v is given by

    $$R = I + rv \otimes u - s(u \otimes u + u \otimes v + v \otimes v).$$

    where $r = (1 + 2u \cdot v)/(1 + u \cdot v)$ and $s = 1/(1 + u \cdot v)$.

23. If w is the axial vector of $W \in \text{Skw}$, then show using Eqns. (1.54) and (1.327) (or Eqns. (1.60) and (1.90)) that $\textbf{cof}\, W = w \otimes w$. Deduce using Eqn. (1.273) that

    $$e^{\textbf{cof}\, W} = I + \frac{(e^{|w|^2} - 1)}{|w|^2} w \otimes w.$$

    The determinant of the above tensor is $e^{\text{tr}\,(\textbf{cof}\, W)} = e^{|w|^2}$.

24. Let W_1 and W_2 be skew-symmetric tensors with axial vectors w_1 and w_2. Show using Eqns. (1.21a) and (1.87) that

    $$W_1 W_2 = w_2 \otimes w_1 - (w_1 \cdot w_2) I, \tag{1.338}$$

    from which it follows that $W_1W_2 - W_2W_1 = w_2 \otimes w_1 - w_1 \otimes w_2$. By using the result of Problem 22a (or, independently, by means of Eqns. (1.21)), we see that $w_1 \times w_2$ is the axial vector of $W_1W_2 - W_2W_1$. If $w_1 = \alpha w_2$, where $\alpha \in \Re$, then, by Eqn. (1.90), $W_1 = \alpha W_2$, and hence $W_1W_2 = W_2W_1$. Conversely, if $W_1W_2 = W_2W_1$, then $w_1 \times w_2$, the axial vector of $W_1W_2 - W_2W_1$, is 0, and by Theorem 1.2.1, it follows that w_1 and w_2 are linearly dependent. Thus, two skew-symmetric tensors W_1 and W_2 commute if and only if their axial vectors are linearly dependent.

It also follows from Eqn. (1.338) that

$$W_1W_2 + W_2W_1 = w_1 \otimes w_2 + w_2 \otimes w_1 - 2(w_1 \cdot w_2)I.$$

The eigenvalues/eigenvectors of the above symmetric tensor are

$$\begin{aligned}&(|w_1|\,|w_2| - w_1 \cdot w_2), \frac{w_1}{|w_1|} + \frac{w_2}{|w_2|},\\ &-(|w_1|\,|w_2| + w_1 \cdot w_2), \frac{w_1}{|w_1|} - \frac{w_2}{|w_2|},\\ &-2(w_1 \cdot w_2), w_1 \times w_2.\end{aligned}$$

One can now easily compute the principal invariants of $W_1W_2 + W_2W_1$ using these eigenvalues.

25. For $W \in \mathrm{Skw}$, find the polar decomposition of $I + W$ in terms of W and its axial vector w.

26. Prove that there is a one-to-one correspondence between Skw and members of Orth^+ with no eigenvalue equal to -1,[5] by showing that

    (a) For every $W \in \mathrm{Skw}$, $Q = (I - W)^{-1}(I + W) \in \mathrm{Orth}^+$. If w is the axial vector of W, then for $n = 3$, show using Eqn. (1.66) and Problem 14 that the above expression simplifies to

$$Q = \frac{1}{1 + w \cdot w}\left[(1 + w \cdot w)I + 2W + 2W^2\right], \tag{1.339}$$

with $w/|w|$ as the axis. In expanded form, if

$$W = \begin{bmatrix} 0 & -\gamma & \beta \\ \gamma & 0 & -\alpha \\ -\beta & \alpha & 0 \end{bmatrix},$$

then

$$Q = \frac{1}{1 + \alpha^2 + \beta^2 + \gamma^2}\begin{bmatrix} 1 + \alpha^2 - \beta^2 - \gamma^2 & 2(\alpha\beta - \gamma) & 2(\beta + \alpha\gamma) \\ 2(\alpha\beta + \gamma) & 1 - \alpha^2 + \beta^2 - \gamma^2 & -2(\alpha - \beta\gamma) \\ 2(-\beta + \alpha\gamma) & 2(\alpha + \beta\gamma) & 1 - \alpha^2 - \beta^2 + \gamma^2 \end{bmatrix}.$$

[5] For members of Orth^+ with at least one eigenvalue equal to -1, the correspondence when the dimension n is three, is given by

$$Q = I + 2W^2,$$

where W is of the form

$$W = \begin{bmatrix} 0 & -\sin\theta & \cos\theta\sin\phi \\ \sin\theta & 0 & -\cos\theta\cos\phi \\ -\cos\theta\sin\phi & \cos\theta\cos\phi & 0 \end{bmatrix}.$$

By comparing Eqn. (1.104) with Eqn. (1.339), we see that $\sin\alpha \equiv 2|w|/(1 + |w|^2)$ and $\cos\alpha = (1 - |w|^2)/(1 + |w|)^2$, which corresponds to letting $\alpha \in [0, \pi)$ in Eqn. (1.104). If $w_1/|w_1|$ and $w_2/|w_2|$ are the axes of R_1 and R_2, respectively, then, from Eqn. (1.107) and the above expressions for $\cos\alpha$ and $\sin\alpha$, the axis of R_2R_1 is $e = w/|w|$, where

$$w = w_1 + w_2 - w_1 \times w_2.$$

(b) For every $Q \in \text{Orth}^+$ such that -1 is not an eigenvalue of Q, $W = (Q - I)(Q + I)^{-1} \in \text{Skw}$.

Thus, the number of parameters required to define an orthogonal tensor is the number of independent components in a skew-symmetric matrix, namely $n(n-1)/2$; e.g., for $n = 3$, the number of parameters is 3.

27. If u is an arbitrary vector, show that there exists $W \in \text{Skw}$ and a vector v (both dependent on u) such that $Wv = u$. Use this result to show that if $WT_1 = WT_2 \ \forall W \in \text{Skw}$, then $T_1 = T_2$.

28. Show that a scalar-valued function $\phi : \text{Skw} \to \Re$ is isotropic if and only if there exists a function $\tilde{\phi} : \text{tr}\,(W^2) \to \Re$ such that

$$\phi(W) = \tilde{\phi}(\text{tr}\, W^2) \quad \forall W \in \text{Skw}.$$

The above result follows directly as a corollary of Theorem 1.6.13, or can be proved independently by using the representation given by Eqn. (1.89), and mimicking the proof of Theorem 1.6.12.

29. For $S \in \text{Sym}$, show using the spectral resolution of S or otherwise that

$$(u, Su) = 0 \quad \forall u \in V,$$

if and only if $S = 0$. It follows that $(u, Tu) = 0 \quad \forall u \in V$, if and only if $T \in \text{Skw}$.

30. A set of Cartesian axes is rotated about the origin to coincide with the unit vectors

$$\bar{e}_1 = (\frac{\sqrt{3}}{4}, \frac{1}{4}, \frac{\sqrt{3}}{2}),$$
$$\bar{e}_2 = (\frac{3}{4}, \frac{\sqrt{3}}{4}, -\frac{1}{2}),$$
$$\bar{e}_3 = (-\frac{1}{2}, \frac{\sqrt{3}}{2}, 0).$$

Write down the rotation matrix corresponding to this rotation, and transform the components of the tensor T with respect to $\{e_i\}$ given by

$$[T] = \begin{bmatrix} 8 & -4 & 0 \\ -4 & -3 & -1 \\ 0 & -1 & -2 \end{bmatrix}.$$

31. Using *direct notation* show that (also verify using indicial notation)

$$(a \otimes b)^T = b \otimes a, \tag{1.340}$$

$$(a \otimes b)(c \otimes d) = (b \cdot c)a \otimes d, \tag{1.341}$$

$$(a \otimes b) : (u \otimes v) = (a \cdot u)(b \cdot v), \tag{1.342}$$

$$T(a \otimes b) = (Ta) \otimes b, \tag{1.343}$$

$$(a \otimes b)T = a \otimes (T^T b), \tag{1.344}$$

$$T : (a \otimes b) = a \cdot Tb. \tag{1.345}$$

Use Eqn. (1.341) to show that for a symmetric S, the spectral decompositions of S^n and S^{-1} (when it exists) are

$$S^n = \lambda_1^n e_1^* \otimes e_1^* + \lambda_2^n e_2^* \otimes e_2^* + \lambda_3^n e_3^* \otimes e_3^*,$$
$$S^{-1} = \lambda_1^{-1} e_1^* \otimes e_1^* + \lambda_2^{-1} e_2^* \otimes e_2^* + \lambda_3^{-1} e_3^* \otimes e_3^*.$$

32. The Fibonacci numbers satisfy the recurrence relation $F_{n+1} = F_n + F_{n-1}$ with $F_1 = F_2 = 1$. In matrix form this recurrence relation can be written as

$$\begin{bmatrix} F_{n+1} \\ F_n \end{bmatrix} = S \begin{bmatrix} F_n \\ F_{n-1} \end{bmatrix},$$

where $S = \begin{bmatrix} 1 & 1 \\ 1 & 0 \end{bmatrix}$. Use the above matrix form to first express $\begin{bmatrix} F_{n+1} \\ F_n \end{bmatrix}$ in terms of $\begin{bmatrix} F_2 \\ F_1 \end{bmatrix}$, and then use this to find an explicit expression for F_n in terms of the eigenvalues λ_1 and λ_2 of S.

33. Let B be an invertible tensor. Show that if n is an eigenvector of A, then Bn is an eigenvector of BAB^{-1} corresponding to the same eigenvalue. It follows that if $Q \in \text{Orth}^+$ and if n is an eigenvector of a tensor T, then Qn is an eigenvector of QTQ^T corresponding to the same eigenvalue.

34. If the eigenvalues of a symmetric tensor S are ordered such that $\lambda_1 \le \lambda_2 \le \lambda_3$, show using the spectral decomposition of S that

$$\lambda_1 = \min_u \frac{u \cdot Su}{u \cdot u}, \quad \lambda_3 = \max_u \frac{u \cdot Su}{u \cdot u}.$$

35. If $T \in \text{Lin}$ and $u \in V$, show that

$$(Tu, Tu) \le (T, T)(u, u).$$

(Hint: $T^T T$ is a symmetric, positive semi-definite tensor.)

36. If $R \in \text{Lin}$, show that $|R^T R| \le |R|^2$ (Hint: $R^T R$ is a symmetric, positive semi-definite tensor). Use this result and the Cauchy–Schwartz inequality to show that if $R, S \in \text{Lin}$, then

$$|RS| \le |R|\,|S|.$$

As a result of the Cauchy–Schwartz inequality, we also have $|(\boldsymbol{R}, \boldsymbol{S})| \leq |\boldsymbol{R}|\,|\boldsymbol{S}|$. However, among the two quantities $|(\boldsymbol{R}, \boldsymbol{S})|$ and $|\boldsymbol{R}\boldsymbol{S}|$, either can be greater. For example, let $\boldsymbol{R} = \begin{bmatrix} 1 & 0 \\ -1 & 2 \end{bmatrix}$; if $\boldsymbol{S} = \begin{bmatrix} 1 & 1 \\ 0 & 1 \end{bmatrix}$, then $|(\boldsymbol{R}, \boldsymbol{S})| = 3$ and $|\boldsymbol{R}\boldsymbol{S}| = 2$, while if $\boldsymbol{S} = \begin{bmatrix} -1 & 1 \\ 0 & 1 \end{bmatrix}$, then $|(\boldsymbol{R}, \boldsymbol{S})| = 1$ and $|\boldsymbol{R}\boldsymbol{S}| = 2$.

37. Show that $\boldsymbol{S}^n$, $\boldsymbol{S}^{-1}$ and $\lambda(\operatorname{tr}\boldsymbol{S})\boldsymbol{I} + 2\mu\boldsymbol{S}$ are isotropic functions.

38. Show that for any orthogonal $\boldsymbol{Q}$, the tensor $\dot{\boldsymbol{Q}}\boldsymbol{Q}^T$ is skew at each t.

39. Show using Eqn. (1.336) that if $\boldsymbol{G}(\boldsymbol{T}) = \operatorname{\mathbf{cof}}\boldsymbol{T}$, then

$$D\boldsymbol{G}(\boldsymbol{T})[\boldsymbol{U}] = [\operatorname{tr}\boldsymbol{T}\operatorname{tr}\boldsymbol{U} - \operatorname{tr}(\boldsymbol{T}\boldsymbol{U})]\,\boldsymbol{I} + [\boldsymbol{T}^T - (\operatorname{tr}\boldsymbol{T})\boldsymbol{I}]\boldsymbol{U}^T + [\boldsymbol{U}^T - (\operatorname{tr}\boldsymbol{U})\boldsymbol{I}]\boldsymbol{T}^T. \tag{1.346}$$

Alternatively, using Eqn. (1.60), show that

$$[D\boldsymbol{G}(\boldsymbol{T})[\boldsymbol{U}]]_{ij} = \epsilon_{imn}\epsilon_{jpq}T_{mp}U_{nq}, \tag{1.347}$$

$$\frac{\partial G_{ij}}{\partial T_{kl}} = \epsilon_{ikm}\epsilon_{jln}T_{mn}. \tag{1.348}$$

Equation (1.348) in direct tensorial notation, obtained using Eqns. (1.55), (1.226a), (1.226b) and (1.233), is given by

$$\frac{\partial}{\partial \boldsymbol{T}}(\operatorname{\mathbf{cof}}\boldsymbol{T}) = \operatorname{tr}\boldsymbol{T}[\boldsymbol{I}\otimes\boldsymbol{I} - \mathbb{T}] - (\boldsymbol{I}\otimes\boldsymbol{T}^T + \boldsymbol{T}^T\otimes\boldsymbol{I}) + (\boldsymbol{I}\boxtimes\boldsymbol{T})\mathbb{T} + (\boldsymbol{T}^T\boxtimes\boldsymbol{I})\mathbb{T}. \tag{1.349}$$

If $\boldsymbol{T} \equiv \boldsymbol{C} \in \text{Sym}$, then, by virtue of Eqn. (1.241), we just pre- and post-multiply the above result with $\mathbb{S}$, and use Eqn. (1.175) to get

$$\frac{\partial}{\partial \boldsymbol{C}}(\operatorname{\mathbf{cof}}\boldsymbol{C}) = \operatorname{tr}\boldsymbol{C}\,[\boldsymbol{I}\otimes\boldsymbol{I} - \mathbb{S}] - (\boldsymbol{I}\otimes\boldsymbol{C} + \boldsymbol{C}\otimes\boldsymbol{I}) + \mathbb{S}\,[\boldsymbol{I}\boxtimes\boldsymbol{C} + \boldsymbol{C}\boxtimes\boldsymbol{I}]\,\mathbb{S}. \tag{1.350}$$

The indicial notation form of the above equation, obtained using Eqns. (1.161) and (1.348), is (for the engineering form, see Eqn. (I.6))

$$\left[\frac{\partial}{\partial \boldsymbol{C}}(\operatorname{\mathbf{cof}}\boldsymbol{C})\right]_{ijkl} = \frac{1}{4}\left[\epsilon_{ikr}\epsilon_{jls} + \epsilon_{ilr}\epsilon_{jks} + \epsilon_{jkr}\epsilon_{ils} + \epsilon_{jlr}\epsilon_{iks}\right]C_{rs}. \tag{1.351}$$

Finally, consider the case when $\boldsymbol{C} \in \text{Sym}$ is invertible, i.e., $\operatorname{\mathbf{cof}}\boldsymbol{C} = (\det\boldsymbol{C})\boldsymbol{C}^{-1}$. Show using Eqns. (1.236) and (1.240) that

$$\frac{\partial}{\partial \boldsymbol{C}}(\operatorname{\mathbf{cof}}\boldsymbol{C}) = \det\boldsymbol{C}\left[\boldsymbol{C}^{-1}\otimes\boldsymbol{C}^{-1} - \mathbb{S}(\boldsymbol{C}^{-1}\boxtimes\boldsymbol{C}^{-1})\mathbb{S}\right]. \tag{1.352}$$

40. Let $\phi_1(\boldsymbol{T}) = \operatorname{tr}(\boldsymbol{T}^{-1}\boldsymbol{T}^{-1}) = \boldsymbol{T}^{-T} : \boldsymbol{T}^{-1}$, $\phi_2(\boldsymbol{T}) = (\det\boldsymbol{T})\boldsymbol{T}^{-1} : \boldsymbol{T}^{-1}$, and $\phi_3(\boldsymbol{T}) = (\operatorname{\mathbf{cof}}\boldsymbol{T}) : (\operatorname{\mathbf{cof}}\boldsymbol{T}) = \left[(\boldsymbol{T}:\boldsymbol{T})^2 - (\boldsymbol{T}^T\boldsymbol{T}) : (\boldsymbol{T}^T\boldsymbol{T})\right]/2$. Using either Eqns. (1.228)–(1.237) or by means of the directional derivative, show that

$$\frac{\partial \phi_1}{\partial \boldsymbol{T}} = -2(\boldsymbol{T}^{-T})^3,$$

$$\frac{\partial \phi_2}{\partial T} = (T^{-1} : T^{-1})\,\mathbf{cof}\, T - 2(\det T) T^{-T} T^{-1} T^{-T},$$

$$\frac{\partial \phi_3}{\partial T} = 2(T : T)T - 2TT^T T.$$

41. Prove that

$$\nabla(\phi v) = \phi \nabla v + v \otimes \nabla \phi,$$

$$\nabla \cdot (\phi v) = \phi(\nabla \cdot v) + v \cdot (\nabla \phi), \tag{1.353}$$

$$\nabla \times (u \times v) = (\nabla \cdot v)u - (\nabla v)u - (\nabla \cdot u)v + (\nabla u)v, \tag{1.354}$$

$$\nabla(u \cdot v) = (\nabla u)^T v + (\nabla v)^T u,$$

$$\nabla[(u \cdot v)w] = (u \cdot v)\nabla w + w \otimes [(\nabla v)^T u] + w \otimes [(\nabla u)^T v],$$

$$\nabla \cdot (u \otimes v) = u \nabla \cdot v + (\nabla u)v,$$

$$\nabla \cdot (T^T v) = T : \nabla v + v \cdot (\nabla \cdot T), \tag{1.355}$$

$$\nabla \cdot (\phi T) = \phi \nabla \cdot T + T \nabla \phi,$$

$$\nabla(\phi T) = \phi \nabla T + T \otimes \nabla \phi,$$

$$\nabla(\phi T v) = \phi \nabla(Tv) + (Tv) \otimes \nabla \phi,$$

$$\nabla^2(u \cdot v) = u \cdot \nabla^2 v + v \cdot \nabla^2 u + 2\nabla u : \nabla v,$$

$$\nabla \cdot [a \times (\nabla \times b)] = (\nabla \times a) \cdot (\nabla \times b) - a \cdot [\nabla \times (\nabla \times b)]. \tag{1.356}$$

By integrating Eqn. (1.356) over a closed volume V with surface S show that

$$-\int_S (\nabla \times b) \cdot (a \times n)\, dS = \int_V \{(\nabla \times a) \cdot (\nabla \times b) - a \cdot [\nabla \times (\nabla \times b)]\}\, dV.$$

42. Show that

$$\nabla \cdot [(\nabla u)u] = \nabla u : (\nabla u)^T + u \cdot [\nabla(\nabla \cdot u)],$$

$$\nabla u : (\nabla u)^T = \nabla \cdot [(\nabla u)u - (\nabla \cdot u)u] + (\nabla \cdot u)^2.$$

43. Show that

$$\int_V \left[u \otimes (\nabla \cdot T^T) + (\nabla u)T\right] dV = \int_S u \otimes (T^T n)\, dS.$$

44. Show that $\Re - \{0\}$ is a commutative group with respect to multiplication.

2 Kinematics

Under the continuum hypothesis, the material being studied, irrespective of whether it is solid, liquid or gas, is assumed to be continuous. In reality, any matter is composed of a large number of discrete atoms or molecules. However, continuum mechanics ignores this discontinuous structure, and treats matter as if it is continuous. This hypothetical continuous material is called a *continuum*. The concept of a continuous medium permits us to define quantities such as density, momentum, energy, etc. at each point in the domain occupied by the material. Since these quantities are assumed to have a continuous distribution in space, we can use all the machinery of calculus, including the concepts of derivatives and integrals, to formulate the governing equations. It has been found that the theories of elasticity, fluid mechanics, etc. based on a continuum model yield to quantitative predictions that agree closely with experience. This is perhaps the best justification for using the continuum hypothesis.

The continuum hypothesis is violated when the mean distance that atoms or molecules move between collisions is of the same order of magnitude as the characteristic length scale of the flow, a situation that occurs in low-density or rarefied flow. Such low-density flows occur only in very special situations such as, for example, flow around an airplane at a very high altitude. We do not consider such special situations in this book.

Although we have said that the continuum hypothesis ignores the molecular structure of matter, it does have to account for the fact that rubber and steel, say, behave very differently under the same loading conditions. This is done by means of constitutive equations, which we shall discuss in Chapter 4.

Kinematics refers to the results that can be obtained about the nature of a continuum without reference to the dynamics of the continuum. Thus, kinematics refers to those results that can be obtained purely from geometrical considerations, without reference to the forces acting on the continuum. These results are subsequently used in formulating the balance laws in Chapter 3.

2.1 Lagrangian and Eulerian Descriptions

Consider a body occupying a region in space, and undergoing motion due to the action of some forces so as to occupy a different region of space as shown in Fig. 2.1. The regions of space occupied by the body are called *configurations*. The initial configuration (also known as the undeformed configuration) is the configuration at time $t = 0$. It is usually chosen as the configuration prior to the application of loads. The reference configuration is one with respect to which the deformations in the current configuration are measured. The choice of the reference configuration is arbitrary, and is usually taken as the initial configuration. The current configuration is also known as the deformed configuration although it may have

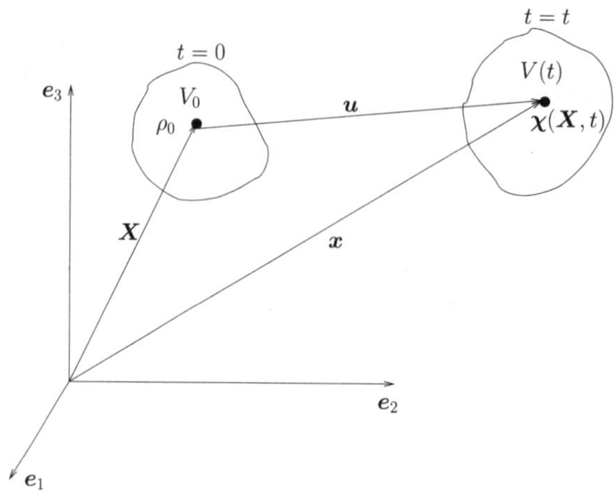

Fig. 2.1 *Reference and deformed configurations.*

resulted due to rigid-body motion. Following the general practice, however imprecise it may be, we shall use the terms reference and deformed configurations.

As shown in Fig. 2.1, the particle occupying a point in the reference configuration V_0 with position vector $\boldsymbol{X}$ occupies a point in the deformed configuration $V(t)$ with position vector $\boldsymbol{x}$. The coordinates $\boldsymbol{X}$ and $\boldsymbol{x}$ are known as the *material* and *spatial* coordinates, respectively. We shall find it convenient to identify points of the body with the positions that they occupy in the reference configuration. Thus, points $\boldsymbol{X} \in V_0$ will often be called as *material points*.

The relation between $\boldsymbol{X}$ and $\boldsymbol{x}$ is expressed as

$$\boldsymbol{x} = \chi(\boldsymbol{X}, t). \tag{2.1}$$

The mapping χ taking $\boldsymbol{X}$ to $\boldsymbol{x}$ is assumed to be one-to-one (except possibly on the boundary) and orientation-preserving (see below), and known as a *deformation*. Physically, the restriction that the mapping characterizing the deformation be one-to-one corresponds to the requirement that the body should not penetrate itself. However, there can be a loss of the one-to-one nature of the mapping at the boundary if self-contact is allowed. Due to the one-to-one nature of the mapping χ, we can invert it to obtain $\boldsymbol{X}$ as a function of $\boldsymbol{x}$ and t, i.e.,

$$\boldsymbol{X} = \chi^{-1}(\boldsymbol{x}, t).$$

A smooth one-parameter family of deformations, with time as the parameter, is known as a *motion*.

The formulation in terms of the reference variable $\boldsymbol{X}$ is called as the *Lagrangian* formulation, while the formulation in terms of the spatial variable $\boldsymbol{x}$ is called the *Eulerian* formulation. We shall develop both these formulations together. The Lagrangian formulation is the one generally preferred in elasticity since the equations of motion in the deformed configuration are expressed in terms of the spatial coordinates $\boldsymbol{x}$, which are themselves unknown. Hence, all the equations are formulated on the reference domain, which is usually taken to be the initial unstressed state. On the other hand, the Eulerian formulation is preferred for problems in fluid mechanics, where we observe the flow in a fixed region of space such as a channel or wind-tunnel.

From Fig. 2.1, we see that the displacement field $\boldsymbol{u}$ is defined by the relation

$$\boldsymbol{x} = \boldsymbol{X} + \boldsymbol{u}. \tag{2.2}$$

At each point in the domain, we define the deformation gradient $\boldsymbol{F}(\boldsymbol{X}, t)$ as $\boldsymbol{\nabla}_X \boldsymbol{\chi}$, with the components given by

$$F_{ij} := \frac{\partial \chi_i}{\partial X_j} = \begin{bmatrix} \frac{\partial \chi_1}{\partial X_1} & \frac{\partial \chi_1}{\partial X_2} & \frac{\partial \chi_1}{\partial X_3} \\ \frac{\partial \chi_2}{\partial X_1} & \frac{\partial \chi_2}{\partial X_2} & \frac{\partial \chi_2}{\partial X_3} \\ \frac{\partial \chi_3}{\partial X_1} & \frac{\partial \chi_3}{\partial X_2} & \frac{\partial \chi_3}{\partial X_3} \end{bmatrix}. \tag{2.3}$$

Since the deformation gradient is defined as the gradient of a vector field, it is a second-order tensor. We shall see later that $\det \boldsymbol{F}$ represents, locally, the volume after deformation per unit original volume; hence it is reasonable to assume $\det \boldsymbol{F} \neq 0$. Further, the deformation gradient in the reference configuration is the identity tensor, so that $\det \boldsymbol{F} = 1$. Thus, a deformation with $\det \boldsymbol{F} < 0$ cannot be reached by a continuous process starting in the reference configuration without $\det \boldsymbol{F}$ becoming zero. Hence, we require the deformation to be orientation-preserving, i.e., (denoting $\det \boldsymbol{F}$ by J)

$$J > 0 \quad \forall \boldsymbol{X} \in V_0,\ t \in [0, \infty). \tag{2.4}$$

Thus, the deformation gradient is invertible at all points in the domain V_0 at any instant of time t, with the inverse given by $\boldsymbol{F}^{-1} = \boldsymbol{\nabla}_x \boldsymbol{\chi}^{-1}$. A deformation with $\boldsymbol{F}$ constant is called *homogeneous*.

Using Eqn. (2.2), we obtain

$$\boldsymbol{F} = \boldsymbol{I} + \boldsymbol{\nabla}_X \boldsymbol{u}, \tag{2.5a}$$

$$\boldsymbol{F}^{-1} = \boldsymbol{I} - \boldsymbol{\nabla}_x \boldsymbol{u}, \tag{2.5b}$$

where $\boldsymbol{\nabla}_X \boldsymbol{u}$ is known as the displacement gradient. In indicial notation, the above equations can be written as

$$F_{ij} = \delta_{ij} + \frac{\partial u_i}{\partial X_j}, \tag{2.6a}$$

$$F_{ij}^{-1} = \delta_{ij} - \frac{\partial u_i}{\partial x_j}. \tag{2.6b}$$

By the polar decomposition theorem discussed in Section 1.6.2, it is possible to decompose the deformation gradient as $\boldsymbol{F} = \boldsymbol{R}\boldsymbol{U} = \boldsymbol{V}\boldsymbol{R}$; $\boldsymbol{U}$ is called the right stretch tensor, $\boldsymbol{V}$ is called the left stretch tensor, and their eigenvalues are known as the principal stretches.

2.2 Length, Area and Volume Elements in the Deformed Configuration

We now find relations between the length, area and volume elements in the deformed configuration, with the corresponding quantities in the reference configuration. The change in length of an element is used for defining the Green and Almansi strain tensors. The relation between the areas in the deformed and reference configurations will be used to define the first and second Piola–Kirchhoff stresses. The relations for the volume transformation are used in the derivation of the conservation laws of mass, momentum, energy, etc. In formulae where indicial notation is used, the indices i, j, etc range from 1 to 3, and summation is assumed over repeated indices.

2.2.1 Length element in the deformed configuration: Strain tensors

At a given instant in time, a length element in the deformed configuration $\boldsymbol{dx}$ is related to a length element in the reference configuration $\boldsymbol{dX}$ by the following relationship:[1]

$$\boldsymbol{dx} = \boldsymbol{F}\boldsymbol{dX}; \qquad dx_i = F_{ij}dX_j = \frac{\partial \chi_i}{\partial X_j}dX_j. \tag{2.7}$$

Since $\boldsymbol{F}$ is invertible, we also have

$$\boldsymbol{dX} = \boldsymbol{F}^{-1}\boldsymbol{dx}; \qquad dX_i = (F^{-1})_{ij}dx_j = \frac{\partial \chi_i^{-1}}{\partial x_j}dx_j. \tag{2.8}$$

Using Eqn. (2.7), the change in length of an element is given by

$$\begin{aligned} ds^2 - dS^2 &= \boldsymbol{dx} \cdot \boldsymbol{dx} - \boldsymbol{dX} \cdot \boldsymbol{dX} \\ &= \boldsymbol{dX} \cdot (\boldsymbol{F}^T\boldsymbol{F} - \boldsymbol{I})\boldsymbol{dX} \\ &= \boldsymbol{dX} \cdot (\boldsymbol{C} - \boldsymbol{I})\boldsymbol{dX} \\ &= 2\boldsymbol{dX} \cdot \boldsymbol{E}\boldsymbol{dX}, \end{aligned} \tag{2.9}$$

where $\boldsymbol{C} = \boldsymbol{F}^T\boldsymbol{F}$ is known as the right Cauchy–Green strain tensor, and

$$\boldsymbol{E} = \frac{1}{2}(\boldsymbol{C} - \boldsymbol{I}) = \frac{1}{2}(\boldsymbol{F}^T\boldsymbol{F} - \boldsymbol{I}), \tag{2.10}$$

is the Lagrangian measure of strain known as the Green strain tensor. In component form, it is given by

$$E_{ij} = \frac{1}{2}\left[\frac{\partial \chi_k}{\partial X_i}\frac{\partial \chi_k}{\partial X_j} - \delta_{ij}\right].$$

The Eulerian measure of strain is obtained by expressing $\boldsymbol{dX}$ in terms of $\boldsymbol{dx}$ as follows:

$$\begin{aligned} ds^2 - dS^2 &= \boldsymbol{dx} \cdot \boldsymbol{dx} - \boldsymbol{dX} \cdot \boldsymbol{dX} \\ &= \boldsymbol{dx} \cdot (\boldsymbol{I} - \boldsymbol{F}^{-T}\boldsymbol{F}^{-1})\boldsymbol{dx} \\ &= \boldsymbol{dx} \cdot (\boldsymbol{I} - \boldsymbol{B}^{-1})\boldsymbol{dx} \\ &= 2\boldsymbol{dx} \cdot \bar{\boldsymbol{E}}\boldsymbol{dx}. \end{aligned}$$

[1] Actually, by the chain rule, $\boldsymbol{dx} = \boldsymbol{F}\boldsymbol{dX} + \boldsymbol{v}dt$, where $\boldsymbol{v}$ is the velocity. However, later, when we derive relations such as Eqns. (2.9) or (2.37), the ds or dV in these relations refers to the length or volume element of the material volume $V(t)$ at a *given* instant in time. Keeping this in mind, we have set $dt = 0$, thereby obtaining Eqn. (2.7).

$B = FF^T$ is known as the left Cauchy–Green strain tensor, and

$$\bar{E} = \frac{1}{2}(I - B^{-1}) = \frac{1}{2}\left(I - F^{-T}F^{-1}\right), \tag{2.11}$$

is known as the Almansi strain tensor. In component form, the Almansi strain tensor is given by

$$\bar{E}_{ij} = \frac{1}{2}\left[\delta_{ij} - \frac{\partial \chi_k^{-1}}{\partial x_i}\frac{\partial \chi_k^{-1}}{\partial x_j}\right].$$

From their definitions, it is immediately obvious that C, B, E and $\bar{E}$ are all symmetric tensors. Since F is assumed to be nonsingular, C and B are positive definite as well. Thus, $B, C \in$ Psym. To see that they are tensors, we first note that F and F^T are tensors. The product of two tensors is a tensor so that F^TF and FF^T are tensors, and, finally, to show that E and $\bar{E}$ are tensors, we use the fact that the sum of two tensors is a tensor. Since $F = RU = VR$, we have $C = U^2$ and $B = V^2$, and

$$C = UU = R^TVRR^TVR = R^TV^2R = R^TBR. \tag{2.12}$$

Thus, B and C are orthogonally equivalent, and hence have the same eigenvalues.

From Eqns. (2.10) and (2.11), we see that the Green strain tensor E, and the Almansi strain tensor $\bar{E}$ are related by

$$E = F^T\bar{E}F. \tag{2.13}$$

To express the strain tensors in terms of the displacement gradients, we substitute Eqns. (2.5) into Eqns. (2.10) and (2.11) to get

$$E = \frac{1}{2}\left[\nabla_X u + (\nabla_X u)^T + (\nabla_X u)^T\nabla_X u\right], \tag{2.14}$$

$$\bar{E} = \frac{1}{2}\left[\nabla_x u + (\nabla_x u)^T - (\nabla_x u)^T\nabla_x u\right], \tag{2.15}$$

or, in component form,

$$E_{ij} = \frac{1}{2}\left[\frac{\partial u_i}{\partial X_j} + \frac{\partial u_j}{\partial X_i} + \frac{\partial u_k}{\partial X_i}\frac{\partial u_k}{\partial X_j}\right], \tag{2.16}$$

$$\bar{E}_{ij} = \frac{1}{2}\left[\frac{\partial u_i}{\partial x_j} + \frac{\partial u_j}{\partial x_i} - \frac{\partial u_k}{\partial x_i}\frac{\partial u_k}{\partial x_j}\right].$$

From Eqn. (2.14), we see that the introduction of the factor $1/2$ in the definition of E is motivated by the requirement that its linear part coincide with the linearized strain tensor (also known as the small strain tensor) $[\nabla_X u + (\nabla_X u)^T]/2$, which we will henceforth denote by ϵ.

To show that the tensor E is a good measure of strain understood in its intuitive sense of 'change in form or size', we need to define a class of deformations called *rigid deformations*. A deformation is said to be rigid if it preserves distances between any two points. A motion is said to be a rigid motion if the distance between any two points is preserved at all times, i.e.,

$$[\chi(X_1,t) - \chi(X_2,t)] \cdot [\chi(X_1,t) - \chi(X_2,t)] = (X_1 - X_2) \cdot (X_1 - X_2). \tag{2.17}$$

for all $X_1, X_2 \in V_0$, and at all t. We have the following important theorem that relates a rigid motion with an absence of strain (we shall present another characterization in Theorem 2.4.1).

Theorem 2.2.1 (Characterization of rigid motions-I). *The following are equivalent:*
(i) The motion is rigid.
(ii) The motion is of the form

$$\chi(X,t) = Q(t)X + c(t) \quad \forall X \in V_0, \tag{2.18}$$

where $Q(t) \in$ Orth$^+$, and $c(t)$ is a vector. (Physically, this corresponds to the idea of rotating the reference configuration around the origin by the rotation $Q(t)$, and translating it by the vector $c(t)$.)
(iii) $E(X,t) = 0$, i.e., $C(X,t) = I$ for all $X \in V_0$, and at all times.

Proof. We shall show that (i) $\implies$ (ii) $\implies$ (iii) $\implies$ (i). Thus, first assuming that the motion is rigid, by Eqn. (2.17) we have

$$[\chi(X_1,t) - \chi(X_2,t)] \cdot [\chi(X_1,t) - \chi(X_2,t)] = (X_1 - X_2) \cdot (X_1 - X_2).$$

Differentiating the above equation with respect to X_2 using Eqn. (1.270), we get

$$F(X_2,t)^T [\chi(X_1,t) - \chi(X_2,t)] = X_1 - X_2.$$

Differentiating, the above equation with respect to X_1, we get

$$F(X_2,t)^T F(X_1,t) = I \quad \forall X_1, X_2 \in V_0.$$

Taking $X_2 = X_1$, we see that $F(X,t)$ is orthogonal at each $X \in V_0$, and at any given instant of time t. Hence, the above equation implies that at any given instant of time

$$F(X_1,t) = F(X_2,t) \quad \forall X_1, X_2 \in V_0,$$

so that F is a function of time alone. By the orientation-preserving condition, $\det F > 0$. Hence, F is a rotation, i.e., $F = Q(t) \in$ Orth$^+$. Integrating the relation $F = \nabla\chi = Q(t)$, we get Eqn. (2.18).

Now we show that (ii) $\implies$ (iii). If Eqn. (2.18) holds then $F = Q(t)$ at all points X in the reference configuration. Hence, $C = F^T F = Q^T Q = I$, which implies that $E = (C - I)/2 = 0$.

Finally, to show that (iii) $\implies$ (i), let $E = 0$ at all times. Then by Eqn. (2.9), $ds = dS$ at each point $X \in V_0$, and at each instant of time t. Hence, the length of each curve in the reference configuration obtained by integrating dS is preserved in the deformed configuration at any given instant of time. Consider the straight line joining two points $\mathbf{X}_1$ and $\mathbf{X}_2$ in the reference configuration[a]. Suppose this straight line of length L gets deformed to some curve (not necessarily a straight line) with endpoints $\mathbf{x}_1 = \chi(\mathbf{X}_1, t)$ and $\mathbf{x}_2 = \chi(\mathbf{X}_2, t)$ in the deformed configuration at time t. Since the total length L of the curve remains unchanged after deformation, L is greater than or equal to the straight line distance between $\mathbf{x}_1$ and $\mathbf{x}_2$. Thus, we have

$$L = |\mathbf{X}_1 - \mathbf{X}_2| \geq |\mathbf{x}_1 - \mathbf{x}_2|.$$

Next, consider the curve in the reference configuration that gets deformed to the straight line joining $\mathbf{x}_1$ and $\mathbf{x}_2$. Now, we have

$$|\mathbf{X}_1 - \mathbf{X}_2| \leq |\mathbf{x}_1 - \mathbf{x}_2|.$$

From the two inequalities above, we have

$$|\mathbf{x}_1 - \mathbf{x}_2| = |\chi(\mathbf{X}_1, t) - \chi(\mathbf{X}_2, t)| = |\mathbf{X}_1 - \mathbf{X}_2|,$$

which is nothing but Eqn. (2.17).

The following shorter proof for showing that (iii) $\implies$ (ii) $\implies$ (i) can be given if we assume that $\chi(\mathbf{X}, t)$ is twice differentiable with respect to $\mathbf{X}$. Since $\mathbf{C} = \mathbf{F}^T\mathbf{F} = \mathbf{I}$, we have $\mathbf{F} = \mathbf{F}^{-T}$. This shows that $\mathbf{F}$ is orthogonal. We now have to prove that it is a *constant* orthogonal tensor over the entire domain. Writing $\mathbf{F}^T\mathbf{F} = \mathbf{I}$ in indicial notation, we have

$$\frac{\partial \chi_m}{\partial X_i}\frac{\partial \chi_m}{\partial X_j} = \delta_{ij}.$$

Differentiating this equation with respect to X_k, we get

$$\frac{\partial^2 \chi_m}{\partial X_i \partial X_k}\frac{\partial \chi_m}{\partial X_j} + \frac{\partial^2 \chi_m}{\partial X_j \partial X_k}\frac{\partial \chi_m}{\partial X_i} = 0. \tag{2.19}$$

By permuting the indices i, j and k, we get the following two equations:

$$\frac{\partial^2 \chi_m}{\partial X_j \partial X_i}\frac{\partial \chi_m}{\partial X_k} + \frac{\partial^2 \chi_m}{\partial X_k \partial X_i}\frac{\partial \chi_m}{\partial X_j} = 0, \tag{2.20}$$

$$\frac{\partial^2 \chi_m}{\partial X_k \partial X_j}\frac{\partial \chi_m}{\partial X_i} + \frac{\partial^2 \chi_m}{\partial X_i \partial X_j}\frac{\partial \chi_m}{\partial X_k} = 0. \tag{2.21}$$

Adding Eqns. (2.20) and (2.21) and subtracting Eqn. (2.19) from it, we get

$$\frac{\partial^2 \chi_m}{\partial X_i \partial X_j}\frac{\partial \chi_m}{\partial X_k} = 0. \tag{2.22}$$

The above equation is of the form $\mathbf{G}\mathbf{F} = \mathbf{0}$, where $\mathbf{G}$ is a third-order tensor. Since $\mathbf{F}$ is invertible, we multiply both sides by $\mathbf{F}^{-1}$ and conclude that $\mathbf{G} = \mathbf{0}$, i.e.,

$$\frac{\partial^2 \chi_m}{\partial X_i \partial X_j} = 0,$$

which proves that the elements of F are independent of X. Integrating the relation $F = \nabla\chi = Q(t)$, we get Eqn. (2.18). If Eqn. (2.18) holds, then

$$[\chi(X_1,t) - \chi(X_2,t)] \cdot [\chi(X_1,t) - \chi(X_2,t)] = Q(t)(X_1 - X_2) \cdot Q(t)(X_1 - X_2)$$
$$= (X_1 - X_2) \cdot (X_1 - X_2),$$

which is Eqn. (2.17). □

[a]The two points can be taken sufficiently close to each other so that the straight line joining them does not lie outside the body.

Thus, $E = (C - I)/2$ is a measure of the 'deviation' between a given deformation and a rigid deformation, since $C = I$ if and only if the deformation is rigid. The above theorem also tells us that, in order to prevent rigid-body motion, six (3 parameters of Q and 3 components of c) and three (1 parameter of Q and 2 components of c) degrees of freedom need to be suppressed in the three and two-dimensional cases, respectively.

A more general version of Theorem 2.2.1 is given by the following theorem that shows that two deformations corresponding to the same tensor E (or, alternatively, the same C) can be obtained from one another by composition with a rigid deformation.

Theorem 2.2.2. *If $\chi_1(X,t)$ and $\chi_2(X,t)$ are two deformations with deformation gradients F_1 and F_2, then the following are equivalent*

(i) $\chi_1(X,t) = Q(t)\chi_2(X,t) + c(t), \forall X \in V_0,$

(ii) $F_1^T F_1 = F_2^T F_2,$

where $Q(t) \in Orth^+$.

Proof. That (i) implies (ii) is obvious since we have $F_1 = Q(t)F_2$. To show that (ii) implies (i), we write it as

$$I = F_2^{-T} F_1^T F_1 F_2^{-1}$$
$$= (F_1 F_2^{-1})^T F_1 F_2^{-1}.$$

Denote $F_1 F_2^{-1}$ by F_3. We see that $F_3^T F_3 = I$, so that $F_3 = Q(X,t)$, where $Q(X,t) \in \text{Orth}^+$. A deformation with F_3 as the deformation gradient exists, and is given by $\chi_1 \circ \chi_2^{-1}$. Hence, applying Theorem 2.2.1, we conclude that Q is constant for all X. Since $F_3 = F_1 F_2^{-1} = Q(t)$, we have $F_1 = Q(t)F_2$. On integrating this equation, we obtain (i). □

Note that by taking $F_2 = Q(t)$ (rigid deformation), Theorem 2.2.2 reduces to Theorem 2.2.1.

The necessary and sufficient compatibility conditions that a tensor field $F \in \text{Lin}^+$ should satisfy, in order that there exists a deformation field $\chi(X,t)$ such that $F = \nabla\chi$ are given by

$$\nabla \times F = 0.$$

The necessity of the above condition follows almost immediately by using the relation $F_{ij} = \partial\chi_i/\partial X_j$, so that

$$\begin{aligned}(\nabla \times F)_{ij} &= \epsilon_{imn}\frac{\partial^2\chi_j}{\partial X_m \partial X_n} \\ &= \epsilon_{inm}\frac{\partial^2\chi_j}{\partial X_n \partial X_m} \quad \text{(since } m \text{ and } n \text{ are dummy indices)} \\ &= -\epsilon_{imn}\frac{\partial^2\chi_j}{\partial X_m \partial X_n} \\ &= -(\nabla \times F)_{ij},\end{aligned}$$

which yields the desired result. Note that we have used the fact that the order of differentiation with respect to X_m and X_n can be interchanged in the second to last step. To prove sufficiency, we need to show that if $\nabla \times F = 0$, there exists a vector function χ such that $F = \nabla\chi$. This result follows directly from Potential theorem (2).

The following discussion uses Christoffel symbols that are discussed in Appendix H. The necessary and sufficient compatibility conditions that a positive definite and symmetric tensor field C should satisfy, in order that there exists a deformation field $\chi(X, t)$ (determined upto rigid deformations) such that $F^T F = C$ are found by using C in place of the metric tensor in Eqn. (H.24), and are thus given by [23, 59]:

$$R_{ijkl} \equiv \frac{\partial \Gamma_{jli}}{\partial X_k} - \frac{\partial \Gamma_{jki}}{\partial X_l} + C^{-1}_{pq}\left(\Gamma_{jkp}\Gamma_{ilq} - \Gamma_{jlp}\Gamma_{ikq}\right) = 0, \qquad 1 \le i, j, k, l \le 3, \tag{2.23}$$

where

$$\Gamma_{ijk} = \Gamma_{jik} = \frac{1}{2}\left(\frac{\partial C_{jk}}{\partial X_i} + \frac{\partial C_{ki}}{\partial X_j} - \frac{\partial C_{ij}}{\partial X_k}\right). \tag{2.24}$$

To directly prove the necessity of the compatibility relations (for a proof of sufficiency, see [59, 358]), write C_{ij} as $\frac{\partial\chi_m}{\partial X_i}\frac{\partial\chi_m}{\partial X_j}$, and differentiate it with respect to X_k to get a relation similar to Eqn. (2.19) with $\partial C_{ij}/\partial X_k$ instead of zero on the right-hand side. Now permute the indices i, j, k in the same manner as was done in obtaining Eqns. (2.20) and (2.21), and subtract the first equation from the sum of the latter two as was done in obtaining Eqn. (2.22), to get

$$\Gamma_{ijk} = \frac{1}{2}\left(\frac{\partial C_{jk}}{\partial X_i} + \frac{\partial C_{ki}}{\partial X_j} - \frac{\partial C_{ij}}{\partial X_k}\right) = \frac{\partial^2\chi_m}{\partial X_i \partial X_j}\frac{\partial\chi_m}{\partial X_k} = \frac{\partial^2\chi_m}{\partial X_i \partial X_j}F_{mk}.$$

Using the above relation, we have

$$C^{-1}_{pq}\Gamma_{jkp}\Gamma_{ilq} = F^{-1}_{pr}F^{-1}_{qr}\frac{\partial^2\chi_m}{\partial X_j \partial X_k}F_{mp}\frac{\partial^2\chi_n}{\partial X_i \partial X_l}F_{nq} = \frac{\partial^2\chi_r}{\partial X_j \partial X_k}\frac{\partial^2\chi_r}{\partial X_i \partial X_l}.$$

Interchanging l and k to obtain an expression for the second term, and simplifying, one can easily verify that Eqn. (2.23) holds. Using Eqn. (H.25), Eqn. (2.23) can also be written as

$$R_{ijkl} \equiv \frac{1}{2}\left(\frac{\partial^2 C_{jk}}{\partial X_i \partial X_l} + \frac{\partial^2 C_{il}}{\partial X_j \partial X_k} - \frac{\partial^2 C_{jl}}{\partial X_i \partial X_k} - \frac{\partial^2 C_{ik}}{\partial X_j \partial X_l}\right) + C^{-1}_{pq}\left(\Gamma_{jkp}\Gamma_{ilq} - \Gamma_{jlp}\Gamma_{ikq}\right) = 0, \quad 1 \le i,j,k,l \le 3.$$

From the above relation, the symmetry of $\boldsymbol{C}$, and Eqn. (2.24), we have

$$\begin{aligned} R_{ijkl} &= -R_{jikl} = -R_{ijlk} = R_{klij}, \\ R_{ijkl} &+ R_{iljk} + R_{iklj} = 0, \end{aligned} \tag{2.25}$$

so that of the 81 equations in Eqn. (2.23), only six are distinct, and these are given by

$$\begin{aligned} &R_{1212} = 0, \quad R_{1313} = 0, \quad R_{2323} = 0, \\ &R_{1323} = 0, \quad R_{2131} = 0, \quad R_{1232} = 0. \end{aligned}$$

By replacing $\boldsymbol{C}$ by $\boldsymbol{B}^{-1}$ and X by x in Eqn. (2.23), we get the compatibility equations for $\boldsymbol{B}^{-1}$. The compatibility conditions for $\boldsymbol{B}$ are discussed in [3] and [23].

2.2.2 Area element in the deformed configuration: The Piola transform

To compute the area element in the deformed configuration in terms of the area element in the reference configuration, it is convenient to introduce the Piola transform that relates tensors defined over the reference and deformed configurations. The Piola transform is also useful for defining the first Piola–Kirchhoff stress tensor (see Section 3.8).

Let $\boldsymbol{T}$ be a second-order tensor defined over the reference configuration V_0. Its divergence with respect to the material variable $\boldsymbol{X}$ is given by

$$\boldsymbol{\nabla}_X \cdot \boldsymbol{T} = \frac{\partial T_{ij}}{\partial X_j}\boldsymbol{e}_i.$$

Similarly, the divergence of a tensor field $\boldsymbol{\tau}$ over the deformed configuration with respect to the spatial variable $\boldsymbol{x}$ is

$$\boldsymbol{\nabla}_x \cdot \boldsymbol{\tau} = \frac{\partial \tau_{ij}}{\partial x_j}\boldsymbol{e}_i.$$

By applying the divergence theorem for tensor fields as given by Eqn. (1.316), we get

$$\int_{V_0} \boldsymbol{\nabla}_X \cdot \boldsymbol{T}\, dV_0 = \int_{S_0} \boldsymbol{T}\boldsymbol{n}_0\, dS_0, \tag{2.26}$$

$$\int_{V(t)} \boldsymbol{\nabla}_x \cdot \boldsymbol{\tau}\, dV = \int_{S(t)} \boldsymbol{\tau}\boldsymbol{n}\, dS, \tag{2.27}$$

where $\boldsymbol{n}_0$ and $\boldsymbol{n}$ denote the unit outward normals along the surfaces of the reference and deformed configurations, respectively. The Piola transform is defined by

$$\boldsymbol{T}(\boldsymbol{X}) := J\boldsymbol{\tau}(\boldsymbol{x})\boldsymbol{F}^{-T} = \boldsymbol{\tau}(\boldsymbol{x})\,\mathbf{cof}\,\boldsymbol{F}. \tag{2.28}$$

The following theorem yields a simple relation between the divergences of $\boldsymbol{T}$ and $\boldsymbol{\tau}$, and as a corollary, the desired relation between the corresponding area elements.

Theorem 2.2.3. *Let the Piola transform be as defined by Eqn. (2.28). Then we have*

$$\nabla_X \cdot T(X) = J\nabla_x \cdot \tau(x) \quad \forall x = \chi(X,t), \tag{2.29}$$

$$T(X)n_0\, dS_0 = \tau(x) n\, dS \quad \forall x = \chi(X,t). \tag{2.30}$$

The area elements dS and dS_0 are related by

$$dS = |(\mathbf{cof}\, F)n_0|\, dS_0 = \sqrt{n_0 \cdot (\mathbf{cof}\, C)n_0}\, dS_0, \tag{2.31}$$

and the normals n and n_0 are related by

$$n = \frac{(\mathbf{cof}\, F)n_0}{|(\mathbf{cof}\, F)n_0|}. \tag{2.32}$$

Proof. We first prove the Piola identity

$$\nabla_X \cdot \mathbf{cof}\, F = \mathbf{0}. \tag{2.33}$$

The elements of the cofactor matrix of F are given by

$$(\mathbf{cof}\, F)_{ij} = \frac{1}{2}\epsilon_{imn}\epsilon_{jpq}\frac{\partial \chi_m}{\partial X_p}\frac{\partial \chi_n}{\partial X_q}.$$

Hence,

$$\begin{aligned}
\frac{\partial(\mathbf{cof}\, F)_{ij}}{\partial X_j} &= \frac{1}{2}\epsilon_{imn}\epsilon_{jpq}\left[\frac{\partial^2 \chi_m}{\partial X_j \partial X_p}\frac{\partial \chi_n}{\partial X_q} + \frac{\partial \chi_m}{\partial X_p}\frac{\partial^2 \chi_n}{\partial X_j \partial X_q}\right] \\
&= \frac{1}{2}\epsilon_{imn}\left[\epsilon_{pjq}\frac{\partial^2 \chi_m}{\partial X_p \partial X_j}\frac{\partial \chi_n}{\partial X_q} + \epsilon_{qpj}\frac{\partial \chi_m}{\partial X_p}\frac{\partial^2 \chi_n}{\partial X_q \partial X_j}\right] \\
&= -\frac{1}{2}\epsilon_{imn}\epsilon_{jpq}\left[\frac{\partial^2 \chi_m}{\partial X_j \partial X_p}\frac{\partial \chi_n}{\partial X_q} + \frac{\partial \chi_m}{\partial X_p}\frac{\partial^2 \chi_n}{\partial X_j \partial X_q}\right] \\
&= -\frac{\partial(\mathbf{cof}\, F)_{ij}}{\partial X_j} \\
\implies \quad \frac{\partial(\mathbf{cof}\, F)_{ij}}{\partial X_j} &= 0.
\end{aligned}$$

Thus,

$$\begin{aligned}
\nabla_X \cdot T &= \nabla_X \cdot (\tau\mathbf{cof}\, F) \\
&= \frac{\partial \tau_{ir}}{\partial X_j}(\mathbf{cof}\, F)_{rj}e_i \\
&= \frac{\partial \tau_{ir}}{\partial x_m}\frac{\partial \chi_m}{\partial X_j}(\mathbf{cof}\, F)_{rj}e_i
\end{aligned}$$

$$
\begin{aligned}
&= \frac{\partial \tau_{ir}}{\partial x_m} F_{mj} (\mathbf{cof}\, \boldsymbol{F})_{rj} \boldsymbol{e}_i \\
&= J \frac{\partial \tau_{ir}}{\partial x_m} \delta_{mr} \boldsymbol{e}_i \qquad \text{(by Eqn. (1.64))} \\
&= J \frac{\partial \tau_{ir}}{\partial x_r} \boldsymbol{e}_i \\
&= J \boldsymbol{\nabla}_x \cdot \boldsymbol{\tau},
\end{aligned}
$$

which proves Eqn. (2.29).

To prove Eqn. (2.30), we use Eqns. (2.26), (2.27) and (2.29), and the relation $dV = J dV_0$:

$$
\begin{aligned}
\int_{S_0} \boldsymbol{T} \boldsymbol{n}_0 \, dS_0 &= \int_{V_0} \boldsymbol{\nabla}_X \cdot \boldsymbol{T} \, dV_0 \\
&= \int_{V(t)} \boldsymbol{\nabla}_x \cdot \boldsymbol{\tau} \, dV \\
&= \int_{S(t)} \boldsymbol{\tau} \boldsymbol{n} \, dS,
\end{aligned}
$$

which proves the relation $\boldsymbol{T}\boldsymbol{n}_0 \, dS_0 = \boldsymbol{\tau}\boldsymbol{n} \, dS$, since the domains S are arbitrary. By taking $\boldsymbol{\tau} = \boldsymbol{I}$, we get

$$
\boldsymbol{n} \, dS = (\mathbf{cof}\, \boldsymbol{F}) \boldsymbol{n}_0 \, dS_0. \tag{2.34}
$$

By taking the Euclidean norm of both sides of the above equation, and noting that $|\boldsymbol{n}| = 1$, we get Eqn. (2.31). From Eqns. (2.31) and (2.34), we get Eqn. (2.32).

An alternative proof of Eqn. (2.34) is as follows:

$$
\begin{aligned}
\boldsymbol{n} \, dS &= \boldsymbol{dx}_1 \times \boldsymbol{dx}_2 \\
&= \boldsymbol{F} \boldsymbol{dX}_1 \times \boldsymbol{F} \boldsymbol{dX}_2 \\
&= \mathbf{cof}\, \boldsymbol{F} (\boldsymbol{dX}_1 \times \boldsymbol{dX}_2) \\
&= \mathbf{cof}\, \boldsymbol{F} (\boldsymbol{n}_0 \, dS_0).
\end{aligned}
$$

□

The spatial version of the Piola identity is given by

$$
\boldsymbol{\nabla}_x \cdot (\mathbf{cof}\, \boldsymbol{F}^{-1}) = \mathbf{0}.
$$

This can be proved by replacing $\boldsymbol{X}$ by $\boldsymbol{x}$ and $\boldsymbol{\chi}$ by $\boldsymbol{\chi}^{-1}$ in the proof of Eqn. (2.33).

Using Eqns. (1.355) and (2.33), we have

$$
\boldsymbol{\nabla}_X \cdot [(\mathbf{cof}\, \boldsymbol{F})^T \boldsymbol{u}] = \mathbf{cof}\, \boldsymbol{F} : \boldsymbol{\nabla}_X \boldsymbol{u} + \boldsymbol{u} \cdot (\boldsymbol{\nabla}_X \cdot \mathbf{cof}\, \boldsymbol{F}) = \mathbf{cof}\, \boldsymbol{F} : \boldsymbol{\nabla}_X \boldsymbol{u} \quad \forall \boldsymbol{u}. \tag{2.35}
$$

In particular, for $\boldsymbol{u} = \boldsymbol{\chi}$, and on using Eqn. (1.334), it follows that

$$
\boldsymbol{\nabla}_X \cdot [(\mathbf{cof}\, \boldsymbol{F})^T \boldsymbol{\chi}] = \mathbf{cof}\, \boldsymbol{F} : \boldsymbol{F} = 3 \det \boldsymbol{F}. \tag{2.36}
$$

2.2.3 Volume element in the deformed configuration

The volume element dV in the deformed configuration is related to the volume element in the reference configuration by the relation

$$dV = J dV_0. \tag{2.37}$$

To prove this, note that the volume of an element in the deformed configuration is

$$\begin{aligned} |[\boldsymbol{dx}_1, \boldsymbol{dx}_2, \boldsymbol{dx}_3]| &= |[\boldsymbol{F}\boldsymbol{dX}_1, \boldsymbol{F}\boldsymbol{dX}_2, \boldsymbol{F}\boldsymbol{dX}_3]| \\ &= |\det \boldsymbol{F}\, [\boldsymbol{dX}_1, \boldsymbol{dX}_2, \boldsymbol{dX}_3]| \quad \text{(by Eqn. (1.43c))} \\ &= J dV_0, \end{aligned}$$

where in the last step, we have used the fact that $|\det \boldsymbol{F}| = \det \boldsymbol{F} = J > 0$. A deformation that is volume-preserving, i.e., one for which $J = 1$, is called *isochoric*.

2.3 Velocity and Acceleration

We first derive the following result that we shall need in the subsequent development:

If $g(t)$ is a scalar-valued function of t, and $\Phi(g(t), t)$ is a scalar-valued function that depends both explicitly and implicitly (through $g(t)$) on t, then its total derivative is given by

$$\frac{d\Phi}{dt} = \left(\frac{\partial \Phi}{\partial t}\right)_{g(t)} + \frac{\partial \Phi}{\partial g}\frac{dg}{dt}. \tag{2.38}$$

To prove this result, consider the Taylor series expansion of $\Phi(g(t+\Delta t), t+\Delta t)$. Denoting dg/dt by $g'(t)$, we have

$$\begin{aligned} \Phi(g(t+\Delta t), t+\Delta t) &= \Phi(g(t+\Delta t), t) + \left(\frac{\partial \Phi}{\partial t}\right)_{g(t+\Delta t)} \Delta t + \mathrm{o}\,(\Delta t) \\ &= \Phi(g(t) + g'(t)\Delta t + \mathrm{o}\,(\Delta t)\,, t) + \left(\frac{\partial \Phi}{\partial t}\right)_{g(t+\Delta t)} \Delta t + \mathrm{o}\,(\Delta t) \\ &= \Phi(g(t), t) + \left(\frac{\partial \Phi}{\partial g}\right)_{g(t)} g'(t)\Delta t + \left(\frac{\partial \Phi}{\partial t}\right)_{g(t+\Delta t)} \Delta t + \mathrm{o}\,(\Delta t) \\ &= \Phi(g(t), t) + \left[\left(\frac{\partial \Phi}{\partial g}\right)_{g(t)} g'(t) + \left(\frac{\partial \Phi}{\partial t}\right)_{g(t+\Delta t)}\right] \Delta t + \mathrm{o}\,(\Delta t)\,. \end{aligned}$$

Hence,

$$\begin{aligned} \frac{d\Phi}{dt} &= \lim_{\Delta t \to 0} \frac{1}{\Delta t}\left[\Phi(g(t+\Delta t), t+\Delta t) - \Phi(g(t), t)\right] \\ &= \left(\frac{\partial \Phi}{\partial t}\right)_{g(t)} + \frac{\partial \Phi}{\partial g}\frac{dg}{dt}, \end{aligned}$$

which proves Eqn. (2.38). The extension of this result to the case when both Φ and g are vector-valued functions is straightforward. In component form, it is given by

$$\frac{d\Phi_i}{dt} = \left(\frac{\partial \Phi_i}{\partial t}\right)_{g(t)} + \frac{\partial \Phi_i}{\partial g_j}\frac{dg_j}{dt}. \tag{2.39}$$

Given a material field $\tilde{\phi}(\boldsymbol{X}, t)$, the particle or material derivative of that field, denoted by $D\tilde{\phi}/Dt$, is defined as the partial derivative of $\tilde{\phi}$ with respect to time, i.e.,

$$\frac{D\tilde{\phi}}{Dt} := \left(\frac{\partial \tilde{\phi}}{\partial t}\right)_{\boldsymbol{X}}. \tag{2.40}$$

It is easy to verify that the material derivative satisfies all the usual rules of differentiation.

In the Lagrangian approach, the velocity and acceleration are defined as the first and second material derivatives, respectively, of the mapping χ in Eqn. (2.1), i.e.,

$$\tilde{\boldsymbol{v}}(\boldsymbol{X}, t) := \frac{D\chi}{Dt} = \left(\frac{\partial \chi}{\partial t}\right)_{\boldsymbol{X}}, \tag{2.41}$$

$$\tilde{\boldsymbol{a}}(\boldsymbol{X}, t) := \frac{D\tilde{\boldsymbol{v}}}{Dt} = \left(\frac{\partial \tilde{\boldsymbol{v}}}{\partial t}\right)_{\boldsymbol{X}} = \left(\frac{\partial^2 \chi}{\partial t^2}\right)_{\boldsymbol{X}}. \tag{2.42}$$

However, in fluid mechanics, the problem with finding the velocities and accelerations from the above formulae is that, in general, at any given time t we do not know the reference position $\boldsymbol{X}$ occupied by a particle now at $\boldsymbol{x}$. In such a case, the computation can be carried out using the Eulerian approach, in which the field quantities are now given as functions of the spatial position $\boldsymbol{x}$ and time t. Since $\boldsymbol{X} = \chi^{-1}(\boldsymbol{x}, t)$, the Eulerian description of the velocity is given by

$$\boldsymbol{v}(\boldsymbol{x}, t) = \tilde{\boldsymbol{v}}(\boldsymbol{X}, t) = \tilde{\boldsymbol{v}}(\chi^{-1}(\boldsymbol{x}, t), t). \tag{2.43}$$

In general, the functions $\tilde{\boldsymbol{v}}$ and $\boldsymbol{v}$ are different, though of course, their values at some material point $\boldsymbol{X}$, and the corresponding spatial point $\boldsymbol{x} = \chi(\boldsymbol{X}, t)$ are the same. As an example, consider the motion

$$x = X(1 + t^2), \quad y = Y(1 + t), \quad z = 0.$$

Using Eqn. (2.41), we have

$$\tilde{v}_x = 2Xt, \quad \tilde{v}_y = Y, \quad \tilde{v}_z = 0.$$

Substituting $X = x/(1 + t^2)$ and $Y = y/(1 + t)$ into the above expressions, we get

$$v_x = \frac{2xt}{1 + t^2}, \quad v_y = \frac{y}{1 + t}, \quad v_z = 0.$$

A comparison of $\tilde{\boldsymbol{v}}$ and $\boldsymbol{v}$ shows that they have different functional forms.

The expression for the acceleration using the spatial description of the velocity $\boldsymbol{v}(\boldsymbol{x}, t)$ is obtained using Eqn. (2.39) as follows:

$$\boldsymbol{a}(\boldsymbol{x}, t) = \tilde{\boldsymbol{a}}(\boldsymbol{X}, t)$$

$$
\begin{aligned}
&= \left(\frac{\partial \tilde{v}}{\partial t}\right)_X \\
&= \left[\frac{\partial}{\partial t} v(\chi(X,t),t)\right]_X \\
&= \left(\frac{\partial v}{\partial t}\right)_x + \frac{\partial v}{\partial x_1}\left(\frac{\partial \chi_1}{\partial t}\right)_X + \frac{\partial v}{\partial x_2}\left(\frac{\partial \chi_2}{\partial t}\right)_X + \frac{\partial v}{\partial x_3}\left(\frac{\partial \chi_3}{\partial t}\right)_X \\
&= \left(\frac{\partial v}{\partial t}\right)_x + \frac{\partial v}{\partial x_j}\left(\frac{\partial \chi_j}{\partial t}\right)_X \\
&= \left(\frac{\partial v}{\partial t}\right)_x + (\nabla_x v)v.
\end{aligned} \tag{2.44}
$$

The above expression for the acceleration shows that one has to carefully distinguish between the acceleration of a particle, and the local rate of change of velocity $(\partial v/\partial t)_x$ at a point. For example, if an instrument that measures velocity were to be immersed in a steady fluid flow, it would show the same reading at a given point at all times, since the local rate of change of velocity is zero. However, this does not mean that the acceleration of a fluid particle passing through that point is zero, since its velocity can change as it moves to a different position.

One can compute the material derivative (also referred to as total, Eulerian or convective derivative) of any spatial field, be it scalar, vector or tensor-valued, in a manner analogous to that above. For example, if $\phi(x,t) = \phi(\chi(X,t),t) = \tilde{\phi}(X,t)$ and $w(x,t) = w(\chi(X,t),t) = \tilde{w}(X,t)$ are scalar and vector-valued fields, then their material derivatives are given by

$$
\frac{D\phi}{Dt} = \left(\frac{\partial \tilde{\phi}}{\partial t}\right)_X = \left(\frac{\partial \phi}{\partial t}\right)_x + \frac{\partial \phi}{\partial x_i}\frac{\partial \chi_i}{\partial t} = \left(\frac{\partial \phi}{\partial t}\right)_x + v \cdot (\nabla_x \phi), \tag{2.45}
$$

$$
\frac{Dw}{Dt} = \left(\frac{\partial \tilde{w}}{\partial t}\right)_X = \left(\frac{\partial w}{\partial t}\right)_x + (\nabla_x w)v. \tag{2.46}
$$

By considering x as a spatial field, we have

$$
v = \frac{Dx}{Dt}, \tag{2.47}
$$

since $(\partial x/\partial t)_x = 0$, and $(\nabla_x x)v = Iv = v$. We shall henceforth write $(\partial/\partial t)_x$ and ∇_x simply as $\partial/\partial t$ and ∇, respectively.

For a discussion on the various kinds of flow lines such as pathlines, streamlines, etc., see Vol. II.

2.4 Rate of Deformation

In this section, we define the rate-of-deformation tensor (also called as stretching tensor) and the vorticity tensor (or the spin tensor) as a function of the velocity gradients.

The velocity gradient L is defined by the relation

$$
L := \nabla_x v.
$$

That $\boldsymbol{L}$ is a second-order tensor follows from the fact that the gradient of a vector is a second-order tensor (see Chapter 1). Equation (2.44) can now be written as

$$\boldsymbol{a} = \frac{D\boldsymbol{v}}{Dt} = \frac{\partial \boldsymbol{v}}{\partial t} + \boldsymbol{L}\boldsymbol{v}. \tag{2.48}$$

The differential $d\boldsymbol{v}$ is given by

$$d\boldsymbol{v} = \boldsymbol{L}d\boldsymbol{x}; \quad dv_i = \frac{\partial v_i}{\partial x_j} dx_j.$$

We have the following relations:

$$\frac{D\boldsymbol{F}}{Dt} = \left(\frac{\partial \boldsymbol{F}}{\partial t}\right)_{\boldsymbol{X}} = \boldsymbol{L}\boldsymbol{F}, \tag{2.49}$$

$$\frac{D\boldsymbol{F}^{-1}}{Dt} = \left(\frac{\partial \boldsymbol{F}^{-1}}{\partial t}\right)_{\boldsymbol{X}} = -\boldsymbol{F}^{-1}\boldsymbol{L}, \tag{2.50}$$

$$\frac{1}{J}\frac{DJ}{Dt} = \frac{1}{J}\left(\frac{\partial J}{\partial t}\right)_{\boldsymbol{X}} = \operatorname{tr} \mathbf{L} = \boldsymbol{\nabla}_x \cdot \boldsymbol{v}, \tag{2.51}$$

$$\frac{\mathrm{D}}{Dt}(\operatorname{\mathbf{cof}} \boldsymbol{F}) = \left[(\operatorname{tr} \mathbf{L})\boldsymbol{I} - \boldsymbol{L}^T\right] \operatorname{\mathbf{cof}} \boldsymbol{F}, \tag{2.52}$$

where $\boldsymbol{L}$ is expressed in material coordinates.

Working with indicial notation,

$$\begin{aligned}
\frac{\partial F_{ij}}{\partial t} &= \frac{\partial}{\partial t}\left(\frac{\partial \chi_i}{\partial X_j}\right) \\
&= \frac{\partial \tilde{v}_i}{\partial X_j} \\
&= \frac{\partial v_i}{\partial x_m}\frac{\partial \chi_m}{\partial X_j} \\
&= L_{im}F_{mj},
\end{aligned}$$

which proves Eqn. (2.49). Taking the material derivative of $\boldsymbol{F}\boldsymbol{F}^{-1} = \boldsymbol{I}$, and using the product rule, we get Eqn. (2.50).

Substituting $\boldsymbol{T} = \boldsymbol{F}$ into Eqn. (1.222), we get

$$\begin{aligned}
\frac{DJ}{Dt} &= \operatorname{\mathbf{cof}} \boldsymbol{F} : \frac{D\boldsymbol{F}}{Dt} \\
&= \operatorname{\mathbf{cof}} \boldsymbol{F} : (\boldsymbol{L}\boldsymbol{F}) && \text{(by Eqn. (2.49))} \\
&= \boldsymbol{F}(\operatorname{\mathbf{cof}} \boldsymbol{F})^T : \boldsymbol{L}^T && \text{(by Eqn. (1.48))} \\
&= J\boldsymbol{I} : \boldsymbol{L}^T && \text{(by Eqn. (1.64))} \\
&= J\boldsymbol{\nabla}_x \cdot \boldsymbol{v},
\end{aligned}$$

which proves Eqn. (2.51).

An alternative proof is obtained by writing $J = \det \boldsymbol{F}$ as

$$J = \epsilon_{ijk} \frac{\partial \chi_1}{\partial X_i} \frac{\partial \chi_2}{\partial X_j} \frac{\partial \chi_3}{\partial X_k}.$$

Taking the material time derivative of this relation, and using the product and chain rules of differentiation, we get

$$\frac{DJ}{Dt} = t_1 + t_2 + t_3, \tag{2.53}$$

where

$$t_1 = \epsilon_{ijk} \frac{\partial v_1}{\partial X_i} \frac{\partial \chi_2}{\partial X_j} \frac{\partial \chi_3}{\partial X_k} = \epsilon_{ijk} \frac{\partial v_1}{\partial x_m} \frac{\partial \chi_m}{\partial X_i} \frac{\partial \chi_2}{\partial X_j} \frac{\partial \chi_3}{\partial X_k},$$

$$t_2 = \epsilon_{ijk} \frac{\partial \chi_1}{\partial X_i} \frac{\partial v_2}{\partial X_j} \frac{\partial \chi_3}{\partial X_k} = \epsilon_{ijk} \frac{\partial \chi_1}{\partial X_i} \frac{\partial v_2}{\partial x_p} \frac{\partial \chi_p}{\partial X_j} \frac{\partial \chi_3}{\partial X_k},$$

$$t_3 = \epsilon_{ijk} \frac{\partial \chi_1}{\partial X_i} \frac{\partial \chi_2}{\partial X_j} \frac{\partial v_3}{\partial X_k} = \epsilon_{ijk} \frac{\partial \chi_1}{\partial X_i} \frac{\partial \chi_2}{\partial X_j} \frac{\partial v_3}{\partial x_q} \frac{\partial \chi_q}{\partial X_k}.$$

Now consider t_1. We can write it as

$$t_1 = \epsilon_{ijk} L_{1m} F_{mi} F_{2j} F_{3k} = \epsilon_{m23} J L_{1m},$$

where the last step follows from Eqn. (1.52). Carrying out the summation over m, we get

$$t_1 = J L_{11} = J \frac{\partial v_1}{\partial x_1}.$$

In an analogous manner, we also obtain

$$t_2 = J L_{22} = J \frac{\partial v_2}{\partial x_2},$$

$$t_3 = J L_{33} = J \frac{\partial v_3}{\partial x_3}.$$

Substituting these expressions into Eqn. (2.53), we obtain Eqn. (2.51).

Equation (2.52) is obtained by writing $\mathbf{cof}\, \boldsymbol{F}$ as $J\boldsymbol{F}^{-T}$, and then using Eqns. (2.50) and (2.51).

By virtue of Eqn. (2.51), we can characterize an isochoric motion ($J = 1$) as one for which

$$\frac{1}{J} \frac{DJ}{Dt} = \boldsymbol{\nabla} \cdot \boldsymbol{v} = 0. \tag{2.54}$$

Using Eqn. (1.33), $\boldsymbol{L}$ can be split into a symmetric and skew-symmetric part as

$$\boldsymbol{L} = \boldsymbol{D} + \hat{\boldsymbol{W}}, \tag{2.55}$$

where $\boldsymbol{D}$ and $\hat{\boldsymbol{W}}$ are referred to as the rate-of-deformation and vorticity (or spin) tensors, respectively, and are given by

$$\boldsymbol{D} = \frac{1}{2}(\boldsymbol{L} + \boldsymbol{L}^T); \quad D_{ij} = \frac{1}{2}\left(\frac{\partial v_i}{\partial x_j} + \frac{\partial v_j}{\partial x_i} \right), \tag{2.56}$$

$$\hat{\boldsymbol{W}} = \frac{1}{2}(\boldsymbol{L} - \boldsymbol{L}^T); \quad \hat{W}_{ij} = \frac{1}{2}\left(\frac{\partial v_i}{\partial x_j} - \frac{\partial v_j}{\partial x_i}\right). \tag{2.57}$$

Since $\boldsymbol{D}$ and $\hat{\boldsymbol{W}}$ are obtained by taking the sum and difference of two second-order tensors, respectively, they are also second-order tensors. The number

$$d(t) = |\boldsymbol{x}_1 - \boldsymbol{x}_2| = |\boldsymbol{\chi}(\boldsymbol{X}_1, t) - \boldsymbol{\chi}(\boldsymbol{X}_2, t)| \tag{2.58}$$

represents the distance at time t between the material points $\boldsymbol{X}_1$ and $\boldsymbol{X}_2$. Recall that a motion is *rigid* if

$$\frac{\partial}{\partial t}|\boldsymbol{\chi}(\boldsymbol{X}_1, t) - \boldsymbol{\chi}(\boldsymbol{X}_2, t)| = 0, \tag{2.59}$$

for all material points $\boldsymbol{X}_1$ and $\boldsymbol{X}_2$, and for all times t.

Theorem 2.4.1 (Characterization of rigid motion-II). *Let $\boldsymbol{\chi}(\boldsymbol{X}, t)$ be a motion, and $\boldsymbol{v}$ the corresponding velocity field. Then the following are equivalent:*
(i) $\boldsymbol{\chi}$ is rigid.
(ii) At each time t, $\boldsymbol{v}$ admits the representation

$$\boldsymbol{v}(\boldsymbol{x}_1, t) = \boldsymbol{v}(\boldsymbol{x}_2, t) + \boldsymbol{W}(t)(\boldsymbol{x}_1 - \boldsymbol{x}_2) \quad \forall \boldsymbol{x}_1, \boldsymbol{x}_2 \in V, \tag{2.60}$$

where $\boldsymbol{W} \in Skw$.
(iii) $\boldsymbol{L}(\boldsymbol{x}, t) \in Skw$, or, alternatively, $\boldsymbol{D}(\boldsymbol{x}, t) = \boldsymbol{0}$ for all $(\boldsymbol{x}, t)$.

Proof. We shall show that (i) $\Longrightarrow$ (ii) $\Longrightarrow$ (iii) $\Longrightarrow$ (i). If the motion is rigid, i.e., if Eqn. (2.59) holds, then

$$(\boldsymbol{x}_1 - \boldsymbol{x}_2) \cdot [\boldsymbol{v}(\boldsymbol{x}_1, t) - \boldsymbol{v}(\boldsymbol{x}_2, t)] = 0 \quad \forall \boldsymbol{x}_1, \boldsymbol{x}_2 \in V.$$

Now differentiating the above equation with respect to $\boldsymbol{x}_1$, we get

$$\boldsymbol{v}(\boldsymbol{x}_1, t) - \boldsymbol{v}(\boldsymbol{x}_2, t) + \boldsymbol{\nabla}\boldsymbol{v}(\boldsymbol{x}_1, t)^T(\boldsymbol{x}_1 - \boldsymbol{x}_2) = \boldsymbol{0},$$

and this result, when differentiated with respect to $\boldsymbol{x}_2$, yields

$$\boldsymbol{\nabla}\boldsymbol{v}(\boldsymbol{x}_2, t) + \boldsymbol{\nabla}\boldsymbol{v}(\boldsymbol{x}_1, t)^T = \boldsymbol{0} \quad \forall \boldsymbol{x}_1, \boldsymbol{x}_2 \in V.$$

Letting $\boldsymbol{x}_2 = \boldsymbol{x}_1$, we see that $\boldsymbol{\nabla}\boldsymbol{v}(\boldsymbol{x}, t)$ is skew at every $\boldsymbol{x} \in V$. Now the above equation implies

$$\boldsymbol{\nabla}\boldsymbol{v}(\boldsymbol{x}_2, t) = \boldsymbol{\nabla}\boldsymbol{v}(\boldsymbol{x}_1, t) \quad \forall \boldsymbol{x}_1, \boldsymbol{x}_2 \in V,$$

or, in other words, $\boldsymbol{\nabla}\boldsymbol{v}$ is a constant skew tensor $\boldsymbol{W}(t)$. Integrating $\boldsymbol{\nabla}\boldsymbol{v} = \boldsymbol{W}(t)$, we get (ii).

To show that (ii) implies (iii), let $\boldsymbol{x}_2$ be the position vector of a fixed point in V, so that, since $\boldsymbol{x}_1$ is arbitrary, Eqn. (2.60) leads to $\boldsymbol{v}(\boldsymbol{x}, t) = \boldsymbol{W}(t)\boldsymbol{x} + \boldsymbol{c}(t)$; hence, $\boldsymbol{L}(\boldsymbol{x}, t) = \boldsymbol{W}(t)$, or, alternatively, $\boldsymbol{D}(\boldsymbol{x}, t) = \boldsymbol{0}$.

Finally, to show that (iii) implies (i), choose $x_1, x_2 \in V$, and let

$$x(s) = x_2 + s(x_1 - x_2)$$

be the parametric equation of the straight line joining x_1 to x_2. We have

$$\begin{aligned} v(x_1,t) - v(x_2,t) &= \int_{x_2}^{x_1} dv \\ &= \int_0^1 \nabla v(s,t)\frac{\partial x}{\partial s}\,ds \\ &= \int_0^1 \nabla v(s,t)(x_1 - x_2)\,ds, \end{aligned}$$

so that

$$(x_1 - x_2)\cdot[v(x_1,t) - v(x_2,t)] = \int_0^1 (x_1 - x_2)\cdot \nabla v(x_1 - x_2)\,ds = 0,$$

since $\nabla v \in$ Skw at each point in the domain.

A simple proof for showing that ∇v is a constant skew tensor when $D = 0$ is obtained if we assume that the velocity field is twice differentiable. Then by differentiating $\partial v_i/\partial x_j$, and using the fact that

$$\frac{\partial v_i}{\partial x_j} = -\frac{\partial v_j}{\partial x_i},$$

we get

$$\begin{aligned} \frac{\partial^2 v_i}{\partial x_k \partial x_j} &= -\frac{\partial^2 v_j}{\partial x_k \partial x_i} = -\frac{\partial^2 v_j}{\partial x_i \partial x_k} = \frac{\partial^2 v_k}{\partial x_i \partial x_j} = \frac{\partial^2 v_k}{\partial x_j \partial x_i} = -\frac{\partial^2 v_i}{\partial x_j \partial x_k} \\ &= -\frac{\partial^2 v_i}{\partial x_k \partial x_j}, \end{aligned}$$

which implies that

$$\frac{\partial^2 v_i}{\partial x_k \partial x_j} = 0,$$

and hence implies that ∇v is a constant skew tensor. □

If w is the axial vector corresponding to $W = \dot{Q}Q^T$, then Eqn. (2.60) can be written as

$$v(x_1,t) = v(x_2,t) + w(t) \times (x_1 - x_2),$$

which is the classical formula for the velocity field of a rigid motion, with w known as the *angular velocity*. Note that for $x_1 - x_2 = \alpha w$, we get $v(x_1,t) = v(x_2,t)$. Thus, for any given x_2, v is the sum of a uniform velocity field $v(x_2)$, and a rigid rotation about the line through x_2 parallel to w. Hence, the span of w is called the spin axis.

Theorem 2.4.1 cannot be applied for static deformations. For example, if $\chi(X) = HX$, where H is a nonorthogonal constant tensor, we see that v, and hence, D is zero even though the deformation is not a rigid one.

In Theorem 2.4.1, we saw the implications of D being zero. To find the necessary and sufficient conditions for $\hat{W}$ being zero everywhere, we see that $\hat{W}$ is zero if and only if $\nabla \times v$ is zero, which in turn implies the existence of a scalar-valued function ϕ such that $v = \nabla\phi$.

In the fluid-mechanics literature, the rate-of-deformation tensor[2] is often referred to as the rate-of-strain tensor. The terminology 'rate-of-strain' seems to imply that D is actually the rate of some strain quantity. In reality, as we will soon show, it is neither the rate of the small strain tensor, nor the rate of the Lagrangian or Eulerian strain tensor, unless the displacements and displacement gradients are extremely small, which is usually not the case in fluid mechanics [207]. It is important to note, however, that the expression for D given by Eqn. (2.56) is valid even when the displacements and displacement gradients are large (even though the *name* 'rate-of-strain tensor' is a misnomer in that case), since no assumption has been made about the magnitude of the displacements or the displacement gradients in deriving the expression for D.

Using Eqns. (2.49) and (2.50), we have

$$\begin{aligned}
\frac{DE}{Dt} &= \frac{1}{2}\frac{DC}{Dt} \\
&= \frac{1}{2}\left[\frac{DF^T}{Dt}F + F^T\frac{DF}{Dt}\right] \\
&= \frac{1}{2}\left[F^TL^TF + F^TLF\right] \\
&= F^TDF, \qquad (2.61)\\
\frac{D\bar{E}}{Dt} &= \frac{DF^{-T}}{Dt}EF^{-1} + F^{-T}\frac{DE}{Dt}F^{-1} + F^{-T}E\frac{DF^{-1}}{Dt} \\
&= -L^TF^{-T}EF^{-1} + F^{-T}F^TDFF^{-1} - F^{-T}EF^{-1}L \\
&= D - L^T\bar{E} - \bar{E}L, \qquad (2.62)\\
\frac{D\epsilon}{Dt} &= \frac{1}{2}\left[LF + F^TL^T\right].
\end{aligned}$$

From Eqn. (2.61) and the relations $F = RU$ and $C = U^2$, we have

$$R^TDR = \frac{1}{2}\left[\dot{U}U^{-1} + U^{-1}\dot{U}\right].$$

From Eqn. (2.62) and (4.39), we see that the convective rate of $\bar{E}$ is equal to D. We now generalize the above results.

If $U^2 = C = \sum_{i=1}^k \lambda_i^2 P_i$, and $V^2 = B = \sum_{i=1}^k \lambda_i^2 G_i$, where λ_i, $i = 1, \ldots, k$, are the distinct eigenvalues of U (known as the principal stretches), P_i are the eigenprojections of C (given by Eqn. (1.242)) and $G_i = RP_iR^T$ are the eigenprojections of B (given by Eqn. (1.242) with B and V in place of C and U, respectively), and if $f(\lambda_i)$ is a strictly increasing function of

[2]For a justification of the terminology 'rate-of-deformation', see Vol. II

λ_i satisfying $f(0) = 0$ and $f'(1) = 1$, then $\boldsymbol{E}_H = \sum_{i=1}^k f(\lambda_i)\boldsymbol{P}_i$ and $\bar{\boldsymbol{E}}_H = \sum_{i=1}^k f(\lambda_i)\boldsymbol{G}_i = \boldsymbol{R}\boldsymbol{E}_H\boldsymbol{R}^T$ are the Lagrangian and Eulerian Hill strain tensors[3]. Their material derivatives, denoted by a superposed dot, are found using Eqn. (1.256) as $f'(\lambda)\boldsymbol{F}^T\boldsymbol{D}\boldsymbol{F}/\lambda$ and $\lambda f'(\lambda)\boldsymbol{D}$, respectively, for the case when all λ_i are equal to λ ($k = 1$), while for $k > 1$,

$$\begin{aligned}
\dot{\boldsymbol{E}}_H &= \mathbb{S}\frac{\partial \boldsymbol{E}_H}{\partial \boldsymbol{C}}\mathbb{S}\dot{\boldsymbol{C}} \\
&= \mathbb{S}\left[\sum_{i=1}^k \frac{f'(\lambda_i)}{2\lambda_i}\boldsymbol{P}_i \boxtimes \boldsymbol{P}_i + \sum_{i=1}^k \sum_{\substack{j=1 \\ j\neq i}}^k \frac{f(\lambda_i) - f(\lambda_j)}{\lambda_i^2 - \lambda_j^2}\boldsymbol{P}_i \boxtimes \boldsymbol{P}_j\right]\mathbb{S}\dot{\boldsymbol{C}} \\
&= \sum_{i=1}^k \frac{f'(\lambda_i)}{2\lambda_i}\boldsymbol{P}_i\dot{\boldsymbol{C}}\boldsymbol{P}_i + \sum_{i=1}^k \sum_{\substack{j=1 \\ j\neq i}}^k \frac{f(\lambda_i) - f(\lambda_j)}{\lambda_i^2 - \lambda_j^2}\boldsymbol{P}_i\dot{\boldsymbol{C}}\boldsymbol{P}_j, && (2.63) \\
&= \sum_{i=1}^k \frac{f'(\lambda_i)}{\lambda_i}\boldsymbol{P}_i\boldsymbol{F}^T\boldsymbol{D}\boldsymbol{F}\boldsymbol{P}_i + \sum_{i=1}^k \sum_{\substack{j=1 \\ j\neq i}}^k \frac{2\left[f(\lambda_i) - f(\lambda_j)\right]}{\lambda_i^2 - \lambda_j^2}\boldsymbol{P}_i\boldsymbol{F}^T\boldsymbol{D}\boldsymbol{F}\boldsymbol{P}_j, && (2.64)
\end{aligned}$$

$$\begin{aligned}
\dot{\bar{\boldsymbol{E}}}_H &= \mathbb{S}\frac{\partial \bar{\boldsymbol{E}}_H}{\partial \boldsymbol{B}}\mathbb{S}\dot{\boldsymbol{B}} \\
&= \sum_{i=1}^k \frac{f'(\lambda_i)}{2\lambda_i}\boldsymbol{G}_i\dot{\boldsymbol{B}}\boldsymbol{G}_i + \sum_{i=1}^k \sum_{\substack{j=1 \\ j\neq i}}^k \frac{f(\lambda_i) - f(\lambda_j)}{\lambda_i^2 - \lambda_j^2}\boldsymbol{G}_i\dot{\boldsymbol{B}}\boldsymbol{G}_j \\
&= \sum_{i=1}^k \lambda_i f'(\lambda_i)\boldsymbol{G}_i\boldsymbol{D}\boldsymbol{G}_i + \sum_{i=1}^k \sum_{\substack{j=1 \\ j\neq i}}^k \left[f(\lambda_i) - f(\lambda_j)\right]\left[\frac{(\lambda_i^2 + \lambda_j^2)}{\lambda_i^2 - \lambda_j^2}\boldsymbol{G}_i\boldsymbol{D}\boldsymbol{G}_j - \boldsymbol{G}_i\hat{\boldsymbol{W}}\boldsymbol{G}_j\right].
\end{aligned} \tag{2.65}$$

where we have used Eqns. (2.61), $\dot{\boldsymbol{B}} = \boldsymbol{L}\boldsymbol{B} + \boldsymbol{B}\boldsymbol{L}^T$ and $\boldsymbol{B}\boldsymbol{G}_i = \lambda_i^2\boldsymbol{G}_i$ (no sum on i) in obtaining the last steps in the expressions for $\dot{\boldsymbol{E}}_H$ and $\dot{\bar{\boldsymbol{E}}}_H$. We now 'invert' Eqn. (2.64). By virtue of Eqns. (1.175), (1.176), and the properties of projections,

$$\left\{\mathbb{S}\left[\sum_{i=1}^k \frac{f'(\lambda_i)}{\lambda_i}\boldsymbol{P}_i \boxtimes \boldsymbol{P}_i + \sum_{i=1}^k \sum_{\substack{j=1 \\ j\neq i}}^k \frac{2\left[f(\lambda_i) - f(\lambda_j)\right]}{\lambda_i^2 - \lambda_j^2}\boldsymbol{P}_i \boxtimes \boldsymbol{P}_j\right]\mathbb{S}\right\}^{-1} =$$

$$\mathbb{S}\left[\sum_{i=1}^k \frac{\lambda_i}{f'(\lambda_i)}\boldsymbol{P}_i \boxtimes \boldsymbol{P}_i + \sum_{i=1}^k \sum_{\substack{j=1 \\ j\neq i}}^k \frac{\lambda_i^2 - \lambda_j^2}{2\left[f(\lambda_i) - f(\lambda_j)\right]}\boldsymbol{P}_i \boxtimes \boldsymbol{P}_j\right]\mathbb{S},$$

so that

$$\boldsymbol{F}^T\boldsymbol{D}\boldsymbol{F} = \mathbb{S}\left[\sum_{i=1}^k \frac{\lambda_i}{f'(\lambda_i)}\boldsymbol{P}_i \boxtimes \boldsymbol{P}_i + \sum_{i=1}^k \sum_{\substack{j=1 \\ j\neq i}}^k \frac{\lambda_i^2 - \lambda_j^2}{2\left[f(\lambda_i) - f(\lambda_j)\right]}\boldsymbol{P}_i \boxtimes \boldsymbol{P}_j\right]\mathbb{S}\dot{\boldsymbol{E}}_H$$

[3]For applications of the Hill strain tensors, see Xiao [355] and references therein.

$$= \sum_{i=1}^{k} \frac{\lambda_i}{f'(\lambda_i)} P_i \dot{E}_H P_i + \sum_{i=1}^{k} \sum_{\substack{j=1 \\ j \neq i}}^{k} \frac{\lambda_i^2 - \lambda_j^2}{[f(\lambda_i) - f(\lambda_j)]} P_i \dot{E}_H P_j. \tag{2.66}$$

Similarly, the 'inverse' of Eqn. (2.65) is

$$D = \sum_{i=1}^{k} \frac{1}{\lambda_i f'(\lambda_i)} G_i \dot{\bar{E}}_H G_i + \sum_{i=1}^{k} \sum_{\substack{j=1 \\ j \neq i}}^{k} \frac{\lambda_i^2 - \lambda_j^2}{\lambda_i^2 + \lambda_j^2} \left[\frac{G_i \dot{\bar{E}}_H G_j}{f(\lambda_i) - f(\lambda_j)} + G_i \hat{W} G_j \right]. \tag{2.67}$$

An alternative expression can be found by substituting

$$\begin{aligned} \dot{E}_H &= R^T \dot{\bar{E}}_H R + R^T \bar{E}_H \dot{R} + \dot{R}^T \bar{E}_H R \\ &= R^T \left[\dot{\bar{E}}_H R + \bar{E}_H \Omega - \Omega \bar{E}_H \right] R, \end{aligned} \tag{2.68}$$

where $\Omega = \dot{R} R^T \in \text{Skw}$, into Eqn. (2.66); we get on using $\bar{E}_H G_i = G_i \bar{E}_H = f(\lambda_i) G_i$ (no sum on i), $P_i R^T = R^T G_i$, and $F = VR$,

$$D = \sum_{i=1}^{k} \frac{1}{\lambda_i f'(\lambda_i)} G_i \dot{\bar{E}}_H G_i + \sum_{i=1}^{k} \sum_{\substack{j=1 \\ j \neq i}}^{k} \frac{(\lambda_i^2 - \lambda_j^2)}{2\lambda_i \lambda_j} \left[\frac{G_i \dot{\bar{E}}_H G_j}{f(\lambda_i) - f(\lambda_j)} + G_i \Omega G_j \right]. \tag{2.69}$$

As an example, if $E_H = U = \sum_{i=1}^{k} \lambda_i P_i$, then using Eqn. (2.63), we have (with $\dot{C} = 2F^T D F$), $\dot{U} = \dot{C}/(2\lambda)$ in case $k = 1$, while for $k > 1$,

$$\begin{aligned} \dot{U} &= \sum_{i=1}^{k} \frac{P_i \dot{C} P_i}{2\lambda_i} + \sum_{i=1}^{k} \sum_{\substack{j=1 \\ j \neq i}}^{k} \frac{1}{\lambda_i + \lambda_j} P_i \dot{C} P_j \\ &= \sum_{i=1}^{k} \sum_{j=1}^{k} \frac{P_i \dot{C} P_j}{\lambda_i + \lambda_j}, \end{aligned} \tag{2.70}$$

which is the same as the first expression in Eqn. (1.246). Similarly, using Eqn. (2.65), we get $\dot{V} = \lambda D$ for $k = 1$, while for $k > 1$,

$$\dot{V} = \sum_{i=1}^{k} \sum_{j=1}^{k} \left[\frac{\lambda_i^2 + \lambda_j^2}{\lambda_i + \lambda_j} G_i D G_j - (\lambda_i - \lambda_j) G_i \hat{W} G_j \right], \tag{2.71}$$

which could also have been obtained from Eqn. (1.248) by substituting $\dot{B} = LB + BL^T$, and using $BG_i = G_i B = \lambda_i^2 G_i$ (no sum on i).

As another example, the material derivatives of $\log U = \sum_{i=1}^{k} \log \lambda_i P_i$ and $\log V = \sum_{i=1}^{k} \log \lambda_i G_i$ are given by $\dot{C}/(2\lambda^2)$ and $\dot{B}/(2\lambda^2)$ when $k = 1$, while for $k > 1$, using Eqns. (2.63) and (2.65), we have

$$\overline{\log U}^{\,\cdot} = \sum_{i=1}^{k} \frac{1}{2\lambda_i^2} P_i \dot{C} P_i + \sum_{i=1}^{k} \sum_{\substack{j=1 \\ j \neq i}}^{k} \frac{\log \lambda_i - \log \lambda_j}{\lambda_i^2 - \lambda_j^2} P_i \dot{C} P_j, \tag{2.72}$$

$$\overline{\log V}^{\cdot} = \sum_{i=1}^{k} G_i D G_i + \sum_{i=1}^{k}\sum_{\substack{j=1\\ j\neq i}}^{k} (\log\lambda_i - \log\lambda_j)\left[\frac{(\lambda_i^2+\lambda_j^2)}{\lambda_i^2-\lambda_j^2} G_i D G_j - G_i \hat{W} G_j\right], \tag{2.73}$$

where $\dot{C} = 2F^T DF$. By Eqns. (2.66), (2.67) and (2.69),

$$F^T DF = \sum_{i=1}^{k} \lambda_i^2 P_i \overline{\log U}^{\cdot} P_i + \sum_{i=1}^{k}\sum_{\substack{j=1\\ j\neq i}}^{k} \frac{\lambda_i^2-\lambda_j^2}{\log\lambda_i^2 - \log\lambda_j^2} P_i \overline{\log U}^{\cdot} P_j,$$

$$D = \sum_{i=1}^{k} G_i \overline{\log V}^{\cdot} G_i + \sum_{i=1}^{k}\sum_{\substack{j=1\\ j\neq i}}^{k} \frac{\lambda_i^2-\lambda_j^2}{\lambda_i^2+\lambda_j^2}\left[\frac{G_i \overline{\log V}^{\cdot} G_j}{\log\lambda_i - \log\lambda_j} + G_i \hat{W} G_j\right] \tag{2.74}$$

$$= \sum_{i=1}^{k} G_i \overline{\log V}^{\cdot} G_i + \sum_{i=1}^{k}\sum_{\substack{j=1\\ j\neq i}}^{k} \frac{\lambda_i^2-\lambda_j^2}{2\lambda_i\lambda_j}\left[\frac{G_i \overline{\log V}^{\cdot} G_j}{\log\lambda_i - \log\lambda_j} + G_i \Omega G_j\right]. \tag{2.75}$$

By taking $E_H \equiv C$ and $\bar{E}_H \equiv B$ in Eqn. (2.68), and using the relations $\dot{C} = 2F^T DF$, $\dot{B} = LB + BL^T = DB + BD + \hat{W}B - B\hat{W}$ and $F = VR$, we get

$$2VDV = (D + \hat{W} - \Omega)B + B(D - \hat{W} + \Omega),$$

or, alternatively,

$$(\hat{W} - \Omega)B - B(\hat{W} - \Omega) = 2VDV - DB - BD.$$

Although the above equation can be written as

$$[I \boxtimes B - B \boxtimes I]\,(\hat{W} - \Omega) = 2VDV - DB - BD,$$

it cannot be inverted to find $\hat{W} - \Omega$ since $\mathbb{A} := I \boxtimes B - B \boxtimes I$ is singular; in fact $\mathbb{A}$ is zero when $B = \lambda I$, so that any skew-symmetric tensor is a solution [355][4]. However, one can find $\hat{W}$ in terms of Ω using the following method. Since $L = \dot{F}F^{-1} = \dot{V}V^{-1} + V\Omega V^{-1}$, we have $VL^T - LV = -(\Omega V + V\Omega)$. Substituting $L = D + \hat{W}$, we get

$$[I \boxtimes V + V \boxtimes I]\,(\hat{W} - \Omega) = (\hat{W} - \Omega)V + V(\hat{W} - \Omega) = VD - DV.$$

The solution of the above equation obtained using Eqn. (1.197) is $\hat{W} = \Omega$ if $k = 1$, while for $k > 1$,

$$\hat{W} = \Omega + \sum_{i=1}^{k}\sum_{j=1}^{k} \frac{G_i(VD - DV)G_j}{\lambda_i + \lambda_j}$$

$$= \Omega + \sum_{i=1}^{k}\sum_{j=1}^{k} \frac{\lambda_i - \lambda_j}{\lambda_i + \lambda_j} G_i D G_j, \tag{2.76}$$

[4]Thus, for example, $\Omega^{\log}$ defined via Eqn. (80) in [220] or Eqn. (55) in [356] or Eqn. (25) in [357], and hence the rate $\dot{\tau} + \tau\Omega^{\log} - \Omega^{\log}\tau$ used in these works, is not unique.

which could also have been obtained using Eqn. (1.249) using $\dot{F} = (D + \hat{W})F = (D + \hat{W})VR$, $VG_i = G_iV = \lambda_i G_i$ (no sum on i) and $\sum_{i=1}^{k} G_i = I$. By substituting Eqn. (2.76) into Eqn. (2.67), and carrying out some manipulation, one gets Eqn. (2.69). By premultiplying Eqn. (2.76) by R^T, and post-multiplying it by R, we get

$$R^T\hat{W}R = R^T\dot{R} + \sum_{i=1}^{k}\sum_{j=1}^{k} \frac{\lambda_i - \lambda_j}{\lambda_i + \lambda_j} P_i R^T D R P_j, \tag{2.77}$$

which could also have been obtained using Eqn. (1.247) using $\dot{F} = (D + \hat{W})F = (D + \hat{W})RU$, $UP_i = P_iU = \lambda_i P_i$ (no sum on i) and $\sum_{i=1}^{k} P_i = I$. By using $F = RU$ and Eqn. (2.49), the following alternative expression to Eqn. (2.76) can also be given:

$$\hat{W} = \Omega + \frac{1}{2}R(\dot{U}U^{-1} - U^{-1}\dot{U})R^T.$$

Thus, $\hat{W} = \Omega$ if and only if $\dot{U}$ and U^{-1} commute, or, alternatively, if $\dot{U}U^{-1} \in \text{Sym}$.

The vorticity vector is defined by

$$\omega := \nabla \times v. \tag{2.78}$$

We now show that $\omega/2$ is the axial vector of $\hat{W}$; it then follows from Eqns. (1.90) that

$$\omega_i = -\epsilon_{ijk}\hat{W}_{jk},$$
$$\hat{W}_{ij} = -\frac{1}{2}\epsilon_{ijk}\omega_k.$$

Since $\omega = \nabla \times v$, we have

$$\begin{aligned}
(\nabla \times v) \times u &= \epsilon_{ijk}\epsilon_{jmn}\frac{\partial v_n}{\partial x_m}u_k e_i \\
&= (\delta_{km}\delta_{in} - \delta_{kn}\delta_{im})\frac{\partial v_n}{\partial x_m}u_k e_i \\
&= \left[\frac{\partial v_i}{\partial x_k} - \frac{\partial v_k}{\partial x_i}\right]u_k e_i \\
&= [\nabla v - (\nabla v)^T]u \\
&= 2\hat{W}u \quad \forall u \in V,
\end{aligned}$$

which is the desired result.

2.5 Examples of Simple Motions

In this section, we consider three simple examples dealing with extension, rigid body motion and shear.

2.5.1 Pure extension

Consider the motion described by

$$x_1 = (1 + \epsilon)X_1$$
$$x_2 = (1 - \nu\epsilon)X_2$$
$$x_3 = (1 - \nu\epsilon)X_3$$

The deformation gradient is given by

$$F = \begin{bmatrix} 1+\epsilon & 0 & 0 \\ 0 & 1-\nu\epsilon & 0 \\ 0 & 0 & 1-\nu\epsilon \end{bmatrix}$$

The right and left Cauchy–Green strain tensor are

$$C = F^T F = \begin{bmatrix} (1+\epsilon)^2 & 0 & 0 \\ 0 & (1-\nu\epsilon)^2 & 0 \\ 0 & 0 & (1-\nu\epsilon)^2 \end{bmatrix} = FF^T = B.$$

The Green and Almansi strain tensors are

$$E = \frac{1}{2}(C - I) = \begin{bmatrix} \epsilon + \frac{\epsilon^2}{2} & 0 & 0 \\ 0 & -\nu\epsilon + \frac{(\nu\epsilon)^2}{2} & 0 \\ 0 & 0 & -\nu\epsilon + \frac{(\nu\epsilon)^2}{2} \end{bmatrix},$$

$$\bar{E} = \frac{1}{2}(I - B^{-1}) = \begin{bmatrix} \frac{2\epsilon+\epsilon^2}{2(1+\epsilon)^2} & 0 & 0 \\ 0 & \frac{-2\nu\epsilon+\nu^2\epsilon^2}{2(1-\nu\epsilon)^2} & 0 \\ 0 & 0 & \frac{-2\nu\epsilon+\nu^2\epsilon^2}{2(1-\nu\epsilon)^2} \end{bmatrix}.$$

The velocity components are

$$v_1 = \dot{\epsilon} X_1 = \frac{\dot{\epsilon} x_1}{1+\epsilon},$$
$$v_2 = -\nu\dot{\epsilon} X_2 = -\frac{\nu\dot{\epsilon} x_2}{1-\nu\epsilon},$$
$$v_3 = -\nu\dot{\epsilon} X_3 = -\frac{\nu\dot{\epsilon} x_3}{1-\nu\epsilon}.$$

The velocity gradient, rate-of-deformation and the vorticity tensors are

$$L = D = \begin{bmatrix} \frac{\dot{\epsilon}}{1+\epsilon} & 0 & 0 \\ 0 & -\frac{\nu\dot{\epsilon}}{1-\nu\epsilon} & 0 \\ 0 & 0 & -\frac{\nu\dot{\epsilon}}{1-\nu\epsilon} \end{bmatrix}, \quad \hat{W} = 0.$$

The rates of the deformation gradient and its determinant are

$$\frac{DF}{Dt} = LF = \begin{bmatrix} \dot{\epsilon} & 0 & 0 \\ 0 & -\nu\dot{\epsilon} & 0 \\ 0 & 0 & -\nu\dot{\epsilon} \end{bmatrix}, \quad \frac{DJ}{Dt} = J \operatorname{tr} D = \dot{\epsilon}(1-\nu\epsilon)(1-2\nu-3\nu\epsilon).$$

The rates of the Green and Almansi tensors are (Eqns. (2.61) and (2.62))

$$\frac{DE}{Dt} = F^T DF = \begin{bmatrix} \dot{\epsilon}(1+\epsilon) & 0 & 0 \\ 0 & -\nu\dot{\epsilon}(1-\nu\epsilon) & 0 \\ 0 & 0 & -\nu\dot{\epsilon}(1-\nu\epsilon) \end{bmatrix},$$

$$\frac{D\bar{E}}{Dt} = D - L^T\bar{E} - \bar{E}L = \begin{bmatrix} \frac{\dot{\epsilon}}{(1+\epsilon)^3} & 0 & 0 \\ 0 & -\frac{\nu\dot{\epsilon}}{(1-\nu\epsilon)^3} & 0 \\ 0 & 0 & -\frac{\nu\dot{\epsilon}}{(1-\nu\epsilon)^3} \end{bmatrix}.$$

2.5.2 Rigid body motion

As already discussed, rigid body motion is characterized by

$$x = Q(t)X + c(t).$$

The deformation gradient is $F = Q(t)$, and hence $C = B = I$, and $E = \bar{E} = 0$. Note that the small-strain tensor $\epsilon = (Q + Q^T - 2I)/2$ is not zero, and in fact, can be quite large at some instant t, thus showing that the use of the small-strain tensor ϵ should be restricted to times (starting from $t = 0$) upto which u and ∇u are 'small' (see beginning of Chapter 6). The velocity and acceleration are

$$\begin{aligned} v &= \dot{Q}X + \dot{c} = \dot{Q}Q^T(x - c) + \dot{c} = w \times (x - c) + \dot{c}, \\ a &= \ddot{Q}X + \ddot{c} = (\dot{W} + W^2)QX + \ddot{c} = (\dot{W} + W^2)(x - c) + \ddot{c} \\ &= \dot{w} \times (x - c) + w \times [w \times (x - c)] + \ddot{c}, \end{aligned} \tag{2.79}$$

where $W := \dot{Q}Q^T \in \text{Skw}$, and w is the axial vector of W. Since W is skew, we have $L = W$ and $D = 0$. Hence, $DE/Dt = D\bar{E}/Dt = 0$.

2.5.3 Simple shear

Consider the motion defined by the following equations (see Fig. 2.2):

$$\begin{aligned} x_1 &= X_1 + \gamma X_2, \\ x_2 &= X_2, \\ x_3 &= X_3, \end{aligned}$$

where $\gamma = \tan\theta$. The deformation gradient, and its cofactor are given by

$$F = \begin{bmatrix} 1 & \gamma & 0 \\ 0 & 1 & 0 \\ 0 & 0 & 1 \end{bmatrix}, \quad \text{cof}\, F = \begin{bmatrix} 1 & 0 & 0 \\ -\gamma & 1 & 0 \\ 0 & 0 & 1 \end{bmatrix}.$$

Since $J = \det F = 1$, the above motion is isochoric. The right and left Cauchy–Green strain tensor are

$$C = F^T F = \begin{bmatrix} 1 & \gamma & 0 \\ \gamma & 1+\gamma^2 & 0 \\ 0 & 0 & 1 \end{bmatrix}, \quad B = FF^T = \begin{bmatrix} 1+\gamma^2 & \gamma & 0 \\ \gamma & 1 & 0 \\ 0 & 0 & 1 \end{bmatrix}.$$

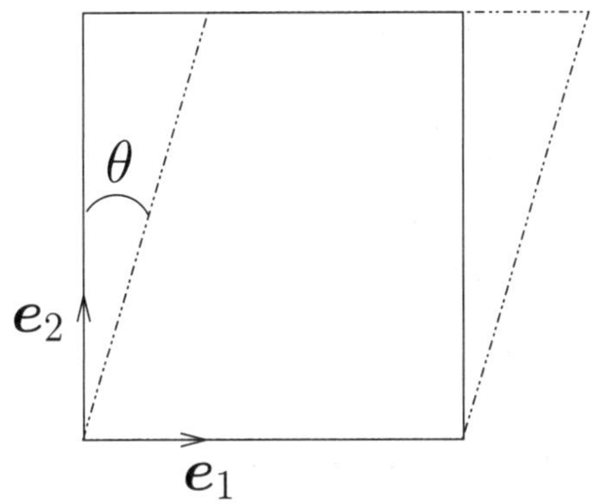

Fig. 2.2 *Simple shear.*

The Green, Almansi and small strain tensors are

$$E = \frac{1}{2}(C - I) = \frac{1}{2}\begin{bmatrix} 0 & \gamma & 0 \\ \gamma & \gamma^2 & 0 \\ 0 & 0 & 0 \end{bmatrix},$$

$$\bar{E} = \frac{1}{2}(I - B^{-1}) = \frac{1}{2}\begin{bmatrix} 0 & \gamma & 0 \\ \gamma & -\gamma^2 & 0 \\ 0 & 0 & 0 \end{bmatrix},$$

$$\epsilon = \frac{1}{2}(F + F^T - 2I) = \frac{1}{2}\begin{bmatrix} 0 & \gamma & 0 \\ \gamma & 0 & 0 \\ 0 & 0 & 0 \end{bmatrix}.$$

The velocity components are

$$v_1 = \dot{\gamma}X_2 = \dot{\gamma}x_2,$$
$$v_2 = 0,$$
$$v_3 = 0,$$

The velocity gradient, rate-of-deformation and vorticity tensors are

$$L = \begin{bmatrix} 0 & \dot{\gamma} & 0 \\ 0 & 0 & 0 \\ 0 & 0 & 0 \end{bmatrix}, \quad D = \frac{1}{2}\begin{bmatrix} 0 & \dot{\gamma} & 0 \\ \dot{\gamma} & 0 & 0 \\ 0 & 0 & 0 \end{bmatrix}, \quad \hat{W} = \frac{1}{2}\begin{bmatrix} 0 & \dot{\gamma} & 0 \\ -\dot{\gamma} & 0 & 0 \\ 0 & 0 & 0 \end{bmatrix}.$$

One can easily verify that $Lv = 0$. Hence, using Eqn. (2.48), we have $Dv/Dt = \partial v/dt$. The rates of the deformation gradient and its determinant are

$$\frac{DF}{Dt} = LF = \begin{bmatrix} 0 & \dot{\gamma} & 0 \\ 0 & 0 & 0 \\ 0 & 0 & 0 \end{bmatrix}, \quad \frac{DJ}{Dt} = J \operatorname{tr} D = 0.$$

The rates of the Green and Almansi tensors are

$$\frac{DE}{Dt} = F^T D F = \frac{1}{2}\begin{bmatrix} 0 & \dot{\gamma} & 0 \\ \dot{\gamma} & 2\gamma\dot{\gamma} & 0 \\ 0 & 0 & 0 \end{bmatrix},$$

$$\frac{D\bar{E}}{Dt} = D - L^T\bar{E} - \bar{E}L = \frac{1}{2}\begin{bmatrix} 0 & \dot{\gamma} & 0 \\ \dot{\gamma} & -2\gamma\dot{\gamma} & 0 \\ 0 & 0 & 0 \end{bmatrix}.$$

Using Eqn. (2.32), we find that the normal e_1 transforms to the vector $\frac{1}{\sqrt{1+\gamma^2}}(1, -\gamma, 0)$.

Having considered the kinematics of a deformable body, we turn to the derivation of the governing equations based on the axioms of conservation of mass, balance of linear and angular momentum, and the laws of thermodynamics, in the following chapter.

EXERCISES

1. Let the χ mapping be given by $x = \chi(X_1, X_2)$, i.e., a flat surface parametrized by (X_1, X_2) is mapped to a curved surface. Using the fact that $n_0 = (0, 0, 1)$, and using Eqns. (1.61) and (2.31) show that

$$dS = \left| \frac{\partial \chi}{\partial X_1} \times \frac{\partial \chi}{\partial X_2} \right|.$$

   This expression is used as the area element for shell geometries with $X_1 \equiv \xi$ and $X_2 \equiv \eta$, where (ξ, η) are the natural coordinates parametrizing the shell mid-surface.

2. Show that a material area element dS_0 remains unchanged in area if and only if $F = Q$, where $Q \in \text{Orth}^+$. It follows that if the material area element remains unchanged for all choices of material surfaces, then $E = 0$ for all X and t, which in turn, by virtue of Theorem 2.2.1, implies that the motion is rigid.

3. Given a vector field $v : V(t) \to \Re^3$ over the deformed configuration $V(t)$, its Piola transform is the vector field $u : V_0 \to \Re^3$ defined over the reference configuration by the relation

$$u(X) := (\mathbf{cof}\, F)^T(X)v(x).$$

   Show that the divergence of the two vector fields are related by

$$\nabla_X \cdot u(X) = J\nabla_x \cdot v(x).$$

   Note that this relation is similar to the relation between the divergences of tensor fields related by the Piola transform.

   If $u(X)$ and $v(x)$ are related by

$$u(X) := F^T v(x),$$

then show using Eqn. (1.52) that the curls of the two vector fields are related by

$$F[\nabla_X \times u(X)] = J\nabla_x \times v(x).$$

If $T(X)$ and $\tau(x)$ are related by

$$T(X) := \tau(x)F,$$

then show using Eqn. (1.52) that the curls of the two tensor fields are related by

$$F[\nabla_X \times T(X)] = J\nabla_x \times \tau(x).$$

4. Show that

$$\begin{aligned} \frac{\partial J}{\partial C} &= \frac{1}{2}JC^{-1}, \\ \frac{\partial J}{\partial E} &= JC^{-1}. \end{aligned} \tag{2.80}$$

It follows from Eqn. (2.80) that $DJ/Dt = JC^{-1} : \dot{C}/2$.

5. Show that

$$\begin{aligned} (I_1)_C &= 3 + 2(I_1)_E, \\ (I_2)_C &= 3 + 4(I_1)_E + 4(I_2)_E, \\ (I_3)_C &= 1 + 2(I_1)_E + 4(I_2)_E + 8(I_3)_E. \end{aligned}$$

6. Using indicial notation, prove that

$$\nabla \times (\omega \times v) = (\nabla\omega)v + (\nabla \cdot v)\omega - (\nabla v)\omega,$$

where $\omega = \nabla \times v$ is the vorticity.

7. Using Eqn. (2.79), show that the acceleration field in a rigid motion has the form

$$a(x_1, t) = a(x_2, t) + \dot{w}(t) \times (x_1 - x_2) + w(t) \times [w(t) \times (x_1 - x_2)],$$

where $w(t)$ is the angular velocity.

8. Show that

$$\int_{V(t)} \left[2\hat{W}v + (\nabla \cdot v)v\right] dV = \int_{S(t)} \left[(v \cdot n)v - \frac{v \cdot v}{2}n\right] dS.$$

9. Show that

$$\begin{aligned} \nabla \cdot \left(\frac{Dv}{Dt}\right) &= \frac{D}{Dt}(\nabla \cdot v) + D : D - \hat{W} : \hat{W} \\ \nabla \times \frac{Dv}{Dt} &= \frac{D\omega}{Dt} + (\nabla \cdot v)\omega - (\nabla v)\omega. \end{aligned}$$

3 Balance Laws

In this chapter, we derive the balance laws by requiring that the material volume, i.e., a volume that comprises a fixed set of particles, obeys the axioms of mass conservation, balance of linear and angular momentum, and the laws of thermodynamics. The important point to note is that the choice of this material volume is arbitrary; for example, any arbitrary subset of a given material volume also constitutes a valid material volume to which the above axioms can be applied. The original statements of all the axioms are in integral form. Using the arbitrariness of the material volume yields a set of differential equations governing the field variables. The solution of these differential equations subject to appropriate boundary and initial conditions then defines the variation of each field variable with space and time.

In fluid mechanics, it is useful to derive balance laws for a control volume, which is a region of space where various flow quantities are observed. The control volume approach has the advantage that, under certain conditions, it is possible to calculate quantities such as the force exerted or the power generated or dissipated, simply by having the appropriate information at the control surface. However, the weakness of this approach is that we do not obtain the details of the various fields at every point within the control volume. For obtaining detailed information, we need to solve the governing differential equations subject to appropriate boundary and initial conditions.

As emphasized in Chapter 1, we write the governing equations in tensorial form, because then it is immediately evident that such equations are independent of the choice of coordinate system. Only while solving specific problems, we *choose* a particular coordinate system, and write the tensor equations in component form in that particular coordinate system. For example, if we were considering flow between two flat surfaces, we would generally prefer to work with a Cartesian coordinate system, and write the incompressibility equation $\boldsymbol{\nabla} \cdot \boldsymbol{v} = 0$ as

$$\frac{\partial v_x}{\partial x} + \frac{\partial v_y}{\partial y} + \frac{\partial v_z}{\partial z} = 0,$$

whereas if we were considering flow through a circular pipe, we would usually prefer working in cylindrical coordinates, and write the incompressibility equation as

$$\frac{1}{r}\frac{\partial (r v_r)}{\partial r} + \frac{1}{r}\frac{\partial v_\theta}{\partial \theta} + \frac{\partial v_z}{\partial z} = 0.$$

The derivation of the governing equations hinges on the transport theorems, which we now derive.

3.1 The First Transport Theorem

We start by deriving the first transport theorem for scalar-valued functions. The corresponding theorems for vector or tensor-valued functions are obtained by applying the result to each component.

Theorem 3.1.1 (Transport theorem-I). *Let $f(x,t)$ be a continuous and differentiable scalar-valued function, and let $V(t)$ be the material volume. Then, obviously, $\int_{V(t)} f(x,t)\,dV$ is a function of time, and its time derivative is given by*

$$\frac{d}{dt}\int_{V(t)} f(x,t)\,dV = \int_{V(t)} \frac{\partial f}{\partial t}\,dV + \int_{S(t)} f(v\cdot n)\,dS. \tag{3.1}$$

Proof. Let $\tilde{f}(X,t) := f(x,t) = f(\chi(X,t),t)$. In order to take the derivative inside the integral sign, we first transform the volume integral over $V(t)$ to one over V_0, where V_0 represents the fixed configuration at $t=0$. Thus,

$$\begin{aligned}
\frac{d}{dt}\int_{V(t)} f(x,t)\,dV &= \frac{d}{dt}\int_{V_0} \tilde{f}J\,dV_0 \\
&= \int_{V_0} \left(\frac{\partial(\tilde{f}J)}{\partial t}\right)_X dV_0 \\
&= \int_{V_0} \left[J\left(\frac{\partial \tilde{f}}{\partial t}\right)_X + \tilde{f}\left(\frac{\partial J}{\partial t}\right)_X\right] dV_0 \\
&= \int_{V_0} \left[\frac{D\tilde{f}}{Dt} + \frac{\tilde{f}}{J}\frac{DJ}{Dt}\right] J\,dV_0.
\end{aligned}$$

Changing the volume integral over V_0 back to one over $V(t)$, we get

$$\begin{aligned}
\frac{d}{dt}\int_{V(t)} f(x,t)\,dV &= \int_{V(t)} \left[\frac{Df}{Dt} + f(\nabla\cdot v)\right] dV && \text{(by Eqn. (2.51))} \\
&= \int_{V(t)} \left[\frac{\partial f}{\partial t} + v\cdot(\nabla f) + f(\nabla\cdot v)\right] dV && \text{(by Eqn. (2.45))} \\
&= \int_{V(t)} \left[\frac{\partial f}{\partial t} + \nabla\cdot(fv)\right] dV && \text{(by Eqn. (1.353))} \\
&= \int_{V(t)} \frac{\partial f}{\partial t}\,dV + \int_{S(t)} f(v\cdot n)\,dS. && \text{(by Eqn. (1.315))}\ \square
\end{aligned}$$

If $w(x,t)$ is a vector-valued function, then applying Eqn. (3.1) to each component of w, we get

$$\frac{d}{dt}\int_{V(t)} w(x,t)\,dV = \int_{V(t)} \frac{\partial w}{\partial t}\,dV + \int_{S(t)} w(v\cdot n)\,dS. \tag{3.2}$$

Note that in Eqns. (3.1) and (3.2), $\boldsymbol{v}$ is the velocity of the surface $S(t)$ as it evolves.

We now derive the Lagrangian and Eulerian forms of the principle of conservation of mass, which in turn are used to derive the second transport theorem. For the subsequent development, we define the density field as

$$\rho(\boldsymbol{x},t) := \lim_{\Delta V \to 0} \frac{\Delta m}{\Delta V},$$

where ΔV is a small volume of material of mass Δm, surrounding the point with position vector $\boldsymbol{x}$.

3.2 Conservation of Mass

3.2.1 Lagrangian version

We denote the position vector of a point and the reference volume at $t = 0$ by $\boldsymbol{X}$ and V_0, respectively, and the corresponding quantities at time t by $\boldsymbol{x}$ and $V(t)$, respectively. As in Chapter 2, we assume that the mapping $\boldsymbol{x} = \boldsymbol{\chi}(\boldsymbol{X},t)$ is one-to-one and orientation-preserving. Let $\tilde{\rho}(\boldsymbol{X},t) := \rho(\boldsymbol{x},t) = \rho(\boldsymbol{\chi}(\boldsymbol{X},t),t)$, and let $\rho_0(\boldsymbol{X})$ denote the density in the reference configuration, i.e., $\rho_0(\boldsymbol{X}) = \tilde{\rho}(\boldsymbol{X},0)$.

Consider a specific mass of the medium enclosed in a material volume $V(t)$, which is arbitrarily chosen. In our non-relativistic setup, the axiom of mass conservation states that as $V(t)$ evolves, its shape and size may change, but its mass remains unchanged (assuming that there are no mass sources within $V(t)$). In particular, the mass in V_0 and $V(t)$ is the same. Thus,

$$\int_{V_0} \rho_0(\boldsymbol{X})\, dV_0 = \int_{V(t)} \rho(\boldsymbol{x},t)\, dV. \tag{3.3}$$

Transforming the element of volume dV to dV_0 using Eqn. (2.37), we get

$$\int_{V_0} \rho_0(\boldsymbol{X})\, dV_0 = \int_{V_0} \tilde{\rho}(\boldsymbol{X},t) J\, dV_0, \tag{3.4}$$

which in turn implies that

$$\int_{V_0} [\rho_0(\boldsymbol{X}) - \tilde{\rho}(\boldsymbol{X},t)J]\, dV_0 = 0.$$

Since V_0 has been arbitrarily chosen, we conclude using the localization theorem (Theorem 1.11.1) that

$$\rho_0(\boldsymbol{X}) = \tilde{\rho}(\boldsymbol{X},t)J = \rho(\boldsymbol{x},t)J. \tag{3.5}$$

Since the first relation in Eqn. (3.5) involves the material coordinates $\boldsymbol{X}$, it is known as the Lagrangian form of the equation of mass conservation.

3.2.2 Eulerian version

We now derive the differential form of the equation of mass conservation involving only the spatial description. Applying the first transport theorem to the mass conservation equation

$$\frac{d}{dt}\int_{V(t)} \rho \, dV = 0, \tag{3.6}$$

we get the integral form of the mass conservation equation as

$$\int_{V(t)} \frac{\partial \rho}{\partial t} \, dV + \int_{S(t)} \rho(\boldsymbol{v}\cdot\boldsymbol{n}) \, dS = 0.$$

Applying the divergence theorem to the second term, we get

$$\int_{V(t)} \left[\frac{\partial \rho}{\partial t} + \boldsymbol{\nabla}\cdot(\rho\boldsymbol{v})\right] dV = 0.$$

Using the fact that $V(t)$ is arbitrary, and invoking the localization theorem, we obtain

$$\frac{\partial \rho}{\partial t} + \boldsymbol{\nabla}\cdot(\rho\boldsymbol{v}) = 0. \tag{3.7}$$

Since by Eqn. (1.353), we have $\boldsymbol{\nabla}\cdot(\rho\boldsymbol{v}) = \boldsymbol{v}\cdot(\boldsymbol{\nabla}\rho) + \rho(\boldsymbol{\nabla}\cdot\boldsymbol{v})$, the above equation can also be written as

$$\frac{D\rho}{Dt} + \rho\boldsymbol{\nabla}\cdot\boldsymbol{v} = 0. \tag{3.8}$$

Another way of deriving the above equation is by taking the material derivative of both sides of Eqn. (3.5), and then using Eqn. (2.51) and the fact that $J > 0$.

An incompressible medium is one that is capable of undergoing only isochoric deformations. Thus, for any motion of an incompressible body, we have $J = 1$. From Eqn. (2.54), the incompressibility condition can be written as

$$\boldsymbol{\nabla}\cdot\boldsymbol{v} = 0. \tag{3.9}$$

From Eqn. (3.8), we observe that the above condition is equivalent to the condition $D\rho/Dt = 0$, while from Eqn. (3.5), we get the condition $\rho(\boldsymbol{x},t) = \rho_0(\boldsymbol{X})$. Thus, the density at any material point remains the same in an incompressible material. If the density is uniform in the reference configuration, then it continues to be uniform in the current configuration. It should be noted that since

$$\frac{D\rho}{Dt} = \frac{\partial \rho}{\partial t} + \boldsymbol{v}\cdot(\boldsymbol{\nabla}\rho),$$

one can have a density field such that $\partial\rho/\partial t = -\boldsymbol{v}\cdot(\boldsymbol{\nabla}\rho)$, so that $D\rho/Dt = 0$, and the incompressibility condition is satisfied. If, however, one assumes the density of an incompressible material to be uniform to begin with, then it remains uniform at all subsequent times.

We now use the equation of conservation of mass to derive the second transport theorem.

3.3 The Second Transport Theorem

Theorem 3.3.1. *Let $f(x,t)$ and $w(x,t)$ be scalar and vector-valued functions defined on $V(t)$. Then,*

$$\frac{d}{dt}\int_{V(t)} \rho f(x,t)\,dV = \int_{V(t)} \rho \frac{Df}{Dt}\,dV = \int_{V(t)} \rho \left\{ \frac{\partial f}{\partial t} + v \cdot (\nabla f) \right\} dV, \tag{3.10a}$$

$$\frac{d}{dt}\int_{V(t)} \rho w(x,t)\,dV = \int_{V(t)} \rho \frac{Dw}{Dt}\,dV = \int_{V(t)} \rho \left\{ \frac{\partial w}{\partial t} + (\nabla w)v \right\} dV. \tag{3.10b}$$

Proof. By letting ρf play the role of f in transport theorem I, we get

$$\begin{aligned}
\frac{d}{dt}\int_{V(t)} \rho f(x,t)\,dV &= \int_{V(t)} \left[\frac{\partial(\rho f)}{\partial t} + \nabla \cdot (\rho f v) \right] dV \\
&= \int_{V(t)} \left[\rho \frac{\partial f}{\partial t} + f\left(\frac{\partial \rho}{\partial t} + \nabla \cdot (\rho v) \right) + (\rho v) \cdot (\nabla f) \right] dV \\
&= \int_{V(t)} \rho \left[\frac{\partial f}{\partial t} + v \cdot (\nabla f) \right] dV \qquad \text{(by Eqn. (3.7))} \\
&= \int_{V(t)} \rho \frac{Df}{Dt}\,dV. \qquad \text{(by Eqn. (2.45))}
\end{aligned}$$

By letting $f \equiv w_i$ in Eqn. (3.10a), we get Eqn. (3.10b). □

We generalize the two transport theorems to include arbitrary time-varying domains V_t that are not restricted to be material volumes in the following section.

3.4 Generalized Transport Theorems

In fluid mechanics, a control volume is a region of space V_t, where we observe the various flow quantities; the subscript t in V_t denotes the possible time dependence that the control volume may have. Unlike the material volume $V(t)$, which comprises the same set of particles as it evolves, the control volume V_t will, in general, have different sets of particles at different times, and even the mass of material enclosed by the control volume may change. An example of such a situation is a control volume attached to a rocket. Let the surface of the control volume be denoted by S_t.

To derive transport theorems analogous to transport theorems I and II for V_t, we introduce the mapping $x = \chi(X,t)$ that maps every X in a stationary reference configuration V_0 to x in V_t. The deformation gradient F now refers to the deformation of the domain (and *not* to the deformation of the material within the domain), so that we have $dV = J\,dV_0$, where $J = \det F$. Analogous to Eqns (2.45) and (2.51), we get

$$\left(\frac{\partial \tilde{f}}{\partial t} \right)_X = \left(\frac{\partial f}{\partial t} \right)_x + v_m \cdot (\nabla_x f), \tag{3.11a}$$

$$\left(\frac{\partial J}{\partial t}\right)_X = J\nabla_x \cdot v_m, \tag{3.11b}$$

where $v_m(x,t)$ denotes the velocity of the domain (with respect to the chosen frame of reference), which is, in general, different than the velocity of the material $v(x,t)$ inside the control volume. In case, the control volume is taken to be the material volume $V(t)$, then, of course, $v(x,t) = v_m(x,t)$.

The generalized first transport theorem can now be stated as

Theorem 3.4.1. *Let $f(x,t)$ be a continuous and differentiable scalar-valued function, and let V_t be the control volume. Then,*

$$\frac{d}{dt}\int_{V_t} f(x,t)\, dV = \int_{V_t}\left[\frac{\partial f}{\partial t} + \nabla\cdot(fv)\right] dV - \int_{S_t} f(v_{rel}\cdot n)\, dS, \tag{3.12}$$

where $v_{rel} = v - v_m$ denotes the relative velocity of the material with respect to the control surface.

Proof. Let $\tilde{f}(X,t) := f(x,t) = f(\chi(X,t),t)$. As in the proof of the first transport theorem, in order to take the derivative inside the integral sign, we first transform the volume integral over V_t to one over V_0. Thus,

$$\begin{aligned}
\frac{d}{dt}\int_{V_t} f(x,t)\, dV &= \frac{d}{dt}\int_{V_0} \tilde{f} J\, dV_0 \\
&= \int_{V_0}\left(\frac{\partial(\tilde{f}J)}{\partial t}\right)_X dV_0 \\
&= \int_{V_0}\left[J\left(\frac{\partial \tilde{f}}{\partial t}\right)_X + \tilde{f}\left(\frac{\partial J}{\partial t}\right)_X\right] dV_0 \\
&= \int_{V_t}\left[\left(\frac{\partial f}{\partial t}\right)_x + v_m\cdot(\nabla f) + f\nabla_x\cdot v_m\right] dV \quad \text{(by Eqns. (3.11))} \\
&= \int_{V_t}\left[\left(\frac{\partial f}{\partial t}\right)_x + \nabla_x\cdot(fv_m)\right] dV \\
&= \int_{V_t}\left[\frac{\partial f}{\partial t} + \nabla\cdot(fv)\right] dV - \int_{S_t} f(v_{\text{rel}}\cdot n)\, dS.
\end{aligned}$$

□

The term $\int_{S_t} f(v_{\text{rel}}\cdot n)\, dS$ represents the rate of efflux of the quantity f from the control volume.

By letting $\rho f(x,t)$ play the role of $f(x,t)$ in the above theorem, we get the following generalized second transport theorem

Theorem 3.4.2. *Let $f(x,t)$ and $w(x,t)$ be scalar and vector fields defined on V_t. Then,*

$$\frac{d}{dt}\int_{V_t} \rho f(x,t)\, dV = \int_{V_t}\rho\left[\frac{\partial f}{\partial t} + v\cdot(\nabla f)\right] dV - \int_{S_t}\rho f(v_{rel}\cdot n)\, dS, \tag{3.13a}$$

$$\frac{d}{dt}\int_{V_t} \rho w(x,t)\,dV = \int_{V_t} \rho\left[\frac{\partial w}{\partial t} + (\nabla w)v\right] dV - \int_{S_t} \rho w(v_{rel}\cdot n)\,dS, \tag{3.13b}$$

where $v_{rel} = v - v_m$ denotes the relative velocity of the material with respect to the control surface.

Proof. Using Eqn. (3.12), we have

$$\begin{aligned}\frac{d}{dt}\int_{V_t} \rho f(x,t)\,dV &= \int_{V_t}\left[\frac{\partial(\rho f)}{\partial t} + \nabla\cdot(\rho f v)\right] dV - \int_{S_t} \rho f(v_{\text{rel}}\cdot n)\,dS \\ &= \int_{V_t} \rho\left[\frac{\partial f}{\partial t} + v\cdot(\nabla f)\right] dV - \int_{S_t} \rho f(v_{\text{rel}}\cdot n)\,dS. \quad \text{(by Eqn. (3.7))}\end{aligned}$$

By letting $f \equiv w_i$, we get Eqn. (3.13b). □

Note that when V_t is chosen to be a material volume $V(t)$, then v_{rel} is zero, and the generalized transport theorems 3.4.1 and 3.4.2 reduce to transport theorems 3.1.1 and 3.3.1, respectively.

It is obviously of interest to find the conditions under which

$$\frac{d}{dt}\int_{V_t} f(x,t)\,dV = \int_{V_t}\left(\frac{\partial f}{\partial t}\right)_x dV. \tag{3.14}$$

From Eqn. (3.12), we get the condition for Eqn. (3.14) to hold as

$$\int_{S_t} f(v_m\cdot n)\,dS = 0,$$

where v_m denotes the velocity of the control surface. If the control volume is fixed, then obviously the above condition is satisfied. Alternatively, if f is a function of time alone, and if the control volume is undergoing rigid motion (i.e., the shape of the control surface does not change), then using the characterization of rigid motion and the divergence theorem, we have

$$\int_{S_t} f(v_m\cdot n)\,dS = f\int_{S_t}(v_m\cdot n)\,dS = f\int_{V_t}\nabla\cdot[w(t)\times x + z(t)]\,dV = 0.$$

Thus, either of the following conditions are sufficient (but not necessary) for Eqn. (3.14) to hold:

1. The control volume is stationary, i.e., $v_m = \mathbf{0}$.
2. The control volume is undergoing rigid motion, *and* f is a function of time alone.

3.5 Balance of Linear Momentum

The axioms of balance of linear and angular momentum in the forms that are used today were first postulated by Euler. The balance of linear momentum states that the rate of change of momentum of the material volume is equal to the net external force acting on it.

The external forces that may act on the given mass can be classed either as body forces, such as gravitational forces or electromagnetic forces, or as surface forces such as pressure or viscous forces that arise from surface contact between bulks of continua. If $\boldsymbol{b}(\boldsymbol{x},t)$ is the body force per unit mass of the medium, then the body force is given by $\int_{V(t)} \rho \boldsymbol{b}\, dV$. We would like to emphasize that the body force $\boldsymbol{b}$ is dependent on the frame of reference (see Section 4.4).

Let $\boldsymbol{n}$ represent the outward normal to the bounding surface $S(t)$ of an arbitrary portion of the body, comprising, as usual, a fixed set of particles. Then, by the *Cauchy hypothesis*, there exists a traction vector field $\boldsymbol{t}(\boldsymbol{x},t,\boldsymbol{n})$ due to the influence of the material outside $S(t)$.[1] The net external force acting on surface $S(t)$ is given by $\int_{S(t)} \boldsymbol{t}(\boldsymbol{x},t,\boldsymbol{n})\, dS$. Note that as a result of the dependence of $\boldsymbol{t}$ on $\boldsymbol{n}$, two surfaces with the same normal but different curvatures, passing through a point $\boldsymbol{x}$ at a given instant of time, have the same traction vector at that point (see Fig. 3.1).

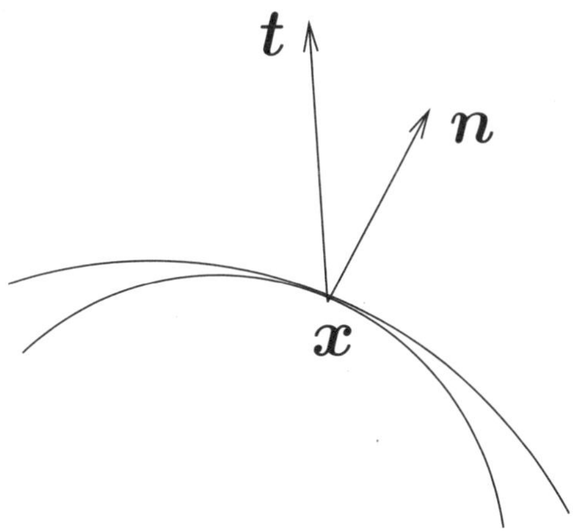

Fig. 3.1 *The traction vector on two different surfaces having the same normal at a point.*

Equating the rate of change of momentum of the mass contained within $V(t)$ to the sum of the body and surface forces, we get

$$\frac{d}{dt}\int_{V(t)} \rho \boldsymbol{v}\, dV = \int_{S(t)} \boldsymbol{t}(\boldsymbol{x},t,\boldsymbol{n})\, dS + \int_{V(t)} \rho \boldsymbol{b}\, dV. \tag{3.15}$$

Note that the choice of $V(t)$ is arbitrary; e.g., any subpart of a given material volume can also be chosen as a material volume for which Eqn. (3.15) holds.

Defining the center of mass of $V(t)$ as

$$\bar{\boldsymbol{x}}(t) = \frac{1}{M}\int_{V(t)} \rho \boldsymbol{x}\, dV,$$

where $M = \int_{V(t)} \rho\, dV$ is the mass contained within $V(t)$, Eqn. (3.15) can be written as

$$M\ddot{\bar{\boldsymbol{x}}} = \int_{S(t)} \boldsymbol{t}(\boldsymbol{x},t,\boldsymbol{n})\, dS + \int_{V(t)} \rho \boldsymbol{b}\, dV, \tag{3.16}$$

[1]The Cauchy hypothesis can be proved under very weak assumptions using the balance of linear momentum; see, e.g., [334]

since by Eqn. (2.47) and transport theorem II

$$\int_{V(t)} \rho \boldsymbol{v}\, dV = \frac{d}{dt}\int_{V(t)} \rho \boldsymbol{x}\, dV = M\dot{\bar{\boldsymbol{x}}}. \tag{3.17}$$

Thus, for any material volume, the force is equal to the mass enclosed within the material volume times the acceleration of its center of mass. Note that the definition of the center of mass is independent of the choice of origin since

$$\bar{\boldsymbol{x}}(t) - \boldsymbol{z} = \frac{1}{M}\left[\int_{V(t)} \rho \boldsymbol{x}\, dV - \int_{V(t)} \rho \boldsymbol{z}\, dV\right] = \frac{1}{M}\int_{V(t)} \rho(\boldsymbol{x}-\boldsymbol{z})\, dV.$$

We now make explicit the dependence of $\boldsymbol{t}$ on $\boldsymbol{n}$.

Theorem 3.5.1 (Cauchy's theorem). *Assuming that the body force $\boldsymbol{b}(\boldsymbol{x},t)$, density $\rho(\boldsymbol{x},t)$ and acceleration $\boldsymbol{a}(\boldsymbol{x},t)$ are continuous functions of $\boldsymbol{x}$ at any time t, and assuming that the traction vector $\boldsymbol{t}(\boldsymbol{x},t,\boldsymbol{n})$ is continuously differentiable with respect to $\boldsymbol{x}$ for each unit vector $\boldsymbol{n}$ and time t, and continuous with respect to $\boldsymbol{n}$ for each $\boldsymbol{x}$ and t, the balance of linear momentum is equivalent to the following two statements:*

1. *There exists a continuously differentiable tensor field $\boldsymbol{\tau}(\boldsymbol{x},t)$ (known as the* Cauchy stress tensor*) such that*

$$\boldsymbol{t}(\boldsymbol{x},t,\boldsymbol{n}) = \boldsymbol{\tau}(\boldsymbol{x},t)\boldsymbol{n}. \tag{3.18}$$

2. *$\boldsymbol{\tau}(\boldsymbol{x},t)$ satisfies the equation of motion*

$$\rho\frac{D\boldsymbol{v}}{Dt} = \nabla_{\boldsymbol{x}} \cdot \boldsymbol{\tau} + \rho \boldsymbol{b}. \tag{3.19}$$

Proof. Consider the tetrahedron shown in Fig. 3.2 with the vertex O having a position vector $\boldsymbol{x}$. Let $\boldsymbol{n} = (n_1, n_2, n_3)$ denote the normal to the face ABC, and let S, S_1, S_2 and S_3 denote the surfaces ΔABC, ΔOBC, ΔOAC and ΔOAB, respectively. To begin with, we assume $n_i > 0$ as shown in the figure. The outward normals of ΔOBC, ΔOAC and ΔOAB are $(-\boldsymbol{e}_1)$, $(-\boldsymbol{e}_2)$ and $(-\boldsymbol{e}_3)$, respectively.

Using transport theorem II (Eqn. (3.10b)) on the left-hand side of Eqn. (3.15), and rearranging terms we get

$$\int_{V(t)} \left(\rho\frac{D\boldsymbol{v}}{Dt} - \rho\boldsymbol{b}\right) dV = \int_{S(t)} \boldsymbol{t}(\boldsymbol{x},t,\boldsymbol{n})\, dS + \int_{S_1(t)} \boldsymbol{t}(\boldsymbol{x},t,-\boldsymbol{e}_1)\, dS_1 + \int_{S_2(t)} \boldsymbol{t}(\boldsymbol{x},t,-\boldsymbol{e}_2)\, dS_2 + \int_{S_3(t)} \boldsymbol{t}(\boldsymbol{x},t,-\boldsymbol{e}_3)\, dS_3.$$

Using the first mean value theorem for integrals (use of this theorem is justified since $\rho(D\boldsymbol{v}/Dt - \boldsymbol{b})$ and $\boldsymbol{t}$ are continuous functions of $\boldsymbol{x}$), and the fact that $dS_i = (\boldsymbol{e}_i \cdot \boldsymbol{n})dS$ so that $\text{area}(S_1) = n_1\,\text{area}(S)$, $\text{area}(S_2) = n_2\,\text{area}(S)$, $\text{area}(S_3) = n_3\,\text{area}(S)$, we get

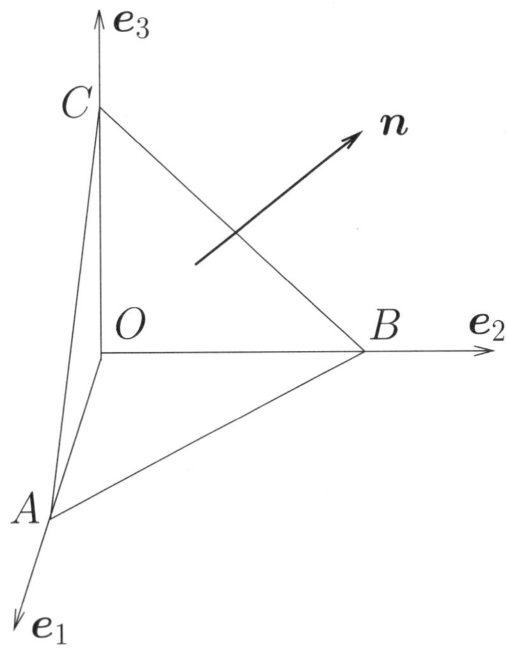

Fig. 3.2 *Tetrahedral element used in Cauchy's proof.*

$$\rho \left[\frac{Dv}{Dt} - b \right]_{x^{**}} \frac{1}{3} Ah = \left[t(x^*, t, n) + t(x_i^*, t, -e_i) n_i \right] A \tag{3.20}$$

where h is the distance of S from the origin O, A is the area of S, x^{**} is in V, x^* is on S, and x_i^* are on S_i. Let h tend to zero in such a manner that the normal to the face ABC continues to coincide with n; note that x_i^* and x^{**} converge to x as $h \to 0$. Again using the continuity of t with respect to x, and the boundedness of $\rho(Dv/Dt - b)$, Eqn. (3.20) reduces to

$$t(x, t, n) = -t(x, t, -e_i) n_i. \tag{3.21}$$

Using next the continuity of t with respect to n, and letting $n \to e_1 = (1, 0, 0)$, we get

$$t(x, t, e_1) = -t(x, t, -e_1).$$

Letting $n \to e_2$ and $n \to e_3$, we get analogous results. Thus,

$$t(x, t, e_i) = -t(x, t, -e_i) \quad i = 1, 2, 3. \tag{3.22}$$

Substituting this result into Eqn. (3.21), we have

$$t(x, t, n) = t(x, t, e_i) n_i. \tag{3.23}$$

We have proved the above result for the case when all $n_i > 0$. The same result holds for all unit normals $\boldsymbol{n}$ even if some or all $n_i < 0$, as we now show. Consider the tetrahedron shown in Fig. 3.3 for which $n_2 < 0$. Proceeding in a similar manner as above, we get instead of Eqn. (3.21) the relation

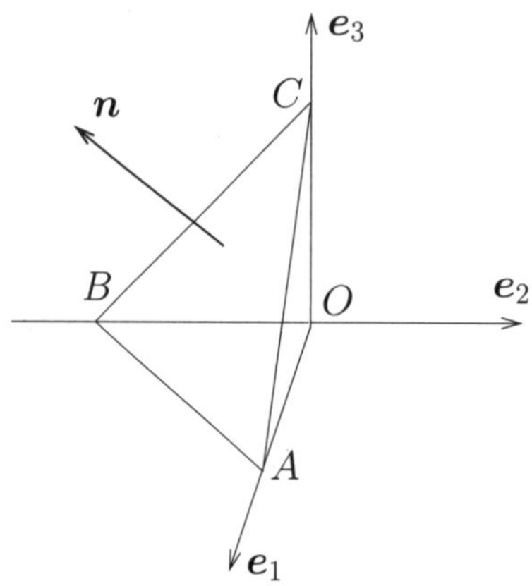

Fig. 3.3 *Tetrahedral element with $n_2 < 0$.*

$$\boldsymbol{t}(\boldsymbol{x},t,\boldsymbol{n}) = -\boldsymbol{t}(\boldsymbol{x},t,-\boldsymbol{e}_1)n_1 + \boldsymbol{t}(\boldsymbol{x},t,\boldsymbol{e}_2)n_2 - \boldsymbol{t}(\boldsymbol{x},t,-\boldsymbol{e}_3)n_3.$$

Letting first $\boldsymbol{n} \to \boldsymbol{e}_1$, and then to $-\boldsymbol{e}_2$ and $\boldsymbol{e}_3$, we again get Eqn. (3.22), and hence Eqn. (3.23). In general, in place of Eqn. (3.21), we have

$$\boldsymbol{t}(\boldsymbol{x},t,\boldsymbol{n}) + \sum_{k=1}^{3} \operatorname{sgn}(n_k)\boldsymbol{t}(\boldsymbol{x},t,-\operatorname{sgn}(n_k)\boldsymbol{e}_k)n_k = 0.$$

If $n_k > 0$, then let $\boldsymbol{n} \to \boldsymbol{e}_k$, else if $n_k < 0$, let $\boldsymbol{n} \to -\boldsymbol{e}_k$ to get Eqn. (3.22), and hence Eqn. (3.23). Thus, Eqn. (3.23) holds for all unit normals $\boldsymbol{n}$.

We now define the Cauchy stress tensor $\boldsymbol{\tau}$ as

$$\boldsymbol{\tau}(\boldsymbol{x},t) := \boldsymbol{t}(\boldsymbol{x},t,\boldsymbol{e}_k) \otimes \boldsymbol{e}_k, \tag{3.24}$$

Since $\boldsymbol{t}(\boldsymbol{x},t,\boldsymbol{n})$ is assumed to be continuously differentiable with respect to $\boldsymbol{x}$, by Eqn. (3.24) it follows that so is $\boldsymbol{\tau}$. Using Eqns. (3.23) and (3.24), we obtain Eqn. (3.18).

We now derive the equation of motion, which is nothing but the differential form of the linear momentum equation. By applying Eqn. (3.10b) to the left side, and substituting Eqn. (3.18) into the right side of Eqn. (3.15), we get

$$\int_{V(t)} \rho \frac{D\boldsymbol{v}}{Dt}\, dV = \int_{S(t)} \boldsymbol{\tau}\boldsymbol{n}\, dS + \int_{V(t)} \rho \boldsymbol{b}\, dV.$$

Using the divergence theorem (Eqn. (1.316)) and grouping terms, we get

$$\int_{V(t)} \left\{ \rho \frac{Dv}{Dt} - \nabla \cdot \tau - \rho b \right\} dV = 0.$$

However, since the volume $V(t)$ is arbitrary, by the localization theorem it follows that the integrand of the above equation is zero, and Eqn. (3.19) follows.

We have shown that balance of linear momentum implies Cauchy's relation given by Eqn. (3.18), and the differential form given by Eqn. (3.19). Conversely, Eqns. (3.18) and (3.19) together imply Eqn. (3.15) since

$$\begin{aligned}\frac{d}{dt}\int_{V(t)} \rho v \, dV &= \int_{V(t)} \rho \frac{Dv}{Dt} \, dV \\ &= \int_{V(t)} (\nabla_x \cdot \tau + \rho b) \, dV \\ &= \int_{S(t)} \tau n \, dS + \int_{V(t)} \rho b \, dV \\ &= \int_{S(t)} t \, dS + \int_{V(t)} \rho b \, dV. \end{aligned}$$ □

In indicial notation, Eqns. (3.18) and (3.19) read

$$\begin{aligned} t_i &= \tau_{ij} n_j, \\ \rho \left\{ \frac{\partial v_i}{\partial t} + \frac{\partial v_i}{\partial x_j} v_j \right\} &= \frac{\partial \tau_{ij}}{\partial x_j} + \rho b_i, \end{aligned}$$

where, by Eqn. (3.24),

$$\tau_{ij} = e_i \cdot \tau e_j = t_i(x, t, e_j).$$

Thus, the stress component τ_{ij} represents the ith component of the traction vector acting on the surface whose normal is in the jth direction, as shown in Fig. 3.4. Note that the element shown in Fig. 3.4 does not have any dimensions, and is simply a convenient device for showing the stress components acting on different planes passing through the point x. Since, by Eqn. (3.18), τ transforms the vector n into the traction vector t by a linear transformation, it is a second-order tensor. Hence, the transformation law for the stress tensor components is as given by Eqn. (1.98).

From Eqn. (3.18) (or Eqn. (3.23)), it follows that

$$t(x, t, -n) = -t(x, t, n). \tag{3.25}$$

Equation (3.25) is the analogue of Newton's law of action and reaction. Figure 3.5 illustrates this principle.

Applying Eqn. (3.2) to the left-hand side of Eqn. (3.15), we get

$$\int_{V(t)} \frac{\partial(\rho v)}{\partial t} \, dV + \int_{S(t)} \rho v (v \cdot n) \, dS = \int_{S(t)} t \, dS + \int_{V(t)} \rho b \, dV,$$

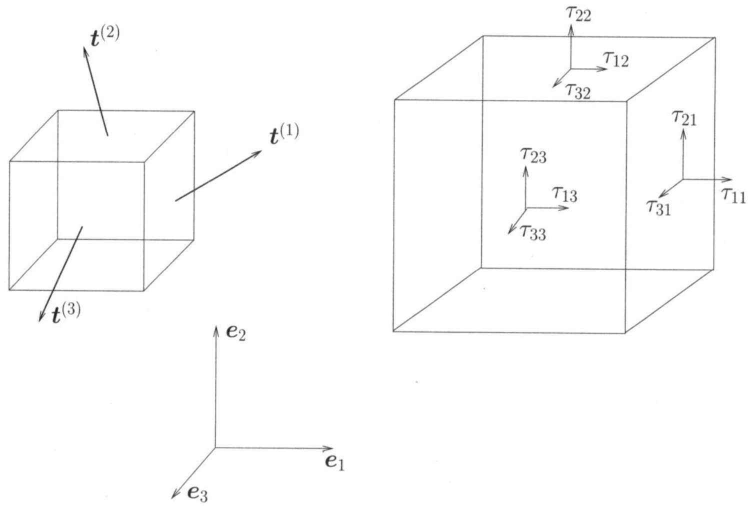

Fig. 3.4 *Traction vectors and the corresponding stress components on an element with $t^{(i)}$ denoting $t(x, t, e_i)$. Tractions and stresses on a hidden face are equal and opposite to those on the corresponding visible face having an opposite normal.*

which is the integral form of the linear momentum balance principle.

By virtue of Eqn. (3.7), Eqn. (3.19) can also be written as

$$\frac{\partial(\rho \boldsymbol{v})}{\partial t} + \boldsymbol{\nabla}_x \cdot [\rho(\boldsymbol{v} \otimes \boldsymbol{v})] = \boldsymbol{\nabla}_x \cdot \boldsymbol{\tau} + \rho \boldsymbol{b}.$$

3.6 Balance of Angular Momentum

This principle states that the rate of change of angular momentum of the material volume is equal to the moments about the origin due to the external forces acting on it, i.e.,

$$\frac{d}{dt} \int_{V(t)} \rho \boldsymbol{x} \times \boldsymbol{v} \, dV = \int_{S(t)} \boldsymbol{x} \times \boldsymbol{t} \, dS + \int_{V(t)} \rho \boldsymbol{x} \times \boldsymbol{b} \, dV. \tag{3.26}$$

Let $\boldsymbol{z}(t)$ be the position vector of a point in motion (e.g., the point could be the center of mass or a material point with $\boldsymbol{z}(t)$ given by the motion $\boldsymbol{\chi}(\boldsymbol{X}, t)$ with $\boldsymbol{X}$ fixed), and let the spin angular momentum $\boldsymbol{h}_{\text{spin}}(t)$, the angular momentum $\boldsymbol{h}_z(t)$ relative to $\boldsymbol{z}(t)$, and the moment $\boldsymbol{m}_z$ about $\boldsymbol{z}(t)$ be defined by

$$\boldsymbol{h}_{\text{spin}}(t) := \int_{V(t)} \rho(\boldsymbol{x} - \bar{\boldsymbol{x}}) \times (\boldsymbol{v} - \dot{\bar{\boldsymbol{x}}}) \, dV, \tag{3.27}$$

$$\boldsymbol{h}_z(t) := \int_{V(t)} \rho(\boldsymbol{x} - \boldsymbol{z}) \times (\boldsymbol{v} - \dot{\boldsymbol{z}}) \, dV = \boldsymbol{h}_{\text{spin}}(t) + M(\bar{\boldsymbol{x}} - \boldsymbol{z}) \times (\dot{\bar{\boldsymbol{x}}} - \dot{\boldsymbol{z}}), \tag{3.28}$$

$$\boldsymbol{m}_z(t) := \int_{S(t)} (\boldsymbol{x} - \boldsymbol{z}) \times \boldsymbol{t}(\boldsymbol{n}) \, dS + \int_{V(t)} \rho(\boldsymbol{x} - \boldsymbol{z}) \times \boldsymbol{b} \, dV = \boldsymbol{m}_{\bar{x}} + M(\bar{\boldsymbol{x}} - \boldsymbol{z}) \times \ddot{\bar{\boldsymbol{x}}}. \tag{3.29}$$

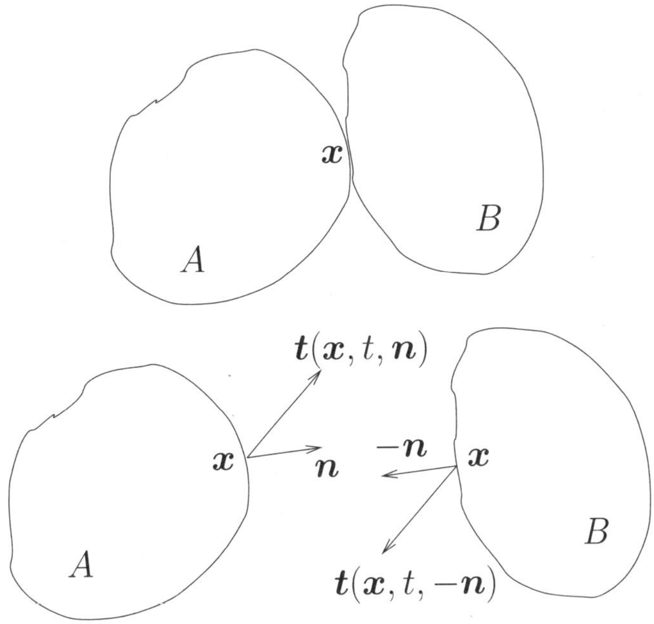

Fig. 3.5 *Principle of action and reaction.*

Then using Eqns. (3.15), (3.26) and transport theorem II, we have

$$
\begin{aligned}
\dot{\boldsymbol{h}}_z(t) &= \int_{V(t)} \rho(\boldsymbol{v} - \dot{\boldsymbol{z}}) \times (\boldsymbol{v} - \dot{\boldsymbol{z}})\, dV + \int_{V(t)} \rho(\boldsymbol{x} - \boldsymbol{z}) \times (\boldsymbol{a} - \ddot{\boldsymbol{z}})\, dV \\
&= \int_{V(t)} \rho(\boldsymbol{x} - \boldsymbol{z}) \times \boldsymbol{a}\, dV - M(\bar{\boldsymbol{x}} - \boldsymbol{z}) \times \ddot{\boldsymbol{z}} \\
&= \int_{S(t)} (\boldsymbol{x} - \boldsymbol{z}) \times \boldsymbol{t}\, dS + \int_{V(t)} \rho(\boldsymbol{x} - \boldsymbol{z}) \times \boldsymbol{b}\, dV - M(\bar{\boldsymbol{x}} - \boldsymbol{z}) \times \ddot{\boldsymbol{z}},
\end{aligned}
$$

so that

$$
\dot{\boldsymbol{h}}_z(t) + M(\bar{\boldsymbol{x}} - \boldsymbol{z}) \times \ddot{\boldsymbol{z}} = \boldsymbol{m}_z(t). \tag{3.30}
$$

If $\boldsymbol{z} = \bar{\boldsymbol{x}}$ or if $\ddot{\boldsymbol{z}} = \boldsymbol{0}$ or if $\ddot{\boldsymbol{z}}$ is parallel to $\bar{\boldsymbol{x}} - \boldsymbol{z}$, then the second term on the left-hand side vanishes. Thus, for the case $\boldsymbol{z} = \bar{\boldsymbol{x}}$, we have

$$
\dot{\boldsymbol{h}}_{\text{spin}} = \boldsymbol{m}_{\bar{x}}(t), \tag{3.31}
$$

while for the case $\boldsymbol{z} = \boldsymbol{x}_0$, where $\boldsymbol{x}_0$ is a fixed point ($\dot{\boldsymbol{x}}_0 = \boldsymbol{0}$), we have

$$
\dot{\boldsymbol{h}}_{x_0} = \boldsymbol{m}_{x_0}(t),
$$

thus showing that the balance of angular momentum holds irrespective of the choice of origin. If the body force $\boldsymbol{b}$ is uniform over the whole body, then we have $\int_{V(t)} \rho \boldsymbol{b}\, dV = M\boldsymbol{b}$, and $\int_{V(t)} \rho(\boldsymbol{x} - \bar{\boldsymbol{x}}) \times \boldsymbol{b}\, dV = M(\bar{\boldsymbol{x}} - \bar{\boldsymbol{x}}) \times \boldsymbol{b} = \boldsymbol{0}$; thus, when the body force is uniform, we can replace the distributed body force by a point force $M\boldsymbol{b}$ acting at the center of mass.

By applying transport theorem II, we get the following result:

Theorem 3.6.1. *Given that balance of linear momentum holds, balance of angular momentum is equivalent to the symmetry of the stress tensor (assuming that there are no extraneous body couples).*

Proof. Applying Eqn. (3.10b) to the left-hand side of Eqn. (3.26), we get

$$\int_{V(t)} \rho \left\{ \frac{D\boldsymbol{x}}{Dt} \times \boldsymbol{v} + \boldsymbol{x} \times \frac{D\boldsymbol{v}}{Dt} \right\} dV = \int_{V(t)} \rho \boldsymbol{x} \times \frac{D\boldsymbol{v}}{Dt}\, dV,$$

since by Eqn. (2.47), $\frac{D\boldsymbol{x}}{Dt} \times \boldsymbol{v} = \boldsymbol{v} \times \boldsymbol{v} = \boldsymbol{0}$. Thus, Eqn. (3.26) can be written as

$$\int_{V(t)} \rho \boldsymbol{x} \times \frac{D\boldsymbol{v}}{Dt}\, dV = \int_{S(t)} \boldsymbol{x} \times \boldsymbol{t}\, dS + \int_{V(t)} \boldsymbol{x} \times \rho \boldsymbol{b}\, dV. \tag{3.32}$$

Substituting for $\rho D\boldsymbol{v}/Dt$ using Eqn. (3.19), and using Eqn. (3.18), Eqn. (3.32) reduces to

$$\int_{V(t)} \boldsymbol{x} \times \boldsymbol{\nabla} \cdot \boldsymbol{\tau}\, dV = \int_{S(t)} \boldsymbol{x} \times \boldsymbol{\tau} \boldsymbol{n}\, dS. \tag{3.33}$$

Let $\hat{\boldsymbol{t}}$ be the axial vector of $\boldsymbol{\tau} - \boldsymbol{\tau}^T$, so that

$$\hat{t}_i = -\frac{1}{2} \epsilon_{ijk} (\tau_{jk} - \tau_{kj}).$$

Using the fact that k and j are dummy indices, and that $\epsilon_{ijk} = -\epsilon_{ikj}$, we get

$$\hat{t}_i = -\epsilon_{ijk} \tau_{jk}.$$

Hence,

$$\begin{aligned}
\boldsymbol{x} \times (\boldsymbol{\nabla} \cdot \boldsymbol{\tau}) + \hat{\boldsymbol{t}} &= \left[\epsilon_{ijk} x_j \frac{\partial \tau_{km}}{\partial x_m} + \hat{t}_i \right] \boldsymbol{e}_i \\
&= \left[\epsilon_{ijk} \frac{\partial (\tau_{km} x_j)}{\partial x_m} + \epsilon_{ikm} \tau_{km} + \hat{t}_i \right] \boldsymbol{e}_i \\
&= \left[\epsilon_{ijk} \frac{\partial (\tau_{km} x_j)}{\partial x_m} \right] \boldsymbol{e}_i,
\end{aligned}$$

which in turn implies that

$$\int_{V(t)} [\boldsymbol{x} \times (\boldsymbol{\nabla} \cdot \boldsymbol{\tau}) + \hat{\boldsymbol{t}}]\, dV = \int_{S(t)} \epsilon_{ijk} \tau_{km} x_j n_m \boldsymbol{e}_i\, dS = \int_{S(t)} \boldsymbol{x} \times (\boldsymbol{\tau} \boldsymbol{n})\, dS. \tag{3.34}$$

Comparing Eqns. (3.33) and (3.34), we get

$$\int_{V(t)} \hat{\boldsymbol{t}}\, dV = \boldsymbol{0}.$$

Since $V(t)$ is arbitrary, by the localization theorem we have $\hat{\boldsymbol{t}} = \boldsymbol{0}$, so that

$$(\boldsymbol{\tau} - \boldsymbol{\tau}^T)\boldsymbol{u} = \hat{\boldsymbol{t}} \times \boldsymbol{u} = \boldsymbol{0} \quad \forall \boldsymbol{u}.$$

Thus, in the absence of body couples, the stress tensor is symmetric, i.e.,

$$\boldsymbol{\tau} = \boldsymbol{\tau}^T. \tag{3.35}$$

Reversing the steps in the above proof, one can show that balance of linear momentum and the symmetry of the stress tensor imply the balance of angular momentum. □

3.7 Properties of the Cauchy Stress Tensor

We now state some properties of the stress tensor.

1. The stress tensor is symmetric.

   Proof: We have proved this in the previous section (see Eqn. (3.35)).

2. At any point in the continuum, there exist planes with unit normal $\boldsymbol{n}$ such that the traction vector on those planes is parallel to the normal, i.e., $\boldsymbol{t} = \lambda\boldsymbol{n}$. The values λ are called as the *principal stresses*, and the corresponding $\boldsymbol{n}$ are called the principal directions. The principal stresses are real. The principal directions $\boldsymbol{n}_1$, $\boldsymbol{n}_2$ and $\boldsymbol{n}_3$ corresponding to the principal stresses λ_1, λ_2 and λ_3 are mutually orthogonal in case the principal stresses are distinct, or can be chosen to be mutually orthogonal in case they are not. If we take a frame $\boldsymbol{e}^*$ such that the unit vectors $\boldsymbol{e}_1^*$, $\boldsymbol{e}_2^*$ and $\boldsymbol{e}_3^*$ coincide with the principal directions $\boldsymbol{n}_1$, $\boldsymbol{n}_2$ and $\boldsymbol{n}_3$, the components of the stress tensor in the frame $\boldsymbol{e}^*$ are

$$\tau^* = \begin{bmatrix} \lambda_1 & 0 & 0 \\ 0 & \lambda_2 & 0 \\ 0 & 0 & \lambda_3 \end{bmatrix}. \tag{3.36}$$

   Proof: We have to determine whether at each point there are surfaces with unit normal $\boldsymbol{n}$ such that

$$\boldsymbol{t} = \boldsymbol{\tau}\boldsymbol{n} = \lambda\boldsymbol{n}. \tag{3.37}$$

   It is important to note that the traction vector is dependent not only on the position and time, but also on the normal to the surface on which the traction vector is being computed. Thus, in general, the traction vector at a point changes as the surface on which it is being computed changes. In this sense, the traction vector is different from say, the velocity vector, which is dependent only on position and time.
   Since $\boldsymbol{\tau}$ is symmetric, and since Eqn. (3.37) is nothing but an eigenvalue problem, all the claims follow from the results of Section 1.6.1.

3. The principal axes correspond to directions for which the normal stress forms an extremum when compared to normal stresses in neighboring directions.

Proof: The normal stress T_n is given by

$$T_n = \boldsymbol{t} \cdot \boldsymbol{n} = t_i n_i = \tau_{ij} n_j n_i.$$

Our goal is to find the unit normal vector $\boldsymbol{n}$ so as to extremize T_n. Thus, using the Lagrange multiplier method, we are interested in extremizing

$$\tau_{ij} n_i n_j + \gamma(1 - n_i n_i),$$

where γ is the Lagrange multiplier associated with the constraint $|\boldsymbol{n}| = 1$. The necessary condition for finding the extremizer $\boldsymbol{n}$ is

$$\begin{aligned}
0 &= \frac{\partial}{\partial n_s}\left[\tau_{ij} n_i n_j + \gamma(1 - n_r n_r)\right] \\
&= \tau_{ij} n_i \delta_{js} + \tau_{ij} \delta_{is} n_j - 2\gamma n_r \delta_{rs} \\
&= \tau_{is} n_i + \tau_{sj} n_j - 2\gamma n_s \\
&= 2(\tau_{is} n_i - \gamma n_s) \\
&= 2(\tau_{is} - \gamma \delta_{is}) n_i,
\end{aligned}$$

which is of the same form as Eqn. (1.112). This proves that the principal directions are also the directions that extremize the normal stress.
If $\lambda_1 > \lambda_2 > \lambda_3$, then $\boldsymbol{n}_1$ and $\boldsymbol{n}_3$ are the principal directions along which the normal stress is a maximum and minimum, respectively. To see this, fix the coordinate system $\{\boldsymbol{e}_i^*\}$ along $\{\boldsymbol{n}_i\}$. Let the normal to a plane in this coordinate frame be denoted by $\boldsymbol{n}^* = (n_1^*, n_2^*, n_3^*)$. Since $\boldsymbol{\tau} = \sum_{k=1}^3 \lambda_k \boldsymbol{e}_k^* \otimes \boldsymbol{e}_k^*$, we have

$$\begin{aligned}
T_{n^*} &= \boldsymbol{n}^* \cdot (\boldsymbol{\tau} \boldsymbol{n}^*) \\
&= \sum_{k=1}^{3} \lambda_k n_k^* n_k^* \\
&\le \sum_{k=1}^{3} \lambda_1 n_k^* n_k^* \\
&= \lambda_1,
\end{aligned}$$

where we have used the fact that $\sum_{k=1}^3 n_k^* n_k^* = 1$ in the last step. We have equality in the above inequality if and only if $\boldsymbol{n}^* = \boldsymbol{e}_1^*$ (or -$\boldsymbol{e}_1^*$). In a similar manner $T_n \ge \lambda_3$ on any plane, with equality if and only if $\boldsymbol{n}^* = \boldsymbol{e}_3^*$. The plane with normal $\boldsymbol{n}_2$ corresponds to a saddle-point since if we perturb the normal in the direction of $\boldsymbol{n}_1$, the normal stress increases, while if we perturb it in the direction of $\boldsymbol{n}_3$, it decreases. If $\lambda_1 > \lambda_2 = \lambda_3$, then $\boldsymbol{n}_1$ corresponds to the direction of maximum normal stress, while any normal in the plane of $\boldsymbol{n}_2$ and $\boldsymbol{n}_3$ corresponds to the direction of minimum normal stress. Similar remarks can be made for the case $\lambda_1 = \lambda_2 > \lambda_3$. If $\lambda_1 = \lambda_2 = \lambda_3 \equiv \lambda$, then the normal stress has a constant value λ on any plane.

4. The normal to the plane on which the maximum shear stress occurs is inclined at 45° to the principal directions $\boldsymbol{n}_1$ and $\boldsymbol{n}_3$, corresponding to the maximum and minimum principal stresses λ_1 and λ_3, respectively (assuming that the principal stresses have

been ordered so that $\lambda_1 \geq \lambda_2 \geq \lambda_3$), and its magnitude is $(\lambda_1 - \lambda_3)/2$.
Proof: Fix the coordinate axes along the principal directions $\boldsymbol{n}_1$, $\boldsymbol{n}_2$ and $\boldsymbol{n}_3$. The stress components in this coordinate frame are given by Eqn. (3.36). The traction vector on a plane whose normal is $\boldsymbol{n}^* = (n_1^*, n_2^*, n_3^*)$ has the components $(\lambda_1 n_1^*, \lambda_2 n_2^*, \lambda_3^* n_2)$. Hence, the magnitude of the shear stress on this plane is

$$\begin{aligned} T_s^2 &= t_i t_i - T_n^2 \\ &= t_i t_i - (t_i n_i^*)^2 \\ &= (\lambda_1 n_1^*)^2 + (\lambda_2 n_2^*)^2 + (\lambda_3 n_3^*)^2 - \left[\lambda_1 (n_1^*)^2 + \lambda_2 (n_2^*)^2 + \lambda_3 (n_3^*)^2\right]^2 \\ &= (n_1^* n_2^*)^2 (\lambda_1 - \lambda_2)^2 + (n_2^* n_3^*)^2 (\lambda_2 - \lambda_3)^2 + (n_1^* n_3^*)^2 (\lambda_1 - \lambda_3)^2, \end{aligned}$$

where the last step has been obtained using the fact that $\sum_{k=1}^3 n_k^* n_k^* = 1$. Using a Lagrange multiplier technique as in the previous proof, it can be proved that T_s is an extremum when $\boldsymbol{n}^*$ is $(1,0,0)$, $(0,1,0)$, $(0,0,1)$, $(1/\sqrt{2}, 1/\sqrt{2}, 0)$, $(0, 1/\sqrt{2}, 1/\sqrt{2})$ and $(1/\sqrt{2}, 0, 1/\sqrt{2})$. The first three cases yield the minimum value of $|T_s|$ (namely, zero), while the last one yields the maximum value. Thus, substituting this value of $\boldsymbol{n}^*$, we get

$$(T_s)_{\max} = \frac{1}{2}(\lambda_1 - \lambda_3).$$

5. If $\operatorname{tr} \boldsymbol{\tau} = 0$, then by Theorem 1.6.3, there exists an orthonormal basis given by Eqn. (1.128) and (1.130) with respect to which the normal stress components are zero. This is illustrated in the following example.

Example: The stress components at the origin in a continuum corresponding to a given basis are given by

$$\boldsymbol{\tau} = \begin{bmatrix} 4 & 0 & 0 \\ 0 & 1 & -3\sqrt{3} \\ 0 & -3\sqrt{3} & -5 \end{bmatrix}$$

Let the unit vector $\boldsymbol{n}$ have components

$$\boldsymbol{n} = (\sin\theta\cos\phi, \sin\theta\sin\phi, \cos\theta).$$

Show that the normal component of stress on a surface whose outer normal is $\boldsymbol{n}$ is given by

$$T_n = 4 - 3(\sin\theta\sin\phi + \sqrt{3}\cos\theta)^2.$$

Hence, find the directions of the normals to surfaces such that the normal stress is a maximum or a minimum. Compare these results with those obtained by solving the eigenvalue problem.
Solution: The traction vector on the plane with normal $\boldsymbol{n}$ is

$$\boldsymbol{t} = \boldsymbol{\tau}\boldsymbol{n} = \begin{bmatrix} 4\sin\theta\cos\phi \\ \sin\theta\sin\phi - 3\sqrt{3}\cos\theta \\ -3\sqrt{3}\sin\theta\sin\phi - 5\cos\theta \end{bmatrix}.$$

Hence the normal stress is

$$\begin{aligned}T_n = \boldsymbol{t}\cdot\boldsymbol{n} &= 4\sin^2\theta\cos^2\phi + \sin^2\theta\sin^2\phi - 6\sqrt{3}\sin\theta\sin\phi\cos\theta - 5\cos^2\theta \\ &= 4 - 3(\sin\theta\sin\phi + \sqrt{3}\cos\theta)^2.\end{aligned}$$

To find the conditions for extremizing T_n, we set $\partial T_n/\partial\theta$ and $\partial T_n/\partial\phi$ to zero, yielding

$$\begin{aligned}&(\sin\theta\sin\phi + \sqrt{3}\cos\theta)(\cos\theta\sin\phi - \sqrt{3}\sin\theta) = 0,\\ &(\sin\theta\sin\phi + \sqrt{3}\cos\theta)(\sin\theta\cos\phi) = 0.\end{aligned}$$

From the above equations, we get the conditions for stationarity of T_n as

$$\theta = \pm\frac{\pi}{6}, \quad \phi = \pm\frac{\pi}{2}$$

or

$$\sin\theta\sin\phi + \sqrt{3}\cos\theta = 0.$$

θ and ϕ satisfying the first condition are the minimizers of T_n, yielding $(T_n)_{\text{min}} = -8$, while the values of (θ,ϕ) satisfying the second condition are the maximizers of T_n, yielding $(T_n)_{\text{max}} = 4$. The eigenvector $\boldsymbol{n} = (\sin\theta\cos\phi, \sin\theta\sin\phi, \cos\theta)$ corresponding to $(T_n)_{\text{min}}$ is, thus, $(0, 1/2, \sqrt{3}/2)$ (discarding the other vector corresponding to the negative values of θ and ϕ, since it is linearly dependent on the above vector). The eigenvector corresponding to $(T_n)_{\text{max}}$ is of the form $\frac{1}{\sqrt{n_1^2+4n_3^2}}(n_1, -\sqrt{3}n_3, n_3)$. Note the orthogonality of the eigenvectors.

Alternatively, one can solve for the principal stresses and the principal directions by considering the eigenvalue problem

$$(\boldsymbol{\tau} - \lambda\boldsymbol{I})\boldsymbol{n} = \boldsymbol{0}.$$

To find the principal stresses we set $\det(\boldsymbol{\tau} - \lambda\boldsymbol{I})$ to zero. Solving the resultant cubic equation in λ yields the principal stresses as $(-8, 4, 4)$. Corresponding to the eigenvalue $\lambda = -8$, we get the eigenvector as $(0, 1/2, \sqrt{3}/2)$, which is the same as that obtained previously. Similarly, corresponding to the eigenvalue $\lambda = 4$, we get $\boldsymbol{n}$ as $\frac{1}{\sqrt{n_1^2+4n_3^2}}(n_1, -\sqrt{3}n_3, n_3)$, which is again the same as that obtained previously. Note that the principal stresses are extremizers of the normal stress T_n, as we had proved. Also note that due to the presence of two repeated eigenvalues, only one eigenvector is determined uniquely; the other two eigenvectors can be chosen arbitrarily in the plane perpendicular to the first one.

Choosing the principal directions corresponding to the eigenvalues $\lambda_1 = 4$, $\lambda_2 = 4$ and $\lambda_3 = -8$ as $\boldsymbol{e}_1^* = (1,0,0)$, $\boldsymbol{e}_2^* = (0, -\sqrt{3}/2, 1/2)$ and $\boldsymbol{e}_3^* = (0, 1/2, \sqrt{3}/2)$, and using Eqns. (1.128) and (1.130), we get the orthogonal basis in which the normal stress components vanish as

$$\boldsymbol{n}_1 = \frac{1}{6}\begin{pmatrix}2\sqrt{3}\\ \sqrt{3}-3\\ \sqrt{3}+3\end{pmatrix}, \quad \boldsymbol{n}_2 = \frac{1}{12}\begin{pmatrix}6+2\sqrt{3}\\ -3+\sqrt{3}\\ -9+\sqrt{3}\end{pmatrix}, \quad \boldsymbol{n}_3 = \frac{1}{12}\begin{pmatrix}2\sqrt{3}-6\\ -3-5\sqrt{3}\\ \sqrt{3}-3\end{pmatrix}.$$

3.8 The Equations of Motion in the Reference Configuration

As we have seen, the equation of motion for the deformed configuration is given by Eqn. (3.19). In elasticity, our final objective is to find the deformation field and the Cauchy stress tensor that arise in a body subjected to a given system of forces. In this respect, the equations of motion in the deformed configuration are not of much help in getting the desired solution, since they are expressed in terms of the spatial variable $\boldsymbol{x}$, which itself is an unknown. Hence, it is desirable to rewrite the equations of motion, and the boundary conditions with respect to the reference configuration, which is considered as given. Thus, we need to transform the term $\boldsymbol{\nabla}_x \cdot \boldsymbol{\tau}$, the body force $\boldsymbol{b}$ and the traction vector $\boldsymbol{t}$.

The term $\boldsymbol{\nabla}_x \cdot \boldsymbol{\tau}$ on the deformed configuration is transformed to a quantity $\boldsymbol{\nabla}_X \cdot \boldsymbol{T}$ on the reference configuration by using the Piola transform. We define the *first Piola–Kirchhoff stress tensor* as the Piola transform of the Cauchy stress tensor (see Eqn. (2.28)), i.e.,

$$\boldsymbol{T}(\boldsymbol{X},t) := J\boldsymbol{\tau}(\boldsymbol{x},t)\boldsymbol{F}^{-T} = \boldsymbol{\tau}(\boldsymbol{x},t)\,\mathbf{cof}\,\boldsymbol{F}. \tag{3.38}$$

By Eqn. (2.29), we have

$$\boldsymbol{\nabla}_X \cdot \boldsymbol{T}(\boldsymbol{X},t) = J\boldsymbol{\nabla}_x \cdot \boldsymbol{\tau}(\boldsymbol{x},t) \quad \forall \boldsymbol{x} = \boldsymbol{\chi}(\boldsymbol{X},t), \tag{3.39}$$

The tractions and body forces in the reference configuration are defined by the relations (see Fig. 3.6)

$$\boldsymbol{t}_0(\boldsymbol{X},t,\boldsymbol{n}_0)dS_0 := \boldsymbol{t}(\boldsymbol{x},t,\boldsymbol{n})dS, \tag{3.40}$$

$$\rho_0(\boldsymbol{X})\boldsymbol{b}_0(\boldsymbol{X},t)dV_0 := \rho(\boldsymbol{x},t)\boldsymbol{b}(\boldsymbol{x},t)dV. \tag{3.41}$$

The vector $\boldsymbol{t}_0(\boldsymbol{X},t,\boldsymbol{n}_0)$ is known as the *first Piola–Kirchhoff traction vector* at the point $\boldsymbol{X}$ of the reference configuration acting on a surface element with normal $\boldsymbol{n}_0$. Although the various terms defined on the reference configuration, such as $\boldsymbol{T}$, $\boldsymbol{t}_0$ and $\boldsymbol{b}_0$, are often given physical interpretations, we prefer to think of them as mathematical entities introduced in order to find the 'true' stress, namely the Cauchy stress tensor.

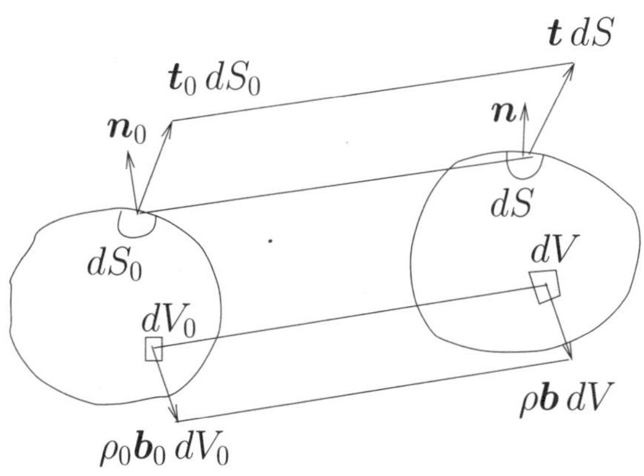

Fig. 3.6 *Applied body force and tractions in the reference and deformed configurations.*

Substituting Eqn. (2.31) into Eqn. (3.40), we get

$$t_0(X,t,n_0) = |(\mathbf{cof}\, F)n_0|\, t(\chi(X,t),t,n). \tag{3.42}$$

As an example, in an inflated balloon, with the deflated state as the reference configuration, we have $t = -p(V)n$, where $p(V)$ is the pressure that is a function of the volume $V(\chi)$ inside the balloon, and hence using Eqn. (2.32), $t_0 = -p(V)(\mathbf{cof}\, F)n_0$. This is an example of a nonlocal boundary condition since the traction at a point X depends on the deformation at other points by virtue of the dependence of p on $V(\chi)$.

The relation between t_0 and T is obtained as follows:

$$\begin{aligned} t_0 dS_0 &= t dS && \text{(by Eqn. (3.40))} \\ &= \tau n\, dS && \text{(by Eqn. (3.18))} \\ &= \tau(\mathbf{cof}\, F)n_0 dS_0 && \text{(by Eqn. (2.34))} \\ &= T n_0 dS_0, && \text{(by Eqn. (3.38))} \end{aligned}$$

which implies that

$$t_0 = T n_0. \tag{3.43}$$

Note that this equation is exactly of the same form as Cauchy's relation in the deformed configuration as given by Eqn. (3.18). From Eqns. (3.42) and (3.43), we get

$$T n_0 = t_0 = |(\mathbf{cof}\, F)n_0|\, t. \tag{3.44}$$

To find the relation between b_0 and b, we use Eqns. (2.37), (3.5) and (3.41) to get

$$b_0(X,t) = b(\chi(X,t),t). \tag{3.45}$$

To find the equation of motion in the reference configuration, we multiply Eqn. (3.19) by J, and use Eqns. (3.5), (3.39) and (3.45) to get

$$\rho_0 \left(\frac{\partial^2 \chi}{\partial t^2}\right)_X = \nabla_X \cdot T + \rho_0 b_0. \tag{3.46}$$

While the Cauchy stress tensor is symmetric, the first Piola–Kirchhoff stress tensor is not; instead we have

$$T^T = F^{-1} T F^T. \tag{3.47}$$

Summarizing, the first Piola–Kirchhoff stress tensor satisfies the relations (see Eqns. (3.44), (3.46) and (3.47))

$$\begin{aligned} &\rho_0 \left(\frac{\partial^2 \chi}{\partial t^2}\right)_X = \nabla_X \cdot T + \rho_0 b_0, \\ &T n_0 = t_0 = |(\mathbf{cof}\, F)n_0|\, t, \\ &F T^T = T F^T. \end{aligned} \tag{3.48}$$

As we saw from Eqn. (3.47), the first Piola–Kirchhoff stress tensor is not symmetric. However, it is desirable to define a symmetric stress tensor on the reference configuration, essentially because the constitutive equation in the reference configuration takes a simpler form. Hence, we define the *second Piola–Kirchhoff stress tensor* $\boldsymbol{S}(\boldsymbol{X}, t)$ as

$$\boldsymbol{S} := \boldsymbol{F}^{-1}\boldsymbol{T} = J\boldsymbol{F}^{-1}\boldsymbol{\tau}\boldsymbol{F}^{-T}. \tag{3.49}$$

In terms of the second Piola–Kirchhoff stress tensor, Eqns. (3.48) take the form

$$\begin{aligned} \rho_0\left(\frac{\partial^2\boldsymbol{\chi}}{\partial t^2}\right)_{\boldsymbol{X}} &= \boldsymbol{\nabla}_X\cdot(\boldsymbol{F}\boldsymbol{S}) + \rho_0\boldsymbol{b}_0, \\ \boldsymbol{F}\boldsymbol{S}\boldsymbol{n}_0 = \boldsymbol{t}_0 &= |(\mathbf{cof}\,\boldsymbol{F})\boldsymbol{n}_0|\,\boldsymbol{t}, \\ \boldsymbol{S}^T &= \boldsymbol{S}. \end{aligned} \tag{3.50}$$

To demonstrate the transformation from the current to the reference configuration, consider the example of a bar rotating with a constant angular velocity ω as shown in Fig. 3.7. The reference configuration is taken to be the state when the bar is stationary, and the bar is assumed to be stress- and strain-free in this configuration. The distance of any point from the axis of rotation with respect to an observer rotating with the bar is given by x (see figure). We approximate the bar as a one-dimensional continuum, i.e., only $\tau_{xx} \equiv \tau$, $E_{xx} \equiv E$, $F_{xx} \equiv F$ etc. are assumed to be nonzero. We also assume the body forces with respect to a stationary observer to be zero, and the constitutive relation to be given by $S = \lambda E$, where λ is a given constant.

We will see in the following chapter that the body force is given by Eqn. (4.18). In the problem under consideration, $\boldsymbol{\Omega} = \omega\boldsymbol{e}_z$, $\boldsymbol{b}^* = \boldsymbol{0}$, $\boldsymbol{c} = \boldsymbol{0}$ and $\boldsymbol{v} = \boldsymbol{0}$ (since the points on the bar are stationary with respect to the rotating frame of reference). Thus, $\boldsymbol{b} = \omega^2 x\boldsymbol{e}_x$. Assuming the deformation to be given by $x = h(X)$, the only nonzero element of the deformation gradient is $F_{xx} = h'(X)$, and thus, $J = \det\boldsymbol{F} = h'(X)$. The governing linear momentum equation for the Cauchy stress is

$$\frac{\partial\tau}{\partial x} + \rho\omega^2 x = 0.$$

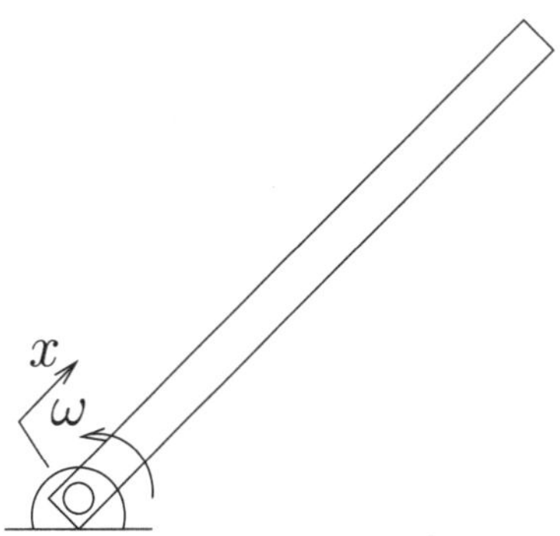

Fig. 3.7 *Rotating bar.*

Using the constitutive relation, we have $S = \lambda[F^T F - 1]/2 = \lambda[h'(X)^2 - 1]/2$, so that $\tau = \lambda[h'(X)^3 - h'(X)]/2$. Substituting this expression into the above equation, we get the governing equation for $h(X)$ as

$$\frac{1}{2}\lambda\left[3(h')^2 h'' - h''\right] + \rho_0 h\omega^2 = 0.$$

The above equation is to be solved subject to the boundary conditions $h(0) = 0$ (since the bar is fixed at $X = 0$), and $h'(L) = 1$ (since $\tau|_{X=L} = 0$). We again emphasize that the above example is meant only to demonstrate the transformation procedure by means of a simple one-dimensional approximation; whether the solution is accurate will depend on the accuracy of the constitutive relation, and the validity of the other assumptions that have been made.

The integral forms of the balance of linear and angular momenta over the reference volume V_0 are obtained by substituting the relations $\rho\, dV = \rho_0\, dV_0$, $\boldsymbol{t}\, dS = \boldsymbol{t}_0\, dS_0$, $\boldsymbol{b}(\boldsymbol{x},t) = \boldsymbol{b}_0(\boldsymbol{X},t)$ and $\dot{\boldsymbol{\chi}} = \dot{\boldsymbol{u}}$ in Eqns. (3.15) and (3.26):

$$\frac{d}{dt}\int_{V_0} \rho_0 \left(\frac{\partial \boldsymbol{u}}{\partial t}\right)_{\boldsymbol{X}} dV_0 = \int_{S_0} \boldsymbol{t}_0\, dS_0 + \int_{V_0} \rho_0 \boldsymbol{b}_0\, dV_0. \tag{3.51}$$

$$\frac{d}{dt}\int_{V_0} \rho_0 \boldsymbol{\chi} \times \left(\frac{\partial \boldsymbol{u}}{\partial t}\right)_{\boldsymbol{X}} dV_0 = \int_{S_0} \boldsymbol{\chi} \times \boldsymbol{t}_0\, dS_0 + \int_{V_0} \boldsymbol{\chi} \times \rho_0 \boldsymbol{b}_0\, dV_0. \tag{3.52}$$

3.9 Variational Formulations

We now discuss variational formulations of initial boundary value problems associated with both the deformed and the reference configurations[2]. We start by presenting the variational form of the following equations over the deformed configuration:

$$\begin{aligned}
&\boldsymbol{\nabla}_x \cdot \boldsymbol{\tau} + \rho \boldsymbol{b} = \rho \frac{D\boldsymbol{v}}{Dt} \text{ in } V,\\
&\boldsymbol{\tau n} = \bar{\boldsymbol{t}} \text{ on } S_t,\\
&\boldsymbol{u} = \bar{\boldsymbol{u}} \text{ on } S_u,
\end{aligned}$$

where $\bar{\boldsymbol{t}}$ and $\bar{\boldsymbol{u}}$ are the prescribed tractions and prescribed displacements on S_t and S_u, respectively, and $\boldsymbol{n}$ is the outward unit normal along S_t. The surface of V is $S = S_t \cup S_u$. For the subsequent development, we find it useful to define the space

$$V_u := \{\boldsymbol{z} :\ \boldsymbol{z} = \boldsymbol{0} \text{ on } S_u\}. \tag{3.53}$$

We obtain the single-field variational formulation using the method of weighted residuals, whereby we set the sum of the weighted residuals based on the equation of motion, and the natural boundary condition to zero, i.e.,

$$\int_{V_t} \boldsymbol{u}_\delta \cdot (\boldsymbol{\nabla}_x \cdot \boldsymbol{\tau} + \rho \boldsymbol{b})\, dV + \int_{S_t} \boldsymbol{u}_\delta \cdot (\bar{\boldsymbol{t}} - \boldsymbol{t})\, dS = \int_{V_t} \rho \boldsymbol{u}_\delta \cdot \frac{D\boldsymbol{v}}{Dt}\, dV \quad \forall \boldsymbol{u}_\delta \in V_u. \tag{3.54}$$

[2] A good introduction to the calculus of variations is presented in [287].

Using Eqn. (1.355), i.e., $\boldsymbol{\nabla}_x \cdot (\boldsymbol{\tau}^T \boldsymbol{u}_\delta) = \boldsymbol{u}_\delta \cdot (\boldsymbol{\nabla}_x \cdot \boldsymbol{\tau}) + \boldsymbol{\tau} : \boldsymbol{\nabla}_x \boldsymbol{u}_\delta$, we get

$$\int_{V_t} \left[\boldsymbol{\nabla}_x \cdot (\boldsymbol{\tau}^T \boldsymbol{u}_\delta) - \boldsymbol{\tau} : \boldsymbol{\nabla}_x \boldsymbol{u}_\delta + \rho \boldsymbol{u}_\delta \cdot \boldsymbol{b} \right] dV + \int_{S_t} \boldsymbol{u}_\delta \cdot (\bar{\boldsymbol{t}} - \boldsymbol{t})\, dS = \int_{V_t} \rho \boldsymbol{u}_\delta \cdot \frac{D\boldsymbol{v}}{Dt}\, dV \quad \forall \boldsymbol{u}_\delta \in V_u.$$

Now applying the divergence theorem to the first term of the above equation, we get

$$\int_{S_t} (\boldsymbol{\tau}^T \boldsymbol{u}_\delta) \cdot \boldsymbol{n}\, dS - \int_{V_t} [\boldsymbol{\tau} : \boldsymbol{\nabla}_x \boldsymbol{u}_\delta - \rho \boldsymbol{u}_\delta \cdot \boldsymbol{b}]\, dV + \int_{S_t} \boldsymbol{u}_\delta \cdot (\bar{\boldsymbol{t}} - \boldsymbol{t})\, dS = \int_{V_t} \rho \boldsymbol{u}_\delta \cdot \frac{D\boldsymbol{v}}{Dt}\, dV \quad \forall \boldsymbol{u}_\delta \in V_u. \tag{3.55}$$

Since

$$\int_{S_t} (\boldsymbol{\tau}^T \boldsymbol{u}_\delta) \cdot \boldsymbol{n}\, dS = \int_{S_t} \boldsymbol{u}_\delta \cdot \boldsymbol{t}\, dS = \int_{S_t} \boldsymbol{u}_\delta \cdot \boldsymbol{t}\, dS \quad \forall \boldsymbol{u}_\delta \in V_u,$$

and since $\boldsymbol{\tau}$ is symmetric, Eqn. (3.55) simplifies to

$$\int_{V_t} \rho \boldsymbol{u}_\delta \cdot \frac{D\boldsymbol{v}}{Dt}\, dV + \frac{1}{2} \int_{V_t} \boldsymbol{\tau} : \left[\boldsymbol{\nabla}_x \boldsymbol{u}_\delta + (\boldsymbol{\nabla}_x \boldsymbol{u}_\delta)^T \right] dV = \int_{V_t} \rho \boldsymbol{u}_\delta \cdot \boldsymbol{b}\, dV + \int_{S_t} \boldsymbol{u}_\delta \cdot \bar{\boldsymbol{t}}\, dS \; \forall \boldsymbol{u}_\delta \in V_u. \tag{3.56}$$

The single-field variational formulation for the reference configuration is found by following a procedure similar to that followed for finding the single-field variational formulation over the deformed configuration. We now define V_u as

$$V_u := \{ \boldsymbol{z} : \; \boldsymbol{z} = \boldsymbol{0} \text{ on } (S_0)_u \}.$$

The governing equations for the first Piola–Kirchhoff stress tensor over the reference configuration are

$$\begin{aligned}
&\boldsymbol{\nabla}_X \cdot \boldsymbol{T} + \rho_0 \boldsymbol{b}_0 = \rho_0 \frac{\partial^2 \boldsymbol{\chi}}{\partial t^2} \text{ in } V_0, \\
&\boldsymbol{T}\boldsymbol{n}_0 = \bar{\boldsymbol{t}}_0 \text{ on } (S_0)_t, \\
&\boldsymbol{F}\boldsymbol{T}^T = \boldsymbol{T}\boldsymbol{F}^T \text{ in } V_0,
\end{aligned}$$

where $\bar{\boldsymbol{t}}_0$ and $\boldsymbol{b}_0$ are defined via Eqns. (3.42) and (3.45). The variational formulation is

$$\int_{V_0} \rho_0 \boldsymbol{u}_\delta \cdot \frac{\partial^2 \boldsymbol{\chi}}{\partial t^2}\, dV + \int_{V_0} \boldsymbol{T} : \boldsymbol{\nabla}_X \boldsymbol{u}_\delta\, dV = \int_{V_0} \rho_0 \boldsymbol{u}_\delta \cdot \boldsymbol{b}_0\, dV + \int_{(S_0)_t} \boldsymbol{u}_\delta \cdot \bar{\boldsymbol{t}}_0\, dS \quad \forall \boldsymbol{u}_\delta \in V_u. \tag{3.57}$$

The second Piola–Kirchoff stress tensor $\boldsymbol{S}$ satisfies the following equations:

$$\begin{aligned}
&\boldsymbol{\nabla}_X \cdot (\boldsymbol{F}\boldsymbol{S}) + \rho_0 \boldsymbol{b}_0 = \rho_0 \frac{\partial^2 \boldsymbol{\chi}}{\partial t^2} \text{ in } V_0, \\
&\boldsymbol{F}\boldsymbol{S}\boldsymbol{n}_0 = \bar{\boldsymbol{t}}_0 \text{ on } (S_0)_t, \\
&\boldsymbol{S}^T = \boldsymbol{S} \text{ in } V_0.
\end{aligned}$$

The single-field variational formulation is obtained simply by substituting $\boldsymbol{T} = \boldsymbol{FS}$ into Eqn. (3.57):

$$\int_{V_0} \rho_0 \boldsymbol{u}_\delta \cdot \frac{\partial^2 \boldsymbol{\chi}}{\partial t^2}\, dV + \int_{V_0} (\boldsymbol{FS}) : \boldsymbol{\nabla}_X \boldsymbol{u}_\delta \, dV = \int_{V_0} \rho_0 \boldsymbol{u}_\delta \cdot \boldsymbol{b}_0 \, dV + \int_{(S_0)_t} \boldsymbol{u}_\delta \cdot \bar{\boldsymbol{t}}_0 \, dS \quad \forall \boldsymbol{u}_\delta \in V_u. \tag{3.58}$$

We can write the above equation in terms of the variation of the Lagrangian strain $\boldsymbol{E}$, rather than in terms of $\boldsymbol{u}_\delta$, which is the variation of $\boldsymbol{u}$. Note that (we write $\boldsymbol{\nabla}_X$ simply as $\boldsymbol{\nabla}$ for brevity)

$$\boldsymbol{E} = \frac{1}{2}\left[\boldsymbol{\nabla u} + (\boldsymbol{\nabla u})^T + (\boldsymbol{\nabla u})^T(\boldsymbol{\nabla u})\right],$$

so that the variation $\boldsymbol{E}_\delta$ is given by

$$\begin{aligned}
\boldsymbol{E}_\delta &= \frac{1}{2}\left[\boldsymbol{\nabla u}_\delta + (\boldsymbol{\nabla u}_\delta)^T + (\boldsymbol{\nabla u})^T \boldsymbol{\nabla u}_\delta + (\boldsymbol{\nabla u}_\delta)^T \boldsymbol{\nabla u}\right] \\
&= \frac{1}{2}\left[\boldsymbol{F}^T \boldsymbol{\nabla u}_\delta + (\boldsymbol{\nabla u}_\delta)^T \boldsymbol{F}\right],
\end{aligned}$$

since $\boldsymbol{F} = \boldsymbol{I} + \boldsymbol{\nabla u}$. Thus,

$$\begin{aligned}
\boldsymbol{FS} : \boldsymbol{\nabla u}_\delta &= \boldsymbol{S} : (\boldsymbol{F}^T \boldsymbol{\nabla u}_\delta) \\
&= \frac{1}{2}\boldsymbol{S} : \left[\boldsymbol{F}^T \boldsymbol{\nabla u}_\delta + (\boldsymbol{\nabla u}_\delta)^T \boldsymbol{F}\right] \quad \text{(since } \boldsymbol{S} \text{ is symmetric)} \\
&= \boldsymbol{S} : (\boldsymbol{E}_\delta).
\end{aligned}$$

Therefore, Eqn. (3.58) can be written as

$$\int_{V_0} \rho_0 \boldsymbol{u}_\delta \cdot \frac{\partial^2 \boldsymbol{\chi}}{\partial t^2}\, dV + \int_{V_0} \boldsymbol{S} : \boldsymbol{E}_\delta \, dV = \int_{V_0} \rho_0 \boldsymbol{u}_\delta \cdot \boldsymbol{b}_0 \, dV + \int_{(S_0)_t} \boldsymbol{u}_\delta \cdot \bar{\boldsymbol{t}}_0 \, dS \quad \forall \boldsymbol{u}_\delta \in V_u. \tag{3.59}$$

A two-field variational formulation involving displacements and stresses is obtained by enforcing the strain–displacement relation also in a weak sense. Thus, if one defines

$$\bar{\boldsymbol{E}}(\boldsymbol{u}) := \frac{1}{2}\left[\boldsymbol{\nabla u} + (\boldsymbol{\nabla u})^T + (\boldsymbol{\nabla u})^T(\boldsymbol{\nabla u})\right],$$

then the two variational statements (which should be compared with Eqns. (6.52)) that we obtain are

$$\int_{V_0} \rho_0 \boldsymbol{u}_\delta \cdot \frac{\partial^2 \boldsymbol{\chi}}{\partial t^2}\, dV + \int_{V_0} \boldsymbol{S} : \bar{\boldsymbol{E}}_\delta \, dV = \int_{V_0} \rho_0 \boldsymbol{u}_\delta \cdot \boldsymbol{b}_0 \, dV + \int_{(S_0)_t} \boldsymbol{u}_\delta \cdot \bar{\boldsymbol{t}}_0 \, dS \;\; \forall \boldsymbol{u}_\delta \in V_u, \tag{3.60a}$$

$$\int_{V_0} \boldsymbol{S}_\delta : \left[\bar{\boldsymbol{E}}(\boldsymbol{u}) - \hat{\boldsymbol{E}}(\boldsymbol{S})\right] dV = 0 \quad \forall \boldsymbol{S}_\delta \in \text{Sym}, \tag{3.60b}$$

where $\boldsymbol{S}_\delta$ denotes the variation of $\boldsymbol{S}$, and $\hat{\boldsymbol{E}}(\boldsymbol{S})$ represents the strain–stress constitutive relation. A numerical implementation of the above two-field formulation may be found in

[156, 159, 160, 161], while a numerical strategy for fluid-structure interaction problems can be found in [162].

The required smoothness on $\boldsymbol{\tau}$ in both the integral form of the balance of linear momentum and the variational formulation is very mild, since it suffices that all integrals make sense. On the contrary, the smoothness required in the classical equation of motion is higher since $\nabla_x \cdot \boldsymbol{\tau}$ has to make sense. Thus, the question naturally arises whether the variational formulation can be proved to be equivalent to the integral form of the linear momentum balance directly, without deriving the classical equations as an intermediate step. Antman and Osborn [11] have shown that the integral forms of the linear and angular impulse-momentum laws (which are generalizations of the linear and angular momentum balance laws presented in this book) are equivalent to the variational formulations involving forces and torques, respectively.

If there are kinematic constraints of the form

$$\mu_i(\boldsymbol{u}) = \boldsymbol{0}, \quad i = 1, \ldots, N, \tag{3.61}$$

they can be incorporated into the variational formulation given by Eqn. (3.59) by using Lagrange multipliers as follows:

$$\int_{V_0} \rho_0 \boldsymbol{u}_\delta \cdot \frac{\partial^2 \boldsymbol{\chi}}{\partial t^2}\, dV + \int_{V_0} \boldsymbol{S} : \boldsymbol{E}_\delta \, dV = \int_{V_0} \rho_0 \boldsymbol{u}_\delta \cdot \boldsymbol{b}_0 \, dV + \int_{(S_0)_t} \boldsymbol{u}_\delta \cdot \bar{\boldsymbol{t}}_0 \, dS + \sum_{i=1}^{N} \lambda_i \delta^{(1)} \mu_i \quad \forall \boldsymbol{u}_\delta \in V_u. \tag{3.62}$$

The Lagrange multipliers may be physically interpreted as being reactions to the constraints; they have to be solved for while solving Eqns. (3.61) and (3.62) (along with appropriate constitutive relations, and initial and boundary conditions). As an example of a kinematic constraint, consider a bead sliding along a rotating wire. This constraint can be expressed as

$$\boldsymbol{v}_r \cdot \boldsymbol{n} = [\boldsymbol{v}_1 - \boldsymbol{v}_2] \cdot \boldsymbol{n} = 0,$$

where $\boldsymbol{v}_r$ denotes the relative velocity of the bead with respect to the wire, and $\boldsymbol{n}$ denotes the unit normal to the wire at the point of contact with the bead. Note that the above constraint is frame-indifferent since $[\boldsymbol{v}_1 - \boldsymbol{v}_2]^* = \boldsymbol{Q}(t)[\boldsymbol{v}_1 - \boldsymbol{v}_2]$ and $\boldsymbol{n}^* = \boldsymbol{Q}(t)\boldsymbol{n}$. For a numerical implementation of such constraints, see, for example, [297].

3.10 Energy Equation

We first derive the integral form of the *mechanical energy balance* equation. This equation is derived solely using the momentum equation, without any thermodynamic considerations. Later we derive a total energy balance equation based on the first law of thermodynamics. Taking the dot product of Eqn. (3.19) with $\boldsymbol{v}$ yields

$$\rho \boldsymbol{v} \cdot \frac{D\boldsymbol{v}}{Dt} = \boldsymbol{v} \cdot (\nabla \cdot \boldsymbol{\tau}) + \rho \boldsymbol{v} \cdot \boldsymbol{b},$$

which implies that

$$\rho \frac{D}{Dt}\left(\frac{\boldsymbol{v} \cdot \boldsymbol{v}}{2}\right) = \boldsymbol{v} \cdot (\nabla \cdot \boldsymbol{\tau}) + \rho \boldsymbol{v} \cdot \boldsymbol{b}$$

$$
\begin{aligned}
&= \nabla \cdot (\tau^T v) - \tau : \nabla v + \rho v \cdot b \\
&= \nabla \cdot (\tau^T v) - \tau : L + \rho v \cdot b \\
&= \nabla \cdot (\tau^T v) - \tau : D + \rho v \cdot b, \qquad (3.63)
\end{aligned}
$$

where the last step follows from Eqn. (1.330). Equation (3.63) is the differential form of the mechanical energy balance equation. Integrating the above expression over the material volume $V(t)$, and applying the divergence theorem (Eqn. (1.315)) to the first term on the right-hand side, we obtain

$$
\begin{aligned}
\int_{V(t)} \rho \frac{D}{Dt}\left(\frac{v \cdot v}{2}\right) dV &= \int_{S(t)} v \cdot (\tau n)\, dS + \int_{V(t)} [-\tau : D + \rho v \cdot b]\, dV \\
&= \int_{S(t)} t \cdot v\, dS + \int_{V(t)} [-\tau : D + \rho b \cdot v]\, dV, \qquad (3.64)
\end{aligned}
$$

where the last step is obtained using Cauchy's relation. Using transport theorem II, we can write Eqn. (3.64) as

$$
\frac{dK}{dt} = \int_{S(t)} t \cdot v\, dS + \int_{V(t)} [\rho b \cdot v - \tau : D]\, dV, \qquad (3.65)
$$

where the kinetic energy K is defined by

$$
K := \frac{1}{2} \int_{V(t)} \rho v \cdot v\, dV.
$$

The relative kinetic energy defined by

$$
K_r := \frac{1}{2} \int_{V(t)} \rho (v - \dot{\bar{x}}) \cdot (v - \dot{\bar{x}})\, dV, \qquad (3.66)
$$

and the kinetic energy K are related by *Konig's theorem* as

$$
K = K_r + \frac{1}{2} M \dot{\bar{x}} \cdot \dot{\bar{x}}. \qquad (3.67)
$$

The proof is as follows:

$$
\begin{aligned}
K_r &= \frac{1}{2} \int_{V(t)} \rho (v - \dot{\bar{x}}) \cdot (v - \dot{\bar{x}})\, dV \\
&= K - \int_{V(t)} \rho \dot{\bar{x}} \cdot v\, dV + \frac{1}{2} \dot{\bar{x}} \cdot \dot{\bar{x}} \int_{V(t)} \rho\, dV \\
&= K - \dot{\bar{x}} \cdot \int_{V(t)} \rho v\, dV + \frac{1}{2} M \dot{\bar{x}} \cdot \dot{\bar{x}} \\
&= K - \frac{1}{2} M \dot{\bar{x}} \cdot \dot{\bar{x}}.
\end{aligned}
$$

Using transport theorem I, we get the integral form of the mechanical energy balance as

$$\int_{V(t)} \frac{\partial}{\partial t}\left\{\rho\left(\frac{\boldsymbol{v}\cdot\boldsymbol{v}}{2}\right)\right\} dV + \int_{S(t)} \rho\left(\frac{\boldsymbol{v}\cdot\boldsymbol{v}}{2}\right)(\boldsymbol{v}\cdot\boldsymbol{n})\, dS = \int_{S(t)} \boldsymbol{t}\cdot\boldsymbol{v}\, dS + \int_{V(t)} [\rho\boldsymbol{b}\cdot\boldsymbol{v} - \boldsymbol{\tau} : \boldsymbol{D}]\, dV. \quad (3.68)$$

The term $\boldsymbol{\tau} : \boldsymbol{D}$ is called the stress power. To see the nontrivial role played by this term, consider the material volume to be an automobile accelerating along a straight line on a horizontal surface, and assume that there is no slip at the wheels. Ignoring wind resistance, the friction force exerted by the road on the wheels is the only external set of horizontal tractions acting on this material volume (since by Eqn. (3.16), the friction force is equal to the mass times the acceleration of the center of mass of the car). Thus, with respect to a stationary observer, since the only body force is that of gravity, and since there is no slip, the first two terms on the right-hand side of Eqn. (3.65) are zero, so that the rate of increase of the kinetic energy of the material volume is solely due to stress power as a result of combustion occurring inside the engine. Running and cycling are other examples similar to the one considered above where the stress power contributes to the rate of change of kinetic energy. If we now choose the material volume so as to exclude the gases in the combustion chamber, and assume it to comprise rigid parts, then the rate of increase of the kinetic energy of the material volume is due to the tractions generated by the gases (which are now *external* to the material volume, in contrast to the previous case where they did not contribute to the energy balance since they were internal), while the contribution due to the stress power is zero. Another point to note is that, in the absence of slip, the friction force on the wheels on which the torque is being applied by the engine has the same direction as the acceleration vector, while on the remaining wheels, it is opposite to the direction of the acceleration vector, the net friction force, of course, being in the direction of the acceleration. Thus, one has to know whether the vehicle is accelerating or decelerating before being able to determine the direction of the friction force. If the wheels slip on the surface, then $\int_{S(t)} \boldsymbol{t}\cdot\boldsymbol{v}\, dS < 0$, since the frictional traction forces are in a direction opposite to the velocity at the contact surface, resulting in a reduction of kinetic energy of the automobile.

To find the integral form of the mechanical energy balance over the volume V_0 in the reference configuration, we take the dot product of Eqn. (3.46) with $\dot{\chi} \equiv \partial\chi/\partial t$, and integrate over V_0 to get

$$\begin{aligned}
\frac{d}{dt}\int_{V_0} \rho_0 \frac{\dot{\chi}\cdot\dot{\chi}}{2}\, dV_0 &= \int_{V_0} \dot{\chi}\cdot(\boldsymbol{\nabla}_X\cdot\boldsymbol{T})\, dV_0 + \int_{V_0} \rho_0\dot{\chi}\cdot\boldsymbol{b}_0\, dV_0 \\
&= \int_{V_0} \boldsymbol{\nabla}_X\cdot(\boldsymbol{T}^T\dot{\chi})\, dV_0 - \int_{V_0} (\boldsymbol{T} : \boldsymbol{\nabla}_X\dot{\chi} - \rho_0\dot{\chi}\cdot\boldsymbol{b}_0)\, dV_0 \\
&= \int_{S_0} \dot{\chi}\cdot(\boldsymbol{T}\boldsymbol{n}^0)\, dS_0 + \int_{V_0} \rho_0\dot{\chi}\cdot\boldsymbol{b}_0\, dV_0 - \int_{V_0} \boldsymbol{T} : \dot{\boldsymbol{F}}\, dV_0 \\
&= \int_{S_0} \dot{\chi}\cdot\boldsymbol{t}_0\, dS_0 + \int_{V_0} \rho_0\dot{\chi}\cdot\boldsymbol{b}_0\, dV_0 - \int_{V_0} \boldsymbol{T} : \dot{\boldsymbol{F}}\, dV_0 \qquad (3.69) \\
&= \int_{S_0} \dot{\chi}\cdot\boldsymbol{t}_0\, dS_0 + \int_{V_0} \rho_0\dot{\chi}\cdot\boldsymbol{b}_0\, dV_0 - \int_{V_0} \boldsymbol{S} : \dot{\boldsymbol{E}}\, dV_0, \qquad (3.70)
\end{aligned}$$

where the last relation follows on using $\dot{\boldsymbol{F}} = \boldsymbol{L}\boldsymbol{F}$, $\boldsymbol{T} = \boldsymbol{F}\boldsymbol{S}$, and Eqns. (1.329) and (2.61).

We now use the first law of thermodynamics to formulate first the integral form, and then the differential form of the energy equation. The first law states that

$$\frac{dU}{dt} = \frac{dW}{dt} + \frac{dQ}{dt}, \tag{3.71}$$

where U is the energy, W is the work done on the system, and Q is the heat supplied to the system. The time derivatives in Eqn. (3.71) denote the rate of change of the respective quantities. We now compute each of these terms.

The total energy U is given by the sum of its kinetic energy and internal energy (we consider the potential energy change as work done by the body forces, and hence include it as part of dW/dt). Thus, if e denotes the *specific internal energy* (internal energy per unit mass), we have

$$U = \int_{V(t)} \rho \left(\frac{\boldsymbol{v} \cdot \boldsymbol{v}}{2} + e \right) dV. \tag{3.72}$$

Work is done on the system by the surface tractions $\boldsymbol{t}$ and the body force $\boldsymbol{b}$. To compute the rate of work done by the surface tractions, consider a small element of surface area dS on the surface S. The total force acting on the element is given by $\boldsymbol{t}\, dS$. Hence, the rate of work done by the tractions on the element is $(\boldsymbol{t} \cdot \boldsymbol{v})\, dS$. Integrating over the surface, the rate of work done by the surface tractions over $S(t)$ is given by $\int_{S(t)} \boldsymbol{t} \cdot \boldsymbol{v}\, dS$. Similarly, the rate of working by the body forces is given by $\int_{V(t)} \rho \boldsymbol{b} \cdot \boldsymbol{v}\, dV$. Hence

$$\frac{dW}{dt} = \int_{S(t)} \boldsymbol{t} \cdot \boldsymbol{v}\, dS + \int_{V(t)} \rho \boldsymbol{b} \cdot \boldsymbol{v}\, dV. \tag{3.73}$$

We now compute the rate of heat received. If Q_h is the heat generated per unit mass per unit time by a heat source, and $\boldsymbol{q}$ is the heat flux vector, then the heat influx rate is given by $\int_{V(t)} \rho Q_h\, dV$, and the rate of heat outflow is given $\int_{S(t)} \boldsymbol{q} \cdot \boldsymbol{n}\, dS$. Hence, the net heat flow rate is

$$\frac{dQ}{dt} = \int_{V(t)} \rho Q_h\, dV - \int_{S(t)} \boldsymbol{q} \cdot \boldsymbol{n}\, dS. \tag{3.74}$$

Substituting Eqns. (3.72), (3.73) and (3.74) into Eqn. (3.71), we get the following statement of the first law of thermodynamics:

$$\frac{d}{dt} \int_{V(t)} \rho \left(\frac{\boldsymbol{v} \cdot \boldsymbol{v}}{2} + e \right) dV = \int_{S(t)} \boldsymbol{t} \cdot \boldsymbol{v}\, dS + \int_{V(t)} \rho \boldsymbol{b} \cdot \boldsymbol{v}\, dV + \int_{V(t)} \rho Q_h\, dV - \int_{S(t)} \boldsymbol{q} \cdot \boldsymbol{n}\, dS. \tag{3.75}$$

Applying Eqn. (3.1) to the left-hand side of the above equation, we get the following integral form of the first law:

$$\begin{aligned} \int_{V(t)} \frac{\partial}{\partial t} \left\{ \rho \left(\frac{\boldsymbol{v} \cdot \boldsymbol{v}}{2} + e \right) \right\} dV + \int_{S(t)} \rho \left(\frac{\boldsymbol{v} \cdot \boldsymbol{v}}{2} + e \right) (\boldsymbol{v} \cdot \boldsymbol{n})\, dS \\ = \int_{S(t)} \boldsymbol{t} \cdot \boldsymbol{v}\, dS + \int_{V(t)} \rho \boldsymbol{b} \cdot \boldsymbol{v}\, dV + \int_{V(t)} \rho Q_h\, dV - \int_{S(t)} \boldsymbol{q} \cdot \boldsymbol{n}\, dS. \end{aligned} \tag{3.76}$$

We now derive the differential form of the energy equation using Eqns. (3.63) and (3.75). Similar to the derivation of the differential form of the momentum equation, we first convert the surface integrals into volume integrals, and then use the fact that $V(t)$ can be chosen arbitrarily to set the integrand to zero. Using the divergence theorem, we have

$$\int_{S(t)} \boldsymbol{q}\cdot\boldsymbol{n}\,dS = \int_{V(t)} \boldsymbol{\nabla}\cdot\boldsymbol{q}\,dV,$$
$$\int_{S(t)} \boldsymbol{t}\cdot\boldsymbol{v}\,dS = \int_{S(t)} (\boldsymbol{\tau}\boldsymbol{n})\cdot\boldsymbol{v}\,dS = \int_{V(t)} \boldsymbol{\nabla}\cdot(\boldsymbol{\tau}^T\boldsymbol{v})\,dV.$$

Substituting these expressions into Eqn. (3.75), and using Eqn. (3.10a) on the left-hand side, we get the following relation:

$$\int_{V(t)} \rho\frac{D}{Dt}\left(\frac{\boldsymbol{v}\cdot\boldsymbol{v}}{2}+e\right)dV = \int_{V(t)} \boldsymbol{\nabla}\cdot(\boldsymbol{\tau}^T\boldsymbol{v})\,dV + \int_{V(t)} \rho\boldsymbol{b}\cdot\boldsymbol{v}\,dV + \int_{V(t)} \rho Q_h\,dV - \int_{V(t)} \boldsymbol{\nabla}\cdot\boldsymbol{q}\,dV.$$

Using the localization theorem, we get

$$\rho\frac{D}{Dt}\left(\frac{\boldsymbol{v}\cdot\boldsymbol{v}}{2}+e\right) = \boldsymbol{\nabla}\cdot(\boldsymbol{\tau}^T\boldsymbol{v}) + \rho\boldsymbol{b}\cdot\boldsymbol{v} + \rho Q_h - \boldsymbol{\nabla}\cdot\boldsymbol{q}. \tag{3.77}$$

From Eqns. (3.63) and (3.77), we get

$$\rho\frac{De}{Dt} = \boldsymbol{\tau}:\boldsymbol{D} - \boldsymbol{\nabla}\cdot\boldsymbol{q} + \rho Q_h, \tag{3.78}$$

which is known as the energy equation. In what follows, we find it convenient to denote the material derivative by a superposed dot, e.g., $\dot{\boldsymbol{F}} \equiv D\boldsymbol{F}/Dt$.

To convert Eqn. (3.78) to one over the reference domain, we use the fact that $\boldsymbol{L} = \dot{\boldsymbol{F}}\boldsymbol{F}^{-1}$, and the Piola transforms for a tensor and vector field (see Problem 3, Chapter 2), to convert the spatial quantities to the material ones. We have

$$\boldsymbol{T}(\boldsymbol{X},t) = \boldsymbol{\tau}(\boldsymbol{x},t)\,\mathbf{cof}\,\boldsymbol{F},$$
$$\boldsymbol{q}_0(\boldsymbol{X},t) = J\boldsymbol{F}^{-1}\boldsymbol{q}(\boldsymbol{x},t),$$

where $\boldsymbol{T}$ is the first Piola–Kirchhoff stress tensor, and $\boldsymbol{q}_0$ is the Piola transform of $\boldsymbol{q}$. Using the relations $\boldsymbol{\nabla}_X\cdot\boldsymbol{q}_0 = J\boldsymbol{\nabla}_x\cdot\boldsymbol{q}$, $J\boldsymbol{\tau}:\boldsymbol{L} = \boldsymbol{T}:\dot{\boldsymbol{F}} = \boldsymbol{S}:\dot{\boldsymbol{E}}$, and $\rho_0 = \rho J$, and defining $e_0(\boldsymbol{X},t) := e(\boldsymbol{x},t)$ and $(Q_h)_0(\boldsymbol{X},t) := Q_h(\boldsymbol{x},t)$, we get the energy equation over the reference configuration as

$$\begin{aligned}\rho_0\left(\frac{\partial e_0}{\partial t}\right)_X &= \boldsymbol{T}:\dot{\boldsymbol{F}} - \boldsymbol{\nabla}_X\cdot\boldsymbol{q}_0 + \rho_0(Q_h)_0 \\ &= \boldsymbol{S}:\dot{\boldsymbol{E}} - \boldsymbol{\nabla}_X\cdot\boldsymbol{q}_0 + \rho_0(Q_h)_0.\end{aligned} \tag{3.79}$$

Note that the stress power $\boldsymbol{\tau}:\boldsymbol{D}$ (which also equals $\boldsymbol{\tau}:\boldsymbol{L}$) in terms of the first Piola–Kirchhoff stress tensor is given by $\boldsymbol{T}:\dot{\boldsymbol{F}}/J$ and $\boldsymbol{S}:\dot{\boldsymbol{E}}/J$. In general, a stress measure $\hat{\boldsymbol{S}}$ is said to be conjugate to a strain measure $\hat{\boldsymbol{E}}$ if

$$\hat{\boldsymbol{S}}:\dot{\hat{\boldsymbol{E}}} = J\boldsymbol{\tau}:\boldsymbol{D} = \frac{1}{2}\boldsymbol{S}:\dot{\boldsymbol{C}}.$$

Thus, the second Piola–Kirchhoff stress is conjugate to the Lagrangian strain. By Eqn. (2.66), the stress conjugate to the Hill strain tensor $\boldsymbol{E}_H = \sum_{i=1}^k f(\lambda_i)\boldsymbol{P}_i$ is (see [355] for an alternative approach)

$$\boldsymbol{S}^{\boldsymbol{E}_H} = \sum_{i=1}^k \frac{\lambda_i}{f'(\lambda_i)}\boldsymbol{P}_i\boldsymbol{S}\boldsymbol{P}_i + \sum_{i=1}^k\sum_{\substack{j=1\\ j\neq i}}^k \frac{\lambda_i^2-\lambda_j^2}{2\left[f(\lambda_i)-f(\lambda_j)\right]}\boldsymbol{P}_i\boldsymbol{S}\boldsymbol{P}_j. \tag{3.80}$$

In particular, for $\boldsymbol{E}_H = \log\boldsymbol{U} = \sum_{i=1}^k \log\lambda_i\boldsymbol{P}_i$, and for $\boldsymbol{E}_H = \boldsymbol{U}-\boldsymbol{I} = \sum_{i=1}^k(\lambda_i-1)\boldsymbol{P}_i$, the conjugate stresses are

$$\boldsymbol{S}^{\log\boldsymbol{U}} = \sum_{i=1}^k \lambda_i^2\boldsymbol{P}_i\boldsymbol{S}\boldsymbol{P}_i + \sum_{i=1}^k\sum_{\substack{j=1\\ j\neq i}}^k \frac{\lambda_i^2-\lambda_j^2}{2\left[\log(\lambda_i)-\log(\lambda_j)\right]}\boldsymbol{P}_i\boldsymbol{S}\boldsymbol{P}_j,$$

$$\boldsymbol{S}^{\boldsymbol{U}-\boldsymbol{I}} = \frac{1}{2}\sum_{i=1}^k\sum_{j=1}^k(\lambda_i+\lambda_j)\boldsymbol{P}_i\boldsymbol{S}\boldsymbol{P}_j = \frac{1}{2}\left[\boldsymbol{U}\boldsymbol{S}+\boldsymbol{S}\boldsymbol{U}\right],$$

where we have used $\sum_{i=1}^k \boldsymbol{P}_i = \boldsymbol{I}$ and $\sum_{i=1}^k \lambda_i\boldsymbol{P}_i = \boldsymbol{U}$ in obtaining the last expression. For an isotropic material, $\boldsymbol{S}$ is of the form $\sum_{i=1}^k S_i\boldsymbol{P}_i$, and using the property $\boldsymbol{P}_i\boldsymbol{P}_j = \delta_{ij}\boldsymbol{P}_i$ (no sum on i), we see that the last term in Eqn. (3.80) is zero. Thus, for an isotropic material, $\boldsymbol{S}^{\log\boldsymbol{U}} = \boldsymbol{U}\boldsymbol{S}\boldsymbol{U} = J\boldsymbol{R}^T\boldsymbol{\tau}\boldsymbol{R}$.

From Eqn. (2.67) or (2.69), we see that, in general, there is no stress that is conjugate to $\bar{\boldsymbol{E}}_H$ (since $\dot{\bar{\boldsymbol{E}}}_H$ is not objective), although, under special circumstances, it may exist. For example, if $\bar{\boldsymbol{E}}_H = \log\boldsymbol{V} = \sum_{i=1}^k \log\lambda_i\boldsymbol{G}_i$, and if the material is isotropic (i.e., $\boldsymbol{\tau}$ is of the form $\sum_{i=1}^k \tau_i\boldsymbol{G}_i$), then on using the property $\boldsymbol{G}_i\boldsymbol{G}_j = \delta_{ij}\boldsymbol{G}_i$ (no sum on i), we see from Eqn. (2.74) or (2.75) that $(J\boldsymbol{\tau}):\boldsymbol{D} = (J\boldsymbol{\tau}):\dot{\overline{\log\boldsymbol{V}}}$. Thus, if the material is isotropic, $J\boldsymbol{\tau}$ is conjugate to $\log\boldsymbol{V}$, a result that we will also prove directly later (see Eqn. (5.46)).

We now discuss the second law of thermodynamics. Consider a hot body brought into contact with a cold one. We observe that heat flows from the hot body to the cold one, reducing the temperature of the hot body and raising that of the cold one. It never happens that the hot body grows hotter and the cold body grows colder, though such a possibility does not violate the first law of thermodynamics. Thus, we conclude that the first law alone is not sufficient to predict whether a particular process is viable or not. Nature imposes another condition on the process, a condition that tells us in which direction the process will take place.

In the classical thermodynamics literature, a state variable η known as the *specific entropy* is defined by the relation

$$d\eta := \frac{dq_{\text{rev}}}{\theta},$$

where dq_{rev} is the incremental amount of specific heat added reversibly to the system, and θ is the system temperature. Since η is a state variable, it can be computed for any kind of process, reversible or irreversible. An effective value of dq_{rev} can always be assigned to an irreversible process, even though the actual heat added dq is different from this amount. Hence, an alternative definition of η is given by the relation

$$d\eta = \frac{dq}{\theta} + d\eta_{\text{irrev}},$$

where $d\eta_{\text{irrev}}$ is the *increase* in entropy caused by the irreversible dissipative phenomena of viscosity and thermal conductivity occurring within the system. Thus, $d\eta_{\text{irrev}} \geq 0$. The equality sign holds for a reversible process, where, by definition, the dissipative phenomena are absent. Hence, we also have

$$d\eta \geq \frac{dq}{\theta},$$

with the equality holding for a reversible process. If the process is adiabatic ($dq = 0$) in addition to being reversible, we see that $d\eta = 0$, i.e., the entropy is constant and the process is known as an isentropic process. The second law states that a process will proceed in a direction such that the change in entropy of the system plus surroundings is always nonnegative. A process that yields a net decrease in entropy is forbidden by the second law of thermodynamics, though that same process might not violate the first law.

One can see that in the classical setup outlined above, the arguments for the existence of the specific entropy function are based on reversible processes. However, such arguments are unsatisfactory when they are extrapolated to irreversible processes. Since most of the real processes that occur in nature involve irreversible changes, one needs to reexamine the basic postulates of thermodynamics. Such a program has been undertaken by several researchers, leading to a more rigorous mathematical basis for the laws of thermodynamics; descriptions can be found in, for example, [295] and [335]. The gist of the findings is that the existence of the specific entropy function can be *proved* using the *Clausius inequality*[3], which states that

$$\int_{t_1}^{t_2} \left[\int_{V(t)} \frac{\rho Q_h}{\theta}\, dV - \int_{S(t)} \frac{\boldsymbol{q} \cdot \boldsymbol{n}}{\theta}\, dS \right] dt \leq 0$$

for all cyclic processes over the time interval $[t_1, t_2]$. The entropy function η is, in general, not unique, although under certain conditions (e.g., for reversible processes), it can be shown to be unique to within a constant. It can be proved that the Clausius inequality is equivalent to the *Clausius–Duhem inequality* given by [295]

$$\frac{d}{dt} \int_{V(t)} \rho\eta\, dV \geq \int_{V(t)} \frac{\rho Q_h}{\theta}\, dV - \int_{S(t)} \frac{\boldsymbol{q} \cdot \boldsymbol{n}}{\theta}\, dS, \tag{3.81}$$

i.e., the rate of entropy increase is greater than or equal to the entropy input rate due to the heat supplied to $V(t)$ from the external world, and the entropy flux at the boundary $S(t)$; equality holds in the above relation only for a reversible process. Equation (3.81) can be said to be integral form of the second law of thermodynamics. Using the divergence theorem, and then invoking the localization theorem, we get the differential form of the Clausius–Duhem inequality as

$$\dot{\eta} \geq \frac{Q_h}{\theta} - \frac{1}{\rho} \boldsymbol{\nabla} \cdot \left(\frac{\boldsymbol{q}}{\theta} \right), \tag{3.82}$$

[3] Under certain hypothesis, the Clausius inequality is shown to be equivalent to various statements of the second law, such as the Carnot, Kelvin and Clausius statements, in [295].

which can also be written as

$$\dot{\eta} \geq \frac{Q_h}{\theta} - \frac{1}{\rho\theta}\boldsymbol{\nabla}\cdot\boldsymbol{q} + \frac{1}{\rho\theta^2}\boldsymbol{q}\cdot\boldsymbol{\nabla}\theta. \tag{3.83}$$

For the example considered of a hot body brought into contact with a cold one, consider the combined volumes of the two bodies as a material volume $V(t)$. Note that the material surface $S(t)$ of this material volume *does not* include the surface of contact between the bodies. Since $\boldsymbol{q}\cdot\boldsymbol{n} = 0$ on $S(t)$ and since $Q_h = 0$, from Eqn. (3.81), we get

$$\frac{d}{dt}\int_{V(t)} \rho\eta\, dV \geq 0,$$

i.e., the rate of change of the entropy of the combined system is non-increasing, although, of course, at points in the interior of the cold and hot body, the entropy increases and decreases, respectively.

Since no explicit formula has been given for the specific entropy, one may wonder as to how one actually computes this quantity. This is done through a constitutive prescription for the specific entropy, which (along with other constitutive relations) has to satisfy the Clausius–Duhem inequality.

We shall find it convenient to introduce the free energy defined by

$$\psi := e - \eta\theta.$$

Taking the material derivative of the above equation, we obtain

$$\dot{e} = \dot{\psi} + \dot{\eta}\theta + \eta\dot{\theta}.$$

Using Eqn. (3.78) to eliminate $\dot{e}$, we get

$$\dot{\eta} = \frac{1}{\theta}\left[Q_h - \frac{1}{\rho}\boldsymbol{\nabla}\cdot\boldsymbol{q} + \frac{1}{\rho}\boldsymbol{\tau} : \boldsymbol{L} - \eta\dot{\theta} - \dot{\psi}\right]. \tag{3.84}$$

Substituting Eqn. (3.84) into Eqn. (3.83), and using the fact that $\theta > 0$, we obtain

$$\dot{\psi} + \eta\dot{\theta} - \frac{1}{\rho}\boldsymbol{\tau} : \boldsymbol{L} + \frac{\boldsymbol{q}\cdot\boldsymbol{g}}{\rho\theta} \leq 0, \tag{3.85}$$

where $\boldsymbol{g} = \boldsymbol{\nabla}_x\theta$ is the temperature gradient.

It is also useful to write Eqns. (3.84) and (3.85) with respect to the reference configuration. Using the fact that

$$\rho_0 = \rho J,$$
$$\boldsymbol{\tau} : \boldsymbol{L} = \frac{1}{J}\boldsymbol{T} : \dot{\boldsymbol{F}},$$
$$\boldsymbol{q}_0(\boldsymbol{X},t) = (\mathbf{cof}\,\boldsymbol{F})^T\boldsymbol{q}(\boldsymbol{x},t),$$
$$\boldsymbol{\nabla}_X\cdot\boldsymbol{q}_0 = J\boldsymbol{\nabla}_x\cdot\boldsymbol{q}, \quad \text{(Problem 3, Chapter 2)}$$

and defining

$$\boldsymbol{g}_0 := \boldsymbol{\nabla}_X\theta = \boldsymbol{F}^T\boldsymbol{\nabla}_x\theta = \boldsymbol{F}^T\boldsymbol{g},$$

$$\psi_0(\boldsymbol{X},t) := \psi(\boldsymbol{x},t),$$
$$\eta_0(\boldsymbol{X},t) := \eta(\boldsymbol{x},t),$$
$$(Q_h)_0(\boldsymbol{X},t) := Q_h(\boldsymbol{x},t),$$
$$\theta_0(\boldsymbol{X},t) := \theta(\boldsymbol{x},t),$$

we get the required versions of Eqn. (3.84) and (3.85) as

$$\left(\frac{\partial \eta_0}{\partial t}\right)_X = \frac{1}{\theta_0}\left[(Q_h)_0 - \frac{1}{\rho_0}\boldsymbol{\nabla}_X \cdot \boldsymbol{q}_0 + \frac{1}{\rho_0}\boldsymbol{T} : \dot{\boldsymbol{F}} - \eta_0\dot{\theta}_0 - \dot{\psi}_0\right]. \tag{3.86}$$

and

$$\left(\frac{\partial \psi_0}{\partial t}\right)_X + \eta_0\left(\frac{\partial \theta_0}{\partial t}\right)_X - \frac{1}{\rho_0}\boldsymbol{T} : \dot{\boldsymbol{F}} + \frac{\boldsymbol{q}_0 \cdot \boldsymbol{g}_0}{\rho_0\theta_0} \le 0. \tag{3.87}$$

By using the relations $\dot{\boldsymbol{F}} = \boldsymbol{L}\boldsymbol{F}$ and $\boldsymbol{T} = \boldsymbol{F}\boldsymbol{S}$, and using Eqns. (1.329) and (2.61), we can also write Eqn. (3.87) as

$$\left(\frac{\partial \psi_0}{\partial t}\right)_X + \eta_0\left(\frac{\partial \theta_0}{\partial t}\right)_X - \frac{1}{2\rho_0}\boldsymbol{S} : \dot{\boldsymbol{C}} + \frac{\boldsymbol{q}_0 \cdot \boldsymbol{g}_0}{\rho_0\theta_0} \le 0. \tag{3.88}$$

Note that, similar to Eqn. (3.40), we have $\boldsymbol{q} \cdot \boldsymbol{n}\, dS = \boldsymbol{q}_0 \cdot \boldsymbol{n}_0\, dS_0$. We will use the above equations to derive restrictions on constitutive relations in Chapter 7.

3.11 Control Volume Form of the Balance Laws

The control volume form of the equation of mass conservation is obtained by letting $f \equiv \rho$ in Eqn. (3.12) and using Eqn. (3.7), and is given by

$$\frac{d}{dt}\int_{V_t} \rho\, dV + \int_{S_t} \rho(\boldsymbol{v}_{\mathrm{rel}} \cdot \boldsymbol{n})\, dS = 0, \tag{3.89}$$

where $\boldsymbol{v}_{\mathrm{rel}}$ represents the velocity of the material with respect to the control surface. If the conditions under which Eqn. (3.14) holds are satisfied, then the following simplification of Eqn. (3.89) results:

$$\int_{V_t} \frac{\partial \rho}{\partial t}\, dV + \int_{S_t} \rho(\boldsymbol{v}_{\mathrm{rel}} \cdot \boldsymbol{n})\, dS = 0. \tag{3.90}$$

By letting $\boldsymbol{w} \equiv \boldsymbol{v}$ in Eqn. (3.13b), and using Eqn. (3.19) and the divergence theorem, we obtain the control volume form of the linear momentum balance principle as

$$\frac{d}{dt}\int_{V_t} \rho\boldsymbol{v}\, dV + \int_{S_t} \rho\boldsymbol{v}(\boldsymbol{v}_{\mathrm{rel}} \cdot \boldsymbol{n})\, dS = \int_{S_t} \boldsymbol{t}\, dS + \int_{V_t} \rho\boldsymbol{b}\, dV. \tag{3.91}$$

Note that $\boldsymbol{v}$ is the velocity of the material as measured in the chosen frame of reference. Again, if the conditions under which Eqn. (3.14) holds are satisfied, then Eqn. (3.91) reduces to the following form:

$$\int_{V_t} \frac{\partial(\rho\boldsymbol{v})}{\partial t}\, dV + \int_{S_t} \rho\boldsymbol{v}(\boldsymbol{v}_{\mathrm{rel}} \cdot \boldsymbol{n})\, dS = \int_{S_t} \boldsymbol{t}\, dS + \int_{V_t} \rho\boldsymbol{b}\, dV. \tag{3.92}$$

By letting $w \equiv x \times v$ in Eqn. (3.13b) and using the facts that $\nabla_x (x \times v) v = x \times [(\nabla v) v]$, $\int_{V_t} x \times \nabla \cdot \tau \, dV = \int_{S_t} x \times (\tau n) \, dS = \int_{S_t} x \times t \, dS$, Eqn. (3.19) and the divergence theorem, we obtain the the following control volume form of the angular momentum balance equation:

$$\frac{d}{dt} \int_{V_t} \rho (x \times v) \, dV + \int_{S_t} \rho (x \times v)(v_{\text{rel}} \cdot n) \, dS = \int_{S_t} x \times t \, dS + \int_{V_t} \rho (x \times b) \, dV.$$

When the conditions under which Eqn. (3.14) are met, the above equation simplifies to

$$\int_{V_t} \frac{\partial}{\partial t} (\rho x \times v) \, dV + \int_{S_t} \rho (x \times v)(v_{\text{rel}} \cdot n) \, dS = \int_{S_t} x \times t \, dS + \int_{V_t} \rho (x \times b) \, dV.$$

By letting $f \equiv (v \cdot v)/2$ in Eqn. (3.13a), and using Eqn. (3.19), we get the control volume form of the mechanical energy balance as

$$\frac{d}{dt} \int_{V_t} \rho \left(\frac{v \cdot v}{2} \right) dV + \int_{S_t} \rho \left(\frac{v \cdot v}{2} \right) (v_{\text{rel}} \cdot n) \, dS = \int_{S_t} t \cdot v \, dS + \int_{V_t} [\rho b \cdot v - \tau : D] \, dV.$$

By letting $f \equiv e$ in Eqn. (3.13a), and using Eqn. (3.78) and the divergence theorem, we get the control volume form of the energy equation as

$$\frac{d}{dt} \int_{V_t} \rho e \, dV + \int_{S_t} \rho e (v_{\text{rel}} \cdot n) \, dS = \int_{V_t} [\rho Q_h + \tau : D] \, dV - \int_{S_t} q \cdot n \, dS.$$

By letting $f \equiv \eta$ in Eqn. (3.13a), and using Eqn. (3.82) and the divergence theorem, we get the control volume form of the Clausius–Duhem inequality as

$$\frac{d}{dt} \int_{V_t} \rho \eta \, dV + \int_{S_t} \rho \eta (v_{\text{rel}} \cdot n) \, dS \geq \int_{V_t} \frac{\rho Q_h}{\theta} dV - \int_{S_t} \frac{1}{\theta} q \cdot n \, dS.$$

EXERCISES

1. Prove the following transport theorems (λ represents the unit tangent vector to the contour C):

$$\frac{d}{dt} \oint_C \phi \, ds = \oint_C \left(\frac{D\phi}{Dt} + \phi \lambda \cdot D \lambda \right) ds,$$

$$\frac{d}{dt} \oint_C \phi \, dx = \oint_C \left(\frac{D\phi}{Dt} + \phi L \right) dx,$$

$$\frac{d}{dt} \int_{S(t)} f \cdot n \, dS = \int_{S(t)} \left(\frac{\partial f}{\partial t} + (\nabla \cdot f) v \right) \cdot n \, dS + \oint_C (f \times v) \cdot \lambda \, ds,$$

$$\frac{d}{dt} \int_{S(t)} H n \, dS = \int_{S(t)} \left[\frac{DH}{Dt} + (\nabla \cdot v) H - H (\nabla v)^T \right] n \, dS.$$

2. By integrating the equations

$$(\nabla_X \cdot T) \otimes X + \rho_0 b_0 \otimes X = 0,$$

over the reference configuration V_0, and with the use of the divergence theorem, show that

$$\int_{V_0} \boldsymbol{T}\, dV = \left[\int_{S_0} \boldsymbol{t}_0 \otimes \boldsymbol{X}\, dS + \int_V \rho_0 \boldsymbol{b}_0 \otimes \boldsymbol{X}\, dV \right]. \tag{3.93}$$

3. The stress components at the origin are given by

$$\tau = \begin{bmatrix} 2 & -1 & 0 \\ -1 & 2 & 0 \\ 0 & 0 & 4 \end{bmatrix}$$

Find the normal stress at the origin on the surface $x^2 + y^2 + z^2 + 2x + 4y + 4z = 0$. Find the principal stresses and the equation of the plane through the origin on which the normal stress is a maximum. Next, align the axes $\bar{e}_1$, $\bar{e}_2$ and $\bar{e}_3$ along the principal directions associated with the maximum, minimum and intermediate principal stresses. Write the rotation matrix $\boldsymbol{Q}$ for this rotated system, and transform the stress matrix to this new coordinate frame. Verify that this transformed stress matrix is a diagonal matrix, with the principal stress values along the diagonal.

4. A rope rests on a semicircular frictionless surface as shown in Fig. 3.8. In the undeformed position, it has a density ρ_0, and subtends an angle θ_0 as shown. Approximating the rope to be a one-dimensional continuum, and assuming the second Piola–Kirchhoff stress to be λ times the Green strain, show that the governing equation for determining the displacement u_θ of the rope under its own weight is given by

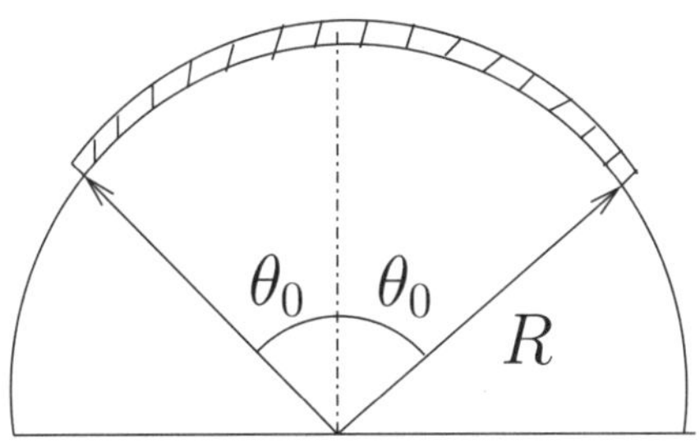

Fig. 3.8 *Problem 4*

$$\lambda \frac{d}{d\theta} \left\{ \left(1 + \frac{1}{R}\frac{du_\theta}{d\theta} \right) \left[\frac{1}{R}\frac{du_\theta}{d\theta} + \frac{1}{2R^2} \left(\frac{du_\theta}{d\theta} \right)^2 \right] \right\} + R\rho_0 g \sin\left(\theta_0 + \frac{u_\theta}{R} \right) = 0.$$

The above equation when solved subject to the boundary conditions

$$u_\theta|_{\theta=0} = 0,$$

$$\left[\frac{1}{R}\frac{du_\theta}{d\theta} + \frac{1}{2R^2} \left(\frac{du_\theta}{d\theta} \right)^2 \right]_{\theta=\theta_0} = 0.$$

yields the displacement u_θ. Show that in the context of the linearized theory of elasticity (which is a good approximation only when the strains are "small"; see Chapter 6), the solution is given by

$$u_\theta = \frac{\rho_0 g R^2}{\lambda} \left[\sin\theta - \theta\cos\theta_0\right].$$

4 Constitutive Equations

The equation of mass conservation and the energy equation yield two equations, while the linear momentum balance equation yields three equations (one for each displacement component), thus giving a total of five equations. The balance of angular momentum is used up implicitly by assuming the stress tensor to be symmetric, and thus having only six independent components. The unknowns to be solved for are the density, displacement field, stress tensor components, temperature, internal energy, entropy and conductivity vector, viz, ρ, $\boldsymbol{u}$ (3 unknowns), $\boldsymbol{\tau}$ (6 unknowns), θ, e, η and $\boldsymbol{q}$ (3 unknowns), giving a total of sixteen unknowns. So we need eleven more equations to solve for all the remaining unknowns. These equations are provided by the *constitutive relations*, an example of which is

$$
\begin{aligned}
\boldsymbol{\tau} &= \tilde{\boldsymbol{\tau}}(\boldsymbol{F}, \theta), && \text{(6 equations)} \\
\boldsymbol{q} &= \tilde{\boldsymbol{q}}(\rho, \theta), && \text{(3 equations)} \\
e &= \tilde{e}(\rho, \theta), && \text{(1 equation)} \\
\eta &= \tilde{\eta}(\rho, \theta). && \text{(1 equation)}
\end{aligned}
$$

The fact that the mathematical model developed so far is incomplete is also evident on physical grounds. While the equations developed so far are valid for any material, irrespective of whether it is solid, liquid or gas, it is clear that the nature of the underlying material should be taken into account. For example, for a given set of forces, the resulting deformations are different if the body is made of steel or of rubber.

In this chapter, we discuss the restrictions imposed on constitutive relations by the *principle of material frame-indifference*, also known as the *principle of material objectivity*. This principle states that constitutive equations must be invariant (in the sense that they have the same form) under changes of frame. As an example, consider an object attached to the end of a spring, and spinning at steady speed. It is tacitly assumed that the spring constant, which has been found in a non-rotating condition, is unaffected by the rigid rotation. The principle of material frame-indifference extends this assumption to general constitutive relations, so that, for example, we postulate that the constitutive relation for the stress in a given body is unchanged even if a rigid-body motion is superimposed on it.

4.1 Frame of Reference

A frame of reference is the background against which observations are made. A change of frame of reference may be interpreted as meaning that two observers who are in arbitrary motion with respect to each other, and have their clocks set differently, but who use the same units of length and time, are equally qualified to describe a given phenomenon. A

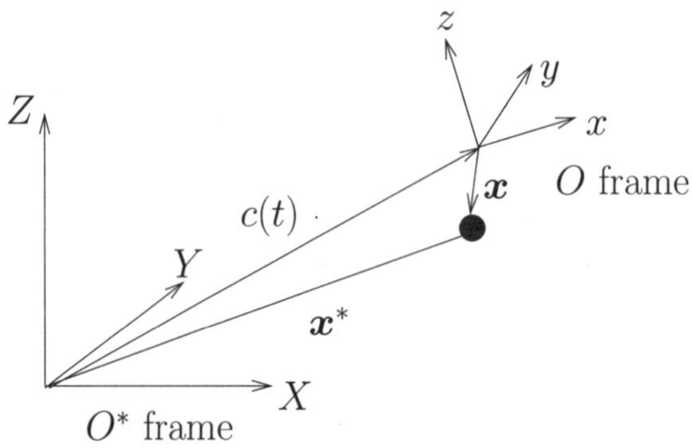

Fig. 4.1 *Frames of reference.*

particular event to which a place and time (x,t) are assigned by the framing O may be assigned another place and timing (x^*,t^*) in another framing O^*, as shown in Fig. 4.1. Let $c(t)$ be the vector connecting the origins of these two coordinate systems as shown. If x, x^* and c are all expressed in terms of the basis vectors of either the O or the O^* frame, then we have the relation $x^* = x + c$; if e_i^* are the (Cartesian) basis vectors associated with the O^* frame, it follows that $x^* \cdot e_i^* = x \cdot e_i^* + c \cdot e_i^*$. However, in practice, it is more convenient to express x in terms of the basis vectors of the O frame. Let e_i denote the basis vectors associated with the O frame, and let $Q(t) = e_k \otimes e_k^*$, with components with respect to the starred frame given by $Q_{ij} = e_i^* \cdot e_j$, i.e., $Q = [e_1 \mid e_2 \mid e_3]$. Then we have $e_i^* = (e_i^* \cdot e_j)e_j = Q_{ij}e_j$, so that $x \cdot e_i^* = x \cdot (Q_{ij}e_j) = Q_{ij}x \cdot e_j$. If we use the symbols x^*, x and c to represent the components of these vectors with respect to the O^*, O and O^* frames, respectively, i.e., if $x^* \equiv (x^* \cdot e_1^*, x^* \cdot e_2^*, x^* \cdot e_3^*)$, $x \equiv (x \cdot e_1, x \cdot e_2, x \cdot e_3)$, and $c \equiv (c \cdot e_1^*, c \cdot e_2^*, c \cdot e_3^*)$, then $x^* = Q(t)x + c(t)$. Assuming that the clock in O is set differently from that in O^* by an amount a, we thus have the following relations:

$$\begin{aligned} x^* &= Q(t)x + c(t), \\ t^* &= t - a. \end{aligned} \tag{4.1}$$

A change of frame of reference should not be confused with a change of coordinate axes. In any given frame, we have an infinite number of coordinate systems, and the components of a vector and a tensor with respect to any two different coordinate axes are related as per Eqns. (1.99) and (1.100). However, the Q matrix in these relations is *time-independent* (although, of course, v or T themselves could be time-dependent), in contrast to the Q matrix in Eqn. (4.1), which is time-dependent. The invariance of the governing equations under change of coordinate axes in a given frame of reference is ensured merely by writing the equations in tensor form. Following Einstein's 'covariance principle', however, we postulate the more extensive principle that there are no preferred reference frames (such as the 'inertial frame' in the classical setup), and that the axioms that we have stated have the same form in reference frames that are in arbitrary motion with respect to each other. We will investigate the implications of this principle in Section 4.4. We now consider a few examples illustrating the first relation in Eqn. (4.1).

Consider the case shown in Fig. 4.2a. The frame O is fixed at the point under observation, and consequently, the position vector of this point is the zero vector with respect

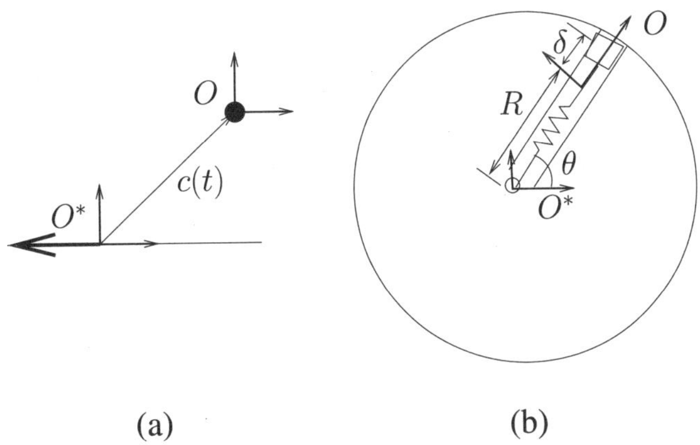

Fig. 4.2 *Two observers in (a) translatory; (b) translatory and rotatory relative motion.*

to O. On the other hand, since the observer in frame O^* is translating to the left, the position vector of the observed point is $\boldsymbol{c}(t)$, as shown in the figure. Thus, the position vectors observed by O and O^* are related as

$$\boldsymbol{x}^* = \boldsymbol{c}(t) = \boldsymbol{I}(\boldsymbol{0}) + \boldsymbol{c}(t),$$

which is a special case of Eqn. (4.1) with $\boldsymbol{Q}(t) = \boldsymbol{I}$.

Now consider the situation shown in Fig. 4.2b. The observer in frame O^* is stationary. Assume that a Cartesian coordinate system in this frame is fixed at the center of the rotating disc as shown. The observer in frame O has the coordinate system fixed at a distance R, which is the distance of the mass from the center before the disc starts spinning. The observer O spins with the disc. The position of the center of the mass with respect to the observer in frame O^* is

$$\begin{aligned} x^* &= (R+\delta)\cos\theta = \delta\cos\theta + R\cos\theta, \\ y^* &= (R+\delta)\sin\theta = \delta\sin\theta + R\sin\theta, \\ z^* &= 0, \end{aligned}$$

where δ is the elongation of the spring, and $\theta(t)$ is the angle made by the groove with the horizontal. The coordinates of the center of the mass with respect to the observer in frame O are $x = \delta$, $y = 0$ and $z = 0$. Thus, once again, we see that the position vectors in the two frames are related via Eqn. (4.1), with $\boldsymbol{c}(t) = (R\cos\theta, R\sin\theta, 0)$, and

$$\boldsymbol{Q}(t) = \begin{bmatrix} \boldsymbol{e}_1 & | & \boldsymbol{e}_2 & | & \boldsymbol{e}_3 \end{bmatrix} = \begin{bmatrix} \cos\theta & -\sin\theta & 0 \\ \sin\theta & \cos\theta & 0 \\ 0 & 0 & 1 \end{bmatrix}.$$

Note that $\boldsymbol{c}(0) = [R, 0, 0]$ is not the zero vector. Some other examples are presented by Malvern [207].

If scalar, vector and tensor functions, ϕ, $\boldsymbol{v}$ and $\boldsymbol{T}$, transform according to the 'expected' transformations,

$$\phi^*(\boldsymbol{x}^*, t^*) = \phi(\boldsymbol{x}, t),$$

$$v^*(x^*, t^*) = Q(t)v(x, t),$$

$$T^*(x^*, t^*) = Q(t)T(x, t)Q(t)^T, \tag{4.2}$$

they are said to be *objective*. We shall soon see that some kinematical quantities are objective, while some others are not. Note that two different observers observe the same value, magnitude and eigenvalues, respectively, of objective scalar, vector and tensor quantities.

4.2 Transformation of Kinematical Quantities

Suppose the motion of a continuum is described by $x = \chi(X, t)$ in frame O, where X are the referential coordinates. We assume that the framings O and O^* coincide at some instant of time, and take the configuration of the body at this instant of time to be the reference configuration. Thus, X as observed from O or O^* is identical[1]. The motion of the continuum as seen from frame O^* is related to that seen from frame O by the relation

$$\chi^*(X, t^*) = Q(t)\chi(X, t) + c(t), \tag{4.3}$$

where $Q \in \text{Orth}^+$, $t^* = t - a$, $Q(0) = I$ and $c(0) = 0$. From Eqn. (4.3), we see that the configurations $\chi^*(X, t^*)$ and $\chi(X, t)$ are related by a rigid deformation (see Fig. 4.3). Two motions related by Eqn. (4.3) will be referred to as equivalent.

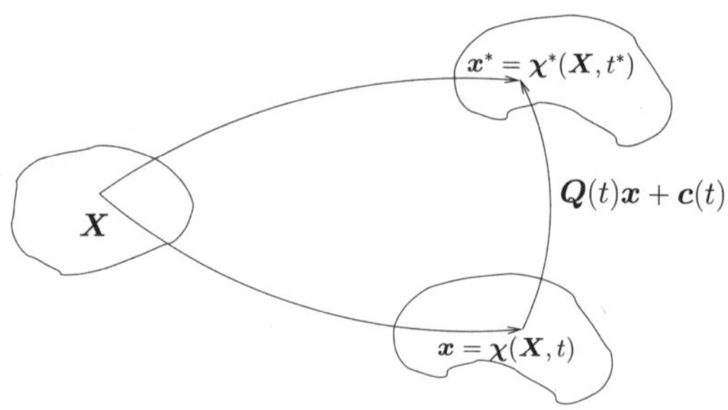

Fig. 4.3 *Change of observer can be given the alternative interpretation of a superposed rigid body motion.*

We now determine how the various kinematical quantities transform under a change of observer. When a quantity is a function of time alone, we use a superposed dot to indicate the time derivative, e.g., $\dot{Q} \equiv dQ/dt$. Note that since $t^* = t - a$, the time derivative of any quantity with respect to either t^* or t is the same.

Since $F = \nabla\chi$, we have

$$F^* = \nabla\chi^* = Q(t)\nabla\chi = QF. \tag{4.4}$$

[1]This assumption can be relaxed by relaxing the constraints on $Q(0)$ and $c(0)$ stated below Eqn. (4.3), i.e., by letting $Q(0) = Q_0$, which is not necessarily the identity tensor, and $c(0) = c_0$, which is not necessarily zero (as in the rotating disc example in the previous section), so that $X^* = Q_0X + c_0$. In this case, some of the relations derived below get modified, e.g., Eqn. (4.4) gets modified to $F^* = QFQ_0^T$; however, whether a quantity is objective or not is not affected by this change. For a more detailed discussion, see [198].

Thus, the deformation gradient transforms like a vector under a change of frame. Since $\det Q = 1$, we have $\det F^* = \det F$. The transformations for the right and left Cauchy–Green strain tensors are

$$C^* = (F^*)^T F^* = F^T Q^T Q F = F^T F = C,$$
$$B^* = F^*(F^*)^T = Q F F^T Q^T = Q B Q^T.$$

Hence, the transformations for the Lagrangian and Eulerian measures of strain are

$$U^* = U,$$
$$V^* = Q V Q^T,$$
$$E^* = E, \tag{4.5}$$
$$\bar{E}^* = Q \bar{E} Q^T. \tag{4.6}$$

Since $F = RU$, it follows that $R^* = QR$. Since $B \in \text{Sym}$, it can be expressed as $\sum_{i=1}^{k} \lambda_i P_i$, where λ_i are the distinct eigenvalues and P_i the associated eigenprojections. Thus, $B^* = \sum_{i=1}^{k} \lambda_i G_i$, where $G_i := Q P_i Q^T$ are also projections. By using Eqn. (1.312), it follows that

$$\log B^* = \sum_{i=1}^{k} (\log \lambda_i) G_i = Q \left(\sum_{i=1}^{k} \lambda_i P_i \right) Q^T = Q (\log B) Q^T. \tag{4.7}$$

The velocity transformation relation follows from material time differentiation of Eqn. (4.3):

$$\begin{aligned} v^* &= \frac{D\chi^*}{Dt} \\ &= Q \frac{D\chi}{Dt} + \dot{Q}\chi + \dot{c} \\ &= Q v + \dot{Q} x + \dot{c}. \end{aligned} \tag{4.8}$$

Consider two points fixed in the unstarred frame, having coordinates x_1^* and x_2^*. Since $v_1 = v_2 = 0$, we get

$$v_1^* - v_2^* = \dot{Q} Q^T (x_1^* - x_2^*) = w \times (x_1^* - x_2^*),$$

where w is the axial vector of the skew tensor $\dot{Q} Q^T$. From the above relation, we see that w represents the angular velocity of the unstarred frame with respect to the starred frame.

To find the relation for the acceleration, we take the material derivative of Eqn. (4.8) to get

$$a^* = Q a + 2\dot{Q} v + \ddot{Q} x + \ddot{c}. \tag{4.9}$$

From Eqns. (4.8) and (4.9), it is apparent that the velocity and acceleration vectors are not objective. Physically, to see that the velocity vector is not indifferent consider a small ball moving outward along a straight radial groove on a spinning disc; with respect to an observer on the spinning disc, the ball moves radially outward with nonzero velocity, while with respect to an observer on the ball, the velocity of the ball is zero. Thus, the two observers observe *different* velocity vectors, and one cannot be obtained from the other by an orthogonal transformation.

One observation that we make at this point is that although physical scalar field variables (such as temperature, density and so on) are objective, *artificially constructed* scalar field variables (e.g., $v \cdot v$) need not be objective.

Using Eqn. (2.49) and the invertibility of F (and hence, also of F^*), we have

$$L^* = \frac{DF^*}{Dt}(F^*)^{-1} = (Q\frac{DF}{Dt} + \dot{Q}F)F^{-1}Q^T,$$

which yields

$$L^* = QLQ^T + \dot{Q}Q^T. \tag{4.10}$$

Using the fact that $\dot{Q}Q^T$ is skew (Problem 38, Chapter 1), we get

$$D^* = \frac{1}{2}\left[L^* + (L^*)^T\right] = QDQ^T \tag{4.11}$$

$$\hat{W}^* = \frac{1}{2}\left[L^* - (L^*)^T\right] = Q\hat{W}Q^T + \dot{Q}Q^T \tag{4.12}$$

Thus, D is objective, whereas $\hat{W}$ is not. Taking the trace of both sides of Eqn. (4.10), and again using the fact that $\dot{Q}Q^T \in$ Skw, we get

$$\nabla_{x^*} \cdot v^* = \operatorname{tr} L = \nabla_x \cdot v. \tag{4.13}$$

From Eqn. (2.61) and (2.62) (or from Eqns. (4.5) and (4.6)), we see that the material rates of the Lagrangian and Eulerian strain measures are not objective. Instead, we have

$$\begin{aligned}
\frac{DE^*}{Dt} &= (F^*)^T D^* F^* \\
&= F^T Q^T Q D Q^T Q F \\
&= F^T D F \\
&= \frac{DE}{Dt} \\
\frac{D\bar{E}^*}{Dt} &= D^* - (L^*)^T \bar{E}^* - \bar{E}^* L^* \\
&= Q\frac{D\bar{E}}{Dt}Q^T + \left[\dot{Q}\bar{E}Q^T + Q\bar{E}\dot{Q}^T\right].
\end{aligned}$$

4.3 Principle of Frame-Indifference

The principle of frame-indifference states that the governing laws must have the same form in reference frames moving arbitrarily with each other, i.e., there is no 'preferred' frame such as the 'inertial' frame in the classical setup. In this section, we shall examine the implications of this principle by applying it to all the balance laws.

Thus, first consider the law of mass conservation. Since $\rho_0(X) = \rho^* J^* = \rho J$, and since $J^* = J$, we have $\rho^* = \rho$. By virtue of Eqn. (4.13), we have $\nabla_{x^*} \cdot v^* = \nabla_x \cdot v$, and thus, the equation of mass conservation given by

$$\frac{D\rho}{Dt} + \rho\nabla \cdot v = \frac{D\rho^*}{Dt^*} + \rho^*\nabla_{x^*} \cdot v^* = 0$$

has the same form in all frames.

In the previous section, we have derived the transformation relations for various kinematical quantities under a change of frame. We saw that certain kinematical quantities were objective, whereas some others were not. We now postulate that the traction vector, and hence the stress tensor, are objective, and analyze the consequences of this postulate. We emphasize that the requirement of objectivity for the stress tensor is an *axiom*, and not a derived relation, as in the case of the kinematical quantities. We shall see that this axiom when combined with the requirement that the linear momentum balance equation has the same form in all frames yields the transformation relation for the body force $\boldsymbol{b}$, which is, in general, not of the form $\boldsymbol{b}^* = \boldsymbol{Q}(t)\boldsymbol{b}$, i.e., $\boldsymbol{b}$ is *not* objective. Note that in contrast to the rather controversial principle of *material* frame-indifference, which is discussed in the following section, the axiom that the stress tensor is objective is universally agreed upon.

To see the physical motivation behind treating contact forces and body forces differently, we consider again the example of a spring mounted on a rotating table. From the viewpoint of an external observer, the body force is merely that of gravity, while from the viewpoint of an observer whose reference frame is fixed to the rotating table, there is an outward force, seemingly exerted by an external field, in addition to the force of gravity. Thus, the body forces observed by the two observers are different. In contrast, both observers observe the same spring constant (which could be a function of the angular velocity of the table), and the same extension, and hence observe the same contact force between the table and spring (or the mass and the spring).

The condition that the traction vector be objective is written as

$$\boldsymbol{t}^*(\boldsymbol{x}^*, t^*, \boldsymbol{n}^*) = \boldsymbol{Q}(t)\boldsymbol{t}(\boldsymbol{x}, t, \boldsymbol{n}) \quad \forall \boldsymbol{x}, t \text{ and } \boldsymbol{n}, \tag{4.14}$$

where

$$\boldsymbol{n}^* = \frac{(\mathbf{cof}\,\boldsymbol{F}^*)\boldsymbol{n}_0}{|(\mathbf{cof}\,\boldsymbol{F}^*)\boldsymbol{n}_0|} = \frac{(\mathbf{cof}\,\boldsymbol{Q})(\mathbf{cof}\,\boldsymbol{F})\boldsymbol{n}_0}{|(\mathbf{cof}\,\boldsymbol{Q})(\mathbf{cof}\,\boldsymbol{F})\boldsymbol{n}_0|} = \boldsymbol{Q}\frac{(\mathbf{cof}\,\boldsymbol{F})\boldsymbol{n}_0}{|(\mathbf{cof}\,\boldsymbol{F})\boldsymbol{n}_0|} = \boldsymbol{Q}(t)\boldsymbol{n},$$

and $\boldsymbol{x}^* = \boldsymbol{Q}(t)\boldsymbol{x} + \boldsymbol{c}(t)$. Since by Cauchy's relation, $\boldsymbol{t} = \boldsymbol{\tau}\boldsymbol{n}$, Eqn. (4.14) can be written in terms of the Cauchy stress tensor as

$$\boldsymbol{\tau}^* = \boldsymbol{Q}(t)\boldsymbol{\tau}\boldsymbol{Q}^T(t), \tag{4.15}$$

i.e., the stress tensor is objective.

To see how the term $\boldsymbol{\nabla}_x \cdot \boldsymbol{\tau}$ in Eqn. (3.19) transforms under a change of frame, we note that

$$\begin{aligned}
\boldsymbol{\nabla}_{x^*} \cdot \boldsymbol{\tau}^* &= \frac{\partial \tau^*_{ij}}{\partial x^*_j}\boldsymbol{e}^*_i \\
&= Q_{ir}Q_{js}\frac{\partial \tau_{rs}}{\partial x^*_j}\boldsymbol{e}^*_i \\
&= Q_{ir}Q_{js}\frac{\partial \tau_{rs}}{\partial x_m}Q_{jm}\boldsymbol{e}^*_i \\
&= Q_{ir}\delta_{ms}\frac{\partial \tau_{rs}}{\partial x_m}\boldsymbol{e}^*_i
\end{aligned}$$

$$= Q_{ir} \frac{\partial \tau_{rs}}{\partial x_s} e_i^*$$
$$= Q \nabla_x \cdot \tau,$$

where we have assumed that the Cauchy stress tensor τ is objective. From the above equation, we see that the term $\nabla \cdot \tau$ transforms as a vector. Thus, if Eqn. (3.19) is to have the same form in all frames, then the term $\rho(b - a)$ must also transform as a vector, i.e., we must have

$$\rho^*(b^* - a^*) = \rho Q(b - a). \tag{4.16}$$

Hence, using Eqn. (4.9), the transformation law for the body force vector is

$$b^* = Qb + 2\dot{Q}v + \ddot{Q}x + \ddot{c},$$

or, alternatively,

$$b = Q^T \left[b^* - (\dot{W} + W^2)(Qx) - 2W(Qv) - \ddot{c} \right], \tag{4.17}$$

where $W = \dot{Q}Q^T \in$ Skw. Since $W \in$ Skw, $Q^T W Q$ and $Q^T \dot{W} Q$ are also skew-symmetric tensors. Note that $Q^T W^2 Q = (Q^T W Q)(Q^T W Q)$. Using $\dot{Q} = WQ$, one can easily verify that $Q^T \dot{W} Q = \overline{Q^T W Q}^{\cdot}$. Hence, the axial vector of $Q^T \dot{W} Q$ is simply the time derivative of the axial vector of $Q^T W Q$. Thus, if we denote the axial vector of $Q^T W Q$ $(= Q^T \dot{Q})$ by Ω, Eqn. (4.17) can be written as

$$b = Q^T [b^* - \ddot{c}] - \dot{\Omega} \times x - \Omega \times (\Omega \times x) - 2\Omega \times v. \tag{4.18}$$

If w denotes the axial vector of W, it can be easily shown (see Problem 17, Chapter 1) that $\Omega = Q^T w$ is the axial vector of $Q^T W Q$, or, in other words, the components of Ω are simply the components of the vector w with respect to the unstarred coordinate frame. Since $Q_{ij} = e_i^* \cdot e_j$, we have (noting that the starred frame is stationary with respect to itself)

$$[Q^T W Q]_{ij} = [Q^T \dot{Q}]_{ij} = (e_k^* \cdot e_i)(e_k^* \cdot \dot{e}_j) = [(e_i \cdot e_k^*) e_k^*] \cdot \dot{e}_j = e_i \cdot \dot{e}_j,$$

so that

$$\Omega = \begin{bmatrix} \dot{e}_2 \cdot e_3 \\ \dot{e}_3 \cdot e_1 \\ \dot{e}_1 \cdot e_2 \end{bmatrix}, \qquad w = Q\Omega. \tag{4.19}$$

In terms of the Euler angle representation of Q given by Eqn. (1.108), the axial vector corresponding to W is

$$w = \begin{bmatrix} \dot{\theta} \cos\psi + \dot{\phi} \sin\theta \sin\psi \\ \dot{\theta} \sin\psi - \dot{\phi} \sin\theta \cos\psi \\ \dot{\psi} + \dot{\phi} \cos\theta \end{bmatrix}. \tag{4.20}$$

The axial vectors of $Q^T W Q$ and $Q^T \dot{W} Q^T$ are given by

$$\Omega = w_{Q^T W Q} = \begin{bmatrix} \dot{\theta}\cos\phi + \dot{\psi}\sin\theta\sin\phi \\ -\dot{\theta}\sin\phi + \dot{\psi}\sin\theta\cos\phi \\ \dot{\phi} + \dot{\psi}\cos\theta \end{bmatrix}, \tag{4.21a}$$

$$\dot{\Omega} = w_{Q^T \dot{W} Q} = \begin{bmatrix} \frac{d}{dt}(\dot{\theta}\cos\phi + \dot{\psi}\sin\theta\sin\phi) \\ \frac{d}{dt}(-\dot{\theta}\sin\phi + \dot{\psi}\sin\theta\cos\phi) \\ \frac{d}{dt}(\dot{\phi} + \dot{\psi}\cos\theta) \end{bmatrix}. \tag{4.21b}$$

If we assume the '$*$' frame to be 'stationary' with a body force $b^* = -ge_z$, and the unstarred frame to be rotating and translating with respect to the '$*$' frame, then we see from Eqn. (4.18) that b now includes among other forces, the centrifugal and Coriolis forces. This is consistent with experience, since we do feel an outward force when sitting in a vehicle going around a curve, or a backward push when a vehicle suddenly accelerates forward. When Q and $\dot{c}$ are constant, these 'extra' forces vanish, i.e., $b^* = Qb$, and such a change of frame is called a *Galilean transformation*.

As a simple example, consider the disc shown in Fig. 4.2b rotating with a constant angular velocity $\Omega = \Omega e_z$, and with the origin of the rotating frame of reference coinciding with the center of the disc. Assume that the mass on the disc has reached a steady-state position with respect to the rotating frame of reference, i.e., $v = 0$. Since c is now zero at all times, and since $\dot{\Omega} = 0$, the only term that contributes to the in-plane body forces in Eqn. (4.18) is the centrifugal body force term. Since $e_1 = (\cos\theta, \sin\theta, 0)$, $e_2 = (-\sin\theta, \cos\theta, 0)$, $e_3 = (0, 0, 1)$, from Eqn. (4.19), we get $\Omega = \dot{\theta} e_z$, and the term $-\Omega \times (\Omega \times x)$ evaluates to $(R + \delta)\Omega^2 e_1$, which is consistent with the outward force that the body experiences on the spinning disc.

We now consider more complicated examples. The spinning disc of radius R shown in Fig. 4.4 spins about its own axis with an angular velocity γ, and the rod of circular cross-section to which it is attached spins about the Y-axis with an angular velocity β. The xyz coordinate frame (assumed to be the unstarred frame) is fixed to the spinning disk, and is given to be parallel to the XYZ coordinate frame (the '$*$'-frame) at $t = 0$. Both γ and β are given to be constant. We are interested in computing the body forces at a point $x = (R, 0, 0)$ with respect to the xyz system, assuming that the body force with respect to XYZ is $b^* = (0, 0, -g)$. The matrix Q has components $Q_{ij} = e_i^* \cdot e_j$, which in the case of this example, can be written as $Q_{ij} = (e_i^* \cdot \hat{e}_m^*)\hat{e}_m^* \cdot e_j = Q_{im}Q_{mj}$, i.e., as a product of the Q matrices corresponding to the rotations β and γ, respectively. The vector c is the position vector of the origin of xyz with respect to XYZ. Thus, we get

$$\begin{aligned} Q &= \begin{bmatrix} \cos\beta t & 0 & \sin\beta t \\ 0 & 1 & 0 \\ -\sin\beta t & 0 & \cos\beta t \end{bmatrix} \begin{bmatrix} \cos\gamma t & -\sin\gamma t & 0 \\ \sin\gamma t & \cos\gamma t & 0 \\ 0 & 0 & 1 \end{bmatrix} \\ &= \begin{bmatrix} \cos\beta t\cos\gamma t & -\cos\beta t\sin\gamma t & \sin\beta t \\ \sin\gamma t & \cos\gamma t & 0 \\ -\sin\beta t\cos\gamma t & \sin\beta t\sin\gamma t & \cos\beta t \end{bmatrix}, \end{aligned} \tag{4.22}$$

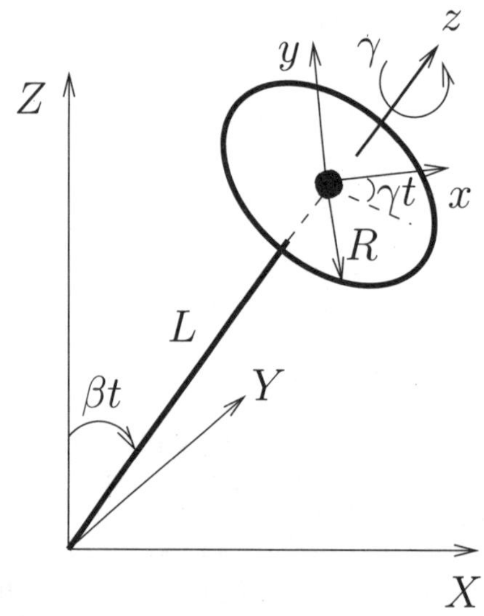

Fig. 4.4 *Body forces experienced by a point on a spinning disc.*

$$c = \begin{bmatrix} L\sin\beta t \\ 0 \\ L\cos\beta t \end{bmatrix}.$$

By means of Eqn. (4.19) (or by comparing the entries of the Q matrix with the Euler angle representation given by Eqn. (1.108), observing that $\theta = \beta t$, $\psi = \pi/2$, and $\phi = \gamma t - \pi/2$, and substituting these values into Eqns. (4.20)–(4.21b)), we get,

$$w = \begin{bmatrix} \gamma\sin\beta t \\ \beta \\ \gamma\cos\beta t \end{bmatrix}, \quad \Omega = \begin{bmatrix} \beta\sin\gamma t \\ \beta\cos\gamma t \\ \gamma \end{bmatrix}, \quad \dot{\Omega} = \begin{bmatrix} \beta\gamma\cos\gamma t \\ -\beta\gamma\sin\gamma t \\ 0 \end{bmatrix}. \tag{4.23}$$

Note that w and Ω are the angular velocities of the xyz frame expressed with respect to the XYZ and xyz frames, respectively. Since the point $x = (R, 0, 0)$ is fixed with respect to xyz, we have $v = 0$. Substituting these quantities into Eqn. (4.18), we get the required body force as

$$b = \begin{bmatrix} g\sin\beta t\cos\gamma t + R\beta^2\cos^2\gamma t + R\gamma^2 \\ -g\sin\beta t\sin\gamma t - R\beta^2\sin\gamma t\cos\gamma t \\ -g\cos\beta t + L\beta^2 - 2R\gamma\beta\sin\gamma t \end{bmatrix}.$$

Note that the body forces experienced by the point are oscillatory!

As another example, consider the flow through a pipe of radius R that is spinning about its own axis with a constant angular speed Ω_a, and about the Z-axis with constant angular speed Ω_p, as shown in Fig. 4.5.

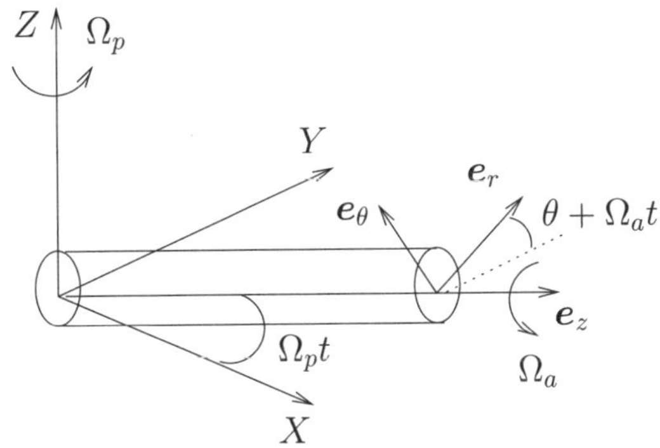

Fig. 4.5 *Flow in a pipe spinning about two axes; the origins of the e_r–e_θ–e_z and XYZ frames are shown separated simply for clarity, although actually they coincide.*

The vector c is zero, and the matrix $Q = [e_1 \mid e_2 \mid e_3]$ is given by

$$Q = \begin{bmatrix} -\sin(\Omega_p t)\cos(\Omega_a t + \theta) & \sin(\Omega_p t)\sin(\Omega_a t + \theta) & \cos(\Omega_p t) \\ \cos(\Omega_p t)\cos(\Omega_a t + \theta) & -\cos(\Omega_p t)\sin(\Omega_a t + \theta) & \sin(\Omega_p t) \\ \sin(\Omega_a t + \theta) & \cos(\Omega_a t + \theta) & 0 \end{bmatrix},$$

so that by means of Eqn. (4.19), we get

$$\Omega = \begin{bmatrix} \Omega_p \sin(\Omega_a t + \theta) \\ \Omega_p \cos(\Omega_a t + \theta) \\ \Omega_a \end{bmatrix}, \quad \dot{\Omega} = \begin{bmatrix} \Omega_p \Omega_a \cos(\Omega_a t + \theta) \\ -\Omega_p \Omega_a \sin(\Omega_a t + \theta) \\ 0 \end{bmatrix}.$$

Note again that Ω is simply the angular velocity of the pipe expressed with respect to the e_r–e_θ–e_z frame. The centrifugal term in the body force, $\Omega \times (\Omega \times x) = -\nabla |\Omega \times x|^2 /2$, can be absorbed into the ∇p term in Eqn. (7.104). Using Eqn. (4.18), the body force components (excluding the centrifugal force) at a point $x = re_r + ze_z$ in the rotating frame are

$$\begin{aligned} b_r &= -\sin(\Omega_a t + \theta)g + \Omega_p \Omega_a z \sin(\Omega_a t + \theta) - 2\Omega_p v_z \cos(\Omega_a t + \theta) + 2\Omega_a v_\theta, \\ b_\theta &= -\cos(\Omega_a t + \theta)g + \Omega_p \Omega_a z \cos(\Omega_a t + \theta) + 2\Omega_p v_z \sin(\Omega_a t + \theta) - 2\Omega_a v_r, \\ b_z &= -\Omega_p \Omega_a r \sin(\Omega_a t + \theta) - 2\Omega_p v_\theta \sin(\Omega_a t + \theta) + 2\Omega_p v_r \cos(\Omega_a t + \theta), \end{aligned}$$

where (v_r, v_θ, v_z) are the velocity components of the fluid particle at x as measured with respect to the rotating frame. Substituting the above body force components into Eqns. (A.2) of Appendix A, Vol. II, and solving these equations in conjunction with the other governing equations (such as the continuity equation), subject to the boundary condition $v|_{r=R} = \mathbf{0}$, and the given initial condition, we can solve for the velocity field of the fluid in the pipe. Note that the Coriolis terms can have a significant influence on the solution.

We have already seen that provided the balance of linear momentum holds, the balance of angular momentum is equivalent to the symmetry of the stress tensor. Thus, the balance of angular momentum is frame-indifferent (in the sense that it has the same form in any arbitrary frame) since the balance of linear momentum is frame-indifferent and since

$$(\boldsymbol{\tau}^*)^T = \boldsymbol{Q}(t)\boldsymbol{\tau}^T\boldsymbol{Q}(t)^T = \boldsymbol{Q}(t)\boldsymbol{\tau}\boldsymbol{Q}(t)^T = \boldsymbol{\tau}^*.$$

We now show that balance of energy coupled with the axiom of frame-indifference is equivalent to all the balance laws that have been derived so far! As usual, let $\rho(\boldsymbol{x},t)$, $\boldsymbol{t}(\boldsymbol{x},t,\boldsymbol{n})$, $\boldsymbol{b}(\boldsymbol{x},t)$, $e(\boldsymbol{x},t)$, $Q_h(\boldsymbol{x},t)$ be the density, surface tractions, body force per unit mass, the internal energy per unit mass, and the heat supply per unit mass, respectively. In addition, let $\gamma(\boldsymbol{x},t,\boldsymbol{n})$ represent the heat flux across a surface with unit normal $\boldsymbol{n}$. Note that $\gamma(\boldsymbol{x},t,\boldsymbol{n})$ depends on the normal to the surface across which the heat conduction is taking place, and is, thus, analogous to the traction term $\boldsymbol{t}(\boldsymbol{x},t,\boldsymbol{n})$.

As we have seen in Eqn. (3.75), the first law of thermodynamics states that the rate of increase of total energy (kinetic and internal) of any portion of the body equals the rate of work done on that portion by the body forces and surface tractions, plus the rate of increase of heat energy from the heat source, and the heat flow across the boundary, i.e.,

$$\frac{d}{dt}\int_{V(t)} \rho\left(e+\frac{1}{2}\boldsymbol{v}\cdot\boldsymbol{v}\right)dV = \int_{S(t)} (\boldsymbol{t}\cdot\boldsymbol{v}+\gamma)\,dS + \int_{V(t)} \rho(\boldsymbol{b}\cdot\boldsymbol{v}+Q_h)\,dV. \tag{4.24}$$

We assume that under a change of observer, the following hold:

$$\rho^*(\boldsymbol{x}^*,t) = \rho(\boldsymbol{x},t), \quad e^*(\boldsymbol{x}^*,t) = e(\boldsymbol{x},t), \quad \gamma^*(\boldsymbol{x}^*,t,\boldsymbol{n}^*) = \gamma(\boldsymbol{x},t),$$
$$Q_h^*(\boldsymbol{x}^*,t) = Q_h(\boldsymbol{x},t), \quad \boldsymbol{t}^*(\boldsymbol{x}^*,t,\boldsymbol{n}^*) = \boldsymbol{Q}\boldsymbol{t}(\boldsymbol{x},t,\boldsymbol{n}), \quad (\boldsymbol{b}^*-\boldsymbol{a}^*) = \boldsymbol{Q}(\boldsymbol{b}-\boldsymbol{a}),$$

with the transformations relations for $\boldsymbol{v}^*$ and $\boldsymbol{a}^*$ given by Eqns. (4.8) and (4.9). Then, we have the following fascinating result:

Theorem 4.3.1. *Let the balance of energy given by Eqn. (4.24) hold. Assume that under a change of observer, the form of Eqn. (4.24) still holds. Then there exist functions* $\boldsymbol{\tau}$ *and* $\boldsymbol{q}$ *such that* $\boldsymbol{t} = \boldsymbol{\tau}\boldsymbol{n}$ *and* $\gamma = -\boldsymbol{q}\cdot\boldsymbol{n}$, *and conservation of mass, balance of linear momentum and balance of angular momentum all hold.*

Conversely, if there exists a function $\boldsymbol{\tau}$ *such that* $\boldsymbol{t} = \boldsymbol{\tau}\boldsymbol{n}$, *and the balance principles all hold, then the balance of energy has the same form under a change of observer.*

Proof. The proof that we present is similar to that in [209]. Let

$$f := \rho\left(e+\frac{1}{2}\boldsymbol{v}\cdot\boldsymbol{v}\right).$$

Then Eqn. (4.24) can be written using transport theorem I (note that transport theorem II cannot be used since conservation of mass is not assumed to hold) as

$$\int_{V(t)}\left(\frac{Df}{Dt}+f\boldsymbol{\nabla}\cdot\boldsymbol{v}\right)dV = \int_{S(t)}(\boldsymbol{t}\cdot\boldsymbol{v}+\gamma)\,dS + \int_{V(t)}\rho(\boldsymbol{b}\cdot\boldsymbol{v}+Q_h)\,dV. \tag{4.25}$$

If Eqn. (4.25) holds under a change of observer, then, since $\rho^* = \rho$, $Q_h^* = Q_h$ and $\gamma^* = \gamma$, $J^* = J$ (so that $dV^* = dV$),

$$\int_{V(t)} \left(\frac{Df^*}{Dt} + f^* \boldsymbol{\nabla}_{\boldsymbol{x}^*} \cdot \boldsymbol{v}^* \right) dV = \int_{S(t)} (\boldsymbol{t}^* \cdot \boldsymbol{v}^* + \gamma)\, dS + \int_{V(t)} \rho(\boldsymbol{b}^* \cdot \boldsymbol{v}^* + Q_h)\, dV, \tag{4.26}$$

where $f^* = \rho^*(e^* + \boldsymbol{v}^* \cdot \boldsymbol{v}^*/2)$. Subtracting Eqn. (4.25) from Eqn. (4.26), and using Eqn. (4.13), we have

$$\begin{aligned}\int_{V(t)} \left[\frac{D}{Dt}(f^* - f) + (f^* - f)\boldsymbol{\nabla} \cdot \boldsymbol{v} \right] dV &= \int_{S(t)} (\boldsymbol{t}^* \cdot \boldsymbol{v}^* - \boldsymbol{t} \cdot \boldsymbol{v})\, dS \\ &+ \int_{V(t)} \rho(\boldsymbol{b}^* \cdot \boldsymbol{v}^* - \boldsymbol{b} \cdot \boldsymbol{v})\, dV. \end{aligned} \tag{4.27}$$

Using the expressions for f^* and f, the left-hand side of the above equation is given by

$$\begin{aligned}\text{LHS} &= \int_{V(t)} \left\{ \frac{1}{2}(\boldsymbol{v}^* \cdot \boldsymbol{v}^* - \boldsymbol{v} \cdot \boldsymbol{v}) \left[\frac{D\rho}{Dt} + \rho \boldsymbol{\nabla} \cdot \boldsymbol{v} \right] + \frac{\rho}{2} \frac{D}{Dt}(\boldsymbol{v}^* \cdot \boldsymbol{v}^* - \boldsymbol{v} \cdot \boldsymbol{v}) \right\} dV \\ &= \int_{V(t)} \left\{ \left[\boldsymbol{Q}\boldsymbol{v} + \frac{1}{2}(\dot{\boldsymbol{Q}}\boldsymbol{x} + \dot{\boldsymbol{c}}) \right] \cdot (\dot{\boldsymbol{Q}}\boldsymbol{x} + \dot{\boldsymbol{c}}) \left[\frac{D\rho}{Dt} + \rho \boldsymbol{\nabla} \cdot \boldsymbol{v} \right] + \rho(\boldsymbol{v}^* \cdot \boldsymbol{a}^* - \boldsymbol{v} \cdot \boldsymbol{a}) \right\} dV. \end{aligned}$$

Substituting $\boldsymbol{t}^* = \boldsymbol{Q}\boldsymbol{t}$, $(\boldsymbol{b}^* - \boldsymbol{a}^*) = \boldsymbol{Q}(\boldsymbol{b} - \boldsymbol{a})$, and the above formula into Eqn. (4.27), we get

$$\begin{aligned}\int_{V(t)} \left\{ \left[\boldsymbol{Q}\boldsymbol{v} + \frac{1}{2}(\dot{\boldsymbol{Q}}\boldsymbol{x} + \dot{\boldsymbol{c}}) \right] \cdot (\dot{\boldsymbol{Q}}\boldsymbol{x} + \dot{\boldsymbol{c}}) \left[\frac{D\rho}{Dt} + \rho \boldsymbol{\nabla} \cdot \boldsymbol{v} \right] + \rho[\boldsymbol{Q}(\boldsymbol{a} - \boldsymbol{b}) \cdot (\dot{\boldsymbol{Q}}\boldsymbol{x} + \dot{\boldsymbol{c}})] \right\} dV \\ = \int_{S(t)} \boldsymbol{Q}\boldsymbol{t} \cdot (\dot{\boldsymbol{Q}}\boldsymbol{x} + \dot{\boldsymbol{c}})\, dS. \end{aligned} \tag{4.28}$$

Eqn. (4.28) holds for all $\boldsymbol{Q}(t) \in \text{Orth}^+$, and $\dot{\boldsymbol{c}} \in V$. Choose $\boldsymbol{Q} \equiv \boldsymbol{I}$ ($\dot{\boldsymbol{Q}} = \boldsymbol{0}$), and $\dot{\boldsymbol{c}} =$ constant. Then, Eqn. (4.28) reduces to

$$\int_{V(t)} \left\{ \left(\boldsymbol{v} \cdot \dot{\boldsymbol{c}} + \frac{1}{2} \dot{\boldsymbol{c}} \cdot \dot{\boldsymbol{c}} \right) \left[\frac{D\rho}{Dt} + \rho \boldsymbol{\nabla} \cdot \boldsymbol{v} \right] + \rho(\boldsymbol{a} - \boldsymbol{b}) \cdot \dot{\boldsymbol{c}} \right\} dV = \int_{S(t)} \boldsymbol{t} \cdot \dot{\boldsymbol{c}}\, dS.$$

Now, using an argument analogous to that in Theorem 3.5.1 yields the existence of a tensor $\boldsymbol{\tau}$ such that $\boldsymbol{t} = \boldsymbol{\tau}\boldsymbol{n}$. Substituting this result into Eqn. (4.25), and again using a similar argument yields the existence of a vector function $\boldsymbol{q}$ such that $\gamma = -\boldsymbol{q} \cdot \boldsymbol{n}$. Since $\boldsymbol{n}^* = \boldsymbol{Q}(t)\boldsymbol{n}$, and since γ is objective, we get $\boldsymbol{q}^* = \boldsymbol{Q}(t)\boldsymbol{q}$.

Using $\boldsymbol{t} = \boldsymbol{\tau}\boldsymbol{n}$ and the divergence theorem, the right-hand side of Eqn. (4.28) can be simplified as

$$\int_{S(t)} \boldsymbol{Q}\boldsymbol{t} \cdot (\dot{\boldsymbol{Q}}\boldsymbol{x} + \dot{\boldsymbol{c}})\, dS = \int_{S(t)} \boldsymbol{Q}\boldsymbol{\tau}\boldsymbol{n} \cdot (\dot{\boldsymbol{Q}}\boldsymbol{x} + \dot{\boldsymbol{c}})\, dS$$

$$
\begin{aligned}
&= \int_{S(t)} \boldsymbol{\tau}^T \boldsymbol{Q}^T (\dot{\boldsymbol{Q}}\boldsymbol{x} + \dot{\boldsymbol{c}}) \cdot \boldsymbol{n}\, dS \\
&= \int_{V(t)} \boldsymbol{\nabla} \cdot [\boldsymbol{\tau}^T \boldsymbol{Q}^T (\dot{\boldsymbol{Q}}\boldsymbol{x} + \dot{\boldsymbol{c}})]\, dV \\
&= \int_{V(t)} (\boldsymbol{\nabla} \cdot \boldsymbol{\tau}) \cdot \boldsymbol{Q}^T (\dot{\boldsymbol{Q}}\boldsymbol{x} + \dot{\boldsymbol{c}})\, dV + \int_{V(t)} \boldsymbol{\tau} : \boldsymbol{Q}^T \dot{\boldsymbol{Q}}\, dV \\
&= \int_{V(t)} \boldsymbol{Q}(\boldsymbol{\nabla} \cdot \boldsymbol{\tau}) \cdot (\dot{\boldsymbol{Q}}\boldsymbol{x} + \dot{\boldsymbol{c}})\, dV + \int_{V(t)} \boldsymbol{\tau} : \boldsymbol{Q}^T \dot{\boldsymbol{Q}}\, dV.
\end{aligned}
$$

Substituting this result into Eqn. (4.28), we get

$$
\begin{aligned}
&\int_{V(t)} \left[\boldsymbol{Q}\boldsymbol{v} + \frac{1}{2}(\dot{\boldsymbol{Q}}\boldsymbol{x} + \dot{\boldsymbol{c}}) \right] \cdot (\dot{\boldsymbol{Q}}\boldsymbol{x} + \dot{\boldsymbol{c}}) \left[\frac{D\rho}{Dt} + \rho \boldsymbol{\nabla} \cdot \boldsymbol{v} \right] dV = \\
&\qquad \int_{V(t)} \boldsymbol{Q}[\boldsymbol{\nabla} \cdot \boldsymbol{\tau} + \rho(\boldsymbol{b} - \boldsymbol{a})] \cdot (\dot{\boldsymbol{Q}}\boldsymbol{x} + \dot{\boldsymbol{c}})\, dV + \int_{V(t)} \boldsymbol{\tau} : \boldsymbol{Q}^T \dot{\boldsymbol{Q}}\, dV. \quad (4.29)
\end{aligned}
$$

Now choosing $\boldsymbol{Q} \equiv \boldsymbol{I}$ ($\dot{\boldsymbol{Q}} = \boldsymbol{0}$), and $\dot{\boldsymbol{c}} = \lambda \boldsymbol{e}$, where $\lambda \in \Re$, and $\boldsymbol{e}$ is a unit vector, we get

$$
\int_{V(t)} \left[\lambda(\boldsymbol{v} \cdot \boldsymbol{e}) + \frac{\lambda^2}{2} \right] \left[\frac{D\rho}{Dt} + \rho \boldsymbol{\nabla} \cdot \boldsymbol{v} \right] dV = \lambda \int_{V(t)} [\boldsymbol{\nabla} \cdot \boldsymbol{\tau} + \rho(\boldsymbol{b} - \boldsymbol{a})] \cdot \boldsymbol{e}\, dV. \tag{4.30}
$$

Differentiating Eqn. (4.30) twice with respect to λ, we get

$$
\int_{V(t)} \left[\frac{D\rho}{Dt} + \rho \boldsymbol{\nabla} \cdot \boldsymbol{v} \right] dV = 0,
$$

which is valid for an arbitrary volume $V(t)$, and hence implies the conservation of mass, i.e.,

$$
\frac{D\rho}{Dt} + \rho \boldsymbol{\nabla} \cdot \boldsymbol{v} = 0. \tag{4.31}
$$

Substituting this result into Eqn. (4.30), and using the arbitrariness of $V(t)$ and $\boldsymbol{e}$, we get

$$
\boldsymbol{\nabla} \cdot \boldsymbol{\tau} + \rho(\boldsymbol{b} - \boldsymbol{a}) = \boldsymbol{0}, \tag{4.32}
$$

which is nothing but the balance of linear momentum.

For $\boldsymbol{W} \in$ Skw, let $\boldsymbol{Q}(t) = e^{\boldsymbol{W}t}$. Then $\boldsymbol{Q}^T\dot{\boldsymbol{Q}} = \boldsymbol{W}$ (since $\dot{\boldsymbol{Q}} = \boldsymbol{W}\boldsymbol{Q} = \boldsymbol{Q}\boldsymbol{W}$). Using Eqns. (4.31) and (4.32), Eqn. (4.29) reduces to

$$
\int_{V(t)} \boldsymbol{\tau} : \boldsymbol{W}\, dV = 0,
$$

which being valid for any arbitrary volume $V(t)$, and any arbitrary skew tensor $\boldsymbol{W}$, yields the fact that $\boldsymbol{\tau}$ is symmetric (see Problem 8, Chapter 1). Thus, the balance of angular momentum holds.

To prove the converse, note that since now we are given that $t = \tau n$, Eqn. (4.29) represents the difference between Eqn. (4.25) and (4.26). Given that Eqn. (4.25) holds, we want to prove that Eqn. (4.26) holds. By the above comment, it suffices to show that Eqn. (4.29) holds. Since by assumption, mass conservation, balance of linear momentum and balance of angular momentum hold, and since $Q^T \dot{Q} \in \text{Skw}$, Eqn. (4.29) holds, and thus the converse is proved. □

A methodology similar to the above is used to derive the governing equations for liquid crystals. See [73], [185]–[187] and [313] for more details.

From Eqn. (3.64), we see that since $\tau : D$ (which is also equal to $T : \dot{F}/J$ and $S : \dot{E}/J$) is frame-indifferent, the quantity

$$\int_{S(t)} t \cdot v \, dS + \int_{V(t)} \rho (b - a) \cdot v \, dV,$$

which is nothing but the mechanical power minus the rate of change of kinetic energy of the material volume (see Eqn. (3.65)), is also frame-indifferent. Thus, the mechanical power itself is *not* frame-indifferent. This is also evident from Eqn. (4.24), since e, γ and Q_h are frame-indifferent.

Though the stress tensor is postulated to be objective, the stress rate is not objective in general, as can be seen by taking the material derivative of Eqn. (4.15):

$$\frac{D\tau^*}{Dt} = \dot{Q}\tau Q^T + Q\dot{\tau}Q^T + Q\tau\dot{Q}^T,$$

which shows that the stress rate is objective if and only if

$$\dot{Q}\tau Q^T + Q\tau\dot{Q}^T = 0 \quad \forall Q \in \text{Orth}^+, \tag{4.33}$$

which in turn is valid if and only if $\tau(x,t) = \lambda(x,t) I$, for all x and t. To see this, let $W \in \text{Skw}$, and let $Q = e^{Wt}$. From the results of Section 1.10, we know that $\dot{Q} = WQ$. Substituting this expression into Eqn. (4.33) and noting that W is arbitrary, we get

$$WQ\tau Q^T = Q\tau Q^T W \quad \forall W \in \text{Skw},$$

By Problem 18, Chapter 1, this proves that $Q\tau Q^T$ is of the form λI, and hence that $\tau = \lambda I$.

Many rate-dependent materials need to be described in terms of stress rates, and the resulting constitutive models must be frame-indifferent. Hence, it is essential to derive stress rates that are objective. The simplest way of generating an objective stress rate uses the fact that the second Piola–Kirchhoff stress tensor remains unchanged under a change of observer, which is proved quite easily as follows:

$$\begin{aligned} S^* &= J^*(F^*)^{-1}\tau^*(F^*)^{-T} \\ &= JF^{-1}Q^{-1}Q\tau Q^T Q^{-T} F^{-T} \\ &= JF^{-1}\tau F^{-T} \\ &= S. \end{aligned}$$

Thus, we also have

$$\frac{DS^*}{Dt} = \frac{DS}{Dt}. \tag{4.34}$$

Taking the material derivative of the relation $S = JF^{-1}\tau F^{-T}$, and using Eqns. (2.50) and (2.51), we get

$$\begin{aligned}\frac{DS}{Dt} &= \dot{J}F^{-1}\tau F^{-T} + J\dot{F}^{-1}\tau F^{-T} + JF^{-1}\dot{\tau}F^{-T} + JF^{-1}\tau\dot{F}^{-T} \\ &= (\operatorname{tr} L)JF^{-1}\tau F^{-T} - JF^{-1}L\tau F^{-T} + JF^{-1}\dot{\tau}F^{-T} - JF^{-1}\tau L^T F^{-T} \\ &= JF^{-1}\left[\dot{\tau} - L\tau - \tau L^T + (\operatorname{tr} L)\tau\right]F^{-T} \\ &= JF^{-1}\tau^{\circ}F^{-T},\end{aligned} \tag{4.35}$$

where

$$\tau^{\circ} := \dot{\tau} - L\tau - \tau L^T + (\operatorname{tr} L)\tau \tag{4.36}$$

is known as the *Truesdell stress rate*. To see that it is objective, we note from Eqn. (4.35) that

$$\tau^{\circ} = \frac{1}{J}F\frac{DS}{Dt}F^T.$$

Now using Eqn. (4.34), we have

$$\begin{aligned}[\tau^{\circ}]^* &= \frac{1}{J^*}F^*\frac{DS^*}{Dt}(F^*)^T \\ &= \frac{1}{J}QF\frac{DS}{Dt}F^TQ^T \\ &= Q\tau^{\circ}Q^T.\end{aligned}$$

The *Jaumann stress rate* (also known as the *corotational stress rate*) is another objective stress rate, which is defined by

$$\tau^{\nabla} := \dot{\tau} - \hat{W}\tau + \tau\hat{W}, \tag{4.37}$$

where $\hat{W}$ is a skew-symmetric part of L. This stress rate is obtained by substituting $L = D + \hat{W}$ into Eqn. (4.36), and shifting all terms with D to the left, i.e.,

$$\tau^{\nabla} = \tau^{\circ} + D\tau + \tau D - (\operatorname{tr} D)\tau.$$

That τ^{∇} is objective is easily seen from the above equation, since τ°, D and τ are all objective.

Writing Eqn. (4.37) as

$$\tau^{\nabla} = \dot{\tau} + \hat{W}^T\tau + \tau\hat{W},$$

substituting $\hat{W} = L - D$, and shifting all terms with D to the left, we get the *convective stress rate* given by

$$\tau^{\diamond} := \dot{\tau} + L^T\tau + \tau L. \tag{4.38}$$

Objectivity of the convective stress rate is a result of the relation

$$\tau^{\diamond} = \tau^{\nabla} + D\tau + \tau D.$$

By replacing $\hat{W}$ with $\dot{R}R^T$ (since $\dot{R}R^T$ is skew-symmetric) into Eqn. (4.37), we obtain the *Green–Naghdi stress rate*:

$$\tau^{\triangle} := \dot{\tau} - \dot{R}R^T\tau + \tau\dot{R}R^T.$$

If $\bar{E}^{\diamond}$ is defined in a manner analogous to $\tau^{\diamond}$, then from Eqns. (2.62) and (4.38), we see that

$$\bar{E}^{\diamond} = D. \tag{4.39}$$

Similar frame-indifferent rates can also be formulated for the heat flux vector. Using the result of Problem 3, Chapter 2, we see that the fluxes in the reference and deformed configurations q_0 and q, respectively, are related by $q_0 = JF^{-1}q$. Since $F^* = QF$ and $q^* = Qq$, we have $q_0^* = q_0$, which in turn implies that $\dot{q}_0^* = \dot{q}_0$. Taking the material time derivative of the relation $q_0 = JF^{-1}q$, and using Eqns. (2.50) and (2.51), we get

$$\dot{q}_0 = JF^{-1}\left[\dot{q} + (\nabla \cdot v)q - Lq\right].$$

Since $(F^{-1})^* = F^{-1}Q^T$, this implies that the bracketed term given by

$$q^{\diamond} := \frac{1}{J}F\dot{q}_0 = \dot{q} + (\nabla \cdot v)q - Lq = \left(\frac{\partial q}{\partial t}\right)_x + (\nabla q)v + (\nabla \cdot v)q - Lq \tag{4.40}$$

is objective. As in the stress case, there are an infinite number of such objective flux rates. For example, other objective rates are

$$q^{\triangle} := \dot{q} - Lq. \tag{4.41}$$

$$q^{\nabla} := \dot{q} - \hat{W}q, \tag{4.42}$$

where $\hat{W} = (L - L^T)/2$. Objective rates such as those given by Eqns. (4.40)–(4.42) can be used in a non-Fourier heat conduction law (see Eqns. (7.14) and (7.64)).

4.4 Principle of Material Frame-Indifference

Ideally, one would like to extend the principle of frame indifference to constitutive relations as well, i.e., we would like to postulate that constitutive relations have the same form in all frames of reference. However, such a postulate has not been universally accepted, and indeed, starting with the works of Muller [228], there have been a number of works that have claimed that the constitutive relations need not have the same form in all frames. This, in turn, has resulted in several works that have tried to clarify the issues involved (see, e.g., [19], [199], [316]).

To see what is at the heart of the controversy, consider again the example of a spring mounted on a rotating table with one end attached to the table, and the other to a mass (see Section 4.1). As noted by Bertram and Svendsen ([19], [316]), the first aspect of material frame-indifference, namely, *observer invariance*, merely requires that the material response function be objective, which, in the context of our example, means that two observers observe the same spring constant. Suppose that the spring constant has been found to be k_0 under stationary conditions. If now the turntable is rotated with an angular speed ω, then although observer invariance states that two observers would observe the same spring constant, that *by itself* does not rule out the possibility that the spring constant is a function of ω, i.e., one could have $k = \hat{k}(\omega)$ with $\hat{k}(0) = k_0$. Muller's example of a constitutive relation obtained from the kinetic theory of gases [228] also depends on the spin of the frame of reference, and is thus, analogous to the dependence of k on ω. From a practical viewpoint, observer invariance by itself is clearly incapable of relating $\hat{k}(\omega)$ to k_0. However, material frame-indifference as traditionally formulated (see, e.g, [108] or [337]) *implicitly assumes* that in addition to observer invariance, *form-invariance* also holds. The hypothesis of form-invariance states that the constitutive relations have the same form in all frames of reference, and hence depend only on variables intrinsic to the material; thus, for example, dependence on the rotation rate of the frame of reference is excluded. In the context of our spring example, form invariance implies that $\hat{k}(\omega) = k_0$. This hypothesis, if valid, is clearly useful from a practical viewpoint, since a designer using the value of the spring constant found under stationary conditions can *analytically* determine the extension of the spring for any ω, which otherwise would have to be determined experimentally for every ω.

If $\hat{\boldsymbol{\tau}}^*$ and $\hat{\boldsymbol{\tau}}$ represent the constitutive relations for $\boldsymbol{\tau}^*$ and $\boldsymbol{\tau}$, respectively, i.e., if $\boldsymbol{\tau}^* = \hat{\boldsymbol{\tau}}^*(\boldsymbol{l}^*)$, and $\boldsymbol{\tau} = \hat{\boldsymbol{\tau}}(\boldsymbol{l})$, where $\boldsymbol{l}$ is a list of kinematical parameters on which $\boldsymbol{\tau}$ is assumed to depend, and $\boldsymbol{l}^*$ is the same list of parameters as observed by O^*, then Eqn. (4.15) implies that

$$\hat{\boldsymbol{\tau}}^*(\boldsymbol{l}^*) = \boldsymbol{Q}(t)\hat{\boldsymbol{\tau}}(\boldsymbol{l})\boldsymbol{Q}(t)^T, \tag{4.43}$$

which is the condition of observer invariance mentioned above. As discussed, $\hat{\boldsymbol{\tau}}^*$ and $\hat{\boldsymbol{\tau}}$ could be *different* functions, so that the above relation by itself does not lead us to any conclusion about the nature of the constitutive relations (indeed, as observed by Liu [199], Eqn. (4.43) can be regarded as the *definition* of $\hat{\boldsymbol{\tau}}^*$, once $\hat{\boldsymbol{\tau}}$ is given). However, if, in addition, we assume form-invariance, i.e.,

$$\hat{\boldsymbol{\tau}}^*(\boldsymbol{l}^*) = \hat{\boldsymbol{\tau}}(\boldsymbol{l}^*), \tag{4.44}$$

then, from Eqn. (4.43), it follows that

$$\hat{\boldsymbol{\tau}}(\boldsymbol{l}^*) = \boldsymbol{Q}(t)\hat{\boldsymbol{\tau}}(\boldsymbol{l})\boldsymbol{Q}(t)^T. \tag{4.45}$$

The above condition, often alluded to as the *invariance of the constitutive relation under superposed rigid-body motion*, or, as the *principle of material frame-indifference* is a restriction that leads us in many cases to the so-called reduced form of the constitutive relations. For example, in the case of our spring example, Eqn. (4.43) does not preclude a dependence of the spring constant on the angular speed ω, but Eqn. (4.45) does. We have shown that granted observer invariance, form-invariance is equivalent to the hypothesis of invariance under superposed rigid body motion, i.e., Eqns. (4.43) and (4.44) together imply Eqn. (4.45). It is also easily seen that Eqns. (4.44) and (4.45) together imply Eqn. (4.43), and that Eqns. (4.43) and (4.45) together imply Eqn. (4.44). Thus, any two hypothesis out of observer invariance, form-invariance and invariance under superposed rigid body motions implies the third.

Note that while observer invariance involves the *same* motion as observed by *two* different observers, invariance under superposed rigid body motions involves two *different* motions as seen by *one* observer. Thus, clearly, these two notions are conceptually quite different. Muller's examples of constitutive relations are observer-invariant, but *not* form-invariant, or, equivalently, not invariant under superposed rigid body motion. Notwithstanding these examples (which, to the best of our knowledge, have not been verified experimentally), form-invariance, or, equivalently, the principle of material frame-indifference, is satisfied by a large class of materials.

Murdoch ([230]–[232]) has claimed that the restrictions on constitutive relations can be deduced by purely objective considerations without the requirement of form-invariance of constitutive relations. Liu [199] remarks that such a claim if true would be a truly remarkable result, since one can easily come up with a counterexample of a constitutive relation that satisfies the observer invariance condition given by Eqn. (4.43), but is not form-invariant, and hence does not satisfy Eqn. (4.45) such as, for example,

$$\hat{\tau}(\boldsymbol{E}) = \alpha \boldsymbol{E}, \tag{4.46}$$

where α is a material constant. By defining $\hat{\tau}^*(\boldsymbol{E}^*) := \alpha \boldsymbol{Q}(t)\boldsymbol{E}^*\boldsymbol{Q}(t)^T$ and recalling that $\boldsymbol{E}^* = \boldsymbol{E}$, we see that

$$\hat{\tau}^*(\boldsymbol{E}^*) = \boldsymbol{Q}(t)\hat{\tau}(\boldsymbol{E})\boldsymbol{Q}(t)^T.$$

Thus, the observer invariance condition given by Eqn. (4.43) is satisfied. On the other hand,

$$\hat{\tau}(\boldsymbol{E}^*) = \alpha\boldsymbol{E}^* = \alpha\boldsymbol{E} = \hat{\tau}(\boldsymbol{E}) \neq \boldsymbol{Q}(t)\hat{\tau}(\boldsymbol{E})\boldsymbol{Q}(t)^T,$$

so that the the restriction of material frame-indifference given by Eqn. (4.45), or, equivalently, that of form-invariance given by Eqn. (4.44) is not satisfied. The above controversy can be resolved if one notes that Murdoch has a condition similar to that of form invariance (condition O.4. on page 311 of [231]; since this condition is not stated in mathematical form, it is not clear whether this statement is weaker than or equivalent to form-invariance) in his definition of objectivity, while objectivity as generally understood and as interpreted by Liu [199, 200] is simply Eqn. (4.43).

Summarizing, notwithstanding claims that in principle material frame-indifference could be violated, in the absence of any *experimental evidence*, we shall henceforth in this book adopt the principle of material frame-indifference as an axiom. This condition is only a necessary condition in the sense that a constitutive relation that violates material frame-indifference such as the one in Eqn. (4.46) is *inadmissible*, but, conversely, a constitutive relation that satisfies material frame-indifference may or may not model the actual physical behavior of a material accurately.

Using Eqns. (3.38), (3.49) and (4.4), the material frame-indifference condition given by Eqn. (4.45) can be written in terms of the first and second Piola–Kirchhoff stress as

$$\hat{\boldsymbol{T}}(\boldsymbol{l}^*) = \boldsymbol{Q}(t)\hat{\boldsymbol{T}}(\boldsymbol{l}), \tag{4.47a}$$

$$\hat{\boldsymbol{S}}(\boldsymbol{l}^*) = \hat{\boldsymbol{S}}(\boldsymbol{l}). \tag{4.47b}$$

We now present some examples of constitutive relations that satisfy material frame-indifference. For a Newtonian fluid, the constitutive relation for the Cauchy stress is

$$\hat{\tau} = -p(\rho,\theta)\boldsymbol{I} + [\lambda(\rho,\theta)\operatorname{tr}\boldsymbol{D}]\boldsymbol{I} + 2\mu(\rho,\theta)\boldsymbol{D}.$$

Since $\rho^* = \rho$, $\theta^* = \theta$ and $D^* = QDQ^T$, we have $p^* = p$, $\lambda^* = \lambda$, $\mu^* = \mu$, so that $\hat{\tau}(\rho^*, \theta^*, D^*) = Q\hat{\tau}(\rho, \theta, D)Q^T$, and material frame-indifference is satisfied. On the other hand, if we were to replace D by L in the above constitutive relation, then since $L^* = QLQ^T + \dot{Q}Q^T$, material frame-indifference is violated, i.e., $\hat{\tau}(\rho^*, \theta^*, L^*) \neq Q\hat{\tau}(\rho, \theta, L)Q^T$, while, by *defining* $\hat{\tau}^*(\rho^*, \theta^*, L^*) := Q\hat{\tau}(\rho, \theta, L)Q^T$, observer invariance is satisfied. As another example, since $E^* = E$, and since the Lamé parameters λ and μ are constants, the constitutive relation $\hat{S} = \lambda(\operatorname{tr} E)I + 2\mu E$ satisfies the constraint given by Eqn. (4.47b), and hence is frame-indifferent.

4.5 Constitutive Relations for Simple Materials

We expect the stress in a body at any given instant of time to depend on what has happened to the body in the past. Also, considering only mechanical influences, we expect the stress at a time t to be determined by the history of the motion of the body upto time t. Since stress represents contact force, we further expect the stress at a point to be determined by the history of the motion in the immediate neighborhood of the point. Since a deformation is characterized locally by its deformation gradient, we are lead to define the following class of materials:

Definition: A material is said to be *simple* if for each particle X and each time t, there exists a functional such that the Cauchy stress at a point $x = \chi(X, t)$ is given by

$$\tau(x, t) = \mathop{\hat{\tau}}_{s=0}^{\infty}(X, F(X, t - s)), \tag{4.48}$$

where F is the deformation gradient, and the variable s, which varies from 0 to ∞, has been introduced to represent the history dependence that the constitutive relation of a material might have. The value $s = 0$ corresponds to the present time t, while large s gives the distant past. This function, which is a *material property* is called the *constitutive equation* of the simple material. Thus, the stress at a point is determined by the history of the deformation gradient relative to some configuration at that point. The adjective 'simple' is used to indicate that the deformation gradient is the only characteristic of the motion involved in the constitutive equation.

Note that we have assumed that the constitutive relation for a simple material is not dependent explicitly on time. Thus, we are assuming that aging does not occur. Also note that the constitutive relation does not depend on the motion χ, since, otherwise, the Cauchy stress would vary if the body is rigidly translated. In addition, the stress at a point x is assumed not to depend on the deformation gradient $F(Y, t)$ evaluated at all other points Y of the domain, as well as on higher-order gradients, e.g., $\partial^2\chi_k/\partial X_i \partial X_j$.

In spite of the simplicity of this model, we shall find that most constitutive models currently in use fall in this category. For example, a material is said to be *elastic* if the history dependence in Eqn. (4.48) is excluded. In this case the Cauchy stress depends only on the current value of the deformation gradient, i.e.,

$$\tau(x, t) = \hat{\tau}(X, F(X, t)), \quad x = \chi(X, t). \tag{4.49}$$

We note, however, that the constitutive relation given by Eqn. (4.49) may be taken to be a good approximation only for 'small' values of F; for large values of F, plasticity or fracture may occur invalidating this relation.

As another example, we shall see that the constitutive relation for a Newtonian fluid depends on the velocity gradient L, which can be expressed as $\dot{F}F^{-1}$. Thus, we see that due to the presence of the term $\dot{F}$, the stress at a given instant of time is influenced by the history of the deformation, even though not a significant amount of such history might need to be considered.

By virtue of the relations of the first and second Piola–Kirchhoff stresses with the Cauchy stress, we also have the following equations corresponding to Eqn. (4.48):

$$T(X,t) = \mathop{\hat{T}}_{s=0}^{\infty}(X, F(X,t-s)),$$
$$S(X,t) = \mathop{\hat{S}}_{s=0}^{\infty}(X, F(X,t-s)).$$

In what follows, we will often make the assumption that the body is homogeneous or isotropic. A *homogeneous* body is one that has identical properties at all points. A body is *isotropic* with respect to certain properties if these properties are identical in all directions. These two properties of homogeneity and isotropy are completely independent of each other, i.e., one could have a non-homogeneous isotropic body or a non-isotropic homogeneous body.

For a homogeneous body, the explicit dependence of the constitutive relations on X can be dropped. Thus, for a homogeneous body, we have

$$\tau(x,t) = \mathop{\hat{\tau}}_{s=0}^{\infty}(F(X,t-s)).$$

Homogeneity is a property satisfied in a given reference configuration, which may no longer hold true if a 'deformed configuration' is chosen as the reference configuration. This same statement holds even for isotropy, as we shall see later.

We now consider the restrictions imposed by material frame-indifference on the constitutive relations of a simple material. By Eqn. (4.3) we know that

$$\chi^*(X,t^*) = Q(t)\chi(X,t) + c(t),$$

where $t^* = t - a$. We need to find $F^*(X, t^* - s)$ associated with this motion. We have

$$\begin{aligned} F^*(X,t^*) &= Q(t)\nabla\chi(X,t) \\ &= Q(t)F(X,t) \\ &= Q(t^* + a)F(X, t^* + a). \end{aligned}$$

Consequently,

$$\begin{aligned} F^*(X,t^*-s) &= Q(t^*+a-s)F(X,t^*+a-s) \\ &= Q(t-s)F(X,t-s). \end{aligned} \tag{4.50}$$

The list of kinematical variables in Eqn. (4.45) comprises only of the deformation gradient, while the response function for the Cauchy stress is as given by Eqn. (4.48). Thus, by virtue of Eqn (4.50), the response function for the Cauchy stress satisfies material frame-indifference if and only if

$$\mathop{\hat{\tau}}_{s=0}^{\infty}(X, F^*(X,t^*-s)) = \mathop{\hat{\tau}}_{s=0}^{\infty}(X, Q(t-s)F(X,t-s))$$

$$= Q(t) \mathop{\hat{\tau}}_{s=0}^{\infty}(X, F(X, t-s))Q(t)^T.$$

Since for any $F \in \text{Lin}^+$, we can find a motion χ such that $\nabla\chi = F$ (the motion $\chi = FX + c(t)$ does the job), we can write the above equation as

$$\mathop{\hat{\tau}}_{s=0}^{\infty}(X, Q(t-s)F(X, t-s)) = Q(t) \mathop{\hat{\tau}}_{s=0}^{\infty}(X, F(X, t-s))Q(t)^T \quad \forall F \in \text{Lin}^+, Q \in \text{Orth}^+, \tag{4.51}$$

We now have the following theorem:

Theorem 4.5.1. *Let $\hat{\tau} : V_0 \times Lin^+ \to Sym$ be the response function for the Cauchy stress. Then the following are equivalent:*

(i) $\hat{\tau}$ satisfies the axiom of material frame-indifference, i.e., Eqn. (4.51) holds for all $X \in V_0$, and for all $t \in \Re$.

(ii) For all $X \in V_0$, and for all $t \in \Re$,

$$\mathop{\hat{\tau}}_{s=0}^{\infty}(X, F(X, t-s)) = R(X, t) \mathop{\hat{\tau}}_{s=0}^{\infty}(X, U(X, t-s))R(X, t)^T \; \forall F = RU \in Lin^+. \tag{4.52}$$

(iii) There exists a mapping $\tilde{S} : V_0 \times Psym \to Sym$, such that, for all $X \in V_0$, and for all $t \in \Re$,

$$\mathop{\hat{S}}_{s=0}^{\infty}(X, F(X, t-s)) = \mathop{\tilde{S}}_{s=0}^{\infty}(X, F^T F(X, t-s)) \quad \forall F \in Lin^+, \tag{4.53}$$

where $\mathop{\hat{S}}_{s=0}^{\infty}$ is the response function for the second Piola–Kirchhoff stress.

Proof. We shall show that (i) $\implies$ (ii) $\implies$ (iii) $\implies$ (i). Let $F(X, t) = R(X, t)U(X, t)$ be the polar decomposition of F. We then have $F(X, t-s) = R(X, t-s)U(X, t-s)$. Since $R(X, t-s) \in \text{Orth}^+$, we choose $Q(t-s) \equiv R(X, t-s)$ (we can make this choice since X is fixed), and since we also have $U \in \text{Lin}^+$, Eqn. (4.51) implies that

$$\begin{aligned} \mathop{\hat{\tau}}_{s=0}^{\infty}(X, F(X, t-s)) &= \mathop{\hat{\tau}}_{s=0}^{\infty}(X, R(X, t-s)U(X, t-s)) \\ &= R(X, t) \mathop{\hat{\tau}}_{s=0}^{\infty}(X, U(X, t-s))R(X, t)^T. \end{aligned}$$

To show that (ii) implies (iii), note that the above equation can be written as

$$\mathop{\hat{\tau}}_{s=0}^{\infty}(X, F(X, t-s)) = FU^{-1}(X, t) \mathop{\hat{\tau}}_{s=0}^{\infty}(X, U(X, t-s))U^{-1}F(X, t)^T,$$

so that

$$\mathop{\hat{S}}_{s=0}^{\infty}(X, F(X, t-s)) = (\det F)F^{-1} \left[\mathop{\hat{\tau}}_{s=0}^{\infty}(X, F(X, t-s)) \right] F^{-T}$$

$$= \overset{\infty}{\underset{s=0}{\tilde{\mathbf{S}}}}(\mathbf{X}, \mathbf{F}^T\mathbf{F}(\mathbf{X}, t-s)),$$

where

$$\overset{\infty}{\underset{s=0}{\tilde{\mathbf{S}}}}(\mathbf{X}, \mathbf{C}(\mathbf{X}, t-s)) := (\det \mathbf{U})\mathbf{U}^{-1}(\mathbf{X}, t) \overset{\infty}{\underset{s=0}{\hat{\boldsymbol{\tau}}}}(\mathbf{X}, \mathbf{U}(\mathbf{X}, t-s))\mathbf{U}^{-1}(\mathbf{X}, t),$$

$$\mathbf{U} = \sqrt{\mathbf{C}}, \ \forall \mathbf{C} \in \text{Psym}.$$

Note that we have written the above relation for all $\mathbf{C} \in$ Psym, since for any $\mathbf{C} \in$ Psym, we have $\mathbf{C} = \mathbf{U}\mathbf{U} = \mathbf{U}^T\mathbf{U}$, where $\mathbf{U} \in \text{Psym} \subset \text{Lin}^+$.

Finally, (iii) implies (i) since

$$\begin{aligned}
\overset{\infty}{\underset{s=0}{\hat{\boldsymbol{\tau}}}}(\mathbf{X}, \mathbf{Q}(t-s)\mathbf{F}(\mathbf{X}, t-s)) &= \frac{1}{\det(\mathbf{Q}\mathbf{F})}\mathbf{Q}(t)\mathbf{F}(\mathbf{X}, t) \overset{\infty}{\underset{s=0}{\hat{\mathbf{S}}}}(\mathbf{X}, \mathbf{Q}(t-s)\mathbf{F}(\mathbf{X}, t-s))\mathbf{F}(\mathbf{X}, t)^T\mathbf{Q}(t)^T \\
&= (\det \mathbf{F})^{-1}\mathbf{Q}(t)\mathbf{F}(\mathbf{X}, t)\left[\overset{\infty}{\underset{s=0}{\tilde{\mathbf{S}}}}(\mathbf{X}, \mathbf{F}^T\mathbf{F}(\mathbf{X}, t-s))\right]\mathbf{F}(\mathbf{X}, t)^T\mathbf{Q}(t)^T \\
&= \mathbf{Q}(t)\left[(\det \mathbf{F})^{-1}\mathbf{F}(\mathbf{X}, t) \overset{\infty}{\underset{s=0}{\hat{\mathbf{S}}}}(\mathbf{X}, \mathbf{F}(\mathbf{X}, t-s))\mathbf{F}(\mathbf{X}, t)^T\right]\mathbf{Q}(t)^T \\
&= \mathbf{Q}(t) \overset{\infty}{\underset{s=0}{\hat{\boldsymbol{\tau}}}}(\mathbf{X}, \mathbf{F}(\mathbf{X}, t-s))\mathbf{Q}(t)^T. \qquad \square
\end{aligned}$$

Using Eqns. (4.47), we get the following restrictions on the constitutive relations for the first and second Piola–Kirchhoff stress:

$$\overset{\infty}{\underset{s=0}{\hat{\mathbf{T}}}}(\mathbf{X}, \mathbf{Q}(t-s)\mathbf{F}(\mathbf{X}, t-s)) = \mathbf{Q}(t) \overset{\infty}{\underset{s=0}{\hat{\mathbf{T}}}}(\mathbf{X}, \mathbf{F}(\mathbf{X}, t-s)) \ \forall \mathbf{F} \in \text{Lin}^+, \mathbf{Q} \in \text{Orth}^+, \tag{4.54}$$

$$\overset{\infty}{\underset{s=0}{\hat{\mathbf{S}}}}(\mathbf{X}, \mathbf{Q}(t-s)\mathbf{F}(\mathbf{X}, t-s)) = \overset{\infty}{\underset{s=0}{\hat{\mathbf{S}}}}(\mathbf{X}, \mathbf{F}(\mathbf{X}, t-s)) \ \forall \mathbf{F} \in \text{Lin}^+, \mathbf{Q} \in \text{Orth}^+. \tag{4.55}$$

It is clear that the response functions $\overset{\infty}{\underset{s=0}{\hat{\mathbf{T}}}}$ and $\overset{\infty}{\underset{s=0}{\hat{\mathbf{S}}}}$ satisfy the axiom of material frame-indifference if and only if the above relations are satisfied. Equation (4.52), known as Richter's theorem, says that the response function $\hat{\boldsymbol{\tau}}$ is completely determined by its restriction to Psym. Another form of Richter's theorem that is useful is obtained by defining

$$\overset{\infty}{\underset{s=0}{\tilde{\boldsymbol{\tau}}}}(\mathbf{X}, \mathbf{C}(\mathbf{X}, t-s)) := \mathbf{U}^{-1}(\mathbf{X}, t) \overset{\infty}{\underset{s=0}{\hat{\boldsymbol{\tau}}}}(\mathbf{X}, \mathbf{U}(\mathbf{X}, t-s))\mathbf{U}^{-1}(\mathbf{X}, t), \quad \mathbf{C} = \mathbf{U}^2.$$

Then, we have the result that the response function $\overset{\infty}{\underset{s=0}{\tilde{\boldsymbol{\tau}}}}(\mathbf{X}, \mathbf{C}(\mathbf{X}, t-s))$ is frame-indifferent if and only if the stress is given by the constitutive relation

$$\boldsymbol{\tau}(\mathbf{x}, t) = \mathbf{F}(\mathbf{X}, t) \overset{\infty}{\underset{s=0}{\tilde{\boldsymbol{\tau}}}}(\mathbf{X}, \mathbf{C}(\mathbf{X}, t-s))\mathbf{F}(\mathbf{X}, t)^T. \tag{4.56}$$

Eqn. (4.53), which relates the right Cauchy–Green tensor (and hence the Lagrangian measure of strain) to the stress tensor, explains why constitutive relations are known as stress–strain relations.

4.6 Material Symmetry

The treatment in this and the following section is based on [35].

In general, a transformation of an object that leaves some property of the object invariant is called a *symmetry transformation*. For example, if a square is rotated by an integer multiple of 90° in its plane, it appears the same as before the rotation. Thus, such a rotation, which leaves the appearance of the square invariant, is a symmetry transformation. We now show by verifying the properties discussed in Section 1.12 that the set of symmetry transformations is a group.

The composition of any of these symmetry transformations is itself a symmetry transformation, e.g., a 180° clockwise rotation followed by a 90° anticlockwise rotation is just a clockwise 90° rotation. Thus, the operation of composition of symmetry transformations of the square is closed.

The composition of these symmetry transformations is also associative, e.g.,

$$[(90^\circ\ \mathrm{CW}) \circ (90^\circ\ \mathrm{CW})] \circ (270^\circ\ \mathrm{CCW}) = 90^\circ\ \mathrm{CW} \circ [(90^\circ\ \mathrm{CW}) \circ (270^\circ\ \mathrm{CCW})].$$

The inverses of these symmetry transformations are also symmetry transformations, e.g., the inverse of the 90° CW rotation is just the 90° CCW rotation.

Finally, the identity transformation, i.e., do nothing to the square, is a symmetry transformation.

Thus, the set of all symmetry transformations of the square is a group with respect to composition.

Now we consider symmetry transformations of material bodies[2]. In the context of continuum mechanics, the material symmetry of a body is characterized by those changes of reference configuration that do not alter the response function. Referring to Fig. 4.6, the stress response with respect to the configuration κ is related to that with respect to the configuration μ by

$$\overset{\infty}{\underset{s=0}{\hat{\boldsymbol{\tau}}_\kappa}}(\boldsymbol{X}, \boldsymbol{\nabla}(f \circ g)(\boldsymbol{X})) = \overset{\infty}{\underset{s=0}{\hat{\boldsymbol{\tau}}_\mu}}(\boldsymbol{Y}, \boldsymbol{\nabla} f(\boldsymbol{Y})),$$

where $\boldsymbol{Y} = g(\boldsymbol{X})$. Let $\boldsymbol{F} = \boldsymbol{\nabla} f(\boldsymbol{Y})$ and $\boldsymbol{G} = \boldsymbol{\nabla} g(\boldsymbol{X})$. Then, by the chain rule

$$\boldsymbol{\nabla}(f \circ g)(\boldsymbol{X}) = \boldsymbol{\nabla} f(\boldsymbol{Y}) \circ \boldsymbol{\nabla} g(\boldsymbol{X}) = \boldsymbol{F}\boldsymbol{G}.$$

Hence,

$$\overset{\infty}{\underset{s=0}{\hat{\boldsymbol{\tau}}_\kappa}}(\boldsymbol{X}, \boldsymbol{F}\boldsymbol{G}) = \overset{\infty}{\underset{s=0}{\hat{\boldsymbol{\tau}}_\mu}}(\boldsymbol{Y}, \boldsymbol{F}) \quad \forall \boldsymbol{F}, \in \mathrm{Lin}^+. \tag{4.57}$$

Since $\boldsymbol{X}$ and $\boldsymbol{Y}$ are the positions occupied by the same material particle in different configurations, the stress response with respect to the configurations is the same if

$$\overset{\infty}{\underset{s=0}{\hat{\boldsymbol{\tau}}_\kappa}}(\boldsymbol{X}, \boldsymbol{A}) = \overset{\infty}{\underset{s=0}{\hat{\boldsymbol{\tau}}_\mu}}(\boldsymbol{Y}, \boldsymbol{A}) \quad \forall \boldsymbol{A} \in \mathrm{Lin}^+, \tag{4.58}$$

[2]We shall write $\boldsymbol{F}(\boldsymbol{X}, t - s)$ simply as $\boldsymbol{F}$.

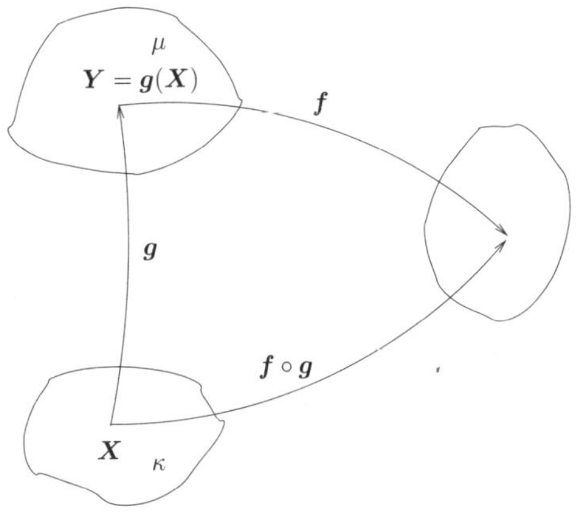

Fig. 4.6 *Stress response with respect to two different reference configurations.*

Thus, from Eqns. (4.57) and (4.58), the condition for the stress response at a point X to be the same is

$$\overset{\infty}{\underset{s=0}{\hat{\tau}_\kappa}}(X, FG) = \overset{\infty}{\underset{s=0}{\hat{\tau}_\kappa}}(X, F) \quad \forall F, G \in \text{Lin}^+. \tag{4.59}$$

We need to impose certain restrictions on G in the above equation in order that Eqn. (4.59) makes sense in a physical setting; more specifically, we need $G \in \text{Unim}^+$. The reason is as follows. By applying Eqn (4.59) repeatedly, we see that

$$\overset{\infty}{\underset{s=0}{\hat{\tau}_\kappa}}(X, FG^n) = \overset{\infty}{\underset{s=0}{\hat{\tau}_\kappa}}(X, F) \quad \forall n \geq 1.$$

Since $F, G \in \text{Lin}^+$, we have $F_n \equiv FG^n \in \text{Lin}^+$. If $\det G > 1$, then we have $\det G_n \to \infty$, but the stress $\overset{\infty}{\underset{s=0}{\hat{\tau}}}(X, F_n) = \overset{\infty}{\underset{s=0}{\hat{\tau}}}(X, F)$ remains the same. Since $\rho_\kappa = \rho_\mu \det G$, we can find a configuration, say μ', with deformation gradient G^n with respect to κ, whose density given by $\rho_\kappa / \det G^n$ can be made arbitrarily close to zero by choosing n large enough, but whose stress is still the same as that in the configuration κ. Clearly, this does not make sense physically. Similarly, we can rule out the case $\det G < 1$, since in that case we can have a configuration whose density approaches infinity, but whose stress remains the same as in the original configuration. Based on the above discussion, we restrict G to lie in the proper unimodular group. Physically, FG with $G \in \text{Unim}^+$ corresponds to a density-preserving deformation ($\det G = 1$), followed by a deformation with gradient F.

We define the *material symmetry group* of a particle X that occupies the position X in the reference configuration κ as the set

$$\partial_\kappa^X := \{H \in \text{Unim}^+ : \overset{\infty}{\underset{s=0}{\hat{\tau}_\kappa}}(X, FH) = \overset{\infty}{\underset{s=0}{\hat{\tau}_\kappa}}(X, F)\}. \tag{4.60}$$

Note that the above definition is for a single particle X. Thus, $H \in \partial_\kappa^X$ is not a field, but rather, it is just the gradient at X of a deformation of κ into a configuration with the same material properties at the particle X.

Two mathematical structures are said to be isomorphic if there is a one-to-one correspondence between them that preserves their structure. Thus, two reference configurations κ and μ are said to be *materially isomorphic* at a particle X if they have the same properties at X, i.e., if

$$\rho_\kappa(\boldsymbol{X}) = \rho_\mu(\boldsymbol{Y}) \text{ and } \boldsymbol{\tau}_\kappa(\boldsymbol{X}) = \boldsymbol{\tau}_\mu(\boldsymbol{Y}).$$

In view of the above considerations, we can say that the two reference configurations κ and μ are materially isomorphic at X if and only if $\boldsymbol{G} = \boldsymbol{\nabla} \boldsymbol{g} \in \partial_\kappa^X$.

As with the plane rotations of the square, the set of symmetry transformations is a group with respect to composition of the symmetry transformations.

Theorem 4.6.1. *The material symmetry group* ∂_κ^X *is a group with respect to tensor multiplication.*

Proof. We shall drop the explicit dependence of $\hat{\boldsymbol{\tau}}$ on $\boldsymbol{X}$ for notational convenience. Let $\boldsymbol{H}_1, \boldsymbol{H}_2 \in \partial_\kappa^X$. Then necessarily $\boldsymbol{H}_1, \boldsymbol{H}_2 \in \text{Unim}^+$, and since Unim^+ is a group with respect to tensor multiplication, $\boldsymbol{H}_1\boldsymbol{H}_2 \in \text{Unim}^+$. Also, for any $\boldsymbol{F} \in \text{Lin}^+$, we have

$$\begin{aligned}
\overset{\infty}{\underset{s=0}{\hat{\boldsymbol{\tau}}_\kappa}}(\boldsymbol{F}(\boldsymbol{H}_1\boldsymbol{H}_2)) &= \overset{\infty}{\underset{s=0}{\hat{\boldsymbol{\tau}}_\kappa}}((\boldsymbol{F}\boldsymbol{H}_1)\boldsymbol{H}_2) \\
&= \overset{\infty}{\underset{s=0}{\hat{\boldsymbol{\tau}}_\kappa}}(\boldsymbol{F}\boldsymbol{H}_1) && (\text{since } \boldsymbol{H}_2 \in \partial_\kappa^X) \\
&= \overset{\infty}{\underset{s=0}{\hat{\boldsymbol{\tau}}_\kappa}}(\boldsymbol{F}) && (\text{since } \boldsymbol{H}_1 \in \partial_\kappa^X).
\end{aligned}$$

Therefore, $\boldsymbol{H}_1\boldsymbol{H}_2 \in \partial_\kappa^X$, i.e., closure is satisfied.

The remaining axioms are easy to check.

1. If $\boldsymbol{H}_1, \boldsymbol{H}_2, \boldsymbol{H}_3 \in \partial_\kappa^X$, then $\boldsymbol{H}_1, \boldsymbol{H}_2, \boldsymbol{H}_3 \in \text{Unim}^+$, and by the associativity of tensor multiplication

$$(\boldsymbol{H}_1\boldsymbol{H}_2)\boldsymbol{H}_3 = \boldsymbol{H}_1(\boldsymbol{H}_2\boldsymbol{H}_3).$$

2. The neutral element is $\boldsymbol{I} \in \partial_\kappa^X$.

3. We need to show that if $\boldsymbol{H} \in \partial_\kappa^X$, then $\boldsymbol{H}^{-1} \in \partial_\kappa^X$. $\boldsymbol{H} \in \partial_\kappa^X$ implies that $\boldsymbol{H} \in \text{Unim}^+$, which in turn implies that there exists $\boldsymbol{H}^{-1} \in \text{Unim}^+$, since Unim^+ is a group. It remains to be shown that $\boldsymbol{H}^{-1} \in \partial_\kappa^X$. We have

$$\begin{aligned}
\overset{\infty}{\underset{s=0}{\hat{\boldsymbol{\tau}}_\kappa}}(\boldsymbol{F}\boldsymbol{H}^{-1}) &= \overset{\infty}{\underset{s=0}{\hat{\boldsymbol{\tau}}_\kappa}}((\boldsymbol{F}\boldsymbol{H}^{-1})\boldsymbol{H}) && (\text{since } \boldsymbol{H} \in \partial_\kappa^X) \\
&= \overset{\infty}{\underset{s=0}{\hat{\boldsymbol{\tau}}_\kappa}}(\boldsymbol{F}(\boldsymbol{H}\boldsymbol{H}^{-1})) && (\text{associativity}) \\
&= \overset{\infty}{\underset{s=0}{\hat{\boldsymbol{\tau}}_\kappa}}(\boldsymbol{F}),
\end{aligned}$$

which proves that $\boldsymbol{H}^{-1} \in \partial_\kappa^X$. □

Since the formulation of material symmetry is in terms of an underlying configuration κ, we now study the effect of using a different reference configuration μ, not necessarily materially isomorphic with κ.

Theorem 4.6.2. *Let the reference configurations κ and μ be connected through a deformation g as shown in Fig. 4.6 with $G = \nabla g$, where G is not necessarily a proper unimodular tensor. Then for any material particle X,*

$$H \in \partial_\kappa^X \Longleftrightarrow G(X)HG^{-1}(X) \in \partial_\mu^X, \tag{4.61}$$

where $X = \kappa(X)$. This rule is referred to as Noll's rule, *and written suggestively as*

$$\partial_\mu^X = G(X)\partial_\kappa^X G^{-1}(X).$$

Proof. We drop the explicit dependence of $\hat{\tau}$ on X for notational convenience. Suppose $H \in \partial_\kappa^X$. Then, $H \in \text{Unim}^+$, and

$$\begin{aligned}
\det[G(X)HG^{-1}(X)] &= (\det G(X))(\det H)(\det G^{-1}(X)) \\
&= \det H \\
&= 1.
\end{aligned}$$

Thus, $G(X)HG^{-1}(X) \in \text{Unim}^+$. We also have

$$\begin{aligned}
\mathop{\hat{\tau}_\mu}_{s=0}^{\infty}(FG(X)HG^{-1}(X)) &= \mathop{\hat{\tau}_\kappa}_{s=0}^{\infty}(FG(X)HG^{-1}(X)G(X)) && \text{(by Eqn. (4.57))} \\
&= \mathop{\hat{\tau}_\kappa}_{s=0}^{\infty}(FG(X)H) \\
&= \mathop{\hat{\tau}_\kappa}_{s=0}^{\infty}(FG(X)) && (H \in \partial_\kappa^X) \\
&= \mathop{\hat{\tau}_\mu}_{s=0}^{\infty}(F) && \text{(by Eqn. (4.57))}.
\end{aligned}$$

Thus, $G(X)HG^{-1}(X) \in \partial_\mu^X$.

Conversely, if $G(X)HG^{-1}(X) \in \partial_\mu^X$, then the following argument using Eqn. (4.57) shows that $H \in \partial_\kappa^X$:

$$\begin{aligned}
\mathop{\hat{\tau}_\kappa}_{s=0}^{\infty}(FH) &= \mathop{\hat{\tau}_\kappa}_{s=0}^{\infty}(FHG^{-1}(X)G(X)) \\
&= \mathop{\hat{\tau}_\mu}_{s=0}^{\infty}(FHG^{-1}(X)) \\
&= \mathop{\hat{\tau}_\mu}_{s=0}^{\infty}(FG^{-1}(X)G(X)HG^{-1}(X)) \\
&= \mathop{\hat{\tau}_\mu}_{s=0}^{\infty}(FG^{-1}(X))
\end{aligned}$$

$$
\begin{aligned}
&= \underset{s=0}{\overset{\infty}{\hat{\tau}_\kappa}}(\boldsymbol{F}\boldsymbol{G}^{-1}(\boldsymbol{X})\boldsymbol{G}(\boldsymbol{X}))\\
&= \underset{s=0}{\overset{\infty}{\hat{\tau}_\kappa}}(\boldsymbol{F}).
\end{aligned}
$$

□

Observe that if the reference configuration μ is obtained from the reference configuration κ by a dilatation, i.e., $\boldsymbol{G} = \lambda \boldsymbol{I}$ where $\lambda > 0$, then $\partial_\kappa^X = \partial_\mu^X$ (see Problem 1). Thus, it is possible to have the same symmetry group with respect to different configurations κ and μ at a given particle X with unequal densities at the particle in the two configurations.

For orthogonal members of the symmetry group, we can get special results by using the principle of material frame-indifference.

Theorem 4.6.3. *Let $\boldsymbol{Q} \in Orth^+$. Then $\boldsymbol{Q} \in \partial_\kappa^X$ if and only if*

$$
\underset{s=0}{\overset{\infty}{\hat{\tau}_\kappa}}(\boldsymbol{X}, \boldsymbol{Q}\boldsymbol{F}\boldsymbol{Q}^T) = \boldsymbol{Q}\,\underset{s=0}{\overset{\infty}{\hat{\tau}_\kappa}}(\boldsymbol{X}, \boldsymbol{F})\boldsymbol{Q}^T. \tag{4.62}
$$

Proof. As usual, we drop the dependence on $\boldsymbol{X}$ for notational convenience. Suppose that $\boldsymbol{Q} \in \partial_\kappa^X$. Then, we have

$$
\underset{s=0}{\overset{\infty}{\hat{\tau}_\kappa}}(\boldsymbol{F}) = \underset{s=0}{\overset{\infty}{\hat{\tau}_\kappa}}(\boldsymbol{F}\boldsymbol{Q}) \quad \forall \boldsymbol{F} \in \mathrm{Lin}^+.
$$

Since $\boldsymbol{Q}\boldsymbol{F}\boldsymbol{Q}^T \in \mathrm{Lin}^+$, we have

$$
\begin{aligned}
\underset{s=0}{\overset{\infty}{\hat{\tau}_\kappa}}(\boldsymbol{Q}\boldsymbol{F}\boldsymbol{Q}^T) &= \underset{s=0}{\overset{\infty}{\hat{\tau}_\kappa}}(\boldsymbol{Q}\boldsymbol{F}\boldsymbol{Q}^T\boldsymbol{Q})\\
&= \underset{s=0}{\overset{\infty}{\hat{\tau}_\kappa}}(\boldsymbol{Q}\boldsymbol{F})\\
&= \boldsymbol{Q}\,\underset{s=0}{\overset{\infty}{\hat{\tau}_\kappa}}(\boldsymbol{F})\boldsymbol{Q}^T \qquad \text{(by material frame-indifference).}
\end{aligned}
$$

Conversely, suppose that

$$
\underset{s=0}{\overset{\infty}{\hat{\tau}_\kappa}}(\boldsymbol{Q}\boldsymbol{F}\boldsymbol{Q}^T) = \boldsymbol{Q}\,\underset{s=0}{\overset{\infty}{\hat{\tau}_\kappa}}(\boldsymbol{F})\boldsymbol{Q}^T \quad \forall \boldsymbol{F} \in \mathrm{Lin}^+.
$$

By using the principle of material frame-indifference, the above equation becomes

$$
\underset{s=0}{\overset{\infty}{\hat{\tau}_\kappa}}(\boldsymbol{Q}\boldsymbol{F}\boldsymbol{Q}^T) = \underset{s=0}{\overset{\infty}{\hat{\tau}_\kappa}}(\boldsymbol{Q}\boldsymbol{F}) \quad \forall \boldsymbol{F} \in \mathrm{Lin}^+.
$$

With $\boldsymbol{Q}^T\boldsymbol{F}\boldsymbol{Q}$ playing the role of $\boldsymbol{F}$, this becomes

$$
\underset{s=0}{\overset{\infty}{\hat{\tau}_\kappa}}(\boldsymbol{F}) = \underset{s=0}{\overset{\infty}{\hat{\tau}_\kappa}}(\boldsymbol{F}\boldsymbol{Q}),
$$

which implies that $\boldsymbol{Q} \in \partial_\kappa^X$, since $\boldsymbol{Q} \in \mathrm{Unim}^+$. □

The two previous results were derived without using a constitutive equation that automatically satisfies material frame-indifference. If we start with the reduced form of the constitutive equation given by Eqn. (4.56), we get the following result:

Theorem 4.6.4. *In terms of the reduced response function* $\underset{s=0}{\overset{\infty}{\tilde{\tau}}}$*, the material group of the particle* X *relative to the configuration* κ *is given by*

$$\partial_\kappa^X = \left\{ \boldsymbol{H} \in Unim^+ : \boldsymbol{H} \underset{s=0}{\overset{\infty}{\tilde{\tau}_\kappa}}(\boldsymbol{H}^T\boldsymbol{C}\boldsymbol{H})\boldsymbol{H}^T = \underset{s=0}{\overset{\infty}{\tilde{\tau}_\kappa}}(\boldsymbol{C}) \quad \forall \boldsymbol{C}(t) \in Psym \right\}.$$

Proof. By definition, a unimodular tensor $\boldsymbol{H}$ belongs to ∂_κ^X if and only if

$$\underset{s=0}{\overset{\infty}{\hat{\tau}_\kappa}}(\boldsymbol{F}\boldsymbol{H}) = \underset{s=0}{\overset{\infty}{\hat{\tau}_\kappa}}(\boldsymbol{F}) \quad \forall \boldsymbol{F} \in \text{Lin}^+.$$

By using Eqn. (4.56), the above requirement for a given particle X can be written as

$$\boldsymbol{F}(t)\boldsymbol{H} \underset{s=0}{\overset{\infty}{\tilde{\tau}_\kappa}}((\boldsymbol{F}\boldsymbol{H})^T(\boldsymbol{F}\boldsymbol{H}))(\boldsymbol{F}(t)\boldsymbol{H})^T = \boldsymbol{F}(t) \underset{s=0}{\overset{\infty}{\tilde{\tau}_\kappa}}(\boldsymbol{C})\boldsymbol{F}^T(t) \quad \forall \boldsymbol{F} \in \text{Lin}^+.$$

Using the facts that $\boldsymbol{F}$ is invertible and $\boldsymbol{F}^T\boldsymbol{F} = \boldsymbol{C}$, we get

$$\boldsymbol{H} \underset{s=0}{\overset{\infty}{\tilde{\tau}_\kappa}}(\boldsymbol{H}^T\boldsymbol{C}\boldsymbol{H})\boldsymbol{H}^T = \underset{s=0}{\overset{\infty}{\tilde{\tau}_\kappa}}(\boldsymbol{C}) \quad \forall \boldsymbol{C}(t) \in \text{Psym}.$$

The requirement that the above condition hold for all $\boldsymbol{C}(t) \in$ Psym follows from the fact that every $\boldsymbol{C} \in$ Psym can be written as $\boldsymbol{U}\boldsymbol{U} = \boldsymbol{U}^T\boldsymbol{U}$, where $\boldsymbol{U}$ belongs to Psym, and hence to Lin$^+$. Thus, every $\boldsymbol{F} \in$ Lin$^+$ generates $\boldsymbol{C} = \boldsymbol{F}^T\boldsymbol{F} \in$ Psym, and every $\boldsymbol{C} \in$ Psym can be written as $\boldsymbol{F}^T\boldsymbol{F}$, where $\boldsymbol{F} \in$ Lin$^+$. □

As a corollary of the above theorem, we have the following companion to Theorem 4.6.3:

Theorem 4.6.5. *Let* $\boldsymbol{Q} \in Orth^+$*. Then* $\boldsymbol{Q} \in \partial_\kappa^X$ *if and only if*

$$\underset{s=0}{\overset{\infty}{\tilde{\tau}_\kappa}}(\boldsymbol{Q}\boldsymbol{C}\boldsymbol{Q}^T) = \boldsymbol{Q} \underset{s=0}{\overset{\infty}{\tilde{\tau}_\kappa}}(\boldsymbol{C})\boldsymbol{Q}^T \quad \forall \boldsymbol{C}(t) \in Psym.$$

Since $\boldsymbol{C} \in$ Psym is a measure of strain, Theorem. 4.6.5 can be expressed as saying: 'Rotating the strain by an element of the symmetry group rotates the stress in the same manner'.

4.7 Classification of Materials

In this section, we give precise definitions of various terms such as isotropy, solid, fluid etc.

If the material properties at a particle X of a body are unaltered by rotations of the reference configuration, we say that the material is *isotropic*. Since the rotation of the reference configuration corresponds to the examination of properties in different directions, this definition corresponds to the intuitive idea of the material properties of an isotropic body being independent of direction. More precisely, we have

Definition: A body is isotropic at a point if there exists a reference configuration κ, called *undistorted*, such that

$$\text{Orth}^+ \subset \partial_\kappa^X,$$

or, in other words,

$$\underset{s=0}{\overset{\infty}{\hat{\boldsymbol{\tau}}_\kappa}}(\boldsymbol{X}, \boldsymbol{F}\boldsymbol{Q}) = \underset{s=0}{\overset{\infty}{\hat{\boldsymbol{\tau}}_\kappa}}(\boldsymbol{X}, \boldsymbol{F}) \quad \forall \boldsymbol{F} \in \text{Lin}^+, \; \boldsymbol{Q} \in \text{Orth}^+. \tag{4.63}$$

Physically, the above equation corresponds to the idea that the stress remains the same even if we carry out an arbitrary rotation of the body about the point $\boldsymbol{X}$ before carrying out the deformation $\boldsymbol{F}$. Since $\chi = \boldsymbol{F}\boldsymbol{Q}\boldsymbol{X} + \boldsymbol{c}(t)$ in this deformed configuration, an alternative interpretation is that the Cauchy stress remains unaltered when the reference configuration is rotated by $\boldsymbol{Q}$ around $\boldsymbol{X}$. Note that while characterizing frame-indifferent materials, we rotate the deformed configuration, so that the product $\boldsymbol{Q}\boldsymbol{F}$ appears in the characterization, while in the case of characterizing isotropy, we rotate the material and then carry out the deformation, which results in the appearance of the product $\boldsymbol{F}\boldsymbol{Q}$. If Eqn. (4.63) is not satisfied for some $\boldsymbol{Q} \in \text{Orth}^+$, then the material is said to be anisotropic at $\boldsymbol{X}$.

It is important to note that isotropy is a property that holds only with respect to special configurations, namely, the undistorted configurations κ in Eqn. (4.63). The material symmetry group relative to any other reference configuration μ is given by Eqn. (4.61) as

$$\partial_\mu^X = \boldsymbol{G}(\boldsymbol{X})\partial_\kappa^X \boldsymbol{G}^{-1}(\boldsymbol{X}),$$

where $\boldsymbol{G} = \boldsymbol{\nabla} \boldsymbol{g}$ is the deformation gradient of the deformation taking κ to μ. It is possible that ∂_μ^X may not contain any orthogonal tensors, let alone all of them. However, if the reference configuration μ is obtained by a rigid deformation of an undistorted configuration κ, then μ is also an undistorted configuration. This follows since $\boldsymbol{G} = \boldsymbol{Q} \in \text{Orth}^+$, and thus, for every $\boldsymbol{R} \in \text{Orth}^+$, we have $\boldsymbol{Q}^T\boldsymbol{R}\boldsymbol{Q} \in \text{Orth}^+ \subset \partial_\kappa^X$, which on applying Noll's rule yields the fact that $\boldsymbol{R} \in \partial_\mu^X$, or, in other words, $\text{Orth}^+ \subset \partial_\mu^X$. Thus, while undistorted configurations are special, they are not unique.

By combining the notions of isotropy and material frame-indifference, we have the following result:

Theorem 4.7.1. *The response functional* $\underset{s=0}{\overset{\infty}{\hat{\boldsymbol{\tau}}}}$ *is isotropic and material frame-indifferent at a particle X with respect to a configuration κ if and only if there exists a mapping* $\underset{s=0}{\overset{\infty}{\bar{\boldsymbol{\tau}}}}$, *such that*

$$\underset{s=0}{\overset{\infty}{\hat{\boldsymbol{\tau}}_\kappa}}(\boldsymbol{X}, \boldsymbol{F}(\boldsymbol{X}, t-s)) = \underset{s=0}{\overset{\infty}{\bar{\boldsymbol{\tau}}_\kappa}}(\boldsymbol{X}, \boldsymbol{F}\boldsymbol{F}^T(\boldsymbol{X}, t-s)) \quad \forall \boldsymbol{F} \in \mathit{Lin}^+, \tag{4.64}$$

and that satisfies

$$\mathop{\bar{\tau}_\kappa}_{s=0}^{\infty}(X, QBQ^T) = Q \mathop{\bar{\tau}_\kappa}_{s=0}^{\infty}(X, B) Q^T \quad \forall B(t) \in Psym,\ Q(t) \in Orth^+. \tag{4.65}$$

Proof. We drop the dependence on X in the proof. Note that in the polar decomposition $F = VR$, $V \in \text{Lin}^+$ and $R \in \text{Orth}^+$. Assuming that the particle is isotropic at X, by Eqn. (4.63) we have

$$\mathop{\hat{\tau}_\kappa}_{s=0}^{\infty}(F) = \mathop{\hat{\tau}_\kappa}_{s=0}^{\infty}(VR) = \mathop{\hat{\tau}_\kappa}_{s=0}^{\infty}(V) = \mathop{\bar{\tau}_\kappa}_{s=0}^{\infty}(B),$$

where $V^2 = B = FF^T$. Thus,

$$\begin{aligned}
Q \mathop{\bar{\tau}_\kappa}_{s=0}^{\infty}(B) Q^T &= Q \mathop{\hat{\tau}_\kappa}_{s=0}^{\infty}(F) Q^T \\
&= \mathop{\hat{\tau}_\kappa}_{s=0}^{\infty}(QFQ^T) \quad \text{(by Theorem 4.6.3)} \\
&= \mathop{\bar{\tau}_\kappa}_{s=0}^{\infty}(QFQ^T QF^T Q^T) \\
&= \mathop{\bar{\tau}_\kappa}_{s=0}^{\infty}(QFF^T Q^T) \\
&= \mathop{\bar{\tau}_\kappa}_{s=0}^{\infty}(QBQ^T).
\end{aligned}$$

Since every $B(t) \in \text{Psym}$ can be expressed as $VV = VV^T$, where $V \in \text{Lin}^+$, the above equation holds for all $B(t) \in \text{Psym}$.

Conversely, assuming Eqns. (4.64) and (4.65) hold

$$\begin{aligned}
\mathop{\hat{\tau}_\kappa}_{s=0}^{\infty}(FQ) &= \mathop{\bar{\tau}_\kappa}_{s=0}^{\infty}(FQQ^T F^T) \\
&= \mathop{\bar{\tau}_\kappa}_{s=0}^{\infty}(FF^T) \\
&= \mathop{\hat{\tau}}_{s=0}^{\infty}(F) \quad \forall Q \in \text{Orth}^+,
\end{aligned}$$

which proves that the particle is isotropic. Substituting $B = FF^T$ into Eqn. (4.65), we have

$$\mathop{\bar{\tau}_\kappa}_{s=0}^{\infty}(QF(QF)^T) = Q \mathop{\bar{\tau}_\kappa}_{s=0}^{\infty}(FF^T) Q^T,$$

which in turn implies that

$$\mathop{\hat{\tau}_\kappa}_{s=0}^{\infty}(QF) = Q \mathop{\hat{\tau}_\kappa}_{s=0}^{\infty}(F) Q^T,$$

thus proving material frame-indifference. □

If we pour a fluid from a container into a test-tube, the fluid undergoes large deformations. Yet, intuitively, it is clear that the density at a fluid particle is preserved, and that the stress response at that particle is the same. Thus, if the stress response at a particle X is unaltered by any density preserving deformation, we say that the material at X is fluid, i.e., a fluid particle has the maximum amount of material symmetry.

Definition: A body is fluid at a particle X if there exists a reference configuration κ such that

$$\partial_\kappa^X = \text{Unim}^+.$$

In contrast to the undistorted configuration in the definition of an isotropic particle, the reference configuration in the definition of a fluid particle is of no special significance as the following result shows.

Theorem 4.7.2. *Let a simple body be fluid at a particle X. Then for any configuration* μ, $\partial_\mu^X = \mathit{Unim}^+$.

Proof. Let $\boldsymbol{H} \in \text{Unim}^+$. Then $\hat{\boldsymbol{H}} := \boldsymbol{G}^{-1}(\boldsymbol{X})\boldsymbol{H}\boldsymbol{G}(\boldsymbol{X}) \in \text{Unim}^+ = \partial_\kappa^X$. By Eqn. (4.61), the corresponding element of ∂_μ^X is

$$\boldsymbol{G}(\boldsymbol{X})\hat{\boldsymbol{H}}\boldsymbol{G}^{-1}(\boldsymbol{X}) = \boldsymbol{G}(\boldsymbol{X})\boldsymbol{G}^{-1}(\boldsymbol{X})\boldsymbol{H}\boldsymbol{G}(\boldsymbol{X})\boldsymbol{G}^{-1}(\boldsymbol{X}) = \boldsymbol{H},$$

i.e., $\boldsymbol{H} \in \partial_\mu^X$. Hence, $\text{Unim}^+ \subset \partial_\mu^X$. By definition, $\partial_\mu^X \subset \text{Unim}^+$, and therefore $\partial_\mu^X = \text{Unim}^+$. □

Thus, a fluid particle has no preferred reference configurations; this is in contrast to the 'undistorted configuration' in the definition of an isotropic particle.

Suppose that a body is a fluid at a particle X. Then $\partial_\kappa^X = \text{Unim}^+$ for any reference configuration κ, and since $\text{Orth}^+ \subset \text{Unim}^+$, $\text{Orth}^+ \subset \partial_\kappa^X$. Thus, we have

Theorem 4.7.3. *Fluid particles of bodies are isotropic with all reference configurations being undistorted.*

If we apply a load to a bar so that it gets deformed, and now choose the deformed bar as the new reference configuration, and apply the same load, then one would expect the response of the bar in the starting and new reference configurations to be different. Thus, we would expect the stress response of a 'solid' to be altered by nonrigid deformation of its reference configuration, although rigid deformations may or may not alter its response. More precisely, we define a solid as follows:

Definition: A body is solid at a particle X if there exists a reference configuration κ, called undistorted, such that

$$\partial_\kappa^X \subset \text{Orth}^+.$$

If $\partial_\kappa^X = \boldsymbol{I}$, then the solid is called *triclinic*.

As in the isotropic particle case, the undistorted reference configuration κ in the definition of a solid particle is of special significance, except when the solid is triclinic. If μ is some other reference configuration, then by Noll's rule

$$\partial_\mu^X = \boldsymbol{G}(\boldsymbol{X})\partial_\kappa^X \boldsymbol{G}^{-1}(\boldsymbol{X}).$$

For $\boldsymbol{Q} \in \text{Orth}^+$, generally $\boldsymbol{G}(\boldsymbol{X})\boldsymbol{Q}\boldsymbol{G}^{-1}(\boldsymbol{X})$ does not belong to Orth^+. So, generally, ∂_μ^X is not a subgroup of Orth^+. However, we have

Theorem 4.7.4. *Let a body be solid at a particle X with κ an undistorted reference configuration. Then any reference configuration μ obtained by a rigid deformation of κ is also undistorted.*

The proof follows from the fact that Orth^+ being a group, if $\boldsymbol{R}, \boldsymbol{Q} \in \text{Orth}^+$, then $\boldsymbol{R}\boldsymbol{Q}\boldsymbol{R}^T \in \text{Orth}^+$, and hence $\partial_\mu^X \subset \text{Orth}^+$.

Suppose that a body is solid at a particle X. Then there exists a reference configuration κ such that $\partial_\kappa^X \subset \text{Orth}^+$. For a particle to be a fluid, we need $\partial_\kappa^X = \text{Unim}^+$. Since by Theorem 1.12.10, Orth^+ is a proper subgroup of Unim^+, we conclude that

Theorem 4.7.5. *A particle of a body cannot be both solid and fluid.*

However, a particle can be both solid and isotropic. To see this, suppose that a body is solid at a particle X. Then there exists a reference configuration κ such that $\partial_\kappa^X \subset \text{Orth}^+$. If the particle X is also isotropic, then there exists a reference configuration $\hat{\kappa}$ such that

$$\text{Orth}^+ \subset \partial_{\hat{\kappa}}^X \subset \text{Unim}^+.$$

Consequently, by the maximality of Orth^+ in Unim^+ (Theorem 1.12.11), either

$$\partial_{\hat{\kappa}}^X = \text{Orth}^+ \text{ or } \partial_{\hat{\kappa}}^X = \text{Unim}^+.$$

The latter possibility implies that X is a fluid particle. However, by hypothesis, it is a solid particle, and it cannot be both solid and fluid (Theorem 4.7.5). Hence,

$$\partial_{\hat{\kappa}}^X = \text{Orth}^+. \tag{4.66}$$

By Eqn. (4.61),

$$\hat{\boldsymbol{Q}} \in \partial_{\hat{\kappa}}^X = \text{Orth}^+ \iff \boldsymbol{Q} = \boldsymbol{G}\hat{\boldsymbol{Q}}\boldsymbol{G}^{-1} \in \partial_\kappa^X \subset \text{Orth}^+,$$

where $\boldsymbol{G}$ is the gradient of the deformation that takes $\hat{\kappa}$ into κ. Since $\boldsymbol{G} = \boldsymbol{Q}\boldsymbol{G}\hat{\boldsymbol{Q}}^T$, we have

$$\boldsymbol{G}^T\boldsymbol{G} = \hat{\boldsymbol{Q}}\boldsymbol{G}^T\boldsymbol{Q}^T\boldsymbol{Q}\boldsymbol{G}\hat{\boldsymbol{Q}}^T = \hat{\boldsymbol{Q}}\boldsymbol{G}^T\boldsymbol{G}\hat{\boldsymbol{Q}}^T \quad \forall \hat{\boldsymbol{Q}} \in \text{Orth}^+,$$

which, by Theorem 1.8.1, implies that $\boldsymbol{G}^T\boldsymbol{G} = \lambda\boldsymbol{I}$ for all possible deformation gradients $\boldsymbol{G}$. Since $\boldsymbol{G}^T\boldsymbol{G} \in \text{Psym}$, we have $\lambda > 0$. Therefore,

$$\left(\frac{1}{\sqrt{\lambda}}\boldsymbol{G}\right)^T\left(\frac{1}{\sqrt{\lambda}}\boldsymbol{G}\right) = \boldsymbol{I}.$$

In terms of $\boldsymbol{R}$ defined by

$$\boldsymbol{R} := \frac{1}{\sqrt{\lambda}}\boldsymbol{G} \in \text{Orth}^+,$$

we can write $\boldsymbol{Q} \in \mathfrak{d}_\kappa^X$ as

$$\boldsymbol{Q} = \boldsymbol{R}\hat{\boldsymbol{Q}}\boldsymbol{R}^T. \tag{4.67}$$

Now let $\tilde{\boldsymbol{Q}} \in \text{Orth}^+$. Then since Orth^+ is a group, $\boldsymbol{R}^T\tilde{\boldsymbol{Q}}\boldsymbol{R} \in \text{Orth}^+ = \mathfrak{d}_{\hat{\kappa}}^X$. Letting $\boldsymbol{R}^T\tilde{\boldsymbol{Q}}\boldsymbol{R}$ play the role of $\hat{\boldsymbol{Q}}$ in Eqn. (4.67), we get

$$\boldsymbol{R}(\boldsymbol{R}^T\tilde{\boldsymbol{Q}}\boldsymbol{R})\boldsymbol{R}^T = \tilde{\boldsymbol{Q}} \in \mathfrak{d}_\kappa^X.$$

Thus, $\text{Orth}^+ \subset \mathfrak{d}_\kappa^X$. But, by hypothesis, $\mathfrak{d}_\kappa^X \subset \text{Orth}^+$. Hence,

$$\mathfrak{d}_\kappa^X = \text{Orth}^+.$$

Summarizing, we have

Theorem 4.7.6. *Suppose that a particle X of a body is both solid, meaning that there exists a reference configuration κ such that $\mathfrak{d}_\kappa^X \subset \text{Orth}^+$, and isotropic, meaning that there exists a reference configuration $\hat{\kappa}$ such that $\text{Orth}^+ \subset \mathfrak{d}_{\hat{\kappa}}^X$. Then*

$$\mathfrak{d}_\kappa^X = \mathfrak{d}_{\hat{\kappa}}^X = \text{Orth}^+.$$

Thus, for an isotropic solid particle, the material symmetry groups relative to the special reference configurations employed in the definition of solid and isotropic (both termed undistorted) each coincide with the orthogonal group. Note that the deformation gradient $\boldsymbol{G}$ is given by $\sqrt{\lambda}\boldsymbol{R}$, i.e., in general, it is a scalar multiple of a rotation matrix. Hence, the undistorted reference configuration κ and $\hat{\kappa}$ need not be the same, and, in general, need not even have the same density. However, by *choosing* $\lambda = 1$ and $\boldsymbol{R} = \boldsymbol{I}$, we can make the undistorted reference configurations coincide, so that the undistorted reference configuration associated with the isotropy is also an undistorted reference configuration for the solidity, and vice versa.

We have seen that an isotropic particle of a body can be fluid (Theorem 4.7.3) or solid (Theorem 4.7.6). A spectacular consequence of the maximality of the orthogonal group in the unimodular group (Theorem 1.12.11) is that these are the only two possibilities for an isotropic material.

Theorem 4.7.7. *An isotropic particle of a body is either solid or fluid.*

Proof. Suppose that a body is isotropic at a particle X. Then there exists a reference configuration κ such that

$$\text{Orth}^+ \subset \mathfrak{d}_\kappa^X \subset \text{Unim}^+.$$

By the maximality of Orth^+ in Unim^+, either

$$\partial_\kappa^X = \text{Orth}^+,$$

in which case $\partial_\kappa^X \subset \text{Orth}^+$, and X is a solid particle, or

$$\partial_\kappa^X = \text{Unim}^+,$$

in which case X is a fluid particle. □

In addition to fluid and solid particles, we can have yet another category of particles, which we refer to as fluid crystals.

Definition: A particle X of a body is a fluid crystal if it is not solid, i.e., X is a fluid crystal if there does not exist a reference configuration κ such that $\partial_\kappa^X \subset \text{Orth}^+$, or, equivalently, if $\partial_\kappa^X \not\subset \text{Orth}^+$ for all configurations κ.

Thus, for a fluid crystal, ∂_κ^X always contains some nonorthogonal elements. While a fluid crystal cannot be solid, it can be fluid or isotropic, as seen from the following result:

Theorem 4.7.8. *A fluid crystal particle of a body is fluid if and only if it is isotropic.*

Proof. Suppose if a fluid crystal particle is isotropic. Then by Theorem 4.7.7, it is either a solid or fluid particle. But a fluid crystal, by definition, cannot be solid. Hence, it is fluid.

Conversely, if a fluid crystal particle is fluid then it has to be isotropic, since all fluid particles are isotropic by Theorem 4.7.3. □

Based on the above discussion, a particle can either be solid or fluid or an anisotropic fluid crystal. We need to show that anisotropic fluid crystals exist (at least mathematically). First, we construct a proper subgroup of ∂_κ^X that contains some nonorthogonal elements.

Let $\boldsymbol{e} \in V$ be a unit vector, and consider

$$\mathcal{G}(\boldsymbol{e}) := \{\boldsymbol{H} \in \text{Unim}^+ : \boldsymbol{He} = \pm\boldsymbol{e}\}.$$

If $\boldsymbol{H}_1, \boldsymbol{H}_2 \in \mathcal{G}(\boldsymbol{e})$, then $\boldsymbol{H}_1\boldsymbol{H}_2 \in \mathcal{G}(\boldsymbol{e})$, since $\boldsymbol{H}_1\boldsymbol{H}_2 \in \text{Unim}^+$, and $\boldsymbol{H}_1\boldsymbol{H}_2\boldsymbol{e} = \pm\boldsymbol{H}_1\boldsymbol{e} = \pm\boldsymbol{e}$. The other properties of associativity, and existence of identity and reverse elements can also be proved easily. Thus, it follows that $\mathcal{G}(\boldsymbol{e})$ is a group.

As an example, let $\boldsymbol{e} = \boldsymbol{e}_1$. Then we see that

$$\begin{bmatrix} 1 & 0 & 0 \\ 1 & 1 & 0 \\ 0 & 0 & 1 \end{bmatrix}$$

defines a proper unimodular tensor that is not in $\mathcal{G}(\boldsymbol{e})$, so $\mathcal{G}(\boldsymbol{e}) \neq \text{Unim}^+$. Similarly,

$$\boldsymbol{H} = \begin{bmatrix} 1 & 0 & 1 \\ 0 & 1 & 0 \\ 0 & 0 & 1 \end{bmatrix}$$

defines an element of $\mathcal{G}(\boldsymbol{e})$, but it is nonorthogonal since $\boldsymbol{H}^T\boldsymbol{H} \neq \boldsymbol{I}$. Thus, $\mathcal{G}(\boldsymbol{e})$ is a group such that

$$\mathcal{G}(\boldsymbol{e}) \neq \text{Unim}^+ \text{ and } \mathcal{G}(\boldsymbol{e}) \not\subset \text{Orth}^+.$$

Next, we try and find a body, some particle of which has $\mathcal{G}(\boldsymbol{e})$ for its material symmetry group. Consider the body defined by the constitutive relation

$$\boldsymbol{\tau}(\boldsymbol{X},t) = \alpha[\boldsymbol{F}(\boldsymbol{X},t)\boldsymbol{e}] \otimes [\boldsymbol{F}(\boldsymbol{X},t)\boldsymbol{e}].$$

Since the constitutive relation involves only the deformation gradient, it is an example of an elastic body (though *not* a solid elastic body, as we shall soon show). A direct calculation shows that the above constitutive relation satisfies the principle of material frame-indifference.

Since α is a constant, this material is homogeneous, and its material symmetry group at any particle X is given by

$$\partial_\kappa^X = \{\boldsymbol{H} \in \text{Unim}^+ : \ (\boldsymbol{FHe}) \otimes (\boldsymbol{FHe}) = (\boldsymbol{Fe}) \otimes (\boldsymbol{Fe}) \quad \forall \boldsymbol{F} \in \text{Lin}^+\}.$$

We now show that $\partial_\kappa^X = \mathcal{G}(\boldsymbol{e})$.

First, suppose that $\boldsymbol{H} \in \mathcal{G}(\boldsymbol{e})$. Then $\boldsymbol{H} \in \text{Unim}^+$, and $\boldsymbol{He} = \pm\boldsymbol{e}$. Hence, we get

$$(\boldsymbol{FHe}) \otimes (\boldsymbol{FHe}) = (\boldsymbol{Fe}) \otimes (\boldsymbol{Fe}) \quad \forall \boldsymbol{F} \in \text{Lin}^+,$$

i.e., $\boldsymbol{H} \in \partial_\kappa^X$, or $\mathcal{G}(\boldsymbol{e}) \subset \partial_\kappa^X$.

Next, suppose that $\boldsymbol{H} \in \partial_\kappa^X$. Then $\boldsymbol{H} \in \text{Unim}^+$, and

$$(\boldsymbol{FHe}) \otimes (\boldsymbol{FHe}) = (\boldsymbol{Fe}) \otimes (\boldsymbol{Fe}) \quad \forall \boldsymbol{F} \in \text{Lin}^+,$$

This can be written as

$$\boldsymbol{FH}(\boldsymbol{e} \otimes \boldsymbol{e})\boldsymbol{H}^T\boldsymbol{F}^T = \boldsymbol{F}(\boldsymbol{e} \otimes \boldsymbol{e})\boldsymbol{F}^T,$$

which, by the invertibility of $\boldsymbol{F}$, implies

$$\boldsymbol{H}(\boldsymbol{e} \otimes \boldsymbol{e})\boldsymbol{H}^T = \boldsymbol{e} \otimes \boldsymbol{e},$$

or,

$$(\boldsymbol{He}) \otimes (\boldsymbol{He}) = \boldsymbol{e} \otimes \boldsymbol{e}.$$

Operating on both sides by $\boldsymbol{e}$, we get

$$(\boldsymbol{e} \cdot \boldsymbol{He})\boldsymbol{He} = \boldsymbol{e}.$$

Clearly, $\boldsymbol{e} \cdot \boldsymbol{He} \neq 0$ since, otherwise, the right-hand side of the above equation would be zero. Taking the cross product of both sides of the above equation with $\boldsymbol{e}$, and using Theorem 1.2.1, we conclude that $\boldsymbol{He} = \beta\boldsymbol{e}$, where β is a constant to be determined. Substituting $\boldsymbol{He} = \beta\boldsymbol{e}$ into the above equation, we get

$$\beta^2 = 1,$$

or, in other words, $\beta = \pm 1$. Therefore, $\boldsymbol{He} = \pm\boldsymbol{e}$, i.e., $\boldsymbol{H} \in \mathcal{G}(\boldsymbol{e})$, or $\partial_\kappa^X \subset \mathcal{G}(\boldsymbol{e})$.

Thus, $\partial_\kappa^X = \mathcal{G}(\boldsymbol{e})$, and we have a theoretical example of an anisotropic fluid crystal.

Summarizing, we have the following classification of materials:

A particle X of a body is

- *isotropic* if $\text{Orth}^+ \subset \partial_\kappa^X$, κ undistorted;
- *solid* if $\partial_\kappa^X \subset \text{Orth}^+$, κ undistorted;
- *a fluid crystal* if $\partial_\kappa^X \not\subset \text{Orth}^+$ $\forall$ configurations κ.
- *fluid* if $\partial_\kappa^X = \text{Unim}^+$ $\forall$ configurations κ.

EXERCISES

1. Show that changes of configuration ($\kappa \to \mu$) defined by deformations of the form

$$g(\boldsymbol{X}) = \boldsymbol{Y}_0 + \alpha(\boldsymbol{X} - \boldsymbol{X}_0), \quad \alpha \in \Re,$$

do not affect the material symmetry group of a body, i.e., $\partial_\kappa^X = \partial_\mu^X$. Deformations of this type are called dilatations if $\alpha > 0$, and central inversions if $\alpha < 0$.

5 Nonlinear Elasticity

An elastic body is one in which the stress depends only on the present configuration of the body, and is independent of how the body attains its present configuration. Thus, the constitutive relation for an elastic material is obtained simply by dropping the history dependence from the constitutive relation for a simple material as in Eqn. (4.49), which we reproduce here:

$$\boldsymbol{\tau}(\boldsymbol{x},t) = \hat{\boldsymbol{\tau}}(\boldsymbol{X},\boldsymbol{F}(\boldsymbol{X},t)), \quad \boldsymbol{x} = \boldsymbol{\chi}(\boldsymbol{X},t).$$

Note from the above equation that the stress at the present time is completely determined by the present value of the deformation gradient at a particle. If all the particles are elastic, the body is said to be elastic. In addition, if the response function is the same at each particle, the body is said to be elastically homogeneous.

The conditions that the response functions for the first and second Piola–Kirchhoff stress be frame-indifferent are obtained from Eqns. (4.54) and (4.55) by dropping the history dependence, i.e.,

$$\hat{\boldsymbol{T}}(\boldsymbol{X},\boldsymbol{Q}(t)\boldsymbol{F}(\boldsymbol{X},t)) = \boldsymbol{Q}(t)\hat{\boldsymbol{T}}(\boldsymbol{X},\boldsymbol{F}(\boldsymbol{X},t)) \ \forall \boldsymbol{F} \in \mathrm{Lin}^+, \boldsymbol{Q} \in \mathrm{Orth}^+, \tag{5.1}$$

$$\hat{\boldsymbol{S}}(\boldsymbol{X},\boldsymbol{Q}(t)\boldsymbol{F}(\boldsymbol{X},t)) = \hat{\boldsymbol{S}}(\boldsymbol{X},\boldsymbol{F}(\boldsymbol{X},t)) \ \forall \boldsymbol{F} \in \mathrm{Lin}^+, \boldsymbol{Q} \in \mathrm{Orth}^+. \tag{5.2}$$

The stress experienced by the body in the reference configuration, i.e., when $\boldsymbol{F} = \boldsymbol{I}$, is known as the *residual stress*. A reference configuration is called a natural state if the residual stress tensor vanishes at all points $\boldsymbol{X} \in V_0$. If a reference state is a natural state, any deformed configuration corresponding to a rigid deformation is also a natural state if it is chosen as the new reference configuration, since by Eqn. (4.64),

$$\hat{\boldsymbol{\tau}}(\boldsymbol{X},\boldsymbol{Q}) = \bar{\boldsymbol{\tau}}(\boldsymbol{X},\boldsymbol{I}) = \boldsymbol{0}.$$

5.1 Isotropic Elastic Particles

Recall from our discussion in Section 4.7 that an elastic particle X is isotropic if

$$\mathrm{Orth}^+ \subset \partial_\kappa^X.$$

or, in other words, if the response function for the Cauchy stress satisfies

$$\hat{\boldsymbol{\tau}}(\boldsymbol{X},\boldsymbol{F}\boldsymbol{Q}) = \hat{\boldsymbol{\tau}}(\boldsymbol{X},\boldsymbol{F}) \quad \forall \boldsymbol{F} \in \mathrm{Lin}^+, \boldsymbol{Q} \in \mathrm{Orth}^+. \tag{5.3}$$

An elastic material is isotropic if it is isotropic at all points of V_0.

Expressed in terms of the response functions for the first and second Piola–Kirchhoff stresses, isotropy at a point X is equivalent to the relations:

$$\hat{T}(X, FQ) = \hat{T}(X, F)Q \quad \forall F \in \text{Lin}^+, Q \in \text{Orth}^+, \tag{5.4}$$

$$\hat{S}(X, FQ) = Q^T\hat{S}(X, F)Q \quad \forall F \in \text{Lin}^+, Q \in \text{Orth}^+. \tag{5.5}$$

Dropping the history dependence from Theorem 4.7.1, we have the following theorem that characterizes an elastic frame-indifferent isotropic material:

Theorem 5.1.1. *The response function of an elastic material is isotropic and frame-indifferent at a point* X *if and only if there exists a mapping* $\bar{\tau} : V_0 \times Psym \to Sym$ *such that*

$$\hat{\tau}(X, F) = \bar{\tau}(X, FF^T) \quad \forall F \in Lin^+, \tag{5.6}$$

and which satisfies

$$\bar{\tau}(X, QBQ^T) = Q\bar{\tau}(X, B)Q^T \quad \forall B \in Psym,\ Q \in Orth^+, \tag{5.7}$$

i.e., the function $\bar{\tau}$ *is isotropic.*

As a consequence of the Rivlin–Ericksen representation theorem for an isotropic function (Theorem 1.6.14), we have

Theorem 5.1.2. *The Cauchy stress at a point* x *in a material that is frame-indifferent and isotropic at* $X \in V_0$ *is given by*

$$\tau(x) = \hat{\tau}(X, F) = \hat{\beta}_0(X, \mathcal{I}_B)I + \hat{\beta}_1(X, \mathcal{I}_B)B + \hat{\beta}_2(X, \mathcal{I}_B)B^2 \quad \forall B = FF^T \in Psym, \tag{5.8}$$

where $\hat{\beta}_0$, $\hat{\beta}_1$, $\hat{\beta}_2$ *are real-valued functions of the principal invariants of* B.

The second Piola–Kirchhoff stress tensor at the point X *is given by*

$$S(X) = \hat{S}(X, C) = \hat{\gamma}_0(X, \mathcal{I}_C)I + \hat{\gamma}_1(X, \mathcal{I}_C)C + \hat{\gamma}_2(X, \mathcal{I}_C)C^2 \tag{5.9}$$

where $C = F^TF \in Psym$.

Conversely, materials that have a constitutive relation of the form given by Eqn. (5.8) (or, equivalently, by Eqn. (5.9)) are frame-indifferent and isotropic.

Proof. Eqn. (5.8) follows directly from the Rivlin–Ericksen theorem. To prove Eqn. (5.9), we first note that $\mathcal{I}_B = \mathcal{I}_C$, since from the polar decomposition theorem, we have $B = RCR^T$. Also note that

$$\det F = I_3(F) = \sqrt{I_3(B)} = \sqrt{I_3(C)}.$$

Thus, we obtain (dropping the explicit dependence on X for notational convenience)

$$\begin{aligned}\hat{S}(F) &= (\det F)F^{-1}\hat{\tau}(F)F^{-T} \\ &= (\det F)(\hat{\beta}_0(\mathcal{I}_B)F^{-1}F^{-T} + \hat{\beta}_1(\mathcal{I}_B)I + \hat{\beta}_2(\mathcal{I}_B)F^TF) \\ &= \sqrt{I_3(C)}(\hat{\beta}_0(\mathcal{I}_C)C^{-1} + \hat{\beta}_1(\mathcal{I}_C)I + \hat{\beta}_2(\mathcal{I}_C)C),\end{aligned}$$

and since by the Cayley–Hamilton theorem

$$C^{-1} = \frac{1}{I_3(C)}(C^2 - I_1(C)C + I_2(C)I),$$

Eqn. (5.9) follows. □

For an anisotropic response at a point, $\hat{S}$ is still a function of C (by virtue of material frame-indifference), for example, $\hat{S} = \mathbf{C}[E]$, where $\mathbf{C}$ is a fourth-order constitutive tensor. However, $\hat{\tau}(F)$ does not reduce to a function of B as in the isotropic case.

By using the Cayley–Hamilton theorem as above, we obtain alternative forms for the response functions $\bar{\tau}(B)$ and $\tilde{S}$ as

$$\bar{\tau}(X,B) = \beta_{-1}(X,\mathcal{I}_B)B^{-1} + \beta_0(X,\mathcal{I}_B)I + \beta_1(X,\mathcal{I}_B)B, \tag{5.10}$$

$$\tilde{S}(X,C) = \gamma_{-1}(X,\mathcal{I}_C)C^{-1} + \gamma_0(X,\mathcal{I}_C)I + \gamma_1(X,\mathcal{I}_C)C. \tag{5.11}$$

For an incompressible elastic material, the constitutive relations given by Eqn. (5.10) is replaced by (for the justification, see the end of Section 7.1, and [39])

$$\bar{\tau}(X,B) = -pI + \beta_{-1}(I_1,I_2)B^{-1} + \beta_1(I_1,I_2)B, \tag{5.12}$$

where the β_0 term has been absorbed in the constitutively 'indeterminate' pressure $p(x,t)$, and β_{-1} and β_1 are functions of $I_1(B)$ and $I_2(B)$ only since $I_3(B) = J^2 = 1$. Experimental data seems to indicate that

$$\beta_{-1} \le 0, \quad \beta_1 > 0. \tag{5.13}$$

The above inequalities are known as the *E-inequalities*. If β_{-1} and β_1 are taken to be constant, then the resulting material model, which is often used as a constitutive equation for rubber-like materials, defines an incompressible *Mooney–Rivlin* material; see Problem 1 for the extension to the compressible case. If β_{-1} is set to zero with β_1 still assumed to be constant, the resulting material model is called (incompressible) *neo-Hookean*. This model is not adequate for rubber, and certain typical effects of nonlinear elasticity are also found wanting in solutions obtained using this model [337]. In contrast, the Mooney–Rivlin model does seem to give at least a qualitative agreement with experimental results.

If a material is isotropic at a point X, the residual stress tensor at X is a pressure since

$$\tau(x) = S(X) = \hat{\tau}(X,I) = \hat{S}(X,I) = -\pi I,$$

where $-\pi = \hat{\beta}_0(X,\mathcal{I}_I) + \hat{\beta}_1(X,\mathcal{I}_I) + \hat{\beta}_2(X,\mathcal{I}_I)$. From this result, it follows that a body loses its isotropy if it undergoes an arbitrary deformation, since $\hat{\tau}(F)$ is not a multiple of the identity tensor for an arbitrary $F \in \text{Lin}^+$. Thus, if we choose an arbitrary deformed configuration as the reference configuration, then we cannot assume that the material is isotropic. In other words, as pointed out in Section 4.7 also, isotropy is a property that holds only in certain reference configurations.

Even though most people have a solid in mind when they speak of an elastic particle, such a restriction is not built into its definition. In fact, we have

Theorem 5.1.3. *If the elastic particle X is a fluid, then there exists a function $p^X : \Re \to \Re$, such that the constitutive relation given by Eqn. (5.8) can be written in the form*

$$\tau(x,t) = -p(X, \rho(x,t))I, \tag{5.14}$$

where $\rho(x,t)$ is the density at the point $x = x(X,t)$ at time t.[a]

Conversely, given any function $p^X : \Re \to \Re$, the constitutive relation Eqn. (5.14) satisfies the principle of material frame-indifference, and defines a fluid elastic particle.

Proof. Suppose that the elastic particle X is a fluid. Then, it is necessarily isotropic (Theorem 4.7.3), and we can use the representation given by Eqn. (5.8). Furthermore, since $\partial_\kappa^X = \text{Unim}^+$ for a fluid particle, we must have

$$\hat{\tau}(x,F) = \hat{\tau}(x,FH) \quad \forall H \in \text{Unim}^+,$$

i.e.,

$$\tau(x,t) = \hat{\beta}_0(X, \mathcal{I}_E)I + \hat{\beta}_1(X, \mathcal{I}_E)E + \hat{\beta}_2(X, \mathcal{I}_E)E^2,$$

where $E = FHH^TF^T$. Choose $H = J^{1/3}F^{-1}$ (note that $\det H = 1$). Then $E = J^{2/3}I$, and $I_1(E) = 3J^{2/3}$, $I_2(E) = 3J^{4/3}$, and $I_3(E) = J^2$. Thus, $\tau(x,t)$ is of the form $\gamma(X, J(X,t))I$. Using the relation $\rho_0 = \rho(x,t)J$, and defining the pressure as

$$p(x,t) = -\gamma(X, \rho_0/\rho),$$

we get the constitutive relation for an elastic fluid as

$$\tau(x,t) = -p(X, \rho(x,t))I,$$

which proves Eqn. (5.14). Note that the above constitutive equation is independent of the choice of reference configuration.

Conversely, if the stress response function is defined by Eqn. (5.14), then noting that

$$\rho^* = \frac{\rho_0}{\det F^*} = \frac{\rho_0}{(\det Q)(\det F)} = \frac{\rho_0}{\det F} = \rho,$$

we have

$$\begin{aligned}\tau^*(x^*,t^*) &= -p(X,\rho^*)I \\ &= -p(X,\rho)Q(t)Q^T(t)\end{aligned}$$

$$= Q(t)[-p(X,\rho)]Q^T(t)$$
$$= Q(t)\tau(x,t)Q^T,$$

thus showing that the stress response is frame-indifferent. For any $H \in \text{Unim}^+$, we have

$$\hat{\tau}(X, FH) = -p(X, \frac{\rho_0}{\det(FH)})I$$
$$= -p(X, \frac{\rho_0}{\det F})I$$
$$= -p(X, \rho(X,t))I$$
$$= \hat{\tau}(X, F),$$

thus proving that $\text{Unim}^+ \subset \partial_\kappa^X$. By its very definition $\partial_\kappa^X \subset \text{Unim}^+$. Hence, $\partial_\kappa^X = \text{Unim}^+$, so that the constitutive relation given by Eqn. (5.14) defines a fluid particle. □

[a] An elastic fluid is also known as barotropic.

The converse assertion in the above proof tells us that additional special choices of elements of the material symmetry group cannot lead to further reductions in the form of the constitutive equation for an elastic fluid. Note that we got this particularly simple relation for the constitutive relation because we did not include the effects of history dependence. We shall include these effects in Section 7.3.

If an isotropic material is also a solid, then its material symmetry group is Orth^+. Thus, its constitutive relations can be taken to have the form given by Eqn. (5.8) (or any other equivalent form); there are no additional reductions possible as in the case of an elastic fluid.

Since an isotropic particle can only be either a solid or a fluid (Theorem 4.7.7), there are no more cases to explore. Henceforth, we shall implicitly assume an isotropic particle to be solid.

5.2 The Constitutive Equation of an Isotropic Solid for 'Small' Deformations

We wish to compute the stress S for deformations that are 'small'. Hence, we carry out an expansion of S in terms of E upto the linear terms. As the following theorem shows, the first-order term involves only two constants.

Theorem 5.2.1. *Assuming that the constants $\hat{\gamma}_i$ that occur in Eqn.* (5.9) *are differentiable at the point $\mathcal{I}_I = (3,3,1)$, the response function of a frame-indifferent, elastic isotropic solid is given by*

$$\tilde{S}(X,C) = -\pi(X)I + \lambda(X)(\operatorname{tr} E)I + 2\mu(X)E + o(E) \quad \forall C = I + 2E \in Psym, \tag{5.15}$$

where $\pi(\boldsymbol{X})$, $\lambda(\boldsymbol{X})$ and $\mu(\boldsymbol{X})$ are constants.

Proof. We use the concept of the directional derivative to find the linearization. We have

$$\tilde{\boldsymbol{S}}(\boldsymbol{I}+2\boldsymbol{E}) = \tilde{\boldsymbol{S}}(\boldsymbol{I}) + 2D\tilde{\boldsymbol{S}}(\boldsymbol{I})[\boldsymbol{E}] + \mathrm{o}\,(\boldsymbol{E})\,.$$

To find $D\tilde{\boldsymbol{S}}(\boldsymbol{I})[\boldsymbol{E}]$, we use the product rule as follows (again dropping the explicit dependence on $\boldsymbol{X}$ for notational convenience):

$$\begin{aligned}
D\tilde{\boldsymbol{S}}(\boldsymbol{I})[\boldsymbol{E}] &= D\hat{\gamma}_0(\mathcal{I}_C)(\boldsymbol{I})[\boldsymbol{E}]\boldsymbol{I} + D\hat{\gamma}_1(\mathcal{I}_C)(\boldsymbol{I})[\boldsymbol{E}]\boldsymbol{I} + D\hat{\gamma}_2(\mathcal{I}_C)(\boldsymbol{I})[\boldsymbol{E}]\boldsymbol{I} \\
&\qquad + \hat{\gamma}_1(\mathcal{I}_I)D\boldsymbol{C}(\boldsymbol{I})[\boldsymbol{E}] + \hat{\gamma}_2(\mathcal{I}_I)D\boldsymbol{C}^2(\boldsymbol{I})[\boldsymbol{E}] \\
&= \left[\frac{\partial\hat{\gamma}_0}{\partial I_1}DI_1(\boldsymbol{I})[\boldsymbol{E}] + \frac{\partial\hat{\gamma}_0}{\partial I_2}DI_2(\boldsymbol{I})[\boldsymbol{E}] + \frac{\partial\hat{\gamma}_0}{\partial I_3}DI_3(\boldsymbol{I})[\boldsymbol{E}]\right]\boldsymbol{I} \\
&\quad + \left[\frac{\partial\hat{\gamma}_1}{\partial I_1}DI_1(\boldsymbol{I})[\boldsymbol{E}] + \frac{\partial\hat{\gamma}_1}{\partial I_2}DI_2(\boldsymbol{I})[\boldsymbol{E}] + \frac{\partial\hat{\gamma}_1}{\partial I_3}DI_3(\boldsymbol{I})[\boldsymbol{E}]\right]\boldsymbol{I} \\
&\quad + \left[\frac{\partial\hat{\gamma}_2}{\partial I_1}DI_1(\boldsymbol{I})[\boldsymbol{E}] + \frac{\partial\hat{\gamma}_2}{\partial I_2}DI_2(\boldsymbol{I})[\boldsymbol{E}] + \frac{\partial\hat{\gamma}_2}{\partial I_3}DI_3(\boldsymbol{I})[\boldsymbol{E}]\right]\boldsymbol{I} \\
&\quad + \hat{\gamma}_1(\mathcal{I}_I)\boldsymbol{E} + 2\hat{\gamma}_2(\mathcal{I}_I)\boldsymbol{E},
\end{aligned}$$

where we have used the facts that $D\boldsymbol{C}(\boldsymbol{I})[\boldsymbol{U}] = \boldsymbol{U}$, and $D\boldsymbol{C}^2(\boldsymbol{I})[\boldsymbol{U}] = 2\boldsymbol{U}$. From the results in Section 1.9, we have

$$\begin{aligned}
DI_1(\boldsymbol{I})[\boldsymbol{E}] &= \operatorname{tr}\boldsymbol{E}, \\
DI_2(\boldsymbol{I})[\boldsymbol{E}] &= 3\operatorname{tr}\boldsymbol{E} - \operatorname{tr}\boldsymbol{E} = 2\operatorname{tr}\boldsymbol{E}, \\
DI_3(\boldsymbol{I})[\boldsymbol{E}] &= \operatorname{tr}\boldsymbol{E},
\end{aligned}$$

which on substituting into the above equation for $D\tilde{\boldsymbol{S}}(\boldsymbol{I})[\boldsymbol{E}]$ yields

$$2D\tilde{\boldsymbol{S}}(\boldsymbol{I})[\boldsymbol{E}] = \lambda(\operatorname{tr}\boldsymbol{E})\boldsymbol{I} + 2\mu\boldsymbol{E},$$

where

$$\begin{aligned}
\lambda &= 2\frac{\partial\hat{\gamma}_0}{\partial I_1}(\mathcal{I}_I) + 4\frac{\partial\hat{\gamma}_0}{\partial I_2}(\mathcal{I}_I) + 2\frac{\partial\hat{\gamma}_0}{\partial I_3}(\mathcal{I}_I) + 2\frac{\partial\hat{\gamma}_1}{\partial I_1}(\mathcal{I}_I) + 4\frac{\partial\hat{\gamma}_1}{\partial I_2}(\mathcal{I}_I) + 2\frac{\partial\hat{\gamma}_1}{\partial I_3}(\mathcal{I}_I) \\
&\qquad + 2\frac{\partial\hat{\gamma}_2}{\partial I_1}(\mathcal{I}_I) + 4\frac{\partial\hat{\gamma}_2}{\partial I_2}(\mathcal{I}_I) + 2\frac{\partial\hat{\gamma}_2}{\partial I_3}(\mathcal{I}_I), \\
\mu &= \hat{\gamma}_1(\mathcal{I}_I) + 2\hat{\gamma}_2(\mathcal{I}_I),
\end{aligned}$$

with $\mathcal{I}_I = (3,3,1)$. Let

$$-\pi = \hat{\gamma}_0(\mathcal{I}_I) + \hat{\gamma}_1(\mathcal{I}_I) + \hat{\gamma}_2(\mathcal{I}_I).$$

Since

$$\tilde{\boldsymbol{S}}(\boldsymbol{I}) = [\hat{\gamma}_0(\mathcal{I}_I) + \hat{\gamma}_1(\mathcal{I}_I) + \hat{\gamma}_2(\mathcal{I}_I)]\,\boldsymbol{I} = -\pi\boldsymbol{I},$$

we get Eqn. (5.15). □

If we make the simultaneous assumptions that the elastic solid is isotropic, homogeneous, and that the reference configuration is a natural state, we obtain the following corollary of Theorem 5.2.1:

Theorem 5.2.2. *Let there be a given homogeneous, isotropic, elastic solid, whose reference configuration is a natural state. If the functions* $\hat{\gamma}_i$, $i = 0, 1, 2$ *in Eqn. (5.9) are differentiable at the point* $\mathcal{I}_I = (3, 3, 1)$, *there exist two constants* λ *and* μ *such that the response function* $\tilde{S} : Psym \to Sym$ *is of the form*

$$\tilde{S}(C) = \lambda(tr\, E)I + 2\mu E + o\,(E) \quad \forall C = F^T F = I + 2E \in Psym,$$

In this case, and only in this case, the constants λ and μ are called the *Lamé constants* of the material under consideration. Note that the existence of the Lamé constants has been derived from an expansion of the constitutive equation with respect to the Green strain tensor E, and not with respect to the linearized strain tensor. If we *assume* that $\tilde{S}$ is a linear function of E, then the result of the above theorem follows directly from Theorem 1.6.15.

5.3 Bounds on the Lamé Constants

By conducting three thought experiments, we can deduce bounds on the Lamé constants, and the corresponding Young modulus and Poisson ratio for an isotropic material whose reference configuration is a natural state. Note that the Cauchy stress tensor in such a case is given by

$$\tau = \frac{1}{J}\left[\lambda(\mathrm{tr}\, E)B + 2\mu F E F^T\right] + o\,(E)\,.$$

The first experiment is one of simple shear (see Section 2.5.3). The Cauchy stress tensor component τ_{12} is given by

$$\tau_{12} = \mu\gamma + o\,(\gamma)\,.$$

Since, intuitively, we expect $\tau_{12} > 0$, we get

$$\mu > 0. \tag{5.16}$$

The second experiment is the uniform compression of a sphere with the deformation given by $\chi = (1 - \epsilon)X$. In this case, the Cauchy stress tensor is given by

$$\tau = -\frac{\epsilon(2-\epsilon)}{2(1-\epsilon)}(3\lambda + 2\mu)I,$$

which agrees with our intuitive idea that the Cauchy stress is a pressure. Since the stress should be compressive, we get

$$3\lambda + 2\mu > 0. \tag{5.17}$$

The third experiment is that of uniform extension (see Section 2.5.1). The components τ_{22} and τ_{33} are given by

$$\tau_{22} = \tau_{33} = \frac{\epsilon}{1+\epsilon}\left[\lambda(1-2\nu) - 2\mu\nu\right] + o\,(\epsilon)\,.$$

Since we expect the first-order terms to vanish, we get

$$\nu = \frac{\lambda}{2(\lambda + \mu)}. \tag{5.18}$$

The component τ_{11} is given by

$$\tau_{11} = \frac{\epsilon(1+\epsilon)}{(1-\nu\epsilon)^2}\left[\lambda(1-2\nu) + 2\mu\right] + \mathrm{o}\left(\epsilon\right).$$

Substituting for ν using Eqn. (5.18), we get

$$\tau_{11} = \frac{\epsilon(1+\epsilon)}{(1-\nu\epsilon)^2}\left[\frac{\mu(3\lambda+2\mu)}{(\lambda+\mu)}\right] + \mathrm{o}\left(\epsilon\right).$$

We expect τ_{11} to be of the form $\epsilon E + \mathrm{o}\left(\epsilon\right)$. Hence,

$$E = \frac{\mu(3\lambda+2\mu)}{(\lambda+\mu)}. \tag{5.19}$$

The constants E and ν are known as the *Young modulus* and *Poisson ratio*, respectively. We summarize the relations between the Lamé constants, and the Young modulus and Poisson ratio:

$$\begin{aligned} E &= \frac{\mu(3\lambda+2\mu)}{(\lambda+\mu)}, \quad & \lambda &= \frac{E\nu}{(1+\nu)(1-2\nu)}, \\ \nu &= \frac{\lambda}{2(\lambda+\mu)}, \quad & \mu &= \frac{E}{2(1+\nu)}. \end{aligned} \tag{5.20}$$

Using Eqns. (5.16), (5.17) and the above relations, we have the following equivalence:

$$\mu > 0 \text{ and } 3\lambda + 2\mu > 0 \iff E > 0 \text{ and } -1 < \nu < 0.5.$$

Note that the bounds on the Lamé constants are obtained from empirical considerations, and not from thermodynamic ones, as in the case of, say, the viscosity coefficients of a Newtonian fluid (see Section 7.3).

5.4 Hyperelastic Solids

An elastic material is hyperelastic if there exists a stored energy function (also called the strain energy density function) $\hat{W} : V_0 \times \mathrm{Lin}^+ \to \Re$ such that

$$\hat{\boldsymbol{T}}(\boldsymbol{X}, \boldsymbol{F}) = \frac{\partial \hat{W}}{\partial \boldsymbol{F}}(\boldsymbol{X}, \boldsymbol{F}) \quad \forall \boldsymbol{X} \in V_0, \boldsymbol{F} \in \mathrm{Lin}^+, \tag{5.21}$$

or, in component form

$$T_{ij}(\boldsymbol{X}, \boldsymbol{F}) = \frac{\partial \hat{W}}{\partial F_{ij}}(\boldsymbol{X}, \boldsymbol{F}).$$

The gradient $\partial \hat{W}/\partial \boldsymbol{F}$ is evaluated as discussed in Section 1.9.4.

Since $\partial \hat{W}/\partial t = (\partial \hat{W}/\partial \boldsymbol{F}) : \dot{\boldsymbol{F}}$, and since V_0 is independent of t, the mechanical energy balance equation for a hyperelastic material can be written using Eqn. (3.69) as

$$\frac{d}{dt}\int_{V_0}\left[\rho_0\frac{\dot{\chi}\cdot\dot{\chi}}{2}+\hat{W}(\boldsymbol{X},\boldsymbol{F}(\boldsymbol{X},t))\right]dV_0=\int_{S_0}\dot{\chi}\cdot(\boldsymbol{t}_0)\,dS_0+\int_{V_0}\rho_0\dot{\chi}\cdot\boldsymbol{b}_0\,dV_0. \tag{5.22}$$

Note that if $\boldsymbol{b}_0 = \boldsymbol{t}_0 = \boldsymbol{0}$, then from Eqns. (3.51), (3.52) and (5.22), it follows that the linear momentum, angular momentum and the total energy (sum of kinetic and strain energies) are conserved, i.e.,

$$\int_{V_0}\rho_0\left(\frac{\partial \boldsymbol{u}}{\partial t}\right)_{\boldsymbol{X}}dV_0=\text{constant}, \tag{5.23a}$$

$$\int_{V_0}\rho_0\chi\times\left(\frac{\partial \boldsymbol{u}}{\partial t}\right)_{\boldsymbol{X}}dV_0=\text{constant}, \tag{5.23b}$$

$$\int_{V_0}\left[\frac{1}{2}\rho_0\left(\frac{\partial \boldsymbol{u}}{\partial t}\right)_{\boldsymbol{X}}\cdot\left(\frac{\partial \boldsymbol{u}}{\partial t}\right)_{\boldsymbol{X}}+\hat{W}\right]dV_0=\text{constant}. \tag{5.23c}$$

A generalization of Eqn. (5.23c) to the case where the loading is conservative is presented in Eqn. (5.36). If $\boldsymbol{b}_0 = \boldsymbol{t}_0 = \boldsymbol{0}$ and $Q_h = \boldsymbol{q}\cdot\boldsymbol{n} = 0$, then from the material version of Eqn. (3.75), we get

$$\int_{V_0}\left[\frac{1}{2}\rho_0\left(\frac{\partial \boldsymbol{u}}{\partial t}\right)_{\boldsymbol{X}}\cdot\left(\frac{\partial \boldsymbol{u}}{\partial t}\right)_{\boldsymbol{X}}+e_0\right]dV_0=\text{constant},$$

where $e_0(\boldsymbol{X},t) := e(\boldsymbol{x},t)$.

The motivation behind defining the first Piola–Kirchhoff stress, and not the Cauchy or the second Piola–Kirchhoff stress, as the gradient of a scalar function, is that the scalar function $\hat{W}$ as defined above transforms in the way an objective scalar function should transform under a change of observer, i.e., $\hat{W}(\boldsymbol{X},\boldsymbol{QF}) = \hat{W}(\boldsymbol{X},\boldsymbol{F})$, as shown in the following theorem:

Theorem 5.4.1. *The stored energy function $\hat{W} : V_0 \times Lin^+ \to \Re$ of a hyperelastic material is frame-indifferent if and only if at all points $\boldsymbol{X} \in V_0$:*

$$\hat{W}(\boldsymbol{X},\boldsymbol{QF})=\hat{W}(\boldsymbol{X},\boldsymbol{F})\quad\forall\boldsymbol{F}\in Lin^+,\boldsymbol{Q}\in Orth^+, \tag{5.24}$$

or, equivalently, if and only if there exists a function $\tilde{W} : V_0 \times Psym \to \Re$ such that

$$\hat{W}(\boldsymbol{X},\boldsymbol{F})=\tilde{W}(\boldsymbol{X},\boldsymbol{F}^T\boldsymbol{F})\quad\forall\boldsymbol{F}\in Lin^+. \tag{5.25}$$

Proof. We drop the dependence on $\boldsymbol{X}$ for notational brevity. From Eqn. (5.1), the response function for the first Piola–Kirchhoff stress is frame-indifferent if and only if

$$\hat{\boldsymbol{T}}(\boldsymbol{QF})=\boldsymbol{Q}\hat{\boldsymbol{T}}(\boldsymbol{F})\quad\forall\boldsymbol{F}\in\text{Lin}^+,\boldsymbol{Q}\in\text{Orth}^+,$$

i.e., if and only if

$$\frac{\partial \hat{W}}{\partial F}(F) = Q^T \frac{\partial \hat{W}}{\partial F}(QF) \quad \forall F \in \mathrm{Lin}^+, Q \in \mathrm{Orth}^+. \tag{5.26}$$

To get the required result, we need to compute the gradient of the mapping $\hat{W}_Q : F \in \mathrm{Lin}^+ \to \hat{W}_Q(F) := \hat{W}(QF)$ for a fixed Q. Using Eqn. (1.48), we have

$$\begin{aligned}\hat{W}_Q(F+U) = \hat{W}(QF+QU) &= \hat{W}(QF) + \frac{\partial \hat{W}}{\partial F}(QF) : (QU) + \mathrm{o}(QU) \\ &= \hat{W}_Q(F) + Q^T \frac{\partial \hat{W}}{\partial F}(QF) : U + \mathrm{o}(U),\end{aligned}$$

since $\mathrm{o}(QU) = \mathrm{o}(U)$ for a fixed Q. Thus,

$$\frac{\partial \hat{W}_Q}{\partial F}(F) = Q^T \frac{\partial \hat{W}}{\partial F}(QF). \tag{5.27}$$

Comparing Eqn. (5.26) and (5.27), we can say that $\hat{W}$ is indifferent if and only if

$$\frac{\partial}{\partial F}\left[\hat{W}_Q(F) - \hat{W}(F)\right] = \mathbf{0} \quad \forall F \in \mathrm{Lin}^+, Q \in \mathrm{Orth}^+. \tag{5.28}$$

If Eqn. (5.24) holds, then clearly Eqn. (5.28) holds. To prove the converse, first note that Eqn. (5.28) implies

$$\hat{W}(QF) = \hat{W}(F) + C(Q) \quad \forall F \in \mathrm{Lin}^+, Q \in \mathrm{Orth}^+. \tag{5.29}$$

We now show that $C(Q) = 0$. Taking $F = I$ in Eqn. (5.29), we get $\hat{W}(Q) = \hat{W}(I) + C(Q)$. Now taking $F = Q$, we get $\hat{W}(Q^2) = \hat{W}(Q) + C(Q) = \hat{W}(I) + 2C(Q)$. Continuing this process, we get

$$\hat{W}(Q^p) = \hat{W}(I) + pC(Q) \quad \forall Q \in \mathrm{Orth}^+, p \geq 1.$$

Since $\hat{W}(I)$ is a bounded quantity independent of p, and since $C(Q)$ is also independent of p, it follows from the above equation that $\left|\hat{W}(Q^p)\right| \to \infty$ as $p \to \infty$. However, since $\hat{W}$ is assumed to be differentiable, and hence continuous, $\hat{W}(Q^p)$ cannot be unbounded. The only way this contradiction can be resolved is if $C(Q) = 0$.

To prove the second equivalence, let $F \in \mathrm{Lin}^+$ be given, and let $F = RU$ be its polar decomposition. Recall that $R \in \mathrm{Orth}^+$ and $U = \sqrt{F^T F} \in \mathrm{Psym}$. Then, from the first equivalence we get

$$\hat{W}(F) = \hat{W}(RU) = \hat{W}(U) = \tilde{W}(F^T F),$$

with

$$\tilde{W}(C) := \hat{W}(\sqrt{C}) \quad \forall C \in \mathrm{Psym}. \tag{5.30}$$

If conversely $\hat{W}(F) = \tilde{W}(F^T F)$ holds, then

$$\hat{W}(QF) = \tilde{W}(F^T Q^T Q F) = \tilde{W}(F^T F) = \hat{W}(F) \quad \forall F \in \mathrm{Lin}^+, Q \in \mathrm{Orth}^+,$$

which shows that Eqn. (5.24) holds. □

Just as the first Piola–Kirchhoff stress is the gradient of $\hat{W}$, the second Piola–Kirchhoff stress is twice the gradient of $\tilde{W}$, as we now show. The gradient $\partial\tilde{W}/\partial\boldsymbol{C}(\boldsymbol{X},\boldsymbol{C})$ is computed as discussed on page 88.

Theorem 5.4.2. *Given a hyperelastic material whose stored energy density function $\hat{W}$ is frame-indifferent, and given the function $\tilde{W}$ defined by*

$$\tilde{W}(\boldsymbol{X},\boldsymbol{C}) := \hat{W}(\boldsymbol{X},\sqrt{\boldsymbol{C}}) \quad \forall \boldsymbol{C} \in Psym, \tag{5.31}$$

if we assume (without loss of generality) that the gradient $\partial\tilde{W}/\partial\boldsymbol{C}(\boldsymbol{X},\boldsymbol{C})$ is a symmetric tensor, then the response function for the second Piola–Kirchhoff stress is given by

$$\hat{\boldsymbol{S}}(\boldsymbol{X},\boldsymbol{F}) = \tilde{\boldsymbol{S}}(\boldsymbol{X},\boldsymbol{C}) = 2\frac{\partial\tilde{W}}{\partial\boldsymbol{C}}(\boldsymbol{X},\boldsymbol{C}), \quad \boldsymbol{C} = \boldsymbol{F}^T\boldsymbol{F}, \quad \forall \boldsymbol{F} \in Lin^+, \tag{5.32}$$

or, equivalently, by

$$\check{\boldsymbol{S}}(\boldsymbol{X},\boldsymbol{E}) = \frac{\partial\check{W}}{\partial\boldsymbol{E}}(\boldsymbol{X},\boldsymbol{E}), \ \textit{with}\ \check{W}(\boldsymbol{X},\boldsymbol{E}) = \tilde{W}(\boldsymbol{X},\boldsymbol{C}),\ \boldsymbol{I}+2\boldsymbol{E} = \boldsymbol{C},\ \forall \boldsymbol{F} \in Lin^+, \tag{5.33}$$

Conversely, an elastic material whose response function $\hat{\boldsymbol{S}}$ is given by Eqn. (5.32) is hyperelastic with the stored energy density function given by

$$\hat{W}(\boldsymbol{X},\boldsymbol{F}) = \tilde{W}(\boldsymbol{X},\boldsymbol{F}^T\boldsymbol{F}) \quad \forall \boldsymbol{F} \in Lin^+.$$

Proof. We drop the dependence on $\boldsymbol{X}$ in the proof. Since

$$\hat{\boldsymbol{S}}(\boldsymbol{F}) = \boldsymbol{F}^{-1}\hat{\boldsymbol{T}}(\boldsymbol{F}) = \boldsymbol{F}^{-1}\frac{\partial\hat{W}}{\partial\boldsymbol{F}}(\boldsymbol{F}),$$

we need to compute the gradient $\partial\hat{W}/\partial\boldsymbol{F}$ in terms of the gradient $\partial\tilde{W}/\partial\boldsymbol{C}$. For $\boldsymbol{U} \in \text{Lin}^+$ with sufficiently small $\|\boldsymbol{U}\|$, the tensor $(\boldsymbol{F}+\boldsymbol{U})^T(\boldsymbol{F}+\boldsymbol{U})$ is also in Psym. Hence,

$$\begin{aligned}
\frac{\partial\hat{W}}{\partial\boldsymbol{F}}(\boldsymbol{F}) : \boldsymbol{U} &= D_{\boldsymbol{F}}\hat{W}(\boldsymbol{F})[\boldsymbol{U}] \\
&= D_{\boldsymbol{F}}\tilde{W}(\boldsymbol{F}^T\boldsymbol{F})[\boldsymbol{U}] \\
&= D_{\boldsymbol{C}}\tilde{W}(\boldsymbol{F}^T\boldsymbol{F})[D_{\boldsymbol{F}}(\boldsymbol{F}^T\boldsymbol{F})[\boldsymbol{U}]] \\
&= D_{\boldsymbol{C}}\tilde{W}(\boldsymbol{F}^T\boldsymbol{F})[\boldsymbol{F}^T\boldsymbol{U}+\boldsymbol{U}^T\boldsymbol{F}] \\
&= \frac{\partial\tilde{W}}{\partial\boldsymbol{C}}(\boldsymbol{F}^T\boldsymbol{F}) : [\boldsymbol{F}^T\boldsymbol{U}+\boldsymbol{U}^T\boldsymbol{F}] \\
&= \boldsymbol{F}\left\{\frac{\partial\tilde{W}}{\partial\boldsymbol{C}}(\boldsymbol{F}^T\boldsymbol{F}) + \left[\frac{\partial\tilde{W}}{\partial\boldsymbol{C}}(\boldsymbol{F}^T\boldsymbol{F})\right]^T\right\} : \boldsymbol{U},
\end{aligned}$$

where we have used Eqn. (1.48) to get the last step. The above equation implies that

$$\begin{aligned}\hat{S}(F) &= F^{-1}\frac{\partial \hat{W}}{\partial F}(F)\\ &= \frac{\partial \tilde{W}}{\partial C}(F^T F) + \left[\frac{\partial \tilde{W}}{\partial C}(F^T F)\right]^T\\ &= 2\frac{\partial \tilde{W}}{\partial C}(F^T F),\end{aligned}$$

where the last step follows from the assumed symmetry of $\partial\tilde{W}/\partial C$. The converse also follows from the above relation.

To prove Eqn. (5.33), note that

$$\begin{aligned}\frac{\partial \tilde{W}}{\partial C} : U &= D_C\tilde{W}(C)[U]\\ &= D_E\check{W}(E)[D_C(E)[U]]\\ &= D_E\check{W}(E)[U/2]\\ &= \frac{1}{2}D_E\check{W}(E)[U]\\ &= \frac{1}{2}\frac{\partial \check{W}}{\partial E} : U.\end{aligned}$$

Since the above equation is valid for all U, we get Eqn. (5.33).

An alternative proof that directly uses the results for fourth-order tensors can be given as in the proof of Theorem 7.1.2. □

Note that the second Piola–Kirchhoff stress is a function of $F^T F$ alone. However, this was true even without the assumption of hyperelasticity (see Eqn. (4.53)).

Loads that are independent of the deformation are called dead loads; loads that are not dead are called live. For example, the body force due to gravity given by $b_0 = b = -ge_z$, where e_z points "upwards" is not dependent on the deformation, and hence a dead load. The body force b_0 is said to be *conservative* if there exists a potential $W_{b_0}(X, u)$ such that the first term on the right-hand side of Eqn. (3.59) can be expressed as the first variation of $\int_{V_0} \rho_0 W_{b_0}\, dV$, i.e.,

$$\int_{V_0} \rho_0 u_\delta \cdot b_0\, dV_0 = \delta^{(1)} \int_{V_0} \rho_0 W_{b_0}\, dV_0. \tag{5.34}$$

Similarly, the traction $\bar{t}_0$ is conservative if there exists a potential function $W_{t_0}(X, u, F)$ such that

$$\int_{(S_0)_t} u_\delta \cdot \bar{t}_0\, dS_0 = \delta^{(1)} \int_{(S_0)_t} W_{t_0}\, dS_0. \tag{5.35}$$

Dead loads are conservative with $W_{b_0} = \boldsymbol{b}_0 \cdot \boldsymbol{u}$ and $W_{t_0} = \bar{\boldsymbol{t}}_0 \cdot \boldsymbol{u}$. An example of a conservative live body force is the centrifugal force since

$$\delta^{(1)} \int_{V_0} \frac{1}{2} |\boldsymbol{\Omega} \times (\boldsymbol{X} + \boldsymbol{u})|^2 \, dV_0 = \int_{V_0} [\boldsymbol{\Omega} \times (\boldsymbol{X} + \boldsymbol{u})] \cdot [\boldsymbol{\Omega} \times \boldsymbol{u}_\delta] \, dV_0$$
$$= \int_{V_0} -\boldsymbol{\Omega} \times [\boldsymbol{\Omega} \times (\boldsymbol{X} + \boldsymbol{u})] \cdot \boldsymbol{u}_\delta \, dV_0,$$

so that $W_{b_0} = |\boldsymbol{\Omega} \times \boldsymbol{x}|^2/2$. An example of a conservative live surface force is uniform pressure loading on $(S_0)_t$ with the remaining boundary $(S_0)_u$ fixed. The function W_{t_0}, in this particular example defined over the entire boundary S_0, is given by $-p[(\mathbf{cof}\,\boldsymbol{F})\boldsymbol{n}_0 \cdot \boldsymbol{\chi}]/3$, since

$$\begin{aligned}
\delta^{(1)} \int_{S_0} -\frac{p}{3} (\mathbf{cof}\,\boldsymbol{F}) \boldsymbol{n}_0 \cdot \boldsymbol{\chi} \, dS_0 &= -\frac{p}{3} \delta^{(1)} \int_{S_0} [(\mathbf{cof}\,\boldsymbol{F})^T \boldsymbol{\chi}] \cdot \boldsymbol{n}_0 \, dS_0 \\
&= -\frac{p}{3} \delta^{(1)} \int_{V_0} \boldsymbol{\nabla}_X \cdot [(\mathbf{cof}\,\boldsymbol{F})^T \boldsymbol{\chi}] \, dV_0 && \text{(by Eqn. (1.315))} \\
&= -p \delta^{(1)} \int_{V_0} \det \boldsymbol{F} \, dV_0 && \text{(by Eqn. (2.36))} \\
&= -p \int_{V_0} (\mathbf{cof}\,\boldsymbol{F}) : \boldsymbol{\nabla}_X \boldsymbol{u}_\delta \, dV_0 && \text{(by Eqn. (1.216))} \\
&= -p \int_{V_0} \boldsymbol{\nabla}_X \cdot [(\mathbf{cof}\,\boldsymbol{F})^T \boldsymbol{u}_\delta] \, dV_0 && \text{(by Eqn. (2.35))} \\
&= -p \int_{S_0} (\mathbf{cof}\,\boldsymbol{F})^T \boldsymbol{u}_\delta \cdot \boldsymbol{n}_0 \, dS_0 && \text{(by Eqn. (1.315))} \\
&= -\int_{(S_0)_t} p (\mathbf{cof}\,\boldsymbol{F}) \boldsymbol{n}_0 \cdot \boldsymbol{u}_\delta \, dS_0. && (\boldsymbol{u}_\delta = \boldsymbol{0} \text{ on } (S_0)_u)
\end{aligned}$$

Assuming the material to be hyperelastic and the loading to be conservative, if one defines the following potential energy functionals

$$\Pi(\boldsymbol{u}) = \int_{V_0} \check{W}(\boldsymbol{E}) \, dV_0 - \int_{V_0} W_{b_0} \, dV_0 - \int_{(S_0)_t} W_{t_0} \, dS_0,$$
$$\Pi_{\mathrm{HR}}(\boldsymbol{u}, \boldsymbol{S}) = \int_{V_0} [\boldsymbol{S} : \bar{\boldsymbol{E}}(\boldsymbol{u}) - W^*(\boldsymbol{S})] \, dV_0 - \int_{V_0} W_{b_0} \, dV - \int_{(S_0)_t} W_{t_0} \, dS_0,$$

where $W^*(\boldsymbol{S})$ is the complementary stored energy function (i.e., $\boldsymbol{E}(\boldsymbol{S}) = \partial W^*/\partial \boldsymbol{S}$), then, under static conditions, the variational formulations given by Eqn. (3.59) and (3.60) are equivalent to the conditions $\delta^{(1)}\Pi = 0$ and $\delta^{(1)}\Pi_{\mathrm{HR}} = 0$, respectively.

Consider a situation where $\tilde{\boldsymbol{v}}$ is zero on $(S_0)_u$. Then by choosing $\boldsymbol{u}_\delta = \tilde{\boldsymbol{v}}$, Eqns. (5.34) and (5.35) can be written as

$$\int_{V_0} \rho_0 \tilde{\boldsymbol{v}} \cdot \boldsymbol{b}_0 \, dV = \frac{d}{dt} \int_{V_0} W_{b_0} \, dV_0,$$
$$\int_{(S_0)_t} \tilde{\boldsymbol{v}} \cdot \bar{\boldsymbol{t}}_0 \, dS = \frac{d}{dt} \int_{(S_0)_t} W_{t_0} \, dS_0.$$

Substituting the above equations into Eqn. (5.22), we get

$$\int_{V_0} \left[\frac{1}{2}\rho_0 \left(\frac{\partial \boldsymbol{u}}{\partial t} \right)_{\boldsymbol{X}} \cdot \left(\frac{\partial \boldsymbol{u}}{\partial t} \right)_{\boldsymbol{X}} + \hat{W} - W_{b_0} \right] dV_0 - \int_{(S_0)_t} W_{t_0} \, dS_0 = \text{constant}, \tag{5.36}$$

which is a generalization of Eqn. (5.23c).

5.5 Isotropic Hyperelastic Solids

By analogy with the definition of isotropy given in Section 5.1, we say that a stored energy density function is isotropic at a point $\boldsymbol{X} \in V_0$ if the corresponding response function $\hat{\boldsymbol{\tau}}$ is isotropic at $\boldsymbol{X}$, i.e., if it satisfies Eqn. (5.3). We have the following necessary and sufficient condition for isotropy:

Theorem 5.5.1. *The stored energy density function $\hat{W} : V_0 \times Lin^+ \to \Re$ of a hyperelastic material is isotropic at* **X** *if and only if*

$$\hat{W}(\boldsymbol{X}, \boldsymbol{F}) = \hat{W}(\boldsymbol{X}, \boldsymbol{F}\boldsymbol{Q}) \quad \forall \boldsymbol{F} \in Lin^+, \boldsymbol{Q} \in Orth^+. \tag{5.37}$$

Proof. The dependence on **X** is dropped in the proof. By Eqn. (5.4), $\hat{W}$ is isotropic if and only if

$$\frac{\partial \hat{W}}{\partial \boldsymbol{F}}(\boldsymbol{F}) = \frac{\partial \hat{W}}{\partial \boldsymbol{F}}(\boldsymbol{F}\boldsymbol{Q})\boldsymbol{Q}^T \quad \forall \boldsymbol{F} \in \text{Lin}^+, \boldsymbol{Q} \in \text{Orth}^+. \tag{5.38}$$

We now find the gradient of the mapping $\hat{W}^Q(\boldsymbol{F}) := \hat{W}(\boldsymbol{F}\boldsymbol{Q})$ in a way analogous to that in Theorem 5.4.1. We find that

$$\frac{\partial \hat{W}^Q}{\partial \boldsymbol{F}}(\boldsymbol{F}) = \frac{\partial \hat{W}}{\partial \boldsymbol{F}}(\boldsymbol{F}\boldsymbol{Q})\boldsymbol{Q}^T \tag{5.39}$$

By comparing Eqns. (5.38) and (5.39), we get

$$\frac{\partial}{\partial \boldsymbol{F}}(\hat{W}^Q(\boldsymbol{F}) - \hat{W}(\boldsymbol{F})) = \boldsymbol{0} \quad \forall \boldsymbol{F} \in \text{Lin}^+, \boldsymbol{Q} \in \text{Orth}^+.$$

If $\hat{W}^Q(\boldsymbol{F}) = \hat{W}(\boldsymbol{F}\boldsymbol{Q}) = \hat{W}(\boldsymbol{F})$, then clearly the above equation is satisfied. Conversely, the above equation implies that

$$\hat{W}(\boldsymbol{F}\boldsymbol{Q}) = \hat{W}(\boldsymbol{F}) + C(\boldsymbol{Q}) \quad \forall \boldsymbol{F} \in \text{Lin}^+, \boldsymbol{Q} \in \text{Orth}^+.$$

As in the proof of Theorem 5.4.1, we get $C(\boldsymbol{Q}) = 0$, and Eqn. (5.37) follows. □

Note again that material frame-indifference involves multiplication on the left, while isotropy involves the multiplication on the right, of $\boldsymbol{F}$ by $\boldsymbol{Q}$. Also note that we have not used the axiom of frame-indifference in deriving the above result. By simultaneous consideration of this axiom, and the property of isotropy, we obtain the following stronger result:

Theorem 5.5.2. *The stored energy density function $\hat{W} : V_0 \times Lin^+ \to \Re$ of a hyperelastic material is frame-indifferent and isotropic at $X \in V_0$ if and only if there exists a function $\breve{W} : \mathcal{I}(Psym) \to \Re$ such that*

$$\hat{W}(X,F) = \breve{W}(X,\mathcal{I}_{F^TF}) = \breve{W}(X,\mathcal{I}_{FF^T}) \quad \forall F \in Lin^+. \tag{5.40}$$

Proof. The dependence on X is dropped in the proof. The invariants of F^TF and FF^T are the same since they are orthogonally equivalent as shown by Eqn. (2.12). Assuming material frame-indifference and isotropy at X, the characterizations established in Theorems 5.4.1 and 5.5.1 imply that

$$\tilde{W}(F^TF) = \hat{W}(F) = \hat{W}(FQ) = \tilde{W}(Q^TF^TFQ) \quad \forall F \in \text{Lin}^+, Q \in \text{Orth}^+.$$

Since every $C \in$ Psym can be written as U^TU, where $U = \sqrt{C} \in \text{Psym} \subset \text{Lin}^+$, we have

$$\tilde{W}(C) = \tilde{W}(Q^TCQ) \quad \forall C \in \text{Psym}, Q \in \text{Orth}^+.$$

By virtue of Theorem 1.6.12, $\tilde{W}$ is only a function of the principal invariants of $C = F^TF$. Conversely, if $\hat{W}(F) = \breve{W}(\mathcal{I}_{F^TF})$, then

$$\hat{W}(F) = \hat{W}(QF) = \hat{W}(FQ) \quad \forall F \in \text{Lin}^+, Q \in \text{Orth}^+,$$

since the principal invariants of $(QF)^T(QF) = F^TF$ and $(FQ)^T(FQ) = Q^TF^TFQ$ are the same as those of F^TF. □

Given a stored energy function expressed in terms of the principal invariants of F^TF, the associated constitutive relations are given by the following result:

Theorem 5.5.3. *Assume that at a point $X \in V_0$, the stored energy function $\hat{W}$ is of the form*

$$\hat{W}(X,F) = \breve{W}(X,\mathcal{I}_{F^TF}) = \breve{W}(X,\mathcal{I}_{FF^T}), \quad F \in Lin^+,$$

and that the function $\breve{W} : \mathcal{I}(Psym) \to \Re$ is differentiable at a point $\mathcal{I}_{F^TF}$ $(= \mathcal{I}_{FF^T})$. Then the associated response functions $\tilde{S}$, $\hat{T}$ and $\hat{\tau}$ are given by

$$\frac{1}{2}\tilde{S}(X,C) = \frac{\partial \breve{W}}{\partial I_1}I + \frac{\partial \breve{W}}{\partial I_2}(I_1 I - C) + \frac{\partial \breve{W}}{\partial I_3}I_3 C^{-1}, \tag{5.41}$$

$$= \left(\frac{\partial \breve{W}}{\partial I_1} + \frac{\partial \breve{W}}{\partial I_2}I_1 + \frac{\partial \breve{W}}{\partial I_3}I_2\right) I - \left(\frac{\partial \breve{W}}{\partial I_2} + \frac{\partial \breve{W}}{\partial I_3}I_1\right) C + \frac{\partial \breve{W}}{\partial I_3}C^2, \tag{5.42}$$

$$\frac{1}{2}\hat{T}(X,F) = \frac{\partial \breve{W}}{\partial I_1}F + \frac{\partial \breve{W}}{\partial I_2}(I_1 I - FF^T)F + \frac{\partial \breve{W}}{\partial I_3}I_3 F^{-T}, \tag{5.43}$$

$$\frac{J}{2}\hat{\tau}(X,B) = \frac{\partial \breve{W}}{\partial I_1}B + \frac{\partial \breve{W}}{\partial I_2}(I_1 B - B^2) + \frac{\partial \breve{W}}{\partial I_3}I_3 I \tag{5.44}$$

$$= \left(\frac{\partial \breve{W}}{\partial I_2}I_2 + \frac{\partial \breve{W}}{\partial I_3}I_3\right)I + \frac{\partial \breve{W}}{\partial I_1}B - \frac{\partial \breve{W}}{\partial I_2}I_3 B^{-1}, \tag{5.45}$$

$$= \frac{\partial \breve{W}}{\partial N}, \tag{5.46}$$

where $J = \sqrt{I_3}$, $N := \log B$, *and*

$$\frac{\partial \breve{W}}{\partial I_k} := \frac{\partial \breve{W}}{\partial I_k}(X, \mathcal{I}_C) = \frac{\partial \breve{W}}{\partial I_k}(X, \mathcal{I}_B), \quad I_k = I_k(B) = I_k(C), \; B = FF^T, \; C = F^T F.$$

Proof. We drop the dependence on X in the proof. We first prove Eqn. (5.41). Using Eqn. (5.32), we get

$$\frac{1}{2}\tilde{S}(C) = \frac{\partial \breve{W}}{\partial I_1}\frac{\partial I_1}{\partial C} + \frac{\partial \breve{W}}{\partial I_2}\frac{\partial I_2}{\partial C} + \frac{\partial \breve{W}}{\partial I_3}\frac{\partial I_3}{\partial C}$$

$$= \frac{\partial \breve{W}}{\partial I_1}I + \frac{\partial \breve{W}}{\partial I_2}[(\operatorname{tr} C)I - C] + \frac{\partial \breve{W}}{\partial I_3}\operatorname{\mathbf{cof}} C, \tag{5.47}$$

where the last step follows from Eqns. (1.226a)–(1.226c), and the symmetry of C. Using the Cayley–Hamilton theorem to eliminate C^{-1}, we get Eqn. (5.42). Multiplying Eqn. (5.41) by F, we get Eqn. (5.43). By using the relation $\tau = TF^T / \sqrt{I_3}$, we get Eqn. (5.44). Using the Cayley–Hamilton theorem leads to Eqn. (5.45).

To show Eqn. (5.46), we use the chain rule as given by Eqn. (1.239) (for an alternative proof, see, e.g., [128]):

$$\frac{\partial \breve{W}}{\partial N} = \mathbb{S}\left(\frac{\partial B}{\partial N}\right)^T \mathbb{S}\frac{\partial \breve{W}}{\partial B}$$

$$= \mathbb{S}\left[\sum_{i=1}^{k}\lambda_i P_i \boxtimes P_i + \sum_{i=1}^{k}\sum_{\substack{j=1\\ j\neq i}}^{k}\frac{\lambda_i - \lambda_j}{\log\lambda_i - \log\lambda_j}P_i \boxtimes P_j\right]$$

$$\left[\frac{\partial \breve{W}}{\partial I_1}I + \frac{\partial \breve{W}}{\partial I_2}(I_1 I - B) + \frac{\partial \breve{W}}{\partial I_3}I_3 B^{-1}\right] \quad \text{since } \mathbb{S}A = A \text{ for } A \in \mathrm{Sym},$$

where (λ_i, P_i) denote the eigenvalues/eigenprojections of B, and the expression for $\partial B/\partial N$ is obtained using Eqn. (1.256), and the fact that $P_i^T = P_i$ (see Eqn. (J.4)). The above expression simplifies to

$$\frac{\partial \breve{W}}{\partial N} = \sum_{i=1}^{k}\lambda_i P_i\left[\frac{\partial \breve{W}}{\partial I_1}I + \frac{\partial \breve{W}}{\partial I_2}(I_1 I - B) + \frac{\partial \breve{W}}{\partial I_3}I_3 B^{-1}\right]P_i$$

$$+ \sum_{i=1}^{k}\sum_{\substack{j=1\\ j\neq i}}^{k}\frac{\lambda_i - \lambda_j}{\log\lambda_i - \log\lambda_j}P_i\left[\frac{\partial \breve{W}}{\partial I_1}I + \frac{\partial \breve{W}}{\partial I_2}(I_1 I - B) + \frac{\partial \breve{W}}{\partial I_3}I_3 B^{-1}\right]P_j$$

$$= \sum_{i=1}^{k} \lambda_i \boldsymbol{P}_i \boldsymbol{P}_i \left[\frac{\partial \breve{W}}{\partial I_1} \boldsymbol{I} + \frac{\partial \breve{W}}{\partial I_2} (I_1 \boldsymbol{I} - \boldsymbol{B}) + \frac{\partial \breve{W}}{\partial I_3} I_3 \boldsymbol{B}^{-1} \right]$$
$$+ \sum_{i=1}^{k} \sum_{\substack{j=1 \\ j \neq i}}^{k} \frac{\lambda_i - \lambda_j}{\log \lambda_i - \log \lambda_j} \boldsymbol{P}_i \boldsymbol{P}_j \left[\frac{\partial \breve{W}}{\partial I_1} \boldsymbol{I} + \frac{\partial \breve{W}}{\partial I_2} (I_1 \boldsymbol{I} - \boldsymbol{B}) + \frac{\partial \breve{W}}{\partial I_3} I_3 \boldsymbol{B}^{-1} \right],$$

where the last step follows from the fact that since $\boldsymbol{P}_i$ is a polynomial in $\boldsymbol{B}$, it commutes with the bracketed terms. Now using the results $\boldsymbol{P}_i \boldsymbol{P}_i = \boldsymbol{P}_i$, $\boldsymbol{P}_i \boldsymbol{P}_j = \boldsymbol{0}$, $\sum_{i=1}^{k} \lambda_i \boldsymbol{P}_i = \boldsymbol{B}$, we get

$$\frac{\partial \breve{W}}{\partial \boldsymbol{N}} = \left[\frac{\partial \breve{W}}{\partial I_1} \boldsymbol{B} + \frac{\partial \breve{W}}{\partial I_2} (I_1 \boldsymbol{B} - \boldsymbol{B}^2) + \frac{\partial \breve{W}}{\partial I_3} I_3 \boldsymbol{I} \right],$$

which is the same as Eqn. (5.44). □

Comparing Eqns. (5.9) and (5.42), we get the necessary and sufficient conditions for a material to be hyperelastic, isotropic and frame-indifferent as

$$\frac{\partial \breve{W}}{\partial I_1} + \frac{\partial \breve{W}}{\partial I_2} I_1 + \frac{\partial \breve{W}}{\partial I_3} I_2 = \frac{\hat{\gamma}_0}{2},$$
$$\frac{\partial \breve{W}}{\partial I_2} + \frac{\partial \breve{W}}{\partial I_3} I_1 = -\frac{\hat{\gamma}_1}{2},$$
$$\frac{\partial \breve{W}}{\partial I_3} = \frac{\hat{\gamma}_2}{2}.$$

From the above equations, we get

$$2\frac{\partial \breve{W}}{\partial I_1} = \hat{\gamma}_0 + \hat{\gamma}_1 I_1 + \hat{\gamma}_2 I_1^2 - \hat{\gamma}_2 I_2,$$
$$2\frac{\partial \breve{W}}{\partial I_2} = -\hat{\gamma}_1 - \hat{\gamma}_2 I_1,$$
$$2\frac{\partial \breve{W}}{\partial I_3} = \hat{\gamma}_2.$$

Now using the relations

$$\frac{\partial^2 \breve{W}}{\partial I_1 \partial I_2} = \frac{\partial^2 \breve{W}}{\partial I_2 \partial I_1}; \quad \frac{\partial^2 \breve{W}}{\partial I_2 \partial I_3} = \frac{\partial^2 \breve{W}}{\partial I_3 \partial I_2}; \quad \frac{\partial^2 \breve{W}}{\partial I_1 \partial I_3} = \frac{\partial^2 \breve{W}}{\partial I_3 \partial I_1},$$

we get the compatibility relations for a material to be hyperelastic, isotropic and frame-indifferent as

$$\frac{\partial \hat{\gamma}_0}{\partial I_2} + \frac{\partial \hat{\gamma}_1}{\partial I_1} + \left(\frac{\partial \hat{\gamma}_1}{\partial I_2} + \frac{\partial \hat{\gamma}_2}{\partial I_1} \right) I_1 + \frac{\partial \hat{\gamma}_2}{\partial I_2} I_1^2 - \frac{\partial \hat{\gamma}_2}{\partial I_2} I_2 = 0,$$
$$\frac{\partial \hat{\gamma}_0}{\partial I_3} + \frac{\partial \hat{\gamma}_1}{\partial I_3} I_1 + \frac{\partial \hat{\gamma}_2}{\partial I_3} \left(I_1^2 - I_2 \right) - \frac{\partial \hat{\gamma}_2}{\partial I_1} = 0,$$

$$\frac{\partial\hat{\gamma}_1}{\partial I_3} + I_1\frac{\partial\hat{\gamma}_2}{\partial I_3} + \frac{\partial\hat{\gamma}_2}{\partial I_2} = 0.$$

In a similar manner, the compatibility relations for the coefficients in Eqn. (5.10) are

$$\begin{aligned}
&\frac{\partial\beta_{-1}}{\partial I_1} + I_3\frac{\partial\beta_1}{\partial I_2} = 0,\\
&\beta_{-1} - I_3\frac{\partial\beta_{-1}}{\partial I_3} - \frac{\partial(I_2\beta_{-1})}{\partial I_2} - I_3\frac{\partial\beta_0}{\partial I_2} = 0,\\
&I_2\frac{\partial\beta_{-1}}{\partial I_1} + I_3\frac{\partial\beta_0}{\partial I_1} - I_3^2\frac{\partial\beta_1}{\partial I_3} = 0,
\end{aligned}$$

while those for the coefficients in Eqn. (5.11) are

$$\begin{aligned}
&\frac{\partial\gamma_0}{\partial I_2} + \frac{\partial\gamma_1}{\partial I_2}I_1 + \frac{\partial\gamma_1}{\partial I_1} = 0,\\
&\frac{1}{I_3}\frac{\partial\gamma_{-1}}{\partial I_2} + \frac{\partial\gamma_1}{\partial I_3} = 0,\\
&-\frac{1}{I_3}\frac{\partial\gamma_{-1}}{\partial I_1} + \frac{\partial\gamma_0}{\partial I_3} + \frac{\partial\gamma_1}{\partial I_3}I_1 = 0.
\end{aligned}$$

5.6 St Venant–Kirchhoff Material

By neglecting the higher-order terms in the expression for the second Piola–Kirchhoff stress given by Eqn. (5.15), and by assuming the material to be homogeneous and the reference configuration to be a natural state, we get the following constitutive relation:

$$\tilde{S}(E) = \lambda(\operatorname{tr} E)I + 2\mu E, \quad E \in \text{Sym}. \tag{5.48}$$

A material whose constitutive relation is given by Eqn. (5.48) is known as a St Venant–Kirchhoff material. Since $\operatorname{tr} C = 3 + 2\operatorname{tr} E$, we have

$$\bar{S}(C) = \left[\frac{\lambda}{2}(I_1(C) - 3) - \mu\right] I + \mu C, \quad C = I + 2E \in \text{Psym},$$

which is a special case of Eqn. (5.9). In component form, Eqn. (5.48) is given by

$$S_{ij}(E) = \lambda E_{kk}\delta_{ij} + 2\mu E_{ij}, \quad E_{ij} = \frac{1}{2}\left(\frac{\partial u_i}{\partial X_j} + \frac{\partial u_j}{\partial X_i} + \frac{\partial u_k}{\partial X_i}\frac{\partial u_k}{\partial X_j}\right),$$

or, equivalently,

$$\begin{aligned}
S &= \mathbb{C}[E], & \mathbb{C} &= \lambda(I \otimes I) + 2\mu\mathbb{S},\\
S_{ij} &= \mathbb{C}_{ijkl}E_{kl}, & \mathbb{C}_{ijkl} &= \lambda\delta_{ij}\delta_{kl} + \mu(\delta_{ik}\delta_{jl} + \delta_{il}\delta_{jk}).
\end{aligned}$$

Regarding hyperelasticity, we have the following result:

Theorem 5.6.1. *A St Venant–Kirchhoff material, whose response is given by Eqn.* (5.48) *is hyperelastic, with a stored energy function given by*

$$\breve{W}(E) = \frac{\lambda}{2}(tr\, E)^2 + \mu tr\, E^2 \tag{5.49}$$

$$= \frac{\lambda}{2}(\gamma_1 + \gamma_2 + \gamma_3)^2 + \mu(\gamma_1^2 + \gamma_2^2 + \gamma_3^2) \tag{5.50}$$

$$= \mu(I_1 - 3) + \frac{\lambda + 2\mu}{8}(I_1 - 3)^2 - \frac{\mu}{2}(I_2 - 3) = \breve{W}(\mathcal{I}_C), \tag{5.51}$$

where γ_i, $i = 1, 2, 3$ *are the eigenvalues of* E, *and* I_1 *and* I_2 *are the first two principal invariants of* C.

Proof. By Eqn. (5.33), it suffices to verify that

$$\breve{S}(E) = \frac{\partial \breve{W}}{\partial E}(E).$$

From Eqn. (5.49), we have

$$\breve{W}(E + U) = \breve{W}(E) + \lambda(\mathrm{tr}\, E)(\mathrm{tr}\, U) + 2\mu\, \mathrm{tr}\,(EU) + \mathrm{o}\,(U)$$

$$= \breve{W}(E) + [\lambda(\mathrm{tr}\, E)I + 2\mu E] : U + \mathrm{o}\,(U).$$

from which Eqn. (5.48) follows. Since $\mathrm{tr}\, E = \gamma_1 + \gamma_2 + \gamma_3$ and $E : E = \gamma_1^2 + \gamma_2^2 + \gamma_3^2$, we get Eqn. (5.50). Substituting $E = (C - I)/2$ into Eqn. (5.49), the relations in terms of the invariants of C follow. □

Equation (5.51) when combined with Theorem 5.5.2 shows that a St Venant–Kirchhoff material is frame-indifferent, hyperelastic and isotropic. However, simply satisfying the condition of frame-indifference does not ensure that it is a good model. We now discuss the shortcomings of this model.

Note that though the mapping $E \to S$ in Eqn. (5.48) is linear, the mapping $u \to E$ is not. The linearity of the mapping between E and S indicates that the stresses are large if and only if the strains are large. However, practical experience with 'eversion problems' shows that large strains are possible even when the stress is small. Another shortcoming is that there is no term in the associated strain energy density function that prevents $J = \det F$ from going to zero (recall that $dV = J dV_0$). Physically speaking, one would expect the stored energy function to increase to infinity as one tries to compress the material to a point, i.e., $\hat{W}(X, F) \to \infty$ as $\det F \to 0^+$. However, as seen from Eqn. (5.51), this condition cannot be satisfied since the stored energy function is independent of $\det C$ (and hence, $\det F$). Due to these shortcomings, a St Venant–Kirchhoff model can be expected to be useful only for 'small strains' E, as it should be since it has been obtained by a linearization of the general response function for isotropic materials; hence it is known as a 'large displacement-small strain' model.

5.7 Examples of Nonlinear Compressible Hyperelastic Models

A small modification of the St Venant–Kirchhoff material model can make it into an effective model even for moderately large elastic strains [5, 6] (for a generalization, see [134]). The modification involves replacing the eigenvalues γ_i in Eqn. (5.50) with $\log \lambda_i$, where λ_i are the principal stretches (the eigenvalues of $\boldsymbol{U}$ and $\boldsymbol{V}$), i.e.,

$$\begin{aligned}
\tilde{W} &= \frac{\lambda}{2}(\log \lambda_1 + \log \lambda_2 + \log \lambda_3)^2 + \mu\left[(\log \lambda_1)^2 + (\log \lambda_2)^2 + (\log \lambda_3)^2\right] \\
&= \frac{\lambda}{8}(\log \det \boldsymbol{C})^2 + \mu(\log \boldsymbol{U}) : (\log \boldsymbol{U}) \\
&= \frac{\lambda}{8}(\operatorname{tr} \boldsymbol{N})^2 + \frac{\mu}{4}\boldsymbol{N} : \boldsymbol{N} \\
&= \frac{\lambda}{8}\left[\log \lambda_1^2 + \log \lambda_2^2 + \log \lambda_3^2\right]^2 + \frac{\mu}{4}\left[(\log \lambda_1^2)^2 + (\log \lambda_2^2)^2 + (\log \lambda_3^2)^2\right],
\end{aligned} \tag{5.52}$$

where $\log \boldsymbol{U}$ is known as the Hencky strain. From Eqn. (1.252), we have for each i, $\partial(\lambda_i^2)/\partial \boldsymbol{C} = \boldsymbol{P}_i$. Thus, using (5.32), we obtain

$$\begin{aligned}
\boldsymbol{S} &= \frac{\lambda}{2}(\log \det \boldsymbol{C})\left[\frac{\operatorname{\mathbf{cof}} \boldsymbol{C}}{\det \boldsymbol{C}}\right] + \mu\left[\frac{(\log \lambda_1^2)\boldsymbol{P}_1}{\lambda_1^2} + \frac{(\log \lambda_2^2)\boldsymbol{P}_2}{\lambda_2^2} + \frac{(\log \lambda_3^2)\boldsymbol{P}_3}{\lambda_3^2}\right] \\
&= \left[\frac{\lambda}{2}(\log \det \boldsymbol{C})\boldsymbol{I} + \mu \log \boldsymbol{C}\right]\boldsymbol{C}^{-1}, \\
J\boldsymbol{\tau} &= \frac{\lambda}{2}\operatorname{tr}(\boldsymbol{N})\boldsymbol{I} + \mu\boldsymbol{N},
\end{aligned} \tag{5.53}$$

where, in obtaining Eqn. (5.53), we have used Eqn. (5.46).

Using Eqns. (1.231), (1.236), (1.241), (1.256) and (J.29), we get

$$\begin{aligned}
\mathbb{C} &= 2\mathbb{S}\frac{\partial \boldsymbol{S}}{\partial \boldsymbol{C}}\mathbb{S} \\
&= \mathbb{D} + 2\mu\mathbb{S}\left[\sum_{i=1}^{3}\frac{(1-\log \lambda_i^2)}{\lambda_i^4}\boldsymbol{P}_i \boxtimes \boldsymbol{P}_i + \left(\frac{\log \lambda_1^2}{\lambda_1^2} - \frac{\log \lambda_2^2}{\lambda_2^2}\right)\left[\frac{\boldsymbol{P}_1 \boxtimes \boldsymbol{P}_2 + \boldsymbol{P}_2 \boxtimes \boldsymbol{P}_1}{\lambda_1^2 - \lambda_2^2}\right]\right. \\
&\quad + \left(\frac{\log \lambda_2^2}{\lambda_2^2} - \frac{\log \lambda_3^2}{\lambda_3^2}\right)\left[\frac{\boldsymbol{P}_2 \boxtimes \boldsymbol{P}_3 + \boldsymbol{P}_3 \boxtimes \boldsymbol{P}_2}{\lambda_2^2 - \lambda_3^2}\right] \\
&\quad \left.+ \left(\frac{\log \lambda_3^2}{\lambda_3^2} - \frac{\log \lambda_1^2}{\lambda_1^2}\right)\left[\frac{\boldsymbol{P}_3 \boxtimes \boldsymbol{P}_1 + \boldsymbol{P}_1 \boxtimes \boldsymbol{P}_3}{\lambda_3^2 - \lambda_1^2}\right]\right]\mathbb{S}, \qquad (\lambda_1 \neq \lambda_2 \neq \lambda_3 \neq \lambda_1) \\
&= \mathbb{D} + 2\mu\mathbb{S}\left[\frac{(1-\log \lambda_1^2)}{\lambda_1^4}\boldsymbol{P}_1 \boxtimes \boldsymbol{P}_1 + \frac{(1-\log \lambda_2^2)}{\lambda_2^4}(\boldsymbol{I}-\boldsymbol{P}_1) \boxtimes (\boldsymbol{I}-\boldsymbol{P}_1)\right. \\
&\quad \left.+ \left(\frac{\log \lambda_1^2}{\lambda_1^2} - \frac{\log \lambda_2^2}{\lambda_2^2}\right)\left[\frac{\boldsymbol{P}_1 \boxtimes (\boldsymbol{I}-\boldsymbol{P}_1) + (\boldsymbol{I}-\boldsymbol{P}_1) \boxtimes \boldsymbol{P}_1}{\lambda_1^2 - \lambda_2^2}\right]\right]\mathbb{S}, \quad (\lambda_1 \neq \lambda_2 = \lambda_3) \\
&= \mathbb{D} + 2\mu\frac{(1-\log \gamma^2)}{\gamma^4}\mathbb{S}, \qquad (\lambda_1 = \lambda_2 = \lambda_3 \equiv \gamma)
\end{aligned}$$

where $\mathbb{D} \equiv \lambda \left[C^{-1} \otimes C^{-1} - (\log \det C)\mathbb{S}(C^{-1} \boxtimes C^{-1})\mathbb{S} \right]$. By approximating $\log C$ as $C - I$ in the constitutive relation for S, we get the compressible neo-Hookean material of Problem 1.

A generalization of the above Hencky model is

$$\tilde{W}(H) = \frac{\lambda}{2}(\operatorname{tr} H)^2 + \mu \operatorname{tr} H^2 + \sum_{n=1}^{\infty} \left[\alpha_n (\operatorname{tr} H^2)^{n+1} + \beta_n (\operatorname{tr} H^3)^{2n} \right], \tag{5.54}$$

where $H = \log V$, and α_n and β_n are material constants. Note that the first two terms are the same as in Eqn. (5.52), and the last term constitutes the generalization. For a better fit to experimental data, the $\lambda(\operatorname{tr} H)^2/2$ term can be replaced by λW_H, where W_H is given by Eqn. (5.56) below. The stress associated with the above generalized Hencky function, obtained using $J\tau = \partial \tilde{W}/\partial H$, is

$$\begin{aligned} J\tau &= \lambda(\operatorname{tr} H)I + 2\mu H + \sum_{n=1}^{\infty} \left[2(n+1)\alpha_n (\operatorname{tr} H^2)^n H + 6n\beta_n (\operatorname{tr} H^3)^{2n-1} H^2 \right] \\ &= \frac{\lambda}{2}(\operatorname{tr} N)I + \mu N + \sum_{n=1}^{\infty} \left[2^{-2n}(n+1)\alpha_n (\operatorname{tr} N^2)^n N + 3n(2^{2-6n})\beta_n (\operatorname{tr} N^3)^{2n-1} N^2 \right], \end{aligned}$$

where $N := \log B$. Note that the above generalization is constructed such that τ is odd in H, i.e., $\tau(-H) = -\tau(H)$.

The Ogden model [240] is another model that is widely used, especially for modeling rubber-type materials. With λ_i denoting the principal stretches as above, and $I_3 \equiv \det C$, the strain energy density function for this model is given by [240]

$$\tilde{W}(C) = \sum_{i=1}^{3} \frac{\mu_i}{\alpha_i} \left[(\lambda_1^2)^{\alpha_i/2} + (\lambda_2^2)^{\alpha_i/2} + (\lambda_3^2)^{\alpha_i/2} - 3 \right] + k_1 \log I_3 + \lambda W_H(J), \tag{5.55}$$

where λ is a Lamé constant. The stresses and constitutive tensor associated with the hydrostatic function $W_H(J)$ are

$$\begin{aligned} S_H &= W_H'(J) J C^{-1}, \\ \tau_H &= W_H'(J) I, \\ \mathbb{C}_H &= \left[W_H''(J) J^2 + W_H'(J) J \right] C^{-1} \otimes C^{-1} - 2W_H'(J) J \mathbb{S}(C^{-1} \boxtimes C^{-1})\mathbb{S}. \end{aligned}$$

As usual, we 'normalize' $W_H(J)$ such that it is zero at the reference configuration ($F = I$ and $J = 1$), i.e., $W_H(1) = 0$. Similarly, we assume the stress to be zero in the reference configuration, i.e., $W_H'(1) = 0$. In order that $\mathbb{C}_H|_{C=I} = I \otimes I$, we need $W_H''(1) = 1$. We expect τ_H to be positive or negative depending on whether $J > 1$ or $J < 1$, so that $W_H'(J) > 0$ for $J > 1$ and $W_H'(J) < 0$ for $J < 1$. Additionally, we expect τ_H to increase as J increases, and to decrease as J decreases, i.e., $W_H''(J) > 0$ for all J. Finally, we expect $W_H(J)$ to tend to infinity as $J \to 0$ and as $J \to \infty$. To summarize, the requirements on $W_H(J)$ are

1. $W_H(1) = 0$.

2. $W_H'(1) = 0$.

3. $W''_H(1) = 1$.
4. $W'_H(J) > 0$ for $J > 1$ and $W'_H(J) < 0$ for $J < 1$.
5. $W''_H(J) > 0$ for all J.
6. $W_H \to \infty$ as $J \to 0$ and as $J \to \infty$.

The W_H function suggested in [21] satisfies all the above requirements except the last one. In order that it does so, we modify it slightly by dividing by J (note that this modification does not modify the stress field substantially for almost incompressible materials since $J \approx 1$), so that

$$W_H(J) = \frac{1}{\alpha^2 J}\left\{\cosh\left[\alpha(J-1)\right] - 1\right\}. \tag{5.56}$$

where α is a positive constant. This function satisfies all the above requirements. Note that several functions used in the literature do not satisfy these requirements. For example, the function

$$W_H = \frac{1}{\beta^2}\left[\beta \log J + J^{-\beta} - 1\right]$$

does not satisfy the fifth requirement above for the value $\beta = 9$ used in [239, 248]. Although all requirements are met by the above function for $\beta < 0$, the function given by Eqn. (5.56) seems to fit the experimental data better [21]. Similarly, the commonly used function $W_H(J) = (\log J)^2/2$ also violates the fifth requirement. For yet another W_H function, see, e.g., [252].

The second Piola–Kirchhoff stress for the Ogden model is given by

$$\begin{aligned}
\boldsymbol{S} &= \sum_{i=1}^{3} \mu_i \left[(\lambda_1^2)^{\alpha_i/2-1}\boldsymbol{P}_1 + (\lambda_2^2)^{\alpha_i/2-1}\boldsymbol{P}_2 + (\lambda_3^2)^{\alpha_i/2-1}\boldsymbol{P}_3\right] + 2k_1\boldsymbol{C}^{-1} \\
&\quad + \frac{\lambda}{\alpha^2}\left[\alpha\sinh\left[\alpha(J-1)\right] - \frac{\cosh\left[\alpha(J-1)\right]-1}{J}\right]\boldsymbol{C}^{-1}, \qquad \lambda_1 \neq \lambda_2 \neq \lambda_3 \neq \lambda_1 \\
&= \sum_{i=1}^{3} \mu_i \left[(\lambda_1^2)^{\alpha_i/2-1}\boldsymbol{P}_1 + (\lambda_2^2)^{\alpha_i/2-1}(\boldsymbol{I}-\boldsymbol{P}_1)\right] + 2k_1\boldsymbol{C}^{-1} \\
&\quad + \frac{\lambda}{\alpha^2}\left[\alpha\sinh\left[\alpha(J-1)\right] - \frac{\cosh\left[\alpha(J-1)\right]-1}{J}\right]\boldsymbol{C}^{-1}, \qquad \lambda_1 \neq \lambda_2 = \lambda_3 \qquad (5.57) \\
&= \sum_{i=1}^{3} \mu_i (\gamma^2)^{\alpha_i/2-1}\boldsymbol{I} + 2k_1\boldsymbol{C}^{-1} \\
&\quad + \frac{\lambda}{\alpha^2}\left[\alpha\sinh\left[\alpha(J-1)\right] - \frac{\cosh\left[\alpha(J-1)\right]-1}{J}\right]\boldsymbol{C}^{-1}, \qquad \lambda_1 = \lambda_2 = \lambda_3 \equiv \gamma
\end{aligned}$$

with $\boldsymbol{P}_i$ given by Eqn. (1.242), while on using Eqns. (1.231), (1.236), (1.241) and (1.256) and (J.29), the constitutive tensor is obtained as

$$\mathbb{C} = 2\mathbb{S}\frac{\partial \boldsymbol{S}}{\partial \boldsymbol{C}}\mathbb{S}$$

$$
\begin{aligned}
&= \sum_{i=1}^{3} 2\mu_i \Bigg[\left(\frac{\alpha_i}{2} - 1\right) (\lambda_1^2)^{\alpha_i/2-2} \mathbb{S}(\boldsymbol{P}_1 \boxtimes \boldsymbol{P}_1)\mathbb{S} + \left(\frac{\alpha_i}{2} - 1\right) (\lambda_2^2)^{\alpha_i/2-2} \mathbb{S}(\boldsymbol{P}_2 \boxtimes \boldsymbol{P}_2)\mathbb{S} \\
&\quad + \left(\frac{\alpha_i}{2} - 1\right) (\lambda_3^2)^{\alpha_i/2-2} \mathbb{S}(\boldsymbol{P}_3 \boxtimes \boldsymbol{P}_3)\mathbb{S} \\
&\quad + (\lambda_1^2)^{\alpha_i/2-1} \mathbb{S} \left[\frac{\boldsymbol{P}_1 \boxtimes \boldsymbol{P}_2 + \boldsymbol{P}_2 \boxtimes \boldsymbol{P}_1}{\lambda_1^2 - \lambda_2^2} + \frac{\boldsymbol{P}_1 \boxtimes \boldsymbol{P}_3 + \boldsymbol{P}_3 \boxtimes \boldsymbol{P}_1}{\lambda_1^2 - \lambda_3^2} \right] \mathbb{S} \\
&\quad + (\lambda_2^2)^{\alpha_i/2-1} \mathbb{S} \left[\frac{\boldsymbol{P}_2 \boxtimes \boldsymbol{P}_3 + \boldsymbol{P}_3 \boxtimes \boldsymbol{P}_2}{\lambda_2^2 - \lambda_3^2} + \frac{\boldsymbol{P}_2 \boxtimes \boldsymbol{P}_1 + \boldsymbol{P}_1 \boxtimes \boldsymbol{P}_2}{\lambda_2^2 - \lambda_1^2} \right] \mathbb{S} \\
&\quad + (\lambda_3^2)^{\alpha_i/2-1} \mathbb{S} \left[\frac{\boldsymbol{P}_3 \boxtimes \boldsymbol{P}_1 + \boldsymbol{P}_1 \boxtimes \boldsymbol{P}_3}{\lambda_3^2 - \lambda_1^2} + \frac{\boldsymbol{P}_3 \boxtimes \boldsymbol{P}_2 + \boldsymbol{P}_2 \boxtimes \boldsymbol{P}_3}{\lambda_3^2 - \lambda_2^2} \right] \mathbb{S} \Bigg] - 4k_1 \mathbb{S}(\boldsymbol{C}^{-1} \boxtimes \boldsymbol{C}^{-1})\mathbb{S} \\
&\quad + \frac{\lambda}{\alpha^2} \left\{ J\alpha^2 \cosh\left[\alpha(J-1)\right] - \alpha \sinh\left[\alpha(J-1)\right] + \frac{\cosh\left[\alpha(J-1)\right] - 1}{J} \right\} \boldsymbol{C}^{-1} \otimes \boldsymbol{C}^{-1} \\
&\quad - \frac{2\lambda}{\alpha^2} \left\{ \alpha \sinh\left[\alpha(J-1)\right] - \frac{\cosh\left[\alpha(J-1)\right] - 1}{J} \right\} \mathbb{S}(\boldsymbol{C}^{-1} \boxtimes \boldsymbol{C}^{-1})\mathbb{S}, \; (\lambda_1 \neq \lambda_2 \neq \lambda_3 \neq \lambda_1) \\
&= \sum_{i=1}^{3} 2\mu_i \Bigg[\left(\frac{\alpha_i}{2} - 1\right) (\lambda_1^2)^{\alpha_i/2-2} \mathbb{S}(\boldsymbol{P}_1 \boxtimes \boldsymbol{P}_1)\mathbb{S} \\
&\quad + \left(\frac{\alpha_i}{2} - 1\right) (\lambda_2^2)^{\alpha_i/2-2} \mathbb{S}\left[(\boldsymbol{I} - \boldsymbol{P}_1) \boxtimes (\boldsymbol{I} - \boldsymbol{P}_1)\right] \mathbb{S} \\
&\quad + \frac{(\lambda_1^2)^{\alpha_i/2-1} - (\lambda_2^2)^{\alpha_i/2-1}}{\lambda_1^2 - \lambda_2^2} \mathbb{S}\left[\boldsymbol{P}_1 \boxtimes (\boldsymbol{I} - \boldsymbol{P}_1) + (\boldsymbol{I} - \boldsymbol{P}_1) \boxtimes \boldsymbol{P}_1\right] \mathbb{S} \Bigg] - 4k_1 \mathbb{S}(\boldsymbol{C}^{-1} \boxtimes \boldsymbol{C}^{-1})\mathbb{S} \\
&\quad + \frac{\lambda}{\alpha^2} \left\{ J\alpha^2 \cosh\left[\alpha(J-1)\right] - \alpha \sinh\left[\alpha(J-1)\right] + \frac{\cosh\left[\alpha(J-1)\right] - 1}{J} \right\} \boldsymbol{C}^{-1} \otimes \boldsymbol{C}^{-1} \\
&\quad - \frac{2\lambda}{\alpha^2} \left\{ \alpha \sinh\left[\alpha(J-1)\right] - \frac{\cosh\left[\alpha(J-1)\right] - 1}{J} \right\} \mathbb{S}(\boldsymbol{C}^{-1} \boxtimes \boldsymbol{C}^{-1})\mathbb{S}, \quad (\lambda_1 \neq \lambda_2 = \lambda_3), \\
&= \sum_{i=1}^{3} 2\mu_i \left[\left(\frac{\alpha_i}{2} - 1\right) (\gamma^2)^{\alpha_i/2-2} \right] \mathbb{S} - 4k_1 \mathbb{S}(\boldsymbol{C}^{-1} \boxtimes \boldsymbol{C}^{-1})\mathbb{S} \\
&\quad + \frac{\lambda}{\alpha^2} \left\{ J\alpha^2 \cosh\left[\alpha(J-1)\right] - \alpha \sinh\left[\alpha(J-1)\right] + \frac{\cosh\left[\alpha(J-1)\right] - 1}{J} \right\} \boldsymbol{C}^{-1} \otimes \boldsymbol{C}^{-1} \\
&\quad - \frac{2\lambda}{\alpha^2} \left\{ \alpha \sinh\left[\alpha(J-1)\right] - \frac{\cosh\left[\alpha(J-1)\right] - 1}{J} \right\} \mathbb{S}(\boldsymbol{C}^{-1} \boxtimes \boldsymbol{C}^{-1})\mathbb{S}, \quad (\lambda_1 = \lambda_2 = \lambda_3 \equiv \gamma).
\end{aligned}
$$

More compact expressions for the above results for $\mathbb{C}$ can be given by using Eqn. (1.174).

The conditions $\boldsymbol{S}(\boldsymbol{I}) = \boldsymbol{0}$ and $\mathbb{C}|_{\boldsymbol{C}=\boldsymbol{I}}\, \boldsymbol{E} = \lambda(\operatorname{tr} \boldsymbol{E})\boldsymbol{I} + 2\mu\boldsymbol{E}$ lead to the constraints

$$
\begin{aligned}
2k_1 &= -\sum_{i=1}^{3} \mu_i, \\
\sum_{i=1}^{3} \mu_i \alpha_i &= 2\mu.
\end{aligned}
$$

The compressible neo-Hookean model of Problem 1 is a special case of the above model when $\mu_1 = 2c_1 = \mu$, $\alpha_1 = 2$ and $\mu_2 = \mu_3 = \alpha_2 = \alpha_3 = 0$.

5.8 The Elasticity Tensors

The first elasticity tensor is defined as

$$\mathbb{A}(\boldsymbol{X}, \boldsymbol{F}) = \frac{\partial \hat{\boldsymbol{T}}}{\partial \boldsymbol{F}}, \tag{5.58}$$

which in component form reads

$$\mathbb{A}_{ijkl} = \frac{\partial \hat{T}_{ij}}{\partial F_{kl}}.$$

Since $\boldsymbol{T}$ and $\boldsymbol{F}$ are both, in general, unsymmetric, $\mathbb{A}$ does not possess minor symmetries. If the material is hyperelastic, then

$$\mathbb{A} = \frac{\partial^2 \hat{W}}{\partial \boldsymbol{F} \partial \boldsymbol{F}} = \mathbb{A}^T, \tag{5.59}$$

i.e., $\mathbb{A}$ possesses major symmetry. Positive definiteness is believed to be too strong a constraint for $\mathbb{A}(\boldsymbol{X}, \boldsymbol{F})$, since it precludes nonuniqueness of solutions as in the case of buckling. From Eqn. (5.59), it follows that $\partial \boldsymbol{T}/\partial \boldsymbol{F} = (\partial \boldsymbol{T}/\partial \boldsymbol{F})^T$ is a necessary condition for a material to be hyperelastic.

The second elasticity tensor is defined as

$$\mathbb{C} = \frac{\partial \check{\boldsymbol{S}}}{\partial \boldsymbol{E}} = 2\frac{\partial \tilde{\boldsymbol{S}}}{\partial \boldsymbol{C}}, \tag{5.60}$$

which in component form reads

$$\mathbb{C}_{ijkl} = 2\frac{\partial \tilde{S}_{ij}}{\partial C_{kl}}.$$

Since $\boldsymbol{S}$ and $\boldsymbol{C}$ are both symmetric, $\mathbb{C}$ has both the minor symmetries, i.e.,

$$\mathbb{C} = \mathbb{SCS},$$
$$\mathbb{C}_{ijkl} = \mathbb{C}_{jikl} = \mathbb{C}_{ijlk}.$$

If the material is assumed to be hyperelastic, then since $\boldsymbol{S} = 2\partial \tilde{W}/\partial \boldsymbol{C}$, we get

$$\mathbb{C} = 4\frac{\partial^2 \tilde{W}}{\partial \boldsymbol{C} \partial \boldsymbol{C}}, \tag{5.61}$$
$$\mathbb{C}_{ijkl} = 4\frac{\partial^2 \tilde{W}}{\partial C_{ij} \partial C_{kl}}.$$

In this case, $\mathbb{C}$ possesses major symmetry as well, i.e.,

$$\mathbb{C} = \mathbb{SCS} = \mathbb{C}^T, \tag{5.62}$$

$$\mathbb{C}_{ijkl} = \mathbb{C}_{jikl} = \mathbb{C}_{ijlk} = \mathbb{C}_{klij}.$$

Note that although, theoretically, we write the expression for $\mathbb{C}$ as in Eqn. (5.61), it should be computed using Eqn. (1.241). From Eqn. (5.62), it follows that $\partial \boldsymbol{S}/\partial \boldsymbol{C} = (\partial \boldsymbol{S}/\partial \boldsymbol{C})^T$ is a necessary condition for the material to be hyperelastic.

For an isotropic, hyperelastic material, from Eqn. (5.41), we have $\boldsymbol{S} = 2(c_{-1}\boldsymbol{C}^{-1} + c_0\boldsymbol{I} + c_1\boldsymbol{C})$, where

$$\begin{aligned} c_{-1} &= \frac{\partial \breve{W}}{\partial I_3} I_3, \\ c_0 &= \frac{\partial \breve{W}}{\partial I_1} + I_1 \frac{\partial \breve{W}}{\partial I_2}, \\ c_1 &= -\frac{\partial \breve{W}}{\partial I_2}. \end{aligned}$$

Using Eqns. (1.231), (1.236), (1.241) and (5.60), we get

$$\frac{1}{4}\mathbb{C} = \boldsymbol{C}^{-1} \otimes \frac{\partial c_{-1}}{\partial \boldsymbol{C}} + \boldsymbol{I} \otimes \frac{\partial c_0}{\partial \boldsymbol{C}} + \boldsymbol{C} \otimes \frac{\partial c_1}{\partial \boldsymbol{C}} - c_{-1}\mathbb{S}(\boldsymbol{C}^{-1} \boxtimes \boldsymbol{C}^{-1})\mathbb{S} + c_1\mathbb{S}.$$

The relation between the first and second elasticity tensors is given by

$$\mathbb{A}(\boldsymbol{X}, \boldsymbol{F}) = \boldsymbol{I} \boxtimes \boldsymbol{S} + (\boldsymbol{F} \boxtimes \boldsymbol{I})\mathbb{S}\mathbb{C}\mathbb{S}(\boldsymbol{F}^T \boxtimes \boldsymbol{I}). \tag{5.63}$$

The proof is obtained directly using Eqns. (1.233), (1.234), (5.58) and (5.60), or by using the relations $T_{ij} = F_{im}S_{mj}$ and $\mathbb{C}_{mjlr} = \mathbb{C}_{mjrl}$, as follows:

$$\begin{aligned} \mathbb{A}_{ijkl} &= \frac{\partial T_{ij}}{\partial F_{kl}} \\ &= \delta_{ik}\delta_{ml}S_{mj} + F_{im}\frac{\partial S_{mj}}{\partial F_{kl}} \\ &= \delta_{ik}S_{jl} + F_{im}\frac{\partial S_{mj}}{\partial C_{rs}}\frac{\partial C_{rs}}{\partial F_{kl}} \\ &= \delta_{ik}S_{jl} + \frac{1}{2}F_{im}\mathbb{C}_{mjrs}\left[F_{ks}\delta_{rl} + F_{kr}\delta_{sl}\right] \\ &= \delta_{ik}S_{jl} + \frac{1}{2}F_{im}\left[\mathbb{C}_{mjls}F_{ks} + \mathbb{C}_{mjrl}F_{kr}\right] \\ &= \delta_{ik}S_{jl} + F_{im}\mathbb{C}_{mjrl}F_{kr}. \end{aligned}$$

If $\boldsymbol{F}$ is symmetric, then $\boldsymbol{T} = \partial \hat{W}/\partial \boldsymbol{F}$ is also symmetric, so that $\mathbb{A}$, similar to $\mathbb{C}$, possesses both minor symmetries; in such a case, the expression on the right-hand side of Eqn. (5.63) should be pre- and post-multiplied by $\mathbb{S}$.

Another elasticity tensor that we need later is defined as

$$\mathbb{Z} := \frac{\partial \boldsymbol{\tau}}{\partial \boldsymbol{N}}, \tag{5.64}$$

where $\boldsymbol{N} := \log \boldsymbol{B}$. Note that $\mathbb{Z}$ has both minor symmetries (but not major symmetry), so that $\mathbb{Z} = \mathbb{S}\mathbb{Z}\mathbb{S}$. Note also that $\mathbb{Z}$ is frame-indifferent. This is proved by noting that since

$\tau^* = \mathbb{Q}\tau$ and $N = \mathbb{Q}^T N^*$ (by Eqn. (4.7)), where $\mathbb{Q} := \mathbb{S}(Q \boxtimes Q)\mathbb{S}$ (writing $Q(t)$ simply as Q for notational brevity), we have

$$\begin{aligned}\frac{\partial \tau^*}{\partial N^*} &= \frac{\partial \tau^*}{\partial N}\frac{\partial N}{\partial N^*} \\ &= \mathbb{Q}\frac{\partial \tau}{\partial N}\mathbb{Q}^T.\end{aligned}$$

It follows from Eqn. (1.194) that the eigenvalues of $\mathbb{Z}$ remain unchanged under a change of observer. By virtue of Eqn. (5.46), for an isotropic material, Eqn. (5.64) can also be written as

$$2J\mathbb{Z} = 4\frac{\partial^2 \breve{W}}{\partial N \partial N} - J\tau \otimes I. \tag{5.65}$$

We have obtained the above formula by writing J as $\sqrt{e^{\operatorname{tr} N}}$.

Let $\mathbb{Z}_1 := \partial\tau/\partial B$ and $\mathbb{Z}_2 := \partial B/\partial N$. By the chain rule, we have

$$\mathbb{S}\mathbb{Z}\mathbb{S} = (\mathbb{S}\mathbb{Z}_1\mathbb{S})(\mathbb{S}\mathbb{Z}_2\mathbb{S}). \tag{5.66}$$

We now evaluate $(\mathbb{S}\mathbb{Z}_1\mathbb{S})$ and $(\mathbb{S}\mathbb{Z}_2\mathbb{S})$. Differentiating the relation $T = J\tau F^{-T}$ and using Eqns. (1.159), (1.173), (1.228), (1.231) and (1.233), we get

$$\begin{aligned}\mathbb{A} &= J(I \boxtimes F^{-1})\mathbb{S}\frac{\partial \tau}{\partial F} + (I \boxtimes F^{-1})[\tau \otimes (JF^{-T})] - (J\tau \boxtimes I)(F^{-T} \boxtimes F^{-1})\mathbb{T}, \\ &= J(I \boxtimes F^{-1})\mathbb{S}\frac{\partial \tau}{\partial F} + (J\tau F^{-T}) \otimes F^{-T} - [(J\tau F^{-T}) \boxtimes F^{-1}]\mathbb{T}.\end{aligned} \tag{5.67}$$

Using the chain rule and Eqn. (1.235), we have

$$J\mathbb{S}\frac{\partial \tau}{\partial F} = J\mathbb{S}\frac{\partial \tau}{\partial B}\mathbb{S}\frac{\partial B}{\partial F} = 2J\mathbb{S}\mathbb{Z}_1\mathbb{S}(I \boxtimes F).$$

Substituting the above expression into Eqn. (5.67), we get

$$\mathbb{A} = 2J(I \boxtimes F^{-1})\mathbb{S}\mathbb{Z}_1\mathbb{S}(I \boxtimes F) + (J\tau F^{-T}) \otimes F^{-T} - [(J\tau F^{-T}) \boxtimes F^{-1}]\mathbb{T}. \tag{5.68}$$

Note that $\mathbb{S}\mathbb{Z}_1\mathbb{S}$, similar to $\mathbb{S}\mathbb{Z}\mathbb{S}$, has both the minor symmetries, but not major symmetry.

If (λ_i, P_i) denote the eigenvalues/eigenprojections of B, then using Eqn. (1.256), we have

$$\begin{aligned}\mathbb{S}\mathbb{Z}_2\mathbb{S} &= \mathbb{S}\left[\sum_{i=1}^{k}\lambda_i P_i \boxtimes P_i + \sum_{i=1}^{k}\sum_{\substack{j=1\\ j\neq i}}^{k}\frac{\lambda_i - \lambda_j}{\log\lambda_i - \log\lambda_j}P_i \boxtimes P_j\right]\mathbb{S} \qquad k > 1, \\ &= \lambda\mathbb{S} \qquad \lambda_1 = \lambda_2 = \lambda_3 \equiv \lambda,\end{aligned} \tag{5.69}$$

where k denotes the number of distinct eigenvalues of B, and P_i is given by Eqn. (J.4). From the above expression, it is evident that $\mathbb{S}\mathbb{Z}_2\mathbb{S}$ has both major and minor symmetries. Note that since $B \in$ Psym, the eigenvalues of B are greater than zero, namely, $\lambda_i > 0$ for all i. Using Eqns. (1.257)–(1.259), the distinct eigenvalues of $\mathbb{S}\mathbb{Z}_2\mathbb{S}$ are

$$\left(\lambda_1, \lambda_2, \lambda_3, \frac{\lambda_1 - \lambda_2}{\log\lambda_1 - \log\lambda_2}, \frac{\lambda_2 - \lambda_3}{\log\lambda_2 - \log\lambda_3}, \frac{\lambda_1 - \lambda_3}{\log\lambda_1 - \log\lambda_3}\right), \quad \lambda_1 \neq \lambda_2 \neq \lambda_3 \neq \lambda_1,$$

$$\left(\lambda_1, \lambda_2, \frac{\lambda_1 - \lambda_2}{\log \lambda_1 - \log \lambda_2}\right), \qquad \lambda_1 \neq \lambda_2 = \lambda_3,$$
$$\lambda, \qquad \lambda_1 = \lambda_2 = \lambda_3 \equiv \lambda.$$

Since all the λ_i are greater than zero, the eigenvalues of $\mathbb{S}\mathbb{Z}_2\mathbb{S}$ are also positive, so that $\mathbb{S}\mathbb{Z}_2\mathbb{S}$ is a symmetric positive definite tensor, and we get $\det(\mathbb{S}\mathbb{Z}_2\mathbb{S}) > 0$ (that $\mathbb{S}\mathbb{Z}_2\mathbb{S}$ is the inverse of $\mathbb{S}(\partial \boldsymbol{N}/\partial \boldsymbol{B})\mathbb{S}$ can be easily verified by multiplying the expression for $\mathbb{S}\mathbb{Z}_2\mathbb{S}$ with the right-hand side of Eqn. (1.313), and using Eqns. (1.175), (1.176) and the properties of projections). At the reference configuration $\boldsymbol{F} = \boldsymbol{I}$, we have

$$2\mathbb{S}\mathbb{Z}\mathbb{S}|_{\boldsymbol{F}=\boldsymbol{I}} = 2\mathbb{S}\mathbb{Z}_1\mathbb{S}|_{\boldsymbol{F}=\boldsymbol{I}} = \mathbb{A}|_{\boldsymbol{F}=\boldsymbol{I}} = \mathbb{C}|_{\boldsymbol{F}=\boldsymbol{I}}. \tag{5.70}$$

It follows that if we assume $\mathbb{C}|_{\boldsymbol{F}=\boldsymbol{I}}$ to be positive definite, then $\mathbb{S}\mathbb{Z}\mathbb{S}$ and $\mathbb{S}\mathbb{Z}_1\mathbb{S}$ are both positive definite at the reference configuration. By Eqn. (5.66),

$$\det(\mathbb{S}\mathbb{Z}\mathbb{S}) = \det(\mathbb{S}\mathbb{Z}_1\mathbb{S})\det(\mathbb{S}\mathbb{Z}_2\mathbb{S}),$$

and we have just shown that $\det(\mathbb{S}\mathbb{Z}_2\mathbb{S}) > 0$. Thus, during the course of a deformation,

$$\mathbb{Z} \text{ becomes singular} \iff \mathbb{Z}_1 \text{ becomes singular.} \tag{5.71}$$

By 'becoming singular', we mean that at least one of the eigenvalues becomes zero. Since only the symmetric parts of $\mathbb{Z}$ and $\mathbb{Z}_1$ decide their positive definiteness (which means that the eigenvalues of the symmetric parts have to be positive), Eqn. (5.71) does not imply that $\mathbb{Z}$ and $\mathbb{Z}_1$ lose positive definiteness simultaneously. However, since $\boldsymbol{B}$ is also a measure of strain, we conjecture that this statement holds. If this conjecture holds, then one can test for the positive definiteness of either $\mathbb{Z}$ or $\mathbb{Z}_1$ in order to decide the positive definiteness of the other. The conjecture is true in the case of equitriaxial loading where $\boldsymbol{\tau}$ is of the form $\gamma \boldsymbol{I}$, since, as evident from Eqns. (5.65), (5.66) and (5.69), both $\mathbb{Z}$ and $\mathbb{Z}_1$ possess major symmetry in this case.

For an isotropic hyperelastic material, the expression for the Cauchy stress is obtained from Eqn. (5.45) as

$$\boldsymbol{\tau} = b_{-1}(I_k)\boldsymbol{B}^{-1} + b_0(I_k)\boldsymbol{I} + b_1(I_k)\boldsymbol{B},$$

where, with I_k, $k = 1, 2, 3$, denoting the principal invariants of $\boldsymbol{B}$,

$$b_{-1} = -2\frac{\partial \breve{W}}{\partial I_2}\sqrt{I_3},$$
$$b_0 = \frac{2}{\sqrt{I_3}}\left(\frac{\partial \breve{W}}{\partial I_2}I_2 + \frac{\partial \breve{W}}{\partial I_3}I_3\right),$$
$$b_1 = \frac{2}{\sqrt{I_3}}\frac{\partial \breve{W}}{\partial I_1}.$$

In this case, using Eqns. (1.168), (1.169), (1.175), (1.176), (1.231), (1.233), and the fact that $\boldsymbol{B}^{-1}\boldsymbol{P}_i = \boldsymbol{P}_i/\lambda_i$, we get

$$\mathbb{S}\mathbb{Z}_1\mathbb{S} = \boldsymbol{B}^{-1} \otimes \left(\frac{\partial b_{-1}}{\partial \boldsymbol{B}}\right) + \boldsymbol{I} \otimes \left(\frac{\partial b_0}{\partial \boldsymbol{B}}\right) + \boldsymbol{B} \otimes \left(\frac{\partial b_1}{\partial \boldsymbol{B}}\right) - b_{-1}\mathbb{S}(\boldsymbol{B}^{-1} \boxtimes \boldsymbol{B}^{-1})\mathbb{S} + b_1\mathbb{S},$$

$$\mathbb{S}\mathbb{Z}\mathbb{S} = \boldsymbol{B}^{-1} \otimes \left(\frac{\partial b_{-1}}{\partial \boldsymbol{B}} \boldsymbol{B} \right) + \boldsymbol{I} \otimes \left(\frac{\partial b_0}{\partial \boldsymbol{B}} \boldsymbol{B} \right) + \boldsymbol{B} \otimes \left(\frac{\partial b_1}{\partial \boldsymbol{B}} \boldsymbol{B} \right)$$
$$- b_{-1}\mathbb{S} \left[\sum_{i=1}^{k} \frac{1}{\lambda_i} \boldsymbol{P}_i \boxtimes \boldsymbol{P}_i + \sum_{i=1}^{k} \sum_{\substack{j=1 \\ j \neq i}}^{k} \frac{\lambda_i - \lambda_j}{\lambda_i \lambda_j (\log \lambda_i - \log \lambda_j)} \boldsymbol{P}_i \boxtimes \boldsymbol{P}_j \right] \mathbb{S} + b_1 \mathbb{S}\mathbb{Z}_2\mathbb{S}.$$

The spatial elasticity tensor is defined via the relation

$$\mathsf{c} := \frac{1}{J} (\boldsymbol{F} \boxtimes \boldsymbol{F}) \mathbb{C} (\boldsymbol{F} \boxtimes \boldsymbol{F})^T,$$
$$\mathsf{c}_{ijkl} = \frac{1}{J} F_{ip} F_{jq} F_{kr} F_{ls} \mathbb{C}_{pqrs}.$$

Obviously, c possesses the same symmetries as $\mathbb{C}$. Given $\mathbb{C}$, the engineering form of c can be obtained using Eqn. (I.13) with $\boldsymbol{J} \equiv \boldsymbol{F}^T$.

The behavior of the constitutive equation $\boldsymbol{T} = \hat{\boldsymbol{T}}(\boldsymbol{F})$ near $\boldsymbol{F} = \boldsymbol{I}$ is governed by the linear transformation $\mathsf{C} : \text{Lin} \to \text{Lin}$ defined by

$$\mathsf{C}(\boldsymbol{X}) := D\hat{\boldsymbol{T}}(\boldsymbol{X}, \boldsymbol{I}) = \mathbb{A}|_{\boldsymbol{F}=\boldsymbol{I}}. \tag{5.72}$$

In indicial notation, it is given by

$$\mathsf{C}_{ijkl}(\boldsymbol{X}) = \frac{\partial \hat{T}_{ij}}{\partial F_{kl}} (\boldsymbol{X}, \boldsymbol{I}).$$

As we shall see in the following chapter, C plays a key role in the derivation of the linearized theory of elasticity. For notational convenience, we drop the explicit dependence of C on $\boldsymbol{X}$ in what follows, and will refer to C as the elasticity tensor.

We assume throughout this section that the residual stress vanishes, i.e.,

$$\hat{\boldsymbol{\tau}}(\boldsymbol{I}) = \hat{\boldsymbol{T}}(\boldsymbol{I}) = \hat{\boldsymbol{S}}(\boldsymbol{I}) = \boldsymbol{0}.$$

We first show that the elasticity tensor is also the derivative of the Cauchy stress and the second Piola–Kirchhoff stress tensors, i.e.,

$$\mathsf{C} = D\hat{\boldsymbol{\tau}}(\boldsymbol{I}) = D\hat{\boldsymbol{T}}(\boldsymbol{I}) = D\hat{\boldsymbol{S}}(\boldsymbol{I}). \tag{5.73}$$

Since $\hat{\boldsymbol{\tau}} = \hat{\boldsymbol{T}}\boldsymbol{F}^T / \det \boldsymbol{F}$, we have

$$D\hat{\boldsymbol{\tau}}(\boldsymbol{F})[\boldsymbol{U}] = -\frac{1}{(\det \boldsymbol{F})^2} D\det \boldsymbol{F}[\boldsymbol{U}] \hat{\boldsymbol{T}}(\boldsymbol{F}) \boldsymbol{F}^T + \frac{1}{\det \boldsymbol{F}} \hat{\boldsymbol{T}}(\boldsymbol{F}) D\boldsymbol{F}^T[\boldsymbol{U}] + \frac{1}{\det \boldsymbol{F}} D\hat{\boldsymbol{T}}(\boldsymbol{F})[\boldsymbol{U}] \boldsymbol{F}^T,$$

for every $\boldsymbol{U} \in \text{Lin}$. Evaluating the above equation at $\boldsymbol{F} = \boldsymbol{I}$, and using the fact that $\hat{\boldsymbol{T}}(\boldsymbol{I}) = \boldsymbol{0}$, we get the first relation in Eqn. (5.73). Similarly using $\hat{\boldsymbol{T}} = \boldsymbol{F}\hat{\boldsymbol{S}}$, we get

$$D\hat{\boldsymbol{T}}(\boldsymbol{F})[\boldsymbol{U}] = D\boldsymbol{F}[\boldsymbol{U}] \hat{\boldsymbol{S}}(\boldsymbol{F}) + \boldsymbol{F} D\hat{\boldsymbol{S}}(\boldsymbol{F})[\boldsymbol{U}],$$

for every $\boldsymbol{U} \in \text{Lin}$. Evaluating the above equation at $\boldsymbol{F} = \boldsymbol{I}$, and using $\hat{\boldsymbol{S}}(\boldsymbol{I}) = \boldsymbol{0}$, we get the second relation in Eqn. (5.73).

The elasticity tensor has the following properties:

1. $\mathsf{C}[U] \in \text{Sym}$ for every $U \in \text{Lin}$.
2. $\mathsf{C}[W] = \mathbf{0}$ for every $W \in \text{Skw}$.

The first assertion follows from the relation

$$\mathsf{C}[U] = \frac{d}{d\alpha}\hat{\tau}(I + \alpha U)\bigg|_{\alpha=0},$$

and the fact that $\hat{\tau}$ assumes symmetric values. To prove the second assertion, we use the fact that $\hat{S}(F) = \check{S}(E)$. Finding the directional derivative of both sides, we get

$$\begin{aligned} D_F\hat{S}(I)[U] &= D_E\check{S}(\mathbf{0})[D_F E(I)[U]] \\ &= \frac{1}{2}D\check{S}(\mathbf{0})[U + U^T], \end{aligned}$$

which for $U = W$ yields the desired relation.

An alternative proof of the second assertion can be given using the properties of the exponential function [108]. Choose $W \in \text{Skw}$ and take $Q(t) = e^{Wt}$, so that $Q \in \text{Orth}^+$ (cf. Section 1.10). Differentiating $\hat{\tau}(QF) = Q\hat{\tau}(F)Q^T$, we have

$$D\hat{\tau}(QF)[\dot{Q}F + Q\dot{F}] = \dot{Q}\hat{\tau}(F)Q^T + QD\hat{\tau}(F)[\dot{F}]Q^T + Q\hat{\tau}(F)\dot{Q}^T.$$

Since $Q(0) = I$, and $\dot{Q}(0) = W$, we get

$$D\hat{\tau}(F)[WF + \dot{F}] = W\hat{\tau}(F) + D\hat{\tau}(F)[\dot{F}] - \hat{\tau}(F)W.$$

Using the linearity of the directional derivative, we obtain

$$D\hat{\tau}(F)[WF] = W\hat{\tau}(F) - \hat{\tau}(F)W.$$

Substituting $F = I$ into the above equation, and using the fact that $\hat{\tau}(I) = \mathbf{0}$, we get $D\hat{\tau}(I)[W] = \mathbf{0}$, which by virtue of Eqn. (5.73) yields the desired result.

Writing U as the sum of its symmetric and skew-symmetric parts U_s and U_{ss}, respectively, result (2) above and the linearity of C imply that

$$\mathsf{C}[U] = \mathsf{C}[U_s]. \tag{5.74}$$

Thus, C is completely determined by its restriction to Sym.

We now establish the invariance properties of C.

Theorem 5.8.1. *If $\hat{\tau}$ is isotropic and frame-indifferent at $\mathbf{X}$, then C is invariant under orthogonal transformations, i.e.,*

$$Q\mathsf{C}[U]Q^T = \mathsf{C}[QUQ^T] \quad \forall Q \in \mathit{Orth},\ U \in \mathit{Sym}. \tag{5.75}$$

Proof. By Theorem 5.1.1, if $\hat{\tau}$ is isotropic and frame-indifferent at $\mathbf{X}$, then there exists an isotropic mapping $\bar{\tau}$ such that $\bar{\tau}(FF^T) = \hat{\tau}(F)$. Since $D\hat{\tau}(I) = 2D\bar{\tau}(I)$, we have $\mathbf{C} = 2D\bar{\tau}(I)$. Let $U \in \text{Sym}$. Then, using Eqn. (1.217), we get

$$\begin{aligned} Q\mathbf{C}[U]Q^T &= 2QD\bar{\tau}(I)[U]Q^T = 2D\bar{\tau}(QIQ^T)[QUQ^T] \\ &= 2D\bar{\tau}(I)[QUQ^T] = \mathbf{C}[QUQ^T]. \end{aligned}$$ □

For a material isotropic at X, it follows from Theorem 5.8.1, and the fact that $\mathbf{C}[U] = \mathbf{C}[U_s] \in \text{Sym}$, that $\mathbf{C}(X)[U_s] : \text{Sym} \to \text{Sym}$ is a linear isotropic function. Thus, using Theorem 1.6.15, we conclude that for a material isotropic at X

$$\mathbf{C}[\epsilon] = \lambda(X)(\operatorname{tr}\epsilon)I + 2\mu(X)\epsilon \quad \forall \epsilon \in \text{Sym}. \tag{5.76}$$

In this case, $\mathbf{C}$ can be written as

$$\begin{aligned} \mathbf{C} &= \lambda(X)I \otimes I + 2\mu(X)\mathbb{S}, \\ \mathsf{C}_{ijkl} &= \lambda(X)\delta_{ij}\delta_{kl} + \mu(X)(\delta_{ik}\delta_{jl} + \delta_{il}\delta_{jk}), \end{aligned} \tag{5.77}$$

where $\mathbb{S} = \frac{1}{2}(\delta_{ik}\delta_{jl} + \delta_{il}\delta_{jk})e_i \otimes e_j \otimes e_k \otimes e_l$ is the symmetrizer tensor. When the material is homogeneous, the constants λ and μ are, of course, independent of X.

We say that $\mathbf{C}$ is symmetric if $\mathbf{C}^T = \mathbf{C}$, or alternatively, if

$$A : \mathbf{C}[B] = B : \mathbf{C}[A],$$

for all tensors A and B. In indicial notation, the above property of symmetry can be written as $\mathsf{C}_{ijkl} = \mathsf{C}_{klij}$. If one assumes that the material is hyperelastic, then $\mathbf{C}$ is symmetric. This follows directly from Eqn. (5.59) and the fact that $\mathbf{C} = \mathbb{A}|_{F=I}$.

Since $\mathbf{C}[W] = \mathbf{0}$ for all $W \in \text{Skw}$, $\mathbf{C}$ can never be positive definite in the usual sense. We call $\mathbf{C}$ *positive definite* (or satisfies the Coleman–Noll inequality) if

$$S : \mathbf{C}[S] > 0 \quad \forall S \in \text{Sym} - \{\mathbf{0}\}.$$

Theorem 5.8.2. *Assume that the material at X is isotropic. Then $\mathbf{C}$ is symmetric. Moreover, $\mathbf{C}$ is positive definite if and only if*

$$\mu > 0, \quad 3\lambda + 2\mu > 0. \tag{5.78}$$

Proof. Since $(I \otimes I)^T = I^T \otimes I^T = I \otimes I$ and $\mathbb{S}^T = \mathbb{S}$, it immediately follows from Eqn. (5.77) that $\mathbf{C}$ is symmetric.

If $\mathbf{C}$ is positive definite then

$$S : \mathbf{C}[S] = \lambda(\operatorname{tr} S)^2 + 2\mu S : S > 0 \quad \forall S \in \text{Sym} - \{\mathbf{0}\}.$$

By choosing $S = I$, and a $S \in \text{Sym}$ whose trace is zero, we get the inequalities in Eqn. (5.78). To prove the converse, let $S \in \text{Sym}$, and let

$$\alpha = \frac{1}{3}\operatorname{tr} S, \quad S_0 = S - \alpha I,$$

so that $\operatorname{tr} S_0 = 0$ and $I : S_0 = 0$. We have

$$S : \mathsf{C}[S] = 3\alpha^2(3\lambda + 2\mu) + 2\mu S_0 : S_0,$$

thus proving that if the inequalities in Eqn. (5.78) are assumed to be satisfied, then C is positive definite. □

For another proof of the above theorem, see Appendix I.

5.9 Elastic and Material Stability

Various restrictions on stored energy functions have been proposed in the literature, such as the Baker–Ericksen and Coleman–Noll inequalities, ellipticity conditions etc.; see [209] for a discussion of these inequalities. First, we shall briefly discuss some of these inequalities, and then in view of shortcomings in them, we propose an alternative approach to characterize elastic and material stability for isotropic hyperelastic materials. The following discussion is based on a corrected version of [164].

The Baker–Ericksen inequality proposes that the greater principal stress occur always in the direction of the greater principal stretch. If τ_i represent the principal (Cauchy) stresses, and λ_i the stretches, then the above inequality can be stated as

$$(\tau_i - \tau_j)(\lambda_i - \lambda_j) > 0 \text{ if } \lambda_i \neq \lambda_j. \tag{5.79}$$

We shall see below when we discuss a material instability in the St Venant–Kirchhoff material model that the Baker–Ericksen inequality is ineffective in detecting this instability.

The tension–extension inequality is the requirement that each principal stress is a *strictly* increasing function of the corresponding principal stretch. Mathematically, this can be stated as $(\tau_\Delta)_i(\lambda_\Delta)_i > 0$, $i = 1,2,3$, where the subscript Δ denotes an increment. Consider a prismatic bar whose lateral surfaces are traction-free and whose end faces have a prescribed traction $\bar{t}$ (see Fig. 5.9). The principal stresses in this case are given by $\tau_1 = \tau_2 = 0$ and $\tau_3 = \bar{t}$. By varying $\bar{t}$, one can get λ_1 to either increase or decrease, while τ_1 just stays constant. Thus, this inequality in incorrect.

The Coleman–Noll inequality states that the first elasticity tensor $\mathbb{A}$ be positive definite, i.e.,

$$H : \mathbb{A}H > 0 \quad \forall H \neq \mathbf{0}. \tag{5.80}$$

Such an inequality precludes nonuniqueness of solutions (as in the case of buckling), and is, hence, not a realistic constitutive assumption, as we discuss below.

If in Eqn. (5.80), H is of the form $H = a \otimes b$, $a \neq \mathbf{0}$, $b \neq \mathbf{0}$, then $\mathbb{A}$ is said to be strongly elliptic. i.e., $\mathbb{A}$ is said to be strongly elliptic if

$$(a \otimes b) : \mathbb{A}(a \otimes b) > 0 \quad \forall a \neq \mathbf{0},\ b \neq \mathbf{0}. \tag{5.81}$$

For an isotropic material at $F = I$, the necessary and sufficient conditions for strong ellipticity are $\mu > 0$ and $\lambda + 2\mu > 0$ (see Problem 4). Thus, the bulk modulus $\lambda + 2\mu/3$ need not be positive, which is an important physical requirement that ensures that a sphere 'compresses' when subjected to pressure loading on its surface. Indeed as Ogden [240] (page 394) remarks: "On the one hand the strong ellipticity condition is not strong enough to ensure 'physically reasonable response' in the sense of $\kappa > 0$, while on the other it is too strong to permit discontinuous solutions of the 'shear band' type. Its status with regard to the conditions it imposes on material properties is therefore uncertain."

The assumption of polyconvexity of the stored energy function has been used to prove existence of solutions in elasticity–see [58, 209] for a detailed discussion. However, this notion is not sufficient to guarantee material stability. For example, one of the two Blatz–Ko material models considered below is polyconvex, but still displays a material instability. Even in the case of a St Venant–Kirchhoff material model that is known to be non-polyconvex, there is no criterion based on the notion of polyconvexity that can predict the exact strain state at which a material instability arises.

Krawietz [180] has proposed a postulate of monotonicity (M-condition). As we show while analyzing specific models below, material models that satisfy the M-condition are not immune to material instabilities.

The shortcomings in the inequalities in the literature motivate the development of a new one to characterize material instabilities as described below. However, we first discuss elastic instabilities, and propose a necessary condition for a class of bifurcations that we term as symmetric bifurcations.

Elastic instabilities correspond to multiple solutions at some critical value of loading. An example is the buckling or barrelling of a column (depending on whether the column is 'thin' or 'thick') under the action of a compressive load along the axis (see Figs. 5.1a and b). Yet another example is that of a square membrane made of an incompressible Mooney–Rivlin material that is loaded by in-plane tensile biaxial loads of equal magnitude (see Fig. 5.1c). At the instability or bifurcation point, a rectangular shape is also a solution in addition to the square one, and in fact, the square solution is unstable [197]. Note that in the buckling/barrelling examples, both the symmetry and homogeneity of $\boldsymbol{F}$ are lost, while in the last example, both of them are retained; hence, we will refer to these instabilities as symmetric bifurcations (note that the 'symmetric' does not refer to the symmetry of the structure, which in fact is retained in barrelling, but, rather, refers to the symmetry of $\boldsymbol{F}$). For various examples and discussions on symmetric bifurcations, see, e.g., [13, 46, 173, 206, 229, 264, 265, 269, 320]. In contrast to an elastic instability, by a material instability we mean non-physical behavior that could be either due to a flaw in the material model, or an improper choice of material constants in an otherwise valid material model. We now try to give characterizations for each of these types of instabilities. We assume throughout the following discussion that the reference configuration is a natural state, and that $\mathbb{S}\mathbb{A}\mathbb{S}|_{F=I}$ (equal to $\mathbb{S}\mathbb{C}\mathbb{S}|_{F=I} = 2\mathbb{S}\mathbb{Z}\mathbb{S}|_{F=I} = 2\mathbb{S}\mathbb{Z}_1\mathbb{S}|_{F=I} = \mathbf{C}$ by virtue of Eqns. (5.70) and (5.72)) is positive definite. For an isotropic material, the latter assumption is equivalent to the restrictions $3\lambda + 2\mu > 0$ and $\mu > 0$. We note that since we have assumed the material to be hyperelastic, both $\mathbb{C} = 4\partial^2\tilde{W}/\partial\boldsymbol{C}\partial\boldsymbol{C}$ and $\mathbb{A} = \partial^2\hat{W}/\partial\boldsymbol{F}\partial\boldsymbol{F}$ possess major symmetry, i.e., $\mathbb{C}^T = \mathbb{C}$ and $\mathbb{A}^T = \mathbb{A}$.

The variational formulation given by Eqn. (3.57) reduces to

$$\int_{V_0} \boldsymbol{T} : \boldsymbol{\nabla}_X \boldsymbol{u}_\delta \, dV = \int_{V_0} \rho_0 \boldsymbol{u}_\delta \cdot \boldsymbol{b}_0 \, dV + \int_{(S_0)_t} \boldsymbol{u}_\delta \cdot \bar{\boldsymbol{t}}_0 \, dS \quad \forall \boldsymbol{u}_\delta \in V_u.$$

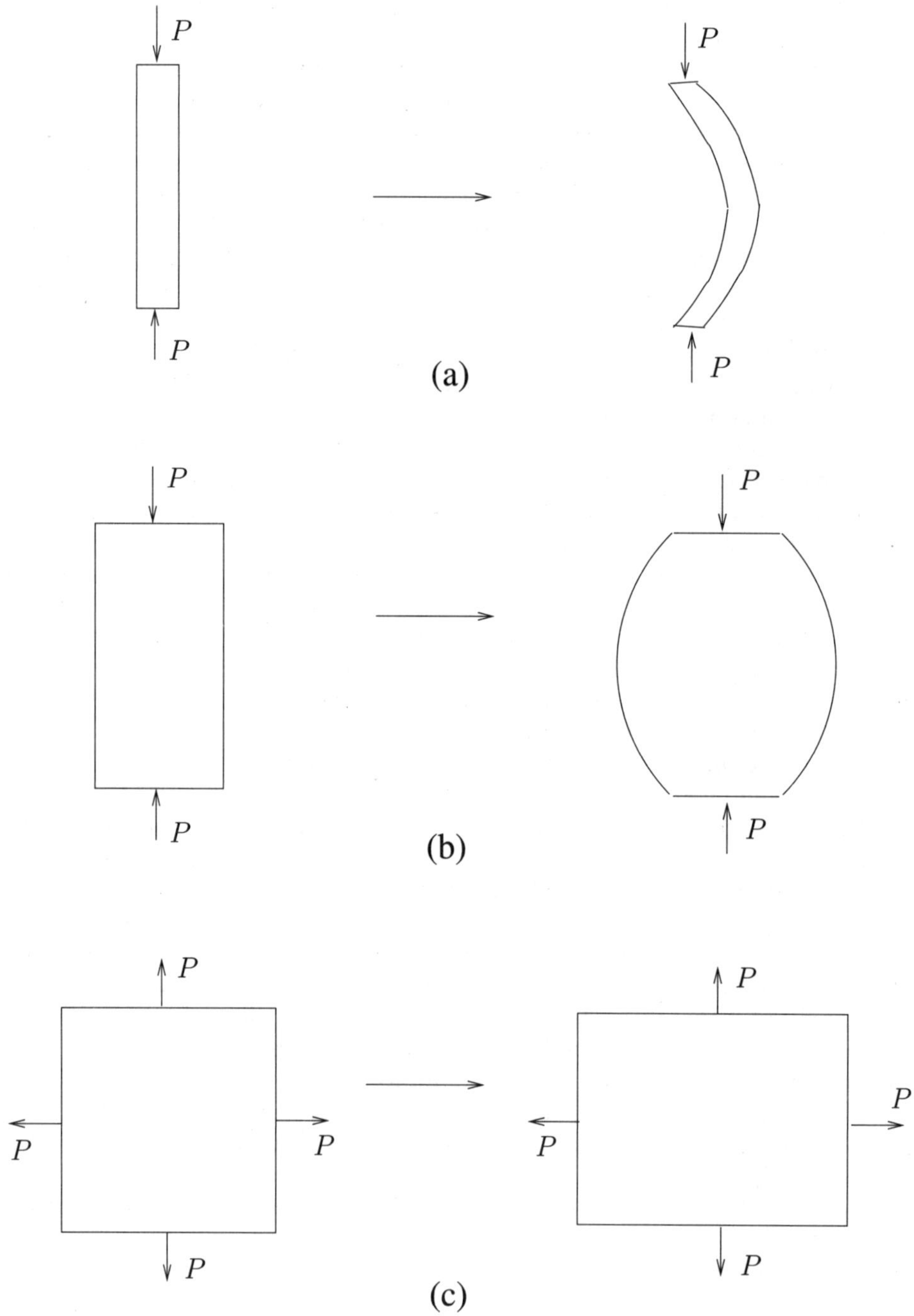

Fig. 5.1 *Examples of elastic instabilities: (a) buckling; (b) barrelling; (c) symmetric bifurcation.*

in the case of static problems. Assuming that the loads are independent of the displacements, the incremental form of the above variational formulation is

$$\int_{V_0} \nabla_X u_\delta : \mathbb{A} \nabla_X u_\Delta \, dV = \int_{V_0} \rho_0 u_\delta \cdot b_0 \, dV + \int_{(S_0)_t} u_\delta \cdot \bar{t}_0 \, dS - \int_{V_0} \nabla_X u_\delta : T_0 \quad \forall u_\delta \in V_u, \tag{5.82}$$

where T_0 is the stress state about which the linearization is carried out. Assuming that the material model is stable, if $\int_{V_0} \nabla_X u_\Delta : \mathbb{A} \nabla_X u_\Delta \, dV$ becomes zero for some nontrivial $u_\Delta \in V_u$, then this indicates the onset of a elastic instability such as buckling, barrelling or a symmetric bifurcation–this condition is equivalent to the stability criterion given by Eqn. (15) of [120]. Note that, if $\mathbb{A}$ is constrained to be positive definite (as in the Coleman–Noll inequality), then $\int_{V_0} \nabla_X u_\Delta : \mathbb{A} \nabla_X u_\Delta \, dV$ can never become zero, which precludes the possibility of elastic instabilities; thus, it is not a realistic constraint.

As mentioned above, buckling or barrelling-type instabilities lead to inhomogeneous deformations and to a loss of symmetry of F, even should such a symmetry exist in the pre-buckling phase. Thus, if one restricts oneself to homogeneous deformations in which $F \in$ Sym, then buckling-type instabilities are automatically excluded. In case F is restricted to be symmetric, T is symmetric (since $T = \partial \hat{W} / \partial F$), so that $\mathbb{A}$ possesses both minor symmetries, i.e., $\mathbb{A} = \mathbb{S}\mathbb{A}\mathbb{S}$. Thus, for a symmetric bifurcation to occur, $\mathbb{S}\mathbb{A}\mathbb{S}$ should lose positive definiteness. Note that in this case, it is possible to give a local criterion since we are assuming the deformation to be homogeneous. Since $\mathbb{S}^T = \mathbb{S}$ and $\mathbb{A}^T = \mathbb{A}$, we have $(\mathbb{S}\mathbb{A}\mathbb{S})^T = \mathbb{S}\mathbb{A}\mathbb{S}$, so that all the six eigenvalues of $\mathbb{S}\mathbb{A}\mathbb{S}$ are real. $\mathbb{S}\mathbb{A}\mathbb{S}$ is positive definite if and only if all these six eigenvalues are positive.

We note that the condition that $\mathbb{S}\mathbb{A}\mathbb{S}$ be positive definite is not the same as the criterion given by Eqn. (17) in [264]. To see this, we use Eqn. (5.63) to get

$$\begin{aligned} \mathbb{S}\mathbb{A}\mathbb{S} &= \mathbb{S}(I \boxtimes S)\mathbb{S} + \mathbb{S}(F \boxtimes I)\mathbb{S}\mathbb{C}\mathbb{S}(F^T \boxtimes I)\mathbb{S} \\ &= \mathbb{S}(F \boxtimes I)(C^{-1} \boxtimes S)(F^T \boxtimes I)\mathbb{S} + \mathbb{S}(F \boxtimes I)\mathbb{S}\mathbb{C}\mathbb{S}(F^T \boxtimes I)\mathbb{S}. \end{aligned} \tag{5.83}$$

Even under the very special case when $C = \lambda I$ and $S = \gamma I$, $C^{-1} \boxtimes S$ *does not* possess minor symmetries, so that checking the eigenvalues of $\mathbb{S}\mathbb{A}\mathbb{S}$ is not equivalent to checking the eigenvalues of

$$\mathbb{S}(C^{-1} \boxtimes S)\mathbb{S} + \mathbb{C}.$$

We now discuss material stability. In the context of hyperelasticity, material stability corresponds to the notion that 'stress increases with strain'. In a one-dimensional setting such a criterion would simply read $\epsilon\tau > 0$, where ϵ and τ are strain and stress measures in the one-dimensional setting. The idea is to generalize this to a three-dimensional framework. Let $N := \log B$, and let $\mathbb{Z}$ be as defined by Eqn. (5.64) (which we have already proved to be objective on page 265). The necessary and sufficient condition that we propose for material stability of an isotropic material (the extension to anisotropic materials would depend on the particular form of the strain energy density function) is that $\mathbb{S}\mathbb{Z}\mathbb{S}$ be positive definite. Since the symmetric part of any tensor decides its positive definiteness, the criterion for material stability can also be stated as the requirement that $\mathbb{S}\mathbb{Z}_{\text{sym}}\mathbb{S}$, where $\mathbb{Z}_{\text{sym}} = (\mathbb{Z} + \mathbb{Z}^T)/2$, be positive definite. Hill [121] (see also [31]) has proposed a similar criterion except that he uses the Kirchhoff stress instead of the Cauchy stress tensor (the disadvantages of doing this are discussed below). If the conjecture stated after Eqn. (5.71) is true, then $\mathbb{SZS}$ loses positive definiteness if and only if $\mathbb{SZ}_1\mathbb{S}$ does, and then, $\mathbb{SZ}_1\mathbb{S}$ can also be used in place of $\mathbb{SZS}$ in the above stability criterion.

Let $\boldsymbol{B}_1 := \boldsymbol{B}(t_1)$, $\boldsymbol{B}_2 := \boldsymbol{B}(t_2)$, $\boldsymbol{N}_1 := \log \boldsymbol{B}_1$, $\boldsymbol{N}_2 := \log \boldsymbol{B}_2$, $\boldsymbol{\tau}_1 := \hat{\boldsymbol{\tau}}(\boldsymbol{N}_1)$, $\boldsymbol{\tau}_2 := \hat{\boldsymbol{\tau}}(\boldsymbol{N}_2)$, $\boldsymbol{Q}_1 := \boldsymbol{Q}(t_1)$, $\boldsymbol{Q}_2 := \boldsymbol{Q}(t_2)$. Note that $\boldsymbol{N}_1, \boldsymbol{N}_2 \in \text{Sym}$. The condition that $\mathbb{Z}$ be positive definite can be stated in the following equivalent form[1]

$$[\boldsymbol{\tau}_2 - \boldsymbol{\tau}_1] : [\boldsymbol{N}_2 - \boldsymbol{N}_1] > 0 \quad \forall \boldsymbol{N}_1, \boldsymbol{N}_2 \in \text{Sym}. \tag{5.84}$$

Since this condition is equivalent to the condition that $\mathbb{Z}$ be positive definite, and since we have shown that $\mathbb{Z}$ is frame-indifferent, the above criterion is also frame-indifferent. One can also prove this directly as follows. We want to show that the criterion given by Eqn. (5.84) holds if and only if

$$[\boldsymbol{\tau}_2^* - \boldsymbol{\tau}_1^*] : [\boldsymbol{N}_2^* - \boldsymbol{N}_1^*] > 0 \quad \forall \boldsymbol{N}_1, \boldsymbol{N}_2 \in \text{Sym}, \quad \forall \boldsymbol{Q}_1, \boldsymbol{Q}_2 \in \text{Orth}^+, \tag{5.85}$$

where $\boldsymbol{N}_1^* := \log \boldsymbol{B}_1^* = \boldsymbol{Q}_1 \boldsymbol{N}_1 \boldsymbol{Q}_1^T$, $\boldsymbol{N}_2^* := \log \boldsymbol{B}_2^* = \boldsymbol{Q}_2 \boldsymbol{N}_2 \boldsymbol{Q}_2^T$, $\boldsymbol{\tau}_1^* := \boldsymbol{Q}_1 \boldsymbol{\tau}_1 \boldsymbol{Q}_1^T$, $\boldsymbol{\tau}_2^* := \boldsymbol{Q}_2 \boldsymbol{\tau}_2 \boldsymbol{Q}_2^T$. First we prove that Eqn. (5.84) implies Eqn. (5.85). Replacing the '$\boldsymbol{N}$' by '$\tilde{\boldsymbol{N}}$' and the '$\boldsymbol{\tau}$' by '$\tilde{\boldsymbol{\tau}}$' in Eqn. (5.84) (note that this is just a change of notation), we get

$$[\tilde{\boldsymbol{\tau}}_2 - \tilde{\boldsymbol{\tau}}_1] : [\tilde{\boldsymbol{N}}_2 - \tilde{\boldsymbol{N}}_1] > 0 \quad \forall \tilde{\boldsymbol{N}}_1, \tilde{\boldsymbol{N}}_2 \in \text{Sym}, \tag{5.86}$$

where $\tilde{\boldsymbol{\tau}}_1 := \hat{\boldsymbol{\tau}}(\tilde{\boldsymbol{N}}_1)$, and $\tilde{\boldsymbol{\tau}}_2 := \hat{\boldsymbol{\tau}}(\tilde{\boldsymbol{N}}_2)$. Now choose $\tilde{\boldsymbol{N}}_1 = \boldsymbol{Q}_1 \boldsymbol{N}_1 \boldsymbol{Q}_1^T = \boldsymbol{N}_1^*$ and $\tilde{\boldsymbol{N}}_2 = \boldsymbol{Q}_2 \boldsymbol{N}_2 \boldsymbol{Q}_2^T = \boldsymbol{N}_2^*$. Note that this choice is admissible, since $\tilde{\boldsymbol{N}}_1, \tilde{\boldsymbol{N}}_2 \in \text{Sym}$, and they are bounded and continuous since $\boldsymbol{N}_1$, $\boldsymbol{N}_2$ are bounded and continuous. Now we have

$$\tilde{\boldsymbol{\tau}}_1 = \hat{\boldsymbol{\tau}}(\tilde{\boldsymbol{N}}_1) = \hat{\boldsymbol{\tau}}(\boldsymbol{Q}_1 \boldsymbol{N}_1 \boldsymbol{Q}_1^T) = \boldsymbol{Q}_1 \hat{\boldsymbol{\tau}}(\boldsymbol{N}_1) \boldsymbol{Q}_1^T = \boldsymbol{Q}_1 \boldsymbol{\tau}_1 \boldsymbol{Q}_1^T = \boldsymbol{\tau}_1^*,$$
$$\tilde{\boldsymbol{\tau}}_2 = \hat{\boldsymbol{\tau}}(\tilde{\boldsymbol{N}}_2) = \hat{\boldsymbol{\tau}}(\boldsymbol{Q}_2 \boldsymbol{N}_2 \boldsymbol{Q}_2^T) = \boldsymbol{Q}_2 \hat{\boldsymbol{\tau}}(\boldsymbol{N}_2) \boldsymbol{Q}_2^T = \boldsymbol{Q}_2 \boldsymbol{\tau}_2 \boldsymbol{Q}_2^T = \boldsymbol{\tau}_2^*,$$

where we have used the result $\hat{\boldsymbol{\tau}}(\boldsymbol{Q}_i \boldsymbol{N}_i \boldsymbol{Q}_i^T) = \boldsymbol{Q}_i \hat{\boldsymbol{\tau}}(\boldsymbol{N}_i) \boldsymbol{Q}_i^T$, $i = 1, 2$ (which follows from material frame-indifference and isotropy). Hence Eqn. (5.86) reduces to Eqn. (5.85).

Conversely, in Eqn. (5.85) choose $\boldsymbol{Q}_2 = \boldsymbol{Q}_1$. This leads to Eqn. (5.84).

Note that Eqn. (5.84) and (5.85) have the same form and hence satisfy form-invariance (although the left-hand sides of the these two equations are not equal).

To describe Hill's criterion for material stability for an isotropic material, let

$$\mathbb{H} := 2\frac{\partial (J\boldsymbol{\tau})}{\partial \boldsymbol{N}}.$$

Note that $\mathbb{H}$ possesses both minor symmetries, so that $\mathbb{H} = \mathbb{SHS}$. For an isotropic hyperelastic material, by Eqn. (5.46), we have $\mathbb{SHS} = 4\mathbb{S}(\partial^2 \breve{W}/\partial \boldsymbol{N} \partial \boldsymbol{N})\mathbb{S}$, so that $\mathbb{H} = \mathbb{H}^T$. Hill proposed the restriction that $\mathbb{SHS}$ should be positive definite [31, 121] (while, as evident from Eqn. (5.65), we require that $\mathbb{H} - J\boldsymbol{\tau} \otimes \boldsymbol{I}$ be positive definite). However, the problem with this restriction is that the Cauchy stress $\boldsymbol{\tau}$, and not the Kirchhoff stress tensor $J\boldsymbol{\tau}$, is a measure of true stress. Thus, $\mathbb{SHS}$ may lose positive definiteness at a point where there is

[1] This equivalent condition was pointed out to the author by Professor Patrizio Neff in a personal communication.

no material instability, as we show below in the case of a St Venant–Kirchhoff material (see Fig. 5.4), or it may remain positive definite even when there is a material instability, as we show below in the case of a model based on the Hencky strain measure.

Based on the above discussion, we summarize the conditions for the onset of elastic and material instabilities:

$$\text{Buckling/Barrelling}: \int_{V_0} \nabla_X u_\Delta : \mathbb{A} \nabla_X u_\Delta \, dV = 0 \text{ for some } u_\Delta \in V_u, \tag{5.87}$$

$$\text{Symmetric Bifurcation}\big|_{F \in \text{Sym}} : \mathbb{S}\mathbb{A}\mathbb{S} \text{ loses positive definiteness}, \tag{5.88}$$

$$\text{Material instability}: \mathbb{S}\mathbb{Z}\mathbb{S} \text{ (equivalently } \mathbb{S}\mathbb{Z}_{\text{sym}}\mathbb{S}) \text{ loses positive definiteness.} \tag{5.89}$$

Conditions given by (5.87) and (5.88) are necessary (but not sufficient, since appropriate constraints can be used to prevent bifurcations), while (5.89) is proposed as a necessary and sufficient condition that characterizes material stability for isotropic materials. In the case of a symmetric bifurcation, the loss of positive definiteness of $\mathbb{S}\mathbb{A}\mathbb{S}$ is due to a loss of positive definiteness of $\mathbb{S}\mathbb{C}\mathbb{S}$. Since, in the case of a St Venant–Kirchhoff material model, $\mathbb{S}\mathbb{C}\mathbb{S}$ is positive definite, a symmetric bifurcation can never occur with the use of this material model.

The practical implication of the above stability criteria is that at a deformation state at which either of them is violated, numerical techniques, such as the finite element method, that are based on some form of the incremental variational formulation given by Eqn. (5.82), either fail to converge or yield absurd results (in the case of material instabilities). The conditions given by Eqns. (5.88) and (5.89) involve only an eigenvalue check of certain matrices (see below), and hence, can either be used to carry out an analytical analysis, or can easily be incorporated in a numerical strategy. Note that the condition $F \in \text{Sym}$ in Eqn. (5.88) is general enough to include uniform uniaxial, biaxial and triaxial tension/compression loading conditions that are usually used for conducting elastic stability analyses in the literature. In case Eqn. (5.89) holds (and we propose that it should hold for material stability), the constitutive relation $\tau = \hat{\tau}(B)$ for an isotropic, hyperelastic material can be inverted to yield $B = \hat{B}(\tau)$, and the latter would not be a 'new class of elastic materials' as stated after Eqn. (2.12) in [33].

Positive definiteness of $\mathbb{A}$ does not imply positive definiteness of $\mathbb{Z}_1$ as seen from the following counterexample. If $W = F : F/2$, then $T = F$ so that $\mathbb{A} = \mathbb{I}$ is positive definite. However, the associated Cauchy stress $\tau = B/J$ does not result in a positive definite $\mathbb{Z}_1$ as can be seen by taking $F = \gamma I$, which results in $\tau = I/\gamma$; thus, τ decreases with increasing γ. Conversely, even if $\mathbb{Z}_1$ is positive definite, $\mathbb{A}$ may lose positive definiteness as noted in the discussion above Eqn. (5.91). Thus, positive definiteness of $\mathbb{A}$ and $\mathbb{Z}_1$ are unrelated.

We check whether $\mathbb{S}\mathbb{A}\mathbb{S}$ and $\mathbb{S}\mathbb{Z}_{\text{sym}}\mathbb{S}$ are positive definite by checking whether all the six eigenvalues are positive. In order to find the eigenvalues of $\mathbb{S}\mathbb{A}\mathbb{S}$ and $\mathbb{S}\mathbb{Z}_{\text{sym}}\mathbb{S}$, we use the χ map of Eqns. (5.83), (5.68) and (5.66), i.e.,

$$\begin{aligned}
\chi(\mathbb{S}\mathbb{A}\mathbb{S}) &= \chi(\mathbb{S}(I \boxtimes S)\mathbb{S}) + \chi(\mathbb{S}(F \boxtimes I)\mathbb{S})H^{-1}\chi(\mathbb{S}\mathbb{C}\mathbb{S})H^{-1}\chi(\mathbb{S}(F^T \boxtimes I)\mathbb{S}),\\
2J\chi(\mathbb{S}\mathbb{Z}_1\mathbb{S}) &= \hat{M}\hat{\Psi}(I \boxtimes F)\hat{\Psi}(\mathbb{A})\hat{\Psi}(I \boxtimes F^{-1})\hat{M}^T + J\chi(\mathbb{S}(I \boxtimes \tau B^{-1})\mathbb{S}) - J\chi(\tau \otimes B^{-1}),\\
2J\chi(\mathbb{S}\mathbb{Z}\mathbb{S}) &= 2J\chi(\mathbb{S}\mathbb{Z}_1\mathbb{S})H^{-1}\chi(\mathbb{S}\mathbb{Z}_2\mathbb{S}).
\end{aligned}$$

The eigenvalues are now found in a manner similar to those for $\mathbb{S}\mathbb{C}\mathbb{S}$ as shown in Appendix I. Note that while $\chi(\mathbb{S}\mathbb{A}\mathbb{S})$ is a symmetric matrix, $\chi(\mathbb{S}\mathbb{Z}\mathbb{S})$ (or $\chi(\mathbb{S}\mathbb{Z}_1\mathbb{S})$) is not, and hence their symmetric parts should be considered while testing for positive definiteness.

We now analyze the elastic and material stability of some popular material models using Eqns. (5.88) and (5.89). We first show that a St Venant–Kirchhoff model is materially unstable under uniaxial compression. As we discuss in Section 5.11.1, and as shown in Fig. 5.2, a St Venant–Kirchhoff material model displays anomalous behavior under uniaxial compression. For any compressive traction $\bar{t}$ below a maximum value of about $0.16554E$, there are two solutions, while above this maximum compressive traction, there are none. At this maximum value of traction which occurs at an axial stretch value λ_a of about 0.606, $\mathbb{S}\mathbb{Z}_{\text{sym}}\mathbb{S}$ loses positive definiteness as seen in Fig. 5.3 (in agreement with the conjecture stated below Eqn. (5.71), $\mathbb{S}(\mathbb{Z}_1)_{\text{sym}}\mathbb{S}$ also loses positive definiteness at this stretch value), and the two solutions at this point are $\lambda_1 = \lambda_2$ and $\lambda_1 = -\lambda_2$; the latter clearly corresponds to a non-physical solution. The path in Fig. 5.2 that lies beyond this material instability point corresponds to non-physical solutions since the stress decreases with increasing strain. Figure 5.4 shows that the peak Cauchy and Kirchhoff stresses occur at different axial stretch values λ_a, namely, 0.606 and 0.707. Thus, Hill's criterion, which is based on the Kirchhoff stress, predicts a material instability at a stretch value where the material is stable, while the Baker–Ericksen inequality fails to detect this instability. We remark that a finite-element-based strategy fails in the vicinity of a stretch of 0.577 (at which point $\mathbb{S}\mathbb{A}\mathbb{S}$ becomes singular), thereby showing that a numerical strategy continues to work even in the presence of a material instability (i.e., even when $\mathbb{S}\mathbb{Z}_{\text{sym}}\mathbb{S}$ loses positive definiteness at a stretch of 0.606), and fails only when $\mathbb{S}\mathbb{A}\mathbb{S}$ loses positive definiteness. The above results indicate a flaw in the St Venant–Kirchhoff material model under compressive loads. In addition to finding shortcomings of existing material models, the stability criterion in Eqn. (5.89) can also be used to find bounds on material constants of newly proposed material models, based on say, uniaxial, biaxial and triaxial loading conditions.

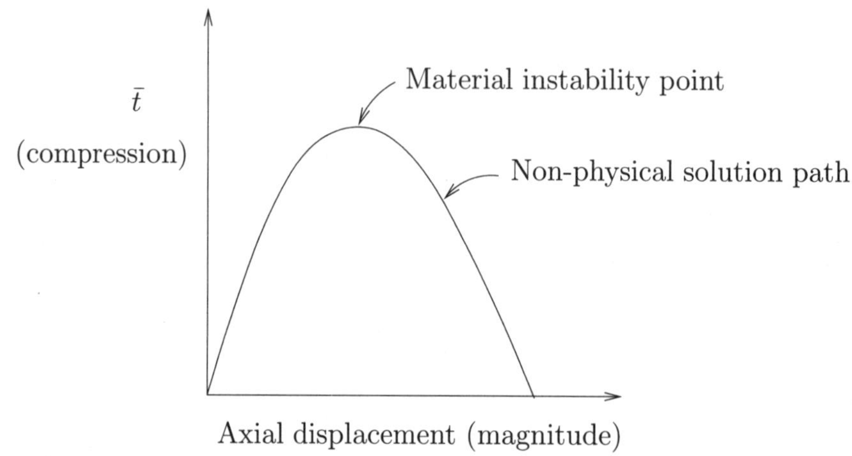

Fig. 5.2 *Anomalous behavior of a St Venant–Kirchhoff material under uniaxial compression.*

Consider now the material stability of the model given by Eqn. (5.53) that is based on the Hencky strain measure. Since the eigenvalues of

$$\mathbb{H} = 2\frac{\partial(J\tau)}{\partial N} = \lambda I \otimes I + 2\mu\mathbb{S},$$

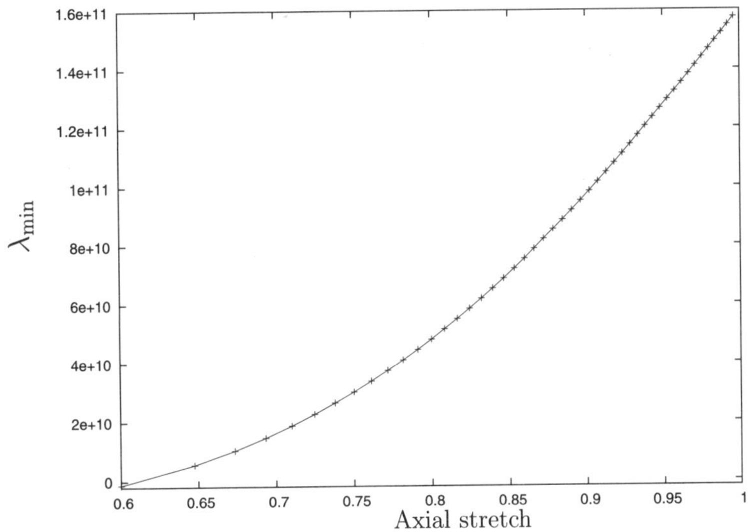

Fig. 5.3 *Lowest eigenvalue of $2J\mathbb{S}\mathbb{Z}_{sym}\mathbb{S}$ as a function of the axial stretch for a St Venant–Kirchhoff material under uniaxial compression ($E = 2 \times 10^{11}$ and $\nu = 0.25$).*

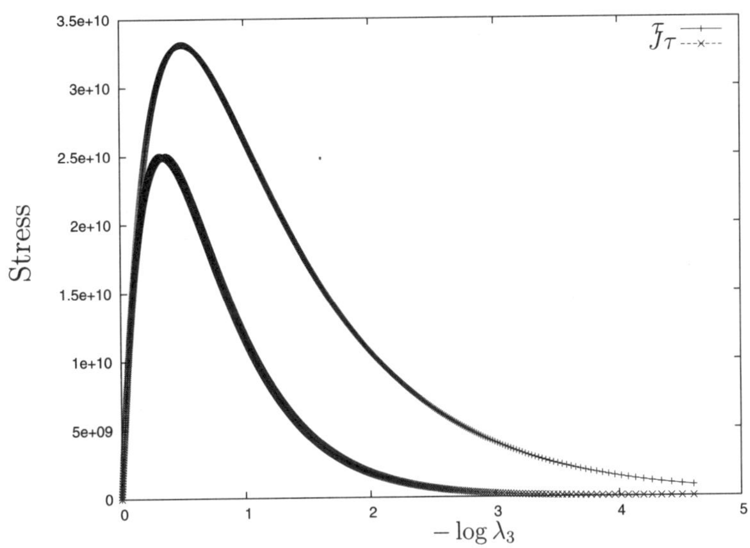

Fig. 5.4 *The longitudinal Cauchy and Kirchhoff stresses as a function of $\log \lambda_3$ for a St Venant–Kirchhoff material under uniaxial compression ($E = 2 \times 10^{11}$ and $\nu = 0.25$).*

are $(2\mu, 2\mu, 2\mu, 2\mu, 3\lambda + 2\mu)$, it follows that Hill's stability criterion leads to the conditions

$$\mu > 0, \quad 3\lambda + 2\mu > 0.$$

Since we assume these bounds on the material constants to hold by virtue of our assumption that $\mathbb{C}|_{F=I}$ be positive definite, according to Hill's criterion, the model given by Eqn. (5.53) is stable irrespective of the strain values. To find the restrictions according to the proposed criterion given by Eqn. (5.89), we need to evaluate $\mathbb{Z}$. Using Eqn. (5.65), we get

$$2J\mathbb{Z} = (\lambda \boldsymbol{I} - J\boldsymbol{\tau}) \otimes \boldsymbol{I} + 2\mu \mathbb{S}. \tag{5.90}$$

We consider the case of equitriaxial loading. Thus, take $\boldsymbol{F}$ to be of the form $\gamma \boldsymbol{I} = J^{1/3}\boldsymbol{I}$. From Eqn. (5.53), we get $J\boldsymbol{\tau} = \tilde{\tau}\boldsymbol{I}$ where $\tilde{\tau} = (\lambda + 2\mu/3)\log J$. The function $\log J/J$ increases with J for $J \in (0, e)$, peaks at $J = e$, and then decreases with increasing J for $J > e$. Thus, $\boldsymbol{\tau}$ increases with increasing γ until $\gamma = e^{1/3}$, and then decreases to zero with increasing γ, demonstrating the onset of a material instability at $J = e$. Of course, in the deformation considered, an elastic instability in the form of a symmetric bifurcation would precede this material instability (i.e., $\mathbb{S}\mathbb{A}\mathbb{S}$ would lose positive definiteness before $\mathbb{S}\mathbb{Z}\mathbb{S}$), but nevertheless shows once again that Hill's criterion fails to detect this instability, while the proposed criterion does, as we now demonstrate. Substituting $J\boldsymbol{\tau} = \tilde{\tau}\boldsymbol{I}$ into Eqn. (5.90), we get

$$2J\mathbb{Z} = (\lambda - \tilde{\tau})\boldsymbol{I} \otimes \boldsymbol{I} + 2\mu \mathbb{S}.$$

Using the χ-map, the eigenvalues of $2J\mathbb{Z}$ are found to be $(2\mu, 2\mu, 2\mu, 2\mu, 2\mu, 3\lambda + 2\mu - 3\tilde{\tau})$. Thus, the conditions for material stability according to the proposed criterion are

$$\mu > 0, \quad 3\lambda + 2\mu - 3\tilde{\tau} > 0.$$

The second condition is equivalent to $\log J = 3\log\gamma < 1$, or equivalently, $\gamma < e^{1/3}$. Quite interestingly, the restriction that we have obtained, namely,

$$\det \boldsymbol{V} = \det \boldsymbol{F} < e,$$

is also one of the restrictions imposed by the strong ellipticity condition (see Eqn. (4.3) in [31]). One can possibly eliminate the above material instability by using a generalized Hencky function such as the one in Eqn. (5.54).

The strain-energy density function and the associated constitutive relation for the Blatz–Ko special I and Blatz–Ko special II material models are given by [295]

$$\tilde{W} = \frac{\mu}{2}\left[I_1 + 2I_3^{-1/2} - 5\right], \tag{5.91a}$$

$$\boldsymbol{\tau} = \frac{\mu}{J^2}(J\boldsymbol{B} - \boldsymbol{I}). \tag{5.91b}$$

and

$$\tilde{W} = \frac{\mu}{2}\left[\frac{I_2}{I_3} + 2I_3^{1/2} - 5\right], \tag{5.92a}$$

$$\tau = \mu \left(I - J^{-1} B^{-1} \right). \tag{5.92b}$$

The model given by Eqns. (5.91) is polyconvex, while that given by Eqns. (5.92) is not.

First consider the model given by Eqns. (5.91). We have

$$\mathbb{Z}_1 = \frac{\partial \tau}{\partial B} = \frac{\mu}{J^2} \left[J\mathbb{S} + \left(I - \frac{J}{2} B \right) \otimes B^{-1} \right].$$

Consider the case of equitriaxial extension where $F = \gamma I$; as pointed out in the discussion below Eqn. (5.71), we can test the positive definiteness of $\mathbb{Z}_1$ in this case. The eigenvalues of $J^2 \mathbb{Z}_1/\mu$ are $\{\gamma^3, \gamma^3, \gamma^3, \gamma^3, \gamma^3, (6-\gamma^5)/(2\gamma^2)\}$. Thus, $\mathbb{Z}_1$ loses positive definiteness at a stretch value of $\gamma = 6^{1/5}$. Exactly consistent with this fact, the Cauchy stress given by Eqn. (5.91b) has a maximum at the stretch value of $6^{1/5}$. Thus, polyconvexity is not sufficient to prevent a material instability.

Now consider the model given by Eqns. (5.92). Note that for this model, $\mathbb{C}|_{F=I} = \mu(I \otimes I + 2\mathbb{S})$, which implies that $\lambda = \mu$. This may not be a good approximation for rubber-type materials. A stability analysis of this model based on strong ellipticity has been carried out by Knowles and Sternberg [174]. Under uniaxial loading along the '1'-direction, we get

$$\tau_{11} = \mu \left(1 - \frac{1}{\lambda_1^{5/2}} \right),$$

with the remaining stress components zero. Thus, even if $\lambda_1 \to \infty$, τ_{11} cannot exceed μ, so that there is no solution in case a traction greater than μ is applied along the longitudinal direction. Again, the proposed stability criterion detects this instability since

$$\frac{2J}{\mu} \mathbb{S}\mathbb{Z}_{\text{sym}}\mathbb{S} = \left\{ \left[2\mathbb{S} + B^{-1} \otimes B \right] \mathbb{S}(B^{-1} \boxtimes B^{-1}) \mathbb{S}\mathbb{Z}_2 \right\}_{\text{sym}}.$$

From the above expression, it is clear that $\mathbb{Z}_{\text{sym}}$ becomes singular as the longitudinal stretch tends to infinity because of the presence of the $\mathbb{S}(B^{-1} \boxtimes B^{-1})\mathbb{S}$ term.

Finally, consider the special class of deformations where $F = U$ (which includes uniaxial, equibiaxial and equitriaxial extensions). Under this condition, a constitutive relation that satisfies the M-condition [180] is

$$\tau = \frac{1}{\det U} \left[\lambda \operatorname{tr}(U - I) U + 2\mu (U - I) U \right].$$

Under equitriaxial loading conditions, the Cauchy stress increases upto a stretch value of 2, and then decreases. Under uniaxial tension, the stress becomes infinite at a stretch value of $3 + 2\mu/\lambda$. Thus, satisfaction of the M-condition *does not* guarantee material stability.

In what follows, we give examples of symmetric bifurcations. Consider the equibiaxial loading of a cube made of a compressible Mooney–Rivlin material (see Eqn. (5.126)) with $c_1 = 0.15 \times 10^6$, $c_2 = 0.015 \times 10^6$, $\lambda = 3.3 \times 10^9$ and $\alpha = 4$. As seen from Fig. 5.5, the lowest eigenvalue of $\mathbb{S}\mathbb{A}\mathbb{S}$ goes to zero at a stretch value of about 3.1677, although $\mathbb{S}\mathbb{Z}\mathbb{S}$ remains positive definite, indicating a bifurcation to a rectangular in-plane shape. This critical stretch value agrees almost perfectly with the value of 3.16854 obtained using either Eqn. (2.8) of [173] or Eqn. (1.5) of [229] (after the correction of a typographical error

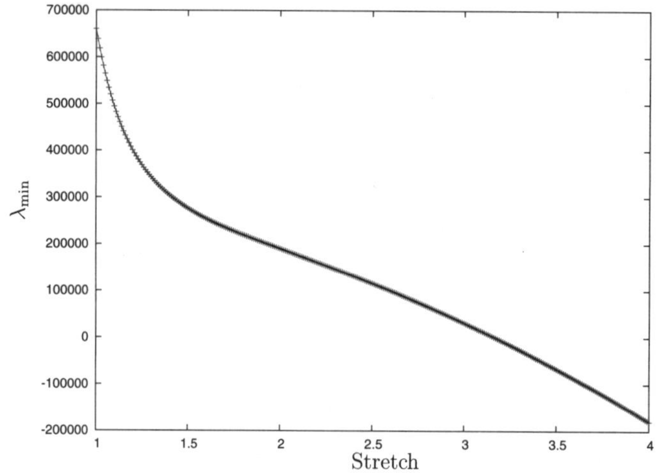

Fig. 5.5 *Lowest eigenvalue of* $\mathbf{S}\mathbb{A}\mathbf{S}$ *as a function of the stretch for a cube made of a compressible Mooney–Rivlin material subjected to equibiaxial extension.*

whereby $\lambda^3\mu^2$ should read $\lambda^3\mu^3$); the small discrepancy is because the aforementioned references consider the material to be perfectly incompressible, while we take it to be almost incompressible. Now consider the same cube and loading as above with the addition of a *compressive* load, of the same magnitude as the in-plane tensile loads (see Fig. 5.6), in the out-of-plane direction. Now the bifurcation occurs at an in-plane stretch value of 3.258, as

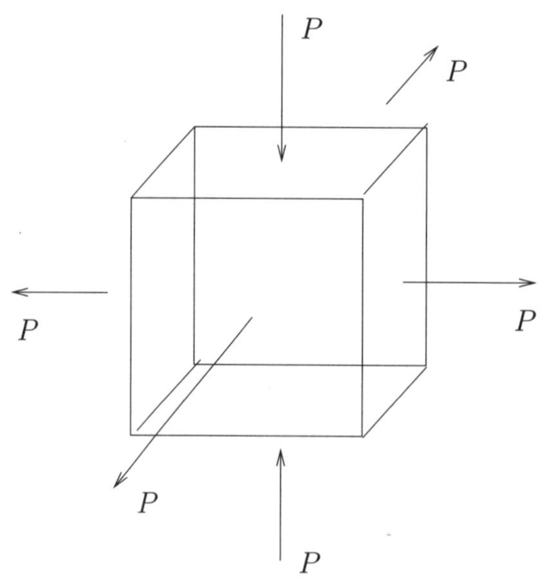

Fig. 5.6 *A cube made of a compressible Mooney–Rivlin material loaded by tensile in-plane and compressive out-of-plane loads.*

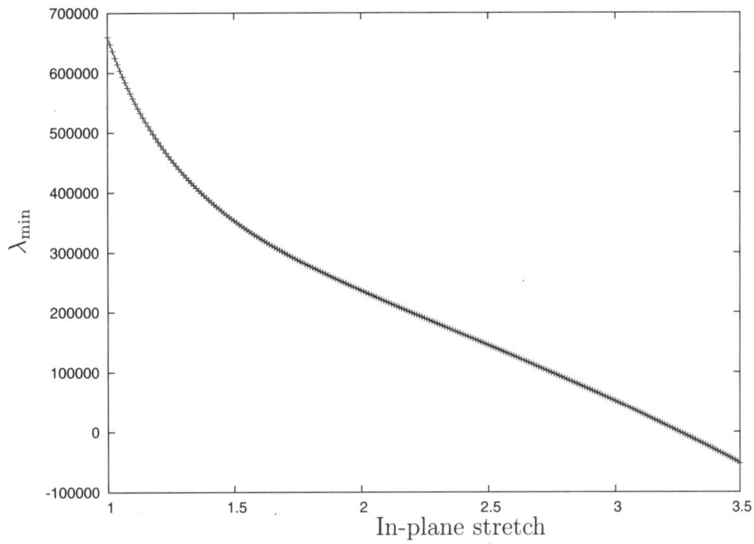

Fig. 5.7 *Lowest eigenvalue of $\mathbb{SAS}$ as a function of the in-plane stretch for a cube made of a compressible Mooney–Rivlin material subjected to the loading shown in Fig. 5.6.*

seen in Fig. 5.7. No bifurcation occurs in both the above load-cases if the cube is made of a neo-Hookean material, i.e., if $c_2 = 0$. The above example illustrates that the statement in [264], that a symmetric bifurcation (referred to there as a 'material instability') arises only when the stress field in the body is positive, is not strictly true. A study of the known cases of symmetric bifurcations does, however, indicate that they do not occur under compressive states of stress, provided, of course, that the material model is stable (this condition is important, since, in the uniaxial compression of a bar made of a St Venant–Kirchhoff material considered above, $\mathbb{SAS}$ loses positive definiteness after $\mathbb{SZS}$ does; however, this loss of positive definiteness of $\mathbb{SAS}$ is clearly due to the material instability, and is not indicative of a symmetric bifurcation). In the first load-case considered above, all eigenvalues of $\mathbb{A}$ are positive upto the bifurcation point, while in the second one, two eigenvalues of $\mathbb{A}$ are zero for any nonzero P, thereby confirming that checking the lowest eigenvalue of $\mathbb{SAS}$, and not that of $\mathbb{A}$, is the correct criterion for finding the onset of a symmetric bifurcation.

We now consider the elastic stability of an Ogden material under equitriaxial extension using Eqn. (5.88). The material parameters that are used are $\mu_1 = 6.1803 \times 10^5$, $\mu_2 = -9.81 \times 10^3$, $\mu_3 = 1.17 \times 10^3$, $\alpha_1 = 1.3$, $\alpha_2 = -2$, $\alpha_3 = 5$, $\lambda = 4.1445 \times 10^9$ and $\alpha = 4$. In this case, five eigenvalues of $\mathbb{SAS}$ are repeated and are lower than the remaining eigenvalue. As can be seen from Fig. 5.8, these five eigenvalues go to zero at a stretch value of about 1.0000667, indicating a symmetric bifurcation at this point. In contrast to the St Venant–Kirchhoff material instability discussed above, this is a physically realizable nonuniqueness, and in fact, the Cauchy stress continues to increase with increasing stretch beyond the bifurcation point, and, consistent with this, $\mathbb{SZS}$ remains positive definite.

Finally, instead of an Ogden material, if a unit cube of a neo-Hookean material with $c_1 = \mu/2$ is subjected to triaxial extension with load T (with $\mu = 0.33 \times 10^6$, $\lambda = 3.3 \times 10^9$, $\alpha = 4$), then we get the symmetric bifurcation point at a load value of $T/\mu = 2$ (again, as in the Ogden material case, the five repeated eigenvalues become zero simultaneously at the

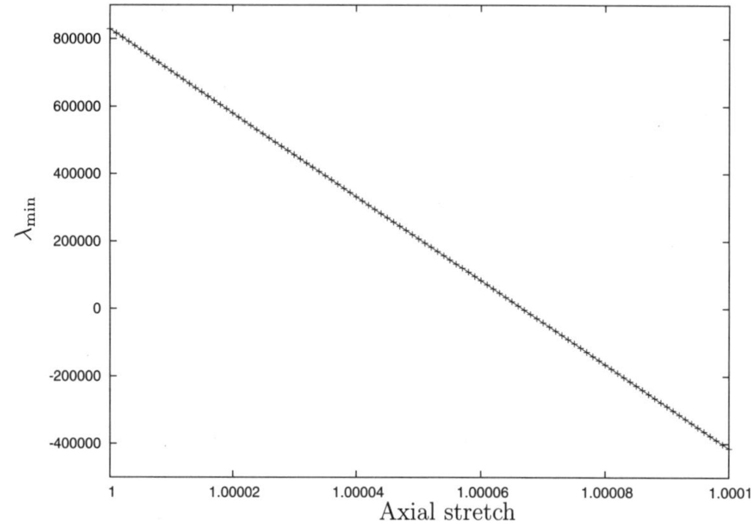

Fig. 5.8 *Lowest eigenvalue of $\mathbb{SAS}$ as a function of the axial stretch for a cube made of Ogden material subjected to triaxial extension.*

critical load), in agreement with the result obtained by Hill [120] (see also the Introduction in [269]).

5.10 Nonuniqueness of Solutions in Elasticity

In this section we present some examples to show that there is lack of uniqueness of solutions to the elasticity problem. Consequently, the mathematical model must be compatible with the existence of several solutions.

As our first example, we consider the eversion problem. When a tennis ball is cut above its equator and pushed hard enough, the larger part of the cut ball snaps into an everted state, which clearly corresponds to a different deformation. The state of stress is also different; if a small cut is made at the bottom of the larger part, the lips of the cut remain together, while after eversion they separate. By contrast if the same experiment is performed on the smaller part, it snaps back to its original state. Another experiment consists in everting a cylindrical tube of rubber. The everted tube is a little longer, its exterior diameter and thickness are slightly smaller, and it has flaring ends.

Another type of nonuniqueness occurs by way of buckling. Consider a plate of small thickness with a traction on the lateral surface given by $p\boldsymbol{n}$, where $\boldsymbol{n}$ is the normal to the surface. When $p > 0$ (tensile loads) or for $-p_c < p < 0$, $p_c > 0$ (small compressive loads), the plate remains in equilibrium at $\boldsymbol{F} = \boldsymbol{I}$, and this is the only solution. However, when compressive loads that are larger in magnitude than the critical load p_c are applied, i.e., when $p < -p_c$, the solution $\boldsymbol{F} = \boldsymbol{I}$ becomes unstable. Any perturbation, however small, causes the plate to buckle, i.e., to occupy one of two symmetric stable solutions. Thus, for a large enough compressive load, we have at least three solutions.

Finally, we consider nonuniqueness for a pure displacement problem. Consider an elastic body, which in its reference configuration occupies the volume enclosed between two concentric spheres, and which is fixed at the outer and inner surfaces. If body forces are neglected, then this configuration is a solution. By rotating either bounding sphere by an angle of $2k\pi$, where k is an integer, we again obtain an infinite number of solutions. More examples of nonuniqueness can be found in [58].

Nonuniqueness of solutions implies that the underlying mathematical model is nonlinear, since the solution to the linearized elastostatics problem is unique (see Theorem 6.4.1). For the linearized model, if one assumes the elasticity tensor to be positive definite, then the stored energy function is convex in $\boldsymbol{F}$. For the general nonlinear case, the stored energy function cannot be strictly convex, since if it were so, the corresponding potential energy functional would have at most one stationary point in contradiction to the counterexamples to uniqueness; thus the notion of convexity has to be replaced with the weaker notion of polyconvexity (see, e.g., [58] for a detailed discussion, and [115] for examples of polyconvex strain energy functions).

We have seen that nonlinearities occur in every aspect of the elasticity problem, whether it be the strain–displacement relation, the equation of motion, the constitutive relation, the traction boundary conditions on the undeformed configuration, and so on. In the next chapter, we discuss how the governing equations and boundary conditions can be linearized to yield what is usually known as the 'linear theory of elasticity'. We shall see that although such a linearized theory is good as an approximation, it cannot serve as a model, since it violates some of the axioms such as material frame-indifference.

5.11 Exact Solutions for Homogeneous, Compressible, Isotropic Elastic Materials

In this section, we consider two examples of homogeneous deformations, i.e., deformations of the type $\boldsymbol{x} = \boldsymbol{F}\boldsymbol{X} + \boldsymbol{c}$, where the deformation gradient $\boldsymbol{F}$ is a constant, and two examples of non-homogeneous deformations. When $\boldsymbol{F}$ is constant, the stress tensor $\boldsymbol{\tau}$ is also a constant, and the equilibrium equations given by $\boldsymbol{\nabla}_x \cdot \boldsymbol{\tau} = \boldsymbol{0}$ are identically satisfied. Thus, if tractions $\boldsymbol{\tau n}$ are applied on the surface, the deformation $\boldsymbol{x} = \boldsymbol{F}\boldsymbol{X} + \boldsymbol{c}$ can be achieved without the application of body forces. A deformation that can be maintained by the application of surface tractions alone is called a *controllable* deformation, and thus, homogeneous deformations are a subset of controllable deformations.

Ericksen [72] and Shield [292] have proved that homogeneous deformations are the only controllable deformations in *every* compressible, homogeneous and isotropic hyperelastic material–i.e., homogeneous deformations are the only universal deformations possible in such materials. Here, we present a slightly different proof. Let (I_1, I_2, I_3) denote the principal invariants of $\boldsymbol{C}$. First choose $\breve{W} = I_1/2$, so that $\boldsymbol{S} = \boldsymbol{I}$. Then the equilibrium equations $\boldsymbol{\nabla}_X \cdot (\boldsymbol{F}\boldsymbol{S}) = \boldsymbol{0}$ reduce to (the subscript X on the del operator is henceforth suppressed)

$$\boldsymbol{\nabla} \cdot \boldsymbol{F} = \boldsymbol{0}, \tag{5.93}$$

or, alternatively,

$$\boldsymbol{\nabla}^2 \boldsymbol{\chi} = \boldsymbol{0}. \tag{5.94}$$

Next choose $\breve{W} = I_3$, so that $\boldsymbol{FS} = I_3\boldsymbol{F}^{-T} = \sqrt{I_3}\,\mathbf{cof}\,\boldsymbol{F}$. By virtue of the Piola identity, the equilibrium equations reduce to

$$\boldsymbol{\nabla}\sqrt{I_3} = \mathbf{0}. \tag{5.95}$$

Now choose $\breve{W} = I_1\sqrt{I_3}$, so that $\boldsymbol{FS} = I_1\mathbf{cof}\,\boldsymbol{F} + 2\sqrt{I_3}\boldsymbol{F}$. Again, by virtue of the Piola identity, and Eqns. (5.93) and (5.95), the equilibrium equations reduce to $\boldsymbol{\nabla} I_1 = \mathbf{0}$. It follows that $\boldsymbol{\nabla}^2 I_1 = 0$, or, on using $I_1 = \boldsymbol{F} : \boldsymbol{F}$,

$$\boldsymbol{\nabla}(\boldsymbol{\nabla}^2\chi) : \boldsymbol{F} + \boldsymbol{\nabla F} \vdots \boldsymbol{\nabla F} = 0, \tag{5.96}$$

where

$$\boldsymbol{\nabla F} \vdots \boldsymbol{\nabla F} = \frac{\partial F_{ij}}{\partial X_k}\frac{\partial F_{ij}}{\partial X_k}.$$

By virtue of Eqn. (5.94), Eqn. (5.96) reduces to $\boldsymbol{\nabla F} \vdots \boldsymbol{\nabla F} = 0$, so that $\boldsymbol{\nabla F} = \mathbf{0}$, which implies that $\boldsymbol{F}$ is a constant.

One can deduce some interesting features about homogeneous controllable deformations from the equilibrium equations. For example, from Eqns. (B.17), we deduce that $J\tau_{\theta\theta} = rT_{rR}/R$, $J\tau_{r\theta} = -rT_{\theta R}/R$ and $T_{zR} = 0$.

If, instead of the class of all constitutive relations, one considers specific constitutive relations, then, controllable, non-homogeneous deformations are possible—some examples of homogeneous and non-homogeneous deformations are presented in the following subsections, and more examples can be found, for example, in [2], [17], [40], [41], [57], [132], [152] and [361], and the references therein. We shall consider examples of non-homogeneous controllable deformations for incompressible materials in the following section.

5.11.1 Uniaxial stretch

We are interested in finding the deformation in a prismatic bar that is subjected to a uniform traction $\bar{t}$ along its axis, and whose lateral surface is traction-free, as shown in Fig. 5.9.

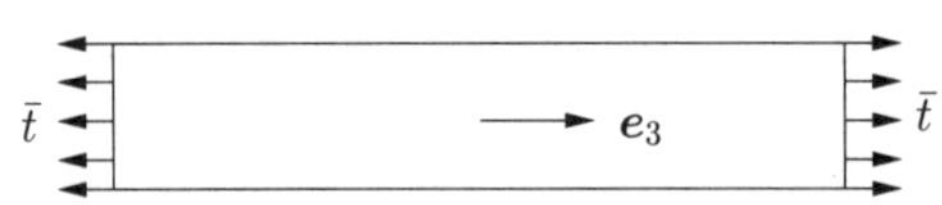

Fig. 5.9 *Prismatic bar undergoing uniaxial stretch.*

We guess the solution to be of the form

$$x_1 = \lambda_1 X_1, \quad x_2 = \lambda_2 X_2, \quad x_3 = \lambda_3 X_3,$$

where the (positive) constants, λ_1, λ_2, and λ_3 are to be determined. We have $\boldsymbol{F} = \mathbf{diag}[\lambda_1, \lambda_2, \lambda_3]$, and $\boldsymbol{B} = \boldsymbol{FF}^T = \mathbf{diag}[\lambda_1^2, \lambda_2^2, \lambda_3^2]$. Using Eqn. (5.8), we get

$$\tau_{11} = \hat{\beta}_0 + \hat{\beta}_1\lambda_1^2 + \hat{\beta}_2\lambda_1^4,$$

$$\tau_{22} = \hat{\beta}_0 + \hat{\beta}_1\lambda_2^2 + \hat{\beta}_2\lambda_2^4,$$

$$\tau_{33} = \hat{\beta}_0 + \hat{\beta}_1\lambda_3^2 + \hat{\beta}_2\lambda_3^4,$$

where $\hat{\beta}_i$ are functions of $(\lambda_1^2, \lambda_2^2, \lambda_3^2)$. The traction boundary conditions yield

$$\hat{\beta}_0 + \hat{\beta}_1\lambda_1^2 + \hat{\beta}_2\lambda_1^4 = 0,$$

$$\hat{\beta}_0 + \hat{\beta}_1\lambda_2^2 + \hat{\beta}_2\lambda_2^4 = 0,$$

$$\hat{\beta}_0 + \hat{\beta}_1\lambda_3^2 + \hat{\beta}_2\lambda_3^4 = \bar{t},$$

The constants $(\lambda_1, \lambda_2, \lambda_3)$ are determined by solving the above three equations, and, thus, obviously depend on the constitutive material of the bar. By subtracting the first equation from the second, we see that $\lambda_1 = \lambda_2$ is an admissible solution.

It is instructive to find the solution for a St Venant–Kirchhoff material for which $\hat{\beta}_0 = 0$, $\hat{\beta}_1 = [(\lambda_1^2 + \lambda_2^2 + \lambda_3^2 - 3)\lambda - 2\mu]/(2\lambda_1\lambda_2\lambda_3)$, and $\hat{\beta}_2 = \mu/(\lambda_1\lambda_2\lambda_3)$. Assuming $\lambda_1 = \lambda_2$, the governing equation for $h \equiv \lambda_1^2$ is obtained as $ah^3 + bh^2 + ch + d = 0$, where, with $\kappa \equiv (3\lambda + 2\mu)$,

$$a = 2(\lambda + \mu) = \frac{E}{(1 - 2\nu)(1 + \nu)},$$

$$b = \frac{\bar{t}^2\lambda^3}{\kappa^2\mu^2} - 7\lambda - 6\mu = \frac{4\nu^3\bar{t}^2 - E^2(3 + \nu)}{E(1 - 2\nu)(1 + \nu)},$$

$$c = 8\lambda + 6\mu = \frac{E(3 + 2\nu)}{(1 - 2\nu)(1 + \nu)},$$

$$d = -\kappa = -\frac{E}{1 - 2\nu}.$$

Once λ_1 is known, λ_3 is determined using the relation $\lambda_3^2 = (1 + \nu - \lambda_1^2)/\nu$. As is evident from the expression for b, the governing equation for λ_1 is the same irrespective of whether the bar is in tension ($\bar{t} > 0$) or compression ($\bar{t} < 0$); however, the different solution values for λ_1 correspond to the tension or compression cases. As an example, for $\nu = 0.25$ and $\bar{t}^2/E^2 = 0.16$ ($\bar{t}/E = \pm 0.4$), we get only one real solution, namely, $\lambda_1 = 0.92828$ and $\lambda_3 = 1.24625$, which clearly corresponds to $\bar{t}/E = 0.4$; there is no solution for $\bar{t}/E = -0.4$. On the other hand for $\bar{t}^2/E^2 = 0.0256$ ($\bar{t}/E = \pm 0.16$), we have the following three real solutions for (λ_1, λ_3): $(0.96623, 1.12497)$, $(1.06369, 0.68866)$ and $(1.08782, 0.51631)$; the first solution corresponds to $\bar{t}/E = 0.16$, while the other two correspond to $\bar{t}/E = -0.16$. One of the two solutions is a non-physical solution that lies beyond a material instability point, as we showed in Section 5.9.

5.11.2 Pure shear

Consider a rectangular block undergoing the deformation (see Fig. 5.10)

$$x_1 = \lambda_1 X_1 + \gamma\lambda_2 X_2, \quad x_2 = \lambda_2 X_2, \quad x_3 = \lambda_3 X_3.$$

Our goal is to determine the constants λ_1, λ_2 and λ_3 such that the normal tractions on all surfaces of the block are zero, i.e., the deformation is being carried out purely by the application of tangential tractions. For the said deformation, we have

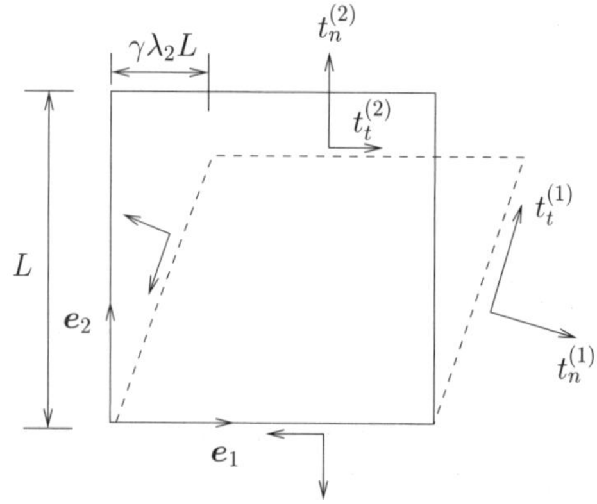

Fig. 5.10 *A block subjected to pure shear.*

$$
\boldsymbol{F} = \begin{bmatrix} \lambda_1 & \gamma\lambda_2 & 0 \\ 0 & \lambda_2 & 0 \\ 0 & 0 & \lambda_3 \end{bmatrix}, \qquad \boldsymbol{B} = \begin{bmatrix} \lambda_1^2+\gamma^2\lambda_2^2 & \gamma\lambda_2^2 & 0 \\ \gamma\lambda_2^2 & \lambda_2^2 & 0 \\ 0 & 0 & \lambda_3^2 \end{bmatrix},
$$

$$
\textbf{cof}\,\boldsymbol{F} = \begin{bmatrix} \lambda_2\lambda_3 & 0 & 0 \\ -\gamma\lambda_2\lambda_3 & \lambda_1\lambda_3 & 0 \\ 0 & 0 & \lambda_1\lambda_2 \end{bmatrix}, \quad \boldsymbol{B}^2 = \begin{bmatrix} (\lambda_1^2+\gamma^2\lambda_2^2)^2+\gamma^2\lambda_2^4 & (\lambda_1^2+\gamma^2\lambda_2^2)\gamma\lambda_2^2+\gamma\lambda_2^4 & 0 \\ (\lambda_1^2+\gamma^2\lambda_2^2)\gamma\lambda_2^2+\gamma\lambda_2^4 & (\gamma^2+1)\lambda_2^4 & 0 \\ 0 & 0 & \lambda_3^4 \end{bmatrix}.
$$

Substituting these matrices into Eqn. (5.8), we get

$$
\begin{aligned}
\tau_{11} &= \hat{\beta}_0 + \hat{\beta}_1(\lambda_1^2+\gamma^2\lambda_2^2) + \hat{\beta}_2[(\lambda_1^2+\gamma^2\lambda_2^2)^2+\gamma^2\lambda_2^4], \\
\tau_{22} &= \hat{\beta}_0 + \hat{\beta}_1\lambda_2^2 + \hat{\beta}_2(\gamma^2+1)\lambda_2^4, \\
\tau_{33} &= \hat{\beta}_0 + \hat{\beta}_1\lambda_3^2 + \hat{\beta}_2\lambda_3^4, \\
\tau_{12} &= \hat{\beta}_1\gamma\lambda_2^2 + \hat{\beta}_2[(\lambda_1^2+\gamma^2\lambda_2^2)\gamma\lambda_2^2+\gamma\lambda_2^4].
\end{aligned}
$$

From the above relations, we see that

$$
\frac{\tau_{11}-\tau_{22}}{\tau_{12}} = \frac{\lambda_1^2+(\gamma^2-1)\lambda_2^2}{\gamma\lambda_2^2}. \tag{5.97}
$$

The above equation is an example of a *universal relation*, so called since such relations hold irrespective of the constitutive equation. More examples of universal relations, and a systematic method to derive them, may be found, for example, in [268] and [275]. Essentially, one eliminates the material constants (such as, say, $\hat{\beta}_0$, $\hat{\beta}_1$, $\hat{\beta}_2$ from Eqn. (5.8)) to obtain relations that involve only the stress and strain variables, and that thus hold irrespective of the constitutive model.

Using Eqn. (2.32) and the expression for $\textbf{cof}\,\boldsymbol{F}$ given above, we find that $\boldsymbol{e}_1 = (1,0,0)$ deforms to $\boldsymbol{n}_1 = (1/\sqrt{1+\gamma^2}, -\gamma/\sqrt{1+\gamma^2}, 0)$, while the normals $\boldsymbol{e}_2$ and $\boldsymbol{e}_3$ remain unaffected. Thus, the normal traction on the inclined face is

$$
t_n^{(1)} = \boldsymbol{n}_1 \cdot \boldsymbol{\tau}\boldsymbol{n}_1 = \frac{\tau_{11}-2\gamma\tau_{12}+\gamma^2\tau_{22}}{(1+\gamma^2)}.
$$

Similarly, the normal tractions on the other surfaces are found to be $t_n^{(2)} = \boldsymbol{n}_2 \cdot \boldsymbol{\tau}\boldsymbol{n}_2 = \tau_{22}$ and $t_n^{(3)} = \boldsymbol{n}_3 \cdot \boldsymbol{\tau}\boldsymbol{n}_3 = \tau_{33}$. The conditions for these normal tractions to vanish are $\tau_{22} = \tau_{33} = 0$ and $\tau_{11} = 2\gamma\tau_{12}$. When these relations are combined with the universal relation given by Eqn. (5.97), we get

$$\lambda_1^2 = (1+\gamma^2)\lambda_2^2. \tag{5.98}$$

Substituting this relation in the expression for τ_{12}, we get

$$\tau_{12} = \frac{\gamma\lambda_1^2}{1+\gamma^2}[\hat{\beta}_1 + 2\hat{\beta}_2\lambda_1^2].$$

Although Eqn. (5.98) is also a universal relation, the manner in which λ_1 and λ_2 depend on γ depends on the constitutive relation of the material.

Since the normal tractions are zero, the traction vectors are tangential to the respective surfaces on which they are acting. Thus,

$$\boldsymbol{t}_t^{(1)} = \boldsymbol{t}^{(1)} = \boldsymbol{\tau}\boldsymbol{n}_1 = \begin{bmatrix} \frac{\tau_{11}-\gamma\tau_{12}}{\sqrt{1+\gamma^2}} \\ \frac{\tau_{12}-\gamma\tau_{22}}{\sqrt{1+\gamma^2}} \end{bmatrix} = \begin{bmatrix} \frac{\gamma\tau_{12}}{\sqrt{1+\gamma^2}} \\ \frac{\tau_{12}}{\sqrt{1+\gamma^2}} \end{bmatrix}$$

$$\boldsymbol{t}_t^{(2)} = \boldsymbol{t}^{(2)} = \boldsymbol{\tau}\boldsymbol{n}_2 = \begin{bmatrix} \tau_{12} \\ 0 \\ 0 \end{bmatrix}.$$

Note that $\left\|\boldsymbol{t}_t^{(1)}\right\| = \left\|\boldsymbol{t}_t^{(2)}\right\| = \tau_{12}$. Problem 5 shows that to maintain simple shear, in contrast to the pure shear case considered here, normal tractions have to be applied on the faces.

5.11.3 Pure bending of a prismatic beam made of a particular St Venant–Kirchhoff material

We now attempt an exact solution to the bending of a prismatic beam of rectangular cross-section with width b and height h into a region bounded by two concentric arcs as shown in Fig. 5.11. The material is assumed to be a St Venant–Kirchhoff material with $\nu = 0$. The neutral plane NP (i.e., plane along which $E_{XX} = 0$) is assumed to deform into an arc of radius R, and each plane $Y =$ constant in the undeformed configuration is assumed to deform into a plane of constant radius $R - f(Y)$, where $f(Y)$ is a function to be determined. Note that the planes $X =$ constant are assumed to remain plane after deformation. Under these assumptions, the deformation is given by

$$r = R - f(Y), \quad \theta = \frac{X}{R}, \quad z = Z.$$

Since the neutral plane deforms into an arc of radius R, we have $f(0) = 0$.

The deformation gradient (obtained using Eqn. (A.27)) and strain tensors are

$$\boldsymbol{F} = \begin{bmatrix} 0 & -f' & 0 \\ 1-\frac{f}{R} & 0 & 0 \\ 0 & 0 & 1 \end{bmatrix}, \quad \boldsymbol{E} = \frac{1}{2}\begin{bmatrix} \left(1-\frac{f}{R}\right)^2 - 1 & 0 & 0 \\ 0 & (f')^2 - 1 & 0 \\ 0 & 0 & 0 \end{bmatrix},$$

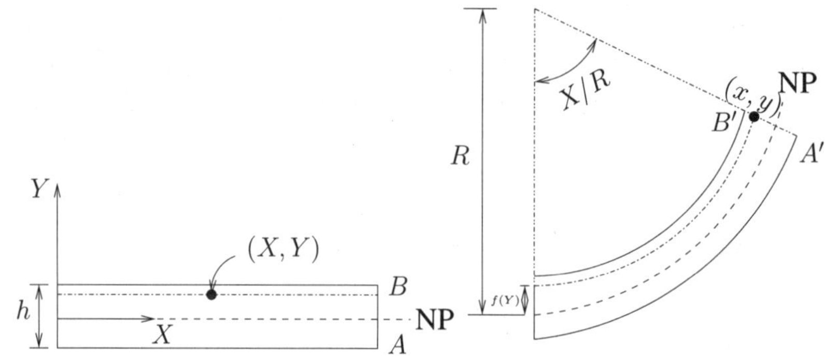

Fig. 5.11 *Bending of a prismatic beam of rectangular cross-section made of a St Venant–Kirchhoff material with $\nu = 0$ into a circular arc; the lower and upper surfaces are assumed to be at distances of h_1 and h_2, respectively, in the undeformed configuration ($h_1 + h_2 = h$).*

where $f' \equiv df/dY$. Since $\nu = 0$, $\lambda = 0$ and $\mu = E/2$, we have $\boldsymbol{S} = E\boldsymbol{E}$. The first Piola–Kirchhoff and Cauchy stresses are given by

$$\boldsymbol{T} = \boldsymbol{F}\boldsymbol{S} = \frac{E}{2}\begin{bmatrix} 0 & -f'[(f')^2 - 1] & 0 \\ \left(1 - \frac{f}{R}\right)\left[\left(1 - \frac{f}{R}\right)^2 - 1\right] & 0 & 0 \\ 0 & 0 & 0 \end{bmatrix},$$

$$J\boldsymbol{\tau} = \frac{E}{2}\begin{bmatrix} (f')^2[(f')^2 - 1] & 0 & 0 \\ 0 & \left(1 - \frac{f}{R}\right)^2\left[\left(1 - \frac{f}{R}\right)^2 - 1\right] & 0 \\ 0 & 0 & 0 \end{bmatrix},$$

where $J = f'(1 - f/R)$. Using Eqn. (B.15), we get the governing differential equation for $f(Y)$ as

$$\frac{1}{R}\left(1 - \frac{f}{R}\right)\left[\left(1 - \frac{f}{R}\right)^2 - 1\right] + \frac{d}{dY}\left[f'((f')^2 - 1)\right] = 0. \tag{5.99}$$

On integrating this equation once, we get

$$3\left[\frac{f^4}{R^4} - \frac{4f^3}{R^3} + 4\frac{f^2}{R^2} + c_1\right] = \left[3(f')^2 - 1\right]^2,$$

where c_1 is a constant of integration. It does not seem possible to integrate this equation to get an analytical expression for $f(Y)$. We have already mentioned the boundary condition $f(0) = 0$. The other conditions are obtained using the fact that the top and bottom surfaces with normals $\pm\boldsymbol{e}_r$ are traction-free. Since $\boldsymbol{t} = \boldsymbol{\tau}\boldsymbol{n}$, this traction-free condition on the top and bottom surfaces yields

$$\left[f'\left((f')^2 - 1\right)\right]_{Y=-h_1} = \left[f'\left((f')^2 - 1\right)\right]_{Y=h_2} = 0. \tag{5.100}$$

The normal to the end surface $A'B'$ is given by e_θ, so that the traction on that surface is $(t_r, t_\theta, t_z) = (E/2)\left[0, (1 - f/R)[(1 - f/R)^2 - 1]/f', 0\right]$. The traction t_θ on the face containing the origin is $-(E/2)(1 - f/R)[(1 - f/R)^2 - 1]/f'$. The condition that the total force acting on the face $A'B'$ is zero yields

$$\int_{-h_1}^{h_2} \left(1 - \frac{f}{R}\right)\left[\left(1 - \frac{f}{R}\right)^2 - 1\right] dY = 0.$$

Using Eqn. (5.99), we see that that this leads us again to the conditions in Eqn. (5.100)! Thus, unlike the linear elastic case, it is not clear whether the condition $h_1 = h_2 = h/2$ holds.

If the bending moment about the neutral plane due to the traction $\boldsymbol{t}$ on face $A'B'$ is $-M_0 e_z$, then $\int \boldsymbol{x} \times \boldsymbol{t}\, dS = -M_0 e_z$, where $\boldsymbol{x} = -f(Y) e_r$ and $dS = b\, dr = -bf'(Y) dY$, which yields

$$\frac{Eb}{2} \int_{-h_1}^{h_2} \left(1 - \frac{f}{R}\right)\left[\left(1 - \frac{f}{R}\right)^2 - 1\right] f(Y)\, dY = -M_0.$$

For thin beams, we have $f(Y) \approx Y$, and $f/R \ll 1$, so that the above equation yields (assuming that the neutral plane coincides with the mid-plane)

$$\frac{M_0}{I} = \frac{E}{R},$$

where $I = bh^3/12$ is the moment of inertia of the cross-section about the Z-axis.

The solution to the same problem as above with a Hencky instead of a St Venant–Kirchhoff material model can be found in [32].

5.11.4 Torsion of a circular shaft made of a St Venant–Kirchhoff material

A circular shaft of initial radius R_0 and length L is subjected to an axial force Ne_z and a torque Te_z by tractions applied to the top and bottom surfaces. The lateral surfaces are traction-free. The Z-axis lies along the axis of the cylinder. The deformation is assumed to be of the form

$$\begin{aligned} r &= g(R), \\ \theta &= \Theta + dZ, \\ z &= fZ. \end{aligned}$$

The use of the scale factors $h_i = (1, r, 1)$ and $h_J^0 = (1, R, 1)$, and Eqn. (A.27) leads to

$$\boldsymbol{F} = \begin{bmatrix} g'(R) & 0 & 0 \\ 0 & \frac{g}{R} & dg \\ 0 & 0 & f \end{bmatrix},$$

$$\boldsymbol{E} = \frac{1}{2}\left(\boldsymbol{F}^T \boldsymbol{F} - \boldsymbol{I}\right) = \begin{bmatrix} \frac{(g')^2 - 1}{2} & 0 & 0 \\ 0 & \frac{1}{2}\left(\frac{g^2}{R^2} - 1\right) & \frac{dg^2}{2R} \\ 0 & \frac{dg^2}{2R} & \frac{d^2 g^2 + f^2 - 1}{2} \end{bmatrix}.$$

Substituting the expression for $\boldsymbol{E}$ into the constitutive relation given by

$$\boldsymbol{\tau} = \frac{1}{\det \boldsymbol{F}} \left[\lambda (\operatorname{tr} \boldsymbol{E}) \boldsymbol{B} + 2\mu \boldsymbol{F}\boldsymbol{E}\boldsymbol{F}^T \right], \tag{5.101}$$

and computing $\boldsymbol{\nabla}_x \cdot \boldsymbol{\tau}$ with the aid of Eqn. (B.7), we find that the second and third equations are identically satisfied, while the first one yields the following governing differential equation for $g(R)$:

$$g\frac{d\tau_{rr}}{dR} + g'(\tau_{rr} - \tau_{\theta\theta}) = 0,$$

with τ_{rr} and $\tau_{\theta\theta}$ obtained via Eqn. (5.101). Note that we could also have used Eqn. (B.17) to obtain the above governing differential equation for $g(R)$. The boundary conditions on $g(R)$ are given by $g(0) = 0$, and $\tau_{rr}|_{R=R_0} = 0$. The constants d and f are evaluated using the fact that the net force and torque in the deformed configuration are N and T, respectively, i.e.,

$$2\pi \int_0^{R_0} g g' \tau_{zz} \, dR = N,$$

$$2\pi \int_0^{R_0} g^2 g' \tau_{\theta z} \, dR = T.$$

A similar solution methodology as above can be used to find solutions to a pressurized thick sphere or thick cylinder under plane strain conditions. In the case of the cylinder we assume $r = g(R)$, $\theta = \Theta$, $z = Z$, while in the case of the sphere, we assume $r = g(R)$, $\theta = \Theta$, $\phi = \Phi$. We then find $\boldsymbol{F}$, substitute it into the constitutive relation, and use Eqn. (B.17) in the case of the cylinder and (C.8) in the case of the sphere to find the governing differential equation for $g(R)$, which is then solved under appropriate boundary conditions on τ_{rr}.

Solutions for inhomogeneous compressible elastic bodies may be found in [280] and [281].

5.12 Exact Solutions for Homogeneous, Incompressible, Isotropic Elastic Materials

In this section, we present non-homogeneous equilibrium solutions for arbitrary homogeneous, incompressible, isotropic elastic materials in the absence of body forces; most of these solutions are due to Rivlin and Ericksen. Numerical solutions for "almost incompressible" materials are presented in [156] and [161]. Since $\boldsymbol{a} = \boldsymbol{b} = \boldsymbol{0}$, Eqn. (3.19) reduces to $\boldsymbol{\nabla}_x \cdot \boldsymbol{\tau} = \boldsymbol{0}$, while the relevant constitutive relation is given by Eqn. (5.12), with β_{-1} and β_1 independent of $\boldsymbol{X}$. In each of the examples considered below, we shall first present the general solution where β_{-1} and β_1 are allowed to be functions of the first two principal invariants of $\boldsymbol{B}$ (recall that since an incompressible material can only undergo isochoric deformations, we have $I_3(\boldsymbol{B}) = J^2 = 1$), and then specialize these results for a Mooney–Rivlin material where they are treated as constants. Before presenting the actual solutions, however, we present a few preliminaries.

Since all the solutions that we discuss here involve orthogonal curvilinear coordinate systems such as the cylindrical or spherical coordinate systems, we shall find it convenient to use *physical* components of all the tensors involved; detailed derivations of all the relevant equations may be found in Appendices B and C. In the development below, we express the desired result, be it stresses, forces or moments, in a manner that does not require the explicit evaluation of the pressure field p. First consider the cylindrical coordinate system r–θ–z. We assume that the extra stress $\boldsymbol{\tau} + p\boldsymbol{I}$ is a function of r only, and we also assume that $\tau_{r\theta} = \tau_{rz} = 0$. Under these assumptions, the equations $\boldsymbol{\nabla}_x \cdot \boldsymbol{\tau} = \boldsymbol{0}$ evaluated using Eqns. (B.7) reduce to

$$\frac{\partial \tau_{rr}}{\partial r} + \frac{1}{r}(\tau_{rr} - \tau_{\theta\theta}) = 0, \tag{5.102}$$
$$\frac{\partial p}{\partial \theta} = 0, \quad \frac{\partial p}{\partial z} = 0.$$

Thus, p, and hence $\boldsymbol{\tau}$, are functions of r only. From the first of the above equations, we have

$$\tau_{rr} = \int \frac{\tau_{\theta\theta} - \tau_{rr}}{r}\, dr, \tag{5.103}$$

$$\tau_{\theta\theta} = \frac{d(r\tau_{rr})}{dr}. \tag{5.104}$$

Consider the annular wedge shown in Fig. 5.12, which is assumed to be loaded such that all the above assumptions hold. Then, using Eqn. (5.104), the resultant normal force R per unit height is given by

$$R = \int_{r_1}^{r_2} \tau_{\theta\theta}\, dr = [r\tau_{rr}]_{r_1}^{r_2}.$$

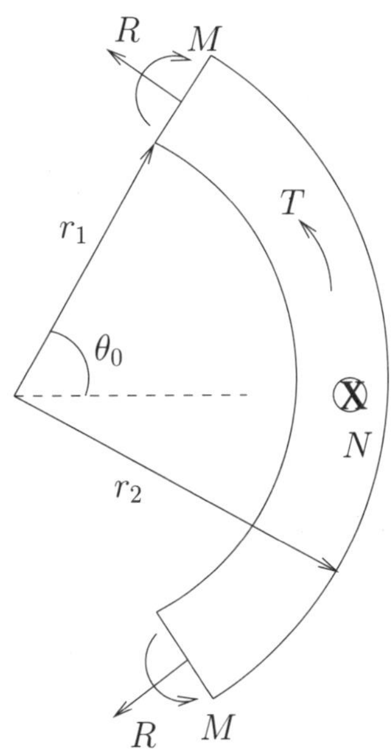

Fig. 5.12 *Forces and moments acting on an annular wedge.*

If a uniform pressure field is superimposed so as to render the surface $r = r_1$ traction-free, from the above equation it follows that $R = 0 \iff \tau_{rr}|_{r=r_2} = 0$. The resultant torque acting on the faces $z =$ const. and the resultant moment per unit height acting on the faces $\theta = \pm\theta_0$, with respect to a point on the z-axis are given by

$$T = 2\theta_0 \int_{r_1}^{r_2} r^2 \tau_{\theta z}\, dr, \tag{5.105}$$

$$\begin{aligned} M &= \int_{r_1}^{r_2} r\tau_{\theta\theta}\, dr \\ &= \frac{1}{2}\int_{r_1}^{r_2} r(\tau_{\theta\theta} + \tau_{rr})\, dr + \frac{1}{2}\int_{r_1}^{r_2} r(\tau_{\theta\theta} - \tau_{rr})\, dr \\ &= \frac{1}{2}\int_{r_1}^{r_2} \left(r^2 \frac{d\tau_{rr}}{dr} + 2r\tau_{rr}\right) dr + \frac{1}{2}\int_{r_1}^{r_2} r(\tau_{\theta\theta} - \tau_{rr})\, dr \quad \text{(by Eqn. (5.102))} \\ &= \frac{1}{2}\int_{r_1}^{r_2} \frac{d(r^2\tau_{rr})}{dr}\, dr + \frac{1}{2}\int_{r_1}^{r_2} r(\tau_{\theta\theta} - \tau_{rr})\, dr \\ &= \frac{1}{2} r^2\tau_{rr}\Big|_{r_1}^{r_2} + \frac{1}{2}\int_{r_1}^{r_2} r(\tau_{\theta\theta} - \tau_{rr})\, dr. \end{aligned} \tag{5.106}$$

The resultant normal force N acting on the surfaces $z =$ const. is given by

$$\begin{aligned} N &= 2\theta_0 \int_{r_1}^{r_2} \tau_{zz} r\, dr \\ &= \theta_0 \int_{r_1}^{r_2} r(\tau_{\theta\theta} + \tau_{rr})\, dr + \theta_0 \int_{r_1}^{r_2} (2\tau_{zz} - \tau_{rr} - \tau_{\theta\theta}) r\, dr \\ &= \theta_0\, r^2\tau_{rr}\Big|_{r_1}^{r_2} + \theta_0 \int_{r_1}^{r_2} (2\tau_{zz} - \tau_{rr} - \tau_{\theta\theta}) r\, dr, \end{aligned} \tag{5.107}$$

where the first term in Eqn. (5.107) is obtained in the same way as the first term in Eqn. (5.106).

Now consider a system of spherical coordinates r–θ–ϕ, and assume that $\boldsymbol{\tau} + p\boldsymbol{I}$ are functions of r only, and that $\tau_{r\theta} = \tau_{r\phi} = \tau_{\theta\phi} = 0$, and $\tau_{\theta\theta} = \tau_{\phi\phi}$. From the relations $\boldsymbol{\nabla}_x \cdot \boldsymbol{\tau} = \boldsymbol{0}$ and Eqns. (C.8), it again follows that p, and hence $\boldsymbol{\tau}$, is a function of r alone, and that

$$\tau_{rr} = 2\int \frac{\tau_{\theta\theta} - \tau_{rr}}{r}\, dr. \tag{5.108}$$

5.12.1 Bending and stretching of a rectangular block

Consider the deformation given by[2]

$$r = \sqrt{2aX}, \quad \theta = bY, \quad z = \frac{Z}{ab}, \tag{5.109}$$

where a and b are nonzero constants. This deformation carries a rectangular block bounded by the planes $X = X_1$, $X = X_2$, $Y = \pm Y_0$, $Z = \pm Z_0$ into the annular wedge bounded by the cylinders $r = r_1 = \sqrt{2aX_1}$ and $r = r_2 = \sqrt{2aX_2}$, and the planes $\theta = \pm\theta_0 = \pm bY_0$,

[2] A generalization of the problem considered here, where r and θ are as in Eqn. (5.109) but z is given by $Z/ab - bcY$, which corresponds to a simple shear in addition to the bending and stretching, has been treated in [337].

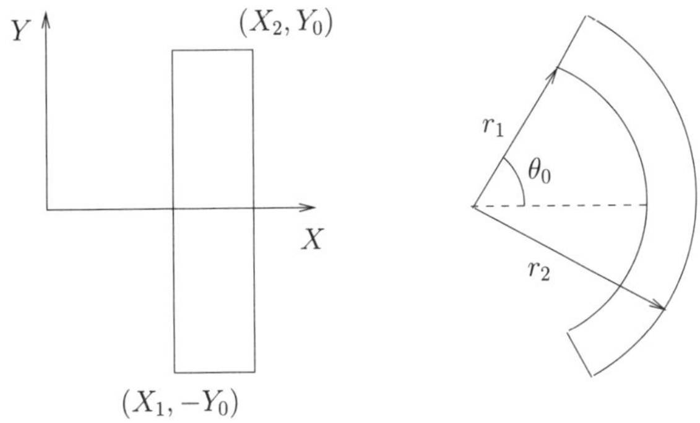

Fig. 5.13 *Bending and stretching of a rectangular block.*

$z = \pm z_0 = \pm Z_0/ab$, as shown in Fig. 5.13. The axial stretch is given by $1/ab$ and the radial stretch is adjusted so as to render the deformation isochoric. The scale factors associated with the coordinate systems used in the deformed and the undeformed coordinate systems are $h_i = (1, r, 1)$ and $h_J^0 = (1, 1, 1)$. Using Eqn. (A.27), we have $\boldsymbol{F} = \mathbf{diag}[a/r, br, 1/(ab)]$, so that $\boldsymbol{B} = \boldsymbol{F}\boldsymbol{F}^T = \mathbf{diag}[a^2/r^2, b^2r^2, 1/(a^2b^2)]$ and $\boldsymbol{B}^{-1} = \mathbf{diag}[r^2/a^2, 1/(b^2r^2), a^2b^2]$. From Eqn. (5.12), it follows that $\boldsymbol{\tau} + p\boldsymbol{I}$ is a function of r only, and in addition, $\tau_{r\theta} = \tau_{rz} = 0$, so that the results derived for the annular wedge shown in Fig. 5.12 can be used directly. Assuming that the surface $r = r_1$ is traction-free, the use of Eqns. (5.12) and (5.103) yields the nonzero stress components as

$$\tau_{rr} = \int_{r_1}^{r} \left(\beta_1 - \frac{\beta_{-1}}{a^2b^2}\right)\left(-\frac{a^2}{r^3} + b^2 r\right) dr, \tag{5.110}$$

$$\begin{aligned} \tau_{\theta\theta} &= \tau_{rr} + (\tau_{\theta\theta} - \tau_{rr}) \\ &= \tau_{rr} + \left[\left(-\frac{r^2}{a^2} + \frac{1}{b^2r^2}\right)\beta_{-1} + \left(-\frac{a^2}{r^2} + b^2r^2\right)\beta_1\right], \end{aligned} \tag{5.111}$$

$$\begin{aligned} \tau_{zz} &= \tau_{rr} + (\tau_{zz} - \tau_{rr}) \\ &= \tau_{rr} + \left[\left(a^2b^2 - \frac{r^2}{a^2}\right)\beta_{-1} + \left(\frac{1}{a^2b^2} - \frac{a^2}{r^2}\right)\beta_1\right], \end{aligned} \tag{5.112}$$

where β_{-1} and β_1 are functions of $I_1(\boldsymbol{B}) = a^2/r^2 + b^2r^2 + 1/(a^2b^2)$ and $I_2(\boldsymbol{B}) = a^2b^2 + r^2/a^2 + 1/(r^2b^2)$. When the surface $r = r_1$ is assumed to be traction-free, we have shown that $R = 0 \iff \tau_{rr}|_{r=r_2} = 0$. Thus, the necessary and sufficient condition that the bending of the block be produced by the couple M (with $R = 0$) follows from Eqn. (5.110), and is given by

$$\int_{r_1}^{r_2} \left(\beta_1 - \frac{\beta_{-1}}{a^2b^2}\right)\left(-\frac{a^2}{r^3} + b^2 r\right) dr = 0. \tag{5.113}$$

For given r_1 and r_2, whether the above equation has a real solution for the constants a and b depends on the nature of β_{-1} and β_1. Note that a nonzero traction field τ_{zz}, which is a function of r, has to be exerted on the faces $z = \pm z_0$ in order to maintain an isochoric deformation.

For a Mooney–Rivlin material, Eqn. (5.113) yields the solution $a^2 = r_1^2 r_2^2 b^2$; thus, it is possible to carry out the bending by a couple M. The expressions for M and N follow from a simplification of Eqns. (5.106) and (5.107) (note that the terms $(\tau_{\theta\theta} - \tau_{rr})$ and $(2\tau_{zz} - \tau_{rr} - \tau_{\theta\theta})$ can be obtained from Eqns. (5.111) and (5.112) without explicitly evaluating τ_{rr}), and are given by

$$\begin{aligned}
M &= \frac{1}{2}\int_{r_1}^{r_2} \left(\beta_1 - \frac{\beta_{-1}}{a^2 b^2}\right)\left(-\frac{a^2}{r} + b^2 r^3\right) dr \\
&= \frac{1}{2}\left(b^2\beta_1 - \frac{\beta_{-1}}{b^2 r_1^2 r_2^2}\right)\left[r_1^2 r_2^2 \ln\left(\frac{r_1}{r_2}\right) + \frac{1}{4}(r_2^4 - r_1^4)\right]. \\
N &= \theta_0 \int_{r_1}^{r_2} (2\tau_{zz} - \tau_{rr} - \tau_{\theta\theta})\, dr \\
&= \theta_0\left[\left(b^4 r_1^2 r_2^2 (r_2^2 - r_1^2) - \frac{r_2^4 - r_1^4}{4b^2 r_1^2 r_2^2} + \frac{1}{b^2}\ln\left(\frac{r_1}{r_2}\right)\right)\beta_{-1}\right. \\
&\quad \left. + \left(\frac{r_2^2 - r_1^2}{b^4 r_1^2 r_2^2} + b^2 r_1^2 r_2^2 \ln\left(\frac{r_1}{r_2}\right) - \frac{b^2(r_2^4 - r_1^4)}{4}\right)\beta_1\right].
\end{aligned}$$

5.12.2 Straightening, stretching and shearing a sector of a cylinder

This problem is in some sense the 'inverse' of the problem considered in the previous subsection. First a sector bounded by the cylinders $R = R_1$ and $R = R_2$, and the planes $\Theta = \pm\Theta_0$, $Z = \pm Z_0$ is straightened to form a rectangular block with faces parallel to the xy, yz and xz planes, followed by a stretching along the x-direction with equal contractions along the y and z directions, in turn followed by a shearing in the yz plane. The deformation is given by

$$x = \frac{1}{2}ab^2R^2, \quad y = \frac{\Theta}{ab}, \quad z = \frac{Z}{b} + \frac{c\Theta}{ab}.$$

The scale factors are $h_i = (1,1,1)$, and $h_J^0 = (1,R,1)$, so that from Eqn. (A.27) it follows that

$$\boldsymbol{F} = \begin{bmatrix} ab^2R & 0 & 0 \\ 0 & \frac{1}{abR} & 0 \\ 0 & \frac{c}{abR} & \frac{1}{b} \end{bmatrix}, \quad \boldsymbol{B} = \begin{bmatrix} 2ab^2x & 0 & 0 \\ 0 & \frac{1}{2ax} & \frac{c}{2ax} \\ 0 & \frac{c}{2abx} & \frac{1}{b^2} + \frac{c^2}{2ax} \end{bmatrix}, \quad \boldsymbol{B}^{-1} = \begin{bmatrix} \frac{1}{2ab^2x} & 0 & 0 \\ 0 & 2ax + b^2c^2 & -b^2c \\ 0 & -b^2c & b^2 \end{bmatrix}.$$

Since $\boldsymbol{B}$ and $\boldsymbol{B}^{-1}$ are functions of x only, it follows from Eqn. (5.12) that $\boldsymbol{\tau} + p\boldsymbol{I}$ is a function of x only, and in addition we observe that $\tau_{xy} = \tau_{xz} = 0$. Thus, the equation $\boldsymbol{\nabla}_x \cdot \boldsymbol{\tau} = \boldsymbol{0}$, reduces to

$$\frac{\partial \tau_{xx}}{\partial x} = 0, \quad \frac{\partial p}{\partial y} = 0, \quad \frac{\partial p}{\partial z} = 0,$$

from which it follows that $\tau_{xx} =$ const. and $p = p(x)$. Assuming the faces $x =$ const. to be free of traction, we have $\tau_{xx} = 0$, which in turn, by virtue of Eqn. (5.12), implies

$$p = \frac{1}{2ab^2x}\beta_{-1} + 2ab^2x\beta_1,$$

$$\tau_{yy} = \left(2ax - \frac{1}{2ab^2x} + b^2c^2\right)\beta_{-1} + \left(\frac{1}{2ax} - 2ab^2x\right)\beta_1,$$
$$\tau_{zz} = \left(b^2 - \frac{1}{2a^2bx}\right)\beta_{-1} + \left(\frac{1}{b^2} + \frac{c^2}{2ax} - 2ab^2x\right)\beta_1,$$
$$\tau_{yz} = -b^2c\beta_{-1} + \frac{c}{2ax}\beta_1,$$

where β_{-1} and β_1 are functions of $I_1(\boldsymbol{B}) = 1/b^2 + 2ab^2x + (1+c^2)/(2ax)$ and $I_2(\boldsymbol{B}) = 2ax + b^2(1+c^2) + 1/(2ab^2x)$; the above results can also be obtained by using Eqn. (B.16). Note that, in general, both normal and tangential tractions must be supplied on the slanted faces $z - cy = \text{const.}$ When $c = 0$, only normal tractions (which are functions of x) need to be supplied. The resultant normal force on the planes $y = \text{const.}$ vanishes if

$$\int_{x_1}^{x_2} \tau_{yy}\,dx = \int_{x_1}^{x_2} \left[\left(2ax - \frac{1}{2ab^2x} + b^2c^2\right)\beta_{-1} + \left(\frac{1}{2ax} - 2ab^2x\right)\beta_1\right] dy = 0.$$

Whether this equation has a real solution for the constants a, b, c again depends on the nature of β_{-1} and β_1.

5.12.3 Torsion, inflation, bending, etc. of an annular wedge

All the problems in this subsection are obtained as special cases of the deformation

$$r = \sqrt{aR^2 + b}, \quad \theta = c\Theta + dZ, \quad z = e\Theta + fZ; \quad a(cf - de) = 1,$$

with the last equation being the constraint $J = 1$. Instead of trying to give a physical interpretation to the deformation above, we shall give physical interpretations to the special cases considered below which are obtained by setting some of the constants to zero. The use of the scale factors $h_i = (1, r, 1)$ and $h_J^0 = (1, R, 1)$, and Eqn. (A.27) leads to

$$\boldsymbol{F} = \begin{bmatrix} \frac{aR}{r} & 0 & 0 \\ 0 & \frac{cr}{R} & dr \\ 0 & \frac{e}{R} & f \end{bmatrix},$$

which in turn leads to

$$\boldsymbol{B} = \begin{bmatrix} \frac{a^2R^2}{r^2} & 0 & 0 \\ 0 & \left(\frac{c^2}{R^2} + d^2\right)r^2 & \frac{cer}{R^2} + dfr \\ 0 & \frac{cer}{R^2} + dfr & \frac{e^2}{R^2} + f^2 \end{bmatrix}, \quad \boldsymbol{B}^{-1} = \begin{bmatrix} \frac{r^2}{a^2R^2} & 0 & 0 \\ 0 & \frac{a^2(e^2+f^2R^2)}{r^2} & -\frac{a^2(ce+dfR^2)}{r} \\ 0 & -\frac{a^2(ce+dfR^2)}{r} & a^2(c^2 + d^2R^2) \end{bmatrix}.$$

From Eqn. (5.12), it follows that $\boldsymbol{\tau} + p\boldsymbol{I}$ is a function of r only, and that $\tau_{r\theta} = \tau_{rz} = 0$. Thus, all the results discussed in connection with Fig. 5.12 can be used here. Using Eqns. (5.12) and (5.103), we get

$$\tau_{rr} = \int \left\{ \left[\frac{a^2(e^2+f^2R^2)}{r^3} - \frac{r}{a^2R^2} \right] \beta_{-1} + \left[\left(\frac{c^2}{R^2} + d^2 \right) r - \frac{a^2R^2}{r^3} \right] \beta_1 \right\} dr,$$

$$\tau_{\theta\theta} = \tau_{rr} + \left[\frac{a^2(e^2+f^2R^2)}{r^2} - \frac{r^2}{a^2R^2} \right] \beta_{-1} + \left[\left(\frac{c^2}{R^2} + d^2 \right) r^2 - \frac{a^2R^2}{r^2} \right] \beta_1,$$

$$\tau_{zz} = \tau_{rr} + \left[a^2(c^2+d^2R^2) - \frac{r^2}{a^2R^2} \right] \beta_{-1} + \left[-\frac{a^2R^2}{r^2} + \frac{e^2}{R^2} + f^2 \right] \beta_1, \tag{5.114}$$

$$\tau_{\theta z} = \left(\frac{ce}{R^2} + df \right) \left(-\frac{a^2R^2}{r}\beta_{-1} + r\beta_1 \right),$$

where β_{-1} and β_1 are functions of $I_1(\boldsymbol{B}) = a^2R^2/r^2 + (c^2/R^2 + d^2)r^2 + e^2/R^2 + f^2$, and $I_2(\boldsymbol{B}) = a^2(c^2 + d^2R^2) + a^2(e^2 + f^2R^2)/r^2 + r^2/(a^2R^2)$. The expressions for the resultant twisting moment T, moment M and normal force N are obtained by substituting the above expressions for the stresses into Eqns. (5.105)–(5.107), and are given by

$$T = 2\theta_0 a \int_{R_1}^{R_2} \left(\frac{ce}{R} + dfR \right) \left[-a^2R^2\beta_{-1} + (aR^2+b)\beta_1 \right] dR, \tag{5.115}$$

$$M = \frac{1}{2}(aR^2+b)\,\tau_{rr}\big|_{R_1}^{R_2} + \frac{a}{2}\int_{R_1}^{R_2} \left\{ \left[\frac{a^2(e^2R+f^2R^3)}{aR^2+b} - \frac{aR^2+b}{a^2R} \right] \beta_{-1} \right.$$
$$\left. + \left[(aR^2+b)\left(\frac{c^2}{R} + d^2R \right) - \frac{a^2R^3}{aR^2+b} \right] \beta_1 \right\} dR, \tag{5.116}$$

$$N = \theta_0(aR^2+b)\,\tau_{rr}\big|_{R_1}^{R_2} + \theta_0 a \int_{R_1}^{R_2} \left\{ \left[2a^2(c^2R+d^2R^3) - \frac{aR^2+b}{a^2R} - \frac{a^2(e^2R+f^2R^3)}{aR^2+b} \right] \beta_{-1} \right.$$
$$\left. + \left[-\frac{a^2R^3}{aR^2+b} - (aR^2+b)\left(\frac{c^2}{R} + d^2R \right) + 2\left(\frac{e^2}{R} + f^2R \right) \right] \beta_1 \right\} dR, \tag{5.117}$$

where $R_1 = \sqrt{(r_1^2 - b)/a}$, $R_2 = \sqrt{(r_2^2 - b)/a}$, and where we have used the fact that $r\,dr = aR\,dR$.

Pure torsion of a hollow circular cylinder

Set $a = 1, b = 0, c = 1, e = 0, f = 1, \theta_0 = \pi$. The constant d now represents the twist. The stresses are obtained from Eqn. (5.114) as

$$\tau_{rr} = d^2 \int \beta_1 r\,dr, \qquad \tau_{\theta\theta} = \tau_{rr} + d^2\beta_1 r^2,$$

$$\tau_{zz} = \tau_{rr} + d^2\beta_{-1}r^2, \qquad \tau_{\theta z} = d(\beta_1 - \beta_{-1})r = d\hat{\mu}r,$$

where β_{-1} and β_1 are functions of $I_1(\boldsymbol{B}) = I_2(\boldsymbol{B}) = 3 + d^2r^2$, and $\hat{\mu} = \beta_1 - \beta_{-1}$, the generalized shear modulus, is the same function as the one that relates τ_{12} to γ in Problem 5. The expression for τ_{rr} shows that if say the outer surface $r = r_2$, is traction-free, and if the E-inequalities given by Eqn. (5.13) hold, then a pressure has to be applied to the inner surface $r = r_1$ in order to effect the torsion. The shear stress is proportional to the generalized shear modulus, which itself, in general, is a function of r^2. A normal stress τ_{zz}, which is a function of r, has to be applied to maintain the length of the cylinder at the same value. The resultant torque and normal force are obtained from Eqns. (5.115) and (5.117) as

$$T = 2\pi d \int_{R_1}^{R_2} \hat{\mu} R^3 \, dR,$$

$$N = \pi[R^2 \tau_{rr}]_{R_1}^{R_2} + \pi d^2 \int_{R_1}^{R_2} (2\beta_{-1} - \beta_1) R^3 \, dR.$$

Note that the torque T is a function of the generalized shear modulus. If the outer surface is assumed to be traction free, then $\tau_{rr} = -d^2 \int_R^{R_2} \beta_1 R \, dR$, so that $\pi[R^2 \tau_{rr}]_{R_1}^{R_2} = \pi d^2 \int_{R_1}^{R_2} \beta_1 R_1^2 R \, dR$. Substituting this value in the above expression for N, we get

$$N = \pi d^2 \int_{R_1}^{R_2} \left[R(R_1^2 - R^2)\beta_1 + 2R^3 \beta_{-1} \right] dR.$$

If the E-inequalities given by Eqn. (5.13) hold, then from the above equation it follows that $N < 0$. Thus, a compressive resultant force proportional to the square of the twist has to be supplied to prevent elongation of the cylinder. For a Mooney–Rivlin material, we have $\tau_{rr} = -d^2 \int_R^{R_2} \beta_1 R \, dR = -d^2 (R_2^2 - R^2)/2$, which leads to

$$\tau_{zz} = -\frac{d^2}{2} \beta_1 (R_2^2 - R^2) + d^2 \beta_{-1} R^2.$$

If the E-inequalities given by Eqn. (5.13) hold, then $\tau_{zz} < 0$. Thus, while in the general case, only the resultant normal force acting on the plane surfaces has to be negative, in the case of a Mooney–Rivlin material, the traction itself has to be a pressure.

Extension and torsion of a solid cylinder

Since in a solid cylinder the line $R_1 = 0$ gets mapped to $r_1 = 0$, we have $b = 0$. In addition, set $c = 1$, $e = 0$ and $\theta_0 = \pi$. The constraint $a(cf - de) = 1$ reduces to $af = 1$. This deformation can be considered as a longitudinal stretch f followed by a torsion with twist d/f. Assuming the outer surface $R = R_2$ to be traction-free, Eqns. (5.115) and (5.117) reduce to

$$T = \frac{2\pi d}{f} \int_0^{R_2} \left(\beta_1 - \frac{1}{f} \beta_{-1} \right) R^3 \, dR,$$

$$N = 2\pi \left(f - \frac{1}{f^2} \right) \int_0^{R_2} \left(\beta_1 - \frac{1}{f} \beta_{-1} \right) R \, dR - \frac{\pi d^2}{f^2} \int_0^{R_2} \left(\beta_1 - \frac{2}{f} \beta_{-1} \right) R^3 \, dR.$$

If the plane ends of the cylinder are free of resultant normal force, i.e., if $N = 0$, then for a given twist d with respect to the undeformed cylinder, the above equation with the left-hand side set to zero is the governing equation for finding the stretch f. If the E-inequalities given by Eqn. (5.13) hold, then the solution f to this equation is such that $f > 1$. Thus, the cylinder elongates when twisted. For a Mooney–Rivlin material, this equation reduces to

$$\left(f - \frac{1}{f^2} \right) \frac{f\beta_1 - \beta_{-1}}{f\beta_1 - 2\beta_{-1}} = \frac{1}{4} \left(\frac{d}{f} \right)^2 R_2^2. \tag{5.118}$$

For a given nonzero value of dR_2/f, it can be shown that the above equation has exactly one real root for f, and that this root has a value greater than unity, which, again, physically means that a twisted cylinder elongates. The expression for the torque T reduces to

$$T = \frac{\pi d}{2f}\left(\beta_1 - \frac{1}{f}\beta_{-1}\right) R_2^4.$$

Thus, for a fixed f, the torque T is proportional to the twist d. However, if the physical condition imposed is $N = 0$, then from Eqn. (5.118), we see that f increases as d/f increases, so that the torsional rigidity, which is proportional to the term in parenthesis in the above equation, decreases. Thus, if the ends of the cylinder are free of resultant force, softening occurs under torsion.

Torsion/Eversion of a hollow circular cylinder

First we consider the problem of eversion without torsion. Set $c = 1, d = 0, e = 0, \theta_0 = \pi$, so that $af = 1$. The condition that the surfaces $R = R_1$ and $R = R_2$ be free of traction is obtained from the first of Eqns. (5.114) as

$$\int_{R_1}^{R_2} \left[-\frac{1}{R^2} + \frac{R^2}{(R^2 + bf)^2} \right] \left(\beta_1 - \frac{1}{a^2}\beta_{-1}\right) R\, dR = 0, \tag{5.119}$$

while, in addition, the condition that the plane ends are free of resultant axial force is obtained from Eqn. (5.117) as

$$0 = \int_{R_1}^{R_2} \left\{ \left[2a^2 R - \frac{aR^2 + b}{a^2 R} - \frac{R^3}{aR^2 + b} \right] \beta_{-1} + \left[\frac{2R}{a^2} - \frac{a^2 R^3}{aR^2 + b} - \frac{aR^2 + b}{R} \right] \beta_1 \right\} dR. \tag{5.120}$$

The above two equations are to be solved for the constants a and b to obtain a solution other than the trivial one $a = 1, b = 0$. Whether a solution $a < 0$, which corresponds to the eversion of a cylindrical tube subject to no applied forces, can be obtained depends on the nature of β_{-1} and β_1. For a Mooney–Rivlin material, Eqn. (5.119) reduces to[3]

$$\ln \frac{1 + \frac{bf}{R_2^2}}{1 + \frac{bf}{R_2^2}} + \frac{1}{1 + \frac{R_2^2}{bf}} - \frac{1}{1 + \frac{R_1^2}{bf}} = 0, \tag{5.121}$$

which is to be solved for bf. Since we seek a solution where $a < 0$, and since $af = 1$, we necessarily need to have $f < 0$, so that the roots of the above equation for bf also need to be negative. It can be shown that Eqn. (5.121) has only one negative root. An application of the mean value theorem to Eqn. (5.119) in the case of a Mooney–Rivlin material shows that $bf = -2\bar{R}^2$, where $R_1 < \bar{R} < R_2$. As an example, if $R_1 = 1$ and $R_2 = 2$, we get $bf \approx -5.46922$, which corresponds to $\bar{R} \approx 1.65366$. Using Eqn. (5.121), Eqn. (5.120) for a Mooney–Rivlin material reduces to

$$\frac{(-\beta_{-1} + \beta_1 f)(f^3 - 1)}{(\beta_{-1} f^2 + \beta_1) f} = H, \tag{5.122}$$

[3]The signs of the last two terms on the left-hand side of Eqn. (5.121) are opposite to those in the corresponding equation (Eqn. (95.8)) in [337].

where, given R_1 and R_2,

$$H = -\frac{1}{2\left(1+\frac{R_1^2}{bf}\right)\left(1+\frac{R_2^2}{bf}\right)},$$

is a known and positive quantity once Eqn. (5.121) has been solved for bf. The solution of Eqn. (5.122) yields the desired stretch f.

In the torsion problem, the solution of determining the lengthening of a cylindrical tube whose surfaces $R = R_1$ and $R = R_2$ are free of traction, and whose plane ends are free of resultant force, for a given twist d, is obtained in the same manner as above by solving the generalized versions of Eqns. (5.119) and (5.120) with d nonzero. For example, for a Mooney–Rivlin material, the equation corresponding to Eqn. (5.121) is

$$\ln\frac{1+\frac{bf}{R_2^2}}{1+\frac{bf}{R_2^2}} + \frac{1}{1+\frac{R_2^2}{bf}} - \frac{1}{1+\frac{R_1^2}{bf}} = \frac{\beta_1(R_2^2-R_1^2)d^2}{\beta_1 - f^2\beta_{-1}}.$$

Solid cylinder spinning about its axis

For solving the problem of steady rotation of a solid cylinder rotating about its axis with angular speed ω, it is convenient to choose a frame of reference that rotates with the cylinder, so that using Eqn. (4.18), the body force components (b_r, b_θ, b_z) are found to be $(r\omega^2, 0, 0)$. With this body force, we expect the cylinder to expand radially outward, and contract along the axis, i.e., we expect $b = d = e = 0$ and $c = 1$. Since ρb_r can be written as $\frac{1}{2}\frac{\partial(\rho r^2\omega^2)}{\partial r}$, the linear momentum balance $\boldsymbol{\nabla}_x \cdot \boldsymbol{\tau} + \rho \boldsymbol{b} = \boldsymbol{0}$ leads to the additional term $-\rho r^2\omega^2/2$ on the right-hand side of the first equation in Eqn. (5.114). With this modification, and setting $b = d = e = 0$, $c = 1$, (so that $r^2 = aR^2$ and $af = 1$) in Eqn. (5.114), we get $\tau_{rr} = \tau_{\theta\theta} = -\rho R^2\omega^2/(2f) + \text{const}$. Assuming the outer surface $r = r_0$ to be traction-free, we get

$$\tau_{rr} = \tau_{\theta\theta} = \frac{\rho\omega^2}{2f}(R_0^2 - R^2),$$

$$\tau_{zz} = \frac{\rho\omega^2}{2f}(R_0^2 - R^2) + \left(f^2 - \frac{1}{f}\right)\left[\beta_1 - \frac{1}{f}\beta_{-1}\right],$$

where R_0 is the radius of the cylinder before the rotation is started, and β_{-1} and β_1 are functions of $I_1(\boldsymbol{B}) = 2/f + f^2$ and $I_2(\boldsymbol{B}) = 1/f^2 + 2f$. Although from the above expression, we see that τ_{zz} is a function of R, the resultant axial force $N = \frac{2\pi}{f}\int_0^{R_0} \tau_{zz} R\,dR$ on the planar ends of cylinder is zero, provided

$$(1-f^3)\left(\beta_1 - \frac{1}{f}\beta_{-1}\right) = \frac{1}{4}\rho\omega^2 R_0^2.$$

If the E-inequalities given by Eqn. (5.13) hold, then we see from the above equation that as expected, $f < 1$, or in other words, the cylinder shortens and thickens as a result of the rotation.

Bending of an annular wedge into another

If we set $d = e = 0$ (so that $acf = 1$), the resulting deformation represents the bending of one annular wedge into another. If $a < 0$, the curvature is reversed during bending. The condition that the curved surfaces $r = r_1$ and $r = r_2$ be free of traction is obtained from the first of Eqns. (5.114) as

$$\int_{r_1}^{r_2} \left[\left(\frac{R^2}{c^2 r^3} - \frac{r}{a^2 R^2} \right) \beta_{-1} + \left(\frac{c^2 r}{R^2} - \frac{a^2 R^2}{r^3} \right) \beta_1 \right] dr = 0,$$

where β_{-1} and β_1 are functions of $I_1(\boldsymbol{B})$ and $I_2(\boldsymbol{B})$. Whether there are real-valued constants a, b, c that satisfy the above equation depends on the nature of β_{-1} and β_1. The resultant moment M and normal force N follow directly from Eqns. (5.116) and (5.117) by setting $\tau_{rr}|_{R_1}^{R_2}$, d and e to zero.

5.12.4 Inflation/eversion of a spherical shell

Consider the deformation that takes a point with spherical coordinates (R, Θ, Φ) to one with spherical coordinates (r, θ, ϕ) given by

$$r = (\pm R^3 + a)^{1/3}, \quad \theta = \pm\Theta, \quad \phi = \Phi,$$

where the '+' and '−' signs in the expression for θ are associated with the '+' and '−' signs in the expression for r. The deformation with the plus sign corresponds to an inflation, while the one with the negative sign corresponds to an eversion. In both cases, the deformation maps the region between two concentric spheres into a region between two other concentric spheres. The scale factors corresponding to the deformed and undeformed configurations are $h_i = (1, r, r\sin\theta)$ and $h_J^0 = (1, R, R\sin\Theta)$. From Eqn. (A.27), if follows that $\boldsymbol{B} = \mathbf{diag}[R^4/r^4, r^2/R^2, r^2/R^2]$, $\boldsymbol{B}^{-1} = \mathbf{diag}[r^4/R^4, R^2/r^2, R^2/r^2]$. By Eqn. (5.12) it follows that $\boldsymbol{\tau} + p\boldsymbol{I}$ is a function of r only, and in addition $\tau_{\theta\theta} = \tau_{\phi\phi}$ and $\tau_{r\theta} = \tau_{r\phi} = \tau_{\theta\phi} = 0$, so that the results obtained for spherical coordinates systems at the beginning of this section can be used for the solution of the problem at hand. Using Eqns. (5.12) and (5.108), we get

$$\tau_{rr} = 2\int \left[\left(\frac{R^2}{r^2} - \frac{r^4}{R^4} \right) \beta_{-1} + \left(\frac{r^2}{R^2} - \frac{R^4}{r^4} \right) \beta_1 \right] \frac{\pm R^2\, dR}{\pm R^3 + a},$$

$$\tau_{\theta\theta} = \tau_{\phi\phi} = \tau_{rr} + \left(\frac{R^2}{r^2} - \frac{r^4}{R^4} \right) \beta_{-1} + \left(\frac{r^2}{R^2} - \frac{R^4}{r^4} \right) \beta_1,$$

where β_{-1} and β_1 are functions of $I_1(\boldsymbol{B}) = R^4/r^4 + 2r^2/R^2$ and $I_2(\boldsymbol{B}) = 2R^2/r^2 + r^4/R^4$. Considering the inflation problem first, we have

$$\tilde{p}(R_1) - \tilde{p}(R_2) = 2\int_{R_1}^{R_2} \frac{(a^2 + 2aR^3)}{(R^3 + a)^{7/3}} \left[\beta_1 - \frac{(R^3 + a)^{2/3}}{R^2} \beta_{-1} \right] dR. \tag{5.123}$$

where $\tilde{p}(R_1) = -\tau_{rr}|_{r=r_1}$ and $\tilde{p}(R_2) = -\tau_{rr}|_{r=r_2}$ are the pressures applied at the surfaces $r_1 = (R_1^3 + a)^{1/3}$ and $r_2 = (R_2^3 + a)^{1/3}$. Given $\tilde{p}(R_1) - \tilde{p}(R_2)$, Eqn. (5.123) can be solved for the constant a, or, alternatively, given the inner radius r_1, the constant a is known, and Eqn. (5.123) can be used to find the difference in pressures that is required to maintain this deformation. As usual, the results for a Mooney–Rivlin material can be obtained by treating β_{-1} and β_1 as constants.

For the eversion problem, if we assume the inner and outer surfaces to be traction-free, we have

$$0 = \int_{R_1}^{R_2} \frac{(a^2 - 2aR^3)}{(a - R^3)^{7/3}} \left[\beta_1 - \frac{(a - R^3)^{2/3}}{R^2} \beta_{-1} \right] dR. \tag{5.124}$$

Whether or not the above equation has a solution a depends on the nature of β_{-1} and β_1; if the E-inequalities are assumed to hold, an application of the mean value theorem to the above equation yields $a = 2\bar{R}^3$, where $R_1 < \bar{R} < R_2$. For a Mooney–Rivlin material, Eqn. (5.124) reduces to[4]

$$\begin{aligned} 2(a - R_1^3)^{-\frac{2}{3}}(2a - R_1^3)\frac{\beta_{-1}}{R_1} + (a - R_1^3)^{-\frac{4}{3}}(4a - 5R_1^3)R_1\beta_1 = \\ 2(a - R_2^3)^{-\frac{2}{3}}(2a - R_2^3)\frac{\beta_{-1}}{R_2} + (a - R_2^3)^{-\frac{4}{3}}(4a - 5R_2^3)R_2\beta_1. \end{aligned} \tag{5.125}$$

For a given R_1 and R_2, it can be shown that the above equation has only one root a such that $a > R_2^3$. Thus, only one everted state is possible for a given spherical shell. For additional discussions, see, e.g., [197].

Although we have considered only incompressible materials in this section, it is possible, even in the case of compressible materials, to reduce the solution of problems that essentially depend on only one spatial variable, such as the inflation of a sphere, or a cylinder subjected to pressure under plane strain conditions, to the solution of a nonlinear ordinary differential equation. For example, for finding the solution to the problem of a hollow sphere that is subjected to pressure on its inner surface, one would assume $r = \chi(R), \theta = \Theta$ and $\phi = \Phi$, then find $\boldsymbol{F}$ and substitute it into the appropriate constitutive relationship for $\boldsymbol{\tau}$ (or $\boldsymbol{S}$), and then use the equilibrium equations and traction conditions to find the governing nonlinear differential equation and boundary conditions for $\chi(R)$.

EXERCISES

1. The stored energy function for a compressible Mooney–Rivlin material is given by [64]

$$\tilde{W}(\boldsymbol{C}) = c_1(I_1 - 3) + c_2(I_2 - 3) + c_3(I_3 - 1) + k_1 \log I_3 + \lambda W_H(J),$$

where $W_H(J)$ is given by Eqn. (5.56), and I_1, I_2 and I_3 are the principal invariants of $\boldsymbol{C}$ (and hence also of $\boldsymbol{B}$). Show using Eqns. (1.226c), (1.231), (1.236) and (1.240), and the fact that $\boldsymbol{C} \in \mathrm{Sym}$, that the constitutive relation and associated constitutive tensor are

$$\begin{aligned} \boldsymbol{S} &= 2c_1\boldsymbol{I} + 2c_2(I_1\boldsymbol{I} - \boldsymbol{C}) + 2c_3I_3\boldsymbol{C}^{-1} + 2k_1\boldsymbol{C}^{-1} \\ &\quad + \frac{\lambda}{\alpha^2}\left[\alpha \sinh\left[\alpha(J-1)\right] - \frac{\cosh\left[\alpha(J-1)\right] - 1}{J}\right]\boldsymbol{C}^{-1}, \end{aligned}$$

[4] Again, similar to the results for the eversion of a cylindrical tube, the signs of certain terms differ in sign from those presented in [337] (compare Eqn. (5.125) with Eqn. (95.13) of [337]).

$$\begin{aligned}\mathbb{C} &= 4c_2\left(\boldsymbol{I}\otimes\boldsymbol{I}-\mathbb{S}\right)+4c_3I_3\left(\boldsymbol{C}^{-1}\otimes\boldsymbol{C}^{-1}-\mathbb{S}(\boldsymbol{C}^{-1}\boxtimes\boldsymbol{C}^{-1})\mathbb{S}\right)-4k_1\mathbb{S}(\boldsymbol{C}^{-1}\boxtimes\boldsymbol{C}^{-1})\mathbb{S}\\
&\quad+\frac{\lambda}{\alpha^2}\left\{J\alpha^2\cosh\left[\alpha(J-1)\right]-\alpha\sinh\left[\alpha(J-1)\right]+\frac{\cosh\left[\alpha(J-1)\right]-1}{J}\right\}\boldsymbol{C}^{-1}\otimes\boldsymbol{C}^{-1}\\
&\quad-\frac{2\lambda}{\alpha^2}\left\{\alpha\sinh\left[\alpha(J-1)\right]-\frac{\cosh\left[\alpha(J-1)\right]-1}{J}\right\}\mathbb{S}(\boldsymbol{C}^{-1}\boxtimes\boldsymbol{C}^{-1})\mathbb{S}\end{aligned}$$

By using the facts that $\boldsymbol{S}|_{\boldsymbol{C}=\boldsymbol{I}}=\boldsymbol{0}$ and that $\mathbb{C}|_{\boldsymbol{C}=\boldsymbol{I}}\boldsymbol{E}=\lambda(\operatorname{tr}\boldsymbol{E})\boldsymbol{I}+2\mu\boldsymbol{E}$, show that

$$\begin{aligned}c_3 &= -c_2,\\ k_1 &= -(c_1+2c_2+c_3)=-(c_1+c_2),\\ c_1+c_2 &= \frac{\mu}{2}.\end{aligned}$$

Thus, the constitutive relations and constitutive tensor for a compressible Mooney–Rivlin material are given by

$$\begin{aligned}\boldsymbol{S} &= 2c_1\left(\boldsymbol{I}-\boldsymbol{C}^{-1}\right)+2c_2\left(I_1\boldsymbol{I}-\boldsymbol{C}-(1+I_3)\boldsymbol{C}^{-1}\right)\\
&\quad+\frac{\lambda}{\alpha^2}\left[\alpha\sinh\left[\alpha(J-1)\right]-\frac{\cosh\left[\alpha(J-1)\right]-1}{J}\right]\boldsymbol{C}^{-1},\\
J\boldsymbol{\tau} &= 2c_1(\boldsymbol{B}-\boldsymbol{I})+2c_2\left[I_1\boldsymbol{B}-\boldsymbol{B}^2-(1+I_3)\boldsymbol{I}\right]\\
&\quad+\frac{\lambda}{\alpha^2}\left[\alpha\sinh\left[\alpha(J-1)\right]-\frac{\cosh\left[\alpha(J-1)\right]-1}{J}\right]\boldsymbol{I}, \qquad (5.126)\\
\mathbb{C} &= 4c_1\mathbb{S}(\boldsymbol{C}^{-1}\boxtimes\boldsymbol{C}^{-1})\mathbb{S}+4c_2\left[\boldsymbol{I}\otimes\boldsymbol{I}-\mathbb{S}-I_3\boldsymbol{C}^{-1}\otimes\boldsymbol{C}^{-1}+(1+I_3)\mathbb{S}(\boldsymbol{C}^{-1}\boxtimes\boldsymbol{C}^{-1})\mathbb{S}\right]\\
&\quad+\frac{\lambda}{\alpha^2}\left\{J\alpha^2\cosh\left[\alpha(J-1)\right]-\alpha\sinh\left[\alpha(J-1)\right]+\frac{\cosh\left[\alpha(J-1)\right]-1}{J}\right\}\boldsymbol{C}^{-1}\otimes\boldsymbol{C}^{-1}\\
&\quad-\frac{2\lambda}{\alpha^2}\left\{\alpha\sinh\left[\alpha(J-1)\right]-\frac{\cosh\left[\alpha(J-1)\right]-1}{J}\right\}\mathbb{S}(\boldsymbol{C}^{-1}\boxtimes\boldsymbol{C}^{-1})\mathbb{S}.\end{aligned}$$

Note that the matrix form of $\mathbb{C}$ is easily found using Eqns. (I.4) and (I.5).

The constitutive relation for an incompressible Mooney–Rivlin material given by Eqn. (5.12) with β_{-1} and β_1 constant, is recovered from Eqn. (5.126) by setting $\det\boldsymbol{F}=\det\boldsymbol{B}=1$, $I_1\boldsymbol{B}-\boldsymbol{B}^2=I_2\boldsymbol{I}-\boldsymbol{B}^{-1}$, and absorbing all terms that are a multiple of $\boldsymbol{I}$ into the pressure term.

The stored energy function for a compressible neo-Hookean material is obtained by setting $c_1=\mu/2$ and $c_2=0$ in the above relations; note that this material model is a special case of the Ogden material model treated in Section 5.7.

Another way that the strain energy density function for a compressible Mooney–Rivlin material can be expressed is [315]

$$\tilde{W}(\boldsymbol{C})=c_1(\bar{I}_1-3)+c_2(\bar{I}_2-3)+\frac{\kappa}{2}(J-1)^2, \qquad (5.127)$$

where $\bar{I}_1=(I_3)^{-1/3}I_1$, $\bar{I}_2=I_3^{-2/3}I_2$ and $\kappa=\lambda+2\mu/3$ is the bulk modulus. Using the conditions $\boldsymbol{S}|_{\boldsymbol{C}=\boldsymbol{I}}=\boldsymbol{0}$, and that $\boldsymbol{S}(\boldsymbol{I})+D\boldsymbol{S}(\boldsymbol{I})[2\boldsymbol{E}]$ agrees with the constitutive relation of a St Venant–Kirchhoff material, we again get $c_1 + c_2 = \mu/2$. The constitutive relations associated with the above strain energy density function are

$$\begin{aligned}
S &= \left[-\frac{2c_1}{3} I_1 (I_3)^{-1/3} - \frac{4c_2}{3} I_2 (I_3)^{-2/3} + \kappa(I_3 - (I_3)^{1/2}) \right] C^{-1} + 2c_1 (I_3)^{-1/3} I + \\
&\quad 2c_2 (I_3)^{-2/3} (I_1 I - C), \\
J\tau &= \left[-\frac{2c_1}{3} I_1 (I_3)^{-1/3} - \frac{4c_2}{3} I_2 (I_3)^{-2/3} + \kappa(I_3 - (I_3)^{1/2}) \right] I + 2c_1 (I_3)^{-1/3} B + \\
&\quad 2c_2 (I_3)^{-2/3} (I_1 B - B^2).
\end{aligned}$$

and the elasticity tensor is

$$\begin{aligned}
\mathbb{C} &= -\frac{4c_1}{3} (I_3)^{-1/3} \left(C^{-1} \otimes I + I \otimes C^{-1} \right) + \\
&\left[\frac{4c_1}{9} I_1 (I_3)^{-1/3} + \frac{16c_2}{9} I_2 (I_3)^{-2/3} + \kappa(2I_3 - I_3^{1/2}) \right] C^{-1} \otimes C^{-1} + \\
&\left[\frac{4c_1}{3} I_1 (I_3)^{-1/3} + \frac{8c_2}{3} I_2 (I_3)^{-2/3} - 2\kappa(I_3 - I_3^{1/2}) \right] \mathbb{S}(C^{-1} \boxtimes C^{-1}) \mathbb{S} + \\
&4c_2 (I_3)^{-2/3} (I \otimes I - \mathbb{S}) - \frac{8c_2}{3} (I_3)^{-2/3} \left[(I_1 I - C) \otimes C^{-1} + C^{-1} \otimes (I_1 I - C) \right].
\end{aligned}$$

An additive split of the strain energy function into an isochoric and a volumetric part as, for example, in Eqn. (5.127) above, although used extensively in numerical strategies, can lead to unphysical results; see, e.g., [21, 71, 248].

2. Hartmann et al. [115] have proposed the following stored energy function for an almost incompressible material:

$$\tilde{W}(C) = \frac{\kappa}{50} \left(I_3^{5/2} + I_3^{-5/2} - 2 \right) + \alpha \left(\frac{I_1^3}{I_3} - 27 \right) + c_{10} \left(\frac{I_1}{I_3^{1/3}} - 3 \right) + c_{01} \left(\frac{I_2^{3/2}}{I_3} - 3\sqrt{3} \right),$$

where I_i, $i = 1, 2, 3$, denote the principal invariants of C. Show using Eqn. (5.32) that the associated constitutive relation is

$$S = \alpha_1 \operatorname{\mathbf{cof}} C + \alpha_2 I + \alpha_3 (I_1 I - C),$$

where

$$\begin{aligned}
\alpha_1 &= \frac{\kappa}{10} \left(I_3^{3/2} - I_3^{-7/2} \right) - 2\alpha I_1^3 I_3^{-2} - \frac{2}{3} c_{10} I_1 I_3^{-4/3} - 2c_{01} I_2^{3/2} I_3^{-2}, \\
\alpha_2 &= 6\alpha I_1^2 I_3^{-1} + 2c_{10} I_3^{-1/3}, \\
\alpha_3 &= 3c_{01} I_2^{1/2} I_3^{-1}.
\end{aligned}$$

Now using Eqns. (1.226a)–(1.226c), (1.240), and the fact that $C \in \mathrm{Sym}$, show that

$$\mathbb{C} = \beta_1 Z + \beta_2 \operatorname{\mathbf{cof}} C \otimes \operatorname{\mathbf{cof}} C + \beta_3 \left[\operatorname{\mathbf{cof}} C \otimes I + I \otimes \operatorname{\mathbf{cof}} C \right] +$$

$$\beta_4\left[\mathbf{cof}\,\boldsymbol{C}\otimes(I_1\boldsymbol{I}-\boldsymbol{C})+(I_1\boldsymbol{I}-\boldsymbol{C})\otimes\mathbf{cof}\,\boldsymbol{C}\right]+\beta_5\left[\boldsymbol{I}\otimes\boldsymbol{I}-\mathbb{S}\right]+$$
$$\beta_6(I_1\boldsymbol{I}-\boldsymbol{C})\otimes(I_1\boldsymbol{I}-\boldsymbol{C})+\beta_7\boldsymbol{I}\otimes\boldsymbol{I},$$

where $\boldsymbol{Z}\equiv\partial(\mathbf{cof}\,\boldsymbol{C})/\partial\boldsymbol{C}$ is given by Eqns. (1.350)–(1.352) and (I.6), and

$$\beta_1=\frac{\kappa}{5}\left(I_3^{3/2}-I_3^{-7/2}\right)-4\alpha I_1^3I_3^{-2}-\frac{4}{3}c_{10}I_1I_3^{-4/3}-4c_{01}I_2^{3/2}I_3^{-2},$$
$$\beta_2=\frac{\kappa}{10}\left(3I_3^{1/2}+7I_3^{-9/2}\right)+8\alpha I_1^3I_3^{-3}+\frac{16}{9}c_{10}I_1I_3^{-7/3}+8c_{01}I_2^{3/2}I_3^{-3},$$
$$\beta_3=-12\alpha I_1^2I_3^{-2}-\frac{4}{3}c_{10}I_3^{-4/3},$$
$$\beta_4=-6c_{01}I_2^{1/2}I_3^{-2},$$
$$\beta_5=6c_{01}I_2^{1/2}I_3^{-1},$$
$$\beta_6=3c_{01}I_2^{-1/2}I_3^{-1},$$
$$\beta_7=24\alpha I_1I_3^{-1}.$$

3. Show that

$$D\hat{\boldsymbol{T}}(\boldsymbol{F})[\boldsymbol{W}\boldsymbol{F}]=\boldsymbol{W}\hat{\boldsymbol{T}}(\boldsymbol{F}).$$

4. $\mathbf{C}$ is strongly elliptic (or satisfies the strong Legendre–Hadamard condition) if

$$\boldsymbol{A}:\mathbf{C}[\boldsymbol{A}]>0,$$

whenever $\boldsymbol{A}$ has the form $\boldsymbol{A}=\boldsymbol{a}\otimes\boldsymbol{c}$, $\boldsymbol{a}\neq\boldsymbol{0}$, $\boldsymbol{c}\neq\boldsymbol{0}$.

(a) Show that if $\mathbf{C}$ is positive definite, then $\mathbf{C}$ is strongly elliptic. (Thus, if one assumes $\mathbf{C}$ to be positive definite, as is usually done, one does not need to check for strong ellipticity.)

(b) Show that for an isotropic material, $\mathbf{C}$ is strongly elliptic if and only if

$$\mu>0,\quad\lambda+2\mu>0.$$

For the necessary and sufficient conditions on the elastic constants for strong ellipticity to hold in a transversely isotropic elastic solid, see, e.g., [218]. These conditions are derived by using the following results:

$$ax^2+bx+c>0\quad\forall x\text{ iff }a>0,\ c>0,\ b^2<4ac,\tag{5.128a}$$
$$ax^4+bx^2+c>0\quad\forall x\text{ iff }a>0,\ c>0,\ b+2\sqrt{ac}>0.\tag{5.128b}$$

To prove the first result, we take $x=0$ and x "large" to conclude that $c>0$ and $a>0$. Next, we divide by a and write the resulting equation as $(x+\lambda_1)(x+\lambda_2)>0\ \forall x$. This inequality holds if and only if either λ_1 and λ_2 are complex or real repeated roots (of any sign), thus leading to the condition $b^2\leq 4ac$.

To prove the second result, we again first prove $a > 0$ and $c > 0$ as in the previous proof, and then write $ax^4 + bx^2 + c > 0$ as

$$(\sqrt{a}x^2 - \sqrt{c})^2 + (b + 2\sqrt{ac})x^2 > 0.$$

If $b + 2\sqrt{ac} > 0$, then clearly, the above inequality holds, while the converse is proved by choosing $x^2 = \sqrt{c}/\sqrt{a}$.

One can extend these results to the case of a sixth-order polynomial:

$$ax^6 + bx^4 + cx^2 + d > 0 \quad \forall x \text{ iff } [a > 0,\ d > 0,\ b > 0,\ c > 0,\ G \leq 0], \\ \text{OR } [a > 0,\ d > 0,\ G > 0].$$

where $G := 4b^3d + 4c^3a - 18abcd + 27a^2d^2 - b^2c^2$. The two conditions are obtained by first proving $a > 0$, $d > 0$, dividing the polynomial by a and writing it as $(x^2 + \lambda_1)(x^2 + \lambda_2)(x^2 + \lambda_3)$, and then considering the cases (i) λ_1, λ_2, λ_3 all positive (including the case when λ_1, λ_2, λ_3 are all equal leading to $b = 3a^{2/3}d^{1/3}$, $c = 3a^{1/3}d^{2/3}$, $G = 0$), (ii) λ_1 positive, and λ_2 and λ_3 complex.

5. Consider a homogeneous, isotropic material subjected to simple shear (Section 2.5.3). Using Eqn. (5.10), show that the Cauchy stress tensor components τ_{13} and τ_{23} are zero, and that τ_{12} is an odd function of γ. Also show that

$$\tau_{11} - \tau_{22} = \gamma\tau_{12}.$$

Note that the above equation is valid regardless of the material properties of the body; it is called Rivlin's universal relation. It shows that

$$\tau_{11} \neq \tau_{22} \text{ for } \gamma \neq 0,$$

which is known as the *Poynting effect*. Using the equation of the slanted faces, $x_1 - \gamma x_2 =$ constant (or Eqn. (2.32)), show that the components of the unit normals and unit tangents on these faces are given by

$$n_1 = \pm\frac{1}{\sqrt{1+\gamma^2}}, \quad n_2 = \mp\frac{\gamma}{\sqrt{1+\gamma^2}}, \quad n_3 = 0,$$

and $(-n_2, n_1, 0)$, respectively. Next, using Cauchy's relation, show that the normal and tangential components of the traction vector on these faces are given by (see Fig. 5.14)

$$t_n = \tau_{22} - \frac{\gamma\tau_{12}}{1+\gamma^2}, \quad t_t = \frac{\tau_{12}}{1+\gamma^2}.$$

Assuming that the sides with normal $(0, 0, \pm 1)$ are traction-free, solve the above problem when the material is incompressible, i.e., when the constitutive relation is given by Eqn. (5.12). In particular, show that τ_{12} is given by the same expression as in the compressible case, and that Rivlin's universal relation still holds.

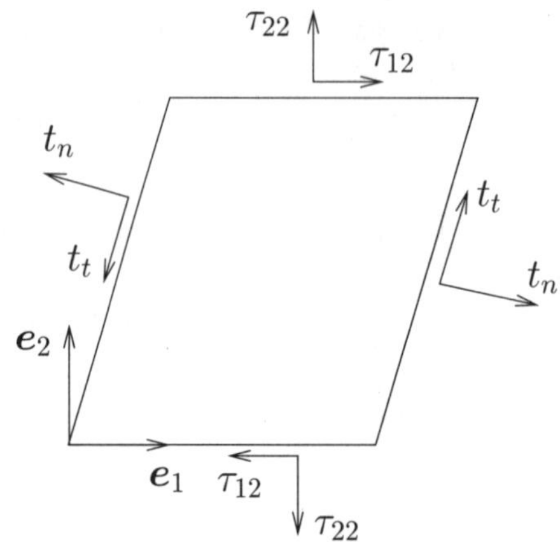

Fig. 5.14 *Tractions on the faces of a block undergoing simple shear.*

6. A circular disc of radius R_0 and thickness h made of an incompressible isotropic homogeneous material is deformed into the shape of a spherical cap as shown in Fig. 5.15. The deformation is given by

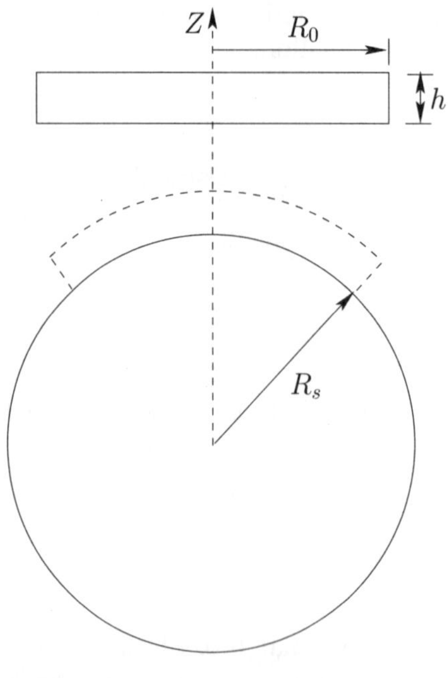

Fig. 5.15 *Deformation of a circular disc to a spherical cap.*

$$r = (R_s^3 + 3b^2 Z)^{1/3}, \quad \theta = \cos^{-1}\left(1 - \frac{R^2}{2b^2}\right), \quad \phi = \Theta,$$

where R_s is the radius of the sphere, b is a constant, and (R, Θ, Z) and (r, θ, ϕ) are cylindrical polar and spherical coordinates, respectively. Note that lines parallel to the Z axis in the undeformed configuration remain straight and are normal to the sphere surface after deformation, while the surfaces $Z =$ const. deform to surfaces $r =$ const. parallel to the spherical surface. Using Eqn. (A.27) and the scale factors $h_i = (1, r, r\sin\theta)$ and $h_J^0 = (1, R, 1)$, show that the deformation gradient is given by

$$\boldsymbol{F} = \begin{bmatrix} 0 & 0 & \frac{b^2}{r^2} \\ \frac{r}{b\cos(\theta/2)} & 0 & 0 \\ 0 & \frac{r\cos(\theta/2)}{b} & 0 \end{bmatrix}.$$

Show that the deformation is isochoric, i.e., $\det \boldsymbol{F} = 1$. Using Eqns. (5.12) and (C.8), show that the above deformation cannot be achieved, in general, in the absence of body forces.

6 Linearized Elasticity

We now discuss the linearized theory of elasticity whose use is appropriate in situations in which displacement $\boldsymbol{u}$ is small (in comparison with $\boldsymbol{X}$), and in which the displacement gradient $\boldsymbol{\nabla u}$ is small ($\|\boldsymbol{\nabla u}\| \ll 1$). The linearization is carried out about the reference configuration (spatially speaking) where $\boldsymbol{F} = \boldsymbol{I}$. Although $\boldsymbol{u}$ can depend on time, the linearized analysis is a good approximation only upto that time (starting from $t = 0$) for which the above-mentioned restrictions on $\boldsymbol{u}$ and $\boldsymbol{\nabla u}$ hold. For example, the use of linearized elasticity is inappropriate if the body undergoes large rotations even though $\boldsymbol{E}$ may be small, as was also pointed out in the special case of rigid motion discussed in Section 2.5.2.

6.1 Kinematics

We first carry out the linearization of the various strain measures. Let $\hat{\boldsymbol{F}}(\boldsymbol{H}) := \boldsymbol{H}$. Then $\boldsymbol{F} = \boldsymbol{I} + \boldsymbol{\nabla u} = \hat{\boldsymbol{F}}(\boldsymbol{I}) + D\hat{\boldsymbol{F}}(\boldsymbol{I})[\boldsymbol{\nabla u}]$. Thus,

$$D\hat{\boldsymbol{F}}(\boldsymbol{I})[\boldsymbol{\nabla u}] = \boldsymbol{\nabla u}.$$

Hence, the linearized Lagrangian strain evaluated at the reference configuration $\boldsymbol{F} = \boldsymbol{I}$ in the direction $\boldsymbol{\nabla u}$ is

$$\begin{aligned} D\boldsymbol{E}(\boldsymbol{I})[\boldsymbol{\nabla u}] &= \frac{1}{2}\left[\hat{\boldsymbol{F}}^T(\boldsymbol{I})D\hat{\boldsymbol{F}}(\boldsymbol{I})[\boldsymbol{\nabla u}] + D\hat{\boldsymbol{F}}^T(\boldsymbol{I})[\boldsymbol{\nabla u}]\hat{\boldsymbol{F}}(\boldsymbol{I})\right] \\ &= \frac{1}{2}\left[\boldsymbol{\nabla}_X \boldsymbol{u} + (\boldsymbol{\nabla}_X \boldsymbol{u})^T\right]. \end{aligned} \tag{6.1}$$

We shall denote $D\boldsymbol{E}(\boldsymbol{I})[\boldsymbol{\nabla u}]$ by $\boldsymbol{\epsilon}$.

Similarly, since $D\hat{\boldsymbol{F}}^{-1}(\boldsymbol{I})[\boldsymbol{\nabla u}] = -\hat{\boldsymbol{F}}^{-1}(\boldsymbol{I})\boldsymbol{\nabla u}\hat{\boldsymbol{F}}^{-1}(\boldsymbol{I}) = -\boldsymbol{\nabla u}$, the linearization of the Almansi strain evaluated at $\boldsymbol{F} = \boldsymbol{I}$ is

$$\begin{aligned} D\bar{\boldsymbol{E}}(\boldsymbol{I})[\boldsymbol{\nabla u}] &= -\frac{1}{2}\left[\hat{\boldsymbol{F}}^{-T}(\boldsymbol{I})D\hat{\boldsymbol{F}}^{-1}(\boldsymbol{I})[\boldsymbol{\nabla u}] + D\hat{\boldsymbol{F}}^{-T}(\boldsymbol{I})[\boldsymbol{\nabla u}]\hat{\boldsymbol{F}}^{-1}(\boldsymbol{I})\right] \\ &= \frac{1}{2}\left[(\boldsymbol{\nabla u})^T + \boldsymbol{\nabla u}\right] = \boldsymbol{\epsilon}. \end{aligned}$$

We shall henceforth denote $D\hat{\boldsymbol{F}}(\boldsymbol{I})[\boldsymbol{\nabla u}]$ as $D\boldsymbol{F}(\boldsymbol{I})[\boldsymbol{\nabla u}]$.

Since $\boldsymbol{U}^2 = \boldsymbol{F}^T\boldsymbol{F}$, we have

$$\boldsymbol{U}|_{\boldsymbol{F}=\boldsymbol{I}}\, D\boldsymbol{U}(\boldsymbol{I})[\boldsymbol{\nabla u}] + D\boldsymbol{U}(\boldsymbol{I})[\boldsymbol{\nabla u}]\, \boldsymbol{U}|_{\boldsymbol{F}=\boldsymbol{I}} = D\boldsymbol{F}^T(\boldsymbol{I})[\boldsymbol{\nabla u}]\boldsymbol{I} + \boldsymbol{I}^T D\boldsymbol{F}(\boldsymbol{I})[\boldsymbol{\nabla u}].$$

Since when $\boldsymbol{F} = \boldsymbol{I}$, we have $\boldsymbol{U} = \boldsymbol{I}$ and $D\boldsymbol{F}(\boldsymbol{I})[\boldsymbol{\nabla u}] = \boldsymbol{\nabla u}$, it follows that $D\boldsymbol{U}(\boldsymbol{I})[\boldsymbol{\nabla u}] = [D\boldsymbol{F}(\boldsymbol{I})[\boldsymbol{\nabla u}] + D\boldsymbol{F}^T(\boldsymbol{I})[\boldsymbol{\nabla u}]]/2 = \boldsymbol{\epsilon}$. Similarly by differentiating the relation $\boldsymbol{V}^2 = \boldsymbol{F}\boldsymbol{F}^T$, we again get $D\boldsymbol{V}(\boldsymbol{I})[\boldsymbol{\nabla u}] = \boldsymbol{\epsilon}$.

By differentiating the relation $F = RU$, we get

$$DF(I)[\nabla u] = R|_{F=I}\, DU(I)[\nabla u] + DR(I)[\nabla u]\, U|_{F=I},$$

leading to the relation

$$DR(I)[\nabla u] = \frac{1}{2}\left[DF(I)[\nabla u] - DF^T(I)[\nabla u]\right] = \frac{1}{2}\left[\nabla u - (\nabla u)^T\right]. \tag{6.2}$$

Thus, $[\nabla u - (\nabla u)^T]/2$ is a measure of the rigid rotation of an element in linearized elasticity.

Since $F = I + \nabla u$, we have

$$\begin{aligned}\det F &= \det I + \nabla u : (\operatorname{cof} I) + o\left(\|\nabla u\|\right)\\ &= 1 + \nabla u : I + o\left(\|\nabla u\|\right).\end{aligned}$$

Thus,

$$DJ(I)[\nabla u] = \operatorname{tr}(\nabla u) = \nabla \cdot u = \operatorname{tr} \epsilon. \tag{6.3}$$

We say that a body undergoes an linearized rigid displacement if the infinitesimal strain ϵ is zero. We have the following result that should be compared with Theorem 2.2.1:

Theorem 6.1.1 (Characterization of linearized rigid displacements). *The displacement field u is a linearized rigid displacement (also usually referred to as an infinitesimal rigid displacement) if and only if it admits the representation*

$$u = WX + c, \tag{6.4}$$

where W is a constant skew tensor. If ω is the axial vector of W, then one can also write the above relation as

$$u = \omega \times X + c.$$

Proof. We note that the results to be proved are very similar to those in Theorem 2.4.1, and hence, the proof is similar too. If the displacement is given by the form Eqn. (6.4), then clearly $\epsilon = 0$ since W is skew. To prove the converse, first note that

$$\nabla u = \epsilon + W = W,$$

where $W = [\nabla u - (\nabla u)^T]/2 \in \text{Skw}$. To show that Eqn. (6.4) follows from this, we have to prove that W is constant on V_0. Towards this end, we first prove that

$$(X_1 - X_2) \cdot [u(X_1) - u(X_2)] = 0 \quad \forall X_1, X_2 \in V_0.$$

Choose $X_1, X_2 \in V_0$, and let

$$X(s) = X_2 + s(X_1 - X_2)$$

be the parametric equation of the straight line joining $\boldsymbol{X}_1$ to $\boldsymbol{X}_2$. We have

$$\boldsymbol{u}(\boldsymbol{X}_1)-\boldsymbol{u}(\boldsymbol{X}_2)=\int_{\boldsymbol{X}_2}^{\boldsymbol{X}_1} d\boldsymbol{u}=\int_0^1 \boldsymbol{\nabla}\boldsymbol{u}(s)\frac{\partial \boldsymbol{X}}{\partial s}ds=\int_0^1 \boldsymbol{\nabla}\boldsymbol{u}(s)(\boldsymbol{X}_1-\boldsymbol{X}_2)ds,$$

so that

$$(\boldsymbol{X}_1-\boldsymbol{X}_2)\cdot[\boldsymbol{u}(\boldsymbol{X}_1)-\boldsymbol{u}(\boldsymbol{X}_2)]=\int_0^1 (\boldsymbol{X}_1-\boldsymbol{X}_2)\cdot\boldsymbol{\nabla}\boldsymbol{u}(\boldsymbol{X}_1-\boldsymbol{X}_2)ds=0,$$

since $\boldsymbol{\nabla}\boldsymbol{u}$ is skew at each point in the domain. Now differentiating the above equation with respect to $\boldsymbol{X}_1$, we get

$$\boldsymbol{u}(\boldsymbol{X}_1)-\boldsymbol{u}(\boldsymbol{X}_2)+\boldsymbol{\nabla}\boldsymbol{u}(\boldsymbol{X}_1)^T(\boldsymbol{X}_1-\boldsymbol{X}_2)=\boldsymbol{0},$$

and this result, when differentiated with respect to $\boldsymbol{X}_2$, yields

$$\boldsymbol{\nabla}\boldsymbol{u}(\boldsymbol{X}_2)+\boldsymbol{\nabla}\boldsymbol{u}(\boldsymbol{X}_1)^T=\boldsymbol{0}\quad \forall \boldsymbol{X}_1,\boldsymbol{X}_2\in V_0.$$

Since $\boldsymbol{\nabla}\boldsymbol{u}$ is skew, the above equation implies

$$\boldsymbol{\nabla}\boldsymbol{u}(\boldsymbol{X}_2)=\boldsymbol{\nabla}\boldsymbol{u}(\boldsymbol{X}_1)\quad \forall \boldsymbol{X}_1,\boldsymbol{X}_2\in V_0,$$

or, in other words, that $\boldsymbol{\nabla}\boldsymbol{u}$ is a constant skew tensor.
Just as in the case of Theorem 2.4.1, an easier proof for the above result can be obtained if we assume that the displacement field is twice differentiable. Then, by differentiating $\partial u_i/\partial X_j$, and using the fact that $\partial u_i/\partial X_j=-\partial u_j/\partial X_i$, we get

$$\begin{aligned}\frac{\partial^2 u_i}{\partial X_k\partial X_j}&=-\frac{\partial^2 u_j}{\partial X_k\partial X_i}=-\frac{\partial^2 u_j}{\partial X_i\partial X_k}=\frac{\partial^2 u_k}{\partial X_i\partial X_j}=\frac{\partial^2 u_k}{\partial X_j\partial X_i}=-\frac{\partial^2 u_i}{\partial X_j\partial X_k}\\&=-\frac{\partial^2 u_i}{\partial X_k\partial X_j},\end{aligned}$$

which implies that

$$\frac{\partial^2 u_i}{\partial X_k\partial X_j}=0,$$

which in turn implies that $\boldsymbol{\nabla}\boldsymbol{u}$ is a constant skew tensor. □

The question arises whether we can relate the results in Theorems 2.2.1 and 6.1.1, both of which deal with rigid deformations, one for the finite deformation case and the other for the small deformation case. Such a connection is presented in [60]. If we consider a rigid deformation as a function of a parameter t, then note that

$$\boldsymbol{Q}(t)=\boldsymbol{Q}(0)+D\boldsymbol{Q}(0)[t]+\mathrm{o}(t)=\boldsymbol{I}+\dot{\boldsymbol{Q}}\big|_{t=0}t+\mathrm{o}(t)\approx \boldsymbol{I}+\boldsymbol{W}t,$$

for small t, where $\boldsymbol{W} = \dot{\boldsymbol{Q}}\boldsymbol{Q}^T\Big|_{t=0}$. Thus, $\boldsymbol{F} \approx \boldsymbol{I} + \boldsymbol{W}t$ (or $\boldsymbol{\nabla u} = \boldsymbol{W}t$) for the small deformation case, which agrees with the results of Theorem 6.1.1 above.

If $\boldsymbol{u}$ is known as a function of $\boldsymbol{X}$, then there is no difficulty in computing $\boldsymbol{\epsilon}$ using Eqn. (6.1). However, if we now consider the reverse problem of determining $\boldsymbol{u}$ given $\boldsymbol{\epsilon}$, then we see that we have a system of six partial differential equations for the three unknown components of $\boldsymbol{u}$. Thus, the strain field needs to satisfy constraints, known as the *compatibility equations*, to ensure that it is integrable. Physically, the compatibility conditions ensure that we obtain a continuous single-valued displacement field (upto linearized rigid displacements) by integrating the strain distribution in case the domain is simply connected. For multiply connected domains, satisfaction of the compatibility relations does not ensure a single-valued displacement field; see the discussion below.

The following strain distributions, which satisfy the compatibility relations given by Eqns. (B.10) and (C.10), respectively, may appear to violate the statement of single-valuedness made above:

$$\epsilon_{rr} = \epsilon_{zz} = c, \qquad \text{[Cylindrical coordinates } (r,\theta,z)], \tag{6.5a}$$

$$\epsilon_{rr} = \epsilon_{\theta\theta} = c, \qquad \text{[Spherical coordinates } (r,\theta,\phi)], \tag{6.5b}$$

with the remaining strain components zero. The displacement field (excluding linearized rigid displacements) associated with this strain field is

$$u_r = cr, \quad u_\theta = -cr\theta, \quad u_z = cz, \qquad \text{[Cylindrical coordinates } (r,\theta,z)], \tag{6.6a}$$

$$u_r = cr, \quad u_\theta = 0, \quad u_\phi = -cr\phi\sin\theta, \qquad \text{[Spherical coordinates } (r,\theta,\phi)]. \tag{6.6b}$$

If the domain is a cylinder or a sphere, then the displacement field given by Eqns. (6.6a) or (6.6b) is not single-valued due to the presence of θ and ϕ, respectively, in these equations. However, strain components in the cylindrical system should also satisfy the *additional* requirement given by Eqns. (6.32) when they are independent of θ, while those in a spherical system should satisfy the additional constraint

$$\epsilon_{\theta\theta}\big|_{\theta=0,\pi} = \epsilon_{\phi\phi}\big|_{\theta=0,\pi}, \tag{6.7}$$

when $\epsilon_{\theta\theta}$ and $\epsilon_{\phi\phi}$ are independent of ϕ, and the additional constraint given by Eqn. (6.34a) when ϵ_{rr}, $\epsilon_{\theta\theta}$ and $\epsilon_{\phi\phi}$ are independent of θ and ϕ. The strain components in Eqns. (6.5a) and (6.5b) fail to meet these constraints (which can be thought of as additional compatibility relations to be satisfied in the cylindrical and spherical coordinate systems to account for the fact that θ and ϕ are not uniquely defined along the axis).

To summarize, when dealing with coordinate systems with angular variables, if the axis where are the angular coordinates are not uniquely defined is part of the domain, then additional constraints need to be imposed on the strain components *besides* those imposed by the conventional compatibility relations, which, if not imposed, can lead to incorrect conclusions about the compatibility of a given strain field.

To show that one need not get a single-valued displacement in the case of a multiply connected domain even when the compatibility conditions are satisfied, consider the following strain field defined in terms of cylindrical coordinates:

$$\epsilon_{rr} = \epsilon_{\theta\theta} = c\log r, \tag{6.8}$$

with the remaining strain components zero. This strain field could possibly be defined on an annular region bounded by two concentric circles (due to the presence of $\log r$ which is singular at the origin). Note that the domain in this case is multiply connected. The compatibility relations given by Eqns. (B.10) are satisfied, but the displacement field (excluding linearized rigid displacements) corresponding to the above strain field

$$\begin{aligned} u_r &= cr(\log r - 1), \\ u_\theta &= cr\theta, \\ u_z &= 0, \end{aligned} \tag{6.9}$$

is neither radially symmetric, nor single-valued on a domain bounded by two concentric circles. The strain field given by Eqn. (6.8) is a potential solution in a radially symmetric thermal loading problem (see Eqn. (6.380)). In Theorem (2.4) of [91], only the existence of a displacement field corresponding to a compatible strain field has been proved (even for multiply connected domains), without any mention of single-valuedness; however, for a multiply connected domain, additional necessary and sufficient conditions such as those given by Eqns. (2.44) and (2.45) (which are both necessary and sufficient) of [358] need to be imposed in order to ensure single-valuedness of the displacements. Quite interestingly, if the domain is an annular wedge with cross-section as shown in Fig. 6.8, then the strain fields given by Eqn. (6.5a) and (6.8) are both compatible (in the sense that the corresponding displacements given by Eqns. (6.6a) and (6.9), respectively, are both single-valued), thus showing how important the topology of the domain is (besides the compatibility equations themselves) in deciding whether a given strain field is compatible.

Theorem 6.1.2 (Compatibility conditions). *The strain field ϵ derived from a displacement field u using Eqn. (6.1) satisfies the conditions*

$$\nabla \times (\nabla \times \epsilon) = \mathbf{0}. \tag{6.10}$$

Conversely, if the compatibility conditions given by Eqn. (6.10) are satisfied and the domain is simply connected, then there exists a single-valued displacement field u that satisfies $\epsilon = [\nabla u + (\nabla u)^T]/2$, and, moreover, this field is unique upto a linearized rigid displacement.

Proof. The first part of the theorem is proved as follows:

$$\begin{aligned} [\nabla \times (\nabla \times \epsilon)]_{ij} &= \epsilon_{imn} \frac{\partial}{\partial X_m} (\nabla \times \epsilon)_{jn} \quad \text{(from Eqn. (1.266))} \\ &= \epsilon_{imn}\epsilon_{jst} \frac{\partial^2 \epsilon_{nt}}{\partial X_m \partial X_s} \\ &= \frac{1}{2}\epsilon_{imn}\epsilon_{jst}\left[\frac{\partial^3 u_n}{\partial X_m \partial X_s \partial X_t} + \frac{\partial^3 u_t}{\partial X_m \partial X_n \partial X_s}\right] \\ &= 0, \end{aligned}$$

where in obtaining the last step, we have used the fact that $\epsilon_{imn} = -\epsilon_{inm}$ and $\epsilon_{jst} = -\epsilon_{jts}$, and also the fact that $\partial^2 u_n/\partial X_s \partial X_t = \partial^2 u_n/\partial X_t \partial X_s$ and $\partial^2 u_t/\partial X_m \partial X_n = \partial^2 u_t/\partial X_n \partial X_m$.

To prove the converse, we first prove that if $S \in \text{Sym}$, then $\text{tr}\,(\nabla \times S) = 0$ as follows:

$$\text{tr}\,(\nabla \times S) = \epsilon_{imn}\frac{\partial S_{in}}{\partial X_m} = \epsilon_{nmi}\frac{\partial S_{ni}}{\partial X_m} = -\epsilon_{imn}\frac{\partial S_{in}}{\partial X_m} = -\text{tr}\,(\nabla \times S).$$

Since $\epsilon \in \text{Sym}$, we get $\text{tr}\,(\nabla \times \epsilon) = 0$, and since $\nabla \times (\nabla \times \epsilon) = \mathbf{0}$, by Potential theorem (3), there exists $W \in \text{Skw}$ such that $\nabla \times \epsilon = \nabla \times W$, or, in other words, $\nabla \times (\epsilon - W) = \mathbf{0}$. Now by Potential theorem (2), we conclude that there exists a vector field u such that

$$\epsilon - W = \nabla u.$$

Adding the transpose of the above equation, we get $\epsilon = [\nabla u + (\nabla u)^T]/2$, as desired.

To show uniqueness of u (modulo linearized rigid displacement), let $\tilde{u}$ be another displacement field that yields the same strain field. Then the strain corresponding to $u - \tilde{u}$ is zero, and by appealing to Theorem 6.1.1, we conclude that u and $\tilde{u}$ can differ only by a linearized rigid displacement. □

Since, as shown while proving Eqn. (6.10),

$$[\nabla \times (\nabla \times \epsilon)]_{ij} = \epsilon_{imn}\epsilon_{jst}\frac{\partial^2 \epsilon_{nt}}{\partial X_m \partial X_s}, \tag{6.11}$$

and since ϵ is symmetric, so is $\nabla \times (\nabla \times \epsilon)$; thus, Eqn. (6.10) has only six independent components that constitute the six compatibility equations.

From Eqns. (6.10), it follows that

$$\text{tr}\,[\nabla \times (\nabla \times \epsilon)] = \nabla^2(\text{tr}\,\epsilon) - \nabla \cdot (\nabla \cdot \epsilon) = 0. \tag{6.12}$$

Using this relation and Eqn. (1.18), Eqns. (6.10) can be stated in the alternate form (with a comma denoting a partial derivative)

$$\nabla^2 \epsilon + \nabla[\nabla(\text{tr}\,\epsilon)] - \nabla(\nabla \cdot \epsilon) - [\nabla(\nabla \cdot \epsilon)]^T = \mathbf{0}, \tag{6.13a}$$

$$\epsilon_{ij,kk} + \epsilon_{kk,ij} - \epsilon_{ik,kj} - \epsilon_{jk,ki} = 0, \quad i,j = 1,2,3. \tag{6.13b}$$

In Cartesian coordinates, the above equations reduce to

$$\frac{\partial^2 \epsilon_{XX}}{\partial Y^2} + \frac{\partial^2 \epsilon_{YY}}{\partial X^2} - 2\frac{\partial^2 \epsilon_{XY}}{\partial X \partial Y} = 0,$$

$$\frac{\partial^2 \epsilon_{YY}}{\partial Z^2} + \frac{\partial^2 \epsilon_{ZZ}}{\partial Y^2} - 2\frac{\partial^2 \epsilon_{YZ}}{\partial Y \partial Z} = 0,$$

$$\frac{\partial^2 \epsilon_{ZZ}}{\partial X^2} + \frac{\partial^2 \epsilon_{XX}}{\partial Z^2} - 2\frac{\partial^2 \epsilon_{ZX}}{\partial Z \partial X} = 0,$$

$$
\begin{aligned}
-\frac{\partial^2 \epsilon_{XX}}{\partial Y \partial Z} + \frac{\partial}{\partial X}\left(-\frac{\partial \epsilon_{YZ}}{\partial X} + \frac{\partial \epsilon_{ZX}}{\partial Y} + \frac{\partial \epsilon_{XY}}{\partial Z}\right) &= 0, \\
-\frac{\partial^2 \epsilon_{YY}}{\partial Z \partial X} + \frac{\partial}{\partial Y}\left(-\frac{\partial \epsilon_{ZX}}{\partial Y} + \frac{\partial \epsilon_{XY}}{\partial Z} + \frac{\partial \epsilon_{YZ}}{\partial X}\right) &= 0, \\
-\frac{\partial^2 \epsilon_{ZZ}}{\partial X \partial Y} + \frac{\partial}{\partial Z}\left(-\frac{\partial \epsilon_{XY}}{\partial Z} + \frac{\partial \epsilon_{YZ}}{\partial X} + \frac{\partial \epsilon_{ZX}}{\partial Y}\right) &= 0.
\end{aligned} \tag{6.14}
$$

Substituting the constitutive relation $E\boldsymbol{\epsilon} = (1+\nu)\boldsymbol{\tau} - \nu(\operatorname{tr}\boldsymbol{\tau})\boldsymbol{I}$ into Eqn. (6.12), and using $\boldsymbol{\nabla} \cdot \boldsymbol{\tau} = -\rho_0 \boldsymbol{b}$, we get

$$
\boldsymbol{\nabla}^2(\operatorname{tr}\boldsymbol{\tau}) = -\frac{1+\nu}{1-\nu}\boldsymbol{\nabla} \cdot (\rho_0 \boldsymbol{b}). \tag{6.15}
$$

Now substituting the constitutive relation into Eqn. (6.13a), and using Eqn. (6.15), we get the following Beltrami–Michell compatibility equations in terms of stresses:

$$
\boldsymbol{\nabla}^2 \boldsymbol{\tau} + \frac{1}{1+\nu}\boldsymbol{\nabla}[\boldsymbol{\nabla}(\operatorname{tr}\boldsymbol{\tau})] = -\left[\boldsymbol{\nabla}(\rho_0 \boldsymbol{b}) + [\boldsymbol{\nabla}(\rho_0 \boldsymbol{b})]^T + \frac{\nu}{1-\nu}\boldsymbol{\nabla} \cdot (\rho_0 \boldsymbol{b})\boldsymbol{I}\right]. \tag{6.16}
$$

We now discuss the issue of the independence of the six compatibility relations, which has been the subject of much discussion in recent times–see, e.g., [9], [25], [208], [242], [243], [302] and [348]. It is often asserted (e.g., [25], [242] with a retraction of some of the claims in [243]) that the six compatibility equations are not independent, since the six 'independent' components of $\boldsymbol{\nabla} \times (\boldsymbol{\nabla} \times \boldsymbol{\epsilon})$ themselves satisfy the three constraint equations $\boldsymbol{\nabla} \cdot [\boldsymbol{\nabla} \times (\boldsymbol{\nabla} \times \boldsymbol{\epsilon})] = \boldsymbol{0}$, *even for a non-compatible strain field*, as can be easily verified from Eqn. (6.11).

Southwell [302] showed that all the six compatibility equations given by Eqn. (6.10) are independent, i.e., it is not sufficient to verify (only) three of these equations in order to conclude whether a strain field is compatible. This finding is, of course, in direct contradiction to the claim by Borodachev [25] that among the six conditions given by Eqn. (6.14), only the first three or the last three conditions need be verified. We now demonstrate, by a method different from that of Southwell, that the six compatibility conditions are indeed independent. Consider the strain field (i) $\epsilon_{XY} = XY$ with the remaining strain components zero. As can be seen from Eqn. (6.14), only the first compatibility condition is violated with the remaining ones satisfied. Thus, for this strain field, only the first compatibility condition yields the conclusion that the given strain field is in fact incompatible. If the last three compatibility conditions are dependent on the first three (or vice versa), both sets should yield the same information about compatibility or incompatibility of a given strain field, which is clearly not the case. We now simply extend this idea to show that each of the remaining compatibility equations are also independent. Thus, successively consider the fields where the only nonzero strain component is given by (ii) $\epsilon_{YZ} = YZ$, (iii) $\epsilon_{XZ} = XZ$, (iv) $\epsilon_{XX} = YZ$, (v) $\epsilon_{YY} = XZ$, (vi) $\epsilon_{ZZ} = XY$. It is easily verified that corresponding to each case, there is one compatibility equation that is violated while the remaining are satisfied. This shows that each equation in Eqn. (6.14) yields information that the remaining five do not yield, thus proving that all six conditions are indeed independent.

Washizu [348] showed that if, say, the first three compatibility conditions are satisfied in the interior, and the remaining on the boundary, then these latter three conditions are also satisfied in the interior. Malyi [208] also reaches to a similar conclusion. The motivation to convert three of the compatibility conditions to boundary conditions comes from the fact that if one were to do so, one would have six equations (three equilibrium and three compatibility equations) in the interior and six boundary conditions (the remaining three compatibility equations and the Cauchy relations) for the six unknown stress components.

6.2 Governing Equations

We now discuss the linearization of the constitutive equation $T = \hat{T}(F)$. We assume that the reference configuration is a natural state, i.e.,

$$\hat{\tau}(I) = \hat{T}(I) = \hat{S}(I) = \mathbf{0}.$$

We have the following result:

Theorem 6.2.1. *Assume that the residual stress vanishes. Then*

$$\hat{T}(F) = \mathbf{C}[\epsilon] + o(\nabla u), \tag{6.17}$$

as $\nabla u \to \mathbf{0}$, *where*

$$\epsilon = \frac{1}{2}\left[\nabla u + (\nabla u)^T\right],$$

is the infinitesimal strain.

Proof. Since the reference configuration is a natural state, we have

$$\begin{aligned}
\hat{T}(F) &= \hat{T}(I + \nabla u) \\
&= \hat{T}(I) + D\hat{T}(I)[\nabla u] + o(\nabla u) \\
&= \mathbf{C}[\nabla u] + o(\nabla u) \\
&= \frac{1}{2}\mathbf{C}[\nabla u + (\nabla u)^T] + o(\nabla u) \quad \text{(by Eqn. (5.74))} \\
&= \mathbf{C}[\epsilon] + o(\nabla u). \quad \square
\end{aligned}$$

Therefore, if the reference configuration is a natural state, then to within terms of $o(\nabla u)$ as $\nabla u \to \mathbf{0}$, the stress T is a linear function of the infinitesimal strain ϵ. Also, since $\mathbf{C}[\epsilon] \in$ Sym, to within the same error, T is symmetric. Since $\mathbf{C} = D\hat{\tau}(I) = D\hat{T}(I) = D\hat{S}(I)$, we also have $\hat{\tau} = \hat{S} = \mathbf{C}[\epsilon]$ to within terms of $o(\nabla u)$. Thus, in the linearized theory of elasticity, the Cauchy, first and second Piola–Kirchhoff stress tensors are identical.

Since $x = \chi(X, t) = X + u$, the equation of motion given by Eqn. (3.46) can be written as

$$\rho_0 \frac{\partial^2 u}{\partial t^2} = \nabla_X \cdot T + \rho_0 b(x, t),$$

Linearizing b, we get

$$b(X + u, t) = b(X, t) + (\nabla b)u + o(u).$$

Since the linearization is about the reference configuration where $\boldsymbol{u} = \boldsymbol{0}$, the body force $\boldsymbol{b}_0(\boldsymbol{X}, t)$ within the linearized framework is

$$\boldsymbol{b}_0(\boldsymbol{X}, t) := \boldsymbol{b}(\boldsymbol{\chi}, t)|_{\boldsymbol{u}=\boldsymbol{0}} = \boldsymbol{b}(\boldsymbol{X}, t).$$

Similarly, the traction condition given by Eqn. (3.44), reduces to $\boldsymbol{T}\boldsymbol{n}_0 = \boldsymbol{t}_0 := \boldsymbol{t}(\boldsymbol{X}, t, \boldsymbol{n}_0)$. For example, the linearization of the 'live' pressure load, $\boldsymbol{t} = -p(\mathbf{cof}\,\boldsymbol{F})\boldsymbol{n}_0$, is $\boldsymbol{t}_0 = -p\boldsymbol{n}_0$.

Summarizing, the linearized theory of elasticity is based on the equations

$$\rho_0 \frac{\partial^2 \boldsymbol{u}}{\partial t^2} = \boldsymbol{\nabla}_X \cdot \boldsymbol{T} + \rho_0 \boldsymbol{b}_0(\boldsymbol{X}, t), \tag{6.18}$$

$$\boldsymbol{T}(\boldsymbol{X}) = \mathbf{C}[\boldsymbol{\epsilon}], \tag{6.19}$$

$$\boldsymbol{\epsilon} = \frac{1}{2}[\boldsymbol{\nabla}\boldsymbol{u} + (\boldsymbol{\nabla}\boldsymbol{u})^T], \tag{6.20}$$

which are to be solved subject to appropriate initial and boundary conditions.

As an example, if one considers a small elastic motion superposed on a possibly large rigid motion, then we choose a frame of reference that is 'attached' to the rigid motion. In this moving frame of reference, the body force vector is given by Eqn. (4.18). The linearization of this body force vector is

$$\boldsymbol{b}_0(\boldsymbol{X}, t) = \boldsymbol{Q}^T[\boldsymbol{b}^* - \ddot{\boldsymbol{c}}] - \dot{\boldsymbol{\Omega}} \times \boldsymbol{X} - \boldsymbol{\Omega} \times (\boldsymbol{\Omega} \times \boldsymbol{X}) - 2\boldsymbol{\Omega} \times \frac{\partial \boldsymbol{u}}{\partial t}. \tag{6.21}$$

Tortorelli [332] proposes that one should use

$$\boldsymbol{b}_0(\boldsymbol{X}, \boldsymbol{u}, t) = \boldsymbol{Q}^T[\boldsymbol{b}^* - \ddot{\boldsymbol{c}}] - \dot{\boldsymbol{\Omega}} \times (\boldsymbol{X} + \boldsymbol{u}) - \boldsymbol{\Omega} \times [\boldsymbol{\Omega} \times (\boldsymbol{X} + \boldsymbol{u})] - 2\boldsymbol{\Omega} \times \frac{\partial \boldsymbol{u}}{\partial t}. \tag{6.22}$$

However, considering the fact that, in the linearized theory, $\|\boldsymbol{u}\|$ is small in comparison with $\|\boldsymbol{X}\|$, this 'correction' hardly serves any purpose (if $\|\boldsymbol{u}\|$ is not small in comparison with $\|\boldsymbol{X}\|$, then one cannot use either of the above theories anyway, and geometric, and possibly material, nonlinearities have to be considered; note that all the classical treatments on linearized elasticity such as [203], [324] etc. use Eqn. (6.21)). We illustrate this by means of the rotating bar example shown in Fig. 3.7, which we now solve within the context of linearized elasticity using both the above approximations. Since $\partial \boldsymbol{u}/\partial t = \boldsymbol{0}$ after steady-state conditions have been reached, Eqn. (6.18), with the body force given by Eqn. (6.21), reduces to

$$0 = E\frac{d^2 u}{dX^2} + \rho_0 X \omega^2.$$

Solving the above differential equation and imposing the boundary conditions, $u|_{X=0} = 0$, and $T|_{X=L} = (E\,du/dX)_{X=L} = 0$, we get

$$\begin{aligned} u(X) &= \frac{1}{6E}\rho_0\omega^2(3L^2X - X^3), \\ T(X) &= \frac{1}{2}\rho_0\omega^2(L^2 - X^2). \end{aligned} \tag{6.23}$$

On the other hand, using Eqn. (6.22), we get the differential equation as

$$0 = E\frac{d^2u}{dX^2} + \rho(X+u)\omega^2.$$

Solving the above differential equation and imposing the boundary conditions, $u|_{X=0} = 0$, and $T|_{X=L} = (Edu/dX)_{X=L} = 0$, we get

$$\begin{aligned} u(X) &= \frac{\sin\alpha X}{\alpha\cos\alpha L} - X, \\ T(X) &= E\left[\frac{\cos\alpha X}{\cos\alpha L} - 1\right], \end{aligned} \tag{6.24}$$

where $\alpha = \omega\sqrt{\rho/E}$. For small α, using the expansions $\sin x \approx x - x^3/3$ and $\cos x \approx 1 - x^2/2$, we see that the solution given by Eqn. (6.24) is approximately the same as the conventional solution given by Eqn. (6.23). For high α, both solutions, being based on a linearized theory, are invalid (in fact, in the solution given by Eqn. (6.24), if ω is such that $\alpha L = \pi/2$, then u is infinity (resonance), similar to the resonance reported in Section 6 of [332]). Thus, it appears that, in lieu of the additional complications that it entails, the new theory offers no advantages over the conventional one. In this regard, we note that a linearized theory is obtained by a systematic linearization of the governing equations at the reference configuration, so that one retains only the dominant part in each nonlinear term. In some cases, the dominant term may involve $\boldsymbol{u}$, while in some other cases it may not, or, in other words, it is not necessary that every term that is linear in $\boldsymbol{u}$ should enter into a linearized theory [165].

Note from Eqn. (6.19) that

$$\hat{\boldsymbol{T}} = \frac{1}{2}\mathsf{C}[\boldsymbol{F} + \boldsymbol{F}^T - 2\boldsymbol{I}],$$

i.e., $\hat{\boldsymbol{T}}(\boldsymbol{QF}) \neq \boldsymbol{Q}\hat{\boldsymbol{T}}(\boldsymbol{F})$, so that material frame-indifference is violated. In addition, we also have [254]

Theorem 6.2.2. *$\boldsymbol{T}$ is frame-indifferent* and *linear in $\boldsymbol{F}$ if and only if $\hat{\boldsymbol{T}}$ is of the form*

$$\hat{\boldsymbol{T}}(\boldsymbol{F}) = \boldsymbol{F}\hat{\boldsymbol{T}}(\boldsymbol{I}) \quad \forall \boldsymbol{F} \in Lin^+. \tag{6.25}$$

Proof. If $\hat{\boldsymbol{T}}$ has the above form, then it is clearly linear in $\boldsymbol{F}$, and since $\hat{\boldsymbol{T}}(\boldsymbol{QF}) = \boldsymbol{Q}\hat{\boldsymbol{T}}(\boldsymbol{F})$, it also satisfies material frame-indifference.

Conversely, let $\boldsymbol{W} \in$ Skw, and let $\boldsymbol{Q}$ be given by the right-hand side of Eqn. (1.104). Substituting this expression into the relation $\hat{\boldsymbol{T}}(\boldsymbol{QF}) = \boldsymbol{Q}\hat{\boldsymbol{T}}(\boldsymbol{F})$, and using the assumed linearity of $\hat{\boldsymbol{T}}$, we get

$$\frac{\sin\alpha}{|\boldsymbol{w}|}\hat{\boldsymbol{T}}(\boldsymbol{WF}) + \frac{(1-\cos\alpha)}{|\boldsymbol{w}|^2}\hat{\boldsymbol{T}}(\boldsymbol{W}^2\boldsymbol{F}) = \frac{\sin\alpha}{|\boldsymbol{w}|}\boldsymbol{W}\hat{\boldsymbol{T}}(\boldsymbol{F}) + \frac{(1-\cos\alpha)}{|\boldsymbol{w}|^2}\boldsymbol{W}^2\hat{\boldsymbol{T}}(\boldsymbol{F}). \tag{6.26}$$

Differentiating the above equation with respect to α, and evaluating the resulting expression at $\alpha = 0$, we get $\hat{\boldsymbol{T}}(\boldsymbol{WF}) = \boldsymbol{W}\hat{\boldsymbol{T}}(\boldsymbol{F})$, which in turn reduces to

$\hat{T}(W) = W\hat{T}(I)$ when $F = I$. Substituting $\hat{T}(WF) = W\hat{T}(F)$ into Eqn. (6.26), and choosing α such that $(1 - \cos\alpha) \neq 0$, we get $\hat{T}(W^2) = W^2\hat{T}(I)$ when $F = I$. Summarizing, we have

$$\hat{T}(W) = W\hat{T}(I) \quad \forall W \in \text{Skw}, \tag{6.27}$$

$$\hat{T}(W^2) = W^2\hat{T}(I) \quad \forall W \in \text{Skw}. \tag{6.28}$$

Let e_1, e_2, e_3 be an orthonormal basis. Choose $W = e_2 \otimes e_3 - e_3 \otimes e_2$. Then,

$$W^2 = -(e_2 \otimes e_2 + e_3 \otimes e_3) = e_1 \otimes e_1 - I,$$

so that, using the linearity of $\hat{T}$ and (6.28), we have

$$\hat{T}(e_1 \otimes e_1) = (e_1 \otimes e_1)\hat{T}(I).$$

Similarly, we also have

$$\hat{T}(e_2 \otimes e_2) = (e_2 \otimes e_2)\hat{T}(I),$$
$$\hat{T}(e_3 \otimes e_3) = (e_3 \otimes e_3)\hat{T}(I).$$

Since any $S \in \text{Sym}$ can be written as $\sum_{i=1}^{3} \lambda_i e_i^* \otimes e_i^*$, where the e_i^* form an orthonormal basis, we have

$$\begin{aligned}\hat{T}(S) &= \sum_{i=1}^{3} \lambda_i \hat{T}(e_i^* \otimes e_i^*) \\ &= \left[\sum_{i=1}^{3} \lambda_i (e_i^* \otimes e_i^*)\right] \hat{T}(I) \\ &= S\hat{T}(I)\end{aligned} \tag{6.29}$$

Since any $F \in \text{Lin}^+$ can be split into a symmetric and skew-symmetric part, Eqns. (6.27), (6.29), and the assumed linearity of $\hat{T}$ yield Eqn. (6.25) □

Since we have assumed the reference configuration to be a natural state, Eqn. (6.25) implies that the stress would have to be zero for all F if material frame-indifference is to be satisfied in the linearized framework. Since this clearly does not hold, the linearized model is *not* frame-indifferent. Thus, although linearized elasticity may serve as a good approximation, it cannot serve as a model consistent with the axioms that we have adopted.

However, failure to satisfy material frame-indifference is not so serious a flaw so as to cause stress singularities or interpenetration of matter (even in problems with smooth loading and no cracks) as claimed in e.g., [81, 82, 83, 88, 131, 146, 184, 319, 328, 329] and references therein. What is perhaps even more intriguing than the singularities is that the hoop stress at the center of a spinning disk is zero if the anisotropy ratio $n = \sqrt{E_\theta / E_r}$ is even slightly greater than one, while it is maximum at the center if the material is isotropic ($n = 1$)! Thus, there is an abrupt transition in the hoop stress behavior as one departs from isotropy [131, 146]. We show that the real cause of such 'absurd' behavior is the use of defective constitutive relations, and not an intrinsic flaw with linearized elasticity [165]. In fact, since $|\nabla u|$ is assumed to be small, it is inconceivable that $\det F = \det(I + \nabla u)$ can become negative, which is the condition for interpenetration to occur. Since these stress singularities arise mostly with the use of a cylindrically aeolotropic or cylindrically orthotropic material model, we first describe this model. For planar problems, the constitutive relations are given by

$$\begin{aligned}\tau_{rr} &= c_{11}(r)\epsilon_{rr} + c_{12}(r)\epsilon_{\theta\theta},\\ \tau_{\theta\theta} &= c_{21}(r)\epsilon_{rr} + c_{22}(r)\epsilon_{\theta\theta},\\ \tau_{r\theta} &= 2c_{66}(r)\epsilon_{r\theta},\end{aligned} \tag{6.30}$$

where $c_{11}(r)$, $c_{22}(r)$, $c_{12}(r) = c_{21}(r)$ and $c_{66}(r)$ are the elastic parameters that depend only on the radial direction r, but not on θ. The inverse relations are

$$\begin{aligned}\epsilon_{rr} &= \frac{1}{E_r(r)}\tau_{rr} - \frac{\nu_r(r)}{E_r(r)}\tau_{\theta\theta},\\ \epsilon_{\theta\theta} &= -\frac{\nu_\theta(r)}{E_\theta(r)}\tau_{rr} + \frac{1}{E_\theta(r)}\tau_{\theta\theta},\\ \epsilon_{r\theta} &= \frac{1}{2G(r)}\tau_{r\theta},\end{aligned}$$

where

$$\begin{aligned}c_{11}(r) &= \frac{E_r(r)}{1-\nu_r(r)\nu_\theta(r)}, & c_{22}(r) &= \frac{E_\theta(r)}{1-\nu_r(r)\nu_\theta(r)},\\ c_{12}(r) &= \frac{E_\theta(r)\nu_r(r)}{1-\nu_r(r)\nu_\theta(r)}, & c_{21}(r) &= \frac{E_r(r)\nu_\theta(r)}{1-\nu_r(r)\nu_\theta(r)},\\ c_{66}(r) &= G(r).\end{aligned}$$

The symmetry requirement $c_{12}(r) = c_{21}(r)$ imposes the constraint

$$\frac{\nu_r(r)}{E_r(r)} = \frac{\nu_\theta(r)}{E_\theta(r)}.$$

In addition, the positive definiteness of the elasticity tensor **C** imposes the restrictions

$$c_{11}(r) > 0, \quad c_{22}(r) > 0 \quad c_{66}(r) > 0, \quad c_{11}(r)c_{22}(r) - c_{21}(r)c_{12}(r) > 0; \tag{6.31a}$$
$$E_r(r) > 0, \quad E_\theta(r) > 0, \quad G(r) > 0, \quad 1 - \nu_r(r)\nu_\theta(r) > 0. \tag{6.31b}$$

Just as, in the case of an isotropic material, one deduces the bounds $\mu > 0$ and $3\lambda + 2\mu > 0$ for the Lamé constants λ and μ based on 'thought experiments' (see page 246), one devises suitable thought experiments to deduce further constraints on the material constants in the above constitutive relations. Consider the radially symmetric deformation of an aeolotropic solid cylinder due to loading by a uniform pressure acting normal to its boundary. As noted by Tarn [321], at $r = 0$, the tangential direction at θ is the radial direction at $\theta + \pi/2$, and vice versa, so that $\tau_{\theta\theta}|_{\substack{r=0\\ \theta}} = \tau_{rr}|_{\substack{r=0\\ \theta+\pi/2}}$. Since the deformation is radially symmetric, τ_{rr} and $\tau_{\theta\theta}$ are independent of θ, so that we get $\tau_{rr}|_{r=0} = \tau_{\theta\theta}|_{r=0}$ (in this regard, we mention that Timoshenko [324] while discussing the problem of a rotating disk states in a footnote 'It can be seen from the definitions of τ_{rr}, $\tau_{\theta\theta}$ that when they are independent of θ, they must be equal at the center'). In a similar manner, when ϵ_{rr} and $\epsilon_{\theta\theta}$ are independent of θ, we should have

$$\epsilon_{rr}|_{r=0} = \epsilon_{\theta\theta}|_{r=0}. \tag{6.32}$$

From Eqns. (6.30), we see that this imposes the constraint

$$c_{11}|_{r=0} = c_{22}|_{r=0} \quad \left(\text{alternatively, } \frac{c_{11}}{c_{22}} = g(r), \text{ where } g(0) = 1\right), \tag{6.33}$$

which in turn implies that $E_r = E_\theta$ and $\nu_r = \nu_\theta$ at $r = 0$. Thus, at the axis $r = 0$, the aeolotropic material model should reduce to an isotropic and to a transversely isotropic model in the planar and three-dimensional cases, respectively. Of course, in case the body is assumed to be homogeneous, the respective models are isotropic and transversely isotropic everywhere; thus, it is *essential* to have inhomogeneity in order to get a cylindrical, or, in general, curvilinear, orthotropic model. In the case of a spherically orthotropic model, using

$$\epsilon_{rr}|_{r=0} = \epsilon_{\theta\theta}|_{r=0} = \epsilon_{\phi\phi}|_{r=0}, \tag{6.34a}$$

$$\tau_{rr}|_{r=0} = \tau_{\theta\theta}|_{r=0} = \tau_{\phi\phi}|_{r=0}, \tag{6.34b}$$

when the origin is part of the domain, and the stress and strain components are independent of θ and ϕ, we get that the material model should reduce to an isotropic one at the center; if the material is constrained to be homogeneous, then it has to be isotropic throughout. Since, similar to the constraints on the Lamé constants, these are also material constraints, they should be imposed even in problems where the axis or center is not a part of the domain, such as, for example, in problems involving hollow cylinders or spheres, or even when the loading is not radially symmetric. The singularities or infinite stress concentration factors in the solutions arise since these constraints are not imposed. In fact, as seen from the solutions presented in Figs. (2) and (3) in [145], there are no singularities in the stresses for the particular choice of material constants that the authors have used, which obey the constraint of Eqn. (6.33). Another benefit of imposing the constraint in Eqn. (6.33) is that it imposes bounds on the individual Poisson ratios ν_r and ν_θ, which, individually, can become unbounded if the only constraint on them is that imposed by the condition $1 - \nu_r(r)\nu_\theta(r) > 0$ in Eqn. (6.31b).

Based on the above discussion, we make the following observations:

1. In the composite isotropic disk considered in Section (3.3) of [83], the linearized stresses do tend to zero as the angle of rigid rotation tends to zero, thus, approaching the correct nonlinear stress limit.

2. For the aeolotropic disk problem considered in Section (3.4) of [83] (with similar constitutive relations being used in [81, 82, 328, 329]), as mentioned above, singularities arise as a result of the constraint given by Eqn. (6.33) not being imposed. Tarn [321] identified that the problem was due to the conflicting definition of anisotropy at $r = 0$, but instead of our constraint $c_{11}|_{r=0} = c_{22}|_{r=0}$, suggested that the inner core of the cylinder be considered as transversely isotropic, while the outer material is anisotropic. This, however, does not resolve the problem since in the limit as the inner core is shrunk to zero, the problem reappears [82].

3. For the functionally graded material considered in Section (3.5) of [83], the Young modulus tends to zero as r tends to zero, leading to a loss of ellipticity at the origin, as noted by the authors themselves.

In conclusion, the highly anomalous behavior of solutions observed with anisotropic linear elastic materials (stress singularities, cavitation, interpenetration etc.) seems more due to flaws in the particular constitutive models being used, rather than to any intrinsic shortcoming of linearized elasticity.

Assuming that $\mathbf{C}$ is symmetric, the expression for the stored energy function is given by

$$\tilde{W}(\epsilon) = \frac{1}{2}\epsilon : \mathbf{C}[\epsilon]. \tag{6.35}$$

To see this note that

$$\begin{aligned}\tilde{W}(\epsilon + \boldsymbol{U}) &= \tilde{W}(\epsilon) + \frac{1}{2}\boldsymbol{U} : \mathbf{C}[\epsilon] + \frac{1}{2}\epsilon : \mathbf{C}[\boldsymbol{U}] + o(\boldsymbol{U}) \\ &= \tilde{W}(\epsilon) + \boldsymbol{U} : \mathbf{C}[\epsilon] + o(\boldsymbol{U}),\end{aligned}$$

which implies that

$$\hat{\boldsymbol{T}} = \frac{\partial \tilde{W}}{\partial \epsilon} = \mathbf{C}[\epsilon].$$

Conversely, if the stored energy function is given by Eqn. (6.35), then from the results of Section 5.8, we know that $\mathbf{C}$ is symmetric. Thus, in the context of linearized elasticity, $\mathbf{C}$ is symmetric if and only if the material is hyperelastic. Throughout the remainder of this chapter, we assume $\mathbf{C}$ to be symmetric.

For a homogeneous, isotropic body, by virtue of Eqn. (5.76), Eqn. (6.19) reduces to

$$\boldsymbol{T} = \lambda(\operatorname{tr}\epsilon)\boldsymbol{I} + 2\mu\epsilon, \tag{6.36}$$

while the expression for the stored energy function given by Eqn. (6.35) reduces to

$$\tilde{W}(\epsilon) = \frac{\lambda}{2}(\operatorname{tr}\epsilon)^2 + \mu\epsilon : \epsilon.$$

By combining Eqns. (6.18), (6.20) and (6.36), we get, for a homogeneous, isotropic body, the following *Navier equations of elasticity*:

$$\rho_0 \frac{\partial^2 \boldsymbol{u}}{\partial t^2} = (\lambda + \mu)\boldsymbol{\nabla}_X(\boldsymbol{\nabla}_X \cdot \boldsymbol{u}) + \mu \boldsymbol{\nabla}_X^2 \boldsymbol{u} + \rho_0 \boldsymbol{b}, \tag{6.37}$$

where ρ_0, λ and μ are all constants since the body is assumed to be homogeneous. In indicial notation, the Navier equations read

$$\rho_0 \frac{\partial^2 u_i}{\partial t^2} = (\lambda + \mu)\frac{\partial \epsilon_{kk}}{\partial X_i} + \mu \frac{\partial^2 u_i}{\partial X_j \partial X_j} + \rho_0 b_i,$$

where $\epsilon_{kk} = \partial u_k / \partial X_k$.

Conducting the same thought experiments as in Section 5.3, we see that Eqns. (5.16), (5.17) and (5.20) are still valid. By substituting Eqns. (6.3), (6.36) and (5.20) into the relation $dV = [1 + DJ(\boldsymbol{I})[\boldsymbol{\nabla u}]]\, dV_0$, we get

$$\begin{aligned} dV &= (1 + \operatorname{tr} \boldsymbol{\epsilon})\, dV_0 \\ &= \left[1 + \frac{\operatorname{tr} \boldsymbol{T}}{3\lambda + 2\mu}\right] dV_0 \\ &= \left[1 + \frac{1 - 2\nu}{E} \operatorname{tr} \boldsymbol{T}\right] dV_0, \end{aligned}$$

so that $dV \to dV_0$ (almost incompressible material in the context of linearized elasticity) if $(3\lambda + 2\mu) \to \infty$, or, alternatively, if $\nu \to 0.5$.

6.3 Energy and Variational Formulations in Linearized Elasticity

First we present a result in elastostatics that relates the work done by the loads to the strain energy, and then proceed to discuss, single-field, two-field and three-field variational formulations in linearized elasticity.

In this section, we consider the following slightly more general form of the constitutive relations to include a prestrain $\boldsymbol{\epsilon}^0$:

$$\boldsymbol{T} = \mathbf{C}[\boldsymbol{\epsilon} - \boldsymbol{\epsilon}^0] \tag{6.38}$$

As an example, the prestrain $\boldsymbol{\epsilon}^0$ could denote thermal strains, which for an isotropic material are given by $\alpha\theta_\Delta \boldsymbol{I}$, where α is the coefficient of thermal expansion and θ_Δ is the increment in temperature. The prestrains and corresponding prestresses are assumed to be zero on the reference configuration, and are thus, not the same as residual stresses.

The strain energy and complementary strain energy density functions are now given by (compare with Eqn. (6.35))

$$W = \frac{1}{2}\left[\boldsymbol{\epsilon}(\boldsymbol{u}) - \boldsymbol{\epsilon}^0\right] : \mathbf{C}[\boldsymbol{\epsilon}(\boldsymbol{u}) - \boldsymbol{\epsilon}^0] = \frac{1}{2}\left[\boldsymbol{\epsilon}(\boldsymbol{u}) - \boldsymbol{\epsilon}^0\right] : \boldsymbol{T}, \tag{6.39a}$$

$$W^* = \frac{1}{2}\left[\boldsymbol{T} + \mathbf{C}[\boldsymbol{\epsilon}^0]\right] : \mathbf{C}^{-1}[\boldsymbol{T} + \mathbf{C}[\boldsymbol{\epsilon}^0]], \tag{6.39b}$$

whose derivatives with respect to $\boldsymbol{\epsilon}$ and $\boldsymbol{T}$ are

$$\begin{aligned} \boldsymbol{T} &= \frac{\partial W}{\partial \boldsymbol{\epsilon}}, \\ \boldsymbol{\epsilon} &= \frac{\partial W^*}{\partial \boldsymbol{T}}. \end{aligned}$$

The integrals of the density functions

$$U(\boldsymbol{\epsilon}) := \int_{V_0} W\, dV, \tag{6.40a}$$

$$U^*(\boldsymbol{T}) := \int_{V_0} W^*\, dV, \tag{6.40b}$$

represent the total strain energy and total complementary strain energy in the body, respectively.

Theorem 6.3.1 (Clayperon's theorem). *Assuming $\partial \boldsymbol{u}/\partial t = \boldsymbol{0}$, if $\boldsymbol{b}$ is the body force, and $\boldsymbol{t} = \boldsymbol{T}\boldsymbol{n}_0$ the traction on the boundary S_0, then*

$$2U(\boldsymbol{\epsilon}) = \int_{V_0} \rho_0 \boldsymbol{b} \cdot \boldsymbol{u}\, dV + \int_{S_0} \boldsymbol{t} \cdot \boldsymbol{u}\, dS - \int_{V_0} \boldsymbol{\epsilon}^0 : \boldsymbol{T}\, dV. \tag{6.41}$$

Proof. The right-hand side of Eqn. (6.41) can be written as

$$\begin{aligned}
\text{RHS} &= \int_{V_0} \rho_0 \boldsymbol{b} \cdot \boldsymbol{u}\, dV + \int_{S_0} (\boldsymbol{T}\boldsymbol{n}_0) \cdot \boldsymbol{u}\, dS - \int_{V_0} \boldsymbol{\epsilon}^0 : \boldsymbol{T}\, dV \\
&= \int_{V_0} \rho_0 \boldsymbol{b} \cdot \boldsymbol{u}\, dV + \int_{V_0} \boldsymbol{\nabla} \cdot (\boldsymbol{T}^T \boldsymbol{u})\, dV - \int_{V_0} \boldsymbol{\epsilon}^0 : \boldsymbol{T}\, dV \\
&\qquad \text{(by the divergence theorem)} \\
&= \int_{V_0} \left[\boldsymbol{u} \cdot (\rho_0 \boldsymbol{b} + \boldsymbol{\nabla} \cdot \boldsymbol{T}) + \boldsymbol{T} : \boldsymbol{\nabla}\boldsymbol{u} \right] dV - \int_{V_0} \boldsymbol{\epsilon}^0 : \boldsymbol{T}\, dV \\
&\qquad \text{(by Eqn. (1.355))} \\
&= \int_{V_0} \boldsymbol{T} : \boldsymbol{\nabla}\boldsymbol{u}\, dV - \int_{V_0} \boldsymbol{\epsilon}^0 : \boldsymbol{T}\, dV \\
&\qquad \text{(by the equilibrium equation)} \\
&= \int_{V_0} \boldsymbol{\epsilon} : \boldsymbol{T}\, dV - \int_{V_0} \boldsymbol{\epsilon}^0 : \boldsymbol{T}\, dV \qquad \text{(by symmetry of } \boldsymbol{T}) \\
&= 2U(\boldsymbol{\epsilon}). \qquad \text{(by Eqn. (6.40a))} \quad \square
\end{aligned}$$

The variational formulations presented below are useful in developing approximate solution methods such as the finite element method.

6.3.1 Single-field variational formulation

The procedure used to obtain the single-field variational formulation is similar to that followed in Section 3.9. Let $\boldsymbol{u}_\delta$ denote the variation of $\boldsymbol{u}$, $\bar{\boldsymbol{t}}$ be the prescribed traction on $(S_0)_t$, $\bar{\boldsymbol{u}}$ be the prescribed displacement on $(S_0)_u$, and let V_u be defined by

$$V_u := \{\boldsymbol{z} : \ \boldsymbol{z} = \boldsymbol{0} \text{ on } (S_0)_u\}.$$

Setting the weighted residual of the equation of motion and the traction boundary condition to zero, and following a procedure exactly analogous to that used in obtaining Eqn. (3.56), we get

$$\int_{V_0} \left(\rho_0 \boldsymbol{u}_\delta \cdot \frac{\partial^2 \boldsymbol{u}}{\partial t^2} + \boldsymbol{T} : \boldsymbol{\nabla}_X \boldsymbol{u}_\delta \right) dV = \int_{V_0} \rho_0 \boldsymbol{u}_\delta \cdot \boldsymbol{b}(\boldsymbol{X}, t)\, dV + \int_{(S_0)_t} \boldsymbol{u}_\delta \cdot \bar{\boldsymbol{t}}(\boldsymbol{X}, t, \boldsymbol{n}_0)\, dS \quad \forall \boldsymbol{u}_\delta \in V_u.$$

Since in our linearized framework, $\boldsymbol{T}$ is symmetric, we have

$$\boldsymbol{T} : \boldsymbol{\nabla}_X \boldsymbol{u}_\delta = \frac{1}{2} \boldsymbol{T} : \left[\boldsymbol{\nabla}_X \boldsymbol{u}_\delta + (\boldsymbol{\nabla}_X \boldsymbol{u}_\delta)^T \right] = \boldsymbol{T} : \boldsymbol{\epsilon}(\boldsymbol{u}_\delta),$$

where $\epsilon(\boldsymbol{u}_\delta) = [\boldsymbol{\nabla}_X \boldsymbol{u}_\delta + (\boldsymbol{\nabla}_X \boldsymbol{u}_\delta)^T]/2$. Hence, the variational formulation can be written as

$$\int_{V_0} \left[\rho_0 \boldsymbol{u}_\delta \cdot \frac{\partial^2 \boldsymbol{u}}{\partial t^2} + \epsilon(\boldsymbol{u}_\delta) : \boldsymbol{T} \right] dV = \int_{V_0} \rho_0 \boldsymbol{u}_\delta \cdot \boldsymbol{b} \, dV + \int_{(S_0)_t} \boldsymbol{u}_\delta \cdot \bar{\boldsymbol{t}} \, dS \quad \forall \boldsymbol{u}_\delta \in V_u. \tag{6.42}$$

Let the strain energy be as given by Eqn. (6.40a), and let

$$K = \int_{V_0} \frac{\rho_0}{2} \frac{\partial \boldsymbol{u}}{\partial t} \cdot \frac{\partial \boldsymbol{u}}{\partial t} \, dV,$$

be the kinetic energy of the body at time t. Then

$$\begin{aligned} \frac{dK}{dt} &= \int_{V_0} \frac{\rho_0}{2} \left[\frac{\partial \boldsymbol{u}}{\partial t} \cdot \frac{\partial^2 \boldsymbol{u}}{\partial t^2} + \frac{\partial^2 \boldsymbol{u}}{\partial t^2} \cdot \frac{\partial \boldsymbol{u}}{\partial t} \right] dV = \int_{V_0} \rho_0 \frac{\partial \boldsymbol{u}}{\partial t} \cdot \frac{\partial^2 \boldsymbol{u}}{\partial t^2} \, dV, \\ \frac{dU}{dt} &= \int_{V_0} \left(\frac{\partial \epsilon}{\partial t} - \frac{\partial \epsilon^0}{\partial t} \right) : \mathsf{C}[\epsilon - \epsilon^0] \, dV = \int_{V_0} \left(\frac{\partial \epsilon}{\partial t} - \frac{\partial \epsilon^0}{\partial t} \right) : \boldsymbol{T} \, dV. \end{aligned} \tag{6.43}$$

Since $\partial \boldsymbol{u}/\partial t = \boldsymbol{0}$ on $(S_0)_u$, we can choose $\boldsymbol{u}_\delta = \partial \boldsymbol{u}/\partial t$ in Eqn. (6.42). Noting using Eqns. (6.43) that

$$\begin{aligned} &\int_{V_0} \rho_0 \frac{\partial \boldsymbol{u}}{\partial t} \cdot \frac{\partial^2 \boldsymbol{u}}{\partial t^2} \, dV = \frac{dK}{dt}, \\ &\int_{V_0} \epsilon \left(\frac{\partial \boldsymbol{u}}{\partial t} \right) : \boldsymbol{T} \, dV = \int_{V_0} \frac{\partial \epsilon}{\partial t} : \boldsymbol{T} \, dV = \frac{dU}{dt} + \int_{V_0} \frac{\partial \epsilon^0}{\partial t} : \boldsymbol{T} \, dV, \end{aligned}$$

we get the following result from Eqn. (6.42), which should be compared to the mechanical energy balance equation given by Eqn. (3.65):

$$\frac{d}{dt}(K + U) + \int_{V_0} \frac{\partial \epsilon^0}{\partial t} : \boldsymbol{T} \, dV = \int_{V_0} \rho_0 \frac{\partial \boldsymbol{u}}{\partial t} \cdot \boldsymbol{b} \, dV + \int_{(S_0)_t} \frac{\partial \boldsymbol{u}}{\partial t} \cdot \bar{\boldsymbol{t}} \, dS.$$

As a corollary it follows that if ϵ^0 is constant with respect to time, $\boldsymbol{b} = \boldsymbol{0}$, and $\frac{\partial \boldsymbol{u}}{\partial t} \cdot \bar{\boldsymbol{t}} = 0$ on $(S_0)_t$, then the sum of the kinetic and strain energy remains constant, i.e.,

$$K + U = \text{constant}. \tag{6.44}$$

Defining the potential energy functional Π as in Eqn. (6.49a) below we see that Eqn. (6.42) can be written as

$$\int_{V_0} \rho_0 \boldsymbol{u}_\delta \cdot \frac{\partial^2 \boldsymbol{u}}{\partial t^2} \, dV + \delta^{(1)} \Pi = 0 \quad \forall \boldsymbol{u}_\delta \in V_u.$$

If we assume that $\boldsymbol{u}$ is prescribed or $\partial \boldsymbol{u}/\partial t = \boldsymbol{0}$ at times t_1 and t_2 ($t_1 < t_2$), then the above equation can be integrated with respect to time to yield Hamilton's principle:

$$\delta^{(1)} \int_{t_1}^{t_2} (K - \Pi) \, dt = 0.$$

We now specialize the above results to the elastostatics case, i.e., $\partial \boldsymbol{u}/\partial t = \boldsymbol{0}$. Equation (6.42) reduces to

$$\int_{V_0} \boldsymbol{\epsilon}(\boldsymbol{u}_\delta) : \mathbf{C}[\boldsymbol{\epsilon}]\, dV = \int_{V_0} \boldsymbol{\epsilon}(\boldsymbol{u}_\delta) : \mathbf{C}[\boldsymbol{\epsilon}^0]\, dV + \int_{V_0} \rho_0 \boldsymbol{u}_\delta \cdot \boldsymbol{b}\, dV + \int_{(S_0)_t} \boldsymbol{u}_\delta \cdot \bar{\boldsymbol{t}}\, dS \quad \forall \boldsymbol{u}_\delta \in V_u. \tag{6.45}$$

The above statement is often referred to as the principle of virtual work. Since $\boldsymbol{u}_\delta = \boldsymbol{0}$ on $(S_0)_u$, the last term can be replaced by $\int_{S_0} \boldsymbol{u}_\delta \cdot \boldsymbol{t}\, dS$.

A complementary variational principle (also called 'complementary virtual work') can similarly be formulated. Let $\boldsymbol{T}_\delta$ denote the variation of the stress field. Since the body forces are prescribed, the variation of $\boldsymbol{\nabla} \cdot \boldsymbol{T} + \rho_0 \boldsymbol{b} = \boldsymbol{0}$ is $\boldsymbol{\nabla} \cdot \boldsymbol{T}_\delta = \boldsymbol{0}$. Taking the dot product of this equation with $\boldsymbol{u}$, and integrating over the domain, we get

$$\int_{V_0} \boldsymbol{u} \cdot [\boldsymbol{\nabla} \cdot \boldsymbol{T}_\delta]\, dV = 0 \quad \forall \boldsymbol{T}_\delta \in V_T.$$

where $V_T := \{\boldsymbol{Z} : \ \boldsymbol{\nabla} \cdot \boldsymbol{Z} = \boldsymbol{0}\}$. Carrying out the usual 'integration by parts', and using the symmetry of $\boldsymbol{T}$, we get

$$\int_{V_0} \boldsymbol{\epsilon}(\boldsymbol{u}) : \boldsymbol{T}_\delta\, dV = \int_{S_0} \boldsymbol{u} \cdot \boldsymbol{t}_\delta\, dS \quad \forall \boldsymbol{T}_\delta \in V_T.$$

where $\boldsymbol{t}_\delta = \boldsymbol{T}_\delta \boldsymbol{n}$. Since $\boldsymbol{t}_\delta = \boldsymbol{0}$ on $(S_t)_0$, the above variational equation reduces to

$$\int_{V_0} \boldsymbol{\epsilon}(\boldsymbol{u}) : \boldsymbol{T}_\delta\, dV = \int_{(S_0)_u} \bar{\boldsymbol{u}} \cdot \boldsymbol{t}_\delta\, dS \quad \forall \boldsymbol{T}_\delta \in V_T.$$

Since $\boldsymbol{\epsilon}(\boldsymbol{u}) = \boldsymbol{\epsilon}^0 + \mathbf{C}^{-1}[\boldsymbol{T}]$, the final form of the complementary virtual work principle is

$$\int_{V_0} [\mathbf{C}^{-1}[\boldsymbol{T}]] : \boldsymbol{T}_\delta\, dV = -\int_{V_0} \boldsymbol{\epsilon}^0 : \boldsymbol{T}_\delta\, dV + \int_{(S_0)_u} \bar{\boldsymbol{u}} \cdot \boldsymbol{t}_\delta\, dS \quad \forall \boldsymbol{T}_\delta \in V_T. \tag{6.46}$$

Since $\boldsymbol{t}_\delta = \boldsymbol{0}$ on $(S_0)_t$, the last term can be replaced by $\int_{S_0} \boldsymbol{u}_\delta \cdot \boldsymbol{t}\, dS$.

Assuming $\mathbf{C}$ to be positive definite in addition to being symmetric, one can also formulate minimization principles associated with the virtual work and complementary virtual work principles. In order to formulate these principles, we first consider the following abstract form of the variational formulations:

Problem (V): Find u such that

$$a(u, v) = L(v) \quad \forall v, \tag{6.47}$$

where u denotes a scalar-, vector- or tensor-valued field, and v denotes its respective variation, $a(.,.)$ is a bilinear, symmetric, positive definite operator, and $L(.)$ is a linear operator, i.e., for all $u, v \in V$ and $\alpha, \beta \in \Re$, we have

$$\begin{aligned}
&a(u, \alpha v + \beta w) = \alpha a(u, v) + \beta a(u, w),\\
&a(u, v) = a(v, u),\\
&a(u, u) > 0 \quad \forall u \neq 0,
\end{aligned}$$

$$L(\alpha u + \beta v) = \alpha L(u) + \beta L(v)$$

We also assume that $a(.,.)$ and $L(.)$ are bounded.

Consider the following minimization problem

Problem (M): Find u such that

$$F(u) = \min_{v} F(v)$$

where

$$F(v) = \frac{1}{2}a(v,v) - L(v). \tag{6.48}$$

We have the following result relating problems (V) and (M) [168]:

Theorem 6.3.2. *The minimization problem (M) is equivalent to problem (V). The solution to Problems (M) and (V) is unique.*

Proof. The Riesz representation theorem from functional analysis can be used to show that a solution to the variational formulation exists; see, e.g., [58].
First we prove that M $\Longrightarrow$ V. Let $v \in V$ and $\epsilon \in \Re$. Since u is a minimum,

$$F(u) \leq F(u + \epsilon v) \quad \forall \epsilon \in \Re, \forall v \in V.$$

Using the notation $g(\epsilon) \equiv F(u + \epsilon v)$, we have

$$g(0) \leq g(\epsilon) \quad \forall \epsilon \in \Re,$$

or, alternatively, $g'(0) = 0$. The function $g(\epsilon)$ can be written as

$$\begin{aligned} g(\epsilon) &= \frac{1}{2}a(u + \epsilon v, u + \epsilon v) - L(u + \epsilon v) \\ &= \frac{1}{2}a(u,u) + \frac{\epsilon}{2}a(u,v) + \frac{\epsilon}{2}a(v,u) + \frac{\epsilon^2}{2}a(v,v) - L(u) - \epsilon L(v) \\ &= \frac{1}{2}a(u,u) - L(u) + \epsilon a(u,v) - \epsilon L(v) + \frac{\epsilon^2}{2}a(v,v), \end{aligned}$$

where we have used the symmetry of $a(.,.)$ in obtaining the last step. The constraint $g'(0) = 0$ implies $a(u,v) = L(v)$ for all $v \in V$.
Now we prove that V $\Longrightarrow$ M. Let $\bar{v} = u + \epsilon v$. Then

$$F(\bar{v}) = \frac{1}{2}a(u,u) - L(u) + \epsilon\,[a(u,v) - L(v)] + \frac{\epsilon^2}{2}a(v,v).$$

Since $a(u,v) = L(v)$, the above equation reduces to

$$F(\bar{v}) = F(u) + \frac{\epsilon^2}{2}a(v,v) \geq F(u).$$

Noting that $F(\bar{v}) = F(u)$ when $\epsilon = 0$, we can write the above equation as

$$F(u) = \min_{\bar{v} \in V} F(\bar{v}),$$

which is nothing but Problem (M).
To prove uniqueness, assume that u_1 and u_2 are two distinct solutions of problem (V). Then

$$a(u_1, v) = L(v) \quad \forall v \in V,$$
$$a(u_2, v) = L(v) \quad \forall v \in V.$$

Using the fact that $a(.,.)$ is a bilinear operator, we get

$$a(u_1 - u_2, v) = 0 \quad \forall v \in V.$$

Choosing $v = u_1 - u_2$, we get

$$a(u_1 - u_2, u_1 - u_2) = 0,$$

which by the assumed positive definiteness of $a(.,.)$ implies that $u_1 = u_2$. □

Note that if $a(.,.)$ is non-symmetric, we still have uniqueness of solutions, but no corresponding minimization problem (M).

By comparing Eqn. (6.45) and Problem (V), it is clear that in the case of the virtual work principle, we have $u := \boldsymbol{u}$ and $v := \boldsymbol{u}_\delta$, so that

$$a(u, v) := \int_{V_0} \boldsymbol{\epsilon}(\boldsymbol{u}_\delta) : \mathbf{C}[\boldsymbol{\epsilon}]\, dV,$$
$$L(v) := \int_{V_0} \boldsymbol{\epsilon}(\boldsymbol{u}_\delta) : \mathbf{C}[\boldsymbol{\epsilon}^0]\, dV + \int_{V_0} \rho_0 \boldsymbol{u}_\delta \cdot \boldsymbol{b}\, dV + \int_{(S_0)_t} \boldsymbol{u}_\delta \cdot \bar{\boldsymbol{t}}\, dS.$$

Similarly, by comparing Eqns. (6.46) and Problem (V), in the case of the complementary virtual work principle, we have $u := \boldsymbol{T}$ and $v := \boldsymbol{T}_\delta$, so that

$$a(u, v) := \int_{V_0} [\mathbf{C}^{-1}[\boldsymbol{T}]] : \boldsymbol{T}_\delta\, dV,$$
$$L(v) := \int_{(S_0)_u} \bar{\boldsymbol{u}} \cdot \boldsymbol{t}_\delta\, dS - \int_{V_0} \boldsymbol{\epsilon}^0 : \boldsymbol{T}_\delta\, dV.$$

Thus, the potential and complementary potential functionals (after adding terms that are independent of $\boldsymbol{u}$ and $\boldsymbol{T}$ to Eqn. (6.48) in order to get it into a convenient form) corresponding to the virtual and complementary virtual work principles are

$$\Pi(\boldsymbol{u}) := \frac{1}{2} \int_{V_0} \left[\boldsymbol{\epsilon}(\boldsymbol{u}) - \boldsymbol{\epsilon}^0\right] : \mathbf{C}[\boldsymbol{\epsilon}(\boldsymbol{u}) - \boldsymbol{\epsilon}^0]\, dV - \int_{V_0} \rho_0 \boldsymbol{u} \cdot \boldsymbol{b}\, dV - \int_{(S_0)_t} \boldsymbol{u} \cdot \bar{\boldsymbol{t}}\, dS, \quad (6.49a)$$

$$\Pi^*(\boldsymbol{T}) := \frac{1}{2} \int_{V_0} \left[\boldsymbol{T} + \mathbf{C}[\boldsymbol{\epsilon}^0]\right] : \mathbf{C}^{-1}[\boldsymbol{T} + \mathbf{C}[\boldsymbol{\epsilon}^0]]\, dV - \int_{(S_0)_u} \bar{\boldsymbol{u}} \cdot \boldsymbol{t}\, dS. \quad (6.49b)$$

Note that by virtue of Eqn. (6.40a), the expression for the potential energy is nothing but the difference in the strain energy and the work done by the loads. The last terms in Eqns. (6.49a) and (6.49b) can be replaced by $\int_{S_0} \boldsymbol{u} \cdot \boldsymbol{t}\, dS$ by virtue of the remarks following Eqns. (6.45) and (6.46). In the absence of body forces and prestrains $\boldsymbol{\epsilon}^0$, the complementary energy Π^* and potential energy Π have the same expressions.

As a result of Theorem 6.3.2, we have

Theorem 6.3.3 (Principle of minimum potential energy). *Assuming* $\mathbf{C}$ *to be symmetric and positive definite, the solution of the variational formulation given by Eqn. (6.45) is also the solution to the following problem: Find* $\boldsymbol{u}$ *such that*

$$\Pi(\boldsymbol{u}) = \min_{\boldsymbol{v}} \Pi(\boldsymbol{v}),$$

where Π *is given by Eqn. (6.49a), and* $\boldsymbol{v}$ *is a field that satisfies the essential boundary condition* $\boldsymbol{v} = \bar{\boldsymbol{u}}$ *on* $(S_0)_u$*, and the strain–displacement relation. Conversely, the solution of the above minimization problem is also the solution of the variational formulation.*

Theorem 6.3.4 (Principle of minimum complementary energy). *Assuming* $\mathbf{C}$ *to be symmetric and positive definite, the solution of the variational formulation given by Eqn. (6.46) is also the solution to the following problem: Find* $\boldsymbol{T}$ *such that*

$$\Pi^*(\boldsymbol{T}) = \min_{\boldsymbol{H}} \Pi^*(\boldsymbol{H}),$$

where Π^* *is given by Eqn. (6.49b), and* $\boldsymbol{H}$ *is a statically admissible field, i.e., one that satisfies* $\boldsymbol{\nabla} \cdot \boldsymbol{H} + \rho_0 \boldsymbol{b} = \boldsymbol{0}$*, and the boundary condition* $\boldsymbol{H}\boldsymbol{n} = \bar{\boldsymbol{t}}$ *on* $(S_0)_t$*. Conversely, the solution of the above minimization problem is also the solution of the variational formulation.*

Consider an unconstrained body (i.e., the entire boundary is traction-free). If one assumes the body to be elastic, then residual stresses cannot exist, since, by the principle of minimum potential energy, the equilibrium state would be one of zero strains (and hence, zero strain energy). Thus, one either needs to consider geometric nonlinearities (e.g., the everted state of a ball), or plastic yielding in order for residual stresses to exist.

Setting the derivative of Π with respect to displacements to zero leads to 'Castigliano's first theorem', while setting the derivative of Π^* with respect to the loads to zero leads to 'Castigliano's second theorem'. The first theorem can be used to find the force corresponding to a given deflection, while the second can be used to find the deflection corresponding to a given load, even in the presence of thermal strains (represented by $\boldsymbol{\epsilon}^0$). In the case of statically indeterminate structures, the derivatives of Π^* with respect to redundant internal forces can be set to zero since they do not contribute to the work term $\int_{S_0} \boldsymbol{u} \cdot \boldsymbol{t}\, dS$. As usual, a dummy load can be applied at a point (which is set to zero after the derivatives have been calculated) where the displacement is desired if no load is acting at that point.

If $\boldsymbol{\epsilon}^0$ represents the thermal strains in an isotropic body given by $\alpha\theta_\Delta \boldsymbol{I}$, where α is the coefficient of thermal expansion and θ_Δ is the increment in temperature, then

$$\mathbf{C}[\boldsymbol{\epsilon}^0] = (3\lambda + 2\mu)\alpha\theta_\Delta \boldsymbol{I} = \frac{E\alpha\theta_\Delta}{1 - 2\nu}\boldsymbol{I}, \tag{6.50a}$$

$$\epsilon^0 : \mathbf{C}[\epsilon(u)] = (3\lambda + 2\mu)(\operatorname{tr}\epsilon)\alpha\theta_\Delta = \frac{E(\operatorname{tr}\epsilon)\alpha\theta_\Delta}{1-2\nu}, \tag{6.50b}$$

$$\epsilon^0 : T = \alpha\theta_\Delta(\operatorname{tr} T). \tag{6.50c}$$

As an application of Castigliano's second theorem, consider the two-bar truss shown in Fig. 6.1, which is subjected to a temperature change of ΔT. In order to find the

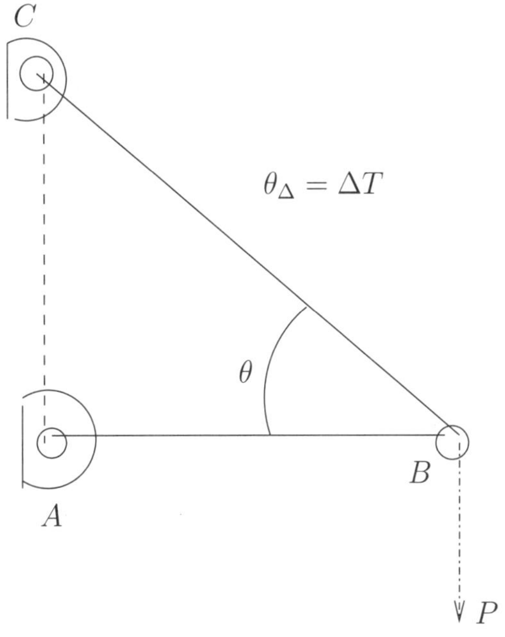

Fig. 6.1 *Two-bar truss subjected to thermal loading.*

vertical deflection at point B due to this thermal loading, we apply the dummy load P as shown. The internal forces are given by $F_{AB} = -P\cot\theta$ and $F_{BC} = P\operatorname{cosec}\theta$. Using a one-dimensional approximation, the complementary potential energy obtained using Eqns. (6.49b) and (6.50c) (after omitting terms not dependent on P) is given by

$$\Pi^* = \frac{(P\cot\theta)^2 L_{AB}}{2E_{AB}A_{AB}} - \alpha\Delta T(P\cot\theta)L_{AB} + \frac{(P\operatorname{cosec}\theta)^2 L_{BC}}{2E_{BC}A_{BC}} + \alpha\Delta T(P\operatorname{cosec}\theta)L_{BC} - P\delta.$$

Setting the derivative of Π^* with respect to P to zero, and then evaluating the resulting expression at $P = 0$, we get the vertical deflection as

$$\delta = \alpha\Delta T L_{BC}\sin\theta.$$

6.3.2 Two-field and three-field variational formulations

The single-field variational formulation given by Eqn. (6.45) was obtained by enforcing the equation of motion and the traction boundary condition in a weak sense. If in addition, we also enforce the strain–displacement relation given by Eqn. (6.20) in a weak sense (with $\epsilon - \epsilon^0$ given by $\mathbf{C}^{-1}[T]$, since the constitutive relation is being enforced in a strong sense), then we obtain a two-field variational formulation as shown below. To keep things simple, we focus on the elastostatics case; the extension to elastodynamics can be obtained by replacing $\boldsymbol{b}$ by $\boldsymbol{b} - \partial^2 \boldsymbol{u}/\partial t^2$. Thus, denoting the variation of $\boldsymbol{T}$ by $\boldsymbol{T}_\delta$, we now have

$$\int_{V_0} \boldsymbol{u}_\delta \cdot (\boldsymbol{\nabla} \cdot \boldsymbol{T} + \boldsymbol{b})\, dV + \int_{(S_0)_t} \boldsymbol{u}_\delta \cdot (\bar{\boldsymbol{t}} - \boldsymbol{t})\, dS + \int_{V_0} \boldsymbol{T}_\delta : \left[\bar{\epsilon}(\boldsymbol{u}) - \epsilon^0 - \mathbf{C}^{-1}[\boldsymbol{T}]\right] dV = 0 \quad \forall (\boldsymbol{u}_\delta, \boldsymbol{T}_\delta),$$

where $\bar{\epsilon}(\boldsymbol{u}) \equiv [\boldsymbol{\nabla u} + (\boldsymbol{\nabla u})^T]/2$. By carrying out the 'integration by parts' as in the single-field formulation, we get

$$\int_{V_0} [\bar{\epsilon}(\boldsymbol{u}_\delta) : \boldsymbol{T} - \boldsymbol{u}_\delta \cdot \boldsymbol{b}]\, dV - \int_{(S_0)_t} \boldsymbol{u}_\delta \cdot \bar{\boldsymbol{t}}\, dS + \int_{V_0} \boldsymbol{T}_\delta : \left[\bar{\epsilon}(\boldsymbol{u}) - \epsilon^0 - \mathbf{C}^{-1}[\boldsymbol{T}]\right] dV = 0 \quad \forall (\boldsymbol{u}_\delta, \boldsymbol{T}_\delta). \tag{6.51}$$

Invoking the arbitrariness of the variations, and considering the two cases $(\boldsymbol{u}_\delta, \boldsymbol{T}_\delta) = (\boldsymbol{u}_\delta, \boldsymbol{0})$, and $(\boldsymbol{u}_\delta, \boldsymbol{T}_\delta) = (\boldsymbol{0}, \boldsymbol{T}_\delta)$, we see that Eqn. (6.51) is equivalent to the following two statements:

$$\int_{V_0} \bar{\epsilon}(\boldsymbol{u}_\delta) : \boldsymbol{T}\, dV = \int_{V_0} \boldsymbol{u}_\delta \cdot \boldsymbol{b}\, dV + \int_{(S_0)_t} \boldsymbol{u}_\delta \cdot \bar{\boldsymbol{t}}\, dS \quad \forall \boldsymbol{u}_\delta, \tag{6.52a}$$

$$\int_{V_0} \boldsymbol{T}_\delta : \left[\bar{\epsilon}(\boldsymbol{u}) - \epsilon^0 - \mathbf{C}^{-1}[\boldsymbol{T}]\right] dV = 0 \quad \forall \boldsymbol{T}_\delta. \tag{6.52b}$$

By reversing the steps in the above proof, it is easy to see that the above two variational statements are equivalent to the original differential equations (given that $\boldsymbol{T}$ is smooth enough so that its spatial derivatives exist).

By setting the first variation of the Hellinger–Reissner functional defined by

$$\Pi_{\text{HR}} := \int_{V_0} \left[\left[\bar{\epsilon}(\boldsymbol{u}) - \epsilon^0\right] : \boldsymbol{T} - \boldsymbol{u} \cdot \boldsymbol{b} - \frac{1}{2}\boldsymbol{T} : \mathbf{C}^{-1}[\boldsymbol{T}]\right] dV - \int_{(S_0)_t} \boldsymbol{u} \cdot \bar{\boldsymbol{t}}\, dS.$$

to zero, we get Eqn. (6.51).

The three-field variational formulation is similar to the two-field variational formulation except in that it enforces the constitutive relation also in a weak sense. If ϵ_δ denotes the variation of the strain field, then, instead of Eqn. (6.51), we now have

$$\begin{aligned}\int_{V_0} [\bar{\epsilon}(\boldsymbol{u}_\delta) : \boldsymbol{T} - \boldsymbol{u}_\delta \cdot \boldsymbol{b}]\, dV - \int_{(S_0)_t} \boldsymbol{u}_\delta \cdot \bar{\boldsymbol{t}}\, dS + \int_{V_0} \boldsymbol{T}_\delta : [\bar{\epsilon}(\boldsymbol{u}) - \epsilon]\, dV \\ + \int_{V_0} \epsilon_\delta : \left[\mathbf{C}[\epsilon - \epsilon^0] - \boldsymbol{T}\right] dV = 0 \quad \forall (\boldsymbol{u}_\delta, \boldsymbol{T}_\delta, \epsilon_\delta).\end{aligned} \tag{6.53}$$

By successively considering $\boldsymbol{u}_\delta$, $\boldsymbol{T}_\delta$ and ϵ_δ to be nonzero with the remaining two fields set to zero, we obtain the three variational statements that are equivalent to the above one as

$$\int_{V_0} \bar{\epsilon}(\boldsymbol{u}_\delta) : \boldsymbol{T}\, dV = \int_{V_0} \boldsymbol{u}_\delta \cdot \boldsymbol{b}\, dV + \int_{(S_0)_t} \boldsymbol{u}_\delta \cdot \bar{\boldsymbol{t}}\, dS \quad \forall \boldsymbol{u}_\delta,$$

$$\int_{V_0} \boldsymbol{T}_\delta : [\bar{\epsilon}(\boldsymbol{u}) - \epsilon]\, dV = 0 \quad \forall \boldsymbol{T}_\delta,$$

$$\int_{V_0} \epsilon_\delta : \left[\mathbf{C}[\epsilon - \epsilon^0] - \boldsymbol{T}\right] dV = 0 \quad \forall \epsilon_\delta.$$

By setting the first variation of the Hu–Washizu functional defined by

$$\Pi_{\mathrm{HW}} := \int_{V_0} \left[\frac{1}{2}\boldsymbol{\epsilon} : \mathbf{C}[\boldsymbol{\epsilon}] - \boldsymbol{\epsilon} : \mathbf{C}[\boldsymbol{\epsilon}^0] - \boldsymbol{u}\cdot\boldsymbol{b} - \boldsymbol{\epsilon} : \boldsymbol{T} + \boldsymbol{T} : \bar{\boldsymbol{\epsilon}}(\boldsymbol{u}) \right] dV - \int_{(S_0)_t} \boldsymbol{u}\cdot\bar{\boldsymbol{t}}\, dS.$$

to zero, we get Eqn. (6.53).

In contrast to the principle of minimization of potential energy or complementary energy, the stationary conditions on the Hellinger–Reissner and Hu–Washizu functionals corresponds to saddle-point problems. Also note that the use of the potential energy, complementary energy, Hellinger–Reissner or Hu–Washizu functionals is not crucial in developing single-field or mixed finite element formulations, since such formulations can be derived more directly and conveniently from the variational forms that have been presented.

6.4 Uniqueness of Solution

We consider the linear elastostatics problem first. If $S_0 = (S_0)_u$, then the problem is said to be a displacement boundary value problem, if $S_0 = (S_0)_t$, then it is said to be a traction boundary value problem, while if $S_0 = (S_0)_u \cup (S_0)_t$, then it is said to be a mixed boundary value problem. In contrast to nonlinear elasticity, we have the following result:

Theorem 6.4.1 (Uniqueness of solution in linear elastostatics). *Assuming* $\mathbf{C}$ *to be symmetric and positive definite, the displacement and mixed boundary value problems have a unique solution for the displacement, strain and stress fields; in the case of the pure traction boundary value problem, the stress and strain solution fields are unique, while the displacement solution field is unique modulo linearized rigid displacements.*

Proof. Let $(\boldsymbol{u}^1, \boldsymbol{\epsilon}^1, \boldsymbol{T}^1)$ and $(\boldsymbol{u}^2, \boldsymbol{\epsilon}^2, \boldsymbol{T}^2)$ be two solutions that satisfy the governing equations and boundary conditions. Let $\tilde{\boldsymbol{u}} = \boldsymbol{u}^1 - \boldsymbol{u}^2$, and $\tilde{\boldsymbol{T}} = \boldsymbol{T}^1 - \boldsymbol{T}^2$. Since by Eqn. (6.18), $\boldsymbol{\nabla}_X \cdot \boldsymbol{T}^1 + \rho_0 \boldsymbol{b} = \boldsymbol{0}$, and $\boldsymbol{\nabla}_X \cdot \boldsymbol{T}^2 + \rho_0 \boldsymbol{b} = \boldsymbol{0}$, we have $\boldsymbol{\nabla}_X \cdot \tilde{\boldsymbol{T}} = \boldsymbol{0}$. Similarly, we have $\tilde{\boldsymbol{t}} = (\boldsymbol{T}^1 - \boldsymbol{T}^2)\boldsymbol{n}_0 = \tilde{\boldsymbol{T}}\boldsymbol{n}_0$, $\tilde{\boldsymbol{\epsilon}} = \boldsymbol{\epsilon}^1 - \boldsymbol{\epsilon}^2 = [\boldsymbol{\nabla}_X \tilde{\boldsymbol{u}} + (\boldsymbol{\nabla}_X \tilde{\boldsymbol{u}})^T]/2$, and $\tilde{\boldsymbol{T}} = \mathbf{C}[\tilde{\boldsymbol{\epsilon}}]$. Thus, $\tilde{\boldsymbol{u}}$, $\tilde{\boldsymbol{\epsilon}}$ and $\tilde{\boldsymbol{T}}$ obey the governing equations given by Eqns. (6.18)–(6.20), with the body force zero. Hence, by Clayperon's theorem, we get

$$\int_{S_0} \tilde{\boldsymbol{t}} \cdot \tilde{\boldsymbol{u}}\, dS = 2\int_{V_0} \tilde{W}(\tilde{\boldsymbol{\epsilon}})\, dV.$$

For the displacement boundary value problem, we have $\tilde{\boldsymbol{u}} = \boldsymbol{0}$, for the traction boundary value problem, we have $\tilde{\boldsymbol{t}} = \boldsymbol{0}$, while for the mixed boundary value problem, we have $\tilde{\boldsymbol{t}} = \boldsymbol{0}$ on $(S_0)_t$ and $\tilde{\boldsymbol{u}} = \boldsymbol{0}$ on $(S_0)_u$. In all three cases, we observe that $\int_{S_0} \tilde{\boldsymbol{t}} \cdot \tilde{\boldsymbol{u}}\, dS = 0$. Thus, the above equation reduces to

$$2\int_{V_0} \tilde{W}(\tilde{\boldsymbol{\epsilon}})\, dV = 0.$$

Since, by assumption, $\mathbf{C}$ is positive definite, $2\tilde{W}(\tilde{\boldsymbol{\epsilon}}) = \tilde{\boldsymbol{\epsilon}} : \mathbf{C}[\tilde{\boldsymbol{\epsilon}}]$ is a positive definite function of $\tilde{\boldsymbol{\epsilon}}$, and hence

$$\tilde{W}(\tilde{\boldsymbol{\epsilon}}) = 0,$$

which in turn implies that

$$\tilde{\epsilon} = \epsilon^1 - \epsilon^2 = \mathbf{0}.$$

From this it immediately follows that $\tilde{T} = \mathsf{C}[\tilde{\epsilon}] = \mathbf{0}$. Thus, the strain and stress fields are unique in all the three kinds of boundary value problems. By Theorem 6.1.1,

$$u^1(X) - u^2(X) = WX + c.$$

For the displacement and mixed boundary value problems, $u^1 - u^2 = \mathbf{0}$ on $(S_0)_u$. Thus, $W = \mathbf{0}$ and $c = \mathbf{0}$ is these cases, and the displacement solution is unique. In the traction boundary value problem, however, the displacements are unique modulo linearized rigid displacements. □

We now prove that the solution to the linear elastodynamics solution is also unique. Let $Tn_0 = \bar{t}(X,t)$ be the prescribed traction on $(S_0)_t$, $\bar{u}(X,t)$ be the prescribed displacement on $(S_0)_u$, and let $u(X,0) = u_0(X)$ and $\dot{u}(X,0) = v_0(X)$ be the initial conditions.

Theorem 6.4.2 (Uniqueness of solution in linear elastodynamics). *Assuming* C *to be symmetric and positive definite, the displacement, strain and stress fields that solve Eqns.* (6.18)–(6.20) *subject to the above initial and boundary conditions are unique.*

Proof. As in the previous proof, let $\tilde{u}$, $\tilde{\epsilon}$ and $\tilde{T}$ denote the difference between two solutions. By substituting the two solutions into the governing equations, and the initial and boundary conditions, we see that $\tilde{u}$, $\tilde{\epsilon}$ and $\tilde{T}$ satisfy the governing equations with the body force zero, and satisfy the initial conditions $\tilde{u}(X,0) = \dot{\tilde{u}}(X,0) = \mathbf{0}$, and the boundary conditions $\tilde{u} = \mathbf{0}$ on $(S_0)_u$ and $\tilde{T}n_0 = \mathbf{0}$ on $(S_0)_t$. Since the conditions used in deriving Eqn. (6.44) are satisfied, we get

$$\int_{V_0} \frac{1}{2}\rho_0 \frac{\partial \tilde{u}}{\partial t} \cdot \frac{\partial \tilde{u}}{\partial t}\, dV + \int_{V_0} \frac{1}{2}\tilde{\epsilon} : \mathsf{C}[\tilde{\epsilon}]\, dV = \text{constant},$$

for all times. Since at time $t = 0$, $\tilde{u}(X,0) = \dot{\tilde{u}}(X,0) = \mathbf{0}$, the constant value in the above equation is zero. Since C is assumed to be positive definite, we deduce from the above equation that both the individual terms in the above equation are zero. This in turn implies that

$$\frac{\partial \tilde{u}}{\partial t}(X,t) = \mathbf{0},$$

i.e., $\tilde{u} = \tilde{u}(X)$ at all times. But at time $t = 0$, $\tilde{u}(X) = \mathbf{0}$. Hence, $\tilde{u} = \mathbf{0}$ at all places and at all times, proving the uniqueness of the solution. □

The uniqueness of solutions for the inverse boundary value problem considered in [75, 234, 235] is discussed briefly in Section 6.5.7.

6.5 Exact Solutions of some Special Problems in Elasticity

In this section, we discuss exact solutions (within the linearized framework) to a few special problems in elasticity. In all the problems discussed below (where the body is suitably constrained to disallow linearized rigid displacements), we will verify that a proposed displacement or stress field is a solution, and by Theorem 6.4.1 it then follows that this is the only possible solution. We assume the reference configuration to be a natural state, and the material to be homogeneous and isotropic, so that by Eqn. (6.36),

$$\boldsymbol{\tau} = \boldsymbol{T} = \mathbf{C}[\boldsymbol{\epsilon}] = \lambda(\operatorname{tr}\boldsymbol{\epsilon})\boldsymbol{I} + 2\mu\boldsymbol{\epsilon},$$

or, equivalently (see Eqn. (5.20)),

$$\boldsymbol{\tau} = \frac{\nu E}{(1+\nu)(1-2\nu)}(\operatorname{tr}\boldsymbol{\epsilon})\boldsymbol{I} + 2G\boldsymbol{\epsilon}, \tag{6.54}$$

where $G = \mu = E/(2(1+\nu))$ is known as the shear modulus. Taking the trace of Eqn. (6.54), we get $\operatorname{tr}\boldsymbol{\tau} = (3\lambda + 2\mu)\operatorname{tr}\boldsymbol{\epsilon} = (E/(1-2\nu))\operatorname{tr}\boldsymbol{\epsilon}$. Substituting this relation into Eqn. (6.54), we get the 'inverse form' of Eqn. (6.54) as

$$\boldsymbol{\epsilon} = \frac{1}{E}\left[(1+\nu)\boldsymbol{\tau} - \nu(\operatorname{tr}\boldsymbol{\tau})\boldsymbol{I}\right]. \tag{6.55}$$

Since here, we are considering problems where $\boldsymbol{a} = \boldsymbol{b} = \boldsymbol{0}$, Eqn. (6.18) reduces to

$$\boldsymbol{\nabla}_X \cdot \boldsymbol{\tau} = \boldsymbol{0}. \tag{6.56}$$

As pointed out in the discussion on page 316, $\boldsymbol{t} = \boldsymbol{t}_0 = \boldsymbol{T}\boldsymbol{n}_0 = \boldsymbol{\tau}\boldsymbol{n}_0$, although subsequently, we shall denote $\boldsymbol{n}_0$ simply as $\boldsymbol{n}$ for notational convenience. In what follows, we shall often use engineering shear strains, which are defined as $\gamma_{ij} := 2\epsilon_{ij}$, $i \neq j$. It should be noted that writing $\boldsymbol{\epsilon}$ and $\boldsymbol{\tau}$ in the form of vectors $\boldsymbol{\epsilon}_c \equiv (\epsilon_{xx}, \epsilon_{yy}, \epsilon_{zz}, \gamma_{xy}, \gamma_{yz}, \gamma_{xz})$[1] and $\boldsymbol{\tau}_c \equiv (\tau_{xx}, \tau_{yy}, \tau_{zz}, \tau_{xy}, \tau_{yz}, \tau_{xz})$ *does not* change their intrinsic tensorial character, and it is done merely for convenience in engineering applications (see Appendix I for a systematic way of converting tensors to matrix form). Equation (6.54) using such a 'vector' notation reads

$$\begin{bmatrix}\tau_{xx}\\ \tau_{yy}\\ \tau_{zz}\\ \tau_{xy}\\ \tau_{yz}\\ \tau_{xz}\end{bmatrix} = \frac{E}{(1+\nu)(1-2\nu)}\begin{bmatrix}1-\nu & \nu & \nu & 0 & 0 & 0\\ \nu & 1-\nu & \nu & 0 & 0 & 0\\ \nu & \nu & 1-\nu & 0 & 0 & 0\\ 0 & 0 & 0 & \frac{1-2\nu}{2} & 0 & 0\\ 0 & 0 & 0 & 0 & \frac{1-2\nu}{2} & 0\\ 0 & 0 & 0 & 0 & 0 & \frac{1-2\nu}{2}\end{bmatrix}\begin{bmatrix}\epsilon_{xx}\\ \epsilon_{yy}\\ \epsilon_{zz}\\ \gamma_{xy}\\ \gamma_{yz}\\ \gamma_{xz}\end{bmatrix}. \tag{6.57}$$

6.5.1 Torsion of a circular cylinder

A circular cylinder of radius a and length L fixed at one end is subjected to a torque T generated by applying suitable tractions on the other end-surface, as shown in Fig. 6.2. The lateral surfaces are assumed to be traction-free. We try and see if the solution derived based

[1]Ideally speaking, we should denote the coordinates and strain component subscripts with capital letters as in Eqn. (6.14), to signify the fact that these quantities are associated with the reference configuration; however, since this can result in very cumbersome notation, we shall henceforth use lower-case letters both for the coordinates and as subscripts for the strain tensor.

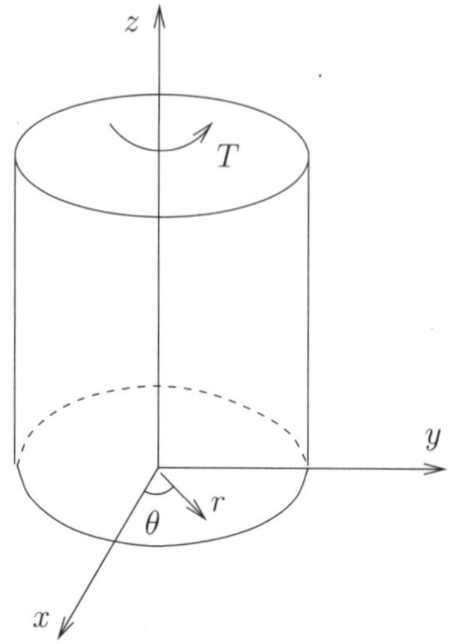

Fig. 6.2 *Torsion of a circular cylinder.*

on the 'engineering theory', where the plane surfaces $z =$ const. are assumed to rotate rigidly about the z axis through an angle αz (where α, known as the twist per unit length, is a constant to be determined), qualifies as a solution. In this solution, the displacements with respect to the cylindrical coordinate system are given by

$$u_r = 0, \quad u_\theta = \alpha r z, \quad u_z = 0. \tag{6.58}$$

Using Eqn. (B.8), we see that the only nonzero strain component is $\gamma_{\theta z} = 2\epsilon_{\theta z} = \alpha r$, which, using Eqn. (6.57), yields the only nonzero stress component as $\tau_{\theta z} = G\gamma_{\theta z} = G\alpha r$; this stress component obviously assumes a maximum value of $G\alpha a$ at $r = a$. The principal stresses at any point on the outer surface are $(-G\alpha a, 0, G\alpha a)$, and the principal directions are inclined at 45° to the horizontal. Thus, if the material is brittle, it is weak in tension, and failure occurs along a helix on the outer surface.

Having derived the strains from the displacement field, we do not need to check the compatibility conditions. Using Eqns. (B.7), it is easily verified that the equilibrium equations given by Eqns. (6.56) are satisfied. On the lateral surface, since $\boldsymbol{n} = (n_r, n_\theta, n_z) = (1, 0, 0)$ the tractions $\boldsymbol{t} = \boldsymbol{\tau n}$ are found to be zero. On the end surface $z = L$, since the normal is $\boldsymbol{n} = (0, 0, 1)$, the traction vector is $\boldsymbol{t} = (t_r, t_\theta, t_z) = (0, G\alpha r, 0)$. Treating the cylindrical and Cartesian coordinate systems as the $\{\bar{\boldsymbol{e}}_i\}$ and $\{\boldsymbol{e}_i\}$ systems, and using Eqn. (1.101), we get $\boldsymbol{t} = (t_x, t_y, t_z) = (-G\alpha r \sin\theta, G\alpha r \cos\theta, 0)$. Thus, the resultant force components are $F_z = 0$, and

$$F_x = \int_0^{2\pi} \int_0^a t_x r \, dr \, d\theta = 0, \quad F_y = \int_0^{2\pi} \int_0^a t_y r \, dr \, d\theta = 0.$$

Similarly, the resultant force components on the face $z = 0$ are also zero. Since the torque about the z-axis is given to be T, we have

$$T\boldsymbol{e}_z = \int_0^{2\pi} \int_0^a [(r\boldsymbol{e}_r) \times \boldsymbol{t}] r \, dr \, d\theta = \int_0^{2\pi} \int_0^a t_\theta r^2 \, dr \, d\theta \boldsymbol{e}_z = GJ\alpha \boldsymbol{e}_z,$$

where $J = \int_A r^2 \, dA = \pi a^4/2$ is the polar moment of inertia, and the ratio $D \equiv T/\alpha$, known as the torsional rigidity, is equal to GJ. Since the stress field associated with the displacement given by Eqn. (6.58) satisfies all the equations of equilibrium, and since these fields also satisfy the traction and displacement boundary conditions, they qualify as a solution to the problem at hand. One can easily verify that this solution is valid even when the cylinder is hollow, with J now given by $\pi(b^4 - a^4)/2$, where a and b are the inner and outer radii of the cylinder. Note that the twisting moment T has to be applied in a manner consistent with the obtained stress distribution, i.e., by applying tractions $t_\theta = G\alpha r$ on the top and bottom surfaces, in order for the solution to qualify as an 'exact' solution. If it is applied through a different traction distribution, then an empirical principle, known as St-Venant's principle, asserts that in regions sufficiently far removed from the region of application of the tractions, the deviation of the actual solution from the exact one is small. The solution for torsion of a circular cylinder with an eccentric circular hole was first presented by Macdonald [205] (see Section 6.5.3).

For future use, we note here the nonzero displacement and stress components of the solution with respect to the Cartesian coordinate system, which have been obtained using Eqns. (1.101) and (1.102):

$$u_x = -\alpha y z, \quad u_y = \alpha x z, \tag{6.59}$$

$$\tau_{xz} = -G\alpha y, \quad \tau_{yz} = G\alpha x. \tag{6.60}$$

6.5.2 Torsion of non-circular bars–Saint-Venant theory of torsion

We now assume the origin to be at the geometrical center of the bar, so that the end surfaces are at $z = \pm L/2$. For circular bars, we assumed that each cross-section remained plane and rotated without distortion during torsion, and obtained the stress distribution given by Eqn. (6.60). Conversely, if the stress distribution is given by Eqn. (6.60), then the cross-section has to be circular. To see this, let s denote the arclength parameter around the contour of a typical cross-section of the bar $z =$ const. In terms of this parameter, the normal to the lateral surface is given by $\boldsymbol{n} = (dy/ds, -dx/ds, 0)$ (see Fig. 6.3), so that absence of tractions on the lateral surface implies that

$$0 = t_z = \tau_{xz} n_x + \tau_{yz} n_y = -G\alpha \left(x\frac{dx}{ds} + y\frac{dy}{ds} \right) = -\frac{G\alpha}{2}\frac{d}{ds}(x^2 + y^2),$$

which in turn implies that the contour has to be a circle. Thus, the stress distribution given by Eqn. (6.60) cannot hold in bars with non-circular cross-sections.

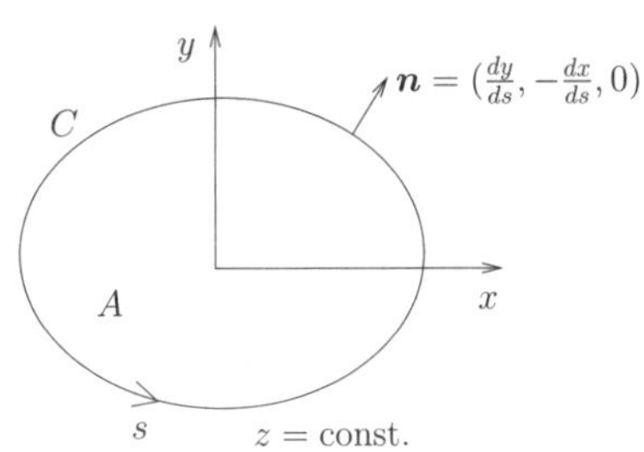

Fig. 6.3 *Cross-section of a non-circular cylinder undergoing torsion.*

Let the displacements, instead of those given by Eqn. (6.59), be given by

$$u_x = -\alpha yz, \quad u_y = \alpha xz, \quad u_z = \alpha\psi(x,y), \tag{6.61}$$

where α, as before, is the twist per unit length, and ψ, which is a function of x and y (so that it is the same for all cross-sections), is known as the warping function (note that now the bar is no longer assumed to be fixed at one end, and the problem is a pure traction boundary value problem). The introduction of this function modifies the assumption made in the case of circular cylinders about plane sections remaining plane, so that plane cross-sections can now warp, with each cross-section warped in the same way. Note that the z-axis is assumed to be the axis of torsion (i.e., the line along which $u_x = u_y = 0$) while writing the above displacement field. If the axis of torsion is different, then the modifications that need to be made are discussed below. We now show that with the above displacement field, we are able to satisfy the equations of equilibrium and the boundary conditions. Again note that the torque T on the ends is to be applied in a manner required by the solution.

Corresponding to the displacement field in Eqn. (6.61), the nonzero strain and stress components obtained using Eqns. (6.20) and (6.57) are

$$\gamma_{xz} = \alpha\left(\frac{\partial\psi}{\partial x} - y\right), \quad \gamma_{yz} = \alpha\left(\frac{\partial\psi}{\partial y} + x\right), \tag{6.62}$$

$$\tau_{xz} = G\alpha\left(\frac{\partial\psi}{\partial x} - y\right), \quad \tau_{yz} = G\alpha\left(\frac{\partial\psi}{\partial y} + x\right). \tag{6.63}$$

Since τ_{xz} and τ_{yz} are the only nonzero stress components, and since they are functions only of x and y, the equilibrium equations given by Eqns. (6.56) reduce to

$$\frac{\partial\tau_{xz}}{\partial x} + \frac{\partial\tau_{yz}}{\partial y} = 0, \tag{6.64}$$

while the condition that the lateral surface C be traction-free reduces to

$$t_z = \tau_{xz}n_x + \tau_{yz}n_y = 0, \tag{6.65}$$

since t_x and t_y are trivially zero. Substituting Eqn. (6.63) into Eqns. (6.64) and (6.65), we get

$$\boldsymbol{\nabla}^2\psi \equiv \frac{\partial^2\psi}{\partial x^2} + \frac{\partial^2\psi}{\partial y^2} = 0 \quad \text{in } A, \tag{6.66}$$

$$(\boldsymbol{\nabla}\psi)\cdot\boldsymbol{n} = yn_x - xn_y \quad \text{on C}. \tag{6.67}$$

Eqn. (6.66) is the governing equation for $\psi(x,y)$, which is to be solved subject to the boundary condition given by Eqn. (6.67). The problem of determining a harmonic function ψ, whose normal derivative is prescribed on the boundary, is referred to as a Neumann problem. The solution to the Neumann problem is unique modulo a constant. The displacements corresponding to different values of the constant differ by a rigid translation of the bar in the z-direction. A necessary condition for the existence of a solution to the Neumann problem is that $\int_C(\boldsymbol{\nabla}\psi)\cdot\boldsymbol{n}\,ds = 0$ since

$$\int_C(\boldsymbol{\nabla}\psi)\cdot\boldsymbol{n}\,ds = \int_A \boldsymbol{\nabla}\cdot(\boldsymbol{\nabla}\psi)\,dA = \int_A \boldsymbol{\nabla}^2\psi\,dA,$$

and $\nabla^2\psi = 0$ by Eqn. (6.66)[2]. This condition is satisfied by the problem at hand since

$$\int_C (\nabla\psi)\cdot \boldsymbol{n}\,ds = \int_C (yn_x - xn_y)\,ds = \int_A \left(\frac{\partial y}{\partial x} - \frac{\partial x}{\partial y}\right) dA = 0.$$

We note that the governing equation and boundary condition for ψ given by Eqn. (6.66) and (6.67) are valid even when the domain is multiply connected.

The resultant force component $F_z = \int_A t_z\,dA$ on the top and bottom surfaces (where $\boldsymbol{n} = (0,0,\pm 1)$) is zero since $t_z = \boldsymbol{\tau n} = 0$ on those surfaces. That the resultant force components F_x and F_y are also zero is proved as follows (we show the proof for the top surface, with the proof for the bottom surface being similar):

$$\begin{aligned}
F_x &= \int_A t_x\,dA \\
&= \int_A \tau_{xz}\,dA \\
&= \int_A \left[\frac{\partial(x\tau_{xz})}{\partial x} + \frac{\partial(x\tau_{yz})}{\partial y}\right] dA && \text{(by Eqn. (6.64))} \\
&= \int_C x(\tau_{xz}n_x + \tau_{yz}n_y)\,ds && \text{(by the divergence theorem)} \\
&= 0, && \text{(by Eqn. (6.65))} \\
F_y &= \int_A \tau_{yz}\,dA \\
&= \int_A \left[\frac{\partial(y\tau_{xz})}{\partial x} + \frac{\partial(y\tau_{yz})}{\partial y}\right] dA \\
&= \int_C y(\tau_{xz}n_x + \tau_{yz}n_y)\,ds \\
&= 0.
\end{aligned}$$

The twisting moment T on the top surface (where we have shown that $t_z = 0$) is given by

$$\begin{aligned}
T\boldsymbol{e}_z &= \int_A [(\boldsymbol{x} - L\boldsymbol{e}_z)] \times \boldsymbol{t}\,dA \\
&= \int_A (x\boldsymbol{e}_x + y\boldsymbol{e}_y) \times (t_x\boldsymbol{e}_x + t_y\boldsymbol{e}_y)\,dA \\
&= \left[\int_A (x\tau_{yz} - y\tau_{xz})\,dA\right] \boldsymbol{e}_z,
\end{aligned}$$

which implies that

$$\begin{aligned}
T &= \int_A (x\tau_{yz} - y\tau_{xz})\,dA && (6.68) \\
&= \int_A (x\tau_{yz} - y\tau_{xz})\,dA \pm \int_C \psi(\tau_{xz}n_x + \tau_{yz}n_y)\,ds && \text{(by Eqn. (6.65))} \\
&= \int_A (x\tau_{yz} - y\tau_{xz})\,dA \pm \int_A \left[\frac{\partial(\psi\tau_{xz})}{\partial x} + \frac{\partial(\psi\tau_{yz})}{\partial y}\right] dA && \text{(by the divergence theorem)}
\end{aligned}$$

[2] This necessary condition holds even for multiply connected domains since the divergence theorem is valid for such domains.

$$= \int_A \left[\left(\pm \frac{\partial \psi}{\partial x} - y \right) \tau_{xz} + \left(\pm \frac{\partial \psi}{\partial y} + x \right) \tau_{yz} \right] dA. \qquad \text{(by Eqn. (6.64))} \tag{6.69}$$

Successively considering first the plus and then the minus sign in Eqn. (6.69), and using Eqn. (6.63), we get the following expressions for the torsional rigidity $D = T/\alpha$:

$$D = \frac{1}{G\alpha^2} \left[\int_A (\tau_{xz}^2 + \tau_{yz}^2) \right] dA \tag{6.70}$$

$$= G \int_A \left\{ (x^2 + y^2) - \left[\left(\frac{\partial \psi}{\partial x} \right)^2 + \left(\frac{\partial \psi}{\partial y} \right)^2 \right] \right\} dA. \tag{6.71}$$

From Eqn. (6.71), it is immediately evident that

$$D \le GJ, \tag{6.72}$$

where $J = \int_A (x^2 + y^2) \, dA$ is the polar moment of inertia of the cross-section. If we have equality in Eqn. (6.72), i.e., if $D = GJ$, then $\partial \psi / \partial x = \partial \psi / \partial y = 0$, which in turn implies that $\psi = \text{const}$. The boundary condition given by Eqn. (6.67) reduces to

$$0 = yn_x - xn_y = x\frac{dx}{ds} + y\frac{dy}{ds} \implies x^2 + y^2 = \text{const}.$$

Thus, the upper bound on torsional rigidity GJ is attained if and only if the cross-section is circular.

If the axis of the torsion is different from the z-axis, say one parallel to the z-axis and intersecting the xy plane at (x_0, y_0), then the displacement field, instead of the one given by Eqn. (6.61), is now given by

$$u_x = -\alpha z(y - y_0), \quad u_y = \alpha z(x - x_0), \quad u_z = \alpha \psi_0(x, y), \tag{6.73}$$

where the warping function $\psi_0(x, y)$ is different from the warping function $\psi(x, y)$. Following the same line of development as before, the counterpart of Eqns. (6.63) is

$$\tau_{xz} = G\alpha \left(\frac{\partial \psi_0}{\partial x} - y + y_0 \right), \quad \tau_{yz} = G\alpha \left(\frac{\partial \psi_0}{\partial y} + x - x_0 \right), \tag{6.74}$$

with the other stress components zero, while the counterparts of Eqns. (6.66) and (6.67) are

$$\boldsymbol{\nabla}^2 \psi_0 \equiv \frac{\partial^2 \psi_0}{\partial x^2} + \frac{\partial^2 \psi_0}{\partial y^2} = 0 \quad \text{in } A,$$

$$[\boldsymbol{\nabla}(\psi_0 + y_0 x - x_0 y)] \cdot \boldsymbol{n} = yn_x - xn_y \quad \text{on } C.$$

Since $\boldsymbol{\nabla}^2 \psi_0 = 0$, we also have $\boldsymbol{\nabla}^2 \tilde{\psi}_0 = 0$, where $\tilde{\psi}_0 := \psi_0 + y_0 x - x_0 y$. Thus, the above two equations can be written as

$$\boldsymbol{\nabla}^2 \tilde{\psi}_0 = 0 \quad \text{in } A,$$

$$(\boldsymbol{\nabla} \tilde{\psi}_0) \cdot \boldsymbol{n} = yn_x - xn_y \quad \text{on } C.$$

Comparing these two equations with Eqns. (6.66) and (6.67), we see that $\tilde{\psi}_0$ satisfies the same governing equation and boundary condition as the function ψ. Since the solution to this problem is unique upto a constant, i.e., $\tilde{\psi}_0 = \psi + \text{const.}$, we have

$$\psi_0 = \psi + x_0 y - y_0 x + \text{const.} \tag{6.75}$$

Substituting the above expression into Eqn. (6.74), we see that the resulting expressions are the same as those given by Eqn. (6.63). Thus, the stress distribution is the same, while the displacements as given by Eqn. (6.73) differ by a linearized rigid displacement, as compared to the corresponding distributions obtained with $x_0 = y_0 = 0$. It immediately follows from Eqn. (6.70) that the torsional rigidity remains unchanged.

Since ψ is harmonic, there exists a conjugate harmonic function g, related to ψ by the Cauchy–Riemann equations

$$\frac{\partial \psi}{\partial x} = \frac{\partial g}{\partial y}, \quad \frac{\partial \psi}{\partial y} = -\frac{\partial g}{\partial x}. \tag{6.76}$$

Using the above equations, the boundary condition given by Eqn. (6.67) can be written as

$$y\frac{dy}{ds} + x\frac{dx}{ds} = \frac{\partial g}{\partial y}\frac{dy}{ds} + \frac{\partial g}{\partial x}\frac{dx}{ds} = \frac{dg}{ds},$$

which implies that $g = (x^2 + y^2)/2 + \text{const.}$ Thus, the governing equation and boundary condition for g are

$$\nabla^2 g \equiv \frac{\partial^2 g}{\partial x^2} + \frac{\partial^2 g}{\partial y^2} = 0 \quad \text{in } A, \tag{6.77}$$

$$g = \frac{1}{2}(x^2 + y^2) + \text{const.} \quad \text{on } C. \tag{6.78}$$

The above problem, where the boundary condition is directly expressed in terms of the variable g rather than in terms of its derivatives as in the Neumann problem, is referred to as a Dirichlet problem. Similar to the Neumann problem, the solution to the Dirichlet problem is unique provided the constant in Eqn. (6.78) is assigned a fixed value. Note that the choice of the constant leaves the values of stresses given by

$$\tau_{xz} = G\alpha\left(\frac{\partial g}{\partial y} - y\right), \quad \tau_{yz} = G\alpha\left(-\frac{\partial g}{\partial x} + x\right) \tag{6.79}$$

unaffected. Hence, in the case of a simply connected domain, we take the constant to be zero on C, while in the case of a multiply connected domain, we can take the constant on one of the free boundaries, say, the outermost one, to be zero.

Since the real and imaginary parts of a complex analytic function $W(z)$, where $z = x + iy$, satisfy the Cauchy–Riemann equations, we can take $W(z) = \psi + ig$; by considering different analytic functions $W(z)$, we get solutions to different problems.

In addition to ψ and g, a third function, known as the Prandtl stress function and denoted by ϕ, can be used to solve the torsion problem in an effective way. This function is defined through the relation

$$g = \frac{1}{2}(x^2 + y^2) + \frac{\phi}{G\alpha}. \tag{6.80}$$

From Eqns. (6.77) and (6.78), we get the governing equation and boundary condition for ϕ as

$$\nabla^2\phi \equiv \frac{\partial^2\phi}{\partial x^2} + \frac{\partial^2\phi}{\partial y^2} = -2G\alpha \quad \text{in } A, \tag{6.81}$$

$$\phi = \text{const.} \quad \text{on } C. \tag{6.82}$$

while from Eqn. (6.79), it follows that

$$\tau_{xz} = \frac{\partial\phi}{\partial y}, \quad \tau_{yz} = -\frac{\partial\phi}{\partial x}. \tag{6.83}$$

The above treatment, which has been carried out with respect to a Cartesian coordinate system, can be generalized to an arbitrary orthogonal curvilinear coordinate system. Let Q be the orthogonal matrix that is used for transforming the components of a vector from the Cartesian to the curvilinear coordinate system, and let x represent the components of the position vector of a point with respect to the curvilinear coordinate system, i.e., $x = Q[x]_{\text{Cartesian}}$. For example, $x = (r, 0)$ for a cylindrical coordinate system. Since in two-dimensions, any Q can be written as $\begin{bmatrix} \cos\gamma & \sin\gamma \\ -\sin\gamma & \cos\gamma \end{bmatrix} = \cos\gamma I + \sin\gamma W$ (for some γ) where $W = \begin{bmatrix} 0 & 1 \\ -1 & 0 \end{bmatrix}$, it follows that W commutes with Q. If (ξ_1, ξ_2) are the in-plane curvilinear coordinates, $u_c \equiv \begin{bmatrix} u_{\xi_1} \\ u_{\xi_2} \end{bmatrix}$, and $\tau_c \equiv \begin{bmatrix} \tau_{\xi_1 z} \\ \tau_{\xi_2 z} \end{bmatrix}$ are the nonzero stresses, then assuming that the axis of torsion passes through the origin of the curvilinear system, and using the fact that Q and W commute, we have

$$\begin{aligned} u_c &= -\alpha z W x, \quad u_z = \alpha\psi(\xi_1, \xi_2), \\ \tau_c &= G\alpha(\nabla\psi - Wx) = G\alpha W(\nabla g - x) = W\nabla\phi. \end{aligned} \tag{6.84}$$

As an example, for a cylindrical coordinate system, we have

$$u_r = 0, \qquad u_\theta = \alpha r z, \qquad u_z = \alpha\psi(r, \theta), \tag{6.85a}$$

$$\tau_{rz} = G\alpha\frac{\partial\psi}{\partial r} = \frac{G\alpha}{r}\frac{\partial g}{\partial\theta} = \frac{1}{r}\frac{\partial\phi}{\partial\theta}, \tag{6.85b}$$

$$\tau_{\theta z} = G\alpha\left[\frac{1}{r}\frac{\partial\psi}{\partial\theta} + r\right] = G\alpha\left[-\frac{\partial g}{\partial r} + r\right] = -\frac{\partial\phi}{\partial r}. \tag{6.85c}$$

The above stress expressions can be verified using the Cartesian formulae and the relations $\tau_{rz} = \tau_{xz}\cos\theta + \tau_{yz}\sin\theta$, $\tau_{\theta z} = \tau_{yz}\cos\theta - \tau_{xz}\sin\theta$.

For an elliptical coordinate system, using Q given by the in-plane part of the Q given by Eqn. (D.3), we get $x = [c^2\sinh 2\xi/(2h_1), -c^2\sin 2\eta/(2h_1)]$, where $h_1 = c\sqrt{(\cosh 2\xi - \cos 2\eta)/2}$. Using Eqn. (6.84), we have

$$\begin{aligned} u_\xi &= \frac{\alpha z c^2}{2h_1}\sin 2\eta, \qquad u_\eta = \frac{\alpha z c^2}{2h_1}\sinh 2\xi, \qquad u_z = \alpha\psi(\xi, \eta), \\ \tau_{\xi z} &= G\alpha\left[\frac{1}{h_1}\frac{\partial\psi}{\partial\xi} + \frac{c^2}{2h_1}\sin 2\eta\right] = G\alpha\left[\frac{1}{h_1}\frac{\partial g}{\partial\eta} + \frac{c^2}{2h_1}\sin 2\eta\right] = \frac{1}{h_1}\frac{\partial\phi}{\partial\eta}, \\ \tau_{\eta z} &= G\alpha\left[\frac{1}{h_1}\frac{\partial\psi}{\partial\eta} + \frac{c^2}{2h_1}\sinh 2\xi\right] = G\alpha\left[-\frac{1}{h_1}\frac{\partial g}{\partial\xi} + \frac{c^2}{2h_1}\sinh 2\xi\right] = -\frac{1}{h_1}\frac{\partial\phi}{\partial\xi}. \end{aligned} \tag{6.86}$$

Similarly, for a bipolar coordinate system, using the in-plane part of the $\boldsymbol{Q}$ given by Eqn. (E.3), and $[\boldsymbol{x}]_{\text{Cartesian}} = (h_1 \sin\eta, h_1 \sinh\xi)$, we get $\boldsymbol{x} = [-h_1 \sinh\xi \cos\eta, -h_1 \cosh\xi \sin\eta]$ where $h_1 = c/(\cosh\xi - \cos\eta)$. From Eqn. (6.84), it follows that

$$
\begin{aligned}
u_\xi &= \alpha z h_1 \cosh\xi \sin\eta, \qquad u_\eta = -\alpha z h_1 \sinh\xi \cos\eta, \qquad u_z = \alpha\psi(\xi,\eta),\\
\tau_{\xi z} &= G\alpha\left[\frac{1}{h_1}\frac{\partial\psi}{\partial\xi} + h_1\cosh\xi\sin\eta\right] = G\alpha\left[\frac{1}{h_1}\frac{\partial g}{\partial\eta} + h_1\cosh\xi\sin\eta\right] = \frac{1}{h_1}\frac{\partial\phi}{\partial\eta},\\
\tau_{\eta z} &= G\alpha\left[\frac{1}{h_1}\frac{\partial\psi}{\partial\eta} - h_1\sinh\xi\cos\eta\right] = G\alpha\left[-\frac{1}{h_1}\frac{\partial g}{\partial\xi} - h_1\sinh\xi\cos\eta\right] = -\frac{1}{h_1}\frac{\partial\phi}{\partial\xi}.
\end{aligned}
\tag{6.87}
$$

The expression for the torque is derived using Eqn. (6.68) and (6.83) as follows:

$$
\begin{aligned}
T &= \int_A \left(-x\frac{\partial\phi}{\partial x} - y\frac{\partial\phi}{\partial y}\right) dA\\
&= \int_A \left[2\phi - \frac{\partial(\phi x)}{\partial x} - \frac{\partial(\phi y)}{\partial y}\right] dA\\
&= 2\int_A \phi\, dA - \int_C \phi\,(x n_x + y n_y)\, ds.
\end{aligned}
\tag{6.88}
$$

For simply connected regions, such as for example, solid bars, we can take $\phi = 0$ on C since the stresses given by Eqn. (6.83) are not affected by the addition of a constant to ϕ; thus, for simply connected domains, Eqn. (6.88) reduces to

$$
T = 2\int_A \phi\, dA. \qquad \text{(for simply connected domains)} \tag{6.89}
$$

In multiply connected regions, although the value of ϕ on any one free boundary, say the outermost one, can still be set to zero, the values that it assumes on the other boundaries are in general nonzero constants. Let C_0 denote the outermost boundary, C_i, $i = 1, 2, \ldots, N$ denote the contours of the interior holes, and let the constant value attained by ϕ on C_i be denoted by k_i. If $\tilde{n}_x = -n_x$ and $\tilde{n}_y = -n_y$ denote the components of the 'inward' normal (where the inward normal is one pointing away from the void region into the domain A, and is thus the outward normal for each void domain), then Eqn. (6.88) can be written as

$$
\begin{aligned}
T &= 2\int_A \phi\, dA + \sum_{i=1}^{N} k_i \int_{C_i} (x\tilde{n}_x + y\tilde{n}_y)\, ds\\
&= 2\int_A \phi\, dA + \sum_{i=1}^{N} k_i \int_{A_i} \left(\frac{\partial x}{\partial x} + \frac{\partial y}{\partial y}\right) dA \quad \text{(divergence theorem)}\\
&= 2\int_A \phi\, dA + 2\sum_{i=1}^{N} k_i A_i,
\end{aligned}
\tag{6.90}
$$

where A_i is the area enclosed by the contour C_i.

Since for the surfaces $\phi = $ const., we have

$$
0 = \frac{d\phi}{ds} = \frac{\partial\phi}{\partial x}\frac{dx}{ds} + \frac{\partial\phi}{\partial y}\frac{dy}{ds} = \tau_{xz}n_x + \tau_{yz}n_y = t_z,
$$

these surfaces are traction-free. On the surface $z = \text{const.}$, with the normal given by $(0, 0, 1)$, the traction vector is

$$\boldsymbol{t}\big|_{z=\text{const.}} = (\tau_{xz}, \tau_{yz}, 0) = \left(\frac{\partial \phi}{\partial y}, -\frac{\partial \phi}{\partial x}, 0\right), \tag{6.91}$$

and its magnitude, denoted by $|\boldsymbol{t}|$, and referred to as the 'shearing stress', is

$$|\boldsymbol{t}| = \sqrt{\tau_{xz}^2 + \tau_{yz}^2} = \sqrt{\left(\frac{\partial \phi}{\partial x}\right)^2 + \left(\frac{\partial \phi}{\partial y}\right)^2} = |\boldsymbol{\nabla}\phi|. \tag{6.92}$$

Note that the traction vector in Eqn. (6.91) is tangential to the curves $\phi = \text{const.}$ (referred to as 'lines of shearing stress') since the unit normal to these curves at a point is given by $\boldsymbol{n} = \boldsymbol{\nabla}\phi / |\boldsymbol{\nabla}\phi|$, so that

$$\boldsymbol{t}\big|_{z=\text{const.}} \cdot \boldsymbol{n} = \frac{1}{|\boldsymbol{\nabla}\phi|}\left(\frac{\partial \phi}{\partial y}\frac{\partial \phi}{\partial x} - \frac{\partial \phi}{\partial x}\frac{\partial \phi}{\partial y}\right) = 0.$$

We now show that the maximum shearing stress $|\boldsymbol{t}|_{\max}$ occurs on the boundary C, so that failure of a bar under torsion is expected to initiate on its lateral surface. Let A be a domain with C as its boundary, and let β be a function defined on A, which is not identically equal to a constant. Then [300] (page 117)

1. if $\boldsymbol{\nabla}^2\beta \geq 0$ (in which case β is said to be subharmonic), then β attains its maximum on the boundary C.
2. if $\boldsymbol{\nabla}^2\beta \leq 0$ (in which case β is said to be superharmonic), then β attains its minimum on the boundary C.

It follows that if $\boldsymbol{\nabla}^2\beta = 0$, then β attains both its maximum and minimum on the boundary C. By virtue of result (1) above, we need to show that $|\boldsymbol{t}|^2$ is subharmonic. By successively differentiating the expression for $|\boldsymbol{t}|^2$, we find that

$$\begin{aligned}
\boldsymbol{\nabla}^2(|\boldsymbol{t}|^2) &= 2\left[\left(\frac{\partial^2 \phi}{\partial x^2}\right)^2 + 2\left(\frac{\partial^2 \phi}{\partial x \partial y}\right)^2 + \left(\frac{\partial^2 \phi}{\partial y^2}\right)^2 + \frac{\partial \phi}{\partial x}\frac{\partial(\boldsymbol{\nabla}^2\phi)}{\partial x} + \frac{\partial \phi}{\partial y}\frac{\partial(\boldsymbol{\nabla}^2\phi)}{\partial y}\right] \\
&= 2\left[\left(\frac{\partial^2 \phi}{\partial x^2}\right)^2 + 2\left(\frac{\partial^2 \phi}{\partial x \partial y}\right)^2 + \left(\frac{\partial^2 \phi}{\partial y^2}\right)^2\right] \geq 0,
\end{aligned}$$

where in obtaining the second step, we have used the fact that $\boldsymbol{\nabla}^2\phi = \text{const.}$ We now discuss the solution of some special problems.

Elliptical cross-section

Consider the torsion of a bar with an elliptical cross-section with semi-major and semi-minor axes dimensions as a and b. Since the cross-section is simply connected, we can choose $\phi = 0$ on the boundary C. A natural 'guess' for the function ϕ satisfying this boundary condition is a scalar multiple of the equation of the boundary, i.e.,

$$\phi = m\left(1 - \frac{x^2}{a^2} - \frac{y^2}{b^2}\right),$$

where m is a constant to be determined. Substituting this equation into Eqn. (6.81), we find that $m = G\alpha a^2b^2/(a^2+b^2)$, so that

$$\phi = \frac{G\alpha a^2 b^2}{a^2+b^2}\left(1 - \frac{x^2}{a^2} - \frac{y^2}{b^2}\right). \tag{6.93}$$

Using Eqn. (6.83), the nonzero stress components are obtained as

$$\tau_{xz} = -\frac{2G\alpha a^2 y}{a^2+b^2}, \quad \tau_{yz} = \frac{2G\alpha b^2 x}{a^2+b^2},$$

from which it follows that

$$|\boldsymbol{t}|^2 = \tau_{xz}^2 + \tau_{yz}^2 = \frac{4G^2\alpha^2}{(a^2+b^2)^2}\left[a^4y^2 + b^4x^2\right].$$

To find $|\boldsymbol{t}|_{\max}$, we use the fact that it occurs on the boundary C, and thus substitute $y^2 = b^2(1 - x^2/a^2)$ into the above equation to get

$$\text{On } C: \quad \tau^2 = \frac{4G^2\alpha^2}{(a^2+b^2)^2}\left[a^4b^2 - x^2b^2(a^2-b^2)\right].$$

Since $a \geq b$, $|\boldsymbol{t}|^2$ is maximum when x^2 is minimum, i.e, when $x = 0$, $y = \pm b$. Substituting $x = 0$ in the above expression, we get

$$|\boldsymbol{t}|_{\max} = \frac{2G\alpha a^2 b}{a^2+b^2}. \tag{6.94}$$

Note that, quite nonintuitively, the maximum shearing stress occurs at the extremities of the minor axis.

Using Eqn. (6.68), we get

$$T = \frac{2G\alpha}{a^2+b^2}\int_A (b^2x^2 + a^2y^2)\, dA.$$

The above integral is conveniently evaluated by making the change of variables $x = ar\cos\theta$, $y = br\sin\theta$. The determinant of the transformation Jacobian matrix $\begin{bmatrix} \frac{\partial x}{\partial r} & \frac{\partial x}{\partial \theta} \\ \frac{\partial y}{\partial r} & \frac{\partial y}{\partial \theta} \end{bmatrix}$ evaluates to abr, so that $dA = abr\, dr\, d\theta$. Thus,

$$T = \frac{2G\alpha}{a^2+b^2}\int_0^{2\pi}\int_0^1 (a^2b^2r^2)abr\, dr\, d\theta = \frac{\pi G\alpha a^3 b^3}{a^2+b^2}.$$

Thus, the torsional rigidity is

$$D = \frac{T}{\alpha} = \frac{\pi G a^3 b^3}{a^2+b^2} = \frac{4a^2b^2}{(a^2+b^2)^2} GJ. \tag{6.95}$$

In order to find the warping and conjugate functions ψ and g, we use the fact that the equation of the boundary C involves quadratic terms in x and y, and hence guess $W(z) \equiv \psi + ig$ to have the form $i(cz^2 + k)$, where c and k are constants to be determined. Since

$z = x + iy$, we get $\psi = -2xyc$ and $g = (x^2 - y^2)c + k$. Since $g = (x^2 + y^2)/2$ on the boundary C, and since $x^2/a^2 + y^2/b^2 = 1$ is the equation of the boundary, we get $c = (a^2 - b^2)/(2(a^2 + b^2))$ and $k = a^2b^2/(a^2 + b^2)$. Thus,

$$\psi = -\frac{(a^2 - b^2)xy}{a^2 + b^2}, \tag{6.96}$$

$$g = \frac{(a^2 - b^2)(x^2 - y^2)}{2(a^2 + b^2)} + \frac{a^2b^2}{a^2 + b^2}.$$

Note that the above expression for g could also have been found directly using Eqn. (6.80).

The displacement field is obtained using Eqn. (6.61) as

$$u_x = -\alpha yz, \quad u_y = \alpha xz, \quad u_z = -\frac{(a^2 - b^2)\alpha xy}{a^2 + b^2}.$$

Note that while it is convenient to find the stress field using the Prandtl stress function, the displacement field is more conveniently found using the warping function. From the expression from u_z, we see that the first and third quadrants become 'concave', while the second and fourth quadrants become 'convex' after deformation.

Upon setting $b = a$, all the above results reduce, as expected, to those for a circular cylinder.

In elliptical coordinates, if the outer boundary is $\xi = \xi_0$, the warping, conjugate and Prandtl stress functions are given by

$$\begin{aligned}
\psi &= -\frac{c^2}{4}\frac{\sinh 2\xi \sin 2\eta}{\cosh 2\xi_0}, \\
g &= \frac{c^2}{4}\left[\frac{\cosh 2\xi \cos 2\eta}{\cosh 2\xi_0} + \cosh 2\xi_0\right], \\
\frac{\phi}{G\alpha} &= \frac{c^2}{4\cosh 2\xi_0}\left[\cosh 2\xi - \cosh 2\xi_0\right]\left[\cos 2\eta - \cosh 2\xi_0\right].
\end{aligned} \tag{6.97}$$

Using Eqns. (6.86), we get

$$u_\xi = \frac{\alpha z c^2}{2h_1}\sin 2\eta, \qquad u_\eta = \frac{\alpha z c^2}{2h_1}\sinh 2\xi, \qquad u_z = \alpha\psi(\xi, \eta),$$

$$\tau_{\xi z} = \frac{\alpha c^2 G}{2h_1}\left[1 - \frac{\cosh 2\xi}{\cosh 2\xi_0}\right]\sin 2\eta,$$

$$\tau_{\eta z} = \frac{\alpha c^2 G}{2h_1}\left[1 - \frac{\cos 2\eta}{\cosh 2\xi_0}\right]\sinh 2\xi.$$

Using the relations $a = c\cosh\xi_0$, $b = c\sinh\xi_0$, and $c^2 = a^2 - b^2$, it follows that $|\boldsymbol{t}|_{\max} = |\tau_{\eta z}|_{\max}$ is given by Eqn. (6.94) at $(\xi, \eta) = (\xi_0, \pm\pi/2)$.

Triangular cross-section

Consider the torsion of a bar with a cross-section which is an equilateral triangle as shown in Fig. 6.4. Similar to the elliptical cross-section case, we propose a scalar multiple of the equation of the boundary as a candidate Prandtl stress function, i.e.,

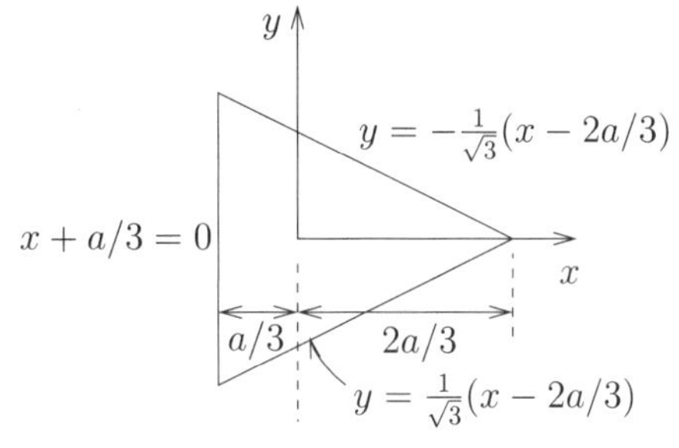

Fig. 6.4 *A bar of triangular cross-section undergoing torsion.*

$$\begin{aligned}\phi &= m\left(x+\frac{a}{3}\right)\left(x-\frac{2a}{3}+\sqrt{3}y\right)\left(x-\frac{2a}{3}-\sqrt{3}y\right)\\ &= m\left[x^3 - ax^2 - 3y^2x + \frac{4a^3}{27} - ay^2\right].\end{aligned}$$

Substituting this equation into Eqn. (6.81), we get $m = G\alpha/(2a)$. Thus,

$$\phi = \frac{G\alpha}{2a}\left[x^3 - ax^2 - 3y^2x + \frac{4a^3}{27} - ay^2\right]. \tag{6.98}$$

The nonzero stress components are obtained using Eqn. (6.83) as

$$\tau_{xz} = -\frac{G\alpha}{a}(3x+a)y, \quad \tau_{yz} = -\frac{G\alpha}{2a}\left[3x^2 - 2ax - 3y^2\right].$$

Note that $\tau_{xz} = \tau_{yz} = 0$ at the corners and at the centroid. Because of symmetry, it suffices to consider the side $x = -a/3$ for finding $|\boldsymbol{t}|_{\text{max}}$; we get $|\boldsymbol{t}|_{\text{max}} = G\alpha a/2$ at $y = 0$ along this side. Using Eqn. (6.89) and (6.98), we find the torsional rigidity to be $D = Ga^4/(15\sqrt{3}) = 3GJ/5$.

Keeping in view the cubic nature of the equation of the boundary, we try to find the warping function by guessing $W(z) \equiv \psi + ig$ to be of the form $i(cz^3 + k)$, where c and k are to be determined using the condition that $g = (x^2+y^2)/2$ on C; we get $c = 1/(2a)$ and $k = 2a^2/27$. Thus,

$$\begin{aligned}\psi &= -\frac{1}{2a}y(3x^2 - y^2),\\ g &= \frac{1}{2a}x(x^2 - 3y^2) + \frac{2}{27}a^2.\end{aligned}$$

As in the case of the elliptical cylinder, g could have been obtained directly from ϕ using Eqn. (6.80). Note that ψ (and hence the warping) is zero on the lines $y = 0$ and $y = \pm\sqrt{3}x$.

The solution for the case when the cross-section is an (arbitrary) isosceles triangle is unknown, although approximate solutions can be found [178, 303]. We now discuss the torsion of a bar whose cross-section is an isosceles right-angled triangle [177, 300] (see Fig. 6.5). Assume the solution to be of the form

$$\phi = G\alpha x(y-x) + \sum_{m=1}^{\infty} A_m \left[\sin\frac{m\pi x}{b} \sinh\frac{m\pi y}{b} - \sin\frac{m\pi y}{b} \sinh\frac{m\pi x}{b} \right].$$

Since the second part of the solution is harmonic, it is easily verified that the above solution satisfies the condition $\nabla^2\phi = -2G\alpha$. We also have $\phi|_{x=0} = \phi|_{y=x} = 0$. Thus, it only remains to determine the coefficients A_m using the boundary condition $\phi|_{y=b} = 0$, i.e.,

$$\sum_{m=1}^{\infty} A_m \sinh m\pi \sin\frac{m\pi x}{b} = -G\alpha x(b-x).$$

Multiplying the above equation by $\sin p\pi x/b$, integrating with respect to x over the interval from 0 to b, and using the orthogonality of the sine functions, we get

$$A_m = -\frac{4G\alpha b^2 \left[1-(-1)^m\right]}{m^3\pi^3 \sinh m\pi}.$$

The corresponding warping and conjugate functions are

$$\psi = \frac{1}{2}(x^2-y^2) + xy - \sum_{m=1}^{\infty} \frac{A_m}{G\alpha} \left[\cos\frac{m\pi x}{b} \cosh\frac{m\pi y}{b} + \cos\frac{m\pi y}{b} \cosh\frac{m\pi x}{b} \right],$$

$$g = xy + \frac{1}{2}(y^2-x^2) + \sum_{m=1}^{\infty} \frac{A_m}{G\alpha} \left[\sin\frac{m\pi x}{b} \sinh\frac{m\pi y}{b} - \sin\frac{m\pi y}{b} \sinh\frac{m\pi x}{b} \right].$$

The solution for the torsion of a cylinder whose cross-section is a 30°-60°-90° triangle is presented by Hay [116].

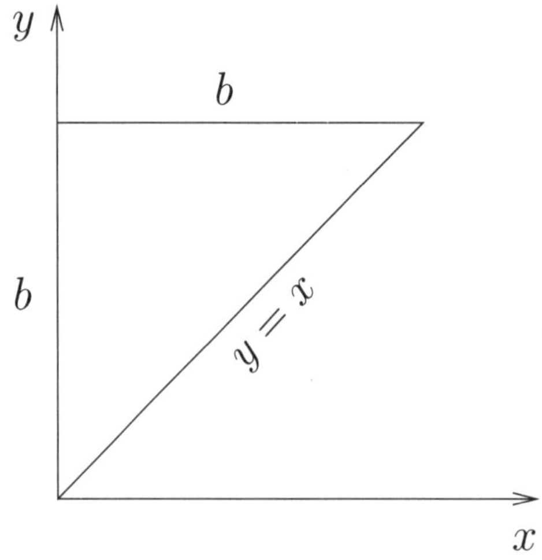

Fig. 6.5 *Torsion of a bar whose cross-section is an isosceles right-angled triangle.*

Circular cut-out in a circular cross-section

The effect of grooves on the maximum shearing stress can be gauged from this example. A groove of radius b is cut out from a shaft of radius a, as shown in Fig. 6.6. Working with

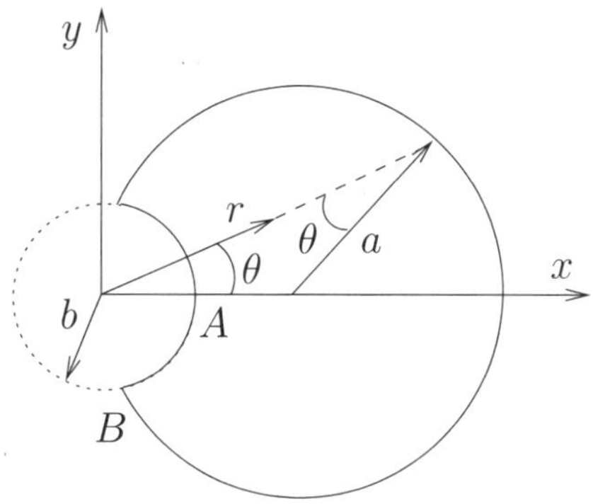

Fig. 6.6 *Groove in a circular shaft undergoing torsion.*

cylindrical coordinates, we try the expression $W(z) = i(az - ab^2/z + b^2/2)$ corresponding to which we get:

$$\psi = -ar\sin\theta - \frac{b^2 a}{r}\sin\theta,$$
$$g = ar\cos\theta\left(1 - \frac{b^2}{r^2}\right) + \frac{1}{2}b^2.$$

We now verify that the condition $g = (x^2+y^2)/2$ that is to be satisfied on C yields the equation of the boundary shown in Fig. 6.6. Since $x^2+y^2 = r^2$, we have on C,

$$a\cos\theta\left(r - \frac{b^2}{r}\right) + \frac{1}{2}b^2 = \frac{1}{2}r^2,$$
$$\implies (r^2 - b^2)\left(1 - \frac{2a\cos\theta}{r}\right) = 0.$$

Thus, the boundary is made of the two circles $r = b$ and $r = 2a\cos\theta$ $(-\pi/2 < \theta < \pi/2)$ as shown in Fig. 6.6. The nonzero stress components τ_{xz} and τ_{yz} are found using the relations $x = r\cos\theta$, $x^2+y^2 = r^2$ and Eqn. (6.79). The shearing stress $|\boldsymbol{t}|$ is zero at the points of intersection of the two circles (point B in Fig. 6.6) The maximum shearing stress occurs at point A, and is given by $|\boldsymbol{t}|_{\max} = G\alpha(2a - b)$, which is approximately equal to $2G\alpha a$ when $b \ll a$. Thus, the maximum shearing stress is almost doubled due to the introduction of a small groove on the periphery of a circular cylinder subjected to torsion, and for this reason it is important to consider stresses in slots and keyways of shafts.

Rectangular cross-section

In the case of a rectangular bar (see Fig. 6.7), we cannot use a multiple of the equation of the boundary $(x^2 - a^2/4)(y^2 - b^2/4)$ as a guess for the Prandtl stress function, since $\boldsymbol{\nabla}^2\phi$

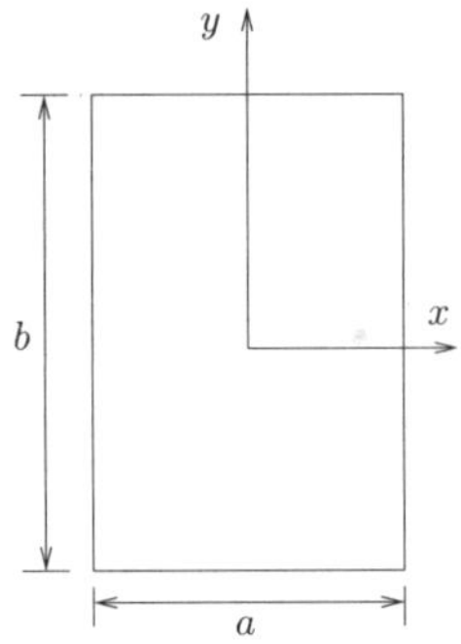

Fig. 6.7 *Rectangular bar undergoing torsion.*

does not evaluate to a constant. Instead, following a methodology similar to that followed by Timoshenko [324] (page 319), we try a double Fourier series of the following form:

$$\phi = G\alpha \sum_{m=1}^{\infty} \sum_{n=1}^{\infty} A_{mn} \cos \frac{(2m-1)\pi x}{a} \cos \frac{(2n-1)\pi y}{b}. \tag{6.99}$$

Note that the condition $\phi = 0$ is satisfied on the boundary $x = \pm a/2$ and $y = \pm b/2$. Substituting the above expression into the governing equation $-\boldsymbol{\nabla}^2 \phi = 2G\alpha$, we get

$$\sum_{m=1}^{\infty} \sum_{n=1}^{\infty} \left[\left(\frac{2m-1}{a} \right)^2 + \left(\frac{2n-1}{b} \right)^2 \right] \pi^2 A_{mn} \cos \frac{(2m-1)\pi x}{a} \cos \frac{(2n-1)\pi y}{b} = 2.$$

Multiplying both sides of the above equation by $\cos \frac{(2p-1)\pi x}{a} \cos \frac{(2q-1)\pi y}{b}$, integrating between the limits $[-\frac{a}{2}, \frac{a}{2}]$ for x and $[-\frac{b}{2}, \frac{b}{2}]$ for y, and using the orthogonality property of cosine functions, we get

$$A_{mn} = \frac{32}{(2m-1)(2n-1)\pi^4} \frac{(-1)^{m+n}}{\left[\left(\frac{2m-1}{a} \right)^2 + \left(\frac{2n-1}{b} \right)^2 \right]}. \tag{6.100}$$

Substituting this result into Eqn. (6.99), we get the required expression for ϕ. The nonzero stress components are obtained immediately using Eqn. (6.83) as

$$\tau_{xz} = -G\alpha \sum_{m=1}^{\infty} \sum_{n=1}^{\infty} A_{mn} \frac{(2n-1)\pi}{b} \cos \frac{(2m-1)\pi x}{a} \sin \frac{(2n-1)\pi y}{b},$$

$$\tau_{yz} = G\alpha \sum_{m=1}^{\infty} \sum_{n=1}^{\infty} A_{mn} \frac{(2m-1)\pi}{a} \sin \frac{(2m-1)\pi x}{a} \cos \frac{(2n-1)\pi y}{b},$$

with A_{mn} given by Eqn. (6.100). Since, from the above expressions, $\tau_{xz}|_{(\pm a/2,0)} = \tau_{yz}|_{(0,\pm b/2)} = 0$, the lateral surfaces are traction-free as required. Assuming that $b \geq a$, the maximum shearing stress occurs at $(\pm a/2, 0)$, and is given by

$$|\boldsymbol{t}|_{\max} = \tau_{yz}|_{(a/2,0)} = G\alpha \sum_{m=1}^{\infty} \sum_{n=1}^{\infty} \frac{(2m-1)\pi}{a} (-1)^{m+1} A_{mn}.$$

The torsional rigidity is obtained from Eqn. (6.89) as

$$D = 256Gab \sum_{m=1}^{\infty} \sum_{n=1}^{\infty} \frac{1}{(2m-1)^2(2n-1)^2\pi^6 \left[\left(\frac{2m-1}{a}\right)^2 + \left(\frac{2n-1}{b}\right)^2 \right]}.$$

With the aid of Eqns. (6.76) and (6.80), the warping and conjugate functions are obtained as

$$\begin{aligned}
\psi &= xy - \sum_{m=1}^{\infty} \sum_{n=1}^{\infty} A_{mn} \frac{(2n-1)/b}{(2m-1)/a} \sin \frac{(2m-1)\pi x}{a} \sin \frac{(2n-1)\pi y}{b} \\
&= -xy + \sum_{m=1}^{\infty} \sum_{n=1}^{\infty} A_{mn} \frac{(2m-1)/a}{(2n-1)/b} \sin \frac{(2m-1)\pi x}{a} \sin \frac{(2n-1)\pi y}{b}, \\
g &= \frac{1}{2}(x^2 + y^2) + \sum_{m=1}^{\infty} \sum_{n=1}^{\infty} A_{mn} \cos \frac{(2m-1)\pi x}{a} \cos \frac{(2n-1)\pi y}{b}.
\end{aligned}$$

Note that $\partial\psi/\partial x$ and $\partial\psi/\partial y$ are obtained by term-by-term differentiation of the first and second expressions for ψ, respectively.

An alternative approach to solving the above problem is to write the conjugate function as a sum of harmonic functions as follows:

$$g = c + k(y^2 - x^2) + \sum_{n=1}^{\infty} A_n \cos \alpha_n x \cosh \alpha_n y,$$

where $\alpha_n = (2n-1)\pi/a$. Writing the boundary condition given by Eqn. (6.78) on the edges $x = \pm a/2$, we get

$$\frac{1}{2}\left(\frac{a^2}{4} + y^2\right) = c + k\left(y^2 - \frac{a^2}{4}\right),$$

which leads to $c = a^2/4$ and $k = 1/2$. Thus, we get

$$g = \frac{a^2}{4} + \frac{1}{2}(y^2 - x^2) + \sum_{n=1}^{\infty} A_n \cos \alpha_n x \cosh \alpha_n y. \tag{6.101}$$

Now considering the boundary conditions on $y = \pm b/2$, we get

$$x^2 - \frac{a^2}{4} = \sum_{n=1}^{\infty} A_n \cos \alpha_n x \cosh \frac{\alpha_n b}{2}.$$

Using the orthogonality of the cosine functions, we obtain

$$A_n = \frac{8(-1)^n a^2}{[(2n-1)\pi]^3 \cosh \frac{\alpha_n b}{2}}.$$

One can interchange x and y, and a and b in Eqn. (6.101) to get an alternative expression. The warping and Prandtl functions corresponding to the conjugate function in Eqn. (6.101), and the corresponding stresses are

$$\psi = xy + \sum_{n=1}^{\infty} A_n \sin \alpha_n x \sinh \alpha_n y,$$

$$\phi = G\alpha \left[\frac{a^2}{4} - x^2 + \sum_{n=1}^{\infty} A_n \cos \alpha_n x \cosh \alpha_n y \right],$$

$$\tau_{xz} = G\alpha \sum_{n=1}^{\infty} A_n \alpha_n \cos \alpha_n x \sinh \alpha_n y,$$

$$\tau_{yz} = G\alpha \left[2x + \sum_{n=1}^{\infty} A_n \alpha_n \sin \alpha_n x \cosh \alpha_n y \right],$$

$$|\boldsymbol{t}|_{\max} = \tau_{yz}\big|_{(a/2,0)} = G\alpha \left[a + \sum_{n=1}^{\infty} A_n \alpha_n \sin \frac{\alpha_n a}{2} \right].$$

The torsional rigidity obtained by using Eqn. (6.89) is given by

$$D = G \left[\frac{a^3 b}{3} - \frac{64}{a} \sum_{n=1}^{\infty} \frac{1}{\alpha_n^5} \tanh \frac{\alpha_n b}{2} \right]. \tag{6.102}$$

The torsion of curved beams of circular and rectangular cross-section has been discussed in [84] and [183], respectively.

Annular Sector

An annular sector, bounded by the arcs $r = a$ and $r = b$ and by the lines $\theta = \pm\theta_0$ (see Fig. 6.8), is subjected to torsion. We assume the Prandtl stress function to be of the form

$$\phi = -\frac{G\alpha}{2} r^2 \left(1 - \frac{\cos 2\theta}{\cos 2\theta_0} \right) - G\alpha \sum_{m=1}^{\infty} \left[A_m r^{\gamma_m} + B_m r^{-\gamma_m} \right] \cos \gamma_m \theta,$$

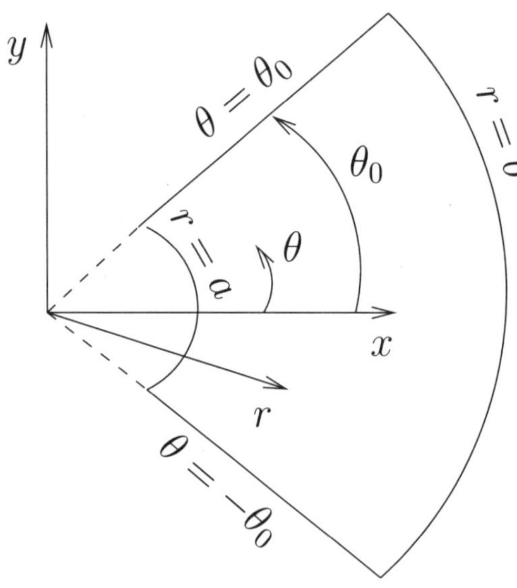

Fig. 6.8 *An annular sector subjected to torsion.*

where $\gamma_m = (2m-1)\pi/(2\theta_0)$. Note that the second term of the first part, and the second part of the above solution are harmonic, while the first term of the first part $\tilde{\phi} = -G\alpha r^2/2$ satisfies $\boldsymbol{\nabla}^2\tilde{\phi} = -2G\alpha$, thus leading to a satisfaction of the governing equation $\boldsymbol{\nabla}^2\phi = -2G\alpha$. The boundary conditions $\phi|_{\theta=\pm\theta_0} = 0$ are also satisfied by the above form. Thus, all that remains is to find the constants A_m and B_m so that the boundary conditions $\phi|_{r=a} = \phi|_{r=b} = 0$ are satisfied, i.e., A_m and B_m are to be determined subject to

$$\sum_{m=1}^{\infty}\left[A_m a^{\gamma_m} + B_m a^{-\gamma_m}\right]\cos\gamma_m\theta = -\frac{a^2}{2}\left(1-\frac{\cos 2\theta}{\cos 2\theta_0}\right),$$
$$\sum_{m=1}^{\infty}\left[A_m b^{\gamma_m} + B_m b^{-\gamma_m}\right]\cos\gamma_m\theta = -\frac{b^2}{2}\left(1-\frac{\cos 2\theta}{\cos 2\theta_0}\right).$$

Multiplying each of the above equations by $\cos\gamma_n\theta$, integrating between the limits $\theta = \pm\theta_0$, and using the orthogonality of the cosine functions, we get $A_m = \tilde{A}_m F_m$, $B_m = \tilde{B}_m F_m$, where

$$\tilde{A}_m = \left[\frac{b^{2+\gamma_m} - a^{2+\gamma_m}}{b^{2\gamma_m} - a^{2\gamma_m}}\right],$$
$$\tilde{B}_m = \left[\frac{b^{\gamma_m}a^2 - a^{\gamma_m}b^2}{b^{2\gamma_m} - a^{2\gamma_m}}\right](ab)^{\gamma_m},$$
$$F_m = \frac{-4(-1)^m}{\theta_0\gamma_m(\gamma_m^2 - 4)}.$$

Note that ϕ can also be written as

$$\phi = G\alpha\sum_{m=1}^{\infty}\left[r^2 - \tilde{A}_m r^{\gamma_m} - \tilde{B}_m r^{-\gamma_m}\right]F_m\cos\gamma_m\theta,$$

from which it follows that the torsional rigidity is

$$D = \frac{T}{\alpha} = 4G\sum_{m=1}^{\infty}\left[\frac{b^4-a^4}{4} - \tilde{A}_m\frac{b^{2+\gamma_m} - a^{2+\gamma_m}}{2+\gamma_m} - \tilde{B}_m\frac{b^{2-\gamma_m} - a^{2-\gamma_m}}{2-\gamma_m}\right]F_m\frac{\sin\gamma_m\theta_0}{\gamma_m}. \tag{6.103}$$

Using Eqn. (6.80), and the fact that g and ψ are conjugate harmonic functions, we get

$$\begin{aligned}\psi &= -\frac{r^2}{2}\frac{\sin 2\theta}{\cos 2\theta_0} + \sum_{m=1}^{\infty}\left[A_m r^{\gamma_m} - B_m r^{-\gamma_m}\right]\sin\gamma_m\theta,\\ g &= \frac{r^2}{2}\frac{\cos 2\theta}{\cos 2\theta_0} - \sum_{m=1}^{\infty}\left[A_m r^{\gamma_m} + B_m r^{-\gamma_m}\right]\cos\gamma_m\theta.\end{aligned} \tag{6.104}$$

Note that the warping is zero on the x-axis. Using Eqns. (6.85), we get

$$\frac{\tau_{rz}}{G\alpha} = -\frac{\sin 2\theta}{\cos 2\theta_0}r + \sum_{m=1}^{\infty}\left[A_m r^{\gamma_m - 1} + B_m r^{-\gamma_m - 1}\right]\gamma_m\sin\gamma_m\theta$$

$$= -\sum_{m=1}^{\infty}\left[r - \tilde{A}_m r^{\gamma_m - 1} - \tilde{B}_m r^{-\gamma_m - 1}\right] F_m \gamma_m \sin\gamma_m\theta,$$

$$\frac{\tau_{\theta z}}{G\alpha} = r\left(1 - \frac{\cos 2\theta}{\cos 2\theta_0}\right) + \sum_{m=1}^{\infty}\left[A_m r^{\gamma_m - 1} - B_m r^{-\gamma_m - 1}\right]\gamma_m \cos\gamma_m\theta$$

$$= -\sum_{m=1}^{\infty}\left[2r - \tilde{A}_m \gamma_m r^{\gamma_m - 1} + \tilde{B}_m \gamma_m r^{-\gamma_m - 1}\right] F_m \cos\gamma_m\theta.$$

The corresponding expressions for the sector of a circle are obtained simply by setting $a = 0$ in the above expressions. For example, from Eqn. (6.103), we get the torsional rigidity for the sector of a circle as

$$D = \frac{64Gb^4\theta_0^3}{\pi^4}\sum_{m=1}^{\infty}\frac{1}{(2m-1)^2\left(2m - 1 + \frac{4\theta_0}{\pi}\right)^2}. \tag{6.105}$$

For a semicircular cross-section, by setting $\theta_0 = \pi/2$, and on using

$$\sum_{m=1}^{\infty}\frac{16}{(4m^2-1)^2} = \pi^2 - 8,$$

we get $D = Gb^4(\pi^2 - 8)/(2\pi)$. For a cracked cylinder of radius b, with the crack extending radially outward from the center to the periphery (i.e., $\theta_0 = \pi$, $a = 0$, $\tilde{B}_m = 0\ \forall m$, $\tilde{A}_m = b^{2-\gamma_m}$, $\gamma_m = (2m-1)/2$), one gets a $1/\sqrt{r}$ singularity as $r \to 0$ in the stress solution. The torsional rigidity as obtained from Eqn. (6.105) on using

$$\sum_{m=1}^{\infty}\frac{576}{(2m-1)^2(2m+3)^2} = 9\pi^2 - 64,$$

is $(9\pi^2 - 64)Gb^4/(9\pi) \approx 0.5589876GJ$, where $J = \pi b^4/2$. Thus, the torsional rigidity is reduced by about 44% due to the presence of the crack.

The solution given by Eqn. (6.104) is not valid when $\theta_0 = \pi/4$ and $\theta_0 = 3\pi/4$. When $\theta_0 = \pi/4$, by considering the real and imaginary parts of $z^2 \log z$ (see Eqn. (6.238)), we get

$$\phi = -\frac{G\alpha r^2}{2}\left[1 + \frac{4}{\pi}(\log r\cos 2\theta - \theta\sin 2\theta)\right] - G\alpha\left(A_0 r^2 + \frac{B_0}{r^2}\right)\cos 2\theta$$

$$- G\alpha\sum_{m=1}^{\infty}\left[A_m r^{2(2m+1)} + B_m r^{-2(2m+1)}\right]\cos 2(2m+1)\theta,$$

$$\psi = \frac{2r^2}{\pi}(\log r\sin 2\theta + \theta\cos 2\theta) + \left(A_0 r^2 - \frac{B_0}{r^2}\right)\sin 2\theta$$

$$+ \sum_{m=1}^{\infty}\left[A_m r^{2(2m+1)} - B_m r^{-2(2m+1)}\right]\sin 2(2m+1)\theta,$$

$$g = -\frac{2r^2}{\pi}(\log r\cos 2\theta - \theta\sin 2\theta) - \left(A_0 r^2 + \frac{B_0}{r^2}\right)\cos 2\theta$$

$$- \sum_{m=1}^{\infty}\left[A_m r^{2(2m+1)} + B_m r^{-2(2m+1)}\right]\cos 2(2m+1)\theta.$$

By considering the orthogonality of the cosine functions, we once again get $A_m = \tilde{A}_m F_m$, $B_m = \tilde{B}_m F_m$, where

$$A_0 = -\frac{3b^4 - 3a^4 + 4b^4 \log b - 4a^4 \log a}{2\pi(b^4 - a^4)},$$

$$B_0 = \frac{2a^4 b^4 \log \frac{b}{a}}{\pi(b^4 - a^4)},$$

$$\tilde{A}_m = \left[\frac{b^{2+\gamma_m} - a^{2+\gamma_m}}{b^{2\gamma_m} - a^{2\gamma_m}}\right],$$

$$\tilde{B}_m = \left[\frac{b^{\gamma_m} a^2 - a^{\gamma_m} b^2}{b^{2\gamma_m} - a^{2\gamma_m}}\right](ab)^{\gamma_m},$$

$$F_m = \frac{(-1)^m}{2\pi m(m+1)(2m+1)},$$

where $\gamma_m = 2(2m+1)$. The torsional rigidity is

$$D = \frac{G}{8\pi(b^4 - a^4)}\left[(12 - \pi^2)(b^4 - a^4)^2 - 32a^4 b^4 \left(\log \frac{b}{a}\right)^2\right] - \frac{G}{2}\sum_{m=1}^{\infty}(-1)^m \left[\frac{b^{4m+4} - a^{4m+4}}{(m+1)(2m+1)} A_m + \frac{a^{-4m} - b^{-4m}}{m(2m+1)} B_m\right],$$

which for $a = 0$, and on using

$$\sum_{m=1}^{\infty} \frac{6}{m(m+1)^2(2m+1)^2} = 36 - 7\pi^2 + 24 \log 4,$$

yields $Gb^4(\pi/6 - \log 4/\pi)$.

For $\theta_0 = 3\pi/4$, we get

$$\phi = -\frac{G\alpha r^2}{2}\left[1 - \frac{4}{3\pi}(\log r \cos 2\theta - \theta \sin 2\theta)\right] - G\alpha\left(A_2 r^2 + \frac{B_2}{r^2}\right)\cos 2\theta - G\alpha \sum_{\substack{m=1\\ m\neq 2}}^{\infty}\left[A_m r^{\frac{2(2m-1)}{3}} + B_m r^{-\frac{2(2m-1)}{3}}\right]\cos \frac{2(2m-1)\theta}{3},$$

$$\psi = -\frac{2r^2}{3\pi}(\log r \sin 2\theta + \theta \cos 2\theta) + \left(A_2 r^2 - \frac{B_2}{r^2}\right)\sin 2\theta + \sum_{\substack{m=1\\ m\neq 2}}^{\infty}\left[A_m r^{\frac{2(2m-1)}{3}} - B_m r^{-\frac{2(2m-1)}{3}}\right]\sin \frac{2(2m-1)\theta}{3},$$

$$g = \frac{2r^2}{3\pi}(\log r \cos 2\theta - \theta \sin 2\theta) - \left(A_2 r^2 + \frac{B_2}{r^2}\right)\cos 2\theta - \sum_{\substack{m=1\\ m\neq 2}}^{\infty}\left[A_m r^{\frac{2(2m-1)}{3}} + B_m r^{-\frac{2(2m-1)}{3}}\right]\cos \frac{2(2m-1)\theta}{3}.$$

where $A_m = \tilde{A}_m F_m$, $B_m = \tilde{B}_m F_m$ $(m \neq 2)$, with

$$A_2 = \frac{3b^4 - 3a^4 + 4b^4 \log b - 4a^4 \log a}{6\pi(b^4 - a^4)},$$

$$B_2 = -\frac{2a^4 b^4 \log \frac{b}{a}}{3\pi(b^4 - a^4)},$$

$$\tilde{A}_m = \frac{a^{4/3} b^{4/3} \left[a^{4m/3} a^{4/3} - b^{4/3} b^{4m/3}\right]}{a^{4/3} b^{8m/3} - a^{8m/3} b^{4/3}},$$

$$\tilde{B}_m = \frac{a^{4m/3} b^{4m/3} \left[a^{4m/3} b^{8/3} - a^{8/3} b^{4m/3}\right]}{a^{4/3} b^{8m/3} - a^{8m/3} b^{4/3}},$$

$$F_m = \frac{9(-1)^m}{2(m-2)(m+1)(2m-1)\pi}.$$

The torsional rigidity is

$$D = \frac{G}{24\pi(b^4 - a^4)} \left[3(b^4 - a^4)^2(4 - 3\pi^2) - 32a^4 b^4 \left(\log \frac{b}{a}\right)^2\right]$$
$$+ \frac{9G}{2} \sum_{\substack{m=1 \\ m \neq 2}}^{\infty} (-1)^m \left[\frac{b^{4/3} b^{4m/3} - a^{4/3} a^{4m/3}}{(m+1)(2m-1)} A_m - \frac{b^{8/3} b^{-4m/3} - a^{8/3} a^{-4m/3}}{(m-2)(2m-1)} B_m\right],$$

which for $a = 0$, and on using

$$\sum_{\substack{m=1 \\ m \neq 2}}^{\infty} \frac{162}{(m-2)(m+1)^2(2m-1)^2} = 18 - 7\pi^2 + 8 \log 4,$$

yields $Gb^4(2\pi^2 - 4\log 4 - 7)/(4\pi)$. Note that in the case when $a = 0$, there is a $r^{-1/3}$ singularity in the stress field as $r \to 0$. Quite interestingly, the formulae for torsional rigidity for the two cases $\theta_0 = \pi/4$ and $3\pi/4$ discussed above also follow from Eqn. (6.105) although the expression for ϕ used for deriving this relation is not valid for these two cases.

For alternative solutions, see [203] (page 319), [301] and [324] (page 313).

Cardioid section

Now consider the torsion of a beam whose cross-section is a cardioid as shown in Fig. 6.9. The equation of the boundary is given by $r = a(1 - \cos\theta)$. Using the fact that $\sin(\theta/2) \geq 0$ for $\theta \in [0, 2\pi]$, so that $\sqrt{1 - \cos\theta} = \sqrt{2}\sin(\theta/2)$, the stress function is given by [289]:

$$\phi = \frac{G\alpha}{2}\left[a^{3/2}\sqrt{2r}\sin\frac{\theta}{2} - ar\cos\theta - r^2\right], \quad \theta \in [0, 2\pi].$$

The warping function and the corresponding stresses are

$$\psi = a^{3/2}\sqrt{\frac{r}{2}}\cos\frac{\theta}{2} + \frac{a}{2} r \sin\theta,$$

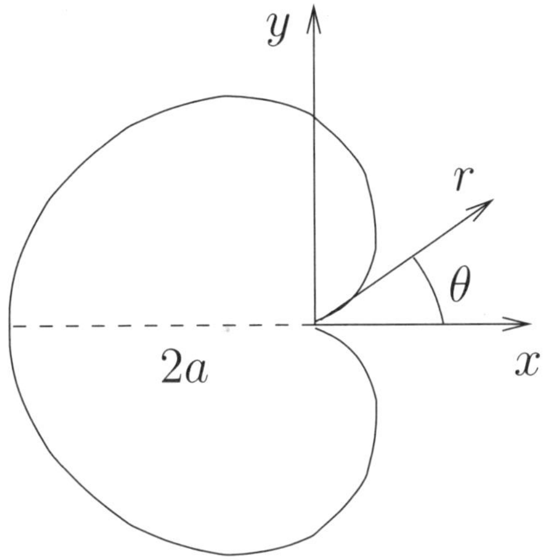

Fig. 6.9 *Torsion of a beam whose cross-section is a cardioid.*

$$\tau_{xz} = G\alpha \left[\frac{a^{3/2}}{2\sqrt{2r}} \cos\frac{\theta}{2} - r\sin\theta \right],$$

$$\tau_{yz} = G\alpha \left[\frac{a^{3/2}}{2\sqrt{2r}} \sin\frac{\theta}{2} + \frac{a}{2} + r\cos\theta \right].$$

The torsional rigidity is given by

$$D = \int_0^{2\pi} \int_0^{a(1-\cos\theta)} 2\phi r\, dr\, d\theta = \frac{17}{16}\pi G a^4.$$

For a comprehensive review of solutions to the Saint-Venant torsion problem, see [118].

Hollow Beams

Polya and Weinstein [255, 256] prove the following result: of all multiply connected cross-sections with given area and with given joint area of holes, the ring bounded by two concentric circles has the maximum torsional rigidity. Thus, if πa^2 is the joint area of all the holes, and $\pi(b^2 - a^2)$ is the area of the cross-section, then

$$D \leq \frac{G\pi}{2}(b^4 - a^4).$$

The special case of a simply connected domain is obtained simply by setting $a = 0$. As an example, if we consider the bar to have a square cross-section of side q with a concentric square hole of side p, then $p^2 = \pi a^2$, $q^2 = \pi b^2$, and

$$D \leq \frac{G\pi}{2}(b^4 - a^4) = \frac{G}{2\pi}(q^4 - p^4).$$

This result is quite difficult to prove since for a given area, the polar moment of inertia of a circular cross-section is minimum, so that obviously the warping function of "all possible cross-sections" plays a key role in the proof. We shall now provide a "partial" proof of the above result. Maximizing the torsional rigidity subject to the cross-section having a constant area can, in principle, be achieved by extremizing the functional

$$
\begin{aligned}
I &= \int_A \left\{ G\left[x^2 + y^2 + x\frac{\partial\psi}{\partial y} - y\frac{\partial\psi}{\partial x}\right] - \lambda \right\} dA \\
&= G\int_A \boldsymbol{\nabla}\cdot \begin{bmatrix} \dfrac{x^3}{3} - \psi y - \dfrac{\lambda x}{2G} \\ \dfrac{y^3}{3} + \psi x - \dfrac{\lambda y}{2G} \end{bmatrix} dA \\
&= G\int_C \left\{ \left[\frac{x^3}{3} - \psi y - \frac{\lambda x}{2G}\right]\frac{dy}{ds} - \left[\frac{y^3}{3} + \psi x - \frac{\lambda y}{2G}\right]\frac{dx}{ds} \right\} ds.
\end{aligned}
$$

Setting the first variation, $\delta^{(1)} I$, to zero, using the fact that $(x(s), y(s))$ are single-valued functions, and invoking the arbitrariness of $(\delta x, \delta y)$, we get the Euler–Lagrange equations

$$
\begin{aligned}
\left[x^2 + y^2 + x\frac{\partial\psi}{\partial y} - y\frac{\partial\psi}{\partial x} - \frac{\lambda}{G}\right]\frac{dx}{ds} &= 0, \\
\left[x^2 + y^2 + x\frac{\partial\psi}{\partial y} - y\frac{\partial\psi}{\partial x} - \frac{\lambda}{G}\right]\frac{dy}{ds} &= 0.
\end{aligned}
$$

Since, both dx/ds and dy/ds are not simultaneously zero on any part of the boundary (these being the components of the normal $\boldsymbol{n}$), we get the condition

$$
x^2 + y^2 + x\frac{\partial\psi}{\partial y} - y\frac{\partial\psi}{\partial x} = \frac{\lambda}{G} \quad \text{on } C. \tag{6.106}
$$

Thus, the extremizing condition is that $x\tau_{yz} - y\tau_{xz} = \tau\left(x\frac{dy}{ds} - y\frac{dx}{ds}\right) = (\boldsymbol{x} \times \boldsymbol{t})_z = \text{constant}$ on the boundary. The warping function for an elliptical cross-section given by Eqn. (6.96) satisfies the above equation, so that the family of ellipses, excepting the extreme case of a degenerate ellipse, is at least a "saddle point". Assuming the semi-major dimension a to be greater than the semi-minor dimension b, the ratio of the torsional rigidity of an ellipse to that of a circle for a given area $A = \pi ab$ is obtained from Eqn. (6.95) as

$$
\frac{D_{\text{ellipse}}}{D_{\text{circle}}} = \frac{2\pi A a^2}{\pi^2 a^4 + A^2},
$$

which is always less than or equal to 1, with equality achieved when the ellipse is a circle ($a = b$). On the other hand, when $a \to \infty$, then the above ratio tends to zero, which is the minimum torsional rigidity possible. Thus, what we have shown is that the minimum possible torsional rigidity is achieved by a "degenerate" ellipse, and that a circular cross-section is a *potential maximizer*. To complete the proof, one has to show that either the warping function of the ellipse is the only one that satisfies Eqn. (6.106), or if it is not, that other warping functions that satisfy Eqn. (6.106) generate a torsional rigidity that does not exceed that of a circle. This is obviously something quite difficult to prove, since all possible cross-sections have to be considered.

6.5.3 Generalization of the Saint-Venant torsion theory to an anisotropic, inhomogeneous bar

Consider the torsion of a prismatic, anisotropic (with at least one symmetry plane normal to the z-axis), inhomogeneous bar whose lateral surfaces are traction-free. The displacement field is still assumed to be given by Eqn. (6.61), while the nonzero stresses are given by [49, 50, 133]

$$\boldsymbol{\tau}_c = \alpha \boldsymbol{C}_c \left[\boldsymbol{\nabla} \psi - \boldsymbol{x}_\perp \right], \tag{6.107}$$

where $\boldsymbol{\nabla}$ is the in-plane gradient operator, and

$$\boldsymbol{\tau}_c = \begin{bmatrix} \tau_{xz} \\ \tau_{yz} \end{bmatrix}, \qquad \boldsymbol{x} = \begin{bmatrix} x \\ y \end{bmatrix}$$

$$\boldsymbol{C}_c = \begin{bmatrix} \mu_{55}(\boldsymbol{x}) & \mu_{45}(\boldsymbol{x}) \\ \mu_{45}(\boldsymbol{x}) & \mu_{44}(\boldsymbol{x}) \end{bmatrix}, \qquad \boldsymbol{x}_\perp = \boldsymbol{W}\boldsymbol{x} = \begin{bmatrix} y \\ -x \end{bmatrix},$$

$$\boldsymbol{W} = \begin{bmatrix} 0 & 1 \\ -1 & 0 \end{bmatrix}.$$

Proofs similar to those in the previous subsection can be given to show that the total resultant force, (F_x, F_y, F_z), on the top and bottom cross-sections is zero. $\boldsymbol{C}_c$ is assumed to be positive definite. For an orthotropic material, we have $\mu_{45} = 0$, while for an isotropic material, we have $\mu_{45} = 0$ and $\mu_{44}(\boldsymbol{x}) = \mu_{55}(\boldsymbol{x})$.

The governing equation on the domain A, and boundary condition on the lateral surface C are given by

$$\boldsymbol{\nabla} \cdot \boldsymbol{\tau}_c = 0 \quad \text{on } A, \tag{6.108a}$$

$$\boldsymbol{\tau}_c \cdot \boldsymbol{n} = \tau_{xz} n_x + \tau_{yz} n_y = 0 \quad \text{on } C. \tag{6.108b}$$

In terms of the warping function, Eqn. (6.108a) reduces to

$$\boldsymbol{\nabla} \cdot (\boldsymbol{C}_c \boldsymbol{\nabla} \psi - \boldsymbol{C}_c \boldsymbol{x}_\perp) = 0, \tag{6.109}$$

which for a homogeneous material reduces to

$$\boldsymbol{\nabla} \cdot (\boldsymbol{C}_c \boldsymbol{\nabla} \psi) = 0. \tag{6.110}$$

Using the relation

$$T = -\int_A \boldsymbol{x}_\perp \cdot \boldsymbol{\tau}_c \, dA = \alpha \int_A \boldsymbol{x}_\perp \cdot \boldsymbol{C}_c \left[\boldsymbol{x}_\perp - \boldsymbol{\nabla} \psi \right], \tag{6.111}$$

one gets the relation for the torsional rigidity as

$$\begin{aligned} D &= \int_A \boldsymbol{x}_\perp \cdot \boldsymbol{C}_c \left[\boldsymbol{x}_\perp - \boldsymbol{\nabla} \psi \right] dA \\ &= \int_A \left[\boldsymbol{x}_\perp \cdot (\boldsymbol{C}_c \boldsymbol{x}_\perp) - \boldsymbol{\nabla} \psi \cdot (\boldsymbol{C}_c \boldsymbol{\nabla} \psi) \right] dA. \end{aligned} \tag{6.112}$$

where the second relation has been derived in a manner analogous to Eqn. (6.71). For a numerical implementation of the above equations, see [166].

Similar to the isotropic case, a stress function ϕ can be introduced such that it automatically satisfies the equation of equilibrium:

$$\boldsymbol{\tau}_c = \boldsymbol{W}\boldsymbol{\nabla}\phi. \tag{6.113}$$

Comparing Eqns. (6.107) and (6.113), we have

$$\alpha \boldsymbol{C}_c \left[\boldsymbol{\nabla}\psi - \boldsymbol{x}_\perp\right] = \boldsymbol{W}\boldsymbol{\nabla}\phi. \tag{6.114}$$

Eliminating ψ from the above equation, we get

$$\boldsymbol{\nabla} \cdot \left(\frac{\boldsymbol{C}_c \boldsymbol{\nabla}\phi}{\det \boldsymbol{C}_c}\right) = -2\alpha. \tag{6.115}$$

As in the isotropic case, the boundary condition is ϕ = constant. As usual, we take $\phi = 0$ on the outer boundary. Equations (6.89) and (6.90) remain valid for simply and multiply connected bars.

If the material is isotropic but inhomogeneous, then $\boldsymbol{C}_c = G(x,y)\boldsymbol{I}$. A varying G can occur either in the case of functionally graded materials or in the case of compound bars that have inclusions of different materials in a matrix (this includes the limiting case of the inclusions being holes when the shear modulus of the inclusions tends to zero). The governing equation given by Eqn. (6.109) reduces to

$$\boldsymbol{\nabla} \cdot \left[G(\boldsymbol{\nabla}\psi - \boldsymbol{x}_\perp)\right] = 0 \text{ in } A, \tag{6.116}$$

The boundary condition either at the lateral surface or on the surface of internal holes (which are also traction-free) is obtained from Eqn. (6.65) as

$$G(\boldsymbol{\nabla}\psi \cdot \boldsymbol{n} - y n_x + x n_y) = 0 \quad \text{on } C_i,\ i = 1, 2, \ldots, N. \tag{6.117}$$

Consider the case of a compound bar that has a material interface separating two regions with different shear moduli G_1 and G_2. At the interface between two materials, the displacements and tractions have to be continuous. This leads to the interface conditions

$$\psi_1 = \psi_2, \tag{6.118a}$$
$$\left[G(\boldsymbol{\nabla}\psi \cdot \boldsymbol{n} - y n_x + x n_y)\right]_1 = \left[G(\boldsymbol{\nabla}\psi \cdot \boldsymbol{n} - y n_x + x n_y)\right]_2. \tag{6.118b}$$

Note that Eqn. (6.118b) reduces to Eqn. (6.117) in case the inclusion, which is, say, material 2, is taken to be a hole ($G_2 \to 0$). The governing equation given by Eqn. (6.116) is to be solved subject to the boundary conditions given by Eqn. (6.118).

Using Eqn. (6.111), we get

$$T = \alpha \int_A G\left[(x^2 + y^2) - \boldsymbol{\nabla}\psi \cdot \boldsymbol{x}_\perp\right] dA,$$

which is the governing equation for α.

An alternative formulation can also be given in terms of a stress function ϕ. If the stresses are given by Eqn. (6.83), then the equilibrium equation given by Eqn. (6.64) is automatically satisfied. Eliminating ψ from Eqns. (6.63) and (6.83), we get the governing equation for ϕ as

$$\nabla \cdot \left(\frac{1}{G}\nabla\phi\right) = -2\alpha. \tag{6.119}$$

On the outer lateral surface or at the surface of holes C_i, $i = 1, 2, \ldots, N$, the traction-free condition $\tau_{xz}n_x + \tau_{yz}n_y = 0$, where $(n_x, n_y) = (dy/ds, -dx/ds)$, leads to the condition

$$\phi = \text{constant on } C_i, \; i = 1, 2, \ldots, N. \tag{6.120}$$

At a material interface between two regions with shear moduli G_1 and G_2, we get the continuity conditions as

$$\phi_1 = \phi_2, \tag{6.121a}$$

$$\frac{1}{G_1}(\nabla\phi \cdot \boldsymbol{n})_1 = \frac{1}{G_2}(\nabla\phi \cdot \boldsymbol{n})_2, \tag{6.121b}$$

In the limit as $G_2 \to 0$, we get $(\nabla^2\phi)_2 = 0$ and $(\nabla\phi \cdot \boldsymbol{n})_2 = 0$, which leads to the fact that $\phi_2 =$ constant. By continuity of ϕ, we get $\phi_1 =$ constant at a traction-free surface, which is the same as the condition given by Eqn. (6.120). Conditions (6.121) are obtained as follows. The continuity of tractions at the interface is given by

$$[\tau_{xz}n_x + \tau_{yz}n_y]_1 = [\tau_{xz}n_x + \tau_{yz}n_y]_2.$$

Substituting for τ_{xz}, τ_{yz} and for (n_x, n_y), we get $(d\phi/ds)_1 = (d\phi/ds)_2$, which leads to $\phi_1 = \phi_2 +$ constant. Since the stresses are unaffected by the constant, it is taken to be zero, leading to Eqn. (6.121a). Since $\psi_1 = \psi_2$ across the entire interface, we have $(d\psi/ds)_1 = (d\psi/ds)_2$, which can be written as

$$\left[-\frac{\partial\psi}{\partial x}n_y + \frac{\partial\psi}{\partial y}n_x\right]_1 = \left[-\frac{\partial\psi}{\partial x}n_y + \frac{\partial\psi}{\partial y}n_x\right]_2.$$

Substituting for ψ in terms of ϕ, we get Eqn. (6.121b).

An alternative condition to Eqn. (6.121b) can be stated in the limit $G_2 \to 0$. If C_i represents the interface with $\boldsymbol{n}$ as the outward normal to region 1, and A represents the domain of the hole whose total area is A_2 (i.e., $\int_A dA = A_2$), then

$$\begin{aligned}
\int_{C_i} \frac{1}{G_1}(\nabla\phi \cdot \boldsymbol{n})_1 \, ds &= \int_{C_i} \frac{1}{G_2}(\nabla\phi \cdot \boldsymbol{n})_2 \, ds \\
&= -\int_{C_i} \frac{1}{G_2}(\nabla\phi \cdot \tilde{\boldsymbol{n}})_2 \, ds \quad (\tilde{\boldsymbol{n}}_2 = -\boldsymbol{n}_2) \\
&= -\int_{A_2} \nabla \cdot \left(\frac{1}{G_2}\nabla\phi\right) dA \\
&= 2\alpha A_2.
\end{aligned} \tag{6.122}$$

Note, however, that this is not a pointwise condition.

A numerical implementation based on ϕ, especially for multiply connected regions, would be cumbersome because of the boundary condition given by Eqn. (6.120). Although

ϕ is a constant at the boundary of each hole, we do not *a-priori* know the value of these various constants. Even the condition given by Eqn. (6.122) is not appropriate since it is not a pointwise condition. Thus, a numerical implementation based on the warping function is much more convenient, since it can routinely handle either compound bars or multiply connected regions.

Isotropic composite beams

As discussed above, in a composite beam, the displacements and tractions have to be continuous across the interface of the dissimilar materials. For example, in a composite beam made up of two materials, if ψ_1 and ψ_2 denote the warping functions, and ϕ_1 and ϕ_2 denote the stress functions in the two regions, then displacement and traction continuity are ensured by having $\psi_1 = \psi_2$ and $\phi_1 = \phi_2$ at the interface. Analogous to Eqn. (6.89), we have

$$T = 2\int_{A_1} \phi_1 \, dA + 2\int_{A_2} \phi_2 \, dA_2. \tag{6.123}$$

We now consider two cases of an example of an elliptical bimaterial composite where, in the first case, the interface between the two materials coincides with a line of shearing stress of the homogeneous isotropic solid beam, so that the same ϕ/G function that is used in the single-material problem can be used, while in the second case, it does not.

Consider a composite beam whose outer boundary and the interface are similar ellipses, i.e., $a_1/a_2 = b_1/b_2 = q$ (see Fig. 6.10). Using the results for a solid elliptical beam, the stress function is taken to be

$$\phi = \begin{cases} \dfrac{G_1 \alpha a_2^2 b_2^2}{a_2^2 + b_2^2}\left(1 - \dfrac{x^2}{a_2^2} - \dfrac{y^2}{b_2^2}\right) + \dfrac{(G_2 - G_1)\alpha a_2^2 b_2^2}{a_2^2 + b_2^2}\left(1 - q^2\right), & \text{in domain 1} \\ \dfrac{G_2 \alpha a_2^2 b_2^2}{a_2^2 + b_2^2}\left(1 - \dfrac{x^2}{a_2^2} - \dfrac{y^2}{b_2^2}\right), & \text{in domain 2.} \end{cases}$$

Note that the above stress function satisfies $\nabla^2\phi = -2G_i\alpha$, $i = 1, 2$, the boundary condition $\phi = 0$ on the outer boundary C_0, and that the second term in the expression for ϕ in domain

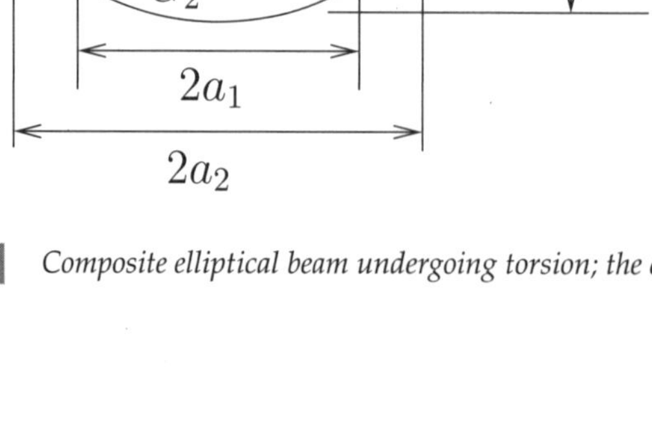

Fig. 6.10 *Composite elliptical beam undergoing torsion; the ellipses are similar.*

1 has been added to ensure continuity of ϕ at the interface. The warping function is the same as that for the solid elliptical beam, namely,

$$\psi_1 = \psi_2 = -\frac{(a_2^2 - b_2^2)xy}{a_2^2 + b_2^2}.$$

Using Eqn. (6.123), the torque is given by

$$\begin{aligned} T &= 2\int_{A_1} \phi\, dA + 2\int_{A_2} \phi\, dA \\ &= 2\int_0^{2\pi}\int_0^q \frac{G_1 \alpha a_2^3 b_2^3}{a_2^2 + b_2^2}(1 - r^2) r\, dr\, d\theta + 2\frac{(G_2 - G_1)\alpha a_2^2 b_2^2}{a_2^2 + b_2^2}\left(1 - q^2\right)\pi a_1 b_1 \\ &\quad + 2\int_0^{2\pi}\int_q^1 \frac{G_2 \alpha a_2^3 b_2^3}{a_2^2 + b_2^2}(1 - r^2) r\, dr\, d\theta \\ &= \frac{\pi \alpha a_2^3 b_2^3}{a_2^2 + b_2^2}\left[q^4 G_1 + (1 - q^4) G_2\right], \end{aligned}$$

so that the torsional rigidity is

$$D = \frac{T}{\alpha} = \frac{\pi a_2^3 b_2^3}{a_2^2 + b_2^2}\left[q^4 G_1 + (1 - q^4) G_2\right].$$

The torsional rigidity for a hollow elliptical beam is obtained by simply setting G_1 to zero, and is given by $4a_2^2 b_2^2 G_2 J/(a_2^2 + b_2^2)^2$, while that for a solid elliptical beam is obtained by setting $G_1 = G_2$. The corresponding results for a composite circular cylinder are obtained by setting $a_1 = b_1$ and $a_2 = b_2$.

Now consider the problem when the ellipses are confocal as shown in Fig. 6.11, and where now the inner ellipse no longer coincides with a line of shearing stress. We shall assume without loss of generality that $a_1 > b_1$. Let $z = x + iy = c\cosh\hat{z} = c\cosh(\xi + i\eta)$,

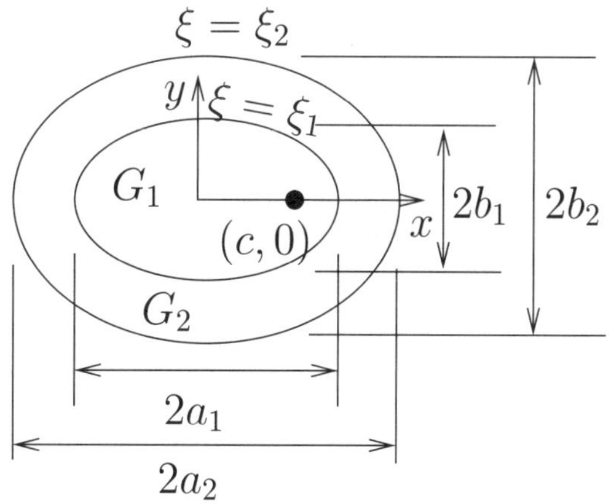

Fig. 6.11 *Composite elliptical beam undergoing torsion; the ellipses are confocal.*

so that $x = c\cosh\xi\cos\eta$ and $y = c\sinh\xi\sin\eta$. The confocal ellipses are defined by the surfaces $\xi = \xi_1$, and $\xi = \xi_2$. We have the relations $a_1 = c\cosh\xi_1$, $b_1 = c\sinh\xi_1$, $a_2 = c\cosh\xi_2$, $b_2 = c\sinh\xi_2$, so that $c^2 = a_1^2 - b_1^2 = a_2^2 - b_2^2$. The foci of both ellipses are located at $(\pm c, 0)$. The relations $a_1^2 + b_1^2 = c^2\cosh 2\xi_1$, $a_2^2 + b_2^2 = c^2\cosh 2\xi_2$, $2a_1b_1 = c^2\sinh 2\xi_1$ and $2a_2b_2 = c^2\sinh 2\xi_2$ will prove to be quite useful in the subsequent development.

The function $\psi + ig$ is assumed to be $i(B_1\cosh 2\hat{z} + k_1)$ and $i(A_2\sinh 2\hat{z} + B_2\cosh 2\hat{z} + k_2)$ in regions 1 and 2, respectively, i.e., [245],

$$
\begin{aligned}
\psi_1 &= -B_1\sinh 2\xi\sin 2\eta,\\
\psi_2 &= -(A_2\cosh 2\xi + B_2\sinh 2\xi)\sin 2\eta,\\
g_1 &= B_1\cosh 2\xi\cos 2\eta + k_1,\\
g_2 &= (A_2\sinh 2\xi + B_2\cosh 2\xi)\cos 2\eta + k_2.
\end{aligned}
$$

Note that there is no term $A_1\cosh 2\xi\cos 2\eta$ in the expression for ψ_1 since the warping is zero at the center of the beam. Using Eqn. (6.80), we have

$$
\begin{aligned}
\frac{\phi_1}{G_1\alpha} &= B_1\cosh 2\xi\cos 2\eta + k_1 - \frac{c^2}{4}(\cosh 2\xi + \cos 2\eta),\\
\frac{\phi_2}{G_2\alpha} &= (A_2\sinh 2\xi + B_2\cosh 2\xi)\cos 2\eta + k_2 - \frac{c^2}{4}(\cosh 2\xi + \cos 2\eta).
\end{aligned}
$$

The conditions $(\psi_1 - \psi_2)_{\xi=\xi_1} = 0$, $(\phi_1 - \phi_2)_{\xi=\xi_1} = 0$ and $\phi_2|_{\xi=\xi_2} = 0$ are used to determine the constants in the above functions. With

$$
D_0 := G_1\cosh 2\xi_1\cosh 2(\xi_1 - \xi_2) - G_2\sinh 2\xi_1\sinh 2(\xi_1 - \xi_2),
$$

we get

$$
\begin{aligned}
k_1 &= \frac{c^2}{4}\left[\frac{G_2}{G_1}(\cosh 2\xi_2 - \cosh 2\xi_1) + \cosh 2\xi_1\right],\\
k_2 &= \frac{c^2}{4}\cosh 2\xi_2,\\
B_1 &= \frac{c^2}{4D_0}\left[(G_1 - G_2)\cosh 2(\xi_1 - \xi_2) + G_2\right],\\
A_2 &= \frac{c^2}{4D_0}\left[(G_1 - G_2)\sinh 2\xi_1(\cosh 2\xi_2 - \cosh 2\xi_1)\right],\\
B_2 &= \frac{c^2}{4D_0}\left[(G_2 - G_1)\sinh 2\xi_1\sinh 2\xi_2 + G_1\cosh^2 2\xi_1 - G_2\sinh^2 2\xi_1\right].
\end{aligned}
$$

The stress field in each domain can be obtained using Eqns. (6.86). Using Eqn. (6.123), the torsional rigidity is obtained as

$$
\begin{aligned}
D &= \frac{c^2}{\alpha}\left[\int_0^{2\pi}\int_0^{\xi_1}\phi_1(\cosh 2\xi - \cos 2\eta)\,d\xi d\eta + \int_0^{2\pi}\int_{\xi_1}^{\xi_2}\phi_2(\cosh 2\xi - \cos 2\eta)\,d\xi d\eta\right]\\
&= \pi G_1\left[-\frac{B_1c^2}{2}\sinh 2\xi_1 + \frac{c^4}{16}\sinh 4\xi_1\right] +
\end{aligned}
$$

$$\pi G_2 \left[\frac{A_2 c^2}{2} (\cosh 2\xi_1 - \cosh 2\xi_2) + \frac{B_2 c^2}{2} (\sinh 2\xi_1 - \sinh 2\xi_2) + \frac{c^4}{16} (\sinh 4\xi_2 - \sinh 4\xi_1) \right].$$

For the case when the section is homogeneous, i.e., $G_1 = G_2 =: G$, we get $D_0 = G \cosh 2\xi_2$, $A_2 = 0$, and $B_1 = B_2 = c^2/(4 \cosh 2\xi_2)$; using these expressions we recover the results given by Eqns. (6.95) and (6.97).

For the case when the inner portion is hollow, i.e., $G_1 = 0$, $G_2 \equiv G$, we obtain

$$A_2 = \frac{c^2(\cosh 2\xi_2 - \cosh 2\xi_1)}{4 \sinh 2(\xi_1 - \xi_2)},$$
$$B_2 = \frac{c^2(\sinh 2\xi_1 - \sinh 2\xi_2)}{4 \sinh 2(\xi_1 - \xi_2)},$$

from which we get the torsional rigidity to be

$$D = \frac{\pi G}{4} \left[a_2 b_2 (a_2^2 + b_2^2) - a_1 b_1 (a_1^2 + b_1^2) - \frac{c^4 (a_1 b_2 - a_2 b_1)}{a_1 a_2 - b_1 b_2} \right].$$

The warping, conjugate and stress functions are given by

$$\psi = \frac{c^2 \left[\cosh 2(\xi_1 - \xi) - \cosh 2(\xi - \xi_2)\right] \sin 2\eta}{4 \sinh 2(\xi_1 - \xi_2)},$$
$$g = \frac{c^2 \left[\sinh 2(\xi_1 - \xi) + \sinh 2(\xi - \xi_2)\right] \cos 2\eta}{4 \sinh 2(\xi_1 - \xi_2)} + \frac{c^2}{4} \cosh 2\xi_2,$$
$$\frac{\phi}{G\alpha} = \frac{c^2}{4} \left[\frac{2 \sinh(\xi - \xi_1) \sinh(\xi - \xi_2) \cos 2\eta}{\cosh(\xi_2 - \xi_1)} + \cosh 2\xi_2 - \cosh 2\xi \right].$$

The associated stresses obtained using Eqns. (6.86) are

$$\tau_{\xi z} = -\frac{\alpha c^2 G}{h_1} \left[\frac{\sinh(\xi - \xi_1) \sinh(\xi - \xi_2) \sin 2\eta}{\cosh(\xi_2 - \xi_1)} \right],$$
$$\tau_{\eta z} = \frac{\alpha c^2 G}{2h_1} \left[\sinh 2\xi - \frac{\sinh(2\xi - \xi_1 - \xi_2) \cos 2\eta}{\cosh(\xi_2 - \xi_1)} \right].$$

The maximum shearing stress occurs at $(\xi, \eta) = (\xi_2, \pm\pi/2)$. Note from the expression for the warping function that there is an ellipse $\xi = (\xi_1 + \xi_2)/2$ confocal with the ellipses $\xi = \xi_1$ and $\xi = \xi_2$ along which there is no warping!

Now consider the torsion of the composite rectangular bar shown in Fig. 6.12. The functions $\psi_j + i g_j$, $j = 1, 2$, for the two regions are taken to be of the form

$$i(A_{jn} \cosh(i\alpha_n z) - B_{jn} \sinh(i\alpha_n z) - \frac{z^2}{2} + k_j),$$

where $\alpha_n = (2n-1)\pi/a$, and k_j are constants, i.e., [245]

$$\psi_1 = xy + \sum_{n=1}^{\infty} (A_{1n} \sinh \alpha_n y + B_{1n} \cosh \alpha_n y) \sin \alpha_n x,$$

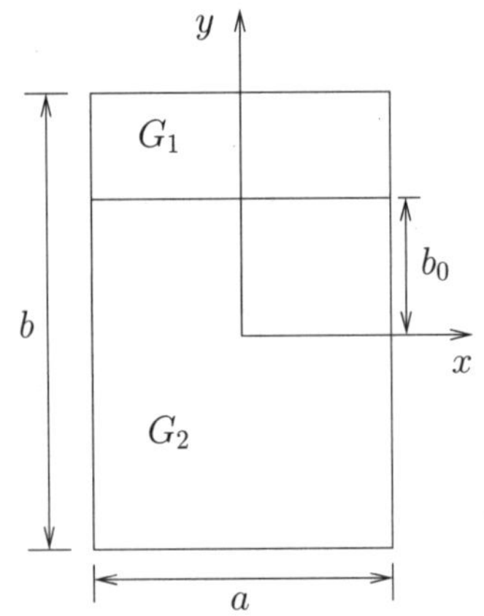

Fig. 6.12 *Composite rectangular bar undergoing torsion.*

$$\psi_2 = xy + \sum_{n=1}^{\infty}(A_{2n}\sinh\alpha_n y + B_{2n}\cosh\alpha_n y)\sin\alpha_n x,$$
$$g_1 = \frac{1}{2}(y^2 - x^2) + k_1 + \sum_{n=1}^{\infty}(A_{1n}\cosh\alpha_n y + B_{1n}\sinh\alpha_n y)\cos\alpha_n x,$$
$$g_2 = \frac{1}{2}(y^2 - x^2) + k_2 + \sum_{n=1}^{\infty}(A_{2n}\cosh\alpha_n y + B_{2n}\sinh\alpha_n y)\cos\alpha_n x,$$
$$\frac{\phi_1}{G_1\alpha} = -x^2 + k_1 + \sum_{n=1}^{\infty}(A_{1n}\cosh\alpha_n y + B_{1n}\sinh\alpha_n y)\cos\alpha_n x,$$
$$\frac{\phi_2}{G_2\alpha} = -x^2 + k_2 + \sum_{n=1}^{\infty}(A_{2n}\cosh\alpha_n y + B_{2n}\sinh\alpha_n y)\cos\alpha_n x.$$

Since ϕ vanishes on the boundary, $\phi_1|_{x=\pm a/2} = \phi_2|_{x=\pm a/2} = 0$, which yields $k_1 = k_2 = a^2/4$. Using the conditions, $(\psi_1 - \psi_2)_{y=b_0} = 0$, $\phi_1|_{y=b/2} = \phi_2|_{y=-b/2} = 0$ and $(\phi_1 - \phi_2)_{y=b_0} = 0$, and using the fact that

$$x^2 - \frac{a^2}{4} = \sum_{n=1}^{\infty}\frac{8a^2(-1)^n}{[(2n-1)\pi]^3}\cos\alpha_n x,$$

yields the equations

$$A_{1n}\sinh\alpha_n b_0 + B_{1n}\cosh\alpha_n b_0 = A_{2n}\sinh\alpha_n b_0 + B_{2n}\cosh\alpha_n b_0,$$
$$A_{1n}\cosh\frac{\alpha_n b}{2} + B_{1n}\sinh\frac{\alpha_n b}{2} = \frac{8a^2(-1)^n}{[(2n-1)\pi]^3},$$
$$A_{2n}\cosh\frac{\alpha_n b}{2} - B_{2n}\sinh\frac{\alpha_n b}{2} = \frac{8a^2(-1)^n}{[(2n-1)\pi]^3},$$
$$A_{1n}\cosh\alpha_n b_0 + B_{1n}\sinh\alpha_n b_0 - \frac{8a^2(-1)^n}{[(2n-1)\pi]^3} =$$

$$\frac{G_2}{G_1}\left[A_{2n}\cosh\alpha_n b_0 + B_{2n}\sinh\alpha_n b_0 - \frac{8a^2(-1)^n}{[(2n-1)\pi]^3}\right],$$

which are to be solved for the constants A_{1n}, A_{2n}, B_{1n} and B_{2n}. We get

$$A_{1n} = \frac{8(-1)^n N_1}{D}, \qquad A_{2n} = \frac{8(-1)^n N_3}{D},$$
$$B_{1n} = \frac{8(-1)^n N_2}{D}, \qquad B_{2n} = \frac{8(-1)^n N_4}{D},$$

where

$$N_1 = (G_1 + 3G_2)\sinh\frac{\alpha_n b}{2} + (G_2 - G_1)\left[\sinh\alpha_n b_0 - \sinh\alpha_n(b+b_0) + \sinh\frac{\alpha_n(b+4b_0)}{2}\right],$$
$$N_2 = (G_2 - G_1)\left[\cosh\alpha_n b_0 + \cosh\alpha_n(b+b_0) - \cosh\frac{\alpha_n b}{2} - \cosh\frac{\alpha_n(b+4b_0)}{2}\right],$$
$$N_3 = (3G_1 + G_2)\sinh\frac{\alpha_n b}{2} + (G_2 - G_1)\left[\sinh\alpha_n b_0 + \sinh\alpha_n(b-b_0) - \sinh\frac{\alpha_n(b-4b_0)}{2}\right],$$
$$N_4 = (G_2 - G_1)\left[\cosh\alpha_n b_0 + \cosh\alpha_n(b-b_0) - \cosh\frac{\alpha_n b}{2} - \cosh\frac{\alpha_n(b-4b_0)}{2}\right],$$
$$D = \alpha_n^3 a\left[(G_2 + G_1)\sinh\alpha_n b + (G_2 - G_1)\sinh 2\alpha_n b_0\right].$$

The torsional rigidity is then obtained by using Eqn. (6.123):

$$\begin{aligned} D &= \frac{2}{\alpha}\left[\int_{-a/2}^{a/2}\int_{b_0}^{b/2}\phi_1\,dydx + \int_{-a/2}^{a/2}\int_{-b/2}^{b_0}\phi_2\,dydx\right], \\ &= \frac{a^3 b(G_1+G_2)}{6} + \frac{a^3 b_0(G_2-G_1)}{3} \\ &\quad + \sum_{n=1}^{\infty}\frac{4(-1)^{n+1}}{\alpha_n^2}\left[h_1\sinh\frac{\alpha_n b}{2} + h_2\sinh\alpha_n b_0 + h_3\left(\cosh\alpha_n b_0 - \cosh\frac{\alpha_n b}{2}\right)\right], \end{aligned}$$

where

$$h_1 = A_{2n}G_2 + A_{1n}G_1,$$
$$h_2 = A_{2n}G_2 - A_{1n}G_1,$$
$$h_3 = B_{2n}G_2 - B_{1n}G_1.$$

For the case, $G_1 = G_2 \equiv G$, we get $B_{1n} = B_{2n} = 0$, and

$$A_{1n} = A_{2n} = \frac{8(-1)^n}{a\alpha_n^3\cosh\frac{\alpha_n b}{2}},$$

so that the above expression for the torsional rigidity reduces to that given by Eqn. (6.102).

Finally, consider the case where the sections are bounded by eccentric circles as shown in Fig. 6.13. We work with bipolar coordinates that are defined through the relation $x + iy = ic\coth\frac{1}{2}(\xi + i\eta)$, which results in

$$x = \frac{c\sin\eta}{\cosh\xi - \cos\eta}, \qquad e^{2\xi} = \frac{x^2 + (y+c)^2}{x^2 + (y-c)^2}, \tag{6.124a}$$

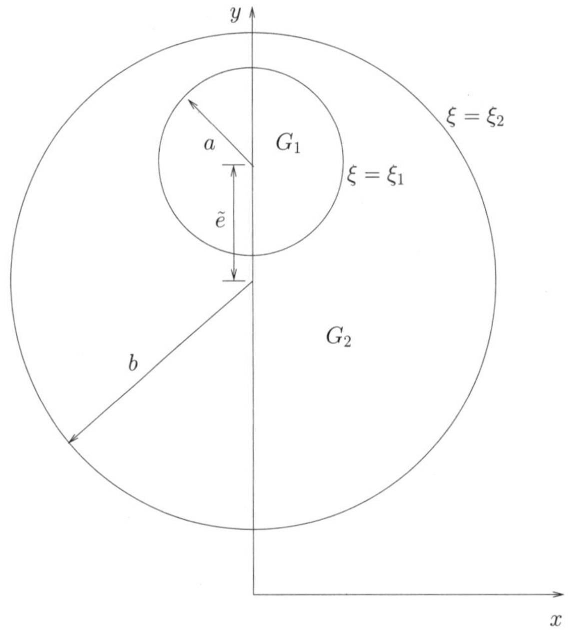

Fig. 6.13 *Torsion of a composite bar where sections are bounded by eccentric circles.*

$$y = \frac{c \sinh \xi}{\cosh \xi - \cos \eta}, \qquad \tan \eta = \frac{2xc}{x^2 + y^2 - c^2}. \tag{6.124b}$$

Eliminating η from the expressions for x and y, we get

$$x^2 + (y - c \coth \xi)^2 = \frac{c^2}{\sinh^2 \xi},$$

which shows that $\xi = \xi_1$ and $\xi = \xi_2$ are circles with centers $(0, c \coth \xi_1)$ and $(0, c \coth \xi_2)$, and with radii $a = c/\sinh \xi_1$ and $b = c/\sinh \xi_2$, respectively. If $\tilde{e} := c(\coth \xi_2 - \coth \xi_1)$ is the eccentricity, then we have the relations

$$c = \frac{1}{2\tilde{e}} \sqrt{[(a+b)^2 - \tilde{e}^2]\,[(b-a)^2 - \tilde{e}^2]},$$
$$\sinh \xi_1 = \frac{c}{a},$$
$$\sinh \xi_2 = \frac{c}{b},$$

which allows one to compute ξ_1 and ξ_2 from the radii and eccentricity. Using Eqn. (6.246) and the fact that $(x^2 + y^2 + c^2)/(2c) = h_1 \cosh \xi$, where $h_1 = c/(\cosh \xi - \cos \eta)$, we assume the respective functions in the two domains to have the following forms

$$\psi_1 = \frac{c^2 \coth \xi_2 \sin \eta}{\cosh \xi - \cos \eta} + \sum_{n=1}^{\infty} A_{1n} \sinh n\xi \sin n\eta,$$

$$g_1 = \frac{c^2 \coth \xi_2 \sinh \xi}{\cosh \xi - \cos \eta} + k_1 - \sum_{n=1}^{\infty} A_{1n} \cosh n\xi \cos n\eta,$$

$$\frac{\phi_1}{G_1 \alpha} = \frac{c^2 \coth \xi_2 \sinh \xi}{\cosh \xi - \cos \eta} + k_1 - \frac{c^2}{2} - \frac{c^2 \cos \eta}{\cosh \xi - \cos \eta} - \sum_{n=1}^{\infty} A_{1n} \cosh n\xi \cos n\eta,$$

$$\psi_2 = \frac{c^2 \coth \xi_2 \sin \eta}{\cosh \xi - \cos \eta} + \sum_{n=1}^{\infty} A_{2n} \cosh n(\xi - \xi_2) \sin n\eta,$$

$$g_2 = c^2 \left[\frac{\coth \xi_2 \sinh \xi}{\cosh \xi - \cos \eta} - \frac{1}{2} \right] - \sum_{n=1}^{\infty} A_{2n} \sinh n(\xi - \xi_2) \cos n\eta,$$

$$\frac{\phi_2}{G_2 \alpha} = -\frac{c^2 \sinh \xi (\coth \xi - \coth \xi_2)}{\cosh \xi - \cos \eta} - \sum_{n=1}^{\infty} A_{2n} \sinh n(\xi - \xi_2) \cos n\eta.$$

Note that the the boundary condition $\phi_2|_{\xi=\xi_2} = 0$ and $\psi_1|_{\eta=0,\pi} = 0$ (warping function zero along the line of symmetry) are automatically satisfied. Thus, the constants k_1, A_{1n} and A_{2n} are to be determined using the constraints $(\psi_1 - \psi_2)_{\xi=\xi_1} = 0$ and $(\phi_1 - \phi_2)_{\xi=\xi_1} = 0$. Using the relations

$$1 + 2\sum_{n=1}^{\infty} e^{-n\xi} \cos n\eta = \frac{\sinh \xi}{(\cosh \xi - \cos \eta)} = \tanh \xi \left[1 + \frac{\cos \eta}{(\cosh \xi - \cos \eta)} \right],$$

$$2\sum_{n=1}^{\infty} e^{-n\xi} \sin n\eta = \frac{\sin \eta}{(\cosh \xi - \cos \eta)} = \tan \eta \left[\frac{\cosh \xi}{\cosh \xi - \cos \eta} - 1 \right],$$

we get

$$k_1 = \frac{c^2}{2G_1} \left[2(G_1 - G_2)(\coth \xi_1 - \coth \xi_2) - G_1 \right],$$

$$A_{1n} = \frac{N}{D_1},$$

$$A_{2n} = \frac{N}{D_2},$$

where

$$N = 2c\tilde{e}e^{-n\xi_1}(G_1 - G_2),$$

$$D_1 = G_1 \cosh n\xi_1 - G_2 \sinh n\xi_1 \tanh n(\xi_1 - \xi_2),$$

$$D_2 = G_1 \cosh n(\xi_1 - \xi_2) \coth n\xi_1 - G_2 \sinh n(\xi_1 - \xi_2).$$

The torsional rigidity is given by

$$D = \frac{T}{\alpha} = \frac{2c^2}{\alpha} \left[\int_0^{2\pi} \int_{\xi_1}^{\infty} \frac{\phi_1 \, d\xi d\eta}{(\cosh \xi - \cos \eta)^2} + \int_0^{2\pi} \int_{\xi_2}^{\xi_1} \frac{\phi_2 \, d\xi d\eta}{(\cosh \xi - \cos \eta)^2} \right].$$

In order to evaluate this integral, we use the series expansion

$$\frac{\cosh\xi}{\sinh^3\xi} - \frac{2}{\sinh\xi}\sum_{n=1}^{\infty}\frac{d}{d\xi}\left(\frac{e^{-n\xi}}{\sinh\xi}\right)\cos n\eta = \frac{1}{(\cosh\xi-\cos\eta)^2}.$$

and the orthogonality of the cosine functions. The final expression is

$$D = \frac{\pi c^4}{2}\left\{G_2\,\mathrm{csch}^4\xi_2 + (G_1-G_2)\,\mathrm{csch}^2\xi_1\left[4(\coth\xi_1-1)(\coth\xi_2+1) - 3\,\mathrm{csch}^2\xi_1\right]\right\}$$
$$-2\pi c^2 G_1(\coth\xi_1-1)\sum_{n=1}^{\infty}A_{1n}e^{-n\xi_1}\left[ne^{n\xi_1}+\cosh n\xi_1(1+\coth\xi_1)\right]$$
$$-2\pi G_2\sum_{n=1}^{\infty}A_{2n}\left[a^2e^{-n\xi_1}\sinh n(\xi_2-\xi_1)+c\tilde{e}ne^{-n\xi_2}\right].$$

The stresses can be calculated using Eqns. (6.87).

For $G_1 = 0$, we get the following solution for a circular cross-section with an eccentric circular hole [205]:

$$\psi_2 = c^2\left[\frac{\coth\xi_2\sin\eta}{\cosh\xi-\cot\eta} - 2(\coth\xi_1-\coth\xi_2)\sum_{n=1}^{\infty}\frac{\cosh n(\xi-\xi_2)}{\sinh n(\xi_1-\xi_2)}e^{-n\xi_1}\sin n\eta\right],$$
$$\frac{\phi_2}{G_2\alpha} = c^2\left[-\frac{\sinh\xi(\coth\xi-\coth\xi_2)}{\cosh\xi-\cos\eta} + 2(\coth\xi_1-\coth\xi_2)\sum_{n=1}^{\infty}\frac{\sinh n(\xi-\xi_2)}{\sinh n(\xi_1-\xi_2)}e^{-n\xi_1}\cos n\eta\right],$$

with

$$\left.\frac{\phi_2}{G_2\alpha}\right|_{\xi=\xi_1} = -c^2(\coth\xi_1-\coth\xi_2).$$

The associated stresses obtained using Eqns. (6.87) are

$$\tau_{\xi z} = -\alpha h_1 G\frac{\sinh(\xi-\xi_2)\sin\eta}{\sinh\xi_2}$$
$$-\frac{2\alpha c^2 G(\coth\xi_1-\coth\xi_2)}{h_1}\sum_{n=1}^{\infty}\frac{n\sinh n(\xi-\xi_2)\sin n\eta}{\sinh n(\xi_1-\xi_2)}e^{-n\xi_1},$$
$$\tau_{\eta z} = \alpha h_1 G\left[(\cosh\xi\cos\eta-1)\coth\xi_2-\sinh\xi\cos\eta\right]$$
$$-\frac{2\alpha c^2 G(\coth\xi_1-\coth\xi_2)}{h_1}\sum_{n=1}^{\infty}\frac{n\cosh n(\xi-\xi_2)\cos n\eta}{\sinh n(\xi_1-\xi_2)}e^{-n\xi_1}.$$

Using the relation

$$2\sum_{n=1}^{\infty}ne^{-n\xi}\sin n\eta = \frac{\sinh\xi\sin\eta}{(\cosh\xi-\cos\eta)^2},$$

it can be verified that $\tau_{\xi z}\big|_{\xi=\xi_1} = 0$. The torsional rigidity is [205]

$$D = \frac{\pi G_2}{2}\left(b^4-a^4\right) - 4\pi c^2\tilde{e}^2 G_2\sum_{n=1}^{\infty}\frac{ne^{-n(\xi_1+\xi_2)}}{\sinh n(\xi_1-\xi_2)},$$

where we have used

$$\sum_{n=1}^{\infty} 2e^{-2n\xi_1} = \coth \xi_1 - 1.$$

For $G_1 = G_2$, we recover the torsional rigidity for a circular bar, and in addition, we get

$$\psi_1 = \psi_2 = (c \coth \xi_2)x,$$
$$\frac{\phi_1}{\alpha G_2} = \frac{\phi_2}{\alpha G_2} = -\frac{1}{2}(x^2 + y^2 - c^2) + cy \coth \xi_2.$$

The warping is of the form $y_0 x$ since the axis of torsion is at the origin, which is at a distance $y_0 = c \coth \xi_2$ from the center of the larger circle (see Eqn. (6.75)).

More examples can be found in [44], and in references therein.

Homogeneous anisotropic beams

Consider the torsion of an anisotropic bar of elliptical cross-section. Since $\phi = 0$ on the outer boundary, similar to the isotropic case, we assume

$$\phi = m\left(1 - \frac{x^2}{a^2} - \frac{y^2}{b^2}\right).$$

Substituting into Eqn. (6.115), we get

$$m = \frac{\alpha a^2 b^2 \det \boldsymbol{C}_c}{\mu_{44} a^2 + \mu_{55} b^2}. \tag{6.125}$$

Note that for the isotropic case where $\mu_{44} = \mu_{55} = G$, and $\det \boldsymbol{C}_c = G^2$, the above solution reduces to that given by Eqn. (6.93). Since Eqn. (6.89) remains valid even for an anisotropic material, the torsional rigidity is given by

$$D = \frac{m}{\alpha} \pi ab.$$

The expressions for the stresses obtained using Eqn. (6.113) are

$$\tau_{xz} = -\frac{2my}{b^2},$$
$$\tau_{yz} = \frac{2mx}{a^2}.$$

From Eqn. (6.114), we get the expression for the warping function as

$$\psi = \frac{1}{\mu_{44} a^2 + \mu_{55} b^2}\left[\mu_{45}(a^2 y^2 - b^2 x^2) + (\mu_{55} b^2 - \mu_{44} a^2) xy\right].$$

Note that for an isotropic material, the above expression reduces to Eqn. (6.96). Unlike the isotropic case, the warping is not zero on the symmetry axes (x and y-axes). For a special orthotropic material with $\mu_{45} = 0$ and $\mu_{55}/\mu_{44} = a^2/b^2$, the warping is zero, a result first derived by Chen [49].

For a hollow elliptical bar with the inner boundary an ellipse that is similar to the outer ellipse with $a_1/a_2 = b_1/b_2 = q$ (see Fig. 6.10), the same warping and stress functions as above are valid. The torsional rigidity is given by

$$D = \frac{m}{\alpha}\pi a_2 b_2(1 - q^4),$$

where m is given by Eqn. (6.125) with (a, b) replaced by (a_2, b_2).

Consider a triangle with sides $y = 0$, $x = ay$ and $x = cy + d$; thus, d is the length of the base of the triangle. Assuming

$$\phi = my(x - ay)(x - cy - d),$$

Eqn. (6.115) is satisfied if

$$a = \frac{\sqrt{3}\mu_{45} + \sqrt{\det \boldsymbol{C}_c}}{\sqrt{3}\mu_{44}},$$

$$c = \frac{\sqrt{3}\mu_{45} - \sqrt{\det \boldsymbol{C}_c}}{\sqrt{3}\mu_{44}},$$

$$m = \frac{\alpha\left[3\det \boldsymbol{C}_c - \sqrt{3\det \boldsymbol{C}_c}\mu_{45}\right]}{d\left[\mu_{45} - \sqrt{3\det \boldsymbol{C}_c}\right]}.$$

The torsional rigidity obtained using Eqn. (6.89) is

$$D = -\frac{\mu_{44}^2 d^5 m}{80\alpha \det \boldsymbol{C}_c},$$

which reduces to the torsional rigidity of an equilateral triangle when the material is isotropic. The warping function can be found using Eqn. (6.114).

Consider the torsion of a bar with cross-section a parallelogram with sides $y = \pm b/2$ and $x + sy = \pm a/2$, where $s = -\mu_{45}/\mu_{44}$. Let the Prandtl stress function be given by

$$\phi = \sum_{m=1}^{\infty}\sum_{n=1}^{\infty} A_{mn} \cos\frac{(2m-1)\pi(x+sy)}{a}\cos\frac{(2n-1)\pi y}{b}.$$

Note that $\phi = 0$ on the boundary of the cross-section. Substituting into Eqn. (6.115), we get

$$\sum_{m=1}^{\infty}\sum_{n=1}^{\infty} A_{mn}\pi^2\left[\det \boldsymbol{C}_c\left(\frac{2m-1}{a}\right)^2 + \mu_{44}^2\left(\frac{2n-1}{b}\right)^2\right]\cos\frac{(2m-1)\pi(x+sy)}{a}\cos\frac{(2n-1)\pi y}{b} = 2\alpha \det \boldsymbol{C}_c.$$

In order to determine A_{mn}, we make the substitution $x = \xi - s\eta$ and $y = \eta$. We now have $dx\,dy = d\xi\,d\eta$. Multiplying both sides of the above equation by $\cos\frac{(2p-1)\pi\xi}{a}\cos\frac{(2q-1)\pi\eta}{b}$, integrating between the limits $[-\frac{a}{2}, \frac{a}{2}]$ for ξ and $[-\frac{b}{2}, \frac{b}{2}]$ for η, and using the orthogonality property of cosine functions, we get

$$A_{mn} = \frac{32\alpha\mu_{44}\det \boldsymbol{C}_c}{(2m-1)(2n-1)\pi^4}\frac{(-1)^{m+n}}{\left[\det \boldsymbol{C}_c\left(\frac{2m-1}{a}\right)^2 + \mu_{44}\left(\frac{2n-1}{b}\right)^2\right]},$$

and the torsional rigidity obtained using Eqn. (6.89) as

$$D = \frac{8ab}{\alpha\pi^2} \sum_{m=1}^{\infty} \sum_{n=1}^{\infty} \frac{(-1)^{m+n} A_{mn}}{(2m-1)(2n-1)}.$$

As in the previous cases, the above results reduce to those of a rectangular bar when the material is isotropic.

6.5.4 Torsion of circular shafts of variable diameter

The torsion of circular shafts of variable diameter can be treated by assuming u_θ to be the only nonzero displacement component. Under this assumption, Eqns. (B.13) reduce to

$$\frac{\partial}{\partial r}\left(\frac{1}{r}\frac{\partial(r u_\theta)}{\partial r}\right) + \frac{\partial^2 u_\theta}{\partial z^2} = 0.$$

A formulation in terms of a stress function ϕ that automatically satisfies the equations of equilibrium (obtained by combining Eqns. (6.56) and (B.7)) can be given by writing the stress distribution as

$$\begin{aligned} \tau_{r\theta} &= Gr\frac{\partial}{\partial r}\left(\frac{u_\theta}{r}\right) = -\frac{1}{r^2}\frac{\partial \phi}{\partial z}, \\ \tau_{\theta z} &= Gr\frac{\partial}{\partial z}\left(\frac{u_\theta}{r}\right) = \frac{1}{r^2}\frac{\partial \phi}{\partial r}. \end{aligned} \tag{6.126}$$

Eliminating u_θ from the above equation, we get the governing equation for ϕ as

$$\frac{\partial^2 \phi}{\partial r^2} - \frac{3}{r}\frac{\partial \phi}{\partial r} + \frac{\partial^2 \phi}{\partial z^2} = 0. \tag{6.127}$$

The boundary condition on ϕ is obtained by imposing the traction-free condition on the lateral surface, i.e.,

$$\begin{bmatrix} t_r \\ t_\theta \\ t_z \end{bmatrix} = \begin{bmatrix} 0 & \tau_{r\theta} & 0 \\ \tau_{r\theta} & 0 & \tau_{\theta z} \\ 0 & \tau_{\theta z} & 0 \end{bmatrix} \begin{bmatrix} n_r \\ 0 \\ n_z \end{bmatrix} = \begin{bmatrix} 0 \\ \tau_{\theta r} n_r + \tau_{\theta z} n_z \\ 0 \end{bmatrix} = \mathbf{0}.$$

Since $(n_r, n_z) = (dz/ds, -dr/ds)$, the above condition in conjunction with Eqn. (6.126) yields $d\phi/ds = 0$, or in other words,

$$\phi = \text{constant} \tag{6.128}$$

on the boundary in the r–z plane. Thus, at least in some cases (see the example of the conical shaft below), the equation of the boundary can be used to guess the form of ϕ that satisfies the governing differential equation Eqn. (6.127). Once ϕ is determined, the displacement u_θ is obtained using Eqn. (6.126). The torque at any cross-section of radius $a(z)$ is given by

$$T = \int_0^{a(z)} 2\pi r^2 \tau_{\theta z}\, dr = 2\pi \int_0^{a(z)} \frac{\partial \phi}{\partial r}\, dr = 2\pi\left[\phi(a(z), z) - \phi(0, z)\right]. \tag{6.129}$$

Since T is a constant at any cross-section, and since $\phi(a(z), z)$ is a constant by virtue of the fact that $(a(z), z)$ lies on the boundary, an important constraint on ϕ is that $\phi(0, z)$ should be a constant. This constant is usually chosen as zero as in the solution given by Eqn. (6.136) below. As is evident from the above derivation, the end surfaces on which the traction is being applied to generate the torque T need not be 'flat' (i.e., need not be of the form $z =$ constant).

The relations corresponding to Eqns. (6.126) and (6.127) for any arbitrary orthogonal curvilinear coordinate system, with '3' corresponding to the circumferential direction (denoted by ϕ below), are

$$\begin{aligned} \tau_{13} &= \frac{Gh_3}{h_1}\frac{\partial}{\partial \xi_1}\left(\frac{u_3}{h_3}\right) = \frac{1}{h_2 h_3^2}\frac{\partial \phi}{\partial \xi_2}, \\ \tau_{23} &= \frac{Gh_3}{h_2}\frac{\partial}{\partial \xi_2}\left(\frac{u_3}{h_3}\right) = -\frac{1}{h_1 h_3^2}\frac{\partial \phi}{\partial \xi_1}, \end{aligned} \tag{6.130}$$

and

$$\frac{\partial}{\partial \xi_1}\left(\frac{h_2}{h_1 h_3^3}\frac{\partial \phi}{\partial \xi_1}\right) + \frac{\partial}{\partial \xi_2}\left(\frac{h_1}{h_2 h_3^3}\frac{\partial \phi}{\partial \xi_2}\right) = 0, \tag{6.131}$$

where (ξ_1, ξ_2) are the orthogonal curvilinear coordinates in the r–z plane. The above relations are derived using Eqns. (A.24). Since $x = (r\cos\phi, r\sin\phi, z)$, the scale factors $h_i = |\partial x/\partial \xi_i|$, $i = 1, 2, 3$ are given by

$$\begin{aligned} h_1 &= \sqrt{\left(\frac{\partial r}{\partial \xi_1}\right)^2 + \left(\frac{\partial z}{\partial \xi_1}\right)^2}, \\ h_2 &= \sqrt{\left(\frac{\partial r}{\partial \xi_2}\right)^2 + \left(\frac{\partial z}{\partial \xi_2}\right)^2}, \\ h_3 &= r. \end{aligned} \tag{6.132}$$

The normal to the surface is given by $(n_1, n_2) = (h_2 \partial\xi_2/\partial s, -h_1 \partial \xi_1/\partial s)$, so that the lateral traction-free condition

$$\tau_{13} n_1 + \tau_{23} n_2 = 0,$$

again leads to Eqn. (6.128) on the boundary. Equations (6.130) and (6.131) can be written in tensorial form as

$$\boldsymbol{\tau}_c = \begin{bmatrix} \tau_{13} \\ \tau_{23} \end{bmatrix} = Gr\boldsymbol{\nabla}\left(\frac{u_3}{r}\right) = \frac{1}{r^2}\boldsymbol{W}\boldsymbol{\nabla}\phi, \tag{6.133a}$$

$$r^2 \boldsymbol{\nabla}^2\left(\frac{\phi}{r^2}\right) - \frac{4\phi}{r^2} = 0, \tag{6.133b}$$

where r is the cylindrical radial coordinate, and $\boldsymbol{W} = \left[\begin{smallmatrix} 0 & 1 \\ -1 & 0 \end{smallmatrix}\right]$.

In particular, for the spherical coordinate system, with R denoting the radial coordinate (so that $r = R\sin\theta$, $z = R\cos\theta$ and $R^2 = r^2 + z^2$), the above relations reduce to

$$\begin{aligned} \tau_{R\phi} &= GR\frac{\partial}{\partial R}\left(\frac{u_\phi}{R}\right) = \frac{1}{R^3 \sin^2\theta}\frac{\partial \phi}{\partial \theta}, \\ \tau_{\theta\phi} &= \frac{G\sin\theta}{R}\frac{\partial}{\partial \theta}\left(\frac{u_\phi}{\sin\theta}\right) = -\frac{1}{R^2\sin^2\theta}\frac{\partial \phi}{\partial R}, \end{aligned} \tag{6.134}$$

and

$$\frac{\partial^2 \phi}{\partial R^2} - \frac{2}{R}\frac{\partial \phi}{\partial R} - \frac{3\cot\theta}{R^2}\frac{\partial \phi}{\partial \theta} + \frac{1}{R^2}\frac{\partial^2 \phi}{\partial \theta^2} = 0. \tag{6.135}$$

The displacement u_ϕ in the relations above is the same as the displacement u_θ in the cylindrical coordinate system. The stresses in the cylindrical system are related to those in the spherical one by the relations

$$\tau_{r\theta} = \sin\theta\,\tau_{R\phi} + \cos\theta\,\tau_{\theta\phi},$$

$$\tau_{\theta z} = \cos\theta\,\tau_{R\phi} - \sin\theta\,\tau_{\theta\phi}.$$

We now discuss some particular solutions; see Higgins [119] for a review, and Abbassi [1], Chattarji [45], Fish [78] and Ling [192] for general solutions to Eqns. (6.127) and (6.135). A general solution of Eqn. (6.127) in cylindrical coordinates (excluding $\tilde{c}_1 + \tilde{c}_2 z$) is

$$\begin{aligned}\phi = k_1 r^4 + k_2 z r^4 + r^2\left[c_1 I_2(kr) + c_2 K_2(kr)\right]\left[c_3\cos(kz) + c_4\sin(kz)\right]\\ + r^2\left[c_5 J_2(hr) + c_6 Y_2(hr)\right]\left[c_7\cosh(hz) + c_8\sinh(hz)\right],\end{aligned}$$

where k and h are arbitrary real or complex-valued constants. Chattarji [45] presents two families of solutions, $\phi_i^{(1)}$ and $\phi_i^{(2)}$, $i = 1, 2, \ldots, \infty$, the first few members of which are as follows:

$$\phi_1^{(1)} = r^4, \qquad \phi_1^{(2)} = r^4 z$$

$$\phi_2^{(1)} = r^4 z^2 - \frac{1}{6}r^6, \qquad \phi_2^{(2)} = r^4 z^3 - \frac{1}{2}r^6 z$$

$$\phi_3^{(1)} = r^4 z^4 - r^6 z^2 + \frac{1}{16}r^8, \qquad \phi_3^{(2)} = r^4 z^5 - \frac{5}{3}r^6 z^3 + \frac{5}{16}r^8 z.$$

These solutions are a subset of the following general solution of Eqn. (6.135) presented by Abbassi [1] and Ling [192]:

$$\begin{aligned}\phi &= (A_0 R^2 + B_0 R)(1-\cos\theta)^2 + (A_1 R^3 + B_1)(1-\cos\theta)^2(2+\cos\theta)\\ &\quad + \sum_{n=2}^{\infty}(A_n R^{n+2} + B_n R^{1-n})\sin^2\theta P_n^{(2)}(\cos\theta),\\ \tau_{R\phi} &= \left(\frac{2A_0}{R} + \frac{2B_0}{R^2}\right)\tan\frac{\theta}{2} + 3\left(A_1 + \frac{B_1}{R^3}\right)\sin\theta\\ &\quad + \sum_{n=2}^{\infty}(n+2)(1-n)(A_n R^{n-1} + B_n R^{-n-2})P_n^{(1)}(\cos\theta),\\ \tau_{\theta\phi} &= -\left(\frac{B_0}{R^2} + \frac{2A_0}{R} + 6A_1 + 3A_1\cos\theta\right)\tan^2\frac{\theta}{2}\\ &\quad - \sum_{n=2}^{\infty}\left[A_n(n+2)R^{n-1} + B_n(1-n)R^{-n-2}\right]P_n^{(2)}(\cos\theta),\end{aligned} \tag{6.136}$$

$$Gu_\phi = -\left(2A_0 + \frac{B_0}{R}\right)\tan\frac{\theta}{2} + 3A_1R\left[(\log R)(\sin\theta) + 2\sin\theta\log\left(\cos\frac{\theta}{2}\right) - \tan\frac{\theta}{2}\right]$$
$$- \frac{B_1}{R^2}\sin\theta - \sum_{n=2}^{\infty}\left[A_n(n+2)R^n - B_n(n-1)R^{-1-n}\right]P_n^{(1)}(\cos\theta),$$

where $P_n^{(m)}(\cos\theta)$, the associated Legendre functions of the first kind of degree n and order m, are solutions of the Legendre equation

$$\frac{d}{d\xi}\left[(1-\xi^2)\frac{dy}{d\xi}\right] + \left[n(n+1) - \frac{m^2}{1-\xi^2}\right]y = 0,$$

where $\xi := \cos\theta$. The rigid displacement $u_\phi = k_1R\sin\theta = k_1r$ has been assumed to be suppressed in Eqn. (6.136). Note that the solution associated with the terms A_0, B_0, A_1 and B_1 in Eqn. (6.136) is valid only for domains with $z \geq 0$, since for $z < 0$, ϕ is not constant along the axis $r = 0$. Similarly, the solution associated with the B_n, $n \geq 2$, terms is valid only when the origin is not part of the domain, since, otherwise, the displacement would be infinite at the origin. The infinite series solution for ϕ in Eqn. (6.136) is obtained by substituting either $\phi = \sin^2\theta R^{n+2}g(\theta)$ or $\phi = \sin^2\theta R^{1-n}g(\theta)$ into Eqn. (6.135), both of which lead to the equation

$$\sin^2\theta g'' + \sin\theta\cos\theta g' + [n(n+1)\sin^2\theta - 4]g = 0,$$

of which $g = P_n^{(2)}(\xi)$ is a solution; the other solution $Q_n^{(2)}(\xi)$ is discarded since the associated stresses are unbounded at $\xi = \pm 1$. If the axis is not part of the domain, then this solution can be included as discussed at the end of this section. The expressions for the stresses and the displacement in Eqn. (6.136) are obtained using Eqn. (6.134).

The associated Legendre polynomials of the first kind $P_n^{(m)}(\xi)$ are defined via the relation

$$P_n^{(m)}(\xi) = (-1)^m(1-\xi^2)^{\frac{m}{2}}\frac{d^mP_n(\xi)}{d\xi^m} = \frac{(-1)^m(1-\xi^2)^{\frac{m}{2}}}{2^nn!}\frac{d^{n+m}}{d\xi^{n+m}}(\xi^2-1)^n. \tag{6.137}$$

The associated Legendre functions of the second kind $Q_n^{(m)}(\xi)$ are defined in an analogous way in terms of derivatives of $Q_n(\xi)$. The associated Legendre polynomials satisfy the following recurrence and orthogonality relations:

$$(n-m+1)P_{n+1}^{(m)}(\xi) - (2n+1)\xi P_n^{(m)}(\xi) + (n+m)P_{n-1}^{(m)}(\xi) = 0,$$

$$\xi P_n^{(2)}(\xi) - P_{n+1}^{(2)}(\xi) = (n+2)\sqrt{1-\xi^2}P_n^{(1)}(\xi),$$

$$\xi P_n^{(2)}(\xi) - P_{n-1}^{(2)}(\xi) = (1-n)\sqrt{1-\xi^2}P_n^{(1)}(\xi),$$

$$\sqrt{1-\xi^2}P_n^{(1)}(\xi) = \frac{n(n+1)}{2n+1}\left[P_{n+1}(\xi) - P_{n-1}(\xi)\right],$$

$$\sqrt{1-\xi^2}P_n^{(1)}(\xi) = (n+1)\left[P_{n+1}(\xi) - \xi P_n(\xi)\right],$$

$$
\begin{aligned}
&\sqrt{1-\xi^2}P_{n+1}^{(1)}(\xi) = (n+1)\left[\xi P_{n+1}(\xi) - P_n(\xi)\right],\\
&\sqrt{1-\xi^2}P_n^{(2)}(\xi) = (n-1)\xi P_n^{(1)}(\xi) - (n+1)P_{n-1}^{(1)}(\xi),\\
&(1-\xi^2)\frac{dP_n^{(m)}(\xi)}{d\xi} = (n+m)P_{n-1}^{(m)}(\xi) - n\xi P_n^{(m)}(\xi),\\
&\int_{-1}^{1} P_n(\xi)\,d\xi = 0, \quad n > 0,\\
&\int_{-1}^{1} P_k^{(m)}(\xi)P_l^{(m)}(\xi)\,d\xi = \frac{2}{2l+1}\frac{(l+m)!}{(l-m)!}\delta_{kl}, \qquad (6.138)\\
&\int_{-1}^{1} \frac{P_l^{(m)}(\xi)P_l^{(n)}(\xi)}{1-\xi^2}\,d\xi = \frac{1}{m}\frac{(l+m)!}{(l-m)!}\delta_{mn},\\
&\int_{-1}^{1} (1-\xi^2)\frac{dP_m(\xi)}{d\xi}\frac{dP_n(\xi)}{d\xi}\,d\xi = \frac{2n(n+1)}{2n+1}\delta_{mn},\\
&\int_{-1}^{1} \frac{P_n^{(2)}(\xi)}{1-\xi^2}\,d\xi = \begin{cases} 0 & \text{when } n \text{ is odd,}\\ n(n+1) & \text{when } n \text{ is even.}\end{cases}
\end{aligned}
$$

The first few functions are

$$
\begin{aligned}
P_1^{(1)}(\cos\theta) &= -\sin\theta,\\
P_2^{(1)}(\cos\theta) &= -3\sin\theta\cos\theta,\\
P_3^{(1)}(\cos\theta) &= -\frac{3}{2}\sin\theta(5\cos^2\theta - 1),\\
P_4^{(1)}(\cos\theta) &= -\frac{5}{2}\sin\theta(7\cos^3\theta - 3\cos\theta),\\
P_5^{(1)}(\cos\theta) &= -\frac{15}{8}\sin\theta(1 - 14\cos^2\theta + 21\cos^4\theta),\\
P_1^{(2)}(\cos\theta) &= 0,\\
P_2^{(2)}(\cos\theta) &= 3\sin^2\theta,\\
P_3^{(2)}(\cos\theta) &= 15\cos\theta\sin^2\theta,\\
P_4^{(2)}(\cos\theta) &= \frac{15}{2}\sin^2\theta(7\cos^2\theta - 1),\\
P_5^{(2)}(\cos\theta) &= \frac{105}{2}\sin^2\theta\cos\theta(3\cos^2\theta - 1).
\end{aligned}
$$

With α denoting the twist per unit length, if $\phi = G\alpha r^4/4 = G\alpha R^4 \sin^4\theta/4$ (which is obtained by setting all constants other than A_2 to zero in Eqn. (6.136)), we recover the solution for a circular shaft of constant diameter with $\tau_{\theta z}$ given by $G\alpha r$. By taking the potential to be

$$\phi = \frac{G\alpha R^4 \sin^4\theta}{4} + \frac{3B_2 \sin^4\theta}{R}, \tag{6.139}$$

and finding B_2 so that $\phi = 0$ on the surface $R = R_0$, we solve the problem of the torsion of a circular shaft with a small spherical hole defined by $R = R_0$ on its axis of rotation, with the diameter of the shaft much larger compared to R_0 so that far away from the hole the shear stress is given by $G\alpha r$, i.e., $\tau_{\theta z}|_{R\to\infty} = G\alpha r$; we get $B_2 = -G\alpha R_0^5/12$. The stresses are recovered using Eqns. (6.134). This problem is similar to that shown in Fig. 6.72 except that here the far-field stress is $\tau_{\theta z} = G\alpha r$. If instead of a void, we had a spherical inclusion of a different material, then we assume the potential for the matrix $\phi^{(m)}$ to be the same as that given by Eqn. (6.139), while we assume that for the inclusion to be $\phi^{(i)} = 3A_2R^4\sin^4\theta$. We evaluate the displacement and stress fields in the inclusion and the matrix by using Eqn. (6.136), and then evaluate the constants A_2 and B_2 by imposing the continuity of the displacement and tractions at the interface, i.e., $u_\phi^{(i)}\Big|_{R=R_0} = u_\phi^{(m)}\Big|_{R=R_0}$ and $\phi^{(i)}\Big|_{R=R_0} = \phi^{(m)}\Big|_{R=R_0}$. We get

$$A_2 = \frac{5\alpha G_i G_m}{12(G_i + 4G_m)},$$
$$B_2 = \frac{\alpha(G_i - G_m)G_m R_0^5}{3(G_i + 4G_m)}.$$

In the limit of $G_i \to 0$, we recover our earlier result for B_2 for a void, while in the limit as $G_i \to \infty$ (rigid inclusion), we get $B_2 = G_m\alpha R_0^5/3$.

A conical shaft of length L is subjected to end twisting moments T as shown in Fig. 6.14. Since the ratio $z/\sqrt{r^2+z^2}$ is constant and equal to $\cos\alpha$, any function of this ratio will satisfy the boundary condition. Thus, we choose ϕ to be of the form

$$\phi = c\left\{\frac{z}{\sqrt{r^2+z^2}} - \frac{1}{3}\left[\frac{z}{\sqrt{r^2+z^2}}\right]^3 - \frac{2}{3}\right\}, \tag{6.140}$$

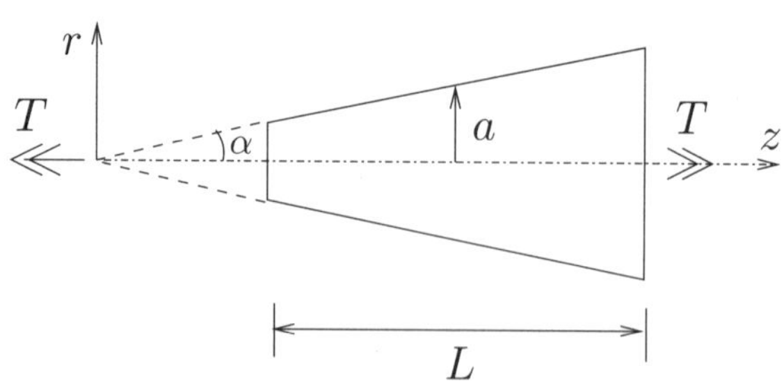

Fig. 6.14 *Conical shaft subjected to end twisting moments.*

where c is a constant. This solution is obtained by setting $B_1 = -c/3$ and all the remaining constants zero in Eqn. (6.136). Using Eqn. (6.129), we get

$$c = \frac{3T}{2\pi(3\cos\alpha - \cos^3\alpha - 2)}.$$

From Eqns. (6.126), we find that

$$\begin{aligned}
\tau_{r\theta} &= -c\left[\frac{1}{(r^2+z^2)^{3/2}} - \frac{z^2}{(r^2+z^2)^{5/2}}\right],\\
\tau_{\theta z} &= -\frac{crz}{(r^2+z^2)^{5/2}},\\
u_\theta &= \frac{cr}{3G(r^2+z^2)^{3/2}}.
\end{aligned}$$

As pointed out after Eqn. (6.129), the above solution is valid even if the flat ends are replaced by curved ones, say, of the form $R =$ constant. If instead of a solid conical shaft, we had a hollow conical shaft whose inner surface is given by $\theta = \beta$, then the solution given by Eqn. (6.140) is still valid except that now $c = 3T/[2\pi(3\cos\alpha - \cos^3\alpha - 3\cos\beta + \cos^3\beta)]$.

The solution for an infinite half-space subjected to a concentrated torque T at $(0,0)$ is also obtained from the above solution simply by setting $\alpha = \pi/2$, so that $c = -3T/(4\pi)$ and $B_1 = T/(4\pi)$. From Eqn. (6.136), we get

$$\begin{aligned}
\phi &= \frac{T}{4\pi}(1-\cos\theta)^2(2+\cos\theta),\\
\tau_{R\phi} &= \frac{3T\sin\theta}{4\pi R^3},\\
\tau_{\theta\phi} &= 0,\\
Gu_\phi &= -\frac{T\sin\theta}{4\pi R^2}.
\end{aligned}$$

The function $\phi = czr^4$, obtained by setting all constants other than A_3 to zero in Eqn. (6.136), is the solution for the torsion of the bar with profile as shown in Fig. 6.15. From Eqns. (6.126), we find that

$$\begin{aligned}
\tau_{r\theta} &= -cr^2,\\
\tau_{\theta z} &= 4crz,\\
u_\theta &= \frac{2cr}{G}\left[z^2 - \frac{r^2}{4}\right],
\end{aligned}$$

while from Eqn. (6.129), we get the torque acting on the top and bottom surfaces to be

$$T = 2\pi ck.$$

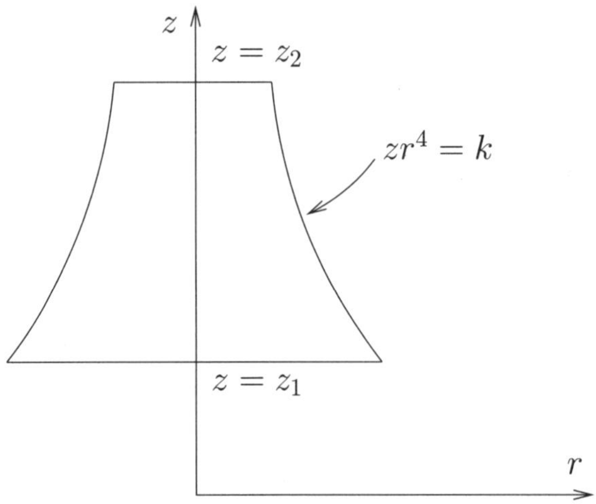

Fig. 6.15 *Circular cylinder with generating surface $zr^4 = k$ subjected to torsion.*

Another solution to Eqn. (6.127), obtained by setting all constants other than A_4 to zero in Eqn. (6.136), is [1, 45]

$$\phi = cr^4(r^2 - 6z^2),$$

which corresponds to a shaft with the generating curve $r^6 - 6r^4z^2 = k$, where k is a constant. From Eqn. (6.129), we get $T = 2\pi ck$. Chattarji [45] also mentions solutions

$$\phi = c(r^4 - az),$$
$$\phi = c(ar^4 - r^4z - z),$$

corresponding to generating curves $r^4 - az = k$ and $ar^4 - r^4z - z = k$, where a and k are constants. However, the above functions violate the constraint $\phi(0, z) =$ constant, and hence are not admissible solutions; they are admissible if the axis is not part of the domain as discussed below.

The solution for a shaft whose generating surface is a paraboloid given by $r^2 = 2zk + k^2$, $z \geq 0$, where k is a constant, obtained by setting all constants other than A_0 to zero in Eqn. (6.136), is

$$\phi = c\left[\sqrt{r^2 + z^2} - z\right]^2.$$

The constraint $z \geq 0$ is required to ensure that $\phi(0, z)$ is a constant. In this case, $T = 2\pi ck^2$, and

$$\tau_{r\theta} = \frac{2c(z - \sqrt{r^2 + z^2})^2}{r^2\sqrt{r^2 + z^2}},$$
$$\tau_{\theta z} = -\frac{2c(z - \sqrt{r^2 + z^2})}{r\sqrt{r^2 + z^2}},$$

$$u_\theta = \frac{2c(z - \sqrt{r^2 + z^2})}{Gr}.$$

Consider the torsion of a bar whose generating surface is expressed in terms of oblate spheroidal coordinates (ξ, η), where $r + iz = c\sin(\eta + i\xi)$, i.e., $r = c\cosh\xi \sin\eta$ and $z = c\sinh\xi\cos\eta$. Using Eqns. (6.132), we obtain $h_1 = h_2 = c\sqrt{\sinh^2\xi + \cos^2\eta}$ and $h_3 = r = c\cosh\xi\sin\eta$, and thus, Eqn. (6.131) reduces to

$$\frac{\partial^2\phi}{\partial\xi^2} - 3\tanh\xi\frac{\partial\phi}{\partial\xi} + \frac{\partial^2\phi}{\partial\eta^2} - 3\cot\eta\frac{\partial\phi}{\partial\eta} = 0.$$

A general solution is given by [54]

$$\begin{aligned}
\phi &= \left\{c_1\left[\sinh^3\xi + 3\sinh\xi\right] + c_2\right\}\left\{c_3\left[3\cos\eta - \cos^3\eta\right] + c_4\right\} \\
&\quad + \cosh^2\xi\sin^2\eta\sum_{n=1}^{\infty}\left[A_nP_n^{(2)}(i\sinh\xi) + B_nQ_n^{(2)}(i\sinh\xi)\right] \\
&\qquad \left[C_nP_n^{(2)}(\cos\eta) + D_nQ_n^{(2)}(\cos\eta)\right], \\
8c^2Gu_\phi &= 3\cosh\xi\sin\eta\left\{8c_1c_3\,\mathrm{sech}^2\xi - 4c_2c_3\left[\tanh\xi\,\mathrm{sech}\,\xi + 2\tan^{-1}\tanh\frac{\xi}{2}\right]\right. \\
&\quad \left. + c_1\left[c_4\left(\csc^2\frac{\eta}{2} - \sec^2\frac{\eta}{2} + 4\log\cot\frac{\eta}{2}\right) + c_3\left(8\csc^2\eta - 8\log(\cosh\xi\sin\eta) - 9\right)\right]\right\} \\
&\quad - 8i\sum_{n=1}^{\infty}(n-1)(n+2)\left[A_nP_n^{(1)}(i\sinh\xi) + B_nQ_n^{(1)}(i\sinh\xi)\right] \\
&\qquad \left[C_nP_n^{(1)}(\cos\eta) + D_nQ_n^{(1)}(\cos\eta)\right],
\end{aligned} \tag{6.141}$$

where the constants A_n and B_n are either real or pure imaginary, so that the final result is real. The first part of the above solution corresponds to the case where the separation-of-variables constant is zero. The stresses $\tau_{\xi\phi}$ and $\tau_{\eta\phi}$ are obtained using Eqn. (6.130) as

$$\begin{aligned}
\tau_{\xi\phi} &= \frac{1}{h_1r^2}\frac{\partial\phi}{\partial\eta}, \\
\tau_{\eta\phi} &= -\frac{1}{h_1r^2}\frac{\partial\phi}{\partial\xi}.
\end{aligned} \tag{6.142}$$

For problems involving infinite domains, $i\pi P_n^{(2)}(i\sinh\xi)/2$ should be subtracted from $Q_n^{(2)}(i\sinh\xi)$ in Eqn. (6.141), so that the combined function decays to zero as $\xi \to \infty$. We illustrate this by considering the torsion of a circular shaft with a small hole defined by $\xi = \xi_0$ on its axis (the setup is similar to that shown in Fig. 6.74 except that the far-field loading is given by $\tau_{\theta z} = G\alpha r$). Similar to the corresponding problem of a spherical hole (see Eqn. (6.139)), we assume the potential ϕ at large ξ to be given by $G\alpha r^4/4$, or, in oblate spheroidal coordinates, $G\alpha c^4\cosh^4\xi\sin^4\eta/4$. Thus, using Eqn. (6.141), we assume the potential to be given by

$$\phi = \frac{1}{4}G\alpha c^4\cosh^4\xi\sin^4\eta + B_2Q_2\cosh^4\xi\sin^4\eta, \tag{6.143}$$

where

$$Q_2 = 6(\tan^{-1}\sinh\xi) + 6\,\mathrm{sech}\,\xi\tanh\xi + 4\,\mathrm{sech}^3\xi\tanh\xi - 3\pi.$$

Setting ϕ to zero on $\xi = \xi_0$, we get

$$B_2 = \frac{G\alpha c^4}{12\left[\pi - 2(\tan^{-1}\sinh\xi_0)\right] - 8\tanh\xi_0\,\mathrm{sech}\,\xi_0(3 + 2\,\mathrm{sech}^2\xi_0)}.$$

The stresses are recovered using Eqns. (6.142). In particular for $\xi = \xi_0$, we get $\tau_{\xi\phi} = 0$ and

$$\tau_{\eta\phi}\big|_{\xi=\xi_0} = -\frac{4G\alpha c^2\sin^2\eta}{h_1\left\{3\left[\pi - 2(\tan^{-1}\sinh\xi_0)\right]\cosh^3\xi_0 - 3\sinh 2\xi_0 - 4\tanh\xi_0\right\}}.$$

By taking the limit as $\xi_0 \to 0$, we get the solution for a penny-shaped crack of radius a lying on the plane $z = 0$ with the far-field loading given by $\tau_{\theta z} = G\alpha r$ as

$$\begin{aligned}
u_\theta &= \frac{2\alpha c^2}{3\pi}\cosh\xi\sin\eta\cos\eta\left[3\sinh\xi(\tan^{-1}\sinh\xi) - \mathrm{sech}^2\xi + 3\right],\\
\tau_{r\theta} &= -\frac{4G\alpha a^3}{3h_1^2\pi}\mathrm{sech}^2\xi\sin^2\eta\cos\eta,\\
\tau_{\theta z} &= \frac{G\alpha a\sin\eta}{3h_1^2\pi}\left\{6\cosh\xi\left[h_1^2(\tan^{-1}\sinh\xi) + a^2\sinh\xi\right] + a^2(1 + 3\cos 2\eta)\tanh\xi\right\}.
\end{aligned}$$

In particular, to obtain the solution along the r-axis, we set $\xi = 0$ and use $a\sin\eta = r$, $a\cos\eta = \pm\sqrt{a^2 - r^2}$ for $r < a$, $z = 0^\pm$ (corresponding to the upper and lower faces of the crack), and set $\eta = \pi/2$ and use $\cosh\xi = r/a$ for $r \geq a$, $z = 0$. The summary of the final results is

$$\left.\begin{aligned}
u_\theta &= \pm\frac{4\alpha}{3\pi}r\sqrt{a^2 - r^2},\\
\tau_{r\theta} &= \mp\frac{4G\alpha}{3\pi}\frac{r^2}{\sqrt{a^2 - r^2}},\\
\tau_{\theta z} &= 0,
\end{aligned}\right\} \quad r < a, z = 0^\pm,$$

$$\left.\begin{aligned}
u_\theta &= \tau_{r\theta} = 0,\\
\tau_{\theta z} &= \frac{2G\alpha}{\pi}\left[\frac{a(3r^2 - a^2)}{3r\sqrt{r^2 - a^2}} + r\cos^{-1}\left(\frac{a}{r}\right)\right]
\end{aligned}\right\} \quad r \geq a, z = 0. \tag{6.144}$$

If in place of an oblate spheroidal void, we had an oblate spheroidal inclusion, then similar to the spherical inclusion case considered earlier, for the outer matrix we still assume the potential to be given by Eqn. (6.143), while in the inclusion we assume the potential to be given by

$$\phi^{(i)} = A_2\cosh^4\xi\sin^4\eta,$$

and determine the constants A_2 and B_2 by using the continuity of ϕ and u_ϕ at $\xi = \xi_0$.

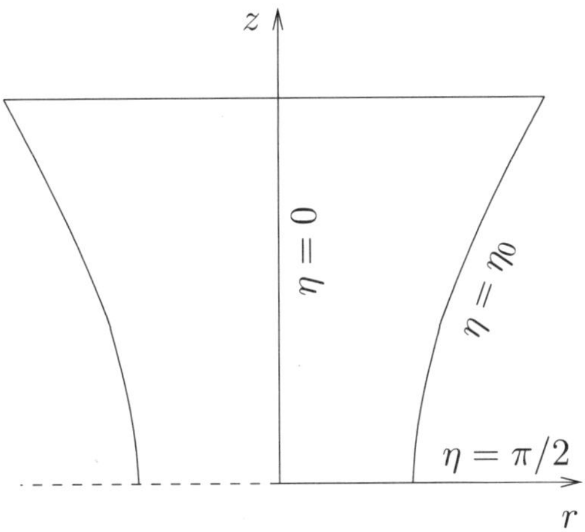

Fig. 6.16 *A shaft whose generating surface is a hyperbola subjected to end twisting moments.*

Taking all constants Eqn. (6.141) except c_2–c_4 to be zero, and noting that $\phi|_{\eta=0} = 0$, we get the following solution to the problem shown in Fig. 6.16 where the generating surface is the hyperbola $\eta = \eta_0$:

$$\phi = c\left[3\cos\eta - \cos^3\eta - 2\right],$$

where c is determined using Eqn. (6.129) as

$$c = \frac{T}{2\pi\left[3\cos\eta_0 - \cos^3\eta_0 - 2\right]}.$$

Similarly, for a prolate spheroid whose generating surface is an ellipse defined by $\xi = \xi_0$, where $r + iz = ci\cos(\eta + i\xi)$, i.e., $r = c\sinh\xi\sin\eta$ and $z = c\cosh\xi\cos\eta$, we have $h_1 = h_2 = c\sqrt{\cosh^2\xi - \cos^2\eta}$, $h_3 = r$, and the governing equation for ϕ is

$$\frac{\partial^2\phi}{\partial\xi^2} - 3\coth\xi\frac{\partial\phi}{\partial\xi} + \frac{\partial^2\phi}{\partial\eta^2} - 3\cot\eta\frac{\partial\phi}{\partial\eta} = 0.$$

A general solution to the above governing equation for ϕ is

$$\begin{aligned}\phi = &\left\{c_1\left[\cosh^3\xi - 3\cosh\xi\right] + c_2\right\}\left\{c_3\left[3\cos\eta - \cos^3\eta\right] + c_4\right\} \\ &+ \sinh^2\xi\sin^2\eta\sum_{n=1}^{\infty}\left[A_nP_n^{(2)}(\cosh\xi) + B_nQ_n^{(2)}(\cosh\xi)\right] \\ &\left[C_nP_n^{(2)}(\cos\eta) + D_nQ_n^{(2)}(\cos\eta)\right],\end{aligned}$$

$$
\begin{aligned}
8c^2Gu_\phi &= 3\sinh\xi\sin\eta\Big\{(c_2-2c_1)c_3\operatorname{csch}^2\frac{\xi}{2}+4c_2c_3\log\tanh\frac{\xi}{2}+(2c_1+c_2)c_3\operatorname{sech}^2\frac{\xi}{2}\\
&\quad +c_1\left[4c_4\cot\eta\csc\eta+4c_4\log\cot\frac{\eta}{2}-9c_3+8c_3\left[\csc^2\eta+\log(2\operatorname{csch}\xi\csc\eta)\right]\right]\Big\}\\
&\quad +8i\sum_{n=1}^{\infty}(n-1)(n+2)\left[A_nP_n^{(1)}(\cosh\xi)+B_nQ_n^{(1)}(\cosh\xi)\right]\\
&\qquad\left[C_nP_n^{(1)}(\cos\eta)+D_nQ_n^{(1)}(\cos\eta)\right].
\end{aligned}
\tag{6.145}
$$

The stresses are given by Eqns. (6.142) with h_1 and r as defined above. In practice, while solving actual boundary value problems, $i\pi P_n^{(2)}(\cosh\xi)/2$ can be added to $Q_n^{(2)}(\cosh\xi)$ in order to keep the coefficients real. Thus, for example, if we consider a prolate spheroidal void defined by $\xi=\xi_0$ subjected to a far-field loading $\tau_{\theta z}|_{\xi\to\infty}=G\alpha r$, then, in place of Eqn. (6.143), we assume the potential to be

$$
\phi=\frac{1}{4}G\alpha c^4\sinh^4\xi\sin^4\eta+B_2Q_2\sinh^4\xi\sin^4\eta,
\tag{6.146}
$$

where

$$
Q_2=\coth\xi\operatorname{csch}\xi(3-2\operatorname{csch}^2\xi)-3\log\coth\frac{\xi}{2}.
$$

Imposing the condition $\phi|_{\xi=\xi_0}=0$, we get

$$
B_2=\frac{G\alpha c^4}{12\log\coth\frac{\xi_0}{2}+4\coth\xi_0\operatorname{csch}\xi_0(2\cosh^2\xi_0-3)}.
$$

Poschl [257] presents the solution for a shaft whose generating surface $\xi=\xi_0$ is a prolate spheroid as

$$
\phi=c\left[\cosh^3\xi-3\cosh\xi+2\right],
$$

where $c=T/[2\pi(\cosh^3\xi_0-3\cosh\xi_0+2)]$, which is a special case of the general solution given by Eqn. (6.145). If c is taken to be $T/[2\pi(\cosh\xi_2(\cosh^2\xi_2-3)-\cosh\xi_1(\cosh^2\xi_1-3))]$, then the above potential solves the problem of a hollow bar with bounding surfaces $\xi=\xi_1$ and $\xi=\xi_2$, similar to the geometry of the spherical shell shown in Fig. 6.18 below. In the former case of the solid prolate cylinder, the top and bottom surfaces should be at a distance less than or equal to c from the $z=0$ plane in order to meet the constraint $\phi=0$ along the z-axis.

Consider the torsion of the truncated sphere shown in Fig. 6.17. Since the domain extends to the region $z<0$, we need to consider only the infinite series part of the solution in Eqn. (6.136), i.e.,

$$
\phi=\sum_{n=2}^{\infty}(A_nR^{n+2}+B_nR^{1-n})(1-\xi^2)P_n^{(2)}(\xi),
$$

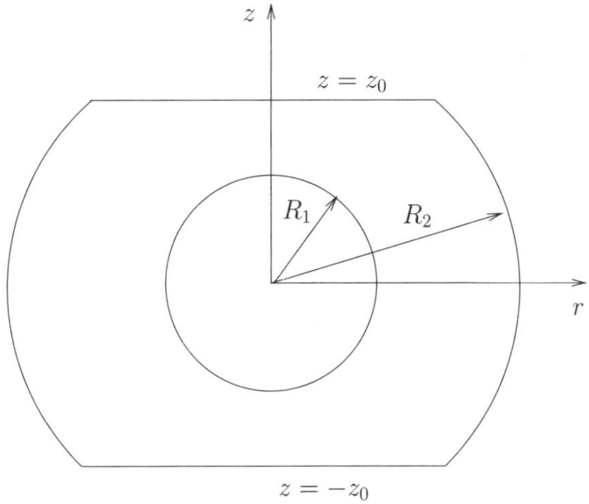

Fig. 6.17 *Truncated hollow sphere subjected to torsion.*

where $\xi = \cos\theta$. Since ϕ is zero on the axis, by continuity, it should be zero on the surface of the inner spherical surface. Thus, $B_n = -R_1^{2n+1} A_n$, and hence

$$\phi = \sum_{n=2}^{\infty} A_n R^{1-n}(R^{2n+1} - R_1^{2n+1})(1-\xi^2)P_n^{(2)}(\xi).$$

Since $u_\phi(z) = -u_\phi(-z)$, the odd n's can be excluded from the above solution, and we get

$$\phi = \sum_{n=1}^{\infty} A_n R^{1-2n}(R^{4n+1} - R_1^{4n+1})(1-\xi^2)P_{2n}^{(2)}(\xi),$$

$$\tau_{R\phi} = \sum_{n=1}^{\infty} 2(n+1)(1-2n)A_n R^{-2n-2}(R^{4n+1} - R_1^{4n+1})P_{2n}^{(1)}(\xi), \tag{6.147}$$

$$\tau_{\theta\phi} = -\sum_{n=1}^{\infty} A_n R^{-2n-2}\left[2(n+1)R^{4n+1} - (1-2n)R_1^{4n+1}\right] P_{2n}^{(2)}(\xi),$$

$$Gu_\phi = -\sum_{n=1}^{\infty} A_n R^{-1-2n}\left[2(n+1)R^{4n+1} + (2n-1)R_1^{4n+1}\right] P_{2n}^{(1)}(\xi),$$

where the constants A_n are evaluated by using Eqn. (6.129):

$$\sum_{n=1}^{\infty} A_n R_2^{1-2n}(R_2^{4n+1} - R_1^{4n+1})P_{2n}^{(2)}(\xi) = \frac{T}{2\pi(1-\xi^2)} \quad \forall \xi \in [z_0/R_2, -z_0/R_2].$$

The above equation can be solved approximately by taking a finite number of terms, and using a weighted residual technique as described on page 485. The solution for a truncated solid sphere is obtained simply by setting R_1 to zero. In the limit as $z_0 \to R_2$, we get the solution for a spherical shell subjected to concentrated torques at the poles. For this case, using the orthogonality properties in Eqn. (6.138), we get (after renaming the A_n) the solution given by Eqn. (6.147) with

$$A_n = \frac{T}{8\pi} \frac{4n+1}{(n+1)(2n-1)R_2^{1-2n}(R_2^{4n+1} - R_1^{4n+1})}.$$

For a solid sphere, by setting R_1 to zero, and renaming R_2 as a, we obtain from Eqn. (6.147) the following solution for a sphere of radius a subjected to concentrated torques at the poles $(r, z) = (0, \pm a)$:

$$\begin{aligned}
\phi &= \sum_{n=1}^{\infty} A_n R^{2n+2}(1-\xi^2)P_{2n}^{(2)}(\xi), \\
\tau_{R\phi} &= \sum_{n=1}^{\infty} 2(n+1)(1-2n)A_n R^{2n-1}P_{2n}^{(1)}(\xi), \\
\tau_{\theta\phi} &= -\sum_{n=1}^{\infty} 2(n+1)A_n R^{2n-1}P_{2n}^{(2)}(\xi), \\
Gu_\phi &= -\sum_{n=1}^{\infty} 2(n+1)A_n R^{2n}P_{2n}^{(1)}(\xi),
\end{aligned} \tag{6.148}$$

where

$$A_n = \frac{T}{8\pi} \frac{4n+1}{(n+1)(2n-1)a^{2n+2}}.$$

We now show how to carry out the summation for the series expression for ϕ in Eqn. (6.148) analytically. We first write the expression for ϕ in the form

$$\phi = \frac{T}{8\pi} \sum_{n=1}^{\infty} \left(\frac{R}{a}\right)^{2n+2} \left[\frac{1}{2n-1} + \frac{1}{2n+2}\right] 2(1-\xi^2)P_{2n}^{(2)}(\xi). \tag{6.149}$$

With $\xi := \cos\theta$, let

$$\begin{aligned}
r_1 &= \sqrt{r^2 + (a-z)^2} = \sqrt{R^2 - 2aR\xi + a^2}, \\
r_2 &= \sqrt{r^2 + (z+a)^2} = \sqrt{R^2 + 2aR\xi + a^2}.
\end{aligned}$$

The generating series for Legendre polynomials is given by

$$\sum_{n=0}^{\infty} \left(\frac{R}{a}\right)^n P_n(\xi) = \frac{a}{r_1}.$$

By using Eqn. (6.137), we get

$$\sum_{n=1}^{\infty} \left(\frac{R}{a}\right)^n P_n^{(2)}(\xi) = \frac{3(1-\xi^2)a^3R^2}{r_1^5}.$$

We write the above relation as

$$\sum_{n=1}^{\infty}\left[\left(\frac{R}{a}\right)^{2n} P_{2n}^{(2)}(\xi)+\left(\frac{R}{a}\right)^{2n-1} P_{2n-1}^{(2)}(\xi)\right]=\frac{3(1-\xi^2)a^3R^2}{r_1^5}. \tag{6.150}$$

Switching the sign on ξ, we get

$$\sum_{n=1}^{\infty}\left[\left(\frac{R}{a}\right)^{2n} P_{2n}^{(2)}(\xi)-\left(\frac{R}{a}\right)^{2n-1} P_{2n-1}^{(2)}(\xi)\right]=\frac{3(1-\xi^2)a^3R^2}{r_2^5}. \tag{6.151}$$

Adding Eqns. (6.150) and (6.151), we get

$$\sum_{n=1}^{\infty}\left(\frac{R}{a}\right)^{2n} 2(1-\xi^2)P_{2n}^{(2)}(\xi)=3(1-\xi^2)^2a^3R^2\left[\frac{1}{r_1^5}+\frac{1}{r_2^5}\right]. \tag{6.152}$$

The above equation can be written as

$$\sum_{n=1}^{\infty}\left(\frac{R^{2n-2}}{a^{2n}}\right) 2(1-\xi^2)P_{2n}^{(2)}(\xi)=3(1-\xi^2)^2a^3\left[\frac{1}{r_1^5}+\frac{1}{r_2^5}\right]. \tag{6.153}$$

On integrating Eqn. (6.153) with respect to R, noting that the integration constant is zero, and multiplying the resulting equation by R^3/a^2, we get

$$\sum_{n=1}^{\infty}\left(\frac{R^{2n+2}}{(2n-1)a^{2n+2}}\right) 2(1-\xi^2)P_{2n}^{(2)}(\xi)=3(1-\xi^2)^2R^3a\int\left[\frac{1}{r_1^5}+\frac{1}{r_2^5}\right]dR. \tag{6.154}$$

Similarly, by multiplying Eqn. (6.152) by R and integrating, and dividing the resulting equation by a^2, we get

$$\sum_{n=1}^{\infty}\left(\frac{R^{2n+2}}{(2n+2)a^{2n+2}}\right) 2(1-\xi^2)P_{2n}^{(2)}(\xi)=3(1-\xi^2)^2a\int R^3\left[\frac{1}{r_1^5}+\frac{1}{r_2^5}\right]dR+4, \tag{6.155}$$

where the last term is the constant of integration that is found by evaluating both sides at $R=0$. From Eqns. (6.149), (6.154) and (6.155), we get the following solution first obtained by Huber [137] using a method different than the one used above (see also Snell [299]; the extension to a spherical shell has been treated by Bramble [29]):

$$\begin{aligned}\frac{8\pi a^3}{T}\phi &= 4a^3-[a^2-(r^2+z^2)]\left[\frac{1}{r_2^3}+\frac{1}{r_1^3}\right]\left[2(r^2+z^2)^2+8a^2z^2+5a^2r^2+2a^4\right]\\ &\quad -6az\left[\frac{1}{r_2^3}-\frac{1}{r_1^3}\right]\left[a^4-(r^2+z^2)^2\right],\\ &= 4a^3-\frac{a^3(a-R\xi)\left[2a^2-4aR\xi-R^2(\xi^2-3)\right]}{r_1^3}-\frac{a^3(a+R\xi)\left[2a^2+4aR\xi-R^2(\xi^2-3)\right]}{r_2^3}\\ &\quad +\frac{R^3(R-a\xi)\left[2R^2-4aR\xi-a^2(\xi^2-3)\right]}{r_1^3}+\frac{R^3(R+a\xi)\left[2R^2+4aR\xi-a^2(\xi^2-3)\right]}{r_2^3},\end{aligned}$$

$$\frac{8\pi a^3 G}{T} u_\theta = 2a^3 r \left[\frac{1}{r_1^3} - \frac{1}{r_2^3} \right] + 3a^2 r \left[\frac{1}{(a-z)r_1} - \frac{1}{(a+z)r_2} \right] + \frac{3z^2}{r} \left[\frac{r_2}{a+z} - \frac{r_1}{a-z} \right] + 3r \log \frac{(a+z+r_2)}{(a-z+r_1)}. \tag{6.156}$$

This solution satisfies the governing equation given by Eqn. (6.127), the boundary condition $\phi|_{R=a} = T/(2\pi)$, and the conditions $u_\theta(0,z) = \phi(0,z) = \tau_{r\theta}(0,z) = \tau_{\theta z}(0,z) = 0$. The stresses on the equatorial plane are $\tau_{r\theta}|_{z=0} = 0$ and

$$\tau_{\theta z}|_{z=0} = \frac{3Tr(2r^4 + 5a^2r^2 + 5a^4)}{4\pi a^3 (r^2 + a^2)^{5/2}},$$

which equals $(-1)\ \tau_{\theta\phi}|_{\theta=\pi/2}$ in the spherical coordinate system. Away from the surface of the sphere $R = a$, there is almost perfect agreement between the numerical results obtained using a *finite* number of terms in the series solution given by Eqn. (6.148) and the exact solution given by Eqn. (6.156), while near to the surface the series solution does not converge. Thus, summing up the series analytically is critical in ensuring an accurate solution.

The solution for the torsion of a circular shaft of finite length L and radius R_2 with a hollow spherical cavity of radius R_1 at the center has been presented by Golovchan [95] and Ling [192]. The solution is given by Eqn. (6.147), with the coefficients A_n now determined using

$$\sum_{n=1}^{\infty} A_n \left(\frac{R_2}{\sin\theta} \right)^{1-2n} \left[\left(\frac{R_2}{\sin\theta} \right)^{4n+1} - R_1^{4n+1} \right] \sin^2\theta P_{2n}^{(2)}(\cos\theta) = \frac{T}{2\pi} \quad \forall \theta \in [\theta_0, \pi - \theta_0],$$

where $\theta_0 = \tan^{-1}(2R_2/L)$. The above equation can again be solved approximately using a weighted residual technique (see page 485). The solution for a circular shaft with a prolate spheroidal cavity has been presented in [342].

If the axis $r = 0$ is not part of the domain, then the constraint that $\phi(0,z)$ be a constant is no longer required (although now, we do require that ϕ be constant on the inner surface as well), and the solution

$$\phi = c_1 R \cos\theta + c_2 R^2 (1 + \cos^2\theta) + c_3 R^3 + \sum_{n=2}^{\infty} (C_n R^{n+2} + D_n R^{1-n}) \sin^2\theta Q_n^{(2)}(\cos\theta),$$

$$\tau_{R\phi} = -\frac{1}{R^2 \sin\theta}(c_1 + 2c_2 R\cos\theta) + \sum_{n=2}^{\infty} (n+2)(1-n)(C_n R^{n-1} + D_n R^{-n-2}) Q_n^{(1)}(\cos\theta),$$

$$\tau_{\theta\phi} = -\frac{c_1 \cos\theta}{R^2 \sin^2\theta} - \frac{c_2}{R\sin^2\theta}(3 + \cos 2\theta) - \frac{3c_3}{\sin^2\theta} - \sum_{n=2}^{\infty} \left[C_n (n+2) R^{n-1} + D_n (1-n) R^{-n-2} \right] Q_n^{(2)}(\cos\theta),$$

$$
\begin{aligned}
Gu_\phi &= \frac{c_1}{2R\sin\theta} + 2c_2\cot\theta + \frac{3c_3 R\cot\frac{\theta}{2}}{2(1+\cos\theta)}\left[\cos\theta + \sin^2\theta\log\left(\cot\frac{\theta}{2}\right)\right] \\
&\quad - \sum_{n=2}^{\infty}\left[C_n(n+2)R^n - D_n(n-1)R^{-1-n}\right]Q_n^{(1)}(\cos\theta),
\end{aligned}
$$

where $Q_n^{(m)}(\xi) = (1-\xi^2)^{m/2} d^m Q_n(\xi)/d\xi^m$ are the associated Legendre polynomials of the second kind, can be added to the solution given by Eqn. (6.136). As an example, the function $\phi = cR^3 = c(r^2+z^2)^{3/2}$, obtained by setting all constants other than c_3 to zero, yields the solution for the torsion of the truncated hollow shell with inner and outer radii R_1 and R_2 as shown in Fig. 6.18. This problem is similar to that shown in Fig. 6.17, except that now $z_0 < R_1$. The stresses and displacements are given by

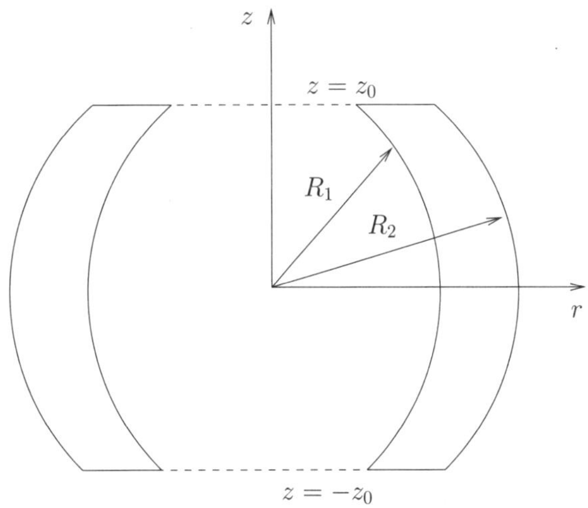

Fig. 6.18 *Truncated hollow shell subjected to torsion.*

$$
\begin{aligned}
\tau_{r\theta} &= -\frac{3c}{r^2} z\sqrt{r^2+z^2}, \\
\tau_{\theta z} &= \frac{3c}{r}\sqrt{r^2+z^2}, \\
Gu_\theta &= \frac{3cz\sqrt{r^2+z^2}}{2r} + \frac{3cr}{2}\log\frac{4(z+\sqrt{r^2+z^2})}{r},
\end{aligned}
$$

and the torque is given by $T = 2\pi c(R_2^3 - R_1^3)$.

6.5.5 Pure bending of prismatic beams

Let $\tilde{x} \equiv (x, y)$ represent the position vector of a point in the planar domain A shown in Fig. 6.19, and define the area moment of inertia tensor as

$$
\mathcal{I} = \int_A \left[(\tilde{x}\cdot\tilde{x})\boldsymbol{I} - \tilde{x}\otimes\tilde{x}\right] dA.
$$

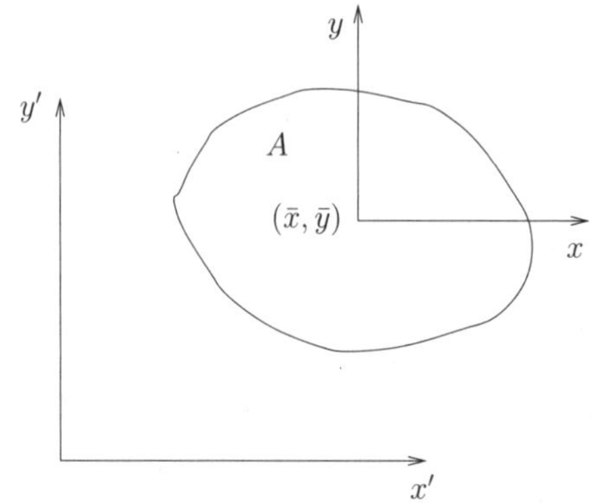

Fig. 6.19 *A planar domain A and parallel set of axes x-y and x'-y'.*

The above second-order tensor is obviously symmetric, so that the results of Section 1.6 can be applied to this tensor. In particular, its principal values are real, and if the coordinate frame is chosen to coincide with the principal axes, then the matrix of its components with respect to this frame is a diagonal matrix with the principal values along the diagonal. The components of $\mathcal{I}$ are given by

$$[\mathcal{I}] \equiv \begin{bmatrix} I_{xx} & I_{xy} \\ I_{xy} & I_{yy} \end{bmatrix} = \begin{bmatrix} \int_A y^2\,dA & -\int_A xy\,dA \\ -\int_A xy\,dA & \int_A x^2\,dA \end{bmatrix}.$$

The transformation law for the components of $\mathcal{I}$ is given by Eqn. (1.100). By choosing the coordinate frame to coincide with the principal axes so that $I_{xy} = 0$, and recalling that the determinant of a tensor is an invariant, we deduce that $\det \mathcal{I} > 0$. Note also that $I_{xy} = 0$, if the domain A is singly or doubly symmetric about the coordinate axes.

Now let the x–y axes be the centroidal axes, so that $\int_A \tilde{\boldsymbol{x}}\,dA = \boldsymbol{0}$, and let x'-y' be a parallel set of axes as shown in Fig. 6.19. Denoting the centroidal coordinates by $\bar{\boldsymbol{x}} \equiv (\bar{x}, \bar{y})$, the area moment of inertia with respect to the x'–y' axes, which we denote by $\mathcal{I}'$ is given by

$$\begin{aligned} \mathcal{I}' &= \int_A \{[(\tilde{\boldsymbol{x}} + \bar{\boldsymbol{x}}) \cdot (\tilde{\boldsymbol{x}} + \bar{\boldsymbol{x}})]\,\boldsymbol{I} - (\tilde{\boldsymbol{x}} + \bar{\boldsymbol{x}}) \otimes (\tilde{\boldsymbol{x}} + \bar{\boldsymbol{x}})\}\,dA \\ &= \mathcal{I}_c + [(\bar{\boldsymbol{x}} \cdot \bar{\boldsymbol{x}})\boldsymbol{I} - \bar{\boldsymbol{x}} \otimes \bar{\boldsymbol{x}}]\,A, \end{aligned} \tag{6.157}$$

where $\mathcal{I}_c$ is the area moment of inertia about the centroidal axes x–y, and where A is the area of cross-section. Equation (6.157) is referred to as the parallel-axis theorem.

Having introduced the area moment of inertia, we are now in a position to consider the bending of a prismatic bar by a couple $\boldsymbol{M} = (M_x, M_y)$ as shown in Fig. 6.20. We choose the z-axis to be the line of centroids[3]. The elementary solution given by

$$\tau_{zz} = -\frac{Ex}{R_x} - \frac{Ey}{R_y}, \tag{6.158}$$

[3] If it is not chosen to be the line of centroids, then replace x and y by $x - \bar{x}$ and $y - \bar{y}$, respectively, in the subsequent development including in the formulae for I_{xx}, I_{xy}, etc.

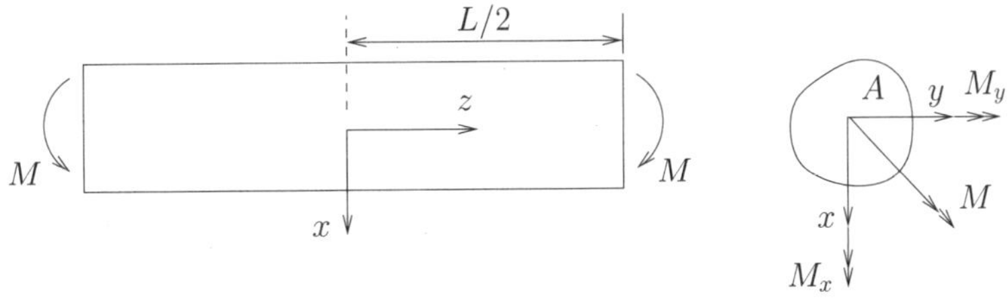

Fig. 6.20 *A prismatic beam subjected to a bending moment.*

where R_x and R_y are constants to be determined, with the other stress components zero, is an exact solution since

1. The above stress field satisfies the equations of equilibrium given by Eqn. (6.56).
2. The strain field obtained using Eqn. (6.55) as
$$\epsilon_{xx} = \epsilon_{yy} = \frac{\nu x}{R_x} + \frac{\nu y}{R_y}, \quad \epsilon_{zz} = -\frac{x}{R_x} - \frac{y}{R_y},$$
with the other components zero, satisfies the equations of compatibility given by Eqns. (6.14); the crucial role played by the compatibility equations is evident from Problem 6.
3. The lateral surface is traction-free as can be easily seen by substituting the stress field $\boldsymbol{\tau}$ and $n_z = 0$ in the relation $\boldsymbol{t} = \boldsymbol{\tau n}$.
4. On the end surfaces, the net forces are zero since $F_x = \int_A \tau_{xz}\,dA = 0$, $F_y = \int_A \tau_{yz}\,dA = 0$, and
$$F_z = \pm \int_A \tau_{zz}\,dA = \mp\frac{E}{R_x}\int_A x\,dA \mp \frac{E}{R_y}\int_A y\,dA = 0,$$
since the z-axis is the line of centroids.

The resultant moment about the x-axis on the right face is given by

$$\begin{aligned} M_x &= \int_A \left[(\boldsymbol{x} - \frac{L}{2}\boldsymbol{e}_z) \times \boldsymbol{t}\right]_x dA = \int_A \left[(x\boldsymbol{e}_x + y\boldsymbol{e}_y) \times (\tau_{zz}\boldsymbol{e}_z)\right]_x dA = \int_A y\tau_{zz}\,dA \\ &= \frac{E}{R_x}I_{xy} - \frac{E}{R_y}I_{xx}. \end{aligned} \tag{6.159}$$

Similarly, the resultant moment about the y-axis is

$$M_y = -\int_A x\tau_{zz}\,dA = \frac{E}{R_x}I_{yy} - \frac{E}{R_y}I_{xy}. \tag{6.160}$$

Solving Eqns. (6.159) and (6.160), we get (with $\boldsymbol{M} \equiv (M_x, M_y)$)

$$\begin{aligned} \frac{1}{R_x} &= \frac{-I_{xy}M_x + I_{xx}M_y}{E(I_{xx}I_{yy} - I_{xy}^2)} = \frac{1}{E}(\boldsymbol{\mathcal{I}}^{-1}\boldsymbol{M})_y, \\ \frac{1}{R_y} &= \frac{-I_{yy}M_x + I_{xy}M_y}{E(I_{xx}I_{yy} - I_{xy}^2)} = -\frac{1}{E}(\boldsymbol{\mathcal{I}}^{-1}\boldsymbol{M})_x, \end{aligned} \tag{6.161}$$

which is known as the Bernoulli–Euler law. Substituting Eqn. (6.161) into Eqn. (6.158) yields

$$\tau_{zz} = \frac{I_{xy}M_x - I_{xx}M_y}{(I_{xx}I_{yy} - I_{xy}^2)}x + \frac{I_{yy}M_x - I_{xy}M_y}{(I_{xx}I_{yy} - I_{xy}^2)}y = [(\mathcal{I}^{-1}\boldsymbol{M}) \times \tilde{\boldsymbol{x}}]_z. \tag{6.162}$$

The displacement field is obtained from the strain field and the boundary conditions $u_x = u_y = u_z = 0$ and $\frac{\partial u_x}{\partial z} = \frac{\partial u_y}{\partial z} = \frac{\partial u_y}{\partial x} = 0$ at $(x, y, z) = (0, 0, 0)$ (which prevent the linearized rigid displacement $\boldsymbol{W}\boldsymbol{X} + \boldsymbol{c}$) as

$$\begin{aligned}
u_x &= \frac{1}{2R_x}\left[z^2 + \nu(x^2 - y^2)\right] + \frac{\nu xy}{R_y},\\
u_y &= \frac{1}{2R_y}\left[z^2 - \nu(x^2 - y^2)\right] + \frac{\nu xy}{R_x},\\
u_z &= -\frac{xz}{R_x} - \frac{yz}{R_y}.
\end{aligned} \tag{6.163}$$

to zero. The deflection of the centroidal line $x = y = 0$ is obtained from the above deflection field as

$$u_x = \frac{z^2}{2R_x}, \quad u_y = \frac{z^2}{2R_y}, \quad u_z = 0.$$

From the above equations, it follows that the radii of curvature of the centroidal line in the x–z and y–z planes are

$$\begin{aligned}
\frac{1}{\kappa_x} &= \frac{\left[1 + \left(\frac{\partial u_x}{\partial z}\right)^2\right]^{\frac{3}{2}}}{\frac{\partial^2 u_x}{\partial z^2}} \approx \frac{1}{\frac{\partial^2 u_x}{\partial z^2}} = R_x,\\
\frac{1}{\kappa_y} &= \frac{\left[1 + \left(\frac{\partial u_y}{\partial z}\right)^2\right]^{\frac{3}{2}}}{\frac{\partial^2 u_y}{\partial z^2}} \approx \frac{1}{\frac{\partial^2 u_y}{\partial z^2}} = R_y.
\end{aligned}$$

This justifies the use of the notation R_x and R_y in Eqn. (6.158). Substituting Eqn. (6.161) into the above equation, we get

$$\begin{aligned}
\frac{\partial^2 u_x}{\partial z^2} &= \frac{-I_{xy}M_x + I_{xx}M_y}{E(I_{xx}I_{yy} - I_{xy}^2)} = \frac{1}{E}(\mathcal{I}^{-1}\boldsymbol{M})_y,\\
\frac{\partial^2 u_y}{\partial z^2} &= \frac{-I_{yy}M_x + I_{xy}M_y}{E(I_{xx}I_{yy} - I_{xy}^2)} = -\frac{1}{E}(\mathcal{I}^{-1}\boldsymbol{M})_x.
\end{aligned}$$

If x–y are the principal axes, then the above equations reduce to

$$EI_{yy}\frac{\partial^2 u_x}{\partial z^2} = M_y, \quad EI_{xx}\frac{\partial^2 u_y}{\partial z^2} = -M_x. \tag{6.164}$$

The above equations indicate that the more the value of EI_{xx} and EI_{yy}, the less is the magnitude of deflection. Hence, these quantities are known as the moduli of flexural rigidity.

The neutral plane is defined as the plane along which there is no extension, and is obtained simply by setting $u_z = 0$. If β is the angle made by the neutral plane with the x-axis (see Fig. 6.21), then

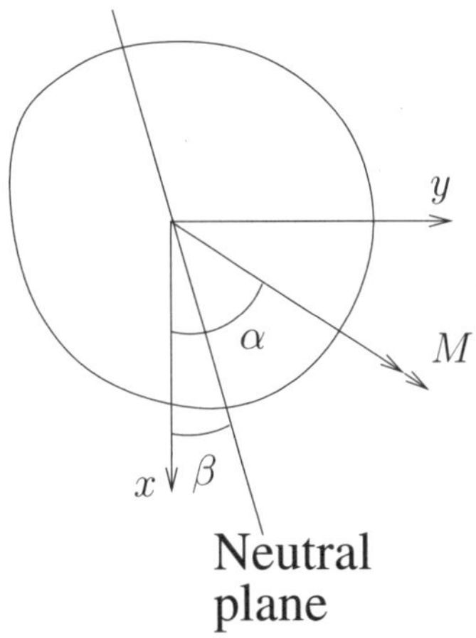

Fig. 6.21 *Neutral plane.*

$$\tan\beta = \frac{y}{x} = \frac{-I_{xy}M_x + I_{xx}M_y}{I_{yy}M_x - I_{xy}M_y} = \frac{(\boldsymbol{\mathcal{I}}^{-1}\boldsymbol{M})_y}{(\boldsymbol{\mathcal{I}}^{-1}\boldsymbol{M})_x}. \tag{6.165}$$

Note that, for the problem at hand, τ_{zz} is zero on the neutral plane. Note also that the angle β need not be the same as the angle $\alpha = \tan^{-1}(M_y/M_x)$ made by the moment vector with the x-axis.

If x and y are the principal axes of the area moment of inertia tensor, i.e., if $I_{xy} = 0$, then Eqns. (6.162) and (6.165) reduce to

$$\tau_{zz} = -\frac{M_y x}{I_{yy}} + \frac{M_x y}{I_{xx}}, \qquad \tan\beta = \frac{I_{xx}M_y}{I_{yy}M_x}. \tag{6.166}$$

As already pointed out, if the x or y axis lies along the line of symmetry in a singly or doubly symmetric beam, then $I_{xy} = 0$. As an example, consider the beam with rectangular cross-section of dimensions $3a \times a$ shown in Fig. 6.22. Choosing the x and y-axes along the principal directions, we have $I_{xy} = 0$, $I_{xx} = a^4/4$, $I_{yy} = 9a^4/4$, $M_x = M/\sqrt{2}$ and $M_y = M/\sqrt{2}$. From Eqn. (6.166), we get

$$\tau_{zz} = -\frac{4Mx}{9\sqrt{2}a^4} + \frac{4My}{\sqrt{2}a^4}, \quad \beta = \tan^{-1}\left(\frac{1}{9}\right).$$

The stresses at points A and B are

$$\tau_{zz}\big|_A = -\frac{4\sqrt{2}M}{3a^3}, \quad \tau_{zz}\big|_B = \frac{2\sqrt{2}M}{3a^3}.$$

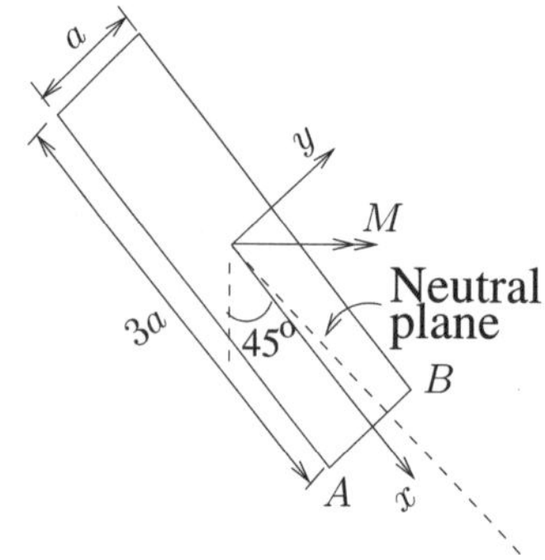

Fig. 6.22 *An inclined rectangular beam undergoing pure bending.*

Note that the normal stress at point A is compressive, while that at point B is tensile.

The pure bending of curved beams in the shape of an incomplete torus has been treated by Sadowsky and Sternberg [277].

6.5.6 Bending of prismatic beams by terminal loads

In this section, we consider the problem of a cantilever beam of uniform cross-section having one end $z = 0$ fixed, and the other end $z = L$ loaded by a traction distribution that is statically equivalent to a force $(W_x, W_y, 0)$ lying in the plane $z = L$, and acting at the point (x_0, y_0, L) (see Fig. 6.23), i.e.,

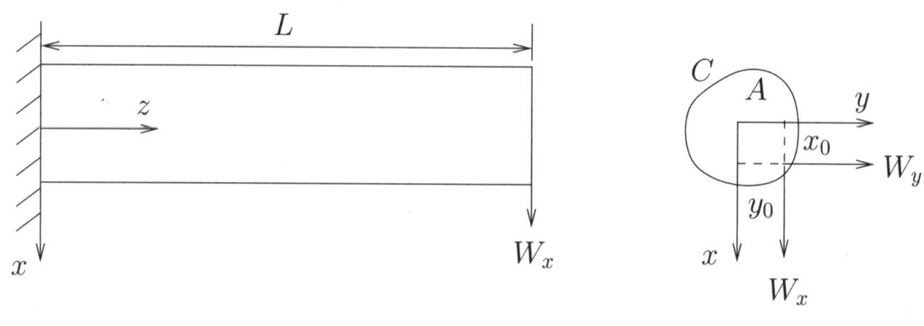

Fig. 6.23 *Bending of a prismatic beam by terminal loads.*

$$\int_A t_x \, dA = W_x, \quad \int_A t_y \, dA = W_y, \quad \int_A [(\boldsymbol{x} - L\boldsymbol{e}_z) \times \boldsymbol{t}]_z \, dA = x_0 W_y - y_0 W_x. \quad (6.167)$$

As in the case of pure bending, we take the z-axis to coincide with the line of centroids[4], but allow the x and y-axes to be arbitrary. The lateral surfaces are traction-free and the body forces are assumed to be negligible.

We make certain assumptions about the stress field that are later justified by showing that all the conditions for an exact solution are met by the resulting solution. Thus, we

[4]As usual, if it is not chosen to be the line of centroids, then replace x, y, x_0, y_0 by $x - \bar{x}$, $y - \bar{y}$, $x_0 - \bar{x}$ and $y_0 - \bar{y}$, respectively, in the subsequent development including in the formulae for I_{xx}, I_{xy}, etc.

assume that $\tau_{xx} = \tau_{xy} = \tau_{yy} = 0$, and guided by our experience of pure bending where the normal stress is proportional to the bending moment, we also assume that

$$\tau_{zz} = -E(L - z)(\kappa_x x + \kappa_y y), \tag{6.168}$$

where the constants κ_x and κ_y are to be determined from the first two conditions in Eqn. (6.167). With the above assumptions the equilibrium equations reduce to

$$\frac{\partial \tau_{xz}}{\partial z} = 0, \tag{6.169a}$$

$$\frac{\partial \tau_{yz}}{\partial z} = 0, \tag{6.169b}$$

$$\frac{\partial \tau_{xz}}{\partial x} + \frac{\partial \tau_{yz}}{\partial y} + E(\kappa_x x + \kappa_y y) = 0, \tag{6.169c}$$

while the traction boundary conditions on the lateral surface reduce to $t_x = t_y = 0$, and

$$t_z = \tau_{xz} n_x + \tau_{yz} n_y = 0 \quad \text{on } C. \tag{6.170}$$

Since the unit normal on the end surface is $(0, 0, 1)$, the first of Eqns. (6.167) can be written as

$$\begin{aligned}
W_x &= \int_A \tau_{xz}\, dA \\
&= \int_A \left[\tau_{xz} + x \left(\frac{\partial \tau_{xz}}{\partial x} + \frac{\partial \tau_{yz}}{\partial y} + E(\kappa_x x + \kappa_y y) \right) \right] dA && \text{(by Eqn. (6.169c))} \\
&= \int_A \left(\frac{\partial (x\tau_{xz})}{\partial x} + \frac{\partial (x\tau_{yz})}{\partial y} \right) dA + \int_A Ex(\kappa_x x + \kappa_y y)\, dA \\
&= \int_C x(\tau_{xz} n_x + \tau_{yz} n_y)\, ds + \int_A Ex(\kappa_x x + \kappa_y y)\, dA && \text{(divergence theorem)} \\
&= \int_A Ex(\kappa_x x + \kappa_y y)\, dA && \text{(by Eqn. (6.170))} \\
&= E\kappa_x I_{yy} - E\kappa_y I_{xy}.
\end{aligned}$$

Similarly,

$$\begin{aligned}
W_y &= \int_A \tau_{yz}\, dA = \int_A \left[\tau_{yz} + y \left(\frac{\partial \tau_{xz}}{\partial x} + \frac{\partial \tau_{yz}}{\partial y} + E(\kappa_x x + \kappa_y y) \right) \right] dA \\
&= -E\kappa_x I_{xy} + E\kappa_y I_{xx}.
\end{aligned}$$

Solving these two equations, we get

$$\kappa_x = \frac{I_{xx} W_x + I_{xy} W_y}{E(I_{xx} I_{yy} - I_{xy}^2)}, \quad \kappa_y = \frac{I_{xy} W_x + I_{yy} W_y}{E(I_{xx} I_{yy} - I_{xy}^2)}. \tag{6.171}$$

The fact that the axial force on any face $z = c$, where c is a constant, is zero follows since, the z-axis being the line of centroids, $\int_A x\,dA = \int_A y\,dA = 0$, which implies that

$$\int_A t_z\,dA = \int_A \tau_{zz}\,dA = E(c-L)\kappa_x \int_A x\,dA + E(c-L)\kappa_y \int_A y\,dA = 0.$$

Similar to the derivation of Eqns. (6.159) and (6.160), we obtain the resultant moments on a plane $z = c$ as

$$M_x = \int_A y\tau_{zz}\,dA = -W_y(L-c), \quad M_y = -\int_A x\tau_{zz}\,dA = W_x(L-c).$$

From Eqn. (6.169), we can write the stresses in terms of a stress function $\phi(x, y)$ as

$$\begin{aligned} \tau_{xz} &= \frac{\partial\phi}{\partial y} + f(y) - \frac{1}{2}E\kappa_x x^2, \\ \tau_{yz} &= -\frac{\partial\phi}{\partial x} - g(x) - \frac{1}{2}E\kappa_y y^2. \end{aligned} \tag{6.172}$$

Although the functions $f(y)$ and $g(x)$ can be absorbed into ϕ, it is convenient to keep them in the above form since it results in a simpler boundary condition for ϕ as explained below. Substituting the corresponding strain field given by $\gamma_{xy} = 0$ and

$$\epsilon_{xx} = \epsilon_{yy} = -\frac{\nu}{E}\tau_{zz}, \quad \epsilon_{zz} = \frac{1}{E}\tau_{zz}, \quad \gamma_{xz} = \frac{1}{G}\tau_{xz}, \quad \gamma_{yz} = \frac{1}{G}\tau_{yz}, \tag{6.173}$$

into the compatibility conditions given by Eqn. (6.14), we get

$$\begin{aligned} \frac{\partial\nabla^2\phi}{\partial x} &= -2G\nu\kappa_y - \frac{d^2 g}{dx^2}, \\ \frac{\partial\nabla^2\phi}{\partial y} &= 2G\nu\kappa_x - \frac{d^2 f}{dy^2}, \end{aligned}$$

which on integrating leads to

$$\nabla^2\phi = -2G\nu\kappa_y x - \frac{dg}{dx} + 2G\nu\kappa_x y - \frac{df}{dy} - 2G\alpha. \tag{6.174}$$

The constant $2G\alpha$ is evaluated by means of Eqn. (6.179) below.

Substituting Eqns. (6.172) into Eqn. (6.170), we get

$$\frac{\partial\phi}{\partial y}\frac{dy}{ds} + \frac{\partial\phi}{\partial x}\frac{dx}{ds} \equiv \frac{d\phi}{ds} = \left[\frac{1}{2}E\kappa_x x^2 - f(y)\right]\frac{dy}{ds} - \left[\frac{1}{2}E\kappa_y y^2 + g(x)\right]\frac{dx}{ds}. \tag{6.175}$$

The functions $g(x)$ and $f(y)$ can be chosen arbitrarily (even as zero), and so, *in general*, ϕ is a function of s on the boundary. Summarizing, ϕ is obtained by solving Eqn. (6.174) subject to the boundary condition given by Eqn. (6.175). Since ϕ is a single-valued function of s on the boundary, we have to ensure that $\int_C \frac{d\phi}{ds}\,ds = 0$. This is easily verified since

$$\int_C \left\{\left[\frac{1}{2}E\kappa_x x^2 - f(y)\right]\frac{dy}{ds} - \left[\frac{1}{2}E\kappa_y y^2 + g(x)\right]\frac{dx}{ds}\right\} ds$$

$$= \int_C \left\{ \left[\frac{1}{2} E\kappa_x x^2 - f(y) \right] n_x + \left[\frac{1}{2} E\kappa_y y^2 + g(x) \right] n_y \right\} ds$$

$$= \int_A E(\kappa_x x + \kappa_y y)\, dA \qquad \text{(divergence theorem)}$$

$$= 0. \qquad \left(\int_A x\, dA = \int_A y\, dA = 0 \right)$$

For certain cross-sections and types of loading, $g(x)$ and $f(y)$ can be chosen such that[5]

$$\left[\frac{1}{2} E\kappa_x x^2 - f(y) \right] \frac{dy}{ds} = 0, \quad \text{on } C, \tag{6.176a}$$

$$\left[\frac{1}{2} E\kappa_y y^2 + g(x) \right] \frac{dx}{ds} = 0, \quad \text{on } C, \tag{6.176b}$$

so that it follows that $d\phi/ds = 0$, or $\phi = \text{const.}$ on C. Since the value of this constant does not affect the stresses, we can take it to be zero. Thus, in such cases, ϕ is to be obtained by solving Eqn. (6.174) subject to the boundary condition

$$\phi = 0 \quad \text{on } C. \tag{6.177}$$

Note that if, say $dy/ds = 0$ on part of the boundary, then $f(y)$ can be prescribed arbitrarily on that part; on the other hand, if $dy/dS \neq 0$ on part of the boundary, then $f(y)$ has to equal $E\kappa_x x^2/2$ on that part. Similar comments apply for $g(x)$. We reiterate that the simplified boundary conditions given by Eqn. (6.176) are applicable only for certain types of cross-sections and loading. In general, ϕ is a function of s on the boundary obtained via Eqn. (6.175) by arbitrarily prescribing $g(x)$ and $f(y)$.

By substituting the expressions for the stresses given by Eqn. (6.172) into the third of Eqns. (6.167) and simplifying, we get an expression for determining the constant term $2G\alpha$ that occurs in Eqn. (6.174) as follows:

$$x_0 W_y - y_0 W_x = \int_A (x\tau_{yz} - y\tau_{xz})\, dA \tag{6.178}$$

$$= \int_A \left[x \left(-\frac{\partial \phi}{\partial x} - g(x) - \frac{1}{2} E\kappa_y y^2 \right) - y \left(\frac{\partial \phi}{\partial y} + f(y) - \frac{1}{2} E\kappa_x x^2 \right) \right] dA$$

$$= \int_A (2\phi - E\kappa_x x^2 y + E\kappa_y x y^2)\, dA + \int_A \left[\frac{\partial}{\partial x} \left(-x\phi - xyf(y) + \frac{1}{2} E\kappa_x x^3 y \right) \right.$$

$$\left. - \frac{\partial}{\partial y} \left(y\phi + xyg(x) + \frac{1}{2} E\kappa_y x y^3 \right) \right] dA$$

$$= \int_A (2\phi - E\kappa_x x^2 y + E\kappa_y x y^2)\, dA + \int_C \left[\left(-x\phi - xyf(y) + \frac{1}{2} E\kappa_x x^3 y \right) \frac{dy}{ds} \right.$$

$$\left. + \left(y\phi + xyg(x) + \frac{1}{2} E\kappa_y x y^3 \right) \frac{dx}{ds} \right] ds \quad \text{(divergence theorem)}$$

[5]We shall show while discussing the case of the hollow circular cylinder below that this may not always be possible.

$$= \int_A (2\phi - E\kappa_x x^2 y + E\kappa_y x y^2)\, dA. \qquad \text{(by Eqns. (6.176) and (6.177))} \tag{6.179}$$

As mentioned earlier, the constant term $2G\alpha$ is determined from the above equation.

To see the physical interpretation of α, recall that by Eqn. (6.2), $\boldsymbol{\nabla u} - (\boldsymbol{\nabla u})^T$, and hence $\boldsymbol{\omega} := \boldsymbol{\nabla} \times \boldsymbol{u}$, which is the axial vector of $\boldsymbol{\nabla u} - (\boldsymbol{\nabla u})^T$, is a measure of the rigid rotation undergone by an element. In particular, the component $\omega_z = \frac{1}{2}\left(\frac{\partial u_y}{\partial x} - \frac{\partial u_x}{\partial y}\right)$ is a measure of the rigid rotation undergone by an element in the x–y plane. Define the local twist at a point (x, y) in the cross-section as

$$\begin{aligned}
\frac{\partial \omega_z}{\partial z} &= \frac{1}{2}\left(\frac{\partial^2 u_y}{\partial x \partial z} - \frac{\partial^2 u_x}{\partial y \partial z}\right) = \frac{1}{2}\left(\frac{\partial \gamma_{yz}}{\partial x} - \frac{\partial \gamma_{xz}}{\partial y}\right) \\
&= \frac{1}{2G}\left(\frac{\partial \tau_{yz}}{\partial x} - \frac{\partial \tau_{xz}}{\partial y}\right).
\end{aligned} \tag{6.180}$$

For the problem of pure torsion, by substituting Eqn. (6.83) into the above equation, we get

$$\frac{\partial \omega_z}{\partial z} = -\frac{\nabla^2 \phi}{2G} = \alpha.$$

This confirms that $\partial \omega_z / \partial z$ is a reliable measure of the local twist. Applying this measure to the problem at hand, by substituting the stresses given by Eqn. (6.172) into Eqn. (6.180), we get

$$\begin{aligned}
\frac{\partial \omega_z}{\partial z} &= -\frac{1}{2G}\left[\nabla^2 \phi + \frac{dg}{dx} + \frac{df}{dy}\right] \\
&= -\frac{1}{2G}\left[-2G\nu\kappa_y x + 2G\nu\kappa_x y - 2G\alpha\right] \\
&= \alpha + \nu\left[\kappa_y x - \kappa_x y\right].
\end{aligned}$$

Thus, since $\int_A x\, dA = \int_A y\, dA = 0$, the mean value of twist $\left(\int_A \frac{\partial \omega_z}{\partial z}\, dA\right)/A$ (or, alternatively, the value of the twist at the origin) is given by α. Thus, in general, there is twisting of the cross-section in addition to bending.

The center of flexure or the shear center is defined as the load point $(x_{\text{cf}}, y_{\text{cf}}, L)$ for which the mean local twist α is zero. To find an expression for $(x_{\text{cf}}, y_{\text{cf}})$, we rewrite Eqn. (6.179) as

$$x_{\text{cf}} W_y - y_{\text{cf}} W_x = \int_A (2\phi - E\kappa_x x^2 y + E\kappa_y x y^2)\, dA,$$

where, now, ϕ is determined from Eqn. (6.174) by setting α to zero. Since the above equation must hold for arbitrary choices of W_x and W_y, by successively setting W_x and then W_y to zero, we get

$$x_{\text{cf}} = \int_A \left[\frac{2\phi}{W_y} - \frac{I_{xy}}{I_{xx} I_{yy} - I_{xy}^2} x^2 y + \frac{I_{yy}}{I_{xx} I_{yy} - I_{xy}^2} x y^2\right] dA, \tag{6.181a}$$

$$y_{\rm cf} = -\int_A \left[\frac{2\phi}{W_x} - \frac{I_{xx}}{I_{xx}I_{yy} - I_{xy}^2} x^2 y + \frac{I_{xy}}{I_{xx}I_{yy} - I_{xy}^2} xy^2 \right] dA. \tag{6.181b}$$

Since ϕ is proportional to W_y in the first case, and proportional to W_x in the second, the shear center is determined by the shape of the cross-section and Poisson ratio only. If the cross-section A is symmetric about the y-axis (so that $I_{xy} = 0$), then one can show that the integrand is an odd function of x, and hence $x_{\rm cf} = 0$. Similarly if the cross-section is symmetric about the x-axis, then $y_{\rm cf} = 0$. For a doubly symmetric cross-section, the center of flexure lies at the centroid. Note that the center of flexure can lie outside the cross-section of the beam.

Although one can resolve the general flexure problem into a bending problem with the load acting through the center of flexure ($\alpha = 0$), and a torsion problem[6] with twist α, due to a couple $W_y(x_0 - x_{\rm cf}) - W_x(y_0 - y_{\rm cf})$, one can also solve a given problem directly by using Eqns. (6.174)–(6.179).

The angle made by the neutral plane, obtained by setting τ_{zz} as given by Eqn. (6.168) to zero, is given by

$$\tan\beta = \frac{y}{x} = -\frac{\kappa_x}{\kappa_y} = -\frac{I_{xx}W_x + I_{xy}W_y}{I_{xy}W_x + I_{yy}W_y},$$

where β is, in general, different from the angle subtended by the resultant load vector, $\gamma = \tan^{-1}(W_y/W_x)$, with the x-axis.

The displacement field is obtained by integrating the strain field given by Eqn. (6.173) subject to the boundary conditions

$$u_x = u_y = u_z = \frac{\partial u_x}{\partial z} = \frac{\partial u_y}{\partial z} = \frac{\partial u_y}{\partial x} = 0 \quad \text{at} \quad (x, y, z) = (0, 0, 0). \tag{6.182}$$

The expression for u_z contains a function $R(x, y)$ that depends on the type of cross-section. For convenience, we write this function in terms of another function $\chi(x, y)$ as $R(x, y) = \chi(x, y) - (\kappa_x x + \kappa_y y)(x^2 + y^2)/4$. Thus, the displacements and corresponding stresses are given by

$$u_x = -\alpha yz + \kappa_x \left[\frac{1}{2}\nu(L - z)(x^2 - y^2) - \frac{1}{6}z^3 + \frac{1}{2}Lz^2 \right] + \kappa_y \nu(L - z)xy, \tag{6.183a}$$

$$u_y = \alpha xz + \kappa_x \nu(L - z)xy + \kappa_y \left[\frac{1}{2}\nu(L - z)(y^2 - x^2) - \frac{1}{6}z^3 + \frac{1}{2}Lz^2 \right], \tag{6.183b}$$

$$u_z = \left(\frac{1}{2}z^2 - Lz \right)(\kappa_x x + \kappa_y y) + \chi(x, y) - \frac{1}{4}(\kappa_x x + \kappa_y y)(x^2 + y^2), \tag{6.183c}$$

$$\frac{\tau_{xz}}{G} = -\alpha y + \frac{\partial \chi}{\partial x} - \frac{\kappa_x}{4}\left[3x^2 + y^2 + 2\nu(x^2 - y^2) \right] - \frac{\kappa_y xy}{2}(1 + 2\nu), \tag{6.183d}$$

$$\frac{\tau_{yz}}{G} = \alpha x + \frac{\partial \chi}{\partial y} - \frac{\kappa_x xy}{2}(1 + 2\nu) - \frac{\kappa_y}{4}\left[x^2 + 3y^2 + 2\nu(y^2 - x^2) \right], \tag{6.183e}$$

$$\tau_{zz} = -E(L - z)(\kappa_x x + \kappa_y y), \tag{6.183f}$$

[6]This is possible since the assumed displacement field for the torsion problem as given by Eqn. (6.59) satisfies the boundary conditions for the flexure problem as given in Eqn. (6.182).

where, using Eqns. (6.169c) and (6.170), the governing equation and boundary condition for determining the function $\chi(x,y)$ are obtained as

$$\nabla^2\chi = 0, \tag{6.184a}$$

$$\boldsymbol{n}\cdot\boldsymbol{\tau}_c = 0, \tag{6.184b}$$

where $\boldsymbol{\tau}_c = (\tau_{xz}, \tau_{yz})$. Equations (6.183a)–(6.183e) can be expressed in tensorial form as

$$\boldsymbol{u}_c = -\alpha z \boldsymbol{W}\boldsymbol{x} + \nu(L-z)\left[\boldsymbol{x}\otimes\boldsymbol{x} - \frac{\boldsymbol{x}\cdot\boldsymbol{x}}{2}\boldsymbol{I}\right]\boldsymbol{\kappa} + \frac{z^2}{2}\left(L-\frac{z}{3}\right)\boldsymbol{\kappa},$$

$$u_z = \left(\frac{1}{2}z^2 - Lz\right)(\boldsymbol{\kappa}\cdot\boldsymbol{x}) + \chi - \frac{1}{4}(\boldsymbol{\kappa}\cdot\boldsymbol{x})(\boldsymbol{x}\cdot\boldsymbol{x}),$$

$$\frac{1}{G}\boldsymbol{\tau}_c = -\alpha\boldsymbol{W}\boldsymbol{x} + \boldsymbol{\nabla}\chi - \frac{1}{4}\left[2(1+2\nu)(\boldsymbol{x}\otimes\boldsymbol{x}) + (1-2\nu)(\boldsymbol{x}\cdot\boldsymbol{x})\boldsymbol{I}\right]\boldsymbol{\kappa},$$

$$\tau_{zz} = -E(L-z)\boldsymbol{\kappa}\cdot\boldsymbol{x},$$

where $\boldsymbol{W} = \left[\begin{smallmatrix}0 & 1\\ -1 & 0\end{smallmatrix}\right]$. A numerical implementation of the above equations may be found in [166].

If $(\bar{x}_c, \bar{y}_c)$ denote the coordinates of the centroid from the origin of the cylindrical system (with θ measured anticlockwise from the y-axis), the displacement and stress components in the cylindrical coordinate system, obtained using the relations $x+\bar{x}_c = -r\sin\theta$, $y+\bar{y}_c = r\cos\theta$, $u_r = u_y\cos\theta - u_x\sin\theta$, $u_\theta = -u_x\cos\theta - u_y\sin\theta$, $\tau_{rz} = -\tau_{xz}\sin\theta + \tau_{yz}\cos\theta$ and $\tau_{\theta z} = -\tau_{xz}\cos\theta - \tau_{yz}\sin\theta$, are

$$\begin{aligned}6u_r &= -6\alpha z(\bar{x}_c\cos\theta + \bar{y}_c\sin\theta) + \kappa_x\Big\{-6\nu\bar{x}_c(L-z)(r-\bar{y}_c\cos\theta)\\ &\quad + \left[3\nu(z-L)(r^2+\bar{x}_c^2-\bar{y}_c^2) + z^3 - 3Lz^2\right]\sin\theta\Big\}\\ &\quad + \kappa_y\Big\{-6\nu\bar{y}_c(L-z)(r+\bar{x}_c\sin\theta)\\ &\quad + \left[3\nu(L-z)(r^2-\bar{x}_c^2+\bar{y}_c^2) - z^3 + 3Lz^2\right]\cos\theta\Big\},\end{aligned} \tag{6.185a}$$

$$\begin{aligned}6u_\theta &= 6\alpha z(r + \bar{x}_c\sin\theta - \bar{y}_c\cos\theta) + \kappa_x\Big\{-6\nu\bar{y}_c(L-z)(r+\bar{x}_c\sin\theta)\\ &\quad + \left[3\nu(L-z)(r^2-\bar{x}_c^2+\bar{y}_c^2) + z^3 - 3Lz^2\right]\cos\theta\Big\}\\ &\quad + \kappa_y\Big\{6\nu\bar{x}_c(L-z)(r-\bar{y}_c\cos\theta)\\ &\quad + \left[3\nu(L-z)(r^2+\bar{x}_c^2-\bar{y}_c^2) + z^3 - 3Lz^2\right]\sin\theta\Big\},\end{aligned} \tag{6.185b}$$

$$\frac{\tau_{rz}}{G} = \alpha\left[-\bar{x}_c\cos\theta - \bar{y}_c\sin\theta\right] + \frac{\partial\chi}{\partial r}$$

$$+\frac{\kappa_x}{4}\Big\{4(1+\nu)r\bar{x}_c+\Big[(3+2\nu)(r^2+\bar{x}_c^2)+(1-2\nu)\bar{y}_c^2\Big]\sin\theta$$

$$-2r\bar{x}_c\cos 2\theta-2(1+2\nu)\bar{x}_c\bar{y}_c\cos\theta-2r\bar{y}_c\sin 2\theta\Big\}$$

$$+\frac{\kappa_y}{4}\Big\{(4+4\nu+2\cos 2\theta)r\bar{y}_c+2(1+2\nu)\bar{x}_c\bar{y}_c\sin\theta-2r\bar{x}_c\sin 2\theta$$

$$-\Big[(3+2\nu)(r^2+\bar{y}_c^2)+(1-2\nu)\bar{x}_c^2\Big]\cos\theta\Big\}, \tag{6.185c}$$

$$\frac{\tau_{\theta z}}{G}=\alpha\left[r+\bar{x}_c\sin\theta-\bar{y}_c\cos\theta\right]+\frac{1}{r}\frac{\partial\chi}{\partial\theta}$$

$$+\frac{\kappa_x}{4}\Big\{2r\bar{y}_c(2\nu-\cos 2\theta)+\Big[(1-2\nu)(r^2+\bar{y}_c^2)+(3+2\nu)\bar{x}_c^2\Big]\cos\theta$$

$$+2\bar{x}_c\bar{y}_c(1+2\nu)\sin\theta+2r\bar{x}_c\sin 2\theta\Big\}$$

$$+\frac{\kappa_y}{4}\Big\{-2r\bar{x}_c(\cos 2\theta+2\nu)+\Big[(1-2\nu)(r^2+\bar{x}_c^2)+(3+2\nu)\bar{y}_c^2\Big]\sin\theta$$

$$+2(1+2\nu)\bar{x}_c\bar{y}_c\cos\theta-2r\bar{y}_c\sin 2\theta\Big\}. \tag{6.185d}$$

The moment condition given by Eqn. (6.178) reduces to

$$x_0W_y-y_0W_x=\int_A\left[r\tau_{\theta z}-\bar{x}_c(\tau_{rz}\cos\theta-\tau_{\theta z}\sin\theta)-\bar{y}_c(\tau_{rz}\sin\theta+\tau_{\theta z}\cos\theta)\right]dA. \tag{6.186}$$

Similarly, for an elliptical coordinate system, using the relations $x=c\cosh\xi\cos\eta$, $y=c\sinh\xi\sin\eta$ (assuming the origin of the elliptical coordinate system to lie at the centroid), $\begin{bmatrix}u_\xi\\u_\eta\end{bmatrix}=Q\begin{bmatrix}u_x\\u_y\end{bmatrix}$, and $\begin{bmatrix}\tau_{\xi z}\\\tau_{\eta z}\end{bmatrix}=Q\begin{bmatrix}\tau_{xz}\\\tau_{yz}\end{bmatrix}$, where Q is the in-plane part of the matrix given in Eqn. (D.3), and with h_1 given by Eqn. (D.2), we get

$$\frac{24h_1}{c}u_\xi=12\alpha cz\sin 2\eta$$

$$+\kappa_x\Big\{3c^2\nu(z-L)\cos 3\eta+\Big[4(3L-z)z^2+3\nu c^2(L-z)(2\cosh 2\xi+3)\Big]\cos\eta\Big\}\sinh\xi$$

$$+\kappa_y\Big\{3c^2\nu(L-z)\left[2\cosh 2\xi\sin\eta-\sin 3\eta\right]+\Big[9c^2\nu(z-L)+4(3L-z)z^2\Big]\sin\eta\Big\}\cosh\xi, \tag{6.187a}$$

$$\frac{24h_1}{c}u_\eta=12\alpha cz\sinh 2\xi$$

$$+\kappa_x\Big\{3c^2\nu(L-z)\left[2\cosh 2\xi\sin\eta-\sin 3\eta\right]+\Big[9c^2\nu(z-L)-4(3L-z)z^2\Big]\sin\eta\Big\}\cosh\xi$$

$$+\kappa_y\Big\{3c^2\nu(L-z)\cos 3\eta+\Big[4(3L-z)z^2+3\nu c^2(z-L)(2\cosh 2\xi+3)\Big]\cos\eta\Big\}\sinh\xi, \tag{6.187b}$$

$$\begin{aligned}\frac{8h_1\tau_{\xi z}}{G} &= 4\alpha c^2\sin 2\eta + 8\frac{\partial\chi}{\partial\xi}\\ &\quad - c^3\kappa_x\left[2(1+2\nu)+(1-2\nu)\cos 2\eta+(3+2\nu)\cosh 2\xi\right]\sinh\xi\cos\eta\\ &\quad + c^3\kappa_y\left[2(1+2\nu)-(1-2\nu)\cos 2\eta-(3+2\nu)\cosh 2\xi\right]\cosh\xi\sin\eta,\end{aligned} \tag{6.187c}$$

$$\begin{aligned}\frac{8h_1\tau_{\eta z}}{G} &= 4\alpha c^2\sinh 2\xi + 8\frac{\partial\chi}{\partial\eta}\\ &\quad + c^3\kappa_x\left[2(1+2\nu)+(1-2\nu)\cosh 2\xi+(3+2\nu)\cos 2\eta\right]\cosh\xi\sin\eta\\ &\quad + c^3\kappa_y\left[2(1+2\nu)-(1-2\nu)\cosh 2\xi-(3+2\nu)\cos 2\eta\right]\sinh\xi\cos\eta.\end{aligned} \tag{6.187d}$$

The moment condition given by Eqn. (6.178) reduces to

$$x_0W_y - y_0W_x = \frac{c^2}{2}\int_A \frac{1}{h_1}(\tau_{\xi z}\sin 2\eta + \tau_{\eta z}\sinh 2\xi)\,dA. \tag{6.188}$$

For a bipolar coordinate system using the relations $x = h_1\sin\eta - \bar{x}_c$, $y = h_1\sinh\xi - \bar{y}_c$, where $h_1 = c/(\cosh\xi - \cos\eta)$, we have

$$\begin{aligned}\begin{bmatrix}u_\xi\\ u_\eta\end{bmatrix} &= Q\begin{bmatrix}u_x\\ u_y\end{bmatrix},\\ \begin{bmatrix}\tau_{\xi z}\\ \tau_{\eta z}\end{bmatrix} &= Q\begin{bmatrix}\tau_{xz}\\ \tau_{yz}\end{bmatrix} + \begin{bmatrix}\frac{1}{h_1}\frac{\partial\chi}{\partial\xi}\\ \frac{1}{h_1}\frac{\partial\chi}{\partial\eta}\end{bmatrix},\end{aligned} \tag{6.189}$$

where Q is given by the in-plane part of the Q in Eqn. (E.3), (u_x, u_y) are given by Eqns. (6.183a), (6.183b), and (τ_{xz}, τ_{yz}) are given by Eqns. (6.183d) and (6.183e) with the χ term excluded (since we have incorporated that separately in Eqn. (6.189)). The moment condition given by Eqn. (6.178) reduces to

$$\begin{aligned}x_0W_y - y_0W_x = \frac{1}{c}\int_A h_1\Big\{&(\bar{x}_c - h_1\sin\eta)[\tau_{\xi z}(\cosh\xi\cos\eta - 1) + \tau_{\eta z}\sinh\xi\sin\eta]\\ &+ (\bar{y}_c - h_1\sinh\xi)[\tau_{\eta z}(\cosh\xi\cos\eta - 1) - \tau_{\xi z}\sinh\xi\sin\eta]\Big\}\,dA.\end{aligned} \tag{6.190}$$

Note that no integration constants need be added while finding χ since the linearized rigid displacements $\boldsymbol{W}\boldsymbol{x}+\boldsymbol{c}$ have already been suppressed by specifying appropriate boundary conditions. From Eqns. (6.183a) and (6.183b), it follows that the displacements of the centroidal line are given by

$$\begin{aligned}u_x &= \frac{1}{6}\kappa_x z^2(3L - z),\\ u_y &= \frac{1}{6}\kappa_y z^2(3L - z),\end{aligned} \tag{6.191}$$

with the maximum deflection occurring at $z = L$, and given by

$$u_x\big|_{z=L} = \frac{1}{3}\kappa_x L^3 = \frac{(I_{xx}W_x + I_{xy}W_y)L^3}{3E(I_{xx}I_{yy} - I_{xy}^2)},$$

$$u_y\big|_{z=L} = \frac{1}{3}\kappa_y L^3 = \frac{(I_{xy}W_x + I_{yy}W_y)L^3}{3E(I_{xx}I_{yy} - I_{xy}^2)}.$$

When the x and y axes coincide with the principal axes, the maximum deflections are

$$u_x\big|_{z=L} = \frac{W_x L^3}{3EI_{yy}}, \quad u_y\big|_{z=L} = \frac{W_y L^3}{3EI_{xx}}.$$

Using Eqn. (6.191), the radii of curvature of the centroidal line in the x–z and y–z planes, R_x and R_y, are given by

$$\frac{1}{R_x} \approx \frac{\partial^2 u_x}{\partial z^2} = (L - z)\kappa_x, \qquad \frac{1}{R_y} \approx \frac{\partial^2 u_y}{\partial z^2} = (L - z)\kappa_y,$$

justifying the use of the terminology κ_x and κ_y used in Eqn. (6.168). Substituting Eqn. (6.171) for κ_x and κ_y, and since we have shown that $M_x = -W_y(L - z)$ and $M_y = W_x(L - z)$, we get

$$\frac{1}{R_x} \approx \frac{\partial^2 u_x}{\partial z^2} = \frac{-I_{xy}M_x + I_{xx}M_y}{E(I_{xx}I_{yy} - I_{xy}^2)} = \frac{1}{E}(\boldsymbol{\mathcal{I}}^{-1}\boldsymbol{M})_y,$$

$$\frac{1}{R_y} \approx \frac{\partial^2 u_y}{\partial z^2} = \frac{-I_{yy}M_x + I_{xy}M_y}{E(I_{xx}I_{yy} - I_{xy}^2)} = -\frac{1}{E}(\boldsymbol{\mathcal{I}}^{-1}\boldsymbol{M})_x,$$

which shows that the Bernoulli–Euler law given by Eqn. (6.161) is valid even when the bending occurs due to terminal loads. For the special case when the x and y axes coincide with the principal axes, i.e., when $I_{xy} = 0$, the above equations reduce to Eqn. (6.164).

The question arises if a generalization similar to that carried out in Section 6.5.3 can be carried out for the problem of flexure of inhomogeneous bars. Reissner [267] has shown that the assumption of vanishing τ_{xx}, τ_{yy} and τ_{xy} is valid if and only if the Poisson ratio ν is constant over the cross-section. The generalization to the case of an orthotropic beam has been carried out by Tolf [331].

We now proceed to the solution of some special problems.

Elliptical cross-section

Consider the bending of a bar of elliptical cross-section shown in Fig. 6.24, by tractions acting on the end $z = L$, whose statically equivalent load has components $W_x = P$, $W_y = 0$ acting through the centroid. Since the cross-section is doubly symmetric about the x–y axes,

1. $I_{xy} = 0$, so that from Eqn. (6.171), we get $\kappa_x = P/(EI_{yy}) = 4P/(E\pi a^3 b)$ and $\kappa_y = 0$. From Eqn. (6.176b), it then follows that $g(x)$ is also zero.

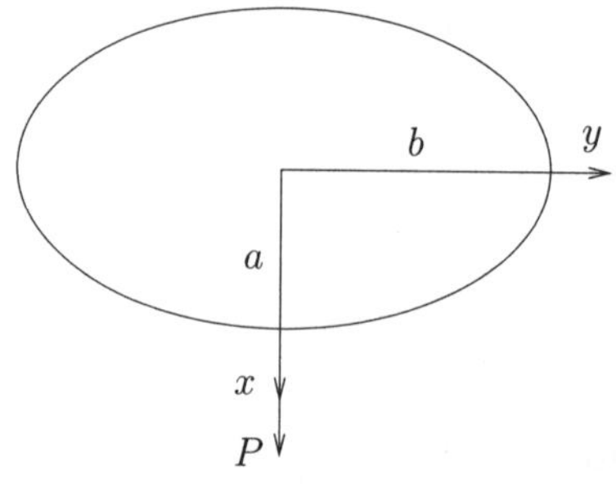

Fig. 6.24 *Bar with elliptical cross-section acted upon by a terminal load.*

2. The shear center lies at the centroid, and since the load also acts through the centroid, no twisting occurs ($\alpha = 0$).

From Eqn. (6.176a) we need $f(y) = Px^2/(2I_{yy})$ on the boundary C whose equation is $\frac{x^2}{a^2} + \frac{y^2}{b^2} - 1 = 0$. If we choose

$$f(y) = \frac{P}{2I_{yy}} \frac{a^2}{b^2}(b^2 - y^2),$$

then this boundary condition is satisfied. Equation (6.174) now becomes

$$\boldsymbol{\nabla}^2\phi = \frac{P}{I_{yy}}\left(\frac{\nu}{1+\nu} + \frac{a^2}{b^2}\right) y.$$

The above equation is to be solved subject to the boundary condition $\phi = 0$ on C. We seek a solution of the form

$$\phi = m\left(\frac{x^2}{a^2} + \frac{y^2}{b^2} - 1\right) y,$$

Note that this form satisfies the boundary condition. We substitute this form into the governing equation, and solve for m; the resulting expression for ϕ is

$$\phi = \frac{a^2[(1+\nu)a^2 + \nu b^2]}{2(1+\nu)(3a^2+b^2)} \frac{P}{I_{yy}} y \left(\frac{x^2}{a^2} + \frac{y^2}{b^2} - 1\right). \tag{6.192}$$

The stresses are obtained using Eqns. (6.168) and (6.172) as

$$\tau_{xz} = \frac{2P}{\pi a^3 b} \frac{[2(1+\nu)a^2 + b^2]}{(1+\nu)(3a^2+b^2)} \left[a^2 - x^2 - \frac{(1-2\nu)a^2y^2}{2(1+\nu)a^2 + b^2}\right],$$

$$\tau_{yz} = -\frac{4P}{\pi a^3 b} \frac{(1+\nu)a^2 + \nu b^2}{(1+\nu)(3a^2+b^2)} xy,$$

$$\tau_{zz} = -\frac{4P}{\pi a^3 b}(L - z)x.$$

Note that the tractions on the face $z = L$, given by $t_x = \tau_{xz}$ and $t_y = \tau_{yz}$, have to be applied as per the above distribution, in order for the solution to qualify as exact. On the line $x = 0$, we have $\tau_{yz} = 0$, and

$$\tau_{xz}\big|_{x=0} = \frac{2P}{\pi ab}\frac{[2(1+\nu)a^2+b^2]}{(1+\nu)(3a^2+b^2)}\left[1 - \frac{(1-2\nu)y^2}{2(1+\nu)a^2+b^2}\right].$$

τ_{xz} assumes the following maximum value at $x = y = 0$:

$$(\tau_{xz})_{\max} = \frac{2P}{A}\frac{[2(1+\nu)a^2+b^2]}{(1+\nu)(3a^2+b^2)},$$

where $A = \pi ab$ is the area of the ellipse. To find $(\tau_{yz})_{\max}$, we need to find $(xy)_{\max}$. Since xy is harmonic, the maximum occurs on the boundary. By putting $x = a\cos\theta$, $y = b\sin\theta$, we find $(xy)_{\max} = (\frac{1}{2}ab\sin 2\theta)_{\max} = \frac{1}{2}ab$. Thus,

$$(\tau_{yz})_{\max} = \frac{2P}{A}\frac{(1+\nu)a^2b+\nu b^3}{(1+\nu)(3a^3+ab^2)}.$$

Note from the above results that when $b \gg a$, τ_{yz} is not negligible, and in such a case, the error in the elementary theory which neglects τ_{yz} can be quite large.

The function χ is of the form $c_1x + c_2(x^3 - 3xy^2)$ and is obtained using Eqns. (6.183d) and (6.183e) as

$$\begin{aligned}\frac{\chi}{\kappa_x} &= \frac{a^2[2(1+\nu)a^2+b^2]}{3a^2+b^2}x + \frac{(b^2-a^2)(1-2\nu)}{12(3a^2+b^2)}(3xy^2-x^3)\\ &= \frac{c^3}{(48+96\cosh 2\xi_0)}\Big\{(1-2\nu)\cosh 3\xi\cos 3\eta\\ &\quad + 3(1+2\cosh 2\xi_0)\left[5+6\nu+(6+4\nu)\cosh 2\xi_0\right]\cosh\xi\cos\eta\Big\},\end{aligned}$$

where $\xi = \xi_0$ represents the outer boundary. The stresses in elliptical coordinates obtained using Eqns. (6.187c) and (6.187d) are

$$\frac{\tau_{\xi z}}{G} = \frac{c^3\kappa_x(\cosh 2\xi_0 - \cosh 2\xi)\sinh\xi\cos\eta}{4h_1(1+2\cosh 2\xi_0)}\left[2-(1-2\nu)\cos 2\eta+(3+2\nu)\cosh 2\xi_0\right],$$

$$\frac{\tau_{\eta z}}{G} = \frac{c^3\kappa_x(\cos 2\eta - \cosh 2\xi_0)\cosh\xi\sin\eta}{4h_1(1+2\cosh 2\xi_0)}\left[2-(1-2\nu)\cosh 2\xi+(3+2\nu)\cosh 2\xi_0\right],$$

where h_1 is given by Eqn. (D.2).

For the case of a section bounded by two confocal ellipses, the inner one $\xi = \xi_1$ and the outer one $\xi = \xi_2$, with the load P acting along the x-direction, the solution for the χ function in terms of elliptic coordinates (ξ, η) is

$$\frac{\chi}{c^3\kappa_x} = (c_1\cosh\xi + c_2\sinh\xi)\cos\eta + (c_3\cosh 3\xi + c_4\sinh 3\xi)\cos 3\eta,$$

where the constants are determined using the boundary conditions $\tau_{\xi z}\big|_{\xi=\xi_1} = \tau_{\xi z}\big|_{\xi=\xi_2} = 0$, and are given by

$$c_1 = \frac{1}{16}\left\{2(1+2\nu) + (3+2\nu)\left[\cosh 2\xi_1 + \cosh 2\xi_2 + \cosh 2(\xi_1+\xi_2)\right]\right\},$$

$$c_2 = -\frac{1}{4}(3+2\nu)\sinh\xi_1 \sinh\xi_2 \sinh(\xi_1+\xi_2),$$

$$c_3 = \frac{(1-2\nu)}{48[1+2\cosh 2(\xi_1-\xi_2)]}\left[\cosh 2\xi_1 + \cosh 2\xi_2 - \cosh 2(\xi_1+\xi_2)\right],$$

$$c_4 = \frac{(1-2\nu)}{12[1+2\cosh 2(\xi_1-\xi_2)]}\sinh\xi_1 \sinh\xi_2 \sinh(\xi_1+\xi_2).$$

The displacements and stresses are now found using Eqn. (6.187a)–(6.187d).

The results for a circular cylinder are recovered from those for a solid ellipse simply by setting $b = a$. We get

$$\tau_{xz}\big|_{x=0} = \frac{(3+2\nu)P}{2\pi a^4(1+\nu)}\left[a^2 - \frac{1-2\nu}{3+2\nu}y^2\right],$$

$$(\tau_{xz})_{\max} = \frac{3+2\nu}{2(1+\nu)}\frac{P}{A},$$

$$(\tau_{yz})_{\max} = \frac{(1+2\nu)P}{2(1+\nu)A}.$$

The elementary beam theory, based on the assumption that τ_{xz} is uniformly distributed along the section $x =$ const., gives $(\tau_{xz})_{\max} = 4P/3A$. Thus, the ratio of the exact to the approximate solution is

$$\frac{[(\tau_{xz})_{\max}]_{\text{exact}}}{[(\tau_{xz})_{\max}]_{\text{approx}}} = \frac{3(3+2\nu)}{8(1+\nu)},$$

which for $\nu = 0.3$ corresponds to an error of 4%.

For a hollow circular cylinder with inner and outer radii given by a and b, respectively (see Fig. 6.25), for the case when $W_y = 0$ and $W_x \equiv P$ passing through the center, we have $\kappa_y = g(x) = 0$ and $\kappa_x = P/(EI_{yy})$. We assume χ to be a superposition of the harmonic functions $r\sin\theta$ and $\sin\theta/r$, i.e., $\chi = c_1 r\sin\theta + c_2\sin\theta/r$. The constants c_1 and c_2 are determined by using the boundary condition given by $\tau_{rz}\big|_{r=a} = \tau_{rz}\big|_{r=b} = 0$, where τ_{rz} is given by Eqn. (6.185c) with $\alpha = \kappa_y = \bar{x}_c = \bar{y}_c = 0$. We obtain

$$\chi = -\frac{(3+2\nu)\kappa_x}{4}\left[(a^2+b^2)r + \frac{a^2b^2}{r}\right]\sin\theta.$$

We simply set $a = 0$ in the above solution to obtain the corresponding result for the solid circular cylinder. Using Eqns. (6.183d), (6.183e) and Eqns. (6.185), we obtain the stresses for the hollow cylinder case as

$$\tau_{xz} = \frac{P}{8(1+\nu)I_{yy}}\left\{(3+2\nu)\left[a^2+b^2-x^2+\frac{a^2b^2}{x^2+y^2} - \frac{2a^2b^2x^2}{(x^2+y^2)^2}\right] - (1-2\nu)y^2\right\},$$

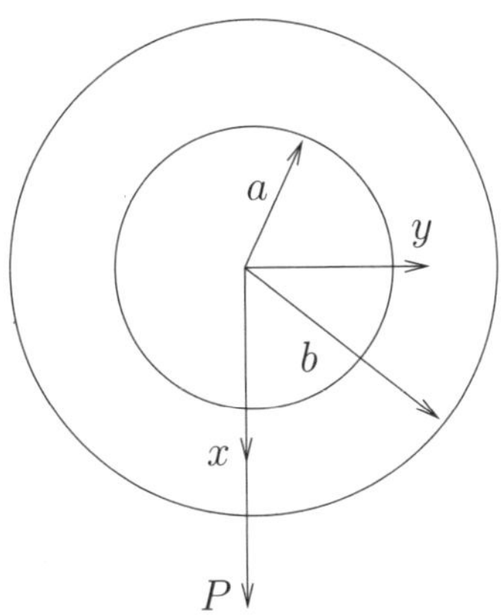

Fig. 6.25 *Bending of a hollow circular cylinder by a terminal load P.*

$$\tau_{yz} = -\frac{Pxy}{4(1+\nu)I_{yy}}\left[1+2\nu+\frac{(3+2\nu)a^2b^2}{(x^2+y^2)^2}\right],$$
$$\tau_{rz} = \frac{P}{8(1+\nu)I_{yy}r^2}(3+2\nu)(r^2-a^2)(r^2-b^2)\sin\theta,$$
$$\tau_{\theta z} = -\frac{P}{8(1+\nu)I_{yy}r^2}\left\{(3+2\nu)\left[a^2b^2+(a^2+b^2)r^2\right]-(1-2\nu)r^4\right\}\cos\theta.$$

The functions ϕ and $f(y)$ are obtained from the expressions for the stresses as

$$\phi = \frac{Py}{8(1+\nu)I_{yy}}\left[(1+2\nu)(x^2+y^2-a^2-b^2)-\frac{(3+2\nu)a^2b^2}{x^2+y^2}\right],$$
$$f(y) = \frac{P}{2I_{yy}}\left[a^2+b^2-y^2\right].$$

Note that ϕ is *not* constant either on the inner or outer boundary, and that is reason that we preferred to obtain the solution using the χ instead of the ϕ function.

The flexure of a hollow shaft with a cavity placed eccentrically has been treated in [286, 312]. First consider the case where $(W_x, W_y) = (P, 0)$. We get $(\kappa_x, \kappa_y) = [P/(EI_{yy}), 0]$, where $I_{yy} = \pi(b^4 - a^4)/2$. Working with bipolar coordinates, let $\xi = \xi_1$ and $\xi = \xi_2$ represent the boundaries of the inner hole and outer surface, respectively, as shown in Fig. 6.26. The radii of the circles are given by $a = c\,\text{csch}\,\xi_1$ and $b = c\,\text{csch}\,\xi_2$. We assume χ to be of the form

$$\chi = c_1 x + c_2 xy + \sum_{n=1}^{\infty} A_n \cosh n(\xi - \xi_2)\sin n\eta.$$

Using Eqns. (6.189) with $(\bar{x}_c, \bar{y}_c) = [0, c(\coth\xi_1 + \text{csch}\,\xi_2\,\text{csch}\,(\xi_1+\xi_2)\sinh\xi_1)]$ (note that $(\bar{x}_c, \bar{y}_c)$ are given here with respect to the bipolar coordinate system shown by the dotted line in Fig. 6.26), and imposing the boundary condition $\tau_{\xi z}\big|_{\xi=\xi_2} = 0$, we get

$$c_1 = (c\coth\xi_2 - \bar{y}_c)\alpha + \frac{\kappa_x}{4}\left\{2c^2\text{csch}^2\,\xi_2 - (1+2\nu)\left[c^2+\bar{y}_c^2 - 2c\bar{y}_c\coth\xi_2\right]\right\},$$

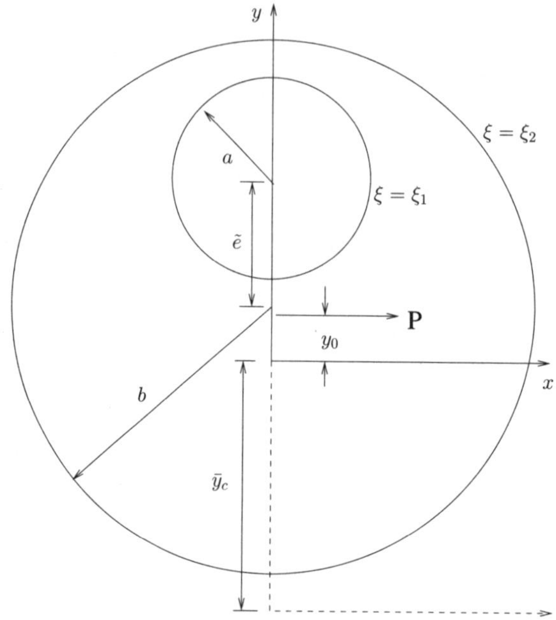

Fig. 6.26 *Flexure of a hollow shaft with a circular cavity placed eccentrically.*

$$c_2 = \frac{\kappa_x}{2}(c \coth \xi_2 - \bar{y}_c).$$

The constants A_n are now to be determined using the boundary condition $\tau_{\xi z}\big|_{\xi=\xi_1} = 0$. We get

$$4 \sum_{n=1}^{\infty} A_n n \sinh n(\xi_1 - \xi_2) \sin n\eta = \frac{k_1 \sin 2\eta}{(\cosh \xi_1 - \cos \eta)^3} + \frac{k_2 \sin \eta}{(\cosh \xi_1 - \cos \eta)^3},$$

where

$$k_1 = -c^2 \operatorname{csch} \xi_2 \sinh(\xi_1 - \xi_2) \left[2(\alpha + \kappa_x \nu \bar{y}_c) + c\kappa_x \coth \xi_2\right],$$

$$k_2 = 2c^2 \operatorname{csch} \xi_2 \sinh(\xi_1 - \xi_2) \left\{\cosh \xi_1 \left[2(\alpha + \kappa_x \nu \bar{y}_c) + c\kappa_x \coth \xi_2\right] + 2c\kappa_x \sinh \xi_1\right\}.$$

Using the relations

$$\frac{1}{\sinh^2 \xi} \sum_{n=1}^{\infty} (n + \coth \xi) e^{-n\xi} n \sin n\eta = \frac{\sin \eta}{(\cosh \xi - \cos \eta)^3},$$

$$\frac{2}{\sinh \xi} \sum_{n=1}^{\infty} (\operatorname{csch}^2 \xi - 1 + n \coth \xi) e^{-n\xi} n \sin n\eta = \frac{\sin 2\eta}{(\cosh \xi - \cos \eta)^3},$$

we get

$$4A_n \sinh n(\xi_1 - \xi_2) = \frac{e^{-n\xi_1}}{\sinh^2 \xi_1} \left[2k_1 \sinh \xi_1 (\operatorname{csch}^2 \xi_1 - 1 + n \coth \xi_1) + k_2 (n + \coth \xi_1)\right].$$

The relation between y_0 and α can now be found using Eqn. (6.190).

Now consider the case when $(W_x, W_y) = (0, P)$. Due to symmetry, we have $\alpha = \kappa_x = 0$, and $\kappa_y = P/(EI_{xx})$, where

$$I_{xx} = \frac{\pi a^4 b^4 \operatorname{csch}(\xi_1+\xi_2)}{64c^4}\Big\{7\left[\sinh(\xi_1-3\xi_2)+\sinh(3\xi_1-\xi_2)\right]-4\sinh(\xi_1-\xi_2)$$

$$-4\sinh 3(\xi_1-\xi_2)+8\left[\sinh(\xi_1+3\xi_2)-\sinh(3\xi_1+\xi_2)\right]+\sinh(5\xi_1+\xi_2)-\sinh(\xi_1+5\xi_2)\Big\}.$$

We assume χ to be

$$\chi = c_0\xi + c_1 y + c_2(x^2-y^2) + \sum_{n=1}^{\infty}\left[A_n^{(1)}\cosh n(\xi-\xi_2) - A_n^{(2)}\cosh n(\xi_1-\xi)\right]\cos n\eta.$$

The constants c_1 and c_2 turn out to be zero, although this is not obvious at this stage. Both conditions $\int_0^{2\pi}(h_1\tau_{\xi z})_{\xi=\xi_1}\,d\eta = 0$ and $\int_0^{2\pi}(h_1\tau_{\xi z})_{\xi=\xi_2}\,d\eta = 0$ lead to

$$c_0 = abc\kappa_y(1+\nu)\operatorname{csch}(\xi_1+\xi_2),$$

on substituting for $\bar{y}_c$! The constants c_1, c_2 and $A_n^{(i)}$, $i=1,2$, are now determined using the conditions $\tau_{\xi z}\big|_{\xi=\xi_1} = \tau_{\xi z}\big|_{\xi=\xi_2} = 0$. We get

$$\frac{p_0^{(i)} + p_1^{(i)}\cos\eta + p_2^{(i)}\cos 2\eta}{(\cosh\xi_i - \cos\eta)^3} + \sum_{n=1}^{\infty} A_n^{(i)} n \sinh n(\xi_1-\xi_2)\cos n\eta$$

$$= \frac{k_0^{(i)} + k_1^{(i)}\cos\eta + k_2^{(i)}\cos 2\eta + k_3^{(i)}\cos 3\eta + k_4^{(i)}\cos 4\eta}{(\cosh\xi_i - \cos\eta)^4},$$

where $p_0^{(i)}, p_1^{(i)}, p_2^{(i)}, k_0^{(i)}, k_1^{(i)}, k_2^{(i)}, k_3^{(i)}$ and $k_4^{(i)}$, $i=1,2$, are given by

$$2p_0^{(i)} = 3c\left[(c_1+2c_2\bar{y}_c)\cosh\xi_i - 2cc_2\sinh\xi_i\right],$$

$$2p_1^{(i)} = -c\left[(c_1+2c_2\bar{y}_c)(3+\cosh 2\xi_i) - 2cc_2\sinh 2\xi_i\right],$$

$$2p_2^{(i)} = c\left[(c_1+2c_2\bar{y}_c)\cosh\xi_i + 2cc_2\sinh\xi_i\right],$$

$$8k_0^{(i)} = -18c_0 - c^3\kappa_y(1+2\nu) + 3c\kappa_y(3+2\nu)\bar{y}_c^2 + 2\left[c\kappa_y\left[c^2+(3+2\nu)\bar{y}_c^2\right] - 8c_0\right]\cosh 2\xi_i$$

$$- 4c^2\kappa_y\bar{y}_c(2+\nu)\sinh 2\xi_i - c_0\cosh 4\xi_i,$$

$$16k_1^{(i)} = 2\left[48c_0 + c^3\kappa_y(1+2\nu) - 7c\kappa_y(3+2\nu)\bar{y}_c^2\right]\cosh\xi_i$$

$$+\left[16c_0 - c\kappa_y(3+2\nu)(c^2+\bar{y}_c^2)\right]\cosh 3\xi_i$$

$$+2c^2\kappa_y\bar{y}_c\left[17+10\nu+(6+4\nu)\cosh 2\xi_i\right]\sin\xi_i,$$

$$8k_2^{(i)} = -16c_0 - 2c^3\kappa_y + 2c\kappa_y\bar{y}_c^2(3+2\nu) + \left[c^3\kappa_y(1+2\nu) + c\kappa_y\bar{y}_c^2(3+2\nu) - 12c_0\right]\cosh 2\xi_i$$
$$- 4c^2\kappa_y\bar{y}_c(1+\nu)\sinh 2\xi_i,$$
$$16k_3^{(i)} = \left[c\kappa_y\left[c^2(1-2\nu) - (3+2\nu)\bar{y}_c^2\right] + 16c_0\right]\cosh\xi_i + 2c^2\kappa_y\bar{y}_c(1+2\nu)\sinh\xi_i,$$
$$8k_4^{(i)} = -c_0.$$

On expressing the functions $\cos l\eta/(\cosh\xi_i - \cos\eta)^3$, $l = 0,1,2$, and $\cos m\eta/(\cosh\xi_i - \cos\eta)^4$, $m = 0,1,2,3,4$, as Fourier series in $\cos n\eta$ (see Appendix G in Vol. II), one can obtain c_1, c_2 and $A_n^{(i)}$. On substituting for c_0 and $\bar{y}_c$, we get $c_1 = c_2 = 0$!

Triangular cross-section

A beam with an equilateral cross-section is loaded by a statically equivalent load P acting through its centroid (which we will show below to be also the position of the shear center), as shown in Fig. 6.27. Since the cross-section is symmetric about the y-axis, both I_{xy} and x_{cf}

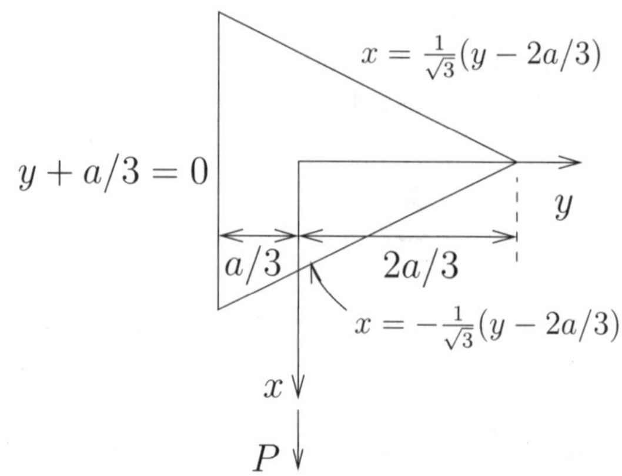

Fig. 6.27 *Bar whose cross-section is an equilateral triangle acted upon by a statically equivalent load P through its centroid.*

are zero. Setting $W_y = 0$ in Eqn. (6.171), we get $\kappa_x = P/(EI_{yy})$, where $I_{yy} = a^4/18\sqrt{3}$, and $\kappa_y = 0$, which in turn, by virtue of Eqn. (6.176b) implies that $g(x)$ is zero. Since $dy/ds = 0$ on the side $y + a/3 = 0$, choosing $f(y)$ such that $f(y)$ is equal to $E\kappa_x x^2/2$ on the inclined surfaces will ensure that Eqn. (6.176a) is satisfied. Thus, we get

$$f(y) = \frac{P}{6I_{yy}}\left(y - \frac{2}{3}a\right)^2.$$

The governing differential equation for ϕ given by Eqn. (6.174) is readily satisfied when the material is incompressible, i.e., when $\nu = 0.5$. Thus, in order to find the position of the shear center in this case, we set $\alpha = 0$ and $\nu = 0.5$ in Eqn. (6.174) to get

$$\nabla^2\phi = \frac{2Pa}{9I_{yy}}. \tag{6.193}$$

Since $\phi = 0$ on the boundary, we assume ϕ to be of the form

$$\phi = m\left(y + \frac{a}{3}\right)\left[x - \frac{1}{\sqrt{3}}\left(y - \frac{2}{3}a\right)\right]\left[x + \frac{1}{\sqrt{3}}\left(y - \frac{2}{3}a\right)\right]$$
$$= m\left(y + \frac{a}{3}\right)\left[x^2 - \frac{1}{3}\left(y - \frac{2}{3}a\right)^2\right].$$

Substituting the above expression into Eqn. (6.193), we find $m = P/6I_{yy}$, so that

$$\phi = \frac{P}{6I_{yy}}\left(y + \frac{a}{3}\right)\left[x^2 - \frac{1}{3}\left(y - \frac{2}{3}a\right)^2\right]. \quad \text{(when } \alpha = 0\text{)} \tag{6.194}$$

We have already mentioned that $x_{\text{cf}} = 0$. From Eqn. (6.181b), we get

$$y_{\text{cf}} = -\frac{1}{3I_{yy}}\int_{-\frac{a}{3}}^{\frac{2a}{3}}\int_{\frac{1}{\sqrt{3}}(y-\frac{2a}{3})}^{-\frac{1}{\sqrt{3}}(y-\frac{2a}{3})}\left\{\left(y + \frac{a}{3}\right)\left[x^2 - \frac{1}{3}\left(y - \frac{2}{3}a\right)^2\right] - 3x^2y\right\}dx\,dy = 0.$$

Thus, the shear center lies at the centroid. Since the load acts through the shear center, the twist α is zero, and thus, the stress function is given by Eqn. (6.194). Using Eqns. (6.168) and (6.172), the stresses are found to be

$$\tau_{xz} = \frac{E\kappa_x}{27}\left(2a^2 - 9x^2 - 3ay\right),$$
$$\tau_{yz} = -\frac{E\kappa_x}{3}\left(y + \frac{a}{3}\right)x,$$
$$\tau_{zz} = -E\kappa_x(L - z)x.$$

The function χ is found using Eqns. (6.183d) and (6.183e), and is given by

$$\chi = \frac{a\kappa_x x}{9}(2a - 3y).$$

Now consider the more general problem of the bending of a bar whose cross-section is an isosceles triangle, by a a statically equivalent load P acting through $y = y_0$ (see Fig. 6.28), where y_0 is determined by the exact solution that we now find. We first find the solution using the same approach as the one used for the equilateral triangle cross-section. We again get $\kappa_x = P/(EI_{yy})$ and $g(x) = \kappa_y = 0$. We choose

$$f(y) = \frac{P}{2I_{yy}}\tan^2\beta\left(y - \frac{2a}{3}\right)^2.$$

The governing equation for ϕ given by Eqn. (6.174) reduces to

$$\nabla^2\phi = \frac{P\nu y}{(1+\nu)I_{yy}} - \frac{P}{I_{yy}}\tan^2\beta\left(y - \frac{2a}{3}\right) - \frac{E\alpha}{(1+\nu)}. \tag{6.195}$$

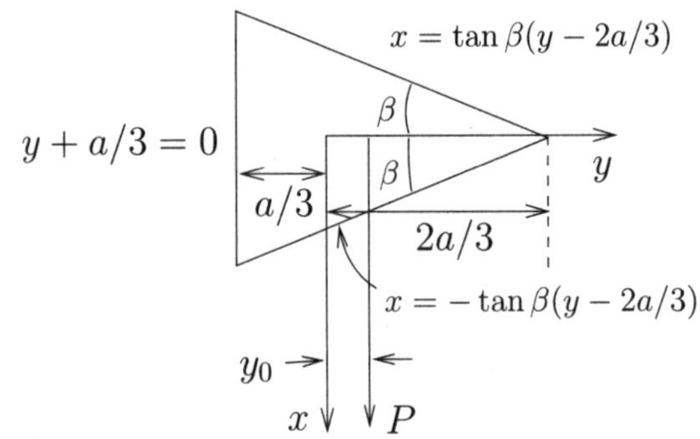

Fig. 6.28 *Bar whose cross-section is an isosceles triangle acted upon by a statically equivalent load P through $y = y_0$.*

Assume ϕ to be of the form

$$\phi = m\left(y - \frac{2a}{3}\right)\left[x^2 - \tan^2\beta\left(y - \frac{2a}{3}\right)^2\right],$$

substitute it into Eqn. (6.195), and solve for m and α. We get

$$m = \frac{P}{2I_{yy}}\frac{\left[\nu - (1+\nu)\tan^2\beta\right]}{(1+\nu)(1-3\tan^2\beta)},$$

$$\alpha = \frac{Pa}{3EI_{yy}}\frac{\left[3\tan^2\beta - \nu - 3(1+\nu)\tan^4\beta\right]}{(1-3\tan^2\beta)}.$$

The stress distribution is

$$\tau_{xz} = G\kappa_x\left[\left(\frac{1+2\nu-(3+2\nu)\cos 2\beta}{4\cos 2\beta - 2}\right)x^2 + \left(\frac{(1-2\nu)\sin^2\beta}{2\cos 2\beta - 1}\right)y^2 + k_1 y + k_2\right],$$

$$\tau_{yz} = G\kappa_x c_1\left(y + \frac{a}{3}\right)x,$$

where

$$k_1 = -\frac{a\tan^2\beta}{3}\left[\frac{1+4\nu-(5+2\nu)\cos 2\beta}{1-2\cos 2\beta}\right],$$

$$k_2 = \frac{4a^2}{9}(1+\nu)\tan^2\beta,$$

$$c_1 = -\frac{1-(1+2\nu)\cos 2\beta}{1-2\cos 2\beta}.$$

By using the condition

$$Py_0 = \int_{-a/3}^{2a/3}\int_{-\tan\beta(2a/3-y)}^{\tan\beta(2a/3-y)}(y\tau_{xz} - x\tau_{yz})\,dx\,dy,$$

and the fact that $\kappa_x = P/(EI_{yy})$, where $I_{yy} = (\tan\beta)^3 a^4/6$, we find that

$$y_0 = \frac{2(1-2\nu)a\cos^2\beta}{15(1+\nu)(1-2\cos 2\beta)}.$$

In general, since α does not vanish at the origin for the above stress distribution, y_0 does not coincide with the center of flexure position y_{cf}. If a load P acts at a point y_p that is different than y_0, then the solution due to the torque $P(y_p - y_0)$ has to be superimposed on the above solution; note, however, that such a torsion solution is not known for an arbitrary isosceles triangle, but only for a right-angled isosceles triangle as discussed in Section 6.5.2. For the case $\beta = \pi/6$ and $\nu = 1/2$, the above solution reduces to the solution for an equilateral triangle presented earlier. For the case $\beta = \pi/6$, $\nu < 1/2$, the above solution is invalid. For the case when $\tan^2\beta = \nu/(1+\nu)$, we get $\phi = 0$ (since $m = 0$), $y_0 = -2a/15$, $\alpha = 2a\nu P/(3EI_{yy})$, $\tau_{yz} = 0$, $\tau_{xz} = G\kappa_x\left[\nu(y-2a/3)^2 - (1+\nu)x^2\right]$.

We can also solve this problem using the formulation based on the function χ by assuming it to be of the form $c_1(3xy^2 - x^3) + c_2xy + c_3x$, and determining the constants using the boundary conditions $\tau_{xz}\cos\beta + \tau_{yz}\sin\beta = 0$ on the lower edge, $-\tau_{xz}\cos\beta + \tau_{yz}\sin\beta = 0$ on the upper one, and $\tau_{yz}\big|_{y=-a/3} = 0$. Using this procedure, we get

$$c_1 = \frac{\kappa_x(1-2\nu)}{12(2\cos 2\beta - 1)},$$

$$c_2 = \frac{\kappa_x a}{6\cos^2\beta}\left[\cos 2\beta - 1 - 2\nu\right],$$

$$c_3 = \frac{4\kappa_x a^2}{9}(1+\nu)\tan^2\beta.$$

The stress solution is the same as obtained above, and the displacements are given by Eqns. (6.183a)–(6.183c).

Now consider the same geometry, but with P acting along the y-axis as shown in Fig. 6.29, and with G equal to $E/3$ (i.e., $\nu = 0.5$; the solution for arbitrary ν is not known). We now have $\alpha = \kappa_x = 0$, and $\kappa_y = P/(EI_{xx})$ with $I_{xx} = (\tan\beta)a^4/18$. We now assume χ

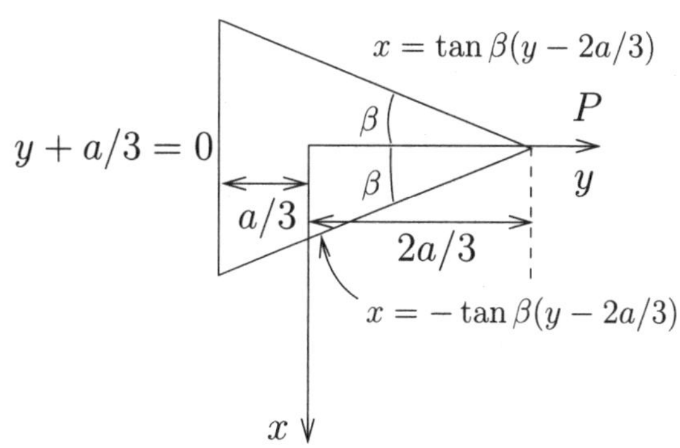

Fig. 6.29 *Beam of triangular cross-section loaded by a statically equivalent load P along the y-axis.*

to be a superposition of $\{y^2 - x^2, y\}$, and determine the constants by using the boundary conditions. The final expressions are

$$\chi = \kappa_y \left[\frac{a}{6}(y^2 - x^2) + \frac{2a^2 y}{9}\right],$$

$$\tau_{xz} = -G\kappa_y x \left(y + \frac{a}{3}\right),$$

$$\tau_{yz} = G\kappa_y \left(y + \frac{a}{3}\right)\left(\frac{2a}{3} - y\right).$$

The displacements are given by Eqns. (6.183a)–(6.183c). Note that the stress function ϕ and the function $g(x)$ corresponding to the above stress distribution are

$$\phi = -\frac{G\kappa_y x}{2}\left(y + \frac{a}{3}\right)^2,$$

$$g(x) = -\frac{1}{6}G\kappa_y a^2.$$

and that, unlike the equilateral triangle case, ϕ is not constant on the boundary.

For a further discussion on flexure of beams with triangular cross-section, see [98, 284, 285].

Rectangular cross-section

If a beam with rectangular cross-section is acted upon by a terminal load $W_x = P$ as shown in Fig. 6.30, then, similar to the elliptical cross-section case, we have $\kappa_y = 0$, $g(x) = 0$ and $\alpha = 0$. Taking $f(y) = E\kappa_x a^2/8$, we see that Eqn. (6.176a) is satisfied since along $x = \pm a/2$, we have $x^2 - a^2/4 = 0$, while along $y = \pm b/2$, we have $dy/ds = 0$. Substituting these values into Eqn. (6.174), we get

$$\nabla^2 \phi = \frac{\nu}{1+\nu}\frac{Py}{I_{yy}},$$

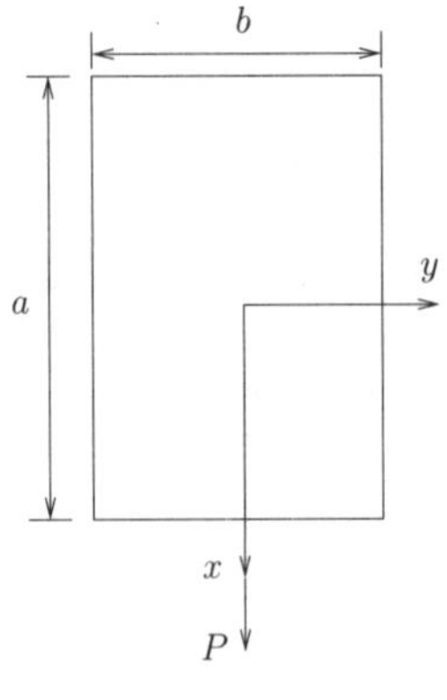

Fig. 6.30 *Bar of rectangular cross-section acted upon by a statically equivalent load P through its centroid.*

which is the governing equation to be solved for ϕ subject to the boundary condition $\phi = 0$ on C. The boundary condition is satisfied if we assume ϕ to be of the form

$$\phi = \sum_{m=1}^{\infty}\sum_{n=1}^{\infty} A_{mn}\cos\frac{(2m-1)\pi x}{a}\sin\frac{2n\pi y}{b}$$

Substituting this expression into the governing equation for ϕ, and using the orthogonality of the sine and cosine functions in the same manner as was done in solving the corresponding torsion problem, we get

$$A_{mn} = -\frac{\nu}{1+\nu}\frac{P}{\pi^4 I_{yy}}\frac{4(-1)^{m+n}b}{(2m-1)n\left[\frac{(2m-1)^2}{a^2}+\frac{4n^2}{b^2}\right]}.$$

The stresses are evaluated using Eqns. (6.172) as

$$\tau_{xz} = \frac{P}{2I_{yy}}\left[\left(\frac{a}{2}\right)^2 - x^2\right] + \frac{\partial\phi}{\partial y},$$

$$\tau_{yz} = -\frac{\partial\phi}{\partial x}.$$

The load P at the end surface $z = L$ has to be applied in a manner consistent with the above stress distribution. The elementary solutions for the stresses are

$$\tau_{xz} = \frac{P}{2I_{yy}}\left[\left(\frac{a}{2}\right)^2 - x^2\right], \qquad \tau_{yz} = 0.$$

Thus, $\tilde{\tau}_{xz} = \partial\phi/\partial y$ and $\tilde{\tau}_{yz} = -\partial\phi/\partial x$ can be considered as corrections to the elementary solution (note that these corrections are zero when $\nu = 0$!). The corrections have the largest values at $(x, y) = (0, \pm b/2)$, where, incidentally, the maximum stress also occurs. Similar to the elliptical cross-section case, the elementary solution is not a good approximation when $b \gg a$, but is a good approximation when $b \ll a$.

To solve the above problem using the χ function, we assume

$$\chi = c_1(x^3 - 3xy^2) + c_2 x + \frac{1}{G}\sum_{m=1}^{\infty}\frac{A_m a}{(2m-1)\pi}\sin\frac{(2m-1)\pi x}{a}\cosh\frac{(2m-1)\pi y}{a}.$$

Evaluating the stresses using Eqns. (6.183d) and (6.183e), and using the boundary conditions $\tau_{xz}\big|_{x=\pm a/2} = \tau_{yz}\big|_{y=\pm b/2} = 0$ along with the orthogonality of the sine functions, we get $c_1 = -\kappa_x(1-2\nu)/12$ and $c_2 = a^2\kappa_x/4$. The final stress distribution is

$$\tau_{xz} = \frac{P}{2I_{yy}(1+\nu)}\left[\left(\frac{a}{2}\right)^2 - x^2\right] + \sum_{m=1}^{\infty} A_m\cos\frac{(2m-1)\pi x}{a}\cosh\frac{(2m-1)\pi y}{a},$$

$$\tau_{yz} = -\frac{\nu Pxy}{I_{yy}(1+\nu)} + \sum_{m=1}^{\infty} A_m\sin\frac{(2m-1)\pi x}{a}\sinh\frac{(2m-1)\pi y}{a},$$

where

$$A_m = \frac{2\nu(-1)^{m+1}abP}{(1+\nu)I_{yy}(2m-1)^2\pi^2\sinh\frac{(2m-1)\pi b}{2a}}.$$

The displacements are given by Eqns. (6.183a)–(6.183c).

Semicircular cross-section

We now consider the problem of bending of a bar whose cross-section is a semicircle as shown in Fig. 6.31. It is convenient to evaluate the integrals in the ensuing formulation by

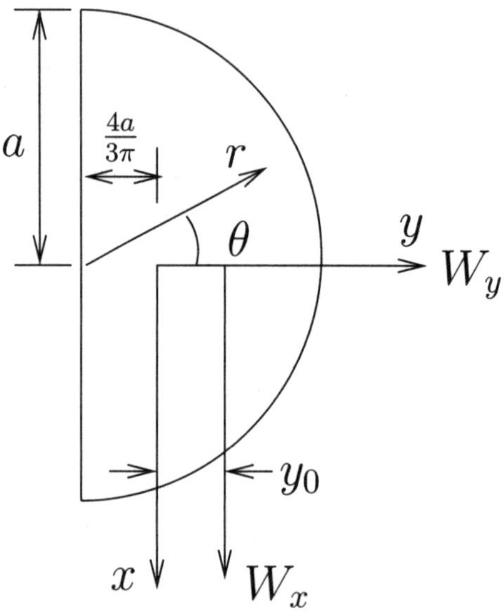

Fig. 6.31 *Bar of semi-circular cross-section.*

making the transformation $x = -r\sin\theta$ and $y + \bar{y}_c = r\cos\theta$, where $\bar{y}_c = 4a/(3\pi)$, and r and θ are as shown in the figure. By symmetry, $I_{xy} = 0$, and the components I_{xx} and I_{yy} are given by

$$I_{xx} = \int_A y^2\,dxdy = \int_0^a \int_{-\pi/2}^{\pi/2} \left(r\cos\theta - \frac{4a}{3\pi}\right)^2 r\,drd\theta = \frac{a^4}{72\pi}(9\pi^2 - 64),$$

$$I_{yy} = \int_A x^2\,dxdy = \int_0^a \int_{-\pi/2}^{\pi/2} r^3 \sin^2\theta\,drd\theta = \frac{\pi a^4}{8}.$$

We first solve for the case when only W_x is applied. From Eqn. (6.171), we get $\kappa_x = W_x/(EI_{yy})$ and $\kappa_y = 0$. Assuming

$$\chi = c_1 r\sin\theta + c_2 r^2 \sin 2\theta + \sum_{m=1}^{\infty} A_m r^{2m-1} \sin(2m-1)\theta,$$

and using Eqns. (6.185), we get

$$\frac{\tau_{rz}}{G} = -\alpha\bar{y}_c \sin\theta + c_1 \sin\theta + 2c_2 r\sin 2\theta + \sum_{m=1}^{\infty} A_m r^{2m-2}(2m-1)\sin(2m-1)\theta$$
$$+ \frac{\kappa_x}{4}\left\{\left[(3+2\nu)r^2 + (1-2\nu)\bar{y}_c^2\right]\sin\theta - 2r\bar{y}_c \sin 2\theta\right\},$$

$$\frac{\tau_{\theta z}}{G} = \alpha\left(r - \bar{y}_c \cos\theta\right) + c_1 \cos\theta + 2c_2 r \cos 2\theta + \sum_{m=1}^{\infty} A_m r^{2m-2}(2m-1)\cos(2m-1)\theta$$
$$+ \frac{\kappa_x}{4}\left[2r\bar{y}_c(2\nu - \cos 2\theta) + (1-2\nu)(r^2 + \bar{y}_c^2)\cos\theta\right].$$

Using the boundary condition $\tau_{\theta z}|_{\theta=\pm\pi/2} = 0$, we get

$$c_2 = \frac{\alpha}{2} + \frac{1}{4}\kappa_x \bar{y}_c(1+2\nu).$$

The boundary condition $\tau_{rz}|_{r=a} = 0$ leads to

$$\sum_{m=1}^{\infty} A_m a^{2m-2}(2m-1)\sin(2m-1)\theta = a_1 \sin\theta + a_2 \sin 2\theta, \tag{6.196}$$

where

$$a_1 = \alpha\bar{y}_c - c_1 - \frac{\kappa_x}{4}\left[(3+2\nu)a^2 + (1-2\nu)\bar{y}_c^2\right],$$
$$a_2 = -(\alpha + \nu\kappa_x \bar{y}_c)a.$$

Since the term $\sin\theta$ on the right-hand side also occur on the left-hand side with the undetermined constant A_1, we choose c_1 such that $a_1 = 0$. Thus, we get

$$c_1 = \alpha\bar{y}_c - \frac{\kappa_x}{4}\left[(3+2\nu)a^2 + (1-2\nu)\bar{y}_c^2\right].$$

Multiplying Eqn. (6.196) by $\sin(2p-1)\theta$, integrating with respect to θ, and using the orthogonality of the cosine functions, we get

$$A_m = \frac{8(-1)^{m+1} a_2}{\pi(2m-3)(4m^2-1)a^{2m-2}}.$$

It remains to determine α. Using Eqn. (6.186), and the series representation

$$\sum_{m=2}^{\infty} \frac{1}{(2m+1)(2m-3)(4m^2-1)} = \frac{\pi^2}{64} - \frac{5}{36},$$

we get

$$y_0 = \frac{16\nu(80-7\pi^2)W_x + \pi^2\left[16W_x - 15\alpha a^3 E(\pi^2-8)\right]}{60(1+\nu)\pi^3 W_x} a. \tag{6.197}$$

The above equation can be used to either (i) determine α given y_0, or (ii) determine y_0 given α. In particular, we obtain the position of the center of flexure y_{cf} by setting $\alpha = 0$ in the above expression:

$$y_{\text{cf}} = \frac{4\left[\pi^2 + \nu(80-7\pi^2)\right]}{15(1+\nu)\pi^3} a,$$

which matches with the expression presented in [300] (page 239; measured there from the origin of the cylindrical system). Although we did find the center of flexure, as evident from the above development and as mentioned earlier, it is possible to solve a flexure problem directly without finding the center of flexure.

Quite interestingly, in the above problem, all the A_m, $m = 1, 2, \ldots \infty$, become zero if $a_2 = 0$, i.e., if $\alpha = -\nu\kappa_x\bar{y}_c$. In this case, from Eqn. (6.197), we get

$$y_0 = \frac{4(1+3\nu)a}{15\pi(1+\nu)},$$

and the stress distribution is given by

$$\begin{aligned}
\tau_{rz} &= -\frac{1}{4}G\kappa_x(3+2\nu)(a^2-r^2)\sin\theta, \\
\tau_{\theta z} &= -\frac{1}{4}G\kappa_x\left[(3+2\nu)a^2-(1-2\nu)r^2\right]\cos\theta, \\
\tau_{xz} &= \frac{1}{4}G\kappa_x\left[(2\nu-1)\left(y+\frac{4a}{3\pi}\right)^2+(3+2\nu)(a^2-x^2)\right], \\
\tau_{yz} &= -\frac{1}{2}G\kappa_x(1+2\nu)\left(y+\frac{4a}{3\pi}\right)x.
\end{aligned}$$

Now consider the case when only the load W_y is applied. We now assume

$$\chi = c_1 r\cos\theta + c_2 r^2\cos 2\theta + c_3 r^3\cos 3\theta + \sum_{m=1}^{\infty} A_m r^{2m}\cos 2m\theta.$$

From Eqn. (6.185), we get

$$\begin{aligned}
\frac{\tau_{rz}}{G} &= c_1\cos\theta + 2c_2 r\cos 2\theta + 3c_3 r^2\cos 3\theta + \sum_{m=1}^{\infty} A_m r^{2m-1}(2m\cos 2m\theta) \\
&\quad + \frac{\kappa_y}{4}\left\{4(1+\nu)r\bar{y}_c + 2r\bar{y}_c\cos 2\theta - (3+2\nu)(r^2+\bar{y}_c^2)\cos\theta\right\}, \\
\frac{\tau_{\theta z}}{G} &= -c_1\sin\theta - 2c_2 r\sin 2\theta - 3c_3 r^2\sin 3\theta - \sum_{m=1}^{\infty} A_m r^{2m-1}(2m\sin 2m\theta) \\
&\quad + \frac{\kappa_y}{4}\left\{\left[(1-2\nu)r^2+(3+2\nu)\bar{y}_c^2\right]\sin\theta - 2r\bar{y}_c\sin 2\theta\right\},
\end{aligned}$$

where $\kappa_y = W_y/(EI_{xx})$. The boundary conditions $\tau_{\theta z}|_{\theta=\pm\pi/2} = 0$ yield

$$\begin{aligned}
c_1 &= \frac{1}{4}\kappa_y(3+2\nu)\bar{y}_c^2, \\
c_3 &= -\frac{1}{12}\kappa_y(1-2\nu),
\end{aligned}$$

while the boundary condition $\tau_{rz}|_{r=a} = 0$ yields

$$\sum_{m=1}^{\infty} A_m a^{2m-1}(2m\cos 2m\theta) = a_0 + a_1\cos\theta + a_2\cos 2\theta + a_3\cos 3\theta, \tag{6.198}$$

where

$$a_0 = -\kappa_y(1+\nu)a\bar{y}_c,$$
$$a_1 = \frac{a^2\kappa_y}{4}(3+2\nu),$$
$$a_2 = -2c_2 a - \frac{1}{2}\kappa_y a\bar{y}_c,$$
$$a_3 = \frac{a^2\kappa_y}{4}(1-2\nu).$$

Since the term $\cos 2\theta$ on the right-hand side also occurs on the left-hand side with the undetermined coefficient A_1, we choose c_2 so that a_2 is zero, i.e.,

$$c_2 = -\frac{1}{4}\kappa_y\bar{y}_c.$$

Multiplying Eqn. (6.198) by $\cos 2p\theta$, integrating with respect to θ between the limits $-\pi/2$ to $\pi/2$, and using the orthogonality of cosine functions, we get

$$A_m = -\frac{2(-1)^m}{m\pi a^{2m-1}}\left[\frac{a_1}{4m^2-1} - \frac{3a_3}{4m^2-9}\right].$$

Now consider the related problem of the bending of an annular sector by loads W_x and W_y as shown in Fig. 6.32. The solution for the loading by W_x was presented in [359], while that under the loading W_y was presented in [283].

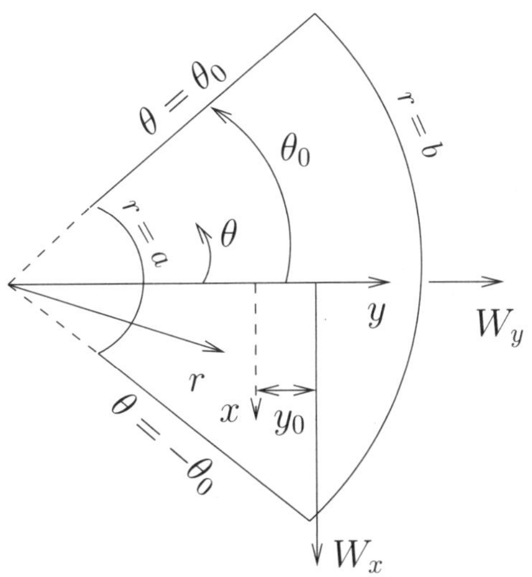

Fig. 6.32 *An annular sector subjected to bending by terminal loads.*

First, let us consider the case when the loading is only due to W_x. We assume χ to be of the form

$$\chi = c_1 r\sin\theta + c_2 r^2\sin 2\theta + c_3 r^3\sin 3\theta + \sum_{m=1}^{\infty}(A_m r^{\gamma_m} + B_m r^{-\gamma_m})\sin\gamma_m\theta, \qquad (6.199)$$

where $\gamma_m = (2m-1)\pi/(2\theta_0)$. Note that the above function is harmonic, and hence satisfies the governing equation given by Eqn. (6.184a). It only remains to determine the constants using the boundary conditions $\tau_{rz}|_{r=a} = \tau_{rz}|_{r=b} = \tau_{\theta z}|_{\theta=\pm\theta_0} = 0$. Using Eqns. (6.185), we get

$$\begin{aligned}\frac{\tau_{rz}}{G} &= -\alpha\bar{y}_c \sin\theta + c_1 \sin\theta + 2c_2 r \sin 2\theta + 3c_3 r^2 \sin 3\theta \\ &\quad + \sum_{m=1}^{\infty} (A_m r^{\gamma_m - 1} - B_m r^{-\gamma_m - 1})\gamma_m \sin\gamma_m\theta \\ &\quad + \frac{\kappa_x}{4}\left\{\left[(3+2\nu)r^2 + (1-2\nu)\bar{y}_c^2\right]\sin\theta - 2r\bar{y}_c \sin 2\theta\right\}, \qquad (6.200a)\\ \frac{\tau_{\theta z}}{G} &= \alpha[r - \bar{y}_c\cos\theta] + c_1\cos\theta + 2c_2 r\cos 2\theta + 3c_3 r^2\cos 3\theta \\ &\quad + \sum_{m=1}^{\infty} (A_m r^{\gamma_m - 1} + B_m r^{-\gamma_m - 1})\gamma_m \cos\gamma_m\theta \\ &\quad + \frac{\kappa_x}{4}\left\{2r\bar{y}_c(2\nu - \cos 2\theta) + (1-2\nu)(r^2 + \bar{y}_c^2)\cos\theta\right\}, \qquad (6.200b)\end{aligned}$$

where $\bar{y}_c = 2\sin\theta_0(b^3 - a^3)/[3\theta_0(b^2 - a^2)]$ and $\kappa_x = W_x/(EI_{yy})$. From the boundary conditions for $\tau_{\theta z}$, we get

$$\begin{aligned} c_1 &= \frac{\bar{y}_c}{4}\left[4\alpha - (1-2\nu)\kappa_x\bar{y}_c\right], \\ c_2 &= \frac{1}{4}\left[(1-2\nu)\kappa_x\bar{y}_c - 2\alpha\sec 2\theta_0\right], \\ c_3 &= -\frac{\kappa_x(1-2\nu)\cos\theta_0}{12\cos 3\theta_0}, \end{aligned}$$

while on enforcing the boundary conditions on τ_{rz}, we get the set of equations

$$\sum_{m=1}^{\infty}(A_m a^{\gamma_m-1} - B_m a^{-\gamma_m-1})\gamma_m \sin\gamma_m\theta = a_1\sin\theta + a_2\sin 2\theta + a_3\sin 3\theta,$$

$$\sum_{m=1}^{\infty}(A_m b^{\gamma_m-1} - B_m b^{-\gamma_m-1})\gamma_m \sin\gamma_m\theta = b_1\sin\theta + b_2\sin 2\theta + b_3\sin 3\theta,$$

where

$$\begin{aligned} a_1 &= -\frac{1}{4}(3+2\nu)\kappa_x a^2, & b_1 &= -\frac{1}{4}(3+2\nu)\kappa_x b^2, \\ a_2 &= a(\alpha + \nu\kappa_x\bar{y}_c)\sec 2\theta_0, & b_2 &= b(\alpha + \nu\kappa_x\bar{y}_c)\sec 2\theta_0, \\ a_3 &= \frac{1}{4}(1-2\nu)a^2\kappa_x\cos\theta_0\sec 3\theta_0, & b_3 &= \frac{1}{4}(1-2\nu)b^2\kappa_x\cos\theta_0\sec 3\theta_0. \end{aligned}$$

Multiplying the above set of equations by $\sin\gamma_p\theta$, integrating with respect to θ, and using the orthogonality of the sine functions, we obtain

$$A_m = \frac{b^{1+\gamma_m}F_2 - a^{1+\gamma_m}F_1}{b^{2\gamma_m} - a^{2\gamma_m}},$$

$$B_m = \frac{a^{2\gamma_m}b^{1+\gamma_m}F_2 - b^{2\gamma_m}a^{1+\gamma_m}F_1}{b^{2\gamma_m} - a^{2\gamma_m}},$$

where

$$F_1 = -\frac{2(-1)^m}{\gamma_m\theta_0}\left[\frac{a_1\cos\theta_0}{\gamma_m^2-1} + \frac{2a_2\cos 2\theta_0}{\gamma_m^2-4} + \frac{3a_3\cos 3\theta_0}{\gamma_m^2-9}\right],$$

$$F_2 = -\frac{2(-1)^m}{\gamma_m\theta_0}\left[\frac{b_1\cos\theta_0}{\gamma_m^2-1} + \frac{2b_2\cos 2\theta_0}{\gamma_m^2-4} + \frac{3b_3\cos 3\theta_0}{\gamma_m^2-9}\right].$$

It remains to determine the constant α. Equation (6.186) reduces to

$$-y_0W_x = \int_a^b\int_{-\theta_0}^{\theta_0}\left[r\tau_{\theta z} - \bar{y}_c(\tau_{rz}\sin\theta + \tau_{\theta z}\cos\theta)\right] r\,dr\,d\theta,$$

where τ_{rz} and $\tau_{\theta z}$ are given by Eqns. (6.200), and on carrying out the integrations, similar to Eqn. (6.197), yields an equation linking α and y_0. Thus, it can be used to find either (i) y_0 given α, or (ii) α given y_0. If α is set to zero, then of course $y_0 \equiv y_{\text{cf}}$ yields the position of the center of flexure. As an example, for $a = 0$, $\theta_0 = \pi$ (cracked cylinder with straight crack running from the periphery to the center), we get $\bar{y}_c = 0$, $I_{yy} = \pi b^4/4$, and

$$Py_0 = \frac{128b(8+9\nu)P}{225(1+\nu)\pi^2} + \alpha Gb^4\left[\frac{64}{9\pi} - \pi\right],$$

so that it follows that

$$y_{\text{cf}} = \frac{128b(8+9\nu)}{225(1+\nu)\pi^2}. \quad (a = 0,\ \theta_0 = \pi)$$

The solution given by Eqn. (6.199) is not valid for $\theta_0 = \pi/6, \pi/4, \pi/2, 3\pi/4$ and $5\pi/6$. Consider the case when $\theta_0 = \pi/4$. Instead of Eqn. (6.199), we now assume (see Eqn. (6.238))

$$\chi = c_1 r\sin\theta + c_2r^2(\log r\sin 2\theta + \theta\cos 2\theta) + c_3r^3\sin 3\theta + \sum_{m=1}^{\infty}(A_mr^{\gamma_m} + B_mr^{-\gamma_m})\sin\gamma_m\theta,$$

where $\gamma_m = 2(2m-1)$. The boundary condition on $\tau_{\theta z}$ leads to

$$c_1 = \frac{\bar{y}_c}{4}\left[4\alpha - (1-2\nu)\kappa_x\bar{y}_c\right],$$

$$c_2 = \frac{2}{\pi}\left[\alpha + \kappa_x\nu\bar{y}_c\right],$$

$$c_3 = \frac{\kappa_x(1-2\nu)}{12}.$$

The constants A_m and B_m are found in exactly the same manner as discussed above by considering the boundary conditions of τ_{rz}.

Now consider the case when the loading is only along the y-direction. In this case, we assume

$$\chi = c\log r + c_1 r\cos\theta + c_2 r^2\cos 2\theta + c_3 r^3\cos 3\theta + \sum_{m=1}^{\infty}(A_m r^{\gamma_m} + B_m r^{-\gamma_m})\cos\gamma_m\theta, \quad (6.201)$$

where $\gamma_m = m\pi/\theta_0$. Again note that the above function is harmonic, and hence only the constants need to be determined using the boundary conditions $\tau_{rz}|_{r=a} = \tau_{rz}|_{r=b} = \tau_{\theta z}|_{\theta=\pm\theta_0} = 0$. Using Eqns. (6.185), we get

$$\begin{aligned}\frac{\tau_{rz}}{G} &= \frac{c}{r} + c_1\cos\theta + 2c_2 r\cos 2\theta + 3c_3 r^2\cos 3\theta + \sum_{m=1}^{\infty}(A_m r^{\gamma_m-1} - B_m r^{-\gamma_m-1})\gamma_m\cos\gamma_m\theta \\ &\quad + \frac{\kappa_y}{4}\left\{4(1+\nu)r\bar{y}_c + 2r\bar{y}_c\cos 2\theta - (3+2\nu)(r^2+\bar{y}_c^2)\cos\theta\right\}, \quad (6.202a)\\ \frac{\tau_{\theta z}}{G} &= -c_1\sin\theta - 2c_2 r\sin 2\theta - 3c_3 r^2\sin 3\theta - \sum_{m=1}^{\infty}(A_m r^{\gamma_m-1} + B_m r^{-\gamma_m-1})\gamma_m\sin\gamma_m\theta \\ &\quad + \frac{\kappa_y}{4}\left\{\left[(1-2\nu)r^2 + (3+2\nu)\bar{y}_c^2\right]\sin\theta - 2r\bar{y}_c\sin 2\theta\right\}, \quad (6.202b)\end{aligned}$$

where $\bar{y}_c = 2\sin\theta_0(b^3 - a^3)/[3\theta_0(b^2 - a^2)]$, $\alpha = \kappa_x = 0$ and $\kappa_y = W_y/(EI_{xx})$. From the boundary conditions for $\tau_{\theta z}$, we get

$$\begin{aligned}c_1 &= \frac{1}{4}(3+2\nu)\kappa_y\bar{y}_c^2,\\ c_2 &= -\frac{1}{4}\kappa_y\bar{y}_c,\\ c_3 &= \frac{\kappa_y(1-2\nu)}{12}\frac{\sin\theta_0}{\sin 3\theta_0},\end{aligned}$$

while on enforcing the boundary conditions on τ_{rz}, we get the set of equations

$$\sum_{m=1}^{\infty}(A_m a^{\gamma_m-1} - B_m a^{-\gamma_m-1})\gamma_m\cos\gamma_m\theta = a_0 + a_1\cos\theta + a_3\cos 3\theta, \quad (6.203a)$$

$$\sum_{m=1}^{\infty}(A_m b^{\gamma_m-1} - B_m b^{-\gamma_m-1})\gamma_m\cos\gamma_m\theta = b_0 + b_1\cos\theta + b_3\cos 3\theta, \quad (6.203b)$$

where

$$\begin{aligned}a_0 &= -(1+\nu)\kappa_y a\bar{y}_c - \frac{c}{a}, & b_0 &= -(1+\nu)\kappa_y b\bar{y}_c - \frac{c}{b},\\ a_1 &= \frac{1}{4}\kappa_y(3+2\nu)a^2, & b_1 &= \frac{1}{4}\kappa_y(3+2\nu)b^2,\\ a_3 &= -\frac{(1-2\nu)a^2\kappa_y\sin\theta_0}{4\sin 3\theta_0}, & b_3 &= -\frac{(1-2\nu)b^2\kappa_y\sin\theta_0}{4\sin 3\theta_0}.\end{aligned}$$

Multiplying the above set of equations by $\cos\gamma_p\theta$, integrating with respect to θ, and using the orthogonality of the cosine functions, we obtain

$$A_m = \frac{b^{1+\gamma_m}F_2 - a^{1+\gamma_m}F_1}{b^{2\gamma_m} - a^{2\gamma_m}},$$

$$B_m = \frac{a^{2\gamma_m}b^{1+\gamma_m}F_2 - b^{2\gamma_m}a^{1+\gamma_m}F_1}{b^{2\gamma_m} - a^{2\gamma_m}},$$

where

$$F_1 = -\frac{2(-1)^m}{\gamma_m\theta_0}\left[\frac{a_1\sin\theta_0}{\gamma_m^2 - 1} + \frac{3a_3\sin 3\theta_0}{\gamma_m^2 - 9}\right],$$

$$F_2 = -\frac{2(-1)^m}{\gamma_m\theta_0}\left[\frac{b_1\sin\theta_0}{\gamma_m^2 - 1} + \frac{3b_3\sin 3\theta_0}{\gamma_m^2 - 9}\right].$$

It remains to determine the constant c. Integrating the equations given by Eqns. (6.203) with respect to θ, we get

$$a_0\theta_0 + a_1\sin\theta_0 + \frac{1}{3}a_3\sin 3\theta_0 = 0,$$

$$b_0\theta_0 + b_1\sin\theta_0 + \frac{1}{3}b_3\sin 3\theta_0 = 0,$$

both of which (amazingly!), on substituting for $\bar{y}_c$, lead to a value of

$$c = -\frac{2}{3}\kappa_y(1+\nu)\frac{a^2b^2}{a+b}\frac{\sin\theta_0}{\theta_0}.$$

The above solution does reduce to the solution presented earlier for a semicircle by setting $\theta_0 = \pi/2$, $a = c = 0$, and by replacing the symbol b by the symbol a.

The solution given by Eqn. (6.201) is not valid for $\theta_0 = \pi/3$ and $2\pi/3$. In this case, we replace the $c_3r^3\cos 3\theta$ term in Eqn. (6.201) by $c_3r^3(\log r\cos 3\theta - \theta\sin 3\theta)$, and then determine the constants in the same way as done above.

Cardioid cross-section

The problem of flexure of a shaft with cardioid cross-section has been discussed in [291, 300, 309], and its generalization to the elliptic limacon cross-section, has been discussed in [129, 310]. Following Shepherd [291], we now discuss the flexure of a beam whose cross-section is a cardioid as shown in Fig. 6.33. Note that $(\bar{x}_c, \bar{y}_c) = (0, -5a/6)$ and $I_{xy} = 0$. First consider the case when W_y is the only nonzero load, so that $\kappa_x = 0$ and $\kappa_y = W_y/(EI_{xx})$, where $I_{xx} = 47a^4\pi/96$. Because of symmetry, we have $\alpha = 0$. We assume

$$\chi = \kappa_y\left[c_1r^2\cos 2\theta + c_2r^{3/2}\sin\frac{3\theta}{2} + c_3r\cos\theta\right],\ \theta\in[0, 2\pi].$$

For $\theta\in[0, 2\pi]$, we have $\sin(\theta/2)\geq 0$ and $\sqrt{1-\cos\theta} = \sqrt{2}\sin(\theta/2)$, so that the normal is given by $(\sin\frac{\theta}{2}, -\cos\frac{\theta}{2})$. The constants c_1, c_2 and c_2 are determined using the boundary condition that

$$\left[\tau_{rz}\sin\frac{\theta}{2} - \tau_{\theta z}\cos\frac{\theta}{2}\right]_{r=a(1-\cos\theta)} = 0, \tag{6.204}$$

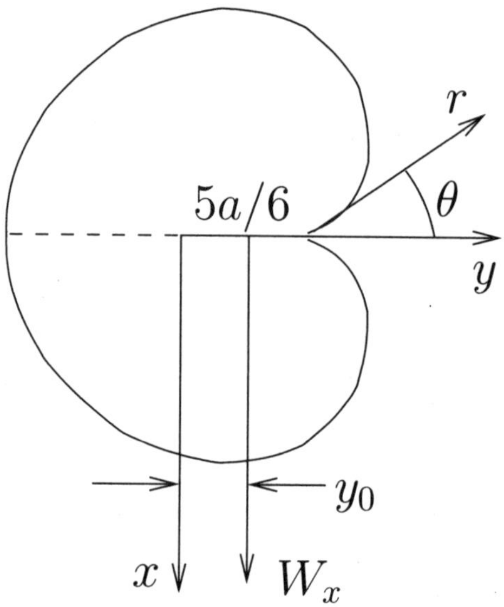

Fig. 6.33 *Flexure of a beam whose cross-section is a cardioid.*

where τ_{rz} and $\tau_{\theta z}$ are given by Eqns. (6.185). We find $c_1 = a/12$, $c_2 = (5+2\nu)a^{3/2}/(12\sqrt{2})$, and $c_3 = (105+62\nu)a^2/144$. The stress distribution is given by

$$\tau_{rz} = \frac{W_y G}{EI_{xx}}\left[-\frac{3+2\nu}{4}r^2\cos\theta - ar\left\{\frac{1}{4}\cos 2\theta + \frac{5}{6}(1+\nu)\right\} + \frac{5+2\nu}{8\sqrt{2}}a^{3/2}\sqrt{r}\sin\frac{3\theta}{2} + \frac{5+2\nu}{24}a^2\cos\theta\right],$$

$$\tau_{\theta z} = \frac{W_y G}{EI_{xx}}\left[\frac{1-2\nu}{4}r^2\sin\theta + \frac{1}{4}ar\sin 2\theta + \frac{5+2\nu}{8\sqrt{2}}a^{3/2}\sqrt{r}\cos\frac{3\theta}{2} - \frac{5+2\nu}{24}a^2\sin\theta\right],$$

while the displacements are given by Eqn. (6.183a)–(6.183c).

Now consider the case when only W_x is applied. We now have $\kappa_y = 0$ and $\kappa_x = W_x/(EI_{yy})$, where $I_{yy} = 21a^4\pi/32$. We assume that

$$\chi = \kappa_x\left[c_1 r^2\sin 2\theta + c_2 r^{3/2}\cos\frac{3\theta}{2} + c_3 r\sin\theta + c_4\sqrt{r}\cos\frac{\theta}{2}\right].$$

The constants are again determined using Eqn. (6.204), and are given by

$$c_1 = -\frac{a}{12},$$

$$c_2 = \frac{a^{3/2}(5+2\nu)}{12\sqrt{2}},$$

$$c_3 = -\frac{48a\alpha + a^2\kappa_x(97+46\nu)}{144\kappa_x},$$

$$c_4 = \frac{a^{3/2}[3\alpha - a\kappa_x(3+4\nu)]}{3\sqrt{2}\kappa_x}.$$

The stresses are given by

$$\tau_{rz} = \frac{G}{48\sqrt{r}}\left[-4\sqrt{2}a^{3/2}\left\{-3\alpha + a\kappa_x(3+4\nu)\right\}\cos\frac{\theta}{2} + \sqrt{r}\left\{3\sqrt{2r}a^{3/2}\kappa_x(5+2\nu)\cos\frac{3\theta}{2}\right.\right.$$

$$\left.\left. +4\left[6a\alpha - 2a^2\kappa_x(3+4\nu) + 3\kappa_x(3+2\nu)r^2 + 6a\kappa_x r\cos\theta\right]\sin\theta\right\}\right],$$

$$\tau_{\theta z} = \frac{G}{48\sqrt{r}}\left[8(6\alpha - 5a\kappa_x\nu)r^{3/2} - 4\sqrt{r}\left\{-6a\alpha + 2a^2\kappa_x(3+4\nu) - 3\kappa_x(1-2\nu)r^2\right\}\cos\theta\right.$$

$$\left. + 12a\kappa_x r^{3/2}\cos 2\theta + \sqrt{2}a^{3/2}\left\{4\left[a\kappa_x(3+4\nu) - 3\alpha\right]\sin\frac{\theta}{2} - 3\kappa_x(5+2\nu)r\sin\frac{3\theta}{2}\right\}\right].$$

Note that there is a stress singularity at $r = 0$ in this case, but not in the case when the loading in along the y-axis.

From Eqn. (6.186), we get

$$y_0 W_x = -\frac{\pi a^4 G}{48}\left[51\alpha + a\kappa_x(3+4\nu)\right].$$

The above equation can be used to find α given y_0 or vice versa. Setting α to zero, we get

$$y_{\text{cf}} = -\frac{a(3+4\nu)}{63(1+\nu)},$$

which agrees with the expression in Sokolnikoff [300] (page 227). On the other hand, if the load is applied along the x-axis, i.e., if $y_0 = 0$, then the angle of twist is

$$\alpha = -\frac{1}{51}a\kappa_x(3+4\nu),$$

which agrees with the corresponding result given by Eqn. (16.16) in [309].

The problem of flexure of a shaft with keyways has been discussed in [290].

Semiparabolic cross-section

We now compute the center of flexure of a bar with the cross-section shown in Fig. 6.34. By symmetry considerations, $x_{\text{cf}} = 0$, so that we simply need to compute y_{cf}. We set W_y to zero and let

$$h(y) := \frac{4a\nu(y+a)}{3(1-\nu)}.$$

It can be easily verified that the origin is at the centroid since $\int_A x\,dA = \int_A y\,dA = 0$. By symmetry $I_{xy} = 0$, and the component I_{yy} is given by

$$I_{yy} = \int_{-a}^{2a/3}\int_{-\sqrt{h(y)}}^{\sqrt{h(y)}} x^2\,dxdy = \frac{160\sqrt{5}}{243}\left[\frac{\nu}{(1-\nu)}\right]^{3/2} a^4.$$

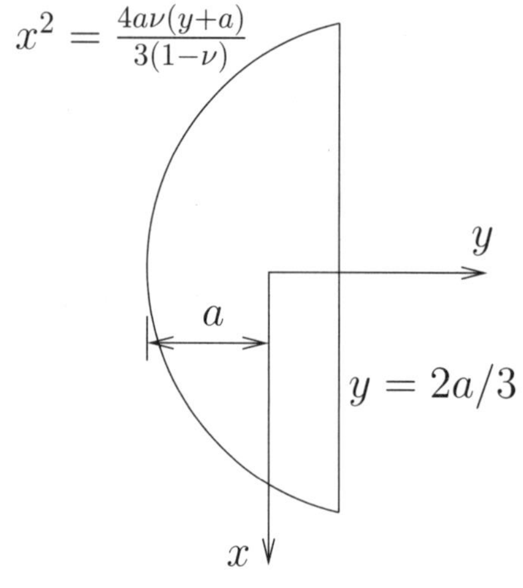

Fig. 6.34 *Bar of semi-parabolic cross-section.*

Setting $W_y = 0$ and $W_x = P$ in Eqn. (6.171), we get $\kappa_x = P/(EI_{yy})$ and $\kappa_y = 0$, which in turn, by virtue of Eqn. (6.176b), implies that $g(x)$ is zero. Since $dy/ds = 0$ on the side $y = 2a/3$, choosing $f(y)$ such that it is equal to $E\kappa_x x^2/2$ on the curved surface will ensure that Eqn. (6.176a) is satisfied. Thus, we get

$$f(y) = \frac{2Pa\nu(y+a)}{3I_{yy}(1-\nu)}.$$

Substituting these values into Eqn. (6.174) with $\alpha = 0$, we get

$$\nabla^2\phi = \frac{\nu P}{I_{yy}}\left[\frac{y}{1+\nu} - \frac{2a}{3(1-\nu)}\right]. \tag{6.205}$$

Since $\phi = 0$ on the boundary, we assume it to be of the form $m[x^2 - h(y)][y - 2a/3]$. Substituting this form into Eqn. (6.205), we get $m = \frac{\nu P}{2(1+\nu)I_{yy}}$. From Eqn. (6.181b), we get

$$y_{cf} = \frac{4a(1-4\nu)}{21(1+\nu)}.$$

Thus, for $\nu = 0.25$, the shear center coincides with the centroid. For this value of Poisson ratio, and for the case $W_x = P$ and $W_y = 0$, the stress distribution is given by

$$\tau_{xz} = \frac{2P}{135I_{yy}}\left[14a^2 + 3(3ay - 9x^2)\right],$$
$$\tau_{yz} = -\frac{Px}{5I_{yy}}\left(y - \frac{2a}{3}\right).$$

For bending of a beam loaded uniformly along its length, including the solution for a circular cylinder bent under its own weight, see [203].

6.5.7 Hollow sphere subjected to uniform pressure/Gravitating sphere

Consider a hollow sphere of inner and outer radii a and b subjected to uniform internal and external pressures p_i and p_o, respectively. By symmetry considerations, we assume that $u_r = u_r(r)$, $u_\theta = u_\phi = 0$. Subject to these assumptions, the Navier equations in Appendix C yield the differential equation

$$\frac{d}{dr}\left[\frac{1}{r^2}\frac{d(r^2 u_r)}{dr}\right] = 0. \tag{6.206}$$

The solution of this differential equation is

$$u_r = c_1 r + \frac{c_2}{r^2},$$

which, on using Eqns. (C.9) and (C.11), yields

$$\tau_{rr} = (3\lambda + 2\mu)c_1 - \frac{4\mu c_2}{r^3},$$

$$\tau_{\theta\theta} = \tau_{\phi\phi} = (3\lambda + 2\mu)c_1 + \frac{2\mu c_2}{r^3}.$$

Using the boundary conditions $\tau_{rr}|_{r=a} = -p_i$ and $\tau_{rr}|_{r=b} = -p_o$, we get the nonzero stress components as

$$\begin{aligned}\tau_{rr} &= \frac{p_i a^3 - p_o b^3}{b^3 - a^3} + \frac{a^3 b^3 (p_o - p_i)}{b^3 - a^3}\frac{1}{r^3},\\ \tau_{\theta\theta} = \tau_{\phi\phi} &= \frac{p_i a^3 - p_o b^3}{b^3 - a^3} - \frac{a^3 b^3 (p_o - p_i)}{b^3 - a^3}\frac{1}{2r^3},\end{aligned} \tag{6.207}$$

and the nonzero displacement component as

$$\begin{aligned}u_r &= \frac{1}{3\lambda + 2\mu}\frac{p_i a^3 - p_o b^3}{b^3 - a^3} r - \frac{1}{4\mu}\frac{a^3 b^3 (p_o - p_i)}{b^3 - a^3}\frac{1}{r^2}\\ &= \frac{(1-2\nu)}{E}\frac{p_i a^3 - p_o b^3}{b^3 - a^3} r - \frac{(1+\nu)}{2E}\frac{a^3 b^3 (p_o - p_i)}{b^3 - a^3}\frac{1}{r^2}.\end{aligned} \tag{6.208}$$

Note that $\tau_{rr} + \tau_{\theta\theta} + \tau_{\phi\phi}$ is a constant throughout the sphere.

Now consider the problem of a gravitating solid sphere of radius R_0, with a body force given by $b_r = -gr/R_0$. Again by symmetry considerations, we have $u_r = u_r(r)$, $u_\theta = u_\phi = 0$. Instead of Eqn. (6.206), we now get

$$(\lambda + 2\mu)\frac{d}{dr}\left[\frac{1}{r^2}\frac{d(r^2 u_r)}{dr}\right] - \frac{\rho_0 g r}{R_0} = 0,$$

whose solution is given by

$$u_r = c_1 r + \frac{c_2}{r^2} + \frac{\rho_0 g r^3}{10(\lambda + 2\mu)R_0}.$$

Since we are considering a solid sphere, in order to prevent the displacements from becoming singular at $r = 0$, we set c_2 to zero. The other constant is determined from the traction-free boundary condition at the outer surface $\tau_{rr}|_{r=R_0} = 0$. The final solution that we get is

$$u_r = \frac{\rho_0 g r R_0}{10(\lambda+2\mu)}\left[\left(\frac{r}{R_0}\right)^2 - \frac{5\lambda+6\mu}{3\lambda+2\mu}\right] = \frac{\rho_0 g r R_0(1+\nu)(1-2\nu)}{10E(1-\nu)}\left[\left(\frac{r}{R_0}\right)^2 - \frac{3-\nu}{1+\nu}\right],$$

$$\tau_{rr} = \frac{(5\lambda+6\mu)\rho_0 g R_0}{10(\lambda+2\mu)}\left[\left(\frac{r}{R_0}\right)^2 - 1\right] = \frac{(3-\nu)\rho_0 g R_0}{(1-\nu)}\left[\left(\frac{r}{R_0}\right)^2 - 1\right],$$

$$\tau_{\theta\theta} = \tau_{\phi\phi} = \frac{(5\lambda+2\mu)\rho_0 g R_0}{10(\lambda+2\mu)}\left[\left(\frac{r}{R_0}\right)^2 - \frac{5\lambda+6\mu}{5\lambda+2\mu}\right] = \frac{(1+3\nu)\rho_0 g R_0}{10(1-\nu)}\left[\left(\frac{r}{R_0}\right)^2 - \frac{3-\nu}{1+3\nu}\right].$$

We now show by means of an example that two possibly inhomogeneous but isotropic bodies having different constitutive properties can have different displacement and stress solutions in the interior but the same traction and displacement distributions on the surface. First consider a solid homogeneous sphere of radius b with Lamé constants given by (λ_0, μ_0), and subjected to a pressure p on the outer surface. From Eqns. (6.207) and (6.208), the displacement and traction vectors on the surface $r = b$ are given by

$$\boldsymbol{u} = \begin{bmatrix} -\frac{pb}{3\lambda_0+2\mu_0} \\ 0 \\ 0 \end{bmatrix}, \quad \boldsymbol{t} = \begin{bmatrix} -p \\ 0 \\ 0 \end{bmatrix}. \tag{6.209}$$

Now consider a sphere of radius b made of a material with properties

$$\lambda = -\frac{16(3\lambda_0+2\mu_0)r/b}{9+4r/b}, \quad \mu = 2(3\lambda_0+2\mu_0). \tag{6.210}$$

Note that, since $3\lambda_0+2\mu_0 > 0$, the material properties in Eqn. (6.210) satisfy the constraints $\mu > 0$ and $3\lambda+2\mu > 0$. The displacement and stress solution is given by

$$u_r = c_1 r + c_2 r^2,$$

$$\tau_{rr} = p\left(\frac{2r}{b} - 3\right),$$

$$\tau_{\theta\theta} = 3p\left(\frac{r}{b} - 1\right),$$

$$\tau_{\phi\phi} = \tau_{\theta\theta},$$

where

$$c_1 = -\frac{3p}{4(3\lambda_0+2\mu_0)},$$

$$c_2 b = -\frac{p}{4(3\lambda_0+2\mu_0)}.$$

Using Eqn. (C.8), it can be easily verified that the above stress distribution satisfies the equations of equilibrium. Moreover, the displacement and traction fields on the surface are the same as given by Eqn. (6.209).

The above demonstration, is, of course, not a counterexample to the uniqueness theorems stated in [75] and [235], since we have shown the surface traction distributions to be the same for *one* choice of the surface displacement field, while the above-mentioned uniqueness theorems require this to be true for *all* choices of surface displacements. A corresponding example for the heat conduction problem is given in Problem 15.

6.6 General Solutions for Elastostatics using Potentials

Consider the equation of equilibrium

$$\boldsymbol{\nabla}_x \cdot \boldsymbol{\tau} + \rho \boldsymbol{b} = \boldsymbol{0}. \tag{6.211}$$

The general solution to the above equation is [109]

$$\boldsymbol{\tau} = \boldsymbol{\nabla} \times (\boldsymbol{\nabla} \times \boldsymbol{H}) + \boldsymbol{\nabla h} + (\boldsymbol{\nabla h})^T - (\boldsymbol{\nabla} \cdot \boldsymbol{h})\boldsymbol{I}, \tag{6.212}$$

where $\boldsymbol{H} \in \text{Sym}$, and $\boldsymbol{h}$ is a vector field that satisfies $\boldsymbol{\nabla}^2 \boldsymbol{h} = -\rho \boldsymbol{b}$. If the body forces are conservative, i.e., there exists a potential W_b such that $\rho \boldsymbol{b} = \boldsymbol{\nabla} W_b$, then using the fact that $\boldsymbol{\nabla} \cdot (W_b \boldsymbol{I}) = \boldsymbol{\nabla} W_b$, one can append the term $-W_b \boldsymbol{I}$ to the right-hand side of Eqn. (6.212), and take $\boldsymbol{h}$ to be a harmonic vector. In what follows, we assume the body forces to be conservative, so that Eqn. (6.211) can be written as

$$\boldsymbol{\nabla} \cdot \tilde{\boldsymbol{\tau}} = \boldsymbol{0}, \tag{6.213}$$

where $\tilde{\boldsymbol{\tau}} := \boldsymbol{\tau} + W_b \boldsymbol{I}$.

An equilibrated stress field $\tilde{\boldsymbol{\tau}}$ is one that satisfies Eqn. (6.213). A self-equilibrated stress field is one whose resultant force and moment corresponding to $\tilde{\boldsymbol{\tau}}$ vanish on every closed regular surface S in V:

$$\int_S \tilde{\boldsymbol{\tau}} \boldsymbol{n} \, dS = \int_S \boldsymbol{x} \times (\tilde{\boldsymbol{\tau}} \boldsymbol{n}) \, dS = \boldsymbol{0}.$$

By using the divergence and localization theorems, it follows that every self-equilibrated stress field satisfies Eqn. (6.213). Conversely, if the boundary of V consists of a single closed surface, then from the divergence theorem, it follows that every $\boldsymbol{\tau}$ that satisfies Eqn. (6.213) is self-equilibrated (equivalently, if the stress field is not self-equilibrated, then V contains more than one closed surface). If the boundary of V contains more than one closed surface, then the stress field may or may not be self-equilibrated.

When the field $\boldsymbol{h}$ in Eqn. (6.212) can be taken to be zero, we get Beltrami's solution:

$$\tilde{\boldsymbol{\tau}} = \boldsymbol{\nabla} \times (\boldsymbol{\nabla} \times \boldsymbol{H}). \tag{6.214}$$

Beltrami's solution always yields a self-equilibrated stress field. Conversely, if $\tilde{\boldsymbol{\tau}}$ is a self-equilibrated field, then there exists $\boldsymbol{H} \in \text{Sym}$ such that Eqn. (6.214) holds. Thus, stress fields that are not self-equilibrated cannot be represented as Beltrami solutions (i.e., $\boldsymbol{h} \neq \boldsymbol{0}$

in Eqn. (6.212)). As mentioned above, if the boundary of V consists of a single-closed surface, then every equilibrated stress field is self-equilibrated, so that the most general solution of Eqn. (6.213) in such a case is given by Eqn. (6.214).

The symmetric tensor field $\boldsymbol{H}$ is not unique. In fact, since, for any arbitrary vector field $\boldsymbol{w}$, $\boldsymbol{\nabla} \times [\boldsymbol{\nabla} \times (\boldsymbol{\nabla w} + (\boldsymbol{\nabla w})^T)] = \boldsymbol{0}$, one can add the field $\boldsymbol{\nabla w} + (\boldsymbol{\nabla w})^T$ to $\boldsymbol{H}$ without affecting the stress field. Gurtin [109] and Rostamian [274] used this technique to construct $\boldsymbol{w}$ such that $\boldsymbol{H}$ has only nonzero off-diagonal or diagonal terms, respectively. Thus, when the stress-field is self-equilibrated, one can, without loss of generality, take $\boldsymbol{H}$ to be either diagonal (Maxwell' solution), or with only the off-diagonal terms nonzero (Morera's solution).[7]

If $H_{11} \equiv \phi_1$, $H_{22} \equiv \phi_2$, $H_{33} \equiv \phi_3$ with the other components of $\boldsymbol{H}$ zero, we can write Maxwell's solution for the stress field as

$$\begin{aligned}
\tau_{11} &= \frac{\partial^2 \phi_2}{\partial x_3^2} + \frac{\partial^2 \phi_3}{\partial x_2^2} - W_b, & \tau_{12} &= -\frac{\partial^2 \phi_3}{\partial x_1 \partial x_2}, \\
\tau_{22} &= \frac{\partial^2 \phi_3}{\partial x_1^2} + \frac{\partial^2 \phi_1}{\partial x_3^2} - W_b, & \tau_{23} &= -\frac{\partial^2 \phi_1}{\partial x_2 \partial x_3}, \\
\tau_{33} &= \frac{\partial^2 \phi_1}{\partial x_2^2} + \frac{\partial^2 \phi_2}{\partial x_1^2} - W_b, & \tau_{31} &= -\frac{\partial^2 \phi_2}{\partial x_3 \partial x_1}.
\end{aligned} \tag{6.215}$$

On the other hand, if $H_{23} \equiv \psi_1$, $H_{13} \equiv \psi_2$ and $H_{12} \equiv \psi_3$ with the other components zero, we can write Morera's solution as

$$\begin{aligned}
\tau_{11} &= -2\frac{\partial^2 \psi_1}{\partial x_2 \partial x_3} - W_b, & \tau_{12} &= \frac{\partial}{\partial x_3}\left(\frac{\partial \psi_1}{\partial x_1} + \frac{\partial \psi_2}{\partial x_2} - \frac{\partial \psi_3}{\partial x_3}\right), \\
\tau_{22} &= -2\frac{\partial^2 \psi_2}{\partial x_3 \partial x_1} - W_b, & \tau_{23} &= \frac{\partial}{\partial x_1}\left(\frac{\partial \psi_2}{\partial x_2} + \frac{\partial \psi_3}{\partial x_3} - \frac{\partial \psi_1}{\partial x_1}\right), \\
\tau_{33} &= -2\frac{\partial^2 \psi_3}{\partial x_1 \partial x_2} - W_b, & \tau_{13} &= \frac{\partial}{\partial x_2}\left(\frac{\partial \psi_3}{\partial x_3} + \frac{\partial \psi_1}{\partial x_1} - \frac{\partial \psi_2}{\partial x_2}\right).
\end{aligned}$$

For a cylindrical coordinate system, since the left-hand side of Eqns. (B.10) are the components of $\boldsymbol{\nabla} \times (\boldsymbol{\nabla} \times \boldsymbol{\epsilon})$, we obtain the components of $\boldsymbol{\nabla} \times (\boldsymbol{\nabla} \times \boldsymbol{H})$ simply by replacing $\boldsymbol{\epsilon}$ by $\boldsymbol{H}$. Following this procedure, the Maxwell and Morera solutions for the stress field with respect to a cylindrical coordinate system are

$$\begin{aligned}
\tau_{rr} &= \frac{\partial^2 \phi_2}{\partial z^2} + \frac{1}{r^2}\frac{\partial^2 \phi_3}{\partial \theta^2} + \frac{1}{r}\frac{\partial \phi_3}{\partial r} - W_b, & \tau_{rr} &= -\frac{2}{r}\frac{\partial^2 \psi_1}{\partial \theta \partial z} - \frac{2}{r}\frac{\partial \psi_2}{\partial z} - W_b, \\
\tau_{\theta\theta} &= \frac{\partial^2 \phi_1}{\partial z^2} + \frac{\partial^2 \phi_3}{\partial r^2} - W_b, & \tau_{\theta\theta} &= -2\frac{\partial^2 \psi_2}{\partial r \partial z} - W_b,
\end{aligned}$$

[7]Bradley [28] claims to be able to generate the stress field, even in the three-dimensional context, using a single stress function. In terms of this single stress function ψ, the expressions for the shear stresses that he obtains are $\tau_{12} = -\frac{\partial^2 \psi}{\partial x_1 \partial x_2}$, $\tau_{23} = -\frac{\partial^2 \psi}{\partial x_2 \partial x_3}$ and $\tau_{13} = -\frac{\partial^2 \psi}{\partial x_1 \partial x_3}$. These expressions impose the constraints $\frac{\partial \tau_{12}}{\partial x_3} = \frac{\partial \tau_{23}}{\partial x_1} = \frac{\partial \tau_{13}}{\partial x_2}$ on the stress field, which will clearly not hold *in general*. Thus, a minimum of three stress functions is required to generate the stress field.

$$\tau_{zz} = \frac{1}{r^2}\frac{\partial^2 \phi_1}{\partial \theta^2} - \frac{1}{r}\frac{\partial \phi_1}{\partial r} + \frac{1}{r}\frac{\partial^2 (r\phi_2)}{\partial r^2} - W_b, \quad \tau_{zz} = -\frac{2}{r^2}\frac{\partial^2 (r\psi_3)}{\partial r \partial \theta} - W_b,$$

$$\tau_{r\theta} = -\frac{\partial^2}{\partial r \partial \theta}\left(\frac{\phi_3}{r}\right), \qquad \tau_{r\theta} = \frac{\partial}{\partial z}\left[-\frac{\partial \psi_3}{\partial z} + r\frac{\partial}{\partial r}\left(\frac{\psi_1}{r}\right) + \frac{1}{r}\frac{\partial \psi_2}{\partial \theta}\right],$$

$$\tau_{\theta z} = -\frac{1}{r}\frac{\partial^2 \phi_1}{\partial \theta \partial z}, \qquad \tau_{\theta z} = -\frac{\partial}{\partial r}\left[\frac{1}{r}\frac{\partial (r\psi_1)}{\partial r} - \frac{1}{r}\frac{\partial \psi_2}{\partial \theta}\right] + \frac{1}{r^2}\frac{\partial^2 (r^2\psi_3)}{\partial r \partial z},$$

$$\tau_{rz} = -\frac{\partial}{\partial z}\left[\frac{\partial \phi_2}{\partial r} + \frac{\phi_2 - \phi_1}{r}\right], \qquad \tau_{rz} = \frac{\partial}{\partial \theta}\left[\frac{1}{r^2}\frac{\partial (r\psi_1)}{\partial r} - \frac{1}{r^2}\frac{\partial \psi_2}{\partial \theta} + \frac{1}{r}\frac{\partial \psi_3}{\partial z}\right].$$

One chooses the stress functions such that they satisfy the equations of compatibility, i.e.,

$$\boldsymbol{\nabla} \times \boldsymbol{\nabla} \times [\mathbf{C}^{-1}(\boldsymbol{\nabla} \times \boldsymbol{\nabla} \times \boldsymbol{H} - W_b \boldsymbol{I})] = \mathbf{0}.$$

For example, for the case of torsionless axisymmetry, one can take $\phi_1 = \phi_r(r,z)$, $\phi_2 = 0$ and $\phi_3 = \phi_z(r,z)$ in the Maxwell solution to get[8]

$$\tau_{rr} = \frac{1}{r}\frac{\partial \phi_z}{\partial r} - W_b, \qquad \tau_{zz} = -\frac{1}{r}\frac{\partial \phi_r}{\partial r} - W_b,$$

$$\tau_{\theta\theta} = \frac{\partial^2 \phi_r}{\partial z^2} + \frac{\partial^2 \phi_z}{\partial r^2} - W_b, \qquad \tau_{rz} = \frac{1}{r}\frac{\partial \phi_r}{\partial z}.$$

The two functions ϕ_r and ϕ_z have to be chosen such that the resulting strain field satisfies the two compatibility relations given by Eqn. (B.11).

Min-zhong and Lu-nan [223] discuss stress functions for torsion, axisymmetric deformation etc. Some examples are

1. For the torsion of multiply connected prismatic beams, since the surfaces are traction-free, the stress field is self-equilibrated ($\boldsymbol{h} = \mathbf{0}$), and one can take $\psi_3 = 0$, and $\frac{\partial \psi_1}{\partial x_1} - \frac{\partial \psi_2}{\partial x_2} = \phi$ in the Morera solution, where ψ_1 and ψ_2 are functions of (x_1, x_2), and ϕ is the Prandtl stress function, to get the Prandtl stress formulation.

2. For the torsion of circular shafts of variable diameter (see page 341 of [324] or Section 6.5.4), one can take $\psi_1 = \psi_2 = 0$ and $r^2 \partial \psi_3/\partial z = \phi$ in the Morera solution, where ϕ is known as the Michell stress function. The governing equation for ϕ is given by Eqn. (6.127).

3. For the case of the hollow cylinder subjected to uniform pressure (which is a special case of the problem treated in Section 6.6.6; see also Problem 10) where the stresses are of the form $\tau_{rr} = c_1 + c_2/r^2$ and $\tau_{zz} = c_3$, we have $\phi_r = -c_3 r^2/2$ and $\phi_z = c_1 r^2/2 + c_2 \log r$.

4. For the "plane-strain" case of the spinning cylinder treated in Section 6.6.11, where the stresses are of the form $\tau_{rr} = c_1 + c_2 r^2$, $\tau_{zz} = c_3 + c_4 r^2$ and $W_b = \rho_0 r^2 \omega^2/2$, we have $\phi_r = -c_3 r^2/2 - (c_4 + \rho_0 \omega^2/2) r^4/2$ and $\phi_z = c_1 r^2/2 + (c_2 + \rho_0 \omega^2/2) r^4/4$.

[8] Choosing $\phi_2 = 0$ is permitted since the resulting expressions for the stresses do not impose any restriction on the stress field other than the equilibrium equations.

We now discuss in detail the Airy stress function approach for planar problems. Under in-plane loading conditions, and assuming that the z-axis is the out-of-plane axis, the plane stress and plane-strain assumptions, which are valid when the thickness is small and large, respectively, compared to the in-plane dimensions, can be written as

$$\tau_{zz} = \tau_{xz} = \tau_{yz} = 0, \qquad \text{(plane stress)} \tag{6.216a}$$

$$\epsilon_{zz} = \epsilon_{xz} = \epsilon_{yz} = 0. \qquad \text{(plane strain)} \tag{6.216b}$$

By assuming $\phi_3(x_1, x_2) =: \phi$ to be the only nonzero component in the Maxwell solution given by Eqn. (6.215), the two-dimensional stress tensor can be expressed as

$$\boldsymbol{\tau} = (\boldsymbol{\nabla}^2\phi)\boldsymbol{I} - \boldsymbol{\nabla}(\boldsymbol{\nabla}\phi) - W_b\boldsymbol{I}, \tag{6.217}$$

so that

$$\operatorname{tr}\boldsymbol{\tau} = \boldsymbol{\nabla}^2\phi - 2W_b. \tag{6.218}$$

In the context of the two-dimensional problems that we are considering $\operatorname{tr}\boldsymbol{\tau}$ denotes $\tau_{xx} + \tau_{yy}$ in the Cartesian system, $\tau_{rr} + \tau_{\theta\theta}$ in the polar coordinate system, etc. Similarly, $\operatorname{tr}\boldsymbol{\epsilon}$ denotes the sum of the normal in-plane strains.

If (ξ, η) denote the in-plane coordinates in an orthogonal curvilinear coordinate system (i.e., (ξ, η) is given by (x, y) in a Cartesian system, by (r, θ) in a polar system, and so on), and if (h_1, h_2) denote the corresponding scale factors, then using Eqns. (A.18) and (A.22), Eqn. (6.217) can be expressed as

$$\begin{aligned}
\tau_{\xi\xi} &= \frac{1}{h_2}\frac{\partial}{\partial\eta}\left(\frac{1}{h_2}\frac{\partial\phi}{\partial\eta}\right) + \frac{1}{h_1^2 h_2}\frac{\partial h_2}{\partial\xi}\frac{\partial\phi}{\partial\xi} - W_b,\\
\tau_{\eta\eta} &= \frac{1}{h_1}\frac{\partial}{\partial\xi}\left(\frac{1}{h_1}\frac{\partial\phi}{\partial\xi}\right) + \frac{1}{h_1 h_2^2}\frac{\partial h_1}{\partial\eta}\frac{\partial\phi}{\partial\eta} - W_b,\\
\tau_{\xi\eta} &= \frac{1}{h_1 h_2}\frac{\partial^2\phi}{\partial\xi\partial\eta} - \frac{1}{h_1}\frac{\partial}{\partial\xi}\left(\frac{1}{h_2}\frac{\partial\phi}{\partial\eta}\right) - \frac{1}{h_2}\frac{\partial}{\partial\eta}\left(\frac{1}{h_1}\frac{\partial\phi}{\partial\xi}\right).
\end{aligned} \tag{6.219}$$

The strain–displacement relations are

$$\begin{aligned}
\epsilon_{\xi\xi} &= \frac{1}{h_1}\frac{\partial u_\xi}{\partial\xi} + \frac{u_\eta}{h_1 h_2}\frac{\partial h_1}{\partial\eta},\\
\epsilon_{\eta\eta} &= \frac{1}{h_2}\frac{\partial u_\eta}{\partial\eta} + \frac{u_\xi}{h_1 h_2}\frac{\partial h_2}{\partial\xi},\\
\gamma_{\xi\eta} &= \frac{h_1}{h_2}\frac{\partial}{\partial\eta}\left(\frac{u_\xi}{h_1}\right) + \frac{h_2}{h_1}\frac{\partial}{\partial\xi}\left(\frac{u_\eta}{h_2}\right).
\end{aligned}$$

Thus, using Eqn. (6.219), with respect to a Cartesian system, we have

$$\begin{aligned}
\tau_{xx} &= \frac{\partial^2\phi}{\partial y^2} - W_b,\\
\tau_{yy} &= \frac{\partial^2\phi}{\partial x^2} - W_b,\\
\tau_{xy} &= -\frac{\partial^2\phi}{\partial x\partial y},
\end{aligned} \tag{6.220}$$

while, in a cylindrical coordinate system, using Eqn. (B.1), we have

$$\tau_{rr} = \frac{1}{r^2}\frac{\partial^2 \phi}{\partial \theta^2} + \frac{1}{r}\frac{\partial \phi}{\partial r} - W_b,$$

$$\tau_{\theta\theta} = \frac{\partial^2 \phi}{\partial r^2} - W_b, \tag{6.221}$$

$$\tau_{r\theta} = -\frac{\partial^2}{\partial r \partial \theta}\left(\frac{\phi}{r}\right).$$

In terms of elliptic coordinates (ξ, η) (see Appendix D), the expressions are

$$\tau_{\xi\xi} = \frac{1}{h_1^2}\frac{\partial^2 \phi}{\partial \eta^2} + \frac{c^2}{2h_1^4}\left[\sinh 2\xi \frac{\partial \phi}{\partial \xi} - \sin 2\eta \frac{\partial \phi}{\partial \eta}\right] - W_b,$$

$$\tau_{\eta\eta} = \frac{1}{h_1^2}\frac{\partial^2 \phi}{\partial \xi^2} - \frac{c^2}{2h_1^4}\left[\sinh 2\xi \frac{\partial \phi}{\partial \xi} - \sin 2\eta \frac{\partial \phi}{\partial \eta}\right] - W_b, \tag{6.222}$$

$$\tau_{\xi\eta} = -\frac{1}{h_1^2}\frac{\partial^2 \phi}{\partial \xi \partial \eta} + \frac{c^2}{2h_1^4}\left[\sinh 2\xi \frac{\partial \phi}{\partial \eta} + \sin 2\eta \frac{\partial \phi}{\partial \xi}\right],$$

while for a bipolar coordinate system that is defined via Eqns. (6.124) (or Eqn. (E.1)), they are given by Eqns. (6.219) with $h_1 = h_2 = c/(\cosh \xi - \cos \eta)$. In the absence of body forces, W_b is set to zero in the above expressions.

For an isotropic material, the two-dimensional constitutive relations for plane stress/ plane strain obtained by using the assumptions in Eqns. (6.216) are given by

$$\boldsymbol{\tau} = \frac{E\nu(\operatorname{tr} \boldsymbol{\epsilon})}{1-\nu^2}\boldsymbol{I} + 2\mu\boldsymbol{\epsilon}, \tag{6.223a}$$

$$\boldsymbol{\tau} = \lambda(\operatorname{tr} \boldsymbol{\epsilon})\boldsymbol{I} + 2\mu\boldsymbol{\epsilon}, \tag{6.223b}$$

The corresponding two-dimensional inverse relations are

$$\boldsymbol{\epsilon} = -\frac{\nu}{E}(\operatorname{tr} \boldsymbol{\tau})\boldsymbol{I} + \frac{1+\nu}{E}\boldsymbol{\tau}, \tag{6.224a}$$

$$\boldsymbol{\epsilon} = -\frac{\nu(1+\nu)}{E}(\operatorname{tr} \boldsymbol{\tau})\boldsymbol{I} + \frac{1+\nu}{E}\boldsymbol{\tau}, \tag{6.224b}$$

Note that the plane strain constitutive relations can be obtained from the plane stress ones, and vice versa using the following transformations

$$E \to \frac{E}{1-\nu^2}, \quad \nu \to \frac{\nu}{1-\nu}, \quad \mu \to \mu, \quad \alpha \to \alpha(1+\nu) \quad \text{(plane stress to plane strain)}, \tag{6.225a}$$

$$E \to \frac{E(1+2\nu)}{(1+\nu)^2}, \quad \nu \to \frac{\nu}{1+\nu}, \quad \mu \to \mu, \quad \alpha \to \frac{\alpha(1+\nu)}{1+2\nu} \quad \text{(plane strain to plane stress)}, \tag{6.225b}$$

where α is the coefficient of thermal expansion[9].

From the above relations, the constitutive matrices that relate the stress components $(\tau_{\alpha\alpha}, \tau_{\beta\beta}, \tau_{\alpha\beta})$ to the engineering strain components $(\epsilon_{\alpha\alpha}, \epsilon_{\beta\beta}, \gamma_{\alpha\beta})$, where (α, β) are orthogonal curvilinear coordinates (e.g., (x, y) in a Cartesian system or (r, θ) in a polar system), in the plane stress and plane strain cases, are obtained as

$$\begin{aligned} C_{\text{plane stress}} &= \frac{E}{1-\nu^2}\begin{bmatrix} 1 & \nu & 0 \\ \nu & 1 & 0 \\ 0 & 0 & \frac{1-\nu}{2} \end{bmatrix}, \\ C_{\text{plane strain}} &= \frac{E}{(1+\nu)(1-2\nu)}\begin{bmatrix} 1-\nu & \nu & 0 \\ \nu & 1-\nu & 0 \\ 0 & 0 & \frac{1-2\nu}{2} \end{bmatrix}. \end{aligned} \tag{6.226}$$

In the plane strain case, the out-of-plane normal stress is given by $\tau_{zz} = \nu(\tau_{\alpha\alpha} + \tau_{\beta\beta}) = \nu(\operatorname{tr} \boldsymbol{\tau})$, with $\operatorname{tr} \boldsymbol{\tau}$ denoting the sum of the in-plane normal stresses within the current context. Hence, using Eqn. (6.15), we get

$$\boldsymbol{\nabla}^2(\operatorname{tr} \boldsymbol{\tau}) = -(1+\nu)\boldsymbol{\nabla}^2 W_b, \qquad \text{(plane stress)} \tag{6.227a}$$

$$\boldsymbol{\nabla}^2(\operatorname{tr} \boldsymbol{\tau}) = -\frac{1}{1-\nu}\boldsymbol{\nabla}^2 W_b, \qquad \text{(plane strain)} \tag{6.227b}$$

where we have used Eqn. (6.225b) to obtain Eqn. (6.227a) from Eqn. (6.227b). Note that although the plane strain assumption satisfies the out-of-plane compatibility relations also, the plane stress assumption satisfies them only if $\operatorname{tr} \boldsymbol{\tau}$ is of the form $A + Bx + Cy$, which, in general, does not hold. Thus, solutions derived under the plane stress assumption have to considered as approximate even in the linearized elasticity framework.

Finally, substituting Eqn. (6.218) into Eqn. (6.227a), and subsequently using Eqn. (6.225a), we get the governing equations for the Airy stress function as

$$\boldsymbol{\nabla}^2(\boldsymbol{\nabla}^2\phi) = (1-\nu)\boldsymbol{\nabla}^2 W_b, \qquad \text{(plane stress)} \tag{6.228a}$$

$$\boldsymbol{\nabla}^2(\boldsymbol{\nabla}^2\phi) = \frac{1-2\nu}{1-\nu}\boldsymbol{\nabla}^2 W_b. \qquad \text{(plane strain)} \tag{6.228b}$$

If $\boldsymbol{\nabla}^2 W_b = 0$, then the above governing equations reduce to $\boldsymbol{\nabla}^4\phi = 0$. A function that satisfies this equation is said to be a biharmonic function. Let Φ be a harmonic function, i.e., a function that satisfies $\boldsymbol{\nabla}^2\Phi = 0$, and let $\boldsymbol{x} := (x, y, z)$ denote the position vector. We now show that the functions $\{\Phi, x\Phi, y\Phi, z\Phi, (\boldsymbol{x}\cdot\boldsymbol{x})\Phi\}$ are biharmonic. If $\beta(\boldsymbol{x})$ is a function of the position vector, then

$$\begin{aligned} \boldsymbol{\nabla}^4(\beta\Phi) &= (\beta\Phi)_{,kkmm} \\ &= \beta_{,kkmm}\Phi + 4\beta_{,kkm}\Phi_{,m} + 4\beta_{,km}\Phi_{,km}, \end{aligned}$$

[9]We consider only isothermal problems in this section with the exception of Section 6.6.5, where we indicate how thermal loading problems can be solved.

where a comma denotes partial differentiation. If $\beta = \{1, x, y, z, \boldsymbol{x} \cdot \boldsymbol{x}\}$, then we see that the right-hand side of the above equation is zero. Summarizing, if Φ is harmonic, then the following functions are biharmonic:

$$
\begin{aligned}
&\Phi, x\Phi, y\Phi, z\Phi, (x^2+y^2+z^2)\Phi, && \text{(3D Cartesian system } (x,y,z)),\\
&\Phi, x\Phi, y\Phi, (x^2+y^2)\Phi, && \text{(2D Cartesian system } (x,y)),\\
&\Phi, (r\cos\theta)\Phi, (r\sin\theta)\Phi, z\Phi, (r^2+z^2)\Phi, && \text{(Cylindrical system } (r,\theta,z)),\\
&\Phi, (r\cos\theta)\Phi, (r\sin\theta)\Phi, r^2\Phi, && \text{(Polar system } (r,\theta)),\\
&\Phi, (r\sin\theta\cos\phi)\Phi, (r\sin\theta\sin\phi)\Phi, (r\cos\theta)\Phi, r^2\Phi, && \text{(Spherical system } (r,\theta,\phi)).
\end{aligned}
\tag{6.229}
$$

Using Eqn. (A.22), if (ξ, η) denote the in-plane curvilinear coordinates, the in-plane Laplacian operator can be expressed as

$$
\boldsymbol{\nabla}^2\phi = \frac{1}{h_1 h_2}\left[\frac{\partial}{\partial\xi}\left(\frac{h_2}{h_1}\frac{\partial\phi}{\partial\xi}\right) + \frac{\partial}{\partial\eta}\left(\frac{h_1}{h_2}\frac{\partial\phi}{\partial\eta}\right)\right]. \tag{6.230}
$$

Assuming constant body forces (i.e., $\boldsymbol{\nabla}^2 W_b = 0$), Eqns. (6.228a) or (6.228b) can be written using Eqn. (6.230) in a Cartesian system as

$$
\left(\frac{\partial^2}{\partial x^2} + \frac{\partial^2}{\partial y^2}\right)\left(\frac{\partial^2\phi}{\partial x^2} + \frac{\partial^2\phi}{\partial y^2}\right) = 0,
$$

or, in the polar coordinate system, as

$$
\left[\frac{1}{r}\frac{\partial}{\partial r}\left(r\frac{\partial}{\partial r}\right) + \frac{1}{r^2}\frac{\partial^2}{\partial\theta^2}\right]\left[\frac{1}{r}\frac{\partial}{\partial r}\left(r\frac{\partial\phi}{\partial r}\right) + \frac{1}{r^2}\frac{\partial^2\phi}{\partial\theta^2}\right] = 0. \tag{6.231}
$$

A similar equation can be written in the elliptic coordinate system as

$$
\left[\frac{\partial^2}{\partial\xi^2} + \frac{\partial^2}{\partial\eta^2}\right]\left\{\frac{1}{(\cosh 2\xi - \cos 2\eta)}\left[\frac{\partial^2\phi}{\partial\xi^2} + \frac{\partial^2\phi}{\partial\eta^2}\right]\right\} = 0, \tag{6.232}
$$

or in terms of bipolar coordinates as

$$
\left[\frac{\partial^2}{\partial\xi^2} + \frac{\partial^2}{\partial\eta^2}\right]\left\{(\cosh\xi - \cos\eta)^2\left[\frac{\partial^2\phi}{\partial\xi^2} + \frac{\partial^2\phi}{\partial\eta^2}\right]\right\} = 0. \tag{6.233}
$$

The general solution to the Laplace equation $\boldsymbol{\nabla}^2\phi = 0$ in Cartesian coordinates is

$$
\begin{aligned}
\phi &= c_0 + c_1 x + c_2 y + c_3 xy\\
&\quad + A\sin k_1 x \sinh k_1 y + B\sin k_2 x\cosh k_2 y\\
&\quad + C\cos k_3 x\sinh k_3 y + D\cos k_y x\cosh k_y y\\
&\quad + E\sinh k_5 x\sin k_5 y + F\sinh k_6 x\cos k_6 y
\end{aligned}
$$

$$
\begin{aligned}
&+ G\cosh k_7 x \sin k_7 y + H \cosh k_8 x \cos k_8 y \\
&+ P(x\cos k_9 x \cosh k_9 y + y \sin k_9 x \sinh k_9 y) \\
&+ Q(x \sin k_{10} x \cosh k_{10} y - y \cos k_{10} x \sinh k_{10} y) \\
&+ R(x \cos k_{11} x \sinh k_{11} y + y \sin k_{11} x \cosh k_{11} y) \\
&+ S(x \sin k_{12} x \sinh k_{12} y - y \cos k_{12} x \cosh k_{12} y) \\
&+ T(x \cosh k_{13} x \cos k_{13} y - y \sinh k_{13} x \sin k_{13} y) \\
&+ U(x \cosh k_{14} x \sin k_{14} y + y \sinh k_{14} x \cos k_{14} y) \\
&+ V(x \sinh k_{15} x \cos k_{15} y - y \cosh k_{15} x \sin k_{15} y) \\
&+ W(x \sinh k_{16} x \sin k_{16} y + y \cosh k_{16} x \cos k_{16} y).
\end{aligned} \tag{6.234}
$$

The polynomial terms in the above solution correspond to the case of zero roots (i.e., the constant that enters into the solution procedure for a separable solution is zero), or, more explicitly, if $\phi = f(x)g(y)$, then the zero root case leads to $f''(x) = g''(y) = 0$ (primes denoting differentiation with respect to the respective variable), which yields the said polynomial solution for ϕ. Noting that if Φ is harmonic, then $\{x\Phi, y\Phi, (x^2+y^2)\Phi\}$ are biharmonic (see Eqn. (6.229)), the general solution to the biharmonic equation $\nabla^4\phi = 0$ in Cartesian coordinates (excluding $\tilde{c}_0 + \tilde{c}_1 x + \tilde{c}_2 y$ since it results in zero stresses) is given by

$$
\begin{aligned}
\phi = {} & c_1 x^2 + c_2 xy + c_3 y^2 \\
& + c_4 x^3 + c_5 x^2 y + c_6 x y^2 + c_7 y^3 \\
& + c_8 x^3 y + c_9 x y^3 \\
& + \sum_{n=1}^{\infty} A_n \sin nx \sinh ny + B_n \sin nx \cosh ny \\
& + C_n \cos nx \sinh ny + D_n \cos nx \cosh ny \\
& + E_n \sinh nx \sin ny + F_n \sinh nx \cos ny \\
& + G_n \cosh nx \sin ny + H_n \cosh nx \cos ny \\
& + x\Big[\hat{A}_n \sin nx \sinh ny + \hat{B}_n \sin nx \cosh ny \\
& + \hat{C}_n \cos nx \sinh ny + \hat{D}_n \cos nx \cosh ny \\
& + \hat{E}_n \sinh nx \sin ny + \hat{F}_n \sinh nx \cos ny \\
& + \hat{G}_n \cosh nx \sin ny + \hat{H}_n \cosh nx \cos ny\Big] \\
& + y\Big[\hat{\hat{A}}_n \sin nx \sinh ny + \hat{\hat{B}}_n \sin nx \cosh ny \\
& + \hat{\hat{C}}_n \cos nx \sinh ny + \hat{\hat{D}}_n \cos nx \cosh ny \\
& + \hat{\hat{E}}_n \sinh nx \sin ny + \hat{\hat{F}}_n \sinh nx \cos ny \\
& + \hat{\hat{G}}_n \cosh nx \sin ny + \hat{\hat{H}}_n \cosh nx \cos ny\Big].
\end{aligned} \tag{6.235}
$$

The first three terms in the solution yield a constant state of stress. Note that the integers n in the above expression can be replaced by complex-valued constants λ_n, where λ_n are the roots of an appropriate characteristic equation. If $z := x + iy$ and $\hat{z} = x - iy$, then the solution given by Eqn. (6.235) corresponds to the real and imaginary parts of $\cos nz$, $\sin nz$, $\cosh nz$, $\sinh nz$, $z \cos nz$, $z \sin nz$, $z \cosh nz$, $z \sinh hz$, $\hat{z} \cos nz$, $\hat{z} \sin nz$, $\hat{z} \cosh nz$ and $\hat{z} \sinh nz$. The displacements corresponding to Eqn. (6.235) in the plane stress case are

$$
\begin{aligned}
Eu_x &= k_1(1+\nu) + k_3(1+\nu)y - 2c_1\nu x - c_2(1+\nu)y + 2c_3 x \\
&\quad - 3c_4(y^2 + \nu x^2) - 2c_5\nu xy + c_6\left[x^2 - (2+\nu)y^2\right] + 6c_7 xy \\
&\quad - c_8 y(3\nu x^2 + y^2) + c_9 y[3x^2 - (2+\nu)y^2] \\
&\quad + \sum_{n=1}^{\infty} -A_n n(1+\nu)\cos nx \sinh ny - B_n n(1+\nu)\cos nx \cosh ny \\
&\quad + C_n n(1+\nu)\sin nx \sinh ny + D_n n(1+\nu)\sin nx \cosh ny \\
&\quad - E_n n(1+\nu)\cosh nx \sin ny - F_n n(1+\nu)\cosh nx \cos ny \\
&\quad - G_n n(1+\nu)\sinh nx \sin ny - H_n n(1+\nu)\sinh nx \cos ny \\
&\quad - \hat{A}_n\left[n(1+\nu)x\cos nx - (1-\nu)\sin nx\right]\sinh ny \\
&\quad - \hat{B}_n\left[n(1+\nu)x\cos nx - (1-\nu)\sin nx\right]\cosh ny \\
&\quad + \hat{C}_n\left[n(1+\nu)x\sin nx + (1-\nu)\cos nx\right]\sinh ny \\
&\quad + \hat{D}_n\left[n(1+\nu)x\sin nx + (1-\nu)\cos nx\right]\cosh ny \\
&\quad - \hat{E}_n\left[n(1+\nu)x\cosh nx - (1-\nu)\sinh nx\right]\sin ny \\
&\quad - \hat{F}_n\left[n(1+\nu)x\cosh nx - (1-\nu)\sinh nx\right]\cos ny \\
&\quad - \hat{G}_n\left[n(1+\nu)x\sinh nx - (1-\nu)\cosh nx\right]\sin ny \\
&\quad - \hat{H}_n\left[n(1+\nu)x\sinh nx - (1-\nu)\cosh nx\right]\cos ny \\
&\quad - \hat{\hat{A}}_n\left[n(1+\nu)y\sinh ny + 2\cosh ny\right]\cos nx \\
&\quad - \hat{\hat{B}}_n\left[n(1+\nu)y\cosh ny + 2\sinh ny\right]\cos nx \\
&\quad + \hat{\hat{C}}_n\left[n(1+\nu)y\sinh ny + 2\cosh ny\right]\sin nx \\
&\quad + \hat{\hat{D}}_n\left[n(1+\nu)y\cosh ny + 2\sinh ny\right]\sin nx \\
&\quad - \hat{\hat{E}}_n\left[n(1+\nu)y\sin ny - 2\cos ny\right]\cosh nx \\
&\quad - \hat{\hat{F}}_n\left[n(1+\nu)y\cos ny + 2\sin ny\right]\cosh nx \\
&\quad - \hat{\hat{G}}_n\left[n(1+\nu)y\sin ny - 2\cos ny\right]\sinh nx \\
&\quad - \hat{\hat{H}}_n\left[n(1+\nu)y\cos ny + 2\sin ny\right]\sinh nx \\
Eu_y &= k_2(1+\nu) - k_3(1+\nu)x + 2c_1 y - c_2(1+\nu)x - 2c_3\nu y \qquad (6.236)
\end{aligned}
$$

$$
\begin{aligned}
&+ 6c_4xy + c_5\left[y^2 - (2+\nu)x^2\right] - 2\nu c_6 xy - 3c_7(x^2 + \nu y^2) \\
&+ c_8 x[3y^2 - (2+\nu)x^2] - c_9 x(x^2 + 3\nu y^2) \\
+ &\sum_{n=1}^{\infty} -A_n n(1+\nu)\sin nx \cosh ny - B_n n(1+\nu)\sin nx \sinh ny \\
&- C_n n(1+\nu)\cos nx \cosh ny - D_n n(1+\nu)\cos nx \sinh ny \\
&- E_n n(1+\nu)\sinh nx \cos ny + F_n n(1+\nu)\sinh nx \sin ny \\
&- G_n n(1+\nu)\cosh nx \cos ny + H_n n(1+\nu)\cosh nx \sin ny \\
&- \hat{A}_n\left[n(1+\nu)x\sin nx - 2\cos nx\right]\cosh ny \\
&- \hat{B}_n\left[n(1+\nu)x\sin nx - 2\cos nx\right]\sinh ny \\
&- \hat{C}_n\left[n(1+\nu)x\cos nx + 2\sin nx\right]\cosh ny \\
&- \hat{D}_n\left[n(1+\nu)x\cos nx + 2\sin nx\right]\sinh ny \\
&- \hat{E}_n\left[n(1+\nu)x\sinh nx + 2\cosh nx\right]\cos ny \\
&+ \hat{F}_n\left[n(1+\nu)x\sinh nx + 2\cosh nx\right]\sin ny \\
&- \hat{G}_n\left[n(1+\nu)x\cosh nx + 2\sinh nx\right]\cos ny \\
&+ \hat{H}_n\left[n(1+\nu)x\cosh nx + 2\sinh nx\right]\sin ny \\
&- \hat{\hat{A}}_n\left[n(1+\nu)y\cosh ny - (1-\nu)\sinh ny\right]\sin nx \\
&- \hat{\hat{B}}_n\left[n(1+\nu)y\sinh ny - (1-\nu)\cosh ny\right]\sin nx \\
&- \hat{\hat{C}}_n\left[n(1+\nu)y\cosh ny - (1-\nu)\sinh ny\right]\cos nx \\
&- \hat{\hat{D}}_n\left[n(1+\nu)y\sinh ny - (1-\nu)\cosh ny\right]\cos nx \\
&- \hat{\hat{E}}_n\left[n(1+\nu)y\cos ny - (1-\nu)\sin ny\right]\sinh nx \\
&+ \hat{\hat{F}}_n\left[n(1+\nu)y\sin ny + (1-\nu)\cos ny\right]\sinh nx \\
&- \hat{\hat{G}}_n\left[n(1+\nu)y\cos ny - (1-\nu)\sin ny\right]\cosh nx \\
&+ \hat{\hat{H}}_n\left[n(1+\nu)y\sin ny + (1-\nu)\cos ny\right]\cosh nx.
\end{aligned}
$$

while in the plane strain case, they are

$$
\begin{aligned}
2\mu u_x = {} & k_1 + k_3 y - 2c_1\nu x - c_2 y + 2(1-\nu)c_3 x - 3c_4[(1-\nu)y^2 + \nu x^2] \\
&- 2c_5\nu xy + c_6\left[(1-\nu)x^2 - (2-\nu)y^2\right] + 6c_7(1-\nu)xy \\
&- c_8 y(3\nu x^2 + (1-\nu)y^2) + c_9 y[3(1-\nu)x^2 - (2-\nu)y^2] \\
+ &\sum_{n=1}^{\infty} -A_n n\cos nx \sinh ny - B_n n\cos nx \cosh ny
\end{aligned}
$$

$$
\begin{aligned}
&+ C_n n \sin nx \sinh ny + D_n n \sin nx \cosh ny \\
&- E_n n \cosh nx \sin ny - F_n n \cosh nx \cos ny \\
&- G_n n \sinh nx \sin ny - H_n n \sinh nx \cos ny \\
&- \hat{A}_n \left[nx \cos nx - (1-2\nu)\sin nx\right] \sinh ny \\
&- \hat{B}_n \left[nx \cos nx - (1-2\nu)\sin nx\right] \cosh ny \\
&+ \hat{C}_n \left[nx \sin nx + (1-2\nu)\cos nx\right] \sinh ny \\
&+ \hat{D}_n \left[nx \sin nx + (1-2\nu)\cos nx\right] \cosh ny \\
&- \hat{E}_n \left[nx \cosh nx - (1-2\nu)\sinh nx\right] \sin ny \\
&- \hat{F}_n \left[nx \cosh nx - (1-2\nu)\sinh nx\right] \cos ny \\
&- \hat{G}_n \left[nx \sinh nx - (1-2\nu)\cosh nx\right] \sin ny \\
&- \hat{H}_n \left[nx \sinh nx - (1-2\nu)\cosh nx\right] \cos ny \\
&- \hat{\hat{A}}_n \left[ny \sinh ny + 2(1-\nu)\cosh ny\right] \cos nx \\
&- \hat{\hat{B}}_n \left[ny \cosh ny + 2(1-\nu)\sinh ny\right] \cos nx \\
&+ \hat{\hat{C}}_n \left[ny \sinh ny + 2(1-\nu)\cosh ny\right] \sin nx \\
&+ \hat{\hat{D}}_n \left[ny \cosh ny + 2(1-\nu)\sinh ny\right] \sin nx \\
&- \hat{\hat{E}}_n \left[ny \sin ny - 2(1-\nu)\cos ny\right] \cosh nx \\
&- \hat{\hat{F}}_n \left[ny \cos ny + 2(1-\nu)\sin ny\right] \cosh nx \\
&- \hat{\hat{G}}_n \left[ny \sin ny - 2(1-\nu)\cos ny\right] \sinh nx \\
&- \hat{\hat{H}}_n \left[ny \cos ny + 2(1-\nu)\sin ny\right] \sinh nx
\end{aligned}
$$

$$
\begin{aligned}
2\mu u_y = {} & k_2 - k_3 x + 2(1-\nu)c_1 y - c_2 x - 2c_3 \nu y + 6c_4(1-\nu)xy \qquad (6.237) \\
&+ c_5\left[(1-\nu)y^2 - (2-\nu)x^2\right] - 2c_6 \nu xy - 3c_7[(1-\nu)x^2 + \nu y^2] \\
&+ c_8 x[3(1-\nu)y^2 - (2-\nu)x^2] - c_9 x[(1-\nu)x^2 + 3\nu y^2] \\
&+ \sum_{n=1}^{\infty} -A_n n \sin nx \cosh ny - B_n n \sin nx \sinh ny \\
&- C_n n \cos nx \cosh ny - D_n n \cos nx \sinh ny \\
&- E_n n \sinh nx \cos ny + F_n n \sinh nx \sin ny \\
&- G_n n \cosh nx \cos ny + H_n n \cosh nx \sin ny \\
&- \hat{A}_n \left[nx \sin nx - 2(1-\nu)\cos nx\right] \cosh ny \\
&- \hat{B}_n \left[nx \sin nx - 2(1-\nu)\cos nx\right] \sinh ny
\end{aligned}
$$

$$
\begin{aligned}
&- \hat{C}_n \left[nx \cos nx + 2(1-\nu) \sin nx \right] \cosh ny \\
&- \hat{D}_n \left[nx \cos nx + 2(1-\nu) \sin nx \right] \sinh ny \\
&- \hat{E}_n \left[nx \sinh nx + 2(1-\nu) \cosh nx \right] \cos ny \\
&+ \hat{F}_n \left[nx \sinh nx + 2(1-\nu) \cosh nx \right] \sin ny \\
&- \hat{G}_n \left[nx \cosh nx + 2(1-\nu) \sinh nx \right] \cos ny \\
&+ \hat{H}_n \left[nx \cosh nx + 2(1-\nu) \sinh nx \right] \sin ny \\
&- \hat{\hat{A}}_n \left[ny \cosh ny - (1-2\nu) \sinh ny \right] \sin nx \\
&- \hat{\hat{B}}_n \left[ny \sinh ny - (1-2\nu) \cosh ny \right] \sin nx \\
&- \hat{\hat{C}}_n \left[ny \cosh ny - (1-2\nu) \sinh ny \right] \cos nx \\
&- \hat{\hat{D}}_n \left[ny \sinh ny - (1-2\nu) \cosh ny \right] \cos nx \\
&- \hat{\hat{E}}_n \left[ny \cos ny - (1-2\nu) \sin ny \right] \sinh nx \\
&+ \hat{\hat{F}}_n \left[ny \sin ny + (1-2\nu) \cos ny \right] \sinh nx \\
&- \hat{\hat{G}}_n \left[ny \cos ny - (1-2\nu) \sin ny \right] \cosh nx \\
&+ \hat{\hat{H}}_n \left[ny \sin ny + (1-2\nu) \cos ny \right] \cosh nx.
\end{aligned}
$$

The above displacements for the infinite series part are obtained by integrating the relations $\partial u_x/\partial x = \epsilon_{xx}$ and $\partial u_y/\partial x = \gamma_{xy} - \partial u_x/\partial y$. As usual, the terms associated with k_1, k_2 and k_3 represent rigid deformations.

In polar coordinates, harmonic functions can be generated by considering the real and imaginary parts of a complex-valued analytic function $W(z)$, where $z = re^{i\theta}$. For example, if $W(z) = z^n$, where n is an integer, then we get the harmonic functions $r^n \cos n\theta$ and $r^n \sin n\theta$, while if $W(z) = r^n \log z$, then we get the harmonic functions $r^n(\log r \cos n\theta - \theta \sin n\theta)$ and $r^n(\log r \sin n\theta + \theta \cos n\theta)$. Thus, a general solution to the Laplace equation $\boldsymbol{\nabla}^2 \phi = 0$ is

$$
\begin{aligned}
\phi &= c_0 + c_1 \log r + c_2\theta + c_3\theta \log r \\
&+ \sum_{\substack{n=-\infty \\ n\neq 0}}^{\infty} r^n \left[A_n \cos n\theta + B_n \sin n\theta + C_n(\log r \cos n\theta - \theta \sin n\theta) + D_n(\log r \sin n\theta + \theta \cos n\theta) \right].
\end{aligned}
\tag{6.238}
$$

Similar to Eqn. (6.234), the first four terms correspond to the zero root case, i.e., if $\phi = f(r)g(\theta)$, then for this case we get $(rf')' = g'' = 0$, which leads to the first four terms. We have already seen in Sections 6.5.2 and 6.5.6 that including the C_n and D_n terms is critical in problems involving an annular sector of a circle ('wedge' geometry) where, for certain wedge angles, the usual separable solution breaks down.

If Φ is a harmonic function, then by virtue of Eqn. (6.229), $(r\cos\theta)\Phi$, $(r\sin\theta)\Phi$ and $r^2\Phi$ are biharmonic functions. Thus, a general solution to Eqn. (6.231) can be written as[10]

$$\begin{aligned}
\phi &= F_1 \log r + F_2\theta + F_3\theta\log r + r^2(G_1\log r + G_2\theta + G_3\theta\log r) + C_1 r^2 \\
&+ r\cos\theta(H_1\log r + H_2\theta + H_3\theta\log r) + r\sin\theta(J_1\log r + J_2\theta + J_3\theta\log r) \\
&+ \sum_{\substack{n=-\infty\\ n\neq 0,-1}}^{\infty} r^{n+1}\left[A_n\cos(n+1)\theta + B_n\sin(n+1)\theta\right] \\
&+ \sum_{\substack{n=-\infty\\ n\neq 0,1}}^{\infty} r^{n+1}\left[C_n\cos(n-1)\theta + D_n\sin(n-1)\theta\right] \qquad (6.239)\\
&+ \sum_{\substack{n=-\infty\\ n\neq 0,-1}}^{\infty} r^{n+1}\left\{\tilde{A}_n\left[\log r\cos(n+1)\theta - \theta\sin(n+1)\theta\right] + \tilde{B}_n\left[\log r\sin(n+1)\theta + \theta\cos(n+1)\theta\right]\right\} \\
&+ \sum_{\substack{n=-\infty\\ n\neq 0,1}}^{\infty} r^{n+1}\left\{\tilde{C}_n\left[\log r\cos(n-1)\theta - \theta\sin(n-1)\theta\right] + \tilde{D}_n\left[\log r\sin(n-1)\theta + \theta\cos(n-1)\theta\right]\right\}.
\end{aligned}$$

As usual, we can replace n by the complex-valued roots of a characteristic equation. Leaving out the terms associated with $\tilde{A}_n$, $\tilde{B}_n$, $\tilde{C}_n$ and $\tilde{D}_n$ can lead to various paradoxes [65] since, similar to the annular sector case mentioned below Eqn. (6.238), for certain wedge angles the usual separable solution breaks down. An exhaustive survey on the two-dimensional biharmonic problem can be found in [216]. Michell [221] presented a solution of the biharmonic equation for a 'cylindrical shell' (by which he probably meant the annular region bounded by two concentric circles). This solution has been misinterpreted in the literature as a general solution of the biharmonic equation in polar coordinates. In fact, the solution presented by Michell [221] agrees with the solution that we present for a region bounded by two concentric circles (see Eqn. (6.351), which is a special case of Eqn. (6.239)) apart from the terms $r\cos\theta$ and $r\sin\theta$, which can be excluded since they result in zero stresses.

The complete solution corresponding to Eqn. (6.239) under conditions of plane stress is

$$\begin{aligned}
2Eu_r &= 2(1+\nu)[k_1\cos\theta + k_2\sin\theta] \\
&\quad - \frac{2(1+\nu)F_1}{r} - \frac{2(1+\nu)\theta F_3}{r} - 2r\left[1+\nu-2(1-\nu)\log r\right]G_1 \\
&\quad + 4(1-\nu)r\theta G_2 - 2r\theta\left[1+\nu-2(1-\nu)\log r\right]G_3 + 4(1-\nu)rC_1 \\
&\quad + 2\left\{\left[1+(1-\nu)\log r\right]\cos\theta + 2\theta\sin\theta\right\}H_1 \\
&\quad + \left[2(1-\nu)\theta\cos\theta - (1-\nu+4\log r)\sin\theta\right]H_2 \\
&\quad + \left\{2\left[(1-\nu)\log r - 1 - \nu\right]\theta\cos\theta + \left[1+\nu+2\theta^2-2\log^2 r\right]\sin\theta\right\}H_3 \\
&\quad + 2\left\{\left[1+(1-\nu)\log r\right]\sin\theta - 2\theta\cos\theta\right\}J_1 \\
&\quad + \left[2(1-\nu)\theta\sin\theta + (1-\nu+4\log r)\cos\theta\right]J_2
\end{aligned}$$

[10] We do not include the constant term (which results in zero stresses), the $n = 0$ term (which result in zero stresses in the first two infinite series), and the $n = -1$ and $n = 1$ terms in the infinite series since those terms are incorporated separately.

$$+\left\{2\left[(1-\nu)\log r-1-\nu\right]\theta\sin\theta-\left[1+\nu+2\theta^2-2\log^2 r\right]\cos\theta\right\}J_3$$
$$-2\sum_{\substack{n=-\infty\\ n\neq 0,-1}}^{\infty}(n+1)(1+\nu)r^n\left\{A_n\cos(n+1)\theta+B_n\sin(n+1)\theta\right\}$$
$$-2\sum_{\substack{n=-\infty\\ n\neq 0,1}}^{\infty}\left[(1+\nu)(n+1)-4\right]r^n\left\{C_n\cos(n-1)\theta+D_n\sin(n-1)\theta\right\}$$
$$-2\sum_{\substack{n=-\infty\\ n\neq 0,-1}}^{\infty}r^n(1+\nu)\left\{\tilde{A}_n\left[\left[1+(n+1)\log r\right]\cos(n+1)\theta-(n+1)\theta\sin(n+1)\theta\right]\right.$$
$$\left.+\tilde{B}_n\left[\left[1+(n+1)\log r\right]\sin(n+1)\theta+(n+1)\theta\cos(n+1)\theta\right]\right\}$$
$$-2\sum_{\substack{n=-\infty\\ n\neq 0,1}}^{\infty}r^n\left\{\tilde{C}_n\left[\left[1+\nu+((n+1)\nu+n-3)\log r\right]\cos(n-1)\theta\right.\right.$$
$$\left.-\left[(n+1)\nu+n-3\right]\theta\sin(n-1)\theta\right]$$
$$+\tilde{D}_n\left[\left[1+\nu+((n+1)\nu+n-3)\log r\right]\sin(n-1)\theta\right.$$
$$\left.\left.+\left[(n+1)\nu+n-3\right]\theta\cos(n-1)\theta\right]\right\},$$

$$2Eu_\theta=2(1+\nu)\left[-k_1\sin\theta+k_2\cos\theta+k_3r\right] \tag{6.240}$$
$$-\frac{2(1+\nu)F_2}{r}-\frac{2(1+\nu)\log rF_3}{r}+8r\theta G_1-8r\log rG_2$$
$$-2r\left[(1+\nu+2\log r)\log r-2\theta^2\right]G_3$$
$$+2\left\{2\theta\cos\theta-\left[2+\nu+(1-\nu)\log r\right]\sin\theta\right\}H_1$$
$$-\left\{2(1-\nu)\theta\sin\theta+\left[3+\nu+4\log r\right]\cos\theta\right\}H_2$$
$$+\left\{\left[1+\nu+2\theta^2-2(1+\nu+\log r)\log r\right]\cos\theta-2(1-\nu)\theta\sin\theta\log r\right\}H_3$$
$$+2\left\{2\theta\sin\theta+\left[2+\nu+(1-\nu)\log r\right]\cos\theta\right\}J_1$$
$$+\left\{2(1-\nu)\theta\cos\theta-\left[3+\nu+4\log r\right]\sin\theta\right\}J_2$$
$$+\left\{\left[1+\nu+2\theta^2-2(1+\nu+\log r)\log r\right]\sin\theta+2(1-\nu)\theta\cos\theta\log r\right\}J_3$$
$$+2\sum_{\substack{n=-\infty\\ n\neq 0,-1}}^{\infty}(n+1)(1+\nu)r^n\left\{A_n\sin(n+1)\theta-B_n\cos(n+1)\theta\right\}$$
$$+2\sum_{\substack{n=-\infty\\ n\neq 0,1}}^{\infty}\left[3+n+(n-1)\nu\right]r^n\left\{C_n\sin(n-1)\theta-D_n\cos(n-1)\theta\right\}$$

$$+2\sum_{\substack{n=-\infty\\n\neq-1}}^{\infty} r^n(1+\nu)\Big\{\tilde{A}_n\,[[1+(n+1)\log r]\sin(n+1)\theta+(n+1)\theta\cos(n+1)\theta]$$

$$-\tilde{B}_n\,[[1+(n+1)\log r]\cos(n+1)\theta-(n+1)\theta\sin(n+1)\theta]\Big\}$$

$$+2\sum_{\substack{n=-\infty\\n\neq1}}^{\infty} r^n\Big\{\tilde{C}_n\Big[[1+\nu+((n-1)\nu+n+3)\log r]\sin(n-1)\theta$$

$$+[(n-1)\nu+n+3]\,\theta\cos(n-1)\theta\Big]$$

$$-\tilde{D}_n\Big[[1+\nu+((n-1)\nu+n+3)\log r]\cos(n-1)\theta$$

$$-[(n-1)\nu+n+3]\,\theta\sin(n-1)\theta\Big]\Big\},$$

$$\tau_{rr}=\frac{F_1}{r^2}+\frac{\theta F_3}{r^2}+(1+2\log r)G_1+2\theta G_2+\theta(1+2\log r)G_3+2C_1$$

$$+\frac{\cos\theta H_1}{r}-\frac{2\sin\theta H_2}{r}+\frac{(\theta\cos\theta-2\sin\theta\log r)H_3}{r}$$

$$+\frac{\sin\theta J_1}{r}+\frac{2\cos\theta J_2}{r}+\frac{(\theta\sin\theta+2\cos\theta\log r)J_3}{r}$$

$$-\sum_{\substack{n=-\infty\\n\neq0,-1}}^{\infty} n(n+1)r^{n-1}\Big\{A_n\cos(n+1)\theta+B_n\sin(n+1)\theta\Big\}$$

$$-\sum_{\substack{n=-\infty\\n\neq0,1}}^{\infty} n(n-3)r^{n-1}\Big\{C_n\cos(n-1)\theta+D_n\sin(n-1)\theta\Big\}$$

$$+\sum_{\substack{n=-\infty\\n\neq0,-1}}^{\infty} r^{n-1}\Big\{\tilde{A}_n\,[n(n+1)\theta\sin(n+1)\theta-[1+2n+n(n+1)\log r]\cos(n+1)\theta]$$

$$-\tilde{B}_n\,[n(n+1)\theta\cos(n+1)\theta+[1+2n+n(n+1)\log r]\sin(n+1)\theta]\Big\}$$

$$+\sum_{\substack{n=-\infty\\n\neq0,1}}^{\infty} r^{n-1}\Big\{\tilde{C}_n\,[n(n-3)\theta\sin(n-1)\theta+[3-2n-n(n-3)\log r]\cos(n-1)\theta]$$

$$-\tilde{D}_n\,[n(n-3)\theta\cos(n-1)\theta-[3-2n-n(n-3)\log r]\sin(n-1)\theta]\Big\},$$

$$\tau_{\theta\theta}=-\frac{F_1}{r^2}-\frac{\theta F_3}{r^2}+(3+2\log r)G_1+2\theta G_2+(3+2\log r)\theta G_3+2C_1$$

$$+\frac{\cos\theta H_1}{r}+\frac{\theta\cos\theta H_3}{r}+\frac{\sin\theta J_1}{r}+\frac{\theta\sin\theta J_3}{r}$$

$$+\sum_{\substack{n=-\infty\\n\neq0,-1}}^{\infty} n(n+1)r^{n-1}\Big\{A_n\cos(n+1)\theta+B_n\sin(n+1)\theta\Big\}$$

$$+\sum_{\substack{n=-\infty\\ n\neq 0,1}}^{\infty} n(n+1)r^{n-1}\left\{C_n\cos(n-1)\theta+D_n\sin(n-1)\theta\right\}$$

$$+\sum_{\substack{n=-\infty\\ n\neq 0,-1}}^{\infty} r^{n-1}\left\{\tilde{A}_n\left[-n(n+1)\theta\sin(n+1)\theta+[1+2n+n(n+1)\log r]\cos(n+1)\theta\right]\right.$$

$$\left.+\tilde{B}_n\left[n(n+1)\theta\cos(n+1)\theta+[1+2n+n(n+1)\log r]\sin(n+1)\theta\right]\right\}$$

$$+\sum_{\substack{n=-\infty\\ n\neq 0,1}}^{\infty} r^{n-1}\left\{\tilde{C}_n\left[-n(n+1)\theta\sin(n-1)\theta+[1+2n+n(n+1)\log r]\cos(n-1)\theta\right]\right.$$

$$\left.+\tilde{D}_n\left[n(n+1)\theta\cos(n-1)\theta+[1+2n+n(n+1)\log r]\sin(n-1)\theta\right]\right\}$$

$$\tau_{r\theta}=\frac{F_2}{r^2}+\frac{(\log r-1)F_3}{r^2}-G_2-(1+\log r)G_3$$

$$+\frac{\sin\theta H_1}{r}+\frac{(\theta\sin\theta-\cos\theta)H_3}{r}-\frac{\cos\theta J_1}{r}-\frac{(\theta\cos\theta+\sin\theta)J_3}{r}$$

$$+\sum_{\substack{n=-\infty\\ n\neq 0,-1}}^{\infty} n(n+1)r^{n-1}\left\{A_n\sin(n+1)\theta-B_n\cos(n+1)\theta\right\}$$

$$+\sum_{\substack{n=-\infty\\ n\neq 0,1}}^{\infty} n(n-1)r^{n-1}\left\{C_n\sin(n-1)\theta-D_n\cos(n-1)\theta\right\}$$

$$+\sum_{\substack{n=-\infty\\ n\neq 0,-1}}^{\infty} r^{n-1}\left\{\tilde{A}_n\left[n(n+1)\theta\cos(n+1)\theta+[1+2n+n(n+1)\log r]\sin(n+1)\theta\right]\right.$$

$$\left.+\tilde{B}_n\left[n(n+1)\theta\sin(n+1)\theta-[1+2n+n(n+1)\log r]\cos(n+1)\theta\right]\right\}$$

$$+\sum_{\substack{n=-\infty\\ n\neq 0,1}}^{\infty} r^{n-1}\left\{\tilde{C}_n\left[n(n-1)\theta\cos(n-1)\theta+[-1+2n+n(n-1)\log r]\sin(n-1)\theta\right]\right.$$

$$\left.+\tilde{D}_n\left[n(n-1)\theta\sin(n-1)\theta+[1-2n-n(n-1)\log r]\cos(n-1)\theta\right]\right\}.$$

To obtain the solution under plane strain conditions, we use the transformation in Eqn. (6.225a) to get

$$2\mu u_r=k_1\cos\theta+k_2\sin\theta-\frac{F_1}{r}-\frac{\theta F_3}{r}-r\left[1-2(1-2\nu)\log r\right]G_1$$

$$+2(1-2\nu)r\theta G_2-r\theta\left[1-2(1-2\nu)\log r\right]G_3+2(1-2\nu)rC_1$$

$$+\left\{[1-\nu+(1-2\nu)\log r]\cos\theta+2(1-\nu)\theta\sin\theta\right\}H_1$$

$$+\left[(1-2\nu)\theta\cos\theta-\frac{1}{2}(1-2\nu+4(1-\nu)\log r)\sin\theta\right]H_2$$

$$+\left\{[(1-2\nu)\log r-1]\,\theta\cos\theta+\left[\frac{1}{2}+(1-\nu)(\theta^2-\log^2 r)\right]\sin\theta\right\}H_3$$

$$+\{[1-\nu+(1-2\nu)\log r]\sin\theta-2(1-\nu)\theta\cos\theta\}\,J_1$$

$$+\left[(1-2\nu)\theta\sin\theta+\frac{1}{2}(1-2\nu+4(1-\nu)\log r)\cos\theta\right]J_2$$

$$+\left\{[(1-2\nu)\log r-1]\,\theta\sin\theta-\left[\frac{1}{2}+(1-\nu)(\theta^2-\log^2 r)\right]\cos\theta\right\}J_3$$

$$-\sum_{\substack{n=-\infty\\ n\neq 0,-1}}^{\infty}(n+1)r^n\left\{A_n\cos(n+1)\theta+B_n\sin(n+1)\theta\right\}$$

$$-\sum_{\substack{n=-\infty\\ n\neq 0,1}}^{\infty}[n-3+4\nu]\,r^n\left\{C_n\cos(n-1)\theta+D_n\sin(n-1)\theta\right\}$$

$$-\sum_{\substack{n=-\infty\\ n\neq 0,-1}}^{\infty}r^n\left\{\tilde{A}_n\left[[1+(n+1)\log r]\cos(n+1)\theta-(n+1)\theta\sin(n+1)\theta\right]\right.$$

$$\left.+\tilde{B}_n\left[[1+(n+1)\log r]\sin(n+1)\theta+(n+1)\theta\cos(n+1)\theta\right]\right\}$$

$$-\sum_{\substack{n=-\infty\\ n\neq 0,1}}^{\infty}r^n\left\{\tilde{C}_n\left[[1+(n-3+4\nu)\log r]\cos(n-1)\theta-[n-3+4\nu]\,\theta\sin(n-1)\theta\right]\right.$$

$$\left.+\tilde{D}_n\left[[1+(n-3+4\nu)\log r]\sin(n-1)\theta+[n-3+4\nu]\,\theta\cos(n-1)\theta\right]\right\},$$

$$2\mu u_\theta=-k_1\sin\theta+k_2\cos\theta+k_3 r \tag{6.241}$$

$$-\frac{F_2}{r}-\frac{\log rF_3}{r}+4(1-\nu)r\theta G_1-4(1-\nu)r\log rG_2$$

$$-r\left[(1+2(1-\nu)\log r)\log r-2(1-\nu)\theta^2\right]G_3$$

$$+\{2(1-\nu)\theta\cos\theta-[2-\nu+(1-2\nu)\log r]\sin\theta\}\,H_1$$

$$-\left\{(1-2\nu)\theta\sin\theta+\frac{1}{2}[3-2\nu+4(1-\nu)\log r]\cos\theta\right\}H_2$$

$$+\left\{(2\nu-1)\theta\sin\theta\log r+\left[\frac{1}{2}-\log r+(1-\nu)(\theta^2-\log^2 r)\right]\cos\theta\right\}H_3$$

$$+\{2(1-\nu)\theta\sin\theta+[2-\nu+(1-2\nu)\log r]\cos\theta\}\,J_1$$

$$+\left\{(1-2\nu)\theta\cos\theta-\frac{1}{2}[3-2\nu+4(1-\nu)\log r]\sin\theta\right\}J_2$$

$$+\left\{(1-2\nu)\theta\cos\theta\log r+\left[\frac{1}{2}-\log r+(1-\nu)(\theta^2-\log^2 r)\right]\sin\theta\right\}J_3$$

$$
\begin{aligned}
&+ \sum_{\substack{n=-\infty \\ n\neq 0,-1}}^{\infty} (n+1)r^n \Big\{ A_n \sin(n+1)\theta - B_n \cos(n+1)\theta \Big\} \\
&+ \sum_{\substack{n=-\infty \\ n\neq 0,1}}^{\infty} [n+3-4\nu]\, r^n \Big\{ C_n \sin(n-1)\theta - D_n \cos(n-1)\theta \Big\} \\
&+ \sum_{\substack{n=-\infty \\ n\neq 0,-1}}^{\infty} r^n \Big\{ \tilde{A}_n \Big[[1+(n+1)\log r] \sin(n+1)\theta + (n+1)\theta \cos(n+1)\theta \Big] \\
&- \tilde{B}_n \Big[[1+(n+1)\log r] \cos(n+1)\theta - (n+1)\theta \sin(n+1)\theta \Big] \Big\} \\
&+ \sum_{\substack{n=-\infty \\ n\neq 0,1}}^{\infty} r^n \Big\{ \tilde{C}_n \Big[[1+(n+3-4\nu)\log r] \sin(n-1)\theta + [n+3-4\nu]\, \theta \cos(n-1)\theta \Big] \\
&- \tilde{D}_n \Big[[1+(n+3-4\nu)\log r] \cos(n-1)\theta - [n+3-4\nu]\, \theta \sin(n-1)\theta \Big] \Big\},
\end{aligned}
$$

with the expressions for the stress components the same. The terms associated with k_1, k_2 and k_3 correspond to linearized rigid displacements. For problems involving sectors (e.g., see Section 6.6.4), the integers n in the above expressions are replaced by the (possibly complex-valued) roots of an appropriate characteristic equation.

As an example of the use of Eqn. (6.239), let $\phi = c\theta$, where c is a constant. Then from Eqns. (6.221), we get

$$\tau_{r\theta} = \frac{c}{r^2},$$

with the remaining stresses zero. Since $\tau_{rr} + \tau_{\theta\theta} = 0$, the plane stress assumption is satisfied, and this solution is in fact the exact solution to the three-dimensional problem of a hollow cylinder of length L and inner and outer radii a and b, which is fixed at $r = a$ and subjected to a uniform circumferential shear at $r = b$ resulting in a net moment M, as shown in Fig. 6.35. By equating the torque generated by $\tau_{r\theta}$ to M, we get $c = M/(2\pi L)$. By using the fact that $\boldsymbol{u}|_{r=a} = \boldsymbol{0}$, the displacements are obtained as

$$
\begin{aligned}
u_r &= u_z = 0, \\
u_\theta &= \frac{M}{4\pi\mu L}\left[\frac{r}{a^2} - \frac{1}{r}\right].
\end{aligned}
$$

As another example, the function ϕ for a vertical line load of magnitude per unit length P, acting on a horizontal straight edge of an infinitely large plate is given by $-Pr\theta\sin\theta/\pi$ ($J_2 \neq 0$, all other coefficients zero in Eqn. (6.239)–this solution is due to Flamant). However, the problem with this solution is that the displacement field does not decay as $r \to \infty$ (see page 103 of [324]; see also the example of an infinite wedge on page 516 where we show that this problem cannot be eliminated simply by excluding these terms in an infinite series solution).

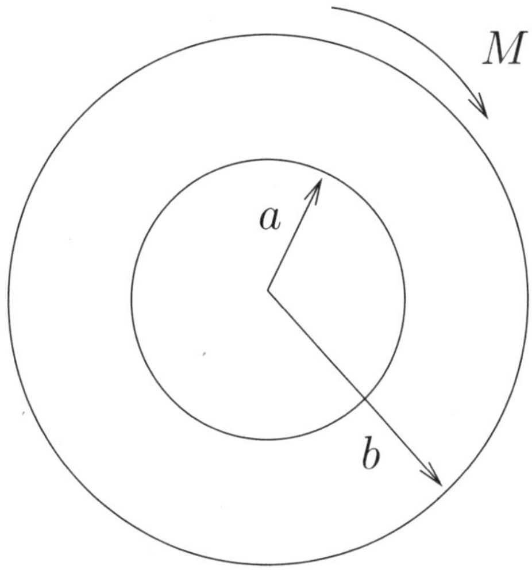

Fig. 6.35 *Hollow cylinder of length L fixed at the inner boundary $r = a$, and subjected to a moment M on the outer boundary $r = b$.*

In terms of elliptical coordinates (ξ, η), a general solution to the Laplace equation $\nabla^2\phi = 0$ is given by

$$\begin{aligned}
\phi &= c_0 + c_1\xi + c_2\eta + c_3\xi\eta \\
&+ \sum_{n=1}^{\infty} (A_n \cosh n\xi \cos n\eta + B_n \cosh n\xi \sin n\eta \\
&+ C_n \sinh n\xi \cos n\eta + D_n \sinh n\xi \sin n\eta) \\
&+ \hat{A}_n(\xi \cosh n\xi \cos n\eta - \eta \sinh n\xi \sin n\eta) \\
&+ \hat{B}_n(\xi \cosh n\xi \sin n\eta + \eta \sinh n\xi \cos n\eta) \\
&+ \hat{C}_n(\xi \sinh n\xi \cos n\eta - \eta \cosh n\xi \sin n\eta) \\
&+ \hat{D}_n(\xi \sinh n\xi \sin n\eta + \eta \cosh n\xi \cos n\eta).
\end{aligned} \tag{6.242}$$

Since $x\beta$, $y\beta$ and $(x^2 + y^2)\beta \equiv (\cosh 2\xi + \cos 2\eta)\beta$ are biharmonic if β is harmonic, a solution to the biharmonic equation given by Eqn. (6.232) (excluding separable terms of the type $\tilde{k}_0 + \tilde{k}_1\xi\eta + (\tilde{c}_0 + \tilde{c}_1\xi + \tilde{c}_2\eta + \tilde{c}_3\xi\eta)x + (\hat{c}_0 + \hat{c}_1\xi + \hat{c}_2\eta + \hat{c}_3\xi\eta)y$) is

$$\begin{aligned}
\phi &= c_1\xi + c_2\eta + \sum_{n=1}^{\infty} (A_n \cosh n\xi \cos n\eta + B_n \cosh n\xi \sin n\eta \\
&+ C_n \sinh n\xi \cos n\eta + D_n \sinh n\xi \sin n\eta) \\
&+ \tilde{A}_n \{\cosh[(n+1)\xi] \cos[(n-1)\eta] + \cosh[(n-1)\xi] \cos[(n+1)\eta]\} \\
&+ \tilde{B}_n \{\cosh[(n+1)\xi] \sin[(n-1)\eta] + \cosh[(n-1)\xi] \sin[(n+1)\eta]\} \\
&+ \tilde{C}_n \{\sinh[(n+1)\xi] \cos[(n-1)\eta] + \sinh[(n-1)\xi] \cos[(n+1)\eta]\}
\end{aligned}$$

$$
\begin{aligned}
&+ \tilde{D}_n \left\{\sinh[(n+1)\xi] \sin[(n-1)\eta] + \sinh[(n-1)\xi] \sin[(n+1)\eta]\right\} \\
&+ \hat{A}_n(\xi \cosh n\xi \cos n\eta - \eta \sinh n\xi \sin n\eta) \\
&+ \hat{B}_n(\xi \cosh n\xi \sin n\eta + \eta \sinh n\xi \cos n\eta) \\
&+ \hat{C}_n(\xi \sinh n\xi \cos n\eta - \eta \cosh n\xi \sin n\eta) \\
&+ \hat{D}_n(\xi \sinh n\xi \sin n\eta + \eta \cosh n\xi \cos n\eta) \qquad (6.243) \\
&+ \hat{\hat{A}}_n\Big\{\xi\left[\cosh(n+1)\xi \cos(n-1)\eta + \cosh(n-1)\xi \cos(n+1)\eta\right] \\
&- \eta\left[\sinh(n+1)\xi \sin(n-1)\eta + \sinh(n-1)\xi \sin(n+1)\eta\right]\Big\} \\
&+ \hat{\hat{B}}_n\Big\{\xi\left[\cosh(n+1)\xi \sin(n-1)\eta + \cosh(n-1)\xi \sin(n+1)\eta\right] \\
&+ \eta\left[\sinh(n+1)\xi \cos(n-1)\eta + \sinh(n-1)\xi \cos(n+1)\eta\right]\Big\} \\
&+ \hat{\hat{C}}_n\Big\{\xi\left[\sinh(n+1)\xi \cos(n-1)\eta + \sinh(n-1)\xi \cos(n+1)\eta\right] \\
&- \eta\left[\cosh(n+1)\xi \sin(n-1)\eta + \cosh(n-1)\xi \sin(n+1)\eta\right]\Big\} \\
&+ \hat{\hat{D}}_n\Big\{\xi\left[\sinh(n+1)\xi \sin(n-1)\eta + \sinh(n-1)\xi \sin(n+1)\eta\right] \\
&+ \eta\left[\cosh(n+1)\xi \cos(n-1)\eta + \cosh(n-1)\xi \cos(n+1)\eta\right]\Big\}.
\end{aligned}
$$

As usual, the integer n in Eqns. (6.242) and (6.243) can be replaced by the (possibly complex-valued) roots of a characteristic equation λ_n.

The displacement field corresponding to the first half of Eqn. (6.243) for the plane stress case is

$$
\begin{aligned}
2\mu h_1 u_\xi &= k_1 \sinh\xi \cos\eta + k_2 \cosh\xi \sin\eta + k_3 \sin 2\eta - c_1 \\
&\quad + \sum_{n=1}^{\infty} -A_n n \sinh n\xi \cos n\eta - B_n n \sinh n\xi \sin n\eta \\
&\quad - C_n n \cosh n\xi \cos n\eta - D_n n \cosh n\xi \sin n\eta \\
&\quad + \frac{\tilde{A}_n}{1+\nu}\left\{a_1 \sinh[(n+1)\xi] \cos[(n-1)\eta] - a_2 \sinh[(n-1)\xi] \cos[(n+1)\eta]\right\} \\
&\quad + \frac{\tilde{B}_n}{1+\nu}\left\{a_1 \sinh[(n+1)\xi] \sin[(n-1)\eta] - a_2 \sinh[(n-1)\xi] \sin[(n+1)\eta]\right\} \\
&\quad + \frac{\tilde{C}_n}{1+\nu}\left\{a_1 \cosh[(n+1)\xi] \cos[(n-1)\eta] - a_2 \cosh[(n-1)\xi] \cos[(n+1)\eta]\right\} \\
&\quad + \frac{\tilde{D}_n}{1+\nu}\left\{a_1 \cosh[(n+1)\xi] \sin[(n-1)\eta] - a_2 \cosh[(n-1)\xi] \sin[(n+1)\eta]\right\}, \\
2\mu h_1 u_\eta &= -k_1 \cosh\xi \sin\eta + k_2 \sinh\xi \cos\eta + k_3 \sinh 2\xi - c_2 \qquad (6.244)
\end{aligned}
$$

$$+ \sum_{n=1}^{\infty} A_n n \cosh n\xi \sin n\eta - B_n n \cosh n\xi \cos n\eta$$
$$+ C_n n \sinh n\xi \sin n\eta - D_n n \sinh n\xi \cos n\eta$$
$$+ \frac{\tilde{A}_n}{1+\nu}\Big\{a_2 \cosh[(n+1)\xi] \sin[(n-1)\eta] - a_1 \cosh[(n-1)\xi] \sin[(n+1)\eta]\Big\}$$
$$+ \frac{\tilde{B}_n}{1+\nu}\Big\{-a_2 \cosh[(n+1)\xi] \cos[(n-1)\eta] + a_1 \cosh[(n-1)\xi] \cos[(n+1)\eta]\Big\}$$
$$+ \frac{\tilde{C}_n}{1+\nu}\Big\{a_2 \sinh[(n+1)\xi] \sin[(n-1)\eta] - a_1 \sinh[(n-1)\xi] \sin[(n+1)\eta]\Big\}$$
$$+ \frac{\tilde{D}_n}{1+\nu}\Big\{-a_2 \sinh[(n+1)\xi] \cos[(n-1)\eta] + a_1 \sinh[(n-1)\xi] \cos[(n+1)\eta]\Big\},$$

where

$$a_1 = 3 - n - (n+1)\nu,$$
$$a_2 = 3 + n + (n-1)\nu,$$

and where the first three terms corresponding to the constants k_1, k_2, k_3 represent rigid-body modes, while for the plane strain case, we have

$$2\mu h_1 u_\xi = k_1 \sinh\xi \cos\eta + k_2 \cosh\xi \sin\eta + k_3 \sin 2\eta - c_1$$
$$+ \sum_{n=1}^{\infty} -A_n n \sinh n\xi \cos n\eta - B_n n \sinh n\xi \sin n\eta$$
$$- C_n n \cosh n\xi \cos n\eta - D_n n \cosh n\xi \sin n\eta$$
$$+ \tilde{A}_n\Big\{b_1 \sinh[(n+1)\xi] \cos[(n-1)\eta] - b_2 \sinh[(n-1)\xi] \cos[(n+1)\eta]\Big\}$$
$$+ \tilde{B}_n\Big\{b_1 \sinh[(n+1)\xi] \sin[(n-1)\eta] - b_2 \sinh[(n-1)\xi] \sin[(n+1)\eta]\Big\}$$
$$+ \tilde{C}_n\Big\{b_1 \cosh[(n+1)\xi] \cos[(n-1)\eta] - b_2 \cosh[(n-1)\xi] \cos[(n+1)\eta]\Big\}$$
$$+ \tilde{D}_n\Big\{b_1 \cosh[(n+1)\xi] \sin[(n-1)\eta] - b_2 \cosh[(n-1)\xi] \sin[(n+1)\eta]\Big\},$$
$$2\mu h_1 u_\eta = -k_1 \cosh\xi \sin\eta + k_2 \sinh\xi \cos\eta + k_3 \sinh 2\xi - c_2 \tag{6.245}$$
$$+ \sum_{n=1}^{\infty} A_n n \cosh n\xi \sin n\eta - B_n n \cosh n\xi \cos n\eta$$
$$+ C_n n \sinh n\xi \sin n\eta - D_n n \sinh n\xi \cos n\eta$$
$$+ \tilde{A}_n\Big\{b_2 \cosh[(n+1)\xi] \sin[(n-1)\eta] - b_1 \cosh[(n-1)\xi] \sin[(n+1)\eta]\Big\}$$
$$+ \tilde{B}_n\Big\{-b_2 \cosh[(n+1)\xi] \cos[(n-1)\eta] + b_1 \cosh[(n-1)\xi] \cos[(n+1)\eta]\Big\}$$
$$+ \tilde{C}_n\Big\{b_2 \sinh[(n+1)\xi] \sin[(n-1)\eta] - b_1 \sinh[(n-1)\xi] \sin[(n+1)\eta]\Big\}$$
$$+ \tilde{D}_n\Big\{-b_2 \sinh[(n+1)\xi] \cos[(n-1)\eta] + b_1 \sinh[(n-1)\xi] \cos[(n+1)\eta]\Big\},$$

where

$$b_1 = 3 - n - 4\nu,$$
$$b_2 = 3 + n - 4\nu.$$

The general solution for Laplace's equation in bipolar coordinates is given by

$$\begin{aligned}\phi &= c_0 + c_1\xi + c_2\eta + c_3\xi\eta \\ &\quad + \sum_{n=0}^{\infty}(A_n \cosh n\xi \cos n\eta + B_n \cosh n\xi \sin n\eta \\ &\quad + C_n \sinh n\xi \cos n\eta + D_n \sinh n\xi \sin n\eta) \\ &\quad + \hat{A}_n(\xi \cosh n\xi \cos n\eta - \eta \sinh n\xi \sin n\eta) \\ &\quad + \hat{B}_n(\xi \cosh n\xi \sin n\eta + \eta \sinh n\xi \cos n\eta) \\ &\quad + \hat{C}_n(\xi \sinh n\xi \cos n\eta - \eta \cosh n\xi \sin n\eta) \\ &\quad + \hat{D}_n(\xi \sinh n\xi \sin n\eta + \eta \cosh n\xi \cos n\eta),\end{aligned} \tag{6.246}$$

while a solution to the biharmonic equation, obtained by using the fact that if β is harmonic and $h_1 = c/(\cosh\xi - \cos\eta)$, then $x\beta = h_1\beta\sin\eta$, $y\beta = h_1\beta\sinh\xi$ and $(x^2 + y^2)\beta = h_1c(\cosh\xi + \cos\eta)\beta$ are biharmonic (see Eqn. (6.229)), is given by [149] (we have again excluded some terms as we did in Eqn. (6.243))

$$\begin{aligned}\phi &= cc_1\xi + cc_2\eta \\ &+ h_1\sum_{n=1}^{\infty}\Big\{[A_n\cosh(n+1)\xi + B_n\sinh(n+1)\xi + C_n\cosh(n-1)\xi + D_n\sinh(n-1)\xi]\cos n\eta \\ &\quad + \big[\tilde{A}_n\cosh(n+1)\xi + \tilde{B}_n\sinh(n+1)\xi + \tilde{C}_n\cosh(n-1)\xi + \tilde{D}_n\sinh(n-1)\xi\big]\sin n\eta\Big\}.\end{aligned} \tag{6.247}$$

The displacements associated with the above stress function for the plane strain case (for the plane stress case use the transformation given by Eqn. (6.225b)) are

$$\begin{aligned}\frac{2\mu u_\xi c}{h_1} &= k_1\cosh\xi\sin\eta + k_2\sinh\xi\sin\eta + k_3(\cosh\xi\cos\eta - 1) - \frac{c_1c^2}{h_1^2} \\ &\quad + \sum_{n=1}^{\infty} A_n\Big\{(2\nu - 1)\sinh\xi\cosh(n+1)\xi\cos n\eta \\ &\quad + \left[\frac{c}{h_1}(1 - n - 2\nu)\cos n\eta + 2(1-\nu)\sin n\eta\sin\eta\right]\sinh(n+1)\xi\Big\} \\ &\quad + B_n\Big\{(2\nu - 1)\sinh\xi\sinh(n+1)\xi\cos n\eta \\ &\quad + \left[\frac{c}{h_1}(1 - n - 2\nu)\cos n\eta + 2(1-\nu)\sin n\eta\sin\eta\right]\cosh(n+1)\xi\Big\}\end{aligned}$$

$$+ C_n\Big\{(2\nu - 1)\sinh\xi\cosh(n-1)\xi\cos n\eta$$

$$+ \left[-\frac{c}{h_1}(1+n-2\nu)\cos n\eta + 2(1-\nu)\sin n\eta\sin\eta\right]\sinh(n-1)\xi\Big\}$$

$$+ D_n\Big\{(2\nu - 1)\sinh\xi\sinh(n-1)\xi\cos n\eta$$

$$+ \left[-\frac{c}{h_1}(1+n-2\nu)\cos n\eta + 2(1-\nu)\sin n\eta\sin\eta\right]\cosh(n-1)\xi\Big\}$$

$$+ \tilde{A}_n\Big\{(2\nu - 1)\sinh\xi\cosh(n+1)\xi\sin n\eta$$

$$+ \left[\frac{c}{h_1}(1-n-2\nu)\sin n\eta - 2(1-\nu)\cos n\eta\sin\eta\right]\sinh(n+1)\xi\Big\}$$

$$+ \tilde{B}_n\Big\{(2\nu - 1)\sinh\xi\sinh(n+1)\xi\sin n\eta$$

$$+ \left[\frac{c}{h_1}(1-n-2\nu)\sin n\eta - 2(1-\nu)\cos n\eta\sin\eta\right]\cosh(n+1)\xi\Big\}$$

$$+ \tilde{C}_n\Big\{(2\nu - 1)\sinh\xi\cosh(n-1)\xi\sin n\eta$$

$$+ \left[-\frac{c}{h_1}(1+n-2\nu)\sin n\eta - 2(1-\nu)\cos n\eta\sin\eta\right]\sinh(n-1)\xi\Big\}$$

$$+ \tilde{D}_n\Big\{(2\nu - 1)\sinh\xi\sinh(n-1)\xi\sin n\eta$$

$$+ \left[-\frac{c}{h_1}(1+n-2\nu)\sin n\eta - 2(1-\nu)\cos n\eta\sin\eta\right]\cosh(n-1)\xi\Big\},$$

$$\frac{2\mu u_\eta c}{h_1} = -k_1\sinh\xi\cos\eta + k_2(1-\cosh\xi\cos\eta) + k_3\sinh\xi\sin\eta - \frac{c_2c^2}{h_1^2} \tag{6.248}$$

$$+ \sum_{n=1}^{\infty} A_n\Big\{2(\nu - 1)\sinh\xi\sinh(n+1)\xi\sin n\eta$$

$$+ \left[\frac{c}{h_1}(n+2-2\nu)\sin n\eta + (2\nu-1)\cos n\eta\sin\eta\right]\cosh(n+1)\xi\Big\}$$

$$+ B_n\Big\{2(\nu - 1)\sinh\xi\cosh(n+1)\xi\sin n\eta$$

$$+ \left[\frac{c}{h_1}(n+2-2\nu)\sin n\eta + (2\nu-1)\cos n\eta\sin\eta\right]\sinh(n+1)\xi\Big\}$$

$$+ C_n\Big\{2(\nu - 1)\sinh\xi\sinh(n-1)\xi\sin n\eta$$

$$+ \left[\frac{c}{h_1}(n-2+2\nu)\sin n\eta + (2\nu-1)\cos n\eta\sin\eta\right]\cosh(n-1)\xi\Big\}$$

$$+ D_n\Big\{2(\nu - 1)\sinh\xi\cosh(n-1)\xi\sin n\eta$$

$$+\left[\frac{c}{h_1}(n-2+2\nu)\sin n\eta+(2\nu-1)\cos n\eta\sin\eta\right]\sinh(n-1)\xi\Big\}$$

$$+\tilde{A}_n\Big\{2(1-\nu)\sinh\xi\sinh(n+1)\xi\cos n\eta$$

$$+\left[-\frac{c}{h_1}(n+2-2\nu)\cos n\eta+(2\nu-1)\sin n\eta\sin\eta\right]\cosh(n+1)\xi\Big\}$$

$$+\tilde{B}_n\Big\{2(1-\nu)\sinh\xi\cosh(n+1)\xi\cos n\eta$$

$$+\left[-\frac{c}{h_1}(n+2-2\nu)\cos n\eta+(2\nu-1)\sin n\eta\sin\eta\right]\sinh(n+1)\xi\Big\}$$

$$+\tilde{C}_n\Big\{2(1-\nu)\sinh\xi\sinh(n-1)\xi\cos n\eta$$

$$+\left[-\frac{c}{h_1}(n-2+2\nu)\cos n\eta+(2\nu-1)\sin n\eta\sin\eta\right]\cosh(n-1)\xi\Big\}$$

$$+\tilde{D}_n\Big\{2(1-\nu)\sinh\xi\cosh(n-1)\xi\cos n\eta$$

$$+\left[-\frac{c}{h_1}(n-2+2\nu)\cos n\eta+(2\nu-1)\sin n\eta\sin\eta\right]\sinh(n-1)\xi\Big\},$$

where the k_1, k_2 and k_3 terms are associated with rigid body motion.

Because of the complications involved in checking whether the compatibility conditions are satisfied, it is perhaps advantageous to start with a representation of the displacement field in terms of potentials. Two well known methods that fall in this category are the Galerkin vector (also referred to as the Cauchy–Kovalevski–Somigliana general solution), and the Boussinesq–Papkovich–Neuber solutions. The completeness of the Galerkin and Boussinesq–Papkovich–Neuber solutions is shown in, e.g., [346]. The treatment of the Galerkin vector approach that we now present is a generalization to the transient case of the method presented in [48].

We shall assume $\mathbf{C}$ to have both the major and minor symmetries, and we also assume the material to be homogeneous. Equations (6.18)–(6.20) can be written as

$$\boldsymbol{M}\boldsymbol{u}+\rho\boldsymbol{b}=\mathbf{0}, \tag{6.249}$$

or, in indicial notation as,

$$M_{ik}u_k+\rho b_i=0, \tag{6.250}$$

where $\boldsymbol{M}$ is a matrix differential operator with components

$$M_{ik}=\mathbb{C}_{ijkl}\frac{\partial^2}{\partial X_j\partial X_l}-\rho\delta_{ik}\frac{\partial^2}{\partial t^2}.$$

Note that since $\mathbf{C}$ is assumed to be symmetric, $\boldsymbol{M}$ is symmetric, so that $(\mathbf{cof}\,\boldsymbol{M})^T=\mathbf{cof}\,(\boldsymbol{M}^T)$ $=\mathbf{cof}\,\boldsymbol{M}$. Take $\boldsymbol{u}=(\mathbf{cof}\,\boldsymbol{M})\boldsymbol{w}$, where $\boldsymbol{w}$ is a vector potential. Substituting into Eqn. (6.249), we get the governing equation for $\boldsymbol{w}$ as

$$(\det\boldsymbol{M})\boldsymbol{w}+\rho\boldsymbol{b}=\mathbf{0}. \tag{6.251}$$

The main advantage of using the Galerkin vector approach is that the three equations for the components of w are uncoupled, although, of course, this is at the cost of increasing the order of the differential operators to six.

When the material is isotropic, one can reduce the order of the differential operators to four as follows. Since $\mathsf{C}_{ijkl} = \lambda\delta_{ij}\delta_{kl} + \mu(\delta_{ik}\delta_{jl} + \delta_{il}\delta_{jk})$, we have

$$M_{ij} = (\lambda + \mu)\frac{\partial^2}{\partial X_i \partial X_j} + \left(\mu\frac{\partial^2}{\partial X_a \partial X_a} - \rho\frac{\partial^2}{\partial t^2}\right)\delta_{ij}.$$

On using Eqn. (1.60), it follows that

$$(\mathbf{cof}\, \boldsymbol{M})_{ij} = (\lambda + 2\mu)\left(\mu\frac{\partial^2}{\partial X_a \partial X_a} - \rho\frac{\partial^2}{\partial t^2}\right)\left\{\left[\frac{\partial^2}{\partial X_b \partial X_b} - \frac{\rho}{\lambda + 2\mu}\frac{\partial^2}{\partial t^2}\right]\delta_{ij} - \frac{1}{2(1-\nu)}\frac{\partial^2}{\partial X_i \partial X_j}\right\}.$$

Using the relation $\det \boldsymbol{M} = \boldsymbol{M} : (\mathbf{cof}\, \boldsymbol{M})/3$, we get

$$\det \boldsymbol{M} = (\lambda + 2\mu)\left(\mu\frac{\partial^2}{\partial X_a \partial X_a} - \rho\frac{\partial^2}{\partial t^2}\right)^2\left[\frac{\partial^2}{\partial X_b \partial X_b} - \frac{\rho}{\lambda + 2\mu}\frac{\partial^2}{\partial t^2}\right].$$

Observing that the factor $(\lambda + 2\mu)\left(\mu\frac{\partial^2}{\partial X_a \partial X_a} - \rho\frac{\partial^2}{\partial t^2}\right)$ is common in both, the expression for $\mathbf{cof}\, \boldsymbol{M}$ and the expression for $\det \boldsymbol{M}$, we define a new vector potential $\boldsymbol{\Psi}$ as

$$\boldsymbol{\Psi} := 2\mu(\lambda + \mu)\left(\mu\frac{\partial^2}{\partial X_a \partial X_a} - \rho\frac{\partial^2}{\partial t^2}\right)\boldsymbol{w}.$$

In terms of this new potential, the displacement field and the governing equation for $\boldsymbol{\Psi}$ are given by

$$u_i = \frac{(1-\nu)}{\mu}\left[\frac{\partial^2 \Psi_i}{\partial X_b \partial X_b} - \frac{\rho}{\lambda + 2\mu}\frac{\partial^2 \Psi_i}{\partial t^2} - \frac{1}{2(1-\nu)}\frac{\partial^2 \Psi_j}{\partial X_i \partial X_j}\right],$$

$$\left(\frac{\partial^2}{\partial X_a \partial X_a} - \frac{\rho}{\mu}\frac{\partial^2}{\partial t^2}\right)\left[\frac{\partial^2 \Psi_i}{\partial X_b \partial X_b} - \frac{\rho}{\lambda + 2\mu}\frac{\partial^2 \Psi_i}{\partial t^2}\right] + \frac{\rho b_i}{1-\nu} = 0,$$

or, in tensorial form, as

$$\boldsymbol{u} = \frac{(1-\nu)}{\mu}\left[\boldsymbol{\nabla}^2\boldsymbol{\Psi} - \frac{1}{c_L^2}\frac{\partial^2 \boldsymbol{\Psi}}{\partial t^2} - \frac{1}{2(1-\nu)}\boldsymbol{\nabla}(\boldsymbol{\nabla}\cdot\boldsymbol{\Psi})\right], \tag{6.252a}$$

$$\left(\boldsymbol{\nabla}^2 - \frac{1}{c_T^2}\frac{\partial^2}{\partial t^2}\right)\left[\boldsymbol{\nabla}^2\boldsymbol{\Psi} - \frac{1}{c_L^2}\frac{\partial^2 \boldsymbol{\Psi}}{\partial t^2}\right] + \frac{\rho \boldsymbol{b}}{1-\nu} = \boldsymbol{0}, \tag{6.252b}$$

where $c_T^2 = \mu/\rho$ and $c_L^2 = (\lambda + 2\mu)/\rho$ are the transverse and longitudinal wave speeds.

For elastostatics problems, Eqns. (6.252a) and (6.252b) reduce to

$$2\mu\boldsymbol{u} = 2(1-\nu)\boldsymbol{\nabla}^2\boldsymbol{\Psi} - \boldsymbol{\nabla}(\boldsymbol{\nabla}\cdot\boldsymbol{\Psi}), \tag{6.253a}$$

$$\boldsymbol{\nabla}^2(\boldsymbol{\nabla}^2\boldsymbol{\Psi}) + \frac{\rho \boldsymbol{b}}{1-\nu} = 0. \tag{6.253b}$$

Thus, in the absence of body forces, the vector $\mathbf{\Psi}$ has to be biharmonic.

As an example, for the problem of torsion of prismatic bars, we can take

$$\Psi_x = -\frac{\alpha\mu x^2 yz}{2(1-\nu)},$$
$$\Psi_y = \frac{\alpha\mu xy^2 z}{2(1-\nu)},$$
$$\Psi_z = \Psi_z(x,y),$$

and the warping function

$$\psi := \frac{1-\nu}{\alpha\mu}\left[\frac{\partial^2\Psi_z}{\partial x^2} + \frac{\partial^2\Psi_z}{\partial y^2}\right].$$

With this choice, we have $\boldsymbol{\nabla}\cdot\mathbf{\Psi} = 0$, and Eqns. (6.253a) and (6.253b) reduce to Eqns. (6.61) and (6.66), respectively. A three-dimensional solution for a rectangular prism under end loads obtained using the Galerkin vector approach is presented in [140]. A related plate solution (obtained without using the Galerkin vector) is presented in [189]. When only one of the components of $\mathbf{\Psi}$, say Ψ_z, is nonzero, it is called as Love's strain function. Of course, such a strain function can only treat the case where the body force components corresponding to the zero components of $\mathbf{\Psi}$ are also zero. The use of Love's strain function to solve axisymmetric problems is discussed below and in [324] (page 380, referred to there as Love's stress function). Although, in this book, we have solved planar problems using the Airy stress function, one can also solve them by either taking $\mathbf{\Psi} = (\Psi(x,y),0,0)$ or $(0,\Psi(x,y),0)$; the $\Psi(x,y)$ function in the latter choice is referred to as the Marguerre displacement function. The governing equation for this function (in the absence of body forces) is the same as for the Airy stress function, namely, $\boldsymbol{\nabla}^4\Psi = 0$. A significant advantage of using the Marguerre displacement function is that the displacement solution in the plane strain case[11] is obtained using the Ψ function as

$$2\mu u_x = -\frac{\partial^2\Psi}{\partial x\partial y},$$
$$2\mu u_y = 2(1-\nu)\left(\frac{\partial^2\Psi}{\partial x^2} + \frac{\partial^2\Psi}{\partial y^2}\right) - \frac{\partial^2\Psi}{\partial y^2},$$

and, then the expressions for the strains and stresses follow directly, while with the Airy stress function approach obtaining the displacements from the strains can be quite cumbersome.

By setting

$$\boldsymbol{\beta} = -\frac{1}{2}\left[\boldsymbol{\nabla}^2\mathbf{\Psi} - \frac{1}{c_L^2}\frac{\partial^2\mathbf{\Psi}}{\partial t^2}\right],$$
$$\alpha = -\boldsymbol{\nabla}\cdot\mathbf{\Psi} - \mathbf{X}\cdot\boldsymbol{\beta}, \tag{6.254}$$

[11]The corresponding expressions for plane stress can then be obtained by using the transformation given by Eqn. (6.225b).

in Eqns. (6.252), we get the Boussinesq–Papkovich–Neuber solution as

$$\begin{aligned} 2\mu \boldsymbol{u} &= \boldsymbol{\nabla}\alpha + (\boldsymbol{\nabla}\boldsymbol{\beta})^T\boldsymbol{X} - (3-4\nu)\boldsymbol{\beta}, \\ \boldsymbol{\nabla}^2\alpha - \frac{1}{c_L^2}\frac{\partial^2\alpha}{\partial t^2} &= -\boldsymbol{X}\cdot\left(\boldsymbol{\nabla}^2\boldsymbol{\beta} - \frac{1}{c_L^2}\frac{\partial^2\boldsymbol{\beta}}{\partial t^2}\right), \\ \boldsymbol{\nabla}^2\boldsymbol{\beta} - \frac{1}{c_T^2}\frac{\partial^2\boldsymbol{\beta}}{\partial t^2} &= \frac{\rho\boldsymbol{b}}{2(1-\nu)}. \end{aligned} \tag{6.255}$$

For elastostatics problems in the absence of body forces, Eqns. (6.254) reduce to

$$\begin{aligned} \boldsymbol{\beta} &= -\frac{1}{2}\boldsymbol{\nabla}^2\boldsymbol{\Psi}, \\ \alpha &= -\boldsymbol{\nabla}\cdot\boldsymbol{\Psi} - \boldsymbol{X}\cdot\boldsymbol{\beta}, \end{aligned} \tag{6.256}$$

while Eqns. (6.255) reduce to

$$2\mu\boldsymbol{u} = \boldsymbol{\nabla}\alpha + (\boldsymbol{\nabla}\boldsymbol{\beta})^T\boldsymbol{X} - (3-4\nu)\boldsymbol{\beta}, \tag{6.257a}$$

$$\boldsymbol{\nabla}^2\alpha = 0, \tag{6.257b}$$

$$\boldsymbol{\nabla}^2\boldsymbol{\beta} = \boldsymbol{0}. \tag{6.257c}$$

One can solve a large class of torsionless axisymmetry problems by assuming the vector $\boldsymbol{\beta}$ to be of the form $(0, 0, \beta)$ in the cylindrical coordinate system, or, equivalently, of the form $(\beta\cos\theta, -\beta\sin\theta, 0)$ in the spherical coordinate system, where β is a function of the cylindrical coordinates (r, z) or spherical coordinates (r, θ), respectively. The functions α and β are known as the Boussinesq potentials. Equations (6.257b) and (6.257c) reduce to

$$\boldsymbol{\nabla}^2\alpha = \boldsymbol{\nabla}^2\beta = 0. \tag{6.258}$$

By letting $\boldsymbol{\beta} = (0, 0, \beta)$ in the cylindrical coordinate system, and using Eqns. (B.3) and (B.5), and the fact that $\boldsymbol{X} = (r, 0, z)$, the components of Eqn. (6.257a) evaluate to

$$\begin{aligned} 2\mu u_r &= \frac{\partial\alpha}{\partial r} + z\frac{\partial\beta}{\partial r}, \\ 2\mu u_z &= \frac{\partial\alpha}{\partial z} + z\frac{\partial\beta}{\partial z} - (3-4\nu)\beta. \end{aligned} \tag{6.259}$$

Substituting these expressions into Eqn. (B.12), and using Eqns. (6.258), we get

$$\begin{aligned} \tau_{rr} &= \frac{\partial^2\alpha}{\partial r^2} + z\frac{\partial^2\beta}{\partial r^2} - 2\nu\frac{\partial\beta}{\partial z}, \\ \tau_{\theta\theta} &= \frac{1}{r}\frac{\partial\alpha}{\partial r} + \frac{z}{r}\frac{\partial\beta}{\partial r} - 2\nu\frac{\partial\beta}{\partial z}, \\ \tau_{zz} &= \frac{\partial^2\alpha}{\partial z^2} + z\frac{\partial^2\beta}{\partial z^2} - 2(1-\nu)\frac{\partial\beta}{\partial z}, \\ \tau_{rz} &= \frac{\partial^2\alpha}{\partial r\partial z} + z\frac{\partial^2\beta}{\partial r\partial z} - (1-2\nu)\frac{\partial\beta}{\partial r}. \end{aligned} \tag{6.260}$$

Similar to the above development in a cylindrical coordinate system, in the case of a spherical coordinate system, by taking $\boldsymbol{\beta} = (\beta\cos\theta, -\beta\sin\theta, 0)$, and using Eqns. (C.3), (C.5) and (C.11), and the fact that $\boldsymbol{X} = r\boldsymbol{e}_r$, the components of Eqn. (6.257a) and the corresponding stresses evaluate to

$$
\begin{aligned}
2\mu u_r &= \frac{\partial\alpha}{\partial r} + r\cos\theta\frac{\partial\beta}{\partial r} - (3-4\nu)\beta\cos\theta, \\
2\mu u_\theta &= \frac{1}{r}\frac{\partial\alpha}{\partial\theta} + \cos\theta\frac{\partial\beta}{\partial\theta} + (3-4\nu)\beta\sin\theta, \\
\tau_{rr} &= \frac{\partial^2\alpha}{\partial r^2} + r\cos\theta\frac{\partial^2\beta}{\partial r^2} - 2(1-\nu)\cos\theta\frac{\partial\beta}{\partial r} + 2\nu\frac{\sin\theta}{r}\frac{\partial\beta}{\partial\theta}, \\
\tau_{\theta\theta} &= \frac{1}{r^2}\frac{\partial^2\alpha}{\partial\theta^2} + \frac{1}{r}\frac{\partial\alpha}{\partial r} + \frac{\cos\theta}{r}\frac{\partial^2\beta}{\partial\theta^2} + 2(1-\nu)\frac{\sin\theta}{r}\frac{\partial\beta}{\partial\theta} + (1-2\nu)\cos\theta\frac{\partial\beta}{\partial r}, \\
\tau_{r\theta} &= \frac{1}{r}\frac{\partial^2\alpha}{\partial r\partial\theta} - \frac{1}{r^2}\frac{\partial\alpha}{\partial\theta} + \cos\theta\frac{\partial^2\beta}{\partial r\partial\theta} + (1-2\nu)\sin\theta\frac{\partial\beta}{\partial r} - 2(1-\nu)\frac{\cos\theta}{r}\frac{\partial\beta}{\partial\theta}, \\
\tau_{\phi\phi} &= \frac{1}{r}\frac{\partial\alpha}{\partial r} + \frac{\cot\theta}{r^2}\frac{\partial\alpha}{\partial\theta} + (1-2\nu)\cos\theta\frac{\partial\beta}{\partial r} + \frac{(\cos^2\theta + 2\nu\sin^2\theta)}{r\sin\theta}\frac{\partial\beta}{\partial\theta}.
\end{aligned}
\tag{6.261}
$$

For a prolate spheroidal coordinate system (see Appendix F), using the fact that

$$
\boldsymbol{X} = \frac{c^2}{h_1}\begin{bmatrix} \sinh\xi\cosh\xi \\ -\sin\eta\cos\eta \\ 0 \end{bmatrix}, \quad \boldsymbol{\beta} = \frac{c\beta}{h_1}\begin{bmatrix} \sinh\xi\cos\eta \\ -\cosh\xi\sin\eta \\ 0 \end{bmatrix},
$$

we get

$$
\begin{aligned}
2\mu u_\xi &= \frac{1}{h_1}\left[\frac{\partial\alpha}{\partial\xi} + c\cosh\xi\cos\eta\frac{\partial\beta}{\partial\xi} + (4\nu-3)c\beta\sinh\xi\cos\eta\right], \\
2\mu u_\eta &= \frac{1}{h_1}\left[\frac{\partial\alpha}{\partial\eta} + c\cosh\xi\cos\eta\frac{\partial\beta}{\partial\eta} - (4\nu-3)c\beta\cosh\xi\sin\eta\right],
\end{aligned}
\tag{6.262}
$$

where α and β are harmonic functions. The expressions for the stresses are then obtained using Eqns. (F.12). Similarly, for an oblate spheroidal coordinate system (see Appendix G), using the fact that

$$
\boldsymbol{X} = \frac{c^2}{h_1}\begin{bmatrix} \sinh\xi\cosh\xi \\ \sin\eta\cos\eta \\ 0 \end{bmatrix}, \quad \boldsymbol{\beta} = \frac{c\beta}{h_1}\begin{bmatrix} \cosh\xi\cos\eta \\ -\sinh\xi\sin\eta \\ 0 \end{bmatrix},
$$

we get

$$
\begin{aligned}
2\mu u_\xi &= \frac{1}{h_1}\left[\frac{\partial\alpha}{\partial\xi} + c\sinh\xi\cos\eta\frac{\partial\beta}{\partial\xi} + (4\nu-3)c\beta\cosh\xi\cos\eta\right], \\
2\mu u_\eta &= \frac{1}{h_1}\left[\frac{\partial\alpha}{\partial\eta} + c\sinh\xi\cos\eta\frac{\partial\beta}{\partial\eta} - (4\nu-3)c\beta\sinh\xi\sin\eta\right],
\end{aligned}
\tag{6.263}
$$

where α and β are harmonic functions. The expressions for the stresses are obtained using Eqns. (G.12).

Corresponding to the choice $\boldsymbol{\beta} = (0, 0, \beta)$ in the cylindrical coordinate system (or, equivalently, $\boldsymbol{\beta} = (\beta\cos\theta, -\beta\sin\theta, 0)$ in the spherical coordinate system), there exists a Galerkin vector $\boldsymbol{\Psi}$ of the form $(0, 0, \Psi)$ with respect to a cylindrical coordinate system (Ψ is thus Love's biharmonic strain function), or of the form $(\Psi\cos\theta, -\Psi\sin\theta, 0)$ with respect to a spherical coordinate system. In the absence of body forces, the governing equation for Ψ is obtained from Eqn. (6.253b) as

$$\boldsymbol{\nabla}^4\Psi = 0, \tag{6.264}$$

where in a cylindrical system

$$\boldsymbol{\nabla}^4\Psi = \left[\frac{1}{r}\frac{\partial}{\partial r}\left(r\frac{\partial}{\partial r}\right) + \frac{\partial^2}{\partial z^2}\right]\left[\frac{1}{r}\frac{\partial}{\partial r}\left(r\frac{\partial \Psi}{\partial r}\right) + \frac{\partial^2\Psi}{\partial z^2}\right].$$

The relation between the harmonic functions (α, β) and the biharmonic Love strain function Ψ is found using Eqns (6.256), and, in the cylindrical coordinate system is given by

$$\alpha = \frac{z}{2}\left[\frac{1}{r}\frac{\partial}{\partial r}\left(r\frac{\partial\Psi}{\partial r}\right) + z^2\frac{\partial}{\partial z}\left(\frac{1}{z^2}\frac{\partial\Psi}{\partial z}\right)\right],$$

$$\beta = -\frac{1}{2}\left[\frac{1}{r}\frac{\partial}{\partial r}\left(r\frac{\partial\Psi}{\partial r}\right) + \frac{\partial^2\Psi}{\partial z^2}\right] = -\frac{1}{2}\boldsymbol{\nabla}^2\Psi,$$

while in the spherical one it is given by

$$\begin{aligned}\alpha &= \frac{1}{2}\left[r\cos\theta\frac{\partial^2\Psi}{\partial r^2} + \frac{\cos\theta}{r}\frac{\partial^2\Psi}{\partial\theta^2} + \frac{1+\sin^2\theta}{r\sin\theta}\frac{\partial\Psi}{\partial\theta}\right],\\ \beta &= -\frac{1}{2}\left[\frac{1}{r^2}\frac{\partial}{\partial r}\left(r^2\frac{\partial\Psi}{\partial r}\right) + \frac{1}{r^2}\frac{\partial^2\Psi}{\partial\theta^2} + \frac{\cot\theta}{r^2}\frac{\partial\Psi}{\partial\theta}\right] = -\frac{1}{2}\boldsymbol{\nabla}^2\Psi.\end{aligned} \tag{6.265}$$

The formulation in terms of Love's strain function in a cylindrical coordinate system is obtained using Eqn. (6.253a) as

$$\begin{aligned}2\mu u_r &= -\frac{\partial^2\Psi}{\partial r\partial z},\\ 2\mu u_z &= 2(1-\nu)\boldsymbol{\nabla}^2\Psi - \frac{\partial^2\Psi}{\partial z^2},\\ \tau_{rr} &= \frac{\partial}{\partial z}\left(\nu\boldsymbol{\nabla}^2\Psi - \frac{\partial^2\Psi}{\partial r^2}\right),\\ \tau_{zz} &= \frac{\partial}{\partial z}\left[(2-\nu)\boldsymbol{\nabla}^2\Psi - \frac{\partial^2\Psi}{\partial z^2}\right],\\ \tau_{rz} &= \frac{\partial}{\partial r}\left[(1-\nu)\boldsymbol{\nabla}^2\Psi - \frac{\partial^2\Psi}{\partial z^2}\right],\\ \tau_{\theta\theta} &= \frac{\partial}{\partial z}\left(\nu\boldsymbol{\nabla}^2\Psi - \frac{1}{r}\frac{\partial\Psi}{\partial r}\right),\end{aligned} \tag{6.266}$$

while in a spherical coordinate system, it is found by substituting Eqns. (6.265) into Eqns. (6.261) (see also [191]).

An axisymmetric solution to $\nabla^2\phi = 0$ in cylindrical coordinates (excluding a constant term) is

$$
\begin{aligned}
\phi &= A_0 I_0(k_1 r)\cos k_1 z + B_0 K_0(k_2 r)\cos k_2 z + C_0 I_0(k_3 r)\sin k_3 z + D_0 K_0(k_4 r)\sin k_4 z \\
&+ P_0 J_0(k_5 r)\cosh k_5 z + Q_0 Y_0(k_6 r)\cosh k_6 z + R_0 J_0(k_7 r)\sinh k_7 z + S_0 Y_0(k_8 r)\sinh k_8 z \\
&+ \tilde{A}\left[z I_0(k_9 r)\cos k_9 z + r I_1(k_9 r)\sin k_9 z\right] \\
&+ \tilde{B}\left[z K_0(k_{10} r)\cos k_{10} z - r K_1(k_{10} r)\sin k_{10} z\right] \\
&+ \tilde{C}\left[z J_0(k_{11} r)\cosh k_{11} z - r J_1(k_{11} r)\sinh k_{11} z\right] \\
&+ \tilde{D}\left[z Y_0(k_{12} r)\cosh k_{12} z - r Y_1(k_{12} r)\sinh k_{12} z\right] \\
&+ \hat{A}\left[z I_0(k_{13} r)\sin k_{13} z - r I_1(k_{13} r)\cos k_{13} z\right] \\
&+ \hat{B}\left[z K_0(k_{14} r)\sin k_{14} z + r K_1(k_{14} r)\cos k_{14} z\right] \\
&+ \hat{C}\left[z J_0(k_{15} r)\sinh k_{15} z - r J_1(k_{15} r)\cosh k_{15} z\right] \\
&+ \hat{D}\left[z Y_0(k_{16} r)\sinh k_{16} z - r Y_1(k_{16} r)\cosh k_{16} z\right] \\
&+ k_0 L_0^{(1)} + k_1 L_1^{(1)} + c_1\phi_1 + c_2\phi_2,
\end{aligned}
\tag{6.267}
$$

where A_0, B_0, C_0, D_0, P_0, Q_0, R_0 and S_0, etc. are constants, J_0 and Y_0 are Bessel functions of the first and second kind, I_0 and K_0 are modified Bessel functions of the first and second kind, and $L_0^{(1)}$, ϕ_1 and ϕ_2 are given by Eqn. (6.276). The ϕ_1 and ϕ_2 terms yield constant and linearly varying values for the stresses. While dealing with solid cylinders, the coefficients of the functions $L_0^{(1)}$, Y_0, K_0, Y_1 and K_1 are set to zero. The (excluded) constant, k_0, k_1 and c_1 terms correspond to the zero root case as in Eqn. (6.238). The c_2 term in Eqn. (6.267) and the c_4 term in Eqn. (6.268) below are not strictly required from a mathematical perspective, but they help in getting accurate solutions to problems involving thin plates (for thick plates, the solutions obtained by including or excluding these terms are almost the same).

An axisymmetric solution to Eqn. (6.264) in cylindrical coordinates (excluding a constant term) obtained using Eqn. (6.229) is

$$
\begin{aligned}
\Psi &= A_0 I_0(k_1 r)\cos k_1 z + B_0 K_0(k_2 r)\cos k_2 z + A r I_1(k_3 r)\cos k_3 z + B r K_1(k_4 r)\cos k_4 z \\
&+ C_0 I_0(k_5 r)\sin k_5 z + D_0 K_0(k_6 r)\sin k_6 z + C r I_1(k_7 r)\sin k_7 z + D r K_1(k_8 r)\sin k_8 z \\
&+ P_0 J_0(k_9 r)\cosh k_9 z + Q_0 Y_0(k_{10} r)\cosh k_{10} z + P r J_1(k_{11} r)\cosh k_{11} z + Q r Y_1(k_{12} r)\cosh k_{12} z \\
&+ R_0 J_0(k_{13} r)\sinh k_{13} z + S_0 Y_0(k_{14} r)\sinh k_{14} z + R r J_1(k_{15} r)\sinh k_{15} z + S r Y_1(k_{16} r)\sinh k_{16} z \\
&+ \left[A_1 I_0(k_{17} r)\cos k_{17} z + B_1 K_0(k_{18} r)\cos k_{18} z + C_1 I_0(k_{19} r)\sin k_{19} z + D_1 K_0(k_{20} r)\sin k_{20} z\right] z \\
&+ \left[P_1 J_0(k_{21} r)\cosh k_{21} z + Q_1 Y_0(k_{22} r)\cosh k_{22} z + R_1 J_0(k_{23} r)\sinh k_{23} z + S_1 Y_0(k_{24} r)\sinh k_{24} z\right] z \\
&+ \frac{2c_1 + c_2}{4(1+\nu)}\phi_2 z + \frac{(2-4\nu)c_1 - (1+4\nu)c_2}{12(1+\nu)}\phi_3 + \frac{\left[2\phi_3 z - (1+2\nu)\phi_4\right]c_4}{24(1+\nu)} \\
&+ \tilde{c}_1 z + \tilde{c}_2 z^2 + \tilde{c}_3 r^2 + \tilde{c}_4 \log r + \tilde{c}_5 r^2 \log r + \tilde{c}_6 r^2 z \log r + \tilde{c}_7 z \log r + \tilde{c}_8 z^2 \log r + \tilde{c}_9 z^3 \log r,
\end{aligned}
\tag{6.268}
$$

where, the terms involving the constants c_1, c_2 and c_4 result in the stresses $\tau_{rr} = \tau_{\theta\theta} = c_1$, $\tau_{zz} = c_2 + c_4 z$, $\tau_{rz} = -c_4 r/2$, and the displacements

$$2\mu u_r = \frac{[(1-\nu)c_1 - \nu c_2 - \nu c_4 z]\, r}{1+\nu},$$

$$2\mu u_z = \frac{2(c_2 - 2\nu c_1)z + c_4(z^2 - r^2)}{2(1+\nu)}.$$

Since $\sin ix = i \sinh x$, $I_0(ir) = J_0(r)$, and so on, one can reduce the number of terms in the above solutions by allowing complex arguments and constants, as we do in Sections 6.6.8–6.6.12.

An axisymmetric solution to $\nabla^2\phi = 0$ with respect to a spherical coordinate system (with $\xi := \cos\theta$) is (see Appendix E of Vol. II for a detailed derivation)

$$\phi = \sum_{n=0}^{\infty} \left\{ [A_n r^n + B_n r^{-(n+1)}] P_n(\xi) + [\hat{A}_n r^n + \hat{B}_n r^{-(n+1)}] Q_n(\xi) + \hat{A}_n^+ L_n^{(1)}(r,\xi) + \hat{B}_n^+ L_n^{(2)}(r,\xi) \right\}, \tag{6.269}$$

where $P_n(\xi)$ and $Q_n(\xi)$, known as the Legendre polynomials of the first kind and the Legendre functions of the second kind, are the solutions of the differential equation

$$\frac{d}{d\xi}\left[(1-\xi^2)\frac{dY}{d\xi}\right] + [n(n+1)]\, Y = 0, \tag{6.270}$$

and where

$$L_n^{(1)}(r,\xi) = r^n \left[P_n(\xi) \log(r \sin\theta) + \hat{a}_n(\xi)\right], \tag{6.271a}$$

$$L_n^{(2)}(r,\xi) = r^{-(2n+1)} \left[r^n P_n(\xi) \log r^2 - L^{(1)}(r,\xi)\right], \tag{6.271b}$$

$$\hat{a}_n(\xi) = \left[\int_\xi^1 \frac{2n[P_n(\alpha) - \alpha P_{n-1}(\alpha)] P_n(\alpha)\, d\alpha}{(1-\alpha^2)}\right] Q_n(\xi) - \left[\int_\xi^1 \frac{2n[P_n(\alpha) - \alpha P_{n-1}(\alpha)] Q_n(\alpha)\, d\alpha}{(1-\alpha^2)}\right] P_n(\xi). \tag{6.271c}$$

Since the functions Q_n are singular at $\xi = \pm 1$, they are useful for solving problems where the axis is not part of the domain such as, for example, hollow cones. Similarly, the functions $L^{(1)}(r,\xi)$ are useful for problems involving hollow cylinders. The Legendre functions of the second kind are given by

$$Q_n(\xi) = P_n(\xi) \int \frac{d\xi}{P_n^2(\xi)(1-\xi^2)}.$$

The P_n and Q_n functions satisfy the following relation (even when n is complex-valued) that can be used to compute the Q_n functions recursively:

$$P_{n+1}(\xi) Q_n(\xi) - P_n(\xi) Q_{n+1}(\xi) = \frac{1}{1+n}.$$

If ϕ is a harmonic function of the form given by Eqn. (6.269), then by virtue of Eqn. (6.229), $r^2\phi$ is a biharmonic function. Thus, we get a general solution to Eqn. (6.264) with respect to a spherical coordinate system as

$$\begin{aligned}\Psi &= \sum_{n=0}^{\infty}(A_n r^{n+2} + B_n r^{1-n} + C_n r^n + D_n r^{-(n+1)})P_n(\xi) \\ &+ \sum_{n=0}^{\infty}(\hat{A}_n r^{n+2} + \hat{B}_n r^{1-n} + \hat{C}_n r^n + \hat{D}_n r^{-(n+1)})Q_n(\xi) \\ &+ \sum_{n=0}^{\infty}[\tilde{A}_n r^2 L_n^{(1)}(r,\xi) + \tilde{B}_n L_n^{(1)}(r,\xi)] \\ &+ \sum_{n=0}^{\infty}[\tilde{C}_n r^2 L_n^{(2)}(r,\xi) + \tilde{D}_n L_n^{(2)}(r,\xi)].\end{aligned} \tag{6.272}$$

For problems involving conical domains (which includes the special case of a hemispherical domain), one either replaces n in the P_n and Q_n terms in Eqns. (6.269) and (6.272) by λ_n, where λ_n are the (complex-valued) roots of a characteristic equation, or one uses the following alternative forms of the solution (see Appendix E of Vol. II; the fact that these two approaches are related is evident from Eqn. (6.388))

$$\phi = \sum_{n=0}^{\infty}\left[A_n r^n + B_n r^{-(n+1)}\right]P_n(\xi) + \left[A_n^+ S_n^+ + A_n^- S_n^-\right], \tag{6.273a}$$

$$\begin{aligned}\Psi &= \sum_{n=0}^{\infty}(A_n r^{n+2} + B_n r^{1-n} + C_n r^n + D_n r^{-(n+1)})P_n(\xi) \\ &+ \left[(A_n^+ + r^2 B_n^+)S_n^+ + (A_n^- + r^2 B_n^-)S_n^-\right],\end{aligned} \tag{6.273b}$$

where

$$\begin{aligned}S_n^+ &= r^n\left[P_n(\xi)\log[r(1+\xi)] + a_n^+(\xi)\right], \\ S_n^- &= r^{-(n+1)}\left[P_n(\xi)\log[r(1+\xi)] + b_n^+(\xi)\right],\end{aligned} \tag{6.274}$$

and the functions $a_n^+(\xi)$ and $b_n^+(\xi)$ are found using the differential equations

$$\begin{aligned}(1-\xi^2)(a_n^+)'' - 2\xi(a_n^+)' + n(n+1)a_n^+ &= -\frac{2n(P_n + P_{n-1})}{1+\xi}, \\ (1-\xi^2)(b_n^+)'' - 2\xi(b_n^+)' + n(n+1)b_n^+ &= \frac{2(n+1+\xi+2n\xi)P_n - 2nP_{n-1}}{1+\xi}.\end{aligned}$$

The solutions to these differential equations are given by

$$\begin{aligned}a_n^+(\xi) &= \left[\int_\xi^1 \frac{2n\left[P_n(\alpha) + P_{n-1}(\alpha)\right]P_n(\alpha)\,d\alpha}{1+\alpha}\right]Q_n(\xi) \\ &- \left[\int_\xi^1 \frac{2n\left[P_n(\alpha) + P_{n-1}(\alpha)\right]Q_n(\alpha)\,d\alpha}{1+\alpha}\right]P_n(\xi), \\ b_n^+(\xi) &= \left[\int_\xi^1 \frac{\left[2nP_{n-1}(\alpha) - 2(n+1+\alpha+2n\alpha)P_n(\alpha)\right]P_n(\alpha)\,d\alpha}{1+\alpha}\right]Q_n(\xi) \\ &- \left[\int_\xi^1 \frac{\left[2nP_{n-1}(\alpha) - 2(n+1+\alpha+2n\alpha)P_n(\alpha)\right]Q_n(\alpha)\,d\alpha}{1+\alpha}\right]P_n(\xi).\end{aligned} \tag{6.275}$$

The solutions in Eqn. (6.273) are singular on the axis $\xi = -1$, which is not part of the conical domain.

The first few Legendre polynomials and Legendre functions are

$$
\begin{aligned}
P_0(\xi) &= 1, & Q_0(\xi) &= \frac{1}{2}\log\left(\frac{1+\xi}{1-\xi}\right),\\
P_1(\xi) &= \xi, & Q_1(\xi) &= \frac{\xi}{2}\log\left(\frac{1+\xi}{1-\xi}\right) - 1,\\
P_2(\xi) &= \frac{1}{2}(3\xi^2-1), & Q_2(\xi) &= \frac{3\xi^2-1}{4}\log\left(\frac{1+\xi}{1-\xi}\right) - \frac{3\xi}{2},\\
P_3(\xi) &= \frac{1}{2}(5\xi^3-3\xi), & Q_3(\xi) &= \frac{5\xi^3-3\xi}{4}\log\left(\frac{1+\xi}{1-\xi}\right) + \frac{4-15\xi^2}{6},\\
P_4(\xi) &= \frac{1}{8}(35\xi^4-30\xi^2+3), & Q_4(\xi) &= \frac{35\xi^4-30\xi^2+3}{16}\log\left(\frac{1+\xi}{1-\xi}\right) + \frac{55\xi-105\xi^3}{24}.
\end{aligned}
$$

The first few terms of the harmonic functions associated with the coefficients A_n, B_n and $\hat{A}_n^+$ in Eqn. (6.269), expressed in terms of cylindrical coordinates (r, z), are

$$
\begin{aligned}
\phi_0 &= 1, & \phi_0^s &= \frac{\phi_0}{(r^2+z^2)^{1/2}}, & L_0^{(1)} &= \log r,\\
\phi_1 &= z, & \phi_1^s &= \frac{\phi_1}{(r^2+z^2)^{3/2}}, & L_1^{(1)} &= z\log r,\\
\phi_2 &= \frac{2z^2-r^2}{2}, & \phi_2^s &= \frac{\phi_2}{(r^2+z^2)^{5/2}}, & L_2^{(1)} &= \frac{1}{2}\left[(2z^2-r^2)\log r + r^2\right],\\
\phi_3 &= \frac{2z^3-3zr^2}{2}, & \phi_3^s &= \frac{\phi_3}{(r^2+z^2)^{7/2}}, & L_3^{(1)} &= \frac{1}{2}\left[(2z^3-3zr^2)\log r + 3zr^2\right],\\
\phi_4 &= \frac{8z^4-24z^2r^2+3r^4}{8}, & \phi_4^s &= \frac{\phi_4}{(r^2+z^2)^{9/2}}, & L_4^{(1)} &= \frac{1}{8}\left[(8z^4-24z^2r^2+3r^4)\log r + 24z^2r^2 - \frac{9r^4}{2}\right].
\end{aligned}
\tag{6.276}
$$

The first few functions $a_n^+(\xi)$ and $b_n^+(\xi)$ in Eqn. (6.274) obtained using Eqns. (6.275) are given by

$$
\begin{aligned}
a_0^+(\xi) &= 0,\\
a_1^+(\xi) &= \xi - 1,\\
a_2^+(\xi) &= \frac{1}{4}(7\xi+1)(\xi-1),\\
a_3^+(\xi) &= \frac{1}{12}(37\xi^2+7\xi-8)(\xi-1),\\
a_4^+(\xi) &= \frac{1}{96}(533\xi^3+113\xi^2-241\xi-21)(\xi-1),\\
b_0^+(\xi) &= -\log(1-\xi^2),\\
b_1^+(\xi) &= -a_1^+ - 2P_1(\xi)\log(1+\xi) + P_1(\xi)\log 4,\\
b_2^+(\xi) &= -a_2^+ - 2P_2(\xi)\log(1+\xi) + P_2(\xi)\log 4,
\end{aligned}
$$

$$b_3^+(\xi) = -a_3^+ - 2P_3(\xi)\log(1+\xi) + P_3(\xi)\log 4,$$
$$b_4^+(\xi) = -a_4^+ - 2P_4(\xi)\log(1+\xi) + P_4(\xi)\log 4.$$

Thus, while solving specific problems, one attempts, by means of the general solution given by Eqn. (6.269), to find harmonic functions $\alpha(r,\theta)$ and $\beta(r,\theta)$, such that the resulting displacement and stress fields obtained via Eqns. (6.261) satisfy the boundary conditions. When the axis is part of the domain, the singular functions can be excluded from Eqn. (6.269), and in this case, α and β can be taken to be

$$\begin{aligned}\alpha &= \sum_{n=0}^{\infty}\left[\hat{A}_n r^n + \hat{C}_n r^{-(n+1)}\right]P_n(\xi),\\ \beta &= \sum_{n=1}^{\infty}\left[\hat{B}_n r^{n-1} + \hat{D}_n r^{-n}\right]P_{n-1}(\xi).\end{aligned} \tag{6.277}$$

By using the first relation in Eqns. (6.138) with $m = 0$, recombining the α and β given above[12], and then using Eqns. (6.261), we get [51]

$$\begin{aligned}\alpha &= \sum_{n=0}^{\infty}\Big[B_n r^n P_n(\xi) - A_n(4\nu+n-2)r^{n+2}P_{n+2}(\xi)\\ &\quad - D_n r^{-n-1}P_n(\xi) + (n+3-4\nu)C_n r^{1-n}P_{n-2}(\xi)\Big],\\ \beta &= \sum_{n=0}^{\infty}\left[A_n(2n+3)r^{n+1}P_{n+1}(\xi) + C_n(1-2n)r^{-n}P_{n-1}(\xi)\right],\\ 2\mu u_r &= \sum_{n=0}^{\infty}\Big[(n+1)(n-2+4\nu)A_n r^{n+1} + nB_n r^{n-1} + n(n+3-4\nu)C_n r^{-n}\\ &\quad + (n+1)D_n r^{-n-2}\Big]P_n(\xi),\\ 2\mu u_\theta &= -\sum_{n=0}^{\infty}\Big[(n+5-4\nu)A_n r^{n+1} + B_n r^{n-1} - (n-4+4\nu)C_n r^{-n}\\ &\quad - D_n r^{-n-2}\Big]\sin\theta\frac{dP_n}{d\xi},\\ \tau_{rr} &= \sum_{n=0}^{\infty}\Big[(n+1)(n^2-n-2-2\nu)A_n r^n + n(n-1)B_n r^{n-2}\\ &\quad - n(n^2+3n-2\nu)C_n r^{-n-1} - (n+1)(n+2)D_n r^{-n-3}\Big]P_n(\xi),\\ \tau_{\theta\theta} &= \sum_{n=0}^{\infty}\Big\{\Big[-(n+1)(n^2+4n+2+2\nu)A_n r^n - n^2B_n r^{n-2}\\ &\quad + n(n^2-2n-1+2\nu)C_n r^{-n-1} + (n+1)^2D_n r^{-n-3}\Big]P_n(\xi)\end{aligned} \tag{6.278}$$

[12] One can derive the displacement and stress fields directly using Eqn. (6.277); the recombination is carried out just so as to bring the resulting displacement and stress expressions in agreement with the expressions found in the literature. One can also derive these equations by substituting the solution given by the first line of Eqn. (6.272) into the relations for the displacements and stresses in terms of Ψ.

$$+\left[(n+5-4\nu)A_n r^n + B_n r^{n-2} - (n-4+4\nu)C_n r^{-n-1} - D_n r^{-n-3}\right]\xi\frac{dP_n}{d\xi}\bigg\},$$

$$\tau_{\phi\phi} = \sum_{n=0}^{\infty}\bigg\{\Big[(n+1)(n-4\nu n-2-2\nu)A_n r^n + nB_n r^{n-2}$$

$$+n(n-4\nu n+3-2\nu)C_n r^{-n-1} + (n+1)D_n r^{-n-3}\Big]P_n(\xi)$$

$$-\left[(n+5-4\nu)A_n r^n + B_n r^{n-2} - (n-4+4\nu)C_n r^{-n-1} - D_n r^{-n-3}\right]\xi\frac{dP_n}{d\xi}\bigg\},$$

$$\tau_{r\theta} = -\sum_{n=0}^{\infty}\Big[(n^2+2n-1+2\nu)A_n r^n + (n-1)B_n r^{n-2}$$

$$+(n^2-2+2\nu)C_n r^{-n-1} + (n+2)D_n r^{-n-3}\Big]\sin\theta\frac{dP_n}{d\xi},$$

where

$$(1-\xi^2)\frac{dP_n(\xi)}{d\xi} = n\left[P_{n-1}(\xi) - \xi P_n(\xi)\right] = (n+1)\left[\xi P_n(\xi) - P_{n+1}(\xi)\right]. \tag{6.279}$$

Eqns. (6.278) can also be written in the following alternative form:

$$\alpha = \sum_{n=0}^{\infty}\left\{\left[B_n - A_n(4\nu+n-2)\right]r^{n+2}P_{n+2}(\xi) + \left[C_n(n+3-4\nu) - D_n\right]r^{1-n}P_{n-2}(\xi)\right\},$$

$$\beta = \sum_{n=0}^{\infty}\left[A_n(2n+3)r^{n+1}P_{n+1}(\xi) + C_n(1-2n)r^{-n}P_{n-1}(\xi)\right],$$

$$2\mu u_r = \sum_{n=0}^{\infty}\Big[(n+1)(n-2+4\nu)A_n r^{n+1}P_n(\xi) + (n+2)B_n r^{n+1}P_{n+2}(\xi)$$

$$+n(n+3-4\nu)C_n r^{-n}P_n(\xi) + (n-1)D_n r^{-n}P_{n-2}(\xi)\Big],$$

$$2\mu u_\theta = -\sum_{n=0}^{\infty}\Big[(n+5-4\nu)A_n r^{n+1}\frac{dP_n}{d\xi} + B_n r^{n+1}\frac{dP_{n+2}}{d\xi} - (n-4+4\nu)C_n r^{-n}\frac{dP_n}{d\xi}$$

$$-D_n r^{-n}\frac{dP_{n-2}}{d\xi}\Big]\sin\theta,$$

$$\tau_{rr} = \sum_{n=0}^{\infty}\Big[(n+1)(n^2-n-2-2\nu)A_n r^n P_n(\xi) + (n+2)(n+1)B_n r^n P_{n+2}(\xi)$$

$$-n(n^2+3n-2\nu)C_n r^{-n-1}P_n(\xi) - (n-1)nD_n r^{-n-1}P_{n-2}(\xi)\Big],$$

$$\tau_{\theta\theta} = \sum_{n=0}^{\infty}\bigg\{\Big[-(n+1)(n^2+4n+2+2\nu)A_n r^n P_n(\xi) - (n+2)^2 B_n r^n P_{n+2}(\xi) \tag{6.280}$$

$$+n(n^2-2n-1+2\nu)C_n r^{-n-1}P_n(\xi) + (n-1)^2 D_n r^{-n-1}P_{n-2}(\xi)\Big]$$

$$+\left[(n+5-4\nu)A_n r^n\frac{dP_n}{d\xi} + B_n r^n\frac{dP_{n+2}}{d\xi} - (n-4+4\nu)C_n r^{-n-1}\frac{dP_n}{d\xi} - D_n r^{-n-1}\frac{dP_{n-2}}{d\xi}\right]\xi\bigg\},$$

$$\tau_{\phi\phi} = \sum_{n=0}^{\infty}\bigg\{\Big[(n+1)(n-4\nu n-2-2\nu)A_n r^n P_n(\xi) + (n+2)B_n r^n P_{n+2}(\xi)$$

$$+ n(n - 4\nu n + 3 - 2\nu)C_n r^{-n-1} P_n(\xi) + (n-1) D_n r^{-n-1} P_{n-2}(\xi)\Big]$$

$$- \left[(n+5-4\nu)A_n r^n \frac{dP_n}{d\xi} + B_n r^n \frac{dP_{n+2}}{d\xi} - (n-4+4\nu)C_n r^{-n-1}\frac{dP_n}{d\xi} - D_n r^{-n-1}\frac{dP_{n-2}}{d\xi}\right]\xi\Bigg\},$$

$$\tau_{r\theta} = -\sum_{n=0}^{\infty}\Bigg[(n^2+2n-1+2\nu)A_n r^n \frac{dP_n}{d\xi} + (n+1)B_n r^n \frac{dP_{n+2}}{d\xi}$$

$$+ (n^2 - 2 + 2\nu)C_n r^{-n-1}\frac{dP_n}{d\xi} + nD_n r^{-n-1}\frac{dP_{n-2}}{d\xi}\Bigg]\sin\theta.$$

Although the same symbols have been used, note that the constants A_n, B_n, C_n and D_n in the above set of equations are not the same as in Eqn. (6.278). The stress distribution in Eqns. (6.278) and (6.280) automatically satisfies the equations of equilibrium (under zero body force and for axisymmetric problems), and thus, the constants need to be determined based only on the boundary conditions. While dealing with problems involving spherical segments, the n in the above equations can be replaced by λ_n, where the λ_n, $n = 1, 2, \ldots, \infty$, are the (complex-valued) roots of an appropriately formulated characteristic equation as discussed towards the end of Section 6.6.7. Note that the $P_\lambda(\xi)$ are no longer linearly independent in this case.

For a prolate spheroidal system defined via the relations $r + iz = ci\cos(\eta + i\xi)$, i.e., $r = c\sinh\xi\sin\eta$ and $z = c\cosh\xi\cos\eta$, we have $h_1 = h_2 = c\sqrt{\cosh^2\xi - \cos^2\eta}$ and $h_3 = r$. The Laplace equation $\nabla^2\phi = 0$ for the axisymmetric case can be written using Eqn. (F.5) as

$$\frac{1}{c^2(\cosh^2\xi - \cos^2\eta)}\left[\frac{\partial^2\phi}{\partial\xi^2} + \coth\xi\frac{\partial\phi}{\partial\xi} + \frac{\partial^2\phi}{\partial\eta^2} + \cot\eta\frac{\partial\phi}{\partial\eta}\right] = 0.$$

A general solution of the above equation is

$$\phi = c_1 + c_2\log\tanh\frac{\xi}{2} + c_3\log\tan\frac{\eta}{2} + c_4\log\tanh\frac{\xi}{2}\log\tan\frac{\eta}{2}$$
$$+ \sum_{n=1}^{\infty}\left[A_n P_n(\cosh\xi) + B_n Q_n(\cosh\xi)\right]\left[C_n P_n(\cos\eta) + D_n Q_n(\cos\eta)\right], \tag{6.281}$$

where P_n and Q_n are Legendre polynomials of the first and second kind. In practice, $i\pi P_n(\cosh\xi)/2$ can be added to $Q_n(\cosh\xi)$ in order to keep the coefficients real.

For an oblate spheroidal system defined using the relations $r + iz = c\sin(\eta + i\xi)$, i.e., $r = c\cosh\xi\sin\eta$ and $z = c\sinh\xi\cos\eta$, we have $h_1 = h_2 = c\sqrt{\sinh^2\xi + \cos^2\eta}$ and $h_3 = r$. The Laplace equation $\nabla^2\phi = 0$ for the axisymmetric case can be written using Eqn. (G.5) as

$$\frac{1}{c^2(\sinh^2\xi + \cos^2\eta)}\left[\frac{\partial^2\phi}{\partial\xi^2} + \tanh\xi\frac{\partial\phi}{\partial\xi} + \frac{\partial^2\phi}{\partial\eta^2} + \cot\eta\frac{\partial\phi}{\partial\eta}\right] = 0.$$

A general solution of the above equation is

$$\phi = c_1 + c_2\tan^{-1}\tanh\frac{\xi}{2} + c_3\log\tan\frac{\eta}{2} + c_4\left(\tan^{-1}\tanh\frac{\xi}{2}\right)\left(\log\tan\frac{\eta}{2}\right)$$
$$+ \sum_{n=1}^{\infty}\left[A_n P_n(i\sinh\xi) + B_n Q_n(i\sinh\xi)\right]\left[C_n P_n(\cos\eta) + D_n Q_n(\cos\eta)\right]. \tag{6.282}$$

For problems involving infinite domains, $i\pi P_n(i\sinh\xi)/2$ should be subtracted from $Q_n(i\sinh\xi)$, so that the combined function decays to zero as $\xi \to \infty$. The constants A_n and B_n are either real or pure imaginary so that the final result is real. One can multiply the function $Q_n(i\sinh\xi) - i\pi P_n(i\sinh\xi)/2$ by the constant $-(-i)^{n+1}$ if one wants to deal only with real coefficients B_n.

We now discuss the solution of several problems using the Airy, Boussinesq/Love and Galerkin potentials.

6.6.1 Cylindrical/elliptical elastic inclusion in an infinite domain with a uniform state of stress at infinity

In the first case that we consider, we shall assume conditions of plane strain to hold with the constitutive relation given by Eqn. (6.226). An isotropic body of infinite extent contains a cylindrical elastic inclusion of radius a as shown in Fig. 6.36. The properties of the materials

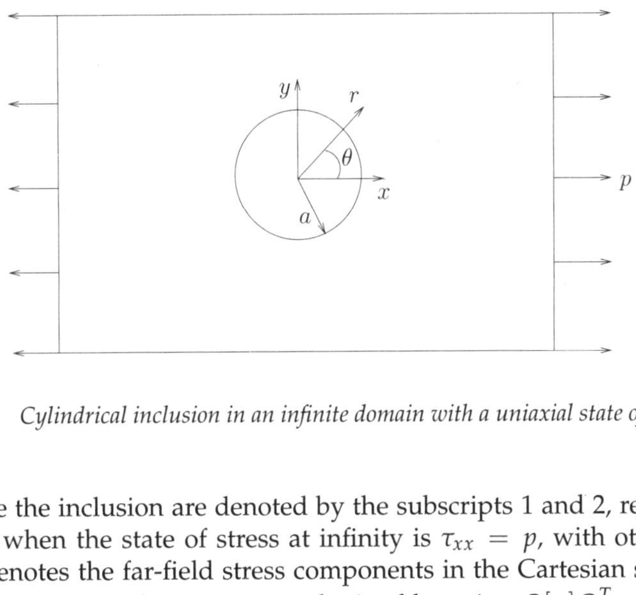

Fig. 6.36 *Cylindrical inclusion in an infinite domain with a uniaxial state of stress at infinity.*

inside and outside the inclusion are denoted by the subscripts 1 and 2, respectively. First consider the case when the state of stress at infinity is $\tau_{xx} = p$, with other components zero [97]. If $[\boldsymbol{\tau}]$ denotes the far-field stress components in the Cartesian system, then the components in the polar coordinate system, obtained by using $\boldsymbol{Q}[\boldsymbol{\tau}]\boldsymbol{Q}^T$, where $\boldsymbol{Q}$ is given by Eqn. (B.2), are given by $\tau_{rr} = p\cos^2\theta$, $\tau_{\theta\theta} = p\sin^2\theta$ and $\tau_{r\theta} = -p\sin\theta\cos\theta$. In the material outside the inclusion, we assume the coefficients F_1, C_0, C_1, A_{-3}, and A_1 to be the only nonzero coefficients in Eqn. (6.239), i.e., we assume ϕ to be of the form

$$\phi = \frac{pr^2}{4}(1-\cos 2\theta) + A\log r + \left(\frac{B}{r^2} + C\right)\cos 2\theta.$$

The stress distribution is obtained using Eqn. (6.221), and the corresponding displacement field is obtained using Eqns. (B.8) and (6.226) as

$$u_r = \frac{1}{2\mu_2}\left\{\frac{pr}{2}(1-2\nu_2+\cos 2\theta) - \frac{A}{r} + \left[\frac{2B}{r^3} + \frac{4C(1-\nu_2)}{r}\right]\cos 2\theta\right\},$$

$$u_\theta = \frac{1}{2\mu_2}\left[-\frac{pr}{2} + \frac{2B}{r^3} - \frac{2C(1-2\nu_2)}{r}\right]\sin 2\theta,$$

$$\tau_{rr} = p\cos^2\theta + \frac{A}{r^2} - \left(\frac{6B}{r^4} + \frac{4C}{r^2}\right)\cos 2\theta,$$
$$\tau_{\theta\theta} = p\sin^2\theta - \frac{A}{r^2} + \frac{6B}{r^4}\cos 2\theta,$$
$$\tau_{r\theta} = -p\sin\theta\cos\theta - \left(\frac{6B}{r^4} + \frac{2C}{r^2}\right)\sin 2\theta. \tag{6.283}$$

Note that the above stress distribution automatically satisfies the far-field conditions as $r \to \infty$.

The Airy stress function for the material within the inclusion is assumed to be of the form

$$\phi = Fr^2 + (Gr^2 + Hr^4)\cos 2\theta,$$

which leads to the following displacement and stress fields:

$$u_r = \frac{1}{\mu_1}\left[(1-2\nu_1)Fr - (Gr + 2\nu_1 Hr^3)\cos 2\theta\right],$$
$$u_\theta = \frac{1}{\mu_1}\left[Gr + (3-2\nu_1)Hr^3\right]\sin 2\theta,$$
$$\tau_{rr} = 2F - 2G\cos 2\theta, \tag{6.284}$$
$$\tau_{\theta\theta} = 2F + 2(G + 6Hr^2)\cos 2\theta,$$
$$\tau_{r\theta} = 2(G + 3Hr^2)\sin 2\theta.$$

The constants A, B, C, F, G and H are determined using the following displacement and traction continuity conditions at the interface:

$$u_r^{(1)}\Big|_{r=a} = u_r^{(2)}\Big|_{r=a}, \quad u_\theta^{(1)}\Big|_{r=a} = u_\theta^{(2)}\Big|_{r=a},$$
$$\tau_{rr}^{(1)}\Big|_{r=a} = \tau_{rr}^{(2)}\Big|_{r=a}, \quad \tau_{r\theta}^{(1)}\Big|_{r=a} = \tau_{r\theta}^{(2)}\Big|_{r=a}. \tag{6.285}$$

We get

$$A = \frac{[\mu_1(1-2\nu_2) - \mu_2(1-2\nu_1)]\,pa^2}{2[(1-2\nu_1)\mu_2 + \mu_1]},$$
$$B = \frac{(\mu_1 - \mu_2)pa^4}{4[\mu_2 + (3-4\nu_2)\mu_1]},$$
$$C = \frac{(\mu_2 - \mu_1)pa^2}{2[\mu_2 + (3-4\nu_2)\mu_1]},$$
$$F = \frac{(1-\nu_2)\mu_1 p}{2[(1-2\nu_1)\mu_2 + \mu_1]},$$
$$G = -\frac{(1-\nu_2)\mu_1 p}{\mu_2 + (3-4\nu_2)\mu_1},$$
$$H = 0.$$

The results for a cylindrical hole are obtained by setting $\mu_1 = 0$:

$$A = -\frac{1}{2}pa^2, \quad B = -\frac{1}{4}pa^4, \quad C = \frac{1}{2}pa^2.$$

At $r = a$, we have $\tau_{\theta\theta} = p(1 - 2\cos 2\theta)$. Thus, at $\theta = 0$ and $\theta = \pi/2$, the values of $\tau_{\theta\theta}$ are $-p$ and $3p$, respectively, demonstrating the stress concentrations that arise due to the presence of the hole.

The results for a rigid inclusion are obtained by letting $\mu_1 \to \infty$:

$$A = \frac{1}{2}(1 - 2\nu_2)pa^2, \quad B = -\frac{1}{4(3 - 4\nu_2)}pa^4, \quad C = -\frac{1}{2(3 - 4\nu_2)}pa^2.$$

The extension of the above results when the cylindrical inclusion is allowed to slide may be found in [135]. The analytical solution for the case of a circular hole in a plate of finite thickness has been presented in [80, 100].

Now consider the case where the state of stress at infinity is one of pure shear (see Fig. 6.37). Outside the inclusion, we assume ϕ to be given by

$$\phi = -\frac{1}{2}pr^2 \sin 2\theta + A \log r + \left(\frac{B}{r^2} + C\right) \sin 2\theta,$$

which leads to

$$u_r = \frac{1}{2\mu_2}\left\{sr \sin 2\theta - \frac{A}{r} - \left[\frac{2B}{r^3} + \frac{4C(1 - \nu_2)}{r}\right] \sin 2\theta\right\},$$

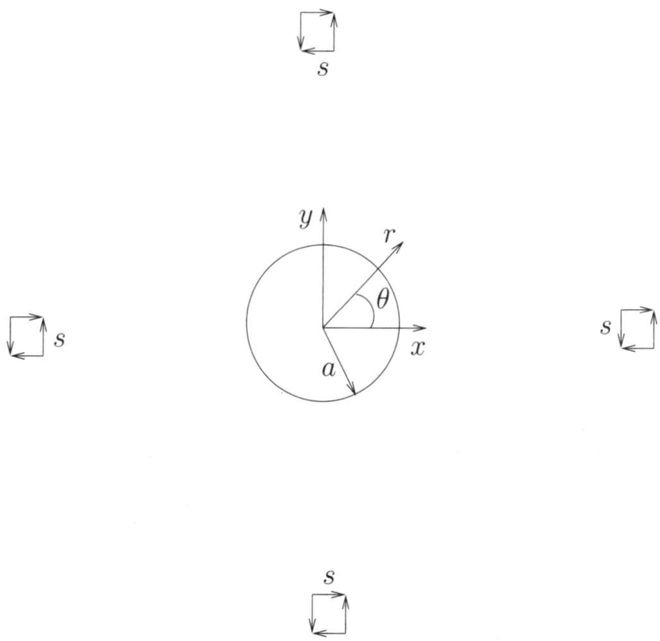

Fig. 6.37 *Cylindrical inclusion in an infinite domain with a state of pure shear at infinity.*

$$
\begin{aligned}
u_\theta &= \frac{1}{2\mu_2}\left[sr + \frac{2B}{r^3} - \frac{2C(1-2\nu_2)}{r}\right]\cos 2\theta, \\
\tau_{rr} &= s\sin 2\theta + \frac{A}{r^2} - \left(\frac{6B}{r^4} + \frac{4C}{r^2}\right)\sin 2\theta, \\
\tau_{\theta\theta} &= -s\sin 2\theta - \frac{A}{r^2} + \frac{6B}{r^4}\sin 2\theta, \\
\tau_{r\theta} &= s\cos 2\theta + \left(\frac{6B}{r^4} + \frac{2C}{r^2}\right)\cos 2\theta.
\end{aligned}
\tag{6.286}
$$

The Airy stress function for the material within the inclusion is assumed to be of the form

$$
\phi = Fr^2 + (Gr^2 + Hr^4)\sin 2\theta,
$$

which leads to the following displacement and stress fields:

$$
\begin{aligned}
u_r &= \frac{1}{\mu_1}\left[(1-2\nu_1)Fr - (Gr + 2\nu_1 Hr^3)\sin 2\theta\right], \\
u_\theta &= -\frac{1}{\mu_1}\left[Gr + (3-2\nu_1)Hr^3\right]\cos 2\theta, \\
\tau_{rr} &= 2F - 2G\sin 2\theta, \\
\tau_{\theta\theta} &= 2F + 2(G + 6Hr^2)\sin 2\theta, \\
\tau_{r\theta} &= -2(G + 3Hr^2)\cos 2\theta.
\end{aligned}
\tag{6.287}
$$

By imposing the boundary conditions given by Eqn. (6.285), we get

$$
\begin{aligned}
A &= F = H = 0, \\
B &= \frac{(\mu_1 - \mu_2)sa^4}{2\left[\mu_2 - (3-4\nu_2)\mu_1\right]}, \\
C &= \frac{(\mu_2 - \mu_1)sa^2}{\mu_2 - (3-4\nu_2)\mu_1}, \\
G &= \frac{(1-2\nu_2)\mu_1 s}{\mu_2 - (3-4\nu_2)\mu_1}.
\end{aligned}
$$

By letting μ_1 tend to zero, we get the results for a cavity as $B = -sa^4/2$ and $C = sa^2$, while by letting μ_1 tend to infinity, we get the results for a rigid inclusion as $B = -sa^4/(6-8\nu_2)$ and $C = sa^2/(3-4\nu_2)$.

Consider the case where a bending moment is applied in the far-field, i.e., $\tau_{xx} \to qy$ as $r \to \infty$, as shown in Fig. 6.38. The Airy stress functions in the inclusion and matrix are assumed to be given by

$$
\begin{aligned}
\phi^{(1)} &= r^3(F_2\sin 3\theta + J_2\sin\theta), \\
\phi^{(2)} &= r^3(B_2\sin 3\theta + D_2\sin\theta) + \frac{1}{r}\left[\tilde{B}_2\sin\theta + \tilde{D}_2\sin 3\theta\right] + \frac{\tilde{B}_4}{r^3}\sin 3\theta,
\end{aligned}
$$

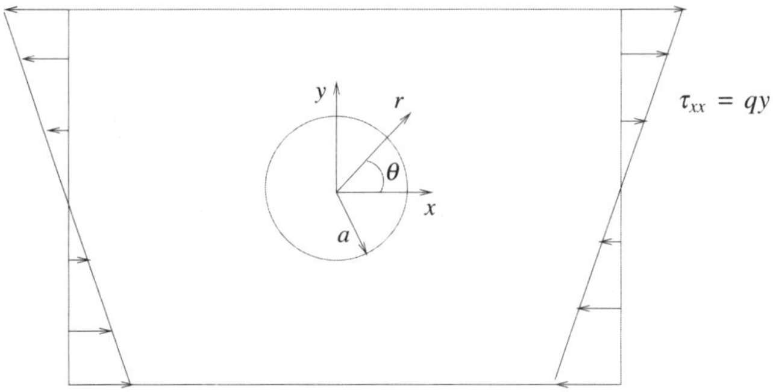

Fig. 6.38 *Cylindrical inclusion in an infinite domain with a bending moment at infinity.*

which lead to the following displacement and stress fields (with $u^{(1)}\big|_{r=0} = 0$):

$$
\begin{aligned}
E_1 u_r^{(1)} &= J_2(1-3\nu_1)r^2 \sin\theta - 3F_2(1+\nu_1)r^2 \sin 3\theta, \\
E_1 u_\theta^{(1)} &= -J_2(5+\nu_1)r^2 \cos\theta - 3F_2(1+\nu_1)r^2 \cos 3\theta, \\
\tau_{rr}^{(1)} &= 2r\left[J_2 \sin\theta - 3F_2 \sin 3\theta\right], \\
\tau_{\theta\theta}^{(1)} &= 6r\left[J_2 \sin\theta + F_2 \sin 3\theta\right], \\
\tau_{r\theta}^{(1)} &= -2r\left[J_2 \cos\theta + 3F_2 \cos 3\theta\right], \\
E_2 u_r^{(2)} &= k\sin\theta + \frac{1}{r^2}\left[\tilde{B}_2(1+\nu_2) + D_2(1-3\nu_2)r^4\right]\sin\theta \\
&\quad + \frac{1}{r^4}\left[3\tilde{B}_4(1+\nu_2) + \tilde{D}_2(5+\nu_2)r^2 - 3B_2(1+\nu_2)r^6\right]\sin 3\theta, \\
E_2 u_\theta^{(2)} &= k\cos\theta - \frac{1}{r^2}\left[\tilde{B}_2(1+\nu_2) + D_2(5+\nu_2)r^4\right]\cos\theta \\
&\quad - \frac{1}{r^4}\left[3\tilde{B}_4(1+\nu_2) + \tilde{D}_2(3\nu_2-1)r^2 + 3B_2(1+\nu_2)r^6\right]\cos 3\theta, \\
\tau_{rr}^{(2)} &= -\frac{2}{r^5}\left[r^2(\tilde{B}_2 - D_2 r^4)\sin\theta + (6\tilde{B}_4 + 5\tilde{D}_2 r^2 + 3B_2 r^6)\sin 3\theta\right], \\
\tau_{\theta\theta}^{(2)} &= \frac{2}{r^5}\left[r^2(\tilde{B}_2 + 3D_2 r^4)\sin\theta + (6\tilde{B}_4 + \tilde{D}_2 r^2 + 3B_2 r^6)\sin 3\theta\right], \\
\tau_{r\theta}^{(2)} &= \frac{2}{r^5}\left[r^2(\tilde{B}_2 - D_2 r^4)\cos\theta + 3(2\tilde{B}_4 + \tilde{D}_2 r^2 - B_2 r^6)\cos 3\theta\right],
\end{aligned}
$$

where the term associated with k represents a rigid displacement. By imposing the boundary conditions given by Eqn. (6.285) and the far-field conditions given by

$$
\begin{aligned}
\lim_{r\to\infty} \tau_{rr}^{(2)} &= qr\sin\theta\cos^2\theta, \\
\lim_{r\to\infty} \tau_{r\theta}^{(2)} &= -qr\sin^2\theta\cos\theta,
\end{aligned}
$$

we get the plane stress solution as

$$B_2 = -\frac{q}{24}, \qquad \tilde{B}_2 = \frac{qa^4}{8}\left[1 - \frac{4E_1}{(1+\nu_2)E_1 + (3-\nu_1)E_2}\right],$$

$$D_2 = \frac{q}{8}, \qquad \tilde{D}_2 = \frac{qa^4}{8}\left[1 - \frac{4E_1}{(3-\nu_2)E_1 + (1+\nu_1)E_2}\right],$$

$$\tilde{B}_4 = \frac{qa^6}{12}\left[\frac{(1+\nu_2)E_1 - (1+\nu_1)E_2}{(3-\nu_2)E_1 + (1+\nu_1)E_2}\right], \quad F_2 = -\frac{qE_1}{6\left[(3-\nu_2)E_1 + (1+\nu_1)E_2\right]},$$

$$J_2 = \frac{qE_1}{2\left[(1+\nu_2)E_1 + (3-\nu_1)E_2\right]}, \qquad k = \frac{qa^2}{4}\left[1 + \nu_2 - \frac{4(1+\nu_1)E_2}{(1+\nu_2)E_1 + (3-\nu_1)E_2}\right].$$

Note that the x-axis ($\theta = 0$) represents the neutral axis along which $u_r^{(1)}, u_r^{(2)}, \tau_{rr}^{(1)}, \tau_{rr}^{(2)}$ are all zero. If the inner inclusion is replaced by a hole, we let $E_1, \nu_1 \to 0$, and add a rigid displacement so that now $u_y\big|_{r=a} = 0$. The plane stress solution in the presence of a hole is thus given by (denoting (E_2, ν_2) by (E, ν))

$$\begin{aligned}
\frac{8Er^4 u_r}{q} &= r^2\left[(1+\nu)a^4 + (1-3\nu)r^4 + 2(1+\nu)a^2r^2\right]\sin\theta \\
&\quad + \left[(5+\nu)a^4r^2 + (1+\nu)r^6 - 2a^6(1+\nu)\right]\sin 3\theta, \\
\frac{8Er^4 u_\theta}{q} &= -r^2\left[(1+\nu)a^4 + (5+\nu)r^4 - 2(1+\nu)a^2r^2\right]\cos\theta \\
&\quad + \left[(1-3\nu)a^4r^2 + (1+\nu)r^6 + 2a^6(1+\nu)\right]\cos 3\theta, \\
\frac{4r^5\tau_{rr}}{q} &= r^2\left(r^4 - a^4\right)\sin\theta + (a^2 - r^2)(4a^4 - a^2r^2 - r^4)\sin 3\theta, \\
\frac{4r^5\tau_{\theta\theta}}{q} &= r^2\left(a^4 + 3r^4\right)\sin\theta - (4a^6 - a^4r^2 + r^6)\sin 3\theta, \\
\frac{4r^5\tau_{r\theta}}{q} &= r^2\left(a^4 - r^4\right)\cos\theta - (a^2 - r^2)(4a^4 + a^2r^2 + r^4)\cos 3\theta.
\end{aligned}$$

Note that $\tau_{\theta\theta}\big|_{r=a} = qa(\sin\theta - \sin 3\theta)$. If the inclusion is rigid, we take the limit as $E_1 \to \infty$ in the general solution. In this case, $\boldsymbol{u}\big|_{r=a} = \mathbf{0}$, and the stresses in the matrix at $r = a$ are

$$\tau_{rr} = qa\left[\frac{\sin\theta}{1+\nu} + \frac{\sin 3\theta}{3-\nu}\right],$$

$$\tau_{\theta\theta} = qa\nu\left[\frac{\sin\theta}{1+\nu} + \frac{\sin 3\theta}{3-\nu}\right],$$

$$\tau_{r\theta} = -qa\left[\frac{\cos\theta}{1+\nu} - \frac{\cos 3\theta}{3-\nu}\right].$$

Now consider an elliptical hole $\xi = \xi_0$ in an infinite plate, with a uniform state of stress at infinity. For the case of simple tension p at infinity (see Fig. 6.39), the boundary conditions are

$$\tau_{\xi\xi}\big|_{\xi\to\infty} = \frac{p}{2}(1 + \cos 2\eta),$$

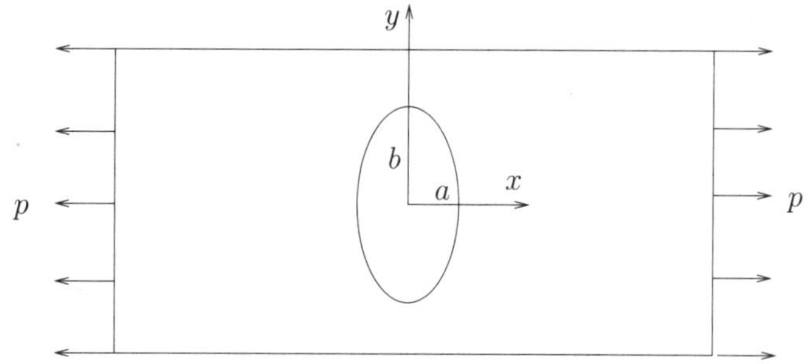

Fig. 6.39 *Elliptical hole in an infinite domain with a state of simple tension at infinity.*

$$\tau_{\eta\eta}\big|_{\xi\to\infty} = \frac{p}{2}(1-\cos 2\eta),$$
$$\tau_{\xi\eta}\big|_{\xi\to\infty} = -\frac{p}{2}\sin 2\eta,$$
$$\tau_{\xi\xi}\big|_{\xi=\xi_0} = \tau_{\xi\eta}\big|_{\xi=\xi_0} = 0.$$

Guided by the boundary conditions, and using Eqn. (6.243) with coefficients c_1, A_2, C_2, $\tilde{A}_1$ and $\tilde{C}_1$ as nonzero, we have (after renaming the constants) [139]

$$\phi = A_1(e^{2\xi}+\cos 2\eta) + A_2(e^{-2\xi}+\cos 2\eta) + A_3 e^{2\xi}\cos 2\eta + A_4 e^{-2\xi}\cos 2\eta + A_5\xi.$$

Using the boundary conditions, we get

$$A_1 = \frac{pc^2}{16}, \qquad A_2 = \frac{pc^2(2e^{2\xi_0}-1)}{16},$$
$$A_3 = -\frac{pc^2}{16}, \qquad A_4 = -\frac{pc^2e^{4\xi_0}}{16},$$
$$A_5 = -\frac{pc^2\sinh^2\xi_0}{2}.$$

The stresses are now recovered using Eqn. (6.222). We find that

$$\tau_{\eta\eta}\big|_{\xi=\xi_0} = \frac{p\left(\sinh 2\xi_0 - e^{2\xi_0}\cos 2\eta + 1\right)}{\cosh 2\xi_0 - \cos 2\eta},$$

which leads to

$$\tau_{\eta\eta}\big|_{\xi=\xi_0,\eta=0} = -p; \quad \tau_{\eta\eta}\big|_{\xi=\xi_0,\eta=\pi/2} = p(1+2\tanh\xi_0) = p\left(1+\frac{2b}{a}\right),$$

where the last relation has been obtained using $a = c\cosh\xi_0$ and $b = c\sinh\xi_0$. Thus, magnitude-wise, at the point $(0,b)$, the stress is more by a factor of $1+2b/a$ compared to the stress at $(a,0)$; for a circle, it yields the stress concentration factor of 3 that we have already derived earlier.

By using the relations $a = c\cosh\xi_0$, $b = c\sinh\xi_0$ and $c^2 = a^2 - b^2$, we get

$$\begin{aligned}\sinh 2\xi_0 &= \frac{2ab}{a^2 - b^2},\\ \cosh 2\xi_0 &= \frac{a^2 + b^2}{a^2 - b^2}.\end{aligned} \tag{6.288}$$

In the limit as $b \to 0$, we get $\xi_0 = 0$, so that by transforming the stresses to the Cartesian system using Eqn. (D.3), we get $\tau_{xx} = p$, and $\tau_{yy} = \tau_{xy} = 0$. Thus the introduction of a crack in the direction of the uniaxial loading does not alter the solution in the uncracked domain.

The displacement field in the plane stress case is given by [30, 87, 311]

$$\begin{aligned}\frac{8Eh_1 e^{2\xi_0} u_\xi}{pc^2} &= (1+\nu)\left[1 - e^{6\xi_0 - 2\xi}\cos 2\eta\right]\\ &\quad + e^{2\xi_0}\left\{\left[(1+\nu)e^{2\xi} - 4\right]\cos 2\eta - 2\left[1 + \nu - (1-\nu)\cosh 2\xi\right]\right\}\\ &\quad + e^{4\xi_0}\left\{1 + \nu + 4\cos 2\eta - 2(1-\nu)e^{-2\xi}\right\},\\ \frac{4Eh_1 e^{-2\xi_0} u_\eta}{pc^2} &= -\left[1 - \nu + (1+\nu)\cosh 2(\xi - \xi_0)\right]\sin 2\eta,\end{aligned} \tag{6.289}$$

where h_1 is given by Eqn. (D.2), while in the plane strain case, it is given by

$$\begin{aligned}\frac{16\mu h_1 e^{2\xi_0} u_\xi}{pc^2} &= 1 - e^{6\xi_0 - 2\xi}\cos 2\eta\\ &\quad + e^{2\xi_0}\left\{\left[e^{2\xi} - 4(1-\nu)\right]\cos 2\eta - 2 + 2(1-2\nu)\cosh 2\xi\right\}\\ &\quad + e^{4\xi_0}\left[1 + 4(1-\nu)\cos 2\eta - 2(1-2\nu)e^{-2\xi}\right],\\ \frac{8\mu h_1 e^{-2\xi_0} u_\eta}{pc^2} &= -\left[1 - 2\nu + \cosh 2(\xi - \xi_0)\right]\sin 2\eta.\end{aligned}$$

The above results may be verified using Eqns. (D.10).

If the far-field loading is equi-biaxial, i.e., $\tau_{xx} = \tau_{yy} = p$ as $\xi \to \infty$, then the stress function can be taken to be of the form

$$\phi = A_1(e^{2\xi} + \cos 2\eta) + A_2(e^{-2\xi} + \cos 2\eta) + A_3\xi.$$

The final solution in the plane stress case is

$$\begin{aligned}\frac{2Eh_1 u_\xi}{pc^2} &= (1+\nu)\cosh 2\xi_0 + (1-\nu)\cosh 2\xi - 2\cos 2\eta,\\ u_\eta &= 0,\end{aligned}$$

$$\tau_{\xi\xi} = \frac{pc^4}{4h_1^4}\left[\cosh 2\xi - \cosh 2\xi_0\right]\sinh 2\xi,$$

$$\tau_{\eta\eta} = \frac{pc^4}{4h_1^4}\left[\cosh 2\xi + \cosh 2\xi_0 - 2\cos 2\eta\right]\sinh 2\xi, \tag{6.290}$$

$$\tau_{\xi\eta} = \frac{pc^4}{4h_1^4}\left[\cosh 2\xi - \cosh 2\xi_0\right]\sin 2\eta,$$

while in the plane strain case, $u_\eta = 0$ and

$$\frac{4\mu h_1 u_\xi}{pc^2} = \cos 2\xi_0 + (1-2\nu)\cosh 2\xi - 2(1-\nu)\cos 2\eta,$$

with the expressions for the stresses the same.

The stresses on the boundary of the elliptical hole $\xi = \xi_0$ are $\tau_{\xi\xi} = \tau_{\xi\eta} = 0$, and

$$\tau_{\eta\eta}\big|_{\xi=\xi_0} = \frac{2p\sinh 2\xi_0}{\cosh 2\xi_0 - \cos 2\eta}.$$

The expressions for $\sinh 2\xi_0$ and $\cosh 2\xi_0$ given by Eqn. (6.288) can be substituted into the above expression for $\tau_{\eta\eta}\big|_{\xi=\xi_0}$ to obtain it in terms of a and b. The maximum value on this boundary is attained at $\eta = 0, \pi$, and is given by $2pa/b$.

By taking the limit as $b \to 0$ (i.e., $\xi_0 \to 0$), and by superimposing the solution corresponding to the uniform stress field $\tau_{xx} = -p$, $\tau_{yy} = \tau_{xy} = 0$, we get the following solution for a crack of length $2a$ along the x-axis subjected to a far-field loading $\tau_{yy} = p$, $\tau_{xx} = \tau_{xy} = 0$ under plane stress conditions (see Fig. 6.40):

$$Eu_x = p\left[\frac{a^3\sinh\xi\cos\eta\left[1+\nu-2\cos 2\eta+(1-\nu)\cosh 2\xi\right]}{2h_1^2} - x\right],$$

$$Eu_y = p\left[\frac{a^3\cosh\xi\sin\eta\left[1+\nu-2\cos 2\eta+(1-\nu)\cosh 2\xi\right]}{2h_1^2} + \nu y\right],$$

$$\tau_{xx} = p\left\{\frac{a^6}{32h_1^6}\left[(9-4\cos 2\eta)\sinh 2\xi + 2(\cos 4\eta - 3\cos 2\eta)\sinh 4\xi + \sinh 6\xi\right] - 1\right\},$$

$$\tau_{yy} = -\frac{pa^6\sinh^3\xi\cosh\xi}{2h_1^6}\left[\cos 2\eta + \cos 4\eta - 2\cosh^2\xi\right],$$

$$\tau_{xy} = -\frac{pa^6\sinh^2\xi}{8h_1^6}\left[2(2+\cosh 2\xi)\sin 2\eta - (1+2\cosh 2\xi)\sin 4\eta\right], \tag{6.291}$$

where $h_1^2 = a^2(\cosh 2\xi - \cos 2\eta)/2$ and $\xi + i\eta = \cosh^{-1}[(x+iy)/a]$.

In particular, to obtain the solution along the x-axis, we set $\xi = 0$ and use $a\sin\eta = \pm\sqrt{x^2-a^2}$ for $|x| < a$, $y = 0^{\pm}$ (corresponding to the upper and lower faces of the crack),

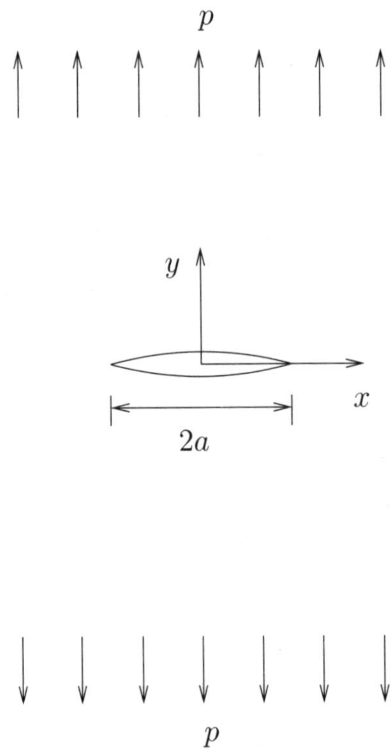

Fig. 6.40 *Crack subjected to a uniform far-field loading orthogonal to the crack.*

and set $\eta = 0, \pi$ and use $\cosh\xi = |x|/a$ for $|x| \geq a$, $y = 0$. With $\operatorname{sgn}(x)$ denoting the sign of x, the summary of the final results is

$$\left.\begin{aligned} Eu_x &= -px, \\ Eu_y &= \pm 2p\sqrt{a^2 - x^2}, \\ \tau_{xx} &= -p, \\ \tau_{yy} &= \tau_{xy} = 0, \end{aligned}\right\} \quad |x| < a, y = 0^{\pm},$$

$$\left.\begin{aligned} u_y &= \tau_{xy} = 0, \\ Eu_x &= \operatorname{sgn}(x)p\left[(1-\nu)\sqrt{x^2 - a^2} - |x|\right], \\ \tau_{xx} &= \frac{p\,|x|}{\sqrt{x^2 - a^2}} - p, \\ \tau_{yy} &= \frac{p\,|x|}{\sqrt{x^2 - a^2}}, \end{aligned}\right\} \quad |x| \geq a, y = 0. \tag{6.292}$$

Note the singularity in the stress solution as $|x| \to a$. Along the y-axis, we get

$$\left.\begin{aligned} u_x &= \tau_{xy} = 0, \\ Eu_y &= \text{sgn}(y)p\left\{\nu\,|y| + \frac{1}{\sqrt{y^2+a^2}}\left[(1+\nu)a^2 + (1-\nu)(y^2+a^2)\right]\right\}, \\ \tau_{xx} &= -\frac{pa}{\sqrt{a^2+y^2}}\left[\frac{a}{\sqrt{y^2+a^2}} - \frac{|y|}{a}\right]\left[1 - \frac{|y|}{\sqrt{y^2+a^2}}\right], \\ \tau_{yy} &= \frac{p\,|y|^3}{(y^2+a^2)^{3/2}}, \end{aligned}\right\} \quad x = 0.$$

By superimposing a uniform state of stress $\tau_{\xi\xi} = -p$, $\tau_{\eta\eta} = -p$ and $\tau_{\xi\eta} = 0$ on the solution given by Eqn. (6.290), we get the following plane-stress solution for an elliptical hole loaded by a uniform pressure p with zero stresses at infinity:

$$\begin{aligned} \frac{2Eh_1u_\xi}{pc^2} &= (1+\nu)\cosh 2\xi_0 + (1-\nu)e^{-2\xi} - 2\cos 2\eta, \\ \frac{2Eh_1u_\eta}{pc^2} &= (1-\nu)\sin 2\eta, \\ \tau_{\xi\xi} &= \frac{pc^4}{4h_1^4}\left[\cosh 2\xi - \cosh 2\xi_0\right]\sinh 2\xi - p, \\ \tau_{\eta\eta} &= \frac{pc^4}{4h_1^4}\left[\cosh 2\xi + \cosh 2\xi_0 - 2\cos 2\eta\right]\sinh 2\xi - p, \\ \tau_{\xi\eta} &= \frac{pc^4}{4h_1^4}\left[\cosh 2\xi - \cosh 2\xi_0\right]\sin 2\eta. \end{aligned}$$

For the plane strain case, we get

$$\begin{aligned} \frac{4\mu h_1u_\xi}{pc^2} &= \cosh 2\xi_0 + (1-2\nu)e^{-2\xi} - 2(1-\nu)\cos 2\eta, \\ \frac{4\mu h_1u_\eta}{pc^2} &= (1-2\nu)\sin 2\eta. \end{aligned}$$

The generalization to the case where the domain is a confocal elliptic ring, which we consider towards the end of this section, may be found in [15].

Now consider the case where the state of stress at infinity is one of pure shear. Based on the boundary conditions given by

$$\begin{aligned} \tau_{\xi\xi}\big|_{\xi\to\infty} &= s\sin 2\eta, \\ \tau_{\eta\eta}\big|_{\xi\to\infty} &= -s\sin 2\eta, \\ \tau_{\xi\eta}\big|_{\xi\to\infty} &= s\cos 2\eta, \\ \tau_{\xi\xi}\big|_{\xi=\xi_0} &= \tau_{\xi\eta}\big|_{\xi=\xi_0} = 0, \end{aligned} \tag{6.293}$$

we choose the Airy stress function to be

$$\phi = A_1 \sin 2\eta + A_2 e^{2\xi} \sin 2\eta + A_3 e^{-2\xi} \sin 2\eta.$$

Using the boundary conditions, we get

$$A_1 = \frac{sc^2 e^{2\xi_0}}{4}, \quad A_2 = -\frac{sc^2}{8}, \quad A_3 = -\frac{sc^2 e^{4\xi_0}}{8}.$$

The stress at the edge of the hole is now given by

$$\tau_{\eta\eta}\big|_{\xi=\xi_0} = -\frac{2se^{2\xi_0} \sin 2\eta}{\cosh 2\xi_0 - \cos 2\eta},$$

so that the value of $\tau_{\eta\eta}\big|_{\xi=\xi_0}$ at $\eta = 0, \pi/2$ is zero.

The displacement field in the plane stress case is

$$\begin{aligned} \frac{2Eh_1 e^{-2\xi_0} u_\xi}{sc^2} &= (1+\nu) \sinh 2(\xi - \xi_0) \sin 2\eta, \\ \frac{2Eh_1 e^{-2\xi_0} u_\eta}{sc^2} &= -2\cosh 2\xi + [1 - \nu + (1+\nu)\cosh 2(\xi - \xi_0)] \cos 2\eta, \end{aligned} \tag{6.294}$$

while in the plane strain case, it is given by

$$\begin{aligned} \frac{4\mu h_1 e^{-2\xi_0} u_\xi}{sc^2} &= \sinh 2(\xi - \xi_0) \sin 2\eta, \\ \frac{4\mu h_1 e^{-2\xi_0} u_\eta}{sc^2} &= -2(1-\nu)\cosh 2\xi + [1 - 2\nu + \cosh 2(\xi - \xi_0)] \cos 2\eta. \end{aligned}$$

By taking the limit as $b \to 0$ (i.e., $\xi_0 \to 0$), we get the following solution for a crack of length $2a$ along the x-axis subjected to a far-field loading $\tau_{xy} = s$, $\tau_{xx} = \tau_{yy} = 0$ under plane stress conditions:

$$\begin{aligned} Eu_x &= -\frac{sa^3 \cosh \xi \sin \eta}{2h_1^2} \left[1 + \nu + 2\cos 2\eta - (3+\nu)\cosh 2\xi\right], \\ Eu_y &= \frac{sa^3 \sinh \xi \cos \eta}{2h_1^2} \left[1 + \nu - 2\nu \cos 2\eta - (1-\nu)\cosh 2\xi\right], \\ \tau_{xx} &= -\frac{sa^6}{16h_1^6} \left[(6 - 2\cosh 2\xi + \cosh 4\xi)\sin 2\eta + (\cosh 4\xi - 5\cosh 2\xi)\sin 4\eta + \sin 6\eta\right], \\ \tau_{yy} &= -\frac{sa^6 \sinh^2 \xi}{8h_1^6} \left[2\left(\cosh 2\xi + 2\right)\sin 2\eta - (1 + 2\cosh 2\xi)\sin 4\eta\right], \\ \tau_{xy} &= \frac{sa^6}{32h_1^6} \left[(9 - 4\cos 2\eta)\sinh 2\xi + 2\left(\cos 4\eta - 3\cos 2\eta\right)\sinh 4\xi + \sinh 6\xi\right], \end{aligned} \tag{6.295}$$

where $h_1^2 = a^2(\cosh 2\xi - \cos 2\eta)/2$ and $\xi + i\eta = \cosh^{-1}[(x+iy)/a]$.

Similar to Eqn. (6.292), we get the solution along the x-axis as

$$\left.\begin{aligned} Eu_x &= \pm 2s\sqrt{a^2 - x^2}, \\ u_y &= \tau_{yy} = \tau_{xy} = 0, \\ \tau_{xx} &= \mp \frac{2s\,|x|}{\sqrt{a^2 - x^2}}, \end{aligned}\right\} \qquad |x| < a,\ y = 0^{\pm},$$

$$\left.\begin{aligned} u_x &= \tau_{xx} = \tau_{yy} = 0, \\ Eu_y &= \operatorname{sgn}(x)s(\nu - 1)\sqrt{x^2 - a^2}, \\ \tau_{xy} &= \frac{s\,|x|}{\sqrt{x^2 - a^2}}, \end{aligned}\right\} \qquad |x| \geq a,\ y = 0. \tag{6.296}$$

Again note the singularity in the stress solution as $|x| \to a$. Along the y-axis, we get

$$\left.\begin{aligned} u_y &= \tau_{xx} = \tau_{yy} = 0, \\ Eu_x &= -\frac{\operatorname{sgn}(y)s}{\sqrt{y^2 + a^2}}\left[(1+\nu)a^2 - (3+\nu)(a^2 + y^2)\right], \\ \tau_{xy} &= \frac{s\,|y|}{\sqrt{y^2 + a^2}}\left[1 + \frac{a^2}{y^2 + a^2}\right], \end{aligned}\right\} \qquad x = 0.$$

We now extend the above results to the case of an elliptic inclusion; for alternative approaches, see [66, 96, 114]. For the case of simple tension at infinity, one again assumes ϕ, for both the inclusion and the matrix, to be given by Eqn. (6.243) with only coefficients c_1, A_2, C_2, $\tilde{A}_1$ and $\tilde{C}_1$ as nonzero. By symmetry considerations, we obviously have $u_\eta\big|_{\eta=0} = u_\eta\big|_{\eta=\pi/2} = 0$. Since $x = c\cosh\xi\cos\eta$ and $y = c\sinh\xi\sin\eta$, the line $\xi = 0$, $\eta \in [0, 2\pi]$ is mapped to the straight line joining $(-c, 0)$ and $(c, 0)$. Using $[\boldsymbol{u}]_{\text{elliptic}} = \boldsymbol{Q}[\boldsymbol{u}]_{\text{Cartesian}}$ and the similar transformation for stresses, where $\boldsymbol{Q}$ is given by Eqn. (D.3), we get $u_\xi\big|_{\xi=0} = u_y\big|_{y=0,x\in[-c,c]} = 0$ and $\tau_{\xi\eta}\big|_{\xi=0} = \tau_{xy}\big|_{y=0,x\in[-c,c]} = 0$, within the inclusion. Enforcing these constraints, we get the form of the displacement field for the plane stress case within the inclusion as

$$\begin{aligned} E_1 h_1 u_\xi^{(1)} &= 2\sinh 2\xi\left[\tilde{A}_1(1-\nu_1) - A_2(1+\nu_1)\cos 2\eta\right], \\ E_1 h_1 u_\eta^{(1)} &= 2\sin 2\eta\left[-\tilde{A}_1(1-\nu_1) + A_2(1+\nu_1)\cosh 2\xi\right]. \end{aligned} \tag{6.297}$$

In the outer matrix, the form of the solution that meets the conditions at infinity is

$$\begin{aligned} 4E_2 h_1 u_\xi^{(2)} &= -4\tilde{c}_1(1+\nu_2) + 2(8\tilde{\tilde{A}}_1 - pc^2)\cos 2\eta - (1-\nu_2)\left[8\tilde{\tilde{A}}_1 e^{-2\xi} - pc^2\cosh 2\xi\right] \\ &\quad + (1+\nu_2)\cos 2\eta\left[pc^2\sinh 2\xi - 8\tilde{C}_2 e^{-2\xi}\right], \\ 4E_2 h_1 u_\eta^{(2)} &= -\sin 2\eta\left\{8\tilde{\tilde{A}}_1(1-\nu_2) + (1+\nu_2)\left[pc^2\cosh 2\xi + 8\tilde{C}_2 e^{-2\xi}\right]\right\}. \end{aligned} \tag{6.298}$$

The constants in the above solution, $\tilde{A}_1$, A_2, $\tilde{c}_1$, $\tilde{C}_2$ and $\tilde{\tilde{A}}_1$, are found so as to satisfy the boundary conditions

$$\begin{aligned} u_\xi^{(1)}\Big|_{\xi=\xi_0} &= u_\xi^{(2)}\Big|_{\xi=\xi_0}, \qquad u_\eta^{(1)}\Big|_{\xi=\xi_0} = u_\eta^{(2)}\Big|_{\xi=\xi_0}, \\ \tau_{\xi\xi}^{(1)}\Big|_{\xi=\xi_0} &= \tau_{\xi\xi}^{(2)}\Big|_{\xi=\xi_0}, \qquad \tau_{\xi\eta}^{(1)}\Big|_{\xi=\xi_0} = \tau_{\xi\eta}^{(2)}\Big|_{\xi=\xi_0}. \end{aligned} \tag{6.299}$$

Let

$$\begin{aligned} a_1 &= (1+\nu_2)E_1 - (1+\nu_1)E_2, \\ a_2 &= (1-\nu_2)E_1 - (1-\nu_1)E_2, \\ a_3 &= (1-\nu_2)E_1 + (1+\nu_1)E_2, \\ a_4 &= (1+\nu_2)E_1 + (1-\nu_1)E_2, \\ a_5 &= (1+\nu_2)E_1 + (3-\nu_1)E_2, \\ D &= 4E_1E_2\cosh 2\xi_0 + \left\{E_2^2 + [\nu_1 E_2 + (3-\nu_2)E_1]\,[(1+\nu_2)E_1 - \nu_1 E_2]\right\}\sinh 2\xi_0. \end{aligned}$$

Then the constants are given by

$$\begin{aligned} 4D\tilde{A}_1 &= pc^2 E_1\left[a_1 + a_3\cosh 2\xi_0 + 2E_1\sinh 2\xi_0\right], \\ 4DA_2 &= pc^2 E_1 e^{\xi_0}\left\{(a_2 - a_4)\sinh\xi_0 - 2E_1\cosh\xi_0\right\}, \\ 4D\tilde{c}_1 &= pc^2 a_2\sinh 2\xi_0\left[a_1 + a_3\cosh 2\xi_0 + 2E_1\sinh 2\xi_0\right], \\ 8D\tilde{C}_2 &= pc^2 e^{2\xi_0}\sinh 2\xi_0\left\{a_1\left[a_2\sinh 2\xi_0 - 2E_1\cosh 2\xi_0\right] - 2E_1 a_2\right\}, \\ 8D\tilde{\tilde{A}}_1 &= pc^2 e^{2\xi_0}\left\{E_1 a_5 + a_1\left[(\sinh 4\xi_0 - \cosh 4\xi_0)E_1 - a_4\sinh 2\xi_0\right]\right\}. \end{aligned}$$

The restriction $\nu_1 = \nu_2$ imposed in [66] is not essential. Using the relation $[\boldsymbol{\tau}]_{\text{Cart}} = \boldsymbol{Q}^T[\boldsymbol{\tau}]_{\text{ellip}}\boldsymbol{Q}$, we get the stresses within the inclusion expressed in Cartesian components to be constant ($\tau_{xx} = 4(\tilde{A}_1 - A_2)/c^2$, $\tau_{yy} = 4(\tilde{A}_1 + A_2)/c^2$ and $\tau_{xy=0}$) as reported in [96, 114]. By setting E_1 to zero, the results in Eqn. (6.289) are recovered, while the results for a rigid inclusion are obtained by letting $E_1 \to \infty$. The results for the plane strain case are recovered from the plane stress one by replacing E_1 and ν_1 by $E_1/(1-\nu_1^2)$ and $\nu_1/(1-\nu_1)$, respectively, with similar replacements for E_2 and ν_2.

If we have an equi-biaxial state of stress at infinity, i.e., $\tau_{xx} = \tau_{yy} = p$ as $\xi \to \infty$, then the form of the displacements within the inclusion is still given Eqn. (6.297), while in the matrix, instead of Eqn. (6.298), we have

$$\begin{aligned} 2E_2h_1u_\xi^{(2)} &= -2\tilde{c}_1(1+\nu_2) + 2(4\tilde{\tilde{A}}_1 + 2\tilde{A}_2(1+\nu_2)e^{-2\xi} - pc^2)\cos 2\eta \\ &\quad - (1-\nu_2)\left[4\tilde{\tilde{A}}_1 e^{-2\xi} - pc^2\cosh 2\xi\right], \\ 2E_2h_1u_\eta^{(2)} &= -4\left[\tilde{\tilde{A}}_1(1-\nu_2) - \tilde{A}_2(1+\nu_2)e^{-2\xi}\right]\sin 2\eta. \end{aligned} \tag{6.300}$$

The constants $\tilde{A}_1$, A_2, $\tilde{c}_1$, $\tilde{\tilde{A}}_1$ and $\tilde{A}_2$ are found subject to Eqns. (6.299), and are given by

$$\begin{aligned}
4De^{2\xi_0}\tilde{A}_1 &= pc^2E_1\left\{(1+\nu_1)(e^{4\xi_0}+1)E_2-\left[1+\nu_2-(3-\nu_2)e^{4\xi_0}\right]E_1\right\}\\
2DA_2 &= -pc^2E_1a_2,\\
2D\tilde{c}_1 &= pc^2a_2\sinh 2\xi_0\left[a_3\cosh 2\xi_0+2E_1\sinh 2\xi_0\right],\\
4De^{2\xi_0}\tilde{\tilde{A}}_1 &= pc^2E_1\left\{(1+\nu_2)(e^{4\xi_0}-1)E_1+\left[1+\nu_1+(3-\nu_1)e^{4\xi_0}\right]E_2\right\},\\
4D\tilde{A}_2 &= pc^2E_1(e^{4\xi_0}-1)a_2.
\end{aligned}$$

The constant stresses within the inclusion are given by $\tau_{xx}=4(\tilde{A}_1-A_2)/c^2$, $\tau_{yy}=4(\tilde{A}_1+A_2)/c^2$ and $\tau_{xy=0}$. The results of Eqn. (6.290) are recovered by setting E_1 to zero, and those for a rigid inclusion are recovered by taking the limit $E_1\to\infty$.

Finally, consider the case when the state of stress at infinity is one of pure shear. We now take only the constants k_3, B_2, D_2 and $\tilde{B}_1$ in Eqn. (6.243) (for both the inclusion and matrix) to be nonzero. Within the inclusion, by imposing the constraints that $u_\eta\big|_{\xi=0}=u_x\big|_{y=0,x\in[-c,c]}=0$ and $\tau_{\xi\xi}\big|_{\xi=0}=\tau_{yy}\big|_{y=0,x\in[-c,c]}=0$, we get for the plane stress case,

$$\begin{aligned}
E_1h_1u_\xi^{(1)} &= (1+\nu_1)(k_3-2D_2\cosh 2\xi)\sin 2\eta,\\
E_1h_1u_\eta^{(1)} &= (1+\nu_1)(k_3-2D_2\cos 2\eta)\sinh 2\xi.
\end{aligned}$$

Similarly, imposing the first three constraints in Eqn. (6.293), we get for the matrix,

$$\begin{aligned}
2E_2h_1u_\xi^{(2)} &= (1+\nu_2)\left[2\tilde{k}_3+sc^2\cosh 2\xi+4\tilde{B}_2e^{-2\xi}\right]\sin 2\eta,\\
2E_2h_1u_\eta^{(2)} &= 4\tilde{\tilde{B}}_1(1-\nu_2)\cos 2\eta-4\left[2\tilde{\tilde{B}}_1+(1+\nu_2)\tilde{B}_2\cos 2\eta\right]\cosh 2\xi\\
&\quad+(1+\nu_2)\left[2\tilde{k}_3+(sc^2+4\tilde{B}_2)\cos 2\eta\right]\sinh 2\xi.
\end{aligned}$$

By choosing $\tilde{k}_3=0$, the remaining constants k_3, $\tilde{B}_2$, D_2 and $\tilde{\tilde{B}}_1$ are obtained using the boundary conditions in Eqn. (6.299), and are given by

$$\begin{aligned}
k_3 &= \frac{sc^2e^{2\xi_0}E_1N\cosh 2\xi_0}{E_2(1+\nu_1)D},\\
\tilde{B}_2 &= \frac{sc^2e^{2\xi_0}N\sinh 4\xi_0}{8D},\\
D_2 &= -\frac{sc^2e^{2\xi_0}E_1}{2D},\\
\tilde{\tilde{B}}_1 &= -\frac{sc^2e^{2\xi_0}N\sinh 2\xi_0}{4D},
\end{aligned}$$

where

$$\begin{aligned}
N &= (1+\nu_2)E_1-(1+\nu_1)E_2,\\
D &= 2E_1\cosh 2\xi_0+\left[E_1(1-\nu_2)+(1+\nu_1)E_2\right]\sinh 2\xi_0.
\end{aligned}$$

The constant stress in the inclusion in the x–y system is given by $\tau_{xx} = \tau_{yy} = 0$ and $\tau_{xy} = -4D_2/c^2$. By setting E_1 to zero, we recover the solution in Eqn. (6.294).

Now consider the problem of a hollow elliptical pressure vessel with inner and outer surfaces $\xi = \xi_i$ and $\xi = \xi_o$ loaded by pressures p_i and p_o, respectively. In this case, we choose only the even A_n and C_n, and odd $\tilde{A}_n$ and $\tilde{C}_n$ in Eqn. (6.243), i.e.,

$$\begin{aligned}\phi = c_1\xi + \sum_{n=1}^{\infty} & A_n \cosh 2n\xi \cos 2n\eta + C_n \sinh 2n\xi \cos 2n\eta \\ & + \tilde{A}_n \left[\cosh 2n\xi \cos(2n-2)\eta + \cosh(2n-2)\xi \cos 2n\eta\right] \\ & + \tilde{C}_n \left[\sinh 2n\xi \cos(2n-2)\eta + \sinh(2n-2)\xi \cos 2n\eta\right].\end{aligned} \tag{6.301}$$

The constants are found using the boundary conditions

$$\tau_{\xi\xi}\big|_{\xi=\xi_1} = -p_i, \qquad \tau_{\xi\xi}\big|_{\xi=\xi_2} = -p_o, \tag{6.302a}$$

$$\tau_{\xi\eta}\big|_{\xi=\xi_1} = 0, \qquad \tau_{\xi\eta}\big|_{\xi=\xi_2} = 0, \tag{6.302b}$$

where the stresses are given by Eqns. (6.222). While it is possible to derive some recurrence relations using Eqns. (6.301) and (6.302b), it is perhaps easier to solve for the constants directly using, say, a weighted residual technique as described on page 485.

If the elliptical domain is simply connected (i.e., it does not have any holes in it), and if the loading on the boundary $\xi = \xi_0$ is symmetric about the x-axis (for example, as in the case of a problem analogous to that shown in Fig. 6.60 below), then keeping in view that $u_\xi\big|_{\xi=0} = u_y\big|_{y=0, x\in[-c,c]} = 0$, the constants c_1, C_n and $\tilde{C}_n$ in the above solution should be set to zero. If the region of loading 2α is small, and the semi-major and semi-minor axes a and b are taken to be almost equal, then one obtains results that are approximately the same as those obtained using Eqn. (6.372), while if one takes $b \ll a$, then the results are approximately the same as those for a rectangular bar subjected to uniaxial tension.

Finally, consider the problem of cylinder whose cross-section is bounded by two non-concentric circles $\xi = \xi_1$ and $\xi = \xi_2$, which is subjected to the boundary conditions given by Eqns. (6.302a) and (6.302b) with ξ and η now denoting bipolar coordinates. We take only the c_1, A_1, B_1 and C_1 terms in Eqn. (6.247) to be nonzero. Using the said boundary conditions, we get

$$\begin{aligned} c_1 &= \frac{c}{D} 2(p_i - p_o), \\ A_1 &= \frac{c}{D}(p_o - p_i)\,\mathrm{sech}\,(\xi_1 - \xi_2) \sinh(\xi_1 + \xi_2), \\ B_1 &= \frac{c}{D}(p_i - p_o)\,\mathrm{sech}\,(\xi_1 - \xi_2) \cosh(\xi_1 + \xi_2), \\ C_1 &= \frac{c}{D}\left[p_i \sinh 2\xi_2 - p_o \sinh 2\xi_1 + (p_i + p_o)\tanh(\xi_1 - \xi_2)\right]. \end{aligned}$$

where

$$D = \sinh 2\xi_1 - \sinh 2\xi_2 - 2\tanh(\xi_1 - \xi_2).$$

The displacements are obtained via Eqns. (6.248), and the stress distribution is [149, 311]

$$
\begin{aligned}
c\tau_{\xi\xi} &= -c_1 \sinh\xi(\cosh\xi - \cos\eta) + A_1(\cosh 2\xi - 2\sinh 2\xi \sinh\xi \cos\eta) \\
&\quad + 2B_1 \sinh\xi(\cosh\xi - \cosh 2\xi \cos\eta) + C_1, \\
c\tau_{\eta\eta} &= c_1 \sinh\xi(\cosh\xi - \cos\eta) + A_1 \left[4\cosh^3\xi \cos\eta - (1 + 2\cos 2\eta)\cosh 2\xi\right] \\
&\quad + B_1 \left[2(2 + \cosh 2\xi)\sinh\xi \cos\eta - (1 + 2\cos 2\eta)\sinh 2\xi\right] + C_1, \\
c\tau_{\xi\eta} &= -(\cosh\xi - \cos\eta)\sin\eta \left[c_1 - 2A_1 \sinh 2\xi - 2B_1 \cosh 2\xi\right].
\end{aligned}
$$

6.6.2 Rectangular domain (e.g., cantilever beam) loaded by tractions on its edges

Consider a rectangular domain that is subjected to some distribution of prescribed tractions and displacements on its edges. A note of caution before we start the discussion. 'Cantilever support' boundary conditions ($\boldsymbol{u} = \boldsymbol{0}$ at a wall) lead to stress singularities at the corner points where the surfaces of the cantilever meet the wall (see, e.g., [101]), although the displacements and the strain energy are bounded (thus, the singular solutions are admissible from a theoretical viewpoint). This singularity arises due to the conflict between the displacement and traction boundary conditions at these points. In such singularity dominated problems, in order to calculate the coefficients in the series solution, the choice of an appropriate strategy such as the weighted residual method described on page 485 becomes absolutely critical in order to obtain a converged displacement solution.

First consider the case where the top and bottom edges are free of traction, while the right edge is subjected to tractions (see Fig. 6.41). Thus, the boundary conditions are

$$
\begin{aligned}
\tau_{xy}\big|_{y=\pm h/2} &= 0, \quad \tau_{xx}\big|_{x=L} = f(y), \\
\tau_{yy}\big|_{y=\pm h/2} &= 0, \quad \tau_{xy}\big|_{x=L} = g(y),
\end{aligned}
\tag{6.303}
$$

in addition to suitable displacement boundary conditions on the left edge (see below). Most of the literature on this subject either considers a semi-infinite rectangular domain (so that boundary conditions on one edge can be avoided), or in case a finite domain is considered, the boundary conditions on some edges edge are satisfied in a weak sense [77, 216].

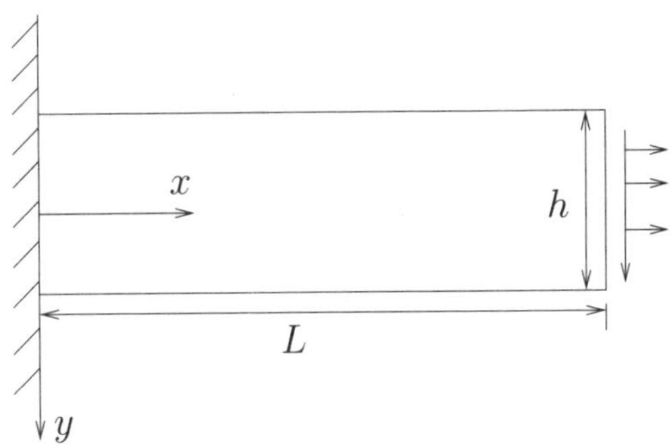

Fig. 6.41 *Rectangular domain with top and bottom edges traction-free.*

We try to remedy this situation in what follows. We do not follow Mathieu's solution method (see [214] for a description) since, although it avoids using complex-valued roots of characteristic equations, the number of terms needed to approximate the solution increases enormously.

From Eqns. (6.220) and (6.235), we see that the homogeneous boundary conditions are satisfied if the Airy stress function is of the form[13]

$$\begin{aligned}\phi = \sum_{n=1}^{\infty} [A_n \sinh \lambda_n x + B_n \cosh \lambda_n x] \left[\left(\frac{h}{2} - y \right) \sin \lambda_n \left(\frac{h}{2} + y \right) \right. \\ \left. - \left(\frac{h}{2} + y \right) \sin \lambda_n \left(\frac{h}{2} - y \right) \right] + c_7 y^3, \qquad [(f(y), g(y)) \text{ odd/even}]\end{aligned} \tag{6.304}$$

provided λ_n are chosen to be the (complex-valued) roots of the characteristic equation

$$\sin \lambda_n h = \lambda_n h, \tag{6.305}$$

and

$$c_7 = \frac{2}{h^3} \int_{-h/2}^{h/2} y f(y)\, dy, \tag{6.306}$$

or, alternatively, if the stress function is of the form

$$\begin{aligned}\phi = \sum_{n=1}^{\infty} [A_n \sinh \gamma_n x + B_n \cosh \gamma_n x] \left[\left(\frac{h}{2} - y \right) \sin \gamma_n \left(\frac{h}{2} + y \right) \right. \\ \left. + \left(\frac{h}{2} + y \right) \sin \gamma_n \left(\frac{h}{2} - y \right) \right] + c_3 y^2, \qquad [(f(y), g(y)) \text{ even/odd}]\end{aligned} \tag{6.307}$$

where γ_n are the (complex-valued) roots of the characteristic equation

$$\sin \gamma_n h = -\gamma_n h, \tag{6.308}$$

and

$$c_3 = \frac{1}{2h} \int_{-h/2}^{h/2} f(y)\, dy. \tag{6.309}$$

Note that the coefficients A_n and B_n in the above equations are also complex-valued[14]. The asymptotic values of the roots of Eqns. (6.305) and (6.308) [123, 270] are obtained as the special case $a = 1$ of the asymptotic complex-valued roots of

$$\sin(\lambda h) = a \lambda h, \tag{6.310a}$$

$$\sin(\gamma h) = -a \gamma h, \tag{6.310b}$$

[13] See page 484 for the case where tractions are applied on both faces $x = \pm L/2$.

[14] The use of complex-valued quantities can be avoided by considering the real and imaginary parts in Eqns. (6.304) and (6.307). However, such a procedure leads to very cumbersome expressions, especially for the displacements.

$a > 0$, which are given by [261][15]

$$\lambda_n \sim \frac{1}{h}\left[\alpha_n - \frac{1}{\alpha_n}\log(2a\alpha_n) + i\log(2a\alpha_n)\right], \tag{6.311a}$$

$$\gamma_n \sim \frac{1}{h}\left[\beta_n - \frac{1}{\beta_n}\log(2a\beta_n) + i\log(2a\beta_n)\right], \tag{6.311b}$$

where $\alpha_n = (4n+1)\pi/2$ and $\beta_n = (4n-1)\pi/2$, $n = 1, 2, \ldots, \infty$. These values can be used as a first guess in the following Newton–Raphson based strategy for finding the roots of Eqns. (6.310):

$$z_n^{\text{new}} = z_n^{\text{prev}} - \left[\frac{\sin z_n - az_n}{\cos z_n - a}\right]_{z_n = z_n^{\text{prev}}},$$

$$z_n^{\text{new}} = z_n^{\text{prev}} - \left[\frac{\sin z_n + az_n}{\cos z_n + a}\right]_{z_n = z_n^{\text{prev}}},$$

where $z_n := \lambda_n h$ and $\gamma_n h$, respectively, in the above two equations. Using this approach, the first twenty roots of $\sin z = z$ and $\sin z = -z$ are presented in Tables 6.1 and 6.2[16]. Since the roots are symmetrically located in the four quadrants of the complex plane, only the roots in the first quadrant are given.

The stress functions in Eqns. (6.304) and (6.307) yield (τ_{xx}, τ_{xy}) that is odd/even and even/odd in y, respectively, and hence should be used when $(f(y), g(y))$ is odd/even and even/odd in y, respectively. The general case of an arbitrary $(f(y), g(y))$ can be tackled by superposing these two results since, for example, the case of $(f(y), g(y))$ being odd/odd can be thought of as a superposition of the cases odd/zero and zero/odd, and so on. The displacements and stresses corresponding to Eqn. (6.304) in the plane stress case are

$$\begin{aligned}
Eu_x &= k_1(1+\nu) + k_3(1+\nu)y + 6c_7xy + 2\sum_{n=1}^{\infty}(A_n\cosh\lambda_n x + B_n\sinh\lambda_n x) \\
&\quad \left\{-(1+\nu)\frac{\lambda_n h}{2}\cos\frac{\lambda_n h}{2}\sin\lambda_n y + [(1+\nu)y\lambda_n\cos\lambda_n y + 2\sin\lambda_n y]\sin\frac{\lambda_n h}{2}\right\}, \\
Eu_y &= k_2(1+\nu) - k_3(1+\nu)x - 3c_7(x^2+\nu y^2) - 2\sum_{n=1}^{\infty}(A_n\sinh\lambda_n x + B_n\cosh\lambda_n x) \\
&\quad \left\{(1+\nu)\frac{\lambda_n h}{2}\cos\frac{\lambda_n h}{2}\cos\lambda_n y + [(1-\nu)\cos\lambda_n y + (1+\nu)y\lambda_n\sin\lambda_n y]\sin\frac{\lambda_n h}{2}\right\}, \\
\tau_{xx} &= 6c_7 y + \sum_{n=1}^{\infty} 2\gamma_n(A_n\sinh\gamma_n x + B_n\cosh\gamma_n x) \\
&\quad \left[(\gamma_n y\cos\gamma_n y + 2\sin\gamma_n y)\sin\frac{\gamma_n h}{2} - \frac{\gamma_n h}{2}\sin\gamma_n y\cos\frac{\gamma_n h}{2}\right],
\end{aligned} \tag{6.312}$$

[15]Note that these asymptotic formulae *cannot* be used to determine the pure real or pure imaginary roots that may exist. For $a = 0$ we have an infinite number of real roots $n\pi/h$, for $a < 1$ Eqns. (6.310) have a finite number of real roots which can be determined using a Newton–Raphson strategy with a first guess $n\pi/h$, and for $a > 1$ Eqn. (6.310b) has no pure imaginary roots, while Eqn. (6.310a) has one pure imaginary root in the first quadrant which can be determined using a first guess $i\log(2a\pi)/h$.

[16]Since $\sin(iz) = i\sinh z$, the roots of $\sinh z = z$ and $\sinh z = -z$ are obtained simply by dividing the roots of $\sin z = z$ and $\sin z = -z$ by i.

n	Roots
1	(7.4976762777763852, 2.7686782829873215)
2	(13.899959713976461, 3.3522098848535071)
3	(20.238517707830020, 3.7167676797524991)
4	(26.554547265491561, 3.9831416403399618)
5	(32.859741005069864, 4.1932514704312087)
6	(39.158816520064988, 4.3667951176706188)
7	(45.454071464355110, 4.5146404494813028)
8	(51.746768302821785, 4.6434279570518964)
9	(58.037662059094266, 4.7575151180816206)
10	(64.327233713285565, 4.8599166478970961)
11	(70.615805061329567, 4.9528053574189448)
12	(76.903600009288397, 5.0377991932918071)
13	(83.190779437837534, 5.1161354659669280)
14	(89.477462085163012, 5.1887816285697888)
15	(95.763737602025358, 5.2565084676014484)
16	(102.04967501274578, 5.3199400450178107)
17	(108.33532837032578, 5.3795887277666505)
18	(114.62074063827144, 5.4358803489774372)
19	(120.90594641796737, 5.4891726599761013)
20	(127.19097390469230, 5.5397691117999068)

Table 6.1 *First few roots of $\sin z = z$ that lie in the first quadrant of the complex plane.*

$$\tau_{yy} = \sum_{n=1}^{\infty} 2\gamma_n^2(A_n \sinh \gamma_n x + B_n \cosh \gamma_n x)\left[\frac{h}{2}\sin\gamma_n y \cos\frac{\gamma_n h}{2} - y\cos\gamma_n y \sin\frac{\gamma_n h}{2}\right],$$

$$\tau_{xy} = \sum_{n=1}^{\infty} 2\gamma_n(A_n \cosh \gamma_n x + B_n \sinh \gamma_n x)$$
$$\left[(\cos\gamma_n y - \gamma_n y \sin\gamma_n y)\sin\frac{\gamma_n h}{2} - \frac{\gamma_n h}{2}\cos\gamma_n y\cos\frac{\gamma_n h}{2}\right],$$

while in the plane strain case, they are

$$\mu u_x = \frac{k_1}{2} + \frac{k_3 y}{2} + 3c_7(1-\nu)xy + \sum_{n=1}^{\infty}(A_n \cosh\lambda_n x + B_n \sinh\lambda_n x)$$
$$\left\{-\frac{\lambda_n h}{2}\cos\frac{\lambda_n h}{2}\sin\lambda_n y + [y\lambda_n\cos\lambda_n y + 2(1-\nu)\sin\lambda_n y]\sin\frac{\lambda_n h}{2}\right\},$$
$$\mu u_y = \frac{k_2}{2} - \frac{k_3 x}{2} - \frac{3c_7}{2}[(1-\nu)x^2 + \nu y^2] - \sum_{n=1}^{\infty}(A_n \sinh\lambda_n x + B_n \cosh\lambda_n x) \tag{6.313}$$
$$\left\{\frac{\lambda_n h}{2}\cos\frac{\lambda_n h}{2}\cos\lambda_n y + [(1-2\nu)\cos\lambda_n y + y\lambda_n \sin\lambda_n y]\sin\frac{\lambda_n h}{2}\right\},$$

n	Roots
1	(4.2123922304906607, 2.2507286116018608)
2	(10.712537397279194, 3.1031487458252616)
3	(17.073364853151830, 3.5510873470220807)
4	(23.398355225651308, 3.8588089931055745)
5	(29.708119825276039, 4.0937049247653343)
6	(36.009866016371603, 4.2837815877750236)
7	(42.306826717639360, 4.4434458303242703)
8	(48.600684124094577, 4.5811045734534375)
9	(54.892405788069240, 4.7020964603617017)
10	(61.182590196833928, 4.8100251374634659)
11	(67.471628634975360, 4.9074384165225551)
12	(73.759788346828003, 4.9962044098711349)
13	(80.047258435889233, 5.0777337322382925)
14	(86.334176690402884, 5.1531177013860292)
15	(92.620646014329395, 5.2232179892477548)
16	(98.906744893767623, 5.2887268570557202)
17	(105.19253428952541, 5.3502088486256767)
18	(111.47806230790954, 5.4081303963803045)
19	(117.76336744566099, 5.4628813161070324)
20	(124.04848089410054, 5.5147907194183361)

Table 6.2 *First few roots of* $\sin z = -z$ *that lie in the first quadrant of the complex plane.*

with the stress expressions the same as in the plane-stress case.

The displacements and stresses corresponding to Eqn. (6.307) in the plane stress case are

$$
\begin{aligned}
Eu_x &= k_1(1+\nu) + k_3(1+\nu)y + 2c_3x + 2\sum_{n=1}^{\infty}(A_n\cosh\gamma_n x + B_n\sinh\gamma_n x) \\
&\quad \left\{[(1+\nu)y\gamma_n\sin\gamma_n y - 2\cos\gamma_n y]\cos\frac{\gamma_n h}{2} - (1+\nu)\frac{\gamma_n h}{2}\sin\frac{\gamma_n h}{2}\cos\gamma_n y\right\}, \\
Eu_y &= k_2(1+\nu) - k_3(1+\nu)x - 2c_3\nu y + 2\sum_{n=1}^{\infty}(A_n\sinh\gamma_n x + B_n\cosh\gamma_n x) \\
&\quad \left\{[(1+\nu)y\gamma_n\cos\gamma_n y - (1-\nu)\sin\gamma_n y]\cos\frac{\gamma_n h}{2} + (1+\nu)\frac{\gamma_n h}{2}\sin\frac{\gamma_n h}{2}\sin\gamma_n y\right\}, \\
\tau_{xx} &= 2c_3 + \sum_{n=1}^{\infty}2\gamma_n(A_n\sinh\gamma_n x + B_n\cosh\gamma_n x) \\
&\quad \left[(\gamma_n y\sin\gamma_n y - 2\cos\gamma_n y)\cos\frac{\gamma_n h}{2} - \frac{\gamma_n h}{2}\cos\gamma_n y\sin\frac{\gamma_n h}{2}\right] \qquad (6.314) \\
\tau_{yy} &= \sum_{n=1}^{\infty}2\gamma_n^2(A_n\sinh\gamma_n x + B_n\cosh\gamma_n x)\left[\frac{h}{2}\cos\gamma_n y\sin\frac{\gamma_n h}{2} - y\sin\gamma_n y\cos\frac{\gamma_n h}{2}\right],
\end{aligned}
$$

$$\tau_{xy} = \sum_{n=1}^{\infty} 2\gamma_n (A_n \cosh \gamma_n x + B_n \sinh \gamma_n x)$$
$$\left[(\gamma_n y \cos \gamma_n y + \sin \gamma_n y) \cos \frac{\gamma_n h}{2} + \frac{\gamma_n h}{2} \sin \gamma_n y \sin \frac{\gamma_n h}{2} \right],$$

while in the plane strain case, they are

$$\mu u_x = \frac{k_1}{2} + \frac{k_3 y}{2} + c_3(1-\nu)x + \sum_{n=1}^{\infty} (A_n \cosh \gamma_n x + B_n \sinh \gamma_n x)$$
$$\left\{ [y\gamma_n \sin \gamma_n y - 2(1-\nu) \cos \gamma_n y] \cos \frac{\gamma_n h}{2} - \frac{\gamma_n h}{2} \sin \frac{\gamma_n h}{2} \cos \gamma_n y \right\},$$
$$\mu u_y = \frac{k_2}{2} - \frac{k_3 x}{2} - c_3 \nu y + \sum_{n=1}^{\infty} (A_n \sinh \gamma_n x + B_n \cosh \gamma_n x) \tag{6.315}$$
$$\left\{ [y\gamma_n \cos \gamma_n y - (1-2\nu) \sin \gamma_n y] \cos \frac{\gamma_n h}{2} + \frac{\gamma_n h}{2} \sin \frac{\gamma_n h}{2} \sin \gamma_n y \right\},$$

with the stresses the same as in the plane stress case. As usual, the terms associated with k_1, k_2 and k_3 in the above solutions represent rigid-body deformations, and the constants c_7 and c_3 are given by Eqns. (6.306) and (6.309). In contrast to the solution where ϕ is taken to be a polynomial (page 55 of [287] or page 41 of [324]), which corresponds to the case where $g(y)$ is parabolic, τ_{yy} is nonzero since the displacement constraint is implemented in a strong sense in the present solution. The complex-valued constants A_n and B_n in the above solutions are found so as to satisfy the boundary conditions $\tau_{xx}|_{x=L} = f(y)$, $\tau_{xy}|_{x=L} = g(y)$, and, if the domain represents a cantilever, say, $\boldsymbol{u}|_{x=0} = \boldsymbol{0}$ using, say, the weighted-residual method described on page 485.

Besides the 'cantilever' type problems considered above, another class of problems that can be solved by the Airy stress functions given by Eqns. (6.304) and (6.307) is the pure traction boundary value problem shown in Fig. 6.42. First consider the case where the loading on the edges $x = \pm L/2$ is symmetric with respect to the y-coordinate (i.e., normal

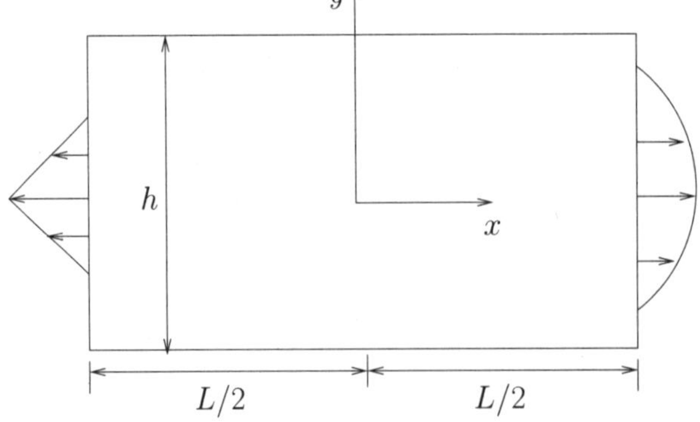

Fig. 6.42 *A traction boundary value problem on a rectangular domain with the edges $y = \pm h/2$ traction-free; shear tractions although possibly applied at $x = \pm L/2$ are not shown.*

tractions are even functions of y), and is such that the body is in equilibrium. Using the condition $u_x|_{(x,y)=(0,0)} = 0$ to suppress rigid deformation, the solution for the Airy stress function and the corresponding displacement and stress field is given by Eqn. (6.307) and by Eqns. (6.314) (plane stress) or (6.315) (plane strain), respectively, with $k_2 = k_3 = 0$, c_3 given by (compare with Eqn. (6.309))

$$c_3 = \frac{1}{2h}\int_{-h/2}^{h/2} f^-(y)\,dy = \frac{1}{2h}\int_{-h/2}^{h/2} f^+(y)\,dy, \tag{6.316}$$

and the constants A_n and B_n found using the boundary conditions

$$\begin{aligned} \tau_{xx}\big|_{x=-L/2} &= f^-(y), \quad \tau_{xx}\big|_{x=L/2} = f^+(y), \\ \tau_{xy}\big|_{x=L/2} &= g^-(y), \quad \tau_{xy}\big|_{x=L/2} = g^+(y). \end{aligned} \tag{6.317}$$

The remaining constant k_1 is then determined using the condition $u_x|_{(x,y)=(0,0)} = 0$. The constants A_n, B_n can be replaced by $\tilde{A}_n/\cosh(\gamma_n L/2)$ and $\tilde{B}_n/\cosh(\gamma_n L/2)$ to prevent ill-conditioning of the matrices in a numerical implementation.

In case uniform normal tractions are applied at both ends, i.e., if $f^-(y) = f^+(y) = c$, and $g^-(y) = g^+(y) = 0$, then $c_3 = c/2$ with the remaining constants zero. An approximate solution for the case where $g^-(y) = g^+(y) = 0$, and $f^-(y)$ and $f^+(y)$ are parabolic and given by $S(1 - 4y^2/h^2)$ has been given by Kantorovich and Krylov [171] (page 286), and by Timoshenko and Goodier [324] (page 258). For this case, we can take k_1, k_2, k_3 and the constants A_n to be zero, and $c_3 = S/3$ (from Eqn. (6.316)). The stress distribution is as given by Eqn. (6.314) with A_n set to zero. The constants B_n are determined by imposing the boundary conditions $\tau_{xx}|_{x=L/2} = f^+(y)$ and $\tau_{xy}|_{x=L/2} = g^+(y)$ using the following weighted residual statement:

$$\int_{\Gamma_t} \boldsymbol{u}_\delta \cdot (\boldsymbol{t} - \bar{\boldsymbol{t}})\,d\Gamma = 0, \tag{6.318}$$

where $\boldsymbol{u}_\delta$ denotes the variation of the displacement, $\boldsymbol{t}$ denotes the tractions, and Γ_t denotes the boundary on which the boundary condition is being enforced, which is this case is the edge $x = L/2$. In the context of the current example, one first writes $\tau_{xx}|_{x=L/2}$ and $\tau_{xy}|_{x=L/2}$ in terms of both B_n and $\bar{B}_n$, where the overbar denotes complex-conjugation. For example, using Eqn. (6.314) one writes

$$g^+(y) = \sum_{n=1}^{\infty} \operatorname{Re}\left\{2B_n\gamma_n \sinh\frac{\gamma_n L}{2}\left[(\gamma_n y\cos\gamma_n y + \sin\gamma_n y)\cos\frac{\gamma_n h}{2} + \frac{\gamma_n h}{2}\sin\gamma_n y \sin\frac{\gamma_n h}{2}\right]\right\}$$

as

$$\begin{aligned} 2g^+(y) = \sum_{n=1}^{\infty} & 2B_n\gamma_n \sinh\frac{\gamma_n L}{2}\left[(\gamma_n y\cos\gamma_n y + \sin\gamma_n y)\cos\frac{\gamma_n h}{2} + \frac{\gamma_n h}{2}\sin\gamma_n y\sin\frac{\gamma_n h}{2}\right] \\ &+ 2\bar{B}_n\bar{\gamma}_n \sinh\frac{\bar{\gamma}_n L}{2}\left[(\bar{\gamma}_n y\cos\bar{\gamma}_n y + \sin\bar{\gamma}_n y)\cos\frac{\bar{\gamma}_n h}{2} + \frac{\bar{\gamma}_n h}{2}\sin\bar{\gamma}_n y\sin\frac{\bar{\gamma}_n h}{2}\right]. \end{aligned}$$

It suffices to consider the roots γ_n that lie in the right half of the complex plane (i.e., with positive real values), since terms obtained with $-\gamma_n$ are linearly dependent on terms with

$+\gamma_n$. Using the above 'expanded' form in terms of constants and their complex-conjugates, one writes the traction vector on the right edge $(t_x, t_y)_{x=L/2}$ in the form

$$\boldsymbol{Z}\begin{bmatrix} B_1 \\ \bar{B}_1 \\ \vdots \\ B_n \\ \bar{B}_n \end{bmatrix} = \begin{bmatrix} 2f^+(y) \\ 2g^+(y) \end{bmatrix}, \tag{6.319}$$

where $\boldsymbol{Z}$ is the matrix of coefficients that are functions of y. Next, in a similar manner, we write the displacement vector $(u_x, u_y)_{x=L/2}$ as a matrix $\boldsymbol{Z}_u$ multiplied by the vector of unknown coefficients. Now Eqn. (6.318) implies that we multiply both sides of Eqn. (6.319) by $\boldsymbol{Z}_u^T$ (multiplying by $\bar{\boldsymbol{Z}}_u^T$ also yields the same final answers), and integrate with respect to y to get

$$\boldsymbol{K}\begin{bmatrix} B_1 \\ \bar{B}_1 \\ \vdots \\ B_n \\ \bar{B}_n \end{bmatrix} = \boldsymbol{f},$$

where $\boldsymbol{K} = \int_{-h/2}^{h/2} \boldsymbol{Z}_u^T \boldsymbol{Z}\, dy$ and $\boldsymbol{f} = \int_{-h/2}^{h/2} \boldsymbol{Z}_u^T \begin{bmatrix} 2f^+(y) \\ 2g^+(y) \end{bmatrix}$ dy. The integrals can be evaluated using Gaussian quadrature. Solving the above linear set of complex-valued equations, we get the coefficients B_i and $\bar{B}_i$, $i = 1, 2, \ldots, n$, which can be substituted into the expressions for the displacements and stresses.

In case displacements are prescribed instead of tractions, say, $\boldsymbol{u} = \bar{\boldsymbol{u}}$ on Γ_u, then, in place of Eqn. (6.318), we use

$$\int_{\Gamma_u} \boldsymbol{t}_\delta \cdot (\boldsymbol{u} - \bar{\boldsymbol{u}})\, d\Gamma = 0, \tag{6.320}$$

For the current example, a converged solution is obtained merely by using seven or eight terms. For $E = 2$, $\nu = 0.25$, $S = 1$, $L = h = 1$, we get the converged values for (u_x, τ_{xx}) at $(0.5, 0)$ as $(0.23084, 1)$, respectively, while at $(0, 0)$, we get (τ_{xx}, τ_{yy}) as $(0.85905, -0.14095)$. For an alternate solution strategy, see [214].

Now consider the case where $f^+(y)$ and $f^-(y)$ are odd, while $g^+(y)$ and $g^-(y)$ are even in y. In Eqn. (6.304) we now replace $c_7 y^3$ by $c_2 xy + c_7 y^3 + c_9 xy^3$, where $c_2 = -3c_9 h^2/4$, and

$$c_7 = \frac{1}{h^3} \int_{-h/2}^{h/2} y[f^+(y) + f^-(y)]\, dy,$$
$$c_9 = \frac{2}{h^3 L} \int_{-h/2}^{h/2} y[f^+(y) - f^-(y)]\, dy.$$

However, for equilibrium, we require that $\int_{-h/2}^{h/2} yf^+(y)\,dy = \int_{-h/2}^{h/2} yf^-(y)\,dy$, which results in $c_2 = c_9 = 0$, and

$$c_7 = \frac{2}{h^3}\int_{-h/2}^{h/2} yf^+(y)\,dy = \frac{2}{h^3}\int_{-h/2}^{h/2} yf^-(y)\,dy. \tag{6.321}$$

The above equation should be compared against Eqn. (6.306). The solution for the Airy stress function and the corresponding displacement field is given by Eqn. (6.304) and by Eqns. (6.312) (plane stress) and (6.313) (plane strain), respectively, with c_7 given by Eqn. (6.321), k_1, k_2 and k_3 found so as to suppress rigid-body motion, and the constants A_n and B_n determined using Eqns. (6.317). Timoshenko and Goodier [324] (page 261) consider an example where $g^-(y) = g^+(y) = 0$, and $f^-(y) = f^+(y) = Ay^3$, and again present an approximate solution. Within our framework, we get k_1, k_3 and the constants A_n as zero, and $c_7 = Ah^2/40$ (using Eqn. (6.321)). The constant k_2 is found in terms of B_n by using the condition $u_y\big|_{x=y=0} = 0$, and the constants B_n are determined using the boundary conditions on either the left or the right edge.

If $g^-(y) = g^+(y) = 0$, and $f^-(y) = f^+(y) = a_0 y/h$, where a_0 is a constant, with the displacement constraints given by $u_x\big|_{x=0} = 0$, and $u_y\big|_{x=y=0} = 0$, then we obtain the case of pure bending where in Eqns. (6.304), (6.312) and (6.313), $c_7 = a_0/(6h)$, and all the remaining constants are zero.

Now consider the case where the top and bottom edges have applied tractions on them, while the right edge of the cantilever is traction-free (see Fig. 6.43), i.e.,

$$\tau_{xx}\big|_{x=L} = 0, \qquad \tau_{xy}\big|_{x=L} = 0, \tag{6.322a}$$

$$\tau_{xy}\big|_{y=-h/2} = s_b(x), \qquad \tau_{xy}\big|_{y=h/2} = s_t(x), \tag{6.322b}$$

$$\tau_{yy}\big|_{y=-h/2} = p_b(x), \qquad \tau_{yy}\big|_{y=h/2} = p_t(x), \tag{6.322c}$$

$$\boldsymbol{u}\big|_{x=0} = 0. \tag{6.322d}$$

Note that the tractions s_t and s_b should obey the restrictions $s_t(L) = s_b(L) = 0$ to maintain continuity of the prescribed tractions at $(L, \pm h/2)$. We assume the Airy stress function to be of the form

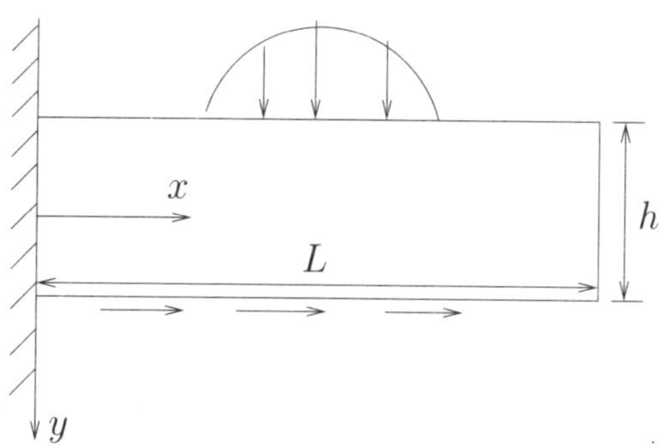

Fig. 6.43 *Rectangular domain with a traction-free right edge.*

$$\phi = \sum_{n=1}^{\infty} \Big\{ [\sin \lambda_n (L-x) - \lambda_n (L-x) \cos \lambda_n (L-x)] \, (A_n \cosh \lambda_n y + B_n \sinh \lambda_n y) + (L-x) \sin \lambda_n (L-x) (C_n \cosh \lambda_n + D_n \sinh \lambda_n) \Big\}, \quad (6.323)$$

which automatically satisfies the homogeneous boundary conditions on the stresses. Imposing the homogeneous boundary conditions on the displacements, we get that λ_n, $n = 1, 2, \ldots, \infty$, are the roots of the characteristic equation

$$(3-\nu)(1+\nu)\cos 2\lambda L - 2(1+\nu)^2(\lambda L)^2 - \nu(2-\nu) + 5 = 0, \qquad \text{(plane stress)}$$
$$(3-4\nu)\cos 2\lambda L - 2(\lambda L)^2 - 4\nu(3-2\nu) + 5 = 0, \qquad \text{(plane strain)}$$

Since the roots are located symmetrically in the four quadrants of the complex plane, it suffices to consider only the roots in the right-half of the complex plane. The asymptotic roots are $(4n-3)\pi/4 + i\log(4n-3)\pi$, which can be used as a first guess for a Newton–Raphson strategy. The first few roots for the plane stress case for $\nu = 0.25$ are listed in Table 6.3. Note that the first root is real.

n	Roots
1	(1.0005595498392583, 0)
2	(2.5911396363741939, 1.3204249219357429)
3	(5.9268510234878367, 2.1245568474626841)
4	(9.1494437322174651, 2.5428397787711807)
5	(12.338652661133029, 2.8332319411534965)
6	(15.512278040818709, 3.0568898717806476)
7	(18.677149616322968, 3.2391190052152994)
8	(21.836544040272113, 3.3930050479870086)
9	(24.992255296348688, 3.5262377490830468)
10	(28.145357350225467, 3.6437341988367522)
11	(31.296536374546903, 3.7488345080549719)
12	(34.446253179349803, 3.8439134827127743)
13	(37.594829703533755, 3.9307210624334652)
14	(40.742498219468928, 4.0105844485537059)
15	(43.889430848513385, 4.0845344317256300)
16	(47.035758053518620, 4.1533877169105455)
17	(50.181580655159166, 4.2178024723955483)
18	(53.326977886060028, 4.2783169314134621)
19	(56.472012935499002, 4.3353769027893074)
20	(59.616736856559896, 4.3893558116058227)

Table 6.3 *First few roots of* $(3-\nu)(1+\nu)\cos 2z - 2(1+\nu)^2 z^2 - \nu(2-\nu) + 5 = 0$ *(for* $\nu = 0.25$*) that lie in the first quadrant of the complex plane.*

The displacements corresponding to Eqn. (6.323) in the plane-stress case (after renaming the constants) are

$$\begin{aligned}
Eu_x &= \sum_{n=1}^{\infty} \Big\{ \lambda_n(\nu-3)(\nu+1)x\cos\lambda_n x + 2\left[2(1-\nu) + (1+\nu)^2\lambda_n^2 L(L-x)\right] \sin\lambda_n x \\
&\quad - (1+\nu)^2\lambda_n x\cos\lambda_n(2L-x)\Big\} \left[A_n\cosh\lambda_n y + B_n\sinh\lambda_n y\right], \\
Eu_y &= \sum_{n=1}^{\infty} \Big\{ \left[5+\nu(\nu-2) + 2(1+\nu)^2\lambda_n^2 L(x-L)\right]\cos\lambda_n x \\
&\quad + (1+\nu)\left[(1+\nu)\lambda_n x\sin\lambda_n(2L-x) + (3-\nu)\left[\cos\lambda_n(2L-x) + \lambda_n(2L-x)\sin\lambda_n x\right]\right]\Big\} \\
&\quad \left[A_n\sinh\lambda_n y + B_n\cosh\lambda_n y\right].
\end{aligned}$$

The complex-valued constants A_n and B_n are now found so as to meet the boundary conditions in Eqns. (6.322b) and (6.322c). For a thin beam, the results for the tip deflection under a uniformly distributed load w on the top edge match with the result $wL^4/(8EI)$ obtained using beam theory.

The above formulation can be used to solve the problem of a cantilever beam loaded under its own weight (see Fig. 6.44). A particular solution that satisfies the equations of equilibrium, and the boundary conditions given by Eqns. (6.322a) and (6.322d) (in both the plane stress and plane strain cases) is

$$\begin{aligned}
u_x &= 0, \\
2\mu u_y &= \rho g(2L-x)x, \\
\tau_{xx} &= \tau_{yy} = 0, \\
\tau_{xy} &= \rho g(L-x).
\end{aligned} \tag{6.324}$$

The complex-valued constants A_n and B_n in Eqn. (6.323) are determined using the boundary conditions in Eqns. (6.322b) and (6.322c) with $p_b(x) = p_t(x) = 0$ and $s_b(x) = s_t(x) = -\rho g(L-x)$, and the resulting solution is superposed on that given by Eqn. (6.324) to obtain the complete solution.

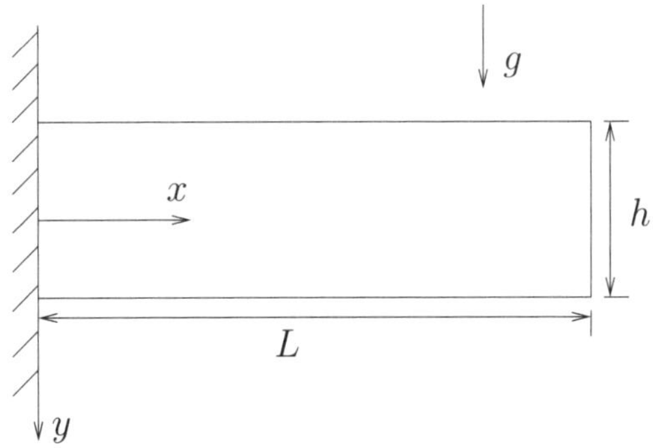

Fig. 6.44 *Cantilever beam loaded under its own weight.*

If instead of the gravity loading being along the y-direction, it is along the negative x direction (e.g., dam under gravity loading), then a particular solution that satisfies the equations of equilibrium, and the boundary conditions given by Eqns. (6.322a) and (6.322d) in the plane stress case is (for plane strain, use the transformation given by Eqn. (6.225a))

$$\begin{aligned}
4\mu u_x &= -\rho g(1-\nu)x(2L-x),\\
4\mu u_y &= 0,\\
\tau_{xx} &= -\rho g(L-x),\\
\tau_{yy} &= -\rho g\nu(L-x),\\
\tau_{xy} &= 0.
\end{aligned} \tag{6.325}$$

The constants are now determined using the boundary conditions in Eqn. (6.322b) and (6.322c) with $p_b(x) = p_t(x) = \rho g\nu(L-x)$ and $s_b(x) = s_t(x) = 0$, and the resulting solution is superposed on that given by Eqn. (6.325) to obtain the complete solution.

Instead of the setup in Fig. 6.43, consider the setup shown in Fig. 6.45 where both ends of the beam are fixed (double cantilever beam), with traction boundary conditions on the top and bottom surfaces given by Eqns. (6.322b) and (6.322c).

We consider first the case where the loading is symmetric about the y-axis. The homogeneous boundary conditions that we use are

$$u_x\big|_{x=0} = 0, \qquad \tau_{xy}\big|_{x=0} = 0, \tag{6.326a}$$

$$\boldsymbol{u}\big|_{x=\pm L/2} = 0. \tag{6.326b}$$

Since the loading is symmetric, we assume the Airy stress function to be given by

$$\phi = \sum_{n=1}^{\infty}\left[\cos\lambda_n x(A_n\sinh\lambda_n y + B_n\cosh\lambda_n y) + x\sin\gamma_n x(C_n\sinh\gamma_n y + D_n\cosh\gamma_n y)\right], \tag{6.327}$$

where λ_n satisfy the characteristic equation

$$\sin\lambda_n L = \left(\frac{1+\nu}{3-\nu}\right)\lambda_n L.$$

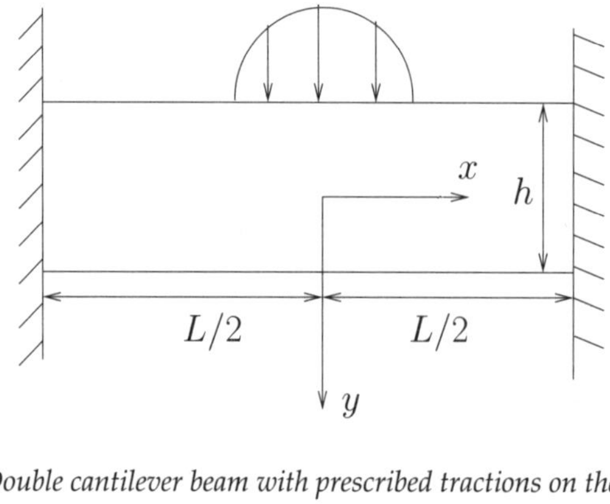

Fig. 6.45 *Double cantilever beam with prescribed tractions on the top and bottom surfaces.*

n	Roots
1	$(2.0002371585994920, 0)$
2	$(7.5930857241008214, 1.9461618205957782)$
3	$(13.954169830647668, 2.5512432770859230)$
4	$(20.276425580195728, 2.9216470489162334)$
5	$(26.583690503962696, 3.1904946104081837)$
6	$(32.883408941561868, 3.4018955219858760)$
7	$(39.178739376476187, 3.5762030297957184)$
8	$(45.471271430546018, 3.7245399074077929)$
9	$(51.761899479863423, 3.8536634417964017)$
10	$(58.051168228449832, 3.9679910537405592)$
11	$(64.339429758943339, 4.0705709110539878)$
12	$(70.626922469859338, 4.1635957394965475)$
13	$(76.913813923647993, 4.2486959673810025)$
14	$(83.200225565480181, 4.3271170636783491)$
15	$(89.486247707359198, 4.3998320040403502)$
16	$(95.771948992615023, 4.4676154254309761)$
17	$(102.05738258535312, 4.5310941434297263)$
18	$(108.34259034040079, 4.5907825380109646)$
19	$(114.62760568607494, 4.6471079425419264)$
20	$(120.91245566260336, 4.7004292456056866)$

Table 6.4 *First few roots of $(3-\nu)\sin z = (1+\nu)z$ for $\nu = 0.25$ that lie in the first quadrant of the complex plane.*

The asymptotic roots of the above characteristic equation are found using Eqn. (6.311a), which can be used as a first guess in a Newton–Raphson strategy. The first few roots are given in Table 6.4; note that the first root is real, and has to be determined separately. The corresponding displacements and stresses for the plane stress case (after renaming the constants) are given by

$$2\mu u_x = \sum_{n=1}^{\infty}\left[\frac{L}{2}\cot\frac{\lambda_n L}{2}\sin\lambda_n x - x\cos\lambda_n x\right]\left[A_n\cosh\lambda_n y + B_n\sinh\lambda_n y\right],$$

$$2\mu u_y = \sum_{n=1}^{\infty}\left[\frac{L}{2}\tan\frac{\lambda_n L}{2}\cos\lambda_n x - x\sin\lambda_n x\right]\left[A_n\sinh\lambda_n y + B_n\cosh\lambda_n y\right],$$

$$\tau_{xx} = \sum_{n=1}^{\infty}\left\{\left[\frac{\nu\lambda_n L}{2}\tan\frac{\lambda_n L}{2} + \left(\frac{\lambda_n L}{2}\cot\frac{\lambda_n L}{2} - 1\right)\right]\frac{\cos\lambda_n x}{(1-\nu)} + \lambda_n x\sin\lambda_n x\right\}$$
$$\left[A_n\cosh\lambda_n y + B_n\sinh\lambda_n y\right],$$

$$\tau_{yy} = \sum_{n=1}^{\infty}\left\{\left[\frac{\lambda_n L}{2}\tan\frac{\lambda_n L}{2} + \nu\left(\frac{\lambda_n L}{2}\cot\frac{\lambda_n L}{2} - 1\right)\right]\frac{\cos\lambda_n x}{(1-\nu)} - \lambda_n x\sin\lambda_n x\right\}$$
$$\left[A_n\cosh\lambda_n y + B_n\sinh\lambda_n y\right],$$

$$\tau_{xy} = \sum_{n=1}^{\infty} \left\{ \left(\frac{\lambda_n L}{2} \cot \lambda_n L - \frac{1}{2} \right) \sin \lambda_n x - \lambda_n x \cos \lambda_n x \right\}$$
$$[A_n \sinh \lambda_n y + B_n \cosh \lambda_n y] .$$

The homogeneous boundary conditions given by Eqns. (6.326a) and (6.326b) are automatically satisfied by the above solution. The constants A_n and B_n are found using the boundary conditions given by Eqns. (6.322b) and (6.322c). As an example, if $E = 2 \times 10^{11}$, $\nu = 0.25$, $L = 1$, $h = 0.2$, and if a uniform compressive normal traction of 10^4 is applied on the top surface with the bottom surface traction-free, i.e., $p_b = -10^4$, $p_t = s_b = s_t = 0$, a converged result for the deflection at the center of the beam is obtained using the weighted residual residual method described on page 485 using just 15 terms from the series solution; the stress solution away from the zone of singularity oscillates about the correct value. The choice of $\boldsymbol{u}_\delta$ as a weighing function is critical in this example since it vanishes at the points of singularity; using, say, $\boldsymbol{t}_\delta$ as a weighing function yields erroneous results. The solution given by 2.855×10^{-7} agrees very closely with a finite element solution. Note that the above solution is valid even for large h/L ratios while a classical beam-theory based solution is not.

For gravity loading, we superpose the particular solution

$$u_x = 0,$$
$$2\mu u_y = \rho g \left(\frac{L^2}{4} - x^2 \right),$$
$$\tau_{xx} = \tau_{yy} = 0,$$
$$\tau_{xy} = -\rho g x,$$

which satisfies the boundary conditions at $x = 0, \pm L/2$, and the equilibrium equations in both the plane stress and plane strain cases.

If the loading is antisymmetric, in place of Eqn. (6.327), we assume ϕ to be

$$\phi = \sum_{n=1}^{\infty} \Big[\sin \lambda_n x (A_n \sinh \lambda_n y + B_n \cosh \lambda_n y) + x \cos \gamma_n x (C_n \sinh \gamma_n y + D_n \cosh \gamma_n y) \Big].$$

where λ_n satisfy the characteristic equation given by

$$\sin \lambda_n L = - \left(\frac{1+\nu}{3-\nu} \right) \lambda_n L,$$

with the asymptotic roots given by Eqn. (6.311b). The corresponding displacements for the plane stress case are

$$E u_x = \sum_{n=1}^{\infty} \left[\frac{\lambda_n L}{2} \tan \frac{\lambda_n L}{2} \cos \lambda_n x - \lambda_n x \sin \lambda_n x \right] [A_n \sinh \lambda_n y + B_n \cosh \lambda_n y] ,$$

$$E u_y = \sum_{n=1}^{\infty} \left\{ \left[\frac{\lambda_n L}{2} \tan \frac{\lambda_n L}{2} + \frac{3-\nu}{1+\nu} \right] \sin \lambda_n x + \lambda_n x \cos \lambda_n x \right\} [A_n \cosh \lambda_n y + B_n \sinh \lambda_n y] .$$

Consider the contact problem shown in Fig. 6.46, where a beam rests on a frictionless rigid surface. The lateral surfaces are traction-free, and a distributed traction acts on the top surface. Thus, the boundary conditions are

$$\tau_{xx}\big|_{x=L} = 0, \qquad \tau_{xy}\big|_{x=L} = 0, \tag{6.328a}$$

$$\tau_{xy}\big|_{y=0} = 0, \qquad u_y\big|_{y=0} = 0, \tag{6.328b}$$

$$\tau_{xy}\big|_{y=h} = s_t(x), \qquad \tau_{yy}\big|_{y=h} = p_t(x), \tag{6.328c}$$

with $s_t(\pm L/2) = 0$ in order to maintain continuity of prescribed tractions at $(\pm L/2, h)$. First consider the case where the loading on the top edge is symmetric. We assume the Airy stress function to be given by

$$\phi = c_1 x^2 + \sum_{n=1}^{\infty} A_n \cosh \gamma_n y \left[\left(\frac{L}{2} - x\right) \sin \gamma_n \left(\frac{L}{2} + x\right) + \left(\frac{L}{2} + x\right) \sin \gamma_n \left(\frac{L}{2} - x\right) \right], \tag{6.329}$$

where γ_n are the roots of the characteristic equation given by (compare with Eqn. (6.308) and see Table 6.2) $\sin \gamma_n L + \gamma_n L = 0$. This form automatically satisfies the conditions in Eqns. (6.328a) and (6.328b), and the conditions

$$u_x\big|_{x=0} = \left.\frac{\partial u_x}{\partial y}\right|_{x=0} = \left.\frac{\partial u_y}{\partial x}\right|_{x=0} = 0.$$

The constant c_1 is given by

$$c_1 = \frac{1}{2L} \int_{-L/2}^{L/2} p_t(x)\, dx.$$

The final solution for the plane stress case is given by

$$Eu_x = -2c_1 \nu x + 2 \sum_{n=1}^{\infty} A_n \cosh \gamma_n y$$

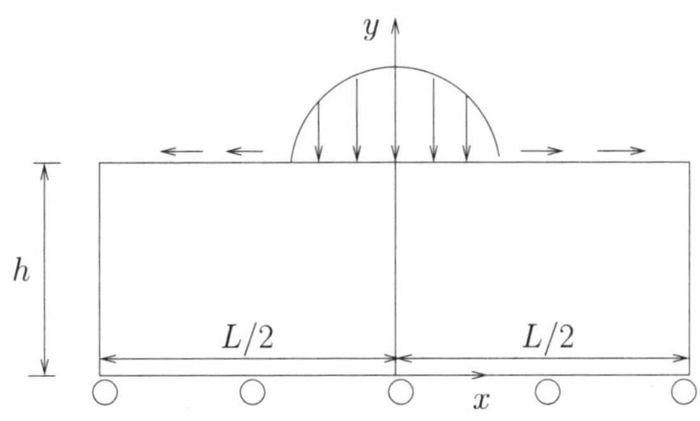

Fig. 6.46 *Beam resting on a frictionless contact surface with prescribed traction on the top, and lateral surfaces traction-free.*

$$\left\{(1+\nu)\frac{\gamma_n L}{2}\sin\frac{\gamma_n L}{2}\sin\gamma_n x + [(1+\nu)x\gamma_n\cos\gamma_n x - (1-\nu)\sin\gamma_n x]\cos\frac{\gamma_n L}{2}\right\},$$

$$Eu_y = 2c_1 y + 2\sum_{n=1}^{\infty} A_n \sinh\gamma_n y$$

$$\left\{[(1+\nu)x\gamma_n\sin\gamma_n x - 2\cos\gamma_n x]\cos\frac{\gamma_n L}{2} - \frac{(1+\nu)\gamma_n L}{2}\cos\gamma_n x\sin\frac{\gamma_n L}{2}\right\},$$

$$\tau_{xx} = 2\sum_{n=1}^{\infty} A_n\gamma_n^2\cosh\gamma_n y\left[\frac{L}{2}\cos\gamma_n x\sin\frac{\gamma_n L}{2} - x\cos\frac{\gamma_n L}{2}\sin\gamma_n x\right],$$

$$\tau_{yy} = 2c_1 + 2\sum_{n=1}^{\infty} A_n\gamma_n\cosh\gamma_n y$$

$$\left[(\gamma_n x\sin\gamma_n x - 2\cos\gamma_n x)\cos\frac{\gamma_n L}{2} - \frac{\gamma_n L}{2}\cos\gamma_n x\sin\frac{\gamma_n L}{2}\right],$$

$$\tau_{xy} = 2\sum_{n=1}^{\infty} A_n\gamma_n\sinh\gamma_n y\left[(\gamma_n x\cos\gamma_n x + \sin\gamma_n x)\cos\frac{\gamma_n L}{2} + \frac{\gamma_n L}{2}\sin\frac{\gamma_n L}{2}\sin\gamma_n x\right],$$

while for the plane strain case, we have

$$2\mu u_x = -2c_1\nu x + 2\sum_{n=1}^{\infty} A_n\cosh\gamma_n y$$

$$\left\{\frac{\gamma_n L}{2}\sin\frac{\gamma_n L}{2}\sin\gamma_n x + [x\gamma_n\cos\gamma_n x - (1-2\nu)\sin\gamma_n x]\cos\frac{\gamma_n L}{2}\right\},$$

$$2\mu u_y = 2(1-\nu)c_1 y + 2\sum_{n=1}^{\infty} A_n\sinh\gamma_n y$$

$$\left\{[x\gamma_n\sin\gamma_n x - 2(1-\nu)\cos\gamma_n x]\cos\frac{\gamma_n L}{2} - \frac{\gamma_n L}{2}\cos\gamma_n x\sin\frac{\gamma_n L}{2}\right\},$$

with the expressions for the stress the same as in the plane stress case. The constants A_n are now determined so as to satisfy the boundary conditions given by Eqn. (6.328c), say, by using a weighted-residual technique. The problems shown in Fig. (iii) of [77] or Figs. (33) and (34) of [324] are identical to the problem considered above, since because of symmetry, we get the conditions given by Eqn. (6.328b) at the plane of symmetry. The problem shown in Fig. (i) of [77] can be solved by replacing $A_n\cosh\gamma_n y$ in Eqn. (6.329) by $A_n\cosh\gamma_n y + B_n\sinh\gamma_n y$, replacing the boundary condition $u_y\big|_{y=0} = 0$ by a traction boundary condition where the two point supports are replaced by reaction tractions (spread over a small area to avoid singularities), which equilibrate the applied traction (which can also be spread over a small area) on the top edge, and then solving for the constants A_n and B_n to satisfy the traction boundary conditions on the top and bottom edges. A very accurate solution is obtained using such a methodology. One can also solve a 'contact problem' where, given the loading on the top surface, the region of contact on the (frictionless) bottom surface is determined by an iterative process of finding the point where the normal stress transitions from compressive to tensile, and then releasing the displacement constraints on the part where the normal stresses are tensile.

If the loading is antisymmetric, then we take the Airy stress function to be

$$\phi = c_4 x^3 + \sum_{n=1}^{\infty} A_n \cosh \lambda_n y \left[\left(\frac{L}{2} - x \right) \sin \lambda_n \left(\frac{L}{2} + x \right) - \left(\frac{L}{2} + x \right) \sin \lambda_n \left(\frac{L}{2} - x \right) \right], \tag{6.330}$$

where λ_n are the roots of the characteristic equation given by (compare with Eqn. (6.305) and see Table 6.1) $\sin \lambda_n L = \lambda_n L$, and

$$c_4 = \frac{2}{L^3} \int_{-L/2}^{L/2} x p_t(x)\, dx.$$

This form automatically satisfies the conditions in Eqns. (6.328a) and (6.328b). The complete solution under plane stress conditions is

$$\begin{aligned}
Eu_x &= -3c_4(y^2 + \nu x^2) - 2 \sum_{n=1}^{\infty} A_n \cosh \lambda_n y \\
&\quad \left\{ (1+\nu) \frac{\lambda_n L}{2} \cos \frac{\lambda_n L}{2} \cos \lambda_n x + [(1+\nu) x \lambda_n \sin \lambda_n x + (1-\nu) \cos \lambda_n x] \sin \frac{\lambda_n L}{2} \right\}, \\
Eu_y &= 6c_4 xy + 2 \sum_{n=1}^{\infty} A_n \sinh \lambda_n y \\
&\quad \left\{ [(1+\nu) x \lambda_n \cos \lambda_n x + 2 \sin \lambda_n x] \sin \frac{\lambda_n L}{2} - \frac{(1+\nu)\lambda_n L}{2} \sin \lambda_n x \cos \frac{\lambda_n L}{2} \right\}, \\
\tau_{xx} &= 2 \sum_{n=1}^{\infty} A_n \lambda_n^2 \cosh \lambda_n y \left[\frac{L}{2} \sin \lambda_n x \cos \frac{\lambda_n L}{2} - x \sin \frac{\lambda_n L}{2} \cos \lambda_n x \right], \\
\tau_{yy} &= 6c_4 x + 2 \sum_{n=1}^{\infty} A_n \lambda_n \cosh \lambda_n y \\
&\quad \left[(\lambda_n x \cos \lambda_n x + 2 \sin \lambda_n x) \sin \frac{\lambda_n L}{2} - \frac{\lambda_n L}{2} \sin \lambda_n x \cos \frac{\lambda_n L}{2} \right], \\
\tau_{xy} &= -2 \sum_{n=1}^{\infty} A_n \lambda_n \sinh \lambda_n y \left[(\lambda_n x \sin \lambda_n x - \cos \lambda_n x) \sin \frac{\lambda_n L}{2} + \frac{\lambda_n L}{2} \cos \frac{\lambda_n L}{2} \cos \lambda_n x \right],
\end{aligned}$$

while for the plane strain case, we have

$$\begin{aligned}
2\mu u_x &= -3c_4[(1-\nu)y^2 + \nu x^2] - 2 \sum_{n=1}^{\infty} A_n \cosh \lambda_n y \\
&\quad \left\{ \frac{\lambda_n L}{2} \cos \frac{\lambda_n L}{2} \cos \lambda_n x + [x \lambda_n \sin \lambda_n x + (1-2\nu) \cos \lambda_n x] \sin \frac{\lambda_n L}{2} \right\}, \\
2\mu u_y &= 6(1-\nu) c_4 xy + 2 \sum_{n=1}^{\infty} A_n \sinh \lambda_n y \\
&\quad \left\{ [x \lambda_n \cos \lambda_n x + 2(1-\nu) \sin \lambda_n x] \sin \frac{\lambda_n L}{2} - \frac{\lambda_n L}{2} \sin \lambda_n x \cos \frac{\lambda_n L}{2} \right\},
\end{aligned}$$

with the expressions for the stress the same as in the plane stress case. The constants A_n are now determined so as to satisfy the boundary conditions given by Eqn. (6.328c), say, by using a weighted-residual technique. The problem shown in Fig. (i) of [77] with anti-symmetric loading instead of symmetric loading on the top can be solved by first replacing the supports by reaction tractions as indicated above, so that the problem now involves traction boundary conditions on the top and bottom edges given by

$$\tau_{yy}\big|_{y=h/2} = p_t(x), \quad \tau_{yy}\big|_{y=-h/2} = p_b(x), \quad \tau_{xy}\big|_{y=h/2} = \tau_{xy}\big|_{y=-h/2} = 0, \tag{6.331}$$

where $p_b(x)$ and $p_t(x)$ are odd functions of x. Replace $A_n \cosh \lambda_n y$ in Eqn. (6.330) by $A_n \cosh \lambda_n y + B_n \sinh \lambda_n y$, find c_4 using

$$c_4 = \frac{2}{L^3} \int_{-L/2}^{L/2} x p_t(x)\, dx = \frac{2}{L^3} \int_{-L/2}^{L/2} x p_b(x)\, dx,$$

and then solve for the constants A_n and B_n using the conditions in Eqn. (6.331).

Finally, consider the case of a simply supported beam of length L and height h, with the origin located at the center of the beam. Let the boundary conditions on the top and bottom surfaces be given by Eqns. (6.322b) and (6.322c). At the end-faces, let $\tau_{xx}\big|_{x=\pm L/2} = 0$[17], but $\tau_{xy}\big|_{x=\pm L/2}$ is not necessarily zero. First consider the case of symmetrical loading, i.e., $p_b(x)$ and $p_t(x)$ are even functions, while $s_b(x)$ and $s_t(x)$ are odd functions of x. In view of this symmetry, we take the Airy stress function to be

$$\phi = c_1 x^2 + c_5 x^2 y + \sum_{n=1}^{\infty} \cos \gamma_n x \left[A_n \sinh \gamma_n y + B_n \cosh \gamma_n y + C_n y \sinh \gamma_n y + D_n y \cosh \gamma_n y \right], \tag{6.332}$$

where $\gamma_n = (2n-1)\pi/L$, and

$$\begin{aligned} c_1 &= \frac{1}{4} \left[p_t\left(\frac{L}{2}\right) + p_b\left(\frac{L}{2}\right) \right], \\ c_5 &= \frac{1}{2h} \left[p_t\left(\frac{L}{2}\right) - p_b\left(\frac{L}{2}\right) \right]. \end{aligned} \tag{6.333}$$

A_n, B_n, C_n and D_n are real-valued constants. The rigid-body-motion constants k_1 and k_3 in Eqn. (6.236) are taken to be zero, and k_2 is adjusted such that $u_y\big|_{x=\pm L/2, y=0} = 0$; we get $k_2 = c_5(2+\nu)L^2/[4(1+\nu)]$. With the use of the above function, the boundary conditions $\tau_{xx}\big|_{x=\pm L/2} = 0$ and $u_x\big|_{x=0} = 0$ are automatically satisfied. The constants A_n, B_n, C_n and D_n are found so as to satisfy the boundary conditions given by Eqns. (6.322b) and (6.322c) by using the orthogonality of the cosine functions. The shear distribution on the end-faces $\tau_{xy}\big|_{x=\pm L/2}$, which is the reaction to the constraint $E\, u_y\big|_{x=\pm L/2} = (2c_1 + c_5 y)y$ (plane stress) and $2\mu\, u_y\big|_{x=\pm L/2} = (1-\nu)(2c_1 + c_5 y)y$ (plane strain) imposed on these faces, is obtained as part of the solution.

[17] Note that solutions in the literature such as those presented by Filon [77] or in Article (22) of [324], satisfy this constraint only in a weak sense. Filon [77] uses only the root λ_n that occurs in Eqn. (6.336), while we use different roots γ_n and λ_n in the even and odd parts of the solution given by Eqns. (6.332) and (6.336), respectively; the use of these different roots is crucial in ensuring that $\tau_{xx}\big|_{x=\pm L/2} = 0$.

The solution corresponding to Eqn. (6.332) in the plane stress case is

$$
\begin{aligned}
Eu_x &= -2\nu x(c_1 + c_5 y) + \sum_{n=1}^{\infty} \sin\gamma_n x\Big\{ [2C_n + \gamma_n(1+\nu)(B_n + D_n y)]\cosh\gamma_n y \\
&\quad + [2D_n + \gamma_n(1+\nu)(A_n + C_n y)]\sinh\gamma_n y\Big\}, \\
Eu_y &= \frac{c_5}{4}(2+\nu)(L^2 - 4x^2) + (2c_1 + c_5 y)y \\
&\quad + \sum_{n=1}^{\infty} \cos\gamma_n x\Big\{ [D_n(1-\nu) - \gamma_n(1+\nu)(A_n + C_n y)]\cosh\gamma_n y \\
&\quad + [C_n(1-\nu) - \gamma_n(1+\nu)(B_n + D_n y)]\sinh\gamma_n y\Big\}, \\
\tau_{xx} &= \sum_{n=1}^{\infty} \gamma_n \cos\gamma_n x\Big\{ [(B_n + D_n y)\gamma_n + 2C_n]\cosh\gamma_n y \\
&\quad + [(A_n + C_n y)\gamma_n + 2D_n]\sinh\gamma_n y\Big\}, \\
\tau_{yy} &= 2(c_1 + c_5 y) - \sum_{n=1}^{\infty} \gamma_n^2 \cos\gamma_n x\,[(B_n + D_n y)\cosh\gamma_n y + (A_n + C_n y)\sinh\gamma_n y], \\
\tau_{xy} &= -2c_5 x + \sum_{n=1}^{\infty} \gamma_n \sin\gamma_n x\Big\{ [(A_n + C_n y)\gamma_n + D_n]\cosh\gamma_n y \\
&\quad + [(B_n + D_n y)\gamma_n + C_n]\sinh\gamma_n y\Big\},
\end{aligned}
\tag{6.334}
$$

while in the plane strain case, it is

$$
\begin{aligned}
2\mu u_x &= -2\nu x(c_1 + c_5 y) + \sum_{n=1}^{\infty} \sin\gamma_n x\Big\{ [2(1-\nu)C_n + \gamma_n(B_n + D_n y)]\cosh\gamma_n y \\
&\quad + [2(1-\nu)D_n + \gamma_n(A_n + C_n y)]\sinh\gamma_n y\Big\}, \\
2\mu u_y &= \frac{c_5}{4}(2-\nu)(L^2 - 4x^2) + (1-\nu)(2c_1 + c_5 y)y \\
&\quad + \sum_{n=1}^{\infty} \cos\gamma_n x\Big\{ [D_n(1-2\nu) - \gamma_n(A_n + C_n y)]\cosh\gamma_n y \\
&\quad + [C_n(1-2\nu) - \gamma_n(B_n + D_n y)]\sinh\gamma_n y\Big\},
\end{aligned}
\tag{6.335}
$$

with the expressions for the stresses the same as in the plane stress case.

In case $p_b(x)$ and $p_t(x)$ are odd functions, and $s_b(x)$ and $s_t(x)$ are even functions of x, ϕ instead of being given by Eqn. (6.332), is now assumed to be

$$
\begin{aligned}
\phi &= c_2 xy + c_4 x^3 + c_8 x^3 y \\
&\quad + \sum_{n=1}^{\infty} \sin\lambda_n x\,[A_n \sinh\lambda_n y + B_n \cosh\lambda_n y + C_n y \sinh\lambda_n y + D_n y\cosh\lambda_n y],
\end{aligned}
\tag{6.336}
$$

where $\lambda_n = 2n\pi/L$ and

$$c_4 = \frac{1}{6L}\left[p_t\left(\frac{L}{2}\right) + p_b\left(\frac{L}{2}\right)\right],$$
$$c_8 = \frac{1}{3hL}\left[p_t\left(\frac{L}{2}\right) - p_b\left(\frac{L}{2}\right)\right]. \tag{6.337}$$

The conditions $\tau_{xx}\big|_{x=\pm L/2} = 0$ and $u_y\big|_{x=0} = 0$ are automatically satisfied. The constant k_2 is set to zero, and k_1 is adjusted so that $u_x\big|_{x=y=0} = 0$; we get $k_1 = \sum_{n=1}^{\infty}[\lambda_n B_n + 2C_n/(1+\nu)]$. The constant k_3 is adjusted so that $u_y\big|_{x=\pm L/2, y=0} = 0$; we get $k_3 = -(2+\nu)L^2 c_8/[4(1+\nu)] - c_2$ for the plane stress case. The stresses $\tau_{xy}\big|_{x=\pm L/2}$ are a reaction to the constraints[18]

$$2E\, u_y\big|_{x=\pm L/2} = \pm 3Ly(2c_4 + c_8 y) \qquad \text{(plane stress)},$$
$$4\mu\, u_y\big|_{x=\pm L/2} = \pm 3(1-\nu)Ly(2c_4 + c_8 y) \qquad \text{(plane strain)},$$

and are obtained as part of the solution. Force balance along the x-direction leads to $\int_{-L/2}^{L/2}[s_t(x) - s_b(x)]\,dx = 0$, while moment balance leads to

$$\frac{L}{2}\int_{-h/2}^{h/2}\left[\tau_{xy}\big|_{x=L/2} + \tau_{xy}\big|_{x=-L/2}\right] dy = \int_{-L/2}^{L/2} x[p_b(x) - p_t(x)]\,dx + \frac{h}{2}\int_{-L/2}^{L/2}[s_t(x) + s_b(x)]\,dx,$$

which in turn leads to

$$\frac{hL}{4}\left(4c_2 + 3c_8 L^2\right) + 2\pi\sum_{n=1}^{\infty}(-1)^n n\left[2A_n \sinh\frac{n\pi h}{L} + D_n h\cosh\frac{n\pi h}{L}\right]$$
$$= \int_{-L/2}^{L/2} x[p_t(x) - p_b(x)]\,dx - \frac{h}{2}\int_{-L/2}^{L/2}[s_t(x) + s_b(x)]\,dx. \tag{6.338}$$

The constant c_2 is found in terms of A_n, B_n, C_n and D_n using the above equation, and these constants in turn are found so as to satisfy the boundary conditions given by Eqns. (6.322b) and (6.322c).

The solution corresponding to Eqn. (6.336) in the plane stress case is

$$Eu_x = -2c_2 y(1+\nu) - 3c_4(\nu x^2 + y^2) - c_8 y\left[3\nu x^2 + y^2 + \frac{(2+\nu)L^2}{4}\right]$$
$$+ \sum_{n=1}^{\infty}(1+\nu)\lambda_n B_n + 2C_n - \sum_{n=1}^{\infty}\cos\lambda_n x\Big\{[2C_n + \lambda_n(1+\nu)(B_n + D_n y)]\cosh\lambda_n y$$
$$+ [2D_n + \lambda_n(1+\nu)(A_n + C_n y)]\sinh\lambda_n y\Big\},$$
$$Eu_y = 6c_4 xy + c_8 x\left[3y^2 + (2+\nu)\left(\frac{L^2}{4} - x^2\right)\right]$$
$$+ \sum_{n=1}^{\infty}\sin\lambda_n x\Big\{[(1-\nu)D_n - (1+\nu)\lambda_n(A_n + C_n y)]\cosh\lambda_n y$$

[18] In a numerical strategy one can set $u_y\big|_{x=\pm L/2}$ to zero in order to get a good approximation to the exact solution obtained here.

$$+ [(1-\nu)C_n - (1+\nu)\lambda_n(B_n + D_n y)] \sinh \lambda_n y \Big\}, \tag{6.339}$$

$$\tau_{xx} = \sum_{n=1}^{\infty} \lambda_n \sin \lambda_n x \Big\{ [(B_n + D_n y)\lambda_n + 2C_n] \cosh \lambda_n y$$

$$+ [(A_n + C_n y)\lambda_n + 2D_n] \sinh \lambda_n y \Big\},$$

$$\tau_{yy} = 6x(c_4 + c_8 y) - \sum_{n=1}^{\infty} \lambda_n^2 \sin \lambda_n x \left[(B_n + D_n y) \cosh \lambda_n y + (A_n + C_n y) \sinh \lambda_n y \right],$$

$$\tau_{xy} = -c_2 - 3c_8 x^2 - \sum_{n=1}^{\infty} \lambda_n \cos \lambda_n x \Big\{ [(A_n + C_n y)\lambda_n + D_n] \cosh \lambda_n y$$

$$+ [(B_n + D_n y)\lambda_n + C_n] \sinh \lambda_n y \Big\},$$

while in the plane strain case, it is

$$2\mu u_x = -2c_2 y - 3c_4 \left[\nu x^2 + (1-\nu) y^2 \right] - c_8 y \left[3\nu x^2 + (1-\nu) y^2 + \frac{(2-\nu)L^2}{4} \right]$$

$$+ \sum_{n=1}^{\infty} (1+\nu)\lambda_n B_n + 2(1-\nu) C_n$$

$$- \sum_{n=1}^{\infty} \cos \lambda_n x \Big\{ [2(1-\nu)C_n + \lambda_n (B_n + D_n y)] \cosh \lambda_n y$$

$$+ [2(1-\nu)D_n + \lambda_n (A_n + C_n y)] \sinh \lambda_n y \Big\},$$

$$2\mu u_y = 6c_4(1-\nu)xy + c_8 x \left[3(1-\nu)y^2 + (2-\nu)\left(\frac{L^2}{4} - x^2 \right) \right] \tag{6.340}$$

$$+ \sum_{n=1}^{\infty} \sin \lambda_n x \Big\{ [(1-2\nu)D_n - \lambda_n (A_n + C_n y)] \cosh \lambda_n y$$

$$+ [(1-2\nu)C_n - \lambda_n (B_n + D_n y)] \sinh \lambda_n y \Big\},$$

with the expressions for the stresses the same as in the plane stress case.

As an illustration of the odd and even loading cases considered above, first consider a simply supported beam with a uniformly distributed load on the top (see Fig. 6.47), i.e., $p_b(x) = -p$, $p_t(x) = s_b(x) = s_t(x) = 0$. Using Eqn. (6.333), we get $c_1 = -p/4$ and $c_5 = p/(2h)$. In Eqn. (6.332), we take $B_n = 0$, $C_n = 0$ and

$$A_n = -\frac{h D_n \coth \frac{\gamma_n h}{2}}{2}.$$

The constants D_n are found using the conditions $\tau_{xy}\big|_{y=\pm h/2} = 0$, and the orthogonality of the sine functions; we get

$$D_n = -\frac{8(-1)^n p \sinh \frac{\gamma_n h}{2}}{\gamma_n^3 h L (\sinh \gamma_n h - \gamma_n h)}.$$

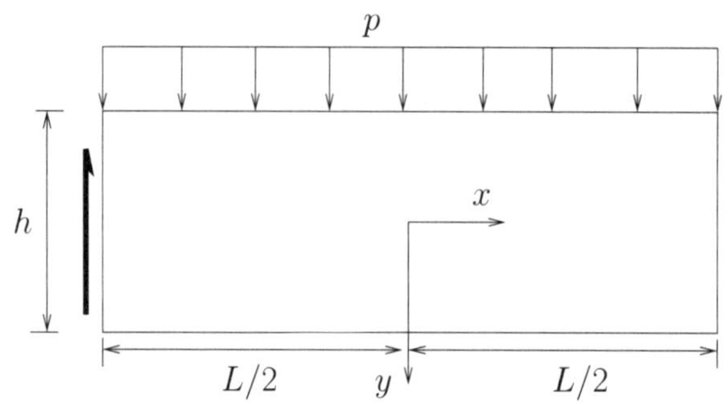

Fig. 6.47 *Simply supported beam subjected to uniform normal traction on its top edge.*

To satisfy the constraint $u_y\big|_{x=\pm L/2, y=0} = 0$, we need $k_2 = pL^2(2+\nu)/[8h(1+\nu)]$. With $\gamma_n := (2n-1)\pi/L$, the final solution for the plane stress case (obtained using Eqn. (6.334)) is

$$\begin{aligned}
2Ehu_x &= p\nu x(h-2y) + \sum_{n=1}^{\infty} D_n h \sin\gamma_n x \Big\{ 2(1+\nu)\gamma_n y \cosh\gamma_n y \\
&\quad + \left[4 - (1+\nu)\gamma_n h \coth\frac{\gamma_n h}{2}\right] \sinh\gamma_n y \Big\}, \\
2Ehu_y &= p\left[(2+\nu)\left(\frac{L^2}{4} - x^2\right) + y(y-h)\right] \\
&\quad + \sum_{n=1}^{\infty} D_n h \cos\gamma_n x \Big\{ \left[2(1-\nu) + (1+\nu)\gamma_n h \coth\frac{\gamma_n h}{2}\right] \cosh\gamma_n y \\
&\quad - 2(1+\nu)\gamma_n y \sinh\gamma_n y \Big\}, \qquad (6.341) \\
\tau_{xx} &= \sum_{n=1}^{\infty} D_n \gamma_n \cos\gamma_n x \left[\gamma_n y \cosh\gamma_n y + \left(2 - \frac{\gamma_n h}{2}\coth\frac{\gamma_n h}{2}\right)\sinh\gamma_n y\right], \\
\tau_{yy} &= p\left(\frac{y}{h} - \frac{1}{2}\right) + \sum_{n=1}^{\infty} D_n \gamma_n^2 \cos\gamma_n x \left[\frac{h}{2}\coth\frac{\gamma_n h}{2}\sinh\gamma_n y - y\cosh\gamma_n y\right], \\
\tau_{xy} &= -\frac{px}{h} + \sum_{n=1}^{\infty} D_n \gamma_n \sin\gamma_n x \left[\left(1 - \frac{\gamma_n h}{2}\coth\frac{\gamma_n h}{2}\right)\cosh\gamma_n y + \gamma_n y \sinh\gamma_n y\right],
\end{aligned}$$

while in the plane strain case (obtained using Eqn. (6.335)), the solution is

$$\begin{aligned}
4\mu h u_x &= p\nu x(h-2y) + \sum_{n=1}^{\infty} D_n h \sin\gamma_n x \Big\{ 2\gamma_n y \cosh\gamma_n y \\
&\quad + \left[4(1-\nu) - \gamma_n h \coth\frac{\gamma_n h}{2}\right] \sinh\gamma_n y \Big\}, \qquad (6.342) \\
4\mu h u_y &= p\left[(2-\nu)\left(\frac{L^2}{4} - x^2\right) + (1-\nu)y(y-h)\right]
\end{aligned}$$

$$+\sum_{n=1}^{\infty} D_n h \cos\gamma_n x \left\{ \left[2(1-2\nu) + \gamma_n h \coth\frac{\gamma_n h}{2} \right] \cosh\gamma_n y - 2\gamma_n y \sinh\gamma_n y \right\},$$

with the expressions for the stresses the same as in the plane stress case. The shear stresses $\tau_{xy}\big|_{x=\pm L/2}$ are a reaction to the imposed constraint $2Eh\ u_y\big|_{x=\pm L/2} = py(y-h)$ in the plane stress case, and $4\mu h\ u_y\big|_{x=\pm L/2} = p(1-\nu)y(y-h)$ in the plane strain case. It can be verified that $\int_{-h/2}^{h/2} \tau_{xy}\big|_{x=\pm L/2}\, dy = \mp pL/2$, $\tau_{xx}\big|_{y=0} = 0$ and $\tau_{yy}\big|_{y=0} = -p/2$. Some solutions in the literature do not have the leading order terms in the expressions for τ_{yy} and τ_{xy}, and thus fail to satisfy the boundary condition on τ_{yy} at $(x, y) = (\pm L/2, -h/2)$.

The deflection at the center of the beam is

$$2E\ u_y\big|_{x=y=0} = \frac{p(2+\nu)L^2}{4h} + \sum_{n=1}^{\infty} D_n \left[2(1-\nu) + (1+\nu)\gamma_n h \coth\frac{\gamma_n h}{2} \right], \quad \text{(plane stress)},$$

$$4\mu\ u_y\big|_{x=y=0} = \frac{p(2-\nu)L^2}{4h} + \sum_{n=1}^{\infty} D_n \left[2(1-2\nu) + \gamma_n h \coth\frac{\gamma_n h}{2} \right], \quad \text{(plane strain)},$$

while the bending stresses at the top and bottom points of the mid-section are

$$\tau_{xx}\big|_{x=0, y=\pm h/2} = \pm 2 \sum_{n=1}^{\infty} D_n \gamma_n \sinh\frac{\gamma_n h}{2}. \tag{6.343}$$

Numerically, for $h = 0.1$, $L = 1$, $\nu = 0.3$, we find from the above result in the plane stress case that $u_y\big|_{(0,0)} = 159.811p/E$, while the displacement computed using beam theory, namely, $(u_y)_{\max} = 5pL^4/(384EI)$ is $156.25p/E$. The bending stress at $(0, h/2)$ from the above exact solution is $75.2p$, while that computed using the beam theory approximation $\tau = My/I\big|_{y=h/2} = 0.75L^2/h^2$ is $75p$. For $h = 0.01$ keeping other things the same, we get from the above exact solution, $u_y\big|_{(0,0)} = 156286p/E$, while using beam theory, we get $(u_y)_{\max} = 156250p/E$; for the bending stress, we get $\tau_{xx}\big|_{(0,h/2)} = 7500.2p$, while the beam theory approximation is $\tau_{xx}\big|_{(0,h/2)} = 7500p$, thus showing that both the displacement and stress exact solutions converge to the corresponding beam theory approximations as the thickness of the beam h is reduced. For beams of large thicknesses, however, the above exact solutions should be used.

The solution for a simply supported beam loaded under its own weight is obtained by superposing, respectively, the solutions in Eqns. (6.341) and (6.342) with $p = \rho g h$ on the solutions $\tau_{xx} = \tau_{xy} = 0$, $\tau_{yy} = \rho g(h/2 - y)$, and

$$2Eu_x = \rho g \nu x(2y-h), \quad 2Eu_y = \rho g \left[\nu\left(\frac{L^2}{4} - x^2\right) + (h-y)y \right], \quad \text{(plane stress)} \tag{6.344a}$$

$$4\mu u_x = \rho g \nu x(2y-h), \quad 4\mu u_y = \rho g \left[\nu\left(\frac{L^2}{4} - x^2\right) + (1-\nu)(h-y)y \right]. \quad \text{(plane strain)} \tag{6.344b}$$

The final solution in the plane strain case is

$$4\mu u_x = \sum_{n=1}^{\infty} D_n \sin\gamma_n x \left\{ 2\gamma_n y \cosh\gamma_n y + \left[4(1-\nu) - \gamma_n h \coth\frac{\gamma_n h}{2} \right] \sinh\gamma_n y \right\},$$

$$
\begin{aligned}
4\mu u_y &= 2\rho g\left[\left(\frac{L^2}{4}-x^2\right)+(1-\nu)y(y-h)\right] \\
&\quad + \sum_{n=1}^{\infty} D_n \cos\gamma_n x\left\{\left[2(1-2\nu)+\gamma_n h\coth\frac{\gamma_n h}{2}\right]\cosh\gamma_n y - 2\gamma_n y\sinh\gamma_n y\right\}, \\
\tau_{xx} &= \sum_{n=1}^{\infty} D_n\gamma_n\cos\gamma_n x\left[\gamma_n y\cosh\gamma_n y+\left(2-\frac{\gamma_n h}{2}\coth\frac{\gamma_n h}{2}\right)\sinh\gamma_n y\right], \qquad (6.345) \\
\tau_{yy} &= \sum_{n=1}^{\infty} D_n\gamma_n^2\cos\gamma_n x\left[\frac{h}{2}\coth\frac{\gamma_n h}{2}\sinh\gamma_n y - y\cosh\gamma_n y\right], \\
\tau_{xy} &= -\rho g x+\sum_{n=1}^{\infty} D_n\gamma_n\sin\gamma_n x\left[\left(1-\frac{\gamma_n h}{2}\coth\frac{\gamma_n h}{2}\right)\cosh\gamma_n y+\gamma_n y\sinh\gamma_n y\right],
\end{aligned}
$$

where

$$
D_n = -\frac{8(-1)^n\rho g\sinh\frac{\gamma_n h}{2}}{\gamma_n^3 L(\sinh\gamma_n h-\gamma_n h)}. \qquad (6.346)
$$

The constraint at the ends is now $u_y\big|_{x=\pm L/2}=0$. Note that $\tau_{xx}\big|_{y=0}=\tau_{yy}\big|_{y=0}=0$ and $u_x\big|_{y=0}=0$. The deflection at the center of the beam is

$$
4\mu\, u_y\big|_{x=y=0} = \frac{\rho g L^2}{2}+\sum_{n=1}^{\infty} D_n\left[2(1-2\nu)+\gamma_n h\coth\frac{\gamma_n h}{2}\right],
$$

while the bending stresses at the top and bottom of the mid-section are given by Eqn. (6.343) with D_n given by Eqn. (6.346). The plane stress solution is obtained by replacing ν by $\nu/(1+\nu)$ on the right-hand side of Eqn. (6.345).

If instead of a uniformly distributed load on the top, we consider a line load $W e_y$ acting at the center of the top edge and acting uniformly along the width (see Fig. 6.48) so that W has units of load per unit length (a similar problem has been considered in Section (14) of [77] and Section (40) of [324]), then the solution is given by Eqns. (6.332), (6.334) (plane stress) and (6.335) (plane strain) with $k_1=k_2=k_3=c_1=c_5=0$, and

$$
A_n = \frac{W}{\gamma_n^2 L(\gamma_n h-\sinh\gamma_n h)}\left[2\cosh\frac{\gamma_n h}{2}+\gamma_n h\sinh\frac{\gamma_n h}{2}\right],
$$

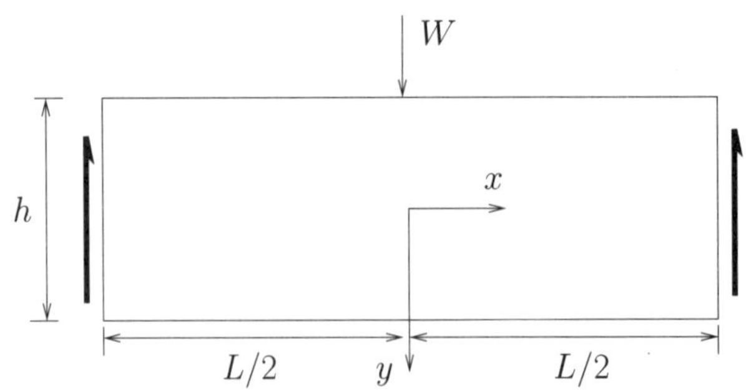

Fig. 6.48 *Simply supported beam subjected to a line load W on its top edge.*

$$B_n = \frac{W}{\gamma_n^2 L(\gamma_n h + \sinh\gamma_n h)}\left[2\sinh\frac{\gamma_n h}{2} + \gamma_n h\cosh\frac{\gamma_n h}{2}\right],$$
$$C_n = -\frac{2W}{\gamma_n L(\gamma_n h + \sinh\gamma_n h)}\sinh\frac{\gamma_n h}{2},$$
$$D_n = -\frac{2W}{\gamma_n L(\gamma_n h - \sinh\gamma_n h)}\cosh\frac{\gamma_n h}{2}.$$

Note that $u_y\big|_{x=\pm L/2} = 0$, $\tau_{yy}\big|_{y=0} = -\sum_{n=1}^{\infty}\gamma_n^2 B_n\cos\gamma_n x$, and $\int_{-h/2}^{h/2}\tau_{xy}\big|_{x=\pm L/2}\,dy = \mp W/2$, where we have used

$$\sum_{n=1}^{\infty}\frac{1}{\gamma_n}\sin\frac{\gamma_n L}{2} = \frac{L}{4}.$$

The vertical displacements and bending stresses at the point of loading, at the center and at the bottom midpoint of the beam in the plane stress case are

$$u_y\big|_{(0,-h/2)} = \frac{2W}{EL}\sum_{n=1}^{\infty}\frac{2\gamma_n h + \sinh 2\gamma_n h}{\gamma_n[\sinh^2\gamma_n h - (\gamma_n h)^2]}, \qquad \text{(does not converge)}$$
$$u_y\big|_{(0,0)} = \frac{W}{EL}\sum_{n=1}^{\infty}\frac{4\cosh\frac{\gamma_n h}{2} + \gamma_n h(1+\nu)\sinh\frac{\gamma_n h}{2}}{\gamma_n(\sinh\gamma_n h - \gamma_n h)},$$
$$u_y\big|_{(0,h/2)} = \frac{4W}{EL}\sum_{n=1}^{\infty}\frac{\gamma_n h\cosh\gamma_n h + \sinh\gamma_n h}{\gamma_n[\sinh^2\gamma_n h - (\gamma_n h)^2]}, \qquad (6.347)$$
$$\tau_{xx}\big|_{(0,-h/2)} = -\frac{W}{L}\sum_{n=1}^{\infty}\frac{\cosh 2\gamma_n h + 2(\gamma_n h)^2 - 1}{\sinh^2\gamma_n h - (\gamma_n h)^2}, \qquad \text{(does not converge)}$$
$$\tau_{xx}\big|_{(0,0)} = \frac{W}{L}\sum_{n=1}^{\infty}\frac{\gamma_n h\cosh\frac{\gamma_n h}{2} - 2\sinh\frac{\gamma_n h}{2}}{\sinh\gamma_n h + \gamma_n h},$$
$$\tau_{xx}\big|_{(0,h/2)} = \frac{4Wh}{L}\sum_{n=1}^{\infty}\frac{\gamma_n\sinh\gamma_n h}{\sinh^2\gamma_n h - (\gamma_n h)^2}.$$

Numerically, we find that the displacements and stresses at the point of loading $(0,-h/2)$ do not converge (as expected), but do converge elsewhere. For $h = 0.1$, $L = 1$, $\nu = 0.3$, we find from the above result that $u_y\big|_{(0,0)} = 256.915W/E$, while the displacement computed using beam theory, namely, $(u_y)_{\max} = WL^3/(48EI)$ is $250W/E$. The bending stress at $(0,h/2)$ from the above exact solution is $147.286W$, while that computed using the beam theory approximation $\tau = My/I|_{y=h/2} = 1.5WL/h^2$ is $150W$. For $h = 0.01$ keeping other things the same, we get from the above exact solution, $u_y\big|_{(0,0)} = 250071W/E$, while using beam theory, we get $(u_y)_{\max} = 250000W/E$; for the bending stress, we get $\tau_{xx}\big|_{(0,h/2)} = 14973.9W$, while the beam theory approximation is $\tau_{xx}\big|_{(0,h/2)} = 15000W$, thus showing that both the displacement and stress exact solutions converge to the corresponding beam theory approximations as the thickness of the beam h is reduced. For beams of large thicknesses, however, the above exact solutions should be used.

Now we consider some examples of antisymmetric loading. If on the top surface of the simply supported beam, we have a linearly varying traction of the form $p_b(x) = -2px/L$, and $p_t(x) = s_b(x) = s_t(x) = 0$, then we assume ϕ to be given by Eqn. (6.336). From Eqn. (6.337), we get $c_4 = -p/(6L)$ and $c_8 = p/(3hL)$. Imposing the boundary conditions, and with $\lambda_n := 2n\pi/L$, the final solutions are given by Eqns. (6.339) and (6.340) in the plane stress and plane strain cases, respectively, with $B_n = C_n = 0$, $c_2 = -pL/(12h)$ (found using Eqn. (6.338)), $k_1 = 0$, $k_3 = -pL/[12h(1+\nu)]$, and

$$A_n = \frac{(-1)^n pL^3 \cosh\frac{n\pi h}{L}}{4(n\pi)^3\left[\frac{L}{2}\sinh\frac{2n\pi h}{L} - n\pi h\right]},$$

$$D_n = -\frac{2A_n}{h}\tanh\frac{\lambda_n h}{2}.$$

As mentioned in the discussion following Eqn. (6.337), the shear distribution on the end-faces $\tau_{xy}\big|_{x=\pm L/2}$ is a reaction to the constraints

$$2Eh\, u_y\big|_{x=\pm L/2} = \mp py(h-y), \qquad \text{(plane stress)},$$

$$4\mu h\, u_y\big|_{x=\pm L/2} = \mp(1-\nu)py(h-y), \qquad \text{(plane strain)},$$

and is obtained as part of the solution. It can be verified that $\int_{-h/2}^{h/2} \tau_{xy}\big|_{x=\pm L/2}\, dy = -pL/6$, $\tau_{xx}\big|_{y=0} = 0$ and $\tau_{yy}\big|_{y=0} = -px/L$.

The problem shown in Fig. (v) of [77] (which is also discussed in Section (24) of [324]) can be split into symmetric and antisymmetric loading cases as shown in Fig. 6.49, which can be solved using the solutions given by Eqns. (6.332) and (6.336). For the symmetric part, we get $k_1 = k_2 = k_3 = 0 = c_1 = c_5 = 0$, $A_n = D_n = 0$, and

$$B_n = \frac{2W\cos\frac{\gamma_n a}{2}}{L\gamma_n^2(\gamma_n h + \sinh\gamma_n h)}\left[\gamma_n h\cosh\frac{\gamma_n h}{2} + 2\sinh\frac{\gamma_n h}{2}\right],$$

$$C_n = -\frac{4W\cos\frac{\gamma_n a}{2}\sinh\frac{\gamma_n h}{2}}{\gamma_n L(\gamma_n h + \sinh\gamma_n h)}.$$

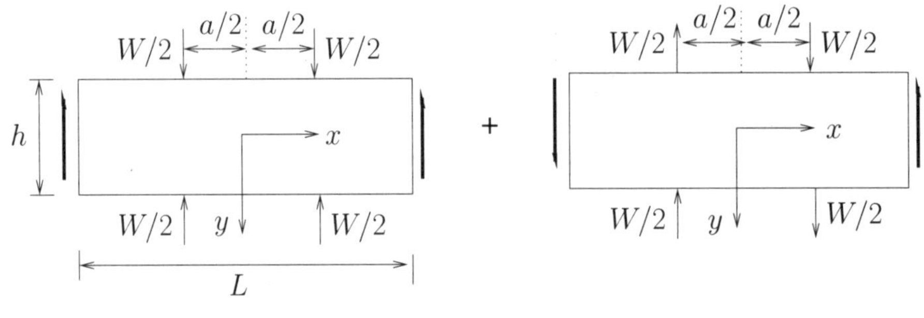

Fig. 6.49 *Simply supported beam subjected to point loads on its top and bottom edges; the loading is split into symmetric and antisymmetric loading cases as shown.*

For the antisymmetric part, we get $k_1 = k_2 = k_3 = c_2 = c_4 = c_8 = 0$, $B_n = C_n = 0$, and

$$A_n = \frac{2W \sin\frac{\lambda_n a}{2}}{\lambda_n^2 L\left[\lambda_n h - \sinh\lambda_n h\right]}\left[2\cosh\frac{\lambda_n h}{2} + \lambda_n h \sinh\frac{\lambda_n h}{2}\right],$$

$$D_n = \frac{4W \sin\frac{\lambda_n a}{2}\cosh\frac{\lambda_n h}{2}}{\lambda_n L\left[\sinh\lambda_n h - \lambda_n h\right]},$$

where we have used

$$\sum_{n=1}^{\infty} \frac{2(-1)^n}{n}\sin\frac{an\pi}{L} = -\frac{a\pi}{L}, \quad (a < L). \tag{6.348}$$

It can be verified that for the symmetric and antisymmetric parts of the solution $u_y\big|_{x=\pm L/2} = 0$, while $\int_{-h/2}^{h/2} \tau_{xy}\big|_{x=\pm L/2}\,dy$ is given by 0 and $-Wa/L$, respectively.

The shear stress at the midsection $x = 0$ for the final combined solution is given by

$$\tau_{xy}\big|_{x=0} = \frac{4W}{L}\sum_{n=1}^{\infty}\frac{\sin\frac{\lambda_n a}{2}}{\sinh\lambda_n h - \lambda_n h}\left[\frac{\lambda_n h}{2}\cosh\lambda_n y \sinh\frac{\lambda_n h}{2} - \lambda_n y \cosh\frac{\lambda_n h}{2}\sinh\lambda_n y\right].$$

The average shear, again obtained using Eqn. (6.348) (by replacing a/L by $1 - a/L$), is $W(1 - a/L)/h$. Thus, the ratio of the midsection shear to the average shear is

$$R = \frac{4h}{L-a}\sum_{n=1}^{\infty}\frac{\sin\frac{\lambda_n a}{2}}{\sinh\lambda_n h - \lambda_n h}\left[\frac{\lambda_n h}{2}\cosh\lambda_n y \sinh\frac{\lambda_n h}{2} - \lambda_n y \cosh\frac{\lambda_n h}{2}\sinh\lambda_n y\right].$$

The variation of R as a function of y/h for $L = 1$, $h = 0.1$ and (i) $a = 0.01$ and (ii) $a = 0.2$, is shown in Fig. 6.50 (similar plots are shown in Fig. (vi) of [77] and Fig. (36) of [324]). As we can see from this figure, for low values of a/L, there are large stresses at the top and bottom of the beam, while the middle part is almost free of shearing stress; thus, this distribution does not match with the parabolic distribution predicted by the elementary theory.

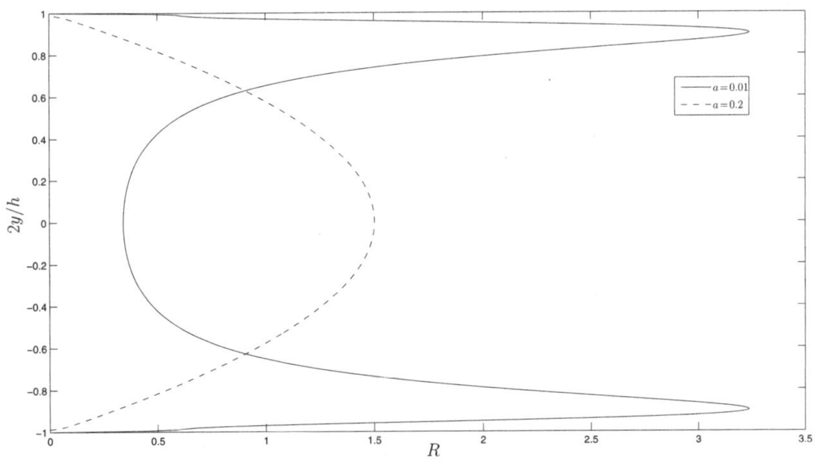

Fig. 6.50 *Variation of R with $2y/h$.*

6.6.3 Circular disc loaded by a traction distribution on its rim

Consider a circular disc of radius a and of unit width that is subjected to the traction loading $\tau_{rr}|_{r=a} = p(\theta)$ and $\tau_{r\theta}|_{r=a} = q(\theta)$. We assume $p(\theta)$ and $q(\theta)$ to be such that the circular disc remains in a state of equilibrium, i.e.,

$$F_x = 2\pi a \int_0^{2\pi} [p(\theta)\cos\theta - q(\theta)\sin\theta]\, d\theta = 0,$$

$$F_y = 2\pi a \int_0^{2\pi} [p(\theta)\sin\theta + q(\theta)\cos\theta]\, d\theta = 0.$$

Since the stresses should be bounded at the origin, and since the displacements should be single-valued, all coefficients in Eqn. (6.239) except $A_n, B_n, C_n, n \geq 1$, and $D_n, n \geq 2$ should be taken to be zero, i.e.,

$$\phi = \sum_{n=1}^{\infty} r^{n+1} \left[A_n \cos(n+1)\theta + B_n \sin(n+1)\theta + C_n \cos(n-1)\theta + D_n \sin(n-1)\theta\right]. \quad (6.349)$$

Using Eqns. (6.240), we get the complete solution under plane stress conditions (after renaming the constants and excluding rigid modes) as

$$\begin{aligned}
2Eu_r &= 4\hat{C}_1(1-\nu)r + 2(1-3\nu)r^2(\hat{C}_2\cos\theta + \hat{D}_2\sin\theta) \\
&\quad - 2\sum_{n=1}^{\infty}\Big[(n+1)(1+\nu)r^n\left[A_n\cos(n+1)\theta + B_n\sin(n+1)\theta\right] \\
&\quad + \left[n-1+(n+3)\nu\right]r^{n+2}\left[C_n\cos(n+1)\theta + D_n\sin(n+1)\theta\right]\Big], \\
2Eu_\theta &= 2(5+\nu)\left[\hat{C}_2\sin\theta - \hat{D}_2\cos\theta\right]r^2 \\
&\quad + 2\sum_{n=1}^{\infty}\Big[(n+1)(1+\nu)r^n\left[A_n\sin(n+1)\theta - B_n\cos(n+1)\theta\right] \\
&\quad + \left[5+n+(n+1)\nu\right]r^{n+2}\left[C_n\sin(n+1)\theta - D_n\cos(n+1)\theta\right]\Big], \\
\tau_{rr} &= 2\hat{C}_1 + 2\left[\hat{C}_2\cos\theta + \hat{D}_2\sin\theta\right]r \\
&\quad - \sum_{n=1}^{\infty}\Big[n(n+1)r^{n-1}\left[A_n\cos(n+1)\theta + B_n\sin(n+1)\theta\right] \\
&\quad + (n-1)(n+2)r^{n+1}\left[C_n\cos(n+1)\theta + D_n\sin(n+1)\theta\right]\Big], \\
\tau_{\theta\theta} &= 2\hat{C}_1 + 6\left[\hat{C}_2\cos\theta + \hat{D}_2\sin\theta\right]r \qquad (6.350) \\
&\quad + \sum_{n=1}^{\infty}\Big[n(n+1)r^{n-1}\left[A_n\cos(n+1)\theta + B_n\sin(n+1)\theta\right] \\
&\quad + (n+2)(n+3)r^{n+1}\left[C_n\cos(n+1)\theta + D_n\sin(n+1)\theta\right]\Big],
\end{aligned}$$

$$\tau_{r\theta} = 2\left[\hat{C}_2 \sin\theta - \hat{D}_2 \cos\theta\right] r$$
$$+ \sum_{n=1}^{\infty} \Big[n(n+1)r^{n-1} \left[A_n \sin(n+1)\theta - B_n \cos(n+1)\theta\right]$$
$$+ (n+1)(n+2)r^{n+1} \left[C_n \sin(n+1)\theta - D_n \cos(n+1)\theta\right]\Big].$$

The corresponding solution under plane strain conditions can be obtained using Eqn. (6.241).

The boundary conditions can be written as

$$2\hat{C}_1 + 2\hat{C}_2 a\cos\theta + 2\hat{D}_2 a\sin\theta$$
$$- \sum_{n=1}^{\infty} \left[n(n+1)a^{n-1}A_n + (n-1)(n+2)a^{n+1}C_n \right] \cos(n+1)\theta$$
$$- \sum_{n=1}^{\infty} \left[n(n+1)a^{n-1}B_n + (n-1)(n+2)a^{n+1}D_n \right] \sin(n+1)\theta = p(\theta),$$
$$2\hat{C}_2 a\sin\theta - 2\hat{D}_2 a\cos\theta + \sum_{n=1}^{\infty} \left[n(n+1)a^{n-1}A_n + (n+1)(n+2)a^{n+1}C_n \right] \sin(n+1)\theta$$
$$- \sum_{n=1}^{\infty} \left[n(n+1)a^{n-1}B_n + (n+1)(n+2)a^{n+1}D_n \right] \cos(n+1)\theta = q(\theta).$$

Using the orthogonality of the sine and cosine functions, we get

$$4\pi C_1 = \int_0^{2\pi} p(\theta)\, d\theta,$$
$$2\pi a C_2 = \int_0^{2\pi} p(\theta)\cos\theta\, d\theta = \int_0^{2\pi} q(\theta)\sin\theta\, d\theta,$$
$$2\pi a D_2 = \int_0^{2\pi} p(\theta)\sin\theta\, d\theta = -\int_0^{2\pi} q(\theta)\cos\theta\, d\theta,$$
$$2\pi n(n+1)a^{n-1}A_n = -\int_0^{2\pi} \left[(n+1)p(\theta)\cos(n+1)\theta + (n-1)q(\theta)\sin(n+1)\theta\right] d\theta,$$
$$2\pi n(n+1)a^{n-1}B_n = -\int_0^{2\pi} \left[(n+1)p(\theta)\sin(n+1)\theta - (n-1)q(\theta)\cos(n+1)\theta\right] d\theta,$$
$$2\pi(n+2)a^{n+1}C_{n+2} = \int_0^{2\pi} \left[p(\theta)\cos(n+1)\theta + q(\theta)\sin(n+1)\theta\right] d\theta,$$
$$2\pi(n+2)a^{n+1}D_{n+2} = \int_0^{2\pi} \left[p(\theta)\sin(n+1)\theta - q(\theta)\cos(n+1)\theta\right] d\theta,$$

for $n \geq 1$.

As an application, consider a circular disc of radius a resting on a flat surface under the influence of gravity. We first approximate the problem as shown in Fig. 6.51. The weight of the disc is balanced by the normal traction of magnitude $p = (\pi\rho a g/2)\operatorname{cosec}\alpha$ acting in

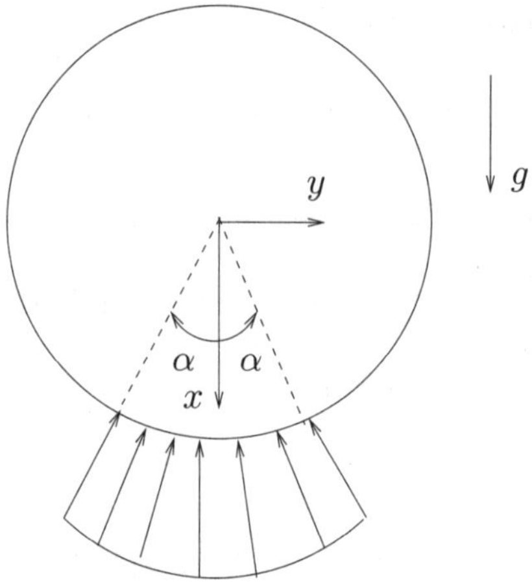

Fig. 6.51 *Circular disc under gravity load.*

the region $-\alpha \leq \theta \leq \alpha$. Because of symmetry considerations and the condition $\tau_{r\theta}|_{r=a} = 0$, the solution given by Eqn. (6.350) reduces to

$$\begin{aligned}
2Eu_r &= 2Ec_1\cos\theta + \rho g(\nu-1)r^2\cos\theta + 4\hat{C}_1(1-\nu)r \\
&\quad + 2\sum_{n=1}^{\infty}\frac{C_n r^n\cos(n+1)\theta}{n}\left[(n+1)(n+2)(1+\nu)a^2 - n\left[n-1+(n+3)\nu\right]r^2\right], \\
2Eu_\theta &= -2Ec_1\sin\theta + \rho g(\nu-1)r^2\sin\theta \\
&\quad - 2\sum_{n=1}^{\infty}\frac{C_n r^n\sin(n+1)\theta}{n}\left[(n+1)(n+2)(1+\nu)a^2 - n\left[n+5+(n+1)\nu\right]r^2\right], \\
\tau_{rr} &= -\rho g r\cos\theta + 2\hat{C}_1 + \sum_{n=1}^{\infty} C_n(n+2)r^{n-1}\cos(n+1)\theta\left[(n+1)a^2-(n-1)r^2\right], \\
\tau_{\theta\theta} &= -\rho g r\cos\theta + 2\hat{C}_1 - \sum_{n=1}^{\infty} C_n(n+2)r^{n-1}\cos(n+1)\theta\left[(n+1)a^2-(n+3)r^2\right], \\
\tau_{r\theta} &= (r^2-a^2)\sum_{n=1}^{\infty} C_n(n+1)(n+2)r^{n-1}\sin(n+1)\theta,
\end{aligned}$$

where the c_1 term corresponds to a rigid displacement that is adjusted to meet the condition $u_r|_{(a,0)} = 0$ (however, under point contact, the displacement at the point of contact is singular, and it is not possible to meet this constraint). Using the condition $\tau_{rr}|_{r=a} = -p$ in the region $-\alpha \leq \theta \leq \alpha$, we get

$$\begin{aligned}
\hat{C}_1 &= -\frac{\alpha\rho a g\,\mathrm{cosec}\,\alpha}{4}, \\
C_n &= -\frac{\rho g\,\mathrm{cosec}\,\alpha\,\sin(n+1)\alpha}{2a^n(n+1)(n+2)}.
\end{aligned}$$

Now taking the limit as $\alpha \to 0$, we get the solution for the case where the disc makes contact with the surface only at a point as

$$8Eu_r = \rho g\left\{\left[3a^2(1-\nu) - 2(1+\nu)r^2 + 8a^2\log J\right]\cos\theta + 2a(1-\nu)(2r + 3a\theta\sin\theta)\right.$$
$$\left. + 4a^2\left[(1-\nu)\tan^{-1}H + \frac{2a(1+\nu)r\sin\theta}{J}\right]\sin\theta\right\} + 8Ec_1\cos\theta,$$
$$8Eu_\theta = \rho g\left\{3a^2(1-\nu)(2\theta\cos\theta + \sin\theta) + 2\left[2a^2(1-\nu)\cos\theta\tan^{-1}H\right.\right.$$
$$+ \frac{1}{J}\left[3a^4(\nu-1) + 2a^2(\nu-2)r^2 + 3(1+\nu)r^4\right.$$
$$\left.\left.\left. - 2ar\left[(\nu-5)a^2 + 3(1+\nu)r^2\right]\cos\theta - 4a^2 J\log J\right]\sin\theta\right]\right\} - 8Ec_1\sin\theta,$$
$$\tau_{rr} = \frac{\rho g(a^2-r^2)\left\{(6a^2+r^2)r\cos\theta - a\left[a^2+3r^2+2(a^2+r^2)\cos 2\theta - ar\cos 3\theta\right]\right\}}{2(a^2+r^2-2ar\cos\theta)^2},$$
$$\tau_{\theta\theta} = -\frac{\rho g(a^2-r^2)\left\{(2a^2+r^2)r\cos\theta - a\left[r^2-a^2+2(a^2+r^2)\cos 2\theta - ar\cos 3\theta\right]\right\}}{2(a^2+r^2-2ar\cos\theta)^2},$$
$$\tau_{r\theta} = \frac{\rho g(a^2-r^2)\sin\theta\left\{4a(a^2+r^2)\cos\theta - r\left[5a^2+r^2+2a^2\cos 2\theta\right]\right\}}{2(a^2+r^2-2ar\cos\theta)^2},$$

where

$$H = \frac{\left[r^2 - 2ar - a^2 + 2a^2\cos\theta\right]\cot\frac{\theta}{2}}{a^2 - 2ar - r^2 + 2a^2\cos\theta},$$
$$J = a^2 + r^2 - 2ar\cos\theta.$$

Note that $\boldsymbol{\tau}|_{r=a} = \mathbf{0}$. The above stress results match with the results presented in [203, 222] (pages 219 and 51). Note, however, that the displacement field is not single-valued due to its dependence on θ; this example does not qualify as a counterexample to Theorem 6.1.2 since the normal strains are unbounded at the point of contact.

For an annular region bounded by two concentric circles of inner and outer radii a and b, respectively, and with prescribed conditions $\tau_{rr}|_{r=a} = p_i(\theta)$, $\tau_{rr}|_{r=b} = p_o(\theta)$, $\tau_{r\theta}|_{r=a} = q_i(\theta)$ and $\tau_{r\theta}|_{r=b} = q_o(\theta)$, instead of the Airy stress function in Eqn. (6.349), we now have

$$\begin{aligned}\phi &= F_1\log r + F_2\theta + G_1 r^2\log r \\ &+ \hat{C}_1 r^2 + \left[\hat{C}_2 r^3 + \frac{A_{-2}}{r}\right]\cos\theta + \left[\hat{D}_2 r^3 + \frac{B_{-2}}{r}\right]\sin\theta \\ &+ r\cos\theta(H_1\log r + H_2\theta) + r\sin\theta(J_1\log r + J_2\theta) \\ &+ \sum_{n=1}^{\infty}\left\{\left[A_n r^{n+1} + C_n r^{n+3}\right]\cos(n+1)\theta + \left[B_n r^{n+1} + D_n r^{n+3}\right]\sin(n+1)\theta\right\} \\ &+ \sum_{n=1}^{\infty}\left\{\left[\tilde{A}_n r^{-n+1} + \tilde{C}_n r^{-n-1}\right]\cos(n+1)\theta + \left[\tilde{B}_n r^{-n+1} + \tilde{D}_n r^{-n-1}\right]\sin(n+1)\theta\right\}.\end{aligned} \tag{6.351}$$

From the displacement field in Eqn. (6.240), we see that in order for the displacements to be single-valued, we need

$$G_1 = 0, \quad 2J_1 = (1-\nu)H_2, \quad 2H_1 = (\nu-1)J_2.$$

The remaining constants are determined using the traction boundary conditions.

6.6.4 Wedge loaded by traction distributions on its edges

For infinite wedges, there is no length scale in the problem, and an infinite number of solutions can exist. For example, see the discussion in [325] on the stress distribution in an infinite elastic wedge with prescribed tractions on its sides. Hence, in this section, we shall treat only the more realistic problem of either finite-sized wedges or semi-infinite wedges in which some length scale is involved. For an alternative treatment to the one given here, see [261, 338].

Consider the wedge shown in Fig. 6.52, which is loaded by the following distribution of tractions on its flat and curved ends:

$$\tau_{\theta\theta} = f(r), \quad \tau_{r\theta} = g(r), \quad a \le r \le b,\ \theta = \pm\theta_0, \tag{6.352a}$$

$$\tau_{rr} = p_i(\theta), \quad \tau_{r\theta} = q_i(\theta), \quad -\theta_0 \le \theta \le \theta_0,\ r = a, \tag{6.352b}$$

$$\tau_{rr} = p_o(\theta), \quad \tau_{r\theta} = q_o(\theta), \quad -\theta_0 \le \theta \le \theta_0,\ r = b. \tag{6.352c}$$

The traction distributions are assumed to be such that the wedge is in equilibrium, i.e., the net resultant forces and moments are zero.

First consider the boundary conditions given by Eqn. (6.352a). These can be satisfied by appropriately determining the constants in the solution given by Eqn. (6.239). In order to determine the characteristic equation under the conditions where the curved surfaces are traction-free, i.e., $p_i = p_o = q_i = q_0 = 0$, we write the first infinite series in Eqn. (6.239) as

$$\phi = \sum_{n=1}^{\infty} \Big[\xi^{\lambda_n}(A_n \cos\lambda_n\theta + B_n \sin\lambda_n\theta) + \xi^{\lambda_n+2}(C_n \cos\lambda_n\theta + D_n \sin\lambda_n\theta)$$
$$+ \xi^{-\lambda_n}(E_n \cos\lambda_n\theta + F_n \sin\lambda_n\theta) + \xi^{2-\lambda_n}(G_n \cos\lambda_n\theta + H_n \sin\lambda_n\theta)\Big], \tag{6.353}$$

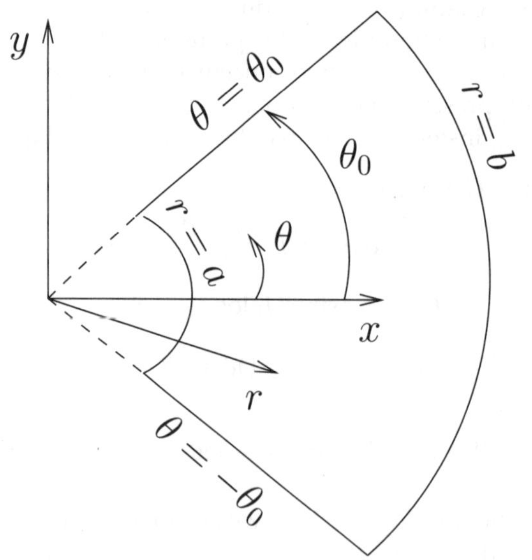

Fig. 6.52 *Wedge subjected to tractions on its edges.*

where $\xi = r/b$. The displacement and stresses corresponding to the above stress function for the plane stress case are

$$\begin{aligned}
bEu_r &= \sum_{n=1}^{\infty} \Big[-(1+\nu)\lambda_n \xi^{\lambda_n - 1}(A_n \cos\lambda_n\theta + B_n \sin\lambda_n\theta) \\
&\quad + [2 - \lambda_n - (2+\lambda_n)\nu]\, \xi^{1+\lambda_n}(C_n \cos\lambda_n\theta + D_n \sin\lambda_n\theta) \\
&\quad + (1+\nu)\lambda_n \xi^{-1-\lambda_n}(E_n \cos\lambda_n\theta + F_n \sin\lambda_n\theta) \\
&\quad + [2 + \lambda_n - (2-\lambda_n)\nu]\, \xi^{1-\lambda_n}(G_n \cos\lambda_n\theta + H_n \sin\lambda_n\theta) \Big], \\
bEu_\theta &= \sum_{n=1}^{\infty} \Big[(1+\nu)\lambda_n \xi^{\lambda_n - 1}(A_n \sin\lambda_n\theta - B_n \cos\lambda_n\theta) \\
&\quad + [4 + (1+\nu)\lambda_n]\, \xi^{1+\lambda_n}(C_n \sin\lambda_n\theta - D_n \cos\lambda_n\theta) \\
&\quad + (1+\nu)\lambda_n \xi^{-1-\lambda_n}(E_n \sin\lambda_n\theta - F_n \cos\lambda_n\theta) \\
&\quad + [-4 + (1+\nu)\lambda_n]\, \xi^{1-\lambda_n}(G_n \sin\lambda_n\theta - H_n \cos\lambda_n\theta) \Big], \\
b^2\tau_{rr} &= \sum_{n=1}^{\infty} \Big[\lambda_n(1-\lambda_n)\xi^{\lambda_n - 2}(A_n \cos\lambda_n\theta + B_n \sin\lambda_n\theta) \\
&\quad + (1+\lambda_n)(2-\lambda_n)\xi^{\lambda_n}(C_n \cos\lambda_n\theta + D_n \sin\lambda_n\theta) \\
&\quad - \lambda_n(1+\lambda_n)\xi^{-\lambda_n-2}(E_n \cos\lambda_n\theta + F_n \sin\lambda_n\theta) \\
&\quad + (1-\lambda_n)(2+\lambda_n)\xi^{-\lambda_n}(G_n \cos\lambda_n\theta + H_n \sin\lambda_n\theta) \Big], \\
b^2\tau_{\theta\theta} &= \sum_{n=1}^{\infty} \Big[\lambda_n(\lambda_n - 1)\xi^{\lambda_n - 2}(A_n \cos\lambda_n\theta + B_n \sin\lambda_n\theta) \\
&\quad + (1+\lambda_n)(2+\lambda_n)\xi^{\lambda_n}(C_n \cos\lambda_n\theta + D_n \sin\lambda_n\theta) \\
&\quad + \lambda_n(1+\lambda_n)\xi^{-\lambda_n-2}(E_n \cos\lambda_n\theta + F_n \sin\lambda_n\theta) \\
&\quad + (1-\lambda_n)(2-\lambda_n)\xi^{-\lambda_n}(G_n \cos\lambda_n\theta + H_n \sin\lambda_n\theta) \Big], \\
b^2\tau_{r\theta} &= \sum_{n=1}^{\infty} \Big[\lambda_n(\lambda_n - 1)\xi^{\lambda_n - 2}(A_n \sin\lambda_n\theta - B_n \cos\lambda_n\theta) \\
&\quad + \lambda_n(1+\lambda_n)\xi^{\lambda_n}(C_n \sin\lambda_n\theta - D_n \cos\lambda_n\theta) \\
&\quad - \lambda_n(1+\lambda_n)\xi^{-\lambda_n-2}(E_n \sin\lambda_n\theta - F_n \cos\lambda_n\theta) \\
&\quad + \lambda_n(1-\lambda_n)\xi^{-\lambda_n}(G_n \sin\lambda_n\theta - H_n \cos\lambda_n\theta) \Big].
\end{aligned} \tag{6.354}$$

Next, imposing the homogeneous boundary conditions on the curved surfaces, and using the fact that $\cos\lambda_n\theta$ and $\sin\lambda_n\theta$ are linearly independent, we get the characteristic equation for λ_n (whose real roots are $(0, \pm 1)$, while the remaining are complex-valued) as[19]

$$\eta^{2-2\lambda_n} + \eta^{2+2\lambda_n} = 2\eta^2 + \lambda_n^2(\eta^2 - 1)^2,$$

[19] In case the inner radius $a = 0$, this characteristic equation yields only the trivial root $\lambda_n = 0$; thus, when $a = 0$, we superpose a particular solution that yields the traction distribution on the flat edges on the solution given by Eqn. (6.360) as illustrated in Section 6.6.16.

where $\eta = a/b$. After adding the non-infinite part of the solution

$$a_1 \log r + a_2 r^2 \log r + a_3 r^2 + \left(b_1 r \log r + b_2 r^3 + \frac{b_3}{r}\right) \sin\theta + \left(c_1 r \log r + c_2 r^3 + \frac{c_3}{r}\right) \cos\theta,$$

and adjusting the constants to satisfy the boundary conditions, the final expression for ϕ (after renaming the constants) is

$$\begin{aligned}
\phi &= \frac{c_0 \eta^2}{4}\left\{2\eta^2 \log\eta \left[\xi^2 - 2\log\xi\right] + (\eta^2-1)\xi^2\left[1 - 2\log\xi\right]\right\} \\
&+ \frac{\eta^2-1}{\xi}\left[\eta^2 - \xi^4 + 2\xi^2(\eta^2+1)\log\xi\right]\left[c_1\cos\theta + c_2\sin\theta\right] \\
&+ \sum_{n=1}^{\infty}\left[\eta^{2+2\lambda_n}(\lambda_n-2)(\lambda_n+1)\left[\xi^{\lambda_n} + \lambda_n(\xi^{-\lambda_n} - \xi^{2-\lambda_n}) - e^{-\lambda_n}\right]\right. \\
&+ \eta^{2\lambda_n}(\lambda_n-1)\lambda_n\left[\xi^{2-\lambda_n} - \xi^{2+\lambda_n} + \lambda_n(\xi^{2-\lambda_n} - e^{-\lambda_n})\right] \\
&- \lambda_n(\lambda_n+1)\left\{\xi^{2-\lambda_n} + \xi^{\lambda_n}\left[(\lambda_n-1)\xi^2 - \lambda_n\right]\right\} \\
&\left. + \eta^2(\lambda_n^2 + \lambda_n - 2)\left\{e^{-\lambda_n} + \xi^{\lambda_n}\left[\lambda_n(\xi^2-1) - 1\right]\right\}\right](A_n\cos\lambda_n\theta + B_n\sin\lambda_n\theta).
\end{aligned} \tag{6.355}$$

Since the real roots have been incorporated in the initial part of the solution, only the complex-valued roots should be considered in the infinite series part. The real-valued constants c_0, c_1 and c_2, and the complex-valued constants A_n and B_n (which occur in complex-conjugate pairs so that the final result is real) are now determined by imposing the traction or displacement boundary conditions at $\theta = \pm\theta_0$. The field distribution corresponding to the non-infinite part of the above Airy stress function for the plane stress case is

$$\begin{aligned}
bE\xi^2 u_r &= -c_0\xi\eta^2\left\{\eta^2\left[(\nu-1)\xi^2 - (1+\nu)\right]\log\eta - \xi^2(\eta^2-1)\left[(\nu-1)\log\xi + 1\right]\right\} \\
&+ (1-\eta^2)\left\{-c_1(1+\nu)\eta^2 - 4(1+\eta^2)(c_1 - \theta c_2)\xi^2 + c_1(1-3\nu)\xi^4\right. \\
&\left. + 2c_1(1+\eta^2)(\nu-1)\xi^2\log\xi\right\}\cos\theta \\
&+ (1-\eta^2)\left\{-c_2(1+\nu)\eta^2 - 4c_1\theta(1+\eta^2)\xi^2 + c_2(1-3\nu)\xi^4\right. \\
&\left. + 2c_2(1+\eta^2)(\nu-1)\xi^2\log\xi\right\}\sin\theta, \\
bE\xi^2 u_\theta &= 2c_0\theta\xi^3\eta^2(1-\eta^2) \\
&+ (1-\eta^2)\left\{c_2(1+\nu)\eta^2 - 2\xi^2(1+\eta^2)\left[(1+\nu)c_2 + 2\theta c_1\right]\right. \\
&\left. - c_2(5+\nu)\xi^4 + 2c_2(1+\eta^2)(\nu-1)\xi^2\log\xi\right\}\cos\theta \\
&+ (\eta^2-1)\left\{c_1(1+\nu)\eta^2 - 2\xi^2(1+\eta^2)\left[(3+\nu)c_1 - 2\theta c_2\right]\right. \\
&\left. - c_1(5+\nu)\xi^4 + 2c_1(1+\eta^2)(\nu-1)\xi^2\log\xi\right\}\sin\theta,
\end{aligned} \tag{6.356}$$

$$
\begin{aligned}
b^2\xi^3\tau_{rr} &= c_0\xi\eta^2\left[\xi^2\eta^2\log\frac{\eta}{\xi}+\xi^2\log\xi-\eta^2\log\eta\right] \\
&\quad +2(\eta^2-\xi^2)(\xi^2-1)(\eta^2-1)(c_1\cos\theta+c_2\sin\theta), \\
b^2\xi^3\tau_{\theta\theta} &= c_0\xi\eta^2\left[\eta^2(1+\xi^2)\log\eta+(1-\eta^2)\xi^2(1+\log\xi)\right] \\
&\quad +2(\eta^2-1)\left[\xi^2-3\xi^4+\eta^2(1+\xi^2)\right](c_1\cos\theta+c_2\sin\theta), \\
b^2\xi^3\tau_{r\theta} &= 2(\eta^2-1)(\eta^2-\xi^2)(\xi^2-1)(c_1\sin\theta-c_2\cos\theta).
\end{aligned}
$$

The field solution corresponding to the infinite series part of Eqn. (6.355) is obtained by comparing it with Eqn. (6.353) to obtain expressions for the constants $A_n, B_n, \ldots, H_n$, and then using Eqn. (6.354).

As an example, the solution to the problem of pure bending of a curved bar by a moment M is obtained by assuming all constants except c_0 in Eqn. (6.355) to be zero. The final result is given by ([324], page 71)

$$
\begin{aligned}
\tau_{rr} &= \frac{4M}{b^2N}\left[\frac{\eta^2}{\xi^2}\log\eta-\log\xi+\eta^2\log\frac{\xi}{\eta}\right], \\
\tau_{\theta\theta} &= -\frac{4M}{b^2N}\left[\frac{\eta^2}{\xi^2}\log\eta+\log\xi+\eta^2\log\frac{\eta}{\xi}+1-\eta^2\right], \\
\tau_{r\theta} &= 0,
\end{aligned}
$$

where

$$
N=(1-\eta^2)^2-4\eta^2\log^2\eta.
$$

Thus, in this case, $g(r)=0$, and $f(r)=\tau_{\theta\theta}$.

As another example, consider the curved cantilever beam in Fig. 6.53. The boundary conditions are

$$
\boldsymbol{u}|_{\theta=0}=0; \quad \tau_{\theta\theta}|_{\theta=\theta_0}=p(r); \quad \tau_{r\theta}|_{\theta=\theta_0}=s(r), \tag{6.357}
$$

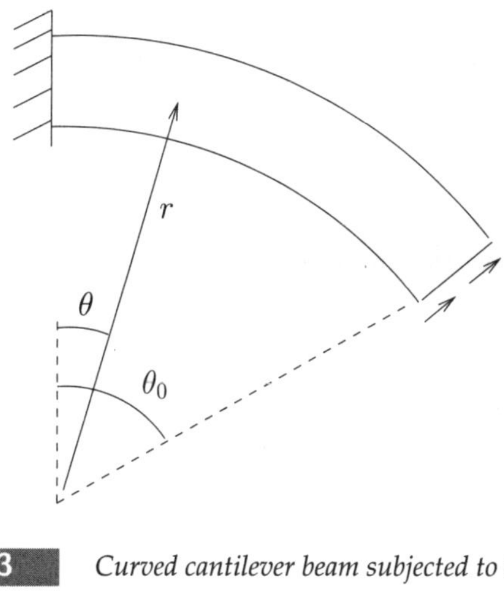

Fig. 6.53 *Curved cantilever beam subjected to terminal loads.*

with the curved surfaces traction-free. We take the constants c_0, c_1 and c_2 in Eqns. (6.355) and (6.356) to be zero, and then determine the complex-valued constants A_n and B_n so as to satisfy the conditions in Eqn. (6.357). The solution for the bending of a curved bar by an arbitrarily directed force at the end given by Timoshenko and Goodier [324] (page 88) is obtained by disregarding the infinite series part of the solution, and by considering a weaker boundary condition than the first condition in Eqn. (6.357). Of course, in this case, the traction distribution that is to be applied at the end is obtained as part of the solution; for a differently prescribed traction, the infinite series part of the solution would need to be considered.

We now look at the problem involving homogeneous traction conditions on the straight edges of the wedge. In order to satisfy the boundary conditions given by Eqns. (6.352b) and (6.352c), one again needs to find the (possibly complex-valued) roots λ_n of a characteristic equation that is derived assuming homogeneous boundary conditions on the edges $\theta = \pm\theta_0$. The loading on the curved edges $r = a$ and $r = b$ can be decomposed into a symmetric and an unsymmetric part. For the symmetric part, the constants B_n and D_n in Eqn. (6.239) can be set to zero, and the solution involving the constants A_n and C_n is to be found so as to satisfy the homogeneous boundary conditions $\tau_{\theta\theta}|_{\theta=\pm\theta_0} = \tau_{r\theta}|_{\theta=\pm\theta_0} = 0$. A nontrivial solution for A_n and C_n will exist if and only if the determinant of the coefficient matrix is zero, i.e.,

$$\det \begin{bmatrix} \cos(\lambda+1)\theta_0 & \cos(\lambda-1)\theta_0 \\ \lambda(\lambda+1)\sin(\lambda+1)\theta_0 & (\lambda-1)\lambda\sin(\lambda-1)\theta_0 \end{bmatrix} = 0,$$

which leads to the characteristic equation

$$\sin 2\lambda\theta_0 + \lambda \sin 2\theta_0 = 0. \tag{6.358}$$

Similarly, for the anti-symmetric part, the constants A_n and C_n can be set to zero, and the characteristic equation based on the constants B_n and D_n turns out to be

$$\sin 2\gamma\theta_0 - \gamma \sin 2\theta_0 = 0. \tag{6.359}$$

Eqns. (6.358) and (6.359) have a finite number of real roots (e.g., $\gamma = 0, -1$) and infinitely many complex roots for all θ_0 except $\theta_0 = \pi/2$ and π. When $\theta_0 = \pi/2$ and π, we have $\lambda_n, \gamma_n \equiv n$ and $n/2$, with n ranging from $-\infty$ to ∞. The asymptotic formulae for the complex-valued roots of Eqns. (6.358) and (6.359) are given by Eqns. (6.311) (note that $\sin 2\theta_0$ can be positive or negative depending on the value of θ_0). These asymptotic formulae can be used as a first guess for a Newton–Raphson iteration to determine the complex-valued roots of the characteristic equations, while the real-valued roots have to be found separately.

Let $\lambda_n \equiv a_n + ib_n$ and $\gamma_n \equiv c_n + id_n$, $n = 1, 2, \ldots, \infty$, denote the roots of Eqns. (6.358) and (6.359). The Airy stress function solution (including the symmetric and anti-symmetric parts) that satisfies the conditions $\tau_{\theta\theta}|_{\theta=\pm\theta_0} = \tau_{r\theta}|_{\theta=\pm\theta_0} = 0$ is given by (the $\hat{D}_2 \sin 2\theta$ term corresponds to the root $\gamma = -1$ in Eqn. (6.239))

$$\begin{aligned} \phi &= \hat{D}_2(\sin 2\theta - 2\theta\cos 2\theta_0) + J_2 r\theta\sin\theta + H_2 r\theta\cos\theta \\ &\quad + \sum_{n=1}^{\infty} r^{\lambda_n+1}\left[A_n\cos(\lambda_n+1)\theta + C_n\cos(\lambda_n-1)\theta\right] \end{aligned} \tag{6.360}$$

$$+\sum_{n=1}^{\infty} r^{\gamma_n+1}\left[B_n \sin(\gamma_n+1)\theta + D_n \sin(\gamma_n-1)\theta\right],$$

where, with F_n and G_n denoting complex-valued constants,

$$A_n = F_n \cos(\lambda_n-1)\theta_0, \qquad B_n = G_n \sin(\gamma_n-1)\theta_0,$$
$$C_n = -F_n \cos(\lambda_n+1)\theta_0, \qquad D_n = -G_n \sin(\gamma_n+1)\theta_0.$$

If $a=0$ and $\theta_0 \le \pi/2$, only those λ_n, γ_n with $a_n > 1$ and $c_n > 1$ should be considered in order to prevent a singularity of the displacements and stresses at the origin, while if $a=0$ and $\theta_0 > \pi/2$, then those λ_n, γ_n with $a_n > 0$ and $c_n > 0$ should be considered since the stresses (but not the displacements) can be singular at the origin.

Using Eqn. (6.240), the complete solution under plane-stress conditions corresponding to the solution in Eqn. (6.360) is obtained as

$$\begin{aligned}
2Eu_r &= 2(1+\nu)[k_1\cos\theta + k_2\sin\theta] + \frac{8\hat{D}_2}{r}\sin 2\theta \\
&\quad + [2(1-\nu)\theta\sin\theta + (1-\nu+4\log r)\cos\theta]\,J_2 \\
&\quad + [2(1-\nu)\theta\cos\theta - (1-\nu+4\log r)\sin\theta]\,H_2 \\
&\quad - 2\sum_{n=1}^{\infty} r^{\lambda_n}\left\{(\lambda_n+1)(1+\nu)A_n\cos(\lambda_n+1)\theta + [(1+\nu)(\lambda_n+1)-4]\,C_n\cos(\lambda_n-1)\theta\right\}, \\
&\quad - 2\sum_{n=1}^{\infty} r^{\gamma_n}\left\{(\gamma_n+1)(1+\nu)B_n\sin(\gamma_n+1)\theta + [(1+\nu)(\gamma_n+1)-4]\,D_n\sin(\gamma_n-1)\theta\right\}, \\
2Eu_\theta &= 2(1+\nu)[-k_1\sin\theta + k_2\cos\theta + k_3 r] \\
&\quad + \frac{4\hat{D}_2}{r}[(1-\nu)\cos 2\theta + (1+\nu)\cos 2\theta_0] + \{2(1-\nu)\theta\cos\theta - [3+\nu+4\log r]\sin\theta\}\,J_2 \\
&\quad - \{2(1-\nu)\theta\sin\theta + [3+\nu+4\log r]\cos\theta\}\,H_2 \\
&\quad + 2\sum_{n=1}^{\infty} r^{\lambda_n}\left\{(\lambda_n+1)(1+\nu)A_n\sin(\lambda_n+1)\theta + [3+\lambda_n+(\lambda_n-1)\nu]\,C_n\sin(\lambda_n-1)\theta\right\} \\
&\quad - 2\sum_{n=1}^{\infty} r^{\gamma_n}\left\{(\gamma_n+1)(1+\nu)B_n\cos(\gamma_n+1)\theta + [3+\gamma_n+(\gamma_n-1)\nu]\,D_n\cos(\gamma_n-1)\theta\right\}, \\
\tau_{rr} &= -\frac{4\hat{D}_2}{r^2}\sin 2\theta + \frac{2J_2}{r}\cos\theta - \frac{2H_2}{r}\sin\theta \\
&\quad - \sum_{n=1}^{\infty} r^{\lambda_n-1}\lambda_n\left\{(\lambda_n+1)A_n\cos(\lambda_n+1)\theta + (\lambda_n-3)C_n\cos(\lambda_n-1)\theta\right\} \\
&\quad - \sum_{n=1}^{\infty} r^{\gamma_n-1}\gamma_n\left\{(\gamma_n+1)B_n\sin(\gamma_n+1)\theta + (\gamma_n-3)D_n\sin(\gamma_n-1)\theta\right\}, \\
\tau_{\theta\theta} &= \sum_{n=1}^{\infty} r^{\lambda_n-1}\lambda_n(\lambda_n+1)\left\{A_n\cos(\lambda_n+1)\theta + C_n\cos(\lambda_n-1)\theta\right\} \\
&\quad + \sum_{n=1}^{\infty} r^{\gamma_n-1}\gamma_n(\gamma_n+1)\left\{B_n\sin(\gamma_n+1)\theta + D_n\sin(\gamma_n-1)\theta\right\},
\end{aligned}$$

$$\tau_{r\theta} = \frac{2\hat{D}_2}{r^2}\left[\cos 2\theta - \cos 2\theta_0\right] + \sum_{n=1}^{\infty} r^{\lambda_n - 1}\lambda_n \left\{(\lambda_n+1)A_n \sin(\lambda_n+1)\theta + (\lambda_n-1)C_n \sin(\lambda_n-1)\theta\right\} - \sum_{n=1}^{\infty} r^{\gamma_n - 1}\gamma_n \left\{(\gamma_n+1)B_n \cos(\gamma_n+1)\theta + (\gamma_n-1)D_n \cos(\gamma_n-1)\theta\right\}.$$

Similarly, the plane strain solution is obtained using Eqn. (6.241). The constants $\hat{D}_2$, J_2, H_2, F_n and G_n (which can be considered as two sets of complex-valued constants corresponding to $a_n, c_n > 1$ and $a_n, c_n \leq 1$) in Eqn. (6.360) are now determined to satisfy the boundary conditions given by Eqns. (6.352b) and (6.352c). For example, if $s_i = s_o = 0$, and $p_i = p\cos\theta/a$ and $p_o = p\cos\theta/b$, then J_2 is the only nonzero coefficient.

As an example, consider the infinite wedge shown in Fig. 6.54 that is subjected to tractions $\tau_{rr} = p(\theta)$ and $\tau_{r\theta} = s(\theta)$ at the surface $r = a$. First consider the case where the

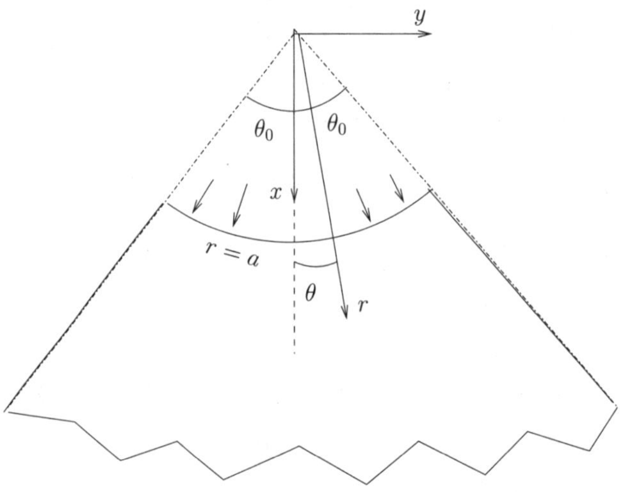

Fig. 6.54 *An infinite wedge subjected to tractions at the surface $r = a$.*

applied tractions $p(\theta)/s(\theta)$ are even/odd in θ. Since the stresses should decay[20] as $r \to \infty$, we consider only those roots λ_n whose real part is negative (if the domain is bounded, then this restriction is removed). All constants except the constant J_2 and the complex-valued constants F_n are taken to be zero. These constants are determined so as to satisfy the traction boundary conditions $\tau_{rr}|_{r=a} = p(\theta)$ and $\tau_{r\theta}|_{r=a} = s(\theta)$. When $\theta_0 = \pi/2$, the solution is given by Eqn. (6.362) below with n replaced by $-n$; the constants A_n and B_n in that solution are determined using the traction boundary conditions at $r = a$. When the applied tractions $p(\theta)/s(\theta)$ are odd/even in θ, then we consider only those roots γ_n whose real part is negative. All constants except $\hat{D}_2$, H_2 and G_n are taken to be zero, where

$$\hat{D}_2 = \frac{a^2}{(2\sin 2\theta_0 - 4\theta_0\cos 2\theta_0)}\int_{-\theta_0}^{\theta_0} s(\theta)\,d\theta.$$

[20]Note that the displacements do not decay due to the presence of the J_2 and H_2 terms; if we exclude these terms, then we are unable to satisfy the traction boundary conditions at $r = a$.

The constants H_2 and G_n are determined so as to satisfy the traction boundary conditions at $r = a$. For $\theta_0 = \pi/2$, the constants C_n and D_n in Eqn. (6.362) are determined using the traction boundary conditions at $r = a$.

If the top edge in Fig. 6.54 is flat and at a perpendicular distance of h from the origin, then the traction boundary conditions on this top flat edge are of the form

$$\tau_{xx}\big|_{x=h} = \left[\tau_{rr}\cos^2\theta + \tau_{\theta\theta}\sin^2\theta - \tau_{r\theta}\sin 2\theta\right]_{r\cos\theta=h} = f(y) = f(r\sin\theta),$$

$$\tau_{xy}\big|_{x=h} = \left[\frac{1}{2}(\tau_{rr} - \tau_{\theta\theta})\sin 2\theta + \tau_{r\theta}\cos 2\theta\right]_{r\cos\theta=h} = g(y) = g(r\sin\theta).$$

The constants in the solution given by Eqn. (6.360) are now determined using the above boundary conditions. One could also have prescribed displacements $u_x\big|_{x=h} = \hat{f}(y)$ and $u_y\big|_{x=h} = \hat{g}(y)$ on the top edge instead of the above prescribed traction boundary conditions.

When $\theta_0 = \pi/2$, we have seen that the roots of the characteristic equation are integers, and thus, it may appear that there are lesser number of unknowns than required to satisfy all boundary conditions. However, as we now show, the series solution can be split into an 'odd' and an 'even' series, with independent sets of constants associated with them. For the case of symmetric loading, we take the infinite series part of the stress function to be of the form

$$\phi = \sum_{n=-\infty}^{\infty} r^{n+1}\left[A_n\cos(n+1)\theta + C_n\cos(n-1)\theta\right]. \tag{6.361}$$

Corresponding to this stress function, we get

$$\tau_{\theta\theta} = \sum_{n=-\infty}^{\infty} r^{n-1}(n+1)n\left[A_n\cos(n+1)\theta + C_n\cos(n-1)\theta\right],$$

$$\tau_{r\theta} = \sum_{n=-\infty}^{\infty} r^{n-1}(n+1)n\left[A_n\sin(n+1)\theta + n(n-1)C_n\sin(n-1)\theta\right].$$

Imposing the conditions $\tau_{\theta\theta}\big|_{\theta=\pm\pi/2} = \tau_{r\theta}\big|_{\theta=\pm\pi/2} = 0$, we get

$$C_{2n} = \frac{(2n+1)A_{2n}}{2n-1},$$

$$C_{2n-1} = A_{2n-1}.$$

One can similarly obtain the results for antisymmetric loading. After renaming the constants, we get the stress function (which includes symmetric and antisymmetric loading) for $\theta_0 = \pi/2$ as

$$\phi = \hat{D}_2(\sin 2\theta + 2\theta) + J_2 r\theta\sin\theta + H_2 r\theta\cos\theta + B_0(1+\cos 2\theta)$$
$$+ \sum_{\substack{n=-\infty \\ n\neq 0}}^{\infty} \Big\{A_n r^{2n+1}\left[(2n-1)\cos(2n+1)\theta + (2n+1)\cos(2n-1)\theta\right]$$

$$+ B_n r^{2n}\left[\cos 2n\theta + \cos 2(n-1)\theta\right] + C_n r^{2n+1}\left[\sin(2n+1)\theta + \sin(2n-1)\theta\right]$$
$$+ D_n r^{2n}\left[(2n-2)\sin 2n\theta + 2n\sin 2(n-1)\theta\right]\Big\}.$$

The complete solution for the plane stress case found using Eqn. (6.221) is given by

$$
\begin{aligned}
Eu_r &= (1+\nu)[k_1\cos\theta + k_2\sin\theta] + \frac{4\hat{D}_2}{r}\sin 2\theta \\
&+ \frac{1}{2}\left[2(1-\nu)\theta\sin\theta + (1-\nu+4\log r)\cos\theta\right]J_2 \\
&+ \frac{1}{2}\left[2(1-\nu)\theta\cos\theta - (1-\nu+4\log r)\sin\theta\right]H_2 + \frac{4B_0\cos 2\theta}{r} \\
&+ \sum_{\substack{n=-\infty \\ n\neq 0}}^{\infty}\Big\{2(2n+1)r^{2n}A_n\left\{2\left[1-n(1+\nu)\right]\cos\theta\cos 2n\theta + (1-\nu)\sin\theta\sin 2n\theta\right\} \\
&+ 2r^{2n-1}B_n\left\{\left[2-n(1+\nu)\right]\cos 2(n-1)\theta - n(1+\nu)\cos 2n\theta\right\} \\
&+ 2r^{2n}C_n\left\{\left[1-\nu-2(1+\nu)n\right]\sin 2n\theta\cos\theta - 2\cos 2n\theta\sin\theta\right\} \\
&- 4nr^{2n-1}D_n\left\{\left[(1+\nu)n-2\right]\sin 2(n-1)\theta + (n-1)(1+\nu)\sin 2n\theta\right\}\Big\}, \\
Eu_\theta &= (1+\nu)[-k_1\sin\theta + k_2\cos\theta + k_3 r] + \frac{2\hat{D}_2}{r}\left[(1-\nu)\cos 2\theta - (1+\nu)\right] \\
&+ \frac{1}{2}\left\{2(1-\nu)\theta\cos\theta - \left[3+\nu+4\log r\right]\sin\theta\right\}J_2 \\
&- \frac{1}{2}\left\{2(1-\nu)\theta\sin\theta + \left[3+\nu+4\log r\right]\cos\theta\right\}H_2 + \frac{2(\nu-1)B_0\sin 2\theta}{r} \\
&+ \sum_{\substack{n=-\infty \\ n\neq 0}}^{\infty}\Big\{2(2n+1)r^{2n}A_n\left\{-2\sin\theta\cos 2n\theta + \left[1-\nu+2n(1+\nu)\right]\cos\theta\sin 2n\theta\right\} \\
&+ 2r^{2n-1}B_n\left\{\left[1+n+(n-1)\nu\right]\sin 2(n-1)\theta + n(1+\nu)\sin 2n\theta\right\} \\
&- 2r^{2n}C_n\left\{2\left[1+(1+\nu)n\right]\cos\theta\cos 2n\theta + (1-\nu)\sin\theta\sin 2n\theta\right\} \\
&- 4nr^{2n-1}D_n\left\{\left[1+n+(n-1)\nu\right]\cos 2(n-1)\theta + (n-1)(1+\nu)\cos 2n\theta\right\}\Big\}, \\
\tau_{rr} &= -\frac{4\hat{D}_2}{r^2}\sin 2\theta + \frac{2J_2}{r}\cos\theta - \frac{2H_2}{r}\sin\theta - \frac{4B_0\cos 2\theta}{r^2} \qquad (6.362) \\
&+ \sum_{\substack{n=-\infty \\ n\neq 0}}^{\infty}\Big\{4n(2n+1)r^{2n-1}A_n\left[\sin\theta\sin 2n\theta - 2(n-1)\cos\theta\cos 2n\theta\right] \\
&- 2(2n-1)r^{2n-2}B_n\left[n\cos 2n\theta + (n-2)\cos 2(n-1)\theta\right] \\
&+ r^{2n-1}C_n\left\{4(1-2n)n\cos\theta\sin 2n\theta - 8n\sin\theta\cos 2n\theta\right\} \\
&- 4n(2n-1)r^{2n-2}D_n\left\{(n-2)\sin 2(n-1)\theta + (n-1)\sin 2n\theta\right\}\Big\},
\end{aligned}
$$

$$\tau_{\theta\theta} = \sum_{\substack{n=-\infty \\ n\neq 0}}^{\infty} \Big\{ 4n(2n+1)r^{2n-1}A_n\left[2n\cos\theta\cos 2n\theta + \sin\theta\sin 2n\theta\right] \\ + 2n(2n-1)r^{2n-2}B_n\left[\cos 2n\theta + \cos 2(n-1)\theta\right] \\ + 4n(2n+1)r^{2n-1}C_n\cos\theta\sin 2n\theta \\ + 4n(2n-1)r^{2n-2}D_n\left\{n\sin 2(n-1)\theta + (n-1)\sin 2n\theta\right\}\Big\},$$

$$\tau_{r\theta} = \frac{2\hat{D}_2}{r^2}\left[\cos 2\theta + 1\right] - \frac{2B_0\sin 2\theta}{r^2} + \sum_{\substack{n=-\infty \\ n\neq 0}}^{\infty} \Big\{ 4n(4n^2-1)A_n r^{2n-1}\cos\theta\sin 2n\theta \\ + 2(2n-1)r^{2n-2}B_n\left[n\sin 2n\theta + (n-1)\sin 2(n-1)\theta\right] \\ + 4nr^{2n-1}C_n\left\{\sin\theta\sin 2n\theta - 2n\cos\theta\cos 2n\theta\right\} \\ - 4n(n-1)(2n-1)r^{2n-2}D_n\left\{\cos 2n\theta + \cos 2(n-1)\theta\right\}\Big\}.$$

Expressions for plane strain can be found using the transformation given by Eqn. (6.225a). The constants $\hat{D}_2$, J_2, H_2, A_n, B_n, C_n and D_n are determined so as to satisfy the boundary conditions given by Eqns. (6.352b) and (6.352c), or, alternatively, to solve the practical problem of a traction-free semi-circular notch of radius a (i.e., $\tau_{rr}|_{r=a} = \tau_{r\theta}|_{r=a} = 0$), subjected to arbitrary displacement or traction boundary conditions on the edges of the domain barring the edges $\theta = \pm\pi/2$ which are constrained to be traction free (see [212] for the solution of a semi-infinite region under tension with a semi-circular notch, and [190] for a generalization to any circular shaped notch). For example, for a semi-circular disc subjected to tractions $\tau_{rr}|_{r=a} = p\sin^2\theta$, $\tau_{r\theta}|_{r=a} = p\sin\theta\cos\theta$ (i.e., $t_x|_{r=a} = 0$, $t_y|_{r=a} = p\sin\theta$), the only nonzero constant is B_1, so that we get $Eu_r = -pr[\nu - 1 + (1+\nu)\cos 2\theta]/2$, $Eu_\theta = (1+\nu)pr\sin\theta\cos\theta$, $\tau_{rr} = p\sin^2\theta$, $\tau_{\theta\theta} = p\cos^2\theta$ and $\tau_{r\theta} = p\sin\theta\cos\theta$; on the other hand, for a semi-circular notch subjected to far-field loading $\tau_{xx} = \tau_{xy} = 0, \tau_{yy} = p$, we get $\tau_{\theta\theta}|_{r=a,\theta=0} = 3.065p$.

When $\theta_0 = \pi$, the solution again separates into even and odd series. We now have

$$\phi = \hat{D}_2(\sin 2\theta - 2\theta) + J_2 r\theta\sin\theta + H_2 r\theta\cos\theta \\ + \sum_{\substack{n=-\infty \\ n\neq 0}}^{\infty} r^{n+1}\Big\{A_n\left[\cos(n+1)\theta - \cos(n-1)\theta\right] \\ + C_n\left[(n-1)\sin(n+1)\theta - (n+1)\sin(n-1)\theta\right]\Big\} \\ + \sum_{n=-\infty}^{\infty} 4r^{\frac{2n+1}{2}}\Big\{B_n\left[(2n-3)\cos\frac{(2n+1)\theta}{2} - (2n+1)\cos\frac{(2n-3)\theta}{2}\right] \\ + D_n\left[\sin\frac{(2n+1)\theta}{2} - \sin\frac{(2n-3)\theta}{2}\right]\Big\}.$$

The complete solution for the plane stress case (for the plane strain case, use the transformation given by Eqn. (6.225a)) under arbitrary loading is given by

$$Eu_r = (1+\nu)\left[k_1\cos\theta + k_2\sin\theta\right] + \frac{4\hat{D}_2}{r}\sin 2\theta \\ + \frac{1}{2}\left[2(1-\nu)\theta\sin\theta + (1-\nu+4\log r)\cos\theta\right]J_2$$

$$+\frac{1}{2}\left[2(1-\nu)\theta\cos\theta-(1-\nu+4\log r)\sin\theta\right]H_2$$
$$+\sum_{\substack{n=-\infty\\n\neq 0}}^{\infty}2r^n\Big\{A_n\left\{[n(1+\nu)+\nu-1]\sin\theta\sin n\theta-2\cos\theta\cos n\theta\right\}$$
$$+(1+n)C_n\left\{[2-(1+\nu)n]\sin\theta\cos n\theta-(1-\nu)\cos\theta\sin n\theta\right\}\Big\}$$
$$+\sum_{n=-\infty}^{\infty}2(2n+1)r^{\frac{2n-1}{2}}B_n\Big\{[2n(1+\nu)+\nu-7]\cos\frac{(2n-3)\theta}{2}$$
$$-(2n-3)(1+\nu)\cos\frac{(2n+1)\theta}{2}\Big\},$$
$$+\sum_{n=-\infty}^{\infty}2r^{\frac{2n-1}{2}}D_n\Big\{[2(1+\nu)n+\nu-7]\sin\frac{(2n-3)\theta}{2}$$
$$-(2n+1)(1+\nu)\sin\frac{(2n+1)\theta}{2}\Big\},$$
$$Eu_\theta=(1+\nu)[-k_1\sin\theta+k_2\cos\theta+k_3r]+\frac{2\hat{D}_2}{r}\left[(1-\nu)\cos 2\theta+(1+\nu)\right]$$
$$+\frac{1}{2}\left\{2(1-\nu)\theta\cos\theta-[3+\nu+4\log r]\sin\theta\right\}J_2$$
$$-\frac{1}{2}\left\{2(1-\nu)\theta\sin\theta+[3+\nu+4\log r]\cos\theta\right\}H_2$$
$$+\sum_{\substack{n=-\infty\\n\neq 0}}^{\infty}2r^n\Big\{A_n\left\{[n(1+\nu)+2]\sin\theta\cos n\theta-(1-\nu)\cos\theta\sin n\theta\right\}$$
$$+(1+n)C_n\left\{2\cos\theta\cos n\theta+[1-\nu+(1+\nu)n]\sin\theta\sin n\theta\right\}\Big\}$$
$$+\sum_{n=-\infty}^{\infty}2(2n+1)r^{\frac{2n-1}{2}}B_n\Big\{(2n-3)(1+\nu)\sin\frac{(2n+1)\theta}{2}$$
$$-[2n(1+\nu)+5-3\nu]\sin\frac{(2n-3)\theta}{2}\Big\},$$
$$+\sum_{n=-\infty}^{\infty}r^{\frac{2n-1}{2}}D_n\Big\{2\left[2n(1+\nu)+5-3\nu\right]\cos\frac{(2n-3)\theta}{2}$$
$$-2(2n+1)(1+\nu)\cos\frac{(2n+1)\theta}{2}\Big\},$$
$$\tau_{rr}=-\frac{4\hat{D}_2}{r^2}\sin 2\theta+\frac{2J_2}{r}\cos\theta-\frac{2H_2}{r}\sin\theta$$
$$+\sum_{\substack{n=-\infty\\n\neq 0}}^{\infty}nr^{n-1}\Big\{A_n\left[2(n-1)\sin\theta\sin n\theta-4\cos\theta\cos n\theta\right]$$
$$-2(n+1)C_n\left[\cos\theta\sin n\theta+(n-2)\sin\theta\cos n\theta\right]\Big\}$$
$$+\sum_{n=-\infty}^{\infty}r^{\frac{2n-3}{2}}\Big\{B_n(4n^2-1)\left[(3-2n)\cos\frac{(2n+1)\theta}{2}+(2n-7)\cos\frac{(2n-3)\theta}{2}\right]$$

$$
\begin{aligned}
&+ D_n \left[(1-4n^2)\sin\frac{(2n+1)\theta}{2} + [4n(n-4)+7]\sin\frac{(2n-3)\theta}{2} \right] \Bigg\}, \\
\tau_{\theta\theta} &= \sum_{\substack{n=-\infty \\ n\neq 0}}^{\infty} 2n(n+1)r^{n-1}\left\{ -A_n \sin\theta\sin n\theta + C_n\left[n\sin\theta\cos n\theta - \cos\theta\sin n\theta\right]\right\} \\
&+ \sum_{n=-\infty}^{\infty} (4n^2-1)r^{\frac{2n-3}{2}} \left\{ B_n \left[(2n-3)\cos\frac{(2n+1)\theta}{2} - (2n+1)\cos\frac{(2n-3)\theta}{2} \right] \right. \\
&\left. + D_n \left[\sin\frac{(2n+1)\theta}{2} - \sin\frac{(2n-3)\theta}{2} \right] \right\}, \\
\tau_{r\theta} &= \frac{2\hat{D}_2}{r^2}\left[\cos 2\theta - 1\right] \\
&+ \sum_{\substack{n=-\infty \\ n\neq 0}}^{\infty} 2nr^{n-1}\left\{ A_n\left[n\sin\theta\cos n\theta + \cos\theta\sin n\theta\right] + C_n(n^2-1)\sin\theta\sin n\theta \right\} \\
&+ \sum_{n=-\infty}^{\infty} r^{\frac{2n-3}{2}} \left\{ B_n(4n^2-1)(2n-3)\left[\sin\frac{(2n+1)\theta}{2} - \sin\frac{(2n-3)\theta}{2} \right] \right. \\
&\left. + D_n \left[(1-4n^2)\cos\frac{(2n+1)\theta}{2} + [4n(n-2)+3]\cos\frac{(2n-3)\theta}{2} \right] \right\}.
\end{aligned}
$$

Once again, the constants A_n, B_n, C_n and D_n are determined so as to satisfy the boundary conditions given by Eqns. (6.352b) and (6.352c). Note the $1/\sqrt{r}$ singularity in the stress solution at the crack tip $r = 0$ when $a = 0$, in which case $\hat{D}_2 = J_2 = H_2 = 0$, and the index n should run from 1 to ∞. With this choice, we can solve the problem shown in Fig. 6.55a with the boundary conditions given by Eqn. (6.352c). However, the above solution can also be used to solve for a domain of any other shape such as the one shown in Fig. 6.55b provided the boundary conditions in Eqn. (6.352c) are replaced by boundary conditions appropriate to that domain in order to solve for A_n, B_n, C_n and D_n, $n = 1, 2, \ldots, \infty$. Keeping in view that the boundary conditions for the domain in Fig. 6.55b may be more appropriately expressed in terms of Cartesian stress components, we express the above solution after setting $\hat{D}_2$, J_2 and H_2 to zero as

$$
\begin{aligned}
Eu_x &= (1+\nu)[k_1 - k_3 r\sin\theta] \\
&+ \sum_{n=1}^{\infty} r^n \Big\{ A_n\left\{(1+\nu)n\cos(n-2)\theta - [4+(1+\nu)n]\cos n\theta\right\} \\
&+ (1+n)C_n\left\{(1+\nu)n\sin(n-2)\theta - [n+2+\nu(n-2)]\sin n\theta\right\} \Big\} \\
&+ \sum_{n=1}^{\infty} 2(2n+1)r^{\frac{2n-1}{2}} B_n \left\{ (2n-1)(1+\nu)\cos\frac{(2n-5)\theta}{2} \right. \\
&\left. - [2n(1+\nu)+3-5\nu]\cos\frac{(2n-1)\theta}{2} \right\}, \\
&+ \sum_{n=1}^{\infty} 2r^{\frac{2n-1}{2}} D_n \left\{ (2n-1)(1+\nu)\sin\frac{(2n-5)\theta}{2} \right.
\end{aligned}
$$

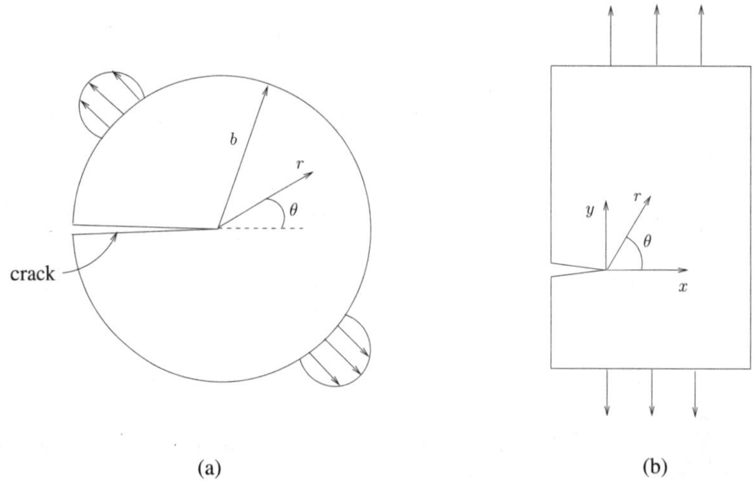

Fig. 6.55 *Domains with crack faces at $\theta = \pm\pi$ and arbitrary loading on the edges.*

$$
\begin{aligned}
&- [7 - \nu + 2(1+\nu)n] \sin\frac{(2n-1)\theta}{2}\Big\}, \\
Eu_y &= (1+\nu)[k_2 + k_3 r\cos\theta] \\
&+ \sum_{n=1}^{\infty} r^n \Big\{ A_n \{[n-2+(n+2)\nu]\sin n\theta - (1+\nu)n\sin(n-2)\theta\} \\
&+ (1+n)C_n\{(1+\nu)n\cos(n-2)\theta + [4-(1+\nu)n]\cos n\theta\}\Big\} \\
&+ \sum_{n=1}^{\infty} 2(2n+1) r^{\frac{2n-1}{2}} B_n \Big\{ [2n(1+\nu) - 9 - \nu]\sin\frac{(2n-1)\theta}{2} \\
&- (2n-1)(1+\nu)\sin\frac{(2n-5)\theta}{2}\Big\}, \\
&+ \sum_{n=1}^{\infty} r^{\frac{2n-1}{2}} D_n \Big\{ 2(2n-1)(1+\nu)\cos\frac{(2n-5)\theta}{2} \\
&- 2[2n(1+\nu) - 5 + 3\nu]\cos\frac{(2n-1)\theta}{2}\Big\}, \\
\tau_{xx} &= \sum_{n=1}^{\infty} n r^{n-1} \Big\{ A_n[(n-1)\cos(n-3)\theta - (n+3)\cos(n-1)\theta] \\
&+ C_n(n+1)[(n-1)\sin(n-3)\theta - (n+1)\sin(n-1)\theta]\Big\} \\
&+ \sum_{n=1}^{\infty} r^{\frac{2n-3}{2}} \Big\{ B_n(4n^2-1)\left[(2n-3)\cos\frac{(2n-7)\theta}{2} - (2n+1)\cos\frac{(2n-3)\theta}{2}\right] \\
&+ D_n(2n-1)\left[(2n-3)\sin\frac{(2n-7)\theta}{2} - (2n+5)\sin\frac{(2n-3)\theta}{2}\right]\Big\},
\end{aligned}
$$

$$
\begin{aligned}
\tau_{yy} = \sum_{n=1}^{\infty} n r^{n-1} \Big\{ & A_n (n-1) \left[\cos(n-1)\theta - \cos(n-3)\theta \right] \\
& - C_n (n+1) \left[(n-1) \sin(n-3)\theta - (n-3) \sin(n-1)\theta \right] \Big\} \\
& + \sum_{n=1}^{\infty} r^{\frac{2n-3}{2}} \left\{ (4n^2-1) B_n \left[(2n-7) \cos \frac{(2n-3)\theta}{2} - (2n-3) \cos \frac{(2n-7)\theta}{2} \right] \right. \\
& \left. + \left[3 + 4n(n-2) \right] D_n \left[\sin \frac{(2n-3)\theta}{2} - \sin \frac{(2n-7)\theta}{2} \right] \right\}, \\
\tau_{xy} = \sum_{n=1}^{\infty} n r^{n-1} \Big\{ & A_n \left[(n+1) \sin(n-1)\theta - (n-1) \sin(n-3)\theta \right] \\
& + C_n (n^2-1) \left[\cos(n-3)\theta - \cos(n-1)\theta \right] \Big\} \\
& + \sum_{n=1}^{\infty} r^{\frac{2n-3}{2}} \left\{ B_n (4n^2-1)(2n-3) \left[\sin \frac{(2n-3)\theta}{2} - \sin \frac{(2n-7)\theta}{2} \right] \right. \\
& \left. + D_n \left[(1-4n^2) \cos \frac{(2n-3)\theta}{2} + \left[4n(n-2) + 3 \right] \cos \frac{(2n-7)\theta}{2} \right] \right\},
\end{aligned}
$$

where $\log(x+iy) = \log r + i\theta$ can be used to find (r, θ) given (x, y). The constants A_n, B_n, C_n and D_n can be found using a weighted-residual method. One can also use the above solution to solve problems where the crack lies completely inside a rectangular domain, and where the placement of the crack and the loading boundary conditions are such that one can analyze it using the domain shown in Fig. 6.55b by imposing suitable symmetry boundary conditions on some of the edges. For example, if a crack was symmetrically placed along the x-axis of a rectangular plate with loading $\tau_{xy} = 0$ and $\tau_{yy} = p$ on the top and bottom edges and the lateral surfaces traction free (this problem is analogous to that shown in Fig. 6.40 except that now the domain, instead of being unbounded, is a rectangle), then we can analyze it by restricting $\theta \geq 0$ in Fig. 6.55b, and by imposing the symmetry boundary conditions $u_x = \tau_{xy} = 0$ and $u_y = \tau_{xy} = 0$ on the left and $\theta = 0$ edges respectively, and by imposing the appropriate traction conditions on the top and right edges. In case the crack of length $2a$ is unsymmetrically placed, then one can establish another local coordinate system $(\tilde{x}, \tilde{y})$ (with associated coordinate system $(\tilde{r}, \tilde{\theta})$ in polar coordinates) at the other crack tip with the relations $\log(\tilde{x} + i\tilde{y}) = \log \tilde{r} + i\tilde{\theta}$, $\tilde{x} + x = -2a$, $\tilde{y} = y$, superpose the two solutions (i.e., use the total solution $(u_x - u_{\tilde{x}}, u_y + u_{\tilde{y}}, \tau_{xx} + \tau_{\tilde{x}\tilde{x}}, \tau_{yy} + \tau_{\tilde{y}\tilde{y}}, \tau_{xy} - \tau_{\tilde{x}\tilde{y}})$), and determine the constants so as to satisfy the boundary conditions at the four edges of the rectangular plate.

Consider the problem of a curved cantilever that is loaded on its curved surfaces as shown in Fig. 6.56, and whose edge $\theta = \theta_0$ is traction-free. The boundary conditions are

$$\tau_{\theta\theta}|_{\theta=\theta_0} = 0, \qquad \tau_{r\theta}|_{\theta=\theta_0} = 0, \tag{6.363a}$$

$$\tau_{rr}|_{r=a} = p_i(\theta), \qquad \tau_{rr}|_{r=b} = p_o(\theta), \tag{6.363b}$$

$$\tau_{r\theta}|_{r=a} = s_i(\theta), \qquad \tau_{r\theta}|_{r=b} = s_o(\theta), \tag{6.363c}$$

$$\boldsymbol{u}|_{\theta=0} = 0. \tag{6.363d}$$

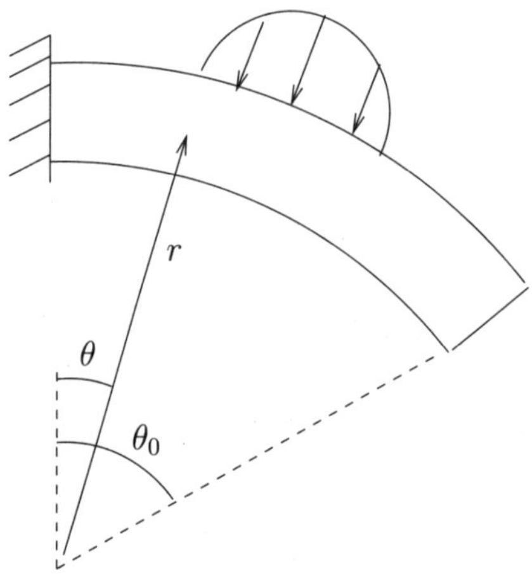

Fig. 6.56 *Curved cantilever beam subjected to tractions on its curved surfaces $r = a$ and $r = b$.*

We take the Airy stress function to be of the form

$$\phi = \sum_{n=1}^{\infty} r^{\lambda_n+1}\Big\{A_n\left[\cos(\lambda_n+1)(\theta-\theta_0) - \cos(\lambda_n-1)(\theta-\theta_0)\right] \\ + B_n\left[(\lambda_n-1)\sin(\lambda_n+1)(\theta-\theta_0) - (\lambda_n+1)\sin(\lambda_n-1)(\theta-\theta_0)\right]\Big\}, \tag{6.364}$$

where $\lambda_n, n = 1, 2, \ldots, \infty$, satisfy the equation

$$(3-\nu)(1+\nu)\cos 2\lambda_n\theta_0 - 2\lambda_n^2(1+\nu)^2\sin^2\theta_0 + \nu(\nu-2) + 5 = 0. \tag{6.365}$$

and where

$$B_n = \frac{1}{1+\lambda_n}\left[\frac{[2+(1+\nu)\lambda_n]\cos\lambda_n\theta_0\sin\theta_0 - (1-\nu)\sin\lambda_n\theta_0\cos\theta_0}{2\cos\lambda_n\theta_0\cos\theta_0 + [1-\nu+(1+\nu)\lambda_n]\sin\lambda_n\theta_0\sin\theta_0}\right]A_n.$$

The above form automatically satisfies the boundary conditions given by Eqns. (6.363a) and (6.363d). The displacement and stress solution corresponding to plane stress conditions is

$$Eu_r = \sum_{n=1}^{\infty} r^{\lambda_n}\Big\{A_n\Big[[(1+\nu)\lambda_n+\nu-3]\cos(\lambda_n-1)(\theta-\theta_0) \\ - (1+\nu)(\lambda_n+1)\cos(\lambda_n+1)(\theta-\theta_0)\Big] \\ + (1+\lambda_n)B_n\Big[[(1+\nu)\lambda_n+\nu-3]\sin(\lambda_n-1)(\theta-\theta_0) \\ - (1+\nu)(\lambda_n-1)\sin(\lambda_n+1)(\theta-\theta_0)\Big]\Big\},$$

$$Eu_\theta = \sum_{n=1}^{\infty} r^{\lambda_n}\Big\{A_n\Big[(1+\nu)(\lambda_n+1)\sin(\lambda_n+1)(\theta-\theta_0) \\ - [3-\nu+(1+\nu)\lambda_n]\sin(\lambda_n-1)(\theta-\theta_0)\Big]$$

$$
\begin{aligned}
&\quad - (1+\lambda_n)B_n\Big[(1+\nu)(\lambda_n-1)\cos(\lambda_n+1)(\theta-\theta_0)\\
&\quad - [3-\nu+(1+\nu)\lambda_n]\cos(\lambda_n-1)(\theta-\theta_0)\Big]\Big\},\\
\tau_{rr} &= \sum_{n=1}^{\infty}\lambda_n r^{\lambda_n-1}\Big\{A_n\left[(\lambda_n-3)\cos(\lambda_n-1)(\theta-\theta_0)-(\lambda_n+1)\cos(\lambda_n+1)(\theta-\theta_0)\right]\\
&\quad + (1+\lambda_n)B_n\left[(\lambda_n-3)\sin(\lambda_n-1)(\theta-\theta_0)-(\lambda_n-1)\sin(\lambda_n+1)(\theta-\theta_0)\right]\Big\},\\
\tau_{\theta\theta} &= \sum_{n=1}^{\infty}\lambda_n(\lambda_n+1)r^{\lambda_n-1}\Big\{A_n\left[\cos(\lambda_n+1)(\theta-\theta_0)-\cos(\lambda_n-1)(\theta-\theta_0)\right]\\
&\quad + B_n\left[(\lambda_n-1)\sin(\lambda_n+1)(\theta-\theta_0)-(\lambda_n+1)\sin(\lambda_n-1)(\theta-\theta_0)\right]\Big\},\\
\tau_{r\theta} &= \sum_{n=1}^{\infty}\lambda_n r^{\lambda_n-1}\Big\{A_n\left[(1+\lambda_n)\sin(\lambda_n+1)(\theta-\theta_0)-(\lambda_n-1)\sin(\lambda_n-1)(\theta-\theta_0)\right]\\
&\quad - (\lambda_n^2-1)B_n\left[\cos(\lambda_n+1)(\theta-\theta_0)-\cos(\lambda_n-1)(\theta-\theta_0)\right]\Big\}.
\end{aligned}
$$

The complex-valued constants A_n in the above stress solution are then found so as to satisfy the boundary conditions in Eqns. (6.363b) and (6.363c). Note that if $a > 0$, then the roots of λ_n with positive and negative real parts should both be considered, while in case the inner radius a is zero, then only those λ_n with a positive real part should be considered. Thus, in case $a > 0$, the complex-valued coefficients A_n corresponding to the positive and negative real parts of λ_n can be considered as four sets of constants that are used for satisfying the four boundary conditions in Eqns. (6.363b) and (6.363c).

The case $\theta_0 = \pi$ and $a = 0$ solves problems where part of a straight edge is fixed and the remaining part is traction free such as the cracked semicircular disc and cantilever beam shown in Figs. 6.57a and 6.57b, respectively. For this case, Eqn. (6.365) reduces to

$$\cos 2\pi\lambda_n + \alpha = 0,$$

where $\alpha = [5+\nu(\nu-2)]/[(3-\nu)(1+\nu)]$. The roots λ_n are given by $n - 1/2 + i(\cosh^{-1}\alpha)/(2\pi)$, $n = 1, 2, \ldots, \infty$, or, equivalently, by $n - 1/2 + (i/(2\pi))\log[(3-\nu)/(1+\nu)]$, which

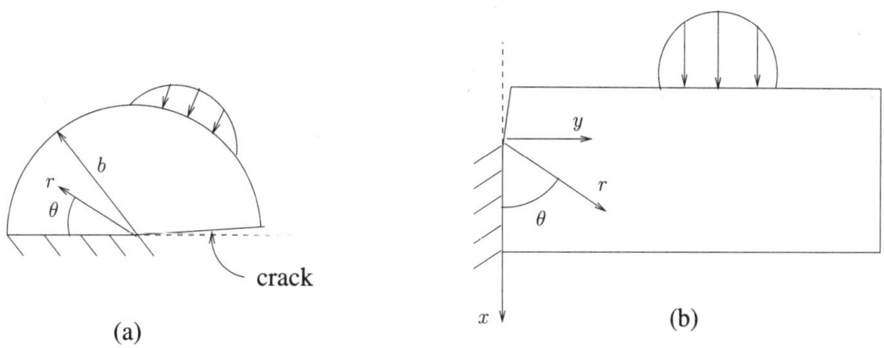

Fig. 6.57 *Domains where part of a straight edge is fixed and the remaining part is traction free ('cracked').*

agrees with the result in [113] (since the complex-conjugate of each λ_n is also a root). The real part of λ_1 leads to a $1/\sqrt{r}$ stress singularity as $r \to 0$, while the imaginary part leads to a bounded oscillatory response since $r^{i\beta} = e^{i\beta \log r} = \cos(\beta \log r) + i \sin(\beta \log r)$. For λ_n given by the above expression, we have

$$B_n = \frac{A_n(\nu - 1)\tan(\pi\lambda_n)}{2(1+\lambda_n)} = -\frac{2i\pi A_n}{(1+2n)\pi + i\log\frac{3-\nu}{1+\nu}}.$$

As in the case of the previous problem, it may be more convenient to use the following Cartesian components while solving problems such as the one shown in Fig. 6.57b:

$$2Eu_x = \sum_{n=1}^{\infty} r^{\lambda_n} A_n \Big\{ 2\big[[(1+\nu)\lambda_n + 4]\cos\lambda_n(\pi-\theta) - (1+\nu)\lambda_n \cos[\lambda_n(\pi-\theta)+2\theta]\big]$$
$$+ (1-\nu)\tan\lambda_n\pi \big[[(1+\nu)\lambda_n + 2(1-\nu)]\sin\lambda_n(\pi-\theta) - (1+\nu)\lambda_n \sin[\lambda_n(\pi-\theta)+2\theta]\big]\Big\},$$

$$2Eu_y = \sum_{n=1}^{\infty} r^{\lambda_n} A_n \Big\{ 2\big[[(1+\nu)\lambda_n - 2(1-\nu)]\sin\lambda_n(\pi-\theta) - (1+\nu)\lambda_n \sin[\lambda_n(\pi-\theta)+2\theta]\big]$$
$$+ (\nu-1)\tan\lambda_n\pi \big[[(1+\nu)\lambda_n - 4]\cos\lambda_n(\pi-\theta) - (1+\nu)\lambda_n \cos[\lambda_n(\pi-\theta)+2\theta]\big]\Big\},$$

$$\tau_{xx} = \sum_{n=1}^{\infty} \lambda_n r^{\lambda_n - 1} A_n \Big\{ \big[(\lambda_n+3)\cos[\lambda_n(\pi-\theta)+\theta] - (\lambda_n-1)\cos[\lambda_n(\pi-\theta)+3\theta]\big]$$
$$+ (\nu-1)\tan\lambda_n\pi \big[(\lambda_n-1)\cos[\lambda_n(\pi-\theta)+2\theta]\sin\theta - \sin[\lambda_n(\pi-\theta)+\theta]\big]\Big\},$$

$$\tau_{yy} = \sum_{n=1}^{\infty} \lambda_n r^{\lambda_n - 1} A_n \Big\{ 2(1-\lambda_n)\sin\theta \sin[\lambda_n(\pi-\theta)+2\theta]$$
$$+ (\nu-1)\tan\lambda_n\pi \big[(1-\lambda_n)\cos[\lambda_n(\pi-\theta)+2\theta]\sin\theta - \sin[\lambda_n(\pi-\theta)+\theta]\big]\Big\},$$

$$\tau_{xy} = \sum_{n=1}^{\infty} \lambda_n r^{\lambda_n - 1} A_n \Big\{ \big[(1+\lambda_n)\sin[\lambda_n(\pi-\theta)+\theta] + (1-\lambda_n)\sin[\lambda_n(\pi-\theta)+3\theta]\big]$$
$$+ (\lambda_n - 1)(\nu-1)\tan\lambda_n\pi \sin\theta \sin[\lambda_n(\pi-\theta)+2\theta]\Big\}.$$

The complex-valued constants A_n are now determined so as to satisfy the traction boundary conditions on the remaining three edges.

Under gravity loading along the vertical direction $\theta = \pi$, a particular solution under plane stress conditions that satisfies the equations of equilibrium is obtained using Eqn. (6.221) with $W_b = -\rho g r \cos\theta$, and the terms associated with A_2, B_2, C_2 and D_2 in Eqn. (6.239) as

$$2Eu_r = (1-\nu)\rho g r^2 \cos\theta + 2r^2\Big[-3(1+\nu)\cos 3\theta A_2 - 3(1+\nu)\sin 3\theta B_2$$
$$+ (1-3\nu)C_2\cos\theta + (1-3\nu)D_2\sin\theta\Big],$$

$$2Eu_\theta = (1-\nu)\rho g r^2 \sin\theta + 2r^2\Big[3(1+\nu)A_2\sin 3\theta - 3(1+\nu)\cos 3\theta B_2$$
$$+ (5+\nu)C_2\sin\theta - (5+\nu)D_2\cos\theta\Big],$$

$$\tau_{rr} = \rho g r\cos\theta - r\left[6A_2\cos 3\theta + 6B_2\sin 3\theta - 2C_2\cos\theta - 2D_2\sin\theta\right],$$
$$\tau_{\theta\theta} = \rho g r\cos\theta + 6r\left[A_2\cos 3\theta + B_2\sin 3\theta + C_2\cos\theta + D_2\sin\theta\right],$$
$$\tau_{r\theta} = r\left[6A_2\sin 3\theta - 6B_2\cos 3\theta + 2C_2\sin\theta - 2D_2\cos\theta\right].$$

The constants A_2, B_2, C_2 and D_2 are now found so as to satisfy Eqns. (6.363a) and (6.363d). We get

$$A_2 = \frac{\rho g N_1}{D}, \quad B_2 = \frac{\rho g N_2}{D}, \quad C_2 = \frac{\rho g N_3}{D}, \quad D_2 = \frac{\rho g N_4}{D},$$

where

$$N_1 = 4(4-\nu-\nu^2)\cos 2\theta_0 + (1+\nu)\left[3\nu - 5 + (5+\nu)\cos 4\theta_0\right],$$
$$N_2 = 2(5+\nu)\left[1-\nu+(1+\nu)\cos 2\theta_0\right]\sin 2\theta_0,$$
$$N_3 = -3\left[9-\nu(3\nu+10)+4\nu(1+\nu)\cos 2\theta_0 + (7-\nu)(1+\nu)\cos 4\theta_0\right],$$
$$N_4 = -6(1+\nu)\left[1-\nu+(1+\nu)\cos 2\theta_0\right]\sin 2\theta_0,$$
$$D = 24\left[1-\nu(3\nu+10)+4(1+\nu)^2\cos 2\theta_0 + (1+\nu)(3-\nu)\cos 4\theta_0\right].$$

The particular solution should be superposed on the solution corresponding to Eqn. (6.364), and the constants adjusted to satisfy the traction boundary conditions on the curved edges.

Now consider the case where a wedge is constrained as shown in Fig. 6.58, with tractions applied on the curved edges. We first consider the case of symmetric loading about $\theta = 0$. The boundary conditions are

$$u_\theta|_{\theta=0} = 0, \qquad \tau_{r\theta}|_{\theta=0} = 0, \tag{6.366a}$$

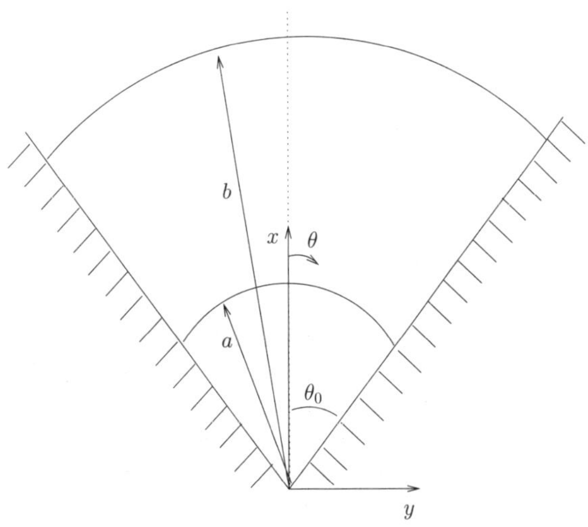

Fig. 6.58 *Constrained wedge subjected to tractions on its curved edges.*

$$\tau_{rr}\big|_{r=a} = p_i(\theta), \qquad \tau_{rr}\big|_{r=b} = p_o(\theta), \tag{6.366b}$$

$$\tau_{r\theta}\big|_{r=a} = s_i(\theta), \qquad \tau_{r\theta}\big|_{r=b} = s_o(\theta), \tag{6.366c}$$

$$\boldsymbol{u}\big|_{\theta=\theta_0} = 0. \tag{6.366d}$$

Since the loading is symmetric, and in view of Eqn. (6.366d), we consider only the infinite series terms that are even in θ in Eqn. (6.239), namely the terms associated with A_n and C_n. The solution for the plane stress case that automatically satisfies Eqns. (6.366a) and (6.366d) is obtained (after renaming the constants) as

$$\begin{aligned}
2\mu u_r &= \sum_{n=1}^{\infty} A_n \left[3 - \nu - (1+\nu)\lambda_n\right] r^{\lambda_n} \Big[\cos(\lambda_n+1)\theta_0 \cos(\lambda_n-1)\theta \\
&\quad - \cos(\lambda_n-1)\theta_0 \cos(\lambda_n+1)\theta\Big], \\
2\mu u_\theta &= \sum_{n=1}^{\infty} A_n r^{\lambda_n} \Big\{\left[3 - \nu + (1+\nu)\lambda_n\right] \cos(\lambda_n+1)\theta_0 \sin(\lambda_n-1)\theta \\
&\quad + \left[3 - \nu - (1+\nu)\lambda_n\right] \cos(\lambda_n-1)\theta_0 \sin(\lambda_n+1)\theta\Big\}, \\
\tau_{rr} &= \sum_{n=1}^{\infty} A_n \lambda_n r^{\lambda_n-1} \Big\{(3-\lambda_n)(1+\nu) \cos(\lambda_n+1)\theta_0 \cos(\lambda_n-1)\theta \\
&\quad + \left[\nu - 3 + (1+\nu)\lambda_n\right] \cos(\lambda_n-1)\theta_0 \cos(\lambda_n+1)\theta\Big\}, \\
\tau_{\theta\theta} &= \sum_{n=1}^{\infty} A_n \lambda_n r^{\lambda_n-1} \Big\{(1+\lambda_n)(1+\nu) \cos(\lambda_n+1)\theta_0 \cos(\lambda_n-1)\theta \\
&\quad + \left[3 - \nu - (1+\nu)\lambda_n\right] \cos(\lambda_n-1)\theta_0 \cos(\lambda_n+1)\theta\Big\}, \\
\tau_{r\theta} &= \sum_{n=1}^{\infty} A_n \lambda_n r^{\lambda_n-1} \Big\{(\lambda_n-1)(1+\nu) \cos(\lambda_n+1)\theta_0 \sin(\lambda_n-1)\theta \\
&\quad + \left[3 - \nu - (1+\nu)\lambda_n\right] \cos(\lambda_n-1)\theta_0 \sin(\lambda_n+1)\theta\Big\},
\end{aligned} \tag{6.367}$$

where λ_n satisfies the characteristic equation (compare with Eqn. (6.359))

$$\sin 2\lambda_n\theta_0 = \left(\frac{1+\nu}{3-\nu}\right) \lambda_n \sin 2\theta_0.$$

The complex-valued constants A_n corresponding to λ_n with positive and negative real parts are found so as to satisfy the conditions given by Eqns. (6.366b)–(6.366c). If $a = 0$, then only those λ_n with positive real parts are considered.

Under gravity loading along the vertical direction $\theta = \pi$, a particular solution under plane stress conditions that satisfies the equations of equilibrium, the boundary conditions $u_\theta\big|_{\theta=0} = \tau_{r\theta}\big|_{\theta=0} = 0$ and Eqn. (6.366d), obtained using Eqn. (6.221) with $W_b = -\rho g r \cos\theta$ and the terms associated with A_2 and C_2 in Eqn. (6.239) is

$$Eu_r = \frac{1}{D}\left[\rho g(1-\nu^2) r^2 (\cos 2\theta - \cos 2\theta_0)\cos\theta\right],$$

$$Eu_\theta = -\frac{1}{D}\left[\rho g(1-\nu^2)r^2(\cos 2\theta - \cos 2\theta_0)\sin\theta\right],$$

$$\tau_{rr} = -\frac{1}{D}\left[2\rho g r(\cos 2\theta_0 - \nu - (1-\nu)\cos 2\theta)\cos\theta\right]$$

$$\tau_{\theta\theta} = \frac{1}{D}\left[2\rho g r(1-\nu\cos 2\theta_0 - \nu - (1-\nu)\cos 2\theta)\cos\theta\right],$$

$$\tau_{r\theta} = -\frac{1}{D}\left[2\rho g(1-\nu)r(\cos 2\theta + \sin^2\theta_0)\sin\theta\right],$$

where

$$D = 1+\nu-(3-\nu)\cos 2\theta_0.$$

The above solution should be superposed on Eqn. (6.367), and the constants adjusted so as to meet the boundary conditions given by Eqns. (6.366b) and (6.366c). The particular solution is invalid for θ_0 that satisfies $\cos 2\theta_0 = (1+\nu)/(3-\nu)$. For this angle, we consider only the C_2, $\tilde{A}_2$ and $\tilde{C}_2$ terms in Eqn. (6.239). The final particular solution that we get is

$$\begin{aligned}
2Eu_r &= r^2\Big\{[(2-6\nu)C_2+\rho g(1-\nu)]\cos\theta + 2\tilde{A}_2(1+\nu)\left[3\theta\sin 3\theta - (1+3\log r)\cos 3\theta\right] \\
&\quad - 2\tilde{C}_2\left[(1-3\nu)\theta\sin\theta + [1+\nu+(3\nu-1)\log r]\cos\theta\right]\Big\}, \\
2Eu_\theta &= r^2\Big\{[(10+2\nu)C_2+\rho g(1-\nu)]\sin\theta + 2\tilde{A}_2(1+\nu)\left[3\theta\cos 3\theta + (1+3\log r)\sin 3\theta\right] \\
&\quad + 2\tilde{C}_2\left[(5+\nu)\theta\cos\theta + [1+\nu+(5+\nu)\log r]\sin\theta\right]\Big\}, \\
\tau_{rr} &= r\Big\{(2C_2+\rho g)\cos\theta + \tilde{A}_2\left[6\theta\sin 3\theta - (5+6\log r)\cos 3\theta\right] \\
&\quad - \tilde{C}_2\left[(1-2\log r)\cos\theta + 2\theta\sin\theta\right]\Big\}, \\
\tau_{\theta\theta} &= r\Big\{(6C_2+\rho g)\cos\theta - \tilde{A}_2\left[6\theta\sin 3\theta - (5+6\log r)\cos 3\theta\right] \\
&\quad + \tilde{C}_2\left[(5+6\log r)\cos\theta - 6\theta\sin\theta\right]\Big\}, \\
\tau_{r\theta} &= r\Big\{2C_2\sin\theta + \tilde{A}_2\left[6\theta\cos 3\theta + (5+6\log r)\sin 3\theta\right] \\
&\quad + \tilde{C}_2\left[(3+2\log r)\sin\theta + 2\theta\cos\theta\right]\Big\},
\end{aligned} \tag{6.368}$$

where the constants C_2, $\tilde{A}_2$ and $\tilde{C}_2$ are found using the boundary condition given by Eqn. (6.366d) and the condition $\cos 2\theta_0 = (1+\nu)/(3-\nu)$ as

$$C_2 = \frac{\rho g(1-\nu)\left[6(3-\nu)\theta_0 + (1+\nu)\sin 2\theta_0\right]}{12\left[-7+\nu(10+\nu)\right]\theta_0 + 6(\nu-3)(1+\nu)\sin 2\theta_0},$$

$$\tilde{A}_2 = \frac{\rho g(3-\nu)(1-\nu)\sin 2\theta_0}{6\left[-7+\nu(10+\nu)\right]\theta_0 + 3(\nu-3)(1+\nu)\sin 2\theta_0},$$

$$\tilde{C}_2 = -\frac{\rho g(1-\nu^2)\sin 2\theta_0}{2\left[-7+\nu(10+\nu)\right]\theta_0 + (\nu-3)(1+\nu)\sin 2\theta_0}.$$

In case the loading on the wedge in Fig. 6.58 is antisymmetric about the x-axis, in place of the solution given by Eqn. (6.367), we now get

$$
\begin{aligned}
2\mu u_r &= \sum_{n=1}^{\infty} A_n r^{\lambda_n}\Big\{[3-\nu-(1+\nu)\lambda_n]\cos(\lambda_n+1)\theta_0\sin(\lambda_n-1)\theta \\
&\quad + [3-\nu+(1+\nu)\lambda_n]\cos(\lambda_n-1)\theta_0\sin(\lambda_n+1)\theta\Big\}, \\
2\mu u_\theta &= -\sum_{n=1}^{\infty}[3-\nu+(1+\nu)\lambda_n]\,A_n r^{\lambda_n}\Big\{\cos(\lambda_n+1)\theta_0\cos(\lambda_n-1)\theta \\
&\quad - \cos(\lambda_n-1)\theta_0\cos(\lambda_n+1)\theta\Big\}, \\
\tau_{rr} &= \sum_{n=1}^{\infty} A_n\lambda_n r^{\lambda_n-1}\Big\{(3-\lambda_n)(1+\nu)\cos(\lambda_n+1)\theta_0\sin(\lambda_n-1)\theta \\
&\quad + [3-\nu+(1+\nu)\lambda_n]\cos(\lambda_n-1)\theta_0\sin(\lambda_n+1)\theta\Big\}, \\
\tau_{\theta\theta} &= \sum_{n=1}^{\infty} A_n\lambda_n r^{\lambda_n-1}\Big\{(1+\lambda_n)(1+\nu)\cos(\lambda_n+1)\theta_0\sin(\lambda_n-1)\theta \\
&\quad - [3-\nu+(1+\nu)\lambda_n]\cos(\lambda_n-1)\theta_0\sin(\lambda_n+1)\theta\Big\}, \\
\tau_{r\theta} &= -\sum_{n=1}^{\infty} A_n\lambda_n r^{\lambda_n-1}\Big\{(\lambda_n-1)(1+\nu)\cos(\lambda_n+1)\theta_0\cos(\lambda_n-1)\theta \\
&\quad - [3-\nu+(1+\nu)\lambda_n]\cos(\lambda_n-1)\theta_0\cos(\lambda_n+1)\theta\Big\},
\end{aligned}
$$

where λ_n satisfies the characteristic equation (compare with Eqn. (6.358))

$$
\sin 2\lambda_n\theta_0 = -\left(\frac{1+\nu}{3-\nu}\right)\lambda_n\sin 2\theta_0.
$$

Similar to the symmetric loading case, the A_n corresponding to λ_n with positive and negative real parts are found so as to satisfy the conditions given by Eqns. (6.366b)–(6.366c); if $a=0$, then only those λ_n with positive real parts are considered.

In case we have roller supports at the ends $\theta=\pm\theta_0$ as shown in Fig. 6.59, i.e.,

$$
u_\theta|_{\theta=\pm\theta_0} = \tau_{r\theta}|_{\theta=\pm\theta_0} = 0, \tag{6.369}
$$

in place of the boundary condition given by Eqn. (6.366d), then the solution for the plane stress case that automatically satisfies Eqns. (6.366a) and (6.369) for the case of symmetric loading about the x-axis is obtained as

$$
\begin{aligned}
2Eu_r &= 4(1-\nu)\hat{C}_1 r - \frac{2(1+\nu)F_1}{r} - 2\sum_{\substack{n=-\infty \\ n\neq 0}}^{\infty}\Big\{r^{\lambda_n}(\lambda_n+1)(1+\nu)A_n\cos(\lambda_n+1)\theta \\
&\quad + r^{\gamma_n}\left[(1+\nu)(\gamma_n+1)-4\right]C_n\cos(\gamma_n-1)\theta\Big\}, \\
2Eu_\theta &= 2\sum_{\substack{n=-\infty \\ n\neq 0}}^{\infty}\Big\{r^{\lambda_n}(\lambda_n+1)(1+\nu)A_n\sin(\lambda_n+1)\theta
\end{aligned}
$$

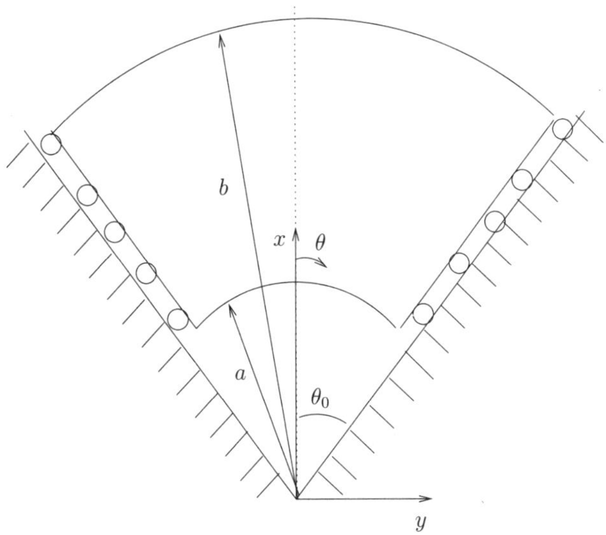

Fig. 6.59 Constrained wedge with roller supports subjected to tractions on its curved edges.

$$+ r^{\gamma_n}\left[3+\gamma_n+(\gamma_n-1)\nu\right]C_n\sin(\gamma_n-1)\theta\Big\},$$

$$\tau_{rr} = 2\hat{C}_1 + \frac{F_1}{r^2} - \sum_{\substack{n=-\infty\\ n\neq 0}}^{\infty}\Big\{r^{\lambda_n-1}\lambda_n(\lambda_n+1)A_n\cos(\lambda_n+1)\theta \tag{6.370}$$

$$+ r^{\gamma_n-1}(\gamma_n-3)\gamma_n C_n\cos(\gamma_n-1)\theta\Big\},$$

$$\tau_{\theta\theta} = 2\hat{C}_1 - \frac{F_1}{r^2} + \sum_{\substack{n=-\infty\\ n\neq 0}}^{\infty}\Big\{r^{\lambda_n-1}\lambda_n(\lambda_n+1)A_n\cos(\lambda_n+1)\theta$$

$$+ r^{\gamma_n-1}\gamma_n(\gamma_n+1)C_n\cos(\gamma_n-1)\theta\Big\},$$

$$\tau_{r\theta} = \sum_{\substack{n=-\infty\\ n\neq 0}}^{\infty}\Big\{r^{\lambda_n-1}\lambda_n(\lambda_n+1)A_n\sin(\lambda_n+1)\theta + r^{\gamma_n-1}(\gamma_n-1)\gamma_n C_n\sin(\gamma_n-1)\theta\Big\},$$

where

$$\lambda_n = \frac{n\pi}{\theta_0} - 1, \tag{6.371a}$$

$$\gamma_n = \frac{n\pi}{\theta_0} + 1, \tag{6.371b}$$

$$4\hat{C}_1\theta_0 + \frac{2F_1\theta_0}{a^2} = \int_{-\theta_0}^{\theta_0} p_i(\theta)\,d\theta, \tag{6.371c}$$

$$4\hat{C}_1\theta_0 + \frac{2F_1\theta_0}{b^2} = \int_{-\theta_0}^{\theta_0} p_o(\theta)\,d\theta. \tag{6.371d}$$

The constants $\hat{C}_1$, F_1, A_n and C_n for positive and negative n (total of four sets) are now found so as to satisfy the conditions given by Eqns. (6.366b)–(6.366c), and Eqns. (6.371c) and (6.371d). If $a = 0$, then the constant F_1 is set to zero, and the index n runs from 1 to ∞.

Timoshenko and Goodier [324] (Article 41) consider the problem of a circular disc of radius a that is subjected to point loads P (actually, P is the load per unit width along the direction perpendicular to the plane of the disc) along two diametrically opposite points on the periphery. We solve this problem by first considering the problem shown in Fig. 6.60, (which by symmetry considerations, is identical to the one shown in Fig. 6.59 with $\theta_0 = \pi/2$), and then taking the limit as $\alpha \to 0$. Thus, we have $p_o = -P/(2\alpha a)$ for $-\alpha \leq \theta \leq \alpha$, $\lambda_n = 2n-1$, $\gamma_n = 2n+1$, $\hat{C}_1 = -P/(2\pi a)$, $F_1 = 0$, and using the fact that $\tau_{r\theta}|_{r=a} = 0$, and the orthogonality of the cosine functions, we get

$$A_n = \frac{a^{1-2n} P \sin 2n\alpha}{2n(2n-1)\pi\alpha},$$

$$C_n = -\frac{a^{-2n-1} P \sin 2n\alpha}{2n(2n+1)\pi\alpha}.$$

The final solution for the problem shown in Fig. 6.60 under plane stress conditions is

$$2Eu_r = \frac{2(\nu-1)Pr}{\pi a} - \frac{2P}{\pi\alpha}\sum_{n=1}^{\infty} \frac{r^{2n-1}\left[n(2n+1)(1+\nu)a^2 - (2n-1)\left[n-1+(1+n)\nu\right]r^2\right]\sin 2n\alpha \cos 2n\theta}{a^{2n+1}n(4n^2-1)},$$

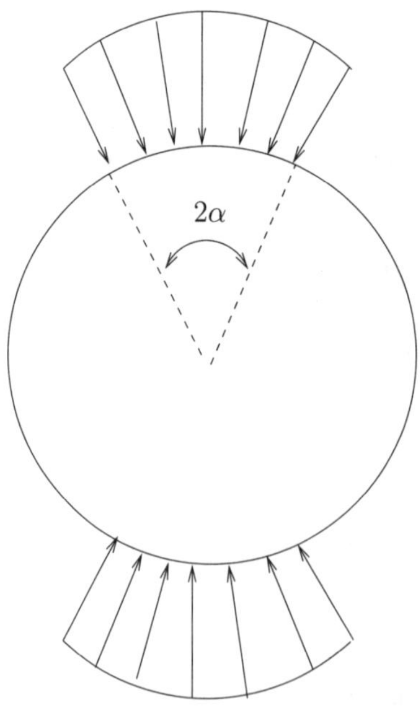

Fig. 6.60 *Circular disc subjected to equal and opposite normal tractions on its curved edge.*

$$2Eu_\theta = \frac{2P}{\pi\alpha}\sum_{n=1}^{\infty}\frac{r^{2n-1}\left[n(2n+1)(1+\nu)a^2-(2n-1)(2+n+n\nu)r^2\right]\sin 2n\alpha\sin 2n\theta}{a^{2n+1}n(4n^2-1)},$$

$$\tau_{rr} = -\frac{P}{\pi a}\left\{1+\frac{1}{\alpha}\sum_{n=1}^{\infty}\frac{r^{2n-2}}{na^{2n}}\left[r^2+n(a^2-r^2)\right]\sin 2n\alpha\cos 2n\theta\right\},$$

$$\tau_{\theta\theta} = -\frac{P}{\pi a}\left\{1+\frac{1}{\alpha}\sum_{n=1}^{\infty}\frac{r^{2n-2}}{na^{2n}}\left[(n+1)r^2-na^2\right]\sin 2n\alpha\cos 2n\theta\right\},$$

$$\tau_{r\theta} = \frac{P(a^2-r^2)}{\pi a\alpha}\sum_{n=1}^{\infty}\frac{r^{2n-2}}{a^{2n}}\sin 2n\alpha\sin 2n\theta.$$

Taking the limit as $\alpha \to 0$, and carrying out the summations, we get the solution under point loads as

$$\begin{aligned}
Eu_r &= \frac{P}{\pi}\left\{\frac{r(1-\nu)}{a} - \log\left(\frac{a^2+r^2+2ar\cos\theta}{a^2+r^2-2ar\cos\theta}\right)\cos\theta - (1-\nu)\tan^{-1}\left(\frac{2ar\sin\theta}{a^2-r^2}\right)\sin\theta\right.\\
&\quad\left.+\frac{2a(1+\nu)r(a^2+r^2)\sin^2\theta}{a^4+r^4-2a^2r^2\cos 2\theta}\right\},\\
Eu_\theta &= \frac{P}{\pi}\left\{\log\left(\frac{a^2+r^2+2ar\cos\theta}{a^2+r^2-2ar\cos\theta}\right)\sin\theta - (1-\nu)\tan^{-1}\left(\frac{2ar\sin\theta}{a^2-r^2}\right)\cos\theta\right.\\
&\quad\left.+\frac{a(1+\nu)r(a^2-r^2)\sin 2\theta}{a^4+r^4-2a^2r^2\cos 2\theta}\right\},\\
\tau_{rr} &= -\frac{P(a^2-r^2)^2\left[a^4-2a^2r^2-r^4+2a^4\cos 2\theta\right]}{\pi a(a^4+r^4-2a^2r^2\cos 2\theta)^2},\\
\tau_{\theta\theta} &= -\frac{P\left[a^8+4a^6r^2-4a^4r^4-r^8+2a^2(2r^6-a^2r^4-a^6)\cos 2\theta\right]}{\pi a(a^4+r^4-2a^2r^2\cos 2\theta)^2},\\
\tau_{r\theta} &= \frac{2aP(a^2-r^2)^2(r^2+a^2)\sin 2\theta}{\pi(a^4+r^4-2a^2r^2\cos 2\theta)^2}.
\end{aligned} \tag{6.372}$$

At $\theta = \pi/2$, we get

$$\tau_{\theta\theta}|_{\theta=\pi/2} = \frac{P}{\pi a}\left[1-\frac{4a^4}{(a^2+r^2)^2}\right],$$

which agrees with the result in [324]. Note that $\boldsymbol{\tau}|_{r=a} = \mathbf{0}$! The strain energy for the above solution (and similar solutions to problems involving point forces and point torques) is unbounded.

An alternative method of derivation (in which care has to be exercised since otherwise it can result in "pseudosolutions" [306]) is to superpose the Flamant solutions for displaced point loads on a half space, i.e., assume the Airy stress function to be (see Fig. 6.61)

$$\phi = -\frac{P}{\pi}\left[r_1\theta_1\sin\theta_1 + r_2\theta_2\sin\theta_2\right]. \tag{6.373}$$

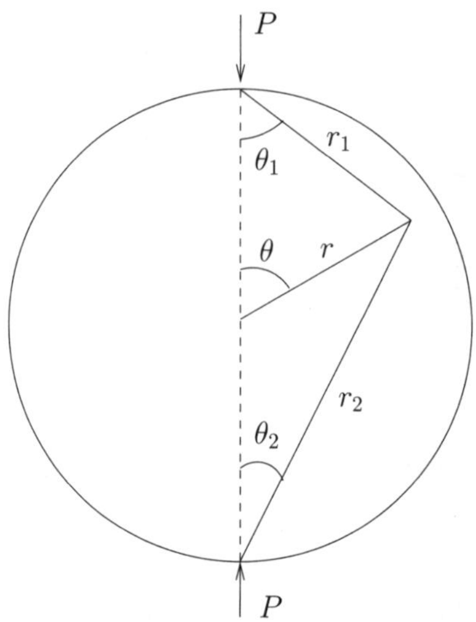

Fig. 6.61 *Circular disc subjected to equal and opposite point loads along its diameter.*

Using the relations

$$r_1 \sin\theta_1 = r\sin\theta, \qquad \tan\theta_1 = \frac{r\sin\theta}{a - r\cos\theta},$$

$$r_2 \sin\theta_2 = r\sin\theta, \qquad \tan\theta_2 = \frac{r\sin\theta}{a + r\cos\theta},$$

Eqn. (6.373) can be written as

$$\phi = -\frac{Pr\sin\theta}{\pi}\tan^{-1}\frac{2ar\sin\theta}{a^2 - r^2}.$$

This leads to the tractions $\tau_{rr}|_{r=a} = -P/(\pi a)$ and $\tau_{r\theta}|_{r=a} = 0$. Since the surface $r = a$ is traction-free (except at the poles), we superpose the uniform stress $\tau_{rr} = \tau_{\theta\theta} = P/(\pi a)$ $\tau_{r\theta} = 0$ on the above solution to obtain the solution given by Eqn. (6.372).

A circular disc of radius b with a concentric hole of radius a, and with point loads acting along the outer diameter can be solved in an analogous manner [330]. Now $\hat{C}_1$ and F_1 are both nonzero, and the constants A_n and C_n for positive and negative n (total of four sets) in Eqn. (6.370) are found so as to satisfy the boundary conditions $\tau_{r\theta}|_{r=a} = \tau_{r\theta}|_{r=b} = \tau_{rr}|_{r=a} = 0$, and $\tau_{rr}|_{r=b} = -P/(2\alpha b)$ in the region $-\alpha \le \theta \le \alpha$. Using Eqns. (6.371c), (6.371d) and these boundary conditions, and noting that $\sin(2n\alpha)/(2n\alpha) \to 1$ as $\alpha \to 0$, we get for $n = 1, 2, \ldots, \infty$,

$$\hat{C}_1 = -\frac{Pb}{2(b^2 - a^2)\pi},$$

$$F_1 = \frac{Pa^2 b}{(b^2 - a^2)\pi},$$

$$A_n = \frac{Pa^2 b^{2n+1}\left\{b^{4n+2} - a^{4n}\left[(1-2n)b^2 + 2na^2\right]\right\}}{(2n-1)D},$$

$$C_n = \frac{Pb^{2n+1}\left\{a^{4n}\left[(2n+1)a^2 - 2nb^2\right] - a^2 b^{4n}\right\}}{(2n+1)D},$$
$$A_{-n} = \frac{Pa^{4n+2}b^{2n+1}\left\{b^{4n}\left[(2n+1)b^2 - 2na^2\right] - a^{4n}b^2\right\}}{(2n+1)D},$$
$$C_{-n} = \frac{Pa^{4n}b^{2n+1}\left\{b^{4n}\left[(2n-1)a^2 - 2nb^2\right] + a^{4n+2}\right\}}{(2n-1)D},$$

where

$$D = \pi\left\{a^2b^2(a^{8n} + b^{8n}) - 2(ab)^{4n}\left[a^2b^2 + 2n^2(b^2 - a^2)^2\right]\right\}.$$

As expected, a summation based on a finite number of terms in Eqn. (6.370) does not converge near to the surface $r = b$.

Under gravity loading along the vertical direction $\theta = \pi$, a particular solution under plane stress conditions that satisfies the equations of equilibrium, the boundary conditions $u_\theta|_{\theta=0} = \tau_{r\theta}|_{\theta=0} = 0$ and Eqns. (6.369), obtained using Eqn. (6.221) with $W_b = -\rho g r\cos\theta$ and the terms associated with A_2 and C_2 in Eqn. (6.239) is

$$\begin{aligned}
2Eu_r(2 + 4\cos 2\theta_0) &= \rho g(1 - \nu^2)r^2\cos\theta\left[2 + 3\cos 2\theta_0 - \cos 2\theta\right],\\
2Eu_\theta(2 + 4\cos 2\theta_0) &= \rho g(1 - \nu^2)r^2\sin\theta\left[\cos 2\theta - \cos 2\theta_0\right],\\
\tau_{rr}(2 + 4\cos 2\theta_0) &= \rho g r\cos\theta\left[2 + (3+\nu)\cos 2\theta_0 - (1-\nu)\cos 2\theta\right],\\
\tau_{\theta\theta}(2 + 4\cos 2\theta_0) &= \rho g r\cos\theta\left[2\nu + (1+3\nu)\cos 2\theta_0 + (1-\nu)\cos 2\theta\right],\\
\tau_{r\theta}(2 + 4\cos 2\theta_0) &= -\rho g(1-\nu)r\sin\theta\left[\cos 2\theta_0 - \cos 2\theta\right].
\end{aligned} \tag{6.374}$$

The above solution is invalid for $\theta_0 = \pi/3$. For this angle, we consider only the C_2 and $\tilde{A}_2$ terms in Eqn. (6.239). The final particular solution that we get is given by Eqns. (6.368) with

$$C_2 = -\frac{1}{8}\rho g(1-\nu),$$
$$\tilde{A}_2 = -\frac{\sqrt{3}}{16\pi}\rho g(1-\nu),$$
$$\tilde{C}_2 = 0.$$

For the semicircular disc ($\theta_0 = \pi/2$) subjected to gravity loading as shown in Fig. 6.62, the solution under plane stress conditions is the sum of the solutions given by Eqns. (6.370) and (6.374) with $F_1 = 0$, and

$$\hat{C}_1 = -\frac{\rho g b(2+\nu)}{3\pi},$$
$$A_n = -\frac{2\rho g(-1)^n b^{3-2n}\left\{\nu - 10 + 4n\left[1 + (n-1)\nu\right]\right\}}{(2n-1)(4n^2-9)(4n^2-1)\pi},$$
$$C_n = \frac{2\rho g(-1)^n b^{1-2n}\left[2 + (2n+1)\nu\right]}{(2n+1)(2n+3)(4n^2-1)\pi},$$

with n ranging from 1 to ∞. It can be easily verified that Eqn. (6.538) is satisfied since the integrals associated with the A_n and C_n terms are zero; the value of each side is $-2\rho g b^3/3$.

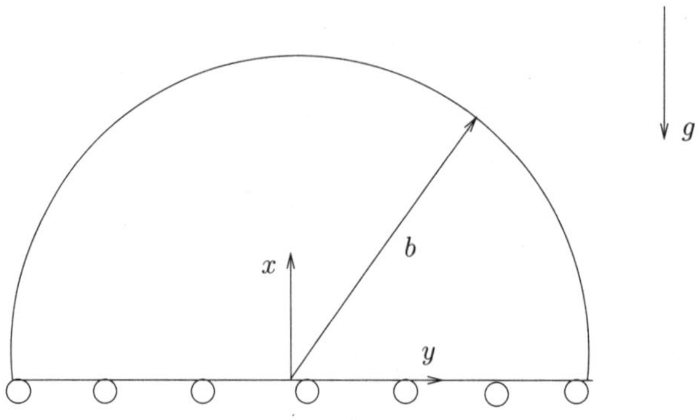

Fig. 6.62 *Semicircular disc under gravity load.*

For the case of antisymmetric loading on the wedge in Fig. 6.59, by considering only the terms associated with B_n and D_n in Eqn. (6.239), we get

$$\begin{aligned}
2Eu_r &= -2 \sum_{\substack{n=-\infty \\ n\neq 0}}^{\infty} \Big\{ r^{\lambda_n}(\lambda_n+1)(1+\nu)B_n \sin(\lambda_n+1)\theta \\
&\quad + r^{\gamma_n}\left[(1+\nu)(\gamma_n+1)-4\right] D_n \sin(\gamma_n-1)\theta \Big\}, \\
2Eu_\theta &= -2 \sum_{\substack{n=-\infty \\ n\neq 0}}^{\infty} \Big\{ r^{\lambda_n}(\lambda_n+1)(1+\nu)B_n \cos(\lambda_n+1)\theta \\
&\quad + r^{\gamma_n}\left[3+\gamma_n+(\gamma_n-1)\nu\right] D_n \cos(\gamma_n-1)\theta \Big\}, \\
\tau_{rr} &= -\sum_{\substack{n=-\infty \\ n\neq 0}}^{\infty} \Big\{ r^{\lambda_n-1}\lambda_n(\lambda_n+1)B_n \sin(\lambda_n+1)\theta + r^{\gamma_n-1}(\gamma_n-3)\gamma_n D_n \sin(\gamma_n-1)\theta \Big\}, \\
\tau_{\theta\theta} &= \sum_{\substack{n=-\infty \\ n\neq 0}}^{\infty} \Big\{ r^{\lambda_n-1}\lambda_n(\lambda_n+1)B_n \sin(\lambda_n+1)\theta + r^{\gamma_n-1}\gamma_n(\gamma_n+1)D_n \sin(\gamma_n-1)\theta \Big\}, \\
\tau_{r\theta} &= \sum_{\substack{n=-\infty \\ n\neq 0}}^{\infty} \Big\{ r^{\lambda_n-1}\lambda_n(\lambda_n+1)B_n \cos(\lambda_n+1)\theta + r^{\gamma_n-1}(\gamma_n-1)\gamma_n D_n \cos(\gamma_n-1)\theta \Big\},
\end{aligned}$$

where

$$\begin{aligned}
\lambda_n &= \frac{(2n-1)\pi}{2\theta_0} - 1, \\
\gamma_n &= \frac{(2n-1)\pi}{2\theta_0} + 1.
\end{aligned}$$

As usual the index n runs from 1 to ∞ in case the inner radius a is zero.

6.6.5 Thermal stresses

In this section, we briefly consider the problem of finding the displacement and stress fields under thermal loading. We assume that the temperature change θ_Δ has been obtained using the heat conduction equation. The constitutive relation is given by Eqn. (6.38), which we rewrite here in terms of $\boldsymbol{\tau}$ (since it is equal to $\boldsymbol{T}$ within the linearized framework)

$$\boldsymbol{\tau} = \mathbf{C}[\boldsymbol{\epsilon} - \boldsymbol{\epsilon}^0].$$

The equilibrium equations can be written as

$$\boldsymbol{\nabla} \cdot [\mathbf{C}[\boldsymbol{\epsilon}]] + \rho_0 \boldsymbol{b} - \boldsymbol{\nabla} \cdot [\mathbf{C}[\boldsymbol{\epsilon}^0]] = \mathbf{0}.$$

Thus, one can obtain a solution to the thermal loading problem by replacing the body force $\rho_0 \boldsymbol{b}$ by $\rho_0 \boldsymbol{b} - \boldsymbol{\nabla} \cdot [\mathbf{C}[\boldsymbol{\epsilon}^0]]$ in the Navier equations of elasticity (written in terms of the displacement field). For an isotropic material and for the case where $\boldsymbol{\epsilon}^0$ represents thermal strains, this transformation is given by the following equation

$$\left.\begin{aligned} \rho_0 \boldsymbol{b} &\to \rho_0 \boldsymbol{b} - \frac{E\alpha \boldsymbol{\nabla} \theta_\Delta}{1 - 2\nu}, \\ W_b &\to W_b - \frac{E\alpha\theta_\Delta}{1 - 2\nu}, \end{aligned}\right\} \quad \text{(3D and plane strain)}, \tag{6.375}$$

where W_b is the body force potential that appears in Eqn. (6.228b). Although W_b also appears in Eqn. (6.219), one should not carry out the above transformation in the case of this equation since the stress field as given there automatically satisfies the equations of equilibrium even under thermal loading. Thus, in the absence of body forces (i.e., $W_b = 0$), Eqns. (6.228a) and (6.228b) get modified to

$$\boldsymbol{\nabla}^4 \phi = -E\alpha \boldsymbol{\nabla}^2 \theta_\Delta, \quad \text{(plane stress)} \tag{6.376a}$$

$$\boldsymbol{\nabla}^4 \phi = -\frac{E\alpha \boldsymbol{\nabla}^2 \theta_\Delta}{1 - \nu}, \quad \text{(plane strain)} \tag{6.376b}$$

where we have used Eqn. (6.225b) in order to obtain Eqn. (6.376a) from Eqn. (6.376b). If we consider steady-state heat conduction with no heat sources, then $\boldsymbol{\nabla}^2 \theta_\Delta = 0$, and both Eqns. (6.376a) and (6.376b) reduce to

$$\boldsymbol{\nabla}^4 \phi = 0. \tag{6.377}$$

If the surface of the body is traction-free, *and the body is simply connected*, then $\phi = 0$ is the unique solution to the above equation that also satisfies the traction boundary conditions. Thus, we have the following

Theorem 6.6.1. *In the absence of surface tractions and body forces, a steady-state (not necessarily uniform) temperature field does not induce any stress in a two-dimensional simply connected isotropic elastic domain.*

The above theorem cannot be extended to multiply connected two-dimensional domains or even to simply connected three-dimensional domains, as the counterexamples of a hollow cylinder and a hollow sphere given below demonstrate.

In the presence of heat sources, we have $\nabla^2\theta_\Delta \neq 0$, and then one has to solve the inhomogeneous biharmonic equations given by Eqns. (6.376a) and (6.376b) in the case of two-dimensional problems. For multiply connected domains, one needs to consider the traction boundary conditions and the single-valuedness of the displacements (even in the case of a pure traction boundary value problem) as we show in the example of a hollow circular cylinder below. For the case of a simply connected domain, however, simply considering the traction boundary conditions in a traction boundary value problem suffices to determine the unknown constants. As an example, consider a rectangular domain $-L/2 \leq x \leq L/2$, $-h/2 \leq y \leq h/2$ under plane stress conditions, and let [217]

$$\theta_\Delta = T_0\left(1 - \frac{4y^2}{h^2}\right).$$

Then using Eqn. (6.376a), we have

$$\nabla^4\phi = \frac{8E\alpha T_0}{h^2}.$$

A particular solution for ϕ is

$$\phi = -\alpha E T_0\left(\frac{y^2}{2} - \frac{y^4}{3h^2}\right),$$

which gives rise to the stresses $\tau_{yy} = \tau_{xy} = 0$ and

$$\tau_{xx} = -\alpha E T_0\left(1 - \frac{4y^2}{h^2}\right).$$

This solution should be superposed on the solution given by Eqn. (6.314) with k_1, k_2, k_3 and A_n set to zero, and the (complex-valued) constants B_n found so as to satisfy the traction-free boundary conditions on the edges $x = \pm L/2$ using, say, the methodology used for solving the analogous problem discussed on page 485 (parabolic load of the form $S(1 - 4y^2/h^2)$ with S now given by $\alpha E T_0$).

To show how the Navier equations of elasticity can be used along with the transformation given by Eqn. (6.375) to solve thermal stress problems, consider a hollow circular cylinder of inner radius a and outer radius b under plane strain conditions subjected to thermal loading $\theta_\Delta(r)$. Because of the axisymmetric nature of the problem we have $u_r = u_r(r)$ and $u_\theta = 0$. Using Eqns. (6.375) and (B.13), we get the governing equation for u_r as

$$\frac{E(1-\nu)}{(1+\nu)(1-2\nu)}\frac{d}{dr}\left[\frac{1}{r}\frac{d(ru_r)}{dr}\right] - \frac{E\alpha}{1-2\nu}\frac{d\theta_\Delta}{dr} = 0, \tag{6.378}$$

which can be written as

$$\frac{d}{dr}\left[\frac{1}{r}\frac{d(ru_r)}{dr}\right] = \frac{(1+\nu)\alpha}{1-\nu}\frac{d\theta_\Delta}{dr}.$$

Thus, the solution is given by

$$u_r = Ar + \frac{B}{r} + \frac{(1+\nu)\alpha}{(1-\nu)r}\int r\theta_\Delta(r)\,dr,$$

$$\tau_{rr} = \lambda\left(\frac{du_r}{dr} + \frac{u_r}{r}\right) + 2\mu\frac{du_r}{dr} - \frac{E\alpha\theta_\Delta}{1-2\nu},$$
$$\tau_{\theta\theta} = \lambda\left(\frac{du_r}{dr} + \frac{u_r}{r}\right) + 2\mu\frac{u_r}{r} - \frac{E\alpha\theta_\Delta}{1-2\nu},$$
$$\tau_{zz} = \lambda\left(\frac{du_r}{dr} + \frac{u_r}{r}\right) - \frac{E\alpha\theta_\Delta}{1-2\nu},$$

where A and B are found subject to traction boundary conditions on τ_{rr} at $r = a$ and $r = b$. For a solid cylinder, the constant B is set to zero. The results under plane stress conditions can be obtained from the above solution using the transformations given by Eqn. (6.225b).

As an example, if $\theta_\Delta = T_i \log(b/r)/\log(b/a)$ (which satisfies $\nabla^2\theta_\Delta = 0$), and the surfaces $r = a$ and $r = b$ are traction-free, then the solution under plane strain conditions is

$$u_r = \frac{\alpha(1+\nu)T_i r}{D}\left[(b^2-a^2)r^2\left(1-\nu+\log\frac{b}{r}\right) - a^2\left[b^2+(1-2\nu)r^2\right]\log\frac{b}{a}\right],$$
$$\tau_{rr} = \frac{\alpha E T_i}{2(1-\nu)}\left[\frac{a^2(b^2-r^2)}{r^2(b^2-a^2)} - \frac{\log\frac{b}{r}}{\log\frac{b}{a}}\right],$$
$$\tau_{\theta\theta} = \frac{\alpha E T_i}{D}\left[(b^2-a^2)r^2\left(1-\log\frac{b}{r}\right) - a^2(b^2+r^2)\log\frac{b}{a}\right],$$
$$\tau_{zz} = \frac{\alpha E T_i r^2}{D}\left[(b^2-a^2)\left(\nu-2\log\frac{b}{r}\right) - 2a^2\nu\log\frac{b}{a}\right],$$

where

$$D = 2(b^2-a^2)(1-\nu)r^2\log\frac{b}{a}.$$

In order to solve the same problem using the Airy stress function approach, we take

$$\phi = F_1\log r + G_1 r^2\log r + C_1 r^2. \tag{6.379}$$

Since $\nabla^2\theta_\Delta = 0$, the above function satisfies the governing equation $\nabla^4\phi = 0$ (Eqn. (6.377)). The stresses are found using Eqns. (6.240), and two of the three constants are eliminated using the conditions $\tau_{rr}|_{r=a} = \tau_{rr}|_{r=b} = 0$. The remaining constant is then found by using the strain–stress relation to first find the strains ϵ_{rr} and $\epsilon_{\theta\theta}$, then finding the displacement field using the relation $u_r = r\epsilon_{\theta\theta}$, and finally substituting into the relation $\epsilon_{rr} = du_r/dr$. As evident from the above procedure, unlike the isothermal case, it is not possible to find the constants simply by considering the boundary conditions even in the case of a traction boundary-value problem. Thus, a combination of the stress boundary conditions and the single-valuedness of the displacement field has to be used in the Airy stress function approach in order to determine the constants in the case of a multiply connected domain.

The 'solution' $\phi = 0$ (i.e., $F_1 = G_1 = C_1 = 0$ in Eqn. (6.379)) satisfies the biharmonic equation and also the traction boundary conditions since the stresses τ_{rr}, $\tau_{\theta\theta}$ and $\tau_{r\theta}$ corresponding to this solution are zero. Thus, we get (for plane strain conditions)

$$\epsilon_{rr} = \epsilon_{\theta\theta} = \frac{1}{\log(b/a)}\alpha(1+\nu)T_i\log\frac{b}{r}, \tag{6.380}$$

with the remaining strains zero. These strains satisfy the compatibility conditions given by Eqn. (B.10). However, as noted in the discussion prior to Theorem 6.1.2, the displacements corresponding to these strains are not single-valued for the domain under consideration. Thus, mere satisfaction of the compatibility relations is not sufficient to ensure single-valuedness of the displacement field in a multiply connected domain. Due to these reasons, it may be more suitable to avoid the Airy stress function approach for multiply connected domains, and work directly with a displacement-based formulation.

For a hollow sphere subjected to thermal loading $\theta_\Delta(r)$, in place of Eqn. (6.378), we get

$$\frac{E(1-\nu)}{(1+\nu)(1-2\nu)}\frac{d}{dr}\left[\frac{1}{r^2}\frac{d(r^2u_r)}{dr}\right]-\frac{E\alpha}{1-2\nu}\frac{d\theta_\Delta}{dr}=0, \tag{6.381}$$

which can be written as

$$\frac{d}{dr}\left[\frac{1}{r^2}\frac{d(r^2u_r)}{dr}\right]=\frac{(1+\nu)\alpha}{1-\nu}\frac{d\theta_\Delta}{dr}.$$

The solution is given by

$$u_r = Ar+\frac{B}{r^2}+\frac{(1+\nu)\alpha}{(1-\nu)r^2}\int r^2\theta_\Delta(r)\,dr,$$

$$\tau_{rr}=\lambda\left(\frac{du_r}{dr}+\frac{2u_r}{r}\right)+2\mu\frac{du_r}{dr}-\frac{E\alpha\theta_\Delta}{1-2\nu},$$

$$\tau_{\theta\theta}=\tau_{\phi\phi}=\lambda\left(\frac{du_r}{dr}+\frac{2u_r}{r}\right)+2\mu\frac{u_r}{r}-\frac{E\alpha\theta_\Delta}{1-2\nu},$$

where A and B are found subject to traction boundary conditions on τ_{rr} at $r=a$ and $r=b$. For a solid sphere, the constant B is set to zero.

As an example, if $\theta_\Delta = T_ia(b-r)/[r(b-a)]$ (which satisfies $\nabla^2\theta_\Delta=0$), and if the inner and outer surfaces are traction-free, then we get

$$u_r=\frac{\alpha a T_i r}{D}\left[b(1+\nu)(a^2+ab+b^2)r^2-2\left[(1-\nu)a^2+\nu b(a+b)\right]r^3-(1+\nu)a^2b^3\right],$$

$$\tau_{rr}=\frac{2\alpha abT_iE}{D}(a-r)(b-r)\left[(a+b)r+ab\right],$$

$$\tau_{\theta\theta}=\tau_{\phi\phi}=\frac{\alpha abT_iE}{D}\left[2(a+b)r^3-(a^2+ab+b^2)r^2-(ab)^2\right],$$

where

$$D=2(b^3-a^3)(1-\nu)r^3.$$

6.6.6 Thick hollow cylinder subjected to a linearly varying pressure on the inner and outer surfaces

A hollow circular cylinder of length L, and inner and outer radii a and b is subjected to a linearly varying pressure on its inner and outer lateral surfaces, while the top and bottom surfaces are subjected to a uniform normal traction T as shown in Fig. 6.63. The Boussinesq

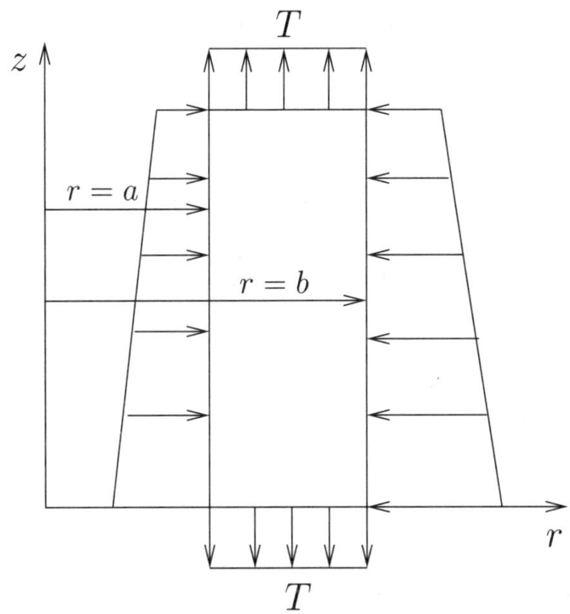

Fig. 6.63 *Thick hollow cylinder subjected to a linearly varying pressure load on the inner and outer surfaces, and a uniform axial traction.*

potentials α and β are assumed to be made up of singular and non-singular harmonic functions as follows:

$$\alpha = A_2\phi_2 + A_3\phi_3 + CL_0^{(1)} + FL_1^{(1)},$$
$$\beta = \hat{A}_1\phi_1 + \hat{A}_2\phi_2 + GL_0^{(1)},$$

where ϕ_1, ϕ_2, ϕ_3, $L_0^{(1)}$ and $L_1^{(1)}$ are given by Eqn. (6.276). Using Eqns. (6.259), and choosing $F = D(1-2\nu)/(2(1-\nu))$ and $G = D/(2(1-\nu))$ (this choice is made to get the shear stress corresponding to the second part to be zero), we get the displacements in the cylindrical coordinate system as

$$2\mu u_r = -A_2 r - 3A_3 rz - \hat{A}_2 rz + \frac{C + Dz}{r},$$
$$2\mu u_z = 2A_2 z + \frac{3A_3}{2}(2z^2 - r^2) + (4\nu - 2)\hat{A}_1 z + \hat{A}_2\left[(4\nu - 1)z^2 + \frac{1}{2}(3 - 4\nu)r^2\right] - D\log r,$$

which on using Eqns. (B.8) and (B.12) lead to

$$\tau_{rr} = -A_2 - 3A_3 z - 2\nu\hat{A}_1 - (1 + 4\nu)\hat{A}_2 z - \frac{C + Dz}{r^2},$$
$$\tau_{\theta\theta} = -A_2 - 3A_3 z - 2\nu\hat{A}_1 - (1 + 4\nu)\hat{A}_2 z + \frac{C + Dz}{r^2},$$
$$\tau_{zz} = 2A_2 + 6A_3 z + 2(\nu - 1)\hat{A}_1 - 2(1 - 2\nu)\hat{A}_2 z,$$
$$\tau_{rz} = \left[-3A_3 + (1 - 2\nu)\hat{A}_2\right] r.$$

Since $\tau_{rz}|_{r=a} = 0$, we get $3A_3 = (1-2\nu)\hat{A}_2$, and since $\tau_{zz}|_{z=L} = T$, we get $2A_2 = T + 2(1-\nu)\hat{A}_1$. Substituting these values, we get

$$
\begin{aligned}
2\mu u_r &= -\left[\frac{T}{2} + (1-\nu)(\hat{A}_1 + 2\hat{A}_2 z)\right] r + \frac{C+Dz}{r},\\
2\mu u_z &= (T + 2\nu\hat{A}_1)z + \left[(1-\nu)r^2 + 2\nu z^2\right]\hat{A}_2 - D\log r,\\
\tau_{rr} &= -\frac{T}{2} - (1+\nu)(\hat{A}_1 + 2\hat{A}_2 z) - \frac{C+Dz}{r^2},\\
\tau_{\theta\theta} &= -\frac{T}{2} - (1+\nu)(\hat{A}_1 + 2\hat{A}_2 z) + \frac{C+Dz}{r^2},\\
\tau_{zz} &= T,\\
\tau_{rz} &= 0.
\end{aligned}
\tag{6.382}
$$

If the pressure distributions on the inner and outer surfaces are given by $p_i + p_i^{(1)}z$ and $p_o + p_o^{(1)}z$, respectively, i.e., if $\tau_{rr}|_{r=a} = -p_i - p_i^{(1)}z$ and $\tau_{rr}|_{r=b} = -p_o - p_o^{(1)}z$, then we get

$$
\begin{aligned}
\hat{A}_1 &= \frac{1}{1+\nu}\left[-\frac{T}{2} + \frac{p_o b^2 - p_i a^2}{b^2 - a^2}\right],\\
\hat{A}_2 &= \frac{1}{2(1+\nu)}\left[\frac{p_o^{(1)} b^2 - p_i^{(1)} a^2}{b^2 - a^2}\right],\\
C &= \frac{(p_i - p_o)a^2 b^2}{b^2 - a^2},\\
D &= \frac{(p_i^{(1)} - p_o^{(1)})a^2 b^2}{b^2 - a^2}.
\end{aligned}
$$

Substituting the above constants into Eqns. (6.382) yields the desired solution. The solution for the case of uniform normal tractions on the inner and outer surfaces is obtained by setting $p_i^{(1)}$ and $p_o^{(1)}$ to zero, while the solution for the case of a solid cylinder whose outer surface $r = b$ is subjected to a linearly varying normal traction $-p_o - p_o^{(1)}z$ is obtained by setting C, D and a to zero in the above solution. For the plane strain case, see Problem 10.

6.6.7 Sphere/spherical segment spinning about its axis

A solid sphere of radius a spins about is axis with angular speed ω. In order to find the displacement and stress fields, we shall find it convenient to work in the cylindrical coordinate system. A particular solution of the equations of equilibrium under a body force $\boldsymbol{b} = (r\omega^2, 0, 0)$ is

$$
\begin{aligned}
2\mu u_r &= \frac{\rho_0\omega^2(2\nu-1)r^3}{8(1-\nu)},\\
\tau_{rr} &= \frac{\rho_0\omega^2(2\nu-3)r^2}{8(1-\nu)},
\end{aligned}
$$

$$\tau_{\theta\theta} = -\frac{\rho_0\omega^2(1+2\nu)r^2}{8(1-\nu)},$$
$$\tau_{zz} = -\frac{\rho_0\omega^2\nu r^2}{2(1-\nu)}, \qquad (6.383)$$
$$\tau_{rz} = 0.$$

We choose the Boussinesq potentials α and β such that the stress solution obtained thereby when superposed with the above solution satisfies the traction-free boundary condition at the surface of the sphere. Thus, we choose

$$\begin{aligned}\alpha &= A_2\phi_2 + A_4\phi_4,\\ \beta &= A_1\phi_1 + A_3\phi_3,\end{aligned} \qquad (6.384)$$

where ϕ_1, ϕ_2, ϕ_3, ϕ_4 are given by Eqn. (6.276). Using Eqns. (6.260) and superposing with the above particular solution, we get

$$\tau_{rr} = -A_2 - 2\nu A_1 + \left(3A_3 + \frac{9}{2}A_4\right)r^2 - (3A_3 + 6A_3\nu + 6A_4)z^2 + \frac{\rho_0\omega^2(2\nu-3)r^2}{8(1-\nu)},$$

$$\tau_{\theta\theta} = -A_2 - 2\nu A_1 + \left(3\nu A_3 + \frac{3}{2}A_4\right)r^2 - (3A_3 + 6A_3\nu + 6A_4)z^2 - \frac{\rho_0\omega^2(1+2\nu)r^2}{8(1-\nu)},$$

$$\tau_{zz} = -2(1-\nu)A_1 + 3(1-\nu)A_3r^2 + 6\nu A_3z^2 - \frac{\rho_0\omega^2\nu r^2}{2(1-\nu)},$$

$$\tau_{rz} = -6(\nu A_3 + 2A_4)rz.$$

The traction-free boundary condition at the surface $r^2 + z^2 = a^2$ is given by

$$\begin{bmatrix}\tau_{rr}r + \tau_{rz}z\\ \tau_{rz}r + \tau_{zz}z\end{bmatrix}_{r^2+z^2=a^2} = \begin{bmatrix}0\\0\end{bmatrix}. \qquad (6.385)$$

Solving for the constants, we get

$$\begin{aligned}A_1 &= \frac{(3-\nu)a^2\rho_0\omega^2}{10(\nu^2-1)}, & A_3 &= \frac{(3+\nu)\rho_0\omega^2}{6(1-\nu)(7+5\nu)},\\ A_2 &= \frac{4(3-2\nu-3\nu^2)a^2\rho_0\omega^2}{5(\nu^2-1)(7+5\nu)}, & A_4 &= \frac{(3-21\nu-10\nu^2)\rho_0\omega^2}{60(1-\nu)(7+5\nu)}.\end{aligned} \qquad (6.386)$$

The final solution is

$$2\mu u_r = \rho_0\omega^2 r\left[\frac{4(3-2\nu-3\nu^2)a^2 + (5\nu^2+3\nu-4)(1+\nu)r^2 + (5\nu^2+8\nu-9)(1+\nu)z^2}{5(1-\nu^2)(7+5\nu)}\right],$$

$$2\mu u_z = \rho_0\omega^2 z\left[\frac{(10\nu^3+3\nu^2-18\nu-3)a^2 + 2(3-\nu)(1+\nu)r^2 + (1+3\nu)(1+\nu)z^2}{5(1-\nu^2)(7+5\nu)}\right],$$

$$\tau_{rr} = \rho_0\omega^2\left[\frac{(5\nu^2-\nu-12)(a^2-r^2) + (9+7\nu)z^2}{5(\nu-1)(7+5\nu)}\right],$$

$$\tau_{\theta\theta} = \rho_0\omega^2 \left[\frac{(5\nu^2 - \nu - 12)a^2 + (4 + 7\nu + 5\nu^2)r^2 + (9 + 7\nu)z^2}{5(\nu - 1)(7 + 5\nu)} \right],$$

$$\tau_{zz} = \rho_0\omega^2 \left[\frac{(3 - 6\nu - 5\nu^2)(a^2 - 2r^2 - z^2)}{5(\nu - 1)(7 + 5\nu)} \right],$$

$$\tau_{rz} = \rho_0\omega^2 \left[\frac{(3 - 6\nu - 5\nu^2)rz}{5(\nu - 1)(7 + 5\nu)} \right].$$

The generalization of the above solution to the case of an ellipsoid may be found in [93].

Now consider the case of a hollow rotating sphere of inner radius a and outer radius b with a fluid of density ρ_f inside the hollow part. Instead of Eqn. (6.384), we now use [94]

$$\alpha = A_2\phi_2 + A_4\phi_4 + \frac{B_0}{\sqrt{r^2 + z^2}} + \frac{B_2(2z^2 - r^2)}{2(r^2 + z^2)^{5/2}},$$

$$\beta = A_1\phi_1 + A_3\phi_3 + \frac{B_1 z}{(r^2 + z^2)^{3/2}},$$

where the functions ϕ_i are given by Eqn. (6.276). The fluid pressure is given by $p = c_0 + \rho_f\omega^2 r^2/2$, where c_0 is a constant, and it exerts a traction $(pr/a, pz/a)$ on the inner surface, while the traction-free boundary condition is enforced at the outer surface $r^2 + z^2 = b^2$. Thus, we have

$$\begin{bmatrix} \tau_{rr}r + \tau_{rz}z + c_0 r + \frac{1}{2}\rho_f\omega^2 r^3 \\ \tau_{rz}r + \tau_{zz}z + c_0 z + \frac{1}{2}\rho_f\omega^2 r^2 z \end{bmatrix}_{r^2+z^2=a^2} = \begin{bmatrix} 0 \\ 0 \end{bmatrix}, \quad \begin{bmatrix} \tau_{rr}r + \tau_{rz}z \\ \tau_{rz}r + \tau_{zz}z \end{bmatrix}_{r^2+z^2=b^2} = \begin{bmatrix} 0 \\ 0 \end{bmatrix}.$$

The seven constants are determined from these boundary conditions as

$$\begin{bmatrix} A_1 \\ A_2 \\ A_3 \\ A_4 \\ B_0 \\ B_1 \\ B_2 \end{bmatrix} = \begin{bmatrix} -2\nu b^5 & -b^5 & -3(1+4\nu)b^7 & -18b^7 & 2b^2 & 2(6-\nu)b^2 & 24 \\ 0 & 0 & 24(1+5\nu)(\nu-1)b^7 & 180(\nu-1)b^7 & 0 & 96(1-\nu)b^2 & 240(1-\nu) \\ 2(\nu-1)b^5 & 2b^5 & 6\nu b^7 & 12b^7 & 2b^2 & (10-4\nu)b^2 & 12 \\ 0 & 0 & 6(1-\nu)(5\nu-1)b^7 & 60(1-\nu)b^7 & 0 & 24(1-\nu)b^2 & 60(1-\nu) \\ -2\nu a^5 & -a^5 & -3(1+4\nu)a^7 & -18a^7 & 2a^2 & 2(6-\nu)a^2 & 24 \\ 0 & 0 & 24(1+5\nu)(\nu-1)a^7 & 180(\nu-1)a^7 & 0 & 96(1-\nu)a^2 & 240(1-\nu) \\ 2(\nu-1)a^5 & 2a^5 & 6\nu a^7 & 12a^7 & 2a^2 & (10-4\nu)a^2 & 12 \end{bmatrix}^{-1} \begin{bmatrix} 0 \\ (2\nu-3)\rho_0\omega^2 b^7 \\ 0 \\ -\nu\rho_0\omega^2 b^7 \\ -c_0 a^5 \\ [(2\nu-3)\rho_0 - 4(\nu-1)\rho_f]\omega^2 a^7 \\ -c_0 a^5 \end{bmatrix}. \tag{6.387}$$

The boundary condition given by

$$6(1-\nu)(5\nu-1)a^7 A_3 + 60(1-\nu)a^7 A_4 + 24(1-\nu)a^2 B_1 + 60(1-\nu)B_2 = -[\nu\rho_0 + (\nu-1)\rho_f]\omega^2 a^7,$$

is linearly dependent on the seven boundary conditions presented above, and hence is automatically satisfied by the solution given by Eqn. (6.387). After determining the constants, the solutions obtained from Eqns. (6.259) and (6.260) are superposed on the particular solution given by Eqn. (6.383) to find the final displacement and stress fields.

It can be easily verified that

$$\lim_{a\to 0} B_0 = \lim_{a\to 0} B_1 = \lim_{a\to 0} B_2 = 0,$$

while the limiting values for A_1, A_2, A_3 and A_4 are the same as those for a solid sphere as given by Eqn. (6.386); thus the solution for the solid sphere is recovered in the limit as $a \to 0$. The solution for a hollow spinning sphere [94] is recovered by setting $c_0 = \rho_f = 0$, while the solution for a sphere subjected to a uniform internal pressure is recovered by setting $\omega = 0$.

Axisymmetric solutions for spherical shell segments are presented in [51, 90, 188]. As an illustration, consider the solution of the problem of a rotating hollow hemisphere. Then, first one considers the solution of the rotating hollow sphere that has been derived above, which we term as the particular solution. This solution will satisfy the governing equations and all boundary conditions except the condition $\tau_{zz}|_{z=0} = 0$. In order to satisfy this boundary condition, we superpose a solution, which we term as the homogeneous solution, that satisfies the governing equations with zero body forces, and the boundary conditions at $r = a$ and $r = b$. The constants in this homogeneous solution are then found such that the complete (i.e., homogeneous + particular) solution satisfies the condition $\tau_{zz}|_{z=0} = 0$. The homogeneous solution is derived using the solution given in Eqn. (6.278), with n no longer restricted to be an integer value, and which we now denote by λ (which, along with the constants A_λ, B_λ, C_λ and D_λ, is complex-valued, and occurs in conjugate pairs). In this case, $P_\lambda(\xi)$ is non-singular at $\xi = 1$, but singular at $\xi = -1$; however, since the negative z-axis is not part of the domain, this does not pose a problem. For simplifying the calculations, we can replace r in Eqn. (6.278) by $\tilde{r} = r/r_m$, where r_m is the mean radius $(a+b)/2$. Let $\delta = (b-a)/(2r_m)$. Imposing the boundary conditions $\tau_{rr}|_{\tilde{r}=1-\delta} = \tau_{rr}|_{\tilde{r}=1+\delta} = \tau_{r\theta}|_{\tilde{r}=1-\delta} = \tau_{r\theta}|_{\tilde{r}=1+\delta} = 0$, results in a set of four homogeneous algebraic equations. The four constants A_λ, B_λ, C_λ and D_λ will have nonzero values only if the determinant of their coefficient matrix is zero. This condition yields a nonlinear characteristic equation for λ (see Eqn. (5) in [51] or Eqn. (10) in [188]). Using the four homogeneous algebraic equations, three of the constants, say B_{λ_n}, C_{λ_n} and D_{λ_n} can be expressed in terms of A_{λ_n}, where the index λ_n corresponds to the root λ_n of the characteristic equation (i.e., the A_{λ_n} are functions of λ_n, ν, and the radii a, b). These complex-valued constants A_{λ_n} are then determined so as to satisfy the boundary conditions $\tau_{r\theta}|_{\theta=\pi/2} = \tau_{\theta\theta}|_{\theta=\pi/2} = 0$[21]. This can be achieved by following a least-squares strategy [51, 188] or a weighted residual method as described on page 485. The functions $P_\lambda(\xi)$, $\lambda \in \mathbb{C}$, can be computed using Eqn. (14) in [169], and are given by

$$P_\lambda(\xi) = \frac{\sin \pi\lambda}{\pi}\left[\log\frac{1+\xi}{2} + 1 + \frac{1}{\lambda(\lambda+1)} - \lambda(\lambda+1)\sum_{n=1}^{\infty}\frac{(-1)^n(2n+1)P_n(\xi)}{n(n+1)\left[n(n+1)-\lambda(\lambda+1)\right]}\right], \tag{6.388}$$

where $\sin \pi\lambda = \sin \pi a \cosh \pi b + i \sinh \pi b \cos \pi a$, when $\lambda = a + ib$.

[21] Of course, although in the above example, we have truncated the hollow sphere at $\theta = \pi/2$, one could truncate it at any $\theta_0 < \pi$; in this case, the term $\pi/2$ in the boundary conditions is replaced by θ_0.

In case the surface $r = b$, say, is loaded by radial and tangential tractions of the form $t_r^0(\theta)$, and $t_\theta^0(\theta)$, then using the orthogonality properties

$$\int_{-1}^{1} P_m(\xi)P_n(\xi)\,d\xi = \frac{2}{2n+1}\delta_{mn},$$
$$\int_{-1}^{1} (1-\xi^2)\frac{dP_m}{d\xi}\frac{dP_n}{d\xi}\,d\xi = \frac{2n(n+1)}{2n+1}\delta_{mn},$$

we can write $t_r^0(\theta) = \sum_{n=1}^{\infty} a_n P_n(\xi)$ and $t_\theta^0(\theta) = \sum_{n=1}^{\infty} b_n \sin\theta (dP_n(\xi)/d\xi)$, where

$$a_n = \frac{2n+1}{2}\int_{-1}^{1} t_r^0(\theta)P_n(\xi)\,d\xi = \frac{2n+1}{2}\int_0^{\pi} t_r^0(\theta)P_n(\cos\theta)\sin\theta\,d\theta,$$
$$b_n = -\frac{2n+1}{2n(n+1)}\int_{-1}^{1} t_\theta^0(\theta)\frac{dP_n(\cos\theta)}{d\theta}\,d\xi = \frac{2n+1}{2n(n+1)}\int_0^{\pi} t_\theta^0(\theta)\frac{dP_n(\cos\theta)}{d\theta}\sin\theta\,d\theta.$$

The constants in Eqn. (6.278) are determined so as to satisfy the boundary conditions $\tau_{rr}|_{r=b} = t_r^0(\theta)$ and $\tau_{r\theta}|_{r=b} = t_\theta^0(\theta)$; thus, we obtain the particular solution. The homogeneous solution is now superposed in order to satisfy all the boundary conditions as described above.

A characteristic equation similar to the above cannot be derived for a segment of a cone ($a = 0$), which includes the case of a solid hemisphere, since the conditions $\tau_{rr}|_{r=b} = \tau_{r\theta}|_{r=b} = 0$ yield only two complex roots, so that it is not possible to satisfy the boundary conditions on the lateral surfaces. In such a case, the characteristic equation is derived using the lateral surfaces as shown in Section 6.6.15. Even the problem of a rotating spherical shell segment that we have considered above can be tackled using this approach (see page 578).

6.6.8 Circular cylinder with loading on its end faces, and lateral surfaces traction-free

Consider a circular cylinder of radius a and length L that has prescribed tractions on its flat surfaces, with the lateral surfaces being traction-free, i.e.,

$$\tau_{zz}|_{z=\pm L/2} = f(r); \qquad \tau_{rz}|_{z=\pm L/2} = \pm g(r), \tag{6.389a}$$
$$\tau_{rr}|_{r=a} = 0; \qquad \tau_{rz}|_{r=a} = 0, \tag{6.389b}$$

with $g(a) = 0$ to have continuity of shear stresses at $(r,z) = (a, \pm L/2)$.

We solve this problem using the Love strain function; see [195] for an alternative approach. One can derive the same solution using the Boussinesq potentials; however, one would need to use the 'nonseparable' parts of the solution given by Eqn. (6.267), and hence, it is more convenient to use the Love strain function. Since the problem is symmetric about the $z = 0$ plane, using Eqn. (6.268), we assume the Love strain function to be of the form

$$\Psi = \sum_{n=1}^{\infty} [A_n J_0(\lambda_n r) + B_n r J_1(\lambda_n r)]\sinh\lambda_n z + \frac{c_2\phi_2 z}{4(1+\nu)} - \frac{c_2(1+4\nu)\phi_3}{12(1+\nu)}, \tag{6.390}$$

where λ_n, $n = 1, 2, \ldots, \infty$, denote the (complex-valued) roots of a characteristic equation that will be derived below, and A_n and B_n are constants that are determined using the boundary conditions in Eqns. (6.389a).

Using Eqns. (6.266), we get

$$\tau_{rr} = \sum_{n=1}^{\infty} A_n \left[\left(\lambda_n B_0(\lambda_n r) - \frac{J_1(\lambda_n r)}{r} \right) \lambda_n^2 \cosh \lambda_n z \right]$$
$$+ \sum_{n=1}^{\infty} B_n \left[(2\nu J_0(\lambda_n r) - J_0(\lambda_n r) + \lambda_n r J_1(\lambda_n r)) \lambda_n^2 \cosh \lambda_n z \right],$$
$$\tau_{rz} = \sum_{n=1}^{\infty} \left\{ A_n \lambda_n^3 J_1(\lambda_n r) \sinh \lambda_n z + B_n \left[(2(\nu - 1) J_1(\lambda_n r) - \lambda_n r J_0(\lambda_n r)) \lambda_n^2 \sinh \lambda_n z \right] \right\}.$$

The conditions in Eqn. (6.389b) lead to a matrix equation of the form

$$\boldsymbol{M} \begin{bmatrix} A_n \\ B_n \end{bmatrix} = \begin{bmatrix} 0 \\ 0 \end{bmatrix}.$$

In order to obtain a nontrivial solution for A_n and B_n, we require that $\det \boldsymbol{M} = 0$. This leads to the characteristic equation

$$a^2 \lambda^2 J_0^2(\lambda a) + \left[a^2 \lambda^2 - 2(1 - \nu) \right] J_1^2(\lambda a) = 0. \tag{6.391}$$

Zero is the only real eigenvalue of Eqn. (6.391), and there are no purely imaginary eigenvalues [258]. Thus, apart from zero, the remaining eigenvalues λ_n, $n = 1, 2, \ldots, \infty$, are complex-valued, and occur in complex-conjugate pairs. The corresponding constants, A_n, B_n also occur in complex-conjugate pairs so that the final result is real. Let $z_n := \lambda_n a$. The asymptotic roots z_n of Eqn. (6.391), which can be used as a first guess in a Newton–Raphson strategy, are $n\pi - \log(4n\pi)/(4n\pi) + i \log(4n\pi)/2$ [317]. The first twenty roots of the characteristic equation obtained using such a Newton–Raphson strategy for $\nu = 0.25$ are listed in Table 6.5 (see also [196]). Since the zero eigenvalues lead to constant stresses, which we have already incorporated in the solution (see the terms involving the constant c_2 in Eqn. (6.390)), it suffices to consider only the complex-valued eigenvalues. By using the condition $\tau_{rz}|_{r=a} = 0$ to eliminate B_n in terms of A_n, we get (after renaming the constants)

$$2\mu u_r = \sum_{n=1}^{\infty} \cosh(\lambda_n z) A_n \left\{ D_1 J_1(\lambda_n r) - \lambda_n r J_0(\lambda_n r) J_1(\lambda_n a) \right\} - \frac{c_2 \nu r}{1 + \nu},$$
$$2\mu u_z = -\sum_{n=1}^{\infty} \sinh(\lambda_n z) A_n \left\{ D_2 J_0(\lambda_n r) + \lambda_n r J_1(\lambda_n r) J_1(\lambda_n a) \right\} + \frac{c_2 z}{1 + \nu},$$
$$\tau_{rr} = \sum_{n=1}^{\infty} \frac{\cosh(\lambda_n z) A_n}{r} \left\{ \lambda_n a J_0(\lambda_n a) \left[\lambda_n r J_0(\lambda_n r) - J_1(\lambda_n r) \right] \right.$$
$$\left. + \left[\lambda_n r J_0(\lambda_n r) - [2(1 - \nu) - \lambda_n^2 r^2] J_1(\lambda_n r) \right] J_1(\lambda_n a) \right\}, \tag{6.392}$$
$$\tau_{zz} = \sum_{n=1}^{\infty} \lambda_n^2 \cosh(\lambda_n z) A_n \left[a J_0(\lambda_n r) J_2(\lambda_n a) - r J_1(\lambda_n a) J_1(\lambda_n r) \right] + c_2,$$
$$\tau_{rz} = \sum_{n=1}^{\infty} \lambda_n^2 \sinh(\lambda_n z) A_n \left[a J_0(\lambda_n a) J_1(\lambda_n r) - r J_0(\lambda_n r) J_1(\lambda_n a) \right],$$

n	Roots
1	(2.6976518048126938, 1.3673569998759254)
2	(6.0512223005490648, 1.6381470848819837)
3	(9.2612734547555835, 1.8285341603000254)
4	(12.438444451168348, 1.9674282971825920)
5	(15.602204445288262, 2.0764211426223569)
6	(18.759055665086660, 2.1660393261524979)
7	(21.911845435020094, 2.2421080309759218)
8	(25.062031889082405, 2.3081756436569010)
9	(28.210443716513613, 2.3665606854539565)
10	(31.357588567275251, 2.4188604004095073)
11	(34.503795629578121, 2.4662214632305690)
12	(37.649288138586861, 2.5094947145580004)
13	(40.794223059136151, 2.5493287796871154)
14	(43.938714110839491, 2.5862293955119156)
15	(47.082845781881687, 2.6205984805165596)
16	(50.226682205087059, 2.6527607128727739)
17	(53.370272973086678, 2.6829821192189875)
18	(56.513657060392369, 2.7114833913729726)
19	(59.656865536276896, 2.7384496282657378)
20	(62.799923483287451, 2.7640375955629906)

Table 6.5 *First twenty roots of $z^2 J_0^2(z) + \left[z^2 - 2(1-\nu)\right] J_1^2(z) = 0$ for $\nu = 0.25$.*

$$
\begin{aligned}
\tau_{\theta\theta} = \sum_{n=1}^{\infty} \frac{\cosh(\lambda_n z) A_n}{r} \Big\{ &\left[\lambda_n a J_0(\lambda_n a) + 2(1-\nu) J_1(\lambda_n a)\right] J_1(\lambda_n r) \\
&- (1-2\nu)\lambda_n r J_0(\lambda_n r) J_1(\lambda_n a) \Big\}.
\end{aligned}
$$

where

$$
\begin{aligned}
D_1 &= \lambda_n a J_0(\lambda_n a) + 2(1-\nu) J_1(\lambda_n a), \\
D_2 &= \lambda_n a J_0(\lambda_n a) - 2(1-\nu) J_1(\lambda_n a).
\end{aligned}
$$

Note that

$$
\lim_{r\to 0} \tau_{rr} = \lim_{r\to 0} \tau_{\theta\theta} = \sum_{n=1}^{\infty} \frac{\lambda_n \cosh(\lambda_n z) A_n}{2} \left[\lambda_n a J_0(\lambda_n a) + 2\nu J_1(\lambda_n a)\right].
$$

The condition $\tau_{rz}|_{z=\pm L/2} = \pm g(r)$ leads to

$$
\sum_{n=1}^{\infty} \lambda_n^2 \sinh\left(\frac{\lambda_n L}{2}\right) A_n \left[a J_0(\lambda_n a) J_1(\lambda_n r) - r J_0(\lambda_n r) J_1(\lambda_n a)\right] = g(r), \tag{6.393}
$$

while integrating the condition $\tau_{zz}|_{z=L/2}\, r = rf(r)$ leads to

$$c_2 = \frac{4}{a}\sum_{n=1}^{\infty}(2-\nu)A_n J_1^2(\lambda_n a)\cosh\frac{\lambda_n L}{2} + \int_{-L/2}^{L/2} rf(r)\,dr.$$

The condition $\tau_{zz}|_{z=\pm L/2} = f(r)$ yields

$$c_2 - \sum_{n=1}^{\infty}\lambda_n \cosh\frac{\lambda_n L}{2} A_n\left[J_0(\lambda_n r)D + \lambda_n r J_1(\lambda_n a)J_1(\lambda_n r)\right] = f(r). \tag{6.394}$$

The constants A_n are determined using Eqns. (6.393) and (6.394), using, say, a weighted residual technique as described on page 485.

We now present an alternative solution. Instead of using two sets of complex constants as in Eqn. (6.390), we can use four sets of real constants as follows[22]:

$$\begin{aligned}\Psi = \sum_{m=1}^{\infty}\left[A_m\sinh\lambda_m z + B_m z\cosh\lambda_m z\right]J_0(\lambda_m r) + \sum_{m=1}^{\infty}\left[C_m I_0(k_m r) + D_m r I_1(k_m r)\right]\sin k_m z \\ + \frac{2c_1 + c_2}{4(1+\nu)}\phi_2 z + \frac{(2-4\nu)c_1 - (1+4\nu)c_2}{12(1+\nu)}\phi_3,\end{aligned} \tag{6.395}$$

where $k_m = 2m\pi/L$, and λ_m, $m = 1, 2, \ldots, \infty$, are the positive roots of $J_1(\lambda a) = 0$. Asymptotically, the mth roots of J_n and Y_n are given by $(4m+2n-1)\pi/4$ and $(4m+n-3)\pi/4$, respectively. The displacement and stress fields corresponding to the above function, obtained using Eqn. (6.266), are

$$\begin{aligned}2\mu u_r &= \frac{[(1-\nu)c_1 - \nu c_2]\,r}{1+\nu} + \sum_{m=1}^{\infty}\Big\{A_m\lambda_m^2 J_1(\lambda_m r)\cosh\lambda_m z \\ &\quad + B_m\lambda_m J_1(\lambda_m r)\left[\cosh\lambda_m z + \lambda_m z\sinh\lambda_m z\right] \\ &\quad - C_m k_m^2 I_1(k_m r)\cos k_m z - D_m k_m^2 r I_0(k_m r)\cos k_m z\Big\}, \\ 2\mu u_z &= \frac{[c_2 - 2\nu c_1]\,z}{1+\nu} + \sum_{m=1}^{\infty}\Big\{-A_m\lambda_m^2 J_0(\lambda_m r)\sinh\lambda_m z \\ &\quad - B_m\lambda_m J_0(\lambda_m r)\left[\lambda_m z\cosh\lambda_m z - 2(1-2\nu)\sinh\lambda_m z\right] \\ &\quad + C_m k_m^2 I_0(k_m r)\sin k_m z + D_m k_m \sin k_m z\left[4(1-\nu)I_0(k_m r) + k_m r I_1(k_m r)\right]\Big\}, \\ \tau_{rr} &= c_1 + \sum_{m=1}^{\infty}\frac{1}{r}\Big\{\lambda_m^2 A_m\left[\lambda_m r J_0(\lambda_m r) - J_1(\lambda_m r)\right]\cosh\lambda_m z \\ &\quad + \lambda_m B_m\Big[\lambda_m r J_0(\lambda_m r)\left[(1+2\nu)\cosh\lambda_m z + \lambda_m z\sinh\lambda_m z\right] \\ &\quad - J_1(\lambda_m r)\left[\cosh\lambda_m z + \lambda_m z\sinh\lambda_m z\right]\Big] + k_m^2 C_m\left[I_1(k_m r) - k_m r I_0(k_m r)\right]\cos k_m z \\ &\quad - k_m^2 r D_m\left[(1-2\nu)I_0(k_m r) + k_m r I_1(k_m r)\right]\cos k_m z\Big\},\end{aligned}$$

[22]This solution procedure is analogous to Mathieu's solution procedure for planar rectangular problems; a similar solution seems to have been presented by Saito [278].

$$
\begin{aligned}
\tau_{zz} &= c_2 + \sum_{m=1}^{\infty}\Big\{-A_m\lambda_m^3 J_0(\lambda_m r)\cosh\lambda_m z \\
&\quad + \lambda_m^2 B_m\left[(1-2\nu)\cosh\lambda_m z - \lambda_m z\sinh\lambda_m z\right] J_0(\lambda_m r) \\
&\quad + C_m k_m^3 I_0(k_m r)\cos k_m z + k_m^2 D_m\left[2(2-\nu)I_0(k_m r) + k_m r I_1(k_m r)\right]\cos k_m z\Big\},
\end{aligned}
\tag{6.396}
$$

$$
\begin{aligned}
\tau_{rz} &= \sum_{m=1}^{\infty}\Big\{A_m\lambda_m^3 J_1(\lambda_m r)\sinh\lambda_m z + \lambda_m^2 B_m J_1(\lambda_m r)\left[\lambda_m z\cosh\lambda_m z + 2\nu\sinh\lambda_m z\right] \\
&\quad + k_m^3 C_m I_1(k_m r)\sin k_m z + k_m^2 D_m\left[k_m r I_0(k_m r) + 2(1-\nu)I_1(k_m r)\right]\sin k_m z\Big\},
\end{aligned}
$$

$$
\begin{aligned}
\tau_{\theta\theta} &= c_1 + \frac{1}{r}\Big\{\lambda_m^2 A_m J_1(\lambda_m r)\cosh\lambda_m z \\
&\quad + \lambda_m B_m\left[2\nu\lambda_m r J_0(\lambda_m r)\cosh\lambda_m z + J_1(\lambda_m r)\left(\cosh\lambda_m z + \lambda_m z\sinh\lambda_m z\right)\right] \\
&\quad - k_m^2 C_m I_1(k_m r)\cos k_m z - k_m^2 r D_m(1-2\nu)I_0(k_m r)\cos k_m z\Big\}.
\end{aligned}
$$

The condition $\tau_{rz}|_{r=a} = 0$ yields

$$
C_m = -\frac{2D_m(1-\nu)}{k_m} - \frac{aD_m I_0(k_m a)}{I_1(k_m a)}.
\tag{6.397}
$$

The condition $\tau_{rz}|_{z=\pm L/2} = \pm g(r)$ yields

$$
\sum_{m=1}^{\infty}\lambda_m^2 J_1(\lambda_m r)\left[B_m\frac{\lambda_m L}{2}\cosh\frac{\lambda_m L}{2} + (\lambda_m A_m + 2\nu B_m)\sinh\frac{\lambda_m L}{2}\right] = g(r).
$$

Multiplying the above expression by $rJ_1(\lambda_n r)$ and integrating over the interval $[0,a]$, we get

$$
B_m\frac{\lambda_m L}{2}\cosh\frac{\lambda_m L}{2} + (\lambda_m A_m + 2\nu B_m)\sinh\frac{\lambda_m L}{2} = \frac{2}{\lambda_m^2 a^2 J_0^2(\lambda_m a)}\int_0^a r g(r) J_1(\lambda_m r)\,dr =: s_m.
\tag{6.398}
$$

The condition $\tau_{rr}|_{r=a} = 0$ yields

$$
\begin{aligned}
c_1 &+ \sum_{m=1}^{\infty}\left\{\frac{k_m a I_0^2(k_m a)}{I_1(k_m a)} - \frac{\left[2(1-\nu) + (k_m a)^2\right] I_1(k_m a)}{k_m a}\right\} D_m k_m^2\cos k_m z \\
&+ \sum_{m=1}^{\infty}\Bigg\{s_m\,\mathrm{cosech}\,\frac{\lambda_m L}{2}\cosh\lambda_m z \\
&+ \left[\left(1 - \frac{\lambda_m L}{2}\coth\frac{\lambda_m L}{2}\right)\cosh\lambda_m z + \lambda_m z\sinh\lambda_m z\right] B_m\Bigg\}\lambda_m^2 J_0(\lambda_m a) = 0.
\end{aligned}
\tag{6.399}
$$

Integrating the above relation over $[-L/2, L/2]$, we get

$$
c_1 = -\frac{2}{L}\sum_{m=1}^{\infty} s_m\lambda_m J_0(\lambda_m a).
$$

Multiplying Eqn. (6.399) by $\cos k_n z$ and integrating over the interval $[-L/2, L/2]$, we get for $n = 1, 2, \ldots, \infty$,

$$\left\{ \frac{k_n a I_0^2(k_n a)}{I_1(k_n a)} - \frac{\left[2(1-\nu) + (k_n a)^2\right] I_1(k_n a)}{k_n a} \right\} \frac{D_n k_n^2 L}{2}$$
$$+ \sum_{m=1}^{\infty} \left\{ \left[s_m \operatorname{cosech} \frac{\lambda_m L}{2} + \left(1 - \frac{\lambda_m L}{2} \coth \frac{\lambda_m L}{2}\right) B_m \right] U_m^{(n)} + \lambda_m V_m^{(n)} B_m \right\} \lambda_m^2 J_0(\lambda_m a) = 0, \tag{6.400}$$

where

$$U_m^{(n)} := \frac{2(-1)^n \lambda_m \sinh \frac{\lambda_m L}{2}}{k_n^2 + \lambda_m^2},$$
$$V_m^{(n)} := \frac{2(-1)^n \left[\frac{\lambda_m L}{2}(k_n^2 + \lambda_m^2)\cosh\frac{\lambda_m L}{2} + (k_n^2 - \lambda_m^2)\sinh\frac{\lambda_m L}{2}\right]}{(k_n^2 + \lambda_m^2)^2}.$$

Finally, the condition $\tau_{zz}|_{z=L/2} = f(r)$ yields

$$c_2 + \sum_{m=1}^{\infty} (-1)^m k_m^3 \left[r I_1(k_m r) - \frac{a I_2(k_m a) I_0(k_m r)}{I_1(k_m a)} \right] D_m$$
$$+ \sum_{m=1}^{\infty} \left\{ (\sinh \lambda_m L + \lambda_m L) \frac{\operatorname{cosech} \frac{\lambda_m L}{2} B_m}{2} - s_m \coth \frac{\lambda_m L}{2} \right\} \lambda_m^2 J_0(\lambda_m r) = f(r). \tag{6.401}$$

By integrating the product of the above expression with r, we get $c_2 = 2\int_0^a r f(r)\,dr/a^2$. Multiplying Eqn. (6.401) by $r J_0(\lambda_n r)$, integrating over the interval $[0, a]$, and using the relations

$$\int_0^a r J_0^2(\lambda_m r)\,dr = \int_0^a r J_1^2(\lambda_m r)\,dr = \frac{a^2 J_0^2(\lambda_m a)}{2},$$
$$\int_0^a r I_0(k_m r) J_0(\lambda_n r)\,dr = \frac{a k_m I_1(k_m a) J_0(\lambda_n a)}{\lambda_n^2 + k_m^2},$$
$$\int r^2 I_1(k_m r) J_0(\lambda_n r)\,dr = \frac{1}{\lambda_n^2 + k_m^2}\left[k_m r^2 I_0(k_m r) J_0(\lambda_n r) - \frac{2k_m^2 r I_1(k_m r) J_0(\lambda_n r)}{\lambda_n^2 + k_m^2}\right.$$
$$\left. - \frac{2k_m \lambda_n r I_0(k_m r) J_1(\lambda_n r)}{\lambda_n^2 + k_m^2} + \lambda_n r^2 I_1(k_m r) J_1(\lambda_n r)\right],$$
$$\int_0^a r^2 I_1(k_m r) J_0(\lambda_n r)\,dr = \frac{k_m a J_0(\lambda_n a)}{(\lambda_n^2 + k_m^2)^2}\left[a(k_m^2 + \lambda_n^2) I_0(k_m a) - 2k_m I_1(k_m a)\right],$$
$$\int_0^a r I_1(k_m r) J_1(\lambda_n r)\,dr = \frac{a \lambda_n I_1(k_m a) J_2(\lambda_n a)}{\lambda_n^2 + k_m^2}, \tag{6.402}$$
$$\int r^2 I_0(k_m r) J_1(\lambda_n r)\,dr = -\frac{\lambda_n r^2}{\lambda_n^2 + k_m^2} J_0(\lambda_n r) I_0(k_m r) + \frac{2\lambda_n k_m r}{(\lambda_n^2 + k_m^2)^2} J_0(\lambda_n r) I_1(k_m r)$$

$$+\frac{2\lambda_n^2 r}{(\lambda_n^2+k_m^2)^2}J_1(\lambda_n r)I_0(k_m r)+\frac{k_m r^2}{\lambda_n^2+k_m^2}J_1(\lambda_n r)I_1(k_m r),$$

$$\int_0^a r^2 I_0(k_m r)J_1(\lambda_n r)\,dr=\frac{\lambda_n a J_0(\lambda_n a)}{(\lambda_n^2+k_m^2)^2}\left[2k_m I_1(k_m a)-a(\lambda_n^2+k_m^2)I_0(k_m a)\right],$$

we get for $n = 1,2,\ldots,\infty$,

$$\sum_{m=1}^{\infty}(-1)^m k_m^3\left[\frac{2a\lambda_n^2 I_1(k_m a)J_0(\lambda_n a)}{(\lambda_n^2+k_m^2)^2}\right]D_m$$
$$+\left\{(\sinh\lambda_n L+\lambda_n L)\,\mathrm{cosech}\,\frac{\lambda_n L}{2}B_n-2s_n\coth\frac{\lambda_n L}{2}\right\}\frac{\lambda_n^2 a^2 J_0^2(\lambda_n a)}{4}=\int_0^a rJ_0(\lambda_n r)f(r)\,dr. \tag{6.403}$$

Eqns. (6.400) and (6.403) are to be solved for the constants B_n and D_n.

Note that

$$\lim_{r\to 0}\tau_{rr}=\lim_{r\to 0}\tau_{\theta\theta}=c_1+\frac{1}{2}\sum_{n=1}^{\infty}\Big\{-\left[k_n C_n+(2-4\nu)D_n\right]k_n^2\cos k_n z$$
$$+\left[\lambda_n A_n+(1+4\nu)B_n\right]\lambda_n^2\cosh\lambda_n z+\lambda_n^3 z B_n\sinh\lambda_n z\Big\}. \tag{6.404}$$

The same strain function used in Eqn. (6.390) can be used to solve the contact problem of a circular plate resting on a rigid frictionless surface, and subjected to tractions on its top surface (see Fig. 6.46, with r in place of x, z in place of y, and a in place of $L/2$), since

$$u_z|_{z=0}=\tau_{rz}|_{z=0}=0.$$

The complete solution is given by Eqn. (6.392), and the constants A_n are determined using the boundary conditions in Eqn. (6.389a) with h in place of $L/2$, i.e., by using Eqns. (6.393) and (6.394) (with, again, h in place of $L/2$).

The solutions presented in Eqns. (6.392) and (6.395) are valid for the case where the tractions (t_r, t_z) are antisymmetric/symmetric or prescribed displacements $(\bar{u}_r, \bar{u}_z)$ are symmetric/antisymmetric at $z=\pm L/2$. If the prescribed tractions are symmetric/antisymmetric or prescribed displacements are antisymmetric/symmetric, then in Eqn. (6.392), c_2 should be set to zero, and $A_n\sinh\lambda_n z$ and $A_n\cosh\lambda_n z$ should be replaced by $B_n\cosh\lambda_n z$ and $B_n\sinh\lambda_n z$, respectively. The constants B_n are found subject to the appropriate boundary conditions at $z = L/2$ (or $z = -L/2$). Similarly, in place of Eqn. (6.395), the solution in terms of real-valued constants is given by

$$\Psi=\sum_{m=1}^{\infty}\left[A_m\cosh\lambda_m z+B_m z\sinh\lambda_m z\right]J_0(\lambda_m r)+\sum_{m=1}^{\infty}\left[C_m I_0(k_m r)+D_m r I_1(k_m r)\right]\cos k_m z, \tag{6.405}$$

where $k_m = (2m-1)\pi/L$, and λ_m, $m = 1,2,\ldots,\infty$, are the positive roots of $J_1(\lambda a) = 0$. A superposition of the symmetric and antisymmetric solutions can be used to solve problems where the loading at the two ends is arbitrary but self-equilibrating as in Problem (1) of [258], or if the prescribed displacements are arbitrary as in the problem considered by Robert and Kerr [271], which we now discuss.

The given material and geometrical parameters are $\nu = 1/3$, $a = 1$, $L = 1.2$, and the prescribed displacements are given by

$$u_r = -8 \times 10^{-3} r, \qquad u_z = 2 \times 10^{-2}, \qquad \text{at } z = -L/2,$$
$$u_r = 8 \times 10^{-3} r, \qquad u_z = -2 \times 10^{-2}, \qquad \text{at } z = L/2.$$

Note that there is a stress singularity at $(r, z) = (a, \pm L/2)$ due to the conflict in the boundary conditions (both displacements and tractions are prescribed at this edge). The (complex-valued) constants A_n and B_n for the symmetric and antisymmetric parts of the solution are obtained using the weighted residual method given by Eqn. (6.320), and the displacement results at $z = \pm L/2$ are presented in Table 6.6; the stress results are not presented due to the aforementioned singularity. As we can see, the accuracy is quite good near to $r = 0$, but

	$z = -0.6$					
	$r = 0$	$r = 0.2$	$r = 0.4$	$r = 0.6$	$r = 0.8$	$r = 1$
u_r	0.000	−1.602E-3	−3.204E-3	−4.803E-3	−6.392E-3	−7.451E-3
u_z	1.999E-2	1.958E-2	1.916E-2	1.875E-2	1.833E-2	1.778E-2
	$z = 0.6$					
	$r = 0$	$r = 0.2$	$r = 0.4$	$r = 0.6$	$r = 0.8$	$r = 1$
u_r	0.000	1.600E-3	3.199E-3	4.799E-3	6.402E-3	8.126E-3
u_z	−2.001E-2	−2.042E-2	−2.084E-3	−2.126E-2	−2.167E-2	−2.206E-2

Table 6.6 *Variation of displacements with respect to r at $z = \pm 0.6$ for the problem in [271].*

deteriorates as one approaches $r = 1$ due to the conflicting boundary condition there. A solution using real constants can be obtained by superposing the solutions in Eqns. (6.395) and (6.405), but is cumbersome due to the large number of constants. Robert and Kerr [271] use a Galerkin approach that yields a more diagonally dominant system, which seems to yield better results for the displacements at and surrounding the region of singularity; however, even this method fails to predict the stresses, for which they devise another method that accounts for the singularity.

Another example that uses the solution given by Eqn. (6.392) is presented in Section 6.6.11.

6.6.9 Circular cylinder with loading on the lateral surface, and end surfaces traction-free ('Filon's problem')

Now consider the problem that is complementary to that considered in Section 6.6.8, namely, one where the traction is prescribed on the lateral surface, and the flat end surfaces are traction-free [76, 215]. Thus, the boundary conditions are of the form

$$\tau_{rr}\big|_{r=a} = f(z); \qquad \tau_{rz}\big|_{r=a} = g(z), \qquad (6.406a)$$
$$\tau_{zz}\big|_{z=\pm L/2} = 0; \qquad \tau_{rz}\big|_{z=\pm L/2} = 0, \qquad (6.406b)$$

with $g(L/2) = g(-L/2) = 0$ in order to maintain continuity of prescribed tractions at $(a, \pm L/2)$.

If $f(z)$ and $g(z)$ are even and odd functions of z, using Eqn. (6.268), we assume the Love strain function to be given by

$$\Psi = \sum_{n=1}^{\infty} \left[A_n \sin(\lambda_n z) + B_n z \cos(\lambda_n z)\right] I_0(\lambda_n r) + \frac{c_1 \phi_2 z}{2(1+\nu)} + \frac{c_1(1-2\nu)\phi_3}{6(1+\nu)}. \tag{6.407}$$

Using Eqns. (6.266), we get

$$\tau_{rz} = \sum_{n=1}^{\infty} \lambda_n^2 I_1(\lambda_n r) \left\{\lambda_n A_n \sin\lambda_n z + B_n \left[\lambda_n z \cos\lambda_n z + 2\nu \sin\lambda_n z\right]\right\},$$

$$\tau_{zz} = \sum_{n=1}^{\infty} \lambda_n^2 I_0(\lambda_n r) \left\{\lambda_n A_n \cos\lambda_n z - B_n \left[(1-2\nu)\cos\lambda_n z + \lambda_n z \sin\lambda_n z\right]\right\}.$$

The conditions in Eqn. (6.406b) lead to a matrix equation of the form

$$\boldsymbol{M} \begin{bmatrix} A_n \\ B_n \end{bmatrix} = \begin{bmatrix} 0 \\ 0 \end{bmatrix}.$$

In order to obtain a nontrivial solution for A_n and B_n, we require that $\det \boldsymbol{M} = 0$. This leads to the characteristic equation

$$\sin\lambda_n L + \lambda_n L = 0. \tag{6.408}$$

Apart from zero, the remaining eigenvalues λ_n, $n = 1, 2, \ldots, \infty$, are complex-valued [262] with an asymptotic approximation given by Eqn. (6.311b); compare with Eqn. (6.308), and see Table 6.2. The corresponding constants, A_n, B_n are also complex-valued, and one takes the real part of the expressions for the stresses while evaluating them. The zero eigenvalues lead to constant stresses. However, since we have already incorporated this by taking the term associated with c_1, it suffices to consider only the complex-valued roots of Eqn. (6.408).

By using the condition $\tau_{rz}|_{\pm z = L/2} = 0$, one can eliminate B_n in terms of A_n to get the solution as (after renaming the constants)

$$2\mu u_r = -\sum_{n=1}^{\infty} I_1(\lambda_n r) A_n \left[(D + 2\nu - 1)\cos\lambda_n z + \lambda_n z \sin\lambda_n z\right] + \frac{(1-\nu)c_1 r}{1+\nu},$$

$$2\mu u_z = \sum_{n=1}^{\infty} I_0(\lambda_n r) A_n \left[(D + 2 - 2\nu)\sin\lambda_n z - \lambda_n z \cos\lambda_n z\right] - \frac{2\nu c_1 z}{1+\nu},$$

$$\begin{aligned} \tau_{rr} = \sum_{n=1}^{\infty} \frac{A_n}{r} \Big\{ &\left[(D + 2\nu - 1)\cos\lambda_n z + \lambda_n z \sin\lambda_n z\right] I_1(\lambda_n r) \\ &- \left[(D-1)\cos\lambda_n z + \lambda_n z \sin\lambda_n z\right] \lambda_n r I_0(\lambda_n r) \Big\} + c_1, \end{aligned} \tag{6.409}$$

$$\tau_{zz} = \sum_{n=1}^{\infty} \lambda_n I_0(\lambda_n r) A_n \left[(D+1)\cos\lambda_n z + \lambda_n z \sin\lambda_n z\right],$$

$$\tau_{rz} = \sum_{n=1}^{\infty} \lambda_n I_1(\lambda_n r) A_n \left[D \sin\lambda_n z - \lambda_n z \cos\lambda_n z\right],$$

$$\tau_{\theta\theta} = \sum_{n=1}^{\infty} \frac{A_n}{r} \left\{2\nu\lambda_n r I_0(\lambda_n r)\cos\lambda_n z - \left[(D+2\nu-1)\cos\lambda_n z + \lambda_n z \sin\lambda_n z\right] I_1(\lambda_n r)\right\} + c_1,$$

where

$$D = \frac{\lambda_n L}{2}\cot\frac{\lambda_n L}{2}.$$

Since $\tau_{rr}|_{r=a} = f(z)$, the constant c_1 is obtained as

$$c_1 = \frac{1}{L}\int_{-L/2}^{L/2} f(z)\,dz - \frac{4\nu}{aL}\sum_{n=1}^{\infty}\frac{A_n I_1(\lambda_n a)\sin\frac{\lambda_n L}{2}}{\lambda_n},$$

where the complex-valued constants A_n are to be taken along with their complex-conjugates to get the real-valued constant c_1. The constants A_n are determined using the conditions in Eqns. (6.406a), using, say, a weighted residual technique as described on page 485. Note that

$$\lim_{r\to 0}\tau_{rr} = \lim_{r\to 0}\tau_{\theta\theta} = c_1 + \frac{1}{2}\sum_{n=1}^{\infty} A_n\lambda_n\left[(1+2\nu-D)\cos\lambda_n z - \lambda_n z\sin\lambda_n z\right]. \tag{6.410}$$

As in the previous section, one can use four sets of real constants instead of two sets of complex-valued constants, and assume the same strain function as given by Eqn. (6.395) with the corresponding displacement and stress fields given by Eqns. (6.396) [215, 278]. The condition $\tau_{rz}|_{z=\pm L/2} = 0$ leads to

$$A_m = -\frac{1}{\lambda_m}\left(2\nu + \frac{\lambda_m L}{2}\coth\frac{\lambda_m L}{2}\right) B_m. \tag{6.411}$$

The condition $\tau_{rz}|_{r=a} = g(z)$ yields

$$\sum_{m=1}^{\infty} k_m^2 \left\{a k_m I_0(k_m a) D_m + \left[k_m C_m + 2(1-\nu) D_m\right] I_1(k_m a)\right\}\sin k_m z = g(z),$$

which in turn yields

$$a k_m I_0(k_m a) D_m + \left[k_m C_m + 2(1-\nu)D_m\right] I_1(k_m a) = \frac{2}{k_m^2 L}\int_{-L/2}^{L/2} g(z)\sin k_m z\,dz =: s_m. \tag{6.412}$$

The condition $\tau_{zz}|_{z=\pm L/2} = 0$ yields

$$\begin{aligned} c_2 + &\sum_{m=1}^{\infty}\frac{\lambda_m^2 J_0(\lambda_m r)\operatorname{cosech}\frac{\lambda_m L}{2} B_m}{2}(\sinh\lambda_m L + \lambda_m L) \\ &= \sum_{m=1}^{\infty}(-1)^{m+1} k_m^3\left\{\frac{I_0(k_m r) s_m}{k_m I_1(k_m a)} + \left[r I_1(k_m r) - \frac{a I_2(k_m a) I_0(k_m r)}{I_1(k_m a)}\right] D_m\right\}. \end{aligned} \tag{6.413}$$

Multiplying by r and integrating between the limits $[0,a]$, we get

$$c_2 = \sum_{m=1}^{\infty} \frac{2(-1)^{m+1} k_m s_m}{a}.$$

Next multiplying Eqn. (6.413) by $rJ_0(\lambda_n r)$, and using Eqns. (6.402), we get

$$\begin{aligned}&\frac{\lambda_n^2 a^2 J_0^2(\lambda_n a)\operatorname{cosech}\frac{\lambda_n L}{2} B_n}{4}(\sinh\lambda_n L + \lambda_n L)\\ &= \sum_{m=1}^{\infty}(-1)^{m+1}k_m^3\left\{\frac{a s_m J_0(\lambda_n a)}{k_m^2+\lambda_n^2} + \left[\frac{2a\lambda_n^2 I_1(k_m a)J_0(\lambda_n a)}{(k_m^2+\lambda_n^2)^2}\right]D_m\right\},\ n=1,2,\ldots,\infty. \qquad (6.414)\end{aligned}$$

Finally, the condition $\tau_{rr}|_{r=a} = f(z)$ yields

$$\begin{aligned}&c_1 + \sum_{m=1}^{\infty}\left\{\left[\frac{1}{k_m a} - \frac{I_0(k_m a)}{I_1(k_m a)}\right]s_m\right.\\ &\left.+\left[\frac{k_m a I_0^2(k_m a)}{I_1(k_m a)} - \frac{\left[2(1-\nu)+(k_m a)^2\right]I_1(k_m a)}{k_m a}\right]D_m\right\}k_m^2\cos k_m z \qquad (6.415)\\ &= f(z) + \sum_{m=1}^{\infty}\left[\left(\frac{\lambda_m L}{2}\coth\frac{\lambda_m L}{2} - 1\right)\cosh\lambda_m z - \lambda_m z\sinh\lambda_m z\right]\lambda_m^2 J_0(\lambda_m a)B_m.\end{aligned}$$

On integrating the above equation between the limits $[-L/2, L/2]$, we get $c_1 = \int_{-L/2}^{L/2} f(z)dz$. Then multiplying Eqn. (6.415) by $\cos k_n z$, and using the relations

$$\int_{-L/2}^{L/2}\cosh\lambda_m z\cos k_n z\,dz = \frac{2(-1)^n\lambda_m\sinh\frac{\lambda_m L}{2}}{k_n^2+\lambda_m^2} =: U_m^{(n)},$$

$$\int_{-L/2}^{L/2} z\sinh\lambda_m z\cos k_n z\,dz = \frac{2(-1)^n\left[\frac{\lambda_m L}{2}(k_n^2+\lambda_m^2)\cosh\frac{\lambda_m L}{2} + (k_n^2-\lambda_m^2)\sinh\frac{\lambda_m L}{2}\right]}{(k_n^2+\lambda_m^2)^2} =: V_m^{(n)},$$

we get

$$\begin{aligned}&\left\{\left[\frac{1}{k_n a} - \frac{I_0(k_n a)}{I_1(k_n a)}\right]s_n + \left[\frac{k_n a I_0^2(k_n a)}{I_1(k_n a)} - \frac{\left[2(1-\nu)+(k_n a)^2\right]I_1(k_n a)}{k_n a}\right]D_n\right\}\frac{k_n^2 L}{2}\\ &= \sum_{m=1}^{\infty}\left[\left(\frac{\lambda_m L}{2}\coth\frac{\lambda_m L}{2} - 1\right)U_m^{(n)} - \lambda_m V_m^{(n)}\right]\lambda_m^2 J_0(\lambda_m a)B_m\\ &\qquad + \int_{-L/2}^{L/2} f(z)\cos k_n z\,dz, \qquad n=1,2,\ldots,\infty. \qquad (6.416)\end{aligned}$$

Eqns. (6.414) and (6.416) are to be solved for the constants B_n and D_n.

The values of τ_{rr} and $\tau_{\theta\theta}$ at the axis are given by Eqn. (6.404), with c_1 given by $\int_{-L/2}^{L/2} f(z)dz$.

If $f(z)$ and $g(z)$ are odd and even functions, respectively, then, in place of Eqn. (6.407), we assume

$$\Psi = \sum_{n=1}^{\infty} [A_n \cos(\lambda_n z) + B_n z \sin(\lambda_n z)]\, I_0(\lambda_n r). \tag{6.417}$$

In place of Eqn. (6.408), we get the characteristic equation

$$\sin \lambda_n L = \lambda_n L,$$

which should be compared with Eqn. (6.305). The asymptotic roots of this equation are given by Eqn. (6.311a); the first few roots of $\sin z = z$ are listed in Table 6.1. In place of Eqns. (6.409), we now get

$$\begin{aligned}
2\mu u_r &= \sum_{n=1}^{\infty} I_1(\lambda_n r) A_n \left[\lambda_n z \cos \lambda_n z + (1 - 2\nu + D) \sin \lambda_n z\right], \\
2\mu u_z &= \sum_{n=1}^{\infty} I_0(\lambda_n r) A_n \left[(D + 2\nu - 2) \cos \lambda_n z - \lambda_n z \sin \lambda_n z\right], \\
\tau_{rr} &= \sum_{n=1}^{\infty} \frac{A_n}{r} \Big\{ [\lambda_n z \cos \lambda_n z + (1 + D) \sin \lambda_n z]\, \lambda_n r I_0(\lambda_n r) \\
&\quad - [\lambda_n z \cos \lambda_n z + (1 - 2\nu + D) \sin \lambda_n z]\, I_1(\lambda_n r) \Big\}, \\
\tau_{zz} &= -\sum_{n=1}^{\infty} \lambda_n I_0(\lambda_n r) A_n \left[\lambda_n z \cos \lambda_n z + (D - 1) \sin \lambda_n z\right], \\
\tau_{rz} &= \sum_{n=1}^{\infty} \lambda_n I_1(\lambda_n r) A_n \left[D \cos \lambda_n z - \lambda_n z \sin \lambda_n z\right], \\
\tau_{\theta\theta} &= \sum_{n=1}^{\infty} \frac{A_n}{r} \Big\{ 2\nu \lambda_n r I_0(\lambda_n r) \sin \lambda_n z + [\lambda_n z \cos \lambda_n z + (1 - 2\nu + D) \sin \lambda_n z]\, I_1(\lambda_n r) \Big\},
\end{aligned}$$

where

$$D = \frac{\lambda_n L}{2} \tan \frac{\lambda_n L}{2}.$$

The constants A_n are found using the conditions in Eqns. (6.406a).

Note that

$$\lim_{r \to 0} \tau_{rr} = \lim_{r \to 0} \tau_{\theta\theta} = \frac{1}{2} \sum_{n=1}^{\infty} \lambda_n A_n \left[\lambda_n z \cos \lambda_n z + (D + 1 + 2\nu) \sin \lambda_n z\right].$$

Using four sets of real constants instead of two sets of complex-valued constants for the case where $f(z)$ and $g(z)$ are odd and even functions, respectively,, we assume

$$\Psi = \sum_{m=1}^{\infty} [A_m \cosh \lambda_m z + B_m z \sinh \lambda_m z]\, J_0(\lambda_m r) + \sum_{m=1}^{\infty} [C_m I_0(k_m r) + D_m r I_1(k_m r)] \cos k_m z, \tag{6.418}$$

where $k_m = (2m-1)\pi/L$, and λ_m, $m = 1,2,\ldots,\infty$, are the positive roots of $J_1(\lambda a) = 0$. The corresponding displacements and stresses are

$$
\begin{aligned}
2\mu u_r &= \sum_{m=1}^{\infty} \Big\{ A_m \lambda_m^2 J_1(\lambda_m r) \sinh \lambda_m z \\
&\quad + B_m \lambda_m J_1(\lambda_m r) \left[\sinh \lambda_m z + \lambda_m z \cosh \lambda_m z\right] \\
&\quad + C_m k_m^2 I_1(k_m r) \sin k_m z + D_m k_m^2 r I_0(k_m r) \sin k_m z \Big\}, \\
2\mu u_z &= \sum_{m=1}^{\infty} \Big\{ -A_m \lambda_m^2 J_0(\lambda_m r) \cosh \lambda_m z \\
&\quad - B_m \lambda_m J_0(\lambda_m r) \left[\lambda_m z \sinh \lambda_m z - 2(1-2\nu) \cosh \lambda_m z\right] \\
&\quad + C_m k_m^2 I_0(k_m r) \cos k_m z + D_m k_m \cos k_m z \left[4(1-\nu) I_0(k_m r) + k_m r I_1(k_m r)\right] \Big\}, \\
\tau_{rr} &= \sum_{m=1}^{\infty} \frac{1}{r} \Big\{ \lambda_m^2 A_m \left[\lambda_m r J_0(\lambda_m r) - J_1(\lambda_m r)\right] \sinh \lambda_m z \\
&\quad + \lambda_m B_m \Big[\lambda_m r J_0(\lambda_m r) \left[(1+2\nu) \sinh \lambda_m z + \lambda_m z \cosh \lambda_m z\right] \\
&\quad - J_1(\lambda_m r) \left[\sinh \lambda_m z + \lambda_m z \cosh \lambda_m z\right] \Big] - k_m^2 C_m \left[I_1(k_m r) - k_m r I_0(k_m r)\right] \sin k_m z \\
&\quad + k_m^2 r D_m \left[(1-2\nu) I_0(k_m r) + k_m r I_1(k_m r)\right] \sin k_m z \Big\}, \\
\tau_{zz} &= \sum_{m=1}^{\infty} \Big\{ -A_m \lambda_m^3 J_0(\lambda_m r) \sinh \lambda_m z \\
&\quad + \lambda_m^2 B_m \left[(1-2\nu) \sinh \lambda_m z - \lambda_m z \cosh \lambda_m z\right] J_0(\lambda_m r) \qquad (6.419) \\
&\quad - C_m k_m^3 I_0(k_m r) \sin k_m z - k_m^2 D_m \left[2(2-\nu) I_0(k_m r) + k_m r I_1(k_m r)\right] \sin k_m z \Big\}, \\
\tau_{rz} &= \sum_{m=1}^{\infty} \Big\{ A_m \lambda_m^3 J_1(\lambda_m r) \cosh \lambda_m z + \lambda_m^2 B_m J_1(\lambda_m r) \left[\lambda_m z \sinh \lambda_m z + 2\nu \cosh \lambda_m z\right] \\
&\quad + k_m^3 C_m I_1(k_m r) \cos k_m z + k_m^2 D_m \left[k_m r I_0(k_m r) + 2(1-\nu) I_1(k_m r)\right] \cos k_m z \Big\}, \\
\tau_{\theta\theta} &= \frac{1}{r} \Big\{ \lambda_m^2 A_m J_1(\lambda_m r) \sinh \lambda_m z \\
&\quad + \lambda_m B_m \left[2\nu \lambda_m r J_0(\lambda_m r) \sinh \lambda_m z + J_1(\lambda_m r) (\sinh \lambda_m z + \lambda_m z \cosh \lambda_m z)\right] \\
&\quad + k_m^2 C_m I_1(k_m r) \sin k_m z + k_m^2 r D_m (1-2\nu) I_0(k_m r) \sin k_m z \Big\}.
\end{aligned}
$$

The condition $\tau_{rz}|_{z=\pm L/2} = 0$ yields

$$
A_m = -\frac{1}{\lambda_m} \left(2\nu + \frac{\lambda_m L}{2} \tanh \frac{\lambda_m L}{2} \right) B_m. \qquad (6.420)
$$

The condition $\tau_{rz}|_{r=a} = g(z)$ yields

$$
\sum_{m=1}^{\infty} k_m^2 \left\{ a k_m I_0(k_m a) D_m + \left[k_m C_m + 2(1-\nu) D_m\right] I_1(k_m a) \right\} \cos k_m z = g(z),
$$

which in turn yields

$$ak_m I_0(k_m a)D_m + [k_m C_m + 2(1-\nu)D_m]\, I_1(k_m a) = \frac{2}{k_m^2 L}\int_{-L/2}^{L/2} g(z)\cos k_m z\, dz =: s_m. \quad (6.421)$$

The condition $\tau_{zz}|_{z=\pm L/2} = 0$ yields

$$\begin{aligned} &c_2 + \sum_{m=1}^{\infty} \frac{\lambda_m^2 J_0(\lambda_m r)\operatorname{sech}\frac{\lambda_m L}{2} B_m}{2}(\sinh\lambda_m L - \lambda_m L) \\ &= \sum_{m=1}^{\infty}(-1)^{m+1}k_m^3\left\{\frac{s_m I_0(k_m r)}{k_m I_1(k_m a)} + \left[rI_1(k_m r) - \frac{aI_0(k_m r)I_2(k_m a)}{I_1(k_m a)}\right]D_m\right\}. \end{aligned} \quad (6.422)$$

Multiplying by r and integrating between the limits $[0, a]$, we get

$$c_2 = \sum_{m=1}^{\infty}\frac{2(-1)^{m+1}k_m s_m}{a}.$$

Next, multiplying Eqn. (6.422) by $rJ_0(\lambda_n r)$, and using Eqns. (6.402), we get

$$\begin{aligned} &\frac{\lambda_n^2 a^2 J_0^2(\lambda_n a)\operatorname{sech}\frac{\lambda_n L}{2} B_n}{4}(\sinh\lambda_n L - \lambda_n L) \\ &= \sum_{m=1}^{\infty}(-1)^{m+1}k_m^3\left\{\frac{as_m J_0(\lambda_n a)}{k_m^2+\lambda_n^2} + \frac{2a\lambda_n^2 I_1(k_m a)J_0(\lambda_n a)}{(k_m^2+\lambda_n^2)^2}D_m\right\},\ n = 1, 2, \ldots, \infty. \end{aligned} \quad (6.423)$$

Finally, the condition $\tau_{rr}|_{r=a} = f(z)$ yields

$$\begin{aligned} &\sum_{m=1}^{\infty}\left\{\left[\frac{1}{k_m a} - \frac{I_0(k_m a)}{I_1(k_m a)}\right]s_m \right.\\ &\left. + \left[\frac{k_m a I_0^2(k_m a)}{I_1(k_m a)} - \frac{[2(1-\nu)+(k_m a)^2]\,I_1(k_m a)}{k_m a}\right]D_m\right\}k_m^2\sin k_m z \\ &= f(z) + \sum_{m=1}^{\infty}\left[\left(\frac{\lambda_m L}{2}\tanh\frac{\lambda_m L}{2} - 1\right)\sinh\lambda_m z - \lambda_m z\cosh\lambda_m z\right]\lambda_m^2 J_0(\lambda_m a)B_m. \end{aligned} \quad (6.424)$$

Multiplying the above relation by $\sin k_n z$, and using the relations

$$\int_{-L/2}^{L/2}\sinh\lambda_m z\sin k_n z\, dz = \frac{2\lambda_m(-1)^{n+1}\cosh\frac{\lambda_m L}{2}}{k_n^2+\lambda_m^2} =: U_m^{(n)},$$

$$\int_{-L/2}^{L/2} z\cosh\lambda_m z\sin k_n z\, dz = \frac{2(-1)^{n+1}\left[\frac{\lambda_m L}{2}(k_n^2+\lambda_m^2)\sinh\frac{\lambda_m L}{2} + (k_n^2-\lambda_m^2)\cosh\frac{\lambda_m L}{2}\right]}{(k_n^2+\lambda_m^2)^2} =: V_m^{(n)},$$

we get for $n = 1, 2, \ldots, \infty$,

$$\left\{\left[\frac{1}{k_n a} - \frac{I_0(k_n a)}{I_1(k_n a)}\right]s_n + \left[\frac{k_n a I_0^2(k_n a)}{I_1(k_n a)} - \frac{[2(1-\nu)+(k_n a)^2]\,I_1(k_n a)}{k_n a}\right]D_n\right\}\frac{k_n^2 L}{2}$$

$$= \int_{-L/2}^{L/2} f(z) \sin k_n z \, dz + \sum_{m=1}^{\infty} \left[\left(\frac{\lambda_m L}{2} \tanh \frac{\lambda_m L}{2} - 1 \right) U_m^{(n)} - \lambda_m V_m^{(n)} \right] \lambda_m^2 J_0(\lambda_m a) B_m. \tag{6.425}$$

Eqns. (6.423) and (6.425) are to be solved for the constants B_n and D_n.

As an example, consider the problem of the following discontinuous tangential stress field applied on the surface [76, 215]:

$$f(z) = 0,$$

$$g(z) = \begin{cases} 0, & |z| < L/6, \ |z| > L/3, \\ S, & L/6 \le z \le L/3, \\ -S, & -L/3 \le z \le -L/6. \end{cases}$$

We denote the solution given by Eqn. (6.409) as Method 1 (obtained using 55 terms), and the immediately following solution based on real coefficients as Method 2 (obtained using 100 terms). From Eqn. (6.412), we get

$$s_m = \frac{SL^2}{m^3 \pi^3} \sin^2 \frac{m\pi}{6} \left(1 + 2 \cos \frac{m\pi}{3} \right).$$

Let $Q := LS/(3a)$. We take $L = \pi a$, and normalize the displacement fields u_r and u_z by $-Q\nu a/E$ and $QL/(2E)$, respectively, and the stress field by Q while presenting the results in the tables below. A finite-element-based solution (FEM) based on [161] is also listed since the analytical solutions do not agree. From Tables 6.7–6.9, we see that the satisfaction of the traction boundary conditions is quite satisfactory with all methods considered (except Filon's method, which yields accurate results only far away from the ends since he satisfies the end conditions only in an approximate sense as seen from the results in Table 6.10), inspite of the discontinuity in the applied τ_{rz} field. Although the stress solution presented by Meleshko [215] is accurate, Tables 6.11 and 6.12 seem to indicate that his method yields a very inaccurate displacement field.

6.6.10 Clamped and simply supported circular cylinders

Consider a circular cylinder of radius a and thickness h that is clamped at the edge $r = a$, i.e. $u_r|_{r=a} = u_z|_{r=a} = 0$, and subjected to the following boundary conditions on the top and bottom surfaces:

$$\begin{aligned} \tau_{zz}|_{z=h/2} &= p_t(r); & \tau_{rz}|_{z=h/2} &= s_t(r), \\ \tau_{zz}|_{z=-h/2} &= p_b(r); & \tau_{rz}|_{z=-h/2} &= s_b(r), \end{aligned} \tag{6.426}$$

with $s_t(0) = s_b(0) = 0$. We consider the Love strain function to be of the form

$$\Psi = \sum_{n=1}^{\infty} [A_n J_0(\lambda_n r) + B_n r J_1(\lambda_n r)] \cosh \lambda_n z + [C_n J_0(\lambda_r) + D_n r J_1(\lambda_n r)] \sinh \lambda_n z, \tag{6.427}$$

where λ_n, $n = 1, 2, \ldots, \infty$, are the roots of the characteristic equation

$$\lambda a \left[J_0^2(\lambda a) + J_1^2(\lambda a) \right] - 4(1 - \nu) J_0(\lambda a) J_1(\lambda a) = 0. \tag{6.428}$$

$r = 0$					
z	Method 1	Method 2	Filon [76]	Meleshko [215]	FEM
0	0.657	0.661	0.689	0.661	0.661
$0.1L$	0.588	0.593	0.631	0.592	0.592
$0.2L$	0.465	0.469	0.539	0.469	0.469
$0.3L$	0.339	0.341	0.457	0.341	0.341
$0.4L$	0.141	0.142	0.270	0.142	0.142
$0.5L$	0.000	−0.005	0.000	−0.001	0.000
$r = a$					
0	1.122	1.115	1.117	1.115	1.127
$0.1L$	1.335	1.345	1.344	1.344	1.361
$0.2L$	1.362	1.435	1.368	1.434	1.409
$0.3L$	−0.270	−0.325	−0.409	−0.325	−0.299
$0.4L$	−0.186	−0.184	−0.357	−0.184	−0.200
$0.5L$	0.000	−0.047	0.000	0.012	0.000

Table 6.7 *Normalized τ_{zz} at $r = 0$ and at $r = a$.*

r	Method 1	Method 2	Filon [76]	Meleshko [215]	FEM
0	0.000	−0.005	0.000	−0.001	0.000
$0.2a$	0.000	0.000	0.000	0.000	0.000
$0.4a$	0.000	0.000	0.000	0.000	0.000
$0.6a$	0.000	0.000	0.000	0.000	0.000
a	0.000	−0.047	0.000	0.012	0.000

Table 6.8 *Normalized τ_{zz} at $z = L/2$.*

z	Method 1	Method 2	Filon [76]	Meleshko [215]	FEM
0	0.000	0.000	0.000	0.000	0.000
$0.1L$	−0.004	−0.002	0.000	0.000	0.000
$0.2L$	0.949	0.964	0.955	0.955	0.950
$0.3L$	0.951	0.964	0.955	0.955	0.950
$0.4L$	0.000	−0.002	0.000	0.000	0.000
$0.5L$	0.000	0.000	0.000	0.000	0.000

Table 6.9 *Normalized τ_{rz} at $r = a$.*

r	Method 1	Method 2	Filon [76]	Meleshko [215]	FEM
0	0.000	0.000	0.000	0.001	0.000
$0.2a$	0.000	0.000	0.092	0.000	0.000
$0.4a$	0.000	0.000	0.172	0.000	0.000
$0.6a$	0.000	0.000	0.210	0.000	0.000
a	0.000	0.000	0.000	0.012	0.000

Table 6.10 *Normalized τ_{rz} at $z = L/2$.*

z	Method 1	Method 2	Filon [76]	Meleshko [215]	FEM
0	0.000	0.000	0.000	0.000	0.000
$0.1L$	0.103	0.103	0.111	0.203	0.103
$0.2L$	0.190	0.191	0.211	0.391	0.191
$0.3L$	0.269	0.271	0.309	0.571	0.271
$0.4L$	0.334	0.336	0.393	0.737	0.336
$0.5L$	0.380	0.383	0.427	0.883	0.382

Table 6.11 *Normalized u_z at $r = 0$.*

z	Method 1	Method 2	Filon [76]	Meleshko [215]	FEM
0	0.551	0.555	0.579	1.056	0.555
$0.1L$	0.428	0.431	0.458	0.932	0.431
$0.2L$	0.316	0.320	0.371	0.821	0.320
$0.3L$	0.596	0.599	0.586	1.100	0.599
$0.4L$	0.560	0.564	0.374	1.064	0.564
$0.5L$	0.654	0.656	0.000	1.136	0.660

Table 6.12 *Normalized u_r at $r = a$.*

The above characteristic equation is derived using the displacement boundary conditions at the clamped edge using a method similar to that used for deriving Eqn. (6.391). The asymptotic roots of Eqn. (6.428) are given by $(2n-1)\pi/2 + (i/2)\log(2n-1)\pi$. The first few roots for $\nu = 0.25$ are $(1.495, 0)$, $(4.529, 1.101)$, $(7.732, 1.373)$ and $(10.901, 1.544)$. The condition $u_z|_{r=a} = 0$ leads to

$$A_n = \left[\frac{4(1-\nu)}{\lambda_n} - \frac{aJ_1(\lambda_n a)}{J_0(\lambda_n a)}\right] B_n,$$

$$C_n = \left[\frac{4(1-\nu)}{\lambda_n} - \frac{aJ_1(\lambda_n a)}{J_0(\lambda_n a)}\right] D_n.$$

The final solution (after renaming the constants) is

$$
\begin{aligned}
2\mu u_r &= -\sum_{n=1}^{\infty} [A_n \sinh \lambda_n z + B_n \cosh \lambda_n z] \\
&\quad \{\lambda_n a J_1(\lambda_n a) J_1(\lambda_n r) + [\lambda_n r J_0(\lambda_n r) - 4(1-\nu) J_1(\lambda_n r)]\, J_0(\lambda_n a)\}, \\
2\mu u_z &= \sum_{n=1}^{\infty} [A_n \cosh \lambda_n z + B_n \sinh \lambda_n z] [\lambda_n a J_1(\lambda_n a) J_0(\lambda_n r) - \lambda_n r J_0(\lambda_n a) J_1(\lambda_n r)], \\
\tau_{rr} &= \sum_{n=1}^{\infty} \frac{1}{r} [A_n \sinh \lambda_n z + B_n \cosh \lambda_n z] \Big\{ [J_1(\lambda_n r) - \lambda_n r J_0(\lambda_n r)]\, \lambda_n a J_1(\lambda_n a) \\
&\quad + \Big[(3-2\nu)\lambda_n r J_0(\lambda_n r) - (4 - 4\nu - \lambda_n^2 r^2) J_1(\lambda_n r) \Big] J_0(\lambda_n a) \Big\}, \\
\tau_{zz} &= \sum_{n=1}^{\infty} \lambda_n [A_n \sinh \lambda_n z + B_n \cosh \lambda_n z] \\
&\quad \{\lambda_n a J_1(\lambda_n a) J_0(\lambda_n r) + [2\nu J_0(\lambda_n r) - \lambda_n r J_1(\lambda_n r)]\, J_0(\lambda_n a)\}, \\
\tau_{rz} &= -\sum_{n=1}^{\infty} \lambda_n [A_n \cosh \lambda_n z + B_n \sinh \lambda_n z] \\
&\quad \{\lambda_n a J_1(\lambda_n a) J_1(\lambda_n r) - [2(1-\nu) J_1(\lambda_n r) - \lambda_n r J_0(\lambda_n r)]\, J_0(\lambda_n a)\}, \\
\tau_{\theta\theta} &= -\sum_{n=1}^{\infty} \frac{1}{r} [A_n \sinh \lambda_n z + B_n \cosh \lambda_n z] \\
&\quad \{\lambda_n a J_1(\lambda_n a) J_1(\lambda_n r) + [(1-2\nu)\lambda_n r J_0(\lambda_n r) - 4(1-\nu) J_1(\lambda_n r)]\, J_0(\lambda_n a)\}.
\end{aligned}
\tag{6.429}
$$

The complex-valued constants A_n and B_n are determined so as to satisfy the boundary conditions in Eqns. (6.426). Note that

$$
\lim_{r\to 0} \tau_{rr} = \lim_{r\to 0} \tau_{\theta\theta} = \sum_{n=1}^{\infty} \lambda_n [A_n \sinh \lambda_n z + B_n \cosh \lambda_n z] \left[J_0(\lambda_n a) - \frac{\lambda_n a}{2} J_1(\lambda_n a) \right].
$$

For example, for a clamped circular plate of radius $a = 1$ and thickness $h = 0.1$ subjected to a uniform pressure loading on the top of magnitude $p = 1$, and with material properties $E = 1$ and $\nu = 0.25$, the maximum midplane displacement and bending stress at $(0, -h/2)$ according to classical plate theory are

$$
\begin{aligned}
u_r\big|_{(0,0)} &= \frac{3pa^4(1-\nu^2)}{16Eh^3} = 175.78, \\
\tau_{rr}\big|_{(a,h/2)} &= \frac{3pa^2(1+\nu)}{8h^2} = 46.875,
\end{aligned}
$$

while the corresponding results that we obtain based on the above solution are 182.67 and 46.75, respectively. If we take $h = 0.5$ keeping all other properties the same, then the results obtained using classical plate theory (which is not valid for such a large h/a ratio) are 1.406 and 1.875, respectively, while those based on our exact solution are 2.84 and 2.000, respectively. Note that there is a singularity in the stresses at $(a, \pm h/2)$ due to the conflict in boundary conditions at these points.

If instead of $u_r|_{r=a} = u_z|_{r=a} = 0$, we had the boundary conditions $u_z|_{r=a} = \tau_{rr}|_{r=a} = 0$, then we again assume the Love strain function to be as given by Eqn. (6.427), resulting in the same solution as in Eqn. (6.429), but where, now, λ_n are the roots of the following characteristic equation that is obtained using the condition $\tau_{rr}|_{r=a} = 0$ (compare with Eqn. (6.428))

$$\lambda a \left[(3-2\nu)J_0^2(\lambda a) + J_1^2(\lambda a)\right] - 4(1-\nu)J_0(\lambda a)J_1(\lambda a) = 0. \tag{6.430}$$

The asymptotic values of the roots of the above equation (for $\nu = 0$) are $(4n-1)\pi/4 + i\log[\sqrt{1.5} + 1/\sqrt{2}]$. The first few roots for $\nu = 0.25$ are $(2.7035, 0.8478)$, $(5.3953, 0.7566)$, $(8.5757, 0.7498)$ and $(11.7346, 0.7478)$. The constants A_n and B_n are again to be determined using the boundary conditions in Eqns. (6.426).

Another possible boundary condition for the simply supported cylinder is that $\tau_{rr}|_{r=a} = 0$ without requiring $u_z|_{r=a}$ to be zero. In this case, we assume the Love strain function to be

$$\begin{aligned}\Psi = \sum_{n=1}^{\infty}\Big\{&[A_n J_0(\lambda_n r) + B_n r J_1(\lambda_n r) + F_n z J_0(\lambda_n r)]\cosh\lambda_n z \\ &+ [C_n J_0(\lambda_n r) + D_n r J_1(\lambda_n r) + E_n z J_0(\lambda_n r)]\sinh\lambda_n z\Big\} \\ &+ \frac{c_2\phi_2 z}{4(1+\nu)} - \frac{(1+4\nu)c_2\phi_3}{12(1+\nu)} + \frac{[2\phi_3 z - (1+2\nu)\phi_4]\,c_4}{24(1+\nu)}.\end{aligned}$$

In order to satisfy the condition $\tau_{rr}|_{r=a} = 0$, we require the λ_n, $n = 1, 2, \ldots, \infty$, to be the (real) roots of the characteristic equation

$$J_0(\lambda a) = J_2(\lambda a), \tag{6.431}$$

(or, equivalently, of $\lambda a J_0(\lambda a) = J_1(\lambda a)$), and

$$E_n = -\frac{\left[2\nu + (\lambda_n a)^2 - 1\right]B_n}{2\nu},$$
$$F_n = -\frac{\left[2\nu + (\lambda_n a)^2 - 1\right]D_n}{2\nu}.$$

A good approximation to the roots λ_n of Eqn. (6.431) is $[\lambda_{n-1}a + \pi]/a$, with $\lambda_1 a \approx 2$. Using this approximation, and a Newton–Raphson strategy based on the equation

$$\lambda_n^{\text{new}} = \lambda_n^{\text{prev}} - \left.\frac{[J_1(\lambda_n) - \lambda_n J_0(\lambda_n)]\lambda_n}{J_1(\lambda_n)(\lambda_n^2 - 1)}\right|_{\lambda_n = \lambda_n^{\text{prev}}},$$

the first few roots obtained for the equation $J_0(\lambda) = J_2(\lambda)$ are presented in Table 6.13. The displacements and stresses are now obtained using Eqns. (6.266). A rigid displacement is added to u_z in order to satisfy the condition $u_z|_{(a,0)} = 0$. The stress $\tau_{rz}|_{r=a}$ is a reaction to the displacement $u_z|_{r=a}$ (which comes out as part of the analytical solution; in a numerical strategy one can get a good approximation to this boundary condition by setting $u_z|_{r=a}$ to zero). The real-valued constants c_2, c_4, A_n, B_n, C_n and D_n are obtained using the boundary

n	Roots	n	Roots
1	1.8411837813406593	11	33.746182898667378
2	5.3314427735250325	12	36.889987409236809
3	8.5363163663462860	13	40.033444053350671
4	11.706004902428946	14	43.176628965448820
5	14.863588633900715	15	46.319597561173907
6	18.015527862681072	16	49.462391139702753
7	21.164369859188696	17	52.605041111556680
8	24.311326857210762	18	55.747571792251001
9	27.457050571059245	19	58.890002299185703
10	30.601922972669094	20	62.032347870661987

Table 6.13 *First few roots of* $J_0(\lambda) = J_2(\lambda)$.

conditions in Eqn. (6.426) and the weighted residual method described on page 485. For example, for a simply supported circular plate of radius $a = 1$ and thickness $h = 0.1$ subjected to a uniform pressure loading on the top of magnitude $p = 1$, and with material properties $E = 1$ and $\nu = 0.25$, the maximum midplane displacement and maximum bending stress according to classical plate theory are

$$\begin{aligned} u_r\big|_{(0,0)} &= \frac{3pa^4(1-\nu)(5+\nu)}{16Eh^3} = 738.38, \\ \tau_{rr}\big|_{(0,-h/2)} &= \frac{3(3+\nu)pa^2}{8h^2} = 121.875, \end{aligned}$$

while the corresponding results that we obtain based on the above solution are 744.52 and 121.99, respectively. If we take $h = 0.5$ keeping all other properties the same, then the results obtained using classical plate theory (which is not valid for such a large h/a ratio) are 5.91 and 4.875, respectively, while those based on our exact solution are 7.149 and 4.99, respectively[23].

6.6.11 Circular cylinder spinning about its axis

A solid cylinder of radius a and length L rotates about the z-axis at a steady angular speed of ω. Similar to the strategy followed in Section 5.12.3, we choose a frame of reference that rotates along with the cylinder, so that the body force components (b_r, b_θ, b_z) are given by $(r\omega^2, 0, 0)$. We obtain the exact solution by first finding a solution assuming that $u_r = u_r(r)$, $u_\theta = 0$ and $u_z = u_z(z)$, and then superimposing another solution in order to satisfy the required boundary conditions. Accordingly, under the above-mentioned assumptions, the Navier equations in Appendix B yield

$$\frac{d}{dr}\left[\frac{1}{r}\frac{d(ru_r)}{dr}\right] = -\frac{\rho_0 r\omega^2}{\lambda + 2\mu},$$

[23] Almost identical results are obtained using the previous analytical formulation (based on the roots of Eqn. (6.430)) and a finite element formulation (based on [153]) both of which use the boundary conditions $u_z\big|_{r=a} = \tau_{rr}\big|_{r=a} = 0$.

$$\frac{d^2u_z}{dz^2} = 0.$$

On solving these equations, we get $u_z = c_3z + c_4$ and

$$u_r = c_1r + \frac{c_2}{r} - \frac{\rho_0\omega^2r^3}{8(\lambda+2\mu)}. \tag{6.432}$$

To prevent an infinite displacement at $r = 0$, we need to choose $c_2 = 0$. Also, using the boundary condition $u_z|_{z=0} = 0$, we get $c_4 = 0$. Now using Eqns. (B.8) and (B.12), we get

$$\begin{aligned}
\tau_{rr} &= \lambda(2c_1+c_3) + 2\mu c_1 - \frac{2\lambda+3\mu}{4(\lambda+2\mu)}\rho_0r^2\omega^2,\\
\tau_{\theta\theta} &= \lambda(2c_1+c_3) + 2\mu c_1 - \frac{2\lambda+\mu}{4(\lambda+2\mu)}\rho_0r^2\omega^2,\\
\tau_{zz} &= \lambda(2c_1+c_3) + 2\mu c_3 - \frac{\lambda}{2(\lambda+2\mu)}\rho_0r^2\omega^2,
\end{aligned}$$

with the other stress components zero. Since the lateral surfaces of the cylinder are traction-free, we need to impose $\tau_{rr}|_{r=a} = 0$, which leads to

$$\lambda(2c_1+c_3) + 2\mu c_1 = \frac{2\lambda+3\mu}{4(\lambda+2\mu)}\rho_0a^2\omega^2. \tag{6.433}$$

Using the fact that $c_3 = \epsilon_{zz}$, we get

$$\begin{aligned}
u_r &= -c_3\nu r + \frac{(1+\nu)(1-2\nu)\rho_0r\omega^2\left[(3-2\nu)a^2-r^2\right]}{8E(1-\nu)},\\
u_z &= c_3z,\\
\tau_{rr} &= \frac{2\lambda+3\mu}{4(\lambda+2\mu)}\rho_0\omega^2(a^2-r^2) = \frac{(3-2\nu)\rho_0\omega^2}{8(1-\nu)}(a^2-r^2),\\
\tau_{\theta\theta} &= \frac{2\lambda+3\mu}{4(\lambda+2\mu)}\rho_0\omega^2a^2 - \frac{2\lambda+\mu}{4(\lambda+2\mu)}\rho_0\omega^2r^2 = \frac{\rho_0\omega^2}{8(1-\nu)}\left[(1+2\nu)(a^2-r^2)+2(1-2\nu)a^2\right],\\
\tau_{zz} &= \frac{\mu(3\lambda+2\mu)}{\lambda+\mu}\epsilon_{zz} + \frac{\rho_0\omega^2\lambda}{4(\lambda+2\mu)}\left[2(a^2-r^2)+\frac{\mu a^2}{\lambda+\mu}\right]\\
&= Ec_3 + \frac{\rho_0\omega^2\nu}{4(1-\nu)}\left[2(a^2-r^2)+(1-2\nu)a^2\right].
\end{aligned} \tag{6.434}$$

By taking $Ec_3 = -\rho_0\omega^2\nu[2a^2+(1-2\nu)a^2]/[4(1-\nu)]$, the above equations reduce to

$$\begin{aligned}
u_r &= \frac{\rho_0r\omega^2\left[(1-\nu)(3-2\nu)a^2-(1+\nu)(1-2\nu)r^2\right]}{8E(1-\nu)},\\
u_z &= -\frac{\rho_0\omega^2\nu z[2a^2+(1-2\nu)a^2]}{4E(1-\nu)},\\
\tau_{rr} &= \frac{(3-2\nu)\rho_0\omega^2}{8(1-\nu)}(a^2-r^2),\\
\tau_{\theta\theta} &= \frac{\rho_0\omega^2}{8(1-\nu)}\left[(1+2\nu)(a^2-r^2)+2(1-2\nu)a^2\right],
\end{aligned} \tag{6.435}$$

	$r = 0$	$r = 0.2a$	$r = 0.4a$	$r = 0.6a$	$r = 0.8a$	$r = a$
u_r	0.000	5.733E-2	0.110	0.151	0.177	0.182
u_z	−0.140	−0.136	−0.125	−0.109	−9.044E-2	−7.493E-2
τ_{rr}	0.390	0.376	0.331	0.255	0.145	0.000
$\tau_{\theta\theta}$	0.390	0.382	0.358	0.317	0.258	0.179
τ_{zz}	7.810E-3	7.548E-3	6.549E-3	4.028E-3	−1.869E-3	−1.292E-2

Table 6.14 *Variation of normalized displacement and stress components with respect to r at* $z/a = 0.8$.

$$\tau_{zz} = -\frac{\rho_0\omega^2\nu r^2}{2(1-\nu)},$$
$$\tau_{rz} = 0.$$

We see that τ_{zz} is of the form βr^2, where $\beta = -\rho_0\omega^2\nu/[2(1-\nu)]$ is a constant. Thus, the traction-free condition at the end surfaces $\tau_{zz}|_{z=\pm L/2} = 0$ can be satisfied by superimposing the solutions discussed in Section 6.6.8 with $f(r) = -\beta r^2$ and $g(r) = 0$. In particular, for the solution given by Eqn. (6.395), we get

$$C_m = -\frac{2D_m(1-\nu)}{k_m} - \frac{aD_m I_0(k_m a)}{I_1(k_m a)},$$
$$A_m = -\frac{B_m}{\lambda_m}\left[2\nu + \frac{\lambda_m L}{2}\coth\frac{\lambda_m L}{2}\right],$$

which leads to $s_m = 0$ for all m, $c_1 = 0$, $c_2 = -\beta a^2/2$, and

$$\int_0^a rJ_0(\lambda_n r)f(r)\,dr = \frac{\beta a^2}{\lambda_n^2}\left[\lambda_n a J_3(\lambda_n a) - 2J_2(\lambda_n a)\right].$$

It can be verified that Eqns. (6.538) and (6.539) are satisfied (the contribution of the infinite series part is zero).

Saito [279] presents the exact solution by means of a different method. We consider the same dimension ratio and Poisson ratio as in this reference, namely, $L = 2a$ and $\nu = 0.25$. Using the method based on complex coefficients, i.e., superposing the solution given by Eqn. (6.392) on the particular solution given by Eqn. (6.435), we get the results shown in Table 6.14 (almost identical results are obtained using the method based on real coefficients), where the presented stresses and displacements have been normalized by $\rho_0\omega^2a^2$ and $\rho_0\omega^2a^3/E$, respectively. By comparing against Figures (3a)–(3c) of Saito [279], we see the almost perfect agreement with his results.

An approximate solution where the traction-free condition is satisfied in an integral sense ($2\pi\int_0^a \tau_{zz} r\,dr = 0$) is presented in [55, 138]. When the Poisson ratio is zero, the solution obtained by setting $\nu = 0$ in Eqn. (6.436) below is an exact solution for the case of a spinning hollow cylinder with all surfaces traction-free, and further by setting $a = 0$, an exact solution for the case of a spinning solid circular cylinder.

An exact solution is also possible for the "plane strain" case when the top and bottom surfaces of a hollow cylinder of inner and outer radii a and b are constrained so that only radial movement is allowed along these surfaces, since then we have $c_3 = \epsilon_{zz} = 0$ in Eqns. (6.434). Since the cylinder is hollow, the constant c_2 is no longer zero, and the solution is given by $\tau_{rz} = 0$ and

$$
\begin{aligned}
u_r &= \frac{(1+\nu)(1-2\nu)\rho_0\omega^2 r}{8E(1-\nu)}\left[(3-2\nu)(a^2+b^2)+\frac{(3-2\nu)}{(1-2\nu)}\frac{a^2b^2}{r^2}-r^2\right],\\
\tau_{rr} &= \frac{(3-2\nu)\rho_0\omega^2}{8(1-\nu)}\left(b^2-r^2\right)\left(1-\frac{a^2}{r^2}\right),\\
\tau_{\theta\theta} &= \frac{\rho_0\omega^2}{8(1-\nu)}\left[(3-2\nu)(a^2+b^2)-(1+2\nu)r^2+\frac{(3-2\nu)a^2b^2}{r^2}\right],\\
\tau_{zz} &= \frac{\rho_0\omega^2\nu}{4(1-\nu)}\left[(3-2\nu)(a^2+b^2)-2r^2\right].
\end{aligned}
\tag{6.436}
$$

Note that in the limit as $a \to 0$, the solution for the solid spinning cylinder under plane strain conditions is recovered.

6.6.12 Circular cylinder on a frictionless surface loaded under its own weight

Consider a cylinder of radius a and length L with its flat surface resting on a frictionless surface, and loaded under its own weight. If z denotes the axis of the cylinder with $z = 0$ being the bottom plane that is in contact with the smooth surface, then the appropriate boundary conditions are $u_z|_{z=0} = \tau_{rz}|_{z=0} = 0$, $\tau_{rr}|_{r=a} = \tau_{rz}|_{r=a} = 0$, and $\tau_{rz}|_{z=L} = \tau_{zz}|_{z=L} = 0$. Solutions for this problem have been presented in [258, 322]. Again, we present a solution using the Love strain function. Since the domain along the z-axis is $[0, L]$, we cannot directly use the results from Section 6.6.9, where the domain was $[-L, L]$.

A particular solution under the body force $\boldsymbol{b} = -g\boldsymbol{e}_z$, which does not satisfy all boundary conditions, is

$$
\begin{aligned}
u_r &= \frac{\nu\rho_0 g r}{E}(L-z),\\
u_z &= \frac{\rho_0 g}{2E}(z^2-2zL+\nu r^2),\\
\tau_{zz} &= \rho_0 g(z-L),
\end{aligned}
\tag{6.437}
$$

with the other stress components zero. Since $u_z|_{z=0} = \rho_0 g\nu r^2/(2E)$ for the above solution, we first superpose another solution that satisfies $u_z|_{z=0} = -\rho_0 g\nu r^2/(2E)$ and $\tau_{rz}|_{z=0} = 0$. Thus, let the Love strain function be of the form

$$\Psi = c_3\phi_3 z + c_4\phi_4.$$

The conditions $u_z|_{z=0} = -\rho_0 g\nu r^2/(2E)$, $\tau_{rz}|_{z=0} = 0$ lead to

$$
\begin{aligned}
c_3 &= \frac{\rho_0\nu g}{6(1-\nu^2)},\\
c_4 &= -\frac{\rho_0\nu^2 g}{12(1-\nu^2)}.
\end{aligned}
$$

Superposing this solution on that given by Eqn. (6.437), we get the following particular solution that satisfies the equations of equilibrium, and the boundary conditions $u_z|_{z=0} = \tau_{zz}|_{z=L} = \tau_{rz}|_{z=0} = \tau_{rz}|_{z=L} = 0$.

$$\begin{aligned}
2\mu u_r &= \frac{\rho_0 \nu g r L}{1+\nu}, \\
2\mu u_z &= -\frac{\rho_0 g z\left[2(1-\nu)L-(1+\nu)(1-2\nu)z\right]}{2(1-\nu^2)}, \\
\tau_{rr} &= \frac{\rho_0 \nu g z}{1-\nu}, \\
\tau_{zz} &= \rho_0 g(z-L), \\
\tau_{rz} &= 0, \\
\tau_{\theta\theta} &= \frac{\rho_0 \nu g z}{1-\nu}.
\end{aligned} \tag{6.438}$$

In order to satisfy all the boundary conditions, we need to superpose on the solution in Eqn. (6.438) a solution that satisfies

$$\begin{aligned}
u_z|_{z=0} &= \tau_{rz}|_{z=0} = 0, & \tau_{zz}|_{z=L} &= \tau_{rz}|_{z=L} = 0, \\
\tau_{rr}|_{r=a} &= -\frac{\rho_0 \nu g z}{1-\nu}, & \tau_{rz}|_{r=a} &= 0.
\end{aligned}$$

In view of this, using Eqn. (6.268), we assume the Love strain function to be of the form

$$\Psi = \sum_{n=1}^{\infty}\left[A_n \sin(\lambda_n z) + B_n z\cos(\lambda_n z)\right] I_0(\lambda_n r) + \frac{c_1\phi_2 z}{2(1+\nu)} + \frac{c_1(1-2\nu)\phi_3}{6(1+\nu)}, \tag{6.439}$$

where λ_n are the roots of the characteristic equation

$$\sin 2\lambda_n L + 2\lambda_n L = 0,$$

that is derived similar to the way Eqn. (6.408) is derived by using the conditions $\tau_{zz}|_{z=L} = \tau_{rz}|_{z=L} = 0$.

By using the condition $\tau_{rz}|_{z=L} = 0$, one can eliminate B_n in terms of A_n to get the same solution as given by Eqn. (6.409) except that now

$$D = \lambda_n L \cot \lambda_n L.$$

The constants A_n are to be determined subject to the boundary conditions $\tau_{rr}|_{r=a} = -\rho_0 \nu g z/(1-\nu)$ and $\tau_{rz}|_{r=a} = 0$. The limiting expression for the stresses for the combined solution is given by the sum of $\rho_0 \nu g z/(1-\nu)$ and the expression in Eqn. (6.410).

Similar to the previous subsection, we now present a method that uses four sets of real constants instead of two sets of complex ones. Thus, let Ψ be as given by Eqn. (6.395), namely

$$\begin{aligned}
\Psi = &\sum_{m=1}^{\infty}\left[A_m \sinh \lambda_m z + B_m z\cosh \lambda_m z\right] J_0(\lambda_m r) + \sum_{m=1}^{\infty}\left[C_m I_0(k_m r) + D_m r I_1(k_m r)\right]\sin k_m z \\
&+ \frac{2c_1+c_2}{4(1+\nu)}\phi_2 z + \frac{(2-4\nu)c_1-(1+4\nu)c_2}{12(1+\nu)}\phi_3,
\end{aligned} \tag{6.440}$$

where $k_m = m\pi/L$, and λ_m, $m = 1, 2, \ldots, \infty$, are the positive roots of $J_1(\lambda a) = 0$. The constants A_n, B_n, C_n, D_n are determined using the boundary conditions $\tau_{rr}|_{r=a} = -\rho_0 \nu g z/(1-\nu)$, $\tau_{rz}|_{r=a} = 0$, and $\tau_{rz}|_{z=L} = \tau_{zz}|_{z=L} = 0$.

The condition $\tau_{rz}|_{r=a} = 0$ yields

$$C_m = -\frac{2D_m(1-\nu)}{k_m} - \frac{aD_m I_0(k_m a)}{I_1(k_m a)}. \tag{6.441}$$

The condition $\tau_{rz}|_{z=L} = 0$ yields

$$A_m = -\frac{(2\nu + \lambda_m L \coth \lambda_m L)B_m}{\lambda_m}.$$

The condition $\tau_{rr}|_{r=a} = -\rho_0 \nu g z/(1-\nu)$ yields

$$\begin{aligned} a\left[c_1 + \frac{\rho_0 g \nu z}{1-\nu}\right] &+ \sum_{m=1}^{\infty} \{[1 - \lambda_m L \coth \lambda_m L] \cosh \lambda_m z + \lambda_m z \sinh \lambda_m z\}\, a\lambda_m^2 J_0(\lambda_m a) B_m \\ &+ \sum_{m=1}^{\infty} \left\{ \frac{k_m^2 a^2 I_0^2(k_m a)}{I_1(k_m a)} - \left[2(1-\nu) + k_m^2 a^2\right] I_1(k_m a) \right\} D_m k_m \cos k_m z = 0. \end{aligned} \tag{6.442}$$

Integrating the above equation over the range $[0, L]$, we get

$$c_1 = -\frac{\nu \rho_0 g L}{2(1-\nu)}.$$

Multiplying Eqn. (6.442) by $\cos k_n z$, and integrating over the interval $[0, L]$, we get

$$\begin{aligned} &-\frac{[1 - (-1)^n]\, \nu \rho_0 g a}{k_n^2(1-\nu)} + \sum_{m=1}^{\infty} \frac{2(-1)^n a k_n^2 \lambda_m^3 \sinh(\lambda_m L) J_0(\lambda_m a)}{(k_n^2 + \lambda_m^2)^2} B_m \\ &+ \left\{ \frac{k_n^2 a^2 I_0^2(k_n a)}{I_1(k_n a)} - \left[2(1-\nu) + k_n^2 a^2\right] I_1(k_n a) \right\} D_n \frac{k_n L}{2} = 0, \quad n = 1, 2, \ldots, \infty. \end{aligned} \tag{6.443}$$

Finally, the condition $\tau_{zz}|_{z=L} = 0$ yields

$$\begin{aligned} c_2 &+ \sum_{m=1}^{\infty} [\cosh \lambda_m L + \lambda_m L \operatorname{cosech} \lambda_m L]\, \lambda_m^2 J_0(\lambda_m r) B_m \\ &+ \sum_{m=1}^{\infty} (-1)^m k_m^3 \left[r I_1(k_m r) - \frac{a I_0(k_m r) I_2(k_m a)}{I_1(k_m a)} \right] D_m = 0. \end{aligned} \tag{6.444}$$

Multiplying Eqn. (6.444) by r, and integrating over the interval $[0, a]$, we get $c_2 = 0$. Multiplying Eqn. (6.444) by $rJ_0(\lambda_n r)$, integrating over the interval $[0, a]$, and using Eqns. (6.402), we get for $n = 1, 2, \ldots, \infty$,

z \ r	0	0.2	0.4	0.6	0.8	1.0
0	0.000	1.232E-2	2.483E-2	3.764E-2	5.075E-2	6.358E-2
0.2	0.000	1.160E-2	2.330E-2	3.510E-2	4.675E-2	5.734E-2
0.4	0.000	9.650E-3	1.925E-2	2.866E-2	3.753E-2	4.552E-2
0.6	0.000	7.015E-3	1.388E-2	2.041E-2	2.637E-2	3.180E-2
0.8	0.000	4.308E-3	8.374E-3	1.200E-2	1.510E-2	1.783E-2
1.0	0.000	2.251E-3	3.985E-3	4.832E-3	4.778E-3	4.391E-3

Table 6.15 *Displacement component u_r.*

z \ r	0	0.2	0.4	0.6	0.8	1.0
0	0.000	0.000	0.000	0.000	0.000	0.000
0.2	−4.739E-2	-4.726E-2	−4.680E-2	−4.572E-2	−4.331E-2	−3.715E-2
0.4	−8.557E-2	-8.528E-2	−8.427E-2	−8.198E-2	−7.724E-2	−6.769E-2
0.6	−0.113	−0.113	−0.111	−0.108	−0.102	−9.057E-2
0.8	−0.130	−0.130	−0.128	−0.124	−0.116	−0.105
1.0	−0.137	−0.136	−0.134	−0.129	−0.122	−0.110

Table 6.16 *Displacement component u_z.*

$$[\cosh \lambda_n L + \lambda_n L \operatorname{cosech} \lambda_n L] \frac{\lambda_n^2 a^2 J_0^2(\lambda_n a) B_n}{2} + \sum_{m=1}^{\infty} (-1)^m k_m^3 \left[\frac{2a\lambda_n^2 I_1(k_m a) J_0(\lambda_n a)}{(k_m^2 + \lambda_n^2)^2} \right] D_m = 0. \quad (6.445)$$

Eqns. (6.443) and (6.445) are to be solved for B_m and D_m. This solution is then superposed on that given by Eqn. (6.438).

Note that for the combined solution

$$\lim_{r\to 0} \tau_{rr} = \lim_{r\to 0} \tau_{\theta\theta} = \frac{\rho_0 g \nu (2z - L)}{2(1-\nu)} + \frac{1}{2} \sum_{n=1}^{\infty} \Big\{ - [k_n C_n + (2-4\nu) D_n] k_n^2 \cos k_n z + [\lambda_n A_n + (1+4\nu) B_n] \lambda_n^2 \cosh \lambda_n z + \lambda_n^3 z B_n \sinh \lambda_n z \Big\}.$$

It can be easily verified that Eqns. (6.538), (6.539) and (6.542) are all satisfied by the solution (the infinite series contribution to all three is zero).

Consider the data given in [258], namely, $L = 1$, $a = 1$, $\nu = 0.3$, $\rho_0 g = 1$, $E = 4$ (this value of E is not given in the mentioned reference). Using the method of complex coefficients (almost identical results are obtained using the method based on real coefficients), we get the results in Tables 6.15–6.18. Once again, there is an almost perfect agreement

z \ r	0	0.2	0.4	0.6	0.8	1.0
0	−1.101	−1.096	−1.080	−1.048	−0.982	−0.797
0.2	−0.892	−0.887	−0.871	−0.839	−0.777	−0.672
0.4	−0.668	−0.664	−0.650	−0.622	−0.579	−0.530
0.6	−0.439	−0.436	−0.427	−0.410	−0.387	−0.371
0.8	−0.212	−0.211	−0.208	−0.202	−0.195	−0.196
1.0	0.000	0.000	0.000	0.000	0.000	0.000

Table 6.17 *Stress component* τ_{zz}.

z \ r	0	0.2	0.4	0.6	0.8	1.0
0	−0.121	−0.117	−0.107	−8.767E-2	−5.519E-2	1.486E-2
0.2	−5.139E-2	−4.857E-2	−3.986E-2	−2.446E-2	−1.235E-3	2.755E-2
0.4	−1.053E-2	−8.912E-3	−4.038E-3	3.917E-3	1.393E-2	2.302E-2
0.6	1.306E-2	1.331E-2	1.410E-2	1.535E-2	1.643E-2	1.590E-2
0.8	3.327E-2	3.195E-2	2.832E-2	2.336E-2	1.822E-2	1.254E-2
1.0	6.692E-2	6.314E-2	5.256E-2	3.775E-2	2.381E-2	1.756E-2

Table 6.18 *Stress component* $\tau_{\theta\theta}$.

with the finite element results based on the formulation in [161]. There is, however, a discrepancy in the current results and those of [258]. For example, $\tau_{zz}|_{(0,0)}$ shown in [258] is much lower than the value we have obtained. Similarly there is a very high discrepancy in the τ_{zz} stress at $(1,0)$. One possible reason is the ill-conditioning caused by the biorthogonality relations that they use [271].

The following problems are solved using a *spherical* coordinate system, so that r now denotes the quantity $\sqrt{x^2+y^2+z^2}$.

6.6.13 Point load in an infinite elastic body (Kelvin problem)

A point load P acts at the origin in the z-direction in an infinite elastic body (see Fig. 6.64). The solution is generated by the Boussinesq potentials

$$\alpha = 0, \quad \beta = -\frac{A}{r},$$

or, alternatively, using the Love strain function (see Eqns. (6.265))

$$\Psi = Ar,$$

where A is a constant to be determined; this solution correspond to the B_0 term in Eqn. (6.272). Note that the solution for β corresponds to B_1 being the only nonzero constant in Eqn. (6.269). From Eqns. (6.261), we get

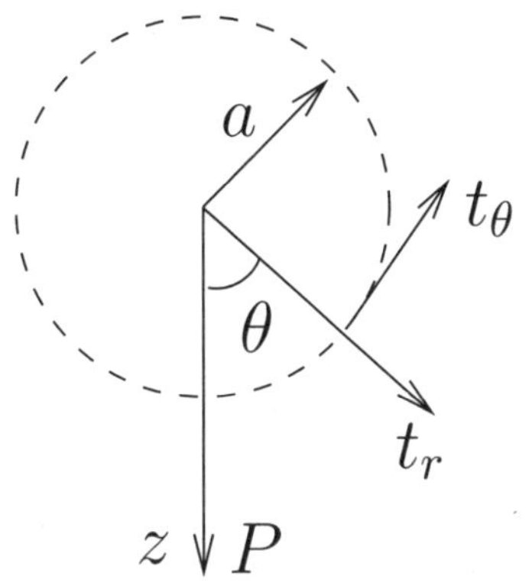

Fig. 6.64 *Point load acting in an infinite elastic body.*

$$
\begin{aligned}
u_r &= \frac{2A(1-\nu)\cos\theta}{\mu r}, \\
u_\theta &= -\frac{A(3-4\nu)\sin\theta}{2\mu r}, \\
\tau_{rr} &= -\frac{A(4-2\nu)\cos\theta}{r^2}, \\
\tau_{\theta\theta} &= \frac{A(1-2\nu)\cos\theta}{r^2}, \\
\tau_{r\theta} = \tau_{\phi\phi} &= \frac{A(1-2\nu)\sin\theta}{r^2}.
\end{aligned}
$$

The constant A is determined by noting that, for equilibrium, the total resultant traction on the surface of the sphere of radius a shown in Fig. 6.64, should be $-Pe_z$. Thus,

$$
\begin{aligned}
P &= -\int_0^\pi t_z(2\pi a^2 \sin\theta)\, d\theta \\
&= -2\pi a^2 \int_0^\pi (\tau_{rr}\cos\theta - \tau_{r\theta}\sin\theta)\sin\theta\, d\theta \\
&= 2\pi \int_0^\pi [A(4-2\nu)\cos^2\theta\sin\theta + A(1-2\nu)\sin^3\theta]\, d\theta \\
&= 8A\pi(1-\nu),
\end{aligned}
$$

leading to

$$
A = \frac{P}{8\pi(1-\nu)}.
$$

It follows from symmetry that the net forces in the x and y directions, and the moments about the origin are zero.

6.6.14 Point load acting normal to the boundary of an infinite half-space (Boussinesq problem)

The solution for the point load P acting at the origin normal to the boundary of a half-space as shown in Fig. 6.65 is generated by the Boussinesq potentials

$$\alpha = -A \log[r(1+\cos\theta)], \quad \beta = -\frac{B}{r}, \tag{6.446}$$

which corresponds to the A_0^+ and B_0 terms in Eqn. (6.273a). The displacements and stresses corresponding to the above potentials, obtained using Eqns. (6.261), are

$$\begin{aligned}
2\mu u_r &= \frac{1}{r}\left[-A + 4(1-\nu)B\cos\theta\right], \\
2\mu u_\theta &= \frac{1}{r}\left[A\tan\frac{\theta}{2} - (3-4\nu)B\sin\theta\right], \\
\tau_{rr} &= \frac{1}{r^2}\left[A + 2B(\nu-2)\cos\theta\right], \\
\tau_{\theta\theta} &= \frac{1}{2r^2}\left[A\left(\sec^2\frac{\theta}{2} - 2\right) + 2(1-2\nu)B\right], \\
\tau_{r\theta} &= \frac{1}{r^2}\left[-A\tan\frac{\theta}{2} + (1-2\nu)B\sin\theta\right], \\
\tau_{\phi\phi} &= \frac{1}{r^2}\left[-\frac{A}{1+\cos\theta} + (1-2\nu)B\cos\theta\right].
\end{aligned} \tag{6.447}$$

Since the boundary $\theta = \pi/2$ is traction-free, we impose

$$\tau_{\theta\theta}|_{\theta=\pi/2} = \tau_{r\theta}|_{\theta=\pi/2} = 0.$$

The first of the above conditions is automatically satisfied, while the second one yields $B = A/(1-2\nu)$. The second relation between A and B is obtained as in the Kelvin problem, i.e.,

$$P = -2\pi a^2 \int_0^{\pi/2} (\tau_{rr}\cos\theta - \tau_{r\theta}\sin\theta)\sin\theta\, d\theta.$$

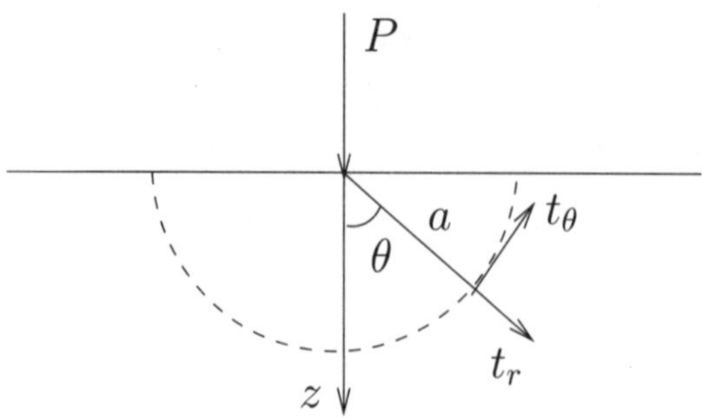

Fig. 6.65 *Point load acting normal to the boundary of a half-space.*

On carrying out the integration, we get $P = -2\pi[A - 2B(1-\nu)]$. On solving for A and B, we get

$$A = \frac{P(1-2\nu)}{2\pi}, \quad B = \frac{P}{2\pi}.$$

As can be easily verified using Eqn. (6.265), the Love strain function corresponding to the Boussinesq potentials given by Eqn. (6.446) is

$$\Psi = Br + A\left[r\xi \log[r(1+\xi)] - r\right], \tag{6.448}$$

which corresponds to the B_0, C_1 and A_1^+ terms in Eqn. (6.273b).

6.6.15 Truncated cone

Consider a truncated cone with radii of the circular faces given by a and b, and cone angle given by θ_0 as shown in Fig. 6.66. First consider the class of problems in which the curved surface is traction-free. We see that the solution given by Eqn. (6.447) serves as a particular solution for this problem provided B is chosen as $A\sec^2(\theta_0/2)/[2(1-2\nu)]$. With this choice, Eqn. (6.447) becomes

$$2\mu u_r = \frac{A}{r}\left[\frac{4(1-\nu)\cos\theta}{(1-2\nu)(1+\cos\theta_0)} - 1\right],$$

$$2\mu u_\theta = \frac{A}{r}\left[\tan\frac{\theta}{2} - \frac{(3-4\nu)\sin\theta}{(1-2\nu)(1+\cos\theta_0)}\right],$$

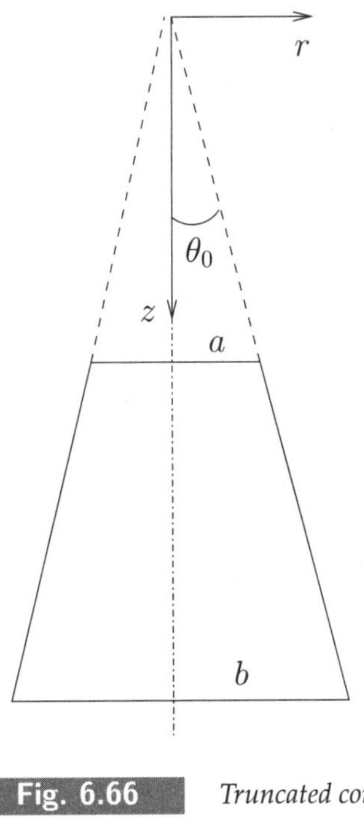

Fig. 6.66 *Truncated cone.*

$$\begin{aligned}
\tau_{rr} &= \frac{A}{r^2}\left[1 - \frac{2(2-\nu)\cos\theta}{(1-2\nu)(1+\cos\theta_0)}\right], \\
\tau_{\theta\theta} &= \frac{A\cos\theta}{2r^2}\left[\sec^2\frac{\theta_0}{2} - \sec^2\frac{\theta}{2}\right], \\
\tau_{r\theta} &= \frac{A\sin\theta}{2r^2}\left[\sec^2\frac{\theta_0}{2} - \sec^2\frac{\theta}{2}\right], \\
\tau_{\phi\phi} &= \frac{A}{2r^2}\left[2\cot\theta\tan\frac{\theta_0}{2} + \cos\theta\sec^2\frac{\theta_0}{2} - 2\right].
\end{aligned} \tag{6.449}$$

The displacement solution in Eqn. (6.449), when expressed in terms of cylindrical coordinates (r,z), with $B = A\sec^2(\theta_0/2)/[2(1-2\nu)]$, is given by

$$\begin{aligned}
2\mu u_r &= \frac{r[-(r^2+z^2)A + Bz(z+\sqrt{r^2+z^2})]}{(r^2+z^2)^{3/2}(z+\sqrt{r^2+z^2})}, \\
2\mu u_z &= \frac{-(r^2+z^2)A + [(3-4\nu)r^2 + 4(1-\nu)z^2]B}{(r^2+z^2)^{3/2}},
\end{aligned}$$

from which τ_{zz} is recovered. Consider the case when the truncated cone in Fig. 6.66 is loaded by an axial load P through a traction distribution on its flat ends. Using the boundary condition $2\pi\int_0^a r\tau_{zz}\,dr = P$, we get

$$\begin{aligned}
A &= -\frac{(1-2\nu)P\cot^2(\theta_0/2)}{\pi(3+4\nu\cos\theta_0+\cos 2\theta_0)}, \\
B &= -\frac{P\csc^2(\theta_0/2)}{2\pi(3+4\nu\cos\theta_0+\cos 2\theta_0)}.
\end{aligned}$$

Since the condition on the flat ends is being satisfied in an integral sense, we get the same solution as above if the flat ends are replaced by curved surfaces, say, $r =$ constant, where r is the spherical radial coordinate.

Now consider problems involving traction distributions on the flat ends different than those in the above problem, but with the curved surface still traction-free. In a manner similar to the development in Section 6.6.7, a characteristic equation can be derived by using the boundary conditions $\tau_{\theta\theta}|_{\theta=\theta_0} = \tau_{r\theta}|_{\theta=\theta_0} = 0$ [195]. Using Eqns. (6.280), we get

$$\begin{aligned}
&\left[-(\lambda+1)(\lambda^2+4\lambda+2+2\nu)P_\lambda(\xi_0) + (\lambda+5-4\nu)\xi_0\frac{dP_\lambda}{d\xi}(\xi_0)\right]A_\lambda \\
&\qquad + \left[-(\lambda+2)^2P_{\lambda+2}(\xi_0) + \xi_0\frac{dP_{\lambda+2}}{d\xi}(\xi_0)\right]B_\lambda = 0,
\end{aligned} \tag{6.450a}$$

$$(\lambda^2+2\lambda-1+2\nu)\frac{dP_\lambda}{d\xi}(\xi_0)A_\lambda + (\lambda+1)\frac{dP_{\lambda+2}}{d\xi}(\xi_0)B_\lambda = 0, \tag{6.450b}$$

$$\begin{aligned}
&\left[\lambda(\lambda^2-2\lambda-1+2\nu)P_\lambda(\xi_0) - (\lambda-4+4\nu)\xi_0\frac{dP_\lambda}{d\xi}(\xi_0)\right]C_\lambda \\
&\qquad + \left[(\lambda-1)^2P_{\lambda-2}(\xi_0) - \xi_0\frac{dP_{\lambda-2}}{d\xi}(\xi_0)\right]D_\lambda = 0,
\end{aligned} \tag{6.450c}$$

$$(\lambda^2 - 2 + 2\nu)\frac{dP_\lambda}{d\xi}(\xi_0)C_\lambda + \lambda\frac{dP_{\lambda-2}}{d\xi}(\xi_0)D_\lambda = 0, \tag{6.450d}$$

where $\xi_0 = \cos\theta_0$. The characteristic equation derived using Eqns. (6.450a) and (6.450b) (obtained by setting the determinant of the matrix of coefficients of A_λ, B_λ to zero), or Eqns. (6.450c) and (6.450d) is identical, and is given by (modulo a factor $(\lambda+1)(\lambda+3/2)$ that multiplies the entire characteristic equation)[24]

$$\begin{aligned}(\lambda+1)^2(\lambda+2)^2\xi_0(1-\xi_0^2)P_{\lambda+1}^2(\xi_0) - (\lambda+1)(\lambda+2)(1-\xi_0^2)^{3/2}P_{\lambda+1}(\xi_0)P_{\lambda+1}^{(1)}(\xi_0)\\ + \xi_0\left[2\nu - 2\xi_0^2 + \lambda(\lambda+3)(1-\xi_0^2)\right][P_{\lambda+1}^{(1)}(\xi_0)]^2 = 0.\end{aligned} \tag{6.451}$$

Note that one would have to include the Q_n terms in the general solution in order to derive the corresponding characteristic equation (the determinant of a 4×4 matrix) for hollow truncated cones. The real roots $\lambda = -1, -3/2$ can be discarded since they either correspond to rigid-body modes or lead to a satisfaction of the traction-free boundary condition only when the associated constants are taken as zero. The roots λ_n of Eqn. (6.451), and the associated constants A_{λ_n}, B_{λ_n}, C_{λ_n} and D_{λ_n}, are complex-valued [323]. The solution is given by Eqn. (6.449) superposed on Eqn. (6.280), with n replaced by λ_n, where λ_n are the roots of the characteristic equation given by Eqn. (6.451). The constants B_{λ_n} can be expressed in terms of A_{λ_n} using either Eqn. (6.450a) or (6.450b), while the constants D_{λ_n} can be expressed in terms of C_{λ_n} using either Eqn. (6.450c) or (6.450d). The constants A_{λ_n} and C_{λ_n} (which being complex-valued can be thought of as four sets of constants) are determined so as to satisfy the given boundary conditions on the flat surfaces $r\cos\theta =$ constant[25]. For instance, if on the flat surfaces $r\cos\theta = a\cot\theta_0$ and $r\cos\theta = b\cot\theta_0$, the shear is specified to be zero, and normal traction is specified by functions $f(r\sin\theta)$ and $g(r\sin\theta)$, then

$$\begin{aligned}
&[\tau_{r\theta}\cos 2\theta + (\tau_{rr} - \tau_{\theta\theta})\sin\theta\cos\theta]_{r\cos\theta = a\cot\theta_0} = 0,\\
&[\tau_{rr}\cos^2\theta + \tau_{\theta\theta}\sin^2\theta - \tau_{r\theta}\sin 2\theta]_{r\cos\theta = a\cot\theta_0} = f(r\sin\theta),\\
&[\tau_{r\theta}\cos 2\theta + (\tau_{rr} - \tau_{\theta\theta})\sin\theta\cos\theta]_{r\cos\theta = b\cot\theta_0} = 0,\\
&[\tau_{rr}\cos^2\theta + \tau_{\theta\theta}\sin^2\theta - \tau_{r\theta}\sin 2\theta]_{r\cos\theta = b\cot\theta_0} = g(r\sin\theta).
\end{aligned}$$

If $a = 0$, then A (in Eqn. (6.449)), and C_n and D_n are set to zero. One could of course have displacement boundary conditions instead of the traction boundary conditions considered above.

If the surfaces $\xi = \xi_0$ have prescribed tractions, $t_r^0(r)$ and $t_\theta^0(r)$ acting on them, then we first express these tractions as $t_r^0(r) = \sum_{n=0}^{\infty} a_n r^n$ and $t_\theta^0(r) = \sum_{n=0}^{\infty} b_n r^n$, and then find the constants in Eqns. (6.280) to satisfy the conditions $\tau_{r\theta}|_{\xi=\xi_0} = t_r^0(r)$ and $\tau_{\theta\theta}|_{\xi=\xi_0} = t_\theta^0(r)$. Equations (6.280) with these determined constants constitute a particular solution that is then added to the solution of the homogeneous problem discussed above; this procedure is illustrated in Section 6.6.16.

Now consider the case when a gravitational body force ge_z acts on the truncated cone. Using Eqns. (6.278) and (6.437) (with $L = a\cot\theta_0$), a particular solution that satisfies the boundary conditions $\tau_{\theta\theta}|_{\theta=\theta_0} = \tau_{r\theta}|_{\theta=\theta_0} = 0$, and the equations of equilibrium, is

[24] The characteristic equation given by Eqn. (6.451), and hence its roots, is different than the one presented in [323], since they have redefined combinations of constants by new constants.

[25] As in Section 6.6.7, this may be achieved by using a weighted residual technique as described on page 485.

$$
\begin{aligned}
12Eu_r &= \rho_0 g r^2 \cos\theta \sec^2\theta_0 \left[(1+\nu)\cos 2\theta - (1+2\nu)\cos 2\theta_0 + \nu - 2\right], \\
12Eu_\theta &= \rho_0 g r^2 \sin\theta \sec^2\theta_0 \left[3\nu - (1+\nu)\cos 2\theta + (3+2\nu)\cos 2\theta_0\right], \\
\tau_{rr} &= -\frac{1}{6}\rho_0 g r \cos\theta \sec^2\theta_0 \left[2 - \cos 2\theta + \cos 2\theta_0\right], \\
\tau_{\theta\theta} &= \frac{1}{6}\rho_0 g r \cos\theta \sec^2\theta_0 \left[\cos 2\theta_0 - \cos 2\theta\right], \\
\tau_{\phi\phi} &= -\frac{1}{3}\rho_0 g r \cos\theta \tan^2\theta_0, \\
\tau_{r\theta} &= \frac{1}{6}\rho_0 g r \sin\theta \sec^2\theta_0 \left[\cos 2\theta_0 - \cos 2\theta\right].
\end{aligned}
$$

The solutions in Eqns. (6.280) with n replaced by λ_n (where λ_n are the roots of the characteristic equation given by Eqn. (6.451)), and (6.449) (in case $a > 0$) are now superposed on the above particular solution to satisfy the boundary conditions on the flat surface. For example, if the bottom surface rests on a frictionless surface, and the top surface is traction-free, then the appropriate boundary conditions are (for $0 \le \theta \le \theta_0$)

$$
\begin{aligned}
&\left[u_r \cos\theta - u_\theta \sin\theta\right]_{r\cos\theta = b\cot\theta_0} = 0, \\
&\left[\tau_{r\theta}\cos 2\theta + (\tau_{rr} - \tau_{\theta\theta})\cos\theta\sin\theta\right]_{r\cos\theta = b\cot\theta_0} = 0, \\
&\left[\tau_{r\theta}\cos 2\theta + (\tau_{rr} - \tau_{\theta\theta})\cos\theta\sin\theta\right]_{r\cos\theta = a\cot\theta_0} = 0, \\
&\left[\tau_{rr}\cos^2\theta + \tau_{\theta\theta}\sin^2\theta - \tau_{r\theta}\sin 2\theta\right]_{r\cos\theta = a\cot\theta_0} = 0.
\end{aligned}
$$

Finally, consider the case when the truncated cone is subjected to centrifugal loading. A particular solution that satisfies the boundary conditions $\tau_{\theta\theta}|_{\theta=\theta_0} = \tau_{r\theta}|_{\theta=\theta_0} = 0$, and the equations of equilibrium, can be found using Eqns. (6.278) and (6.383), and is given by

$$
\begin{aligned}
\frac{24E(\nu-1)A}{(1+\nu)\rho_0\omega^2 r^3} u_r &= 3(1-2\nu)A\sin^4\theta - 4\sin^2\theta_0\Big[9 + 33\nu - 30\nu^2 \\
&\quad + (-9 + 63\nu + 30\nu^2)\cos 2\theta_0 + 5\cos^4\theta\left[21 + \nu(5+34\nu) + 3(2\nu-1)(7+5\nu)\cos 2\theta_0\right] \\
&\quad - 6\cos^2\theta\left[15 + \nu(2\nu+19) + 15[-1+\nu(2\nu+3)]\cos 2\theta_0\right]\Big], \\
\frac{6E(\nu-1)A}{(1+\nu)\rho_0\omega^2 r^3} u_\theta &= \sin 2\theta\Big\{2\left[-9+\nu(4\nu-19)\right]\cos 2\theta_0 \\
&\quad + 2\cos 2\theta\Big[3 + 7\nu(1+2\nu) + 5(2\nu^2+\nu-3)\cos 2\theta_0\Big] \\
&\quad + (1-\nu)\left[12 + 41\nu + 15(2+\nu)\cos 4\theta_0\right]\Big\},
\end{aligned}
\tag{6.452}
$$

where

$$
A = 73 + 43\nu + 20(1+3\nu)\cos 2\theta_0 + 5(7+5\nu)\cos 4\theta_0.
$$

Again, solutions given by Eqns. (6.280) (with n replaced by the roots of the characteristic equation given by Eqn. (6.451)) and (6.449) (in case $a > 0$) are now superposed on the above particular solution to satisfy the boundary conditions on the flat surfaces $r\cos\theta = a\cot\theta_0$ and $r\cos\theta = b\cot\theta_0$.

For the case when $\theta_0 = \pi/2$ (e.g., rotating hemisphere of radius a), the only solutions of Eqn. (6.451) are integers. Imposing the conditions $\tau_{\theta\theta}|_{\theta=\pi/2} = \tau_{r\theta}|_{\theta=\pi/2} = 0$ on the solution given by Eqn. (6.280) with $C_n = D_n = 0$ (since there are no singularities), we get the following solutions for even and odd integers:

$$\begin{aligned} B_{2n} &= \frac{(2n^2+4n+1+\nu)A_{2n}}{n+1}, \\ B_{2n-1} &= \frac{(4n^2-2+2\nu)A_{2n-1}}{2n+1}. \end{aligned}$$

Substituting these values into Eqn. (6.280), and renaming the constants associated with the even and odd series as A_n and B_n, we get

$$\begin{aligned} 2\mu u_r &= \sum_{n=1}^{\infty} 2r^{2n-1}A_n\left[(2n-1)(n-2+2\nu)P_{2n-2}(\xi)+(2n^2-1+\nu)P_{2n}(\xi)\right] \\ &\quad + \sum_{n=1}^{\infty} 2r^{2n}B_n\left[n(2n+4\nu-3)P_{2n-1}(\xi)+(2n^2-1+\nu)P_{2n+1}(\xi)\right], \\ 2\mu u_\theta &= \sum_{n=1}^{\infty}(4n-1)r^{2n-1}A_n\left[(1-2\nu)\sin\theta P_{2n-1}(\xi)+\frac{(1+n-\nu)\xi P^{(1)}_{2n-1}(\xi)}{n}\right] \\ &\quad + \sum_{n=1}^{\infty}\frac{r^{2n}B_n}{\sin\theta}\Bigg[4n(n+2-2\nu)\left[P_{2n}(\xi)-\xi P_{2n-1}(\xi)\right] \\ &\quad - \frac{4(n+1)(2n^2+\nu-1)\left[\xi P_{2n+1}(\xi)-P_{2n+2}(\xi)\right]}{2n+1}\Bigg], \\ \tau_{rr} &= \sum_{n=1}^{\infty}(2n-1)r^{2n-2}A_n\left[(4n^2-10n+4-2\nu)P_{2n-2}(\xi)+2(2n^2-1+\nu)P_{2n}(\xi)\right] \\ &\quad + \sum_{n=1}^{\infty}2nr^{2n-1}B_n\left[(4n^2-6n-2\nu)P_{2n-1}(\xi)+2(2n^2+\nu-1)P_{2n+1}(\xi)\right], \\ \tau_{\theta\theta} &= \sum_{n=1}^{\infty}(1-4n)r^{2n-2}\xi A_n\left[(4n^2-1)P_{2n-1}(\xi)+\frac{(1+n-\nu)\xi P^{(1)}_{2n-1}(\xi)}{n\sin\theta}\right] \\ &+\sum_{n=1}^{\infty}r^{2n-1}B_n\left[-2n(2n+1)(4n+1)\xi P_{2n}(\xi)+\frac{(4n+1)\left[2n+1+(2\nu-4n-3)\xi^2\right]P^{(1)}_{2n}(\xi)}{(2n+1)\sin\theta}\right], \\ \tau_{\phi\phi} &= \sum_{n=1}^{\infty}(4n-1)r^{2n-2}A_n\Bigg[(2n-1)(1-2\nu)\xi P_{2n-1}(\xi) \\ &\quad + \frac{\left[1+\nu+2(n-1)\nu+n(1-2\nu)\xi^2\right]P^{(1)}_{2n-1}(\xi)}{n\sin\theta}\Bigg] \end{aligned}$$

$$+\sum_{n=1}^{\infty}(4n+1)r^{2n-1}B_n\left[\left[1+4n\nu+(4n+1)(1-2\nu)\xi^2\right]P_{2n+1}(\xi)\right.$$

$$\left.+\frac{\xi\left[1-2n+8n\nu+(4n+1)(1-2\nu)\xi^2\right]P^{(1)}_{2n+1}(\xi)}{(2n+1)\sin\theta}\right],$$

$$\tau_{r\theta}=\sum_{n=1}^{\infty}(4n-1)r^{2n-2}A_n\left[(1-2n)\sin\theta P_{2n-1}(\xi)+\frac{[n(2n-1)+\nu-1]\,\xi P^{(1)}_{2n-1}(\xi)}{n}\right]$$

$$+\sum_{n=1}^{\infty}\frac{r^{2n-1}B_n}{2n+1}\left[2(4n+1)(2n^2+\nu-1)\xi P^{(1)}_{2n}(\xi)\right].$$

Along the axis, we have $u_\theta|_{\theta=0}=0$, and

$$\lim_{\theta\to 0}\frac{P^{(1)}_{2n-1}(\xi)}{n\sin\theta}=1-2n,$$

$$\lim_{\theta\to 0}\frac{P^{(1)}_{2n}(\xi)}{(2n+1)\sin\theta}=-n,$$

$$\lim_{\theta\to 0}\frac{P^{(1)}_{2n+1}(\xi)}{(2n+1)\sin\theta}=-1-n.$$

The above solution is superposed on the solution generated using Eqn. (6.452) with $\theta_0=\pi/2$, i.e.,

$$\frac{24E(\nu-1)(11+\nu)u_r}{\rho_0\omega^2r^3(1+\nu)}=9\left[1-\nu(3+2\nu)\right]-4\left[6+\nu(11\nu-5)\right]\cos 2\theta$$

$$-(9+\nu+2\nu^2)\cos 4\theta,$$

$$\frac{12E(\nu-1)(11+\nu)u_\theta}{\rho_0\omega^2r^3(1+\nu)}=\left[15+(13-16\nu)\nu+(9+\nu+2\nu^2)\cos 2\theta\right]\sin 2\theta,$$

$$\frac{4(\nu-1)(11+\nu)\tau_{rr}}{\rho_0\omega^2r^2}=4(\nu-1)(3+\nu)\cos 2\theta+(9+\nu+2\nu^2)\sin^2 2\theta,$$

$$\frac{2(\nu-1)(11+\nu)\tau_{\theta\theta}}{\rho_0\omega^2r^2}=\left[-3+9\nu+6\nu^2+(9+\nu+2\nu^2)\cos 2\theta\right]\cos^2\theta,$$

$$\frac{2(\nu-1)(11+\nu)\tau_{\phi\phi}}{\rho_0\omega^2r^2}=4+\nu(7+\nu)+\left[2+\nu(3+7\nu)\right]\cos 2\theta,$$

$$\frac{(\nu-1)(11+\nu)\tau_{r\theta}}{\rho_0\omega^2r^2}=(9+\nu+2\nu^2)\sin\theta\cos^3\theta,$$

and the constants A_n and B_n determined so as to satisfy the boundary conditions $\tau_{rr}|_{r=a}=\tau_{r\theta}|_{r=a}=0$. The analytical and finite element results (based on the formulation in [161]) for $E=a=\rho_0\omega^2=1$ and $\nu=0.25$ are presented in Table 6.19.

	$r = a/2, \theta = \pi/2$					
	u_r	u_θ	τ_{rr}	$\tau_{\theta\theta}$	$\tau_{r\theta}$	$\tau_{\phi\phi}$
Exact	0.1458	−0.0169	0.3210	0.0000	0.0000	0.3719
FEM	0.1458	−0.0169	0.3212	0.0000	0.0000	0.3720
	$r = a/2, \theta = 0$					
	u_r	u_θ	τ_{rr}	$\tau_{\theta\theta}$	$\tau_{r\theta}$	$\tau_{\phi\phi}$
Exact	−0.0946	0.0000	−0.0081	0.2937	0.0000	0.2937
FEM	−0.0946	0.0000	−0.0081	0.2940	0.0000	0.2937
	$r = a, \theta = \pi/4$					
	u_r	u_θ	τ_{rr}	$\tau_{\theta\theta}$	$\tau_{r\theta}$	$\tau_{\phi\phi}$
Exact	0.0259	0.0697	0.0000	0.0184	0.0000	0.1002
FEM	0.0259	0.0697	0.0001	0.0184	0.0000	0.1002

Table 6.19 *Displacements and stresses at various points for the rotating hemisphere problem.*

Consider the problem shown in Fig. 6.67. A particular solution for $\boldsymbol{b} = -g\boldsymbol{e}_z$, which satisfies the conditions $u_\theta|_{\theta=\theta_0} = \tau_{r\theta}|_{\theta=\theta_0} = 0$, and the equations of equilibrium is

$$\begin{aligned}
2EAu_r &= \rho_0 g r^2(2\nu^2 + \nu - 1)\cos\theta(\cos 2\theta - 3\cos 2\theta_0 - 2), \\
2EAu_\theta &= -\rho_0 g r^2(2\nu^2 + \nu - 1)\sin\theta(\cos 2\theta - \cos 2\theta_0), \\
A\tau_{rr} &= \rho_0 g r\cos\theta\left[2 - \nu - (1-2\nu)\cos 2\theta + (3-\nu)\cos 2\theta_0\right], \\
A\tau_{\theta\theta} &= \rho_0 g r\cos\theta\left[3\nu + (1-2\nu)\cos 2\theta + (1+3\nu)\cos 2\theta_0\right], \\
A\tau_{\phi\phi} &= \rho_0 g r\cos\theta\left[1 + \nu + (1+3\nu)\cos 2\theta_0\right], \\
A\tau_{r\theta} &= \rho_0 g r(1-2\nu)\sin\theta\left[\cos 2\theta - \cos 2\theta_0\right],
\end{aligned} \tag{6.453}$$

where $A = (1-\nu)(3 + 5\cos 2\theta_0)$. Using Eqns. (6.280), the boundary conditions $u_\theta|_{\theta=\theta_0} = \tau_{r\theta}|_{\theta=\theta_0} = 0$ lead to the equations

$$(\lambda + 5 - 4\nu)\left.\frac{dP_\lambda}{d\xi}\right|_{\xi=\xi_0} A_\lambda + \left.\frac{dP_{\lambda+2}}{d\xi}\right|_{\xi=\xi_0} B_\lambda = 0, \tag{6.454a}$$

$$(\lambda^2 + 2\lambda - 1 + 2\nu)\left.\frac{dP_\lambda}{d\xi}\right|_{\xi=\xi_0} A_\lambda + (\lambda+1)\left.\frac{dP_{\lambda+2}}{d\xi}\right|_{\xi=\xi_0} B_\lambda = 0, \tag{6.454b}$$

$$(\lambda - 4 + 4\nu)\left.\frac{dP_\lambda}{d\xi}\right|_{\xi=\xi_0} C_\lambda + \left.\frac{dP_{\lambda-2}}{d\xi}\right|_{\xi=\xi_0} D_\lambda = 0, \tag{6.454c}$$

$$(\lambda^2 - 2 + 2\nu)\left.\frac{dP_\lambda}{d\xi}\right|_{\xi=\xi_0} C_\lambda + \lambda\left.\frac{dP_{\lambda-2}}{d\xi}\right|_{\xi=\xi_0} D_\lambda = 0. \tag{6.454d}$$

Similar to Eqn. (6.451), the characteristic equation based on Eqns. (6.454a) and (6.454b) (or Eqns. (6.454c) and (6.454d)) (apart from a factor $(1+\lambda)(3+\lambda)(3+2\lambda)$ which can be discarded since these roots are either rigid-body modes or lead to constant stress states

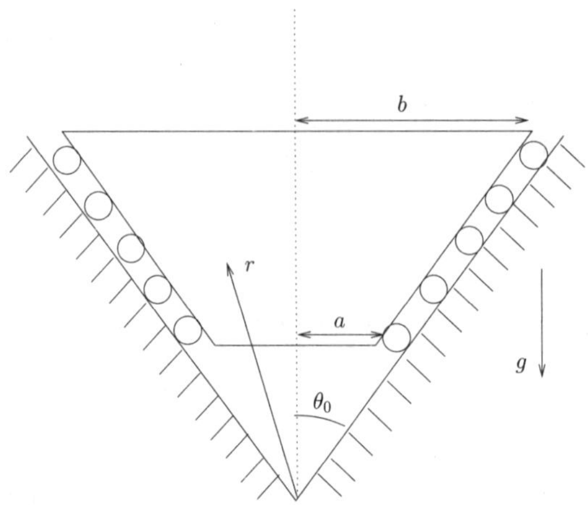

Fig. 6.67 *Truncated cone supported on its lateral surface subjected to gravity loading and to prescribed tractions on its free surfaces.*

(which are already part of the solution) or lead to a violation of the boundary conditions unless the associated constants are set to zero), is given by

$$[\xi_0 P_\lambda(\xi_0) - P_{\lambda+1}(\xi_0)]\,[\xi_0 P_{\lambda+2}(\xi_0) - P_{\lambda+3}(\xi_0)] = 0,$$

which can be written as

$$\left.\frac{dP_\lambda}{d\xi}\right|_{\xi=\xi_0} \left.\frac{dP_{\lambda+2}}{d\xi}\right|_{\xi=\xi_0} = 0. \tag{6.455}$$

The roots of the above equation are real [16]. We substitute these roots for a given θ_0 into Eqn. (6.280), and superpose this solution on the one given by Eqn. (6.453). The constants A_n, B_n, C_n, D_n are determined so as to satisfy the boundary conditions

$$\begin{aligned}
\left[\tau_{r\theta}\cos 2\theta + (\tau_{rr} - \tau_{\theta\theta})\sin\theta\cos\theta\right]_{r\cos\theta=a\cot\theta_0} &= s_i(\theta),\\
\left[\tau_{rr}\cos^2\theta + \tau_{\theta\theta}\sin^2\theta - \tau_{r\theta}\sin 2\theta\right]_{r\cos\theta=a\cot\theta_0} &= p_i(\theta),\\
\left[\tau_{r\theta}\cos 2\theta + (\tau_{rr} - \tau_{\theta\theta})\sin\theta\cos\theta\right]_{r\cos\theta=b\cot\theta_0} &= s_o(\theta),\\
\left[\tau_{rr}\cos^2\theta + \tau_{\theta\theta}\sin^2\theta - \tau_{r\theta}\sin 2\theta\right]_{r\cos\theta=b\cot\theta_0} &= p_o(\theta),
\end{aligned} \tag{6.456}$$

where $s_i(\theta)$, $s_o(\theta)$, $p_i(\theta)$, $p_o(\theta)$ are the prescribed tangential and normal tractions on the free surfaces. For the case of a full cone (i.e., $a = 0$), all constants that lead to singular displacements at the origin such as $C_n, D_n, n = 1, 2, \ldots, \infty$, are set to zero, and the remaining constants are found so as to satisfy the boundary conditions on the top surface. If instead of a truncated cone, we have a segment of a spherical shell, i.e., the top and bottom surfaces are given by $r = a$ and $r = b$ (see Fig. 6.68), then the boundary conditions instead of those given by Eqns. (6.456) are now given by $\tau_{rr}|_{r=a} = p_i(\theta)$, $\tau_{r\theta}|_{r=a} = s_i(\theta)$, $\tau_{rr}|_{r=b} = p_o(\theta)$ and $\tau_{r\theta}|_{r=b} = s_o(\theta)$.

For the special case of a gravity-loaded hemisphere of radius a whose flat end rests on a frictionless surface, and whose curved surface is subjected to tractions $\tau_{rr}|_{r=a} = p(\theta)$ and

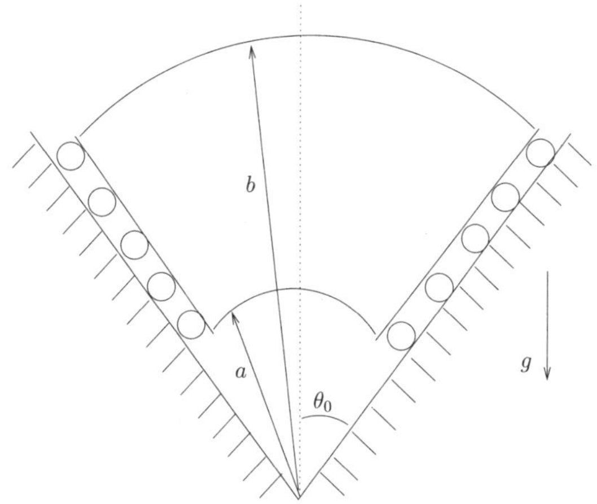

Fig. 6.68 *Segment of a spherical shell supported on its lateral surface subjected to gravity loading and to prescribed tractions on its free surfaces.*

$\tau_{r\theta}|_{r=a} = s(\theta)$ with $s(\pi/2) = 0$, the constants C_n and D_n can be set to zero since both the displacements and stresses are nonsingular. Since $\xi_0 = 0$, the roots of the characteristic equation given by Eqn. (6.455) are even integers. Thus, by considering only even integers in the solution given by Eqn. (6.280), we get (after renaming the constants)

$$\begin{aligned}
\alpha &= A_0(1-2\nu)(3\xi^2-1)r^2 + \sum_{n=1}^{\infty}\left[B_n r^{2n}P_{2n}(\xi) - 2A_n(2\nu+n-1)r^{2n+2}P_{2n+2}(\xi)\right],\\
\beta &= 3A_0 r\xi + \sum_{n=1}^{\infty} A_n(4n+3)r^{2n+1}P_{2n+1}(\xi),\\
2\mu u_r &= 2(2\nu-1)rA_0 + \sum_{n=1}^{\infty}\left[2(2n+1)(n-1+2\nu)A_n r^{2n+1} + 2nB_n r^{2n-1}\right]P_{2n}(\xi),\\
2\mu u_\theta &= -\sum_{n=1}^{\infty}\left[(2n+5-4\nu)A_n r^{2n+1} + B_n r^{2n-1}\right]\frac{dP_{2n}}{d\xi}\sin\theta,\\
\tau_{rr} &= -2(1+\nu)A_0\\
&\quad + \sum_{n=1}^{\infty}\left[2(2n+1)(2n^2-n-1-\nu)A_n r^{2n} + 2n(2n-1)B_n r^{2n-2}\right]P_{2n}(\xi),\\
\tau_{\theta\theta} &= -2(1+\nu)A_0 \qquad\qquad (6.457)\\
&\quad + \sum_{n=1}^{\infty}\left\{\left[-2(2n+1)(2n^2+4n+1+\nu)A_n r^{2n} - 4n^2B_n r^{2n-2}\right]P_{2n}(\xi)\right.\\
&\quad \left. + \left[(2n+5-4\nu)A_n r^{2n} + B_n r^{2n-2}\right]\xi\frac{dP_{2n}}{d\xi}\right\},\\
\tau_{\phi\phi} &= -2(1+\nu)A_0\\
&\quad + \sum_{n=1}^{\infty}\left\{\left[2(2n+1)(n-4\nu n-1-\nu)A_n r^{2n} + 2nB_n r^{2n-2}\right]P_{2n}(\xi)\right.
\end{aligned}$$

$$- \left[(2n+5-4\nu) A_n r^{2n} + B_n r^{2n-2} \right] \xi \frac{dP_{2n}}{d\xi} \right\},$$

$$\tau_{r\theta} = - \sum_{n=1}^{\infty} \left[(4n^2 + 4n - 1 + 2\nu) A_n r^{2n} + (2n-1) B_n r^{2n-2} \right] \frac{dP_{2n}}{d\xi} \sin\theta,$$

where

$$\frac{dP_{2n}}{d\xi} = \frac{(2n+1)}{(1-\xi^2)} \left[\xi P_{2n}(\xi) - P_{2n+1}(\xi) \right],$$

with the limiting value of the above expression as $\xi \to 1$ given by $n(2n+1)$. Superposing the above solution on the solution given by Eqn. (6.453), namely,

$$\begin{aligned}
2EAu_r &= \rho_0 g r^2 (2\nu^2 + \nu - 1) \cos\theta (1 + \cos 2\theta), \\
2EAu_\theta &= -\rho_0 g r^2 (2\nu^2 + \nu - 1) \sin\theta (1 + \cos 2\theta), \\
A\tau_{rr} &= -\rho_0 g r \cos\theta \left[1 + (1-2\nu) \cos 2\theta \right], \\
A\tau_{\theta\theta} &= \rho_0 g r \cos\theta \left[(1-2\nu) \cos 2\theta - 1 \right], \\
A\tau_{\phi\phi} &= -2\nu \rho_0 g r \cos\theta, \\
A\tau_{r\theta} &= \rho_0 g r (1-2\nu) \sin\theta (1 + \cos 2\theta),
\end{aligned}$$

where $A = -2(1-\nu)$, the equilibrium equations and boundary conditions $u_\theta|_{\theta=\pi/2} = \tau_{r\theta}|_{\theta=\pi/2} = 0$, are automatically satisfied. The constants A_n and B_n are now found so as to satisfy the boundary conditions on the curved surface, namely, $\tau_{rr}|_{r=a} = p(\theta)$ and $\tau_{r\theta}|_{r=a} = s(\theta)$. The constant A_0 is given by $\int_0^{\pi/2} p(\theta)\, d\theta / [2(1+\nu)] + \rho_0 g a / [8(1-\nu^2)]$. One can use a weighted residual technique to determine these constants.

The case of a sphere subjected to point loads along its diameter [124, 306] (extensions to an ellipsoid, hollow ellipsoid and a hollow sphere are treated in [181], [182] and [343], respectively) can be solved by first considering a uniformly distributed normal stress $-P/[2\pi a^2(1-\cos\alpha)]$ acting in the region $0 \le \theta \le \alpha$ in Fig. 6.68 with $\theta_0 = \pi/2$ (analogous to the disc problem shown in Fig. 6.60). Using the condition $\tau_{r\theta}|_{r=a} = 0$ in Eqn. (6.457), we get a relation between A_n and B_n. Next, considering the condition $\tau_{rr}|_{r=a} = -P/[2\pi a^2(1-\cos\alpha)]$ in the region $0 \le \theta \le \alpha$, and zero elsewhere, and using the relations

$$\int_{\cos\alpha}^{1} P_{2n}(\xi)\, d\xi = \frac{P_{2n-1}(\cos\alpha) - P_{2n+1}(\cos\alpha)}{4n+1},$$

$$\int_0^1 P_{2m}(\xi) P_{2n}(\xi)\, d\xi = \frac{1}{4n+1} \delta_{mn} \quad \text{(no sum on } n\text{)},$$

we get the solution given by Eqn. (6.457) with

$$A_0 = \frac{P}{4\pi a^2 (1+\nu)},$$

$$A_n = \frac{P \operatorname{cosec}^2 \frac{\alpha}{2} \left[P_{2n-1}(\cos\alpha) - P_{2n+1}(\cos\alpha) \right]}{8\pi a^{2(1+n)} \left[1 + \nu + 2n(1 + 2n + 2\nu) \right]},$$

$$B_n = \frac{P \operatorname{cosec}^2 \frac{\alpha}{2} \left[1 - 2\nu - 4n(1+n) \right] \left[P_{2n-1}(\cos\alpha) - P_{2n+1}(\cos\alpha) \right]}{8\pi a^{2n} (2n-1) \left[1 + \nu + 2n(1 + 2n + 2\nu) \right]}.$$

The solution for a point load is obtained by taking the limit as $\alpha \to 0$. In this limit, we have

$$A_0 = \frac{P}{4\pi a^2(1+\nu)},$$

$$A_n = \frac{P(4n+1)}{4\pi a^{2+2n}\left[1+\nu+2n(1+2n+2\nu)\right]},$$

$$B_n = \frac{P(4n+1)\left[1-2\nu-4n(1+n)\right]}{4\pi a^{2n}(2n-1)\left[1+\nu+2n(1+2n+2\nu)\right]}.$$

In the limit as $r \to 0$, only a few terms of the infinite series solution are nonzero, and we get $\boldsymbol{u} = \boldsymbol{0}$ and

$$\tau_{rr} = \frac{3P\left[7-5(7+2\nu)\xi^2\right]}{4\pi a^2(7+5\nu)},$$

$$\tau_{\theta\theta} = \frac{3P\left[5(7+2\nu)\xi^2-2(14+5\nu)\right]}{4\pi a^2(7+5\nu)},$$

$$\tau_{r\theta} = \frac{15P(7+2\nu)\xi\sin\theta}{4\pi a^2(7+5\nu)},$$

$$\tau_{\phi\phi} = \frac{21P}{4\pi a^2(7+5\nu)}.$$

For arbitrary r, one could attempt an approximate solution by taking a finite number of terms in the infinite series. However, such a methodology yields an accurate solution only away from the surface $r = a$. To obtain an accurate solution, we sum up the series solutions for α and β in Eqn. (6.457). Let

$$r_1(x,\xi) = \sqrt{x^2 - 2ax\xi + a^2},$$

$$r_2(x,\xi) = \sqrt{x^2 + 2ax\xi + a^2},$$

$$h = \sqrt{4\nu^2 - 3},$$

$$\gamma_1 = \frac{1}{2}\left[-1-2\nu-h\right],$$

$$\bar{\gamma}_1 = \frac{1}{2}\left[-1-2\nu+h\right],$$

where a superposed bar denotes the complex conjugate, e.g., $\bar{\gamma}_1$ is the complex conjugate of γ_1. The terms $r_1(r,\xi)$ and $r_2(r,\xi)$ are denoted simply as r_1 and r_2, respectively (see Fig. 6.69). We use the following relations in the subsequent development:

$$\sum_{n=1}^{\infty}\left(\frac{r}{a}\right)^{2n} P_{2n}(\xi) = \frac{a}{2}\left(\frac{1}{r_1}+\frac{1}{r_2}\right) - 1, \tag{6.458a}$$

$$\sum_{n=1}^{\infty}\left(\frac{r}{a}\right)^{2n+2} P_{2n+2}(\xi) = \frac{a}{2}\left(\frac{1}{r_1}+\frac{1}{r_2}\right) - \left(\frac{r}{a}\right)^2\frac{(3\xi^2-1)}{2} - 1, \tag{6.458b}$$

$$\sum_{n=1}^{\infty}\left(\frac{r}{a}\right)^{2n+1} P_{2n+1}(\xi) = \frac{a}{2}\left(\frac{1}{r_1}-\frac{1}{r_2}\right) - \left(\frac{r}{a}\right)\xi, \tag{6.458c}$$

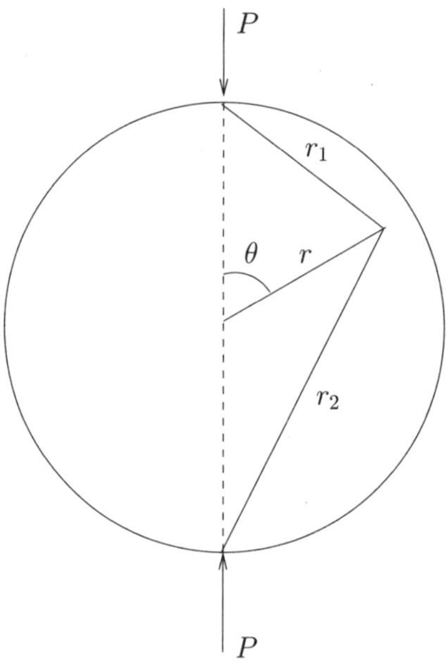

Fig. 6.69 *Sphere subjected to equal and opposite point loads along its diameter.*

The expressions for α and β in Eqns. (6.457) can be written as

$$\alpha = A_0(1-2\nu)(3\xi^2-1)r^2 - \frac{P}{\pi}\sum_{n=1}^{\infty}\left(\frac{r}{a}\right)^{2n} P_{2n}(\xi)\left[\frac{1}{2}+\frac{1}{2(2n-1)}+\frac{c_2}{2n-\gamma_1}+\frac{\bar{c}_2}{2n-\bar{\gamma}_1}\right]$$
$$-\frac{P}{\pi}\sum_{n=1}^{\infty}\left(\frac{r}{a}\right)^{2n+2} P_{2n+2}(\xi)\left[\frac{1}{2}+\frac{c_4}{2n-\gamma_1}+\frac{\bar{c}_4}{2n-\bar{\gamma}_1}\right], \tag{6.459}$$
$$\beta = 3A_0 r\xi + \frac{P}{a\pi}\sum_{n=1}^{\infty}\left(\frac{r}{a}\right)^{2n+1} P_{2n+1}(\xi)\left[1+\frac{c_6}{2n-\gamma_1}+\frac{\bar{c}_6}{2n-\bar{\gamma}_1}\right],$$

where

$$c_2 = \frac{1}{8h}\{1+3h+2\nu[3-4\nu-2h]\}, \qquad \bar{c}_2 = \frac{1}{8h}\{-1+3h+2\nu[4\nu-3-2h]\},$$
$$c_4 = \frac{1}{8h}\{3-5h+2\nu[4\nu-5+2h]\}, \qquad \bar{c}_4 = \frac{1}{8h}\{-3-5h+2\nu[5-4\nu+2h]\},$$
$$c_6 = \frac{1}{4h}\{3+2h-4\nu[2\nu-1+h]\}, \qquad \bar{c}_6 = \frac{1}{4h}\{-3+2h-4\nu[1-2\nu+h]\}.$$

Multiplying Eqn. (6.458a) by $r^{-\gamma-1}$, where γ is a real or complex-valued constant, we get

$$\sum_{n=1}^{\infty}\left(\frac{r^{2n-\gamma-1}}{a^{2n}}\right) P_{2n}(\xi) = r^{-\gamma-1}\left[\frac{a}{2}\left(\frac{1}{r_1}+\frac{1}{r_2}\right)-1\right].$$

Integrating the above equation, and multiplying by r^γ, we get

$$\sum_{n=1}^{\infty} \frac{1}{2n-\gamma}\left(\frac{r}{a}\right)^{2n} P_{2n}(\xi) = r^\gamma \int_0^r x^{-\gamma-1}\left[\frac{a}{2}\left(\frac{1}{r_1(x,\xi)} + \frac{1}{r_2(x,\xi)}\right) - 1\right] dx.$$

In a similar way, by using Eqns. (6.458b) and (6.458c), we can obtain expressions for the remaining terms in Eqn. (6.459). The final expressions for α and β are

$$\alpha = A_0(1-2\nu)(3\xi^2-1)r^2 - \frac{P}{\pi}\left\{\frac{a}{4}\left[\frac{1}{r_1}+\frac{1}{r_2}\right] - \frac{r_1+r_2}{4a} + \frac{r\xi}{4a}\log\frac{a+r\xi+r_2}{a-r\xi+r_1}\right.$$

$$+ c_2 r^{\gamma_1}\int_0^r x^{-\gamma_1-1} g_1(x,\xi)\,dx + \bar{c}_2 r^{\bar{\gamma}_1}\int_0^r x^{-\bar{\gamma}_1-1} g_1(x,\xi)\,dx \tag{6.460}$$

$$\left. + \frac{1}{2}g_2(r,\xi) + c_4 r^{\gamma_1+2}\int_0^r x^{-\gamma_1-3} g_2(x,\xi)\,dx + \bar{c}_4 r^{\bar{\gamma}_1+2}\int_0^r x^{-\bar{\gamma}_1-3} g_2(x,\xi)\,dx\right\},$$

$$\beta = 3A_0 r\xi + \frac{P}{a\pi}\left\{g_3(r,\xi) + c_6 r^{\gamma_1+1}\int_0^r x^{-\gamma_1-2} g_3(x,\xi)\,dx + \bar{c}_6 r^{\bar{\gamma}_1+1}\int_0^r x^{-\bar{\gamma}_1-2} g_3(x,\xi)\,dx\right\},$$

where

$$g_1(x,\xi) = \frac{a}{2}\left[\frac{1}{r_1(x,\xi)} + \frac{1}{r_2(x,\xi)}\right] - 1,$$

$$g_2(x,\xi) = \frac{a}{2}\left[\frac{1}{r_1(x,\xi)} + \frac{1}{r_2(x,\xi)}\right] - \left(\frac{x}{a}\right)^2\frac{(3\xi^2-1)}{2} - 1,$$

$$g_3(x,\xi) = \frac{a}{2}\left[\frac{1}{r_1(x,\xi)} - \frac{1}{r_2(x,\xi)}\right] - \left(\frac{x}{a}\right)\xi.$$

The displacement and stress fields are now recovered using Eqns. (6.261) and the Leibnitz formula. The integrals in the above solution can be carried out using standard Gaussian quadrature. Now, in contrast to the series solution based on a finite number of terms that does not converge in the vicinity of $r = a$, we get a solution that is accurate everywhere (except at the points of singularity). For example, for $P = 1$, $a = 1$, $E = 1$, $\nu = 0.25$, we get the results given in Table 6.20.

Sternberg and Rosenthal [306] use an alternative method where they consider the singular terms S_0^+, S_1^+ and S_2^+ in Eqn. (6.273a), and adjust the associated coefficients such that $\tau_{r\theta}|_{\theta=0} = 0$ [306]; this is in contrast to the analogous two-dimensional problem of a disc

(r,θ)	u_r	u_θ	τ_{rr}	$\tau_{\theta\theta}$	$\tau_{r\theta}$	$\tau_{\phi\phi}$
$(1/2, 0)$	−0.6624	0	−2.0872	0.2892	0	0.2892
$(1/2, \pi/6)$	−0.2916	0.3067	−0.4720	−0.5801	0.6931	0.2162
$(1, \pi/6)$	−0.2850	0.1333	0.0	−7.056E-2	0.0	−7.1737E-2

Table 6.20 *Displacement and stress fields at specific points for the sphere under point loads problem.*

subjected to point loads where no higher-order singular functions over and above those given by Eqn. (6.373) needed to be considered.

To solve the problem of a sphere under its own weight and contacting a rigid surface at a point, we follow the same approach that we followed for solving the corresponding disc problem (see Fig. 6.51). Thus, we superpose the particular solution (assuming that the z-axis points downwards, and θ is measured as usual from the z-axis) given by

$$\begin{aligned}
2Eu_r &= -\rho_0 g r^2(1-2\nu)\cos\theta, \\
2Eu_\theta &= -\rho_0 g r^2(1-2\nu)\sin\theta, \\
\tau_{rr} &= \tau_{\theta\theta} = \tau_{\phi\phi} = -\rho_0 g r\cos\theta, \\
\tau_{r\theta} &= 0,
\end{aligned}$$

on the solution given by Eqn. (6.278) with $C_n = D_n = 0$. We then determine the constants A_n and B_n so that the combined solution satisfies the traction boundary conditions $\tau_{r\theta}\big|_{r=a} = 0$,

$$\tau_{rr}\big|_{r=a} = -\frac{4\rho_0 g a \operatorname{cosec}^2\alpha}{3}, \quad 0 \le \theta \le \alpha,$$

and zero on the remaining part of the boundary. After taking the limit $\alpha \to 0$, and on using Eqn. (6.138), we get $B_0 = A_1 = B_1 = 0$, $A_0 = \rho_0 a g/[6(1+\nu)]$. Renumbering the constants as $A_n \to A_{n-1}$ and $B_n \to B_{n-1}$, the complete solution obtained using Eqn. (6.278) can be written as

$$\begin{aligned}
2\mu u_r &= 2\mu c_3\xi + \frac{\rho g r(2\nu-1)(2a+3r\xi)}{6(1+\nu)} \\
&\quad + \sum_{n=1}^{\infty}\left[(n+2)(n-1+4\nu)A_n r^{n+2} + (n+1)B_n r^n\right]P_{n+1}(\xi), \\
2\mu u_\theta &= -2\mu c_3\sin\theta + \frac{\rho g r^2(2\nu-1)\sin\theta}{2(1+\nu)} - \sum_{n=1}^{\infty}\left[(n+6-4\nu)A_n r^{n+2} + B_n r^n\right]\sin\theta\frac{dP_{n+1}}{d\xi}, \\
\tau_{rr} &= -\frac{\rho g}{3}(a+3r\xi) + \sum_{n=1}^{\infty}\left[(n+2)(n^2+n-2-2\nu)A_n r^{n+1} + n(n+1)B_n r^{n-1}\right]P_{n+1}(\xi),
\end{aligned} \tag{6.461}$$

$$\begin{aligned}
\tau_{\theta\theta} &= -\frac{\rho g}{3}(a+3r\xi) + \sum_{n=1}^{\infty}\left\{\left[-(n+2)(n^2+6n+7+2\nu)A_n r^{n+1} - (n+1)^2 B_n r^{n-1}\right]P_{n+1}(\xi)\right. \\
&\quad \left. + \left[(n+6-4\nu)A_n r^{n+1} + B_n r^{n-1}\right]\xi\frac{dP_{n+1}}{d\xi}\right\}, \\
\tau_{\phi\phi} &= -\frac{\rho g}{3}(a+3r\xi) + \sum_{n=1}^{\infty}\left\{\left[(n+2)(n-1-4\nu n-6\nu)A_n r^{n+1} + (n+1)B_n r^{n-1}\right]P_{n+1}(\xi)\right. \\
&\quad \left. - \left[(n+6-4\nu)A_n r^{n+1} + B_n r^{n-1}\right]\xi\frac{dP_{n+1}}{d\xi}\right\}, \\
\tau_{r\theta} &= -\sum_{n=1}^{\infty}\left[(n^2+4n+2+2\nu)A_n r^{n+1} + nB_n r^{n-1}\right]\sin\theta\frac{dP_{n+1}}{d\xi},
\end{aligned}$$

where the c_3 term is a rigid body displacement added to meet the constraint $u_r|_{(a,0)} = 0$ (in this particular problem though, due to the singularity in the displacements, this constraint cannot be met; note, however, that the stresses are not affected by this term), and

$$A_n = \frac{\rho_0 g(2n+3)}{6a^n\left[n^2+3n+3+(2n+3)\nu\right]},$$
$$B_n = -\frac{\rho_0 g a^{2-n}(2n+3)\left[n^2+4n+2+2\nu\right]}{6n\left[n^2+3n+3+(2n+3)\nu\right]}.$$

As $r \to 0$, the series terminates, and in this limit we get

$$\tau_{rr} = \frac{\rho a g\left[7-5(7+2\nu)\xi^2\right]}{2(7+5\nu)},$$
$$\tau_{\theta\theta} = \frac{\rho a g\left[5(7+2\nu)\xi^2-2(14+5\nu)\right]}{2(7+5\nu)},$$
$$\tau_{\phi\phi} = \frac{7\rho a g}{2(7+5\nu)},$$
$$\tau_{r\theta} = \frac{5\rho a g(7+2\nu)\xi\sin\theta}{2(7+5\nu)}.$$

As usual, due to the presence of the singularity at the pole, an approximation based on a finite number of terms of the series solution is reasonably accurate only away from the surface $r = a$. We find that though the stresses obtained using a finite number of terms in the series solution converge at least at some points in the interior, the displacements do not seem to converge anywhere if the bottom pole is constrained to have zero displacement using the c_3 term (this was also verified using a finite element code). Because of this non-convergence of the displacement field throughout the domain, we have not proceeded to sum the series analytically as we did for the problem shown in Fig. 6.69, although the same procedure followed there would work here too. Recall also that we found an even more severe pathology (a multi-valued displacement field) in the two-dimensional analogue of this problem, namely, a circular disc loaded under its own weight.

6.6.16 Contact problems on a finite domain

Consider the contact problem shown in Fig. 6.70, where a semicircular disk of radius b is subjected to either symmetric prescribed normal tractions or prescribed normal displacements on the region $0 \le r \le a$ and $\theta = \pm\pi/2$; physically, the prescribed displacements can be imposed by a rigid indenter indenting the flat surface, and hence we refer to this problem as a contact problem. The surface $r = b$ is supported using roller supports or is fixed. Thus, the boundary conditions (considering only the right half because of symmetry) are

$$\tau_{r\theta}|_{\theta=\pm\pi/2} = 0, \tag{6.462a}$$
$$u_r|_{r=b} = \tau_{r\theta}|_{r=b} = 0 \quad \text{or} \quad u_r|_{r=b} = u_\theta|_{r=b} = 0, \tag{6.462b}$$
$$\tau_{\theta\theta}|_{\theta=\pm\pi/2} = g_t(r) \quad \text{or} \quad u_\theta|_{\theta=\pm\pi/2} = g_u(r), \quad 0 \le r \le a, \tag{6.462c}$$
$$\tau_{\theta\theta}|_{\theta=\pi/2} = 0, \quad a \le r \le b. \tag{6.462d}$$

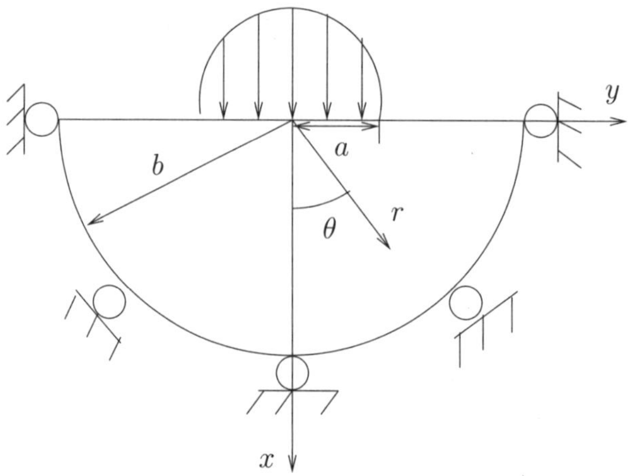

Fig. 6.70 *Contact problem on a semicircular domain.*

Since the domain is finite, the problems associated with solutions derived using the Flamant solution (i.e., $\boldsymbol{u}|_{r\to\infty} \to \infty$) do not arise. The Airy stress function and the associated displacements (under plane stress conditions) that automatically satisfy Eqn. (6.462a) are[26]

$$\phi = C_0 r^2 + \sum_{n=1}^{\infty}\left\{\left[r^{2n}A_n + r^{2n+2}C_n\right]\cos 2n\theta + r^{2n+1}\left[\hat{A}_n\cos(2n+1)\theta + \hat{C}_n\cos(2n-1)\theta\right]\right\}$$

$$\begin{aligned}2Eu_r &= k_1\cos\theta + 4(1-\nu)rC_0\\ &\quad -2\sum_{n=1}^{\infty}\left\{\left[2n(1+\nu)r^{2n-1}A_n + ((1+\nu)(2n+2)-4)r^{2n+1}C_n\right]\cos 2n\theta\right.\\ &\quad \left.+ r^{2n}\left[(2n+1)(1+\nu)\hat{A}_n\cos(2n+1)\theta + \left[(1+\nu)(2n+1)-4\right]\hat{C}_n\cos(2n-1)\theta\right]\right\}\\ 2Eu_\theta &= -k_1\sin\theta + 2\sum_{n=1}^{\infty}\left\{\left[2n(1+\nu)r^{2n-1}A_n + (4+2n+2n\nu)r^{2n+1}C_n\right]\sin 2n\theta\right.\\ &\quad \left.+ r^{2n}\left[(2n+1)(1+\nu)\hat{A}_n\sin(2n+1)\theta + \left[3+2n+(2n-1)\nu\right]\hat{C}_n\sin(2n-1)\theta\right]\right\}\end{aligned}$$

where

$$\hat{C}_n = \frac{2n+1}{2n-1}\hat{A}_n. \tag{6.463}$$

The terms associated with k_1 represent a rigid-body displacement. The constants are determined so as to satisfy the boundary conditions given by Eqns. (6.462b)–(6.462d) using, for example, a weighted-residual method as described on page 485. In case $g_u(r)$ is prescribed on part of the top edge, the displacement field on the top edge is used as a weighting function in the weighted-residual statement for that edge.

[26] If the prescribed tangential tractions on the top surface are nonzero, then the constraint given by Eqn. (6.463) is relaxed.

The function $g_u(r)$ in Eqn. (6.462c) depends on the shape of the indenter. Suppose that the rigid indenter has a flat surface of width $2a$, and its axis is coincident with the x-axis. Let the flat surface of the indenter be brought into contact with the surface $x = 0$, and then let the indenter be moved along the x-axis by a prescribed amount δ. Then we have $g_u(r) = \delta$ for $0 \le r \le a$ and $g_t(r) = 0$ for $r \ge a$. If the indenter is wedge-shaped with wedge-angle 2α, and it is given a prescribed displacement of δ, then we have $g_u(r) = (1 - r/a)\delta$ where $2a = 2\delta \tan\alpha$ is the contact width along the y-axis. Finally, if the indenter is circular with radius R, then in the zone $r \le a$, we have $r^2 + [R - \delta + g_u(r)]^2 = R^2$, which leads to $g_u(r) = \delta - R + \sqrt{R^2 - r^2}$. Noting that $g_u(a) = 0$, we get $a = \sqrt{2\delta R - \delta^2}$, and $g_u(r) = \sqrt{R^2 - r^2} - \sqrt{R^2 - a^2}$.

If instead of the displacement, the total load per unit length P exerted by the indenter is specified, then the width of contact $2a$ has to be determined using the equation

$$
\begin{aligned}
P &= 2\int_0^a \tau_{\theta\theta}|_{\theta=\pi/2}\, dr \\
&= 4aC_0 + 4\sum_{n=1}^{\infty}(-1)^n a^{2n-1}\left[A_n n + a^2(n+1)C_n\right].
\end{aligned}
$$

Since the constants C_0, A_n and C_n could themselves be functions of a, a Newton–Raphson procedure may not be feasible, and a non-gradient method such as the bisection method would probably have to be used.

As an example, for $a = 1$, $b = 5$, $E = 1$, $\nu = 0.25$, $g_t(r) = -(1 - r/a)$, and with the second set of boundary conditions in Eqn. (6.462b), we get[27] at $(r,\theta) = (1,0)$, the solution for the nonzero components as $u_r = 0.969$, $\tau_{rr} = -0.5\ (-0.5)$, $\tau_{\theta\theta} = -0.003\ (-0.0587)$, $P = 1\ (1)$, where the bracketed values show the predictions based on the solution for an infinite half-space. As another example, for the same geometry and material properties as in the previous example, but now with $g_u(r) = -(1 - r^2/a^2)$, we get at $(r,\theta) = (1/2, \pi/2)$ the solution $u_r = -0.19$, $u_\theta = -0.74$, $\tau_{rr} = -0.36$, $\tau_{\theta\theta} = -0.40$, $\tau_{r\theta} = 0$ and $P = 0.52$.

Now consider the three-dimensional analogue of the above problem as shown in Fig. 6.71 with the boundary conditions given by

$$\tau_{r\theta}|_{\theta=\pi/2} = 0 \tag{6.464a}$$

$$u_r|_{r=b} = \tau_{r\theta}|_{r=b} = 0 \quad \text{or} \quad u_r|_{r=b} = u_\theta|_{r=b} = 0, \tag{6.464b}$$

$$\tau_{\theta\theta}|_{\theta=\pi/2} = g_t(r) \quad \text{or} \quad u_\theta|_{\theta=\pi/2} = g_u(r), \quad 0 \le r \le a, \tag{6.464c}$$

$$\tau_{\theta\theta}|_{\theta=\pi/2} = 0, \quad a \le r \le b. \tag{6.464d}$$

This problem can be solved by splitting the Boussinesq potentials give in Eqns. (6.278) and (6.280) into even and odd parts. The potentials and the corresponding field variables are given by

$$\alpha = A_0(1-2\nu)(3\xi^2 - 1)r^2 + \sum_{n=1}^{\infty}\left\{\left[B_n r^{2n}P_{2n}(\xi) - 2A_n(2\nu + n - 1)r^{2n+2}P_{2n+2}(\xi)\right]\right.$$

[27] Quadruple precision was used to obtain these results. The matrices in the case of the mixed problem are ill-conditioned because of the displacement and traction boundary conditions that have to be met on the top edge; relatively speaking, the prescribed traction condition on the top edge is much easier to handle.

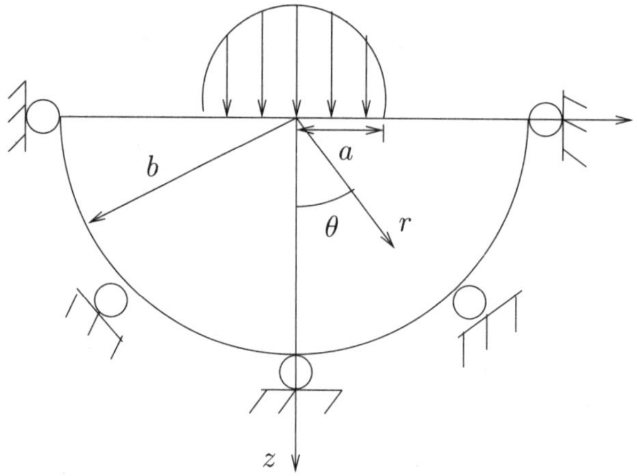

Fig. 6.71 Contact problem on a hemispherical domain.

$$+ \left[\tilde{B}_n - \tilde{A}_n(4\nu + 2n - 3)\right] r^{2n+1} P_{2n+1}(\xi) \Big\},$$

$$\beta = 3A_0 r\xi + \sum_{n=1}^{\infty} \Big\{ A_n(4n+3) r^{2n+1} P_{2n+1}(\xi) + \tilde{A}_n(4n+1) r^{2n} P_{2n}(\xi) \Big\},$$

$$2\mu u_r = k_3 \cos\theta + 2(2\nu - 1) r A_0$$
$$+ \sum_{n=1}^{\infty} \Big\{ 2(2n+1)(n-1+2\nu) A_n r^{2n+1} P_{2n}(\xi) + 2n B_n r^{2n-1} P_{2n}(\xi)$$
$$+ 2n(2n-3+4\nu)\tilde{A}_n r^{2n} P_{2n-1}(\xi) + (2n+1)\tilde{B}_n r^{2n} P_{2n+1}(\xi) \Big\},$$

$$2\mu u_\theta = -k_3 \sin\theta - \sum_{n=1}^{\infty} \Big\{ (2n+5-4\nu) A_n r^{2n+1} \frac{dP_{2n}}{d\xi} + B_n r^{2n-1} \frac{dP_{2n}}{d\xi}$$
$$+ (2n+4-4\nu)\tilde{A}_n r^{2n} \frac{dP_{2n-1}}{d\xi} + \tilde{B}_n r^{2n} \frac{dP_{2n+1}}{d\xi} \Big\} \sin\theta,$$

$$\tau_{rr} = -2(1+\nu) A_0$$
$$+ \sum_{n=1}^{\infty} \Big\{ 2(2n+1)(2n^2 - n - 1 - \nu) A_n r^{2n} P_{2n}(\xi) + 2n(2n-1) B_n r^{2n-2} P_{2n}(\xi)$$
$$+ 2n(4n^2 - 6n - 2\nu)\tilde{A}_n r^{2n-1} P_{2n-1}(\xi) + 2n(2n+1)\tilde{B}_n r^{2n-1} P_{2n+1}(\xi) \Big\},$$

$$\tau_{\theta\theta} = -2(1+\nu) A_0$$
$$+ \sum_{n=1}^{\infty} \Big\{ -2(2n+1)(2n^2 + 4n + 1 + \nu) A_n r^{2n} P_{2n}(\xi) - 4n^2 B_n r^{2n-2} P_{2n}(\xi)$$
$$- 2n(4n^2 + 4n - 1 + 2\nu)\tilde{A}_n r^{2n-1} P_{2n-1}(\xi) - (2n+1)^2 \tilde{B}_n r^{2n-1} P_{2n+1}(\xi)$$

$$+\left[(2n+5-4\nu)A_n r^{2n}\frac{dP_{2n}}{d\xi}+B_n r^{2n-2}\frac{dP_{2n}}{d\xi}\right.$$
$$\left.+(2n+4-4\nu)\tilde{A}_n r^{2n-1}\frac{dP_{2n-1}}{d\xi}+\tilde{B}_n r^{2n-1}\frac{dP_{2n+1}}{d\xi}\right]\xi\Big\},$$
$$\tau_{\phi\phi}=-2(1+\nu)A_0+\sum_{n=1}^{\infty}\Big\{2(2n+1)(n-4\nu n-1-\nu)A_n r^{2n}P_{2n}(\xi)+2nB_n r^{2n-2}P_{2n}(\xi)$$
$$+2n(2n-3-8\nu n+2\nu)\tilde{A}_n r^{2n-1}P_{2n-1}(\xi)+(2n+1)\tilde{B}_n r^{2n-1}P_{2n+1}(\xi)$$
$$-\left[(2n+5-4\nu)A_n r^{2n}\frac{dP_{2n}}{d\xi}+B_n r^{2n-2}\frac{dP_{2n}}{d\xi}\right.$$
$$\left.+(2n+4-4\nu)\tilde{A}_n r^{2n-1}\frac{dP_{2n-1}}{d\xi}+\tilde{B}_n r^{2n-1}\frac{dP_{2n+1}}{d\xi}\right]\xi\Big\},$$
$$\tau_{r\theta}=-\sum_{n=1}^{\infty}\Big\{(4n^2+4n-1+2\nu)A_n r^{2n}\frac{dP_{2n}}{d\xi}+(2n-1)B_n r^{2n-2}\frac{dP_{2n}}{d\xi}$$
$$+(4n^2-2+2\nu)\tilde{A}_n r^{2n-1}\frac{dP_{2n-1}}{d\xi}+2n\tilde{B}_n r^{2n-1}\frac{dP_{2n+1}}{d\xi}\Big\}\sin\theta.$$

with

$$\tilde{B}_n=\frac{2(2n^2-1+\nu)}{(2n+1)}\tilde{A}_n.$$

The terms associated with k_3 yield rigid displacements. The above choice of potentials automatically satisfy the boundary conditions given by Eqns. (6.464a), so that the constants are determined numerically using the conditions in Eqns. (6.464b)–(6.464d).

As in the two-dimensional case, the function $g_u(r)$ in Eqn. (6.464c) depends on the shape of the indenter. Suppose that the rigid indenter is cylindrical with radius a, and its axis is coincident with the z-axis. Let the flat surface of the indenter be brought into contact with the surface $z=0$, and then let the indenter be moved along the z-axis by a prescribed amount δ. Then we have $g_u(r)=\delta$ for $0\le r\le a$ and $g_t(r)=0$ for $a\le r\le b$. If the indenter is conical with cone-angle α, and it is given a prescribed displacement of δ, then we have $g_u(r)=(1-r/a)\delta$ where $a=\delta\tan\alpha$ is the radius of the contact zone on the $z=0$ plane. Finally, if the indenter is spherical with radius R, then in the zone $r\le a$, we have $r^2+[R-\delta+g_u(r)]^2=R^2$, which leads to $g_u(r)=\delta-R+\sqrt{R^2-r^2}$. Noting that $g_u(a)=0$, we get $a=\sqrt{2\delta R-\delta^2}$, and $g_u(r)=\sqrt{R^2-r^2}-\sqrt{R^2-a^2}$.

If instead of the displacement, the total load P exerted by the indenter is specified, then the radius of the contact zone a has to be determined using the equation

$$P=\int_0^a \tau_{\theta\theta}|_{\theta=\pi/2}\,2\pi r\,dr$$
$$=-2(1+\nu)\pi a^2A_0+\sum_{n=1}^{\infty}\left[\frac{4a^{2n+2}\left[2n^2+4n+1+\nu\right]\pi^{3/2}}{\Gamma\left(-\frac{1}{2}-n\right)\Gamma(n+2)}A_n-\frac{4a^{2n}\pi^{3/2}}{\Gamma\left(\frac{1}{2}-n\right)\Gamma(n)}B_n\right],$$

where Γ denotes the gamma function. As in the two-dimensional case, an iterative method would have to be used to determine a.

As an example, for $a = 1$, $b = 5$, $E = 1$, $\nu = 0.25$, $g_t(r) = -\sqrt{1 - r^2/a^2}$, and with the second set of boundary conditions in Eqn. (6.464b), we get at $(r, \theta) = (1/2, \pi/2)$ the solution $u_r = -0.14\ (-0.15)$, $u_\theta = -1.09\ (-1.29)$, $\tau_{rr} = -0.63\ (-0.63)$, $\tau_{\theta\theta} = -0.88\ (-0.87)$, $\tau_{r\theta} = 0\ (0)$, $\tau_{\phi\phi} = -0.66\ (-0.67)$, and the total force $P = 2.06\ (2.09)$, with again the bracketed values showing the solution for an infinite half-space. As another example, for the same geometry and material properties as in the previous example, but now with $g_u(r) = -1$, we get at $(r, \theta) = (1/2, \pi/2)$, the solution $u_r = -0.09$, $u_\theta = -1.07\ (-1)$, $\tau_{rr} = -0.74$, $\tau_{\theta\theta} = -0.91\ (-0.39)$, $\tau_{\phi\phi} = -0.6$, and the total force $P = 2.08\ (2.13)$, while at $(r, \theta) = (3, \pi/2)$, we get $u_r = -0.04$, $u_\theta = -0.04\ (-0.34)$, $\tau_{rr} = 0.04$, $\tau_{\theta\theta} = 0.01\ (0)$ and $\tau_{\phi\phi} \approx 0$.

6.6.17 Spherical cavity/inclusion in an infinite domain with a uniaxial state of stress at infinity

An extensive review on three-dimensional stress concentrations in the theory of elasticity can be found in [307]. Here, we consider a few of such problems. An isotropic body of infinite extent contains a spherical cavity of radius a, as shown in Fig. 6.72. The state of stress at infinity is $\tau_{zz} = p$, with other components zero. If $[\boldsymbol{u}]$ and $[\boldsymbol{\tau}]$ are the Cartesian components of the displacement and stress under simple tension, the corresponding components in the spherical coordinate system are given by $\boldsymbol{Q}[\boldsymbol{u}]$ and $\boldsymbol{Q}[\boldsymbol{\tau}]\boldsymbol{Q}^T$, where $\boldsymbol{Q}$ is given by Eqn. (C.2). Carrying out this computation, we get

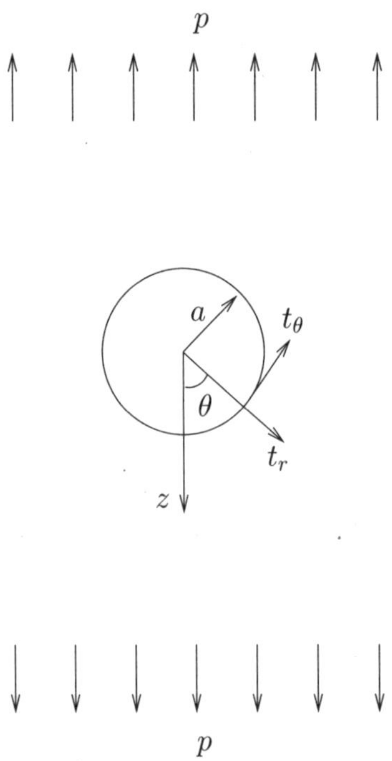

Fig. 6.72 *Spherical cavity in an infinite domain with a uniaxial state of stress at infinity.*

$$2\mu u_r = \frac{pr}{(1+\nu)}(\cos^2\theta - \nu\sin^2\theta), \quad 2\mu u_\theta = -\frac{pr}{2}\sin(2\theta),$$

$$\tau_{rr} = p\cos^2\theta, \quad \tau_{\theta\theta} = p\sin^2\theta, \quad \tau_{r\theta} = -\frac{p}{2}\sin(2\theta), \quad \tau_{\phi\phi} = 0.$$

The above simple tension solution is generated by the following Boussinesq potentials and Love strain function:

$$\alpha = \frac{p\nu}{2(1+\nu)}r^2(3\cos^2\theta - 1), \quad \beta = -\frac{p}{2(1+\nu)}r\cos\theta,$$

$$\Psi = \frac{p\nu}{2(1+\nu)}r^3\cos\theta + \frac{p(1-5\nu)}{6(1+\nu)}r^3\cos^3\theta.$$

The solution to the spherical cavity problem is obtained by combining the solution for simple tension with the solution generated by the Boussinesq potentials/Love strain function [191]

$$\alpha = \frac{A}{r} + \frac{B(3\cos^2\theta - 1)}{2r^3}, \quad \beta = \frac{C\cos\theta}{r^2},$$

$$\Psi = 2\left[\frac{(C-A)}{4}\log\left(\frac{1+\cos\theta}{1-\cos\theta}\right) - \frac{C}{2}\cos\theta + \frac{B}{4}\frac{\cos\theta}{r^2}\right].$$

Carrying out the required superposition, we get

$$\begin{aligned}
2\mu u_r &= \frac{pr}{(1+\nu)}(\cos^2\theta - \nu\sin^2\theta) + \frac{1}{2r^4}\left\{3B - 2Ar^2 + [-9B + (8\nu - 10)Cr^2]\cos^2\theta\right\},\\
2\mu u_\theta &= -\frac{pr}{2}\sin(2\theta) - \frac{\sin(2\theta)}{2r^4}\left[3B + (4\nu - 2)Cr^2\right],\\
\tau_{rr} &= p\cos^2\theta + \frac{1}{r^5}\left\{2Ar^2 - 6B + \left[18B + (10 - 4\nu)Cr^2\right]\cos^2\theta - 2C\nu r^2\sin^2\theta\right\},\\
\tau_{\theta\theta} &= p\sin^2\theta + \frac{1}{4r^5}\left\{-3B - (4A + 10C - 12\nu C)r^2 + [-21B + (4\nu - 2)Cr^2]\cos 2\theta\right\},\\
\tau_{r\theta} &= -\frac{p}{2}\sin 2\theta + \frac{\sin 2\theta}{r^5}\left[6B + (1+\nu)Cr^2\right],\\
\tau_{\phi\phi} &= \frac{1}{2r^5}\left[3B - 2Ar^2 + [-15B + (8\nu - 6)Cr^2]\cos^2\theta - 4\nu Cr^2\sin^2\theta\right].
\end{aligned} \tag{6.465}$$

Using the boundary conditions

$$\tau_{rr}\big|_{r=a} = \tau_{r\theta}\big|_{r=a} = 0,$$

we obtain

$$A = \frac{a^3p(6-5\nu)}{2(7-5\nu)}, \quad B = \frac{a^5p}{(7-5\nu)}, \quad C = -\frac{5a^3p}{2(7-5\nu)}.$$

The stresses $\tau_{\theta\theta}$ and $\tau_{\phi\phi}$ at the cavity for $\theta = 0$ and $\theta = \pi/2$ are

$$\tau_{\theta\theta}|_{\theta=0} = -\frac{3p(1+5\nu)}{2(7-5\nu)}, \qquad \tau_{\phi\phi}|_{\theta=0} = -\frac{3p(1+5\nu)}{2(7-5\nu)},$$

$$\tau_{\theta\theta}|_{\theta=\frac{\pi}{2}} = \frac{p(27-15\nu)}{2(7-5\nu)}, \qquad \tau_{\phi\phi}|_{\theta=\frac{\pi}{2}} = -\frac{3p(1-5\nu)}{2(7-5\nu)}.$$

These expressions give an idea of the stress concentrations that occur due to the introduction of a spherical cavity in a bar under simple tension.

If instead of a spherical cavity, we had a rigid spherical inclusion of radius a, then the constants A, B and C are determined by using the boundary conditions

$$u_r|_{r=a} = u_\theta|_{r=a} = 0.$$

We obtain

$$A = -\frac{\nu p a^3}{(1+\nu)} - \frac{3pa^3}{4(4-5\nu)}, \quad B = -\frac{pa^5}{2(4-5\nu)}, \quad C = \frac{5a^3 p}{4(4-5\nu)}.$$

Now consider the case when the spherical cavity is in an infinite domain with a state of stress $\tau_{xz} = p$ at infinity (see Fig. 6.73b). We first convert the displacement solution in Eqn. (6.465) to the Cartesian system by multiplying it by Q^T (with Q given by Eqn. (C.2)), and then find the solution to the problem shown in Fig. 6.73a by superposing the solutions for simple tension and simple compression at infinity along the x and z directions, respectively. The axes are then rotated by $\pi/4$ to obtain the solution for the problem shown in Fig. 6.73b. The final solution obtained is

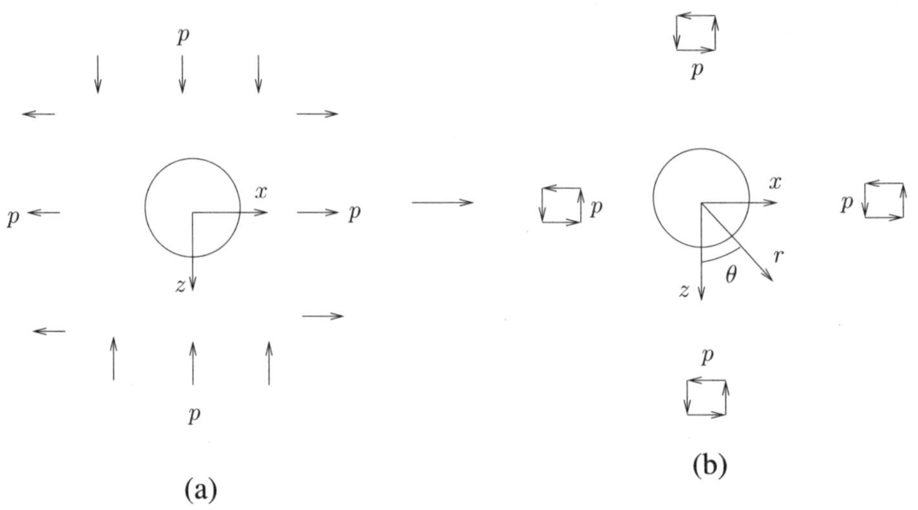

Fig. 6.73 *Spherical cavity in an infinite domain with a state of pure shear at infinity.*

$$2\mu u_x = p\left[\frac{h_1 z + h_2 x}{h_3}\right],$$
$$2\mu u_y = p\frac{h_2 y}{h_3}, \tag{6.466}$$
$$2\mu u_z = p\left[\frac{h_1 x + h_2 z}{h_3}\right],$$

where (with $r = \sqrt{x^2 + y^2 + z^2}$),

$$h_1 = 12a^5 + 20(1-2\nu)a^3r^2 + 4(7-5\nu)r^5,$$
$$h_2 = -60a^3(a^2 - r^2)\frac{xz}{r^2},$$
$$h_3 = 4(7-5\nu)r^5.$$

With respect to the spherical coordinate system, the solution is

$$2\mu u_r = \frac{p}{2(7-5\nu)r^4}\left[-9a^5 + 5(5-4\nu)a^3r^2 + 2(7-5\nu)r^5\right]\sin 2\theta\cos\phi,$$
$$2\mu u_\theta = \frac{p}{(7-5\nu)r^4}\left[3a^5 + 5(1-2\nu)a^3r^2 + (7-5\nu)r^5\right]\cos 2\theta\cos\phi,$$
$$2\mu u_\phi = \frac{p}{(7-5\nu)r^4}\left[-3a^5 - 5(1-2\nu)a^3r^2 - (7-5\nu)r^5\right]\cos\theta\sin\phi,$$
$$\tau_{rr} = \frac{p}{(7-5\nu)r^5}\left[18a^5 - 5(5-\nu)a^3r^2 + (7-5\nu)r^5\right]\sin 2\theta\cos\phi,$$
$$\tau_{\theta\theta} = -\frac{p}{2(7-5\nu)r^5}\left[21a^5 - 5(1-2\nu)a^3r^2 + 2(7-5\nu)r^5\right]\sin 2\theta\cos\phi,$$
$$\tau_{\phi\phi} = -\frac{15pa^3}{2(7-5\nu)r^5}\left[a^2 - (1-2\nu)r^2\right]\sin 2\theta\cos\phi,$$
$$\tau_{r\theta} = \frac{p(r-a)}{(7-5\nu)r^5}\left\{12a^4 + 12a^3r + (7-5\nu)\left[a^2r^2 + ar^3 + r^4\right]\right\}\cos 2\theta\cos\phi,$$
$$\tau_{\theta\phi} = \frac{p}{(7-5\nu)r^5}\left[3a^5 + 5(1-2\nu)a^3r^2 + (7-5\nu)r^5\right]\sin\theta\sin\phi,$$
$$\tau_{r\phi} = \frac{p}{(7-5\nu)r^5}\left[12a^5 - 5(1+\nu)a^3r^2 - (7-5\nu)r^5\right]\cos\theta\sin\phi.$$

It is easily verified that the above solution satisfies the boundary conditions

$$\tau_{rr}\big|_{r=a} = \tau_{r\theta}\big|_{r=a} = \tau_{r\phi}\big|_{r=a} = 0.$$

Finally, consider a generalization of the problem shown in Fig. 6.72, where we now have a spherical elastic inclusion of radius a instead of a cavity or a rigid inclusion [97]. The properties of the materials outside and inside the inclusion are denoted by subscripts 1 and 2, respectively. The solution outside the inclusion is still assumed to be given by

Eqns. (6.465) with μ and ν replaced by μ_1 and ν_1. The solution inside the inclusion is assumed to be generated by the Boussinesq potentials and Love strain functions given by

$$\alpha^{(2)} = \frac{Dr^2(3\cos^2\theta - 1)}{2}, \quad \beta^{(2)} = Gr\cos\theta,$$
$$\Psi^{(2)} = \frac{1}{2}Dr^3\cos\theta - \frac{1}{6}(5D + 2G)r^3\cos^3\theta.$$

Corresponding to the above potentials, we get from Eqns. (6.261)

$$\begin{aligned}
2\mu_2 u_r^{(2)} &= r\left\{-D + [3D - (2 - 4\nu_2)G]\cos^2\theta\right\},\\
2\mu_2 u_\theta^{(2)} &= -\frac{r}{2}[3D - 2G + 4\nu_2 G]\sin 2\theta,\\
\tau_{rr}^{(2)} &= \frac{1}{2}[D - 2G + (3D - 2G + 4\nu_2 G)\cos 2\theta],\\
\tau_{\theta\theta}^{(2)} &= \frac{1}{2}\{D - 2G - [3D - (2 - 4\nu_2)G]\cos 2\theta\},\\
\tau_{r\theta}^{(2)} &= -\frac{1}{2}[3D - 2(1 - 2\nu_2)G]\sin 2\theta,\\
\tau_{\phi\phi}^{(2)} &= -D - 2\nu_2 G.
\end{aligned}$$

The constants A, B, C, D and G are determined using the conditions

$$u_r^{(1)}\Big|_{r=a} = u_r^{(2)}\Big|_{r=a}, \quad u_\theta^{(1)}\Big|_{r=a} = u_\theta^{(2)}\Big|_{r=a}, \quad \tau_{rr}^{(1)}\Big|_{r=a} = \tau_{rr}^{(2)}\Big|_{r=a},$$

and leads to the equations

$$\begin{bmatrix} A\\ B\\ C\\ D\\ G \end{bmatrix} = \begin{bmatrix} -\frac{\gamma}{a^2} & -\frac{3\gamma}{a^4} & \frac{(4\nu_1-5)\gamma}{a^2} & -2a & (2-4\nu_2)a\\ -\frac{\gamma}{a^2} & \frac{3\gamma}{2a^4} & 0 & a & 0\\ 0 & -\frac{3\gamma}{a^4} & \frac{2(1-2\nu_1)\gamma}{a^2} & 3a & 2(2\nu_2-1)a\\ \frac{2}{a^2} & \frac{12}{a^4} & \frac{10-4\nu_1}{a^2} & -2a & 2(1-\nu_2)a\\ \frac{2}{a^2} & -\frac{6}{a^4} & -\frac{2\nu_1}{a^2} & a & 2\nu_2 a \end{bmatrix}^{-1} \begin{bmatrix} -\frac{pa\gamma}{1+\nu_1}\\ \frac{p\nu_1 a\gamma}{1+\nu_1}\\ pa\gamma\\ -pa\\ 0 \end{bmatrix}.$$

where $\gamma = \mu_2/\mu_1$. The boundary condition $\tau_{r\theta}^{(1)}\Big|_{r=a} = \tau_{r\theta}^{(2)}\Big|_{r=a}$ which leads to the condition

$$\frac{12}{a^4}B + \frac{2(1+\nu_1)}{a^2}C + 3aD - 2(1 - 2\nu_2)aG = pa,$$

is automatically satisfied by the above solution. The results for the hollow cavity and rigid inclusion are recovered in the limits $\gamma \to 0$ and $\gamma \to \infty$, respectively.

The extension of the above results to the case of a spherical or ellipsoidal inclusion or cavity under various far-field loading conditions can be found in [70, 74, 172, 193, 236, 276, 339, 340, 341]. Although the solution presented in [193] is correct, the stress concentration factor for the limiting case when the radius of the cavity tends to the radius of the cylinder, deduced based on 'physical reasoning', is incorrect [307]. If b represents the radius of the circular cylinder, this solution can be found by superposing the particular solution given by Eqn. (6.465) on the general solution given by Eqn. (6.278) (with only even $n \geq 4$ considered due to symmetry, and the form of the particular solution), and then finding the constants using the boundary conditions $\tau_{rr}|_{r=a} = \tau_{r\theta}|_{r=a} = 0$, and $\tau_{rr}|_{r\sin\theta=b} = \tau_{r\theta}|_{r\sin\theta=b} = 0$.

6.6.18 Prolate or oblate spheroidal cavity in an infinite domain with a uniform stress state at infinity

This problem is similar to that shown in Fig. 6.72 except that the shape of the cavity is that of a prolate or oblate spheroid, and the stress state at infinity is given by $\tau_{rr} = \tau_{\theta\theta} = p_r$ and $\tau_{zz} = p_z$. The prolate spheroidal cavity is obtained by rotating the ellipse $\xi = \xi_0$ around the z-axis. The semi-major and semi-minor axes (along the z- and r-axes, respectively) are given by $a = c\cosh\xi_0$ and $b = c\sinh\xi_0$, so that $c^2 = a^2 - b^2$. Since the problem is axisymmetric, we assume one part of the solution to be given by Eqn. (6.262) with the harmonic potentials α and β given by (see Eqn. (6.281))

$$\alpha = AQ_0(\cosh\xi)P_0(\cos\eta) + BQ_2(\cosh\xi)P_2(\cos\eta),$$
$$\beta = CQ_1(\cosh\xi)P_1(\cos\eta).$$

The functions Q_0, Q_1 and Q_2 are given by

$$Q_0(\cosh\xi) = \frac{1}{2}\log\left(\frac{\cosh\xi+1}{\cosh\xi-1}\right),$$
$$Q_1(\cosh\xi) = \frac{\cosh\xi}{2}\log\left(\frac{\cosh\xi+1}{\cosh\xi-1}\right) - 1,$$
$$Q_2(\cosh\xi) = \frac{3\cosh^2\xi-1}{4}\log\left(\frac{\cosh\xi+1}{\cosh\xi-1}\right) - \frac{3}{2}\cosh\xi.$$

This solution is superposed on the stress solution

$$\boldsymbol{Q}\text{diag}[p_r, p_r, p_z]\boldsymbol{Q}^T, \tag{6.467}$$

with $\boldsymbol{Q}$ given by Eqn. (F.3). The constants A, B and C are found by imposing the boundary conditions

$$\tau_{\xi\xi}\big|_{\xi=\xi_0} = \tau_{\xi\eta}\big|_{\xi=\xi_0} = 0, \tag{6.468}$$

on the total stress field, and are given by

$$A = \frac{N_1}{24D},$$
$$B = \frac{N_2}{3D},$$
$$C = \frac{N_3}{D},$$

where

$$N_1 = c^2\Big\{4\left[(8\nu-6)p_r + (17-24\nu)p_z\right] - 2\left[6p_r + (35-64\nu)p_z\right]\cosh 2\xi_0$$

$$
\begin{aligned}
&- 4\left[(8\nu-6)p_r + (1+8\nu)p_z\right]\cosh 4\xi_0 + 6(2p_r+p_z)\cosh 6\xi_0 \\
&+ \left[(32\nu-50)p_r + (32\nu-9)p_z\right]\cosh\xi_0 \log\left(\coth\frac{\xi_0}{2}\right) \\
&+ 3\left[(22-16\nu)p_r + (3-16\nu)p_z\right]\cosh 3\xi_0 \log\left(\coth\frac{\xi_0}{2}\right) \\
&+ \left[2(8\nu-5)p_r + (3+16\nu)p_z\right]\cosh 5\xi_0 \log\left(\coth\frac{\xi_0}{2}\right) \\
&- 3(2p_r+p_z)\cosh 7\xi_0 \log\left(\coth\frac{\xi_0}{2}\right)\Bigg\},
\end{aligned}
$$

$$
\begin{aligned}
N_2 &= 4c^2\sinh^2\xi_0\Bigg\{4\nu(p_r-p_z) + 4\left[\nu p_r + (\nu-1)p_z\right]\cosh 2\xi_0 \\
&\quad + 8\left[(1-\nu)p_r - \nu p_z\right]\sinh^2\xi_0\cosh\xi_0 \log\left(\coth\frac{\xi_0}{2}\right)\Bigg\}, \\
N_3 &= -4c\sinh^2\xi_0\left\{2p_r - 3p_z + (2p_r+p_z)\left[\cosh 2\xi_0 + 2\sinh^2\xi_0\cosh\xi_0 \log\left(\tanh\frac{\xi_0}{2}\right)\right]\right\}, \\
D &= 2\left[2\nu - 5 + 4\cosh 2\xi_0 + (1-2\nu)\cosh 4\xi_0\right]\log\left(\coth\frac{\xi_0}{2}\right) \\
&\quad + 8\cosh\xi_0\left[\nu + (\nu-2)\cosh 2\xi_0 + 2(1+\nu)\sinh^4\xi_0 \log^2\left(\coth\frac{\xi_0}{2}\right)\right].
\end{aligned}
$$

In the limit as $\xi_0 \to 0$, the constants A, B and H all tend to zero; thus, the limiting process to obtain the solution for a line crack of length $2a$ along the z-axis does not seem to work.

For the oblate spheroidal cavity, the semi-major and semi-minor axes are given by $a = c\cosh\xi_0$ and $b = c\sinh\xi_0$ (see Fig. 6.74), so that $c^2 = a^2 - b^2$. We assume one part of the solution to be given by Eqn. (6.263) with the harmonic potentials given by (see Eqn. (6.282))

$$
\begin{aligned}
\alpha &= AQ_0(i\sinh\xi)P_0(\cos\eta) + BQ_2(i\sinh\xi)P_2(\cos\eta), \\
\beta &= CQ_1(i\sin\xi)P_1(\cos\eta),
\end{aligned}
$$

where the functions Q_0, Q_1 and Q_2 are

$$
\begin{aligned}
Q_0 &= \cot^{-1}\sinh\xi, \\
Q_1 &= \sinh\xi(\cot^{-1}\sinh\xi) - 1, \\
Q_2 &= \frac{1}{2}\left[(3\sinh^2\xi + 1)(\cot^{-1}\sinh\xi) - 3\sinh\xi\right].
\end{aligned}
$$

This solution is superposed on the solution given by Eqn. (6.467) with Q now given by Eqn. (G.3). The constants A, B and C are found using the boundary conditions given by Eqn. (6.468), and are given by

$$
\begin{aligned}
A &= \frac{N_1}{6D}, \\
B &= \frac{N_2}{3D}, \\
C &= \frac{N_3}{D},
\end{aligned}
$$

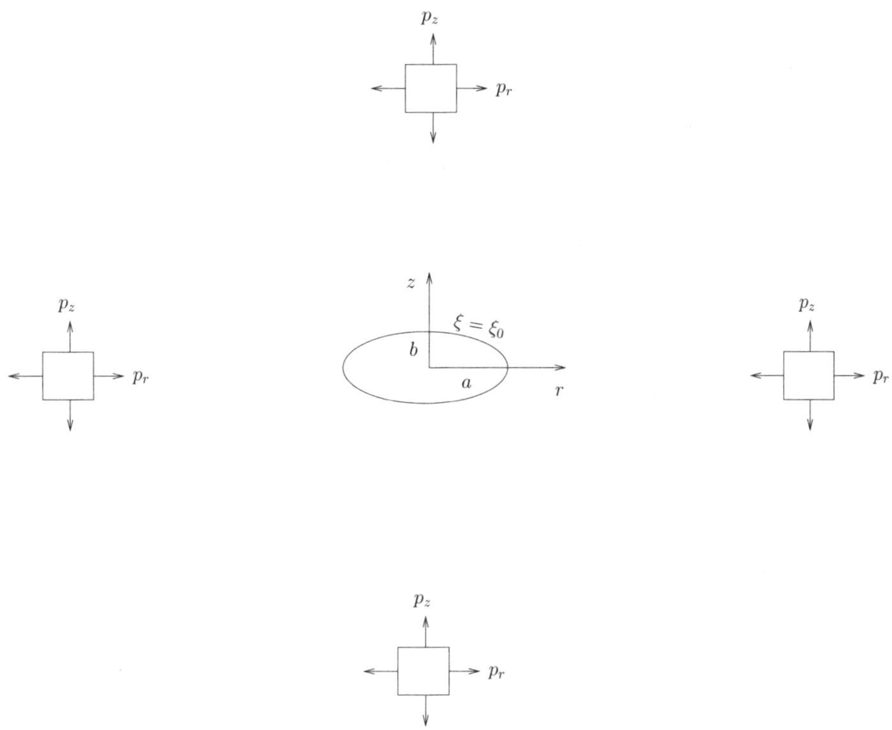

Fig. 6.74 *Oblate spheroidal cavity in an infinite domain with uniform normal stresses at infinity.*

where

$$N_1 = c^2 \cosh^2 \xi_0 \Big\{ (36 - 32\nu)p_r + (32\nu - 30)p_z + 8\left[(4\nu - 6)p_r + (4\nu - 1)p_z\right] \cosh 2\xi_0$$

$$-8 \sinh \xi_0 \cosh^2 \xi_0 (\cot^{-1} \sinh \xi_0)[2(4\nu - 7)p_r + (8\nu - 3)p_z + 3(2p_r + p_z)\cosh 2\xi_0]$$

$$+ 6(2p_r + p_z)\cosh 4\xi_0 \Big\},$$

$$N_2 = 16c^2 \cosh^2 \xi_0 \Big\{ \nu(p_z - p_r) + [\nu p_r + (\nu - 1)p_z] \cosh 2\xi_0$$

$$+ 2 \sinh \xi_0 \cosh^2 \xi_0 (\cot^{-1} \sinh \xi_0)\, [(1 - \nu)p_r - \nu p_z] \Big\},$$

$$N_3 = -4c \cosh^2 \xi_0 \Big\{ 3p_z - 2p_r + (2p_r + p_z) \left[\cosh 2\xi_0 - 2 \sinh \xi_0 \cosh^2 \xi_0 (\cot^{-1} \sinh \xi_0)\right] \Big\},$$

$$D = 2(2\nu - 5)(\cot^{-1} \sinh \xi_0) - 8 \cosh 2\xi_0 (\cot^{-1} \sinh \xi_0) + 2(1 - 2\nu)\cosh 4\xi_0 (\cot^{-1} \sinh \xi_0)$$

$$+ 2\left[4 - 6\nu + (1 + \nu)(\cot^{-1} \sinh \xi_0)^2\right] \sinh \xi_0$$

$$+ \left[4(\nu - 2) + 3(1 + \nu)(\cot^{-1} \sinh \xi_0)^2\right] \sinh 3\xi_0 + (1 + \nu)(\cot^{-1} \sinh \xi_0)^2 \sinh 5\xi_0.$$

The stress distribution in the cylindrical coordinate system is obtained by using $Q^T[\boldsymbol{\tau}]_{\text{oblate}}$ Q, with Q given by Eqn. (G.3). In the limit as $\xi_0 \to 0$, we get

$$
\begin{aligned}
A = B &= \frac{2c^2(1-2\nu)p_z}{3\pi}, \\
H &= \frac{2cp_z}{\pi},
\end{aligned}
$$

which yield the results for a penny-shaped crack of radius a lying in the plane $z = 0$. For this case, the relation $r + iz = a\sin(\eta + i\xi)$ can be used to find (ξ, η) given (r, z). In particular, to obtain the solution along the r-axis, we set $\xi = 0$ and use $a\sin\eta = r$, $a\cos\eta = \pm\sqrt{a^2 - r^2}$ for $r < a$, $z = 0^\pm$ (corresponding to the upper and lower faces of the crack), and set $\eta = \pi/2$ and use $\cosh\xi = r/a$ for $r \geq a$, $z = 0$. The summary of the final results is

$$
\left.
\begin{aligned}
2\mu u_r &= \frac{(1-\nu)\left[2p_r - (1+2\nu)p_z\right] r}{2(1+\nu)}, \\
2\mu u_z &= \pm\frac{4}{\pi}(1-\nu)p_z\sqrt{a^2 - r^2}, \\
\tau_{rr} = \tau_{\theta\theta} &= p_r - \frac{1}{2}(1+2\nu)p_z, \\
\tau_{zz} = \tau_{rz} &= 0,
\end{aligned}
\right\} \quad r < a, z = 0^\pm,
$$

$$
\left.
\begin{aligned}
u_z = \tau_{rz} &= 0, \\
2\mu u_r &= -\frac{1}{(1+\nu)\pi r}\Big\{\pi\left[(\nu-1)p_r + \nu p_z\right] r^2 \\
&\quad + (1+\nu)(2\nu-1)p_z\left[a\sqrt{r^2 - a^2} - r^2\sin^{-1}\left(\frac{a}{r}\right)\right]\Big\}, \\
\tau_{rr} &= p_r + \frac{p_z}{\pi}\Big\{\frac{a}{r^2\sqrt{r^2-a^2}}\left[a^2(1-2\nu) + (1+2\nu)r^2\right] \\
&\quad - (1+2\nu)\sin^{-1}\left(\frac{a}{r}\right)\Big\}, \\
\tau_{zz} &= \frac{2p_z}{\pi}\left[\frac{a}{\sqrt{r^2-a^2}} + \cos^{-1}\left(\frac{a}{r}\right)\right], \\
\tau_{\theta\theta} &= p_r + \frac{p_z}{\pi r^2}\Big\{(1-2\nu)a\sqrt{r^2-a^2} + \frac{4\nu a r^2}{\sqrt{r^2-a^2}} \\
&\quad - (1+2\nu)r^2\sin^{-1}\left(\frac{a}{r}\right)\Big\},
\end{aligned}
\right\} \quad r \geq a, z = 0. \tag{6.469}
$$

Note the singularity in the stress solution as $r \to a$.

Along the z-axis, using the fact that $\sinh\xi = |z|/a$, we get

$$\left.\begin{aligned}
u_r &= \tau_{rz} = 0,\\
2\mu u_z &= \frac{\operatorname{sgn}(z)}{(1+\nu)\pi(a^2+z^2)}\left\{4a^3(1-\nu^2)p_z + (p_z - 2\nu p_r)\pi|z|(a^2+z^2)\right.\\
&\quad \left.+\,2(1+\nu)(1-2\nu)p_z\left[az^2 - |z|(a^2+z^2)\cot^{-1}\left(\frac{|z|}{a}\right)\right]\right\},\\
\tau_{rr} &= \tau_{\theta\theta} = p_r + \frac{a|z|p_z}{\pi(a^2+z^2)^2}\left[(3+2\nu)a^2 + (1+2\nu)z^2\right]\\
&\quad - \frac{(1+2\nu)p_z}{\pi}\cot^{-1}\left(\frac{|z|}{a}\right),\\
\tau_{zz} &= p_z\left[1 - \frac{2a|z|(a^2-z^2)}{\pi(a^2+z^2)^2} - \frac{2}{\pi}\cot^{-1}\left(\frac{|z|}{a}\right)\right],
\end{aligned}\right\}\ r = 0.$$

6.6.19 Point load acting tangential to the boundary of an infinite half-space (Cerruti problem)

The solution for the point load P acting at the origin tangential to the boundary of a half-space along the x-direction as shown in Fig. 6.75 is generated by the Galerkin vector (compare with Eqn. (6.448))

$$\begin{aligned}
\Psi_x &= Ar + B[r - z\log(r+z)],\\
\Psi_y &= 0,\\
\Psi_z &= Cx\log(r+z),
\end{aligned} \tag{6.470}$$

where $r = \sqrt{x^2+y^2+z^2}$, or, alternatively, by the Papkovich–Neuber potentials (see Eqn. (6.256))

$$\begin{aligned}
\alpha &= -\frac{(B+C)x}{r+z},\\
\beta_x &= -\frac{A}{r},\\
\beta_y &= 0,\\
\beta_z &= -\frac{Cx}{r(r+z)}.
\end{aligned}$$

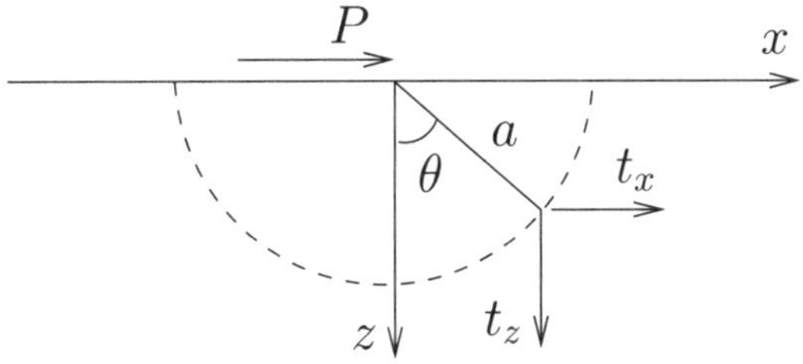

Fig. 6.75 *Point load acting tangential to the boundary of a half-space.*

Quite interestingly, instead of the Galerkin vector given by Eqn. (6.470), one can also use the vector

$$\Psi_x = Ar,$$
$$\Psi_y = 0,$$
$$\Psi_z = Cx\log(r+z) + Dx\left[\log(r+z) + \frac{z}{r+z}\right],$$

where, now, A, C and D are the constants to be determined (D turns out to be $B/2$).

Substituting Eqns. (6.470) into Eqn. (6.253a), we get

$$\begin{aligned}
u_x &= \frac{1}{2\mu}\left\{A\left[\frac{3-4\nu}{r} + \frac{x^2}{r^3}\right] + B\left[-\frac{1}{r+z} + \frac{x^2}{r(r+z)^2}\right] + C\left[-\frac{1}{r} + \frac{x^2}{r^3}\right]\right\},\\
u_y &= \frac{1}{2\mu}\left[A\frac{xy}{r^3} + B\frac{xy}{r(r+z)^2} + C\frac{xy}{r^3}\right],\\
u_z &= \frac{A}{2\mu}\frac{xz}{r^3} + \frac{B}{2\mu}\frac{x}{r(r+z)} + \frac{(1-\nu)C}{\mu}\left[\frac{2x}{r(r+z)} + \frac{xz}{2(1-\nu)r^3}\right].
\end{aligned} \tag{6.471}$$

Finding the stresses corresponding to the above displacement field, and imposing the boundary conditions,

$$\tau_{xz}\big|_{\substack{z=0\\ r\neq 0}} = \tau_{yz}\big|_{\substack{z=0\\ r\neq 0}} = \tau_{zz}\big|_{\substack{z=0\\ r\neq 0}} = 0,$$

we get the relations

$$\begin{aligned}
B &= 2C(\nu-1),\\
A &= \frac{1}{1-2\nu}\left[B + C(3-2\nu)\right].
\end{aligned} \tag{6.472}$$

The third equation is obtained by considering the equilibrium of the hemisphere shown in Fig. 6.75, which leads to the conditions $\int_S \boldsymbol{t}\,dS = -P\boldsymbol{e}_x$ and $\int_S \boldsymbol{x}\times\boldsymbol{t}\,dS = \boldsymbol{0}$, with $\boldsymbol{x}$ given by Eqn. (C.1). In expanded form, these conditions read

$$\int_0^{\pi/2}\int_0^{2\pi}\begin{bmatrix} t_x \\ t_y \\ t_z \\ a\sin\theta\sin\phi t_z - a\cos\theta t_y \\ a\cos\theta t_x - a\sin\theta\cos\phi t_z \\ a\sin\theta\cos\phi t_y - a\sin\theta\sin\phi t_x \end{bmatrix}_{r=a} a^2\sin\theta\, d\phi d\theta = \begin{bmatrix} -P \\ 0 \\ 0 \\ 0 \\ 0 \\ 0 \end{bmatrix}, \tag{6.473}$$

where

$$\begin{bmatrix} t_x \\ t_y \\ t_z \end{bmatrix} = \begin{bmatrix} \tau_{xx} & \tau_{xy} & \tau_{xz} \\ \tau_{xy} & \tau_{yy} & \tau_{yz} \\ \tau_{xz} & \tau_{yz} & \tau_{zz} \end{bmatrix}\begin{bmatrix} \sin\theta\cos\phi \\ \sin\theta\sin\phi \\ \cos\theta \end{bmatrix}.$$

Note that in the above equation, the stress components are expressed in terms of spherical coordinates (again by using Eqn. (C.1)), and that the expression for the normal has been obtained by using $[\boldsymbol{n}]_{\text{Cartesian}} = \boldsymbol{Q}^T \begin{bmatrix} 1 \\ 0 \\ 0 \end{bmatrix}$, with $\boldsymbol{Q}$ given by Eqn. (C.2). One finds that the last five equations in Eqn. (6.473) are identically satisfied, while the first one yields the third required equation. Solving these three equations, we get

$$A = \frac{P}{4\pi(1-\nu)}, \quad B = -2(1-2\nu)\frac{P}{4\pi}, \quad C = \frac{P(1-2\nu)}{4\pi(1-\nu)}.$$

For the sake of completeness, we list the entire solution:

$$\begin{aligned}
u_x &= \frac{P}{4\pi\mu}\left\{\frac{1}{r} + \frac{x^2}{r^3} + (1-2\nu)\left[\frac{1}{r+z} - \frac{x^2}{r(r+z)^2}\right]\right\},\\
u_y &= \frac{P}{4\pi\mu}\left[\frac{xy}{r^3} - \frac{(1-2\nu)xy}{r(r+z)^2}\right],\\
u_z &= \frac{P}{4\pi\mu}\left[\frac{xz}{r^3} + \frac{(1-2\nu)x}{r(r+z)}\right],\\
\tau_{xx} &= \frac{Px}{2\pi r^3}\left\{-\frac{3x^2}{r^2} + \frac{(1-2\nu)}{(r+z)^2}\left[r^2 - y^2 - \frac{2ry^2}{r+z}\right]\right\},\\
\tau_{yy} &= \frac{Px}{2\pi r^3}\left\{-\frac{3y^2}{r^2} + \frac{(1-2\nu)}{(r+z)^2}\left[3r^2 - x^2 - \frac{2rx^2}{r+z}\right]\right\},\\
\tau_{zz} &= -\frac{3P}{2\pi}\frac{xz^2}{r^5},\\
\tau_{xy} &= \frac{Py}{2\pi r^3}\left\{-\frac{3x^2}{r^2} + \frac{(1-2\nu)}{(r+z)^2}\left[-r^2 + x^2 + \frac{2rx^2}{r+z}\right]\right\},\\
\tau_{yz} &= -\frac{3P}{2\pi}\frac{xyz}{r^5},\\
\tau_{xz} &= -\frac{3P}{2\pi}\frac{x^2 z}{r^5}.
\end{aligned}$$

6.7 Elastodynamics

6.7.1 Progressive waves

We look for a solution of the type

$$\boldsymbol{u}(\boldsymbol{X},t) = \hat{\boldsymbol{a}}\sin(\boldsymbol{p}\cdot\boldsymbol{X} - ct), \quad |\boldsymbol{p}| = 1, \tag{6.474}$$

to the Navier equations given by Eqn. (6.37) (which hold for a homogeneous, isotropic body), when body forces are absent, i.e., we find conditions under which Eqn. (6.474), which is referred to as a sinusoidal progressive wave, solves

$$\rho_0 \frac{\partial^2 \boldsymbol{u}}{\partial t^2} = (\lambda+\mu)\boldsymbol{\nabla}_X(\boldsymbol{\nabla}_X\cdot\boldsymbol{u}) + \mu\boldsymbol{\nabla}_X^2\boldsymbol{u}. \tag{6.475}$$

$\hat{a}$, p and c denote the amplitude, direction and velocity of the wave, respectively. The wave is said to be longitudinal if $\hat{a}$ and p are parallel, and transverse if $\hat{a}$ and p are perpendicular. Evaluating the various terms in Eqn. (6.475), we get

$$\begin{aligned}\frac{\partial^2 u}{\partial t^2} &= -c^2 \hat{a} \sin(p \cdot X - ct),\\ \nabla \cdot u &= (\hat{a} \cdot p) \cos(p \cdot X - ct),\\ \nabla(\nabla \cdot u) &= -(\hat{a} \cdot p) p \sin(p \cdot X - ct),\\ \nabla^2 u &= -\hat{a} \sin(p \cdot X - ct). \qquad (\text{since } p \cdot p = 1)\end{aligned}$$

Substituting these quantities into Eqn. (6.475), and using the fact that for some X and t, $\sin(p \cdot X - ct)$ is nonzero, we get the necessary and sufficient condition that u given by Eqn. (6.474) is a solution of Eqn. (6.475) as

$$A(p)\hat{a} = c^2 \hat{a}, \tag{6.476}$$

where $A(p)$, known as the acoustic tensor, is defined by

$$A(p) := \frac{1}{\rho_0} \left[\mu I + (\lambda + \mu) p \otimes p \right].$$

Equation (6.476) is simply an eigenvalue problem with c^2 an eigenvalue, and $\hat{a}$ an eigenvector of the symmetric second-order tensor $A(p)$. Rewriting the above expression for $A(p)$ as

$$A(p) = \frac{\lambda + 2\mu}{\rho_0} p \otimes p + \frac{\mu}{\rho_0} (I - p \otimes p),$$

and comparing with Eqn. (1.123), we see that the eigenvalues of $A(p)$ are $(\lambda + 2\mu)/\rho_0$ and μ/ρ_0, with the corresponding eigenvectors given by p, and any vector in the plane perpendicular to p. Thus, the sinusoidal progressive wave given by Eqn. (6.474) is a solution of Eqn. (6.475) if and only if

1. $c^2 = (\lambda + 2\mu)/\rho_0 =: c_L^2$ and $\hat{a}$ is parallel to p (Longitudinal wave), or
2. $c^2 = \mu/\rho_0 =: c_T^2$ and $\hat{a}$ is perpendicular to p (Transverse wave).

Note from the above result that for a homogeneous, isotropic material, only two types of sinusoidal waves, longitudinal and transverse, are possible. Since by virtue of Eqns. (5.16) and (5.17), $\lambda + 2\mu > 0$ and $\mu > 0$, the longitudinal and transverse wave speeds c_L and c_T are real.

For an anisotropic material, the governing equation in place of Eqn. (6.475) is now

$$\rho_0 \frac{\partial^2 u}{\partial t^2} = \nabla_X \cdot [\mathsf{C}[\nabla u]]. \tag{6.477}$$

Assuming C to be independent of position, the above equation in indicial notation reads

$$\rho_0 \frac{\partial^2 u_i}{\partial t^2} = \mathsf{C}_{ijkl} \frac{\partial^2 u_k}{\partial X_j \partial X_l}. \tag{6.478}$$

Assuming the solution form to be still given by Eqn. (6.474), and substituting it into Eqn. (6.478), we get

$$\rho_0 c^2 \hat{a}_i = \mathsf{C}_{ijkl} p_j a_k p_l, \tag{6.479}$$

which in tensorial form reads

$$\rho_0 c^2 \hat{\boldsymbol{a}} = \mathsf{C}[\hat{\boldsymbol{a}} \otimes \boldsymbol{p}]\boldsymbol{p}.$$

Defining the acoustic tensor $A_{ik}(\boldsymbol{p}) := \frac{1}{\rho_0}\mathsf{C}_{ijkl} p_j p_l$, we get the same propagation condition given by Eqn. (6.476). However, the waves are neither longitudinal or transverse. The acoustic tensor is positive definite for all directions $\boldsymbol{p}$ (i.e., $\boldsymbol{q} \cdot \boldsymbol{A}(\boldsymbol{p})\boldsymbol{q} > 0$ for all $\boldsymbol{p}, \boldsymbol{q}$) if and only if C is strongly elliptic (see Problem 4 in Chapter 5). For the isotropic case, we have $\mathsf{C} = \lambda \boldsymbol{I} \otimes \boldsymbol{I} + 2\mu\mathsf{S}$, so that

$$\begin{aligned}
[\boldsymbol{p} \otimes \boldsymbol{q}] : \mathsf{C}[\boldsymbol{p} \otimes \boldsymbol{q}] &= [\boldsymbol{p} \otimes \boldsymbol{q}] : [\lambda(\boldsymbol{p} \cdot \boldsymbol{q})\boldsymbol{I} + 2\mu(\boldsymbol{p} \otimes \boldsymbol{q} + \boldsymbol{q} \otimes \boldsymbol{p})] \\
&= (\lambda + 2\mu)(\boldsymbol{p} \cdot \boldsymbol{q})^2 + 2\mu(\boldsymbol{p} \cdot \boldsymbol{p})(\boldsymbol{q} \cdot \boldsymbol{q}).
\end{aligned}$$

By making appropriate choices of $\boldsymbol{p}$ and $\boldsymbol{q}$, we deduce that strong ellipticity holds in the isotropic case if and only if $\lambda + 2\mu > 0$ and $\mu > 0$.

It can be shown rather easily that if ϕ and $\boldsymbol{f}$ are scalar and vector fields that satisfy

$$\boldsymbol{\nabla}^2 \phi = \frac{1}{c_L^2}\frac{\partial^2 \phi}{\partial t^2}, \tag{6.480a}$$

$$\boldsymbol{\nabla}^2 \boldsymbol{f} = \frac{1}{c_T^2}\frac{\partial^2 \boldsymbol{f}}{\partial t^2}, \tag{6.480b}$$

then the displacement field given by

$$\boldsymbol{u} = \boldsymbol{\nabla}\phi + \boldsymbol{\nabla} \times \boldsymbol{f}, \tag{6.481}$$

satisfies the Navier equations of elasticity given by

$$\frac{\partial^2 \boldsymbol{u}}{\partial t^2} = c_L^2 \boldsymbol{\nabla}(\boldsymbol{\nabla} \cdot \boldsymbol{u}) - c_T^2 \boldsymbol{\nabla} \times (\boldsymbol{\nabla} \times \boldsymbol{u}),$$

where we have used Eqn. (1.271b) to obtain the above form from Eqn. (6.475). Conversely, by the Clebsch completeness theorem [308], if $\boldsymbol{u}$ satisfies the Navier equations of elasticity, then there exist scalar and vector fields ϕ and $\boldsymbol{f}$ that satisfy Eqns. (6.480), and such that Eqn. (6.481) holds.

6.7.2 Solution to special problems

To gain some insight into wave phenomena, we now solve a number of one-dimensional problems using the method of Laplace transforms. For a numerical strategy to solve linear elastodynamics problems, see [167].

One-dimensional wave propagation problem in an elastic solid

We assume that $\nu = 0$, so that $\lambda + 2\mu = E$. Consider the setup in Fig. 6.76 that is unbounded in the y and z directions. From symmetry considerations, it is clear that only the displacement component u_x is nonzero, and that it is a function of (x, t). Thus, the Navier equations reduce to

$$\frac{\partial^2 u}{\partial x^2} = \frac{1}{c^2}\frac{\partial^2 u}{\partial t^2}, \tag{6.482}$$

where $c \equiv c_L = \sqrt{E/\rho_0}$. The longitudinal stress is $\tau_{xx} = \rho_0 c^2 \partial u/\partial x$. We assume that the initial displacement and velocity are both zero.

First consider the case when $h \to \infty$. Taking the Laplace transform of Eqn. (6.482), we get

$$\frac{\partial^2 \bar{u}}{\partial x^2} - \tilde{q}^2 \bar{u} = 0, \tag{6.483}$$

where $\tilde{q} = s/c$. The general solution of this equation is $\bar{u}(x,s) = A(s)e^{\tilde{q}x} + B(s)e^{-\tilde{q}x}$. In order for the displacement solution to stay bounded as $x \to \infty$, we set $A(s)$ to zero. Using the boundary condition that $\bar{\tau}_{xx}(0,s) = \bar{T}(s)$, we get

$$\bar{u}(x,s) = -\frac{\bar{T}(s)}{\rho_0 c s} e^{-xs/c}.$$

Inverting the above relation using the convolution theorem, and using the constitutive relation, we get

$$\begin{aligned} u(x,t) &= -\frac{1}{\rho_0 c} H\left(t - \frac{x}{c}\right) \int_0^{t-\frac{x}{c}} T(\tau)\, d\tau \\ &= \begin{cases} 0 & \text{for } t < x/c \\ -\frac{1}{\rho_0 c}\int_0^{t-\frac{x}{c}} T(\tau)\, d\tau & \text{for } t \geq x/c, \end{cases} \end{aligned}$$

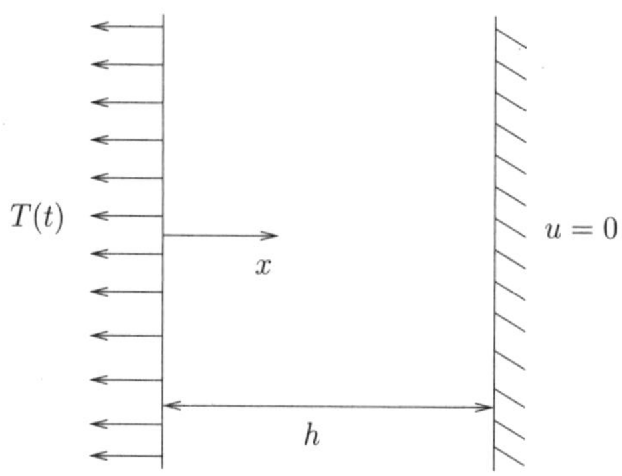

Fig. 6.76 *Elastic solid of thickness h subjected to a uniform time-varying traction at one end, and fixed at the other.*

$$v(x,t) = -\frac{1}{\rho_0 c} H\left(t - \frac{x}{c}\right) T\left(t - \frac{x}{c}\right),$$
$$\tau_{xx}(x,t) = H\left(t - \frac{x}{c}\right) T\left(t - \frac{x}{c}\right),$$

where $H(.)$ denotes the Heaviside function. Thus, at a point x_0, the displacement and stress remain zero for times less than x_0/c, or, in other words, a disturbance propagates at finite speed.

Now consider the case when h is finite. Instead of Eqn. (6.483), we now use

$$\frac{\partial^2 \bar{u}}{\partial x^2} + q^2 \bar{u} = 0,$$

where $q^2 = -s^2/c^2$. The general solution of this equation is $\bar{u}(x,s) = A(s) \sin qx + B(s) \cos qx$. Using the boundary conditions $\bar{\tau}_{xx}(0,s) = \bar{T}(s)$ and $\bar{u}(h,s) = 0$, we get

$$\bar{u}(x,s) = \frac{\sin q(x-h)}{\rho_0 c^2 q \cos qh} \bar{T}(s).$$

The roots of $\cos qh = 0$ are $qh = (2n-1)\pi/2$, $n = 1, 2, \ldots, \infty$. Thus, with a view towards using the convolution theorem, we write

$$\frac{\sin q(x-h)}{\rho_0 c^2 q \cos qh} = \sum_{n=1}^{\infty} \frac{c_n}{s^2 + \lambda_n^2 c^2},$$

where $\lambda_n := (2n-1)\pi/(2h)$. Using the Heaviside formula (see Appendix F in Vol. II), we get

$$c_n = -\frac{2}{\rho_0 h} \cos \lambda_n x.$$

Thus, on applying the convolution theorem, we get

$$\begin{aligned} u(x,t) &= -\frac{4}{\rho_0 c \pi} \sum_{n=1}^{\infty} \frac{\cos \lambda_n x}{2n-1} \int_0^t \sin(\lambda_n c \tau) T(t-\tau)\, d\tau \\ &= -\frac{4}{\rho_0 c \pi} \sum_{n=1}^{\infty} \frac{\cos \lambda_n x}{2n-1} \int_0^t \sin \lambda_n c(t-\tau) T(\tau)\, d\tau. \end{aligned} \tag{6.484}$$

The velocity and stress are found using the relations $v = \partial u/\partial t$ and $\tau_{xx} = \rho_0 c^2 \partial u/\partial x$.

As an example, first consider the case when a constant stress T is impulsively applied at the left wall, i.e., $T(t) = T$. Then, using the above solution, we get

$$u(x,t) = -\frac{Th}{\rho_0 c^2} \left[1 - \frac{x}{h} - \sum_{n=1}^{\infty} \frac{8 \cos \lambda_n x \cos \lambda_n ct}{(2n-1)^2 \pi^2}\right],$$
$$\tau_{xx}(x,t) = T \left[1 - \sum_{n=1}^{\infty} \frac{4 \sin \lambda_n x \cos \lambda_n ct}{(2n-1)\pi}\right],$$

where we have used the relation

$$\sum_{n=1}^{\infty} \frac{8\cos\lambda_n x}{(2n-1)^2\pi^2} = 1 - \frac{x}{h}.$$

As another example, consider the case when $T(t) = T\sin\omega t$. We get (with $k \equiv \omega/c$)

$$u(x,t) = -\frac{T}{\rho_0 c\omega}\left[\frac{\sin k(h-x)}{\cos kh}\sin\omega t - \sum_{n=1}^{\infty}\frac{4\cos\lambda_n x \sin\lambda_n ct}{(2n-1)\pi\left[\left(\frac{\lambda_n}{k}\right)^2 - 1\right]}\right],$$

$$\tau_{xx}(x,t) = T\left[\frac{\cos k(h-x)}{\cos kh}\sin\omega t - \frac{2}{kh}\sum_{n=1}^{\infty}\frac{\sin\lambda_n x \sin\lambda_n ct}{\left(\frac{\lambda_n}{k}\right)^2 - 1}\right],$$

where we have used the fact that

$$\sum_{n=1}^{\infty}\frac{\cos\lambda_n x}{\left[\left(\frac{\lambda_n}{k}\right)^2 - 1\right]} = \frac{kh\sin k(h-x)}{2\cos kh}.$$

For the data $\lambda = 0$, $c = \rho_0 = 2\mu = T = 1$, $\omega = 3\pi$ and $h = 4$, we get the displacement and stress responses shown in Fig. 6.77. Note the doubling of the stress at the right end at $t = 5.5$ (after reflection of the wave has occurred) in order to maintain zero displacement at that end.

Similarly, for the case when $T(t) = T\cos\omega t$, we get

$$u(x,t) = -\frac{T}{\rho_0 c\omega}\left[\frac{\sin k(h-x)}{\cos kh}\cos\omega t - \frac{2}{kh}\sum_{n=1}^{\infty}\frac{\cos\lambda_n x \cos\lambda_n ct}{\left[\left(\frac{\lambda_n}{k}\right)^2 - 1\right]}\right],$$

$$\tau_{xx}(x,t) = T\left[\frac{\cos k(h-x)}{\cos kh}\cos\omega t - \frac{2}{k^2h}\sum_{n=1}^{\infty}\frac{\lambda_n \sin\lambda_n x \cos\lambda_n ct}{\left(\frac{\lambda_n}{k}\right)^2 - 1}\right],$$

while for the case $T(t) = Te^{-\omega t}$, we get

$$u(x,t) = -\frac{T}{\rho_0 c\omega}\left[\frac{\sinh k(h-x)}{\cosh kh}e^{-\omega t} - 4\sum_{n=1}^{\infty}\frac{\left[\frac{\lambda_n}{k}\cos\lambda_n ct - \sin\lambda_n ct\right]\cos\lambda_n x}{(2n-1)\pi\left[\left(\frac{\lambda_n}{k}\right)^2 + 1\right]}\right],$$

$$\tau_{xx}(x,t) = T\left[\frac{\cosh k(h-x)}{\cosh kh}e^{-\omega t} - \frac{2}{kh}\sum_{n=1}^{\infty}\frac{\left[\frac{\lambda_n}{k}\cos\lambda_n ct - \sin\lambda_n ct\right]\sin\lambda_n x}{\left(\frac{\lambda_n}{k}\right)^2 + 1}\right],$$

where we have used the fact that

$$\sum_{n=1}^{\infty}\frac{\cos\lambda_n x}{\left[\left(\frac{\lambda_n}{k}\right)^2 + 1\right]} = \frac{kh\sinh k(h-x)}{2\cosh kh}.$$

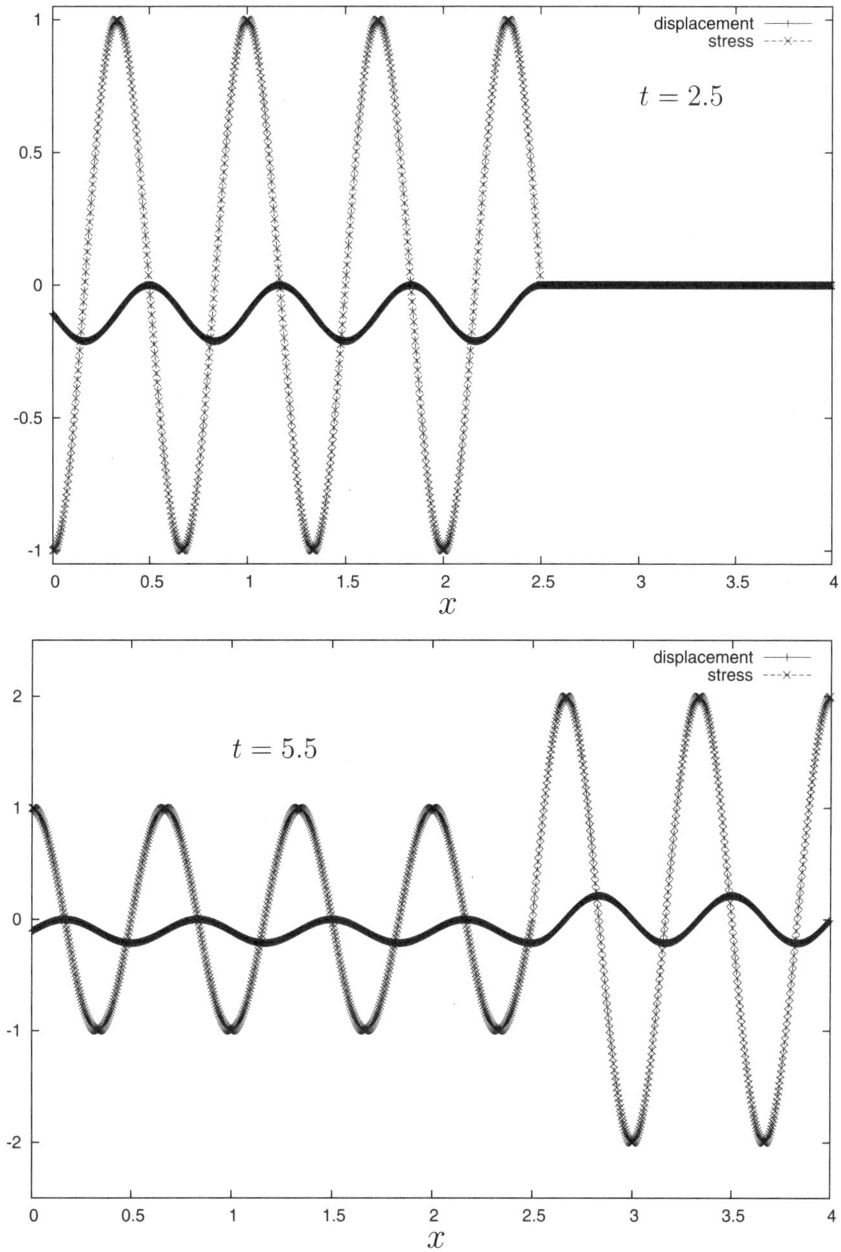

Fig. 6.77 *Displacement and stress plots for the case $T = \sin 3\pi t$ at times $t = 2.5$ and $t = 5.5$.*

While considering the case, $T(t) = T\sin\omega t$, we implicitly assumed that $k \neq \lambda_m$ for any integer m. If, however, $k = \lambda_m$, i.e., $\omega = c\lambda_m$, then resonance occurs. We now find the solution for the case $\omega = c\lambda_m$. Using Eqn. (6.484), we get

$$u = -\frac{T}{\rho_0 c\omega\pi}\left[\frac{g_1(\xi)\sin\omega t}{2m-1} + \frac{2(\sin\omega t - \omega t\cos\omega t)\cos\lambda_m x}{2m-1} - \sum_{\substack{n=1\\ n\neq m}}^{\infty}\frac{4\cos\lambda_n x\sin\lambda_n ct}{(2n-1)\left[\left(\frac{2n-1}{2m-1}\right)^2 - 1\right]}\right],$$

$$\tau_{xx} = \frac{T}{(2m-1)\pi}\left[\frac{g_2(\xi)\sin\omega t}{2m-1} + 2(\sin\omega t - \omega t\cos\omega t)\sin\lambda_m x - \sum_{\substack{n=1\\ n\neq m}}^{\infty}\frac{4\sin\lambda_n x\sin\lambda_n ct}{\left(\frac{2n-1}{2m-1}\right)^2 - 1}\right],$$

where, with $\xi := x/h$,

$$g_1(\xi) = \cos\frac{(2m-1)\pi\xi}{2} - (1-\xi)(2m-1)\pi\sin\frac{(2m-1)\pi\xi}{2},$$

$$g_2(\xi) = (1-\xi)(2m-1)^2\pi\cos\frac{(2m-1)\pi\xi}{2} - (2m-1)\sin\frac{(2m-1)\pi\xi}{2},$$

and where we have used

$$\sum_{\substack{n=1\\ n\neq m}}^{\infty}\frac{4\cos\frac{(2n-1)\pi\xi}{2}}{\left(\frac{2n-1}{2m-1}\right)^2 - 1} = g_1(\xi),$$

$$\sum_{\substack{n=1\\ n\neq m}}^{\infty}\frac{4(2n-1)\sin\frac{(2n-1)\pi\xi}{2}}{\left(\frac{2n-1}{2m-1}\right)^2 - 1} = g_2(\xi), \quad \xi \in (0,1].$$

Similarly, for the case $T(t) = T\cos\omega t$, where $\omega = c\lambda_m$, we get

$$u = -\frac{T}{\rho_0 c\omega(2m-1)\pi}\left[g_1(\xi)\cos\omega t + 2\omega t\sin\omega t\cos\lambda_m x - \sum_{\substack{n=1\\ n\neq m}}^{\infty}\frac{4\cos\lambda_n x\cos\lambda_n ct}{\left(\frac{2n-1}{2m-1}\right)^2 - 1}\right],$$

$$\tau_{xx} = \frac{T}{(2m-1)^2\pi}\left[g_2(\xi)\cos\omega t + 2(2m-1)\omega t\sin\omega t\sin\lambda_m x \right.$$
$$\left. - \sum_{\substack{n=1\\ n\neq m}}^{\infty}\frac{4(2n-1)\sin\lambda_n x\cos\lambda_n ct}{\left(\frac{2n-1}{2m-1}\right)^2 - 1}\right].$$

The function $g_1(\xi)$ when $T(t) = T\sin\omega t$ is determined as follows. First, the general solution for $g(\xi)$ is found so that the function

$$\tilde{u} = g(\xi)\sin\omega t + \frac{(\sin\omega t - \omega t\cos\omega t)\cos\lambda_m x}{2\omega(2m-1)}$$

satisfies the wave equation $\partial^2\tilde{u}/\partial x^2 - (1/c^2)\partial^2\tilde{u}/\partial t^2 = 0$. Then the two integration constants in the solution for $g(\xi)$ are found using the conditions $\int_0^h g(\xi)\cos\lambda_m x\,dx = 0$ and $g(\xi)|_{\xi=1} = 0$.

Viscoelastic bar impacting a rigid wall

As in the previous sub-section, we assume that $\nu = 0$, so that $\lambda + 2\mu = E$. Consider a bar of length h (which we again approximate as a one-dimensional continuum) traveling with velocity V_0 impacting a rigid wall as shown in Fig. 6.78. We now assume the material

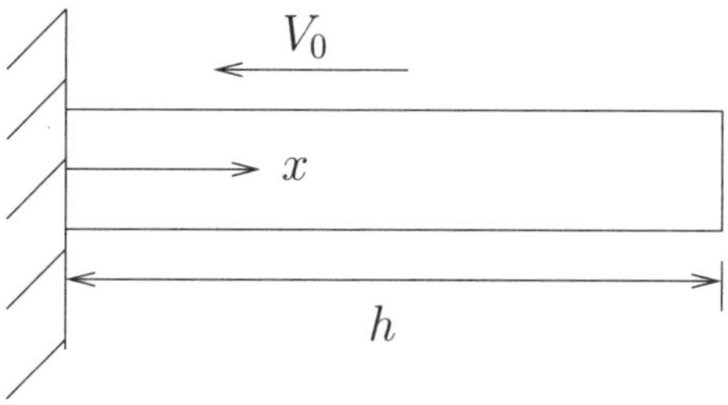

Fig. 6.78 *Elastic bar of length h impacting a rigid wall with velocity V_0.*

to be viscoelastic with the constitutive relation give by the Kelvin–Voight model, namely, $\tau_{xx} = E(\partial u/\partial x + \alpha \partial \dot{u}/\partial x)$, $\alpha \geq 0$. Note that the elastic model is recovered simply by setting α to zero. Instead of Eqn. (6.482), we now have

$$\frac{\partial^2 u}{\partial x^2} + \alpha \frac{\partial^2 \dot{u}}{\partial x^2} = \frac{1}{c^2}\frac{\partial^2 u}{\partial t^2}. \tag{6.485}$$

The time $t = 0$ is assumed to be the instant when the left end of the bar just touches the wall. Taking the Laplace transform of Eqn. (6.485), we get

$$\frac{\partial^2 \bar{u}}{\partial x^2} + q^2 \bar{u} = \frac{V_0}{c^2(1+\alpha s)},$$

where $q^2 = -s^2/[c^2(1+\alpha s)]$. The general solution of this equation is

$$\bar{u}(x,s) = A \sin qx + B \cos qx - \frac{V_0}{s^2}. \tag{6.486}$$

Using the boundary conditions $\bar{u}|_{x=0} = 0$ and $\partial \bar{u}/\partial x|_{x=h} = 0$, we get

$$\bar{u} = -\frac{V_0}{s^2} + \frac{V_0 \cos q(x-h)}{s^2 \cos qh}. \tag{6.487}$$

Since the roots of $\cos qh = 0$, are $q = (2n-1)\pi/(2h)$, $n = 1, 2, \ldots, \infty$, we write the above equation as

$$\bar{u} = -\frac{V_0}{s^2} + \frac{A}{s^2} + \frac{B}{s} + \sum_{n=1}^{\infty} \frac{c_n}{(s+\zeta_n)^2 + \omega_n^2},$$

where

$$\zeta_n = \frac{(2n-1)^2\pi^2c^2\alpha}{8h^2},$$

$$\omega_n^2 = \frac{(2n-1)^2\pi^2c^2}{4h^2}\left[1 - \frac{(2n-1)^2\pi^2c^2\alpha^2}{16h^2}\right].$$

The constants A and B are found as

$$A = \lim_{s\to 0} \frac{V_0 \cos q(x-h)}{\cos qh} = V_0,$$

$$B = \lim_{s\to 0} \frac{d}{ds}\left[\frac{V_0 \cos q(x-h)}{\cos qh}\right] = 0.$$

In order to find c_n, we write Eqn. (6.487) as

$$\bar{u} = \sum_{n=1}^{\infty}\left[\frac{c_n^{(1)}}{s+\zeta_n+i\omega_n} + \frac{c_n^{(2)}}{s+\zeta_n-i\omega_n}\right],$$

where $i = \sqrt{-1}$. We find $c_n^{(1)}$ and $c_n^{(2)}$ using the Heaviside formula, namely,

$$c_n^{(1)} = \left.\frac{P(q)}{\frac{dQ(q)}{dq}\frac{dq}{ds}}\right|_{s=-\zeta_n-i\omega_n},$$

$$c_n^{(2)} = \left.\frac{P(q)}{\frac{dQ(q)}{dq}\frac{dq}{ds}}\right|_{s=-\zeta_n+i\omega_n},$$

where $P(q) = V_0 \cos q(x-h)$, $Q(q) = s^2 \cos qh$ and $dq/ds = -(2c^2s + \alpha c^2 s^2)/[2qc^4(1+\alpha s)^2]$. After some simplification, we find that

$$\begin{aligned} u(x,t) &= \sum_{n=1}^{\infty} e^{-\zeta_n t}\left[c_n^{(1)} e^{-i\omega_n t} + c_n^{(2)} e^{i\omega_n t}\right] \\ &= \sum_{n=1}^{\infty} -\frac{4V_0 e^{-\zeta_n t}}{(2n-1)\pi\omega_n} \sin\frac{(2n-1)\pi x}{2h} \sin\omega_n t. \end{aligned}$$

For a given $\alpha > 0$, there is a value of n beyond which ω becomes imaginary, but such that $\sin\omega t/\omega$ is still real; we call this value of n as N, and it is given by

$$N = \mathrm{Int}\left[\frac{2h}{\pi c\alpha} + \frac{1}{2}\right], \tag{6.488}$$

where $\mathrm{Int}(x)$ denotes the largest integer less than or equal to x. Thus, the final solution on using the relation $\sin\theta = (e^{i\theta} - e^{-i\theta})/(2i)$ is given by

$$\begin{aligned} u(x,t) = &-\sum_{n=1}^{N} \frac{4V_0 e^{-\zeta_n t}}{(2n-1)\pi\omega_n} \sin\frac{(2n-1)\pi x}{2h} \sin\omega_n t \\ &- \sum_{n=N+1}^{\infty} \frac{2V_0}{(2n-1)\pi\hat{\omega}_n} \sin\frac{(2n-1)\pi x}{2h}\left[e^{(-\zeta_n+\hat{\omega}_n)t} - e^{(-\zeta_n-\hat{\omega}_n)t}\right], \end{aligned}$$

$$v(x,t) = -\sum_{n=1}^{N} \frac{4V_0 e^{-\zeta_n t}}{(2n-1)\pi} \sin\frac{(2n-1)\pi x}{2h} \cos\omega_n t$$
$$+ \sum_{n=N+1}^{\infty} \frac{2V_0}{(2n-1)\pi\hat{\omega}_n} \sin\frac{(2n-1)\pi x}{2h}$$
$$\left[(\zeta_n - \hat{\omega}_n)e^{(-\zeta_n+\hat{\omega}_n)t} - (\zeta_n + \hat{\omega}_n)e^{(-\zeta_n-\hat{\omega}_n)t}\right],$$
$$\tau_{xx}(x,t) = -\sum_{n=1}^{N} \frac{2V_0\rho_0 c^2 e^{-\zeta_n t}}{\omega_n h} \cos\frac{(2n-1)\pi x}{2h} \sin\omega_n t$$
$$- \sum_{n=N+1}^{\infty} \frac{V_0\rho_0 c^2}{\hat{\omega}_n h} \cos\frac{(2n-1)\pi x}{2h} \left[e^{(-\zeta_n+\hat{\omega}_n)t} - e^{(-\zeta_n-\hat{\omega}_n)t}\right],$$

where

$$\omega_n = \frac{(2n-1)\pi c}{2h}\sqrt{1 - \frac{(2n-1)^2\pi^2c^2\alpha^2}{16h^2}},$$
$$\hat{\omega}_n = \frac{(2n-1)\pi c}{2h}\sqrt{\frac{(2n-1)^2\pi^2c^2\alpha^2}{16h^2} - 1}.$$

Of course, when $\alpha = 0$, we take N to be infinity, and exclude the second part in the expressions for $u(x,t)$ and $\tau_{xx}(x,t)$. If the value of α is such that ω_n becomes exactly zero for some $n = N$, then we replace the term $\sin(\omega_N t)/\omega_N$ by t. In case α is zero, the total energy is conserved, which leads to an expression for the stress at the wall valid in the time interval $t \in (0, 2h/c)$, and given by $\tau_w = -\rho_0 c V_0$. At time $2h/c$, the stress at the wall becomes zero, and the bar bounces off the wall with speed V_0, and in a stress-free state.

From the solution, we see that modes that are higher than N as given by Eqn. (6.488) are damped out due to the presence of damping. The stress and velocity solutions for the values $\lambda = 0$, $c = \rho_0 = 2\mu = 1$, $h = 4$, $V_0 = 10^{-3}$ and (i) $\alpha = 0$, (ii) $\alpha = 0.01$, at times $t = 3$ and $t = 7$ (after reflection of the stress wave has occurred) are shown in Figs. 6.79, 6.80 and 6.81, respectively. Note that the shock front gets diffused due to the presence of viscous dissipation. The above solution is a strong solution for $\alpha > 0$, but weak (in the sense that $\partial\tau_{xx}/\partial x$ and the acceleration are infinity at the shock front, so that the governing differential equation is satisfied only in its variational form) when $\alpha = 0$. While computing the above solution, a large number of terms are required to capture the solution in Fig. 6.79, while relatively fewer terms are required to capture the one shown in Fig. 6.80. This explains why numerical strategies (e.g., [160]) that try to capture the solution using few elements show oscillations in the undamped case, but relatively smoother solutions in the damped one.

Elastic bar impacting another elastic bar

Instead of an elastic bar impacting a rigid wall, now consider an elastic bar impacting another elastic bar whose other end is constrained by a rigid wall as shown in Fig. 6.82. Both bars are assumed to have the same cross-sectional area. Similar to Eqn. (6.486), we now have

$$\bar{u}^{(1)}(x,s) = A_1 \sin q_1 x + B_1 \cos q_1 x,$$

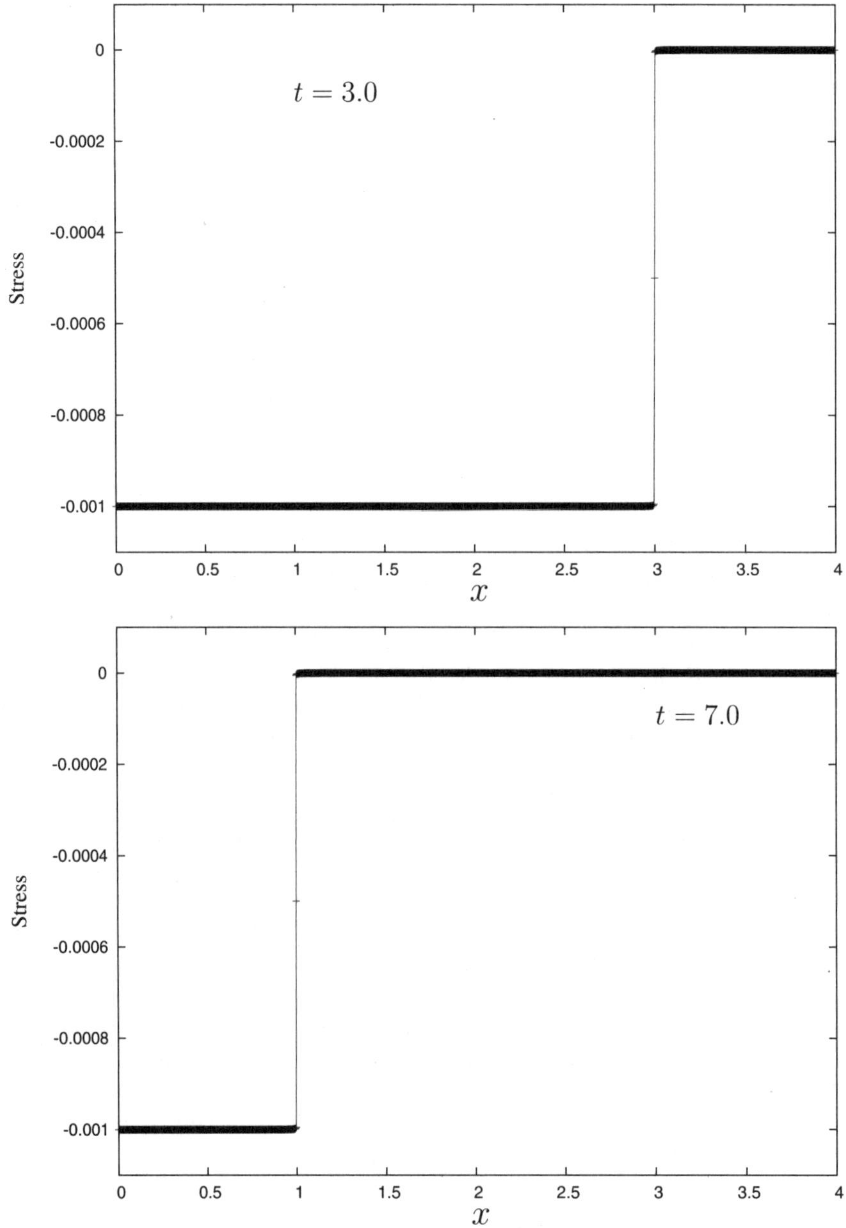

Fig. 6.79 *Stress plots for the case $\alpha = 0$ at times $t = 3$ and $t = 7$.*

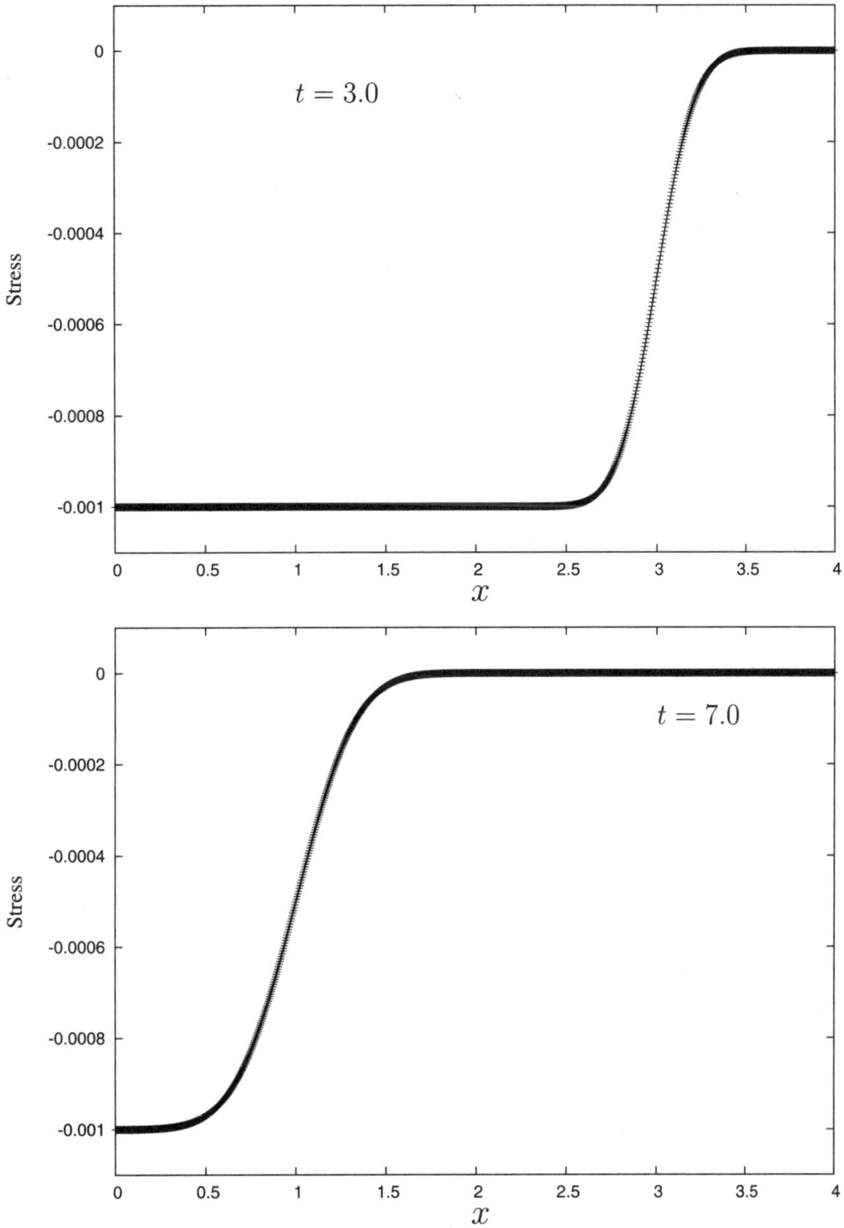

Fig. 6.80 *Stress plots for the case $\alpha = 0.01$ at times $t = 3$ and $t = 7$.*

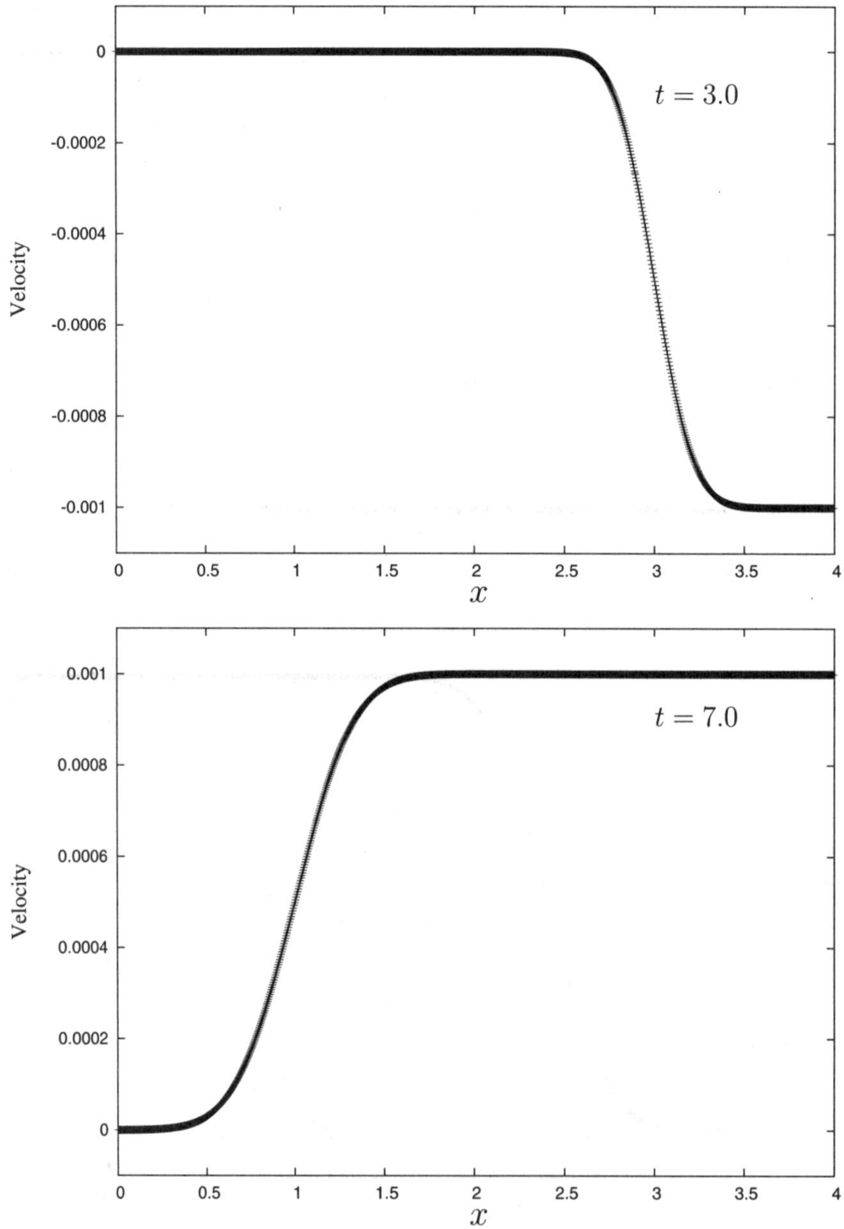

Fig. 6.81 *Velocity plots for the case* $\alpha = 0.01$ *at times* $t = 3$ *and* $t = 7$.

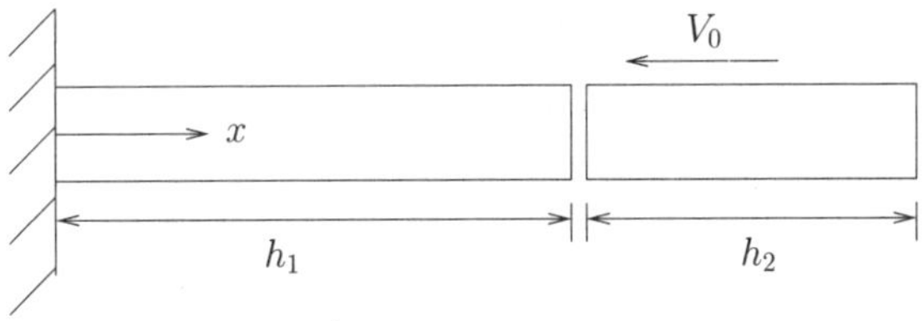

Fig. 6.82 *Elastic bar of length h_2 impacting a stationary elastic bar of length h_1 constrained by a rigid wall.*

$$\bar{u}^{(2)}(x,s) = A_2 \sin q_2 x + B_2 \cos q_2 x - \frac{V_0}{s^2},$$

as the solutions in the two bars, where $q_1^2 = -s^2/c_1^2$ and $q_2^2 = -s^2/c_2^2$. By imposing the displacement and traction continuity boundary conditions at the interface besides the displacement and traction conditions at the two ends, i.e., $\bar{u}^{(1)}(0,s) = 0$, $\bar{u}^{(1)}(h_1,s) = \bar{u}^{(2)}(h_1,s)$, $\rho_1 c_1^2 \partial \bar{u}^{(1)}(h_1,s)/\partial x = \rho_2 c_2^2 \partial \bar{u}^{(2)}(h_1,s)/\partial x$, and $\partial \bar{u}^{(2)}(h_1+h_2,s)/\partial x = 0$, we get

$$\bar{u}^{(1)}(x,s) = \frac{\rho_2 c_2 V_0 \sin(q_1 c_1 h_2/c_2)\sin(q_1 x)}{s^2\left[\rho_1 c_1 \cos(q_1 h_1)\cos(q_1 c_1 h_2/c_2) - \rho_2 c_2 \sin(q_1 h_1)\sin(q_1 c_1 h_2/c_2)\right]},$$
$$\bar{u}^{(2)}(x,s) = \frac{\rho_1 c_1 V_0 \cos(q_1 h_1)\cos(q_1 c_1 (h_1+h_2-x)/c_2)}{s^2\left[\rho_1 c_1 \cos(q_1 h_1)\cos(q_1 c_1 h_2/c_2) - \rho_2 c_2 \sin(q_1 h_1)\sin(q_1 c_1 h_2/c_2)\right]} - \frac{V_0}{s^2},$$

where we have used the relation $q_1 c_1 = q_2 c_2$. The inversion is carried out as in the previous subsection. Let λ_n, $n = 1,2,\ldots,\infty$, be the positive roots of

$$\rho_1 c_1 \cos(h_1 x)\cos\frac{c_1 h_2 x}{c_2} - \rho_2 c_2 \sin(h_1 x)\sin\frac{c_1 h_2 x}{c_2} = 0.$$

Then the solution is given by

$$\begin{aligned} u^{(1)}(x,t) &= \sum_{n=1}^{\infty} \frac{2\rho_2 c_2 V_0}{c_1 \lambda_n^2 D'} \sin\frac{\lambda_n c_1 h_2}{c_2} \sin(\lambda_n x)\sin\lambda_n c_1 t, \\ u^{(2)}(x,t) &= \sum_{n=1}^{\infty} \frac{2\rho_1 V_0}{\lambda_n^2 D'} \cos(\lambda_n h_1)\cos\frac{\lambda_n c_1 (h_1+h_2-x)}{c_2} \sin\lambda_n c_1 t, \end{aligned} \tag{6.489}$$

where

$$\begin{aligned} D' = &-\rho_1 c_1 h_1 \sin(\lambda_n h_1)\cos\frac{\lambda_n c_1 h_2}{c_2} - \frac{\rho_1 c_1^2 h_2}{c_2}\cos(\lambda_n h_1)\sin\frac{\lambda_n c_1 h_2}{c_2} \\ &- \rho_2 c_2 h_1 \cos(\lambda_n h_1)\sin\frac{\lambda_n c_1 h_2}{c_2} - \rho_2 c_1 h_2 \sin(\lambda_n h_1)\cos\frac{\lambda_n c_1 h_2}{c_2}. \end{aligned}$$

In order to find the value of the velocity V_i and stress τ_i at the interface just after impact takes place, we use the principles of balance of linear momentum, and conservation of the total (i.e., strain+kinetic) energy. Consider an instant of time t_0 just after impact when the impact wave has not yet reached the wall, so that the reaction exerted by the wall on the bar is zero. Thus, both linear momentum and energy are conserved. From the balance of linear momentum, we get

$$\rho_2 h_2 V_0 = \rho_1 V_i c_1 t_0 + \rho_2 V_i c_2 t_0 + \rho_2 V_0 (h_2 - c_2 t_0),$$

which leads to

$$V_i = \frac{\rho_2 V_0 c_2}{\rho_1 c_1 + \rho_2 c_2}.$$

From the balance of energy, we get

$$\frac{1}{2}\rho_2 h_2 V_0^2 = \frac{1}{2}\rho_1 c_1 t_0 V_i^2 + \frac{1}{2}\rho_2 c_2 t_0 V_i^2 + \frac{1}{2}\rho_2 (h_2 - c_2 t_0) V_0^2 + \frac{1}{2}\tau_i^2 t_0 \left(\frac{1}{\rho_1 c_1} + \frac{1}{\rho_2 c_2}\right),$$

which leads to

$$\tau_i = -\frac{\rho_1 \rho_2 c_1 c_2 V_0}{\rho_1 c_1 + \rho_2 c_2}.$$

If both bars are assumed to have the same material properties, then $\lambda_n = (2n-1)\pi/[2(h_1 + h_2)]$, and the solution given by Eqn. (6.489) reduces to

$$u^{(1)}(x,t) = u^{(2)}(x,t) = -\sum_{n=1}^{\infty} \frac{8(h_1+h_2)V_0}{(2n-1)^2\pi^2 c} \cos\frac{(2n-1)\pi h_1}{2(h_1+h_2)} \sin\frac{(2n-1)\pi x}{2(h_1+h_2)} \sin\frac{(2n-1)\pi c t}{2(h_1+h_2)},$$

$$v^{(1)}(x,t) = v^{(2)}(x,t) = -\sum_{n=1}^{\infty} \frac{4V_0}{(2n-1)\pi} \cos\frac{(2n-1)\pi h_1}{2(h_1+h_2)} \sin\frac{(2n-1)\pi x}{2(h_1+h_2)} \cos\frac{(2n-1)\pi c t}{2(h_1+h_2)},$$

$$\tau_{xx}^{(1)}(x,t) = \tau_{xx}^{(2)}(x,t) = -\sum_{n=1}^{\infty} \frac{4\rho c V_0}{(2n-1)\pi} \cos\frac{(2n-1)\pi h_1}{2(h_1+h_2)} \cos\frac{(2n-1)\pi x}{2(h_1+h_2)} \sin\frac{(2n-1)\pi c t}{2(h_1+h_2)}.$$

The velocity and stress solutions for the values $\lambda = 0$, $c = \rho = 2\mu = 1$, $h_1 = 1$, $h_2 = 3$, $V_0 = 10^{-3}$ at times 0, 0.5, 2, 6 and 8 are shown in Figs. 6.83 and 6.84. Note that the impacting bar separates from the other bar at $t = 8$, which is the same time at which an impacting bar of length 4 would separate after colliding with a rigid wall.

Now consider the problem shown in Fig. 6.85 where $V_1 > V_2$. Applying the balance of linear momentum and energy, we get the velocity and stress at the interface immediately after the impact as

$$v_i = \frac{\rho_1 V_1 c_1 + \rho_2 V_2 c_2}{\rho_1 c_1 + \rho_2 c_2},$$

$$\tau_i = -\frac{\rho_1 \rho_2 c_1 c_2 (V_1 - V_2)}{\rho_1 c_1 + \rho_2 c_2}.$$

The complete solution can be found in a manner similar to the development above. We have

$$\bar{u}^{(1)}(x,s) = -\frac{\rho_2 c_2 (V_1 - V_2) \sin\frac{q_1 c_1 h_2}{c_2} \cos(q_1 x)}{s^2 \left[\rho_1 c_1 \sin(q_1 h_1) \cos\frac{q_1 c_1 h_2}{c_2} + \rho_2 c_2 \sin\frac{q_1 c_1 h_2}{c_2} \cos(q_1 h_1)\right]} + \frac{V_1}{s^2},$$

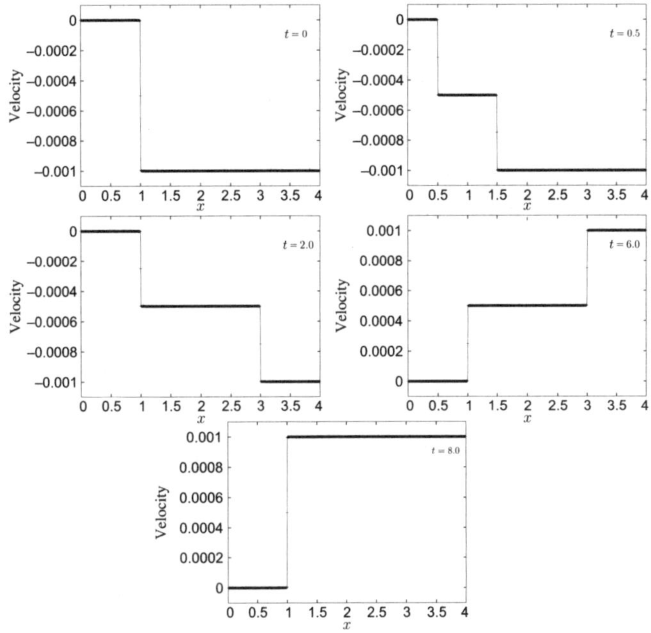

Fig. 6.83 *Velocity plots at times 0, 0.5, 2, 6 and 8 for the problem shown in Fig. 6.82.*

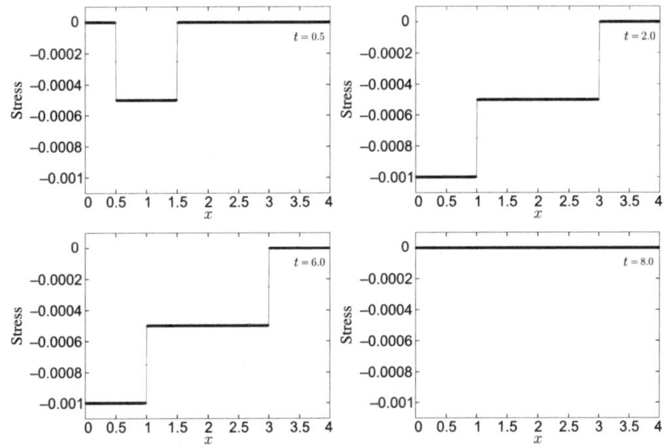

Fig. 6.84 *Stress plots at times 0.5, 2, 6 and 8 for the problem shown in Fig. 6.82.*

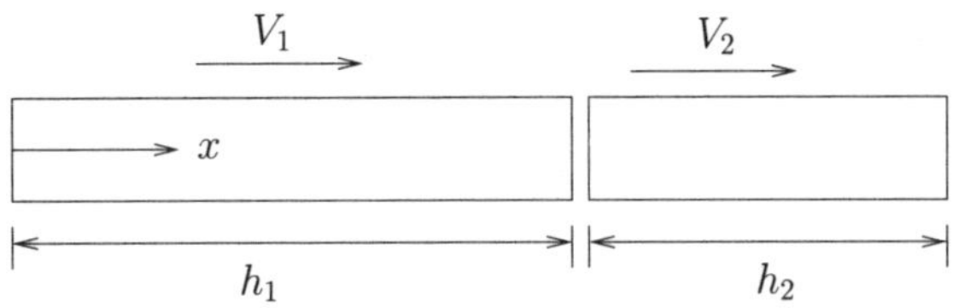

Fig. 6.85 *Elastic bar of length h_1 impacting a moving elastic bar of length h_2.*

$$\bar{u}^{(2)}(x,s) = \frac{\rho_1 c_1 (V_1 - V_2)\sin(q_1 h_1)\cos\frac{q_1 c_1 (h_1+h_2-x)}{c_2}}{s^2\left[\rho_1 c_1 \sin(q_1 h_1)\cos\frac{q_1 c_1 h_2}{c_2} + \rho_2 c_2 \sin\frac{q_1 c_1 h_2}{c_2}\cos(q_1 h_1)\right]} + \frac{V_2}{s^2},$$

Let λ_n, $n = 1, 2, \ldots, \infty$, be the positive roots of

$$\rho_1 c_1 \sin(h_1 x)\cos\frac{c_1 h_2 x}{c_2} + \rho_2 c_2 \cos(h_1 x)\sin\frac{c_1 h_2 x}{c_2} = 0.$$

Then the solution is given by

$$\begin{aligned} u^{(1)}(x,t) &= \frac{\rho_1 V_1 h_1 + \rho_2 V_2 h_2}{\rho_1 h_1 + \rho_2 h_2} t - \sum_{n=1}^{\infty} \frac{2\rho_2 c_2 (V_1 - V_2)}{c_1 \lambda_n^2 D'} \sin\frac{\lambda_n c_1 h_2}{c_2}\cos(\lambda_n x)\sin\lambda_n c_1 t, \\ u^{(2)}(x,t) &= \frac{\rho_1 V_1 h_1 + \rho_2 V_2 h_2}{\rho_1 h_1 + \rho_2 h_2} t + \sum_{n=1}^{\infty} \frac{2\rho_1 (V_1 - V_2)}{\lambda_n^2 D'} \sin(\lambda_n h_1)\cos\frac{\lambda_n c_1 (h_1 + h_2 - x)}{c_2}\sin\lambda_n c_1 t, \end{aligned} \tag{6.490}$$

where

$$\begin{aligned} D' = \rho_1 c_1 h_1 \cos(\lambda_n h_1)\cos\frac{\lambda_n c_1 h_2}{c_2} &- \frac{\rho_1 c_1^2 h_2}{c_2}\sin(\lambda_n h_1)\sin\frac{\lambda_n c_1 h_2}{c_2} \\ &- \rho_2 c_2 h_1 \sin(\lambda_n h_1)\sin\frac{\lambda_n c_1 h_2}{c_2} + \rho_2 c_1 h_2 \cos(\lambda_n h_1)\cos\frac{\lambda_n c_1 h_2}{c_2}. \end{aligned}$$

If the material properties of the two bars are the same, then $\lambda_n = n\pi/(h_1 + h_2)$, and the solution given by Eqn. (6.490) reduces to

$$\begin{aligned} u^{(1)}(x,t) = u^{(2)}(x,t) &= \frac{V_1 h_1 + V_2 h_2}{h_1 + h_2} t \\ &+ \sum_{n=1}^{\infty} \frac{2(h_1+h_2)(V_1 - V_2)}{n^2 \pi^2 c}\sin\frac{n\pi h_1}{h_1+h_2}\cos\frac{n\pi x}{h_1+h_2}\sin\frac{n\pi c t}{h_1+h_2}, \\ v^{(1)}(x,t) = v^{(2)}(x,t) &= \frac{V_1 h_1 + V_2 h_2}{h_1 + h_2} + \sum_{n=1}^{\infty} \frac{2(V_1 - V_2)}{n\pi}\sin\frac{n\pi h_1}{h_1+h_2}\cos\frac{n\pi x}{h_1+h_2}\cos\frac{n\pi c t}{h_1+h_2}, \\ \tau_{xx}^{(1)}(x,t) = \tau_{xx}^{(2)}(x,t) &= -\sum_{n=1}^{\infty} \frac{2\rho_0 c (V_1 - V_2)}{n\pi}\sin\frac{n\pi h_1}{h_1+h_2}\sin\frac{n\pi x}{h_1+h_2}\sin\frac{n\pi c t}{h_1+h_2}. \end{aligned}$$

The velocity and stress plots for the values $\lambda = 0$, $c = \rho_0 = 2\mu = 1$, $h_1 = 1$, $h_2 = 3$, $V_1 = 10^{-3}$, $V_2 = 0$ at times 0.5, 2, 4 and 6 are shown in Figs. 6.86 and 6.87. Note that the bars separate at $t = 6$ when the calculated stress at the interface becomes greater than zero.

Hollow circular cylinder under plane strain conditions subjected to time-varying pressures on its inner and outer surfaces

To find the solution to the problem of a hollow circular cylinder of length L and with inner and outer radii a and b that is initially at rest, and that is subjected to a pressure $p_i(t)$ on

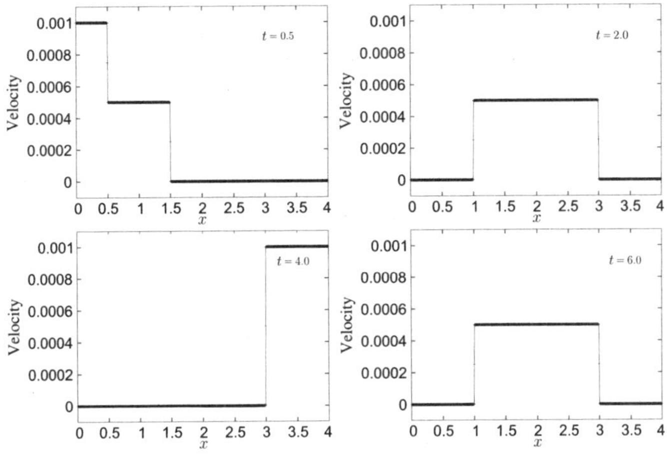

Fig. 6.86 *Velocity plots at times* 0.5, 2, 4 *and* 6 *for the problem shown in Fig. 6.85.*

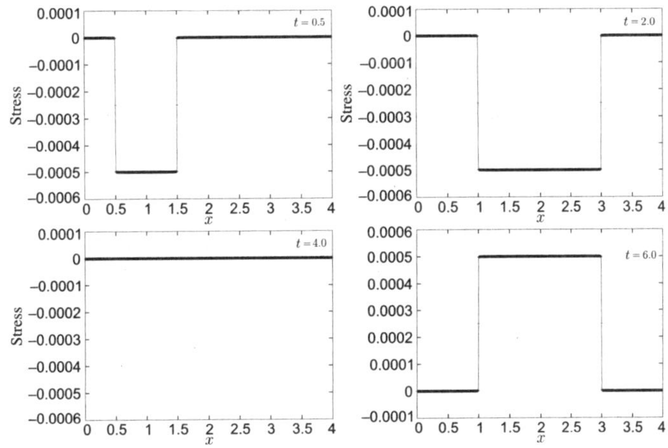

Fig. 6.87 *Stress plots at times* 0.5, 2, 4 *and* 6 *for the problem shown in Fig. 6.85.*

the inner surface and $p_o(t)$ on the outer one under plane strain conditions (for alternate solutions to the one presented here, see [92, 333, 347]), we take the Laplace transform of the governing equation

$$\frac{\partial}{\partial r}\left[\frac{1}{r}\frac{\partial(ru_r)}{\partial r}\right] = \frac{1}{c^2}\frac{\partial^2 u_r}{\partial t^2},$$

where $c = \sqrt{(\lambda + 2\mu)/\rho_0}$, to get

$$\frac{\partial}{\partial r}\left[\frac{1}{r}\frac{\partial(r\bar{u}_r)}{\partial r}\right] + q^2\bar{u}_r = 0,$$

where $q^2 = -s^2/c^2$. The general solution of this equation is $\bar{u}_r = c_1(s)J_1(qr) + c_2(s)Y_1(qr)$, where J_1 and Y_1 are Bessel functions of the first and second kinds. The constants $c_1(s)$

and $c_2(s)$ are determined using the conditions $\bar{\tau}_{rr}|_{r=a} = -\bar{p}_i(s)$ and $\bar{\tau}_{rr}|_{r=b} = -\bar{p}_o(s)$. On evaluating these constants, the solution that we get is

$$\bar{u}_r = \frac{1}{qD(q)}\{[\mu Y_2(qb) - (\lambda+\mu)Y_0(qb)]\,J_1(qr) + [(\lambda+\mu)J_0(qb) - \mu J_2(qb)]\,Y_1(qr)\}\,\bar{p}_i$$
$$+\frac{1}{qD(q)}\{[\mu J_2(qa) - (\lambda+\mu)J_0(qa)]\,Y_1(qr) + [(\lambda+\mu)Y_0(qa) - \mu Y_2(qa)]\,J_1(qr)\}\,\bar{p}_o,$$

where

$$D(x) = \det\begin{bmatrix}(\lambda+\mu)J_0(xa) - \mu J_2(xa) & (\lambda+\mu)Y_0(xa) - \mu Y_2(xa)\\ (\lambda+\mu)J_0(xb) - \mu J_2(xb) & (\lambda+\mu)Y_0(xb) - \mu Y_2(xb)\end{bmatrix}.$$

This can be written as

$$\bar{u}_r = \bar{p}_i(s)\sum_{n=1}^{\infty}\left[\frac{c_n^{(1)}}{s+i\lambda_n c} + \frac{c_n^{(2)}}{s-i\lambda_n c}\right] + \bar{p}_o(s)\sum_{n=1}^{\infty}\left[\frac{d_n^{(1)}}{s+i\lambda_n c} + \frac{d_n^{(2)}}{s-i\lambda_n c}\right],$$

where λ_n, $n = 1,2,\ldots,\infty$, denote the positive roots of $D(x) = 0$. The constants $c_n^{(1)}$, $c_n^{(2)}$, $d_n^{(1)}$ and $d_n^{(2)}$ are determined using the Heaviside formula as in the previous problem. The final solution is given by

$$u_r(r,t) = -\sum_{n=1}^{\infty}\frac{2cP_n^{(i)}}{\lambda_n D'}\int_0^t \sin(\lambda_n c\tau)p_i(t-\tau)\,d\tau - \sum_{n=1}^{\infty}\frac{2cP_n^{(o)}}{\lambda_n D'}\int_0^t \sin(\lambda_n c\tau)p_o(t-\tau)\,d\tau,$$

where

$$P_n^{(i)} = (\lambda+\mu)\left[J_0(\lambda_n b)Y_1(\lambda_n r) - Y_0(\lambda_n b)J_1(\lambda_n r)\right] + \mu\left[Y_2(\lambda_n b)J_1(\lambda_n r) - J_2(\lambda_n b)Y_1(\lambda_n r)\right],$$
$$P_n^{(o)} = (\lambda+\mu)\left[Y_0(\lambda_n a)J_1(\lambda_n r) - J_0(\lambda_n a)Y_1(\lambda_n r)\right] + \mu\left[J_2(\lambda_n a)Y_1(\lambda_n r) - Y_2(\lambda_n a)J_1(\lambda_n r)\right],$$
$$D' = \left[\frac{\mu a}{2}J_3(\lambda_n a) - \left(\lambda+\frac{3}{2}\mu\right)aJ_1(\lambda_n a)\right]\left[(\lambda+\mu)Y_0(\lambda_n b) - \mu Y_2(\lambda_n b)\right]$$
$$+\left[\frac{\mu b}{2}Y_3(\lambda_n b) - \left(\lambda+\frac{3}{2}\mu\right)bY_1(\lambda_n b)\right]\left[(\lambda+\mu)J_0(\lambda_n a) - \mu J_2(\lambda_n a)\right]$$
$$-\left[\frac{\mu b}{2}J_3(\lambda_n b) - \left(\lambda+\frac{3}{2}\mu\right)bJ_1(\lambda_n b)\right]\left[(\lambda+\mu)Y_0(\lambda_n a) - \mu Y_2(\lambda_n a)\right]$$
$$-\left[\frac{\mu a}{2}Y_3(\lambda_n a) - \left(\lambda+\frac{3}{2}\mu\right)aY_1(\lambda_n a)\right]\left[(\lambda+\mu)J_0(\lambda_n b) - \mu J_2(\lambda_n b)\right].$$

Once the displacement is known, the stresses are recovered using

$$\tau_{rr} = (\lambda+2\mu)\frac{\partial u_r}{\partial r} + \lambda\frac{u_r}{r},$$
$$\tau_{\theta\theta} = \lambda\frac{\partial u_r}{\partial r} + (\lambda+2\mu)\frac{u_r}{r}, \tag{6.491}$$

along with the relations

$$\frac{P_n^{(i)}}{r}+\frac{\partial P_n^{(i)}}{\partial r}=(\lambda+\mu)\lambda_n\left[J_0(\lambda_n b)Y_0(\lambda_n r)-Y_0(\lambda_n b)J_0(\lambda_n r)\right]$$
$$+\mu\lambda_n\left[Y_2(\lambda_n b)J_0(\lambda_n r)-J_2(\lambda_n b)Y_0(\lambda_n r)\right],$$
$$\frac{P_n^{(o)}}{r}+\frac{\partial P_n^{(o)}}{\partial r}=(\lambda+\mu)\lambda_n\left[Y_0(\lambda_n a)J_0(\lambda_n r)-J_0(\lambda_n a)Y_0(\lambda_n r)\right]$$
$$+\mu\lambda_n\left[J_2(\lambda_n a)Y_0(\lambda_n r)-Y_2(\lambda_n a)J_0(\lambda_n r)\right],$$
$$2\frac{\partial P_n^{(i)}}{\partial r}=(\lambda+\mu)\lambda_n\left[J_0(\lambda_n b)[Y_0(\lambda_n r)-Y_2(\lambda_n r)]-Y_0(\lambda_n b)[J_0(\lambda_n r)-J_2(\lambda_n r)]\right]$$
$$+\mu\lambda_n\left[Y_2(\lambda_n b)[J_0(\lambda_n r)-J_2(\lambda_n r)]-J_2(\lambda_n b)[Y_0(\lambda_n r)-Y_2(\lambda_n r)]\right],$$
$$2\frac{\partial P_n^{(o)}}{\partial r}=(\lambda+\mu)\lambda_n\left[Y_0(\lambda_n a)[J_0(\lambda_n r)-J_2(\lambda_n r)]-J_0(\lambda_n a)[Y_0(\lambda_n r)-Y_2(\lambda_n r)]\right]$$
$$+\mu\lambda_n\left[J_2(\lambda_n a)[Y_0(\lambda_n r)-Y_2(\lambda_n r)]-Y_2(\lambda_n a)[J_0(\lambda_n r)-J_2(\lambda_n r)]\right].$$

If the pressure is impulsively applied on the inner surface with the outer surface being traction-free, i.e., $p_i(t)=p_i$ and $p_o(t)=0$, then we get

$$u_r(r,t)=\frac{a^2p_i}{2(b^2-a^2)}\left[\frac{b^2}{\mu r}+\frac{1}{\lambda+\mu}r\right]+\sum_{n=1}^{\infty}\frac{2p_iP_n^{(i)}}{\lambda_n^2D'}\cos(\lambda_n ct),$$

$$\tau_{rr}(r,t)=\frac{p_ia^2}{(b^2-a^2)}\left[1-\frac{b^2}{r^2}\right]+\sum_{n=1}^{\infty}\frac{2p_i\left[\lambda\left(\frac{P_n^{(i)}}{r}+\frac{\partial P_n^{(i)}}{\partial r}\right)+2\mu\frac{\partial P_n^{(i)}}{\partial r}\right]}{\lambda_n^2D'}\cos(\lambda_n ct),$$

where we have used

$$\sum_{n=1}^{\infty}\frac{2P_n^{(i)}}{\lambda_n^2D'}=-\frac{a^2}{2(b^2-a^2)}\left[\frac{b^2}{\mu r}+\frac{1}{\lambda+\mu}r\right],$$
$$\sum_{n=1}^{\infty}\frac{2\left[\lambda\left(\frac{P_n^{(i)}}{r}+\frac{\partial P_n^{(i)}}{\partial r}\right)+2\mu\frac{\partial P_n^{(i)}}{\partial r}\right]}{\lambda_n^2D'}=-\frac{a^2}{(b^2-a^2)}\left[1-\frac{b^2}{r^2}\right],\quad r\in(a,b].$$

If $p_i(t)=p_i\sin\omega t$ and $p_o(t)=0$, then (with $k\equiv\omega/c$)

$$u_r(r,t)=\frac{p_i}{kD_1(k)}\Big\{(\lambda+\mu)\left[J_0(kb)Y_1(kr)-Y_0(kb)J_1(kr)\right]$$
$$+\mu\left[Y_2(kb)J_1(kr)-J_2(kb)Y_1(kr)\right]\Big\}\sin\omega t+\sum_{n=1}^{\infty}\frac{2p_iP_n^{(i)}}{\lambda_nk\left[\left(\frac{\lambda_n}{k}\right)^2-1\right]D'}\sin(\lambda_n ct),$$

$$\tau_{rr}(r,t) = -\frac{p_i N_1(k)}{D_1(k)} \sin \omega t + \sum_{n=1}^{\infty} \frac{2p_i \left[\lambda \left(\frac{P_n^{(i)}}{r} + \frac{\partial P_n^{(i)}}{\partial r}\right) + 2\mu \frac{\partial P_n^{(i)}}{\partial r}\right]}{\lambda_n k \left[\left(\frac{\lambda_n}{k}\right)^2 - 1\right] D'} \sin(\lambda_n ct),$$

where

$$N_1(k) = \det \begin{bmatrix} (\lambda+\mu)J_0(kr) - \mu J_2(kr) & (\lambda+\mu)Y_0(kr) - \mu Y_2(kr) \\ (\lambda+\mu)J_0(kb) - \mu J_2(kb) & (\lambda+\mu)Y_0(kb) - \mu Y_2(kb) \end{bmatrix},$$

$$D_1(k) = \det \begin{bmatrix} (\lambda+\mu)J_0(ka) - \mu J_2(ka) & (\lambda+\mu)Y_0(ka) - \mu Y_2(ka) \\ (\lambda+\mu)J_0(kb) - \mu J_2(kb) & (\lambda+\mu)Y_0(kb) - \mu Y_2(kb) \end{bmatrix},$$

and where we have used

$$\sum_{n=1}^{\infty} \frac{2P_n^{(i)}}{\left[\left(\frac{\lambda_n}{k}\right)^2 - 1\right] D'} = -\frac{k}{D_1(k)} \Big\{ (\lambda+\mu) \left[J_0(kb)Y_1(kr) - Y_0(kb)J_1(kr)\right]$$
$$+ \mu \left[Y_2(kb)J_1(kr) - J_2(kb)Y_1(kr)\right] \Big\},$$

$$\sum_{n=1}^{\infty} \frac{2\left[\lambda \left(\frac{P_n^{(i)}}{r} + \frac{\partial P_n^{(i)}}{\partial r}\right) + 2\mu \frac{\partial P_n^{(i)}}{\partial r}\right]}{\left[\left(\frac{\lambda_n}{k}\right)^2 - 1\right] D'} = \frac{k^2 N_1(k)}{D_1(k)} \quad r \in (a,b].$$

If $p_i(t) = p_i \cos \omega t$ and $p_o(t) = 0$, then

$$u_r(r,t) = \frac{p_i}{kD_1(k)} \Big\{ (\lambda+\mu) \left[J_0(kb)Y_1(kr) - Y_0(kb)J_1(kr)\right]$$
$$+ \mu \left[Y_2(kb)J_1(kr) - J_2(kb)Y_1(kr)\right] \Big\} \cos \omega t + \sum_{n=1}^{\infty} \frac{2p_i P_n^{(i)}}{k^2 \left[\left(\frac{\lambda_n}{k}\right)^2 - 1\right] D'} \cos(\lambda_n ct),$$

$$\tau_{rr}(r,t) = -\frac{p_i N_1(k)}{D_1(k)} \cos \omega t + \sum_{n=1}^{\infty} \frac{2p_i \left[\lambda \left(\frac{P_n^{(i)}}{r} + \frac{\partial P_n^{(i)}}{\partial r}\right) + 2\mu \frac{\partial P_n^{(i)}}{\partial r}\right]}{k^2 \left[\left(\frac{\lambda_n}{k}\right)^2 - 1\right] D'} \cos(\lambda_n ct).$$

Finally, if $p_i(t) = p_i e^{-\omega t}$ and $p_o(t) = 0$, then, with $k := \omega / c$, we have

$$u_r(r,t) = \frac{p_i}{kD_2(k)} \Big\{ (\lambda+\mu) \left[I_0(kb)K_1(kr) + K_0(kb)I_1(kr)\right]$$
$$+ \mu \left[I_2(kb)K_1(kr) + K_2(kb)I_1(kr)\right] \Big\} e^{-\omega t}$$
$$+ \sum_{n=1}^{\infty} \frac{2p_i P_n^{(i)}}{\lambda_n k \left[\left(\frac{\lambda_n}{k}\right)^2 + 1\right] D'} \left[\frac{\lambda_n}{k} \cos \lambda_n ct - \sin \lambda_n ct\right],$$

$$\tau_{rr}(r,t) = -\frac{p_i N_2(k)}{D_2(k)} e^{-\omega t} + \sum_{n=1}^{\infty} \frac{2p_i \left[\lambda \left(\frac{P_n^{(i)}}{r} + \frac{\partial P_n^{(i)}}{\partial r}\right) + 2\mu \frac{\partial P_n^{(i)}}{\partial r}\right]}{\lambda_n k \left[\left(\frac{\lambda_n}{k}\right)^2 + 1\right] D'} \left[\frac{\lambda_n}{k} \cos \lambda_n ct - \sin \lambda_n ct\right],$$

where

$$N_2(k) = \det \begin{bmatrix} (\lambda+\mu)K_0(kr) + \mu K_2(kr) & (\lambda+\mu)I_0(kr) + \mu I_2(kr) \\ (\lambda+\mu)K_0(kb) + \mu K_2(kb) & (\lambda+\mu)I_0(kb) + \mu I_2(kb) \end{bmatrix},$$

$$D_2(k) = \det \begin{bmatrix} (\lambda+\mu)K_0(ka) + \mu K_2(ka) & (\lambda+\mu)I_0(ka) + \mu I_2(ka) \\ (\lambda+\mu)K_0(kb) + \mu K_2(kb) & (\lambda+\mu)I_0(kb) + \mu I_2(kb) \end{bmatrix},$$

and where we have used

$$\sum_{n=1}^{\infty} \frac{2P_n^{(i)}}{\left[\left(\frac{\lambda_n}{k}\right)^2 + 1\right] D'} = -\frac{k}{D_2(k)} \Big\{ (\lambda+\mu) \left[I_0(kb)K_1(kr) + K_0(kb)I_1(kr)\right]$$

$$+ \mu \left[I_2(kb)K_1(kr) + K_2(kb)I_1(kr)\right] \Big\},$$

$$\sum_{n=1}^{\infty} \frac{2\left[\lambda \left(\frac{P_n^{(i)}}{r} + \frac{\partial P_n^{(i)}}{\partial r}\right) + 2\mu \frac{\partial P_n^{(i)}}{\partial r}\right]}{\left[\left(\frac{\lambda_n}{k}\right)^2 + 1\right] D'} = \frac{k^2 N_2(k)}{D_2(k)}, \quad r \in (a,b].$$

If the region outside the inner cylindrical cavity is unbounded, i.e., if $b \to \infty$, then we get

$$\bar{u}_r(r,s) = \frac{K_1(qr)\bar{p}_i(s)}{q\left[\lambda K_0(qa) + \mu(K_0(qa) + K_2(qa))\right]},$$

where $q = s/c$, and K_0, K_1 and K_2 are modified Bessel functions; this transform seems difficult to invert.

For a solid cylinder with pressure $p_0(t)$ applied on the outer radius b, we get

$$\bar{u}_r = -\frac{\bar{p}_0 b J_1(qr)}{(\lambda+2\mu)qbJ_0(qb) - 2\mu J_1(qb)}.$$

Let λ_n, $n = 1, 2, \ldots, \infty$, be the positive roots of

$$(\lambda + 2\mu)xbJ_0(xb) - 2\mu J_1(xb) = 0.$$

Then the solution is given by

$$\begin{aligned} u_r &= c \sum_{n=1}^{\infty} \lambda_n c_n \int_0^t \sin(\lambda_n c\tau) p_o(t-\tau)\, d\tau, \\ \tau_{rr} &= c \sum_{n=1}^{\infty} \lambda_n d_n \int_0^t \sin(\lambda_n c\tau) p_o(t-\tau)\, d\tau, \end{aligned} \tag{6.492}$$

where with

$$D = (\lambda + \mu)J_0(\lambda_n b) + \mu J_2(\lambda_n b) - (\lambda + 2\mu)\lambda_n b J_1(\lambda_n b),$$

we have

$$c_n = \frac{2J_1(\lambda_n r)}{\lambda_n D},$$
$$d_n = \frac{2\left[(\lambda + 2\mu)\lambda_n r J_0(\lambda_n r) - 2\mu J_1(\lambda_n r)\right]}{\lambda_n r D}.$$

If the pressure is applied impulsively, i.e., $p_0(t) = p_0$, then from Eqn. (6.492), we get

$$u_r = -p_0 \left[\frac{r}{2(\lambda + \mu)} + \sum_{n=1}^{\infty} c_n \cos(\lambda_n c t)\right],$$
$$\tau_{rr} = -p_0 \left[1 + \sum_{n=1}^{\infty} d_n \cos(\lambda_n c t)\right],$$

where we have used

$$\sum_{n=1}^{\infty} c_n = -\frac{r}{2(\lambda + \mu)},$$
$$\sum_{n=1}^{\infty} d_n = -1, \quad r \in [0, b).$$

If $p_0(t) = p_0 \sin \omega t$, then we get, with $k := \omega / c$,

$$u_r = -\frac{p_0}{k}\left[\frac{kbJ_1(kr)\sin \omega t}{(\lambda + 2\mu)kbJ_0(kb) - 2\mu J_1(kb)} + \sum_{n=1}^{\infty} \frac{\lambda_n c_n \sin(\lambda_n c t)}{\left(\frac{\lambda_n}{k}\right)^2 - 1}\right],$$

$$\tau_{rr} = -p_0 \left[\frac{b\left[(\lambda + 2\mu)krJ_0(kr) - 2\mu J_1(kr)\right]}{r\left[(\lambda + 2\mu)kbJ_0(kb) - 2\mu J_1(kb)\right]}\sin \omega t + \sum_{n=1}^{\infty} \frac{\lambda_n d_n \sin(\lambda_n c t)}{k\left[\left(\frac{\lambda_n}{k}\right)^2 - 1\right]}\right],$$

where we have used

$$\sum_{n=1}^{\infty} \frac{\lambda_n^2 c_n}{\left(\frac{\lambda_n}{k}\right)^2 - 1} = -\frac{k^2 b J_1(kr)}{(\lambda + 2\mu)kbJ_0(kb) - 2\mu J_1(kb)},$$
$$\sum_{n=1}^{\infty} \frac{\lambda_n^2 d_n}{\left(\frac{\lambda_n}{k}\right)^2 - 1} = -\frac{k^2 b\left[(\lambda + 2\mu)krJ_0(kr) - 2\mu J_1(kr)\right]}{r\left[(\lambda + 2\mu)kbJ_0(kb) - 2\mu J_1(kb)\right]}, \quad r \in [0, b).$$

If $p_0(t) = p_0 \cos \omega t$, then we get

$$u_r = -\frac{p_0}{k}\left[\frac{kbJ_1(kr)\cos \omega t}{(\lambda + 2\mu)kbJ_0(kb) - 2\mu J_1(kb)} + \sum_{n=1}^{\infty} \frac{\lambda_n^2 c_n \cos(\lambda_n c t)}{k\left[\left(\frac{\lambda_n}{k}\right)^2 - 1\right]}\right],$$

$$\tau_{rr} = -p_0 \left[\frac{b\left[(\lambda+2\mu)krJ_0(kr) - 2\mu J_1(kr)\right]}{r\left[(\lambda+2\mu)kbJ_0(kb) - 2\mu J_1(kb)\right]} \cos\omega t + \sum_{n=1}^{\infty} \frac{\lambda_n d_n \cos(\lambda_n ct)}{k^2 \left[\left(\frac{\lambda_n}{k}\right)^2 - 1\right]} \right].$$

Finally, if $p_0(t) = p_0 e^{-\omega t}$, then we get with $k := \omega/c$,

$$u_r = -\frac{p_0}{k} \left[\frac{kbI_1(kr)e^{-\omega t}}{(\lambda+2\mu)kbI_0(kb) - 2\mu I_1(kb)} + \sum_{n=1}^{\infty} \frac{\lambda_n c_n \left[\frac{\lambda_n}{k}\cos\lambda_n ct - \sin\lambda_n ct\right]}{\left(\frac{\lambda_n}{k}\right)^2 + 1} \right],$$

$$\tau_{rr} = -p_0 \left[\frac{b\left[(\lambda+2\mu)krI_0(kr) - 2\mu I_1(kr)\right]}{r\left[(\lambda+2\mu)kbI_0(kb) - 2\mu I_1(kb)\right]} e^{-\omega t} + \sum_{n=1}^{\infty} \frac{\lambda_n d_n \left[\frac{\lambda_n}{k}\cos\lambda_n ct - \sin\lambda_n ct\right]}{k\left[\left(\frac{\lambda_n}{k}\right)^2 + 1\right]} \right],$$

where we have used

$$\sum_{n=1}^{\infty} \frac{\lambda_n^2 c_n}{\left(\frac{\lambda_n}{k}\right)^2 + 1} = -\frac{k^2 bI_1(kr)}{(\lambda+2\mu)kbI_0(kb) - 2\mu I_1(kb)},$$

$$\sum_{n=1}^{\infty} \frac{\lambda_n^2 d_n}{\left(\frac{\lambda_n}{k}\right)^2 + 1} = -\frac{k^2 b\left[(\lambda+2\mu)krI_0(kr) - 2\mu I_1(kr)\right]}{r\left[(\lambda+2\mu)kbI_0(kb) - 2\mu I_1(kb)\right]}, \quad r \in [0, b).$$

If instead of a traction, a displacement field $u_0(t)$ is prescribed on the outer surface of the cylinder of radius b, then we get

$$\frac{\bar{u}}{\bar{u}_0} = \frac{J_1(qr)}{J_1(qb)}.$$

where $q^2 = -s^2/c^2$. Let λ_n, $n = 1, 2, \ldots, \infty$, be the roots of $J_1(xb) = 0$. Then, on inverting the above transform, we get

$$u = \frac{c}{b} \sum_{n=1}^{\infty} c_n \int_0^t \sin(\lambda_n c\tau) u_0(t-\tau)\, d\tau, \tag{6.493}$$

where $c_n = -2J_1(\lambda_n r)/J_0(\lambda_n b)$.

For the case of an impulsively applied displacement, i.e., $u_0(t) = u_0$, we get from Eqn. (6.493),

$$u_r = u_0 \left[\frac{r}{b} - \sum_{n=1}^{\infty} \frac{c_n \cos\lambda_n ct}{b\lambda_n} \right],$$

where we have used

$$\sum_{n=1}^{\infty} \frac{c_n}{\lambda_n} = r.$$

If $u_0(t) = u_0 \sin \omega t$, then we get, with $k := \omega/c$,

$$u_r = u_0 \left[\frac{J_1(kr)}{J_1(kb)} \sin \omega t - \frac{1}{kb} \sum_{n=1}^{\infty} \frac{c_n \sin \lambda_n ct}{\left(\frac{\lambda_n}{k}\right)^2 - 1} \right],$$

where we have used

$$\sum_{n=1}^{\infty} \frac{c_n \lambda_n}{\left(\frac{\lambda_n}{k}\right)^2 - 1} = k^2 b \frac{J_1(kr)}{J_1(kb)}.$$

If $u_0(t) = u_0 \cos \omega t$, then we get

$$u_r = u_0 \left[\frac{J_1(kr)}{J_1(kb)} \cos \omega t - \frac{1}{k^2 b} \sum_{n=1}^{\infty} \frac{c_n \lambda_n \cos \lambda_n ct}{\left(\frac{\lambda_n}{k}\right)^2 - 1} \right],$$

Finally, if $u_0(t) = u_0 e^{-\omega t}$, then we get, with $k := \omega/c$,

$$u_r = u_0 \left[\frac{I_1(kr)}{I_1(kb)} e^{-\omega t} - \frac{1}{kb} \sum_{n=1}^{\infty} \frac{c_n \left[\frac{\lambda_n}{k} \cos \lambda_n ct - \sin \lambda_n ct\right]}{\left(\frac{\lambda_n}{k}\right)^2 + 1} \right],$$

where we have used

$$\sum_{n=1}^{\infty} \frac{c_n \lambda_n}{\left(\frac{\lambda_n}{k}\right)^2 + 1} = k^2 b \frac{I_1(kr)}{I_1(kb)}.$$

The stresses in all the above cases can be recovered using Eqns. (6.491).

To find the periodic steady-state response under harmonic loading and under conditions of plane strain, let $\phi = \tilde{\phi} e^{i\omega t}$ and $f = (0, 0, f_z) e^{i\omega t}$. By virtue of Eqn. (6.480), the functions $\tilde{\phi}$ and f_z are subject to the following governing equations:

$$\frac{1}{r} \frac{\partial}{\partial r} \left(r \frac{\partial \tilde{\phi}}{\partial r} \right) + \frac{1}{r^2} \frac{\partial^2 \tilde{\phi}}{\partial \theta^2} + k_L^2 \tilde{\phi} = 0, \tag{6.494a}$$

$$\frac{1}{r} \frac{\partial}{\partial r} \left(r \frac{\partial f_z}{\partial r} \right) + \frac{1}{r^2} \frac{\partial^2 f_z}{\partial \theta^2} + k_T^2 f_z = 0, \tag{6.494b}$$

where

$$k_L^2 = \frac{\omega^2}{c_L^2}, \qquad k_T^2 = \frac{\omega^2}{c_T^2}.$$

A separable solution for the above equations is given by

$$\tilde{\phi} = \sum_{n=0}^{\infty} \left[A_n J_n(k_L r) + B_n Y_n(k_L r) \right] (C_n \cos n\theta + D_n \sin n\theta),$$

$$f_z = \sum_{n=0}^{\infty} [E_n J_n(k_T r) + F_n Y_n(k_T r)] (G_n \cos n\theta + H_n \sin n\theta),$$

where J_n and Y_n are Bessel functions of the first and second kind. Substituting the above expressions into Eqn. (6.481), we get

$$\begin{aligned} u_r &= \frac{k_L e^{i\omega t}}{2} \sum_{n=0}^{\infty} \{[J_{n-1}(k_L r) - J_{n+1}(k_L r)] A_n + [Y_{n-1}(k_L r) - Y_{n+1}(k_L r)] B_n\} \\ &\quad (C_n \cos n\theta + D_n \sin n\theta) + \frac{e^{i\omega t}}{r} \sum_{n=0}^{\infty} n [E_n J_n(k_T r) + F_n Y_n(k_T r)] (H_n \cos n\theta - G_n \sin n\theta), \\ u_\theta &= \frac{e^{i\omega t}}{r} \sum_{n=0}^{\infty} n [A_n J_n(k_L r) + B_n Y_n(k_L r)] (D_n \cos n\theta - C_n \sin n\theta) \\ &\quad - \frac{k_T e^{i\omega t}}{2} \sum_{n=0}^{\infty} \{[J_{n-1}(k_T r) - J_{n+1}(k_T r)] E_n + [Y_{n-1}(k_T r) - Y_{n+1}(k_T r)] F_n\} \\ &\quad (G_n \cos n\theta + H_n \sin n\theta). \end{aligned}$$

The stresses are now found using Eqns. (B.12). The constants are determined using appropriate boundary conditions at $r = a$ and at $r = b$. The solution for a solid cylinder is found by setting a, B_n and F_n to zero. The radially symmetric solution is found by considering only the $n = 0$ term excluding G_0, which is taken to be zero. Note that this solution agrees with the first term (the steady-state part) of the transient solutions that have been presented for the radially symmetric case with either prescribed tractions or prescribed displacements proportional to $\cos \omega t$ or $\sin \omega t$.

We have presented separable solutions above to the Helmholtz equations given by Eqns. (6.494). A nonseparable solution, which may be useful in the solution of wedge-type problems, can be found by assuming[28]

$$\begin{aligned} \tilde{\phi} = \sum_{n=1}^{\infty} \Big\{ & \tilde{A}_n [J_n(k_L r)\theta \cos n\theta + f(r) \sin n\theta] + \tilde{B}_n [J_n(k_L r)\theta \sin n\theta - f(r) \cos n\theta] \\ & + \tilde{C}_n [Y_n(k_L r)\theta \cos n\theta + g(r) \sin n\theta] + \tilde{D}_n [Y_n(k_L r)\theta \sin n\theta - g(r) \cos n\theta] \Big\}, \end{aligned} \tag{6.495}$$

where $f(r)$ and $g(r)$ are the particular solutions of the inhomogeneous Bessel equations

$$r^2 f''(r) + r f'(r) + (k_L^2 r^2 - n^2) f(r) = 2n J_n(k_L r), \tag{6.496a}$$

$$r^2 g''(r) + r g'(r) + (k_L^2 r^2 - n^2) g(r) = 2n Y_n(k_L r). \tag{6.496b}$$

We use the method of variation of parameters to find $f(r)$ and $g(r)$. Thus, let

$$f(r) = f_1(r) J_n(k_L r) + f_2(r) Y_n(k_L r).$$

so that

$$f'(r) = f_1'(r) J_n(k_L r) + f_2'(r) Y_n(k_L r) + f_1(r) J_n'(k_L r) + f_2(r) Y_n'(k_L r).$$

[28] As usual, n in Eqn. (6.495) can be replaced by the (possibly complex-valued) roots of a characteristic equation.

Set

$$f_1'(r)J_n(k_Lr) + f_2'(r)Y_n(k_Lr) = 0, \tag{6.497}$$

from which it follows that

$$\begin{aligned}
f'(r) &= f_1(r)J_n'(k_Lr) + f_2(r)Y_n'(k_Lr),\\
f''(r) &= f_1'(r)J_n'(k_Lr) + f_2'(r)Y_n'(k_Lr) + f_1(r)J_n''(k_Lr) + f_2(r)Y_n''(k_Lr).
\end{aligned}$$

Substituting into Eqn. (6.496a), we get

$$f_1'(r)J_n'(k_Lr) + f_2'(r)Y_n'(r) = \frac{2nJ_n(k_Lr)}{r^2}. \tag{6.498}$$

Solving Eqns. (6.497) and (6.498), and using the fact that

$$J_n(k_Lr)Y_n'(k_Lr) - Y_n(k_Lr)J_n'(k_Lr) = \frac{2}{\pi r},$$

we get

$$\begin{aligned}
f_1'(r) &= -\frac{n\pi J_n(k_Lr)Y_n(k_Lr)}{r},\\
f_2'(r) &= \frac{n\pi J_n^2(k_Lr)}{r}.
\end{aligned}$$

Thus,

$$\begin{aligned}
f(r) &= -n\pi J_n(k_Lr)\int \frac{J_n(k_Lr)Y_n(k_Lr)}{r}\,dr + n\pi Y_n(k_Lr)\int \frac{J_n^2(k_Lr)}{r}\,dr,\\
g(r) &= -n\pi J_n(k_Lr)\int \frac{Y_n^2(k_Lr)}{r}\,dr + n\pi Y_n(k_Lr)\int \frac{J_n(k_Lr)Y_n(k_Lr)}{r}\,dr,
\end{aligned}$$

where [349] (page 135)

$$\begin{aligned}
\int \frac{J_n(k_Lr)Y_n(k_Lr)}{r}\,dr &= \frac{k_Lr}{2n}\left[J_{n+1}(k_Lr)\frac{\partial Y_n(k_Lr)}{\partial n} - J_n(k_Lr)\frac{\partial Y_{n+1}(k_Lr)}{\partial n}\right] + \frac{J_n(k_Lr)Y_n(k_Lr)}{2n},\\
\int \frac{J_n^2(k_Lr)}{r}\,dr &= \frac{k_Lr}{2n}\left[J_{n+1}(k_Lr)\frac{\partial J_n(k_Lr)}{\partial n} - J_n(k_Lr)\frac{\partial J_{n+1}(k_Lr)}{\partial n}\right] + \frac{J_n^2(k_Lr)}{2n},\\
\int \frac{Y_n^2(k_Lr)}{r}\,dr &= \frac{k_Lr}{2n}\left[Y_{n+1}(k_Lr)\frac{\partial Y_n(k_Lr)}{\partial n} - Y_n(k_Lr)\frac{\partial Y_{n+1}(k_Lr)}{\partial n}\right] + \frac{Y_n^2(k_Lr)}{2n}.
\end{aligned}$$

When the origin is part of the domain, then $\tilde{C}_n$ and $\tilde{D}_n$ should be set to zero in Eqn. (6.495). The functions $f(r)$ and $f'(r)$, inspite of having terms such as $Y_n(k_Lr)$ in their expressions, tend to zero in the limit as $r \to 0$, at least for noninteger n (thus, the displacement field is bounded at the origin).

Hollow sphere subjected to time-varying pressures (or prescribed zero displacement) on its inner and outer surfaces

A hollow sphere with inner and outer radii a and b, which is initially at rest, is subjected to a pressure $p_i(t)$ on the inner surface and $p_o(t)$ on the outer one. Due to the symmetry inherent in the geometry and loading, we can solve the problem in terms of the potential ϕ;[29] the governing equation for ϕ in spherical coordinates assumes the simple form

$$\frac{\partial^2(r\phi)}{\partial r^2} = \frac{1}{c^2}\frac{\partial^2(r\phi)}{\partial t^2}, \tag{6.499}$$

where c denotes the longitudinal wave speed. The displacement is then recovered using the relation $u_r = \partial\phi/\partial r$. Note that the static solution $u_r = c_1 r + c_2/r^2$ can also be recovered by using $\phi = c_1 r^2/2 - c_2/r - 3c_1c^2t^2/2$ in this approach. First let us consider the case when $b \to \infty$. Taking the Laplace transform of the above equation and solving, we get

$$r\bar{\phi} = A(s)e^{-sr/c} + B(s)e^{sr/c}$$

from which we get

$$\bar{u}_r = -A(s)\frac{s}{cr}e^{-sr/c} - A(s)\frac{1}{r^2}e^{-sr/c} + B(s)\frac{s}{cr}e^{sr/c} - B(s)\frac{1}{r^2}e^{sr/c}.$$

Using the condition that the displacement is zero as $r \to \infty$, we get $B(s) = 0$, while on using the condition that $\bar{\tau}_{rr}|_{r=a} = -\bar{p}_i$, we can find $A(s)$. The final result is

$$\bar{u}_r = \frac{\bar{p}_i a c^2}{r^2(\lambda+2\mu)} \frac{e^{s(a-r)/c}(1+sr/c)}{s^2 + \frac{4\mu}{\lambda+2\mu}\frac{sc}{a} + \frac{4\mu}{\lambda+2\mu}\frac{c^2}{a^2}}.$$

Let $\zeta = \sqrt{\mu/(\lambda+2\mu)}$, $\omega = 2\zeta c/a$ and $h = r/c$. Then on inverting the above transform, we get $u_r(r,t) = 0$ for $t < (r-a)/c$, and

$$\begin{aligned} u_r(r,t) &= \frac{ac^2}{(\lambda+2\mu)\omega r^2}\int_0^{t_s} e^{-\zeta\omega\tau}\left[\frac{1-h\zeta\omega}{\sqrt{1-\zeta^2}}\sin\left[\omega\sqrt{1-\zeta^2}\,\tau\right]\right. \\ &\quad \left. + h\omega\cos\left[\omega\sqrt{1-\zeta^2}\,\tau\right]\right] p_i(t_s-\tau)\,d\tau \\ &= \frac{ac^2}{(\lambda+2\mu)\omega r^2}\int_0^{t_s} e^{-\zeta\omega(t_s-\tau)}\left[\frac{1-h\zeta\omega}{\sqrt{1-\zeta^2}}\sin\left[\omega\sqrt{1-\zeta^2}\,(t_s-\tau)\right]\right. \\ &\quad \left. + h\omega\cos\left[\omega\sqrt{1-\zeta^2}\,(t_s-\tau)\right]\right] p_i(\tau)\,d\tau \qquad t \ge \frac{r-a}{c}, \end{aligned}$$

[29] If one works directly with the Laplace transform of the governing equation for u_r, namely,

$$\frac{d}{dr}\left(\frac{1}{r^2}\frac{d(r^2\bar{u}_r)}{dr}\right) + q^2\bar{u}_r = 0,$$

then we get the solution as $\bar{u}_r = c_1(s)j_1(qr) + c_2(s)y_1(qr)$, where j_1 and y_1 are spherical Bessel functions of the first and second kind. Since the spherical Bessel functions can be expressed in terms of sine and cosine functions, the final solution is the same as the one presented; for an alternative solution methodology, see [288].

where the 'shifted time' t_s is given by

$$t_s := t - \frac{r-a}{c}. \tag{6.500}$$

The velocity and stresses are found using the relations

$$\begin{aligned} v_r &= \frac{\partial u_r}{\partial t}, \\ \tau_{rr} &= (\lambda + 2\mu)\frac{\partial u_r}{\partial r} + \lambda\frac{2u_r}{r}, \\ \tau_{\theta\theta} &= \tau_{\phi\phi} = 2(\lambda+\mu)\frac{u_r}{r} + \lambda\frac{\partial u_r}{\partial r}. \end{aligned} \tag{6.501}$$

For the case when the pressure is impulsively applied, i.e., $p(t) = p$, we get the solution for $t \geq (r-a)/c$ as

$$u_r(r,t) = \frac{a^3 p_i \left[e^{\zeta\omega t_s} - \cos(\omega t_s\sqrt{1-\zeta^2}) + \left(\frac{2r}{a} - 1\right)\frac{\zeta}{\sqrt{1-\zeta^2}}\sin(\omega t_s\sqrt{1-\zeta^2})\right]}{4\mu r^2 e^{\zeta\omega t_s}}.$$

Thus, as $t \to \infty$, the above displacement solution tends to the static solution $a^3 p_i/(4\mu r^2)$. Note that the transient part of the solution at any given r is attenuated although damping is not present in the model!

For the case when b is finite, one can write the solution for $r\bar{\phi}$ as $c_1(s)\sin qr + c_2(s)\cos qr$, where $q^2 = -s^2/c^2$. Using the boundary conditions $\bar{\tau}_{rr}|_{r=a} = -\bar{p}_i$ and $\bar{\tau}_{rr}|_{r=b} = -\bar{p}_o$, we find that

$$\bar{u}_r = \frac{a^3\bar{p}_i h_1(q) + b^3\bar{p}_o h_2(q)}{r^2 D(q)}, \tag{6.502}$$

where

$$\begin{aligned} h_1(x) &= x\left[4\mu(b-r) + (\lambda+2\mu)rb^2x^2\right]\cos(b-r)x \\ &\quad + \left[\lambda b^2x^2 + \mu\left(2bx^2(b-2r) - 4\right)\right]\sin(b-r)x, \\ h_2(x) &= -x\left[4a\mu - 4\mu r + ra^2x^2(\lambda+2\mu)\right]\cos(r-a)x \\ &\quad + \left[\lambda a^2x^2 - \mu\left(2ax^2(2r-a) + 4\right)\right]\sin(r-a)x, \\ g(x) &= -4\mu(b-r)x\left[4\mu + (\lambda+2\mu)brx^2\right]\cos(b-r)x \\ &\quad + \left\{16\mu^2 - 4\mu\left[(\lambda+2\mu)(b^2+r^2) - 4\mu br\right]x^2 + (\lambda+2\mu)^2b^2r^2x^4\right\}\sin(b-r)x, \\ D(x) &= 4\mu(b-a)x\left[4\mu + (\lambda+2\mu)abx^2\right]\cos(b-a)x \\ &\quad - \left\{16\mu^2 - 4\mu\left[(\lambda+2\mu)(a^2+b^2) - 4\mu ab\right]x^2 + (\lambda+2\mu)^2a^2b^2x^4\right\}\sin(b-a)x. \end{aligned}$$

Let λ_n, $n = 1, 2, \ldots, \infty$, denote the positive roots of $D(x) = 0$. Then we can write the above equation as

$$\bar{u} = \bar{p}_i(s)\left[\sum_{n=1}^{\infty}\frac{c_n^{(1)}}{s + ic\lambda_n} + \frac{c_n^{(2)}}{s - ic\lambda_n}\right] + \bar{p}_o(s)\left[\sum_{n=1}^{\infty}\frac{d_n^{(1)}}{s + ic\lambda_n} + \frac{d_n^{(2)}}{s - ic\lambda_n}\right],$$

where the constants $c_n^{(1)}$, $c_n^{(2)}$, $d_n^{(1)}$ and $d_n^{(2)}$ are found using the Heaviside formula. The final result is

$$u_r(r,t) = -\sum_{n=1}^{\infty} \frac{2ca^3 h_1(\lambda_n)}{\lambda_n r^2 D'} \int_0^t \sin(\lambda_n c\tau) p_i(t-\tau)\, d\tau - \sum_{n=1}^{\infty} \frac{2cb^3 h_2(\lambda_n)}{\lambda_n r^2 D'} \int_0^t \sin(\lambda_n c\tau) p_o(t-\tau)\, d\tau,$$

where

$$D' = (b-a)\lambda_n \left[4(a^2+3ab+b^2)\lambda\mu + 8(a^2+ab+b^2)\mu^2 - a^2b^2(\lambda+2\mu)^2\lambda_n^2\right]\cos(b-a)\lambda_n$$
$$+ 4\left[2(a^2+b^2)\lambda\mu - ab(\lambda+2\mu)[ab\lambda + (a^2+b^2)\mu]\lambda_n^2\right]\sin(b-a)\lambda_n.$$

If a pressure $p_i(t) = p_i$ is impulsively applied to the inner surface with the outer surface traction-free, then we get

$$u_r = \frac{1}{3\lambda+2\mu}\frac{p_i a^3 r}{(b^3-a^3)} + \frac{1}{4\mu}\frac{a^3b^3 p_i}{(b^3-a^3)r^2} + p_i \sum_{n=1}^{\infty} \frac{2a^3 h_1(\lambda_n)\cos(\lambda_n ct)}{\lambda_n^2 r^2 D'},$$
$$\tau_{rr} = -\frac{p_i a^3(b^3-r^3)}{r^3(b^3-a^3)} + p_i \sum_{n=1}^{\infty} \frac{2a^3 g(\lambda_n)\cos(\lambda_n ct)}{\lambda_n^2 r^3 D'},$$

where we have used the fact that

$$-\sum_{n=1}^{\infty} \frac{2h_1(\lambda_n)}{\lambda_n^2 D'} = \frac{1}{(3\lambda+2\mu)}\frac{r^3}{(b^3-a^3)} + \frac{1}{4\mu}\frac{b^3}{(b^3-a^3)},$$
$$\sum_{n=1}^{\infty} \frac{2g(\lambda_n)}{\lambda_n^2 D'} = \frac{b^3-r^3}{b^3-a^3}, \quad r \in (a,b].$$

If $p_i(t) = p_i \sin\omega t$ and $p_o(t) = 0$, then we get (with $k \equiv \omega/c$),

$$u_r = \frac{p_i a^3}{r^2}\left[\frac{h_1(k)\sin\omega t}{D(k)} + \sum_{n=1}^{\infty} \frac{2h_1(\lambda_n)\sin(\lambda_n ct)}{\lambda_n k\left[\left(\frac{\lambda_n}{k}\right)^2 - 1\right] D'}\right],$$

$$\tau_{rr} = \frac{p_i a^3}{r^3}\left[\frac{g(k)\sin\omega t}{D(k)} + \sum_{n=1}^{\infty} \frac{2g(\lambda_n)\sin(\lambda_n ct)}{\lambda_n k\left[\left(\frac{\lambda_n}{k}\right)^2 - 1\right] D'}\right],$$

where we have used

$$\sum_{n=1}^{\infty} \frac{2h_1(\lambda_n)}{\left[\left(\frac{\lambda_n}{k}\right)^2 - 1\right] D'} = -\frac{k^2 h_1(k)}{D(k)},$$
$$\sum_{n=1}^{\infty} \frac{2g(\lambda_n)}{\left[\left(\frac{\lambda_n}{k}\right)^2 - 1\right] D'} = -\frac{k^2 g(k)}{D(k)}, \quad r \in (a,b],$$

while if $p_i(t) = p_i \cos\omega t$ and $p_o(t) = 0$, then we get

$$u_r = \frac{p_i a^3}{r^2}\left[\frac{h_1(k)\cos\omega t}{D(k)} + \sum_{n=1}^{\infty}\frac{2h_1(\lambda_n)\cos(\lambda_n ct)}{k^2\left[\left(\frac{\lambda_n}{k}\right)^2 - 1\right]D'}\right],$$

$$\tau_{rr} = \frac{p_i a^3}{r^3}\left[\frac{g(k)\cos\omega t}{D(k)} + \sum_{n=1}^{\infty}\frac{2g(\lambda_n)\cos(\lambda_n ct)}{k^2\left[\left(\frac{\lambda_n}{k}\right)^2 - 1\right]D'}\right].$$

Finally, for the case $p_i(t) = p_i e^{-\omega t}$ and $p_o(t) = 0$, we get (with $k \equiv \omega/c$),

$$u_r = \frac{p_i a^3}{r^2}\left[\frac{\hat{h}_1(k)e^{-\omega t}}{\hat{D}(k)} + \sum_{n=1}^{\infty}\frac{2h_1(\lambda_n)\left[\frac{\lambda_n}{k}\cos(\lambda_n ct) - \sin(\lambda_n ct)\right]}{\lambda_n k\left[\left(\frac{\lambda_n}{k}\right)^2 + 1\right]D'}\right],$$

$$\tau_{rr} = \frac{p_i a^3}{r^3}\left[\frac{\hat{g}(k)e^{-\omega t}}{\hat{D}(k)} + \sum_{n=1}^{\infty}\frac{2g(\lambda_n)\left[\frac{\lambda_n}{k}\cos(\lambda_n ct) - \sin(\lambda_n ct)\right]}{\lambda_n k\left[\left(\frac{\lambda_n}{k}\right)^2 + 1\right]D'}\right],$$

where,

$$\begin{aligned}
\hat{h}_1(k) &= k\left[4\mu(b-r) - (\lambda+2\mu)rb^2k^2\right]\cosh(b-r)k \\
&\quad - \left[\lambda b^2k^2 + \mu\left(2bk^2(b-2r) + 4\right)\right]\sinh(b-r)k, \\
\hat{g}(k) &= -4\mu(b-r)k\left[4\mu - (\lambda+2\mu)brk^2\right]\cosh(b-r)k \\
&\quad + \left\{16\mu^2 + 4\mu\left[(\lambda+2\mu)(b^2+r^2) - 4\mu br\right]k^2 + (\lambda+2\mu)^2b^2r^2k^4\right\}\sinh(b-r)k, \\
\hat{D}(k) &= 4\mu(b-a)k\left[4\mu - (\lambda+2\mu)abk^2\right]\cosh(b-a)k \\
&\quad - \left\{16\mu^2 + 4\mu\left[(\lambda+2\mu)(a^2+b^2) - 4\mu ab\right]k^2 + (\lambda+2\mu)^2a^2b^2k^4\right\}\sinh(b-a)k.
\end{aligned}$$

and, where we have used

$$\sum_{n=1}^{\infty}\frac{2\hat{h}_1(\lambda_n)}{\left[\left(\frac{\lambda_n}{k}\right)^2 + 1\right]D'} = -\frac{k^2\hat{h}_1(k)}{\hat{D}(k)},$$

$$\sum_{n=1}^{\infty}\frac{2\hat{g}(\lambda_n)}{\left[\left(\frac{\lambda_n}{k}\right)^2 + 1\right]D'} = -\frac{k^2\hat{g}(k)}{\hat{D}(k)}, \quad r \in (a, b].$$

The results for a solid sphere are recovered simply by setting p_i and a to zero. We get

$$u_r(r,t) = -\sum_{n=1}^{\infty}\frac{2ch_s(\lambda_n)}{\lambda_n r^2 D_s'}\int_0^t \sin(\lambda_n c\tau)p_o(t-\tau)\,d\tau,$$

where λ_n are the roots of $D_s(x) = 0$, with

$$b^3 D_s(x) = 4\mu bx \cos bx + \left[-4\mu + (\lambda + 2\mu)x^2 b^2\right] \sin bx,$$

and

$$\begin{aligned}
h_s(x) &= rx \cos rx - \sin rx, \\
g_s(x) &= -4\mu rx \cos rx - \left[-4\mu + (\lambda + 2\mu)x^2 r^2\right] \sin rx, \\
D_s' &= (\lambda + 2\mu)\lambda_n \cos(\lambda_n b) + \frac{2\lambda}{b} \sin(\lambda_n b).
\end{aligned}$$

If the pressure is impulsively applied, i.e., $p_o(t) = p_0$, then we get

$$\begin{aligned}
u_r &= -\frac{p_o r}{3\lambda + 2\mu} + p_o \sum_{n=1}^{\infty} \frac{2h_s(\lambda_n)\cos(\lambda_n ct)}{\lambda_n^2 r^2 D_s'}, \\
\tau_{rr} &= -p_o + p_o \sum_{n=1}^{\infty} \frac{2g_s(\lambda_n)\cos(\lambda_n ct)}{\lambda_n^2 r^3 D_s'},
\end{aligned}$$

where we have used the fact that

$$\begin{aligned}
\sum_{n=1}^{\infty} \frac{2h_s(\lambda_n)}{\lambda_n^2 D_s'} &= \frac{r^3}{3\lambda + 2\mu}, \\
\sum_{n=1}^{\infty} \frac{2g_s(\lambda_n)}{\lambda_n^2 D_s'} &= r^3, \quad r \in [0, b).
\end{aligned}$$

If $p_o(t) = p_o \sin \omega t$, then we get (with $k \equiv \omega/c$)

$$\begin{aligned}
u_r &= \frac{p_o}{r^2}\left[\frac{h_s(k)\sin \omega t}{D_s(k)} + \sum_{n=1}^{\infty} \frac{2h_s(\lambda_n)\sin(\lambda_n ct)}{\lambda_n k\left[\left(\frac{\lambda_n}{k}\right)^2 - 1\right] D_s'}\right], \\
\tau_{rr} &= \frac{p_o}{r^3}\left[\frac{g_s(k)\sin \omega t}{D_s(k)} + \sum_{n=1}^{\infty} \frac{2g_s(\lambda_n)\sin(\lambda_n ct)}{\lambda_n k\left[\left(\frac{\lambda_n}{k}\right)^2 - 1\right] D_s'}\right],
\end{aligned}$$

where we have used

$$\begin{aligned}
\sum_{n=1}^{\infty} \frac{2h_s(\lambda_n)}{\left[\left(\frac{\lambda_n}{k}\right)^2 - 1\right] D_s'} &= -\frac{k^2 h_s(k)}{D_s(k)}, \\
\sum_{n=1}^{\infty} \frac{2g_s(\lambda_n)}{\left[\left(\frac{\lambda_n}{k}\right)^2 - 1\right] D_s'} &= -\frac{k^2 g_s(k)}{D_s(k)}, \quad r \in [0, b),
\end{aligned}$$

while if $p_o(t) = p_o \cos \omega t$, then we get

$$u_r = \frac{p_o}{r^2}\left[\frac{h_s(k)\cos\omega t}{D_s(k)} + \sum_{n=1}^{\infty}\frac{2h_s(\lambda_n)\cos(\lambda_n ct)}{k^2\left[\left(\frac{\lambda_n}{k}\right)^2 - 1\right]D_s'}\right],$$

$$\tau_{rr} = \frac{p_o}{r^3}\left[\frac{g_s(k)\cos\omega t}{D_s(k)} + \sum_{n=1}^{\infty}\frac{2g_s(\lambda_n)\cos(\lambda_n ct)}{k^2\left[\left(\frac{\lambda_n}{k}\right)^2 - 1\right]D_s'}\right].$$

Finally, if $p_o(t) = p_o e^{-\omega t}$, then we get (with $k \equiv \omega/c$)

$$u_r = \frac{p_o}{r^2}\left[\frac{\hat{h}_s(k)e^{-\omega t}}{\hat{D}_s(k)} + \sum_{n=1}^{\infty}\frac{2h_s(\lambda_n)\left[\frac{\lambda_n}{k}\cos(\lambda_n ct) - \sin(\lambda_n ct)\right]}{\lambda_n k\left[\left(\frac{\lambda_n}{k}\right)^2 + 1\right]D_s'}\right],$$

$$\tau_{rr} = \frac{p_o}{r^3}\left[\frac{\hat{g}_s(k)e^{-\omega t}}{\hat{D}_s(k)} + \sum_{n=1}^{\infty}\frac{2g_s(\lambda_n)\left[\frac{\lambda_n}{k}\cos(\lambda_n ct) - \sin(\lambda_n ct)\right]}{\lambda_n k\left[\left(\frac{\lambda_n}{k}\right)^2 + 1\right]D_s'}\right],$$

where

$$\hat{h}_s(k) = rk\cosh rk - \sinh rk,$$

$$\hat{g}_s(k) = -4\mu rk\cosh rk + \left[4\mu + (\lambda + 2\mu)k^2r^2\right]\sinh rk,$$

$$b^3\hat{D}_s(k) = 4\mu bk\cosh bk - \left[4\mu + (\lambda + 2\mu)k^2b^2\right]\sinh bk,$$

and where we have used

$$\sum_{n=1}^{\infty}\frac{2h_s(\lambda_n)}{\left[\left(\frac{\lambda_n}{k}\right)^2 + 1\right]D_s'} = -\frac{k^2\hat{h}_s(k)}{\hat{D}_s(k)},$$

$$\sum_{n=1}^{\infty}\frac{2g_s(\lambda_n)}{\left[\left(\frac{\lambda_n}{k}\right)^2 + 1\right]D_s'} = -\frac{k^2\hat{g}_s(k)}{\hat{D}_s(k)}, \quad r \in [0, b).$$

If, in the hollow sphere case, instead of the outer surface being subjected to a pressure p_o, it is subjected to a zero displacement condition, then instead of Eqn. (6.502), we get

$$\bar{u}_r = \frac{a^3\bar{p}_i h(q)}{r^2 D(q)},$$

where

$$h(x) = (1 + brx^2)\sin(b - r)x - x(b - r)\cos(b - r)x,$$

$$g(x) = -x\left[4\mu(r-b) + (\lambda+2\mu)r^2x^2b\right]\cos(b-r)x$$
$$- \left[4\mu(1+rb) - (\lambda+2\mu)r^2x^2\right]\sin(b-r)x,$$
$$D(x) = x\left[4\mu(a-b) + (\lambda+2\mu)a^2x^2b\right]\cos(b-a)x$$
$$+ \left[4\mu(1+ab) - (\lambda+2\mu)a^2x^2\right]\sin(b-a)x.$$

Let λ_n, $n = 1, 2, \ldots, \infty$, be the positive roots of $D(x) = 0$. Then inverting the above transform, we get

$$u_r(r,t) = -\sum_{n=1}^{\infty} \frac{2ca^3h(\lambda_n)}{\lambda_n r^2 D'} \int_0^t \sin(\lambda_n c\tau) p_i(t-\tau)\, d\tau,$$

where

$$D' = a\left[2ab\lambda + 4b^2\mu + a^2(\lambda+2\mu)\right]\lambda_n \cos\lambda_n(b-a)$$
$$+ \left[4b^2\mu - 2\lambda a^2 - a^2b(b-a)(\lambda+2\mu)\lambda_n^2\right]\sin\lambda_n(b-a).$$

If a pressure $p_i(t) = p_i$ is impulsively applied, then we get

$$u_r = \frac{a^3p_i(b^3-r^3)}{r^2\left[(3\lambda+2\mu)a^3 + 4\mu b^3\right]} + p_i \sum_{n=1}^{\infty} \frac{2a^3h(\lambda_n)\cos(\lambda_n ct)}{\lambda_n^2 r^2 D'},$$
$$\tau_{rr} = -\frac{a^3p_i\left[(3\lambda+2\mu)r^3 + 4\mu b^3\right]}{r^3\left[(3\lambda+2\mu)a^3 + 4\mu b^3\right]} + p_i \sum_{n=1}^{\infty} \frac{2a^3g(\lambda_n)\cos(\lambda_n ct)}{\lambda_n^2 r^3 D'},$$

where we have used the fact that

$$-\sum_{n=1}^{\infty} \frac{2a^3h(\lambda_n)}{\lambda_n^2 D'} = \frac{a^3(b^3-r^3)}{[(3\lambda+2\mu)a^3 + 4\mu b^3]},$$
$$\sum_{n=1}^{\infty} \frac{2g(\lambda_n)}{\lambda_n^2 D'} = \frac{(3\lambda+2\mu)r^3 + 4\mu b^3}{(3\lambda+2\mu)a^3 + 4\mu b^3}, \quad r \in (a,b].$$

Unlike the infinite domain case ($b \to \infty$) treated earlier, it is not evident whether the transients get damped out in this case; however, the λ_n do increase as b increases.

If $p_i(t) = p_i \sin\omega t$, then we get (with $k \equiv \omega/c$)

$$u_r = \frac{p_i a^3}{r^2}\left[\frac{h(k)\sin\omega t}{D(k)} + \sum_{n=1}^{\infty} \frac{2h(\lambda_n)\sin(\lambda_n ct)}{\lambda_n k\left[\left(\frac{\lambda_n}{k}\right)^2 - 1\right]D'}\right],$$

$$\tau_{rr} = \frac{p_i a^3}{r^3}\left[\frac{g(k)\sin\omega t}{D(k)} + \sum_{n=1}^{\infty} \frac{2g(\lambda_n)\sin(\lambda_n ct)}{\lambda_n k\left[\left(\frac{\lambda_n}{k}\right)^2 - 1\right]D'}\right],$$

where we have used

$$\sum_{n=1}^{\infty} \frac{2h(\lambda_n)}{\left[\left(\frac{\lambda_n}{k}\right)^2 - 1\right] D'} = -\frac{k^2 h(k)}{D(k)},$$

$$\sum_{n=1}^{\infty} \frac{2g(\lambda_n)}{\left[\left(\frac{\lambda_n}{k}\right)^2 - 1\right] D'} = -\frac{k^2 g(k)}{D(k)}, \quad r \in (a, b],$$

while for $p_i(t) = p_i \cos \omega t$, we get

$$u_r = \frac{p_i a^3}{r^2} \left[\frac{h(k) \cos \omega t}{D(k)} + \sum_{n=1}^{\infty} \frac{2h(\lambda_n) \cos(\lambda_n ct)}{k^2 \left[\left(\frac{\lambda_n}{k}\right)^2 - 1\right] D'} \right],$$

$$\tau_{rr} = \frac{p_i a^3}{r^3} \left[\frac{g(k) \cos \omega t}{D(k)} + \sum_{n=1}^{\infty} \frac{2g(\lambda_n) \cos(\lambda_n ct)}{k^2 \left[\left(\frac{\lambda_n}{k}\right)^2 - 1\right] D'} \right].$$

Finally, for $p_i(t) = p_i e^{-\omega t}$, we get (with $k \equiv \omega / c$)

$$u_r = \frac{p_i a^3}{r^2} \left[\frac{\hat{h}(k) e^{-\omega t}}{\hat{D}(k)} + \sum_{n=1}^{\infty} \frac{2h(\lambda_n) \left[\frac{\lambda_n}{k} \cos(\lambda_n ct) - \sin(\lambda_n ct)\right]}{\lambda_n k \left[\left(\frac{\lambda_n}{k}\right)^2 + 1\right] D'} \right],$$

$$\tau_{rr} = \frac{p_i a^3}{r^3} \left[\frac{\hat{g}(k) e^{-\omega t}}{\hat{D}(k)} + \sum_{n=1}^{\infty} \frac{2g(\lambda_n) \left[\frac{\lambda_n}{k} \cos(\lambda_n ct) - \sin(\lambda_n ct)\right]}{\lambda_n k \left[\left(\frac{\lambda_n}{k}\right)^2 + 1\right] D'} \right],$$

where

$$\begin{aligned}
\hat{h}(k) &= (1 - brk^2) \sinh(b - r)k - k(b - r) \cosh(b - r)k, \\
\hat{g}(k) &= -k \left[4\mu(r - b) - (\lambda + 2\mu) r^2 k^2 b\right] \cosh(b - r)k \\
&\quad - \left[4\mu(1 + rb) + (\lambda + 2\mu) r^2 k^2\right] \sinh(b - r)k, \\
\hat{D}(k) &= k \left[4\mu(a - b) - (\lambda + 2\mu) a^2 k^2 b\right] \cosh(b - a)k \\
&\quad + \left[4\mu(1 + ab) + (\lambda + 2\mu) a^2 k^2\right] \sinh(b - a)k,
\end{aligned}$$

and where we have used

$$\sum_{n=1}^{\infty} \frac{2h(\lambda_n)}{\left[\left(\frac{\lambda_n}{k}\right)^2 + 1\right] D'} = -\frac{k^2 \hat{h}(k)}{\hat{D}(k)},$$

$$\sum_{n=1}^{\infty} \frac{2g(\lambda_n)}{\left[\left(\frac{\lambda_n}{k}\right)^2+1\right] D'} = -\frac{k^2\hat{g}(k)}{\hat{D}(k)}, \quad r \in (a,b].$$

Now consider the problem of determining the 'periodic steady-state' solution for the hollow sphere when the loading is harmonic. We consider only the axisymmetric case here. Thus, let $\phi = \tilde{\phi}e^{i\omega t}$ and $f = (0,0,f_\phi)e^{i\omega t}$. From Eqn. (6.480), we get

$$\frac{1}{r^2}\frac{\partial}{\partial r}\left(r^2\frac{\partial\tilde{\phi}}{\partial r}\right) + \frac{1}{r^2}\frac{\partial}{\partial\xi}\left[(1-\xi^2)\frac{\partial\tilde{\phi}}{\partial\xi}\right] + k_L^2\tilde{\phi} = 0,$$
$$\frac{1}{r^2}\frac{\partial}{\partial r}\left(r^2\frac{\partial f_\phi}{\partial r}\right) + \frac{1}{r^2}\frac{\partial}{\partial\xi}\left[(1-\xi^2)\frac{\partial f_\phi}{\partial\xi}\right] - \frac{f_\phi}{r^2(1-\xi^2)} + k_T^2 f_\phi = 0,$$

(thus, $\tilde{\phi}$ satisfies the Helmholtz equation) where $\xi := \cos\theta$, and

$$k_L^2 = \frac{\omega^2}{c_L^2}, \qquad k_T^2 = \frac{\omega^2}{c_T^2}.$$

The separable solutions for the above equations are

$$\tilde{\phi} = \sum_{n=0}^{\infty}\left[A_n j_n(k_L r) + B_n y_n(k_L r)\right] P_n(\xi),$$
$$f_\phi = \sum_{n=0}^{\infty}\left[C_n j_n(k_T r) + D_n y_n(k_T r)\right] P_n^{(1)}(\xi),$$

where $P_n(\xi)$ are the Legendre functions of the first kind of degree n, and $P_n^{(1)}(\xi)$ are the associated Legendre functions of the first kind of degree n and order 1. The displacements and now recovered using Eqn. (6.481). We get

$$\begin{aligned} u_r &= \frac{e^{i\omega t}}{r}\sum_{n=0}^{\infty}\Big\{A_n\left[n j_n(k_L r) - k_L r j_{n+1}(k_L r)\right]P_n(\xi) + B_n\left[n y_n(k_L r) - k_L r y_{n+1}(k_L r)\right]P_n(\xi) \\ &+ \frac{nC_n}{\sqrt{1-\xi^2}}\left[P_{n+1}^{(1)}(\xi) - \xi P_n^{(1)}(\xi)\right] j_n(k_T r) + \frac{nD_n}{\sqrt{1-\xi^2}}\left[P_{n+1}^{(1)}(\xi) - \xi P_n^{(1)}(\xi)\right] y_n(k_T r)\Big\}, \\ u_\theta &= \frac{e^{i\omega t}}{r}\sum_{n=0}^{\infty}\Big\{C_n\left[k_T r j_{n+1}(k_T r) - (n+1) j_n(k_T r)\right]P_n^{(1)}(\xi) \\ &+ D_n\left[k_T r y_{n+1}(k_T r) - (n+1) y_n(k_T r)\right]P_n^{(1)}(\xi)\Big\}. \end{aligned}$$

The stresses are now found using Eqns. (C.11). The constants A_n, B_n, C_n and D_n are found using the imposed boundary conditions on the surfaces $r = a$ and $r = b$. For a solid sphere, we set a, B_n and D_n to zero. The radially symmetric solution is recovered by considering only $n = 0$. Note that this solution agrees with the first term (the steady-state part) of the transient solutions that have been presented for the radially symmetric case with either prescribed tractions or prescribed displacements proportional to $\cos\omega t$ or $\sin\omega t$ since $j_1(k_L r) = [\sin(k_L r) - k_L r\cos(k_L r)]/(k_L r)^2$ and $y_1(k_L r) = -[\cos(k_L r) + k_L r\sin(k_L r)]/(k_L r)^2$.

Infinite waveguide

Consider a waveguide that is infinite in the z-direction, and which has a rectangular cross-section in the x–y plane as shown in Fig. 6.7. The displacements are of the form $u_x = u_y = 0$, and $u_z = u_z(x, y, t)$, and the boundary condition is $\boldsymbol{u} = \boldsymbol{0}$ on the lateral boundaries. The motion is initiated by the initial displacement $u_0(x, y)$ for u_z. We assume that $u_0(x, y)$ is symmetric about the x and y axis, and that its Fourier representation is given by

$$u_0(x, y) = \sum_{m=1}^{\infty} \sum_{n=1}^{\infty} H_{mn} \cos \frac{(2m-1)\pi x}{a} \cos \frac{(2n-1)\pi y}{b}.$$

Then the displacement u_z is given by

$$u_z(x, y, t) = \sum_{m=1}^{\infty} \sum_{n=1}^{\infty} H_{mn} \cos \frac{(2m-1)\pi x}{a} \cos \frac{(2n-1)\pi y}{b} \cos \sqrt{\frac{(2m-1)^2\pi^2}{a^2} + \frac{(2n-1)^2\pi^2}{b^2}}\, ct,$$

where $c \equiv c_T$. It can be easily verified that $u_z(x, y, 0) = u_0(x, y)$, and that the above solution satisfies the wave equation and boundary conditions.

For a waveguide of circular cross-section of radius R, with the initial displacement prescribed as $u_0(r) = f(r)$ and boundary condition $\boldsymbol{u}(R) = \boldsymbol{0}$, by following a method similar to that used for finding the Poiseuille flow solution in circular pipes (see Chapter 5 in Vol. II), we get

$$\bar{u}(r, s) = \frac{sJ_0(qr)}{c^2} \left[\int_r^R \frac{1}{\xi J_0^2(q\xi)} \left[\int_0^\xi \eta J_0(q\eta) f(\eta)\, d\eta \right] d\xi \right],$$

where $q^2 = -s^2/c^2$.

Torsional oscillations of a circular cylinder

Consider the problem of the torsional oscillations of a circular cylinder of radius a and length L that is fixed at $z = 0$, and that is given an initial twist corresponding to the static solution given by Eqn. (6.58), namely $(u_\theta)_0 = \alpha_0 rz$, where α_0 is the initial twist per unit length, and then released to allow it to oscillate about its axis. To find the solution, we assume that $u_r = u_z = 0$, so that Eqns. (B.13) reduce to

$$\frac{\partial}{\partial r}\left(\frac{1}{r} \frac{\partial (r u_\theta)}{\partial r} \right) + \frac{\partial^2 u_\theta}{\partial z^2} = \frac{1}{c^2} \frac{\partial^2 u_\theta}{\partial t^2},$$

where $c \equiv c_T = \sqrt{\mu/\rho_0}$. We further assume that $u_\theta(r, z, t) = rw(z, t)$, so that the above equation reduces to

$$\frac{\partial^2 w}{\partial z^2} = \frac{1}{c^2} \frac{\partial^2 w}{\partial t^2}.$$

Taking the Laplace transform of this equation, and noting that the initial value w_0 is $\alpha_0 z$, we get

$$\frac{\partial^2 \bar{w}}{\partial z^2} + q^2 \bar{w} = -\frac{\alpha_0 z s}{c^2}.$$

where $q^2 = -s^2/c^2$. The solution of the above equation is

$$\bar{w}(z,s) = A(s)\cos qz + B(s)\sin qz + \frac{\alpha_0 z}{s}.$$

Since $\bar{w}(0,s) = 0$, we get $A(s) = 0$. Hence,

$$\bar{u}_\theta(r,z,s) = r\left[B(s)\sin qz + \frac{\alpha_0 z}{s}\right].$$

The only nonzero stress is

$$\bar{\tau}_{\theta z}(r,z,s) = \rho_0 c^2 \bar{\gamma}_{\theta z} = \rho_0 c^2 r\left[B(s) q\cos qz + \frac{\alpha_0}{s}\right].$$

Since the top surface is traction-free, we get $B(s) = -\alpha_0/(qs\cos qL)$. Thus,

$$\bar{u}_\theta(r,z,s) = \frac{\alpha_0 r}{s}\left[z - \frac{\sin qz}{q\cos qL}\right].$$

Since $qL = (2n-1)\pi/2$, $n = 1,2,\ldots,\infty$, are the roots of $\cos qL = 0$, we can write the second term on the right-hand side as

$$-\frac{\alpha_0 r\sin qz}{sq\cos qL} = \frac{c_0}{s} + \sum_{n=1}^{\infty}\frac{c_n^{(1)}}{s+i\omega_n} + \frac{c_n^{(2)}}{s-i\omega_n},$$

where $\omega_n = (2n-1)\pi c/(2L)$, and

$$c_0 = \lim_{s\to 0} -\frac{\alpha_0 r\sin qz}{q\cos qL} = -\alpha_0 rz.$$

The constants $c_n^{(1)}$ and $c_n^{(2)}$ are determined using the Heaviside formula as

$$c_n^{(1)} = \left.\frac{P(s)}{\frac{dQ(q)}{dq}\frac{dq}{ds}}\right|_{s=-i\omega_n} = -\frac{(-1)^n\alpha_0 rc^2}{\omega_n^2 L}\sin\frac{(2n-1)\pi z}{2L},$$

$$c_n^{(2)} = \left.\frac{P(s)}{\frac{dQ(q)}{dq}\frac{dq}{ds}}\right|_{s=i\omega_n} = -\frac{(-1)^n\alpha_0 rc^2}{\omega_n^2 L}\sin\frac{(2n-1)\pi z}{2L},$$

where $dq/ds = -s/(c^2 q)$. Thus, the final solution can be written as

$$\begin{aligned}
u_\theta(r,z,t) &= \sum_{n=1}^{\infty}\left[c_n^{(1)} + c_n^{(2)}\right]\cos\omega_n t + i\left[c_n^{(1)} - c_n^{(2)}\right]\sin\omega_n t\\
&= -\frac{8\alpha_0 rL}{\pi^2}\sum_{n=1}^{\infty}\frac{(-1)^n}{(2n-1)^2}\sin\frac{(2n-1)\pi z}{2L}\cos\frac{(2n-1)\pi ct}{2L},\\
\tau_{\theta z}(r,z,t) &= -\frac{4\rho_0 c^2\alpha_0 r}{\pi}\sum_{n=1}^{\infty}\frac{(-1)^n}{(2n-1)}\cos\frac{(2n-1)\pi z}{2L}\cos\frac{(2n-1)\pi ct}{2L}.
\end{aligned}$$

Using the result

$$\sum_{n=1}^{\infty} \frac{8(-1)^{n+1}}{(2n-1)^2\pi^2} \sin\frac{(2n-1)\pi z}{2L} = \frac{z}{L},$$

one can easily verify that the initial condition is satisfied. The above solution is also valid for a hollow cylinder. The scalar ϕ and the vector f in Eqn. (6.481) are given by zero and $(f_r, 0, 0)$ where

$$f_r = \frac{16\alpha_0 r L^2}{\pi^3} \sum_{n=1}^{\infty} \frac{(-1)^n}{(2n-1)^3} \cos\frac{(2n-1)\pi z}{2L} \cos\frac{(2n-1)\pi c t}{2L}.$$

If instead of an initial angular displacement, the torsional motion is caused by a traction $\boldsymbol{t} = rT(t)\boldsymbol{e}_\theta$ applied on the face $z = L$, then using the solution given by Eqn. (6.484), we get (with $\lambda_n \equiv (2n-1)\pi/(2L)$)

$$u_\theta(r,z,t) = \frac{4r}{\rho_0 c \pi} \sum_{n=1}^{\infty} \frac{(-1)^{n+1} \sin\lambda_n z}{2n-1} \int_0^t \sin\lambda_n c(t-\tau) T(\tau)\, d\tau. \tag{6.503}$$

For the case when $T(t)$ is impulsively applied, i.e., when $T(t) = T_0$, we get

$$u_\theta = \frac{T_0 r}{\rho_0 c^2} \left[z - \sum_{n=1}^{\infty} \frac{4(-1)^{n+1} \sin\lambda_n z \cos\lambda_n c t}{(2n-1)\pi\lambda_n} \right],$$

$$\tau_{\theta z} = T_0 r \left[1 - \sum_{n=1}^{\infty} \frac{4(-1)^{n+1} \cos\lambda_n z \cos\lambda_n c t}{(2n-1)\pi} \right],$$

where we have used

$$\sum_{n=1}^{\infty} \frac{4(-1)^{n+1} \sin\lambda_n z}{(2n-1)\pi\lambda_n} = z,$$

$$\sum_{n=1}^{\infty} \frac{4(-1)^{n+1} \cos\lambda_n z}{(2n-1)\pi} = 1, \quad z \in [0, L).$$

For the case when $T(t) = T\sin\omega t$, we get (with $k \equiv \omega/c$)

$$u_\theta = \frac{Tr}{\rho_0 c\omega} \left[\frac{\sin kz \sin\omega t}{\cos kL} - \sum_{n=1}^{\infty} \frac{4(-1)^{n+1} \sin\lambda_n z \sin\lambda_n c t}{(2n-1)\pi \left[\left(\frac{\lambda_n}{k}\right)^2 - 1 \right]} \right],$$

$$\tau_{\theta z} = Tr \left[\frac{\cos kz \sin\omega t}{\cos kL} - \sum_{n=1}^{\infty} \frac{2(-1)^{n+1} \cos\lambda_n z \sin\lambda_n c t}{kL \left[\left(\frac{\lambda_n}{k}\right)^2 - 1 \right]} \right],$$

while for the case when $T(t) = T\cos\omega t$, we get

$$u_\theta = \frac{Tr}{\rho_0 c\omega} \left[\frac{\sin kz \cos\omega t}{\cos kL} - \sum_{n=1}^{\infty} \frac{2(-1)^{n+1} \sin\lambda_n z \cos\lambda_n c t}{kL \left[\left(\frac{\lambda_n}{k}\right)^2 - 1 \right]} \right],$$

$$\tau_{\theta z} = Tr\left[\frac{\cos kz \cos \omega t}{\cos kL} - \sum_{n=1}^{\infty} \frac{(-1)^{n+1}(2n-1)\pi \cos \lambda_n z \cos \lambda_n ct}{k^2L^2\left[\left(\frac{\lambda_n}{k}\right)^2 - 1\right]}\right].$$

In deriving the above relations, we have used

$$\sum_{n=1}^{\infty} \frac{2(-1)^{n+1} \sin \lambda_n z}{\left(\frac{\lambda_n}{k}\right)^2 - 1} = \frac{kL \sin kz}{\cos kL},$$

$$\sum_{n=1}^{\infty} \frac{(-1)^{n+1}(2n-1)\pi \cos \lambda_n z}{\left(\frac{\lambda_n}{k}\right)^2 - 1} = \frac{k^2L^2 \cos kz}{\cos kL}, \quad z \in [0, L).$$

Finally, for the case when $T(t) = Te^{-\omega t}$, we get (with $k \equiv \omega/c$)

$$u_\theta = \frac{Tr}{\rho_0 c\omega}\left[\frac{\sinh kze^{-\omega t}}{\cosh kL} - \sum_{n=1}^{\infty} \frac{4(-1)^{n+1} \sin \lambda_n z \left[\frac{\lambda_n}{k} \cos \lambda_n ct - \sin \lambda_n ct\right]}{(2n-1)\pi\left[\left(\frac{\lambda_n}{k}\right)^2 + 1\right]}\right],$$

$$\tau_{\theta z} = Tr\left[\frac{\cosh kze^{-\omega t}}{\cosh kL} - \sum_{n=1}^{\infty} \frac{2(-1)^{n+1} \cos \lambda_n z \left[\frac{\lambda_n}{k} \cos \lambda_n ct - \sin \lambda_n ct\right]}{kL\left[\left(\frac{\lambda_n}{k}\right)^2 + 1\right]}\right],$$

where we have used

$$\sum_{n=1}^{\infty} \frac{2(-1)^{n+1} \sin \lambda_n z}{\left(\frac{\lambda_n}{k}\right)^2 + 1} = \frac{kL \sinh kz}{\cosh kL},$$

$$\sum_{n=1}^{\infty} \frac{(-1)^{n+1}(2n-1)\pi \cos \lambda_n z}{\left(\frac{\lambda_n}{k}\right)^2 + 1} = \frac{k^2L^2 \cosh kz}{\cosh kL}, \quad z \in [0, L).$$

While considering the cases $T(t) = T\sin \omega t$ and $T(t) = T\cos \omega t$ above, we implicitly assumed while using Eqn. (6.503) that $k = \omega/c \neq \lambda_m$. If, however, $k = \lambda_m = (2m-1)\pi/(2L)$, i.e., $\omega = (2m-1)\pi c/(2L)$, then 'resonance', where the solution becomes unbounded as $t \to \infty$, occurs. First consider the case where $T(t) = T\sin \omega t$. From Eqn. (6.503), we get

$$u_\theta = \frac{Tr}{\rho_0 c\omega}\left[\frac{g_1(\xi) \sin \omega t}{(2m-1)\pi} + \frac{2(-1)^{m+1}(\sin \omega t - \omega t \cos \omega t) \sin \lambda_m z}{(2m-1)\pi}\right.$$
$$\left. - \sum_{\substack{n=1\\n\neq m}}^{\infty} \frac{4(-1)^{n+1} \sin \lambda_n z \sin \lambda_n ct}{(2n-1)\pi\left[\left(\frac{2n-1}{2m-1}\right)^2 - 1\right]}\right],$$

$$\tau_{\theta z} = \frac{Tr}{(2m-1)\pi}\left[g_2(\xi) \sin \omega t + 2(-1)^{m+1}(\sin \omega t - \omega t \cos \omega t) \cos \lambda_m z\right.$$

$$- \sum_{\substack{n=1\\n\neq m}}^{\infty} \frac{4(-1)^{n+1}\cos\lambda_n z \sin\lambda_n ct}{\left[\left(\frac{2n-1}{2m-1}\right)^2 - 1\right]}\Bigg],$$

where, with $\xi := z/L$,

$$g_1(\xi) = (-1)^{m+1}\left[\sin\frac{(2m-1)\pi\xi}{2} - (2m-1)\pi\xi\cos\frac{(2m-1)\pi\xi}{2}\right],$$

$$g_2(\xi) = (-1)^{m}\left[\cos\frac{(2m-1)\pi\xi}{2} - (2m-1)\pi\xi\sin\frac{(2m-1)\pi\xi}{2}\right],$$

and where we have used

$$\sum_{\substack{n=1\\n\neq m}}^{\infty} \frac{4(-1)^{n+1}\sin\frac{(2n-1)\pi\xi}{2}}{\left[\left(\frac{2n-1}{2m-1}\right)^2 - 1\right]} = g_1(\xi),$$

$$\sum_{\substack{n=1\\n\neq m}}^{\infty} \frac{4(-1)^{n+1}(2n-1)\cos\frac{(2n-1)\pi\xi}{2}}{\left[\left(\frac{2n-1}{2m-1}\right)^2 - 1\right]} = (2m-1)g_2(\xi), \quad \xi \in [0,1).$$

Similarly, for $T(t) = T\cos\omega t$ with $\omega = (2m-1)\pi c/(2L)$, we get

$$u_\theta = \frac{Tr}{(2m-1)\pi\rho_0 c\omega}\Bigg[g_1(\xi)\cos\omega t + 2(-1)^{m+1}\omega t\sin\omega t\sin\lambda_m z$$

$$- \sum_{\substack{n=1\\n\neq m}}^{\infty} \frac{4(-1)^{n+1}\sin\lambda_n z\cos\lambda_n ct}{\left[\left(\frac{2n-1}{2m-1}\right)^2 - 1\right]}\Bigg],$$

$$\tau_{\theta z} = \frac{Tr}{(2m-1)\pi}\Bigg[g_2(\xi)\cos\omega t + 2(-1)^{m+1}\omega t\sin\omega t\cos\lambda_m z$$

$$- \sum_{\substack{n=1\\n\neq m}}^{\infty} \frac{4(-1)^{n+1}(2n-1)\cos\lambda_n z\sin\lambda_n ct}{(2m-1)\left[\left(\frac{2n-1}{2m-1}\right)^2 - 1\right]}\Bigg].$$

Beam vibrations

Before we discuss beam vibrations, we briefly review the method of variation of parameters for finding the solution to a nonhomogeneous ordinary differential equation. Let the solutions of the homogeneous differential equation be given by $\{y_1(x), y_2(x), \ldots, y_n(x)\}$, and let $g(x)$ be the forcing function. Then the general solution to the nonhomogeneous equation is given by $y = h_1y_1 + h_2y_2 + \ldots + h_ny_n$, where

$$h_k(x) = \int_0^x \frac{g(\xi)W_k(\xi)}{W(\xi)}\,d\xi + c_k.$$

In the above equation, c_k is a constant (dependent only on the Laplace transform parameter s), while $W(\xi)$ is the Wronskian of $\{y_1(\xi), y_2(\xi), \ldots, y_n(\xi)\}$ and

$$W_k(\xi) = (-1)^{n-k}W[y_1, \ldots, y_{k-1}, y_{k+1}, \ldots, y_n](\xi),$$

is the Wronskian of the homogeneous solution set excluding y_k. Note that $\sum_{k=1}^{n} c_k y_k$ is the homogeneous part, while the remaining is the particular part of the solution. We now apply these ideas to the problem of beam vibrations. For alternative approaches, see [7, 69, 79, 99, 224].

The equation of motion of an Euler–Bernoulli beam model is

$$\rho_0 A \frac{\partial^2 w}{\partial t^2} + EI \frac{\partial^4 w}{\partial x^4} = q(x,t), \tag{6.504}$$

where w, ρ_0, A, E, I denote the transverse deflection, density, cross-sectional area, Young modulus, and moment of inertia of the cross-section, respectively, and $q(x,t)$ is the distributed load per unit length applied to the beam. The above equation is to be solved subject to the initial conditions $w(x,0) = w_0(x)$ and $\dot{w}(x,0) = v_0(x)$. Taking the Laplace transform of the above equation, we get

$$\frac{d^4 \bar{w}}{dx^4} - \alpha^4 \bar{w} = \frac{1}{EI} \left[\bar{q}(x) + \rho_0 A s w_0(x) + \rho_0 A v_0(x) \right]. \tag{6.505}$$

where $\alpha^4 = -\rho_0 A s^2/(EI)$. The homogeneous solution set is $\{\sin\alpha x, \cos\alpha x, \sinh\alpha x, \cosh\alpha x\}$. By the method of variation of parameters described above, the general solution is given by

$$\bar{w} = \bar{c}_1(x)\sin\alpha x + \bar{c}_2(x)\cos\alpha x + \bar{c}_3(x)\sinh\alpha x + \bar{c}_4(x)\cosh\alpha x, \tag{6.506}$$

where

$$\begin{aligned}
\bar{c}_1(x) &= -\frac{1}{2\alpha^3 EI} \int_0^x \left[\bar{q}(\xi) + \rho_0 A s w_0(\xi) + \rho_0 A v_0(\xi) \right] \cos\alpha\xi \, d\xi + c_1, \\
\bar{c}_2(x) &= \frac{1}{2\alpha^3 EI} \int_0^x \left[\bar{q}(\xi) + \rho_0 A s w_0(\xi) + \rho_0 A v_0(\xi) \right] \sin\alpha\xi \, d\xi + c_2, \\
\bar{c}_3(x) &= \frac{1}{2\alpha^3 EI} \int_0^x \left[\bar{q}(\xi) + \rho_0 A s w_0(\xi) + \rho_0 A v_0(\xi) \right] \cosh\alpha\xi \, d\xi + c_3, \\
\bar{c}_4(x) &= -\frac{1}{2\alpha^3 EI} \int_0^x \left[\bar{q}(\xi) + \rho_0 A s w_0(\xi) + \rho_0 A v_0(\xi) \right] \sinh\alpha\xi \, d\xi + c_4.
\end{aligned}$$

The constants c_1, c_2, c_3, c_4 are found from the appropriate boundary conditions, and then the Laplace transform in Eqn. (6.506) is inverted to obtain the solution. We now illustrate this procedure by means of several examples.

First consider the case when a prescribed initial displacement $w_0(x) = \hat{w}_0 \xi(1-\xi)$, where $\xi := x/L$, is given to a simply supported beam of length L supported at the ends $x = 0$ and $x = L$, and then the beam is left free to vibrate. Using Eqn. (6.506) and the boundary conditions

$$\begin{aligned}
\bar{w}(0) &= 0, & \bar{w}(L) &= 0, \\
\bar{w}''(0) &= 0, & \bar{w}''(L) &= 0,
\end{aligned}$$

we get

$$\bar{w} = \frac{\hat{w}_0}{s} \xi (1-\xi) + \frac{\hat{w}_0 \rho_0 A s}{EI\alpha^6 L^2} \left[\frac{\cos\alpha(x - L/2)}{\cos\alpha L/2} - \frac{\cosh\alpha(x - L/2)}{\cosh\alpha L/2} \right].$$

Let $\beta := \sqrt{EI/(\rho_0 A)}$. Then the above equation can be written as

$$\bar{w} = \frac{c_0}{s^2} + \frac{d_0}{s} + \sum_{s=1}^{\infty} \frac{c_n^{(1)}}{s + \lambda_n^2 \beta i} + \frac{c_n^{(2)}}{s - \lambda_n^2 \beta i}. \tag{6.507}$$

where $\lambda_n := (2n-1)\pi/L$. We get $c_0 = \lim_{s\to 0} s^2 \bar{w} = 0$. It then follows that $d_0 = \lim_{s\to 0} s\bar{w} = 0$. The constants $c_n^{(1)}$ and $c_n^{(2)}$ are evaluated using the Heaviside formula (see Appendix F, Vol. II–if $Q(s)$ denotes the denominator, then the derivative dQ/ds in the Heaviside formula is found using the chain rule as $(dQ/d\alpha)(d\alpha/ds)$, where $d\alpha/ds = -s/(2\beta^2\alpha^3)$). We get

$$c_n^{(1)} = c_n^{(2)} = \frac{4\hat{w}_0}{(2n-1)^3\pi^3} \sin \lambda_n x.$$

Thus, the final solution is given by

$$w(x,t) = \hat{w}_0 \sum_{n=1}^{\infty} \frac{8}{(2n-1)^3\pi^3} \sin \lambda_n x \cos \lambda_n^2 \beta t.$$

Note that at $t = 0$, we recover the original shape since

$$\sum_{n=1}^{\infty} \frac{8 \sin(2n-1)\pi\xi}{(2n-1)^3\pi^3} = \xi\,(1-\xi)\,.$$

Now consider the case when the above simply supported beam is loaded by a uniform load $q(x,t) = \hat{q}(t)$. For simplicity, we assume $w_0 = v_0 = 0$. Using Eqn. (6.506), we get

$$\frac{\bar{w}}{\bar{q}} = \frac{P}{Q},$$

where

$$\begin{aligned} P &= (-2 + \cos\alpha x + \cosh\alpha x)\cos\frac{\alpha L}{2}\cosh\frac{\alpha L}{2} + \sin\alpha x \sin\frac{\alpha L}{2}\cosh\frac{\alpha L}{2} \\ &\quad - \sinh\alpha x \sinh\frac{\alpha L}{2}\cos\frac{\alpha L}{2}, \\ Q &= 2EI\alpha^4 \cos\frac{\alpha L}{2}\cosh\frac{\alpha L}{2}. \end{aligned}$$

We can again write $\bar{w}$ as given in Eqn. (6.507), and, as in the previous example, we get $c_0 = d_0 = 0$. Using the Heaviside formula, we get $c_n^{(1)} = -c_n^{(2)}$, and with $\lambda_n := (2n-1)\pi/L$,

$$w(x,t) = \frac{\beta}{EI} \sum_{n=1}^{\infty} c_n \int_0^t \sin(\lambda_n^2 \beta\tau)\hat{q}(t-\tau)\, d\tau, \tag{6.508}$$

where

$$c_n = -\frac{2iEIc_n^{(1)}}{\beta} = \frac{4 \sin\lambda_n x}{\lambda_n^3 L}.$$

If $\hat{q} = q_0$, where q_0 is a constant, then Eqn. (6.508) yields

$$w(x,t) = \frac{q_0}{24EI}x(L^3 - 2Lx^2 + x^3) - \frac{q_0}{EI}\sum_{n=1}^{\infty}\frac{c_n \cos\lambda_n^2\beta t}{\lambda_n^2},$$

where we have used

$$\sum_{n=1}^{\infty}\frac{1}{(2n-1)^5\pi^5}\sin\lambda_n x = \frac{x}{96L}\left[1 - \frac{2x^2}{L^2} + \frac{x^3}{L^3}\right].$$

If $\hat{q}(t) = q_0 \sin\hat{\omega}t$, then from Eqn. (6.508), we get

$$w(x,t) = w_{ss} - \frac{q_0}{k^2EI}\sum_{n=1}^{\infty}\frac{c_n \sin\lambda_n^2\beta t}{\left(\frac{\lambda_n}{k}\right)^4 - 1},$$

where $k := \sqrt{\hat{\omega}/\beta}$, and the periodic steady-state solution w_{ss} is given by

$$w_{ss} = \frac{q_0}{2k^4EI}\left[-2 + \cos kx + \cosh kx + \sin kx \tan\frac{kL}{2} - \sinh kx \tanh\frac{kL}{2}\right]\sin\hat{\omega}t.$$

In deriving the above solution, we have used the fact that

$$\sum_{n=1}^{\infty}\frac{8\sin\lambda_n x}{\lambda_n L\left[\left(\frac{\lambda_n}{k}\right)^4 - 1\right]} = -2 + \cos kx + \cosh kx + \sin kx \tan\frac{kL}{2} - \sinh kx \tanh\frac{kL}{2}. \tag{6.509}$$

If $\hat{q}(t) = q_0 \cos\hat{\omega}t$, then from Eqn. (6.508), we get

$$w(x,t) = w_{ss} - \frac{q_0}{k^4EI}\sum_{n=1}^{\infty}\frac{c_n\lambda_n^2 \cos\lambda_n^2\beta t}{\left(\frac{\lambda_n}{k}\right)^4 - 1},$$

where

$$w_{ss} = \frac{q_0}{2k^4EI}\left[-2 + \cos kx + \cosh kx + \sin kx \tan\frac{kL}{2} - \sinh kx \tanh\frac{kL}{2}\right]\cos\hat{\omega}t.$$

Finally, if $\hat{q}(t) = q_0 e^{-\hat{\omega}t}$, then

$$w(x,t) = \frac{q_0}{k^4EI}\left[\frac{N_2}{D_2}e^{-\hat{\omega}t} - \sum_{n=1}^{\infty}\frac{c_n(\lambda_n^2\cos\lambda_n^2\beta t - k^2\sin\lambda_n^2\beta t)}{\left(\frac{\lambda_n}{k}\right)^4 + 1}\right],$$

where, with $k_1 := \sqrt{\hat{\omega}/(2\beta)} = k/\sqrt{2}$,

$$N_2 = \cos(k_1L) + \cosh(k_1L) - \cos(k_1x)\cosh[k_1(L-x)] - \cos[k_1(L-x)]\cosh(k_1x),$$
$$D_2 = \cos(k_1L) + \cosh(k_1L).$$

In deriving the above, we have used the fact that

$$\sum_{n=1}^{\infty} \frac{4\sin\lambda_n x}{\lambda_n L\left[\left(\frac{\lambda_n}{k}\right)^4+1\right]} = \frac{N_2}{D_2}.$$

While considering the cases $q(t) = q_0 \sin\hat{\omega}t$ and $q(t) = q_0\cos\hat{\omega}t$ above, we implicitly assumed while using Eqn. (6.508) that $k = \sqrt{\hat{\omega}/\beta} \neq \lambda_m$. If, however, $k = \lambda_m = (2m-1)\pi/L$, i.e., $\hat{\omega} = (2m-1)^2\pi^2\beta/L^2$, then 'resonance', where the solution becomes unbounded as $t \to \infty$, occurs. First consider the case where $q(t) = q_0\sin\hat{\omega}t$. From Eqn. (6.508), we get

$$w = \frac{q_0 L^4}{EI(\lambda_m L)^5}\left[g(\xi)\sin\hat{\omega}t + 2(\sin\hat{\omega}t - \hat{\omega}t\cos\hat{\omega}t)\sin(2m-1)\pi\xi - \lambda_m^3 L\sum_{\substack{n=1\\n\neq m}}^{\infty}\frac{c_n\sin\lambda_n^2\beta t}{\left(\frac{2n-1}{2m-1}\right)^4-1}\right],$$

where,

$$g(\xi) = \frac{(2m-1)\pi}{2}\left[(1-2\xi)\cos(2m-1)\pi\xi + \frac{\cosh\frac{(1-2\xi)(2m-1)\pi}{2}}{\cosh\frac{(2m-1)\pi}{2}} + \frac{5}{(2m-1)\pi}\sin(2m-1)\pi\xi - 2\right].$$

and where we have used the fact that

$$\sum_{\substack{n=1\\n\neq m}}^{\infty}\frac{c_n\lambda_n^2}{\left(\frac{2n-1}{2m-1}\right)^4-1} = \frac{g(\xi)}{(2m-1)\pi}.$$

Similarly, when $q(t) = q_0\cos\hat{\omega}t$, we get

$$w = \frac{q_0 L^4}{EI(\lambda_m L)^5}\left[g(\xi)\cos\hat{\omega}t + 2\hat{\omega}t\sin\hat{\omega}t\sin(2m-1)\pi\xi - \lambda_m L\sum_{\substack{n=1\\n\neq m}}^{\infty}\frac{c_n\lambda_n^2\cos\lambda_n^2\beta t}{\left(\frac{2n-1}{2m-1}\right)^4-1}\right].$$

The function $g(\xi)$ is determined as follows. First, we find the general solution for $g(\xi)$ such that

$$\bar{w} = \frac{q_0 L^4}{EI(\lambda_m L)^5}\left[g(\xi)\sin\hat{\omega}t + 2(\sin\hat{\omega}t - \hat{\omega}t\cos\hat{\omega}t)\sin(2m-1)\pi\xi\right]$$

satisfies the governing equation given by Eqn. (6.504). Then the four integration constants are found using the conditions $g|_{\xi=0} = g|_{x=L} = [\partial^2 g/\partial x^2]_{\xi=0} = 0$, and the condition $\int_0^L g\sin(2m-1)\pi\xi\,dx = 0$. The condition $[\partial^2 g/\partial x^2]_{\xi=L} = 0$ is automatically satisfied.

Now consider the vibrations of a simply supported beam of length L that is subjected to a time-varying moment $M(t)$ at one end. Assuming that $w_0(x) = v_0(x) = 0$, from Eqn. (6.506), we obtain

$$\bar{w} = c_1(s)\sin\alpha x + c_2(s)\cos\alpha x + c_3(s)\sinh\alpha x + c_4(s)\cosh\alpha x. \tag{6.510}$$

Using the boundary conditions

$$\bar{w}(0) = 0, \qquad \bar{w}''(0) = 0,$$
$$\bar{w}(L) = 0 \qquad EI\bar{w}''(L) = \bar{M}(s),$$

we get

$$EI\frac{\bar{w}}{\bar{M}} = \frac{N}{D},$$

where

$$N = \sinh \alpha x \sin \alpha L - \sin \alpha x \sinh \alpha L,$$
$$D = 2\alpha^2(\sin \alpha L \sinh \alpha L).$$

Let $\lambda_n := n\pi/L, n = 1, 2, \ldots, \infty$. Since $\lim_{\alpha \to 0} EI\alpha\bar{w}/\bar{M} = 0$, there are no terms of the form $1/s, 1/s^2$, etc. in the expansion for $EI\bar{w}/\bar{M}$. Thus, we can write

$$EI\frac{\bar{w}}{\bar{M}} = \sum_{n=1}^{\infty} \frac{c_n^{(1)}}{s + \beta\lambda_n^2 i} + \frac{c_n^{(2)}}{s - \beta\lambda_n^2 i}.$$

Using the Heaviside formula, we get $c_n^{(1)} = -c_n^{(2)}$, which leads to the final solution given by

$$w(x,t) = \frac{\beta}{EI} \sum_{n=1}^{\infty} c_n \int_0^t \sin(\lambda_n^2 \beta\tau) M(t - \tau)\, d\tau, \tag{6.511}$$

where, the expression for c_n is

$$c_n = -\frac{2ic_n^{(1)}}{\beta} = \frac{2(-1)^n \sin \lambda_n x}{\lambda_n L}.$$

Now we find the solution for some special choices of $M(t)$. If $M(t) = M_0$, where M_0 is a constant, then from Eqn. (6.511), we get

$$w(x,t) = \frac{M_0}{EI}\left[-\frac{xL}{6}\left(1 - \frac{x^2}{L^2}\right) - \sum_{n=1}^{\infty} \frac{c_n \cos \lambda_n^2 \beta t}{\lambda_n^2}\right],$$

where we have used

$$\sum_{n=1}^{\infty} \frac{c_n}{\lambda_n^2} = -\frac{xL}{6}\left(1 - \frac{x^2}{L^2}\right).$$

If $M(t) = M_0 \sin \hat{\omega} t$, then from Eqn. (6.511), we get

$$w(x,t) = \frac{M_0}{k^2 EI}\left[\frac{N_1}{D_1} \sin \hat{\omega} t - \sum_{n=1}^{\infty} \frac{c_n \sin \lambda_n^2 \beta t}{\left(\frac{\lambda_n}{k}\right)^4 - 1}\right],$$

where $k := \sqrt{\hat{\omega}/\beta}$, and

$$N_1 = \sinh kx \sin kL - \sin kx \sinh kL,$$
$$D_1 = 2\sin kL \sinh kL.$$

In deriving the above expression, we have used

$$\sum_{n=1}^{\infty} \frac{c_n\lambda_n^2}{\left(\frac{\lambda_n}{k}\right)^4 - 1} = \frac{k^2 N_1}{D_1}.$$

If $M(t) = M_0 \cos\hat{\omega}t$, then we get

$$w(x,t) = \frac{M_0}{k^4 EI}\left[\frac{k^2 N_1}{D_1}\cos\hat{\omega}t - \sum_{n=1}^{\infty}\frac{c_n\lambda_n^2\cos\lambda_n^2\beta t}{\left(\frac{\lambda_n}{k}\right)^4 - 1}\right].$$

Finally, if $M(t) = M_0 e^{-\hat{\omega}t}$, then

$$w(x,t) = \frac{M_0}{k^4 EI}\left[\frac{2k^2 N_2}{D_2}e^{-\hat{\omega}t} - \sum_{n=1}^{\infty}\frac{c_n(\lambda_n^2\cos\lambda_n^2\beta t - k^2\sin\lambda_n^2\beta t)}{\left(\frac{\lambda_n}{k}\right)^4 + 1}\right],$$

where, with $k_1 := k/\sqrt{2}$,

$$N_2 = \sin k_1x\cosh k_1x\cos k_1L\sinh k_1L - \cos k_1x\sinh k_1x\sin k_1L\cosh k_1L,$$
$$D_2 = \cosh 2k_1L - \cos 2k_1L.$$

In deriving the above expression, we have used

$$\sum_{n=1}^{\infty} \frac{c_n\lambda_n^2}{\left(\frac{\lambda_n}{k}\right)^4 + 1} = \frac{2k^2 N_2}{D_2}.$$

While considering the cases $M(t) = M_0\sin\hat{\omega}t$ and $M(t) = M_0\cos\hat{\omega}t$ above, we implicitly assumed while using Eqn. (6.511) that $k = \sqrt{\hat{\omega}/\beta} \neq \lambda_m$ for some integer m. If, however, $k = \lambda_m$, i.e., $\hat{\omega} = \lambda_m^2\beta$, then 'resonance', where the solution becomes unbounded as $t \to \infty$, occurs. We now present the solution for this case, which is a generalization of the one in [69, 79]. First consider the case where $M(t) = M_0\sin\lambda_m^2\beta t$. From Eqn. (6.511), we get

$$w = \frac{M_0}{\lambda_m^2 EI}\left[g(\xi)\sin\hat{\omega}t + (\sin\hat{\omega}t - \hat{\omega}t\cos\hat{\omega}t)\frac{(-1)^m\sin\lambda_m x}{m\pi} - \sum_{\substack{n=1\\n\neq m}}^{\infty}\frac{c_n\sin\lambda_n^2\beta t}{\left(\frac{n}{m}\right)^4 - 1}\right],$$

where, with $\xi := x/L$,

$$g(\xi) = \frac{1}{4m\pi}\left[\frac{2m\pi\sinh m\pi\xi}{\sinh m\pi} + (-1)^m(\sin m\pi\xi - 2m\pi\xi\cos m\pi\xi)\right].$$

and where we have used the fact that $\lambda_n/\lambda_m = n/m$, and

$$\sum_{\substack{n=1\\ n\neq m}}^{\infty} \frac{8(-1)^n n \sin n\pi\xi}{\left(\frac{n}{m}\right)^4 - 1} = m\left[\frac{2m\pi \sinh m\pi\xi}{\sinh m\pi} + (-1)^m\left(\sin m\pi\xi - 2m\pi\xi\cos m\pi\xi\right)\right].$$

Similarly, when $M(t) = M_0 \cos \lambda_m^2 \beta t$, we get

$$w = \frac{M_0}{\lambda_m^4 EI}\left[\lambda_m^2 g(\xi)\cos\hat{\omega}t + \beta\lambda_m^4 t \sin\hat{\omega}t \frac{(-1)^m \sin\lambda_m x}{m\pi} - \sum_{\substack{n=1\\ n\neq m}}^{\infty} \frac{c_n \lambda_n^2 \cos\lambda_n^2\beta t}{\left(\frac{n}{m}\right)^4 - 1}\right].$$

Instead of a time-varying moment $M(t)$ considered above, if the slope $\theta(t)$ is prescribed at one end, then from Eqn. (6.510) and the boundary conditions

$$\bar{w}(0) = 0, \qquad \bar{w}''(0) = 0,$$
$$\bar{w}(L) = 0 \qquad \bar{w}'(L) = \bar{\theta}(s),$$

we get

$$\frac{\bar{w}}{\bar{\theta}} = \frac{N}{D},$$

where

$$N = \sinh\alpha x \sin\alpha L - \sin\alpha x \sinh\alpha L,$$
$$D = \alpha(\sin\alpha L\cosh\alpha L - \sinh\alpha L\cos\alpha L).$$

Let λ_n, $n = 1, 2, \ldots, \infty$, be the positive roots of $\tan Lx = \tanh Lx$. Since $\lim_{\alpha\to 0} \alpha\bar{w}/\bar{\theta} = 0$, there are no terms of the form $1/s$, $1/s^2$, etc. in the expansion for $\bar{w}/\bar{\theta}$. Thus, we can write

$$\frac{\bar{w}}{\bar{\theta}} = \sum_{n=1}^{\infty} \frac{c_n^{(1)}}{s + \beta\lambda_n^2 i} + \frac{c_n^{(2)}}{s - \beta\lambda_n^2 i}.$$

Using the Heaviside formula, we get $c_n^{(1)} = -c_n^{(2)}$, which leads to the final solution given by

$$w(x,t) = \frac{\beta}{L}\sum_{n=1}^{\infty} c_n \int_0^t \sin(\lambda_n^2\beta\tau)\theta(t-\tau)\,d\tau, \tag{6.512}$$

where, the expression for c_n is

$$c_n = -\frac{2iLc_n^{(1)}}{\beta} = \frac{2(\sin\lambda_n x \sinh\lambda_n L - \sinh\lambda_n x \sin\lambda_n L)}{\sin\lambda_n L \sinh\lambda_n L}.$$

Now we find the solution for some special choices of $\theta(t)$. If $\theta(t) = \theta_0$, where θ_0 is a constant, then from Eqn. (6.512), we get

$$w(x,t) = \frac{\theta_0}{L}\left[-\frac{xL}{2}\left(1 - \frac{x^2}{L^2}\right) - \sum_{n=1}^{\infty} \frac{c_n \cos\lambda_n^2\beta t}{\lambda_n^2}\right],$$

where we have used

$$\sum_{n=1}^{\infty} \frac{c_n}{\lambda_n^2} = -\frac{xL}{2}\left(1 - \frac{x^2}{L^2}\right).$$

If $\theta(t) = \theta_0 \sin \hat{\omega} t$, then from Eqn. (6.512), we get

$$w(x,t) = \frac{\theta_0}{k^2 L}\left[\frac{N_1}{D_1}\sin \hat{\omega} t - \sum_{n=1}^{\infty} \frac{c_n \sin \lambda_n^2 \beta t}{\left(\frac{\lambda_n}{k}\right)^4 - 1}\right],$$

where $k := \sqrt{\hat{\omega}/\beta}$, and

$$N_1 = kL(\sinh kx \sin kL - \sin kx \sinh kL),$$
$$D_1 = \sin kL \cosh kL - \sinh kL \cos kL.$$

In deriving the above expression, we have used

$$\sum_{n=1}^{\infty} \frac{c_n \lambda_n^2}{\left(\frac{\lambda_n}{k}\right)^4 - 1} = \frac{k^2 N_1}{D_1}.$$

If $\theta(t) = \theta_0 \cos \hat{\omega} t$, then we get

$$w(x,t) = \frac{\theta_0}{k^4 L}\left[\frac{k^2 N_1}{D_1}\cos \hat{\omega} t - \sum_{n=1}^{\infty} \frac{c_n \lambda_n^2 \cos \lambda_n^2 \beta t}{\left(\frac{\lambda_n}{k}\right)^4 - 1}\right].$$

Finally, if $\theta(t) = \theta_0 e^{-\hat{\omega} t}$, then

$$w(x,t) = \frac{\theta_0}{k^4 L}\left[\frac{2\sqrt{2}k^3 L N_2}{D_2} e^{-\hat{\omega} t} - \sum_{n=1}^{\infty} \frac{c_n(\lambda_n^2 \cos \lambda_n^2 \beta t - k^2 \sin \lambda_n^2 \beta t)}{\left(\frac{\lambda_n}{k}\right)^4 + 1}\right],$$

where, with $k_1 := k/\sqrt{2}$,

$$N_2 = \sin k_1 x \cosh k_1 x \cos k_1 L \sinh k_1 L - \cos k_1 x \sinh k_1 x \sin k_1 L \cosh k_1 L,$$
$$D_2 = \sinh 2k_1 L - \sin 2k_1 L.$$

In deriving the above expression, we have used

$$\sum_{n=1}^{\infty} \frac{c_n \lambda_n^2}{\left(\frac{\lambda_n}{k}\right)^4 + 1} = \frac{2\sqrt{2}k^3 L N_2}{D_2}.$$

While considering the cases $\theta(t) = \theta_0 \sin \hat{\omega} t$ and $\theta(t) = \theta_0 \cos \hat{\omega} t$ above, we implicitly assumed while using Eqn. (6.512) that $k = \sqrt{\hat{\omega}/\beta} \neq \lambda_m$ for some integer m. If, however,

$k = \lambda_m$, i.e., $\hat{\omega} = \lambda_m^2 \beta$, then 'resonance', where the solution becomes unbounded as $t \to \infty$, occurs. First consider the case where $\theta(t) = \theta_0 \sin \lambda_m^2 \beta t$. From Eqn. (6.512), we get

$$w = \frac{\theta_0}{\lambda_m^2 L} \left[g(x) \sin \hat{\omega} t + (\sin \hat{\omega} t - \hat{\omega} t \cos \hat{\omega} t) \left(\frac{\sin \lambda_m x}{\sin \lambda_m L} - \frac{\sinh \lambda_m x}{\sinh \lambda_m L} \right) - \sum_{\substack{n=1 \\ n \neq m}}^{\infty} \frac{c_n \sin \lambda_n^2 \beta t}{\left(\frac{\lambda_n}{\lambda_m} \right)^4 - 1} \right],$$

where,

$$g(x) = \frac{1}{4} \left[\frac{\sinh \lambda_m x + 2\lambda_m x \cosh \lambda_m x}{\sinh \lambda_m L} - \frac{\sin \lambda_m x + 2\lambda_m x \cos \lambda_m x}{\sin \lambda_m L} \right],$$

and where we have used the fact that

$$\sum_{\substack{n=1 \\ n \neq m}}^{\infty} \frac{c_n \lambda_n^2}{\left(\frac{\lambda_n}{\lambda_m} \right)^4 - 1} = \lambda_m^2 g(x).$$

Similarly, when $\theta(t) = \theta_0 \cos \lambda_m^2 \beta t$, we get

$$w = \frac{\theta_0}{\lambda_m^4 L} \left[\lambda_m^2 g(x) \cos \hat{\omega} t + \beta \lambda_m^4 t \sin \hat{\omega} t \left(\frac{\sin \lambda_m x}{\sin \lambda_m L} - \frac{\sinh \lambda_m x}{\sinh \lambda_m L} \right) - \sum_{\substack{n=1 \\ n \neq m}}^{\infty} \frac{c_n \lambda_n^2 \cos \lambda_n^2 \beta t}{\left(\frac{\lambda_n}{\lambda_m} \right)^4 - 1} \right].$$

Next, consider the oscillations of a cantilever beam of length L fixed at $x = 0$, due to an initial deflection $w_0(x) = \hat{w}_0 x^2 / L^2$. Using Eqn. (6.506) and the boundary conditions

$$\bar{w}(0) = 0, \qquad \bar{w}''(L) = 0,$$
$$\bar{w}'(0) = 0, \qquad \bar{w}'''(L) = 0,$$

we get

$$\frac{L^2 \bar{w}}{\hat{w}_0} = \frac{N}{D},$$

where

$$N = -s \Big[\alpha^2 x^2 + \cos \alpha (L - x) + \big[\alpha^2 x^2 \cos \alpha L + \cos \alpha x \big] \cosh \alpha L$$
$$- (\cos \alpha L + \cosh \alpha L) \cosh \alpha x - \sin \alpha L \sinh \alpha x + (\sinh \alpha x - \sin \alpha x) \sinh \alpha L \Big],$$
$$D = \alpha^6 \beta^2 (1 + \cos \alpha L \cosh \alpha L).$$

Let λ_n, $n = 1, 2, \ldots, \infty$, be the positive roots of $1 + \cos Lx \cosh Lx = 0$. Since $\lim_{\alpha \to 0} L^2 \alpha \bar{w} / \hat{w}_0 = 0$, there are no terms of the form $1/s$, $1/s^2$, etc. in the expansion for $L^2 \bar{w} / \hat{w}_0$. Thus, we can write

$$\frac{L^2 \bar{w}}{\hat{w}_0} = \sum_{n=1}^{\infty} \frac{c_n^{(1)}}{s + \beta \lambda_n^2 i} + \frac{c_n^{(2)}}{s - \beta \lambda_n^2 i}.$$

Using the Heaviside formula, we get $c_n^{(1)} = c_n^{(2)}$, which leads to the final solution given by

$$w(x,t) = \hat{w}_0 \sum_{n=1}^{\infty} c_n \cos \lambda_n^2 \beta t,$$

where, with

$$N_1 = 4\Big[\lambda_n^2 x^2 + \cos \lambda_n (L-x) + \Big[\lambda_n^2 x^2 \cos \lambda_n L + \cos \lambda_n x\Big] \cosh \lambda_n L$$
$$- (\cos \lambda_n L + \cosh \lambda_n L) \cosh \lambda_n x - \sin \lambda_n L \sinh \lambda_n x + (\sinh \lambda_n x - \sin \lambda_n x) \sinh \lambda_n L\Big],$$
$$D_1 = \lambda_n^3 L^3 (\cos \lambda_n L \sinh \lambda_n L - \cosh \lambda_n L \sin \lambda_n L),$$

we have

$$c_n = \frac{N_1}{D_1}.$$

At time $t = 0$, we get $w(x,0) = \hat{w}_0 x^2 / L^2$, since

$$\sum_{n=1}^{\infty} c_n = \frac{x^2}{L^2}.$$

Now consider the vibrations of a cantilever beam of length L that is subjected to a time-varying end-moment $M(t)$. Assuming that $w_0(x) = v_0(x) = 0$, from Eqn. (6.506), we again obtain Eqn. (6.510). Using the boundary conditions

$$\bar{w}(0) = 0, \qquad \bar{w}'(0) = 0,$$
$$EI\bar{w}''(L) = \bar{M}(s), \qquad \bar{w}'''(L) = 0,$$

we get

$$EI\frac{\bar{w}}{\bar{M}} = \frac{N}{D},$$

where

$$N = (\cosh \alpha x - \cos \alpha x)(\cos \alpha L + \cosh \alpha L) + (\sinh \alpha x - \sin \alpha x)(\sin \alpha L - \sinh \alpha L),$$
$$D = 2\alpha^2 (1 + \cos \alpha L \cosh \alpha L).$$

Let λ_n, $n = 1, 2, \ldots, \infty$, be the positive roots of $1 + \cos Lx \cosh Lx = 0$. Then, we can write

$$EI\frac{\bar{w}}{\bar{M}} = \frac{c_0}{s} + \sum_{n=1}^{\infty} \frac{c_n^{(1)}}{s + \beta \lambda_n^2 i} + \frac{c_n^{(2)}}{s - \beta \lambda_n^2 i}.$$

The constant $c_0 = EI \lim_{s \to 0} s\bar{w}/\bar{M} = 0$. Using the Heaviside formula, we get $c_n^{(1)} = -c_n^{(2)}$, which leads to the final solution given by

$$w(x,t) = \frac{\beta}{EI} \sum_{n=1}^{\infty} c_n \int_0^t \sin(\lambda_n^2 \beta \tau) M(t-\tau)\, d\tau, \tag{6.513}$$

where, with

$$\hat{N} = -2\Big[(\cosh\lambda_n x - \cos\lambda_n x)(\cos\lambda_n L + \cosh\lambda_n L) + (\sinh\lambda_n x - \sin\lambda_n x)(\sin\lambda_n L - \sinh\lambda_n L)\Big],$$

$$\hat{D} = \lambda_n L(\cos\lambda_n L \sinh\lambda_n L - \cosh\lambda_n L \sin\lambda_n L),$$

the expression for c_n is

$$c_n = -\frac{2ic_n^{(1)}}{\beta} = \frac{\hat{N}}{\hat{D}}.$$

Now we find the solution for some special choices of $M(t)$. If $M(t) = M_0$, where M_0 is a constant, then from Eqn. (6.513), we get

$$w(x,t) = \frac{M_0}{EI}\left[\frac{x^2}{2} - \sum_{n=1}^{\infty}\frac{c_n\cos\lambda_n^2\beta t}{\lambda_n^2}\right],$$

where we have used

$$\sum_{n=1}^{\infty}\frac{c_n}{\lambda_n^2} = \frac{x^2}{2}.$$

If $M(t) = M_0\sin\hat{\omega}t$, then from Eqn. (6.513), we get

$$w(x,t) = \frac{M_0}{k^2EI}\left[\frac{N_1}{D_1}\sin\hat{\omega}t - \sum_{n=1}^{\infty}\frac{c_n\sin\lambda_n^2\beta t}{\left(\frac{\lambda_n}{k}\right)^4 - 1}\right],$$

where $k := \sqrt{\hat{\omega}/\beta}$, and

$$N_1 = (\cosh kx - \cos kx)(\cos kL + \cosh kL) + (\sinh kx - \sin kx)(\sin kL - \sinh kL),$$

$$D_1 = 2(1 + \cos kL\cosh kL).$$

In deriving the above expression, we have used

$$\sum_{n=1}^{\infty}\frac{c_n\lambda_n^2}{\left(\frac{\lambda_n}{k}\right)^4 - 1} = \frac{k^2N_1}{D_1}.$$

If $M(t) = M_0\cos\hat{\omega}t$, then we get

$$w(x,t) = \frac{M_0}{k^4EI}\left[\frac{k^2N_1}{D_1}\cos\hat{\omega}t - \sum_{n=1}^{\infty}\frac{c_n\lambda_n^2\cos\lambda_n^2\beta t}{\left(\frac{\lambda_n}{k}\right)^4 - 1}\right].$$

Finally, if $M(t) = M_0e^{-\hat{\omega}t}$, then

$$w(x,t) = \frac{M_0}{k^4EI}\left[\frac{2k^2N_2}{D_2}e^{-\hat{\omega}t} - \sum_{n=1}^{\infty}\frac{c_n(\lambda_n^2\cos\lambda_n^2\beta t - k^2\sin\lambda_n^2\beta t)}{\left(\frac{\lambda_n}{k}\right)^4 + 1}\right],$$

where, with $k_1 := k/\sqrt{2}$,

$$\begin{aligned} N_2 &= 2\sin k_1x \sinh k_1x \cos k_1L \cosh k_1L \\ &\quad + [\cos k_1x \sinh k_1x - \cosh k_1x \sin k_1x]\, [\cos k_1L \sinh k_1L - \cosh k_1L \sin k_1L]\,, \\ D_2 &= 2 + \cos 2k_1L + \cosh 2k_1L. \end{aligned}$$

In deriving the above expression, we have used

$$\sum_{n=1}^{\infty} \frac{c_n\lambda_n^2}{\left(\frac{\lambda_n}{k}\right)^4 + 1} = \frac{2k^2N_2}{D_2}.$$

If $M = M_0 \sin \hat{\omega}t$ or $M = M_0 \cos \hat{\omega}t$ with $\hat{\omega} = \beta\lambda_m^2$ (i.e., $k = \lambda_m$) for some integer m, then resonance occurs, with the displacement increasing linearly with t.

If instead of a time-varying moment, we have a time-varying force $P(t)$ at the end of the cantilever, then imposing the boundary conditions

$$\begin{aligned} \bar{w}(0) &= 0, & \bar{w}'(0) &= 0, \\ \bar{w}''(L) &= 0, & EI\bar{w}'''(L) &= -\bar{P}(s), \end{aligned}$$

on the solution in Eqn. (6.510), we get

$$EI\frac{\bar{w}}{\bar{P}} = \frac{N}{D},$$

where

$$\begin{aligned} N &= (\cosh \alpha x - \cos \alpha x)(\sin \alpha L + \sinh \alpha L) + (\sin \alpha x - \sinh \alpha x)(\cos \alpha L + \cosh \alpha L), \\ D &= 2\alpha^3(1 + \cos \alpha L \cosh \alpha L). \end{aligned}$$

Let λ_n, $n = 1, 2, \ldots, \infty$, be the positive roots of $1 + \cos Lx \cosh Lx = 0$. Then, since $\lim_{\alpha \to 0} \alpha EI\bar{w}/\bar{P} = 0$, there are no terms of the type $1/s$, $1/s^2$ etc. in the expansion for $\bar{w}/\bar{P}$, and we can write

$$EI\frac{\bar{w}}{\bar{P}} = \sum_{n=1}^{\infty} \frac{d_n^{(1)}}{s + \beta\lambda_n^2 i} + \frac{d_n^{(2)}}{s - \beta\lambda_n^2 i}.$$

Using the Heaviside formula, we get $d_n^{(1)} = -d_n^{(2)}$, which leads to the final solution given by

$$w(x,t) = \frac{\beta}{EI}\sum_{n=1}^{\infty} d_n \int_0^t \sin(\lambda_n^2\beta\tau)P(t-\tau)\,d\tau, \tag{6.514}$$

where, with

$$\begin{aligned} \hat{N} = 2\Big[&(\cos \lambda_n x - \cosh \lambda_n x)(\sin \lambda_n L + \sinh \lambda_n L) \\ &+ (\sinh \lambda_n x - \sin \lambda_n x)(\cos \lambda_n L + \cosh \lambda_n L)\Big], \end{aligned}$$

$$\hat{D} = \lambda_n^2 L(\cos\lambda_n L \sinh\lambda_n L - \cosh\lambda_n L \sin\lambda_n L),$$

the expression for d_n is

$$d_n = -\frac{2id_n^{(1)}}{\beta} = \frac{\hat{N}}{\hat{D}}.$$

Now we find the solution for some special choices of $P(t)$. If $P(t) = P_0$, where P_0 is a constant, then from Eqn. (6.514), we get

$$w(x,t) = \frac{P_0}{EI}\left[\frac{1}{6}x^2(3L-x) - \sum_{n=1}^{\infty}\frac{d_n\cos\lambda_n^2\beta t}{\lambda_n^2}\right],$$

where we have used

$$\sum_{n=1}^{\infty}\frac{d_n}{\lambda_n^2} = \frac{1}{6}x^2(3L-x).$$

If $P(t) = P_0\sin\hat{\omega}t$, then from Eqn. (6.514), we get

$$w(x,t) = \frac{P_0}{k^3EI}\left[\frac{N_1}{D_1}\sin\hat{\omega}t - \sum_{n=1}^{\infty}\frac{d_n k\sin\lambda_n^2\beta t}{\left(\frac{\lambda_n}{k}\right)^4 - 1}\right],$$

where $k := \sqrt{\hat{\omega}/\beta}$, and

$$N_1 = (\cosh kx - \cos kx)(\sin kL + \sinh kL) + (\sin kx - \sinh kx)(\cos kL + \cosh kL),$$
$$D_1 = 2(1 + \cos kL\cosh kL).$$

In deriving the above expression, we have used

$$\sum_{n=1}^{\infty}\frac{d_n\lambda_n^2}{\left(\frac{\lambda_n}{k}\right)^4 - 1} = \frac{kN_1}{D_1}.$$

If $P(t) = P_0\cos\hat{\omega}t$, then we get

$$w(x,t) = \frac{P_0}{k^4EI}\left[\frac{kN_1}{D_1}\cos\hat{\omega}t - \sum_{n=1}^{\infty}\frac{d_n\lambda_n^2\cos\lambda_n^2\beta t}{\left(\frac{\lambda_n}{k}\right)^4 - 1}\right].$$

Finally, if $P(t) = P_0e^{-\hat{\omega}t}$, then

$$w(x,t) = \frac{P_0}{k^4EI}\left[\frac{4k_1N_2}{D_2}e^{-\hat{\omega}t} - \sum_{n=1}^{\infty}\frac{d_n(\lambda_n^2\cos\lambda_n^2\beta t - k^2\sin\lambda_n^2\beta t)}{\left(\frac{\lambda_n}{k}\right)^4 + 1}\right],$$

where, with $k_1 := k/\sqrt{2}$,

$$N_2 = \cos k_1(L-x)\cosh k_1 L \sinh k_1 x - \cos k_1 L \cosh k_1(L-x)\sin k_1 x,$$
$$D_2 = 2 + \cos 2k_1 L + \cosh 2k_1 L.$$

In deriving the above expression, we have used

$$\sum_{n=1}^{\infty} \frac{d_n \lambda_n^2}{\left(\frac{\lambda_n}{k}\right)^4 + 1} = \frac{4k_1 N_2}{D_2}.$$

If $P = P_0 \sin \hat{\omega} t$ or $P = P_0 \cos \hat{\omega} t$ with $\hat{\omega} = \beta \lambda_m^2$ (i.e., $k = \lambda_m$) for some integer m, then resonance occurs, with the displacement increasing linearly with t.

Now we consider a prescribed time-varying displacement $W(t)$ at the free end of the cantilever. Imposing the boundary conditions

$$\bar{w}(0) = 0, \qquad \bar{w}'(0) = 0,$$
$$\bar{w}(L) = \bar{W}(s), \qquad \bar{w}''(L) = 0,$$

on Eqn. (6.510), we get

$$\frac{\bar{w}}{\bar{W}} = \frac{N}{D},$$

where

$$N = (\cosh \alpha x - \cos \alpha x)(\sin \alpha L + \sinh \alpha L) - (\sinh \alpha x - \sin \alpha x)(\cos \alpha L + \cosh \alpha L),$$
$$D = 2(\cosh \alpha L \sin \alpha L - \cos \alpha L \sinh \alpha L).$$

Let λ_n, $n = 1, 2, \ldots, \infty$, be the positive roots of $\tan Lx = \tanh Lx$. Then, since $\lim_{\alpha \to 0} \alpha \bar{w}/\bar{W} = 0$, we can write

$$\frac{\bar{w}}{\bar{W}} = \sum_{n=1}^{\infty} \frac{c_n^{(1)}}{s + \beta \lambda_n^2 i} + \frac{c_n^{(2)}}{s - \beta \lambda_n^2 i}.$$

Using the Heaviside formula, we get $c_n^{(1)} = -c_n^{(2)}$, which leads to the final solution given by

$$w(x,t) = \frac{\beta}{L} \sum_{n=1}^{\infty} c_n \int_0^t \sin(\lambda_n^2 \beta \tau) W(t-\tau)\, d\tau, \tag{6.515}$$

where, with

$$\hat{N} = \lambda_n \Big[(\sinh \lambda_n x - \sin \lambda_n x)(\cos \lambda_n L + \cosh \lambda_n L)$$
$$- (\cosh \lambda_n x - \cos \lambda_n x)(\sin \lambda_n L + \sinh \lambda_n L) \Big],$$
$$\hat{D} = \sin \lambda_n L \sinh \lambda_n L,$$

the expression for c_n is

$$c_n = -\frac{2iLc_n^{(1)}}{\beta} = \frac{\hat{N}}{\hat{D}}.$$

Now we find the solution for some special choices of $W(t)$. If $W(t) = w_0$, where w_0 is a constant, then from Eqn. (6.515), we get

$$w(x,t) = w_0 \left[\frac{3x^2}{L^2} - \frac{x^3}{L^3} - \frac{1}{L} \sum_{n=1}^{\infty} \frac{c_n \cos \lambda_n^2 \beta t}{\lambda_n^2} \right],$$

where we have used

$$\sum_{n=1}^{\infty} \frac{c_n}{\lambda_n^2} = \frac{L}{2} \left[\frac{3x^2}{L^2} - \frac{x^3}{L^3} \right], \quad x \in [0, L). \tag{6.516}$$

If $W(\tau) = p\tau^2$, then

$$w(x,t) = \frac{pt^2}{2} \left[\frac{3x^2}{L^2} - \frac{x^3}{L^3} \right] + \frac{pL^4}{840\beta^2} \left[\frac{x^7}{L^7} - \frac{7x^6}{L^6} + \frac{39x^3}{L^3} - \frac{33x^2}{L^2} \right] + \frac{2p}{\beta^2 L} \sum_{n=1}^{\infty} \frac{c_n \cos \lambda_n^2 \beta t}{\lambda_n^6},$$

where we have used Eqn. (6.516) and

$$\sum_{n=1}^{\infty} \frac{-2c_n}{\lambda_n^6} = \frac{L^5}{840} \left[\frac{x^7}{L^7} - \frac{7x^6}{L^6} + \frac{39x^3}{L^3} - \frac{33x^2}{L^2} \right].$$

If $W(t) = w_0 \sin \hat{\omega} t$, then from Eqn. (6.515), we get

$$w(x,t) = w_0 \left[\frac{N_1}{D_1} \sin \hat{\omega} t - \frac{1}{k^2 L} \sum_{n=1}^{\infty} \frac{c_n \sin \lambda_n^2 \beta t}{\left(\frac{\lambda_n}{k}\right)^4 - 1} \right],$$

where $k := \sqrt{\hat{\omega}/\beta}$, and

$$N_1 = (\cosh kx - \cos kx)(\sin kL + \sinh kL) - (\sinh kx - \sin kx)(\cos kL + \cosh kL),$$
$$D_1 = 2(\cosh kL \sin kL - \cos kL \sinh kL).$$

In deriving the above expression, we have used

$$\sum_{n=1}^{\infty} \frac{c_n \lambda_n^2}{\left(\frac{\lambda_n}{k}\right)^4 - 1} = \frac{k^4 L N_1}{D_1}, \quad x \in [0, L).$$

If $W(t) = w_0 \cos \hat{\omega} t$, then we get

$$w(x,t) = w_0 \left[\frac{N_1}{D_1} \cos \hat{\omega} t - \frac{1}{k^4 L} \sum_{n=1}^{\infty} \frac{c_n \lambda_n^2 \cos \lambda_n^2 \beta t}{\left(\frac{\lambda_n}{k}\right)^4 - 1} \right].$$

Finally, if $W(t) = w_0 e^{-\hat{\omega}t}$, then

$$w(x,t) = w_0 \left[\frac{N_2}{D_2} e^{-\hat{\omega}t} - \frac{1}{k^4 L} \sum_{n=1}^{\infty} \frac{c_n(\lambda_n^2 \cos\lambda_n^2\beta t - k^2 \sin\lambda_n^2\beta t)}{\left(\frac{\lambda_n}{k}\right)^4 + 1} \right],$$

where, with $k_1 := k/\sqrt{2}$,

$$N_2 = 2\left[\cos k_1 L \cosh k_1(L-x) \sin k_1 x - \cos k_1(L-x) \cosh k_1 L \sinh k_1 x\right],$$
$$D_2 = \sin 2k_1 L - \sinh 2k_1 L.$$

In deriving the above expression, we have used

$$\sum_{n=1}^{\infty} \frac{c_n \lambda_n^2}{\left(\frac{\lambda_n}{k}\right)^4 + 1} = \frac{k^4 L N_2}{D_2}, \quad x \in [0, L).$$

In the cases where $W(t) = w_0 \sin\hat{\omega}t$ or $W(t) = w_0 \cos\hat{\omega}t$, if $\hat{\omega} = \beta\lambda_m^2$ for some integer m, then resonance occurs, with the displacement increasing linearly with t.

Now consider a prescribed time-varying displacement $W(t)$ at the fixed end of the cantilever instead of the free end. Imposing the boundary conditions

$$\bar{w}(0) = \bar{W}(s), \qquad \bar{w}'(0) = 0,$$
$$\bar{w}''(L) = 0, \qquad \bar{w}'''(L) = 0,$$

on the solution in Eqn. (6.510), we get

$$\frac{\bar{w}}{\bar{W}} = \frac{N}{D},$$

where

$$N = \left[\cosh\alpha(L-x) + \cos\alpha(L-x)\right]\left[\cosh\alpha L + \cos\alpha L\right] + \left[\sinh\alpha(L-x) + \sin\alpha(L-x)\right]\left[\sin\alpha L - \sinh\alpha L\right],$$
$$D = 2(1 + \cos\alpha L \cosh\alpha L).$$

Let λ_n, $n = 1, 2, \ldots, \infty$, be the positive roots of $1 + \cos Lx \cosh Lx = 0$. Then, since $\lim_{\alpha\to 0} \alpha\bar{w}/\bar{W} = 0$, we can write

$$\frac{\bar{w}}{\bar{W}} = \sum_{n=1}^{\infty} \frac{d_n^{(1)}}{s + \beta\lambda_n^2 i} + \frac{d_n^{(2)}}{s - \beta\lambda_n^2 i}.$$

Using the Heaviside formula, we get $d_n^{(1)} = -d_n^{(2)}$, which leads to the final solution given by

$$w(x,t) = \frac{\beta}{L} \sum_{n=1}^{\infty} d_n \int_0^t \sin(\lambda_n^2 \beta\tau) W(t-\tau)\, d\tau, \tag{6.517}$$

where, with

$$\hat{N} = 2\lambda_n \Big\{ [\cosh \lambda_n (L - x) + \cos \lambda_n (L - x)] [\cosh \lambda_n L + \cos \lambda_n L] + [\sinh \lambda_n (L - x) + \sin \lambda_n (L - x)] [\sin \lambda_n L - \sinh \lambda_n L] \Big\},$$

$$\hat{D} = \cosh \lambda_n L \sin \lambda_n L - \cos \lambda_n L \sinh \lambda_n L,$$

the expression for d_n is

$$d_n = -\frac{2iLd_n^{(1)}}{\beta} = \frac{\hat{N}}{\hat{D}}.$$

Now we find the solution for some special choices of $W(t)$. If $W(t) = w_0$, where w_0 is a constant, then from Eqn. (6.517), we get

$$w(x,t) = w_0 \left[1 - \frac{1}{L} \sum_{n=1}^{\infty} \frac{d_n \cos \lambda_n^2 \beta t}{\lambda_n^2} \right],$$

where we have used

$$\sum_{n=1}^{\infty} \frac{d_n}{\lambda_n^2} = L, \quad x \in (0, L].$$

If $W(t) = w_0 \sin \hat{\omega} t$, then from Eqn. (6.517), we get

$$w(x,t) = w_0 \left[\frac{N_1}{D_1} \sin \hat{\omega} t - \frac{1}{k^2 L} \sum_{n=1}^{\infty} \frac{d_n \sin \lambda_n^2 \beta t}{\left(\frac{\lambda_n}{k}\right)^4 - 1} \right],$$

where $k := \sqrt{\hat{\omega}/\beta}$, and

$$N_1 = [\cosh k(L - x) + \cos k(L - x)] [\cosh kL + \cos kL] + [\sinh k(L - x) + \sin k(L - x)] [\sin kL - \sinh kL],$$

$$D_1 = 2(1 + \cos kL \cosh kL).$$

In deriving the above expression, we have used

$$\sum_{n=1}^{\infty} \frac{d_n \lambda_n^2}{\left(\frac{\lambda_n}{k}\right)^4 - 1} = \frac{k^4 L N_1}{D_1}, \quad x \in (0, L].$$

If $W(t) = w_0 \cos \hat{\omega} t$, then we get

$$w(x,t) = w_0 \left[\frac{N_1}{D_1} \cos \hat{\omega} t - \frac{1}{k^4 L} \sum_{n=1}^{\infty} \frac{d_n \lambda_n^2 \cos \lambda_n^2 \beta t}{\left(\frac{\lambda_n}{k}\right)^4 - 1} \right].$$

Finally, if $W(t) = w_0 e^{-\hat{\omega}t}$, then

$$w(x,t) = w_0 \left[\frac{N_2}{D_2} e^{-\hat{\omega}t} - \frac{1}{k^4 L} \sum_{n=1}^{\infty} \frac{d_n(\lambda_n^2 \cos \lambda_n^2 \beta t - k^2 \sin \lambda_n^2 \beta t)}{\left(\frac{\lambda_n}{k}\right)^4 + 1} \right],$$

where, with $k_1 := k/\sqrt{2}$,

$$\begin{aligned} N_2 &= 4 \cos k_1(L-x) \cosh k_1(L-x) \cos k_1 L \cosh k_1 L \\ &\quad + 2 \left[\cosh k_1(L-x) \sin k_1(L-x) + \cos k_1(L-x) \sinh k_1(L-x)\right] \\ &\quad \left[\cosh k_1 L \sin k_1 L - \cos k_1 L \sinh k_1 L\right], \\ D_2 &= 2 + \cos 2k_1 L + \cosh 2k_1 L. \end{aligned}$$

In deriving the above expression, we have used

$$\sum_{n=1}^{\infty} \frac{d_n \lambda_n^2}{\left(\frac{\lambda_n}{k}\right)^4 + 1} = \frac{k^4 L N_2}{D_2}, \quad x \in (0, L].$$

In the cases where $W(t) = w_0 \sin \hat{\omega}t$ or $W(t) = w_0 \cos \hat{\omega}t$, if $\hat{\omega} = \beta \lambda_m^2$ for some integer m, then resonance occurs, with the displacement increasing linearly with t.

Axisymmetric motion of circular plates

The governing equation for a thin plate is

$$D \boldsymbol{\nabla}^4 w + \rho_0 h \ddot{w} = q,$$

where $D = Eh^3/[12(1-\nu^2)]$, w represents the transverse displacement, and q is the distributed load (force per unit area) acting on the plate. If C represents the boundary of Ω, and $\boldsymbol{n}$ is the unit normal to C, the boundary conditions to be imposed on C are

w is prescribed or Vertical shear force is prescribed,

$(\boldsymbol{\nabla} w) \cdot \boldsymbol{n}$ is prescribed or $D\left\{\left[\nu(\boldsymbol{\nabla}^2 w)\boldsymbol{I} + (1-\nu)\boldsymbol{\nabla}(\boldsymbol{\nabla} w)\right]\right\}$ is prescribed.

As an example, we have

$$\boldsymbol{\nabla}(\boldsymbol{\nabla} w) = \begin{bmatrix} \frac{\partial^2 w}{\partial x^2} & \frac{\partial^2 w}{\partial x \partial y} \\ \frac{\partial^2 w}{\partial x \partial y} & \frac{\partial^2 w}{\partial y^2} \end{bmatrix}, \qquad \text{(Cartesian system)}$$

$$\boldsymbol{\nabla}(\boldsymbol{\nabla} w) = \begin{bmatrix} \frac{\partial^2 w}{\partial r^2} & \frac{1}{r}\left(\frac{\partial^2 w}{\partial r \partial \theta} - \frac{1}{r}\frac{\partial w}{\partial \theta}\right) \\ \frac{1}{r}\left(\frac{\partial^2 w}{\partial r \partial \theta} - \frac{1}{r}\frac{\partial w}{\partial \theta}\right) & \frac{1}{r}\left(\frac{1}{r}\frac{\partial^2 w}{\partial \theta^2} + \frac{\partial w}{\partial r}\right) \end{bmatrix}, \qquad \text{(Polar system)}$$

where we have used Eqn. (B.5) in obtaining the latter equation.

For the axisymmetrical motion of a circular plate, the governing equation is given by

$$D \left[\frac{1}{r}\frac{\partial}{\partial r}\left(r \frac{\partial}{\partial r}\right)\right] \left[\frac{1}{r}\frac{\partial}{\partial r}\left(r \frac{\partial w}{\partial r}\right)\right] + \rho_0 h \ddot{w} = q. \tag{6.518}$$

while the permissible boundary conditions are

$$w \text{ is prescribed} \quad \text{or} \quad D\frac{\partial}{\partial r}\left[\frac{1}{r}\frac{\partial}{\partial r}\left(r\frac{\partial w}{\partial r}\right)\right] \text{ is prescribed,}$$
$$\frac{\partial w}{\partial r} \text{ is prescribed} \quad \text{or} \quad D\left[\frac{\partial^2 w}{\partial r^2}+\frac{\nu}{r}\frac{\partial w}{\partial r}\right] \text{ is prescribed.}$$

Taking the Laplace transform of Eqn. (6.518), we get

$$\left[\frac{1}{r}\frac{d}{dr}\left(r\frac{d}{dr}\right)\right]\left[\frac{1}{r}\frac{d}{dr}\left(r\frac{d\bar{w}}{dr}\right)\right]-\alpha^4\bar{w}=\frac{1}{D}\left[\bar{q}+\rho_0 hsw_0+\rho_0 hv_0\right], \tag{6.519}$$

where $w_0(r)$ is the initial deflection of the plate, $v_0(r)$ is the initial velocity, and $\alpha^4 = -\rho_0 hs^2/D$. The homogeneous solution of Eqn. (6.519), after excluding the singular solutions $\{Y_0(\alpha r), K_0(\alpha r)\}$ (which would have to included if there were a circular hole at the center of the circular plate) is the set $\{J_0(\alpha r), I_0(\alpha r)\}$. We use the method of variation of parameters to find the general solution. Thus, assume the solution to be of the form

$$\bar{w}=h_1(r,s)J_0(\alpha r)+h_2(r,s)I_0(\alpha r),$$

where $h_1(r,s)$ and $h_2(r,s)$ are functions to be determined. We get

$$\begin{aligned}\nabla^2\bar{w}&=\alpha^2\left[I_0(\alpha r)h_2(r,s)-J_0(\alpha r)h_1(r,s)\right]+\frac{1}{r}\left[J_0(\alpha r)h_1'(r,s)+I_0(\alpha r)h_2'(r,s)\right]\\&\quad+2\alpha\left[I_1(\alpha r)h_2'(r,s)-J_1(\alpha r)h_1'(r,s)\right]+J_0(\alpha r)h''(r,s)+I_0(\alpha r)h_2''(r,s),\end{aligned}$$

where primes denote differentiation with respect to r. If we set all the terms barring the first term on the right-hand side to zero, i.e., if

$$\frac{[rh_1'(r,s)J_0^2(\alpha r)]'}{J_0(\alpha r)}+\frac{[rh_2'(r,s)I_0^2(\alpha r)]'}{I_0(\alpha r)}=0, \tag{6.520}$$

then we have

$$\nabla^2\bar{w}=\alpha^2\left[I_0(\alpha r)h_2(r,s)-J_0(\alpha r)h_1(r,s)\right],$$

so that Eqn. (6.519) reduces to

$$-\frac{[rh_1'(r,s)J_0^2(\alpha r)]'}{J_0(\alpha r)}+\frac{[rh_2'(r,s)I_0^2(\alpha r)]'}{I_0(\alpha r)}=\frac{rg(r,s)}{\alpha^2 D}, \tag{6.521}$$

where

$$g(r,s)=\bar{q}(r,s)+\rho_0 hsw_0(r)+\rho_0 hv_0(r).$$

From Eqns. (6.520) and (6.521), we get

$$\frac{[rh_1'(r,s)J_0^2(\alpha r)]'}{J_0(\alpha r)}=-\frac{rg(r,s)}{2\alpha^2 D},$$

$$\frac{[rh_2'(r,s)I_0^2(\alpha r)]'}{I_0(\alpha r)} = \frac{rg(r,s)}{2\alpha^2 D}.$$

Solving the above two equations, we get

$$\begin{aligned} h_1(r,s) &= -\int_0^r \left[\frac{1}{\eta J_0^2(\alpha\eta)} \int_0^\eta \frac{\xi g(\xi,s) J_0(\alpha\xi)\, d\xi}{2\alpha^2} \right] d\eta + c_1(s), \\ h_2(r,s) &= \int_0^r \left[\frac{1}{\eta I_0^2(\alpha\eta)} \int_0^\eta \frac{\xi g(\xi,s) I_0(\alpha\xi)\, d\xi}{2\alpha^2} \right] d\eta + c_2(s), \end{aligned} \tag{6.522}$$

where $c_1(s)$ and $c_2(s)$ are constants dependent only on s.

Consider the case where a uniform load $q(r,t) = \hat{q}(t)$ acts on the circular plate. Then using Eqn. (6.522), and absorbing some constant terms into $c_1(s)$ and $c_2(s)$, we get

$$\begin{aligned} h_1(r,s) &= -\frac{\bar{q}}{2\alpha^4 D J_0(\alpha r)} + c_1(s), \\ h_2(r,s) &= -\frac{\bar{q}}{2\alpha^4 D I_0(\alpha r)} + c_2(s), \end{aligned}$$

so that

$$\bar{w} = c_1(s) J_0(\alpha r) + c_2(s) I_0(\alpha r) - \frac{\bar{q}}{\alpha^4 D}.$$

For a simply supported plate, using the boundary conditions

$$\bar{w}\big|_{r=a} = \left. \frac{d^2\bar{w}}{dr^2} + \frac{\nu}{r}\frac{d\bar{w}}{dr} \right|_{r=a} = 0,$$

we get

$$\bar{w} = \frac{\bar{q} N_1(\alpha)}{D D_1(\alpha)}, \tag{6.523}$$

where

$$\begin{aligned} N_1(\alpha) &= (1-\nu)\left\{ [J_0(\alpha a) - J_0(\alpha r)]\, I_1(\alpha a) + [I_0(\alpha a) - I_0(\alpha r)]\, J_1(\alpha a) \right\} \\ &\quad + \alpha a \left[J_0(\alpha a) I_0(\alpha r) + I_0(\alpha a) J_0(\alpha r) - 2 J_0(\alpha a) I_0(\alpha a) \right], \\ D_1(\alpha) &= \alpha^4 \left\{ 2\alpha a J_0(\alpha a) I_0(\alpha a) - (1-\nu)\left[J_0(\alpha a) I_1(\alpha a) + I_0(\alpha a) J_1(\alpha a) \right] \right\}. \end{aligned}$$

Let $\beta := \sqrt{D/(\rho_0 h)}$, and let λ_n, $n = 1,2,\ldots,\infty$, be the roots of

$$2\lambda a J_0(a\lambda) I_0(a\lambda) - (1-\nu)\left[J_0(a\lambda) I_1(a\lambda) + I_0(a\lambda) J_1(a\lambda) \right] = 0. \tag{6.524}$$

We can write Eqn. (6.523) as

$$\frac{\bar{w}}{\bar{q}} = \frac{c_0}{s^2} + \frac{d_0}{s} + \sum_{s=1}^{\infty} \frac{c_n^{(1)}}{s + \lambda_n^2 \beta i} + \frac{c_n^{(2)}}{s - \lambda_n^2 \beta i}.$$

Similar to the beam case, we can show that $c_0 = d_0 = 0$. Using the Heaviside formula with $d\alpha/ds = -s/(2\alpha^3\beta^2)$, we get

$$c_n^{(1)} = -c_n^{(2)} = \frac{\beta N_2 i}{D D_2},$$

where

$$N_2 = \lambda_n a\left[J_0(\lambda_n a)I_0(\lambda_n r) + I_0(\lambda_n a)J_0(\lambda_n r)\right] - (1-\nu)\left[J_0(\lambda_n r)I_1(\lambda_n a) + J_1(\lambda_n a)I_0(\lambda_n r)\right],$$
$$D_2 = \left\{\lambda_n^2 a^2\left[J_1(\lambda_n a)I_0(\lambda_n a) - J_0(\lambda_n a)I_1(\lambda_n a)\right] - (1+\nu)\lambda_n a J_0(\lambda_n a)I_0(\lambda_n a)\right\}\lambda_n^2.$$

Using the convolution theorem, we get

$$w(r,t) = \frac{\beta}{D}\sum_{n=1}^{\infty} c_n \int_0^t \sin(\lambda_n^2\beta\tau)\hat{q}(t-\tau)\,d\tau, \tag{6.525}$$

where $c_n = 2N_2/D_2$.

If $\hat{q} = q_0$, then from Eqn. (6.525), we get

$$w(r,t) = \frac{q_0}{D}\left\{\frac{(a^2-r^2)}{64}\left[\frac{(5+\nu)a^2}{1+\nu} - r^2\right] - \sum_{n=1}^{\infty}\frac{c_n\cos\lambda_n^2\beta t}{\lambda_n^2}\right\},$$

where we have used the fact that

$$\sum_{n=1}^{\infty}\frac{c_n}{\lambda_n^2} = \frac{(a^2-r^2)}{64}\left[\frac{(5+\nu)a^2}{1+\nu} - r^2\right].$$

If $\hat{q} = q_0\sin\hat{\omega}t$, then from Eqn. (6.525), we get

$$w(r,t) = \frac{q_0}{D}\left[\frac{N_1(k)}{D_1(k)}\sin\hat{\omega}t - \frac{1}{k^2}\sum_{n=1}^{\infty}\frac{c_n\sin\lambda_n^2\beta t}{\left(\frac{\lambda_n}{k}\right)^4 - 1}\right],$$

where $k := \sqrt{\hat{\omega}/\beta}$, and where we have used the fact that

$$\sum_{n=1}^{\infty}\frac{c_n\lambda_n^2}{\left(\frac{\lambda_n}{k}\right)^4 - 1} = \frac{k^4 N_1(k)}{D_1(k)}.$$

If $\hat{q} = q_0\cos\hat{\omega}t$, then we get

$$w(r,t) = \frac{q_0}{D}\left[\frac{N_1(k)}{D_1(k)}\cos\hat{\omega}t - \frac{1}{k^4}\sum_{n=1}^{\infty}\frac{c_n\lambda_n^2\cos\lambda_n^2\beta t}{\left(\frac{\lambda_n}{k}\right)^4 - 1}\right].$$

Finally, if $\hat{q} = q_0 e^{-\hat{\omega}t}$, then

$$w(r,t) = \frac{q_0}{D}\left\{\frac{N_3(k)}{D_3(k)}e^{-\hat{\omega}t} - \frac{1}{k^2}\sum_{n=1}^{\infty}\frac{c_n\left[\frac{\lambda_n^2}{k^2}\cos\lambda_n^2\beta t - \sin\lambda_n^2\beta t\right]}{\left(\frac{\lambda_n}{k}\right)^4 + 1}\right\},$$

where

$$N_3(k) = \sqrt{2}\nu\Big\{[\text{ber}_0(ka) - \text{ber}_0(kr)]\,[\text{bei}_1(ka) - \text{ber}_1(ka)]$$
$$+ [\text{bei}_1(ka) + \text{ber}_1(ka)]\,[\text{bei}_0(kr) - \text{bei}_0(ka)]\Big\}$$
$$+ ka\Big\{[\text{bei}_0(ka) - \text{bei}_0(kr)]\,[\text{bei}_0(ka) - \text{bei}_2(ka)]$$
$$+ [\text{ber}_0(ka) - \text{ber}_0(kr)]\,[\text{ber}_0(ka) - \text{ber}_2(ka)]\Big\},$$
$$\frac{D_3(k)}{k^4} = ka\left\{\text{bei}_0^2(ka) - \text{bei}_0(ka)\text{bei}_2(ka) + [\text{ber}_0(ka) - \text{ber}_2(ka)]\,\text{ber}_0(ka)\right\}$$
$$+ \sqrt{2}\nu\left\{[\text{bei}_1(ka) - \text{ber}_1(ka)]\,\text{ber}_0(ka) - [\text{bei}_1(ka) + \text{ber}_1(ka)]\,\text{bei}_0(ka)\right\},$$

where $\text{ber}_n(.)$ and $\text{bei}_n(.)$ are the nth order Kelvin functions that obey

$$\text{ber}_0(ka) + \text{ber}_2(ka) = -\frac{\sqrt{2}}{ka}\,[\text{ber}_1(ka) - \text{bei}_1(ka)],$$
$$\text{bei}_0(ka) + \text{bei}_2(ka) = -\frac{\sqrt{2}}{ka}\,[\text{ber}_1(ka) + \text{bei}_1(ka)].$$

In deriving the above result for $w(r,t)$, we have used

$$\sum_{n=1}^{\infty} \frac{c_n \lambda_n^2}{\left(\frac{\lambda_n}{k}\right)^4 + 1} = \frac{k^4 N_3(k)}{D_3(k)}.$$

Now consider the case of a simply supported circular plate with a time-varying bending moment per unit length $M(t)$ applied at $r = a$. For simplicity, we assume that the initial displacement and velocity is zero. In this case, from Eqn. (6.522), we get $h_1 = c_1(s)$ and $h_2 = c_2(s)$, so that

$$\bar{w} = c_1(s) J_0(\alpha r) + c_2(s) I_0(\alpha r).$$

Applying the boundary conditions

$$w\big|_{r=a} = 0; \quad D\left[\frac{d^2\bar{w}}{dr^2} + \frac{\nu}{r}\frac{d\bar{w}}{dr}\right] = \bar{M},$$

we get

$$\frac{\bar{w}}{\bar{M}} = \frac{N_1(\alpha)}{D D_1(\alpha)},$$

where

$$N_1(\alpha) = a\left[J_0(\alpha a) I_0(\alpha r) - J_0(\alpha r) I_0(\alpha a)\right],$$
$$D_1(\alpha) = \alpha\left\{2\alpha a J_0(\alpha a) I_0(\alpha a) - (1-\nu)\left[J_0(\alpha a) I_1(\alpha a) + I_0(\alpha a) J_1(\alpha a)\right]\right\}.$$

Let $\beta := \sqrt{D/(\rho_0 h)}$, and let λ_n, $n = 1, 2, \ldots, \infty$, be the roots of Eqn. (6.524). Then, similar to the previous case, we have

$$\frac{\bar{w}}{\bar{M}} = \sum_{s=1}^{\infty} \frac{c_n^{(1)}}{s + \lambda_n^2 \beta i} + \frac{c_n^{(2)}}{s - \lambda_n^2 \beta i}.$$

Using the Heaviside formula, we get

$$c_n^{(1)} = -c_n^{(2)} = \frac{\beta N_2 i}{D D_2},$$

where

$$N_2 = \lambda_n a \left[J_0(\lambda_n a) I_0(\lambda_n r) - J_0(\lambda_n r) I_0(\lambda_n a) \right],$$
$$D_2 = \left\{ \lambda_n^2 a^2 \left[J_1(\lambda_n a) I_0(\lambda_n a) - J_0(\lambda_n a) I_1(\lambda_n a) \right] - (1+\nu) \lambda_n a J_0(\lambda_n a) I_0(\lambda_n a) \right\}.$$

Using the convolution theorem, we get

$$w(r,t) = \frac{\beta}{D} \sum_{n=1}^{\infty} c_n \int_0^t \sin(\lambda_n^2 \beta \tau) M(t-\tau)\, d\tau, \tag{6.526}$$

where $c_n = 2N_2/D_2$.

If $M(t) = M_0$, then from Eqn. (6.526), we get

$$w(r,t) = \frac{M_0}{D} \left[\frac{(r^2 - a^2)}{2(1+\nu)} - \sum_{n=1}^{\infty} \frac{c_n \cos \lambda_n^2 \beta t}{\lambda_n^2} \right],$$

where we have used the fact that

$$\sum_{n=1}^{\infty} \frac{c_n}{\lambda_n^2} = \frac{(r^2 - a^2)}{2(1+\nu)}.$$

If $M(t) = M_0 \sin \hat{\omega} t$, then from Eqn. (6.526), we get

$$w(r,t) = \frac{M_0}{D} \left[\frac{N_1(k)}{D_1(k)} \sin \hat{\omega} t - \frac{1}{k^2} \sum_{n=1}^{\infty} \frac{c_n \sin \lambda_n^2 \beta t}{\left(\frac{\lambda_n}{k}\right)^4 - 1} \right],$$

where $k := \sqrt{\hat{\omega}/\beta}$, and where we have used the fact that

$$\sum_{n=1}^{\infty} \frac{c_n \lambda_n^2}{\left(\frac{\lambda_n}{k}\right)^4 - 1} = \frac{k^4 N_1(k)}{D_1(k)}.$$

If $M(t) = M_0 \cos \hat{\omega} t$, then we get

$$w(r,t) = \frac{M_0}{D} \left[\frac{N_1(k)}{D_1(k)} \cos \hat{\omega} t - \frac{1}{k^4} \sum_{n=1}^{\infty} \frac{c_n \lambda_n^2 \cos \lambda_n^2 \beta t}{\left(\frac{\lambda_n}{k}\right)^4 - 1} \right].$$

Finally, if $M(t) = M_0 e^{-\hat{\omega}t}$, then

$$w(r,t) = \frac{M_0}{D}\left\{\frac{N_3(k)}{D_3(k)}e^{-\hat{\omega}t} - \frac{1}{k^2}\sum_{n=1}^{\infty}\frac{c_n\left[\frac{\lambda_n^2}{k^2}\cos\lambda_n^2\beta t - \sin\lambda_n^2\beta t\right]}{\left(\frac{\lambda_n}{k}\right)^4+1}\right\},$$

where

$$N_3(k) = 2a\left[\mathrm{ber}_0(ka)\mathrm{bei}_0(kr) - \mathrm{ber}_0(kr)\mathrm{bei}_0(ka)\right],$$

$$\frac{D_3(k)}{k} = ka\left[\mathrm{bei}_0^2(ka) - \mathrm{bei}_0(ka)\mathrm{bei}_2(ka) + 2\mathrm{ber}_0^2(ka)\right],$$

$$+\sqrt{2}\Big\{(1-\nu)\left[\mathrm{ber}_1(ka) - \mathrm{bei}_1(ka)\right]\mathrm{ber}_0(ka) - \nu\left[\mathrm{ber}_1(ka) + \mathrm{bei}_1(ka)\right]\mathrm{bei}_0(ka)\Big\},$$

and where we have used

$$\sum_{n=1}^{\infty}\frac{c_n\lambda_n^2}{\left(\frac{\lambda_n}{k}\right)^4+1} = \frac{k^4 N_3(k)}{D_3(k)}.$$

For a clamped plate subjected to uniform loading $\hat{q}(t)$, using the boundary conditions

$$\bar{w}|_{r=a} = \left.\frac{d\bar{w}}{dr}\right|_{r=a} = 0,$$

we get

$$\bar{w} = \frac{\bar{q}N_1(\alpha)}{DD_1(\alpha)}, \tag{6.527}$$

where

$$N_1(\alpha) = \left[J_0(\alpha r) - J_0(\alpha a)\right]I_1(\alpha a) + \left[I_0(\alpha r) - I_0(\alpha a)\right]J_1(\alpha a),$$

$$D_1(\alpha) = \alpha^4\left[J_0(\alpha a)I_1(\alpha a) + J_1(\alpha a)I_0(\alpha a)\right].$$

Let λ_n, $n = 1, 2, \ldots, \infty$, be the roots of

$$J_0(\lambda a)I_1(\lambda a) + J_1(\lambda a)I_0(\lambda a) = 0.$$

The first root of the above equation is given by approximately 3.1962/a. Approximations to the subsequent roots are obtained by adding π/a to the previous root. These approximations can be used as a first guess in a Newton–Raphson strategy.

We can write Eqn. (6.527) as

$$\frac{\bar{w}}{\bar{q}} = \sum_{s=1}^{\infty}\frac{c_n^{(1)}}{s+\lambda_n^2\beta i} + \frac{c_n^{(2)}}{s-\lambda_n^2\beta i}.$$

Using the Heaviside formula, we get

$$c_n^{(1)} = -c_n^{(2)} = \frac{\beta N_2 i}{DD_2},$$

where

$$N_2 = -\left[J_0(\lambda_n r) I_1(\lambda_n a) + J_1(\lambda_n a) I_0(\lambda_n r)\right],$$

$$D_2 = \lambda_n^3 a J_0(\lambda_n a) I_0(\lambda_n a).$$

Using the convolution theorem, we get

$$w(r,t) = \frac{\beta}{D} \sum_{n=1}^{\infty} c_n \int_0^t \sin(\lambda_n^2 \beta \tau) \hat{q}(t-\tau)\, d\tau, \tag{6.528}$$

where $c_n = 2N_2/D_2$.

If $\hat{q} = q_0$, then from Eqn. (6.528), we get

$$w(r,t) = \frac{q_0}{D} \left\{ \frac{(a^2 - r^2)^2}{64} - \sum_{n=1}^{\infty} \frac{c_n \cos \lambda_n^2 \beta t}{\lambda_n^2} \right\},$$

where we have used the fact that

$$\sum_{n=1}^{\infty} \frac{c_n}{\lambda_n^2} = \frac{(a^2 - r^2)^2}{64}.$$

If $\hat{q} = q_0 \sin \hat{\omega} t$, then from Eqn. (6.528), we get

$$w(r,t) = \frac{q_0}{D} \left[\frac{N_1(k)}{D_1(k)} \sin \hat{\omega} t - \frac{1}{k^2} \sum_{n=1}^{\infty} \frac{c_n \sin \lambda_n^2 \beta t}{\left(\frac{\lambda_n}{k}\right)^4 - 1} \right],$$

where $k := \sqrt{\hat{\omega}/\beta}$, and where we have used the fact that

$$\sum_{n=1}^{\infty} \frac{c_n \lambda_n^2}{\left(\frac{\lambda_n}{k}\right)^4 - 1} = \frac{k^4 N_1(k)}{D_1(k)}.$$

If $\hat{q} = q_0 \cos \hat{\omega} t$, then we get

$$w(r,t) = \frac{q_0}{D} \left[\frac{N_1(k)}{D_1(k)} \cos \hat{\omega} t - \frac{1}{k^4} \sum_{n=1}^{\infty} \frac{c_n \lambda_n^2 \cos \lambda_n^2 \beta t}{\left(\frac{\lambda_n}{k}\right)^4 - 1} \right].$$

Finally, if $\hat{q} = q_0 e^{-\hat{\omega} t}$, then

$$w(r,t) = \frac{q_0}{D} \left\{ \frac{N_3(k)}{D_3(k)} e^{-\hat{\omega} t} - \frac{1}{k^2} \sum_{n=1}^{\infty} \frac{c_n \left[\frac{\lambda_n^2}{k^2} \cos \lambda_n^2 \beta t - \sin \lambda_n^2 \beta t \right]}{\left(\frac{\lambda_n}{k}\right)^4 + 1} \right\},$$

where

$$N_3(k) = [\mathrm{ber}_0(ka) - \mathrm{ber}_0(kr)]\,[\mathrm{ber}_1(ka) - \mathrm{bei}_1(ka)] + [\mathrm{ber}_1(ka) + \mathrm{bei}_1(ka)]\,[\mathrm{bei}_0(ka) - \mathrm{bei}_0(kr)]\,,$$

$$\frac{D_3(k)}{k^4} = [\mathrm{ber}_1(ka) - \mathrm{bei}_1(ka)]\,\mathrm{ber}_0(ka) + [\mathrm{ber}_1(ka) + \mathrm{bei}_1(ka)]\,\mathrm{bei}_0(ka),$$

and where we have used

$$\sum_{n=1}^{\infty} \frac{c_n \lambda_n^2}{\left(\frac{\lambda_n}{k}\right)^4 + 1} = \frac{k^4 N_3(k)}{D_3(k)}.$$

Kinetic analogue of the Kelvin problem

For a discussion of the kinetic analogue of the Kelvin problem, see, e.g., [10], [109], [202], [203], [351], and references therein. If $\boldsymbol{g}(t)$ is the load acting at a point $\boldsymbol{z}$, and if $\boldsymbol{y} = \boldsymbol{x} - \boldsymbol{z}$ and $y = |\boldsymbol{y}|$, then the solution to this problem for the case when the solid is initially at rest is given by

$$4\pi\rho_0\phi(\boldsymbol{x},t) = \int_0^{y/c_L} \frac{\alpha}{y^3}\boldsymbol{g}(t-\alpha)\cdot\boldsymbol{y}\,d\alpha,$$

$$4\pi\rho_0\boldsymbol{f}(\boldsymbol{x},t) = \int_0^{y/c_T} \frac{\alpha}{y^3}\boldsymbol{g}(t-\alpha)\times\boldsymbol{y}\,d\alpha,$$

$$\begin{aligned}4\pi\rho_0\boldsymbol{u}(\boldsymbol{x},t) &= \boldsymbol{\nabla}\phi + \boldsymbol{\nabla}\times\boldsymbol{f}\\ &= \frac{\boldsymbol{A}}{y^2}\int_{y/c_L}^{y/c_T} \alpha\boldsymbol{g}(t-\alpha)\,d\alpha + \boldsymbol{B}\left[\frac{1}{c_L^2}\boldsymbol{g}\left(t-\frac{y}{c_L}\right) - \frac{1}{c_T^2}\boldsymbol{g}\left(t-\frac{y}{c_T}\right)\right] + \frac{1}{yc_T^2}\boldsymbol{g}\left(t-\frac{y}{c_T}\right)\\ &= \boldsymbol{A}\int_{1/c_L}^{1/c_T} \lambda\boldsymbol{g}(t-\lambda y)\,d\lambda + \boldsymbol{B}\left[\frac{1}{c_L^2}\boldsymbol{g}\left(t-\frac{y}{c_L}\right) - \frac{1}{c_T^2}\boldsymbol{g}\left(t-\frac{y}{c_T}\right)\right] + \frac{1}{yc_T^2}\boldsymbol{g}\left(t-\frac{y}{c_T}\right),\end{aligned}$$

$$4\pi\rho_0\boldsymbol{\nabla}\cdot\boldsymbol{u} = -\frac{1}{c_L^2 y^3}\boldsymbol{y}\cdot\boldsymbol{g}(t-y/c_L) - \frac{1}{c_L^3 y}[\boldsymbol{B}\boldsymbol{g}'(t-y/c_L)]\cdot\boldsymbol{y},$$

$$\begin{aligned}4\pi\rho_0\boldsymbol{\epsilon} &= \frac{3}{y^5}\left[(\boldsymbol{y}\cdot\boldsymbol{h}_1)\boldsymbol{I} + \boldsymbol{y}\otimes\boldsymbol{h}_1 + \boldsymbol{h}_1\otimes\boldsymbol{y}\right] - \frac{15}{y^7}(\boldsymbol{y}\cdot\boldsymbol{h}_1)\boldsymbol{y}\otimes\boldsymbol{y}\\ &\quad + \frac{1}{2c_T^2}\left[\frac{12}{y^5}[\boldsymbol{y}\cdot\boldsymbol{g}(t-y/c_T)]\boldsymbol{y}\otimes\boldsymbol{y} - \frac{3}{y^3}\boldsymbol{g}(t-y/c_T)\otimes\boldsymbol{y} - \frac{3}{y^3}\boldsymbol{y}\otimes\boldsymbol{g}(t-y/c_T)\right]\\ &\quad - \frac{1}{c_L^2}\left[\frac{6}{y^5}[\boldsymbol{y}\cdot\boldsymbol{g}(t-y/c_L)]\boldsymbol{y}\otimes\boldsymbol{y} - \frac{1}{y^3}\boldsymbol{g}(t-y/c_L)\otimes\boldsymbol{y} - \frac{1}{y^3}\boldsymbol{y}\otimes\boldsymbol{g}(t-y/c_L)\right]\\ &\quad + \frac{1}{y^3}\left[\frac{1}{c_L^2}\boldsymbol{y}\cdot\boldsymbol{g}(t-y/c_L) - \frac{1}{c_T^2}\boldsymbol{y}\cdot\boldsymbol{g}(t-y/c_T)\right]\boldsymbol{I}\\ &\quad + \frac{1}{2c_T^3 y}\left[\boldsymbol{B}\boldsymbol{g}'(t-y/c_T)\otimes\boldsymbol{y} + \boldsymbol{y}\otimes[\boldsymbol{B}\boldsymbol{g}'(t-y/c_T)]\right]\\ &\quad - \frac{1}{2c_L^3 y}\left[\boldsymbol{B}\boldsymbol{g}'(t-y/c_L)\otimes\boldsymbol{y} + \boldsymbol{y}\otimes[\boldsymbol{B}\boldsymbol{g}'(t-y/c_L)]\right]\\ &\quad - \frac{1}{2c_T^3 y^2}\left[\boldsymbol{g}'(t-y/c_T)\otimes\boldsymbol{y} + \boldsymbol{y}\otimes\boldsymbol{g}'(t-y/c_T)\right],\end{aligned}$$

$$4\pi\boldsymbol{\tau} = c_L^2(\boldsymbol{\nabla}\cdot\boldsymbol{u})\boldsymbol{I} + 2c_T^2\left[\boldsymbol{\epsilon} - (\boldsymbol{\nabla}\cdot\boldsymbol{u})\boldsymbol{I}\right],$$

where

$$
\begin{aligned}
A &= \frac{3}{y^3}\boldsymbol{y} \otimes \boldsymbol{y} - \frac{1}{y}\boldsymbol{I}, \\
B &= \frac{1}{y^3}\boldsymbol{y} \otimes \boldsymbol{y}, \\
h_1 &= \int_{y/c_L}^{y/c_T} \alpha g(t-\alpha)\, d\alpha, \\
h_2 &= \frac{1}{c_L^2} g\left(t - \frac{y}{c_L}\right) - \frac{1}{c_T^2} g\left(t - \frac{y}{c_T}\right).
\end{aligned}
$$

It can be shown that the above is a strong solution for all $\boldsymbol{x} \neq \boldsymbol{z}$ at all times, and that

$$
\begin{aligned}
&\boldsymbol{u}(\boldsymbol{x},.) = \mathrm{o}\left(y^{-1}\right) \text{ as } \boldsymbol{y} \to \boldsymbol{0}, \\
&\boldsymbol{\tau}(\boldsymbol{x},.) = \mathrm{o}\left(y^{-2}\right) \text{ as } \boldsymbol{y} \to \boldsymbol{0}, \\
&\lim_{r\to 0} \int_{S_r} \boldsymbol{\tau}(\boldsymbol{x},t)\boldsymbol{n}\, dA = -\boldsymbol{g}(t), \\
&\lim_{r\to 0} \int_{S_r} (\boldsymbol{x}-\boldsymbol{z}) \times [\boldsymbol{\tau}(\boldsymbol{x},t)\boldsymbol{n}]\, dA = \boldsymbol{0}, \\
&\lim_{r\to 0} \int_{B_r} \rho_0 \ddot{\boldsymbol{u}}(\boldsymbol{x},t)\, dV = \boldsymbol{0}, \\
&\lim_{r\to 0} \int_{B_r} (\boldsymbol{x}-\boldsymbol{z}) \times [\rho_0 \ddot{\boldsymbol{u}}(\boldsymbol{x},t)]\, dV = \boldsymbol{0},
\end{aligned}
$$

where $\boldsymbol{n}$ is the outward unit normal to the surface S_r of the spherical ball B_r of radius r and center $\boldsymbol{z}$. Thus, the point force is carried entirely by the surface tractions.

For a discussion of the kinetic analogue of the Boussinesq (the so-called 'dynamic Boussinesq problem'; we will consider the special case of harmonic loading in the following subsection on Rayleigh waves), Cerruti and related problems, see, e.g., [14], [89], [201], and references therein.

Rayleigh waves

Consider the propagation of waves in a homogeneous, isotropic, elastic half-space with $z = 0$ denoting the surface, and the material occupying the region $z \geq 0$. The derivation that we present is alternate to the one presented by Rayleigh [263], and is based on potentials. Let the potentials in Eqn. (6.480) be given by

$$
\begin{aligned}
\phi &= \tilde{\phi} e^{i(\omega t + \boldsymbol{k}\cdot\boldsymbol{x})} e^{-k_z z}, \\
\boldsymbol{f} &= \begin{bmatrix} f_x \\ f_y \\ f_z \end{bmatrix} e^{i(\omega t + \boldsymbol{k}\cdot\boldsymbol{x})} e^{-c_z z},
\end{aligned}
\tag{6.529}
$$

where ω is the frequency of the waves (which we are interested in finding), $\boldsymbol{k} := (k_x, k_y)$ are the wave numbers of the waves in the x and y-directions, and $\boldsymbol{x} := (x, y)$ is the position vector in the x–y plane. The quantities k_x, k_y, k_z, c_z and ω are all real, and in addition, k_z and c_z are both positive. Because of the exponential decay with respect to z, these waves are confined to a layer near to the surface. Note that, here, we are interested in finding only the 'periodic steady-state' solution, and hence, have assumed a dependence of $e^{i\omega t}$ for the potentials. Initial conditions play no role in the solution process since we assume that the initial transients are damped out (although damping is not explicitly included in the model), and the solution reaches the 'periodic steady-state' mentioned above. The surface $z = 0$ is assumed to be traction-free, so that the boundary conditions are

$$\tau_{xz}\big|_{z=0} = \tau_{yz}\big|_{z=0} = \tau_{zz}\big|_{z=0} = 0. \tag{6.530}$$

Substituting Eqns. (6.529) into Eqns. (6.480), we get the constraint equations

$$\begin{aligned} k_x^2 + k_y^2 - k_z^2 &= k_L^2, \\ k_x^2 + k_y^2 - c_z^2 &= k_T^2, \end{aligned} \tag{6.531}$$

where

$$k_L^2 = \frac{\omega^2}{c_L^2}, \qquad k_T^2 = \frac{\omega^2}{c_T^2}.$$

Substituting Eqns. (6.529) into Eqn. (6.481), we get the expressions for the displacement field. Using the first two boundary conditions in Eqn. (6.530), we can eliminate f_x and f_y. The displacement field (after renaming $\tilde{\phi}$) is now given by

$$\begin{aligned} u_x &= \tilde{\phi} i e^{i(\omega t + \boldsymbol{k}\cdot\boldsymbol{x})} k_x \left[(k_x^2 + k_y^2 + c_z^2) e^{-k_z z} - 2c_z k_z e^{-c_z z} \right], \\ u_y &= \tilde{\phi} i e^{i(\omega t + \boldsymbol{k}\cdot\boldsymbol{x})} k_y \left[(k_x^2 + k_y^2 + c_z^2) e^{-k_z z} - 2c_z k_z e^{-c_z z} \right], \\ u_z &= -\tilde{\phi} e^{i(\omega t + \boldsymbol{k}\cdot\boldsymbol{x})} k_z \left[(k_x^2 + k_y^2 + c_z^2) e^{-k_z z} - 2(k_x^2 + k_y^2) e^{-c_z z} \right]. \end{aligned} \tag{6.532}$$

Now imposing the third condition in Eqn. (6.530), and eliminating k_z^2 and c_z^2 using Eqns. (6.531), we get

$$k_L^2 \left[k_T^2 - 2(k_x^2 + k_y^2) \right] (\lambda + 2\mu) - 2(k_x^2 + k_y^2) \left[k_T^2 - 2(k_x^2 + k_y^2 - c_z k_z) \right] \mu = 0.$$

After finding an expression for $c_z k_z$ from the above equation, squaring it, again using Eqns. (6.531) to eliminate c_z and k_z, and using the fact that $k_T^2/k_L^2 = (\lambda + 2\mu)/\mu$, we get

$$k_T^2 - k_x^2 - k_y^2 + \frac{\left[k_T^2 - 2(k_x^2 + k_y^2) \right]^4}{16(k_x^2 + k_y^2)^2 (k_x^2 + k_y^2 - k_L^2)} = 0. \tag{6.533}$$

Let

$$\tilde{k}_L^2 = \frac{k_L^2}{k_x^2 + k_y^2}, \qquad \tilde{k}_T^2 = \frac{k_T^2}{k_x^2 + k_y^2}.$$

Equation (6.533) simplifies to (and matches with Eqn. (24) of [263])

$$\tilde{k}_T^8 - 8\tilde{k}_T^6 + 24\tilde{k}_T^4 - 16(1 + \tilde{k}_L^2)\tilde{k}_T^2 + 16\tilde{k}_L^2 = 0.$$

Dividing by $\tilde{k}_T^2$, and using the relation $\tilde{k}_L^2/\tilde{k}_T^2 = \mu/(\lambda + 2\mu)$, we get

$$\tilde{k}_T^6 - 8\tilde{k}_T^4 + \frac{8(2-\nu)}{1-\nu}\tilde{k}_T^2 - \frac{8}{1-\nu} = 0. \tag{6.534}$$

One can solve the above cubic equation in $\tilde{k}_T^2$. First one notes that only the real roots of $\tilde{k}_T^2$ should be considered since ω, k_x and k_y are real. Secondly, since

$$1 - \tilde{k}_T^2 = \frac{c_z^2}{k_x^2 + k_y^2},$$

and since c_z, k_x and k_y are all real, only those real roots whose magnitude is less than one should be considered. These relevant roots for some typical values of ν are listed in Table 6.21. To get the results for the surface, set $z = 0$ in the above results.

ν	$\tilde{k}_L^2$	$\tilde{k}_T^2$
0	$\frac{3-\sqrt{5}}{2}$	$3-\sqrt{5}$
0.25	0.281766	0.8453
0.3	0.2457	0.8601
0.5	0	0.9126

Table 6.21 *Roots of Eqn. (6.534) for typical values of ν.*

If the motion is assumed to take place in two-dimensions only, then we can set $k_y = 0$ in the above solution, so that $u_y = 0$, $k_x^2 - k_z^2 = k_L^2$, $k_x^2 - c_z^2 = k_T^2$, $\tilde{k}_T^2 = k_T^2/k_x^2$, and

$$u_x = \tilde{\phi} i e^{i(\omega t + k_x x)} k_x \left[(k_x^2 + c_z^2)e^{-k_z z} - 2c_z k_z e^{-c_z z} \right],$$

$$u_z = -\tilde{\phi} e^{i(\omega t + k_x x)} k_z \left[(k_x^2 + c_z^2)e^{-k_z z} - 2k_x^2 e^{-c_z z} \right].$$

The wave speed is given by $c = \omega/k_x = c_T \tilde{k}_T$. Since $\tilde{k}_T < 1$, we have $c < c_T$. If instead of the boundary conditions $\tau_{xz}|_{z=0} = \tau_{zz}|_{z=0} = 0$, we had $\tau_{xz}|_{z=0} = X e^{ik_x x} e^{i\omega t}$ and $\tau_{zz}|_{z=0} = Z e^{ik_x x} e^{i\omega t}$, then we get $u_y = 0$, and

$$u_x = c_z f_y e^{i(\omega t + k_x x)} e^{-c_z z} + i k_x \tilde{\phi} e^{i(\omega t + k_x x)} e^{-k_z z},$$

$$u_z = i k_x f_y e^{i(\omega t + k_x x)} e^{-c_z z} - k_z \tilde{\phi} e^{i(\omega t + k_x x)} e^{-k_z z},$$

where, with $D = (2k_x^2 - k_T^2)^2 - 4c_z k_z k_x^2$, we have

$$\tilde{\phi} = \frac{1}{\mu D} \left[(2k_x^2 - k_T^2) Z - 2i k_x c_z X \right],$$

$$f_y = \frac{1}{\mu D} \left[(k_T^2 - 2k_x^2) X - 2i k_x k_z Z \right].$$

For concentrated forces $Pe^{i\omega t}$ and $Qe^{i\omega t}$ acting tangential and normal to the surface at $x = z = 0$, we write $X = -P\,d\xi/(2\pi)$ and $Z = -Q\,d\xi/(2\pi)$, and integrate from $-\infty$ to ∞ to get the surface displacements as

$$u_x^0 = -\frac{e^{i\omega t}}{2\pi\mu}\int_{-\infty}^{\infty}\frac{1}{D(\xi)}\left[c_z k_T^2 P + iQ\xi(2\xi^2 - k_T^2 - 2c_z k_z)\right]e^{i\xi x}\,d\xi,$$

$$u_z^0 = \frac{e^{i\omega t}}{2\pi\mu}\int_{-\infty}^{\infty}\frac{1}{D(\xi)}\left[iP\xi(2\xi^2 - k_T^2 - 2c_z k_z) - k_z k_T^2 Q\right]e^{i\xi x}\,d\xi,$$

where, with $k_z = \sqrt{\xi^2 - k_L^2}$ and $c_z = \sqrt{\xi^2 - k_T^2}$,

$$D(\xi) = (2\xi^2 - k_T^2)^2 - 4c_z k_z \xi^2.$$

To get the axisymmetric Rayleigh wave solution, we work with cylindrical coordinates. Let $\boldsymbol{f} = (0, f, 0)$, and

$$\phi = \tilde{\phi}e^{i\omega t}J_0(k_r r)e^{-k_z z},$$

$$f = f_\theta e^{i\omega t}J_1(k_r r)e^{-c_z z},$$

where k_r, ω, k_z and c_z are real, and, in addition, k_z and c_z are positive. From Eqn. (6.480), we get the constraint equations

$$k_r^2 - k_z^2 = k_L^2,$$

$$k_r^2 - c_z^2 = k_T^2.$$

The displacement field is obtained using the relations

$$u_r = \frac{\partial\phi}{\partial r} - \frac{\partial f}{\partial z},$$

$$u_z = \frac{\partial\phi}{\partial z} + \frac{1}{r}\frac{\partial(rf)}{\partial r},$$

as

$$u_r = e^{i\omega t}\left[-k_r\tilde{\phi}e^{-k_z z} + c_z f_\theta e^{-c_z z}\right]J_1(k_r r),$$

$$u_z = e^{i\omega t}\left[-k_z\tilde{\phi}e^{-k_z z} + k_r f_\theta e^{-c_z z}\right]J_0(k_r r).$$

The condition $\tau_{rz}|_{z=0} = 0$ leads to

$$f_\theta = \frac{2k_r k_z\tilde{\phi}}{k_r^2 + c_z^2},$$

so that (after renaming $\tilde{\phi}$)

$$u_r = e^{i\omega t}k_r\tilde{\phi}\left[-(k_r^2 + c_z^2)e^{-k_z z} + 2k_z c_z e^{-c_z z}\right]J_1(k_r r),$$

$$u_z = e^{i\omega t}k_z\tilde{\phi}\left[-(k_r^2 + c_z^2)e^{-k_z z} + 2k_r^2 e^{-c_z z}\right]J_0(k_r r).$$

Let

$$\tilde{k}_L^2 = \frac{k_L^2}{k_r^2}, \qquad \tilde{k}_T^2 = \frac{k_T^2}{k_r^2}.$$

Imposing the boundary condition $\tau_{zz}|_{z=0} = 0$, we again get Eqn. (6.534).

If instead of the boundary condition $\tau_{zz}|_{z=0} = 0$, we had a surface loading condition of the type $\tau_{zz}|_{z=0} = ZJ_0(k_r r)e^{i\omega t}$, then we get

$$\tilde{\phi} = \frac{Z}{\mu\left[(k_T^2 - 2k_r^2)^2 - 4c_z k_z k_r^2\right]},$$

and the surface displacements are

$$\begin{aligned} u_r^0 &= \tilde{\phi} k_r e^{i\omega t}\left[2c_z k_z - c_z^2 - k_r^2\right] J_1(k_r r), \\ u_z^0 &= \tilde{\phi} k_z e^{i\omega t} k_T^2 J_0(k_r r). \end{aligned} \tag{6.535}$$

If a concentrated force $Re^{i\omega t}$ acts at the surface, then we replace Z by $-R\xi\, d\xi/(2\pi)$, and integrate from 0 to ∞ [201] to get the surface displacements as

$$\begin{aligned} u_r^0 &= \frac{Re^{i\omega t}}{2\pi\mu}\int_0^\infty \frac{\xi^2\left[2\xi^2 - k_T^2 - 2c_z k_z\right] J_1(\xi r)\, d\xi}{(2\xi^2 - k_T^2)^2 - 4c_z k_z \xi^2}, \\ u_z^0 &= -\frac{Re^{i\omega t}}{2\pi\mu}\int_0^\infty \frac{k_z k_T^2 \xi J_0(\xi r)\, d\xi}{(2\xi^2 - k_T^2)^2 - 4c_z k_z \xi^2}, \end{aligned} \tag{6.536}$$

where $k_z = \sqrt{\xi^2 - k_L^2}$ and $c_z = \sqrt{\xi^2 - k_T^2}$.

For a discussion of additional theorems, and for the solution of additional special problems, see, e.g., [109], [195], [300] and [324].

EXERCISES

1. Show that

$$\begin{aligned} \boldsymbol{\epsilon} : \boldsymbol{\epsilon} + \boldsymbol{W} : \boldsymbol{W} &= \boldsymbol{\nabla u} : (\boldsymbol{\nabla u}), \\ \boldsymbol{\epsilon} : \boldsymbol{\epsilon} - \boldsymbol{W} : \boldsymbol{W} &= \boldsymbol{\nabla u} : (\boldsymbol{\nabla u})^T. \end{aligned}$$

2. Let $\boldsymbol{u} = \boldsymbol{0}$ on the surface S_0 of a volume V_0. Using the results of Problem 42, Chapter 1, prove *Korn's inequality*:

$$\int_{V_0} \boldsymbol{\nabla u} : \boldsymbol{\nabla u}\, dV \le 2\int_{V_0} \boldsymbol{\epsilon} : \boldsymbol{\epsilon}\, dV.$$

3. A body in equilibrium occupies a volume V with surface S. By integrating the equations

$$(\boldsymbol{\nabla}_X \cdot \boldsymbol{\tau}) \otimes \boldsymbol{X} + \rho_0 \boldsymbol{b} \otimes \boldsymbol{X} = \boldsymbol{0},$$

over the domain V, and with the use of the divergence theorem and the symmetry of the stress tensor, show that (compare against Eqn. (3.93))

$$\begin{aligned}\int_V \boldsymbol{\tau}\, dV &= \left[\int_S \boldsymbol{t}\otimes\boldsymbol{X}\, dS + \int_V \rho_0 \boldsymbol{b}\otimes\boldsymbol{X}\, dV\right] \\ &= \left[\int_S \boldsymbol{X}\otimes\boldsymbol{t}\, dS + \int_V \rho_0 \boldsymbol{X}\otimes\boldsymbol{b}\, dV\right].\end{aligned} \tag{6.537}$$

The above equation obviously holds only for Cartesian components. Results for other curvilinear systems can be derived using this result. For example, for the cylindrical coordinate system (r, θ, z), we get

$$\int_V (\tau_{rr} + \tau_{\theta\theta})\, dV = \int_S t_r r\, dS + \int_V \rho_0 b_r r\, dV, \tag{6.538}$$

$$\int_V \tau_{zz}\, dV = \int_C t_z z\, dS + \int_A \rho_0 b_z z\, dV. \tag{6.539}$$

while for the spherical coordinate system (r, θ, ϕ), we get

$$\int_V (\tau_{rr} + \tau_{\theta\theta} + \tau_{\phi\phi})\, dV = \int_S t_r r\, dS + \int_V \rho_0 b_r r\, dV. \tag{6.540}$$

Show using Eqn. (6.537) that for a prismatic bar with uniform density ρ_0 standing vertically on a smooth horizontal surface, which coincides with the xy plane as shown in Fig. 6.88, the average value of τ_{zz} is $-\rho_0 g L/2$, while the average value of the other

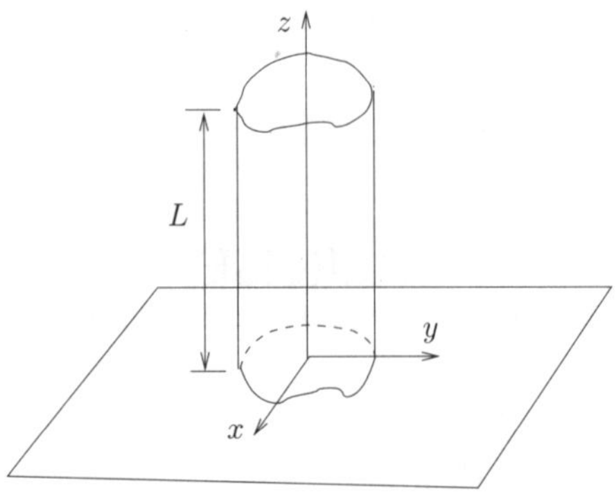

Fig. 6.88 *Prismatic bar standing on a smooth horizontal surface.*

stress components is zero, where the average value of τ_{ij} is $\int_V \tau_{ij}\, dV/V$.
For the case of $\nu = 0$, the exact solution is given by $w = \rho_0 g(z^2 - 2zL)/(2E)$, $\tau_{zz} = \rho_0 g(z - L)$, with the other displacement and stress components zero. Verify in this case that the average value of the stresses is as given by the above result.

4. A prismatic bar of length L hangs from a fixed support. If the origin is located at the centroid of the fixed cross-section with the z-axis pointing downwards, show that the stress distribution $\tau_{zz} = \rho_0 g(L - z)$, other stress components zero, satisfies the equations of equilibrium and the traction boundary conditions at the free surfaces of

the bar. Show that the corresponding strains satisfy the equations of compatibility, and hence can be integrated using the boundary conditions $u = v = w = \partial u/\partial z = \partial v/\partial z = \partial u/\partial y = 0$ at $(0, 0, 0)$ to yield the following displacement field:

$$u = -\frac{1}{E}\nu\rho_0 g x(L - z),$$
$$v = -\frac{1}{E}\nu\rho_0 g y(L - z),$$
$$w = \frac{1}{2E}\rho_0 g\left[2zL - z^2 - \nu(x^2 + y^2)\right].$$

5. In Section 6.5.1, we considered the problem of a circular cylinder subjected to a torque applied through tractions acting on the end surfaces, with the lateral surfaces being free of traction. In this problem, we consider a circular cylinder subjected to a torque

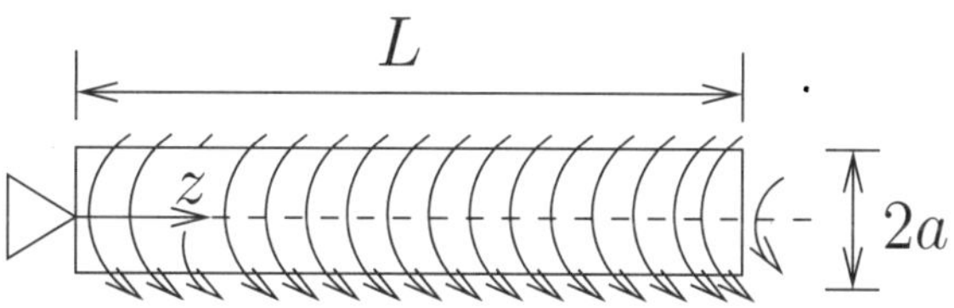

Fig. 6.89 *Torque applied on the lateral and end surfaces.*

through a traction distribution on the lateral and end surface ($z = L$) as shown in Fig. 6.89. The bar is fixed at $(r, z) = (0, 0)$ as shown, and we assume that $u_r = u_z = 0$ along the entire bar, i.e., all particles only move in tangential directions. Based on this assumption, it follows from Eqns. (6.36) and (B.8) that $\tau_{r\theta}$ and $\tau_{\theta z}$ are the only nonzero components of stress, and that if they are expressed in terms of a function ϕ through Eqns. (6.126), then the equations of equilibrium are automatically satisfied, and the governing equation for ϕ is once again given by Eqn. (6.127).

(a) We have seen in Section 6.5.4 that $\phi = czr^4$ satisfies Eqn. (6.127). Determine the traction distributions on the lateral and end surfaces corresponding to this ϕ, and find the *total* resultant torque exerted about the z-axis due to these distributions.

(b) Using Eqns. (6.126) and the given boundary conditions, find an expression for u_θ.

6. A prismatic beam of rectangular cross-section is subjected to a bending moment M at its ends (see Fig. 6.90). It is proposed that the stress distribution instead of being given by $\tau_{zz} = -Ex/R_x$ is given by

$$\tau_{zz} = -\frac{Ex^3}{R_x},$$

with the other stress components zero.

(a) Verify that the equilibrium equations given by Eqn. (6.56) are satisfied.

(b) Verify that the lateral surfaces are traction-free.

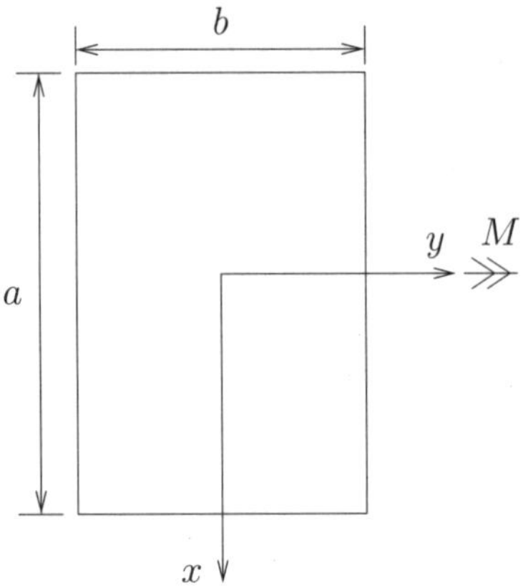

Fig. 6.90 *Bending of a rectangular beam.*

(c) Verify that the forces on the end surfaces are zero.

(d) Find an expression for R_x in terms of M.

The above 'solution' is different from the solution presented in Section 6.5.5. However, by the uniqueness of the stress solution in linear elasticity problems, this cannot be the case. How does one resolve this contradiction?

7. For the problem shown in Fig. 6.90, show using Eqn. (6.163) that the radius of curvature in the x–z plane is R_x, while that in the x–y plane is $-R_x/\nu$. Thus, the beam deforms into a saddle-shape.

8. A beam of the doubly symmetric cross-section comprising two vertical sides $y = \pm a$, and two hyperbolas $(1+\nu)x^2 - \nu y^2 = a^2$ is loaded by a statically equivalent load of

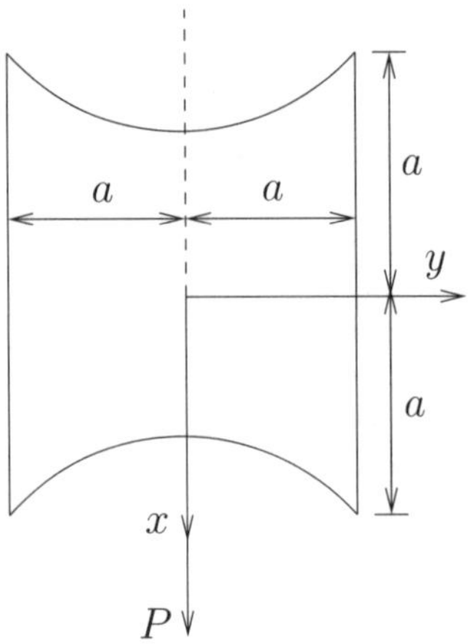

Fig. 6.91 *Problem 8.*

P acting at the centroid and directed along the x-axis as shown in Fig. 6.91.

(a) State with justification (but without mathematical proof), the value of twist at the centroid α.

(b) State the value of I_{xy}, and use it to find the 'curvatures' κ_x and κ_y.

(c) Find the functions $g(x)$ and $f(y)$ such that the boundary conditions are satisfied.

(d) Using the governing differential equation and the appropriate boundary conditions, show that $\phi = 0$.

(e) Find τ_{xz} and τ_{yz}, and using these expressions find $(\tau_{xz})_{\max}$, and the location where it occurs.

(f) Show that

$$\frac{\chi}{\kappa_x} = -a^2 x + \frac{1}{6}(2+\nu)(x^3 - 3xy^2).$$

9. A beam of the doubly symmetric cross-section shown in Fig. 6.92, and described by the equation

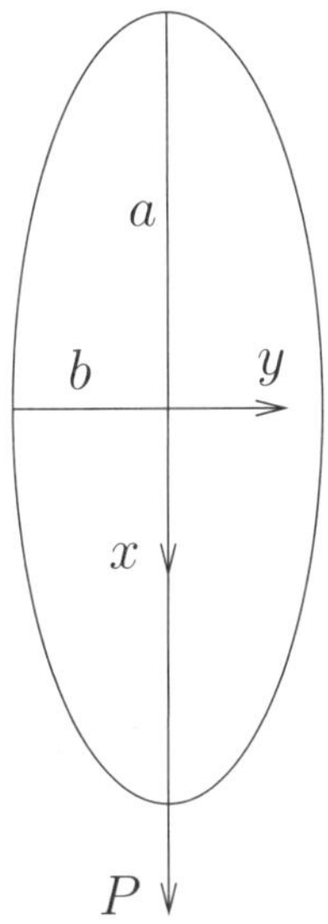

Fig. 6.92 *Problem 9.*

$$\left(1 - \frac{x^2}{a^2}\right)^{2\nu} = \left(\frac{y}{b}\right)^2,$$

is loaded by a statically equivalent load of P acting at the centroid and directed along the x-axis. Show that the function

$$\frac{\chi}{\kappa_x} = -a^2 x + \frac{1}{6}(2-\nu)(x^3 - 3xy^2),$$

with the corresponding stresses

$$\tau_{xz} = \frac{P}{2(1+\nu)I_{yy}}(a^2 - x^2),$$
$$\tau_{yz} = -\frac{\nu P}{(1+\nu)I_{yy}}xy,$$

is the solution to this problem.

10. Show that the displacements and stresses for the cylindrical pressure vessel problem under "plane strain" conditions (i.e., when the top and bottom surfaces of the cylinder are constrained so that they can move only radially), and when $p_i^{(1)} = p_o = p_o^{(1)} = 0$ (using the notation of Section 6.6.6), are given by

$$u_r = \frac{(1+\nu)p_i a^2}{E(b^2 - a^2)}\left[\frac{b^2}{r} + (1-2\nu)r\right],$$
$$\tau_{rr} = \frac{p_i a^2}{(b^2 - a^2)}\left[1 - \frac{b^2}{r^2}\right],$$
$$\tau_{\theta\theta} = \frac{p_i a^2}{(b^2 - a^2)}\left[1 + \frac{b^2}{r^2}\right],$$
$$\tau_{zz} = \frac{2\nu p_i a^2}{(b^2 - a^2)}.$$

11. A hollow cylinder of inner radius $a - \Delta$ and outer radius b is shrink-fitted around a solid cylinder of radius a as shown in Fig. 6.93 (the individual cylinders *before* shrink fitting are shown in the upper part of the figure, and the assembly *after* shrink fitting is shown in the lower part). Both cylinders are made of the same material. Plane strain conditions are maintained during the shrink fitting process by constraining the motion along the z-direction (but *not* along the other directions) as shown in the figure, i.e., $u_z = \epsilon_{zz} = 0$. Assuming the process to be axisymmetric, i.e., $u_r = u_r(r)$, the Navier equations given by Eqn. (B.13) reduce to

$$\frac{d}{dr}\left[\frac{1}{r}\frac{d(ru_r)}{dr}\right] = 0.$$

(a) Find a general solution to the above equation.

(b) Write the appropriate displacement and stress boundary conditions. Assuming Δ to be small in comparison to a and b, you may state the boundary conditions at $r = a$ and $r = b$. (Hint: Be careful of signs for boundary conditions on u_r.)

(c) Use the strain–displacement relations $\epsilon_{rr} = \partial u_r/\partial r$ and $\epsilon_{\theta\theta} = u_r/r$ (you may assume the other strain components to be zero), and the constitutive relation to find expressions for the displacement and stress fields in the solid and hollow cylinders as a function of Δ.

(d) Show that the *total* axial force F_z exerted by the top wall on the top surface of the total cylinder is zero.

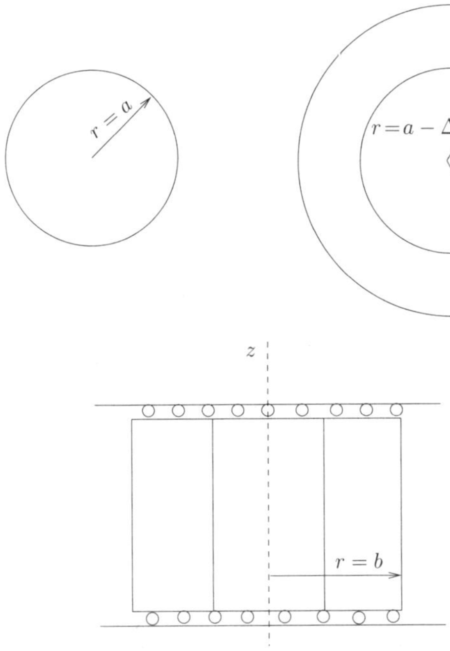

Fig. 6.93 *Problem 11.*

12. A uniform radial traction p is applied to the lateral surface of a circular cylinder of radius a and length $2L$ made of an isotropic material as shown in Fig. 6.94a. By

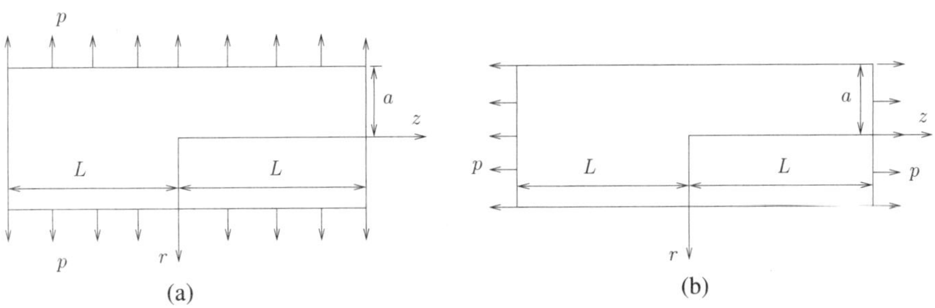

Fig. 6.94 *Problem 12.*

assuming the body forces to be zero, and the displacement field to be given by

$$u_r = c_1 r + d_1 r^2,$$
$$u_z = c_2 z,$$

use the relevant differential equations and boundary conditions to find the displacement field in terms of E and ν.

13. [*Reciprocal theorem*] Let $(\boldsymbol{u}^{(1)}, \boldsymbol{\epsilon}^{(1)}, \boldsymbol{\tau}^{(1)}, \boldsymbol{t}^{(1)})$ and $(\boldsymbol{u}^{(2)}, \boldsymbol{\epsilon}^{(2)}, \boldsymbol{\tau}^{(2)}, \boldsymbol{t}^{(2)})$, where $\boldsymbol{\epsilon}^{(1)} \equiv \boldsymbol{\epsilon}(\boldsymbol{u}^{(1)})$ and $\boldsymbol{\epsilon}^{(2)} \equiv \boldsymbol{\epsilon}(\boldsymbol{u}^{(2)})$, denote the static displacement, strain, stress and traction fields (which satisfy all the equations of equilibrium and boundary conditions) under two different loading conditions on the *same domain* V with surface S. Let the body forces in these two loading conditions be denoted by $\boldsymbol{b}^{(1)}$ and $\boldsymbol{b}^{(2)}$, respectively. Assume that $\mathbf{C}$ is symmetric.
   (a) Show that

$$\int_S \boldsymbol{t}^{(1)} \cdot \boldsymbol{u}^{(2)}\, dS + \int_V \rho_0 \boldsymbol{b}^{(1)} \cdot \boldsymbol{u}^{(2)}\, dV = \int_V \boldsymbol{\tau}^{(1)} : \boldsymbol{\epsilon}^{(2)}\, dV.$$

   Interchanging the indices (1) and (2), we get

$$\int_S \boldsymbol{t}^{(2)} \cdot \boldsymbol{u}^{(1)}\, dS + \int_V \rho_0 \boldsymbol{b}^{(2)} \cdot \boldsymbol{u}^{(1)}\, dV = \int_V \boldsymbol{\tau}^{(2)} : \boldsymbol{\epsilon}^{(1)}\, dV.$$

   (b) Using the symmetry of $\mathbf{C}$, show that

$$\int_S \boldsymbol{t}^{(1)} \cdot \boldsymbol{u}^{(2)}\, dS + \int_V \rho_0 \boldsymbol{b}^{(1)} \cdot \boldsymbol{u}^{(2)}\, dV = \int_S \boldsymbol{t}^{(2)} \cdot \boldsymbol{u}^{(1)}\, dS + \int_V \rho_0 \boldsymbol{b}^{(2)} \cdot \boldsymbol{u}^{(1)}\, dV. \tag{6.541}$$

   (c) Consider the prismatic bar shown in Fig. 6.88 with now a uniform normal traction p applied on its top surface instead of the body force loading considered in that problem. In this case, the axial displacement and stress field is given by

$$\begin{aligned} u_z &= -\frac{pz}{E}, \\ \tau_{zz} &= -p. \end{aligned}$$

   Applying the reciprocal theorem given by Eqn. (6.541) to the prismatic bar shown in Fig. 6.88 with gravity loading, and the above-mentioned normal traction loading, show that under gravity loading

$$\int_A u_z|_{z=L}\, dA = -\frac{\rho_0 g A L^2}{2E}, \tag{6.542}$$

   where A is the area of cross-section of the prismatic bar.

14. A hollow cylinder of length L and inner and outer radii a and b, is fixed at $r = a$, and subjected to a uniform shear stress at $r = b$ resulting in a net shear force T, as shown in Fig. 6.95. By assuming the body forces to be zero, and the displacement field (generated by using the Boussinesq potentials $\alpha = k_2 z \log r$ and $\beta = k_1 - k_2 \log r$) to be given by

$$\begin{aligned} u_r &= 0, \\ u_\theta &= 0, \\ u_z &= c_1 + c_2 \log r, \end{aligned}$$

   find the displacement, strain and stress fields in terms of the Lamé constants and T. Note that although $t_r \neq 0$ on both the top and bottom surfaces, the resultant force on each of these surfaces is still equal to zero due to symmetry.

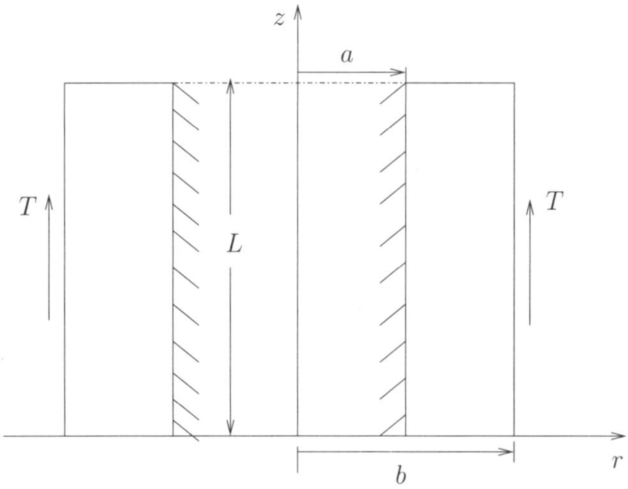

Fig. 6.95 *Hollow cylinder of length L fixed at the inner boundary $r = a$, and subjected to a shear force T at the outer boundary $r = b$.*

15. This problem is the thermal counterpart of the inverse boundary value problem discussed in Section 6.5.7. First consider a hollow homogeneous sphere of inner radius a and outer radius b, and with thermal conductivity $k_0 > 0$. Let the temperature distribution be given by $T = c_1 + c_2/r$, where c_1 and c_2 are constants. The heat flux is given by $\boldsymbol{q} = -k_0 \boldsymbol{\nabla} T = k_0 c_2/r^2 \boldsymbol{e}_r$. Now consider a hollow sphere of the same dimensions as above but made of a material with conductivity $k = abk_0/r^2$. Show using Eqns. (C.3) and (C.7) that the temperature distribution

$$T = c_1 + c_2 \left(\frac{1}{a} + \frac{1}{b} \right) - \frac{c_2 r}{ab}$$

satisfies the governing equation $\boldsymbol{\nabla} \cdot (k \boldsymbol{\nabla} T) = 0$, and in addition yields the same temperature and flux distribution at the surfaces $r = a$ and $r = b$.
Again, this is not a counterexample to the uniqueness result of [318] for the same reason as stated in Section 6.5.7.

7 Thermomechanics

In this chapter, we deal with the restrictions imposed by the Clausius–Duhem inequality on solids and fluids. We first introduce some terminology. The specific volume v is defined as $1/\rho$. Equation (3.8) can be written in terms of v as

$$-\frac{Dv}{Dt} + v\nabla \cdot \boldsymbol{v} = 0. \tag{7.1}$$

The specific enthalpy h is defined by

$$h := e + \frac{p}{\rho} = e + pv. \tag{7.2}$$

We now examine the restrictions imposed on constitutive relations by the second law of thermodynamics. The momentum and energy balance equations do not pose any restrictions, because any motion is consistent with them for a suitable choice of body force $\boldsymbol{b}$, and heat source Q_h, i.e., for a given motion, balance of momentum and energy *define* what $\boldsymbol{b}$ and Q_h have to be. Since the restrictions imposed by the second law on constitutive relations are in the form of necessary and sufficient conditions (as we show in the treatment that follows), by simply adopting constitutive relations that satisfy these restrictions, *we no longer have to consider the second law of thermodynamics explicitly in numerical implementations.*

Throughout this chapter, we *assume* that the constitutive relations do not depend explicitly on time, i.e., there is no aging. We shall also adopt the principle of *equipresence* (first proposed by Truesdell), which states that *an independent variable present in one constitutive equation of a material should be present in all, unless its presence is in direct contradiction with the assumed symmetry of the material, the principle of material-objectivity or the laws of thermodynamics.* We begin by considering the restrictions imposed by the Clausius–Duhem inequality on thermoelastic materials. We follow the treatment in [36].

7.1 Thermoelastic materials

Let the elastic material be defined by the following constitutive relations:

$$\begin{aligned}
\psi_0 &= \hat{\psi}(\boldsymbol{X}, \boldsymbol{F}, \theta_0, \boldsymbol{g}_0),\\
\boldsymbol{T} &= \hat{\boldsymbol{T}}(\boldsymbol{X}, \boldsymbol{F}, \theta_0, \boldsymbol{g}_0),\\
\eta_0 &= \hat{\eta}(\boldsymbol{X}, \boldsymbol{F}, \theta_0, \boldsymbol{g}_0),\\
\boldsymbol{q}_0 &= \hat{\boldsymbol{q}}_0(\boldsymbol{X}, \boldsymbol{F}, \theta_0, \boldsymbol{g}_0),
\end{aligned}$$

where $\boldsymbol{g}_0 = \nabla_X \theta = \boldsymbol{F}^T \boldsymbol{g}$. A thermodynamic process is said to be *admissible* if it is compatible with the above constitutive equations. As a consequence of the Clausius–Duhem inequality, we have the following result:

Theorem 7.1.1. *A necessary and sufficient condition that every admissible thermodynamic process obey Eqn.* (3.87) *is that the following three conditions hold:*

1. *The response function* $\hat{\psi}$, $\hat{\boldsymbol{T}}$ *and* $\hat{\eta}$ *are independent of the temperature gradient* $\boldsymbol{g}_0$*:*

$$\psi_0 = \hat{\psi}(\boldsymbol{X}, \boldsymbol{F}, \theta_0), \quad \boldsymbol{T} = \hat{\boldsymbol{T}}(\boldsymbol{X}, \boldsymbol{F}, \theta_0), \quad \eta_0 = \hat{\eta}(\boldsymbol{X}, \boldsymbol{F}, \theta_0).$$

2. $\hat{\psi}$ *determines* $\hat{\boldsymbol{T}}$ *and* $\hat{\eta}$ *through the relations*

$$\hat{\boldsymbol{T}}(\boldsymbol{X}, \boldsymbol{F}, \theta_0) = \rho_0 \frac{\partial}{\partial \boldsymbol{F}} \hat{\psi}(\boldsymbol{X}, \boldsymbol{F}, \theta_0), \tag{7.3}$$

$$\hat{\eta}(\boldsymbol{X}, \boldsymbol{F}, \theta_0) = -\frac{\partial}{\partial \theta_0} \hat{\psi}(\boldsymbol{X}, \boldsymbol{F}, \theta_0). \tag{7.4}$$

3. $\hat{\boldsymbol{q}}_0$ *obeys the heat conduction inequality*

$$\hat{\boldsymbol{q}}_0(\boldsymbol{F}, \theta_0, \boldsymbol{g}_0) \cdot \boldsymbol{g}_0 \le 0. \tag{7.5}$$

Proof. We shall drop the explicit dependence on $\boldsymbol{X}$ for notational convenience. Substituting the constitutive relations into Eqn. (3.87), we get the necessary and sufficient condition for an admissible thermodynamic process as

$$\left[\frac{\partial \hat{\psi}}{\partial \boldsymbol{F}}(\boldsymbol{F}, \theta_0, \boldsymbol{g}_0) - \frac{1}{\rho_0} \hat{\boldsymbol{T}}(\boldsymbol{F}, \theta_0, \boldsymbol{g}_0)\right] : \dot{\boldsymbol{F}} + \left[\frac{\partial}{\partial \theta_0} \hat{\psi}(\boldsymbol{F}, \theta_0, \boldsymbol{g}_0) + \hat{\eta}(\boldsymbol{F}, \theta_0, \boldsymbol{g}_0)\right] \dot{\theta}_0$$
$$+ \left[\frac{\partial}{\partial \boldsymbol{g}_0} \hat{\psi}(\boldsymbol{F}, \theta_0, \boldsymbol{g}_0)\right] \cdot \dot{\boldsymbol{g}}_0 + \frac{1}{\rho_0 \theta_0} \hat{\boldsymbol{q}}_0(\boldsymbol{F}, \theta_0, \boldsymbol{g}_0) \cdot \boldsymbol{g}_0 \le 0. \tag{7.6}$$

The sufficiency of conditions (1)–(3) stated in the theorem follows from Eqn. (7.6). To prove the necessity, we first prove that given $(\boldsymbol{F}^\diamond, \theta^\diamond, \boldsymbol{g}^\diamond, \boldsymbol{X}^\diamond, t^\diamond)$, a tensor $\boldsymbol{A}$, a scalar α and a vector $\boldsymbol{a}$, with $\det \boldsymbol{F}^\diamond > 0$ and $\theta^\diamond > 0$, there exists an admissible thermodynamic process with

$$\boldsymbol{F}(\boldsymbol{X}^\diamond, t^\diamond) = \boldsymbol{F}^\diamond, \qquad \dot{\boldsymbol{F}}(\boldsymbol{X}^\diamond, t^\diamond) = \boldsymbol{A},$$
$$\theta_0(\boldsymbol{X}^\diamond, t^\diamond) = \theta^\diamond, \qquad \dot{\theta}_0(\boldsymbol{X}^\diamond, t^\diamond) = \alpha,$$
$$\boldsymbol{g}_0(\boldsymbol{X}^\diamond, t^\diamond) = \boldsymbol{g}^\diamond, \qquad \dot{\boldsymbol{g}}_0(\boldsymbol{X}^\diamond, t^\diamond) = \boldsymbol{a}.$$

Clearly, there exists an open ball about $\boldsymbol{X}^\diamond$, and an open interval about $t^\diamond$ such that

$$\det[\boldsymbol{F}^\diamond + (t - t^\diamond)\boldsymbol{A}] > 0,$$
$$\theta^\diamond + \alpha(t - t^\diamond) + [\boldsymbol{g}^\diamond + (t - t^\diamond)\boldsymbol{a}] \cdot (\boldsymbol{X} - \boldsymbol{X}^\diamond) > 0.$$

Let $\chi(\boldsymbol{X}, t)$ and $\theta_0(\boldsymbol{X}, t)$ be defined by

$$\chi(\boldsymbol{X}, t) = [\boldsymbol{F}^\diamond + (t - t^\diamond)\boldsymbol{A}](\boldsymbol{X} - \boldsymbol{X}^\diamond),$$
$$\theta_0(\boldsymbol{X}, t) = \theta^\diamond + \alpha(t - t^\diamond) + [\boldsymbol{g}^\diamond + (t - t^\diamond)\boldsymbol{a}] \cdot (\boldsymbol{X} - \boldsymbol{X}^\diamond).$$

Then, the above process has the desired properties. Applying the inequality given by Eqn. (7.6) to the process constructed above yields

$$\left[\frac{\partial \hat{\psi}}{\partial F}(F^\diamond,\theta^\diamond,g^\diamond) - \frac{1}{\rho_0}\hat{T}(F^\diamond,\theta^\diamond,g^\diamond)\right] : A + \left[\frac{\partial}{\partial \theta_0}\hat{\psi}(F^\diamond,\theta^\diamond,g^\diamond) + \hat{\eta}(F^\diamond,\theta^\diamond,g^\diamond)\right]\alpha$$
$$+ \left[\frac{\partial}{\partial g_0}\hat{\psi}(F^\diamond,\theta^\diamond,g^\diamond)\right]\cdot a + \frac{1}{\rho_0\theta^\diamond}\hat{q}_0(F^\diamond,\theta^\diamond,g^\diamond)\cdot g^\diamond \leq 0. \quad (7.7)$$

Note that $F^\diamond$, $\theta^\diamond$, $g^\diamond$, A, α and a are arbitrary. Observe that Eqn. (7.7) can be written as

$$\left[\frac{\partial}{\partial g_0}\hat{\psi}(F^\diamond,\theta^\diamond,g^\diamond)\right]\cdot a + f_1(F^\diamond,\theta^\diamond,g^\diamond,A,\alpha) \leq 0.$$

If we fix the values of $F^\diamond$, $\theta^\diamond$, $g^\diamond$, A and α, then $f_1(F^\diamond,\theta^\diamond,g^\diamond,A,\alpha)$ is a finite constant, say c. Thus, we need the inequality

$$\left[\frac{\partial}{\partial g_0}\hat{\psi}(F^\diamond,\theta^\diamond,g^\diamond)\right]\cdot a + c \leq 0.$$

to be satisfied for all values of a. This is possible only if

$$\frac{\partial}{\partial g_0}\hat{\psi}(F^\diamond,\theta^\diamond,g^\diamond) = \mathbf{0}, \quad (7.8)$$

or, in other words, $\hat{\psi}$ is independent of g_0. Using Eqn. (7.8), Eqn. (7.7) can be written as

$$\left[\frac{\partial}{\partial \theta_0}\hat{\psi}(F^\diamond,\theta^\diamond,g^\diamond) + \hat{\eta}(F^\diamond,\theta^\diamond,g^\diamond)\right]\alpha + f_2(F^\diamond,\theta^\diamond,g^\diamond,A) \leq 0.$$

Fixing $F^\diamond$, $\theta^\diamond$, $g^\diamond$ and A, the necessary condition for the above inequality to hold for any arbitrary α is that the term in the square brackets disappears, i.e.,

$$\hat{\eta}(F^\diamond,\theta^\diamond) = -\frac{\partial}{\partial \theta_0}\hat{\psi}(F^\diamond,\theta^\diamond).$$

By following the above procedure repeatedly, the necessity of all the conditions stated in the theorem follows. □

Note that by using the same procedure as above, we can rule out the dependence of $\hat{\psi}$ on second and higher-order gradients of the deformation and the temperature, even if we assume such a dependence at the outset.

Since by Eqns. (7.3) and (7.4),

$$\left(\frac{\partial \psi_0}{\partial t}\right)_X = \frac{1}{\rho_0}T : \dot{F} - \eta_0\dot{\theta}_0,$$

Eqn. (3.86) reduces to

$$\rho_0\theta_0\left(\frac{\partial\eta_0}{\partial t}\right)_X = \rho_0(Q_h)_0 - \boldsymbol{\nabla}_X\cdot\boldsymbol{q}_0. \tag{7.9}$$

Thus, an admissible thermodynamic process in a thermoelastic material is adiabatic, i.e., $\rho_0(Q_h)_0 = \boldsymbol{\nabla}_X\cdot\boldsymbol{q}_0 = 0$, if and only if it is isentropic ($\dot{\eta}_0 = 0$). This observation essentially says that there is no dissipation in a thermoelastic process, since we know that a process is isentropic if and only if it is adiabatic and reversible. From Eqns. (7.3) and (7.4), it also follows that

$$\frac{\partial}{\partial\theta_0}\left(\frac{\hat{\boldsymbol{T}}(\boldsymbol{F},\theta_0)}{\rho_0}\right) = -\frac{\partial}{\partial\boldsymbol{F}}\hat{\eta}(\boldsymbol{F},\theta_0).$$

We now consider the restrictions imposed by the principle of material frame-indifference on the constitutive relations. First note that the quantities $\boldsymbol{g}_0$ and $\boldsymbol{q}_0$ remain unchanged under a change of observer since

$$\begin{aligned}(\boldsymbol{q}_0)^* &= J^*(\boldsymbol{F}^*)^{-1}\boldsymbol{q}^* = J\boldsymbol{F}^{-1}\boldsymbol{Q}^T\boldsymbol{Q}\boldsymbol{q} = J\boldsymbol{F}^{-1}\boldsymbol{q} = \boldsymbol{q}_0,\\ (\boldsymbol{g}_0)^* &= \boldsymbol{\nabla}_X\theta^* = \boldsymbol{g}_0.\end{aligned}$$

The result for $\boldsymbol{q}_0$ is analogous to the case of the second Piola–Kirchoff stress, which is defined over the reference configuration, and which remains unchanged under a change of observer (see Eqn. (5.2)). The transformation law for the first Piola–Kirchoff stress is given by Eqn. (5.1). Thus, we have

$$\begin{aligned}\hat{\psi}(\boldsymbol{F},\theta_0) &= \hat{\psi}(\boldsymbol{QF},\theta_0),\\ \hat{\boldsymbol{T}}(\boldsymbol{F},\theta_0) &= \boldsymbol{Q}^T\hat{\boldsymbol{T}}(\boldsymbol{QF},\theta_0),\\ \hat{\eta}(\boldsymbol{F},\theta_0) &= \hat{\eta}(\boldsymbol{QF},\theta_0),\\ \hat{\boldsymbol{q}}_0(\boldsymbol{F},\theta_0,\boldsymbol{g}_0) &= \hat{\boldsymbol{q}}_0(\boldsymbol{QF},\theta_0,\boldsymbol{g}_0),\end{aligned}$$

for all $\boldsymbol{Q}\in\text{Orth}^+$, and for all $\boldsymbol{F}\in\text{Lin}^+$.

By the polar decomposition theorem, $\boldsymbol{F} = \boldsymbol{RU}$. Choosing $\boldsymbol{Q} = \boldsymbol{R}^T$, we arrive at the relations

$$\begin{aligned}\psi &= \hat{\psi}(\boldsymbol{U},\theta_0),\\ \boldsymbol{T} &= \boldsymbol{FU}^{-1}\hat{\boldsymbol{T}}(\boldsymbol{U},\theta_0),\\ \eta_0 &= \hat{\eta}(\boldsymbol{U},\theta_0),\\ \boldsymbol{q}_0 &= \hat{\boldsymbol{q}}_0(\boldsymbol{U},\theta_0,\boldsymbol{g}_0),\end{aligned}$$

Since $\boldsymbol{U} = \sqrt{\boldsymbol{F}^T\boldsymbol{F}}$, we have the existence of a function $\tilde{\psi}$ defined by

$$\tilde{\psi}(\boldsymbol{X},\boldsymbol{E},\theta_0) := \hat{\psi}(\boldsymbol{X},\boldsymbol{F},\theta_0)\quad\forall\boldsymbol{F}\in\text{Lin}^+,\ \theta_0>0. \tag{7.10}$$

We have the following result that should be compared with Theorem 5.4.2:

Theorem 7.1.2. *Given a thermoelastic material with the free energy function $\hat{\psi}$, and given the function $\tilde{\psi}$ defined by Eqn. (7.10), and if we assume without loss of generality that the derivative $\partial\tilde{\psi}/\partial\mathbf{E}(\mathbf{X},\mathbf{E},\theta_0)$ is a symmetric tensor (see discussion on page 88), then the principle of material frame-indifference is satisfied if and only if the response functions for the second Piola–Kirchoff stress tensor and the entropy are given by*

$$\tilde{\mathbf{S}}(\mathbf{X},\mathbf{E},\theta_0) = \rho_0 \frac{\partial\tilde{\psi}}{\partial\mathbf{E}}(\mathbf{X},\mathbf{E},\theta_0) \quad \forall \boldsymbol{F} \in Lin^+, \theta_0 > 0. \tag{7.11}$$

$$\tilde{\eta}(\mathbf{X},\mathbf{E},\theta_0) = -\frac{\partial\tilde{\psi}}{\partial\theta_0}(\mathbf{X},\mathbf{E},\theta_0) \quad \forall \boldsymbol{F} \in Lin^+, \theta_0 > 0. \tag{7.12}$$

Proof. Using the facts that $\mathbb{S}^T = \mathbb{T}$ and $(\boldsymbol{A} \boxtimes \boldsymbol{B})^T = \boldsymbol{A}^T \boxtimes \boldsymbol{B}^T$, we have

$$\begin{aligned}
\frac{\partial\hat{\psi}}{\partial\boldsymbol{F}}(\boldsymbol{F},\theta_0) &= \left(\frac{\partial\mathbf{E}}{\partial\boldsymbol{F}}\right)^T \frac{\partial\tilde{\psi}}{\partial\mathbf{E}} && \text{(by Eqn. (1.238))} \\
&= \left[\boldsymbol{F}^T \boxtimes \boldsymbol{I}\right]^T \mathbb{S}\frac{\partial\tilde{\psi}}{\partial\mathbf{E}} && \text{(by Eqn. (1.234))} \\
&= (\boldsymbol{F} \boxtimes \boldsymbol{I})\mathbb{S}\frac{\partial\tilde{\psi}}{\partial\mathbf{E}} \\
&= \frac{\boldsymbol{F}}{2}\left[\left(\frac{\partial\tilde{\psi}}{\partial\mathbf{E}}\right)^T + \frac{\partial\tilde{\psi}}{\partial\mathbf{E}}\right].
\end{aligned}$$

Now using the fact that $\frac{\partial\tilde{\psi}}{\partial\mathbf{E}}(\mathbf{X},\mathbf{E},\theta_0)$ is a symmetric tensor, we get

$$\hat{\boldsymbol{T}}(\boldsymbol{F},\theta_0) = \rho_0\frac{\partial\hat{\psi}}{\partial\boldsymbol{F}}(\boldsymbol{F},\theta_0) = \rho_0\boldsymbol{F}\frac{\partial\tilde{\psi}}{\partial\mathbf{E}}(\mathbf{E},\theta_0).$$

Hence,

$$\begin{aligned}
\tilde{\mathbf{S}}(\mathbf{E},\theta_0) &= \boldsymbol{F}^{-1}\hat{\boldsymbol{T}}(\boldsymbol{F},\theta_0) \\
&= \rho_0\frac{\partial\tilde{\psi}}{\partial\mathbf{E}}(\mathbf{E},\theta_0).
\end{aligned}$$

The entropy relation follows immediately since the θ_0 dependence of the functions $\hat{\psi}$ and $\tilde{\psi}$ is the same. The converse also follows from these relations since $\mathbf{E}^* = \mathbf{E}$, $\theta^* = \theta_0$, and $\rho_0^* = \rho_0$. □

As in the hyperelastic case, note from Eqn. (7.11) that the second Piola–Kirchhoff stress is symmetric, i.e., balance of angular momentum is automatically satisfied by a thermoelastic material.

We now examine the consequences of the heat conduction inequality given by Eqn. (7.5). The conductivity tensor is defined by

$$[\boldsymbol{K}(\mathbf{E},\theta_0)]_{ij} := -\left.\frac{\partial(\hat{q}_0)_i}{\partial(g_0)_j}(\mathbf{E},\theta_0,g_0)\right|_{g_0=\mathbf{0}}.$$

Note that K is not necessarily symmetric. Fixing E and θ_0, and expanding $\hat{q}_0$ about $g_0 = 0$, we get

$$\hat{q}_0(g_0) = \hat{q}_0(0) - Kg_0 + o(|g_0|),$$

as $g_0 \to 0$. Substituting this linearization into Eqn. (7.5), we get

$$\hat{q}_0(0) \cdot g_0 - g_0 \cdot Kg_0 + o\left(|g_0|^2\right) \le 0 \quad \forall g_0.$$

This inequality holds if and only if $\hat{q}_0(0) = 0$, and $g_0 \cdot Kg_0 \ge 0$ for every g_0. Thus, we have the following result:

Theorem 7.1.3. *The heat flux vanishes whenever the temperature gradient vanishes, i.e.,*

$$\hat{q}_0(E, \theta_0, 0) = 0,$$

and the conductivity tensor $K(E, \theta_0)$ is positive semi-definite.

As a result of the above theorem, we can write to a first approximation,

$$\hat{q}_0(E, \theta_0, g_0) = -K(E, \theta_0)g_0. \tag{7.13}$$

Equation (7.13) is known as the Fourier law for heat conduction in solids.

Under a rotation of the reference configuration of the form $X = Q(t)\bar{X}$ (which leads to $\bar{F} = FQ$ and hence to $\bar{B} = B$ and $\bar{C} = Q^T CQ$), we see that $\bar{g}_0 = Q^T g_0$. The quantity $g = \nabla_x \theta$ remains unaffected by this rotation of the reference configuration, i.e., $\bar{g} = g$. In order for the flux response to be isotropic, we need that $\hat{q}(\bar{F}, \theta, \bar{g}) = \hat{q}(F, \theta, g)$, or, alternatively, on using the polar decomposition,

$$\hat{q}(F, \theta, g) = \tilde{q}(B, \theta, g).$$

Using the relation $q_0 = (\mathbf{cof}\, F)^T q$, we see that the requirement for isotropy in the Lagrangian framework is that

$$\tilde{q}_0(\bar{C}, \theta_0, \bar{g}_0) = (\mathbf{cof}\, \bar{F})^T \tilde{q}(\bar{B}, \theta, \bar{g}) = Q^T (\mathbf{cof}\, F)^T \tilde{q}(B, \theta, g) = Q^T \tilde{q}_0(C, \theta_0, g_0).$$

Thus, the relations for an isotropic frame-indifferent material are given by

$$\begin{aligned}
\tilde{S}(X, \theta_0, C) &= \gamma_0(X, \theta_0, \mathcal{I}_C)I + \gamma_1(X, \theta_0, \mathcal{I}_C)C + \gamma_2(X, \theta_0, \mathcal{I}_C)C^2, \\
\tilde{q}_0(X, \theta_0, C, g_0) &= -(\beta_0 I + \beta_1 C + \beta_2 C^2)g_0,
\end{aligned}$$

where

$$\beta_\alpha = \tilde{\beta}_\alpha(\theta_0, \operatorname{tr} C, \operatorname{tr} C^2, \operatorname{tr} C^3, g_0 \cdot g_0, g_0 \cdot Cg_0, g_0 \cdot C^2 g_0).$$

The conductivity tensor is given by $K = \beta_0 I + \beta_1 C + \beta_2 C^2$. As an example, if we have $q = -k(\theta)g$, then since $q_0 = JF^{-1}q$ and $g = F^{-T}g_0$, we get $K = k(\theta_0)\sqrt{\det C}C^{-1}$.

To obtain finite wave speeds, the Fourier law of heat conduction can be generalized as [56]

$$\tau_0 \left(\frac{\partial \hat{q}_0}{\partial t} \right)_X + \hat{q}_0 = -K g_0, \tag{7.14}$$

where τ_0 is the thermal relaxation time. However, as pointed out in [293, 314] and references therein, some solutions to the hyperbolic heat conduction equation obtained using the above constitutive relation display heat conduction from cold to hot bodies, in violation of the second law of thermodynamics.

The number

$$c(E, \theta_0) := \frac{\partial}{\partial \theta_0} \hat{e}(E, \theta_0),$$

is called the *specific heat*. Since $\hat{e} = \hat{\psi} + \hat{\eta}\theta_0$, we have

$$c = \frac{\partial \hat{\psi}}{\partial \theta_0} + \hat{\eta} + \theta_0 \frac{\partial \hat{\eta}}{\partial \theta_0}.$$

By Eqn. (7.4), the above equation reduces to

$$c = \theta_0 \frac{\partial \hat{\eta}}{\partial \theta_0}. \tag{7.15}$$

The latent heat tensor $\boldsymbol{H}$ is defined as

$$\boldsymbol{H} := \theta_0 \frac{\partial \hat{\eta}}{\partial \boldsymbol{E}} = -\frac{\theta_0}{\rho_0} \frac{\partial \tilde{\boldsymbol{S}}}{\partial \theta_0}. \tag{7.16}$$

Combining Eqns. (7.9), (7.13), (7.15) and (7.16) we get the *heat equation*:

$$\rho_0 \left[c \left(\frac{\partial \theta_0}{\partial t} \right)_X + \boldsymbol{H} : \dot{\boldsymbol{E}} \right] = \boldsymbol{\nabla} \cdot (\boldsymbol{K} \boldsymbol{g}_0) + \rho_0 (Q_h)_0.$$

As an example, if $\hat{\alpha}$ denotes the (linear) coefficient of thermal expansion, and c is assumed to be constant, then, on combining with a St Venant–Kirchhoff model, we get

$$\begin{aligned}
\rho_0 \hat{\psi} &= \frac{1}{2} \lambda (\operatorname{tr} \boldsymbol{E})^2 + \mu \boldsymbol{E} : \boldsymbol{E} - 3\kappa\hat{\alpha} (\operatorname{tr} \boldsymbol{E}) \theta_\Delta + \rho_0 c \left(\theta_\Delta - \theta_0 \log \frac{\theta_0}{\theta_R} \right), \\
\boldsymbol{S} &= \lambda (\operatorname{tr} \boldsymbol{E}) \boldsymbol{I} + 2\mu \boldsymbol{E} - 3\kappa\hat{\alpha}\theta_\Delta \boldsymbol{I}, \\
\rho_0 \boldsymbol{H} &= -\theta_0 \frac{\partial \tilde{\boldsymbol{S}}}{\partial \theta_0} = 3\kappa\hat{\alpha}\theta_0 \boldsymbol{I}, \\
\rho_0 \hat{\eta} &= 3\kappa\hat{\alpha} (\operatorname{tr} \boldsymbol{E}) + \rho_0 c \log \frac{\theta_0}{\theta_R}, \\
\rho_0 \hat{e} &= \frac{1}{2} \left[\lambda (\operatorname{tr} \boldsymbol{E})^2 + 2\mu \boldsymbol{E} : \boldsymbol{E} \right] + 3\kappa\hat{\alpha} (\operatorname{tr} \boldsymbol{E}) \theta_R + \rho_0 c \theta_\Delta,
\end{aligned}$$

where θ_R represents the reference temperature, $\theta_\Delta := \theta_0 - \theta_R$, and $\kappa = \lambda + 2\mu/3$ is the bulk modulus, so that $3\kappa = 3\lambda + 2\mu = E/(1 - 2\nu)$. Note that the last term in the expression for S is such that $S = 0$ when $E = \hat{\alpha}(\theta_0 - \theta_R)I$. Substituting for H, the heat conduction equation for the St Venant–Kirchhoff model takes the form

$$\rho_0 c \left(\frac{\partial \theta_0}{\partial t}\right)_X + 3\kappa\hat{\alpha}\theta_0 \operatorname{tr}\dot{E} = \nabla \cdot (K g_0) + \rho_0 (Q_h)_0.$$

For rubber-type materials, the elastic constants vary linearly with temperature. Thus, in the Ogden model given by Eqn. (5.55), the elastic coefficients are given by $\mu_i(\theta_0) = \theta_0 \mu_i(\theta_R)/\theta_R$ and $\lambda(\theta_0) = \theta_0 \lambda(\theta_R)/\theta_R$, where $\mu_i(\theta_R)$ and $\lambda(\theta_R)$ are the values of the elastic coefficients at the reference temperature θ_R. Thus, the thermoelastic extension of the Ogden model is given by [42]

$$\begin{aligned}\rho_0 \hat{\psi} &= \sum_{i=1}^{3} \frac{\mu_i}{\alpha_i} \left[(\lambda_1^2)^{\alpha_i/2} + (\lambda_2^2)^{\alpha_i/2} + (\lambda_3^2)^{\alpha_i/2} - \frac{\alpha_i}{2} \log I_3 - 3 \right] \\ &\quad + \lambda W_H(J) - \frac{3}{\gamma} \kappa_0 (J^\gamma - 1) \alpha \theta_\Delta + \rho_0 c \left(\theta_\Delta - \theta_0 \log \frac{\theta_0}{\theta_R} \right),\end{aligned}$$

where $W_H(J)$ is given by Eqn. (5.56), $\gamma = 2.5$, and $3\kappa_0 = 3\lambda_0 + \sum_{i=1}^{3} (\mu_0)_i \alpha_i$. The associated constitutive relations based on this model are

$$\begin{aligned} S &= S_{\text{mech}} - 3\kappa_0 \hat{\alpha} \theta_\Delta (\det C)^{\gamma/2} C^{-1}, \\ \rho_0 H &= -S_{\text{mech}} + 3\kappa_0 \hat{\alpha} \theta_0 (\det C)^{\gamma/2} C^{-1}, \\ \rho_0 \hat{e} &= \frac{3}{\gamma} \kappa_0 \hat{\alpha} (J^\gamma - 1) \theta_R + \rho_0 c \theta_\Delta, \end{aligned}$$

where S_{mech} is given by Eqn. (5.57) with the elastic moduli now dependent on temperature. Identical approximations as those given above result for the compressible Mooney–Rivlin material model (see Problem 1, Chapter 5) if, again, the elastic moduli are assumed to depend linearly on the temperature.

If the temperature θ_0 is maintained constant throughout the deformation process, then we see that $\hat{\psi}$ is just a function of F. Hence, for an *isothermal* deformation, a thermoelastic material reduces to a hyperelastic material with $\hat{W}(F) = \hat{\psi}(F)$. Similarly, for an *isentropic* deformation, using $\hat{e} = \hat{\psi} + \hat{\eta}\theta_0$, we have

$$\frac{\partial \hat{e}}{\partial \theta_0} = \frac{\partial \hat{\psi}}{\partial \theta_0} + \hat{\eta} = 0,$$

i.e., $e = \hat{e}(F)$. Hence, for an isentropic deformation, a thermoelastic material reduces to a hyperelastic material with $\hat{W}(F) = \hat{e}(F)$. Summarizing, for an isothermal or isentropic process, a thermoelastic material may be considered as hyperelastic with the stored-energy function given either by $\hat{\psi}$ or $\hat{e}$.

Besides the above conventional approach, there is an alternative approach based on a multiplicative decomposition of the deformation gradient into an elastic part and a thermal part; e.g., see [130], [194], [204] and [345]. There is an assumption in this theory that the free energy can be decomposed into a part that depends on the elastic part of the strain $\hat{\psi}_e$ (with the material parameters dependent on the temperature), and a part that depends on the temperature alone $\hat{\psi}_{\theta_0}$, e.g., see Eqn. (52) in [345]. The part $\hat{\psi}_e$ is chosen to be one of the strain energy functions of isothermal finite-strain elasticity. It is not clear whether the above assumption is backed by *experimental* evidence at *large* strains and temperature increments.

The results of Theorem 7.1.1 were derived in the absence of any constraints on $\boldsymbol{F}$ or θ. However, now consider the presence of scalar-valued kinematic or internal constraints such as, for example,

$$\mu_1(\boldsymbol{F}) = 0, \quad \mu_2(\theta_0) = 0,$$

which lead to the restrictions

$$\frac{\partial \mu_1}{\partial \boldsymbol{F}} : \dot{\boldsymbol{F}} = 0, \quad \frac{\partial \mu_2}{\partial \theta_0}\dot{\theta}_0 = 0.$$

In view of these constraints, Eqn. (7.6) can be written as

$$\left[\frac{\partial \hat{\psi}}{\partial \boldsymbol{F}}(\boldsymbol{F}, \theta_0, \boldsymbol{g}_0) - \frac{1}{\rho_0}\hat{\boldsymbol{T}}(\boldsymbol{F}, \theta_0, \boldsymbol{g}_0) + \frac{\lambda_1}{\rho_0}\frac{\partial \mu_1}{\partial \boldsymbol{F}}\right] : \dot{\boldsymbol{F}} + \left[\frac{\partial}{\partial \theta_0}\hat{\psi}(\boldsymbol{F}, \theta_0, \boldsymbol{g}_0) + \hat{\eta}(\boldsymbol{F}, \theta_0, \boldsymbol{g}_0) + \lambda_2 \frac{\partial \mu_2}{\partial \theta_0}\right]\dot{\theta}_0$$
$$+ \left[\frac{\partial}{\partial \boldsymbol{g}_0}\hat{\psi}(\boldsymbol{F}, \theta_0, \boldsymbol{g}_0)\right] \cdot \dot{\boldsymbol{g}}_0 + \frac{1}{\rho_0 \theta_0}\hat{\boldsymbol{q}}_0(\boldsymbol{F}, \theta_0, \boldsymbol{g}_0) \cdot \boldsymbol{g}_0 \leq 0,$$

where λ_1 and λ_2 are arbitrary scalar-valued parameters. Now following the same procedure as in the proof of Theorem 7.1.1, we get (compare with Eqns. (7.3) and (7.4))

$$\hat{\boldsymbol{T}}(\boldsymbol{X}, \boldsymbol{F}, \theta_0) = \rho_0 \frac{\partial}{\partial \boldsymbol{F}}\hat{\psi}(\boldsymbol{X}, \boldsymbol{F}, \theta_0) + \lambda_1 \frac{\partial \mu_1}{\partial \boldsymbol{F}},$$
$$\hat{\eta}(\boldsymbol{X}, \boldsymbol{F}, \theta_0) = -\frac{\partial}{\partial \theta_0}\hat{\psi}(\boldsymbol{X}, \boldsymbol{F}, \theta_0) - \lambda_2 \frac{\partial \mu_2}{\partial \theta_0}.$$

The extra term $\boldsymbol{N} \equiv \lambda_1 \partial \mu_1 / \partial \boldsymbol{F}$ is termed as the reaction stress. The corresponding reaction stress that is part of the Cauchy stress tensor is $\boldsymbol{N}_\tau = \boldsymbol{N}\boldsymbol{F}^T/J = \frac{\lambda_1}{J}\frac{\partial \mu_1}{\partial \boldsymbol{F}}\boldsymbol{F}^T$. Since the reaction stress is a part of the total stress tensor, we subject it to the same restrictions such as material frame-indifference, symmetry, etc. as the total stress tensor itself. In the above development, we have considered only one constraint on the deformation gradient. If there are multiple constraints, $\mu_i(\boldsymbol{F}) = 0$, $i = 1, \ldots, n$, then the reaction stress is given by $\sum_{i=1}^n \lambda_i \partial \mu_i / \partial \boldsymbol{F}$. Note that the stress power, $\boldsymbol{N} : \dot{\boldsymbol{F}}/J = \boldsymbol{N}_\tau : \boldsymbol{L}$, corresponding to the reaction stress is zero. Also note that since the λ_i's are arbitrary, the reaction stress and 'reaction entropy' are indeterminate. However, they can be determined as a part of the total stress or entropy field when appropriate boundary conditions are prescribed.

As an example, consider the incompressibility constraint $\mu(\boldsymbol{F}) = \det \boldsymbol{F} - 1 = 0$. Using Eqn. (1.226c), the reaction stress is obtained as $\boldsymbol{N} = \lambda_1 \operatorname{\mathbf{cof}} \boldsymbol{F}$, which implies that $\boldsymbol{N}_\tau = \lambda_1 \boldsymbol{I} = -p\boldsymbol{I}$, where, since λ_1 is arbitrary, $p(\boldsymbol{x}, t) = -\lambda_1$ denotes the 'indeterminate' pressure. As another example, consider a rigid body that is capable of undergoing only rigid motions for which, as we have seen, the deformation gradient is $\boldsymbol{F} = \boldsymbol{Q}(t) \in \text{Orth}^+$. This can be thought of as nine constraint equations; writing the corresponding nine λ_i's in the form of a matrix with components λ_{ij}, we have $N_{ij} = \lambda_{kl}\partial F_{kl}/\partial F_{ij} = \lambda_{ij}$, from which it follows that the entire stress tensor is indeterminate.

7.2 Viscoplastic materials

In this section, we derive a finite strain viscoplasticity theory under isothermal conditions that is based on the following hypothesis:

1. The deformation gradient $\boldsymbol{F}(\boldsymbol{X},t)$ can be multiplicatively decomposed into an elastic part $\boldsymbol{F}^e(\boldsymbol{X},t)$ and a plastic part $\boldsymbol{F}^p(\boldsymbol{X},t)$, i.e., $\boldsymbol{F}(\boldsymbol{X},t) = \boldsymbol{F}^e(\boldsymbol{X},t)\boldsymbol{F}^p(\boldsymbol{X},t)$. Since we have assumed that $J := \det \boldsymbol{F} > 0$, it follows that $J^e := \det \boldsymbol{F}^e$ and $J^p := \det \boldsymbol{F}^p$ are both nonzero, so that $\boldsymbol{F}^e$ and $\boldsymbol{F}^p$ can both be assumed to be invertible. Note that $\boldsymbol{F}^e$ and $\boldsymbol{F}^p$ are *not* the gradients of a deformation field, and are thus to be viewed as internal variables that are determined through appropriate evolution laws.

2. We assume the free energy ψ to be dependent on $\boldsymbol{C}^e := (\boldsymbol{F}^e)^T\boldsymbol{F}^e$, but independent of $\boldsymbol{C}^p := (\boldsymbol{F}^p)^T\boldsymbol{F}^p$–this is a reasonable assumption for viscoplastic solids, but not for amorphous solids where a dependence on $\boldsymbol{C}^p$ also needs to be assumed [112]. Since $(\boldsymbol{F}^e)^* = \boldsymbol{Q}\boldsymbol{F}^e$ and $(\boldsymbol{F}^p)^* = \boldsymbol{F}^p$ [107, 112], we have $(\boldsymbol{C}^e)^* = \boldsymbol{C}^e$–thus, the free energy function ψ is automatically frame-indifferent. If one assumes isotropic response, then ψ is an isotropic function of $\boldsymbol{C}^e$, i.e., a function of the principal invariants of $\boldsymbol{C}^e$.

3. To derive the evolution laws, we adopt the principle of maximum dissipation that is discussed in greater detail below.

Quite remarkably, these three hypotheses are sufficient to yield the governing equations for the physical and internal variables, which in turn yield solutions that are in good agreement with experiments. In essence, the theory that is discussed below is based on the treatment in [296] with the following minor differences:

- We develop the theory in the material framework that is suitable for numerical implementations of the total Lagrangian approach.

- As stated above, the free energy ψ is assumed to depend on $\boldsymbol{C}^e$ instead of $\boldsymbol{B}^e$, so that the material behavior is not restricted to be isotropic.

- The yield function is formulated in terms of the Cauchy (which is the 'true' stress) rather than the Kirchhoff stress tensor [219].

Using the decomposition $\boldsymbol{F} = \boldsymbol{F}^e\boldsymbol{F}^p$, we get $\boldsymbol{C} = (\boldsymbol{F}^p)^T\boldsymbol{C}^e\boldsymbol{F}^p$, where $\boldsymbol{C}^e := (\boldsymbol{F}^e)^T\boldsymbol{F}^e$, and

$$\dot{\boldsymbol{C}} = (\boldsymbol{F}^p)^T\dot{\boldsymbol{C}}^e\boldsymbol{F}^p + (\boldsymbol{F}^p)^T\boldsymbol{C}^e\dot{\boldsymbol{F}}^p + [(\boldsymbol{F}^p)^T\boldsymbol{C}^e\dot{\boldsymbol{F}}^p]^T. \tag{7.17}$$

The free energy function is assumed to be of the form

$$\rho_0\psi_0 = \hat{\psi}(\boldsymbol{C}^e, \boldsymbol{\alpha}_K^0, \alpha),$$

where $\boldsymbol{\alpha}_K^0$ and α are strain-like internal variables that describe kinematic and isotropic hardening. Substituting the above free energy form into the isothermal version of Eqn. (3.88) and using Eqns. (1.48) and (1.329), we get

$$\left[\frac{1}{2}\boldsymbol{F}^p\boldsymbol{S}(\boldsymbol{F}^p)^T - \frac{\partial\hat{\psi}}{\partial\boldsymbol{C}^e}\right] : \dot{\boldsymbol{C}}^e + \boldsymbol{S} : (\boldsymbol{F}^p)^T\boldsymbol{C}^e\dot{\boldsymbol{F}}^p - \frac{\partial\hat{\psi}}{\partial\boldsymbol{\alpha}_K^0} : \dot{\boldsymbol{\alpha}}_K^0 - \frac{\partial\hat{\psi}}{\partial\alpha}\dot{\alpha} \geq 0. \tag{7.18}$$

In the absence of yielding, i.e., when $\dot{\boldsymbol{\alpha}}_K^0 = \mathbf{0}$, $\dot{\alpha} = 0$ and $\boldsymbol{C}^e = \boldsymbol{C}$, we have equality in the above equation for arbitrary $\dot{\boldsymbol{C}}^e$. We thus conclude that

$$\boldsymbol{S} = 2(\boldsymbol{F}^p)^{-1}\frac{\partial\hat{\psi}}{\partial\boldsymbol{C}^e}(\boldsymbol{F}^p)^{-T}, \tag{7.19}$$

which is the same as Eqn. (110) in [12]. Defining the internal stress-like hardening variables[1],

$$\boldsymbol{q}_K^0 := -\frac{\partial\hat{\psi}}{\partial\boldsymbol{\alpha}_K^0}, \quad q := -\frac{\partial\hat{\psi}}{\partial\alpha}, \tag{7.20}$$

the dissipation function (the expression on the left-hand side of Eqn. (7.18)) is now given by

$$\mathcal{D} = \boldsymbol{S} : (\boldsymbol{F}^p)^T\boldsymbol{C}^e\dot{\boldsymbol{F}}^p + \boldsymbol{q}_K^0 : \dot{\boldsymbol{\alpha}}_K^0 + q\dot{\alpha} \geq 0 \tag{7.21}$$

For an elastic deformation (i.e., the plastic internal variables $\boldsymbol{F}^p$, $\boldsymbol{\alpha}_K^0$ and α are constant), the dissipation is zero, and using Eqn. (3.70) and assuming conservative loading, we get, similar to Eqn. 5.36,

$$\int_{V_0}\left[\frac{1}{2}\rho_0\left(\frac{\partial\boldsymbol{u}}{\partial t}\right)_{\boldsymbol{X}}\cdot\left(\frac{\partial\boldsymbol{u}}{\partial t}\right)_{\boldsymbol{X}} + \hat{\psi}(\boldsymbol{C}^e) - W_{b_0}\right]dV_0 - \int_{(S_0)_t} W_{t_0}\,dS_0 = \text{constant}. \tag{7.22}$$

Since $\boldsymbol{q}_K^0$ is a stress-like material variable, we assume that it is related to its spatial counterpart $\boldsymbol{q}_K$ via the relation

$$\boldsymbol{q}_K = \frac{1}{J}\boldsymbol{F}\boldsymbol{q}_K^0\boldsymbol{F}^T. \tag{7.23}$$

Assuming that $\boldsymbol{q}_K : \boldsymbol{\alpha}_K = \boldsymbol{q}_K^0 : \boldsymbol{\alpha}_K^0$, the corresponding relation for the strain-like hardening variable is

$$\boldsymbol{\alpha}_K^0 = \frac{1}{J}\boldsymbol{F}^T\boldsymbol{\alpha}_K\boldsymbol{F}. \tag{7.24}$$

Under a change of frame of reference, we assume $(\boldsymbol{q}_K^0)^* = \boldsymbol{q}_K^0$ and $(\boldsymbol{\alpha}_K^0)^* = \boldsymbol{\alpha}_K^0$, thus leading to

$$\boldsymbol{q}_K^* = \boldsymbol{Q}\boldsymbol{q}_K\boldsymbol{Q}^T, \tag{7.25a}$$

$$\boldsymbol{\alpha}_K^* = \boldsymbol{Q}\boldsymbol{\alpha}_K\boldsymbol{Q}^T. \tag{7.25b}$$

The elastic domain is assumed to be defined via the following restriction on a yield function $\phi(\boldsymbol{\tau}, \boldsymbol{q}_K, q)$:

$$\phi(\boldsymbol{\tau}, \boldsymbol{q}_K, q) \leq 0. \tag{7.26}$$

As mentioned above, physically speaking, it is more reasonable to assume the yield function to depend on the Cauchy stress tensor rather than on the Kirchhoff stress tensor. Since

[1] Although α and q are material variables, for notational convenience, we do not put the superscript '0', since these being scalar variables, their values in the spatial or material settings are the same.

$\tau^* = Q\tau Q^T$, $q_K^* = Qq_K Q^T$, we have $\phi(Q\tau Q^T, Qq_K Q^T, q) = \phi(\tau, q_K, q)$. It follows that ϕ is an isotropic function of τ and q_K, and thus a function of the principal invariants of τ (or alternatively, by virtue of Eqn. (J.21), the principal invariants of CS) and of q_K (note that the isotropy of ϕ does not imply that the material response as given by Eqn. (7.19) is isotropic).

The principle of maximum dissipation states that among all possible admissible states $(\hat{S}, \hat{q}_K^0, \hat{q})$ that satisfy Eqn. (7.26), the actual state (S, q_K^0, q) is the one that maximizes $\mathcal{D}$ for a given set of kinematical variables $(F^e, F^p, \dot{F}^p, \dot{\alpha}_K^0, \dot{\alpha})$. Thus, in terms of the Lagrangian

$$\mathcal{L} := -S : \frac{1}{2}\left[(F^p)^T C^e \dot{F}^p + (\dot{F}^p)^T C^e F^p\right] - q_K^0 : \dot{\alpha}_K^0 - q\dot{\alpha} + \gamma\phi(\tau, q_k, q),$$

where γ denotes the Lagrange multiplier, the evolution equations are given by

$$\begin{aligned}
\frac{\partial \mathcal{L}}{\partial S} &= -\frac{1}{2}\left[(F^p)^T C^e \dot{F}^p + (\dot{F}^p)^T C^e F^p\right] + \gamma\frac{\partial \phi}{\partial S} = \mathbf{0},\\
\frac{\partial \mathcal{L}}{\partial q_K^0} &= -\dot{\alpha}_K^0 + \gamma\frac{\partial \phi}{\partial q_K^0} = \mathbf{0},\\
\frac{\partial \mathcal{L}}{\partial q} &= -\dot{\alpha} + \gamma\frac{\partial \phi}{\partial q} = 0.
\end{aligned}$$

Let $L^p := \dot{F}^p(F^p)^{-1}$ be the plastic distortion rate. Similar to Eqn. (106) of [107], we *assume* that the plastic spin given by skew$(C^e L^p)$ vanishes, or, alternatively, $C^e L^p \in$ Sym. This implies that $(F^p)^T C^e \dot{F}^p \in$ Sym, so that the first of the above equations can be written as

$$(F^p)^T C^e \dot{F}^p = \gamma\frac{\partial \phi}{\partial S}.$$

Using the fact that

$$\begin{aligned}
\frac{\partial \phi}{\partial S} &= \frac{1}{J}F^T\frac{\partial \phi}{\partial \tau}F,\\
\frac{\partial \phi}{\partial q^0} &= \frac{1}{J}F^T\frac{\partial \phi}{\partial q_K}F,
\end{aligned}$$

and using the invertibility of F^e and F^p, the evolution equations can be written as

$$\dot{F}^p = \frac{\gamma}{J}\left[(F^e)^{-1}\frac{\partial \phi}{\partial \tau}F^e\right]F^p, \tag{7.27a}$$

$$\dot{\alpha}_K^0 = \frac{\gamma}{J}F^T\frac{\partial \phi}{\partial q_K}F, \tag{7.27b}$$

$$\dot{\alpha} = \gamma\frac{\partial \phi}{\partial q}. \tag{7.27c}$$

For an isotropic material, Eqn. (7.27a) leads to Eqn. (2.14) in [296] (apart from the factor J due to their use of the Kirchhoff versus our use of the Cauchy stress tensor). Note that Eqns. (7.27a)–(7.27c) satisfy the axiom of frame-indifference, i.e., they have the same form in all frames of reference (this property would not hold had we written the principle of maximum dissipation in the spatial framework, and derived a relation of the form $\dot{\alpha}_K = \gamma\partial\phi/\partial q_K$, as can be seen by using Eqn. (7.25b)).

From Eqns. (1.223) and (7.27a), we have

$$\begin{aligned}\dot{J}^p &= J^p \operatorname{tr}(\dot{F}^p (F^p)^{-1}) \\ &= \frac{\gamma}{J^e} \operatorname{tr}\left[(F^e)^{-1} \frac{\partial \phi}{\partial \tau} F^e\right] \\ &= \frac{\gamma}{J^e} \operatorname{tr}\left(\frac{\partial \phi}{\partial \tau}\right).\end{aligned}$$

A pressure-insensitive yield condition is one for which ϕ is independent of the first invariant I_1 of τ. Pressure-insensitive yield conditions (such as the von Mises yield condition) are formulated in terms of the deviatoric stress τ_{dev}, so that

$$\operatorname{tr}\left(\frac{\partial \phi}{\partial \tau}\right) = 0,$$

which implies plastic volume preservation, i.e., since $J^p = 1$ at time $t = 0$, we have $J^p = 1$ at all times for materials with a pressure-insensitive yield condition. The rate-dependent viscoplasticity formulation can be obtained by a penalty regularization of the rate-independent model presented above, i.e., by taking

$$\gamma = \frac{1}{\eta} \langle g(\phi) \rangle := \frac{1}{2\eta} \left[g(\phi) + |g(\phi)| \right],$$

where $\eta > 0$ is known as a viscosity parameter, and where the function $g(x)$ is a continuous, monotonically increasing function of x with $g(0) = 0$, e.g., $g(x) = x$.

If one assumes the material to be isotropic, i.e., $\hat{\psi}$ is a function of the principal invariants of C^e, then from Eqn. (7.19), it follows that τ is of the form $\sum_{i=0}^{2} \gamma_i (B^e)^i$, with $(B^e)^0 \equiv I$, and since ϕ is a function of the principal invariants of τ, $\partial\phi/\partial\tau$ is of the form $\sum_{i=0}^{2} \beta_i \tau^i$, or alternatively, by substituting for τ and using the Cayley–Hamilton theorem, of the form $\sum_{i=0}^{2} \alpha_i (B^e)^i$. Thus, from Eqn. (7.27a), one now gets $L^p \in \text{Sym}$, which is consistent with Eqn. (13.1) of [112].

Usually, $\hat{\psi}(C^e, \alpha_K^0, \alpha)$ is assumed to be separable into parts that depend on C^e, α_K^0 and α, i.e.,

$$\hat{\psi}(C^e, \alpha_K^0, \alpha) = \hat{W}(C^e) + \hat{K}_1(\alpha_K^0) + \hat{K}_2(\alpha),$$

where $\hat{W}(C^e)$ is analogous to a stored-energy function for an elastic material, e.g., see Eqns. (5.50) or (5.52). Examples of $\hat{K}_1$ and $\hat{K}_2$ are $H_K \alpha_K^0 : \alpha_K^0/2$ and $H\alpha^2/2 + (\sigma_Y^\infty - \sigma_Y)(\alpha + e^{-\delta\alpha}/\delta)$, where $H_K > 0$ and $H > 0$ are kinematic and isotropic hardening coefficients, σ_Y is the uniaxial yield stress, and where $\sigma_Y^\infty \geq \sigma_Y > 0$, $\delta \geq 0$ are also material constants. A typical example for the yield function is the Drucker–Prager criterion, which is given by

$$\begin{aligned}\phi &= \|\operatorname{dev}(\tau - q_K)\| + \beta \operatorname{tr}(\tau - q_K) - \sqrt{\frac{2}{3}}(\sigma_Y - q) \\ &= \frac{1}{\sqrt{\det C}} \sqrt{C(S - q_K^0) : (S - q_K^0)C - \frac{1}{3}[C : (S - q_K^0)]^2} + \frac{\beta}{\sqrt{\det C}} C : (S - q_K^0) \\ &\quad - \sqrt{\frac{2}{3}}(\sigma_Y - q),\end{aligned}$$

where β is a material parameter. The von Mises yield criterion is obtained by simply setting β to zero. Note that the von Mises yield criterion is pressure-insensitive. Using Eqns. (7.19), (7.20), and (7.27), we have for the above example

$$\begin{aligned}
S &= 2(F^p)^{-1}\frac{\partial \hat{W}}{\partial C^e}(F^p)^{-T}, \\
q_K^0 &= -H_K \alpha_K^0, \\
q &= -\left[H\alpha + (\sigma_Y^\infty - \sigma_Y)\left(1 - e^{-\delta\alpha}\right)\right], \\
\dot{F}^p &= \frac{\gamma}{J}\left[(F^e)^{-1} N F^e\right] F^p = \frac{\gamma}{J} F^p \left[F^{-1} N F\right], \\
\dot{\alpha}_K^0 &= -\frac{\gamma}{J} F^T N F, \\
\dot{\alpha} &= \sqrt{\frac{2}{3}}\gamma,
\end{aligned}$$

where

$$N := \frac{\partial \phi}{\partial \tau} = \frac{\operatorname{dev}(\tau - q_K)}{\|\operatorname{dev}(\tau - q_K)\|} + \beta I = \frac{\operatorname{dev}[F(S - q_K^0)F^T]}{\|\operatorname{dev}[F(S - q_K^0)F^T]\|} + \beta I.$$

By eliminating α_K^0, the above set of equations can be written as

$$\begin{aligned}
S &= 2(F^p)^{-1}\frac{\partial \hat{W}}{\partial C^e}(F^p)^{-T}, \\
\dot{F}^p &= \frac{\gamma}{J} F^p \left[F^{-1} N F\right], \\
\dot{q}_K^0 &= \frac{H_K \gamma}{J} F^T N F.
\end{aligned}$$

A numerical implementation of the above set of equations can be found in [159].

7.3 Restrictions on the constitutive relations for fluids

In this section, we consider a special kind of nonlinear fluid response that includes viscous effects but not long-range memory. For notational convenience, we drop the dependence of the constitutive relations on X; however, all the arguments are valid for materially homogeneous and materially inhomogeneous bodies. The constitutive relations are assumed to be

$$\begin{aligned}
\psi &= \hat{\psi}(F, \theta, g, \dot{F}), \\
\tau &= \hat{\tau}(F, \theta, g, \dot{F}), \\
\eta &= \hat{\eta}(F, \theta, g, \dot{F}), \\
q &= \hat{q}(F, \theta, g, \dot{F}).
\end{aligned} \tag{7.28}$$

The above constitutive relations define a simple material as given by Eqn. (4.48), since the history of $F(X, t)$ upto time t determines $\dot{F}$. Since we are considering only $\dot{F}$, and not higher-order derivatives such as $\ddot{F}$, we use the the terminology viscous fluid 'without long-range memory'.

The symmetry group of a fluid is the set of all proper unimodular tensors H for which the following identities hold:

$$\hat{\psi}(F, \theta, g, \dot{F}) = \hat{\psi}(FH, \theta, g, \dot{F}H), \tag{7.29}$$

$$\hat{\tau}(F, \theta, g, \dot{F}) = \hat{\tau}(FH, \theta, g, \dot{F}H), \tag{7.30}$$

$$\hat{\eta}(F, \theta, g, \dot{F}) = \hat{\eta}(FH, \theta, g, \dot{F}H), \tag{7.31}$$

$$\hat{q}(F, \theta, g, \dot{F}) = \hat{q}(FH, \theta, g, \dot{F}H), \tag{7.32}$$

for all scalars θ, all vectors g, and all tensors F and L. If $H = J^{1/3}F^{-1}$, we see that $\det H = 1$. Substituting this expression for H into the above expressions for $\hat{\psi}$, $\hat{\eta}$ and $\hat{q}$, we get

$$\hat{\psi}(F, \theta, g, \dot{F}) = \hat{\psi}(J^{1/3}I, \theta, g, J^{1/3}L) = \tilde{\psi}(\rho, \theta, g, L),$$
$$\hat{\tau}(F, \theta, g, \dot{F}) = \hat{\tau}(J^{1/3}I, \theta, g, J^{1/3}L) = \tilde{\tau}(\rho, \theta, g, L),$$
$$\hat{\eta}(F, \theta, g, \dot{F}) = \hat{\eta}(J^{1/3}I, \theta, g, J^{1/3}L) = \tilde{\eta}(\rho, \theta, g, L),$$
$$\hat{q}(F, \theta, g, \dot{F}) = \hat{q}(J^{1/3}I, \theta, g, J^{1/3}L) = \tilde{q}(\rho, \theta, g, L),$$

since $\rho_0 = \rho J$. Generalizing earlier work of Coleman and Noll [61], Coleman and Mizel [62] derived the following result:

Theorem 7.3.1. *Assuming that $\tilde{\tau}$ is continuous with respect to L, the necessary and sufficient conditions that every admissible thermodynamic process obey Eqn. (3.85) are*

1. *The response functions $\tilde{\psi}$ and $\tilde{\eta}$ are independent of the temperature gradient g and L, i.e.,*

$$\psi = \tilde{\psi}(\rho, \theta), \quad \eta = \tilde{\eta}(\rho, \theta).$$

2. *$\tilde{\psi}$ determines $\tilde{\eta}$ through the relation*

$$\tilde{\eta}(\rho, \theta) = -\frac{\partial}{\partial \theta}\tilde{\psi}(\rho, \theta). \tag{7.33}$$

3. *The equilibrium stress defined by*

$$\tilde{\tau}_0(\rho, \theta) := \tilde{\tau}(\rho, \theta, \mathbf{0}, \mathbf{0}),$$

is given by

$$\tilde{\tau}_0(\rho, \theta) = -p(\rho, \theta)I, \tag{7.34}$$

where

$$p(\rho, \theta) = \rho^2 \frac{\partial}{\partial \rho}\tilde{\psi}(\rho, \theta), \tag{7.35}$$

or, alternatively,

$$p(v,\theta) = -\frac{\partial}{\partial v}\tilde{\tilde{\psi}}(v,\theta), \tag{7.36}$$

where $\tilde{\tilde{\psi}}(v,\theta) := \tilde{\psi}(\rho,\theta)$.

4. *The extra stress (or the viscous stress) defined by*

$$\tilde{\sigma}(\rho,\theta,g,L) := \tilde{\tau}(\rho,\theta,g,L) - \tilde{\tau}_0(\rho,\theta),$$

and the heat flux vector satisfy the general dissipation inequality

$$-\tilde{\sigma}(\rho,\theta,g,L) : D + \frac{1}{\theta}\tilde{q}(\rho,\theta,g,L)\cdot g \le 0. \tag{7.37}$$

Proof. Substituting the constitutive equations into Eqn. (3.85), we get the necessary and sufficient condition for an admissible thermodynamic process as

$$\left[\frac{\partial\tilde{\psi}}{\partial F}F^T - \frac{\tilde{\tau}}{\rho}\right] : L + \left[\frac{\partial\tilde{\psi}}{\partial\theta} + \tilde{\eta}\right]\dot{\theta} + \frac{\partial\tilde{\psi}}{\partial g}\cdot\dot{g} + \frac{\partial\tilde{\psi}}{\partial L} : \dot{L} + \frac{q\cdot g}{\rho\theta} \le 0. \tag{7.38}$$

If the conditions enumerated in the statement of the theorem hold, then clearly Eqn. (7.38) is satisfied. To prove the necessity, we follow steps similar to those used in proving Theorem 7.1.1, and first prove that given $(F^\diamond,\theta^\diamond,g^\diamond,L^\diamond)$, a tensor A, a scalar α and a vector a, with $\det F^\diamond > 0$ and $\theta^\diamond > 0$, there exists an admissible thermodynamic process with

$$\begin{aligned}
\theta(X^\diamond,t^\diamond) &= \theta^\diamond, & \dot{\theta}(X^\diamond,t^\diamond) &= \alpha,\\
g(X^\diamond,t^\diamond) &= g^\diamond, & \dot{g}(X^\diamond,t^\diamond) &= a,\\
L(X^\diamond,t^\diamond) &= L^\diamond, & \ddot{F}(X^\diamond,t^\diamond) &= A,\\
F(X^\diamond,t^\diamond) &= F^\diamond.
\end{aligned}$$

Clearly, there exists an open ball about $X^\diamond$, and an open interval about $t^\diamond$ such that

$$\det[F^\diamond + (t-t^\diamond)L^\diamond F^\diamond + \frac{1}{2}(t-t^\diamond)^2 A] > 0$$
$$\theta^\diamond + \alpha(t-t^\diamond) + (F^\diamond)^T[g^\diamond + (t-t^\diamond)(a + (L^\diamond)^T g^\diamond)]\cdot(X - X^\diamond) > 0.$$

Let $\chi(X,t)$ and $\theta(X,t)$ be defined by

$$\chi(X,t) = [F^\diamond + (t-t^\diamond)L^\diamond F^\diamond + \frac{1}{2}(t-t^\diamond)^2 A](X - X^\diamond),$$
$$\theta(X,t) = \theta^\diamond + \alpha(t-t^\diamond) + (F^\diamond)^T[g^\diamond + (t-t^\diamond)(a + (L^\diamond)^T g^\diamond)]\cdot(X - X^\diamond).$$

Then the above process has the desired properties. Applying the inequality given by Eqn. (7.38) to the process constructed above yields

$$\left\{\left[\frac{\partial}{\partial F}\tilde{\psi}(\rho^\diamond,\theta^\diamond,g^\diamond,L^\diamond)\right](F^\diamond)^T - \frac{1}{\rho^\diamond}\tilde{\tau}(\rho^\diamond,\theta^\diamond,g^\diamond,L^\diamond)\right\} : L^\diamond +$$
$$\left[\frac{\partial}{\partial\theta}\tilde{\psi}(\rho^\diamond,\theta^\diamond,g^\diamond,L^\diamond) + \tilde{\eta}(\rho^\diamond,\theta^\diamond,g^\diamond,L^\diamond)\right]\alpha + \frac{\partial}{\partial\dot{F}}\tilde{\psi}(\rho^\diamond,\theta^\diamond,g^\diamond,L^\diamond) : A$$
$$+\left[\frac{\partial}{\partial g}\tilde{\psi}(\rho^\diamond,\theta^\diamond,g^\diamond,L^\diamond)\right]\cdot a + \frac{1}{\rho^\diamond\theta^\diamond}\tilde{q}(\rho^\diamond,\theta^\diamond,g^\diamond,L^\diamond)\cdot g^\diamond \le 0, \quad (7.39)$$

where $\rho^\diamond = \rho_0/\det F^\diamond$. Note that $F^\diamond$, $\theta^\diamond$, $g^\diamond$, $L^\diamond$, A , α and a are arbitrary.

Equation (7.39) can be written as

$$\frac{\partial}{\partial\dot{F}}\tilde{\psi}(\rho^\diamond,\theta^\diamond,g^\diamond,L^\diamond) : A + f_1(F^\diamond,\theta^\diamond,g^\diamond,L^\diamond,\alpha,a) \le 0.$$

If we fix the values of $F^\diamond$, $\theta^\diamond$, $g^\diamond$, $L^\diamond$, α and a, then $f_1(F^\diamond,\theta^\diamond,g^\diamond,L^\diamond,\alpha,a)$ is a fixed constant, say c. Hence, we have

$$\frac{\partial}{\partial\dot{F}}\tilde{\psi}(\rho^\diamond,\theta^\diamond,g^\diamond,L^\diamond) : A + c \le 0.$$

The above inequality is satisfied for any arbitrary choice of A only if

$$\frac{\partial}{\partial\dot{F}}\tilde{\psi}(\rho^\diamond,\theta^\diamond,g^\diamond,L^\diamond) = \mathbf{0},$$

which, by virtue of the relation $\dot{F} = LF$, leads to

$$\frac{\partial}{\partial L}\tilde{\psi}(\rho^\diamond,\theta^\diamond,g^\diamond,L^\diamond) = \mathbf{0}. \quad (7.40)$$

Using Eqn. (7.40), Eqn. (7.39) reduces to

$$\frac{\partial}{\partial g}\tilde{\psi}(\rho^\diamond,\theta^\diamond,g^\diamond,L^\diamond)\cdot a + f_2(F^\diamond,\theta^\diamond,g^\diamond,L^\diamond,\alpha) \le 0.$$

Now fixing $F^\diamond$, $\theta^\diamond$, $g^\diamond$, $L^\diamond$ and α, and invoking the arbitrariness of a, we get

$$\frac{\partial}{\partial g}\tilde{\psi}(\rho^\diamond,\theta^\diamond,g^\diamond,L^\diamond) = \mathbf{0}.$$

Thus, $\tilde{\psi}$ is independent of g and L, i.e.,

$$\psi = \tilde{\psi}(\rho,\theta). \quad (7.41)$$

Using the above results, Eqn. (7.39) reduces to

$$\left[\frac{\partial}{\partial\theta}\tilde{\psi}(\rho^\diamond,\theta^\diamond) + \tilde{\eta}(\rho^\diamond,\theta^\diamond,g^\diamond,L^\diamond)\right]\alpha + f_3(F^\diamond,\theta^\diamond,g^\diamond,L^\diamond) \le 0,$$

which on using our usual argument yields

$$\tilde{\eta}(\rho,\theta) = -\frac{\partial}{\partial\theta}\tilde{\psi}(\rho,\theta).$$

Splitting $\tilde{\tau}$ into the equilibrium stress and the extra stress as given in the statement of the theorem, Eqn. (7.39) can be written as

$$\left\{\left[\frac{\partial}{\partial F}\tilde{\psi}(\rho^\diamond,\theta^\diamond)\right](F^\diamond)^T - \frac{1}{\rho^\diamond}\tilde{\tau}_0(\rho^\diamond,\theta^\diamond)\right\} : L^\diamond - \frac{1}{\rho^\diamond}\tilde{\sigma}(\rho^\diamond,\theta^\diamond,g^\diamond,L^\diamond) : L^\diamond + \frac{1}{\rho^\diamond\theta^\diamond}\tilde{q}(\rho^\diamond,\theta^\diamond,g^\diamond,L^\diamond)\cdot g^\diamond \le 0 \quad (7.42)$$

Choosing $g^\diamond = \mathbf{0}$ in the above equation, we get

$$\left\{\left[\frac{\partial}{\partial F}\tilde{\psi}(\rho^\diamond,\theta^\diamond)\right](F^\diamond)^T - \frac{1}{\rho^\diamond}\tilde{\tau}_0(\rho^\diamond,\theta^\diamond)\right\} : L^\diamond - \frac{1}{\rho^\diamond}\tilde{\sigma}(\rho^\diamond,\theta^\diamond,\mathbf{0},L^\diamond) : L^\diamond \le 0. \quad (7.43)$$

Since $\tilde{\tau}$ is continuous with respect to L, we have

$$\tilde{\tau}(\rho,\theta,\mathbf{0},L) \to \tilde{\tau}_0(\rho,\theta) \text{ as } L \to \mathbf{0},$$

or, alternatively,

$$\tilde{\sigma}(\rho,\theta,\mathbf{0},L) \to \mathbf{0} \text{ as } L \to \mathbf{0}.$$

Replacing $L^\diamond$ by $\beta L^\diamond$, and choosing β sufficiently small, so that the first term in Eqn. (7.43) dominates the second, and then replacing β by $-\beta$, we conclude that the term in curly parenthesis must disappear, i.e.,

$$\tilde{\tau}_0(\rho,\theta) = \rho\left[\frac{\partial}{\partial F}\tilde{\psi}(\rho,\theta)\right]F^T.$$

Substituting this result into Eqn. (7.42), and using the symmetry of σ to replace $\tilde{\sigma} : L$ by $\tilde{\sigma} : D$, the general dissipation inequality given by Eqn. (7.37) follows.

Summarizing the above results, we have

$$\psi = \tilde{\psi}(\rho,\theta),$$
$$\tilde{\eta} = -\frac{\partial\tilde{\psi}}{\partial\theta},$$
$$\tilde{\tau}_0 = \rho\frac{\partial\hat{\psi}}{\partial F}F^T,$$
$$-\tilde{\sigma} : D + \frac{1}{\theta}\tilde{q}\cdot g \le 0.$$

Using Eqn. (7.41), and the fact that

$$\frac{\partial\rho}{\partial F} = -\rho F^{-T},$$

we get

$$\tilde{\tau}_0(\rho,\theta) = \rho\frac{\partial\hat{\psi}}{\partial F}F^T$$

$$= \rho \frac{\partial \tilde{\psi}}{\partial \rho} \frac{\partial \rho}{\partial \boldsymbol{F}} \boldsymbol{F}^T$$
$$= -\rho^2 \frac{\partial \tilde{\psi}}{\partial \rho} \boldsymbol{I}.$$

Since, by definition, $\tilde{\boldsymbol{\tau}}_0(\rho, \theta) = -p(\rho, \theta)\boldsymbol{I}$, we get Eqn. (7.35), which in turn leads to Eqn. (7.36). □

If we assume that the relation $\eta = \tilde{\eta}(v, \theta)$ is invertible in θ, i.e., if we assume, $(\partial \eta / \partial \theta)_v \neq 0$, then we have $\theta = \tilde{\theta}(v, \eta)$. Taking (v, η) as the independent variables, we now derive the following expressions for the pressure and temperature (temporarily using the same symbol for the function and its value):

$$\theta = \left(\frac{\partial e}{\partial \eta} \right)_v, \tag{7.44}$$
$$p = -\left(\frac{\partial e}{\partial v} \right)_\eta. \tag{7.45}$$

Since $\psi = e - \eta\theta$, we have

$$\eta = -\left(\frac{\partial \psi}{\partial \theta} \right)_v$$
$$= -\left(\frac{\partial e}{\partial \theta} \right)_v + \theta \left(\frac{\partial \eta}{\partial \theta} \right)_v + \eta,$$

which implies that

$$\theta \left(\frac{\partial \eta}{\partial \theta} \right)_v = \left(\frac{\partial e}{\partial \theta} \right)_v = \left(\frac{\partial e}{\partial \eta} \right)_v \left(\frac{\partial \eta}{\partial \theta} \right)_v.$$

Since we have assumed that $(\partial \eta / \partial \theta)_v \neq 0$, we get Eqn. (7.44). Using Eqn. (7.36), we get

$$p = -\left(\frac{\partial \psi}{\partial v} \right)_\theta$$
$$= -\left(\frac{\partial e}{\partial v} \right)_\theta + \left(\frac{\partial \eta}{\partial v} \right)_\theta \theta$$
$$= -\left(\frac{\partial e}{\partial v} \right)_\eta - \left(\frac{\partial e}{\partial \eta} \right)_v \left(\frac{\partial \eta}{\partial v} \right)_\theta + \left(\frac{\partial \eta}{\partial v} \right)_\theta \theta$$
$$= -\left(\frac{\partial e}{\partial v} \right)_\eta,$$

where the last step follows from Eqn. (7.44).

Theorem 7.3.1 shows that in the special class of fluids whose constitutive relations can be assumed to be of the form given by Eqns. (7.28), the stress tensor can be decomposed into a viscous stress and an 'equilibrium stress', namely the thermodynamic pressure, which is related to the temperature and density via a 'state equation' given by Eqn. (7.35). Thus,

$$\boldsymbol{\tau} = -p\boldsymbol{I} + \boldsymbol{\sigma}. \tag{7.46}$$

Substituting Eqn. (7.46) into Eqn. (3.19) yields

$$\rho \left\{ \frac{\partial \boldsymbol{v}}{\partial t} + (\boldsymbol{\nabla} \boldsymbol{v})\boldsymbol{v} \right\} = -\boldsymbol{\nabla} p + \boldsymbol{\nabla} \cdot \boldsymbol{\sigma} + \rho \boldsymbol{b}, \tag{7.47}$$

while substituting it into Eqn. (3.78) yields

$$\rho \frac{De}{Dt} = -p\boldsymbol{\nabla} \cdot \boldsymbol{v} + \boldsymbol{\sigma} : \boldsymbol{D} - \boldsymbol{\nabla} \cdot \boldsymbol{q} + \rho Q_h. \tag{7.48}$$

In Eqn. (7.48), the first term on the right-hand side represents the work done by the pressure against volume dilatation, while the second term represents viscous dissipation in the form of heat. The third term represents the heat being lost due to conduction, and the final term represents the energy being added due to a heat source. The term $\boldsymbol{\sigma} : \boldsymbol{D}$ is called a dissipation function because it is a measure of the rate at which mechanical energy is converted into thermal energy. In the absence of thermal effects, it can be shown to be positive for nonzero $\boldsymbol{D}$ (see Eqn. (7.58)), and thus increases irreversibly the internal energy of a fluid. Using Eqns. (7.1) and (7.48), and the fact that $h = e + pv$, we get

$$\rho \frac{Dh}{Dt} = \frac{Dp}{Dt} + \boldsymbol{\sigma} : \boldsymbol{D} + \rho Q_h - \boldsymbol{\nabla} \cdot \boldsymbol{q}. \tag{7.49}$$

If we assume a fluid to be incompressible and inviscid, and if in addition, we assume that there is no heat addition, heat conduction and external work, then from Eqn. (7.48) we observe that $De/Dt = 0$. In such a situation, the mechanical and total energy balance equations given by Eqns. (3.68) and (3.76) both reduce to the same equation given by

$$\int_{V(t)} \frac{\partial}{\partial t} \left\{ \rho \left(\frac{\boldsymbol{v} \cdot \boldsymbol{v}}{2} \right) \right\} dV + \int_{S(t)} \rho \left(\frac{\boldsymbol{v} \cdot \boldsymbol{v}}{2} \right) (\boldsymbol{v} \cdot \boldsymbol{n}) \, dS = \int_{S(t)} \boldsymbol{t} \cdot \boldsymbol{v} \, dS + \int_{V(t)} \rho \boldsymbol{b} \cdot \boldsymbol{v} \, dV.$$

Equation (7.46) is one way of decomposing the stress tensor. Alternatively, the stress tensor is sometimes expressed as

$$\boldsymbol{\tau} = -\hat{p}\boldsymbol{I} + \boldsymbol{S}, \tag{7.50}$$

where $\boldsymbol{S}$ is the deviatoric stress tensor, and

$$\hat{p} = -\frac{1}{3} \operatorname{tr} \boldsymbol{\tau} = -\frac{1}{3} \tau_{ii} \tag{7.51}$$

is the mean pressure. Note that the trace of the deviatoric stress tensor is zero. In general, the thermodynamic pressure p is different from the mean pressure $\hat{p}$ (for example, see Eqn. (7.62)). If the fluid is assumed to be incompressible, then, as discussed towards the end of the previous section, the pressure p is constitutively indeterminate (i.e., there is no constitutive relation between p, ρ and θ), and has to be found from the governing equations and boundary conditions.

Using Eqns. (3.18) and (7.46), the expression for the traction vector is given by

$$t = \tau n = -pn + \sigma n. \tag{7.52}$$

To see the consequences of material frame-indifference, note that $\tilde{\psi}$, $\tilde{\eta}$ and $\tilde{\tau}_0$, being only functions of the scalar fields ρ and θ, are frame-indifferent. The constitutive equations for the viscous stress and the heat flux vector are frame-indifferent if and only if

$$\tilde{\sigma}(\rho, \theta, Qg, QLQ^T + \dot{Q}Q^T) = Q\tilde{\sigma}(\rho, \theta, g, L)Q^T \quad \forall Q \in \text{Orth}^+, \tag{7.53a}$$

$$\tilde{q}(\rho, \theta, Qg, QLQ^T + \dot{Q}Q^T) = Q\tilde{q}(\rho, \theta, g, L) \quad \forall Q \in \text{Orth}^+, \tag{7.53b}$$

since $g^* = \nabla_{x^*}\theta^* = Qg$, and $L^* = QLQ^T + \dot{Q}Q^T$ (by Eqn. (4.10)). Fix the time t and choose Q to be the function

$$Q(\tau) = e^{(t-\tau)\hat{W}},$$

where $\hat{W}$ is the skew-symmetric part of L. Then $Q(t) = I$, and

$$\dot{Q} = -\hat{W}e^{(t-\tau)\hat{W}},$$

so that $\dot{Q}(t) = -\hat{W}$. With this choice of Q, we get $L^* = D$, so that from Eqns. (7.53), we get $\tilde{\sigma}(\rho, \theta, g, L) = \tilde{\sigma}(\rho, \theta, g, D)$ and $\tilde{q}(\rho, \theta, g, L) = \tilde{q}(\rho, \theta, g, D)$– i.e., the dependence of the constitutive relations on L reduces to a dependence on D. If we had assumed a dependence of the constitutive relations on the velocity, then material frame indifference can be used to rule out such a dependence. To see this, note that

$$v^* = Qv + \dot{Q}x + \dot{c}(t).$$

Choosing Q as above, and $c(\tau) = -Qx(\tau)$, so that

$$\dot{c}(t) = -Q(t)v - \dot{Q}(t)x(t) = -v + \hat{W}x.$$

we get $v^* = 0$.

From the above discussion, we conclude that $\tilde{\sigma}$ and $\tilde{q}$ are isotropic functions of D and g, i.e.,

$$\tilde{\sigma}(\rho, \theta, Qg, QDQ^T) = Q\tilde{\sigma}(\rho, \theta, g, D)Q^T, \tag{7.54}$$

$$\tilde{q}(\rho, \theta, Qg, QDQ^T) = Q\tilde{q}(\rho, \theta, g, D). \tag{7.55}$$

In Eqn. (7.54) we can replace g by $g \otimes g$, and then use the representation theorem for isotropic, symmetric, tensor-valued functions of two symmetric tensors [179, 249]. Representing $g \otimes g$ by G, we get

$$\begin{aligned}\tilde{\sigma}(\rho, \theta, G, D) = \gamma_0 I + \gamma_1 D + \gamma_2 D^2 + \gamma_3 G + \gamma_4 G^2 + \gamma_5(DG + GD) \\ + \gamma_6(D^2G + GD^2) + \gamma_7(DG^2 + G^2D) + \gamma_8(D^2G^2 + G^2D^2),\end{aligned} \tag{7.56}$$

where the constants $\gamma_i(\rho, \theta, G, D)$ are of the form

$$\gamma_i = \tilde{\gamma}_i(\rho, \theta, \text{tr}\, D, \text{tr}\, D^2, \text{tr}\, D^3, \text{tr}\, G, \text{tr}\, G^2, \text{tr}\, G^3,$$

$$\operatorname{tr}(\boldsymbol{DG}), \operatorname{tr}(\boldsymbol{DG}^2), \operatorname{tr}(\boldsymbol{D}^2\boldsymbol{G}), \operatorname{tr}(\boldsymbol{D}^2\boldsymbol{G}^2)).$$

By a representation theorem for isotropic vector-valued functions [179, 249], the necessary and sufficient condition for Eqn. (7.55) to be satisfied is that $\tilde{q}$ be of the form

$$\tilde{q} = (\beta_0 \boldsymbol{I} + \beta_1 \boldsymbol{D} + \beta_2 \boldsymbol{D}^2)\boldsymbol{g}, \tag{7.57}$$

where

$$\beta_\alpha = \tilde{\beta}_\alpha(\rho, \theta, \operatorname{tr}\boldsymbol{D}, \operatorname{tr}\boldsymbol{D}^2, \operatorname{tr}\boldsymbol{D}^3, \boldsymbol{g}\cdot\boldsymbol{g}, \boldsymbol{g}\cdot\boldsymbol{D}\boldsymbol{g}, \boldsymbol{g}\cdot\boldsymbol{D}^2\boldsymbol{g}).$$

From Eqn. (7.57), one immediately observes that the heat flux vanishes when the temperature gradient vanishes. From Eqns. (7.56) and (7.57), we see the way $\tilde{\sigma}$ depends on $\boldsymbol{g}$, and $\tilde{q}$ depends on $\boldsymbol{D}$. Perhaps, we would have missed observing these coupling effects had we not adopted the principle of equipresence at the outset. The classical (uncoupled) constitutive relations

$$\tilde{\sigma} = \lambda(\operatorname{tr}\boldsymbol{D})\boldsymbol{I} + 2\mu\boldsymbol{D},$$
$$\tilde{q} = -k\boldsymbol{g},$$

are special cases of Eqns. (7.56) and (7.57) as we shall soon show.

It follows from Eqn. (7.37) that when $\boldsymbol{g} = \boldsymbol{0}$, we have the *mechanical dissipation inequality*

$$\tilde{\sigma}(\rho, \theta, \boldsymbol{0}, \boldsymbol{D}) : \boldsymbol{D} \geq 0 \quad \forall \boldsymbol{D} \in \mathrm{Sym}, \tag{7.58}$$

and when $\boldsymbol{D} = \boldsymbol{0}$, we have the *heat conduction inequality*

$$\tilde{q}(\rho, \theta, \boldsymbol{g}, \boldsymbol{0}) \cdot \boldsymbol{g} \leq 0 \quad \forall \boldsymbol{g} \in V. \tag{7.59}$$

We note, however, that in general, a resolution of Eqn. (7.37) into a mechanical dissipation inequality for nonzero $\boldsymbol{g}$, and a heat conduction inequality for nonzero $\boldsymbol{D}$ is not valid.

For the case when $\boldsymbol{g} = \boldsymbol{0}$, Eqn. (7.56) reduces to

$$\tilde{\sigma}(\rho, \theta, \boldsymbol{0}, \boldsymbol{D}) = \gamma_0(\rho, \theta, \boldsymbol{0}, \mathcal{I}_{\boldsymbol{D}})\boldsymbol{I} + \gamma_1(\rho, \theta, \boldsymbol{0}, \mathcal{I}_{\boldsymbol{D}})\boldsymbol{D} + \gamma_2(\rho, \theta, \boldsymbol{0}, \mathcal{I}_{\boldsymbol{D}})\boldsymbol{D}^2.$$

The coefficients γ_i in the above equation are subject to restrictions imposed by Eqn. (7.58). Fluids that obey the above constitutive relation are called Reiner–Rivlin fluids. If the fluid is stationary or uniformly moving (no relative motion between fluid elements) then $\boldsymbol{D} = \boldsymbol{0}$, and Eqn. (7.46) reduces to

$$\boldsymbol{\tau}(\boldsymbol{x}, t) = \beta_0(\theta, \rho)\boldsymbol{I}.$$

Thus, we see that a fluid at rest or in uniform motion cannot support a shear stress, or, alternatively, if a fluid at rest is subjected to a shear stress, then it must flow. Some solids also flow in this sense, and, therefore, these statements do not suffice to characterize a fluid. For example, by setting $\beta_0 = -K(3 - \operatorname{tr}\boldsymbol{U})$, $\beta_1 = \beta_2 = 0$, in the constitutive equation given by Eqn. (5.8), we get $\hat{\boldsymbol{\tau}}(\boldsymbol{X}, \boldsymbol{F}) = -K(3 - \operatorname{tr}\boldsymbol{U})\boldsymbol{I}$, which describes an isotropic, homogeneous, elastic solid whose reference configuration is a natural state, and which never experiences a nonzero shear stress no matter how it is deformed.

If one restricts the constitutive relation for the viscous stress to be a linear relation, then by Theorem 1.6.15, we have

$$\hat{\sigma}(\rho,\theta,\boldsymbol{D}) = \lambda(\rho,\theta)(\operatorname{tr}\boldsymbol{D})\boldsymbol{I} + 2\mu(\rho,\theta)\boldsymbol{D}. \tag{7.60}$$

λ is known as the coefficient of dilatational viscosity, and μ is known as the coefficient of shear viscosity or dynamic viscosity. The coefficient of kinematic viscosity is defined by the relation

$$\nu := \frac{\mu}{\rho}.$$

If the coefficients λ and μ vanish, then we obtain an elastic fluid particle with the constitutive relation of the same form as that given by Eqn. (5.14), but now with the temperature dependence added. Thus, elastic fluids are also referred to as *inviscid*, because the shear stress on any plane is always zero.

A fluid that obeys Eqn. (7.60) is said to be *Newtonian*. Most fluids, including air and water, are modeled quite well as Newtonian fluids, and, unfortunately, those which are not, such as polymer solutions, blood, etc., are not modeled well by the more general Reiner–Rivlin constitutive model either, since long-range memory has not been taken into account in deriving this model. In other words, the only known Reiner–Rivlin fluids are Newtonian [138]. The constitutive relations for a Newtonian fluid are obtained by substituting Eqn. (7.60) into Eqn. (7.46); we get

$$\boldsymbol{\tau} = -p(\rho,\theta)\boldsymbol{I} + \lambda(\rho,\theta)(\operatorname{tr}\boldsymbol{D})\boldsymbol{I} + 2\mu(\rho,\theta)\boldsymbol{D}. \tag{7.61}$$

From Eqn. (7.52), the expression for the traction vector is

$$\boldsymbol{t} = (-p + \lambda \operatorname{tr}\boldsymbol{D})\boldsymbol{n} + 2\mu\boldsymbol{D}\boldsymbol{n},$$

which for an incompressible fluid reduces to

$$\boldsymbol{t} = -p\boldsymbol{n} + 2\mu\boldsymbol{D}\boldsymbol{n}.$$

In order to find a relation between the thermodynamic pressure and the mean pressure for a Newtonian fluid, we note from Eqns. (7.51) and (7.61) that

$$\begin{aligned}\hat{p} &= -\frac{1}{3}\operatorname{tr}\boldsymbol{\tau} \\ &= p - \lambda\operatorname{tr}\boldsymbol{D} - \frac{2}{3}\mu\operatorname{tr}\boldsymbol{D} \\ &= p - \left(\lambda + \frac{2}{3}\mu\right)\boldsymbol{\nabla}\cdot\boldsymbol{v}\end{aligned} \tag{7.62}$$

The factor $\kappa = \lambda + 2\mu/3$ is known as the coefficient of bulk viscosity. From Eqn. (7.62), we note that for an incompressible fluid ($\boldsymbol{\nabla}\cdot\boldsymbol{v} = 0$), we have $\hat{p} = p$.

Substituting Eqn. (7.60) into Eqn. (7.58), we get

$$\lambda(\operatorname{tr}\boldsymbol{D})^2 + 2\mu\boldsymbol{D}:\boldsymbol{D} \geq 0,$$

which is valid for all $D \in$ Sym. We have seen while proving Theorem 5.8.2, that the above inequality is equivalent to the inequalities

$$\mu \geq 0, \quad \lambda + \frac{2}{3}\mu \geq 0.$$

Thus, the Clausius–Duhem inequality leads us to the conclusion that the dynamic viscosity μ, and the bulk viscosity $\lambda + 2\mu/3$, must be nonnegative.

For the case when $D = 0$, Eqn. (7.57) reduces to (compare with Theorem 1.6.9, Chapter 1),

$$\tilde{q}(\rho, \theta, g, 0) = -k(\rho, \theta, |g|)g \quad \forall g \in V, \tag{7.63}$$

where $k(\rho, \theta, |g|) = -\tilde{\beta}_0(\rho, \theta, g \cdot g)$. Equation (7.63) is known as the Fourier law for heat conduction in fluids[2]. Substituting this relation into Eqn. (7.59), we get

$$k(\rho, \theta, |g|) \geq 0.$$

Summarizing, we have the following result that should be compared with Theorem 7.1.3:

Theorem 7.3.2. *The heat flux vanishes whenever the temperature gradient vanishes, i.e.,*

$$\tilde{q}(\rho, \theta, 0, D) = 0. \quad \textit{(by Eqn. (7.57))}$$

When $D = 0$, *the thermal conductivity* $k(\rho, \theta, |g|)$ *is nonnegative.*

Note that the results in this section are easily generalizable to viscoelastic materials if one does not replace F by ρ at the outset–for a more detailed discussion, see, e.g., [266]. Using the same technique as in section 5.4 to incorporate the effects of material frame-indifference, we get

$$\tilde{\tau}_0 = \rho \frac{\partial \hat{\psi}}{\partial F} F^T = 2\rho F \frac{\partial \tilde{\psi}}{\partial C} F^T,$$

where $\tilde{\psi}(C, \theta) := \hat{\psi}(\sqrt{C}, \theta)$.

One of the important conclusions that follows from Theorem 7.3.1 is that Eqn. (7.33) guarantees the existence of an equation of state for the entropy as a function of the density and temperature, something that is *assumed* in customary treatments on thermodynamics. We would like to reiterate, however, that this equation of state could be different for different material points unless the fluid is homogeneous. Since the pressure is a state function of the density and temperature, and since the specific entropy is a state function of the density and temperature, it follows that the pressure is a state function of the density and specific entropy. As is the case with the previously discussed state functions, this state function also could be different for different material points.

[2] Again, similar to Eqn. (7.14), this law can be generalized as [56]

$$\tau_0 q_{\text{obj}} + \tilde{q} = -kg, \tag{7.64}$$

where τ_0 is the thermal relaxation time, and q_{obj} is an appropriately chosen objective flux rate (see, e.g., Eqns. (4.40)–(4.42)), the actual choice being made based on experimental data.

Using Eqns. (7.33) and (7.36), we have

$$\frac{D\psi}{Dt} = \frac{\partial\tilde{\psi}}{\partial v}\frac{Dv}{Dt} + \frac{\partial\tilde{\psi}}{\partial\theta}\frac{D\theta}{Dt} = -p\frac{Dv}{Dt} - \eta\frac{D\theta}{Dt}. \tag{7.65}$$

Since $\psi = e - \eta\theta$, we also have

$$\frac{D\psi}{Dt} = \frac{De}{Dt} - \theta\frac{D\eta}{Dt} - \eta\frac{D\theta}{Dt}. \tag{7.66}$$

Comparing Eqn. (7.65) and (7.66), we get the Gibbs relation

$$\theta\frac{D\eta}{Dt} = \frac{De}{Dt} + p\frac{Dv}{Dt} = \frac{De}{Dt} + \frac{p}{\rho}\boldsymbol{\nabla}\cdot\boldsymbol{v}. \tag{7.67}$$

This equation provides the justification for traditional thermodynamic treatments that start with an equation of the type '$\theta d\eta = de + pdv$' based on 'reversible' processes. The corresponding equation in terms of the enthalpy $h = e + pv$ is

$$\theta\frac{D\eta}{Dt} = \frac{Dh}{Dt} - v\frac{Dp}{Dt}. \tag{7.68}$$

Multiplying Eqn. (7.67) by ρ, we get

$$\rho\theta\frac{D\eta}{Dt} = \rho\frac{De}{Dt} + \rho p\frac{Dv}{Dt}.$$

Using Eqns. (7.1) and (7.48), we get

$$\rho\theta\frac{D\eta}{Dt} = \boldsymbol{\sigma} : \boldsymbol{D} + \rho Q_h - \boldsymbol{\nabla}\cdot\boldsymbol{q}. \tag{7.69}$$

Note that the above equation could also have been obtained by multiplying Eqn. (7.68) by ρ, and then using Eqn. (7.49). Equation (7.48) is an equation for the material time rate of internal energy, and is based on the first law of thermodynamics, whereas Eqn. (7.69) yields the material rate of change of the entropy, and is based on the second law of thermodynamics. A flow is said to be *isentropic* if $D\eta/Dt = 0$, while it is said to be *homentropic* is η is uniform over the fluid. Thus, an isentropic process essentially assumes that the viscosity coefficients, thermal conductivity and heat input are negligible, i.e., Q_h, $\boldsymbol{q}$ and $\boldsymbol{\sigma}$ are all zero. In this case, the energy equation given by Eqn. (7.48) reduces to

$$\rho\frac{De}{Dt} = -p\boldsymbol{\nabla}\cdot\boldsymbol{v}.$$

Setting $D\eta/Dt$ to zero in Gibbs relation given by Eqn. (7.67), we observe that the above relation is automatically satisfied.

We have seen that the free energy ψ and the specific entropy η are both state variables of the absolute temperature θ and the specific volume v, only. Hence, it follows that the internal energy $e = \psi + \eta\theta$ is a state variable of the absolute temperature and the specific volume, i.e., $e = \tilde{e}(\theta, v)$. Similarly, the specific enthalpy $h = e + pv$ is a state variable of the temperature and pressure, since the pressure is related to the volume and temperature through the state equation, i.e., $h = \tilde{h}(\theta, p)$. Note that we have used the absolute temperature and specific volume as the independent variables in defining the internal energy, while we have used the absolute temperature and pressure as the independent variables in defining the specific enthalpy.

We define the specific heat at constant volume, and the specific heat at constant pressure by

$$c_v := \left(\frac{\partial \tilde{e}}{\partial \theta}\right)_v, \tag{7.70}$$

$$c_p := \left(\frac{\partial \tilde{h}}{\partial \theta}\right)_p. \tag{7.71}$$

Since $e = \tilde{e}(\theta, v)$, we have

$$\frac{De}{Dt} = \left(\frac{\partial \tilde{e}}{\partial \theta}\right)_v \frac{D\theta}{Dt} + \left(\frac{\partial \tilde{e}}{\partial v}\right)_\theta \frac{Dv}{Dt}. \tag{7.72}$$

Substituting Eqn. (7.72) into Eqn. (7.67), we get

$$\frac{D\eta}{Dt} = \frac{1}{\theta}\left(\frac{\partial \tilde{e}}{\partial \theta}\right)_v \frac{D\theta}{Dt} + \frac{1}{\theta}\left[\left(\frac{\partial \tilde{e}}{\partial v}\right)_\theta + p\right] \frac{Dv}{Dt}. \tag{7.73}$$

Since $\eta = \tilde{\eta}(\theta, v)$, we also have

$$\frac{D\eta}{Dt} = \frac{\partial \tilde{\eta}}{\partial \theta}\frac{D\theta}{Dt} + \frac{\partial \tilde{\eta}}{\partial v}\frac{Dv}{Dt}. \tag{7.74}$$

From Eqns. (7.73) and (7.74), we get

$$\begin{aligned} \frac{\partial \tilde{\eta}}{\partial \theta} &= \frac{1}{\theta}\left(\frac{\partial \tilde{e}}{\partial \theta}\right)_v, \\ \frac{\partial \tilde{\eta}}{\partial v} &= \frac{1}{\theta}\left[\left(\frac{\partial \tilde{e}}{\partial v}\right)_\theta + p\right]. \end{aligned} \tag{7.75}$$

Since $\frac{\partial}{\partial v}\left(\frac{\partial \tilde{\eta}}{\partial \theta}\right) = \frac{\partial}{\partial \theta}\left(\frac{\partial \tilde{\eta}}{\partial v}\right)$, we obtain

$$\frac{\partial}{\partial v}\left[\frac{1}{\theta}\left(\frac{\partial \tilde{e}}{\partial \theta}\right)_v\right] = \frac{\partial}{\partial \theta}\left\{\frac{1}{\theta}\left[\left(\frac{\partial \tilde{e}}{\partial v}\right)_\theta + p\right]\right\}, \tag{7.76}$$

which, on simplifying, yields

$$\left(\frac{\partial \tilde{e}}{\partial v}\right)_\theta = \theta\left(\frac{\partial p}{\partial \theta}\right)_v - p. \tag{7.77}$$

Substituting Eqns. (7.70) and (7.77) into Eqn. (7.72), we get

$$\frac{De}{Dt} = c_v \frac{D\theta}{Dt} + \left[\theta\left(\frac{\partial p}{\partial \theta}\right)_v - p\right]\frac{Dv}{Dt} \tag{7.78}$$

Substituting Eqn. (7.78) into Eqn. (7.48), and using Eqn. (7.1), we get

$$\rho c_v \frac{D\theta}{Dt} = -\theta \left(\frac{\partial p}{\partial \theta}\right)_\rho (\boldsymbol{\nabla} \cdot \boldsymbol{v}) + \boldsymbol{\sigma} : \boldsymbol{D} - \boldsymbol{\nabla} \cdot \boldsymbol{q} + \rho Q_h, \tag{7.79}$$

while substituting it into Eqn. (7.67) yields

$$\theta \frac{D\eta}{Dt} = c_v \frac{D\theta}{Dt} + \theta \left(\frac{\partial p}{\partial \theta}\right)_v \frac{Dv}{Dt}. \tag{7.80}$$

We follow a similar procedure to find an expression for Dh/Dt. By Eqn. (7.68), we have

$$\frac{D\eta}{Dt} = \frac{1}{\theta}\left(\frac{Dh}{Dt} - v\frac{Dp}{Dt}\right) = \frac{1}{\theta}\frac{\partial \tilde{h}}{\partial \theta}\frac{D\theta}{Dt} + \frac{1}{\theta}\left(\frac{\partial \tilde{h}}{\partial p} - v\right)\frac{Dp}{Dt}. \tag{7.81}$$

Since $\eta = \hat{\eta}(\theta, p)$ is a state function, we get

$$\frac{D\eta}{Dt} = \frac{\partial \hat{\eta}}{\partial \theta}\frac{D\theta}{Dt} + \frac{\partial \hat{\eta}}{\partial p}\frac{Dp}{Dt}. \tag{7.82}$$

Comparing Eqns. (7.81) and (7.82), we get

$$\frac{\partial \hat{\eta}}{\partial \theta} = \frac{1}{\theta}\frac{\partial \tilde{h}}{\partial \theta},$$
$$\frac{\partial \hat{\eta}}{\partial p} = \frac{1}{\theta}\left(\frac{\partial \tilde{h}}{\partial p} - v\right).$$

Now, using the fact that $\frac{\partial}{\partial p}(\frac{\partial \hat{\eta}}{\partial \theta}) = \frac{\partial}{\partial \theta}(\frac{\partial \hat{\eta}}{\partial p})$, we obtain

$$\left(\frac{\partial \tilde{h}}{\partial p}\right)_\theta = v - \theta\left(\frac{\partial v}{\partial \theta}\right)_p. \tag{7.83}$$

Thus, using Eqns. (7.71) and (7.83), we have

$$\begin{aligned}\frac{Dh}{Dt} &= \left(\frac{\partial \tilde{h}}{\partial \theta}\right)_p \frac{D\theta}{Dt} + \left(\frac{\partial \tilde{h}}{\partial p}\right)_\theta \frac{Dp}{Dt} \\ &= c_p \frac{D\theta}{Dt} + \left[v - \theta\left(\frac{\partial v}{\partial \theta}\right)_p\right]\frac{Dp}{Dt}.\end{aligned} \tag{7.84}$$

Substituting Eqn. (7.84) into Eqn. (7.49), and using the fact that $v = 1/\rho$, we get

$$\rho c_p \frac{D\theta}{Dt} = -\frac{\theta}{\rho}\left(\frac{\partial \rho}{\partial \theta}\right)_p \frac{Dp}{Dt} + \boldsymbol{\sigma} : \boldsymbol{D} - \boldsymbol{\nabla} \cdot \boldsymbol{q} + \rho Q_h. \tag{7.85}$$

Substituting Eqn. (7.84) into Eqn. (7.68), we get

$$\theta \frac{D\eta}{Dt} = c_p \frac{D\theta}{Dt} - \theta\left(\frac{\partial v}{\partial \theta}\right)_p \frac{Dp}{Dt}. \tag{7.86}$$

Equation (7.80), or, alternatively, Eqn. (7.86) is the governing differential equation for the specific entropy.

From the state equation given by Eqn. (7.35), we know that v, θ and p are related. Hence,

$$\left(\frac{\partial \tilde{e}}{\partial \theta}\right)_p = \left(\frac{\partial \tilde{e}}{\partial \theta}\right)_v + \left(\frac{\partial \tilde{e}}{\partial v}\right)_\theta \left(\frac{\partial v}{\partial \theta}\right)_p. \tag{7.87}$$

Since $h = e + pv$,

$$\begin{aligned} c_p &= \left(\frac{\partial \tilde{h}}{\partial \theta}\right)_p \\ &= \left(\frac{\partial \tilde{e}}{\partial \theta}\right)_p + p\left(\frac{\partial v}{\partial \theta}\right)_p \\ &= \left(\frac{\partial \tilde{e}}{\partial \theta}\right)_v + \left(\frac{\partial \tilde{e}}{\partial v}\right)_\theta \left(\frac{\partial v}{\partial \theta}\right)_p + p\left(\frac{\partial v}{\partial \theta}\right)_p && \text{(by Eqn. (7.87))} \\ &= c_v + \left[\left(\frac{\partial \tilde{e}}{\partial v}\right)_\theta + p\right]\left(\frac{\partial v}{\partial \theta}\right)_p && \text{(by Eqn. (7.70))} \\ &= c_v + \theta\left(\frac{\partial p}{\partial \theta}\right)_v \left(\frac{\partial v}{\partial \theta}\right)_p. && \text{(by Eqn. (7.77))} \end{aligned}$$

Thus, the difference in the specific heats at constant pressure and constant volume is given by

$$c_p - c_v = \theta\left(\frac{\partial p}{\partial \theta}\right)_v \left(\frac{\partial v}{\partial \theta}\right)_p. \tag{7.88}$$

Before we conclude, we note that the above analysis is more useful in the case of gases than liquids. This is because liquids can be assumed to be nearly incompressible ($(\partial v/\partial \theta)_p \approx 0$), and hence, from Eqn. (7.88) it follows that they have a single reasonably constant specific heat. Thus, for liquids (and especially for water), we have $\rho \approx$ constant, $c_p \approx c_v$, and $\theta(D\eta/Dt) \approx De/Dt \approx c_p D\theta/Dt$.

7.3.1 Thermodynamic relations for a perfect gas

We specialize the results that we have obtained above for the case of a perfect gas. For such a gas, the state relation is given by

$$p = \rho R\theta, \tag{7.89}$$

where R is the specific gas constant. The value of R is different for different gases. For air at standard conditions, $R = 287$ J/(kg-K). Define the ratio

$$\gamma = \frac{c_p}{c_v}. \tag{7.90}$$

Substituting Eqn. (7.89) into (7.88), we get

$$c_p - c_v = R. \tag{7.91}$$

Solving Eqns. (7.90) and (7.91), we get

$$c_v = \frac{\gamma}{\gamma - 1},$$
$$c_p = \frac{\gamma R}{\gamma - 1}.$$

Assuming that γ is a constant dependent only on $\boldsymbol{X}$, and since R is also only a function of $\boldsymbol{X}$, we observe that c_p and c_v are constants dependent only on $\boldsymbol{X}$ for a perfect gas. Equations (7.78) and (7.84) reduce to the relations $De/Dt = c_v(\boldsymbol{X})D\theta/Dt$ and $Dh/Dt = c_p(\boldsymbol{X})D\theta/Dt$, which on integrating yield $e = c_v(\boldsymbol{X})T + c_1(\boldsymbol{X})$ and $h = c_p(\boldsymbol{X})T + c_2(\boldsymbol{X})$, where $c_1(\boldsymbol{X})$ and $c_2(\boldsymbol{X})$ are constants dependent only on $\boldsymbol{X}$. Assuming $h = e = 0$ at $T = 0$, we get (suppressing the dependence of c_v and c_p on $\boldsymbol{X}$ for notational convenience).

$$e = c_v\theta,$$
$$h = c_p\theta.$$

The energy equations given by Eqns. (7.79) and (7.85) now simplify to

$$\rho c_v \frac{D\theta}{Dt} = -p\boldsymbol{\nabla}\cdot\boldsymbol{v} + \boldsymbol{\sigma} : \boldsymbol{D} - \boldsymbol{\nabla}\cdot\boldsymbol{q} + \rho Q_h, \tag{7.92a}$$
$$\rho c_p \frac{D\theta}{Dt} = \frac{Dp}{Dt} + \boldsymbol{\sigma} : \boldsymbol{D} - \boldsymbol{\nabla}\cdot\boldsymbol{q} + \rho Q_h. \tag{7.92b}$$

Eqns. (7.80) and (7.86) can be integrated to yield the change in entropy along a pathline (i.e., the trajectory of a particle as it flows):

$$\eta_2 - \eta_1 = c_v \ln\frac{\theta_2}{\theta_1} + R\ln\frac{v_2}{v_1}, \tag{7.93}$$
$$= c_p \ln\frac{\theta_2}{\theta_1} - R\ln\frac{p_2}{p_1}. \tag{7.94}$$

The free energy corresponding to the above constitutive relations for p and η is

$$\tilde{\psi} = R\theta\log\frac{\rho}{\rho_0} + c_v\left(\theta - \theta\log\frac{\theta}{\theta_0}\right) + c(\boldsymbol{X})\theta$$
$$= R\theta\left(\log\frac{p}{p_0} - 1\right) + c_p\left(\theta - \theta\log\frac{\theta}{\theta_0}\right) + c(\boldsymbol{X})\theta,$$

where p_0, ρ_0 and θ_0 are reference values, and $c(\boldsymbol{X})$ is a constant dependent on $\boldsymbol{X}$. The dependence of R, c_v, c_p and c on $\boldsymbol{X}$ reflects the fact that the constitutive relation can be different at different material points.

The relation between θ and v for an isentropic process can be obtained by setting $\eta_1 = \eta_2$ in Eqn. (7.93). After some algebraic manipulation, we get

$$\theta v^{\gamma-1} = \frac{\theta}{\rho^{\gamma-1}} = \text{constant along a pathline.} \tag{7.95}$$

Now substituting $\theta = pv/R$, we get

$$pv^{\gamma} = \frac{p}{\rho^{\gamma}} = \text{constant along a pathline.} \tag{7.96}$$

From Eqns. (7.95) and (7.96), we get

$$\frac{p}{\theta^{\gamma/(\gamma-1)}} = \text{constant along a pathline.} \tag{7.97}$$

As already mentioned, if one uses the above relations, the energy equation given by either Eqn. (7.92a) or (7.92b) with $Q_h = \boldsymbol{\nabla} \cdot \boldsymbol{q} = \boldsymbol{\sigma} : \boldsymbol{D} = 0$ is automatically satisfied.

7.3.2 The Navier–Stokes and energy equations

By substituting the constitutive relations for a Newtonian fluid into the momentum equation, i.e., substituting Eqn. (7.61) into Eqn. (3.19), we get

$$\rho \left\{ \frac{\partial \boldsymbol{v}}{\partial t} + (\boldsymbol{\nabla v})\boldsymbol{v} \right\} = -\boldsymbol{\nabla} p + (\lambda + \mu)\boldsymbol{\nabla}(\boldsymbol{\nabla} \cdot \boldsymbol{v}) + \mu \boldsymbol{\nabla} \cdot (\boldsymbol{\nabla v}) + \rho \boldsymbol{b}, \tag{7.98}$$

which are the celebrated Navier–Stokes equations. In indicial notation, they read

$$\begin{aligned} \rho \frac{Dv_i}{Dt} &= \rho \left\{ \frac{\partial v_i}{\partial t} + v_j \frac{\partial v_i}{\partial x_j} \right\} \\ &= -\frac{\partial p}{\partial x_i} + (\lambda + \mu) \frac{\partial D_{kk}}{\partial x_i} + \mu \frac{\partial^2 v_i}{\partial x_j \partial x_j} + \rho b_i. \end{aligned} \tag{7.99}$$

Note that the Navier–Stokes equations are nonlinear in $\boldsymbol{v}$ due to the presence of the $(\boldsymbol{\nabla v})\boldsymbol{v}$ term. This term poses the greatest difficulty in finding analytical solutions.

By substituting the constitutive equations into the energy equation, we get a differential equation that relates the temperature, density and velocity fields (which we continue to refer to as the 'energy equation'). Thus, substituting Eqns. (7.60) and (7.63) into Eqns. (7.79) and (7.85), we get

$$\rho c_v \frac{D\theta}{Dt} = -\theta \left(\frac{\partial p}{\partial \theta} \right)_\rho (\boldsymbol{\nabla} \cdot \boldsymbol{v}) + \lambda (\boldsymbol{\nabla} \cdot \boldsymbol{v})^2 + 2\mu \boldsymbol{D} : \boldsymbol{D} + k \boldsymbol{\nabla}^2 \theta + \rho Q_h, \tag{7.100}$$

$$\rho c_p \frac{D\theta}{Dt} = -\frac{\theta}{\rho} \left(\frac{\partial \rho}{\partial \theta} \right)_p \frac{Dp}{Dt} + \lambda (\boldsymbol{\nabla} \cdot \boldsymbol{v})^2 + 2\mu \boldsymbol{D} : \boldsymbol{D} + k \boldsymbol{\nabla}^2 \theta + \rho Q_h. \tag{7.101}$$

Finally, substituting Eqn. (7.60) into Eqn. (7.69), we get

$$\rho \theta \frac{D\eta}{Dt} = \lambda (\boldsymbol{\nabla} \cdot \boldsymbol{v})^2 + 2\mu \boldsymbol{D} : \boldsymbol{D} + \rho Q_h - \boldsymbol{\nabla} \cdot \boldsymbol{q}. \tag{7.102}$$

7.3.3 Summary of the governing equations for a Newtonian fluid

We summarize the governing equations for a Newtonian fluid (not necessarily incompressible) with coefficient of dynamic viscosity μ, coefficient of thermal conductivity k, specific heats c_v and c_p, and obeying the Fourier law of heat conduction.

Continuity equation:

$$\frac{\partial \rho}{\partial t} + \boldsymbol{\nabla} \cdot (\rho \boldsymbol{v}) = 0, \tag{7.103}$$

The Navier–Stokes equations:

$$\rho \left[\frac{\partial \boldsymbol{v}}{\partial t} + (\boldsymbol{\nabla v})\boldsymbol{v} \right] = -\boldsymbol{\nabla} p + (\lambda + \mu)\boldsymbol{\nabla}(\boldsymbol{\nabla} \cdot \boldsymbol{v}) + \mu \boldsymbol{\nabla} \cdot (\boldsymbol{\nabla v}) + \rho \boldsymbol{b}, \tag{7.104}$$

Energy equation:

$$\rho c_v \left[\frac{\partial \theta}{\partial t} + \boldsymbol{v} \cdot (\boldsymbol{\nabla} \theta) \right] = -\theta \left(\frac{\partial p}{\partial \theta} \right)_\rho (\boldsymbol{\nabla} \cdot \boldsymbol{v}) + \lambda (\boldsymbol{\nabla} \cdot \boldsymbol{v})^2 + k \boldsymbol{\nabla}^2 \theta + \rho Q_h + 2\mu \boldsymbol{D} : \boldsymbol{D}, \tag{7.105}$$

or, alternatively,

$$\rho c_p \left[\frac{\partial \theta}{\partial t} + \boldsymbol{v} \cdot (\boldsymbol{\nabla} \theta) \right] = -\frac{\theta}{\rho} \left(\frac{\partial \rho}{\partial \theta} \right)_p \frac{Dp}{Dt} + \lambda (\boldsymbol{\nabla} \cdot \boldsymbol{v})^2 + k \boldsymbol{\nabla}^2 \theta + \rho Q_h + 2\mu \boldsymbol{D} : \boldsymbol{D}, \tag{7.106}$$

Entropy equation

$$\rho \theta \frac{D\eta}{Dt} = \lambda (\boldsymbol{\nabla} \cdot \boldsymbol{v})^2 + 2\mu \boldsymbol{D} : \boldsymbol{D} + \rho Q_h - \boldsymbol{\nabla} \cdot \boldsymbol{q}.$$

Equation of state:

$$f(p, \rho, \theta) = 0, \tag{7.107}$$

where

$$\boldsymbol{v} \equiv (v_x, v_y, v_z),$$

$$\boldsymbol{\nabla}^2 \equiv \frac{\partial^2}{\partial x^2} + \frac{\partial^2}{\partial y^2} + \frac{\partial^2}{\partial z^2},$$

$$\boldsymbol{D} : \boldsymbol{D} \equiv D_{xx}^2 + D_{yy}^2 + D_{zz}^2 + 2(D_{xy}^2 + D_{yz}^2 + D_{xz}^2),$$

$$D_{xx} = \frac{\partial v_x}{\partial x}, \quad D_{xy} = \frac{1}{2}\left(\frac{\partial v_x}{\partial y} + \frac{\partial v_y}{\partial x} \right),$$

$$D_{yy} = \frac{\partial v_y}{\partial y}, \quad D_{yz} = \frac{1}{2}\left(\frac{\partial v_y}{\partial z} + \frac{\partial v_z}{\partial y} \right),$$

$$D_{zz} = \frac{\partial v_z}{\partial z}, \quad D_{xz} = \frac{1}{2}\left(\frac{\partial v_x}{\partial z} + \frac{\partial v_z}{\partial x} \right).$$

In general, Eqns (7.103)–(7.107) have to be solved simultaneously for the variables ρ, p, $\boldsymbol{v}$ and θ. However, a great simplification is achieved if we assume the fluid to be incompressible, and assume the viscosity coefficients λ and μ, and the density ρ, to be independent of the temperature. In such a case the velocity and pressure fields can be obtained simply by solving $\boldsymbol{\nabla} \cdot \boldsymbol{v} = 0$ and (7.104). The velocity field can then be used in Eqn. (7.105) (or

Eqn. (7.106)) to determine the temperature field. Thus, the energy equation is uncoupled from the continuity and momentum equations.

EXERCISES

Using the fact that $\nabla_X \tilde{v} = \dot{F}$, show that the Lagrangian governing equations for a Newtonian fluid are

$$
\begin{aligned}
&\rho_0 = \tilde{\rho} J \quad \text{or} \quad J\left(\frac{\partial \tilde{\rho}}{\partial t}\right)_X + \tilde{\rho}\nabla_X \tilde{v} : \mathbf{cof}\, F = 0, \\
&\rho_0 \left(\frac{\partial^2 u}{\partial t^2}\right)_X = \nabla_X \cdot (FS) + \rho_0 b_0, \\
&S = -\tilde{p} J C^{-1} + \lambda \dot{J} C^{-1} + \mu\left[(\mathbf{cof}\, F)^T \dot{F} C^{-1} + C^{-1} \dot{F}^T \mathbf{cof}\, F\right], \\
&\rho_0 c_v \left(\frac{\partial \theta_0}{\partial t}\right)_X = \left[\tilde{p} - \theta_0 \left(\frac{\partial \tilde{p}}{\partial \theta_0}\right)_X\right] \dot{J} + S : \dot{E} - \nabla_X \cdot q_0 + \rho_0 (Q_h)_0, \\
&q_0 = -kJC^{-1}(\nabla_X \theta_0), \\
&\tilde{p} = \hat{p}(\tilde{\rho}, \theta_0),
\end{aligned}
$$

where, by virtue of the Piola identity, $\nabla_X \cdot (\tilde{p}\,\mathbf{cof}\, F) = (\mathbf{cof}\, F)\nabla \tilde{p}$, $S : \dot{E} = (FS) : \dot{F}$, $\theta_0(X,t) := \theta(x,t)$, $\tilde{p}(X,t) := p(x,t)$ and $\tilde{\rho}(X,t) := \rho(x,t)$. For a perfect gas, the first term on the right-hand side of the energy equation is zero. Specialize the above formulation to the case of an incompressible Newtonian fluid (which is useful in dealing with problems such as sloshing).

With T denoting the first Piola–Kirchhoff stress tensor, the surface tension condition $[\tau^{(1)} - \tau^{(2)}]n = \alpha(\nabla \cdot n)n$ when expressed in Lagrangian form is given by

$$
[T^{(1)} - T^{(2)}]n_0 = \alpha\left[\nabla_X \left[\frac{(\mathbf{cof}\, F)n_0}{|(\mathbf{cof}\, F)n_0|}\right] : F^{-T}\right](\mathbf{cof}\, F)n_0.
$$

8 Rigid-Body Dynamics

A rigid body is one capable of undergoing only rigid motions. Of course, this is an idealization, but one that works quite well for analyzing a variety of motions ranging from the spinning of a top to the motion of planets around the sun.

Recall that the center of mass of a body at time t is given by the relation

$$\bar{x}(t) = \frac{1}{M}\int_{V(t)} \rho x \, dV, \tag{8.1}$$

where $M = \int_{V(t)} \rho \, dV$ is the mass of the body. Also recall that by Eqn. (3.17), the linear momentum $\boldsymbol{l}(t)$ can be expressed as

$$\boldsymbol{l}(t) = \int_{V(t)} \rho \boldsymbol{v} \, dV = M\dot{\bar{\boldsymbol{x}}}. \tag{8.2}$$

In Chapter 2, we have seen that the position, velocity and acceleration in a rigid motion have the forms

$$\boldsymbol{x} = \boldsymbol{\chi}(\boldsymbol{X},t) = \boldsymbol{Q}(t)\boldsymbol{X} + \boldsymbol{c}(t) \quad \forall \boldsymbol{X} \in V_0, \tag{8.3}$$
$$\boldsymbol{v}(\boldsymbol{x}_1,t) = \boldsymbol{v}(\boldsymbol{x}_2,t) + \boldsymbol{w} \times (\boldsymbol{x}_1 - \boldsymbol{x}_2), \tag{8.4}$$
$$\boldsymbol{a}(\boldsymbol{x}_1,t) = \boldsymbol{a}(\boldsymbol{x}_2,t) + \dot{\boldsymbol{w}}(t) \times (\boldsymbol{x}_1 - \boldsymbol{x}_2) + \boldsymbol{w}(t) \times [\boldsymbol{w}(t) \times (\boldsymbol{x}_1 - \boldsymbol{x}_2)], \tag{8.5}$$

where $\boldsymbol{x}_1, \boldsymbol{x}_2 \in V$, $\boldsymbol{Q}(t) \in \text{Orth}^+$, $\boldsymbol{Q}(0) = \boldsymbol{I}$, $\boldsymbol{c}(0) = \boldsymbol{0}$, and $\boldsymbol{w}$, the angular velocity, is the axial vector of $\dot{\boldsymbol{Q}}\boldsymbol{Q}^T$. Using Eqns. (8.1), (8.3), and the relation $\rho \, dV = \rho_0 \, dV_0$, the position of the center of mass is given by

$$\bar{\boldsymbol{x}}(t) = \frac{1}{M}\boldsymbol{Q}(t)\int_{V_0} \rho_0 \boldsymbol{X} \, dV_0 + \boldsymbol{c}(t) = \boldsymbol{Q}(t)\bar{\boldsymbol{x}}(0) + \boldsymbol{c}(t) = \boldsymbol{\chi}(\bar{\boldsymbol{x}}(0),t).$$

Thus, the center of mass obeys the same law of motion as the other particles in the body, although, it may physically not be a part of the body. From Eqn. (8.4), it follows that

$$\boldsymbol{v} - \dot{\bar{\boldsymbol{x}}} = \boldsymbol{w} \times (\boldsymbol{x} - \bar{\boldsymbol{x}}(t)). \tag{8.6}$$

Substituting Eqn. (8.6) into Eqn. (3.27), and using Eqn. (1.21a), we get

$$\begin{aligned} \boldsymbol{h}_{\text{spin}}(t) &= \int_{V(t)} \rho(\boldsymbol{x} - \bar{\boldsymbol{x}}(t)) \times [\boldsymbol{w} \times (\boldsymbol{x} - \bar{\boldsymbol{x}}(t))] \, dV \\ &= \boldsymbol{J}\boldsymbol{w}, \end{aligned} \tag{8.7}$$

where

$$J(t) = \int_{V(t)} \rho\,[(x - \bar{x}(t)) \cdot (x - \bar{x}(t))I - (x - \bar{x}(t)) \otimes (x - \bar{x}(t))]\,dV, \tag{8.8}$$

is the *inertia tensor* relative to the center of mass. The above expression can also be written as $J = (\operatorname{tr} M)I - M$, where

$$M = \int_{V(t)} \rho(x - \bar{x}(t)) \otimes (x - \bar{x}(t))\,dV,$$

is known as the *Euler tensor*.

Let

$$J_z(t) = \int_{V(t)} \rho\,[(x - z) \cdot (x - z)I - (x - z) \otimes (x - z)]\,dV \tag{8.9}$$

be the inertia tensor relative to a point $z(t)$ 'fixed' to the body (i.e., although $z(t)$ may not be a physical point on the body, it obeys $z(t) = Q(t)z(0) + c(t)$). Then using the fact that $\bar{x}$ and z are functions of time only, we have

$$J_z(t) = J(t) + M\,[(\bar{x} - z) \cdot (\bar{x} - z)I - (\bar{x} - z) \otimes (\bar{x} - z)]. \tag{8.10}$$

Eqn. (8.10), known as the parallel-axis theorem, can be used for the computation of the moment of inertia about z knowing the moment of inertia about $\bar{x}$. By Eqns. (3.28), (8.6) and (8.7), we have

$$\begin{aligned} h_z &= Jw + M(\bar{x} - z) \times [w \times (\bar{x} - z)] \\ &= Jw + M\,[(\bar{x} - z) \cdot (\bar{x} - z)w - [(\bar{x} - z) \cdot w](\bar{x} - z)] \\ &= Jw + M\,[(\bar{x} - z) \cdot (\bar{x} - z)I - (\bar{x} - z) \otimes (\bar{x} - z)]\,w \\ &= J_z w. \end{aligned} \tag{8.11}$$

Since $x - \bar{x}(t) = Q\,(X - \bar{x}(0))$, $\bar{x} - z(t) = Q[\bar{x}(0) - z(0)]$, and $\rho_0\,dV_0 = \rho\,dV$, Eqns. (8.8), and (8.10) can be written as

$$J(t) = Q(t)J(0)Q(t)^T, \tag{8.12}$$

$$J_z(t) = Q(t)J_z(0)Q(t)^T, \tag{8.13}$$

where

$$J(0) = \int_{V_0} \rho_0\,[(X - \bar{x}(0)) \cdot (X - \bar{x}(0))I - (X - \bar{x}(0)) \otimes (X - \bar{x}(0))]\,dV_0,$$

$$J_z(0) = J(0) + M\,[(\bar{x}(0) - z(0)) \cdot (\bar{x}(0) - z(0))I - (\bar{x}(0) - z(0)) \otimes (\bar{x}(0) - z(0))].$$

Assuming $\bar{x}(0) = \mathbf{0}$, we have

$$J(0) = \begin{bmatrix} \int_{V_0} \rho_0(Y^2 + Z^2)\,dV_0 & -\int_{V_0} \rho_0 XY\,dV_0 & -\int_{V_0} \rho_0 XZ\,dV_0 \\ -\int_{V_0} \rho_0 XY\,dV_0 & \int_{V_0} \rho_0(X^2 + Z^2)\,dV_0 & -\int_{V_0} \rho_0 YZ\,dV_0 \\ -\int_{V_0} \rho_0 XZ\,dV_0 & -\int_{V_0} \rho_0 YZ\,dV_0 & \int_{V_0} \rho_0(X^2 + Y^2)\,dV_0 \end{bmatrix}$$

From Eqn. (8.8), it is clear that $\boldsymbol{J}(t) \in \text{Sym}$ at all times. Hence, $\boldsymbol{J}(0) \in \text{Sym}$, and, by Theorem. 1.6.1, it follows that it is diagonalizable with the principal values J_1, J_2 and J_3 along the diagonal. By Eqn. (8.9), $\boldsymbol{J}_z(0) \in \text{Sym}$; let J_{z1}, J_{z2} and J_{z3} denote the principal values of $\boldsymbol{J}_z(0)$. If $\bar{\boldsymbol{x}} - \boldsymbol{z}$ is parallel to a principal axis of $\boldsymbol{J}(0)$, say $\boldsymbol{e}_2^*(0)$, then from Eqn. (8.10), it is clear that the $\boldsymbol{e}_i^*(0)$ are eigenvectors of $\boldsymbol{J}_z(0)$, and the corresponding eigenvalues are

$$J_{z1} = J_1 + M\left|\bar{\boldsymbol{x}} - \boldsymbol{z}\right|^2, \quad J_{z2} = J_2, \quad J_{z3} = J_3 + M\left|\bar{\boldsymbol{x}} - \boldsymbol{z}\right|^2.$$

As an example, consider a circular cylinder of radius R, length h and mass M, with the origin at the center of mass, and the 3-axis along the length. We have

$$J_1 = J_2 = \frac{M}{4}\left[R^2 + \frac{h^2}{3}\right], \quad J_3 = \frac{1}{2}MR^2. \tag{8.14}$$

If $\boldsymbol{z}$ is chosen along the 2-axis on the periphery of the cylinder, then

$$J_{z1} = J_1 + MR^2, \quad J_{z2} = J_2, \quad J_{z3} = J_3 + MR^2. \tag{8.15}$$

For a rectangular parallelepiped of dimensions $a \times b \times h$, with the origin at the center of mass, and the 3-axis along the height h, we have

$$J_3 = \frac{M}{12}(a^2 + b^2),$$

with similar expressions for J_1 and J_2.

Since $\boldsymbol{J}(0) \in \text{Sym}$, using the spectral resolution, we can write

$$\boldsymbol{J}(0) = J_1\boldsymbol{e}_1^*(0) \otimes \boldsymbol{e}_1^*(0) + J_2\boldsymbol{e}_2^*(0) \otimes \boldsymbol{e}_2^*(0) + J_3\boldsymbol{e}_3^*(0) \otimes \boldsymbol{e}_3^*(0).$$

Now assume that the three vectors $\boldsymbol{e}_i^*$ are fixed to the body, i.e., $\boldsymbol{e}_i^*(t) = \boldsymbol{Q}(t)\boldsymbol{e}_i^*(0)$ so that $\boldsymbol{Q}(t) = \boldsymbol{e}_k^*(t) \otimes \boldsymbol{e}_k^*(0)$ and the components with respect to $\{\boldsymbol{e}_i^*(t)\}$ are $Q_{ij} = \boldsymbol{e}_i^*(0) \cdot \boldsymbol{e}_j^*(t)$ (note that this is the same $\boldsymbol{Q}$ that appears in Eqn. (8.3) since $\boldsymbol{e}_i^*(t)$ and $\boldsymbol{e}_i^*(0)$ can be expressed as $\boldsymbol{x}_2^{(i)} - \boldsymbol{x}_1^{(i)}$ and $\boldsymbol{X}_2^{(i)} - \boldsymbol{X}_1^{(i)}$, respectively). Then, using Eqn. (8.12), we have

$$\begin{aligned}\boldsymbol{J}(t) &= \boldsymbol{Q}(t)\boldsymbol{J}(0)\boldsymbol{Q}(t)^T \\ &= J_1\boldsymbol{e}_1^*(t) \otimes \boldsymbol{e}_1^*(t) + J_2\boldsymbol{e}_2^*(t) \otimes \boldsymbol{e}_2^*(t) + J_3\boldsymbol{e}_3^*(t) \otimes \boldsymbol{e}_3^*(t).\end{aligned}$$

Thus, the principal values of $\boldsymbol{J}(t)$ with respect to the basis $\{\boldsymbol{e}_i^*(t)\}$ that is fixed to the body, are independent of time. The basis $\{\boldsymbol{e}_i^*(t)\}$ is called the principal basis, and the corresponding J_i are called the moments of inertia. Using the above spectral resolution, Eqn. (8.7) can be written as

$$\boldsymbol{h}_{\text{spin}}(t) = J_1 w_1^*(t)\boldsymbol{e}_1^*(t) + J_2 w_2^*(t)\boldsymbol{e}_2^*(t) + J_3 w_3^*(t)\boldsymbol{e}_3^*(t),$$

where $w_i^*(t) = \boldsymbol{w} \cdot \boldsymbol{e}_i^*(t)$ are the components of $\boldsymbol{w}$ with respect to the principal basis. Since $\boldsymbol{e}_i^*(t) = \boldsymbol{Q}(t)\boldsymbol{e}_i^*(0)$, we can also write the above expression in terms of the basis $\boldsymbol{e}_i^*(0)$ as

$$\boldsymbol{h}_{\text{spin}}(t) = \boldsymbol{Q}\left[J_1 w_1^*(t)\boldsymbol{e}_1^*(0) + J_2 w_2^*(t)\boldsymbol{e}_2^*(0) + J_3 w_3^*(t)\boldsymbol{e}_3^*(0)\right]$$

$$= Q \begin{bmatrix} J_1 w_1^*(t) \\ J_2 w_2^*(t) \\ J_3 w_3^*(t) \end{bmatrix}.$$

Similarly, the components of $\boldsymbol{h}_z$ with respect to $\boldsymbol{e}_i^*(0)$ are

$$\boldsymbol{h}_z = \boldsymbol{Q} \begin{bmatrix} J_{z1} w_1^*(t) \\ J_{z2} w_2^*(t) \\ J_{z3} w_3^*(t) \end{bmatrix}. \tag{8.16}$$

For future reference, we rewrite Eqns. (3.16), (3.30) and (3.31) here as

$$\hat{\boldsymbol{f}} = \int_{S(t)} \boldsymbol{t}(\boldsymbol{x}, t, \boldsymbol{n})\, dS + \int_{V(t)} \rho \boldsymbol{b}\, dV = \dot{\boldsymbol{l}} = M\ddot{\bar{\boldsymbol{x}}}. \tag{8.17}$$

$$\boldsymbol{m}_z(t) = \dot{\boldsymbol{h}}_z + M(\bar{\boldsymbol{x}} - \boldsymbol{z}) \times \ddot{\boldsymbol{z}} \tag{8.18}$$

$$\boldsymbol{m}_{\bar{x}}(t) = \dot{\boldsymbol{h}}_{\text{spin}}. \tag{8.19}$$

Differentiating $\boldsymbol{h}_{\text{spin}} = \boldsymbol{J}\boldsymbol{w} = \boldsymbol{Q}(t)\boldsymbol{J}(0)\boldsymbol{Q}(t)^T\boldsymbol{w}$, and using $\dot{\boldsymbol{Q}} = \boldsymbol{W}\boldsymbol{Q}$ and $\boldsymbol{W}\boldsymbol{w} = \boldsymbol{0}$, we get

$$\begin{aligned} \dot{\boldsymbol{h}}_{\text{spin}} &= \boldsymbol{J}(t)\dot{\boldsymbol{w}}(t) + \dot{\boldsymbol{Q}}\boldsymbol{J}(0)\boldsymbol{Q}^T\boldsymbol{w} + \boldsymbol{Q}\boldsymbol{J}(0)\dot{\boldsymbol{Q}}^T\boldsymbol{w} \\ &= \boldsymbol{J}(t)\dot{\boldsymbol{w}}(t) + \boldsymbol{W}\boldsymbol{Q}\boldsymbol{J}(0)\boldsymbol{Q}^T\boldsymbol{w} - \boldsymbol{Q}\boldsymbol{J}(0)\boldsymbol{Q}^T\boldsymbol{W}\boldsymbol{w} \\ &= \boldsymbol{J}(t)\dot{\boldsymbol{w}}(t) + \boldsymbol{W}\boldsymbol{J}(t)\boldsymbol{w} \\ &= \boldsymbol{J}(t)\dot{\boldsymbol{w}}(t) + \boldsymbol{w}(t) \times [\boldsymbol{J}(t)\boldsymbol{w}(t)]. \end{aligned} \tag{8.20}$$

From Eqn. (8.19) and the above expression, we get the Euler equations

$$\boldsymbol{m}_{\bar{x}}(t) = \boldsymbol{J}(t)\dot{\boldsymbol{w}}(t) + \boldsymbol{w}(t) \times [\boldsymbol{J}(t)\boldsymbol{w}(t)]. \tag{8.21}$$

If $m_i(t)$ denote the components of $\boldsymbol{m}_{\bar{x}}$ expressed with respect to the principal basis $\{\boldsymbol{e}_i^*(t)\}$, then from Eqn. (8.21), we get

$$\begin{aligned} m_1(t) &= J_1\dot{w}_1^*(t) + (J_3 - J_2)w_2^* w_3^*, \\ m_2(t) &= J_2\dot{w}_2^*(t) + (J_1 - J_3)w_1^* w_3^*, \\ m_3(t) &= J_3\dot{w}_3^*(t) + (J_2 - J_1)w_1^* w_2^*. \end{aligned} \tag{8.22}$$

We now derive analogous expressions with respect to the point $\boldsymbol{z}(t)$. Equation (8.18) is especially useful when either point $\boldsymbol{z}(t)$ is stationary at all times so that $\ddot{\boldsymbol{z}} = \boldsymbol{0}$, or when its acceleration is directed towards $\bar{\boldsymbol{x}}$, so that $(\bar{\boldsymbol{x}} - \boldsymbol{z}) \times \ddot{\boldsymbol{z}} = \boldsymbol{0}$. In either case, Eqn. (8.18) reduces to

$$\boldsymbol{m}_z(t) = \dot{\boldsymbol{h}}_z. \tag{8.23}$$

If $m_{zi}(t)$ denote the components of $\boldsymbol{m}_z$ expressed with respect to $\boldsymbol{e}_i^*(t)$, then analogous to Eqn. (8.22), we have

$$\begin{aligned} m_{z1}(t) &= J_{z1}\dot{w}_1^*(t) + (J_{z3} - J_{z2})w_2^* w_3^*, \\ m_{z2}(t) &= J_{z2}\dot{w}_2^*(t) + (J_{z1} - J_{z3})w_1^* w_3^*, \\ m_{z3}(t) &= J_{z3}\dot{w}_3^*(t) + (J_{z2} - J_{z1})w_1^* w_2^*. \end{aligned} \tag{8.24}$$

The components of a vector $\boldsymbol{a}$ with respect to $\boldsymbol{e}_i^*(t) = [\boldsymbol{e}_i^*(t) \cdot \boldsymbol{e}_j^*(0)]\boldsymbol{e}_j^*(0) = Q_{ji}\boldsymbol{e}_j^*(0)$ are $\boldsymbol{a} \cdot \boldsymbol{e}_i^*(t) = \boldsymbol{a} \cdot [Q_{ji}\boldsymbol{e}_j^*(0)] = Q_{ji}[\boldsymbol{a} \cdot \boldsymbol{e}_j^*(0)]$, i.e., $[\boldsymbol{a}]_{\boldsymbol{e}_i^*(t)} = \boldsymbol{Q}^T[\boldsymbol{a}]_{\boldsymbol{e}_i^*(0)}$. Similarly, for a tensor $\boldsymbol{T}$, we have $[\boldsymbol{T}]_{\boldsymbol{e}_i^*(t)} = \boldsymbol{Q}^T[\boldsymbol{T}]_{\boldsymbol{e}_i^*(0)}\boldsymbol{Q}$. It follows that the components of $\boldsymbol{w}$ and $\dot{\boldsymbol{w}}$ with respect to the principal basis $\boldsymbol{e}_i^*(t)$ are the axial vectors of $\boldsymbol{Q}^T\boldsymbol{W}\boldsymbol{Q}$ and $\boldsymbol{Q}^T\dot{\boldsymbol{W}}\boldsymbol{Q}$. Thus, using Eqn. (4.19), we get

$$\boldsymbol{w}_{\boldsymbol{e}_i^*(t)} = \begin{bmatrix} \dot{\boldsymbol{e}}_2^*(t) \cdot \boldsymbol{e}_3^*(t) \\ \dot{\boldsymbol{e}}_3^*(t) \cdot \boldsymbol{e}_1^*(t) \\ \dot{\boldsymbol{e}}_1^*(t) \cdot \boldsymbol{e}_2^*(t) \end{bmatrix}.$$

In terms of the Euler angle representation of $\boldsymbol{Q}$ given by (see Eqn. (1.108))

$$\begin{bmatrix} \cos\phi\cos\psi - \cos\theta\sin\psi\sin\phi & -\sin\phi\cos\psi - \cos\theta\sin\psi\cos\phi & \sin\theta\sin\psi \\ \cos\phi\sin\psi + \cos\theta\cos\psi\sin\phi & -\sin\phi\sin\psi + \cos\theta\cos\psi\cos\phi & -\sin\theta\cos\psi \\ \sin\phi\sin\theta & \cos\phi\sin\theta & \cos\theta \end{bmatrix}, \tag{8.25}$$

$\boldsymbol{w}_{\boldsymbol{e}_i^*(t)}$ and $\dot{\boldsymbol{w}}_{\boldsymbol{e}_i^*(t)}$ are given by Eqns. (4.21), i.e.,

$$\boldsymbol{w}_{\boldsymbol{e}_i^*(t)} = \begin{bmatrix} \dot\theta\cos\phi + \dot\psi\sin\theta\sin\phi \\ -\dot\theta\sin\phi + \dot\psi\sin\theta\cos\phi \\ \dot\phi + \dot\psi\cos\theta \end{bmatrix} \tag{8.26a}$$

$$\dot{\boldsymbol{w}}_{\boldsymbol{e}_i^*(t)} = \begin{bmatrix} \frac{d}{dt}(\dot\theta\cos\phi + \dot\psi\sin\theta\sin\phi) \\ \frac{d}{dt}(-\dot\theta\sin\phi + \dot\psi\sin\theta\cos\phi) \\ \frac{d}{dt}(\dot\phi + \dot\psi\cos\theta) \end{bmatrix}. \tag{8.26b}$$

Note that when $\theta \neq 0$, $\boldsymbol{w}_{\boldsymbol{e}_i^*(t)}$, as given by Eqn. (8.26a), is oriented differently from the axis $\boldsymbol{e}$, as given by Eqn. (1.109).

For a rigid motion, one can find a simple expression for the relative kinetic energy K_r. Substituting Eqn. (8.6) into Eqn. (3.66) and using Eqn. (1.20), we get

$$\begin{aligned} K_r &= \frac{1}{2}\int_{V(t)} \rho\left[(\boldsymbol{w}\cdot\boldsymbol{w})[(\boldsymbol{x}-\bar{\boldsymbol{x}})\cdot(\boldsymbol{x}-\bar{\boldsymbol{x}})] - [\boldsymbol{w}\cdot(\boldsymbol{x}-\bar{\boldsymbol{x}})]^2\right] dV \\ &= \frac{1}{2}\boldsymbol{w}\cdot\left\{\int_{V(t)} \rho\,[(\boldsymbol{x}-\bar{\boldsymbol{x}})\cdot(\boldsymbol{x}-\bar{\boldsymbol{x}})\boldsymbol{I} - (\boldsymbol{x}-\bar{\boldsymbol{x}})\otimes(\boldsymbol{x}-\bar{\boldsymbol{x}})]\,dV\right\}\boldsymbol{w} \\ &= \frac{1}{2}\boldsymbol{w}\cdot\boldsymbol{J}\boldsymbol{w}. \end{aligned} \tag{8.27}$$

Using the principal basis, we have

$$K_r = \frac{1}{2}\left[J_1(w_1^*)^2 + J_2(w_2^*)^2 + J_3(w_3^*)^2\right].$$

Thus, by Konig's theorem (Eqn. (3.67)), the total kinetic energy of a rigid body is

$$K = \frac{1}{2}M\dot{\bar{\boldsymbol{x}}}\cdot\dot{\bar{\boldsymbol{x}}} + \frac{1}{2}\boldsymbol{w}\cdot\boldsymbol{J}\boldsymbol{w} \tag{8.28}$$

$$= \frac{1}{2} M \dot{\bar{x}} \cdot \dot{\bar{x}} + \frac{1}{2} \left[J_1 (w_1^*)^2 + J_2 (w_2^*)^2 + J_3 (w_3^*)^2 \right].$$

The mechanical energy balance for rigid motion is given by the second of the following equations:

$$\frac{dK_r}{dt} = \boldsymbol{m}_{\bar{x}} \cdot \boldsymbol{w}, \tag{8.29a}$$

$$\frac{dK}{dt} = \hat{\boldsymbol{f}} \cdot \dot{\bar{\boldsymbol{x}}} + \boldsymbol{m}_{\bar{x}} \cdot \boldsymbol{w}, \tag{8.29b}$$

$$\frac{1}{2} \frac{d \left| \boldsymbol{h}_{\text{spin}} \right|^2}{dt} = \boldsymbol{m}_{\bar{x}} \cdot \boldsymbol{h}_{\text{spin}}. \tag{8.29c}$$

The first equation is obtained using Eqns. (8.20), (8.21) and (8.27) as follows:

$$\begin{aligned} \frac{dK_r}{dt} &= \frac{1}{2} \frac{d}{dt} (\boldsymbol{w} \cdot \boldsymbol{h}_{\text{spin}}) \\ &= \frac{1}{2} \left[\dot{\boldsymbol{w}} \cdot (\boldsymbol{J} \boldsymbol{w}) + \boldsymbol{w} \cdot \dot{\boldsymbol{h}}_{\text{spin}} \right] \\ &= \frac{1}{2} \left[\dot{\boldsymbol{w}} \cdot (\boldsymbol{J} \boldsymbol{w}) + \boldsymbol{w} \cdot (\boldsymbol{J} \dot{\boldsymbol{w}}) \right] \\ &= \boldsymbol{w} \cdot (\boldsymbol{J} \dot{\boldsymbol{w}}) \quad (\text{since } \boldsymbol{J} \in \text{Sym}) \\ &= \boldsymbol{w} \cdot \boldsymbol{m}_{\bar{x}}. \end{aligned}$$

The second equation is obtained using Konig's theorem and Eqn. (8.17). Equation (8.29b) can also be obtained from Eqn. (3.65) by setting $\boldsymbol{\tau} : \boldsymbol{D} = 0$, substituting for $\boldsymbol{v}$ using Eqn. (8.6), and using the properties of the scalar triple product. The third equation follows directly from Eqn. (8.19).

If $\boldsymbol{z}(t)$ is a point on the body for which $\dot{\boldsymbol{z}} = \boldsymbol{0}$ (but not necessarily $\ddot{\boldsymbol{z}} = \boldsymbol{0}$), then $\dot{\bar{\boldsymbol{x}}} = \boldsymbol{w} \times (\bar{\boldsymbol{x}} - \boldsymbol{z})$, so that the kinetic energy is

$$\begin{aligned} K &= \frac{1}{2} M \dot{\bar{\boldsymbol{x}}} \cdot \dot{\bar{\boldsymbol{x}}} + \frac{1}{2} \boldsymbol{w} \cdot \boldsymbol{J} \boldsymbol{w} \\ &= \frac{1}{2} M [\boldsymbol{w} \times (\bar{\boldsymbol{x}} - \boldsymbol{z})] \cdot [\boldsymbol{w} \times (\bar{\boldsymbol{x}} - \boldsymbol{z})] + \frac{1}{2} \boldsymbol{w} \cdot \boldsymbol{J} \boldsymbol{w} \\ &= \frac{1}{2} M \left[(\boldsymbol{w} \cdot \boldsymbol{w}) [(\bar{\boldsymbol{x}} - \boldsymbol{z}) \cdot (\bar{\boldsymbol{x}} - \boldsymbol{z})] - [\boldsymbol{w} \cdot (\bar{\boldsymbol{x}} - \boldsymbol{z})]^2 \right] + \frac{1}{2} \boldsymbol{w} \cdot \boldsymbol{J} \boldsymbol{w} \\ &= \frac{1}{2} \boldsymbol{w} \cdot \boldsymbol{J}_z \boldsymbol{w}, \end{aligned} \tag{8.30}$$

where the last step follows from Eqn. (8.10). Also, from Eqn. (3.29) and the properties of the scalar triple product, it follows that

$$\hat{\boldsymbol{f}} \cdot \dot{\bar{\boldsymbol{x}}} + \boldsymbol{m}_{\bar{x}} \cdot \boldsymbol{w} = \hat{\boldsymbol{f}} \cdot [\boldsymbol{w} \times (\bar{\boldsymbol{x}} - \boldsymbol{z})] + \left[\boldsymbol{m}_z + (\boldsymbol{z} - \bar{\boldsymbol{x}}) \times \hat{\boldsymbol{f}} \right] \cdot \boldsymbol{w} = \boldsymbol{m}_z \cdot \boldsymbol{w}.$$

Thus, from Eqn. (8.29b), we get for a point $\boldsymbol{z}$ with $\dot{\boldsymbol{z}} = \boldsymbol{0}$, the relation

$$\boldsymbol{m}_z \cdot \boldsymbol{w} = \frac{d}{dt} \left[\frac{1}{2} \boldsymbol{w} \cdot \boldsymbol{J}_z \boldsymbol{w} \right]. \tag{8.31}$$

Also analogous to Eqn. (8.29c), we have

$$\frac{1}{2}\frac{d\left|\boldsymbol{h}_z\right|^2}{dt} = \boldsymbol{m}_z \cdot \boldsymbol{h}_z.$$

If the loads are conservative, i.e., there exist potentials W_b and W_t such that

$$\frac{d}{dt}\int_{V(t)} \rho W_b\, dV = \int_{V(t)} \rho \boldsymbol{b} \cdot \boldsymbol{v}\, dV,$$
$$\frac{d}{dt}\int_{S(t)} W_t\, dS = \int_{S(t)} \boldsymbol{t} \cdot \boldsymbol{v}\, dS,$$

then analogous to Eqn. (5.36), we get

$$K - \int_{V(t)} \rho W_b\, dV - \int_{S(t)} W_t\, dS = \text{constant}. \tag{8.32}$$

As noted in the discussion on page 252, if the body forces are constant, then $\int_{V(t)} \rho W_b\, dV = M\boldsymbol{b} \cdot \bar{\boldsymbol{x}}$, while if it is of centrifugal type, the potential W_b is given by $\left|\boldsymbol{\Omega} \times \boldsymbol{x}\right|^2/2$. Similarly, if the traction $\boldsymbol{t}$ acting on $S(t)$ is a point dead load not varying with time, then $\int_{S(t)} W_t\, dS = \boldsymbol{t} \cdot \boldsymbol{x}$, where $\boldsymbol{x}$ is the position vector of the point of application of the load. Of course, if we have $\int_{S(t)} \boldsymbol{t} \cdot \boldsymbol{v}\, dV = 0$ at all times, then the W_t term can be dropped from the above equation; a similar comment applies to the body force term also.

Equations (8.17) and (8.22) (or Eqn. (8.24)) constitute the governing equations for the unknowns $\boldsymbol{w}$ and $\bar{\boldsymbol{x}}$. Equations (8.22) (or Eqn. (8.24)) can be quite difficult to integrate. Hence, we shall often use the fact that in some specially chosen coordinate system(s), some (or all) components of $\boldsymbol{m}_{\bar{x}}$ (or $\boldsymbol{m}_z$) are constant, and then by virtue of Eqn. (8.19) (or Eqn. (8.23)), we get the corresponding components of $\boldsymbol{h}_{\text{spin}}$ (or $\boldsymbol{h}_z$) as integrals of these constant components. In particular, if some components of $\boldsymbol{m}_{\bar{x}}$ (or $\boldsymbol{m}_z$) are zero, then the corresponding components of $\boldsymbol{h}_{\text{spin}}$ (or $\boldsymbol{h}_z$) are just constant. If the problem is two-dimensional, and if the loads are conservative, then the single unknown component of $\boldsymbol{w}$ can be found using the energy equation given by either of Eqns. (8.29b), (8.31) or (8.32). In three-dimensional problems where the forces are conservative, we can again use the energy equation, and in addition, attempt to find two directions along which the components of $\boldsymbol{m}_{\bar{x}}$ (or $\boldsymbol{m}_z$) are constant; this is the procedure we use to find the solution to the spinning top problem below. If the loads are not conservative, then of course, one may need to solve the Euler equations in the form given by Eqn. (8.22) (or (8.24)).

The above formulation can be termed as Eulerian since it has been carried out on the deformed configuration. We now develop the Lagrangian formulation on the reference configuration V_0. Using the second transport theorem, Eqn. (3.31) can be written as

$$\int_{V(t)} \rho(\boldsymbol{x} - \bar{\boldsymbol{x}}) \times (\boldsymbol{a} - \ddot{\bar{\boldsymbol{x}}})\, dV = \boldsymbol{m}_{\bar{x}}.$$

Since $\boldsymbol{x} - \bar{\boldsymbol{x}} = \boldsymbol{Q}(\boldsymbol{X} - \bar{\boldsymbol{x}}(0))$, $\boldsymbol{a} - \ddot{\bar{\boldsymbol{x}}} = \ddot{\boldsymbol{Q}}(\boldsymbol{X} - \bar{\boldsymbol{x}}(0))$, $\rho = \rho_0$ and $dV = dV_0$, we can write the above equation as

$$\int_{V_0} \rho_0 \boldsymbol{Q}(\boldsymbol{X} - \bar{\boldsymbol{x}}(0)) \times \ddot{\boldsymbol{Q}}(\boldsymbol{X} - \bar{\boldsymbol{x}}(0))\, dV_0 = \int_{S_0} \boldsymbol{Q}(\boldsymbol{X} - \bar{\boldsymbol{x}}(0)) \times \tilde{\boldsymbol{t}}\, dS_0 + \int_{V_0} \rho_0 \boldsymbol{Q}(\boldsymbol{X} - \bar{\boldsymbol{x}}(0)) \times \tilde{\boldsymbol{b}}\, dV_0,$$

where $\tilde{t}(X, t, n_0) := t(x, t, n)$ and $\tilde{b}(X, t) := b(x, t)$ are the tractions and body forces using the material description. Since $u \times v$ is the axial vector of $v \otimes u - u \otimes v = 2\mathbb{W}(v \otimes u)$, and since $(T_1 u) \otimes (T_2 v) = T_1(u \otimes v)T_2^T$, the above equation can be written as

$$\mathbb{W}\left[\ddot{Q}M_0 - \int_{S_0} \tilde{t} \otimes (X - \bar{x}(0))\, dS_0 - \int_{V_0} \rho_0 \tilde{b} \otimes (X - \bar{x}(0))\, dV_0\right] Q^T = 0. \tag{8.33}$$

where $M_0 = \int_{V_0} \rho_0 (X - \bar{x}(0)) \otimes (X - \bar{x}(0))\, dV_0 = (\operatorname{tr} J(0)/2)I - J(0)$ is the Euler tensor in the reference configuration. Thus, the term following $\mathbb{W}$ in the above equation is a symmetric tensor. Since, if $S \in \text{Sym}$, then $Q^T S Q \in \text{Sym}$, it follows that Eqn. (8.33) can also be written as

$$\mathbb{W} Q^T \left[\ddot{Q}M_0 - \int_{S_0} \tilde{t} \otimes (X - \bar{x}(0))\, dS_0 - \int_{V_0} \rho_0 \tilde{b} \otimes (X - \bar{x}(0))\, dV_0\right] = 0.$$

Since $\dot{Q}M_0\dot{Q}^T \in \text{Sym}$, yet another way in which Eqn. (8.33) can be written is

$$\frac{d}{dt}[\mathbb{W}(\dot{Q}M_0 Q^T)] - \mathbb{W}\left[\int_{S_0} \tilde{t} \otimes (X - \bar{x}(0))\, dS_0 + \int_{V_0} \rho_0 \tilde{b} \otimes (X - \bar{x}(0))\, dV_0\right] Q^T = 0.$$

If Q is represented as $e^{W(t)}$, then the above equation can be written as

$$\frac{d}{dt}[\tilde{W} e^{W(t)} M_0 e^{-W(t)} + e^{W(t)} M_0 e^{-W(t)} \tilde{W}] - \tilde{F} e^{-W(t)} + e^{W(t)} \tilde{F}^T = 0,$$

where $\tilde{F} = \int_{S_0} \tilde{t} \otimes (X - \bar{x}(0))\, dS_0 + \int_{V_0} \rho_0 \tilde{b} \otimes (X - \bar{x}(0))\, dV_0$, and $\tilde{W} \in \text{Skw}$ is the right-hand side of Eqn. (1.290a). On simplification, the above equation yields

$$\left[\dot{\tilde{W}} + \tilde{W}^2\right] e^{W(t)} M_0 e^{-W(t)} + e^{W(t)} M_0 e^{-W(t)} \left[\dot{\tilde{W}} - \tilde{W}^2\right] - \tilde{F} e^{-W(t)} + e^{W(t)} \tilde{F}^T = 0.$$

By setting the axial vector of the above equation to zero, one can solve for the three independent components of $W(t)$.

We now present a few examples solved using the Eulerian approach.

Example 1: Central-force motion

Consider the motion of a planet of mass m relative to a star of mass M, as shown in Fig. 8.1. We assume that the distance r between the star and the planet is large compared to the size of the planet, and that the body force exerted by the star with respect to the xy frame fixed to it, at a point which is at a radial distance $\hat{r}$ from the center of the star is $b = -(GM/\hat{r}^2)e_r$. Since we have assumed r to be large compared to the size of the planet, the body force may be assumed to be uniform over the volume occupied by the planet, so that the total force acting on the planet is fe_r, where $f = -GMm/r^2$. We now show that the motion of the planet is in a plane at all instants of time. Since there is no contact between the bodies, the traction force acting on the planet is zero. Thus, the force exerted on the planet is entirely due to the body force. Applying Eqn. (8.17) to the planet, we get

$$m\ddot{\bar{x}} = fe_r. \tag{8.34}$$

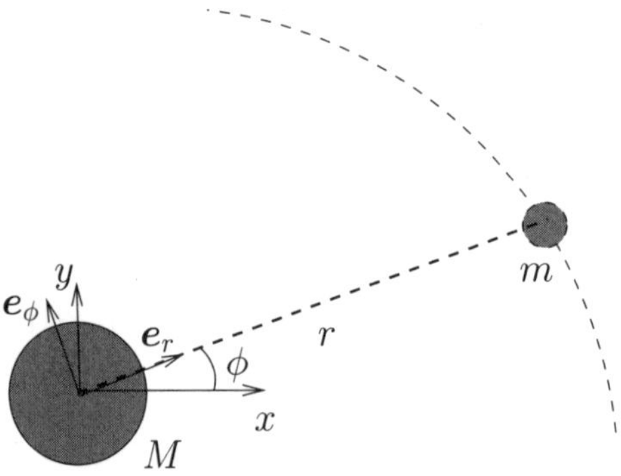

Fig. 8.1 *Central-force motion.*

Since $\bar{x} = re_r$, we have

$$\bar{x} \times \ddot{\bar{x}} = \frac{rf}{m} e_r \times e_r = \mathbf{0}.$$

However, since $\bar{x} \times \ddot{\bar{x}} = \frac{d}{dt}(\bar{x} \times \dot{\bar{x}})$, we get

$$\bar{x} \times \dot{\bar{x}} = c,$$

where c is a constant vector. Thus, the central force motion occurs in the plane of $\bar{x}$ and $\dot{\bar{x}}$.

To solve Eqn. (8.34), we write it as

$$\ddot{\bar{x}} = -\frac{GM}{r^2} e_r. \tag{8.35}$$

It is easier to solve the above equation using polar rather than Cartesian coordinates. Since $e_r = \cos\phi\, e_x + \sin\phi\, e_y$, we have $\bar{x} = re_r = r\cos\phi\, e_x + r\sin\phi\, e_y$. Differentiating this equation twice (or directly from Eqn. (B.14)), we get

$$\ddot{\bar{x}} = (\ddot{r} - r\dot{\phi}^2)e_r + (r\ddot{\phi} + 2\dot{r}\dot{\phi})e_\phi,$$

where $e_\phi = -\sin\phi\, e_x + \cos\phi\, e_y$. Substituting this expression into Eqn. (8.35), we get

$$\ddot{r} - r\dot{\phi}^2 = -\frac{GM}{r^2}, \tag{8.36}$$

$$r\ddot{\phi} + 2\dot{r}\dot{\phi} = 0. \tag{8.37}$$

Multiplying Eqn. (8.37) by r, we get

$$r^2\ddot{\phi} + 2r\dot{r}\dot{\phi} = \frac{d(r^2\dot{\phi})}{dt} = 0,$$

which implies that

$$r^2\dot{\phi} = c, \tag{8.38}$$

where c is a constant. Since the rate of area swept by the trajectory dA/dt is given by $r^2\dot{\phi}/2$, we get

$$\frac{dA}{dt} = \text{constant}.$$

This is Kepler's second law of planetary motion.

Using Eqn. (8.38), Eqn. (8.36) can be integrated to yield

$$\frac{1}{r} = \frac{GM}{c^2} + d\cos(\phi - \alpha),$$

where d and α are constants of integration. This is the equation of a circle, ellipse, parabola or hyperbola depending upon whether the eccentricity dc^2/GM is 0, less than 1, equal to 1 or greater than 1, respectively. Of course, the above solution is valid for any gravitational central-force, e.g., the motion of a satellite about the earth.

Example 2: Cylinder rolling down a plane

Consider a cylinder that starts from rest, and that rolls down a plane without slipping as shown in Fig. 8.2. Let the principal basis $e_i^*(t)$ be denoted by 123 (Fig. 8.2). The principal

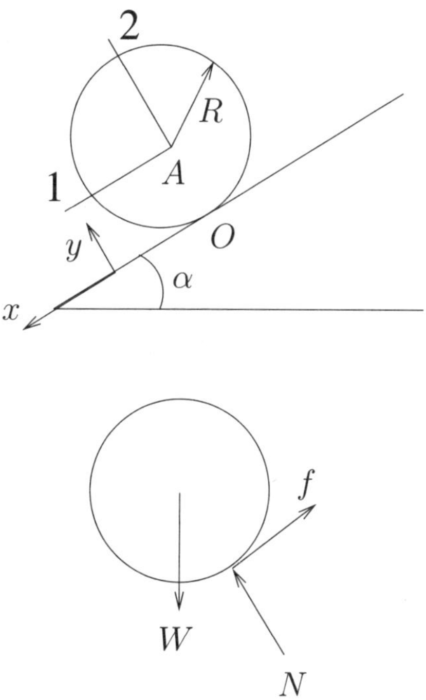

Fig. 8.2 *Cylinder rolling down an inclined plane.*

values of inertia of the cylinder with respect to 123 are given by Eqn. (8.14). Equation (8.17) reduces to

$$-f + W\sin\alpha = M\ddot{x}.$$

Since $\omega_x^* = \omega_y^* = 0$, the only nontrivial component of Eqn. (8.22) is the z-component; we get (dropping the '$\star$' superscript)

$$-fR = J_3\dot{\omega}_z.$$

Note that we have three unknowns (f, ω_z and $\bar{x}$), but only two equations. The third equation is obtained from the 'constraint' that the cylinder is rolling on the plane. Using the fact that the velocity of the contact point O is zero, and applying Eqn. (8.6), we get

$$\dot{\bar{x}}\boldsymbol{e}_x = \omega_z\boldsymbol{e}_z \times (R\boldsymbol{e}_y) = -R\omega_z\boldsymbol{e}_x,$$

which implies $\dot{\bar{x}} = -R\omega_z$. Note that this equation is valid at *all* times. Hence,

$$\ddot{\bar{x}} = -R\dot{\omega}_z.$$

From the three governing equations, we now get

$$\dot{\omega}_z = -\frac{WR\sin\alpha}{J_3 + MR^2}. \tag{8.39}$$

Integrating the above relation using the initial condition $\omega_z(0) = 0$, we get

$$\omega_z = -\frac{WRt\sin\alpha}{J_3 + MR^2}. \tag{8.40}$$

Applying Eqn. (8.5) to the points A and O, we get

$$\begin{aligned}\boldsymbol{a}_A &= \boldsymbol{a}_O + \dot{\omega}_z\boldsymbol{e}_z \times (R\boldsymbol{e}_y) + \omega_z\boldsymbol{e}_z \times (\omega_z\boldsymbol{e}_z \times R\boldsymbol{e}_y)\\ &= \boldsymbol{a}_O - R\dot{\omega}_z\boldsymbol{e}_x - \omega_z^2 R\boldsymbol{e}_y.\end{aligned}$$

Since $\boldsymbol{a}_A = \ddot{\bar{x}}\boldsymbol{e}_x = -R\dot{\omega}_z\boldsymbol{e}_x$, we get $\boldsymbol{a}_O = \omega_z^2 R\boldsymbol{e}_y$, i.e., the acceleration of the contact point O is along the radial direction, and hence Eqn. (8.23) is valid. Thus, using Eqn. (8.15) and either of Eqns. (8.24) or (8.31), we get

$$-WRw_z\sin\alpha = \frac{1}{2}\frac{d}{dt}\left[(J_3 + MR^2)w_z^2\right] = (J_3 + MR^2)w_z\dot{w}_z,$$

which is nothing but Eqn. (8.39). From Eqn. (8.32), we get

$$\frac{1}{2}(J_3 + MR^2)\omega_z^2 - (W\sin\alpha)\bar{x} = \text{constant},$$

which on using Eqn. (8.40) yields $\bar{x}$ as a function of time.

Example 3: Sliding rod

Consider a rigid rod sliding against a frictionless wall and floor as shown in Fig. 8.3. From Eqn. (8.17), we get

$$\begin{aligned}F - N_b &= M\ddot{\bar{x}},\\ N_a - W &= M\ddot{\bar{y}}.\end{aligned} \tag{8.41}$$

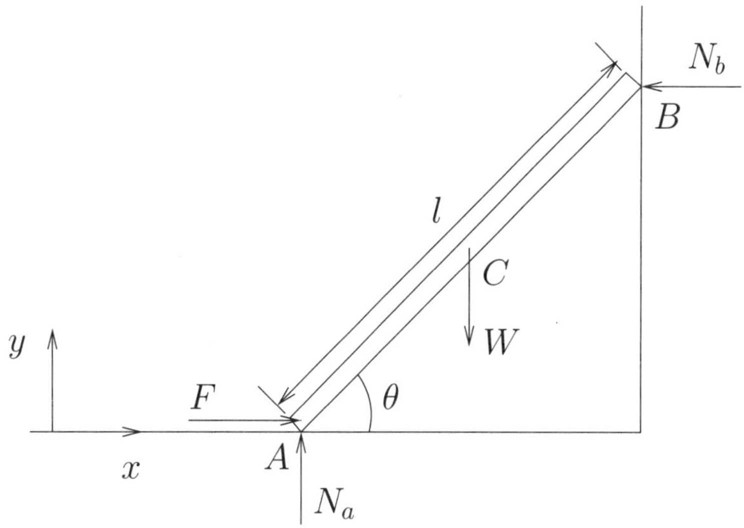

Fig. 8.3 *Rod sliding against a wall.*

From Eqn. (8.22), we get

$$\frac{1}{2}Fl\sin\theta + \frac{1}{2}N_b l\sin\theta - \frac{1}{2}N_a l\cos\theta = J_3\dot{\omega}_z. \tag{8.42}$$

Since we have three equations, and six unknowns (N_a, N_b, θ, $\ddot{\bar{x}}$ and $\ddot{\bar{y}}$, ω_z), we need to consider the constraint equations. From the figure, we see that

$$\bar{x} = c - \frac{l}{2}\cos\theta,$$
$$\bar{y} = \frac{l}{2}\sin\theta,$$

where c is the distance of the vertical wall from the origin. This relation is valid at all times. Thus,

$$\begin{aligned} \dot{\bar{x}} &= \frac{1}{2}l\dot{\theta}\sin\theta, \qquad \ddot{\bar{x}} = \frac{1}{2}l\left(\dot{\theta}^2\cos\theta + \ddot{\theta}\sin\theta\right), \\ \dot{\bar{y}} &= \frac{1}{2}l\dot{\theta}\cos\theta, \qquad \ddot{\bar{y}} = \frac{1}{2}l\left(-\dot{\theta}^2\sin\theta + \ddot{\theta}\cos\theta\right). \end{aligned} \tag{8.43}$$

Applying Eqn. (8.6) to points A and C, and then to B and C, and noting that $\boldsymbol{v}_A = v_A\boldsymbol{e}_x$, $\boldsymbol{v}_B = v_B\boldsymbol{e}_y$, we get

$$\dot{\bar{x}}\boldsymbol{e}_x + \dot{\bar{y}}\boldsymbol{e}_y = v_A\boldsymbol{e}_x + \omega_z\boldsymbol{e}_z \times \left(\frac{l}{2}\cos\theta\boldsymbol{e}_x + \frac{l}{2}\sin\theta\boldsymbol{e}_y\right),$$

$$\dot{\bar{x}}\boldsymbol{e}_x + \dot{\bar{y}}\boldsymbol{e}_y = v_B\boldsymbol{e}_y + \omega_z\boldsymbol{e}_z \times \left(-\frac{l}{2}\cos\theta\boldsymbol{e}_x - \frac{l}{2}\sin\theta\boldsymbol{e}_y\right),$$

which implies that

$$\dot{x} = v_A - \frac{1}{2}\omega_z l \sin\theta = \frac{1}{2}\omega_z l \sin\theta,$$
$$\dot{y} = v_B - \frac{1}{2}\omega_z l \cos\theta = \frac{1}{2}\omega_z l \cos\theta.$$

Comparing these equations with Eqns. (8.43), we get $\dot{\theta} = \omega_z$, $v_A = \omega_z l \sin\theta$, and $v_B = \omega_z l \cos\theta$. Substituting these results into Eqn. (8.41) and (8.42), and using (8.43) and the fact that $J_3 = Ml^2/12$, we get

$$\ddot{\theta} = \frac{3}{2Ml}\left(2F\sin\theta - W\cos\theta\right).$$

Multiplying by $\dot{\theta}$ and integrating, we get the governing equation for θ as

$$\frac{\dot{\theta}^2}{2} + \frac{3}{2Ml}\left(2F\cos\theta + W\sin\theta\right) = c_1, \tag{8.44}$$

where c_1 is a constant. The solution to the above equation can be given in terms of the Appell hypergeometric function.

Since this is a two-dimensional problem and since the forces are conservative, Eqn. (8.44) can be obtained directly using Eqns. (8.28) and (8.32) as follows:

$$\frac{1}{2}M(\dot{x}^2 + \dot{y}^2) + \frac{1}{2}J\dot{\theta}^2 - F(c - l\cos\theta) + \frac{1}{2}Wl\sin\theta = \text{constant}.$$

On substituting $\dot{x} = l\dot{\theta}\sin\theta/2$, $\dot{y} = l\dot{\theta}\cos\theta/2$ and $J = Ml^2/12$, we get Eqn. (8.44).

The accelerations of points A and B are given by

$$\boldsymbol{a}_A = l(\ddot{\theta}\sin\theta + \dot{\theta}^2\cos\theta)\boldsymbol{e}_x,$$
$$\boldsymbol{a}_B = l(\ddot{\theta}\cos\theta - \dot{\theta}^2\sin\theta)\boldsymbol{e}_y.$$

Example 4: Motion of a sleigh

The motion of a sleigh is often modeled as shown in Fig. 8.4. The point O is constrained such that it cannot move along the η direction, i.e., if v denotes the velocity component along the η direction, then $v = \dot{v} = \ddot{v} = 0$. This constraint gives rise to a force F as shown. However, the sleigh is free to move in the ξ direction, and has a velocity component u in that direction (in general, $\dot{u} \neq 0$). The sleigh is also free to rotate about the point O–its angular velocity and angular acceleration are given by $\omega = \dot{\phi}$, and $\dot{\omega} = \ddot{\phi}$, respectively. The center of mass G is at a distance of a from O along the ξ axis. Assuming the mass of the sleigh to be M, the moment of inertia about an axis perpendicular to the plane of the sleigh and passing through G to be I, and the body forces with respect to the stationary frame xy to be negligible, find the equations of motion with respect to the xy frame. Using Eqn. (8.32), show that the kinetic energy K remains a constant.

Since $\dot{\boldsymbol{x}} = \boldsymbol{v}_O + \boldsymbol{w} \times (a\boldsymbol{e}_\xi)$, and since $v = 0$, we have

$$\dot{x} = u\cos\phi - a\omega\sin\phi,$$
$$\dot{y} = u\sin\phi + a\omega\cos\phi.$$

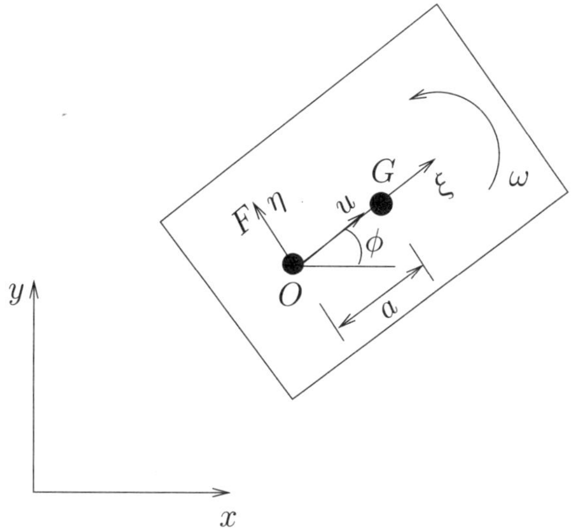

Fig. 8.4 *Model of a sleigh.*

The linear momentum equations, $M\ddot{x} = -F\sin\phi$, and $M\ddot{y} = F\cos\phi$, lead to

$$M\left[\dot{u}\cos\phi - a\dot{\omega}\sin\phi - u\omega\sin\phi - a\omega^2\cos\phi\right] = -F\sin\phi,$$
$$M\left[\dot{u}\sin\phi + a\dot{\omega}\cos\phi + u\omega\cos\phi - a\omega^2\sin\phi\right] = F\cos\phi,$$

which being valid for arbitrary ϕ yield

$$\dot{u} - a\omega^2 = 0,$$
$$a\dot{\omega} + u\omega = \frac{F}{M}.$$

The third governing equation is obtained from Eqn. (8.22) as

$$I\dot{\omega} = -Fa.$$

The same equations as above can be obtained if we use the $\xi\eta$ instead of the xy coordinate frame. Since the axis of rotation does not change with time, we use the relation

$$\boldsymbol{b} = \boldsymbol{Q}^T\left[\boldsymbol{b}^* - \ddot{\boldsymbol{c}}\right] - \dot{\boldsymbol{w}} \times \boldsymbol{x} - \boldsymbol{w} \times (\boldsymbol{w} \times \boldsymbol{x}) - 2\boldsymbol{w} \times \boldsymbol{v},$$

to compute the body forces in the moving frame of reference. We have $\dot{x} = \dot{y} = 0$, and

$$\boldsymbol{Q} = \begin{bmatrix} \cos\phi & -\sin\phi \\ \sin\phi & \cos\phi \end{bmatrix}, \qquad \dot{\boldsymbol{c}} = \begin{bmatrix} u\cos\phi \\ u\sin\phi \end{bmatrix},$$

so that

$$\boldsymbol{Q}^T\ddot{\boldsymbol{c}} = \begin{bmatrix} \dot{u} \\ u\omega \end{bmatrix}.$$

Using $\boldsymbol{b}^* = \boldsymbol{0}$, $\boldsymbol{v} = \boldsymbol{0}$, $\boldsymbol{x} = a\boldsymbol{e}_x$, $\boldsymbol{w} = \omega\boldsymbol{e}_z$, $\int \rho\boldsymbol{x}\,dV = M\bar{\boldsymbol{x}}$, and the fact that the contact force F remains the same in the rotating frame by the axiom of objectivity of the traction vector, we get the first two governing equations; since the body force $\boldsymbol{b}$ in the rotating frame acts through $\bar{\boldsymbol{x}}$, it exerts no moment about $\bar{\boldsymbol{x}}$, so that the third governing equation again follows directly from Eqn. (8.22).

To show that the kinetic energy K remains constant, observe that with respect to the stationary frame of reference, the body forces are negligible, and $\int_{S(t)} \boldsymbol{t} \cdot \boldsymbol{v}\,dS = 0$ so that W_t in Eqn. (8.32) can be taken to be zero, from which it directly follows that

$$K = \frac{1}{2}M[u^2 + a^2\omega^2] + \frac{1}{2}I\omega^2 = \text{constant.}$$

Example 5: Spinning disc

Consider again the example of the spinning disc considered in Section 4.4. The setup is as shown in Fig. 8.5. Remember that in this case, we are considering the motion of the rigid disc in a single frame of reference. The axes $\boldsymbol{e}_i^*(0)$ and $\boldsymbol{e}_i^*(t)$ are the principal axes of inertia at $t = 0$ and at $t = t$, respectively, at the center of mass of the disc. Using the fact that $Q_{ij} = \boldsymbol{e}_i^*(0) \cdot \boldsymbol{e}_j^*(t)$, we get the same $\boldsymbol{Q}$ matrix as in Eqn. (4.22); hence, $\boldsymbol{w}_{\boldsymbol{e}_i^*(t)}$ and $\dot{\boldsymbol{w}}_{\boldsymbol{e}_i^*(t)}$ are the same as $\boldsymbol{\Omega}$ and $\dot{\boldsymbol{\Omega}}$ in Eqn. (4.23). By symmetry $J_1 = J_2$. Let this common value be denoted by J, and let $J_3 = J'$. The position of the center of mass with respect to XYZ is the

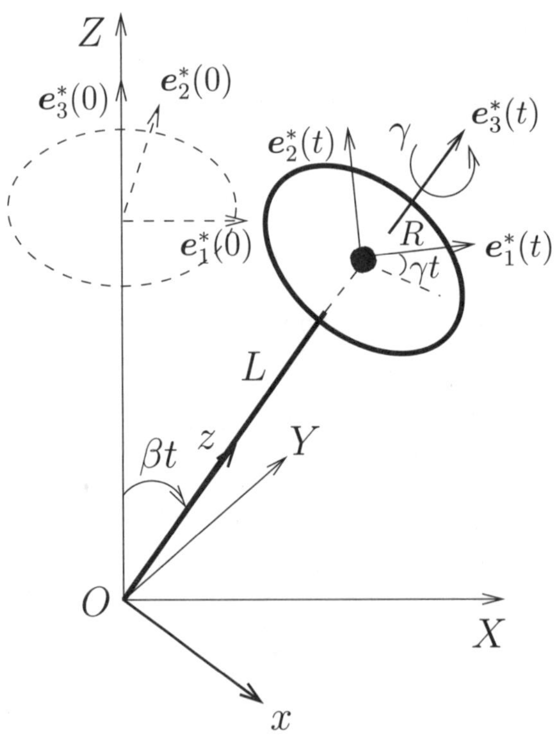

Fig. 8.5 *Spinning disc.*

same as c in Eqn. (4.22). Hence, if M is the mass of the disc, and if $\hat{f}$ represents the force due to the tractions acting on the disc (expressed with respect to XYZ), then by Eqn. (8.17),

$$\hat{f} = M \begin{bmatrix} -L\beta^2 \sin\beta t \\ 0 \\ g - L\beta^2 \cos\beta t \end{bmatrix}.$$

We can take an alternative viewpoint, and fix our coordinate system to the disc; the most convenient choice is, of course, to let it coincide with $e_i^*(t)$. Now, although the acceleration, $\ddot{x} = \mathbf{0}$, the body force is given by Eqn. (4.18) with $v = \mathbf{0}$. Since $\boldsymbol{\Omega}$ is a function of time alone, and since now $\bar{x} = \mathbf{0}$, we get from Eqn. (8.17) the tractions acting on the disc (expressed with respect to $e_i^*(t)$) as

$$\begin{aligned}
\hat{f}_{e_i^*(t)} &= -\int_{V(t)} \rho b \, dV \\
&= -MQ^T(b^* - \ddot{c}) + \dot{\boldsymbol{\Omega}} \times \int_{V(t)} \rho x \, dV + \boldsymbol{\Omega} \times \boldsymbol{\Omega} \times \int_{V(t)} \rho x \, dV \\
&= -MQ^T(b^* - \ddot{c}) \\
&= M \begin{bmatrix} -L\beta^2 \sin\beta t \\ 0 \\ g - L\beta^2 \cos\beta t \end{bmatrix},
\end{aligned}$$

which agrees with the result obtained above.

By Eqn. (8.22), the net moment that needs to be exerted about the center of mass to maintain the motion (expressed with respect to $e_i^*(t)$) is

$$\begin{bmatrix} J'\beta\gamma \cos\gamma t \\ -J'\beta\gamma \sin\gamma t \\ 0 \end{bmatrix}.$$

Since the moment exerted by the body force about the center of mass is zero, this net moment is entirely due to the tractions acting on the disc.

Finally, consider the case when we are interested in finding the forces and moments at the point O with respect to xyz as shown. Note that now we have to consider the forces due to the rotation of the bar and the disc. Since we have two rigid bodies in relative motion with each other, we cannot use the Euler equations directly. In such a case, it is better to go back to the original integral form of the linear and angular momentum balance equations. Since the angular velocity $\boldsymbol{\Omega}$ of the xyz system is βe_y, and since $c = \mathbf{0}$, the body force with respect to the xyz frame is

$$b = Q^T b^* - \beta e_y \times (\beta e_y \times x) - 2\beta e_y \times v,$$

where

$$Q = \begin{bmatrix} \cos\beta t & 0 & \sin\beta t \\ 0 & 1 & 0 \\ -\sin\beta t & 0 & \cos\beta t \end{bmatrix}.$$

Note that the velocity of the rod with respect to xyz is zero, and that of a point $\boldsymbol{x} = (x, y, z)$ on the disc is $\gamma\boldsymbol{e}_z \times \boldsymbol{x} = -\gamma y\boldsymbol{e}_x + \gamma x\boldsymbol{e}_y$. Since the positions of the center of mass of the rod and disc are also fixed with respect to xyz, $\ddot{\bar{\boldsymbol{x}}}_1 = \ddot{\bar{\boldsymbol{x}}}_2 = \boldsymbol{0}$. Denoting the forces due to traction at O by $\hat{\boldsymbol{f}}_{xyz}$, the mass of the rod by $\bar{M}$, the volume occupied by the rod and disc by $V(t)$, and using Eqn. (8.17), we get

$$\begin{aligned}\hat{\boldsymbol{f}}_{xyz} &= -\int_{V(t)} \rho\boldsymbol{b}\, dV \\ &= (M+\bar{M})\begin{bmatrix} -g\sin\beta t \\ 0 \\ g\cos\beta t \end{bmatrix} - \begin{bmatrix} 0 \\ 0 \\ \frac{1}{2}\beta^2\bar{M}L + \beta^2 ML \end{bmatrix}.\end{aligned}$$

Denoting the moment due to the tractions at O by $\hat{\boldsymbol{m}}_{xyz}$, and using Eqn. (3.26), we get

$$\begin{aligned}\hat{\boldsymbol{m}}_{xyz} &= \frac{d}{dt}\int_{V(t)} \boldsymbol{x} \times \rho\boldsymbol{v}\, dV - \int_{V(t)} \boldsymbol{x} \times \rho\boldsymbol{b}\, dV \\ &= -\int_{V(t)} \rho\boldsymbol{x} \times \boldsymbol{b}\, dV,\end{aligned}$$

since γ, and hence $\int_{V(t)} \boldsymbol{x} \times \rho\boldsymbol{v}\, dV$, is a constant. Using

$$\boldsymbol{x} \times (\boldsymbol{\Omega} \times \boldsymbol{\Omega} \times \boldsymbol{x}) = (\boldsymbol{\Omega}\cdot\boldsymbol{x})(\boldsymbol{x} \times \boldsymbol{\Omega}) = \beta^2(-yz\boldsymbol{e}_x + xy\boldsymbol{e}_z),$$

we get $\int_{V_{\text{disc}}} \rho\boldsymbol{x} \times (\boldsymbol{\Omega} \times \boldsymbol{\Omega} \times \boldsymbol{x})\, dV = \boldsymbol{0}$. The same result holds for the rod. Also,

$$\begin{aligned}2\int_{V_{\text{disc}}} \rho\boldsymbol{x} \times (\boldsymbol{\Omega} \times \boldsymbol{v})\, dV &= 2\int_{V_{\text{disc}}} \rho\boldsymbol{x} \times [\beta\boldsymbol{e}_y \times (-\gamma y\boldsymbol{e}_x + \gamma x\boldsymbol{e}_y)]\, dV \\ &= 2\int_{V_{\text{disc}}} \rho(\beta\gamma y^2\boldsymbol{e}_x - \beta\gamma xy\boldsymbol{e}_y)\, dV = J'\beta\gamma\boldsymbol{e}_x.\end{aligned}$$

Thus, we get

$$\hat{\boldsymbol{m}}_{xyz} = J'\beta\gamma\,\boldsymbol{e}_x - \left(\frac{\bar{M}}{2} + M\right) Lg\sin\beta t\,\boldsymbol{e}_y.$$

Example 6: Spinning top

Consider the symmetric spinning top shown in Fig. 8.6, which is fixed at the point of contact with the ground. The $\boldsymbol{e}_i^*(0)$ and $\boldsymbol{e}_i^*(t)$ axes are parallel to the XYZ and xyz axes, respectively. The linear momentum equations do not yield any useful information about the motion since the reactions at the contact point are unknown. Hence, we use the angular momentum and energy equations to solve this problem. Since the origin is at the point of contact, $\int_{S(t)} \boldsymbol{x} \times \boldsymbol{t}\, dS = \boldsymbol{0}$, and since the body force $\boldsymbol{b}$ is constant, $\int_{V(t)} \rho\boldsymbol{x} \times \boldsymbol{b}\, dV = M\bar{\boldsymbol{x}} \times \boldsymbol{b}$. Thus, the components of the moment along the $\bar{\boldsymbol{x}}$ and $\boldsymbol{b}$ directions are zero, which implies that z and Z components of $\boldsymbol{h}_z$ are constant. The third equation is obtained using the energy equation.

To find the moment components due to the weight about the point of contact with respect to xyz, we note that the components of the weight in the xyz frame are given by (with $\boldsymbol{Q}$ given by Eqn. (8.25))

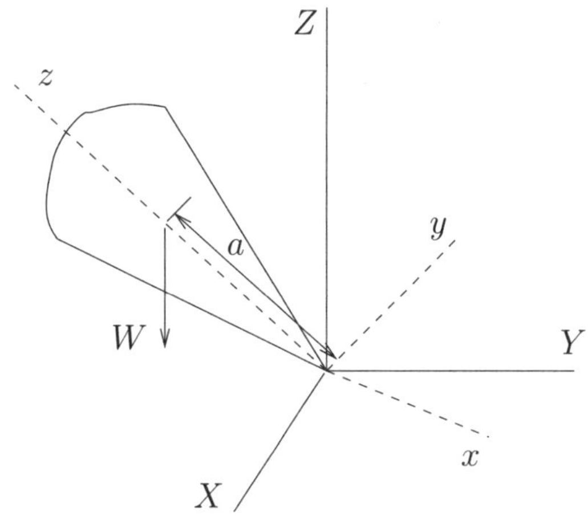

Fig. 8.6 Spinning top.

$$f_W = Q^T \begin{bmatrix} 0 \\ 0 \\ -W \end{bmatrix} = \begin{bmatrix} -W \sin\phi \sin\theta \\ -W \cos\phi \sin\theta \\ -W \cos\theta. \end{bmatrix}. \tag{8.45}$$

Hence, the moment about the point of contact with respect to xyz is

$$\begin{aligned} \boldsymbol{m}_z &= a\boldsymbol{e}_z \times (-W \sin\phi \sin\theta \boldsymbol{e}_x - W\cos\phi \sin\theta \boldsymbol{e}_y - W\cos\theta \boldsymbol{e}_z) \\ &= Wa(\sin\theta \cos\phi \boldsymbol{e}_x - \sin\theta \sin\phi \boldsymbol{e}_y). \end{aligned}$$

Due to the symmetry of the top, we have $J_{z1} = J_{z2}$; we denote these components by J', and denote J_{z3} by J. Using Eqn. (8.26b), the z-component of Eqn. (8.24) reduces to

$$J\frac{d}{dt}(\dot{\phi} + \dot{\psi}\cos\theta) = 0,$$

which implies that

$$\dot{\phi} + \dot{\psi}\cos\theta = \alpha, \tag{8.46}$$

where α is a constant.

Now substituting this result into the x and y-components of Eqn. (8.24), and using Eqn. (8.26a), we get

$$Wa\sin\theta\cos\phi = J'\frac{dw_1}{dt} + (J - J')\alpha w_2, \tag{8.47}$$

$$-Wa\sin\theta\sin\phi = J'\frac{dw_2}{dt} + (J' - J)\alpha w_1, \tag{8.48}$$

where $w_1 = \dot{\theta}\cos\phi + \dot{\psi}\sin\theta\sin\phi$, and $w_2 = -\dot{\theta}\sin\phi + \dot{\psi}\sin\theta\cos\phi$. Equations (8.46)–(8.48) are the governing equations to be solved for θ, ϕ and ψ, subject to appropriate initial conditions. We now try and simplify these equations. Multiplying Eqn. (8.47) by w_1, and Eqn. (8.48) by w_2, and adding the resulting equations, we get

$$\frac{J'}{2}\frac{d}{dt}(w_1^2 + w_2^2) + Wa\frac{d}{dt}(\cos\theta) = 0,$$

or, alternatively,

$$\frac{1}{2}J'(w_1^2 + w_2^2) + Wa\cos\theta = \beta, \tag{8.49}$$

where β is a constant. To see that this is the same equation as the mechanical energy balance equation given by Eqn. (8.32), we observe that $\int_{S(t)} \boldsymbol{t}\cdot\boldsymbol{v}\,dS = 0$, and that by using Eqn. (8.45), we have

$$M\boldsymbol{b}\cdot\bar{\boldsymbol{x}} = -W(\sin\phi\sin\theta\boldsymbol{e}_x + \cos\phi\sin\theta\boldsymbol{e}_y + \cos\theta\boldsymbol{e}_z)\cdot(a\boldsymbol{e}_z) = -Wa\cos\theta.$$

Thus, by virtue of Eqn. (8.30), Eqn. (8.32) reduces to

$$\frac{1}{2}\boldsymbol{w}\cdot\boldsymbol{J}_z\boldsymbol{w} + Wa\cos\theta = \text{constant}.$$

Since w_3 is a constant, we get Eqn. (8.49).

To find the last 'integrated form' of the governing differential equations, as already noted, we use the fact that the Z component of the right-hand side of Eqn. (8.16) is constant, i.e.,

$$J'(\dot\theta\cos\phi + \dot\psi\sin\theta\sin\phi)\sin\phi\sin\theta + J'(-\dot\theta\sin\phi + \dot\psi\sin\theta\cos\phi)\cos\phi\sin\theta + J\alpha\cos\theta = \text{constant},$$

which on simplifying leads us to the relation

$$J'\dot\psi\sin^2\theta + J\alpha\cos\theta = \gamma, \tag{8.50}$$

where γ is a constant.

Equations (8.46), (8.49) and (8.50) are the integrated forms of the governing differential equations; the constants α, β, γ are determined from the initial conditions.

Example 7: Force on a bar

Consider the stationary rigid bar shown in Fig. 8.7 that, at time $t = 0$, is subjected to a point force $F\boldsymbol{e}_y$ at $x = L/2$, and a uniform friction force given by $-\mu Mg\boldsymbol{e}_y$ so that $\int \boldsymbol{t}\,dS = (F - \mu Mg)\boldsymbol{e}_y$, where μ is the coefficient of friction, M is the mass and g the gravitational acceleration. Approximating the bar in its initial position (as shown in the figure) as a one-dimensional body with $\boldsymbol{x} = x\boldsymbol{e}_x$ and $\rho\,dV = M\,dx/L$, we are interested in finding the acceleration at the point of application of the force F at time $t = 0$.

Using Eqns. (8.17) and (8.22) and the fact that $\omega|_{t=0} = 0$, we have

$$M\ddot{y} = F - \mu Mg,$$

$$\frac{ML^2}{12}\dot\omega|_{t=0} = \frac{1}{2}FL,$$

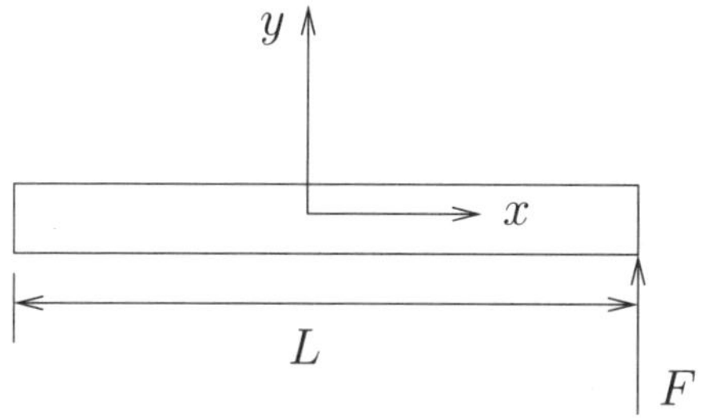

Fig. 8.7 *Bar subjected to point load F and total frictional force μMg.*

from which we get

$$\ddot{y}|_{t=0} = \frac{F-\mu Mg}{M},$$
$$\dot{\omega}|_{t=0} = \frac{6F}{ML}.$$

Using Eqn. (8.5), we get at time $t=0$, $a_x|_{x=L/2}=0$, and

$$a_y\big|_{x=L/2} = \frac{4F}{M} - \mu g.$$

Example 8: Rotating bar

This example shows that "energy methods" should be used with some caution, since even if $\int_{S(t)} \boldsymbol{t}\cdot\boldsymbol{v}\,dS$ is zero under steady-state conditions (but not zero at all times), the total energy would not be conserved. Consider a bar of length L and mass per unit length of m_0 hinged to a spinning disk of radius R as shown in Fig. 8.8. The local axes $\{\boldsymbol{e}_i\}$ can be expressed in terms of the stationary axes $\{\boldsymbol{e}_i^*\}$ as

$$\begin{aligned}
\boldsymbol{e}_1 &= \cos\omega t\boldsymbol{e}_1^* - \sin\omega t\boldsymbol{e}_2^*,\\
\boldsymbol{e}_2 &= \cos\theta(\sin\omega t\boldsymbol{e}_1^* + \cos\omega t\boldsymbol{e}_2^*) - \sin\theta\boldsymbol{e}_3^*,\\
\boldsymbol{e}_3 &= \sin\theta(\sin\omega t\boldsymbol{e}_1^* + \cos\omega t\boldsymbol{e}_2^*) + \cos\theta\boldsymbol{e}_3^*.
\end{aligned} \tag{8.51}$$

In order to find the angle θ made by the bar under steady-state conditions, we apply the balance of angular momentum to the bar:

$$\frac{d}{dt}\int_{V(t)} \boldsymbol{x}\times(\rho\boldsymbol{v})\,dV = \int_{V(t)} \boldsymbol{x}\times(\rho\boldsymbol{b})\,dV.$$

Using the fact that $\boldsymbol{x} = x\boldsymbol{e}_3$, $\boldsymbol{v} = (R + x\sin\theta)\omega\boldsymbol{e}_1$, $\rho\,dV = m_0\,dx$, $\boldsymbol{b} = g\boldsymbol{e}_3^*$, and using Eqns. (8.51), we get after simplification

$$\omega^2\left(R + \frac{2L}{3}\sin\theta\right)\cos\theta = g\sin\theta,$$

which is the governing equation for determining θ. In the limit as $\omega\to 0$, $\theta\to 0$, and in the limit as $\omega\to\infty$, $\tan\theta\to\infty$, or $\theta\to\pi/2$, which agrees with intuition.

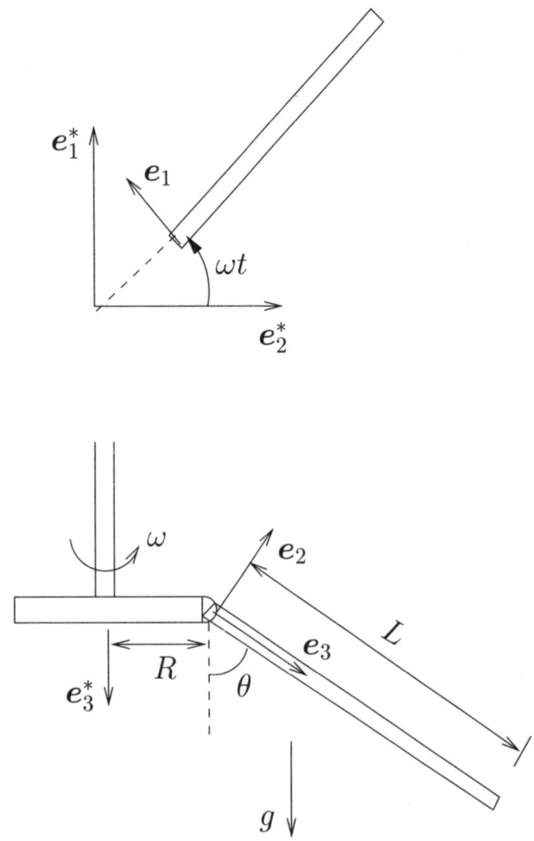

Fig. 8.8 *Rotating bar.*

By the balance of linear momentum, we get the reaction at the pin as

$$f = -m_0 L\ddot{x} + m_0 g L e_3^* = \frac{1}{2} m_0 L^2 \omega^2 \sin\theta \left[\sin\omega t e_1^* + \cos\omega t e_2^*\right] + m_0 g L e_3^*.$$

Note that each of the individual terms in Eqn. (3.65) is zero. Note also that the total energy (sum of kinetic and potential energies) in the steady state is higher compared to the energy in the initial state (assuming that the bar hangs vertically downward at time $t = 0$). Thus, although $\int_{S(t)} \boldsymbol{t} \cdot \boldsymbol{v}$ is zero at steady state, the integral of this term with respect to time from time $t = 0$ to the time that it reaches steady state provides this increase in energy.

Example 9: Slider-crank mechanism

Consider the slider-crank mechanism shown in Fig. 8.9, which is initially at rest. At $t = 0$, the bottom end of the connecting rod is given an acceleration $a_0 e_x$. Assuming the connecting rod to be rigid and massless, and the disc to be rigid with density ρ, and of unit width, we want to find the force exerted by the connecting rod on the disc at the pin joint 1 at the instant the acceleration $a_0 e_x$ is applied.

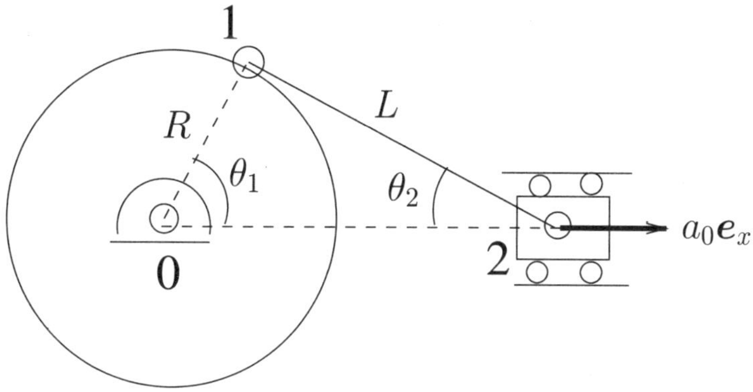

Fig. 8.9 *Slider-crank mechanism.*

Let $\dot{\omega}_1$ and $\dot{\omega}_2$ be the angular accelerations of the disk and the connecting rod. By applying the relation $\boldsymbol{a}_2 = \boldsymbol{a}_1 + \dot{\boldsymbol{\omega}} \times (\boldsymbol{x}_2 - \boldsymbol{x}_1) + \boldsymbol{\omega} \times [\boldsymbol{\omega} \times (\boldsymbol{x}_2 - \boldsymbol{x}_1)]$, between 0 and 1, we get $\boldsymbol{a}_1 = \dot{\omega}_1 R \left[\cos\theta_1 \boldsymbol{e}_y - \sin\theta_1 \boldsymbol{e}_x\right]$. Then applying it between 1 and 2, we get $a_0\boldsymbol{e}_x = \left[-\dot{\omega}_1 R \sin\theta_1 + \dot{\omega}_2 L \sin\theta_2\right]\boldsymbol{e}_x + \left[\dot{\omega}_1 R \cos\theta_1 + \dot{\omega}_2 L \cos\theta_2\right]\boldsymbol{e}_y$, from which it follows that

$$\dot{\omega}_1 = -\frac{a_0\cos\theta_2}{R\sin(\theta_1+\theta_2)},$$
$$\dot{\omega}_2 = \frac{a_0\cos\theta_1}{L\sin(\theta_1+\theta_2)}.$$

Applying the balance of angular momentum about 0 for the crank 0–1, we get

$$2\pi\int_0^R [\rho(r\boldsymbol{e}_r)\times(\dot{\omega}_1 r)\boldsymbol{e}_\theta]\, r\, dr = (R\cos\theta_1\boldsymbol{e}_x + R\sin\theta_1\boldsymbol{e}_y)\times(F_x\boldsymbol{e}_x + F_y\boldsymbol{e}_y),$$

where F_x and F_y are the forces acting on the pin due to the action of the connecting rod. The above equation reduces to

$$\frac{1}{2}\pi\rho\dot{\omega}_1 R^4 = (R\cos\theta_1)F_y - (R\sin\theta_1)F_x.$$

Applying the angular momentum balance to the connecting rod about point 2, we get

$$(L\sin\theta_2)F_x + (L\cos\theta_2)F_y = 0.$$

Solving for F_x and F_y, we get

$$F_x = \frac{a_0\pi R^2\rho\cos^2\theta_2}{2\sin^2(\theta_1+\theta_2)},$$
$$F_y = -\frac{a_0\pi R^2\rho\sin\theta_2\cos\theta_2}{2\sin^2(\theta_1+\theta_2)}.$$

EXERCISES

1. A rod of length l and mass per unit length m, which is pinned at one end, oscillates about the x^* axis as shown in Fig. 8.10. Using the transformation law for the body force field given by Eqn. (4.18) with $Q_{ij} = e_i^* \cdot e_j$, find the body force field in the x–y–z frame. Approximating the rod as a one-dimensional rigid body ($\rho dV \equiv m\,dx$), and applying the principles of linear and angular momenta in integral form in the x–y–z frame, show that the governing equation for the angle θ is

$$\ddot{\theta} + \frac{3}{2}\frac{g \sin\theta}{l} = 0,$$

which on multiplying with $\dot{\theta}$ and integrating yields

$$\dot{\theta}^2 - \frac{3g\cos\theta}{l} = c_1, \tag{8.52}$$

where c_1 is a constant. Verify Eqn. (8.52) using Eqns. (8.30) and (8.32). Assuming that the initial conditions are such that the pendulum does not cross the $\theta = \pi$ position, we have $\dot{\theta} = 0$ at $\theta = \theta_{\text{max}}$, so that $c_1 = -3g\cos\theta_{\text{max}}/l$. Thus,

$$\sqrt{\frac{3g}{l}}t = \int_0^{\theta} \frac{d\hat{\theta}}{\sqrt{\cos\hat{\theta} - \cos\theta_{\text{max}}}}.$$

Carrying out the change of variable $\sin(\theta/2) = k\sin\phi$, where $k = \sin(\theta_{\text{max}}/2)$, we get

$$\sqrt{\frac{3g}{2l}}t = \int_0^{\phi} \frac{d\hat{\phi}}{\sqrt{1 - k^2\sin^2\hat{\phi}}} = \mathcal{F}(\phi, k), \tag{8.53}$$

where $\mathcal{F}(\phi, k)$ is the elliptic integral of the first kind. Since when $\theta = \theta_{\text{max}}$, we have $\phi = \pi/2$, the time period T is given by

$$\sqrt{\frac{3g}{2l}}T = 4\mathcal{K}(k),$$

where $\mathcal{K}(.)$ is the complete elliptic integral of the first kind.

Since $\theta = 2\sin^{-1}(k\sin\phi)$, using the definition of the elliptic sine function sn(.,.) and Eqn. (8.53), we get the desired solution as

$$\theta(t) = 2\sin^{-1}\left[k\,\text{sn}\left(\sqrt{\frac{3g}{2l}}t, k\right)\right].$$

If the rod is given an initial angular speed $\dot{\theta}(0)$ at the position $\theta = 0$, then from Eqn. (8.52)

$$\dot{\theta}(0) = \sqrt{\frac{6g}{l}}k,$$

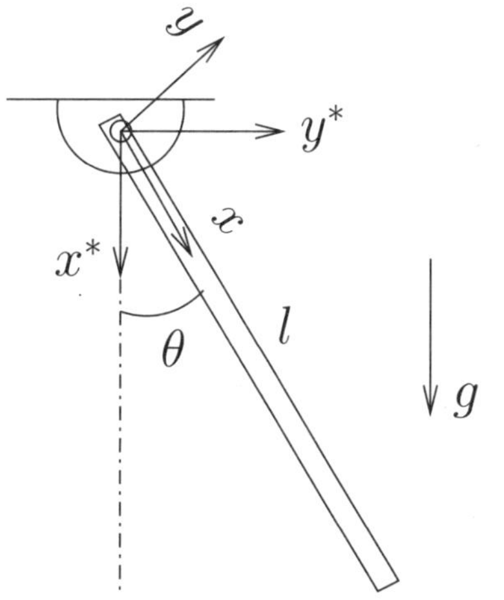

Fig. 8.10 *Oscillating rod.*

from which the maximum angle $\theta_{\max}$ can be determined. Alternatively, the above equation can be used to find the angular speed $\dot{\theta}$ at the position $\theta = 0$ if the rod is released with zero angular speed from a position $\theta = \theta_{\max}$.

Find the reactions exerted on the rod at the pinned-support with respect to the x–y–z axes. Linearize the governing equation for θ, and present an explicit solution to this linearized equation.

A Orthogonal Curvilinear Coordinate Systems

In this appendix, we present a new method[1] for deriving the component forms of various differential operators when referred to an orthogonal curvilinear coordinate system. Throughout this treatment, we use *physical components* of vectors or tensors, which are simply the components relative to a local Cartesian system aligned with the curvilinear system. Truesdell [334] recommends first deriving the equations in covariant form, and then converting them to physical coordinates. However, since this can be cumbersome, here we shall derive these equations directly.

Let $\{\boldsymbol{e}_i\}$ and $\{\boldsymbol{i}_i\}$ denote the (unit) basis vectors of the orthogonal curvilinear and Cartesian coordinate systems, respectively[2]. A vector $\boldsymbol{v}$ and tensor $\boldsymbol{T}$ referred to the curvilinear basis is written as $\boldsymbol{v} = v_i \boldsymbol{e}_i$ and $T_{ij}\boldsymbol{e}_i \otimes \boldsymbol{e}_j$, where v_i and T_{ij} are the physical components of $\boldsymbol{v}$ and $\boldsymbol{T}$, respectively. Since the $\{\boldsymbol{e}_i\}$ coordinate system is analogous to the $\{\bar{\boldsymbol{e}}_i\}$ coordinate system in Fig. 1.1, by Eqn. (1.95), we have $\boldsymbol{e}_i = \boldsymbol{Q}^T \boldsymbol{i}_i$, where $\boldsymbol{Q} = \boldsymbol{i}_1 \otimes \boldsymbol{e}_1 + \boldsymbol{i}_2 \otimes \boldsymbol{e}_2 + \boldsymbol{i}_3 \otimes \boldsymbol{e}_3$. If $[\boldsymbol{v}]_i$ and $[\boldsymbol{v}]_e$, and $[\boldsymbol{T}]_i$ and $[\boldsymbol{T}]_e$ denote the components of the vector $\boldsymbol{v}$ and tensor $\boldsymbol{T}$ with respect to $\{\boldsymbol{i}_i\}$ and $\{\boldsymbol{e}_i\}$, respectively, then by Eqns. (1.99) and (1.100), we have $[\boldsymbol{v}]_e = \boldsymbol{Q}[\boldsymbol{v}]_i$ and $[\boldsymbol{T}]_e = \boldsymbol{Q}[\boldsymbol{T}]_i\boldsymbol{Q}^T$. Thus, finding the components of a vector or tensor in the curvilinear system, knowing the components in a Cartesian system (or vice versa) is quite straightforward. However, when partial derivatives with respect to the curvilinear coordinates are involved, one needs to compute the spatial derivatives of the basis vectors themselves, since they change direction from point to point.

Let $\boldsymbol{x}$ represent the position vector of a point with Cartesian coordinates (x_1, x_2, x_3), and curvilinear coordinates (ξ_1, ξ_2, ξ_3). The position vector can be expressed in terms of the curvilinear coordinates as $\boldsymbol{x} = \hat{\boldsymbol{x}}(\xi_1, \xi_2, \xi_3)$. Its differential is given by

$$d\boldsymbol{x} = \sum_{i=1}^{3} \frac{\partial \hat{\boldsymbol{x}}}{\partial \xi_i} d\xi_i = \sum_{i=1}^{3} h_i \, d\xi_i \boldsymbol{e}_i,$$

where

$$h_i = \left| \frac{\partial \hat{\boldsymbol{x}}}{\partial \xi_i} \right|, \tag{A.1}$$

are known as the scale factors, and

$$\boldsymbol{e}_i = \frac{1}{h_i} \frac{\partial \hat{\boldsymbol{x}}}{\partial \xi_i}, \quad \text{(no sum on i)} \tag{A.2}$$

[1]Actually, Malvern ([207], page 645) does derive Eqn. (A.4) for cylindrical and spherical coordinate systems, but does not generalize the method to arbitrary orthogonal curvilinear coordinate systems–here we carry out this generalization.

[2]We shall see in Appendix H that one need not distinguish between covariant and contravariant basis vectors when the coordinate system in orthonormal, and hence one can simply denote the orthonormal basis vectors of the curvilinear system as $\{\boldsymbol{e}_i\}$.

are the orthogonal basis vectors of unit magnitude. The element of length, element of surface area on a surface $\xi_3 =$ constant, and element of volume are given by

$$ds^2 = \boldsymbol{dx} \cdot \boldsymbol{dx} = h_1^2(d\xi_1)^2 + h_2^2(d\xi_2)^2 + h_3^2(d\xi_3)^2,$$

$$dS = \left| \frac{\partial \hat{\boldsymbol{x}}}{\partial \xi_1} \times \frac{\partial \hat{\boldsymbol{x}}}{\partial \xi_2} \right| d\xi_1 d\xi_2 = h_1 h_2 \left| \boldsymbol{e}_1 \times \boldsymbol{e}_2 \right| d\xi_1 d\xi_2 = h_1 h_2 d\xi_1 d\xi_2,$$

$$dV = \left[\frac{\partial \hat{\boldsymbol{x}}}{\partial \xi_1}, \frac{\partial \hat{\boldsymbol{x}}}{\partial \xi_2}, \frac{\partial \hat{\boldsymbol{x}}}{\partial \xi_3} \right] d\xi_1 d\xi_2 d\xi_3 = h_1 h_2 h_3 \left[\boldsymbol{e}_1, \boldsymbol{e}_2, \boldsymbol{e}_3 \right] d\xi_1 d\xi_2 d\xi_3 = h_1 h_2 h_3 d\xi_1 d\xi_2 d\xi_3.$$

Differentiating the relation $\boldsymbol{e}_i = \boldsymbol{Q}^T \boldsymbol{i}_i$ with respect to ξ_j, we get

$$\frac{\partial \boldsymbol{e}_i}{\partial \xi_j} = \frac{\partial \boldsymbol{Q}^T}{\partial \xi_j} \boldsymbol{i}_i = \frac{\partial \boldsymbol{Q}^T}{\partial \xi_j} \boldsymbol{Q} \boldsymbol{e}_i. \tag{A.3}$$

Since $\boldsymbol{Q}^T \boldsymbol{Q} = \boldsymbol{I}$, we have

$$\frac{\partial \boldsymbol{Q}^T}{\partial \xi_j} \boldsymbol{Q} + \boldsymbol{Q}^T \frac{\partial \boldsymbol{Q}}{\partial \xi_j} = \boldsymbol{0},$$

so that

$$\left[\frac{\partial \boldsymbol{Q}^T}{\partial \xi_j} \boldsymbol{Q} \right]^T = \boldsymbol{Q}^T \frac{\partial \boldsymbol{Q}}{\partial \xi_j} = -\frac{\partial \boldsymbol{Q}^T}{\partial \xi_j} \boldsymbol{Q},$$

which implies that $\frac{\partial \boldsymbol{Q}^T}{\partial \xi_j} \boldsymbol{Q} \in \text{Skw}$. If $\boldsymbol{\omega}_j$ denotes its axial vector, then Eqn. (A.3) can be written as

$$\frac{\partial \boldsymbol{e}_i}{\partial \xi_j} = \boldsymbol{\omega}_j \times \boldsymbol{e}_i, \quad i, j = 1, 2, 3. \tag{A.4}$$

The above equation plays a key role in the subsequent derivations.

Expressions for the axial vectors in terms of the scale factors are as follows:

$$\begin{aligned} \boldsymbol{\omega}_1 &= \frac{1}{h_3} \frac{\partial h_1}{\partial \xi_3} \boldsymbol{e}_2 - \frac{1}{h_2} \frac{\partial h_1}{\partial \xi_2} \boldsymbol{e}_3, \\ \boldsymbol{\omega}_2 &= \frac{1}{h_1} \frac{\partial h_2}{\partial \xi_1} \boldsymbol{e}_3 - \frac{1}{h_3} \frac{\partial h_2}{\partial \xi_3} \boldsymbol{e}_1, \\ \boldsymbol{\omega}_3 &= \frac{1}{h_2} \frac{\partial h_3}{\partial \xi_2} \boldsymbol{e}_1 - \frac{1}{h_1} \frac{\partial h_3}{\partial \xi_1} \boldsymbol{e}_2. \end{aligned} \tag{A.5}$$

Note from the above equations that

$$\boldsymbol{\omega}_1 \cdot \boldsymbol{e}_1 = \boldsymbol{\omega}_2 \cdot \boldsymbol{e}_2 = \boldsymbol{\omega}_3 \cdot \boldsymbol{e}_3 = 0. \tag{A.6}$$

To prove Eqns. (A.5), we first prove that when $i \neq j \neq k \neq i$,

$$\boldsymbol{e}_k \cdot \frac{\partial^2 \hat{\boldsymbol{x}}}{\partial \xi_i \partial \xi_j} = 0. \tag{A.7}$$

Since we have assumed our curvilinear coordinate system to be orthogonal, $e_i \cdot e_j = 0$. Using Eqn. (A.2), we get

$$\frac{\partial \hat{x}}{\partial \xi_i} \cdot \frac{\partial \hat{x}}{\partial \xi_j} = 0.$$

Differentiating the above equation with respect to ξ_k, and again using Eqn. (A.2), we get

$$h_j e_j \cdot \frac{\partial^2 \hat{x}}{\partial \xi_i \partial \xi_k} + h_i e_i \cdot \frac{\partial^2 \hat{x}}{\partial \xi_j \partial \xi_k} = 0. \tag{A.8}$$

By considering $e_j \cdot e_k = 0$ and $e_i \cdot e_k = 0$, we get the following two additional equations similar to Eqn. (A.8):

$$h_k e_k \cdot \frac{\partial^2 \hat{x}}{\partial \xi_j \partial \xi_i} + h_j e_j \cdot \frac{\partial^2 \hat{x}}{\partial \xi_k \partial \xi_i} = 0, \tag{A.9}$$

$$h_i e_i \cdot \frac{\partial^2 \hat{x}}{\partial \xi_k \partial \xi_j} + h_k e_k \cdot \frac{\partial^2 \hat{x}}{\partial \xi_i \partial \xi_j} = 0. \tag{A.10}$$

Adding Eqns. (A.9) and (A.10), and subtracting Eqn. (A.8) from it yields Eqn. (A.7).

From Eqn. (A.2), we have

$$h_i \frac{\partial e_i}{\partial \xi_j} + \frac{\partial h_i}{\partial \xi_j} e_i = \frac{\partial^2 x}{\partial \xi_i \partial \xi_j} = \frac{\partial h_j}{\partial \xi_i} e_j + h_j \frac{\partial e_j}{\partial \xi_i}. \tag{A.11}$$

Since $e_i \cdot e_i = e_j \cdot e_j = 1$ (no sum on i, j),

$$e_i \cdot \frac{\partial e_i}{\partial \xi_j} = e_j \cdot \frac{\partial e_j}{\partial \xi_i} = 0.$$

In other words, $\partial e_i / \partial \xi_j$ is orthogonal to e_i, and $\partial e_j / \partial \xi_i$ is orthogonal to e_j (this is also evident from Eqn. (A.4)). From Eqn. (A.7), $\partial^2 x / \partial \xi_i \partial \xi_j$ lies in the plane spanned by e_i and e_j. Hence, from Eqn. (A.11), we get

$$\frac{\partial e_j}{\partial \xi_i} = \frac{1}{h_j} \frac{\partial h_i}{\partial \xi_j} e_i, \quad i \neq j,$$

$$\frac{\partial e_i}{\partial \xi_j} = \frac{1}{h_i} \frac{\partial h_j}{\partial \xi_i} e_j, \quad i \neq j. \tag{A.12}$$

Now we have

$$\begin{aligned}
\omega_1 &= (\omega_1 \cdot e_1) e_1 + (\omega_1 \cdot e_2) e_2 + (\omega_1 \cdot e_3) e_3 \\
&= [\omega_1 \cdot (e_2 \times e_3)]\, e_1 + [\omega_1 \cdot (e_3 \times e_1)]\, e_2 + [\omega_1 \cdot (e_1 \times e_2)]\, e_3 \\
&= [e_3 \cdot (\omega_1 \times e_2)]\, e_1 + [e_1 \cdot (\omega_1 \times e_3)]\, e_2 - [e_1 \cdot (\omega_1 \times e_2)]\, e_3 \qquad \text{(by Eqn. (1.14))} \\
&= \left[e_3 \cdot \frac{\partial e_2}{\partial \xi_1}\right] e_1 + \left[e_1 \cdot \frac{\partial e_3}{\partial \xi_1}\right] e_2 - \left[e_1 \cdot \frac{\partial e_2}{\partial \xi_1}\right] e_3 \qquad \text{(by Eqn. (A.4))}
\end{aligned}$$

$$= \frac{1}{h_3}\frac{\partial h_1}{\partial \xi_3}e_2 - \frac{1}{h_2}\frac{\partial h_1}{\partial \xi_2}e_3, \qquad \text{(by Eqn. (A.12))}$$

which proves the first relation in Eqns. (A.5). The other two relations are proved similarly.

Let $\xi = \hat{\xi}(x) = \hat{\xi}(\chi(X,t)) \equiv \tilde{\xi}(X,t)$. Then using Eqns. (A.2) and (A.4), the expressions for the velocity and acceleration are given by

$$\begin{aligned}
\tilde{v} &= \left(\frac{\partial \chi}{\partial t}\right)_X = \sum_{i=1}^{3} \frac{\partial \hat{x}}{\partial \xi_i}\left(\frac{\partial \tilde{\xi}_i}{\partial t}\right)_X = \sum_{i=1}^{3}\left(\frac{\partial \tilde{\xi}_i}{\partial t}\right)_X h_i e_i,\\
\tilde{a} &= \left(\frac{\partial^2 \chi}{\partial t^2}\right)_X = \sum_{i=1}^{3}\left(\frac{\partial^2 \tilde{\xi}_i}{\partial t^2}\right)_X h_i e_i + \sum_{i=1}^{3}\sum_{j=1}^{3}\left(\frac{\partial \tilde{\xi}_i}{\partial t}\right)_X \left(\frac{\partial \tilde{\xi}_j}{\partial t}\right)_X \left[\frac{\partial h_i}{\partial \xi_j}e_i + h_i \omega_j \times e_i\right].
\end{aligned} \qquad (A.13)$$

The gradient of a scalar function ϕ is given by

$$(\nabla\phi, e_i) = \frac{1}{h_i}\frac{\partial \phi}{\partial x_m}\frac{\partial \hat{x}_m}{\partial \xi_i} = \frac{1}{h_i}\frac{\partial \phi}{\partial \xi_i}, \quad \text{no sum on } i$$

so that we can write

$$\nabla\phi = \sum_{i=1}^{3}\frac{e_i}{h_i}\frac{\partial \phi}{\partial \xi_i}. \qquad (A.14)$$

Similarly,

$$(\nabla v)e_j = \frac{1}{h_j}\frac{\partial v}{\partial x_m}\frac{\partial \hat{x}_m}{\partial \xi_j} = \frac{1}{h_j}\frac{\partial v}{\partial \xi_j}, \quad \text{no sum on } j. \qquad (A.15)$$

Using $v = v_i e_i$ and Eqn. (A.4), we have

$$\frac{\partial v}{\partial \xi_j} = \frac{\partial v_i}{\partial \xi_j}e_i + v_i(\omega_j \times e_i).$$

Substituting this relation into Eqn. (A.15), we get

$$\nabla v = \sum_{i=1}^{3}\sum_{j=1}^{3}\frac{1}{h_j}\left[\frac{\partial v_i}{\partial \xi_j}e_i \otimes e_j + v_i(\omega_j \times e_i)\otimes e_j\right], \qquad (A.16)$$

with its components with respect to the curvilinear system given by

$$(\nabla v)_{ij} = \frac{1}{h_j}\frac{\partial v_i}{\partial \xi_j} + \sum_{k=1}^{3} v_k \hat{\omega}_j \cdot (e_k \times e_i), \quad \text{(no sum on } j) \qquad (A.17)$$

where $\hat{\omega}_1 \equiv \omega_1/h_1$, $\hat{\omega}_2 \equiv \omega_2/h_2$, and $\hat{\omega}_3 \equiv \omega_3/h_3$. In matrix form (on using Eqn. (A.6))

$$[\nabla v] = \begin{bmatrix}
\frac{1}{h_1}\frac{\partial v_1}{\partial \xi_1} + \hat{\omega}_1 \cdot (v_3 e_2 - v_2 e_3) & \frac{1}{h_2}\frac{\partial v_1}{\partial \xi_2} - v_2\hat{\omega}_2 \cdot e_3 & \frac{1}{h_3}\frac{\partial v_1}{\partial \xi_3} + v_3\hat{\omega}_3 \cdot e_2 \\
\frac{1}{h_1}\frac{\partial v_2}{\partial \xi_1} + v_1\hat{\omega}_1 \cdot e_3 & \frac{1}{h_2}\frac{\partial v_2}{\partial \xi_2} + \hat{\omega}_2 \cdot (v_1 e_3 - v_3 e_1) & \frac{1}{h_3}\frac{\partial v_2}{\partial \xi_3} - v_3\hat{\omega}_3 \cdot e_1 \\
\frac{1}{h_1}\frac{\partial v_3}{\partial \xi_1} - v_1\hat{\omega}_1 \cdot e_2 & \frac{1}{h_2}\frac{\partial v_3}{\partial \xi_2} + v_2\hat{\omega}_2 \cdot e_1 & \frac{1}{h_3}\frac{\partial v_3}{\partial \xi_3} + \hat{\omega}_3 \cdot (v_2 e_1 - v_1 e_2)
\end{bmatrix}. \qquad (A.18)$$

The vector $(\boldsymbol{\nabla v})\boldsymbol{v}$ is obtained simply by multiplying the above matrix with the vector $\boldsymbol{v}$. $\boldsymbol{\nabla}\times\boldsymbol{v}$, which is the axial vector of $\boldsymbol{\nabla v}-(\boldsymbol{\nabla v})^T$, i.e., $\boldsymbol{\nabla}\times\boldsymbol{v}=\epsilon_{ijk}(\boldsymbol{\nabla v})_{kj}\boldsymbol{e}_i$, is given by

$$\begin{aligned}(\boldsymbol{\nabla}\times\boldsymbol{v})_1&=\frac{1}{h_2h_3}\left[\frac{\partial(h_3v_3)}{\partial\xi_2}-\frac{\partial(h_2v_2)}{\partial\xi_3}\right],\\(\boldsymbol{\nabla}\times\boldsymbol{v})_2&=\frac{1}{h_3h_1}\left[\frac{\partial(h_1v_1)}{\partial\xi_3}-\frac{\partial(h_3v_3)}{\partial\xi_1}\right],\\(\boldsymbol{\nabla}\times\boldsymbol{v})_3&=\frac{1}{h_1h_2}\left[\frac{\partial(h_2v_2)}{\partial\xi_1}-\frac{\partial(h_1v_1)}{\partial\xi_2}\right].\end{aligned}\tag{A.19}$$

Taking the trace of both sides of Eqn. (A.16), we obtain

$$\boldsymbol{\nabla}\cdot\boldsymbol{v}=\sum_{i=1}^{3}\left[\frac{1}{h_i}\frac{\partial v_i}{\partial\xi_i}+\sum_{j=1}^{3}v_i\hat{\boldsymbol{\omega}}_j\cdot(\boldsymbol{e}_i\times\boldsymbol{e}_j)\right]\tag{A.20}$$

$$=\frac{1}{h_1h_2h_3}\left[\frac{\partial}{\partial\xi_1}(h_2h_3v_1)+\frac{\partial}{\partial\xi_2}(h_3h_1v_2)+\frac{\partial}{\partial\xi_3}(h_1h_2v_3)\right].\tag{A.21}$$

Letting $\boldsymbol{v}=\boldsymbol{\nabla}\phi$ in the above expression, and using Eqn. (A.14), we get

$$\nabla^2\phi=\frac{1}{h_1h_2h_3}\left[\frac{\partial}{\partial\xi_1}\left(\frac{h_2h_3}{h_1}\frac{\partial\phi}{\partial\xi_1}\right)+\frac{\partial}{\partial\xi_2}\left(\frac{h_3h_1}{h_2}\frac{\partial\phi}{\partial\xi_2}\right)+\frac{\partial}{\partial\xi_3}\left(\frac{h_1h_2}{h_3}\frac{\partial\phi}{\partial\xi_3}\right)\right].\tag{A.22}$$

On the other hand, letting $\phi=\boldsymbol{\nabla}\cdot\boldsymbol{v}$ in Eqn. (A.14), we get an expression for $\boldsymbol{\nabla}(\boldsymbol{\nabla}\cdot\boldsymbol{v})$.

Similar to the above development, we can get an expression for $\boldsymbol{\nabla T}$ as follows:

$$\begin{aligned}(\boldsymbol{\nabla T})\boldsymbol{e}_k&=\frac{1}{h_k}\frac{\partial\boldsymbol{T}}{\partial\xi_k}\quad\text{(no sum on }k\text{)}\\&=\frac{1}{h_k}\frac{\partial}{\partial\xi_k}(T_{ij}\boldsymbol{e}_i\otimes\boldsymbol{e}_j)\\&=\frac{1}{h_k}\left[\frac{\partial T_{ij}}{\partial\xi_k}\boldsymbol{e}_i\otimes\boldsymbol{e}_j+T_{ij}(\boldsymbol{\omega}_k\times\boldsymbol{e}_i)\otimes\boldsymbol{e}_j+T_{ij}\boldsymbol{e}_i\otimes(\boldsymbol{\omega}_k\times\boldsymbol{e}_j)\right],\end{aligned}$$

so that

$$\boldsymbol{\nabla T}=\sum_{i=1}^{3}\sum_{j=1}^{3}\sum_{k=1}^{3}\frac{1}{h_k}\left[\frac{\partial T_{ij}}{\partial\xi_k}\boldsymbol{e}_i\otimes\boldsymbol{e}_j\otimes\boldsymbol{e}_k+T_{ij}(\boldsymbol{\omega}_k\times\boldsymbol{e}_i)\otimes\boldsymbol{e}_j\otimes\boldsymbol{e}_k+T_{ij}\boldsymbol{e}_i\otimes(\boldsymbol{\omega}_k\times\boldsymbol{e}_j)\otimes\boldsymbol{e}_k\right].$$

To find an expression for $\boldsymbol{\nabla}\cdot\boldsymbol{T}$, we use Eqn. (1.263). Since $\boldsymbol{T}^T\boldsymbol{u}=T_{ij}(\boldsymbol{e}_j\otimes\boldsymbol{e}_i)\boldsymbol{u}=T_{ij}(\boldsymbol{u}\cdot\boldsymbol{e}_i)\boldsymbol{e}_j$, we can choose $\boldsymbol{v}=v_j\boldsymbol{e}_j\equiv\boldsymbol{T}^T\boldsymbol{u}$, with $v_j=T_{ij}(\boldsymbol{u}\cdot\boldsymbol{e}_i)$, in Eqn. (A.20). Using the fact that $\boldsymbol{u}$ is a *constant* vector, we get

$$\boldsymbol{\nabla}\cdot\boldsymbol{T}=\sum_{i=1}^{3}\sum_{j=1}^{3}\left[\frac{1}{h_j}\frac{\partial T_{ij}}{\partial\xi_j}\boldsymbol{e}_i+T_{ij}(\hat{\boldsymbol{\omega}}_j\times\boldsymbol{e}_i)+\sum_{k=1}^{3}T_{ij}[\hat{\boldsymbol{\omega}}_k\cdot(\boldsymbol{e}_j\times\boldsymbol{e}_k)]\boldsymbol{e}_i\right],$$

$$(\boldsymbol{\nabla}\cdot\boldsymbol{T})_i=\sum_{j=1}^{3}\left[\frac{1}{h_j}\frac{\partial T_{ij}}{\partial\xi_j}+\sum_{k=1}^{3}\hat{\boldsymbol{\omega}}_k\cdot(T_{jk}\boldsymbol{e}_j\times\boldsymbol{e}_i+T_{ij}\boldsymbol{e}_j\times\boldsymbol{e}_k)\right].\tag{A.23}$$

In expanded form

$$
\begin{aligned}
(\nabla\cdot T)_1 &= \sum_{j=1}^{3}\frac{1}{h_j}\frac{\partial T_{1j}}{\partial\xi_j} + (T_{13}+T_{31})\hat{\omega}_1\cdot e_2 - (T_{12}+T_{21})\hat{\omega}_1\cdot e_3 \\
&\quad - T_{13}\hat{\omega}_2\cdot e_1 + (T_{11}-T_{22})\hat{\omega}_2\cdot e_3 \\
&\quad + T_{12}\hat{\omega}_3\cdot e_1 + (T_{33}-T_{11})\hat{\omega}_3\cdot e_2, \\
(\nabla\cdot T)_2 &= \sum_{j=1}^{3}\frac{1}{h_j}\frac{\partial T_{2j}}{\partial\xi_j} + T_{23}\hat{\omega}_1\cdot e_2 + (T_{11}-T_{22})\hat{\omega}_1\cdot e_3 \\
&\quad - (T_{23}+T_{32})\hat{\omega}_2\cdot e_1 + (T_{12}+T_{21})\hat{\omega}_2\cdot e_3 \\
&\quad + (T_{22}-T_{33})\hat{\omega}_3\cdot e_1 - T_{21}\hat{\omega}_3\cdot e_2, \\
(\nabla\cdot T)_3 &= \sum_{j=1}^{3}\frac{1}{h_j}\frac{\partial T_{3j}}{\partial\xi_j} + (T_{33}-T_{11})\hat{\omega}_1\cdot e_2 - T_{32}\hat{\omega}_1\cdot e_3 \\
&\quad + (T_{22}-T_{33})\hat{\omega}_2\cdot e_1 + T_{31}\hat{\omega}_2\cdot e_3 \\
&\quad + (T_{23}+T_{32})\hat{\omega}_3\cdot e_1 - (T_{13}+T_{31})\hat{\omega}_3\cdot e_2.
\end{aligned}
\tag{A.24}
$$

Letting $T_{ij} \equiv (\nabla v)_{ij}$ in the above equations, we get expressions for $(\nabla^2 v)_i$.

Let $v \equiv T^T u$, where u is a constant vector, in the expression $\nabla\times v = \epsilon_{imn}(\nabla v)_{mn}e_i$. Then using Eqn. (A.17) for obtaining $(\nabla v)_{mn}$, substituting into Eqn. (1.265), and using the fact that u is a constant vector, $\nabla\times T$ and its components are obtained as

$$
\nabla\times T = \sum_{i=1}^{3}\sum_{j=1}^{3}\sum_{m=1}^{3}\sum_{n=1}^{3}\epsilon_{imn}\left[\frac{1}{h_m}\frac{\partial T_{jn}}{\partial\xi_m}e_i\otimes e_j + T_{jn}e_i\otimes(\hat{\omega}_m\times e_j) + \sum_{k=1}^{3}T_{jk}[\hat{\omega}_m\cdot(e_k\times e_n)]e_i\otimes e_j\right], \tag{A.25}
$$

$$
(\nabla\times T)_{ij} = \sum_{m=1}^{3}\sum_{n=1}^{3}\epsilon_{imn}\left\{\frac{1}{h_m}\frac{\partial T_{jn}}{\partial\xi_m} + \hat{\omega}_m\cdot\left[\sum_{k=1}^{3}T_{kn}(e_k\times e_j) + T_{jk}(e_k\times e_n)\right]\right\}. \tag{A.26}
$$

Note from the above results that

$$
\begin{aligned}
\nabla\cdot v &= \sum_{i=1}^{3}\frac{1}{h_i}\left(\frac{\partial v}{\partial\xi_i}\right)_i, \\
(\nabla\times v)_i &= \sum_{j=1}^{3}\sum_{k=1}^{3}\epsilon_{ijk}\frac{1}{h_j}\left(\frac{\partial v}{\partial\xi_j}\right)_k, \\
(\nabla\cdot T)_i &= \sum_{j=1}^{3}\frac{1}{h_j}\left(\frac{\partial T}{\partial\xi_j}\right)_{ij}, \\
(\nabla\times T)_{ij} &= \sum_{m=1}^{3}\sum_{n=1}^{3}\epsilon_{imn}\frac{1}{h_m}\left(\frac{\partial T}{\partial\xi_m}\right)_{jn},
\end{aligned}
$$

where $\boldsymbol{v} = v_m \boldsymbol{e}_m$ and $\boldsymbol{T} = T_{rs}\boldsymbol{e}_r \otimes \boldsymbol{e}_s$.

Similarly, using Eqn. (1.268), we get

$$\boldsymbol{\nabla}^2 \boldsymbol{T} = \sum_{k=1}^{3} \frac{1}{h_k}\frac{\partial}{\partial \xi_k}\left[\frac{1}{h_k}\frac{\partial \boldsymbol{T}}{\partial \xi_k}\right],$$

where $\boldsymbol{T} = T_{ij}\boldsymbol{e}_i \otimes \boldsymbol{e}_j$.

Let $\boldsymbol{X}$ be the position vector of a point in the reference configuration with Cartesian coordinates (X_1, X_2, X_3) and curvilinear coordinates (η_1, η_2, η_3). Let $\boldsymbol{\xi} \equiv (\xi_1, \xi_2, \xi_3) = \chi(\eta_1, \eta_2, \eta_3, t)$ represent the mapping from curvilinear material to curvilinear spatial coordinates, and let $\boldsymbol{X} = \hat{\boldsymbol{X}}(\eta_1, \eta_2, \eta_3)$ and $\boldsymbol{x} = \hat{\chi}(\xi_1, \xi_2, \xi_3) = \hat{\chi} \circ \chi \circ \hat{\boldsymbol{X}}^{-1}(\boldsymbol{X}, t) \equiv \tilde{\chi}(\boldsymbol{X}, t)$. Then, analogous to Eqn. (A.2), we have

$$\boldsymbol{e}_I^0 = \frac{1}{h_I^0}\frac{\partial \hat{\boldsymbol{X}}}{\partial \eta_I}, \quad \boldsymbol{e}_i = \frac{1}{h_i}\frac{\partial \hat{\chi}}{\partial \xi_i} \quad I, i = 1, 2, 3, \quad \text{no sum on } I, i,$$

where capital letters as subscripts and a superscript '0' are used to distinguish the variables associated with the reference configuration from those in the current configuration. Thus,

$$\begin{aligned} \boldsymbol{F}\boldsymbol{e}_J^0 &= \frac{1}{h_J^0}\frac{\partial \tilde{\chi}}{\partial X_m}\frac{\partial \hat{X}_m}{\partial \eta_J} \\ &= \frac{1}{h_J^0}\frac{\partial \tilde{\chi}}{\partial \eta_J} \\ &= \sum_{i=1}^{3} \frac{1}{h_J^0}\frac{\partial \hat{\chi}}{\partial \xi_i}\frac{\partial \chi_i}{\partial \eta_J} \\ &= \sum_{i=1}^{3} \frac{h_i}{h_J^0}\frac{\partial \chi_i}{\partial \eta_J}\boldsymbol{e}_i, \end{aligned}$$

so that

$$F_{iJ} = \frac{h_i}{h_J^0}\frac{\partial \chi_i}{\partial \eta_J}, \quad \text{no sum on } i,\, J. \tag{A.27}$$

Once $\boldsymbol{F}$ is determined using the above relation, the various measures of strain, or other kinematical quantities can be determined using the relations in Chapter 2. The deformation gradient relates differential elements of length in the deformed and undeformed configurations:

$$(h\,d\xi)_i = F_{iJ}(h^0 d\eta)_J. \tag{A.28}$$

For a rigid angular motion expressed in cylindrical coordinates as $r = R, \theta = \Theta + \phi, z = Z$, we get, on using Eqn. (A.27), $\boldsymbol{F} = \boldsymbol{I}$. This is consistent with Eqn. (A.28) since $dr = dR$, $r d\theta = R d\Theta$ and $dz = dZ$.

Care has to be exercised in evaluating gradients of tensors of 'mixed' type, such as the first Piola–Kirchhoff stress $\boldsymbol{T}$ whose first and second indices are associated with the deformed and undeformed configurations. One cannot directly use Eqn. (A.23) for evaluating $\nabla \cdot \boldsymbol{T}$ since there the tensor is expressed with respect to one curvilinear basis[3]. We now derive such an expression (the derivation has been carried out in a slightly different way in [163]).

$$
\begin{aligned}
(\nabla \boldsymbol{T})\boldsymbol{e}_K^0 &= \frac{1}{h_K^0}\frac{\partial \boldsymbol{T}}{\partial \eta_K} \quad \text{(no sum on } K\text{)} \\
&= \sum_{i=1}^{3}\sum_{J=1}^{3}\frac{1}{h_K^0}\frac{\partial}{\partial \eta_K}(T_{iJ}\boldsymbol{e}_i \otimes \boldsymbol{e}_J^0) \\
&= \sum_{i=1}^{3}\sum_{J=1}^{3}\frac{1}{h_K^0}\left[\frac{\partial T_{iJ}}{\partial \eta_K}\boldsymbol{e}_i \otimes \boldsymbol{e}_J^0 + T_{iJ}\frac{\partial \chi_m}{\partial \eta_K}\frac{\partial \boldsymbol{e}_i}{\partial \xi_m} \otimes \boldsymbol{e}_J^0 + T_{iJ}\boldsymbol{e}_i \otimes (\boldsymbol{\omega}_K^0 \times \boldsymbol{e}_J^0)\right] \\
&= \sum_{i=1}^{3}\sum_{J=1}^{3}\frac{1}{h_K^0}\left[\frac{\partial T_{iJ}}{\partial \eta_K}\boldsymbol{e}_i \otimes \boldsymbol{e}_J^0\right] + T_{iJ}F_{mK}(\hat{\boldsymbol{\omega}}_m \times \boldsymbol{e}_i) \otimes \boldsymbol{e}_J^0 + T_{iJ}\boldsymbol{e}_i \otimes (\hat{\boldsymbol{\omega}}_K^0 \times \boldsymbol{e}_J^0),
\end{aligned}
$$

so that

$$
\nabla \boldsymbol{T} = \sum_{i=1}^{3}\sum_{J=1}^{3}\sum_{K=1}^{3}\frac{1}{h_K^0}\frac{\partial T_{iJ}}{\partial \eta_K}\boldsymbol{e}_i \otimes \boldsymbol{e}_J^0 \otimes \boldsymbol{e}_K^0 + T_{iJ}F_{mK}(\hat{\boldsymbol{\omega}}_m \times \boldsymbol{e}_i) \otimes \boldsymbol{e}_J^0 \otimes \boldsymbol{e}_K^0 + T_{iJ}\boldsymbol{e}_i \otimes (\hat{\boldsymbol{\omega}}_K^0 \times \boldsymbol{e}_J^0) \otimes \boldsymbol{e}_K^0.
$$

It follows that

$$
\nabla \cdot \boldsymbol{T} = \sum_{i=1}^{3}\left[\sum_{J=1}^{3}\frac{1}{h_J^0}\frac{\partial T_{iJ}}{\partial \eta_J}\boldsymbol{e}_i + \sum_{j=1}^{3}J\tau_{ij}(\hat{\boldsymbol{\omega}}_j \times \boldsymbol{e}_i) + \sum_{J=1}^{3}\sum_{K=1}^{3}T_{iJ}[\hat{\boldsymbol{\omega}}_K^0 \cdot (\boldsymbol{e}_J^0 \times \boldsymbol{e}_K^0)]\boldsymbol{e}_i\right],
$$

$$
(\nabla \cdot \boldsymbol{T})_i = \sum_{J=1}^{3}\left[\frac{1}{h_J^0}\frac{\partial T_{iJ}}{\partial \eta_J} + \sum_{K=1}^{3}T_{iJ}\hat{\boldsymbol{\omega}}_K^0 \cdot (\boldsymbol{e}_J^0 \times \boldsymbol{e}_K^0)\right] + \sum_{j=1}^{3}\sum_{k=1}^{3}J\tau_{jk}\hat{\boldsymbol{\omega}}_k \cdot (\boldsymbol{e}_j \times \boldsymbol{e}_i). \tag{A.29}
$$

In expanded form, with $J \equiv \det \boldsymbol{F}$,

$$
\begin{aligned}
(\nabla \cdot \boldsymbol{T})_1 &= \sum_{J=1}^{3}\frac{1}{h_J^0}\frac{\partial T_{1J}}{\partial \eta_J} + T_{11}\hat{\boldsymbol{\omega}}_2^0 \cdot \boldsymbol{e}_3^0 - T_{11}\hat{\boldsymbol{\omega}}_3^0 \cdot \boldsymbol{e}_2^0 \\
&\quad - T_{12}\hat{\boldsymbol{\omega}}_1^0 \cdot \boldsymbol{e}_3^0 + T_{12}\hat{\boldsymbol{\omega}}_3^0 \cdot \boldsymbol{e}_1^0 + T_{13}\hat{\boldsymbol{\omega}}_1^0 \cdot \boldsymbol{e}_2^0 - T_{13}\hat{\boldsymbol{\omega}}_2^0 \cdot \boldsymbol{e}_1^0 \\
&\quad - J\tau_{21}\hat{\boldsymbol{\omega}}_1 \cdot \boldsymbol{e}_3 - J\tau_{22}\hat{\boldsymbol{\omega}}_2 \cdot \boldsymbol{e}_3 + J\tau_{31}\hat{\boldsymbol{\omega}}_1 \cdot \boldsymbol{e}_2 + J\tau_{33}\hat{\boldsymbol{\omega}}_3 \cdot \boldsymbol{e}_2, \\
(\nabla \cdot \boldsymbol{T})_2 &= \sum_{J=1}^{3}\frac{1}{h_J^0}\frac{\partial T_{2J}}{\partial \eta_J} + T_{22}\hat{\boldsymbol{\omega}}_3^0 \cdot \boldsymbol{e}_1^0 - T_{22}\hat{\boldsymbol{\omega}}_1^0 \cdot \boldsymbol{e}_3^0 \\
&\quad - T_{23}\hat{\boldsymbol{\omega}}_2^0 \cdot \boldsymbol{e}_1^0 + T_{23}\hat{\boldsymbol{\omega}}_1^0 \cdot \boldsymbol{e}_2^0 + T_{21}\hat{\boldsymbol{\omega}}_2^0 \cdot \boldsymbol{e}_3^0 - T_{21}\hat{\boldsymbol{\omega}}_3^0 \cdot \boldsymbol{e}_2^0 \\
&\quad - J\tau_{32}\hat{\boldsymbol{\omega}}_2 \cdot \boldsymbol{e}_1 - J\tau_{33}\hat{\boldsymbol{\omega}}_3 \cdot \boldsymbol{e}_1 + J\tau_{12}\hat{\boldsymbol{\omega}}_2 \cdot \boldsymbol{e}_3 + J\tau_{11}\hat{\boldsymbol{\omega}}_1 \cdot \boldsymbol{e}_3, \\
(\nabla \cdot \boldsymbol{T})_3 &= \sum_{J=1}^{3}\frac{1}{h_J^0}\frac{\partial T_{3J}}{\partial \eta_J} + T_{33}\hat{\boldsymbol{\omega}}_1^0 \cdot \boldsymbol{e}_2^0 - T_{33}\hat{\boldsymbol{\omega}}_2^0 \cdot \boldsymbol{e}_1^0 \\
&\quad - T_{31}\hat{\boldsymbol{\omega}}_3^0 \cdot \boldsymbol{e}_2^0 + T_{31}\hat{\boldsymbol{\omega}}_2^0 \cdot \boldsymbol{e}_3^0 + T_{32}\hat{\boldsymbol{\omega}}_3^0 \cdot \boldsymbol{e}_1^0 - T_{32}\hat{\boldsymbol{\omega}}_1^0 \cdot \boldsymbol{e}_3^0 \\
&\quad - J\tau_{13}\hat{\boldsymbol{\omega}}_3 \cdot \boldsymbol{e}_2 - J\tau_{11}\hat{\boldsymbol{\omega}}_1 \cdot \boldsymbol{e}_2 + J\tau_{23}\hat{\boldsymbol{\omega}}_3 \cdot \boldsymbol{e}_1 + J\tau_{22}\hat{\boldsymbol{\omega}}_2 \cdot \boldsymbol{e}_1.
\end{aligned} \tag{A.30}
$$

[3]This is the reason that analytical solutions to nonlinear elasticity problems in curvilinear coordinates are generally worked out in terms of the Cauchy stress tensor (both of whose indices are associated with the deformed configuration), rather than in terms of the first Piola–Kirchhoff stress–the divergence $\nabla \cdot \boldsymbol{\tau}$ can be directly computed using Eqn. (A.23) with $\boldsymbol{\tau}$ in place of $\boldsymbol{T}$.

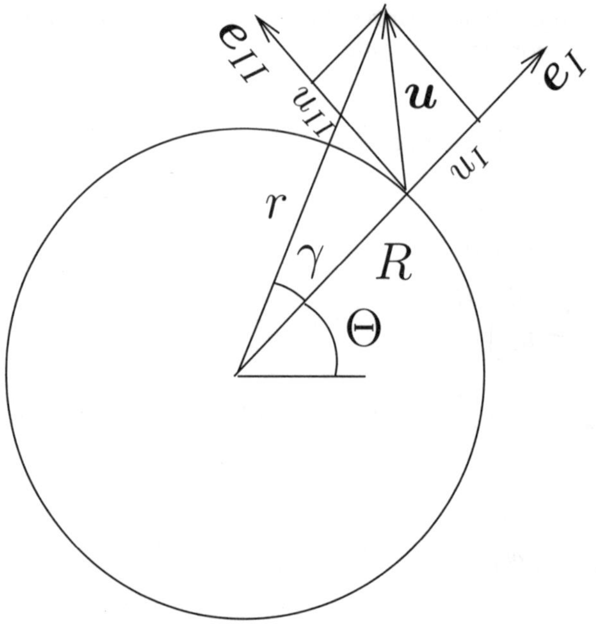

Fig. A.1 *Physical components of the displacement vector with respect to the undeformed configuration basis vectors.*

Note that the Kirchhoff stress tensor $J\boldsymbol{\tau}$ enters into the above equations, and also note the similarity of the above equation with the spatial counterpart given by Eqn. (A.24).

Physical displacements are defined as projections of the displacement vector on the basis vectors of the reference or deformed configuration, as shown by means of an example in Fig. A.1, where

$$u_I = r\cos\gamma - R,$$
$$u_{II} = r\sin\gamma.$$

Thus, even for a rigid angular motion with $r = R$, u_r is nonzero. In the linearized elasticity context, however, γ is constrained to be small, and then $u_r = 0$ and $u_\theta = R\gamma$, which also follows as a special case of Eqn. (B.9).[4] The physical components of velocity, acceleration, etc. are defined in a manner similar to that of the displacement. In general, we have

$$\boldsymbol{u} = \sum_{i=1}^{3} u_i \boldsymbol{e}_i = \sum_{I=1}^{3} u_I \boldsymbol{e}_I. \tag{A.31}$$

[4]Similarly, in the case of torsion of a circular cylinder, the u_r and u_θ displacements given by Eqn. (6.58) are approximations to the actual displacements $u_R = R[\cos(\alpha Z) - 1]$, $u_\Theta = R\sin(\alpha Z)$, when αZ is small.

B Cylindrical Coordinate System

The cylindrical coordinates are $\xi_1 = r$, $\xi_2 = \theta$, $\xi_3 = z$, and the position vector is given by

$$\boldsymbol{x} = (r\cos\theta, r\sin\theta, z).$$

Note that $\log(x+iy) = \log r + i\theta$, which is useful for finding (r,θ) given (x,y). Using Eqn. (A.1), the scale factors are found to be

$$h_1 = 1, \;\; h_2 = r, \;\; h_3 = 1. \tag{B.1}$$

From Eqn. (A.2), the basis vectors are $\boldsymbol{e}_r = (\cos\theta, \sin\theta, 0)$, $\boldsymbol{e}_\theta = (-\sin\theta, \cos\theta, 0)$, and $\boldsymbol{e}_z = (0,0,1)$. The rotation matrix that occurs in Eqns. (1.99), (1.100), etc. (i.e., $[\boldsymbol{u}]_{\text{cylindrical}} = \boldsymbol{Q}[\boldsymbol{u}]_{\text{Cartesian}}$, etc.) is

$$\boldsymbol{Q} = \begin{bmatrix} \boldsymbol{e}_r \\ \boldsymbol{e}_\theta \\ \boldsymbol{e}_z \end{bmatrix} = \begin{bmatrix} \cos\theta & \sin\theta & 0 \\ -\sin\theta & \cos\theta & 0 \\ 0 & 0 & 1 \end{bmatrix}. \tag{B.2}$$

The axial vectors $\boldsymbol{\omega}_j$ and the corresponding $\hat{\boldsymbol{\omega}}_j \equiv \boldsymbol{\omega}_j/h_j$ are obtained using Eqns. (A.5) as

$$\begin{aligned} \boldsymbol{\omega}_r &= \boldsymbol{0}, \quad \boldsymbol{\omega}_\theta = \boldsymbol{e}_z, \quad \boldsymbol{\omega}_z = \boldsymbol{0}, \\ \hat{\boldsymbol{\omega}}_r &= \boldsymbol{0}, \quad \hat{\boldsymbol{\omega}}_\theta = \frac{1}{r}\boldsymbol{e}_z, \quad \hat{\boldsymbol{\omega}}_z = \boldsymbol{0}. \end{aligned}$$

Let Φ, $\boldsymbol{v}$ (with physical components v_r, v_θ and v_z) and $\boldsymbol{T}$ (with physical components T_{rr}, $T_{r\theta}$, T_{rz} etc.) be scalar, vector and tensor-valued fields. Then

$$\boldsymbol{\nabla}\Phi = \frac{\partial\Phi}{\partial r}\boldsymbol{e}_r + \frac{1}{r}\frac{\partial\Phi}{\partial\theta}\boldsymbol{e}_\theta + \frac{\partial\Phi}{\partial z}\boldsymbol{e}_z, \tag{B.3}$$

$$\boldsymbol{\nabla}^2\Phi = \frac{1}{r}\frac{\partial}{\partial r}\left(r\frac{\partial\Phi}{\partial r}\right) + \frac{1}{r^2}\frac{\partial^2\Phi}{\partial\theta^2} + \frac{\partial^2\Phi}{\partial z^2}. \tag{B.4}$$

The components of $\boldsymbol{\nabla}\boldsymbol{v}$ and associated quantities are

$$[\boldsymbol{\nabla}\boldsymbol{v}] = \begin{bmatrix} \dfrac{\partial v_r}{\partial r} & \dfrac{1}{r}\left(\dfrac{\partial v_r}{\partial\theta} - v_\theta\right) & \dfrac{\partial v_r}{\partial z} \\ \dfrac{\partial v_\theta}{\partial r} & \dfrac{1}{r}\left(\dfrac{\partial v_\theta}{\partial\theta} + v_r\right) & \dfrac{\partial v_\theta}{\partial z} \\ \dfrac{\partial v_z}{\partial r} & \dfrac{1}{r}\dfrac{\partial v_z}{\partial\theta} & \dfrac{\partial v_z}{\partial z} \end{bmatrix}, \tag{B.5}$$

$$\nabla \times v = \left[\frac{1}{r}\frac{\partial v_z}{\partial \theta} - \frac{\partial v_\theta}{\partial z}\right] e_r + \left[\frac{\partial v_r}{\partial z} - \frac{\partial v_z}{\partial r}\right] e_\theta + \left[\frac{1}{r}\frac{\partial (r v_\theta)}{\partial r} - \frac{1}{r}\frac{\partial v_r}{\partial \theta}\right] e_z,$$
$$\nabla \cdot v = \frac{1}{r}\frac{\partial (r v_r)}{\partial r} + \frac{1}{r}\frac{\partial v_\theta}{\partial \theta} + \frac{\partial v_z}{\partial z}. \tag{B.6}$$

The components of $\nabla \cdot T$ are given by

$$\begin{aligned}
(\nabla \cdot T)_r &= \frac{\partial T_{rr}}{\partial r} + \frac{1}{r}\frac{\partial T_{r\theta}}{\partial \theta} + \frac{\partial T_{rz}}{\partial z} + \frac{T_{rr} - T_{\theta\theta}}{r}, \\
(\nabla \cdot T)_\theta &= \frac{\partial T_{\theta r}}{\partial r} + \frac{1}{r}\frac{\partial T_{\theta\theta}}{\partial \theta} + \frac{\partial T_{\theta z}}{\partial z} + \frac{T_{r\theta} + T_{\theta r}}{r}, \\
(\nabla \cdot T)_z &= \frac{\partial T_{zr}}{\partial r} + \frac{1}{r}\frac{\partial T_{z\theta}}{\partial \theta} + \frac{\partial T_{zz}}{\partial z} + \frac{T_{zr}}{r}.
\end{aligned} \tag{B.7}$$

Letting $T_{ij} \equiv (\nabla v)_{ij}$ in the above equations, we get expressions for $(\nabla^2 v)_i$ (see the bracketed terms multiplying μ in Eqns. (B.13) below).

The small strain tensor components obtained using the relations $\epsilon = [\nabla u + (\nabla u)^T]/2$ and Eqn. (B.5) are given by (as stated on page 333, to avoid the notation from becoming cumbersome, we shall use lower-case letters to denote the coordinates and the strain components, although, to be consistent with the notation used so far, we should use capital letters.)

$$\begin{aligned}
\epsilon_{rr} &= \frac{\partial u_r}{\partial r}, & \epsilon_{r\theta} &= \frac{1}{2}\left[\frac{1}{r}\frac{\partial u_r}{\partial \theta} + r\frac{\partial}{\partial r}\left(\frac{u_\theta}{r}\right)\right], \\
\epsilon_{\theta\theta} &= \frac{1}{r}\left(\frac{\partial u_\theta}{\partial \theta} + u_r\right), & \epsilon_{\theta z} &= \frac{1}{2}\left(\frac{\partial u_\theta}{\partial z} + \frac{1}{r}\frac{\partial u_z}{\partial \theta}\right), \\
\epsilon_{zz} &= \frac{\partial u_z}{\partial z}, & \epsilon_{rz} &= \frac{1}{2}\left(\frac{\partial u_r}{\partial z} + \frac{\partial u_z}{\partial r}\right).
\end{aligned} \tag{B.8}$$

From the above relations,

$$\operatorname{tr}\epsilon = \frac{1}{r}\frac{\partial (r u_r)}{\partial r} + \frac{1}{r}\frac{\partial u_\theta}{\partial \theta} + \frac{\partial u_z}{\partial z}.$$

The 'linearized rigid displacements' obtained by using $Q(Wx + c)$, where $W \in$ Skw, and x are the Cartesian coordinates, are

$$\begin{aligned}
u_r &= (b\cos\theta + c\sin\theta)z + c_1\cos\theta + c_2\sin\theta, \\
u_\theta &= ar + (c\cos\theta - b\sin\theta)z - c_1\sin\theta + c_2\cos\theta, \\
u_z &= -(b\cos\theta + c\sin\theta)r + c_3,
\end{aligned} \tag{B.9}$$

where a, b, c, c_1, c_2 and c_3 are constants.

The compatibility conditions, obtained by applying Eqn. (A.26) twice, are given by

$$
\begin{aligned}
&\frac{1}{r^2}\frac{\partial^2 \epsilon_{rr}}{\partial \theta^2} - \frac{1}{r}\frac{\partial \epsilon_{rr}}{\partial r} + \frac{1}{r}\frac{\partial^2 (r\epsilon_{\theta\theta})}{\partial r^2} - \frac{2}{r^2}\frac{\partial^2 (r\epsilon_{r\theta})}{\partial r \partial \theta} = 0,\\
&\frac{\partial^2 \epsilon_{\theta\theta}}{\partial z^2} + \frac{1}{r^2}\frac{\partial^2 \epsilon_{zz}}{\partial \theta^2} + \frac{1}{r}\frac{\partial \epsilon_{zz}}{\partial r} - \frac{2}{r}\frac{\partial^2 \epsilon_{\theta z}}{\partial \theta \partial z} - \frac{2}{r}\frac{\partial \epsilon_{rz}}{\partial z} = 0,\\
&\frac{\partial^2 \epsilon_{rr}}{\partial z^2} + \frac{\partial^2 \epsilon_{zz}}{\partial r^2} - 2\frac{\partial^2 \epsilon_{rz}}{\partial r \partial z} = 0,\\
&-\frac{\partial}{\partial r}\left[\frac{1}{r}\frac{\partial (r\epsilon_{\theta z})}{\partial r} - \frac{1}{r}\frac{\partial \epsilon_{rz}}{\partial \theta}\right] + \frac{\partial}{\partial z}\left[\frac{1}{r^2}\frac{\partial (r^2\epsilon_{r\theta})}{\partial r} - \frac{1}{r}\frac{\partial \epsilon_{rr}}{\partial \theta}\right] = 0,\\
&\frac{1}{r}\frac{\partial}{\partial \theta}\left[\frac{1}{r}\frac{\partial (r\epsilon_{\theta z})}{\partial r} - \frac{1}{r}\frac{\partial \epsilon_{rz}}{\partial \theta}\right] - \frac{\partial}{\partial z}\left[\frac{\partial \epsilon_{\theta\theta}}{\partial r} - \frac{1}{r}\frac{\partial \epsilon_{r\theta}}{\partial \theta} + \frac{\epsilon_{\theta\theta} - \epsilon_{rr}}{r}\right] = 0,\\
&-\frac{\partial^2}{\partial r \partial \theta}\left(\frac{\epsilon_{zz}}{r}\right) + \frac{\partial}{\partial z}\left[-\frac{\partial \epsilon_{r\theta}}{\partial z} + r\frac{\partial}{\partial r}\left(\frac{\epsilon_{\theta z}}{r}\right) + \frac{1}{r}\frac{\partial \epsilon_{rz}}{\partial \theta}\right] = 0.
\end{aligned}
\tag{B.10}
$$

By using the strain–displacement relations, it can be easily verified that the above relations are satisfied. In case the z-axis is part of the domain, the strain components have to satisfy the additional constraint given by Eqn. (6.32) provided ϵ_{rr} and $\epsilon_{\theta\theta}$ are independent of θ.

For the case of torsionless axisymmetry, the two independent compatibility relations from among Eqns. (B.10) are

$$
\begin{aligned}
\epsilon_{rr} &= \frac{\partial (r\epsilon_{\theta\theta})}{\partial r},\\
\frac{\partial^2 (r\epsilon_{\theta\theta})}{\partial z^2} + \frac{\partial \epsilon_{zz}}{\partial r} &= \frac{\partial \gamma_{rz}}{\partial z}.
\end{aligned}
\tag{B.11}
$$

The constitutive relations given by Eqns. (6.36) become

$$
\begin{aligned}
\tau_{rr} &= \lambda(\operatorname{tr}\epsilon) + 2\mu\epsilon_{rr}, & \tau_{r\theta} &= 2\mu\epsilon_{r\theta},\\
\tau_{\theta\theta} &= \lambda(\operatorname{tr}\epsilon) + 2\mu\epsilon_{\theta\theta}, & \tau_{\theta z} &= 2\mu\epsilon_{\theta z},\\
\tau_{zz} &= \lambda(\operatorname{tr}\epsilon) + 2\mu\epsilon_{zz}, & \tau_{rz} &= 2\mu\epsilon_{rz},
\end{aligned}
\tag{B.12}
$$

while the Navier equations of elasticity (Eqns. (6.37)) are given by

$$
\begin{aligned}
\rho_0\frac{\partial^2 u_r}{\partial t^2} &= (\lambda+\mu)\frac{\partial (\operatorname{tr}\epsilon)}{\partial r} + \mu\left[\frac{\partial}{\partial r}\left(\frac{1}{r}\frac{\partial (ru_r)}{\partial r}\right) + \frac{1}{r^2}\frac{\partial^2 u_r}{\partial \theta^2} + \frac{\partial^2 u_r}{\partial z^2} - \frac{2}{r^2}\frac{\partial u_\theta}{\partial \theta}\right] + \rho_0(\boldsymbol{b}_0)_r,\\
\rho_0\frac{\partial^2 u_\theta}{\partial t^2} &= (\lambda+\mu)\frac{1}{r}\frac{\partial (\operatorname{tr}\epsilon)}{\partial \theta} + \mu\left[\frac{\partial}{\partial r}\left(\frac{1}{r}\frac{\partial (ru_\theta)}{\partial r}\right) + \frac{1}{r^2}\frac{\partial^2 u_\theta}{\partial \theta^2} + \frac{\partial^2 u_\theta}{\partial z^2} + \frac{2}{r^2}\frac{\partial u_r}{\partial \theta}\right] + \rho_0(\boldsymbol{b}_0)_\theta,\\
\rho_0\frac{\partial^2 u_z}{\partial t^2} &= (\lambda+\mu)\frac{\partial (\operatorname{tr}\epsilon)}{\partial z} + \mu\left[\frac{1}{r}\frac{\partial}{\partial r}\left(r\frac{\partial u_z}{\partial r}\right) + \frac{1}{r^2}\frac{\partial^2 u_z}{\partial \theta^2} + \frac{\partial^2 u_z}{\partial z^2}\right] + \rho_0(\boldsymbol{b}_0)_z,
\end{aligned}
\tag{B.13}
$$

where the Lamé constants (λ,μ) are related to (E,ν) by means of Eqns. (5.20).

If $r=\tilde{r}(\boldsymbol{X},t)$, $\theta=\tilde{\theta}(\boldsymbol{X},t)$, and $z=\tilde{z}(\boldsymbol{X},t)$, using Eqns. (A.13) we have

$$\tilde{v}(\boldsymbol{X},t)=\dot{\tilde{r}}\boldsymbol{e}_r+\tilde{r}\dot{\tilde{\theta}}\boldsymbol{e}_\theta+\dot{\tilde{z}}\boldsymbol{e}_z,$$

$$\tilde{a}(\boldsymbol{X},t)=(\ddot{\tilde{r}}-\tilde{r}\dot{\tilde{\theta}}^2)\boldsymbol{e}_r+(\tilde{r}\ddot{\tilde{\theta}}+2\dot{\tilde{r}}\dot{\tilde{\theta}})\boldsymbol{e}_\theta+\ddot{\tilde{z}}\boldsymbol{e}_z. \tag{B.14}$$

Let $\boldsymbol{T}$ and $\boldsymbol{\tau}$ denote the first Piola–Kirchhoff and Cauchy stress tensors, and let $J\equiv\det\boldsymbol{F}$. If the material points are expressed in terms of Cartesian coordinates (X,Y,Z), and spatial points in terms of cylindrical coordinates (r,θ,z), then using the fact that $\hat{\omega}_1^0=\hat{\omega}_2^0=\hat{\omega}_3^0=\boldsymbol{0}$, and $\hat{\omega}_1=\boldsymbol{0}$, $\hat{\omega}_2=\boldsymbol{e}_z/r$, $\hat{\omega}_3=\boldsymbol{0}$, we get on using Eqn. (A.30):

$$\begin{aligned}
(\boldsymbol{\nabla}\cdot\boldsymbol{T})_r&=\frac{\partial T_{rX}}{\partial X}+\frac{\partial T_{rY}}{\partial Y}+\frac{\partial T_{rZ}}{\partial Z}-\frac{J\tau_{\theta\theta}}{r},\\
(\boldsymbol{\nabla}\cdot\boldsymbol{T})_\theta&=\frac{\partial T_{\theta X}}{\partial X}+\frac{\partial T_{\theta Y}}{\partial Y}+\frac{\partial T_{\theta Z}}{\partial Z}+\frac{J\tau_{r\theta}}{r},\\
(\boldsymbol{\nabla}\cdot\boldsymbol{T})_z&=\frac{\partial T_{zX}}{\partial X}+\frac{\partial T_{zY}}{\partial Y}+\frac{\partial T_{zZ}}{\partial Z}.
\end{aligned}\tag{B.15}$$

If the material points are expressed in terms of cylindrical coordinates (R,Θ,Z), and spatial points in terms of Cartesian coordinates (x,y,z), then using the fact that $\hat{\omega}_1^0=\boldsymbol{0}$, $\hat{\omega}_2^0=\boldsymbol{e}_Z^0/R$, $\hat{\omega}_3^0=\boldsymbol{0}$, and $\hat{\omega}_1=\hat{\omega}_2=\hat{\omega}_3=\boldsymbol{0}$, we get on using Eqn. (A.30):

$$\begin{aligned}
(\boldsymbol{\nabla}\cdot\boldsymbol{T})_x&=\frac{\partial T_{xR}}{\partial R}+\frac{1}{R}\frac{\partial T_{x\Theta}}{\partial\Theta}+\frac{\partial T_{xZ}}{\partial Z}+\frac{T_{xR}}{R},\\
(\boldsymbol{\nabla}\cdot\boldsymbol{T})_y&=\frac{\partial T_{yR}}{\partial R}+\frac{1}{R}\frac{\partial T_{y\Theta}}{\partial\Theta}+\frac{\partial T_{yZ}}{\partial Z}+\frac{T_{yR}}{R},\\
(\boldsymbol{\nabla}\cdot\boldsymbol{T})_z&=\frac{\partial T_{zR}}{\partial R}+\frac{1}{R}\frac{\partial T_{z\Theta}}{\partial\Theta}+\frac{\partial T_{zZ}}{\partial Z}+\frac{T_{zR}}{R}.
\end{aligned}\tag{B.16}$$

If the material and spatial points are expressed in terms of cylindrical coordinates (R,Θ,Z) and (r,θ,z), respectively, then using the fact that $\hat{\omega}_1^0=\boldsymbol{0}$, $\hat{\omega}_2^0=\boldsymbol{e}_Z^0/R$, $\hat{\omega}_3^0=\boldsymbol{0}$, and $\hat{\omega}_1=\boldsymbol{0}$, $\hat{\omega}_2=\boldsymbol{e}_z/r$, $\hat{\omega}_3=\boldsymbol{0}$, we get on using Eqn. (A.30):

$$\begin{aligned}
(\boldsymbol{\nabla}\cdot\boldsymbol{T})_r&=\frac{\partial T_{rR}}{\partial R}+\frac{1}{R}\frac{\partial T_{r\Theta}}{\partial\Theta}+\frac{\partial T_{rZ}}{\partial Z}+\frac{T_{rR}}{R}-\frac{J\tau_{\theta\theta}}{r},\\
(\boldsymbol{\nabla}\cdot\boldsymbol{T})_\theta&=\frac{\partial T_{\theta R}}{\partial R}+\frac{1}{R}\frac{\partial T_{\theta\Theta}}{\partial\Theta}+\frac{\partial T_{\theta Z}}{\partial Z}+\frac{T_{\theta R}}{R}+\frac{J\tau_{r\theta}}{r},\\
(\boldsymbol{\nabla}\cdot\boldsymbol{T})_z&=\frac{\partial T_{zR}}{\partial R}+\frac{1}{R}\frac{\partial T_{z\Theta}}{\partial\Theta}+\frac{\partial T_{zZ}}{\partial Z}+\frac{T_{zR}}{R}.
\end{aligned}\tag{B.17}$$

If (R,Θ,Z) and (r,θ,z) denote the cylindrical coordinates of a point in the reference and deformed configurations, the physical components of the displacement with respect to the reference configuration basis are given by $u_R=r\cos(\theta-\Theta)-R$, $u_\Theta=r\sin(\theta-\Theta)$

and $u_Z = z - Z$. In the special case of axisymmetry ($u_\Theta = 0$, $u_r = u_R$, $u_z = u_Z$, and $\partial(.)/\partial\Theta = 0$), the deformation gradient and nonzero strain components are

$$\boldsymbol{F} = \begin{bmatrix} 1 + \frac{\partial u_r}{\partial R} & 0 & \frac{\partial u_r}{\partial Z} \\ 0 & 1 + \frac{u_r}{R} & 0 \\ \frac{\partial u_z}{\partial R} & 0 & 1 + \frac{\partial u_z}{\partial Z} \end{bmatrix},$$

$$E_{RR} = \frac{\partial u_r}{\partial R} + \frac{1}{2}\left[\left(\frac{\partial u_r}{\partial R}\right)^2 + \left(\frac{\partial u_z}{\partial R}\right)^2\right],$$

$$E_{ZZ} = \frac{\partial u_z}{\partial Z} + \frac{1}{2}\left[\left(\frac{\partial u_r}{\partial Z}\right)^2 + \left(\frac{\partial u_z}{\partial Z}\right)^2\right],$$

$$2E_{RZ} = \frac{\partial u_z}{\partial R} + \frac{\partial u_r}{\partial Z} + \frac{\partial u_r}{\partial R}\frac{\partial u_r}{\partial Z} + \frac{\partial u_z}{\partial R}\frac{\partial u_z}{\partial Z},$$

$$E_{\Theta\Theta} = \frac{u_r}{R} + \frac{1}{2}\left(\frac{u_r}{R}\right)^2.$$

C Spherical Coordinate System

The spherical coordinates are $\xi_1 = r, \xi_2 = \theta, \xi_3 = \phi$, and the position vector is given by

$$x = (r \sin\theta \cos\phi, r \sin\theta \sin\phi, r \cos\theta). \tag{C.1}$$

The scale factors are

$$h_1 = 1, \;\; h_2 = r, \;\; h_3 = r \sin\theta.$$

The basis vectors are $e_r = (\sin\theta\cos\phi, \sin\theta\sin\phi, \cos\theta)$, $e_\theta = (\cos\theta\cos\phi, \cos\theta\sin\phi, -\sin\theta)$, and $e_\phi = (-\sin\phi, \cos\phi, 0)$. The rotation matrix that occurs in Eqns. (1.99), (1.100), etc., (i.e., $[u]_{\text{spherical}} = Q[u]_{\text{Cartesian}}$, etc.) is

$$Q = \begin{bmatrix} e_r \\ e_\theta \\ e_\phi \end{bmatrix} = \begin{bmatrix} \sin\theta\cos\phi & \sin\theta\sin\phi & \cos\theta \\ \cos\theta\cos\phi & \cos\theta\sin\phi & -\sin\theta \\ -\sin\phi & \cos\phi & 0 \end{bmatrix}, \tag{C.2}$$

and the vectors ω_j and $\hat{\omega}_j$ are obtained using Eqns. (A.5) as

$$\omega_r = \mathbf{0}, \quad \omega_\theta = e_\phi, \quad \omega_\phi = \cos\theta e_r - \sin\theta e_\theta,$$
$$\hat{\omega}_r = \mathbf{0}, \quad \hat{\omega}_\theta = \frac{1}{r} e_\phi, \quad \hat{\omega}_\phi = \frac{1}{r}(\cot\theta e_r - e_\theta).$$

Let Φ, v (with physical components v_r, v_θ and v_ϕ) and T (with physical components T_{rr}, $T_{r\theta}$, $T_{r\phi}$ etc.) be scalar, vector and tensor-valued fields. Then

$$\nabla\Phi = \frac{\partial\Phi}{\partial r} e_r + \frac{1}{r}\frac{\partial\Phi}{\partial\theta} e_\theta + \frac{1}{r\sin\theta}\frac{\partial\Phi}{\partial\phi} e_\phi, \tag{C.3}$$

$$\nabla^2\Phi = \frac{1}{r^2}\frac{\partial}{\partial r}\left(r^2\frac{\partial\Phi}{\partial r}\right) + \frac{1}{r^2\sin\theta}\frac{\partial}{\partial\theta}\left(\sin\theta\frac{\partial\Phi}{\partial\theta}\right) + \frac{1}{r^2\sin^2\theta}\frac{\partial^2\Phi}{\partial\phi^2}. \tag{C.4}$$

The components of ∇v and associated quantities are

$$[\nabla v] = \begin{bmatrix} \dfrac{\partial v_r}{\partial r} & \dfrac{1}{r}\left(\dfrac{\partial v_r}{\partial\theta} - v_\theta\right) & \dfrac{1}{r\sin\theta}\dfrac{\partial v_r}{\partial\phi} - \dfrac{v_\phi}{r} \\ \dfrac{\partial v_\theta}{\partial r} & \dfrac{1}{r}\left(\dfrac{\partial v_\theta}{\partial\theta} + v_r\right) & \dfrac{1}{r\sin\theta}\dfrac{\partial v_\theta}{\partial\phi} - \dfrac{v_\phi}{r}\cot\theta \\ \dfrac{\partial v_\phi}{\partial r} & \dfrac{1}{r}\dfrac{\partial v_\phi}{\partial\theta} & \dfrac{1}{r\sin\theta}\dfrac{\partial v_\phi}{\partial\phi} + \dfrac{v_\theta}{r}\cot\theta + \dfrac{v_r}{r} \end{bmatrix}, \tag{C.5}$$

$$\nabla \times v = \left[\frac{1}{r\sin\theta}\frac{\partial}{\partial\theta}(v_\phi \sin\theta) - \frac{1}{r\sin\theta}\frac{\partial v_\theta}{\partial\phi}\right] e_r + \left[\frac{1}{r\sin\theta}\frac{\partial v_r}{\partial\phi} - \frac{1}{r}\frac{\partial(rv_\phi)}{\partial r}\right] e_\theta$$
$$+ \left[\frac{1}{r}\frac{\partial(rv_\theta)}{\partial r} - \frac{1}{r}\frac{\partial v_r}{\partial\theta}\right] e_\phi, \tag{C.6}$$

$$\nabla \cdot v = \frac{1}{r^2}\frac{\partial(r^2 v_r)}{\partial r} + \frac{1}{r\sin\theta}\frac{\partial}{\partial\theta}(v_\theta\sin\theta) + \frac{1}{r\sin\theta}\frac{\partial v_\phi}{\partial\phi}. \tag{C.7}$$

The components of $\nabla \cdot T$ are given by

$$(\nabla \cdot T)_r = \frac{\partial T_{rr}}{\partial r} + \frac{1}{r}\frac{\partial T_{r\theta}}{\partial\theta} + \frac{1}{r\sin\theta}\frac{\partial T_{r\phi}}{\partial\phi} + \frac{1}{r}\left(2T_{rr} - T_{\theta\theta} - T_{\phi\phi} + \cot\theta\, T_{r\theta}\right),$$

$$(\nabla \cdot T)_\theta = \frac{\partial T_{\theta r}}{\partial r} + \frac{1}{r}\frac{\partial T_{\theta\theta}}{\partial\theta} + \frac{1}{r\sin\theta}\frac{\partial T_{\theta\phi}}{\partial\phi} + \frac{1}{r}\left(T_{r\theta} + 2T_{\theta r} + \cot\theta(T_{\theta\theta} - T_{\phi\phi})\right), \tag{C.8}$$

$$(\nabla \cdot T)_\phi = \frac{\partial T_{\phi r}}{\partial r} + \frac{1}{r}\frac{\partial T_{\phi\theta}}{\partial\theta} + \frac{1}{r\sin\theta}\frac{\partial T_{\phi\phi}}{\partial\phi} + \frac{1}{r}\left(T_{r\phi} + 2T_{\phi r} + \cot\theta(T_{\theta\phi} + T_{\phi\theta})\right).$$

Letting $T_{ij} \equiv (\nabla v)_{ij}$ in the above equations, we get expressions for $(\nabla^2 v)_i$ (see the bracketed terms multiplying μ in Eqns. (C.12) below).

The small strain tensor components are given by (changing notation, as in the previous appendix, from (R, Θ, Φ) to (r, θ, ϕ))

$$\epsilon_{rr} = \frac{\partial u_r}{\partial r}, \qquad \epsilon_{r\theta} = \frac{1}{2}\left[\frac{1}{r}\frac{\partial u_r}{\partial\theta} + r\frac{\partial}{\partial r}\left(\frac{u_\theta}{r}\right)\right],$$

$$\epsilon_{\theta\theta} = \frac{1}{r}\left(\frac{\partial u_\theta}{\partial\theta} + u_r\right), \qquad \epsilon_{\theta\phi} = \frac{1}{2}\left[\frac{1}{r\sin\theta}\frac{\partial u_\theta}{\partial\phi} + \frac{\sin\theta}{r}\frac{\partial}{\partial\theta}\left(\frac{u_\phi}{\sin\theta}\right)\right], \tag{C.9}$$

$$\epsilon_{\phi\phi} = \frac{1}{r\sin\theta}\frac{\partial u_\phi}{\partial\phi} + \frac{u_r}{r} + \frac{u_\theta\cot\theta}{r}, \qquad \epsilon_{r\phi} = \frac{1}{2}\left[\frac{1}{r\sin\theta}\frac{\partial u_r}{\partial\phi} + r\frac{\partial}{\partial r}\left(\frac{u_\phi}{r}\right)\right].$$

From the above relations, it follows that

$$\operatorname{tr}\epsilon = \frac{1}{r^2}\frac{\partial(r^2 u_r)}{\partial r} + \frac{1}{r\sin\theta}\frac{\partial}{\partial\theta}(u_\theta\sin\theta) + \frac{1}{r\sin\theta}\frac{\partial u_\phi}{\partial\phi}.$$

The 'linearized rigid displacements' obtained by using $Q(Wx + c)$, where $W \in \text{Skw}$, and x are the Cartesian coordinates, are

$$u_r = c_1\sin\theta\cos\phi + c_2\sin\theta\sin\phi + c_3\cos\theta,$$

$$u_\theta = (b\cos\phi + c\sin\phi)r + c_1\cos\theta\cos\phi + c_2\cos\theta\sin\phi - c_3\sin\theta,$$

$$u_\phi = (c\cos\theta\cos\phi - b\cos\theta\sin\phi - a\sin\theta)r - c_1\sin\phi + c_2\cos\phi,$$

where a, b, c, c_1, c_2 and c_3 are constants.

The compatibility conditions, obtained by applying Eqn. (A.26) twice, are given by

$$\begin{aligned}
&2(\epsilon_{\theta\theta} - \epsilon_{rr} - \cot\theta\epsilon_{r\theta}) + \operatorname{cosec}^2\theta\frac{\partial}{\partial\phi}\left[\frac{\partial\epsilon_{\theta\theta}}{\partial\phi} - 2\frac{\partial}{\partial\theta}\left(\sin\theta\epsilon_{\theta\phi}\right) - 2\sin\theta\epsilon_{r\phi}\right] \\
&+\cot\theta\frac{\partial}{\partial\theta}\left[2\epsilon_{\phi\phi} - \epsilon_{\theta\theta}\right] - 2\frac{\partial\epsilon_{r\theta}}{\partial\theta} + \frac{\partial^2\epsilon_{\phi\phi}}{\partial\theta^2} + r\frac{\partial}{\partial r}(\epsilon_{\theta\theta} + \epsilon_{\phi\phi}) = 0, \\
&\cot\theta\left[\frac{\partial\epsilon_{rr}}{\partial\theta} - 2\epsilon_{r\theta}\right] + r\frac{\partial}{\partial r}(2\epsilon_{\phi\phi} - \epsilon_{rr} - 2\cot\theta\epsilon_{r\theta}) \\
&+\operatorname{cosec}^2\theta\frac{\partial}{\partial\phi}\left[\frac{\partial\epsilon_{rr}}{\partial\phi} - 2\sin\theta\frac{\partial(r\epsilon_{r\phi})}{\partial r}\right] + r^2\frac{\partial^2\epsilon_{\phi\phi}}{\partial r^2} = 0, \\
&-2\frac{\partial\epsilon_{r\theta}}{\partial\theta} + \frac{\partial^2\epsilon_{rr}}{\partial\theta^2} - r\frac{\partial}{\partial r}\left[\epsilon_{rr} + 2\frac{\partial\epsilon_{r\theta}}{\partial\theta}\right] + \frac{\partial}{\partial r}\left(r^2\frac{\partial\epsilon_{\theta\theta}}{\partial r}\right) = 0, \\
&-2\epsilon_{r\theta} + \operatorname{cosec}^2\theta\frac{\partial}{\partial\phi}\left[\frac{\partial}{\partial\theta}\left(\sin\theta\epsilon_{r\phi}\right) + r\sin\theta\frac{\partial\epsilon_{\theta\phi}}{\partial r} - \frac{\partial\epsilon_{r\theta}}{\partial\phi}\right] \\
&+\frac{\partial}{\partial\theta}\left[\epsilon_{rr} - r\frac{\partial\epsilon_{\phi\phi}}{\partial r}\right] + r\cot\theta\frac{\partial}{\partial r}\left(\epsilon_{\theta\theta} - \epsilon_{\phi\phi}\right) = 0, \\
&\operatorname{cosec}\theta\frac{\partial}{\partial\phi}\left[\epsilon_{r\theta} + r\frac{\partial\epsilon_{r\theta}}{\partial r} - \sin\theta\frac{\partial}{\partial\theta}\left(\frac{\epsilon_{rr}}{\sin\theta}\right)\right] - \cot\theta\epsilon_{r\phi} \\
&+r\sin\theta\frac{\partial^2}{\partial r\partial\theta}\left(\frac{\epsilon_{r\phi}}{\sin\theta}\right) + \frac{\partial\epsilon_{r\phi}}{\partial\theta} - \frac{\partial}{\partial r}\left(r^2\frac{\partial\epsilon_{\theta\phi}}{\partial r}\right) = 0, \\
&(\operatorname{cosec}^2\theta - 2)\epsilon_{r\phi} + \operatorname{cosec}\theta\frac{\partial}{\partial\phi}\left[\epsilon_{rr} + \sin\theta\frac{\partial}{\partial\theta}\left(\frac{\epsilon_{r\theta}}{\sin\theta}\right) - r\frac{\partial\epsilon_{\theta\theta}}{\partial r}\right] \\
&-\operatorname{cosec}\theta\frac{\partial}{\partial\theta}\left(\sin\theta\frac{\partial\epsilon_{r\phi}}{\partial\theta}\right) + r\operatorname{cosec}^2\theta\frac{\partial^2}{\partial r\partial\theta}\left(\sin^2\theta\epsilon_{\theta\phi}\right) = 0.
\end{aligned} \tag{C.10}$$

By using the strain–displacement relations, it can be easily verified that the above relations are satisfied. In case the axis $\theta = 0$ or $\theta = \pi$ is part of the domain, the strain components have to satisfy the additional constraint given by Eqn. (6.7) provided $\epsilon_{\theta\theta}$ and $\epsilon_{\phi\phi}$ are independent of ϕ, and the additional constraint given by Eqn. (6.34a) provided ϵ_{rr}, $\epsilon_{\theta\theta}$ and $\epsilon_{\phi\phi}$ are independent of θ and ϕ.

The constitutive relations given by Eqns. (6.36) become

$$\begin{aligned}
\tau_{rr} &= \lambda(\operatorname{tr}\epsilon) + 2\mu\epsilon_{rr}, & \tau_{r\theta} &= 2\mu\epsilon_{r\theta}, \\
\tau_{\theta\theta} &= \lambda(\operatorname{tr}\epsilon) + 2\mu\epsilon_{\theta\theta}, & \tau_{\theta\phi} &= 2\mu\epsilon_{\theta\phi}, \\
\tau_{\phi\phi} &= \lambda(\operatorname{tr}\epsilon) + 2\mu\epsilon_{\phi\phi}, & \tau_{r\phi} &= 2\mu\epsilon_{r\phi},
\end{aligned} \tag{C.11}$$

while the Navier equations of elasticity (Eqns. (6.37)) are given by

$$\rho_0 \frac{\partial^2 u_r}{\partial t^2} = (\lambda+\mu)\frac{\partial(\operatorname{tr}\boldsymbol{\epsilon})}{\partial r} + \mu\left[\frac{\partial}{\partial r}\left(\frac{1}{r^2}\frac{\partial(r^2 u_r)}{\partial r}\right) + \frac{1}{r^2\sin\theta}\frac{\partial}{\partial\theta}\left(\sin\theta\frac{\partial u_r}{\partial\theta}\right)\right.$$
$$\left. + \frac{1}{r^2\sin^2\theta}\frac{\partial^2 u_r}{\partial\phi^2} - \frac{2}{r^2}\frac{\partial u_\theta}{\partial\theta} - \frac{2u_\theta\cot\theta}{r^2} - \frac{2}{r^2\sin\theta}\frac{\partial u_\phi}{\partial\phi}\right] + \rho_0(\boldsymbol{b}_0)_r,$$

$$\rho_0 \frac{\partial^2 u_\theta}{\partial t^2} = (\lambda+\mu)\frac{1}{r}\frac{\partial(\operatorname{tr}\boldsymbol{\epsilon})}{\partial \theta} + \mu\left[\frac{1}{r^2}\frac{\partial}{\partial r}\left(r^2\frac{\partial u_\theta}{\partial r}\right) + \frac{1}{r^2\sin\theta}\frac{\partial}{\partial\theta}\left(\sin\theta\frac{\partial u_\theta}{\partial\theta}\right)\right.$$
$$\left. + \frac{1}{r^2\sin^2\theta}\frac{\partial^2 u_\theta}{\partial\phi^2} - \frac{u_\theta}{r^2\sin^2\theta} + \frac{2}{r^2}\frac{\partial u_r}{\partial\theta} - \frac{2\cos\theta}{r^2\sin^2\theta}\frac{\partial u_\phi}{\partial\phi}\right] + \rho_0(\boldsymbol{b}_0)_\theta,$$

$$\rho_0 \frac{\partial^2 u_\phi}{\partial t^2} = (\lambda+\mu)\frac{1}{r\sin\theta}\frac{\partial(\operatorname{tr}\boldsymbol{\epsilon})}{\partial \phi} + \mu\left[\frac{1}{r^2}\frac{\partial}{\partial r}\left(r^2\frac{\partial u_\phi}{\partial r}\right) + \frac{1}{r^2\sin\theta}\frac{\partial}{\partial\theta}\left(\sin\theta\frac{\partial u_\phi}{\partial\theta}\right)\right.$$
$$\left. + \frac{1}{r^2\sin^2\theta}\frac{\partial^2 u_\phi}{\partial\phi^2} + \frac{2}{r^2\sin\theta}\frac{\partial u_r}{\partial\phi} + \frac{2\cos\theta}{r^2\sin^2\theta}\frac{\partial u_\theta}{\partial\phi} - \frac{u_\phi}{r^2\sin^2\theta}\right] + \rho_0(\boldsymbol{b}_0)_\phi, \tag{C.12}$$

where the Lamé constants (λ, μ) are related to (E, ν) by means of Eqns. (5.20).

If $r = \tilde{r}(\boldsymbol{X}, t)$, $\theta = \tilde{\theta}(\boldsymbol{X}, t)$, and $\phi = \tilde{\phi}(\boldsymbol{X}, t)$, using Eqns. (A.13) we have

$$\tilde{\boldsymbol{v}}(\boldsymbol{X}, t) = \dot{\tilde{r}}\boldsymbol{e}_r + \tilde{r}\dot{\tilde{\theta}}\boldsymbol{e}_\theta + \tilde{r}\dot{\tilde{\phi}}\sin\theta\boldsymbol{e}_\phi,$$

$$\tilde{\boldsymbol{a}}(\boldsymbol{X}, t) = (\ddot{\tilde{r}} - \tilde{r}\dot{\tilde{\theta}}^2 - \tilde{r}\dot{\tilde{\phi}}^2\sin^2\theta)\boldsymbol{e}_r + (\tilde{r}\ddot{\tilde{\theta}} + 2\dot{\tilde{r}}\dot{\tilde{\theta}} - \tilde{r}\dot{\tilde{\phi}}^2\sin\tilde{\theta}\cos\tilde{\theta})\boldsymbol{e}_\theta +$$
$$(\tilde{r}\ddot{\tilde{\phi}}\sin\tilde{\theta} + 2\dot{\tilde{r}}\dot{\tilde{\phi}}\sin\tilde{\theta} + 2\tilde{r}\dot{\tilde{\theta}}\dot{\tilde{\phi}}\cos\tilde{\theta})\boldsymbol{e}_\phi.$$

Let $\boldsymbol{T}$ and $\boldsymbol{\tau}$ denote the first Piola–Kirchhoff and Cauchy stress tensors, and let $J \equiv \det \boldsymbol{F}$. If the material and spatial points are expressed in terms of spherical coordinates (R, Θ, Z) and (r, θ, z), respectively, then using the fact that $\hat{\omega}_1^0 = \boldsymbol{0}$, $\hat{\omega}_2^0 = \boldsymbol{e}_\Phi^0/R$, $\hat{\omega}_3^0 = (\cot\Theta\boldsymbol{e}_R^0 - \boldsymbol{e}_\Theta^0)/R$, and $\hat{\omega}_1 = \boldsymbol{0}$, $\hat{\omega}_2 = \boldsymbol{e}_\phi/r$, $\hat{\omega}_3 = (\cot\theta\boldsymbol{e}_r - \boldsymbol{e}_\theta)/r$, we get on using Eqn. (A.30):

$$(\boldsymbol{\nabla}\cdot\boldsymbol{T})_r = \frac{\partial T_{rR}}{\partial R} + \frac{1}{R}\frac{\partial T_{r\Theta}}{\partial\Theta} + \frac{1}{R\sin\Theta}\frac{\partial T_{r\Phi}}{\partial\Phi} + \frac{1}{R}(2T_{rR} + \cot\Theta T_{r\Theta}) - \frac{J}{r}(\tau_{\theta\theta} + \tau_{\phi\phi}),$$

$$(\boldsymbol{\nabla}\cdot\boldsymbol{T})_\theta = \frac{\partial T_{\theta R}}{\partial R} + \frac{1}{R}\frac{\partial T_{\theta\Theta}}{\partial\Theta} + \frac{1}{R\sin\Theta}\frac{\partial T_{\theta\Phi}}{\partial\Phi} + \frac{1}{R}(2T_{\theta R} + \cot\Theta T_{\theta\Theta}) + \frac{J}{r}(\tau_{r\theta} - \cot\theta\tau_{\phi\phi}),$$

$$(\boldsymbol{\nabla}\cdot\boldsymbol{T})_\phi = \frac{\partial T_{\phi R}}{\partial R} + \frac{1}{R}\frac{\partial T_{\phi\Theta}}{\partial\Theta} + \frac{1}{R\sin\Theta}\frac{\partial T_{\phi\Phi}}{\partial\Phi} + \frac{1}{R}(2T_{\phi R} + \cot\Theta T_{\phi\Theta}) + \frac{J}{r}(\tau_{r\phi} + \cot\theta\tau_{\theta\phi}). \tag{C.13}$$

D Elliptic Cylindrical Coordinate System

The elliptic cylindrical coordinates are $\xi_1 = \xi$, $\xi_2 = \eta$, $\xi_3 = z$, and the position vector is given by

$$\boldsymbol{x} = (c\cosh\xi\cos\eta, c\sinh\xi\sin\eta, z). \tag{D.1}$$

Note that $x + iy = c\cosh(\xi + i\eta)$, which is useful for finding (ξ, η) given (x, y). The scale factors are

$$h_1 = h_2 = c\sqrt{\cosh^2\xi - \cos^2\eta} = \frac{c}{\sqrt{2}}\sqrt{\cosh 2\xi - \cos 2\eta}, \;\; h_3 = 1. \tag{D.2}$$

The in-plane basis vectors are

$$\begin{aligned}
\boldsymbol{e}_\xi &= \frac{c}{h_1}\left[\sinh\xi\cos\eta\,\boldsymbol{e}_x + \cosh\xi\sin\eta\,\boldsymbol{e}_y\right], & \boldsymbol{e}_x &= \frac{c}{h_1}\left[\sinh\xi\cos\eta\,\boldsymbol{e}_\xi - \cosh\xi\sin\eta\,\boldsymbol{e}_\eta\right], \\
\boldsymbol{e}_\eta &= \frac{c}{h_1}\left[-\cosh\xi\sin\eta\,\boldsymbol{e}_x + \sinh\xi\cos\eta\,\boldsymbol{e}_y\right], & \boldsymbol{e}_y &= \frac{c}{h_1}\left[\cosh\xi\sin\eta\,\boldsymbol{e}_\xi + \sinh\xi\cos\eta\,\boldsymbol{e}_\eta\right].
\end{aligned}$$

The rotation matrix that occurs in Eqns. (1.99), (1.100), etc., (i.e., $[\boldsymbol{u}]_{\text{elliptic}} = \boldsymbol{Q}[\boldsymbol{u}]_{\text{Cartesian}}$, etc.) is

$$\boldsymbol{Q} = \begin{bmatrix} \boldsymbol{e}_\xi \\ \boldsymbol{e}_\eta \\ \boldsymbol{e}_z \end{bmatrix} = \begin{bmatrix} \frac{c}{h_1}\sinh\xi\cos\eta & \frac{c}{h_1}\cosh\xi\sin\eta & 0 \\ -\frac{c}{h_1}\cosh\xi\sin\eta & \frac{c}{h_1}\sinh\xi\cos\eta & 0 \\ 0 & 0 & 1 \end{bmatrix}, \tag{D.3}$$

and the vectors $\boldsymbol{\omega}_j$ and $\hat{\boldsymbol{\omega}}_j$ are obtained using Eqns. (A.5) as

$$\begin{aligned}
\boldsymbol{\omega}_\xi &= -\frac{1}{h_1}\frac{\partial h_1}{\partial \eta}\boldsymbol{e}_z, & \boldsymbol{\omega}_\eta &= \frac{1}{h_1}\frac{\partial h_1}{\partial \xi}\boldsymbol{e}_z, & \boldsymbol{\omega}_z &= \boldsymbol{0}, \\
\hat{\boldsymbol{\omega}}_\xi &= -\frac{1}{h_1^2}\frac{\partial h_1}{\partial \eta}\boldsymbol{e}_z, & \hat{\boldsymbol{\omega}}_\eta &= \frac{1}{h_1^2}\frac{\partial h_1}{\partial \xi}\boldsymbol{e}_z, & \hat{\boldsymbol{\omega}}_z &= \boldsymbol{0}.
\end{aligned}$$

Let Φ, $\boldsymbol{v}$ (with physical components v_ξ, v_η and v_z) and $\boldsymbol{T}$ (with physical components $T_{\xi\xi}$, $T_{\xi\eta}$, $T_{\xi z}$ etc.) be scalar, vector and tensor-valued fields. Then

$$\boldsymbol{\nabla}\Phi = \frac{1}{h_1}\frac{\partial\Phi}{\partial\xi}\boldsymbol{e}_\xi + \frac{1}{h_1}\frac{\partial\Phi}{\partial\eta}\boldsymbol{e}_\eta + \frac{\partial\Phi}{\partial z}\boldsymbol{e}_z, \tag{D.4}$$

$$\boldsymbol{\nabla}^2\Phi = \frac{1}{h_1^2}\left[\frac{\partial^2\Phi}{\partial\xi^2} + \frac{\partial^2\Phi}{\partial\eta^2}\right] + \frac{\partial^2\Phi}{\partial z^2}. \tag{D.5}$$

The components of ∇v and associated quantities are

$$[\nabla v] = \begin{bmatrix} \frac{1}{h_1}\frac{\partial v_\xi}{\partial \xi} + \frac{v_\eta}{h_1^2}\frac{\partial h_1}{\partial \eta} & \frac{1}{h_1}\frac{\partial v_\xi}{\partial \eta} - \frac{v_\eta}{h_1^2}\frac{\partial h_1}{\partial \xi} & \frac{\partial v_\xi}{\partial z} \\ \frac{1}{h_1}\frac{\partial v_\eta}{\partial \xi} - \frac{v_\xi}{h_1^2}\frac{\partial h_1}{\partial \eta} & \frac{1}{h_1}\frac{\partial v_\eta}{\partial \eta} + \frac{v_\xi}{h_1^2}\frac{\partial h_1}{\partial \xi} & \frac{\partial v_\eta}{\partial z} \\ \frac{1}{h_1}\frac{\partial v_z}{\partial \xi} & \frac{1}{h_1}\frac{\partial v_z}{\partial \eta} & \frac{\partial v_z}{\partial z} \end{bmatrix}, \tag{D.6}$$

$$\nabla \times v = \left[\frac{1}{h_1}\frac{\partial v_z}{\partial \eta} - \frac{\partial v_\eta}{\partial z}\right] e_\xi + \left[\frac{\partial v_\xi}{\partial z} - \frac{1}{h_1}\frac{\partial v_z}{\partial \xi}\right] e_\eta + \frac{1}{h_1^2}\left[\frac{\partial}{\partial \xi}(h_1 v_\eta) - \frac{\partial}{\partial \eta}(h_1 v_\xi)\right] e_z, \tag{D.7}$$

$$\nabla \cdot v = \frac{1}{h_1^2}\left[\frac{\partial}{\partial \xi}(h_1 v_\xi) + \frac{\partial}{\partial \eta}(h_1 v_\eta)\right] + \frac{\partial v_z}{\partial z}, \tag{D.8}$$

where $\partial h_1/\partial \xi = c^2 \sinh \xi \cosh \xi / h_1$ and $\partial h_1/\partial \eta = c^2 \sin \eta \cos \eta / h_1$. The components of $\nabla \cdot T$ are given by

$$\begin{aligned} (\nabla \cdot T)_\xi &= \frac{1}{h_1}\frac{\partial T_{\xi\xi}}{\partial \xi} + \frac{1}{h_1}\frac{\partial T_{\xi\eta}}{\partial \eta} + \frac{\partial T_{\xi z}}{\partial z} + \frac{(T_{\xi\eta} + T_{\eta\xi})}{h_1^2}\frac{\partial h_1}{\partial \eta} + \frac{(T_{\xi\xi} - T_{\eta\eta})}{h_1^2}\frac{\partial h_1}{\partial \xi}, \\ (\nabla \cdot T)_\eta &= \frac{1}{h_1}\frac{\partial T_{\eta\xi}}{\partial \xi} + \frac{1}{h_1}\frac{\partial T_{\eta\eta}}{\partial \eta} + \frac{\partial T_{\eta z}}{\partial z} + \frac{(T_{\xi\eta} + T_{\eta\xi})}{h_1^2}\frac{\partial h_1}{\partial \xi} - \frac{(T_{\xi\xi} - T_{\eta\eta})}{h_1^2}\frac{\partial h_1}{\partial \eta}, \\ (\nabla \cdot T)_z &= \frac{1}{h_1^2}\frac{\partial}{\partial \xi}(h_1 T_{z\xi}) + \frac{1}{h_1^2}\frac{\partial}{\partial \eta}(h_1 T_{z\eta}) + \frac{\partial T_{zz}}{\partial z}. \end{aligned} \tag{D.9}$$

Letting $T_{ij} \equiv (\nabla v)_{ij}$ in the above equations, we get expressions for $(\nabla^2 v)_i$ (see the bracketed terms multiplying μ in Eqns. (D.13) below).

The small strain tensor components are given by

$$\begin{aligned} \epsilon_{\xi\xi} &= \frac{1}{h_1}\frac{\partial u_\xi}{\partial \xi} + \frac{u_\eta}{h_1^2}\frac{\partial h_1}{\partial \eta}, & \epsilon_{\xi\eta} &= \frac{1}{2}\left[\frac{\partial}{\partial \eta}\left(\frac{u_\xi}{h_1}\right) + \frac{\partial}{\partial \xi}\left(\frac{u_\eta}{h_1}\right)\right], \\ \epsilon_{\eta\eta} &= \frac{1}{h_1}\frac{\partial u_\eta}{\partial \eta} + \frac{u_\xi}{h_1^2}\frac{\partial h_1}{\partial \xi}, & \epsilon_{\eta z} &= \frac{1}{2}\left[\frac{\partial u_\eta}{\partial z} + \frac{1}{h_1}\frac{\partial u_z}{\partial \eta}\right], \\ \epsilon_{zz} &= \frac{\partial u_z}{\partial z}, & \epsilon_{\xi z} &= \frac{1}{2}\left[\frac{\partial u_\xi}{\partial z} + \frac{1}{h_1}\frac{\partial u_z}{\partial \xi}\right]. \end{aligned} \tag{D.10}$$

From the above relations, it follows that

$$\operatorname{tr} \epsilon = \frac{1}{h_1^2}\frac{\partial (h_1 u_\xi)}{\partial \xi} + \frac{1}{h_1^2}\frac{\partial (h_1 u_\eta)}{\partial \eta} + \frac{\partial u_z}{\partial z}.$$

The 'linearized rigid displacements' obtained by using $Q(Wx+c)$, where $W \in \text{Skw}$, and x are the Cartesian coordinates, are

$$\begin{aligned}
u_\xi &= \frac{c}{h_1}\left[(c_2-\alpha z)\cosh\xi\sin\eta + c\gamma\sin\eta\cos\eta + (c_1+\beta z)\sinh\xi\cos\eta\right],\\
u_\eta &= \frac{c}{h_1}\left[(c_2-\alpha z)\sinh\xi\cos\eta + c\gamma\sinh\xi\cosh\xi - (c_1+\beta z)\cosh\xi\sin\eta\right],\\
u_z &= c(\alpha\sinh\xi\sin\eta - \beta\cosh\xi\cos\eta) + c_3,
\end{aligned} \tag{D.11}$$

where $\alpha, \beta, \gamma, c_1, c_2$ and c_3 are constants.

The constitutive relations given by Eqns. (6.36) become

$$\begin{aligned}
\tau_{\xi\xi} &= \lambda(\operatorname{tr}\epsilon) + 2\mu\epsilon_{\xi\xi}, & \tau_{\xi\eta} &= 2\mu\epsilon_{\xi\eta},\\
\tau_{\eta\eta} &= \lambda(\operatorname{tr}\epsilon) + 2\mu\epsilon_{\eta\eta}, & \tau_{\eta z} &= 2\mu\epsilon_{\eta z},\\
\tau_{zz} &= \lambda(\operatorname{tr}\epsilon) + 2\mu\epsilon_{zz}, & \tau_{\xi z} &= 2\mu\epsilon_{\xi z},
\end{aligned} \tag{D.12}$$

while the Navier equations of elasticity (Eqns. (6.37)) are given by

$$\begin{aligned}
\rho_0\frac{\partial^2 u_\xi}{\partial t^2} &= \frac{(\lambda+\mu)}{h_1}\frac{\partial(\operatorname{tr}\epsilon)}{\partial\xi} + \mu\left\{\frac{1}{h_1^2}\left(\frac{\partial^2 u_\xi}{\partial\xi^2} + \frac{\partial^2 u_\xi}{\partial\eta^2}\right) + \frac{\partial^2 u_\xi}{\partial z^2}\right.\\
&\quad\left. + \frac{1}{h_1^3}\left(h_1^3\frac{\partial^2 u_\xi}{\partial z^2} + 2\frac{\partial u_\eta}{\partial\xi}\frac{\partial h_1}{\partial\eta} - 2\frac{\partial u_\eta}{\partial\eta}\frac{\partial h_1}{\partial\xi}\right) - \frac{u_\xi}{h_1^4}\left[\left(\frac{\partial h_1}{\partial\xi}\right)^2 + \left(\frac{\partial h_1}{\partial\eta}\right)^2\right]\right\} + \rho_0(b_0)_\xi,\\
\rho_0\frac{\partial^2 u_\eta}{\partial t^2} &= \frac{(\lambda+\mu)}{h_1}\frac{\partial(\operatorname{tr}\epsilon)}{\partial\eta} + \mu\left\{\frac{1}{h_1^2}\left(\frac{\partial^2 u_\eta}{\partial\xi^2} + \frac{\partial^2 u_\eta}{\partial\eta^2}\right) + \frac{\partial^2 u_\eta}{\partial z^2}\right.\\
&\quad\left. + \frac{1}{h_1^3}\left(h_1^3\frac{\partial^2 u_\eta}{\partial z^2} + 2\frac{\partial u_\xi}{\partial\eta}\frac{\partial h_1}{\partial\xi} - 2\frac{\partial u_\xi}{\partial\xi}\frac{\partial h_1}{\partial\eta}\right) - \frac{u_\eta}{h_1^4}\left[\left(\frac{\partial h_1}{\partial\xi}\right)^2 + \left(\frac{\partial h_1}{\partial\eta}\right)^2\right]\right\} + \rho_0(b_0)_\eta,\\
\rho_0\frac{\partial^2 u_z}{\partial t^2} &= (\lambda+\mu)\frac{\partial(\operatorname{tr}\epsilon)}{\partial z} + \mu\left[\frac{1}{h_1^2}\left(\frac{\partial^2 u_z}{\partial\xi^2} + \frac{\partial^2 u_z}{\partial\eta^2}\right) + \frac{\partial^2 u_z}{\partial z^2}\right] + \rho_0(b_0)_z,
\end{aligned} \tag{D.13}$$

where the Lamé constants (λ, μ) are related to (E, ν) by means of Eqns. (5.20).

If $\xi = \tilde{\xi}(X, t)$, $\eta = \tilde{\eta}(X, t)$ and $z = \tilde{z}(X, t)$, using Eqns. (A.13) we have

$$\begin{aligned}
\tilde{v}(X,t) &= h_1\dot{\tilde{\xi}}e_\xi + h_1\dot{\tilde{\eta}}e_\eta + \dot{\tilde{z}}e_z,\\
\tilde{a}(X,t) &= \left[\ddot{\tilde{\xi}}h_1 + \left(\dot{\tilde{\xi}}^2 - \dot{\tilde{\eta}}^2\right)\frac{\partial h_1}{\partial\xi} + 2\dot{\tilde{\xi}}\dot{\tilde{\eta}}\frac{\partial h_1}{\partial\eta}\right]e_\xi\\
&\quad + \left[\ddot{\tilde{\eta}}h_1 + \left(\dot{\tilde{\eta}}^2 - \dot{\tilde{\xi}}^2\right)\frac{\partial h_1}{\partial\eta} + 2\dot{\tilde{\xi}}\dot{\tilde{\eta}}\frac{\partial h_1}{\partial\xi}\right]e_\eta + \ddot{\tilde{z}}e_z.
\end{aligned}$$

Let T and τ denote the first Piola–Kirchhoff and Cauchy stress tensors, and let $J \equiv \det F$. If the material and spatial points are expressed in terms of elliptic cylindrical coordinates (Ξ, Γ, Z) and (ξ, η, z), respectively, then using the fact that $\hat{\omega}_1^0 = -\frac{1}{(h_1^0)^2}\frac{\partial h_1^0}{\partial\Gamma}e_Z$,

$\hat{\omega}_2^0 = \frac{1}{(h_1^0)^2}\frac{\partial h_1^0}{\partial \Xi}\boldsymbol{e}_Z$, $\hat{\omega}_3^0 = \boldsymbol{0}$, and $\hat{\omega}_1 = -\frac{1}{h_1^2}\frac{\partial h_1}{\partial \eta}\boldsymbol{e}_z$, $\hat{\omega}_2 = \frac{1}{h_1^2}\frac{\partial h_1}{\partial \xi}\boldsymbol{e}_z$, $\hat{\omega}_3 = \boldsymbol{0}$, we get on using Eqn. (A.30):

$$
\begin{aligned}
(\boldsymbol{\nabla}\cdot\boldsymbol{T})_\xi &= \frac{1}{h_1^0}\frac{\partial T_{\xi\Xi}}{\partial \Xi} + \frac{1}{h_1^0}\frac{\partial T_{\xi\Gamma}}{\partial \Gamma} + \frac{\partial T_{\xi Z}}{\partial Z} + \frac{T_{\xi\Gamma}}{(h_1^0)^2}\frac{\partial h_1^0}{\partial \Gamma} + \frac{J\tau_{\xi\eta}}{h_1^2}\frac{\partial h_1}{\partial \eta} + \frac{T_{\xi\Xi}}{(h_1^0)^2}\frac{\partial h_1^0}{\partial \Xi} - \frac{J\tau_{\eta\eta}}{h_1^2}\frac{\partial h_1}{\partial \xi},\\
(\boldsymbol{\nabla}\cdot\boldsymbol{T})_\eta &= \frac{1}{h_1^0}\frac{\partial T_{\eta\Xi}}{\partial \Xi} + \frac{1}{h_1^0}\frac{\partial T_{\eta\Gamma}}{\partial \Gamma} + \frac{\partial T_{\eta Z}}{\partial Z} + \frac{T_{\eta\Xi}}{(h_1^0)^2}\frac{\partial h_1^0}{\partial \Xi} + \frac{J\tau_{\xi\eta}}{h_1^2}\frac{\partial h_1}{\partial \xi} + \frac{T_{\eta\Gamma}}{(h_1^0)^2}\frac{\partial h_1^0}{\partial \Gamma} - \frac{J\tau_{\xi\xi}}{h_1^2}\frac{\partial h_1}{\partial \eta},\\
(\boldsymbol{\nabla}\cdot\boldsymbol{T})_z &= \frac{1}{(h_1^0)^2}\frac{\partial}{\partial \Xi}(h_1^0 T_{z\Xi}) + \frac{1}{(h_1^0)^2}\frac{\partial}{\partial \Gamma}(h_1^0 T_{z\Gamma}) + \frac{\partial T_{zZ}}{\partial Z}.
\end{aligned}
\tag{D.14}
$$

E Bipolar Cylindrical Coordinate System

The bipolar cylindrical coordinates are $\xi_1 = \xi$, $\xi_2 = \eta$, $\xi_3 = z$, and the position vector is given by (see Fig. 6.13)

$$x = \left(\frac{c \sin \eta}{\cosh \xi - \cos \eta}, \frac{c \sinh \xi}{\cosh \xi - \cos \eta}, z \right). \tag{E.1}$$

Note that $z = x + iy = ic \coth \frac{1}{2}(\xi + i\eta)$, or, alternatively, $\xi + i\eta = \log[(z + ic)/(z - ic)]$, which is useful for finding (ξ, η) given (x, y).

$$h_1 = h_2 = \frac{c}{\cosh \xi - \cos \eta}, \quad h_3 = 1. \tag{E.2}$$

The in-plane basis vectors are

$$\begin{aligned}
e_\xi &= -\frac{h_1}{c} \left[\sinh \xi \sin \eta \, e_x + (\cosh \xi \cos \eta - 1) \, e_y \right], \\
e_\eta &= \frac{h_1}{c} \left[(\cosh \xi \cos \eta - 1) \, e_x - \sinh \xi \sin \eta \, e_y \right], \\
e_x &= -\frac{h_1}{c} \left[\sinh \xi \sin \eta \, e_\xi - (\cosh \xi \cos \eta - 1) \, e_\eta \right], \\
e_y &= -\frac{h_1}{c} \left[(\cosh \xi \cos \eta - 1) \, e_\xi + \sinh \xi \sin \eta \, e_\eta \right].
\end{aligned}$$

The rotation matrix that occurs in Eqns. (1.99), (1.100), etc., (i.e., $[u]_{\text{bipolar}} = Q[u]_{\text{Cartesian}}$, etc.) is

$$Q = \begin{bmatrix} e_\xi \\ e_\eta \\ e_z \end{bmatrix} = \begin{bmatrix} -\frac{h_1}{c} \sinh \xi \sin \eta & -\frac{h_1}{c} (\cosh \xi \cos \eta - 1) & 0 \\ \frac{h_1}{c} (\cosh \xi \cos \eta - 1) & -\frac{h_1}{c} \sinh \xi \sin \eta & 0 \\ 0 & 0 & 1 \end{bmatrix}, \tag{E.3}$$

and the vectors ω_j and $\hat{\omega}_j$ are obtained using Eqns. (A.5) as

$$\begin{aligned}
\omega_\xi &= -\frac{1}{h_1} \frac{\partial h_1}{\partial \eta} e_z, \quad \omega_\eta = \frac{1}{h_1} \frac{\partial h_1}{\partial \xi} e_z, \quad \omega_z = \mathbf{0}, \\
\hat{\omega}_\xi &= -\frac{1}{h_1^2} \frac{\partial h_1}{\partial \eta} e_z, \quad \hat{\omega}_\eta = \frac{1}{h_1^2} \frac{\partial h_1}{\partial \xi} e_z, \quad \hat{\omega}_z = \mathbf{0}.
\end{aligned}$$

Let Φ, v (with physical components v_ξ, v_η and v_z) and T (with physical components $T_{\xi\xi}$, $T_{\xi\eta}$, $T_{\xi z}$ etc.) be scalar, vector and tensor-valued fields. Then

$$\nabla \Phi = \frac{1}{h_1} \frac{\partial \Phi}{\partial \xi} e_\xi + \frac{1}{h_1} \frac{\partial \Phi}{\partial \eta} e_\eta + \frac{\partial \Phi}{\partial z} e_z, \tag{E.4}$$

$$\nabla^2\Phi = \frac{1}{h_1^2}\left[\frac{\partial^2\Phi}{\partial\xi^2} + \frac{\partial^2\Phi}{\partial\eta^2}\right] + \frac{\partial^2\Phi}{\partial z^2}. \tag{E.5}$$

The components of ∇v and associated quantities are

$$[\nabla v] = \begin{bmatrix} \frac{1}{h_1}\frac{\partial v_\xi}{\partial\xi} + \frac{v_\eta}{h_1^2}\frac{\partial h_1}{\partial\eta} & \frac{1}{h_1}\frac{\partial v_\xi}{\partial\eta} - \frac{v_\eta}{h_1^2}\frac{\partial h_1}{\partial\xi} & \frac{\partial v_\xi}{\partial z} \\ \frac{1}{h_1}\frac{\partial v_\eta}{\partial\xi} - \frac{v_\xi}{h_1^2}\frac{\partial h_1}{\partial\eta} & \frac{1}{h_1}\frac{\partial v_\eta}{\partial\eta} + \frac{v_\xi}{h_1^2}\frac{\partial h_1}{\partial\xi} & \frac{\partial v_\eta}{\partial z} \\ \frac{1}{h_1}\frac{\partial v_z}{\partial\xi} & \frac{1}{h_1}\frac{\partial v_z}{\partial\eta} & \frac{\partial v_z}{\partial z} \end{bmatrix}, \tag{E.6}$$

$$\nabla\times v = \left[\frac{1}{h_1}\frac{\partial v_z}{\partial\eta} - \frac{\partial v_\eta}{\partial z}\right]e_\xi + \left[\frac{\partial v_\xi}{\partial z} - \frac{1}{h_1}\frac{\partial v_z}{\partial\xi}\right]e_\eta + \frac{1}{h_1^2}\left[\frac{\partial}{\partial\xi}(h_1 v_\eta) - \frac{\partial}{\partial\eta}(h_1 v_\xi)\right]e_z, \tag{E.7}$$

$$\nabla\cdot v = \frac{1}{h_1^2}\left[\frac{\partial}{\partial\xi}(h_1 v_\xi) + \frac{\partial}{\partial\eta}(h_1 v_\eta)\right] + \frac{\partial v_z}{\partial z}, \tag{E.8}$$

where $\partial h_1/\partial\xi = -h_1^2\sinh\xi/c$ and $\partial h_1/\partial\eta = -h_1^2\sin\eta/c$. The components of $\nabla\cdot T$ are given by

$$\begin{aligned}(\nabla\cdot T)_\xi &= \frac{1}{h_1}\frac{\partial T_{\xi\xi}}{\partial\xi} + \frac{1}{h_1}\frac{\partial T_{\xi\eta}}{\partial\eta} + \frac{\partial T_{\xi z}}{\partial z} + \frac{(T_{\xi\eta}+T_{\eta\xi})}{h_1^2}\frac{\partial h_1}{\partial\eta} + \frac{(T_{\xi\xi}-T_{\eta\eta})}{h_1^2}\frac{\partial h_1}{\partial\xi},\\ (\nabla\cdot T)_\eta &= \frac{1}{h_1}\frac{\partial T_{\eta\xi}}{\partial\xi} + \frac{1}{h_1}\frac{\partial T_{\eta\eta}}{\partial\eta} + \frac{\partial T_{\eta z}}{\partial z} + \frac{(T_{\xi\eta}+T_{\eta\xi})}{h_1^2}\frac{\partial h_1}{\partial\xi} - \frac{(T_{\xi\xi}-T_{\eta\eta})}{h_1^2}\frac{\partial h_1}{\partial\eta},\\ (\nabla\cdot T)_z &= \frac{1}{h_1^2}\frac{\partial}{\partial\xi}(h_1T_{z\xi}) + \frac{1}{h_1^2}\frac{\partial}{\partial\eta}(h_1T_{z\eta}) + \frac{\partial T_{zz}}{\partial z}.\end{aligned} \tag{E.9}$$

Letting $T_{ij} \equiv (\nabla v)_{ij}$ in the above equations, we get expressions for $(\nabla^2 v)_i$ (see the bracketed terms multiplying μ in Eqns. (E.13) below).

The small strain tensor components are given by

$$\begin{aligned}\epsilon_{\xi\xi} &= \frac{1}{h_1}\frac{\partial u_\xi}{\partial\xi} + \frac{u_\eta}{h_1^2}\frac{\partial h_1}{\partial\eta}, & \epsilon_{\xi\eta} &= \frac{1}{2}\left[\frac{\partial}{\partial\eta}\left(\frac{u_\xi}{h_1}\right) + \frac{\partial}{\partial\xi}\left(\frac{u_\eta}{h_1}\right)\right],\\ \epsilon_{\eta\eta} &= \frac{1}{h_1}\frac{\partial u_\eta}{\partial\eta} + \frac{u_\xi}{h_1^2}\frac{\partial h_1}{\partial\xi}, & \epsilon_{\eta z} &= \frac{1}{2}\left[\frac{\partial u_\eta}{\partial z} + \frac{1}{h_1}\frac{\partial u_z}{\partial\eta}\right],\\ \epsilon_{zz} &= \frac{\partial u_z}{\partial z}, & \epsilon_{\xi z} &= \frac{1}{2}\left[\frac{\partial u_\xi}{\partial z} + \frac{1}{h_1}\frac{\partial u_z}{\partial\xi}\right].\end{aligned} \tag{E.10}$$

From the above relations, it follows that

$$\operatorname{tr}\epsilon = \frac{1}{h_1^2}\frac{\partial(h_1u_\xi)}{\partial\xi} + \frac{1}{h_1^2}\frac{\partial(h_1u_\eta)}{\partial\eta} + \frac{\partial u_z}{\partial z}.$$

The 'linearized rigid displacements' obtained by using $Q(Wx + c)$, where $W \in \text{Skw}$, and x are the Cartesian coordinates, are

$$
\begin{aligned}
u_\xi &= \frac{h_1}{c}\left[(c_2 - \alpha z)(1 - \cosh\xi\cos\eta) + c\gamma\cosh\xi\sin\eta - (c_1 + \beta z)\sinh\xi\sin\eta\right],\\
u_\eta &= -\frac{h_1}{c}\left[(c_2 - \alpha z)\sinh\xi\sin\eta + c\gamma\sinh\xi\cos\eta + (c_1 + \beta z)(1 - \cosh\xi\cos\eta)\right],\\
u_z &= h_1(\alpha\sinh\xi - \beta\sin\eta) + c_3,
\end{aligned}
\tag{E.11}
$$

where α, β, γ, c_1, c_2 and c_3 are constants.

The constitutive relations given by Eqns. (6.36) become

$$
\begin{aligned}
\tau_{\xi\xi} &= \lambda(\operatorname{tr}\epsilon) + 2\mu\epsilon_{\xi\xi}, &\quad \tau_{\xi\eta} &= 2\mu\epsilon_{\xi\eta},\\
\tau_{\eta\eta} &= \lambda(\operatorname{tr}\epsilon) + 2\mu\epsilon_{\eta\eta}, &\quad \tau_{\eta z} &= 2\mu\epsilon_{\eta z},\\
\tau_{zz} &= \lambda(\operatorname{tr}\epsilon) + 2\mu\epsilon_{zz}, &\quad \tau_{\xi z} &= 2\mu\epsilon_{\xi z},
\end{aligned}
\tag{E.12}
$$

while the Navier equations of elasticity (Eqns. (6.37)) are given by

$$
\begin{aligned}
\rho_0\frac{\partial^2 u_\xi}{\partial t^2} &= \frac{(\lambda+\mu)}{h_1}\frac{\partial(\operatorname{tr}\epsilon)}{\partial\xi} + \mu\left\{\frac{1}{h_1^2}\left(\frac{\partial^2 u_\xi}{\partial\xi^2} + \frac{\partial^2 u_\xi}{\partial\eta^2}\right) + \frac{\partial^2 u_\xi}{\partial z^2}\right.\\
&\quad\left. + \frac{1}{h_1^3}\left(h_1^3\frac{\partial^2 u_\xi}{\partial z^2} + 2\frac{\partial u_\eta}{\partial\xi}\frac{\partial h_1}{\partial\eta} - 2\frac{\partial u_\eta}{\partial\eta}\frac{\partial h_1}{\partial\xi}\right) - \frac{u_\xi}{h_1^4}\left[\left(\frac{\partial h_1}{\partial\xi}\right)^2 + \left(\frac{\partial h_1}{\partial\eta}\right)^2\right]\right\} + \rho_0(\boldsymbol{b}_0)_\xi,\\
\rho_0\frac{\partial^2 u_\eta}{\partial t^2} &= \frac{(\lambda+\mu)}{h_1}\frac{\partial(\operatorname{tr}\epsilon)}{\partial\eta} + \mu\left\{\frac{1}{h_1^2}\left(\frac{\partial^2 u_\eta}{\partial\xi^2} + \frac{\partial^2 u_\eta}{\partial\eta^2}\right) + \frac{\partial^2 u_\eta}{\partial z^2}\right.\\
&\quad\left. + \frac{1}{h_1^3}\left(h_1^3\frac{\partial^2 u_\eta}{\partial z^2} + 2\frac{\partial u_\xi}{\partial\eta}\frac{\partial h_1}{\partial\xi} - 2\frac{\partial u_\xi}{\partial\xi}\frac{\partial h_1}{\partial\eta}\right) - \frac{u_\eta}{h_1^4}\left[\left(\frac{\partial h_1}{\partial\xi}\right)^2 + \left(\frac{\partial h_1}{\partial\eta}\right)^2\right]\right\} + \rho_0(\boldsymbol{b}_0)_\eta,\\
\rho_0\frac{\partial^2 u_z}{\partial t^2} &= (\lambda+\mu)\frac{\partial(\operatorname{tr}\epsilon)}{\partial z} + \mu\left[\frac{1}{h_1^2}\left(\frac{\partial^2 u_z}{\partial\xi^2} + \frac{\partial^2 u_z}{\partial\eta^2}\right) + \frac{\partial^2 u_z}{\partial z^2}\right] + \rho_0(\boldsymbol{b}_0)_z,
\end{aligned}
\tag{E.13}
$$

where the Lamé constants (λ, μ) are related to (E, ν) by means of Eqns. (5.20).

If $\xi = \tilde{\xi}(\boldsymbol{X}, t)$, $\eta = \tilde{\eta}(\boldsymbol{X}, t)$, and $z = \tilde{z}(\boldsymbol{X}, t)$, using Eqns. (A.13) we have

$$
\begin{aligned}
\tilde{\boldsymbol{v}}(\boldsymbol{X}, t) &= h_1\dot{\tilde{\xi}}\boldsymbol{e}_\xi + h_1\dot{\tilde{\eta}}\boldsymbol{e}_\eta + \dot{\tilde{z}}\boldsymbol{e}_z,\\
\tilde{\boldsymbol{a}}(\boldsymbol{X}, t) &= \left[\ddot{\tilde{\xi}}h_1 + \left(\dot{\tilde{\xi}}^2 - \dot{\tilde{\eta}}^2\right)\frac{\partial h_1}{\partial\xi} + 2\dot{\tilde{\xi}}\dot{\tilde{\eta}}\frac{\partial h_1}{\partial\eta}\right]\boldsymbol{e}_\xi\\
&\quad + \left[\ddot{\tilde{\eta}}h_1 + \left(\dot{\tilde{\eta}}^2 - \dot{\tilde{\xi}}^2\right)\frac{\partial h_1}{\partial\eta} + 2\dot{\tilde{\xi}}\dot{\tilde{\eta}}\frac{\partial h_1}{\partial\xi}\right]\boldsymbol{e}_\eta + \ddot{\tilde{z}}\boldsymbol{e}_z.
\end{aligned}
$$

Let $\boldsymbol{T}$ and $\boldsymbol{\tau}$ denote the first Piola–Kirchhoff and Cauchy stress tensors, and let $J \equiv \det\boldsymbol{F}$. If the material and spatial points are expressed in terms of bipolar cylindrical coordinates (Ξ, Γ, Z) and (ξ, η, z), respectively, then using the fact that $\hat{\omega}_1^0 = -\frac{1}{(h_1^0)^2}\frac{\partial h_1^0}{\partial\Gamma}\boldsymbol{e}_Z$,

$\hat{\omega}_2^0 = \frac{1}{(h_1^0)^2}\frac{\partial h_1^0}{\partial \Xi}\boldsymbol{e}_Z$, $\hat{\omega}_3^0 = \boldsymbol{0}$, and $\hat{\omega}_1 = -\frac{1}{h_1^2}\frac{\partial h_1}{\partial \eta}\boldsymbol{e}_z$, $\hat{\omega}_2 = \frac{1}{h_1^2}\frac{\partial h_1}{\partial \xi}\boldsymbol{e}_z$, $\hat{\omega}_3 = \boldsymbol{0}$, we get on using Eqn. (A.30):

$$\begin{aligned}
(\boldsymbol{\nabla}\cdot\boldsymbol{T})_\xi &= \frac{1}{h_1^0}\frac{\partial T_{\xi\Xi}}{\partial \Xi} + \frac{1}{h_1^0}\frac{\partial T_{\xi\Gamma}}{\partial \Gamma} + \frac{\partial T_{\xi Z}}{\partial Z} + \frac{T_{\xi\Gamma}}{(h_1^0)^2}\frac{\partial h_1^0}{\partial \Gamma} + \frac{J\tau_{\xi\eta}}{h_1^2}\frac{\partial h_1}{\partial \eta} + \frac{T_{\xi\Xi}}{(h_1^0)^2}\frac{\partial h_1^0}{\partial \Xi} - \frac{J\tau_{\eta\eta}}{h_1^2}\frac{\partial h_1}{\partial \xi}, \\
(\boldsymbol{\nabla}\cdot\boldsymbol{T})_\eta &= \frac{1}{h_1^0}\frac{\partial T_{\eta\Xi}}{\partial \Xi} + \frac{1}{h_1^0}\frac{\partial T_{\eta\Gamma}}{\partial \Gamma} + \frac{\partial T_{\eta Z}}{\partial Z} + \frac{T_{\eta\Xi}}{(h_1^0)^2}\frac{\partial h_1^0}{\partial \Xi} + \frac{J\tau_{\xi\eta}}{h_1^2}\frac{\partial h_1}{\partial \xi} + \frac{T_{\eta\Gamma}}{(h_1^0)^2}\frac{\partial h_1^0}{\partial \Gamma} - \frac{J\tau_{\xi\xi}}{h_1^2}\frac{\partial h_1}{\partial \eta}, \\
(\boldsymbol{\nabla}\cdot\boldsymbol{T})_z &= \frac{1}{(h_1^0)^2}\frac{\partial}{\partial \Xi}(h_1^0 T_{z\Xi}) + \frac{1}{(h_1^0)^2}\frac{\partial}{\partial \Gamma}(h_1^0 T_{z\Gamma}) + \frac{\partial T_{zZ}}{\partial Z}.
\end{aligned} \tag{E.14}$$

F Prolate Spheroidal Coordinate System

The prolate spheroidal coordinates are $\xi_1 = \xi$, $\xi_2 = \eta$, $\xi_3 = \phi$, and the position vector is given by

$$\boldsymbol{x} = (c \sinh\xi \sin\eta \cos\phi, c \sinh\xi \sin\eta \sin\phi, c \cosh\xi \cos\eta). \tag{F.1}$$

In what follows, r–θ–z denote cylindrical coordinates; the ϕ-direction of the prolate spheroidal coordinate system is identical to the θ-direction of the cylindrical system. Note that $r + iz = ci\cos(\eta + i\xi)$, which is useful for finding (ξ, η) given (r, z). The scale factors are

$$h_1 = h_2 = c\sqrt{\cosh^2\xi - \cos^2\eta} = \frac{c}{\sqrt{2}}\sqrt{\cosh 2\xi - \cos 2\eta}, \; h_3 = r = c\sinh\xi\sin\eta. \tag{F.2}$$

The basis vectors in the r-z plane are

$$\begin{aligned}
\boldsymbol{e}_\xi &= \frac{c}{h_1}\left[\cosh\xi\sin\eta\,\boldsymbol{e}_r + \sinh\xi\cos\eta\,\boldsymbol{e}_z\right], \quad \boldsymbol{e}_r = \frac{c}{h_1}\left[\cosh\xi\sin\eta\,\boldsymbol{e}_\xi + \sinh\xi\cos\eta\,\boldsymbol{e}_\eta\right],\\
\boldsymbol{e}_\eta &= \frac{c}{h_1}\left[\sinh\xi\cos\eta\,\boldsymbol{e}_r - \cosh\xi\sin\eta\,\boldsymbol{e}_z\right], \quad \boldsymbol{e}_z = \frac{c}{h_1}\left[\sinh\xi\cos\eta\,\boldsymbol{e}_\xi - \cosh\xi\sin\eta\,\boldsymbol{e}_\eta\right].
\end{aligned}$$

The rotation matrix that relates components of vectors and tensors in the prolate spheroidal and the cylindrical systems (i.e., $[\boldsymbol{u}]_{\text{prolate}} = \boldsymbol{Q}[\boldsymbol{u}]_{\text{cylindrical}}$, etc.) is

$$\boldsymbol{Q} = \begin{bmatrix} \boldsymbol{e}_\xi \\ \boldsymbol{e}_\eta \\ \boldsymbol{e}_\phi \end{bmatrix} = \begin{bmatrix} \frac{c}{h_1}\cosh\xi\sin\eta & 0 & \frac{c}{h_1}\sinh\xi\cos\eta \\ \frac{c}{h_1}\sinh\xi\cos\eta & 0 & -\frac{c}{h_1}\cosh\xi\sin\eta \\ 0 & 1 & 0 \end{bmatrix}, \tag{F.3}$$

and the vectors $\boldsymbol{\omega}_j$ and $\hat{\boldsymbol{\omega}}_j$ are obtained using Eqns. (A.5) as

$$\begin{aligned}
\boldsymbol{\omega}_\xi &= -\frac{1}{h_1}\frac{\partial h_1}{\partial\eta}\boldsymbol{e}_\phi, \quad \boldsymbol{\omega}_\eta = \frac{1}{h_1}\frac{\partial h_1}{\partial\xi}\boldsymbol{e}_\phi, \quad \boldsymbol{\omega}_\phi = \boldsymbol{e}_z,\\
\hat{\boldsymbol{\omega}}_\xi &= -\frac{1}{h_1^2}\frac{\partial h_1}{\partial\eta}\boldsymbol{e}_\phi, \quad \hat{\boldsymbol{\omega}}_\eta = \frac{1}{h_1^2}\frac{\partial h_1}{\partial\xi}\boldsymbol{e}_\phi, \quad \hat{\boldsymbol{\omega}}_\phi = \frac{1}{r}\boldsymbol{e}_z.
\end{aligned}$$

Let Φ, $\boldsymbol{v}$ (with physical components v_ξ, v_η and v_ϕ) and $\boldsymbol{T}$ (with physical components $T_{\xi\xi}$, $T_{\xi\eta}$, $T_{\xi\phi}$ etc.) be scalar, vector and tensor-valued fields. Then

$$\boldsymbol{\nabla}\Phi = \frac{1}{h_1}\frac{\partial\Phi}{\partial\xi}\boldsymbol{e}_\xi + \frac{1}{h_1}\frac{\partial\Phi}{\partial\eta}\boldsymbol{e}_\eta + \frac{1}{r}\frac{\partial\Phi}{\partial\phi}\boldsymbol{e}_\phi, \tag{F.4}$$

$$\nabla^2\Phi = \frac{1}{h_1^2}\left[\frac{\partial^2\Phi}{\partial\xi^2} + \coth\xi\frac{\partial\Phi}{\partial\xi} + \frac{\partial^2\Phi}{\partial\eta^2} + \cot\eta\frac{\partial\Phi}{\partial\eta} + \frac{h_1^2}{r^2}\frac{\partial^2\Phi}{\partial\phi^2}\right]. \tag{F.5}$$

The components of ∇v and associated quantities are

$$[\nabla v] = \begin{bmatrix} \frac{1}{h_1}\frac{\partial v_\xi}{\partial\xi} + \frac{v_\eta}{h_1^2}\frac{\partial h_1}{\partial\eta} & \frac{1}{h_1}\frac{\partial v_\xi}{\partial\eta} - \frac{v_\eta}{h_1^2}\frac{\partial h_1}{\partial\xi} & \frac{1}{r}\frac{\partial v_\xi}{\partial\phi} - \frac{cv_\phi}{rh_1}\cosh\xi\sin\eta \\ \frac{1}{h_1}\frac{\partial v_\eta}{\partial\xi} - \frac{v_\xi}{h_1^2}\frac{\partial h_1}{\partial\eta} & \frac{1}{h_1}\frac{\partial v_\eta}{\partial\eta} + \frac{v_\xi}{h_1^2}\frac{\partial h_1}{\partial\xi} & \frac{1}{r}\frac{\partial v_\eta}{\partial\phi} - \frac{cv_\phi}{rh_1}\sinh\xi\cos\eta \\ \frac{1}{h_1}\frac{\partial v_\phi}{\partial\xi} & \frac{1}{h_1}\frac{\partial v_\phi}{\partial\eta} & \frac{1}{r}\frac{\partial v_\phi}{\partial\phi} + \frac{c}{rh_1}\left(v_\eta\sinh\xi\cos\eta + v_\xi\cosh\xi\sin\eta\right) \end{bmatrix}, \tag{F.6}$$

$$\nabla\times v = \left[\frac{1}{h_1 r}\frac{\partial(rv_\phi)}{\partial\eta} - \frac{1}{r}\frac{\partial v_\eta}{\partial\phi}\right]e_\xi + \left[\frac{1}{r}\frac{\partial v_\xi}{\partial\phi} - \frac{1}{h_1 r}\frac{\partial(rv_\phi)}{\partial\xi}\right]e_\eta + \frac{1}{h_1^2}\left[\frac{\partial}{\partial\xi}(h_1 v_\eta) - \frac{\partial}{\partial\eta}(h_1 v_\xi)\right]e_z, \tag{F.7}$$

$$\nabla\cdot v = \frac{1}{h_1^2 r}\left[\frac{\partial}{\partial\xi}(h_1 r v_\xi) + \frac{\partial}{\partial\eta}(h_1 r v_\eta)\right] + \frac{1}{r}\frac{\partial v_\phi}{\partial\phi}, \tag{F.8}$$

where $\partial h_1/\partial\xi = c^2\sinh\xi\cosh\xi/h_1$ and $\partial h_1/\partial\eta = c^2\sin\eta\cos\eta/h_1$. The components of $\nabla\cdot T$ are given by

$$\begin{aligned}
(\nabla\cdot T)_\xi &= \frac{1}{h_1}\left[\frac{\partial T_{\xi\xi}}{\partial\xi} + \frac{\partial T_{\xi\eta}}{\partial\eta}\right] + \frac{1}{r}\frac{\partial T_{\xi\phi}}{\partial\phi} + \frac{(T_{\xi\eta}+T_{\eta\xi})}{h_1^2}\frac{\partial h_1}{\partial\eta} + \frac{(T_{\xi\xi}-T_{\eta\eta})}{h_1^2}\frac{\partial h_1}{\partial\xi} \\
&\quad + \frac{c}{rh_1}\left[(T_{\xi\xi}-T_{\phi\phi})\cosh\xi\sin\eta + T_{\xi\eta}\sinh\xi\cos\eta\right], \\
(\nabla\cdot T)_\eta &= \frac{1}{h_1}\left[\frac{\partial T_{\eta\xi}}{\partial\xi} + \frac{\partial T_{\eta\eta}}{\partial\eta}\right] + \frac{1}{r}\frac{\partial T_{\eta\phi}}{\partial\phi} + \frac{(T_{\xi\eta}+T_{\eta\xi})}{h_1^2}\frac{\partial h_1}{\partial\xi} + \frac{(T_{\eta\eta}-T_{\xi\xi})}{h_1^2}\frac{\partial h_1}{\partial\eta} \\
&\quad + \frac{c}{rh_1}\left[(T_{\eta\eta}-T_{\phi\phi})\sinh\xi\cos\eta + T_{\eta\xi}\cosh\xi\sin\eta\right], \\
(\nabla\cdot T)_\phi &= \frac{1}{h_1}\left[\frac{\partial T_{\phi\xi}}{\partial\xi} + \frac{\partial T_{\phi\eta}}{\partial\eta}\right] + \frac{1}{r}\frac{\partial T_{\phi\phi}}{\partial\phi} + \frac{T_{\phi\eta}}{h_1^2}\frac{\partial h_1}{\partial\eta} + \frac{T_{\phi\xi}}{h_1^2}\frac{\partial h_1}{\partial\xi} \\
&\quad + \frac{c}{rh_1}\left[(T_{\eta\phi}+T_{\phi\eta})\sinh\xi\cos\eta + (T_{\xi\phi}+T_{\phi\xi})\cosh\xi\sin\eta\right].
\end{aligned} \tag{F.9}$$

Letting $T_{ij} \equiv (\nabla v)_{ij}$ in the above equations, we get expressions for $(\nabla^2 v)_i$ (see the see the bracketed terms multiplying μ in Eqns. (F.13) below).

The small strain tensor components are given by

$$\begin{aligned}
\epsilon_{\xi\xi} &= \frac{1}{h_1}\frac{\partial u_\xi}{\partial\xi} + \frac{u_\eta}{h_1^2}\frac{\partial h_1}{\partial\eta}, & \epsilon_{\xi\eta} &= \frac{1}{2}\left[\frac{\partial}{\partial\eta}\left(\frac{u_\xi}{h_1}\right) + \frac{\partial}{\partial\xi}\left(\frac{u_\eta}{h_1}\right)\right], \\
\epsilon_{\eta\eta} &= \frac{1}{h_1}\frac{\partial u_\eta}{\partial\eta} + \frac{u_\xi}{h_1^2}\frac{\partial h_1}{\partial\xi}, & \epsilon_{\eta\phi} &= \frac{1}{2}\left[\frac{1}{r}\frac{\partial u_\eta}{\partial\phi} + \frac{\sin\eta}{h_1}\frac{\partial(\csc\eta\, u_\phi)}{\partial\eta}\right], \\
\epsilon_{\phi\phi} &= \frac{1}{r}\frac{\partial u_\phi}{\partial\phi} + \frac{c}{rh_1}\left(u_\eta\sinh\xi\cos\eta + u_\xi\cosh\xi\sin\eta\right), & \epsilon_{\xi\phi} &= \frac{1}{2}\left[\frac{1}{r}\frac{\partial u_\xi}{\partial\phi} + \frac{\sinh\xi}{h_1}\frac{\partial(\operatorname{csch}\xi\, u_\phi)}{\partial\xi}\right].
\end{aligned} \tag{F.10}$$

From the above relations, it follows that

$$\operatorname{tr}\epsilon = \frac{1}{h_1^2 r}\frac{\partial(h_1 r u_\xi)}{\partial \xi} + \frac{1}{h_1^2 r}\frac{\partial(h_1 r u_\eta)}{\partial \eta} + \frac{1}{r}\frac{\partial u_\phi}{\partial \phi}.$$

The 'linearized rigid displacements' obtained by using $Q_{\text{Cart}}(\boldsymbol{W}\boldsymbol{x}+\boldsymbol{c})$, where Q_{Cart} is the orthogonal transformation matrix from Cartesian to prolate spheroidal coordinates, $\boldsymbol{W} \in \text{Skw}$, and $\boldsymbol{x}$ are the Cartesian coordinates, are

$$\begin{aligned}
u_\xi &= \frac{c}{h_1}\left[\cosh\xi \sin\eta(c_4\cos\phi + c_5\sin\phi) + c_6\sinh\xi\cos\eta + \sin\eta\cos\eta(c_2\cos\phi - c_1\sin\phi)\right],\\
u_\eta &= \frac{c}{h_1}\left[\sinh\xi\cos\eta(c_4\cos\phi + c_5\sin\phi) - c_6\cosh\xi\sin\eta + \sinh\xi\cosh\xi(c_2\cos\phi - c_1\sin\phi)\right],\\
u_\phi &= c_5\cos\phi - c_4\sin\phi - \cosh\xi\cos\eta(c_1\cos\phi + c_2\sin\phi) + c_3\sinh\xi\sin\eta,
\end{aligned} \tag{F.11}$$

where c_1–c_6 are constants.

The constitutive relations given by Eqns. (6.36) become

$$\begin{aligned}
\tau_{\xi\xi} &= \lambda(\operatorname{tr}\epsilon) + 2\mu\epsilon_{\xi\xi}, & \tau_{\xi\eta} &= 2\mu\epsilon_{\xi\eta},\\
\tau_{\eta\eta} &= \lambda(\operatorname{tr}\epsilon) + 2\mu\epsilon_{\eta\eta}, & \tau_{\eta\phi} &= 2\mu\epsilon_{\eta\phi},\\
\tau_{\phi\phi} &= \lambda(\operatorname{tr}\epsilon) + 2\mu\epsilon_{\phi\phi}, & \tau_{\xi\phi} &= 2\mu\epsilon_{\xi\phi},
\end{aligned} \tag{F.12}$$

while the Navier equations of elasticity (Eqns. (6.37)) are given by

$$\begin{aligned}
\rho_0\frac{\partial^2 u_\xi}{\partial t^2} &= \frac{(\lambda+\mu)}{h_1}\frac{\partial(\operatorname{tr}\epsilon)}{\partial \xi}\\
&+ \mu\left\{\frac{1}{h_1^2}\left[\frac{\partial^2 u_\xi}{\partial \xi^2} + \coth\xi\frac{\partial u_\xi}{\partial \xi} - \coth^2\xi u_\xi + \frac{\partial^2 u_\xi}{\partial \eta^2} + \cot\eta\frac{\partial u_\xi}{\partial \eta} - \coth\xi\cot\eta u_\eta\right]\right.\\
&+ \frac{1}{h_1^3}\left[u_\eta\left[\coth\xi\frac{\partial h_1}{\partial \eta} - \cot\eta\frac{\partial h_1}{\partial \xi}\right] - 2\frac{\partial h_1}{\partial \xi}\frac{\partial u_\eta}{\partial \eta} + 2\frac{\partial h_1}{\partial \eta}\frac{\partial u_\eta}{\partial \xi}\right]\\
&\left.- \frac{u_\xi}{h_1^4}\left[\left(\frac{\partial h_1}{\partial \xi}\right)^2 + \left(\frac{\partial h_1}{\partial \eta}\right)^2\right] - \frac{2}{ch_1}\coth\xi\operatorname{csch}\xi\csc\eta\frac{\partial u_\phi}{\partial \phi} + \frac{1}{c^2}\operatorname{csch}^2\xi\csc^2\eta\frac{\partial^2 u_\xi}{\partial \phi^2}\right\}\\
&+ \rho_0(\boldsymbol{b}_0)_\xi,\\
\rho_0\frac{\partial^2 u_\eta}{\partial t^2} &= \frac{(\lambda+\mu)}{h_1}\frac{\partial(\operatorname{tr}\epsilon)}{\partial \eta}\\
&+ \mu\left\{\frac{1}{h_1^2}\left[\frac{\partial^2 u_\eta}{\partial \xi^2} + \coth\xi\frac{\partial u_\eta}{\partial \xi} - \cot^2\eta u_\eta + \frac{\partial^2 u_\eta}{\partial \eta^2} + \cot\eta\frac{\partial u_\eta}{\partial \eta} - \coth\xi\cot\eta u_\xi\right]\right.
\end{aligned}$$

$$+\frac{1}{h_1^3}\left[u_\xi\left[\cot\eta\frac{\partial h_1}{\partial\xi}-\coth\xi\frac{\partial h_1}{\partial\eta}\right]+2\frac{\partial h_1}{\partial\xi}\frac{\partial u_\xi}{\partial\eta}-2\frac{\partial h_1}{\partial\eta}\frac{\partial u_\xi}{\partial\xi}\right]$$

$$-\frac{u_\eta}{h_1^4}\left[\left(\frac{\partial h_1}{\partial\xi}\right)^2+\left(\frac{\partial h_1}{\partial\eta}\right)^2\right]-\frac{2}{ch_1}\operatorname{csch}\xi\cot\eta\csc\eta\frac{\partial u_\phi}{\partial\phi}+\frac{1}{c^2}\operatorname{csch}^2\xi\csc^2\eta\frac{\partial^2 u_\eta}{\partial\phi^2}\Bigg\}$$

$$+\rho_0(\boldsymbol{b}_0)_\eta,$$

$$\rho_0\frac{\partial^2 u_\phi}{\partial t^2}=\frac{(\lambda+\mu)}{r}\frac{\partial(\operatorname{tr}\boldsymbol{\epsilon})}{\partial\phi}+\mu\left\{\frac{1}{h_1^2}\left[\frac{\partial^2 u_\phi}{\partial\xi^2}+\frac{\partial^2 u_\phi}{\partial\eta^2}+\coth\xi\frac{\partial u_\phi}{\partial\xi}+\cot\eta\frac{\partial u_\phi}{\partial\eta}\right]\right.$$

$$+\frac{1}{c^2}\operatorname{csch}^2\xi\csc^2\eta\frac{\partial^2 u_\phi}{\partial\phi^2}+\frac{2}{ch_1}\operatorname{csch}\xi\csc\eta\left[\coth\xi\frac{\partial u_\xi}{\partial\phi}+\cot\eta\frac{\partial u_\eta}{\partial\phi}\right]$$

$$\left.-\frac{1}{h_1^2}(\coth^2\xi+\cot^2\eta)u_\phi\right\}+\rho_0(\boldsymbol{b}_0)_\phi,$$

(F.13)

where the Lamé constants (λ,μ) are related to (E,ν) by means of Eqns. (5.20).

If $\xi=\tilde{\xi}(\boldsymbol{X},t)$, $\eta=\tilde{\eta}(\boldsymbol{X},t)$ and $\phi=\tilde{\phi}(\boldsymbol{X},t)$, using Eqns. (A.13) we have

$$\tilde{\boldsymbol{v}}(\boldsymbol{X},t)=h_1\dot{\tilde{\xi}}\boldsymbol{e}_\xi+h_1\dot{\tilde{\eta}}\boldsymbol{e}_\eta+r\dot{\tilde{\phi}}\boldsymbol{e}_\phi,$$

$$\tilde{\boldsymbol{a}}(\boldsymbol{X},t)=\left[\ddot{\tilde{\xi}}h_1+\left(\dot{\tilde{\xi}}^2-\dot{\tilde{\eta}}^2\right)\frac{\partial h_1}{\partial\xi}+2\dot{\tilde{\xi}}\dot{\tilde{\eta}}\frac{\partial h_1}{\partial\eta}-\frac{rc}{h_1}\dot{\tilde{\phi}}^2\cosh\xi\sin\eta\right]\boldsymbol{e}_\xi$$

$$+\left[\ddot{\tilde{\eta}}h_1+\left(\dot{\tilde{\eta}}^2-\dot{\tilde{\xi}}^2\right)\frac{\partial h_1}{\partial\eta}+2\dot{\tilde{\xi}}\dot{\tilde{\eta}}\frac{\partial h_1}{\partial\xi}-\frac{rc}{h_1}\dot{\tilde{\phi}}^2\sinh\xi\cos\eta\right]\boldsymbol{e}_\eta$$

$$+\left[r\ddot{\tilde{\phi}}+2c\dot{\tilde{\phi}}\left(\dot{\tilde{\xi}}\cosh\tilde{\xi}\sin\tilde{\eta}+\dot{\tilde{\eta}}\sinh\tilde{\xi}\cos\tilde{\eta}\right)\right]\boldsymbol{e}_\phi.$$

G Oblate Spheroidal Coordinate System

The oblate spheroidal coordinates are $\xi_1 = \xi$, $\xi_2 = \eta$, $\xi_3 = \phi$, and the position vector is given by

$$x = (c \cosh \xi \sin \eta \cos \phi, c \cosh \xi \sin \eta \sin \phi, c \sinh \xi \cos \eta). \tag{G.1}$$

In what follows, r–θ–z denote cylindrical coordinates; the ϕ-direction of the oblate spheroidal coordinate system is identical to the θ-direction of the cylindrical system. Note that $r + iz = c \sin(\eta + i\xi)$, which is useful for finding (ξ, η) given (r, z). The scale factors are

$$h_1 = h_2 = c\sqrt{\sinh^2 \xi + \cos^2 \eta} = \frac{c}{\sqrt{2}}\sqrt{\cosh 2\xi + \cos 2\eta}, \quad h_3 = r = c \cosh \xi \sin \eta. \tag{G.2}$$

The basis vectors in the r-z plane are

$$\begin{aligned} e_\xi &= \frac{c}{h_1}\left[\sinh \xi \sin \eta \, e_r + \cosh \xi \cos \eta \, e_z\right], \quad & e_r &= \frac{c}{h_1}\left[\sinh \xi \sin \eta \, e_\xi + \cosh \xi \cos \eta \, e_\eta\right], \\ e_\eta &= \frac{c}{h_1}\left[\cosh \xi \cos \eta \, e_r - \sinh \xi \sin \eta \, e_z\right], \quad & e_z &= \frac{c}{h_1}\left[\cosh \xi \cos \eta \, e_\xi - \sinh \xi \sin \eta \, e_\eta\right]. \end{aligned}$$

The rotation matrix that relates components of vectors and tensors in the oblate spheroidal and the cylindrical systems (i.e., $[u]_{\text{oblate}} = Q[u]_{\text{cylindrical}}$, etc.) is

$$Q = \begin{bmatrix} e_\xi \\ e_\eta \\ e_\phi \end{bmatrix} = \begin{bmatrix} \frac{c}{h_1} \sinh \xi \sin \eta & 0 & \frac{c}{h_1} \cosh \xi \cos \eta \\ \frac{c}{h_1} \cosh \xi \cos \eta & 0 & -\frac{c}{h_1} \sinh \xi \sin \eta \\ 0 & 1 & 0 \end{bmatrix}, \tag{G.3}$$

and the vectors ω_j and $\hat{\omega}_j$ are obtained using Eqns. (A.5) as

$$\begin{aligned} \omega_\xi &= -\frac{1}{h_1}\frac{\partial h_1}{\partial \eta} e_\phi, \quad & \omega_\eta &= \frac{1}{h_1}\frac{\partial h_1}{\partial \xi} e_\phi, \quad & \omega_\phi &= e_z, \\ \hat{\omega}_\xi &= -\frac{1}{h_1^2}\frac{\partial h_1}{\partial \eta} e_\phi, \quad & \hat{\omega}_\eta &= \frac{1}{h_1^2}\frac{\partial h_1}{\partial \xi} e_\phi, \quad & \hat{\omega}_\phi &= \frac{1}{r} e_z. \end{aligned}$$

Let Φ, v (with physical components v_ξ, v_η and v_ϕ) and T (with physical components $T_{\xi\xi}$, $T_{\xi\eta}$, $T_{\xi\phi}$ etc.) be scalar, vector and tensor-valued fields. Then

$$\nabla \Phi = \frac{1}{h_1}\frac{\partial \Phi}{\partial \xi} e_\xi + \frac{1}{h_1}\frac{\partial \Phi}{\partial \eta} e_\eta + \frac{1}{r}\frac{\partial \Phi}{\partial \phi} e_\phi, \tag{G.4}$$

$$\nabla^2\Phi = \frac{1}{h_1^2}\left[\frac{\partial^2\Phi}{\partial\xi^2} + \tanh\xi\frac{\partial\Phi}{\partial\xi} + \frac{\partial^2\Phi}{\partial\eta^2} + \cot\eta\frac{\partial\Phi}{\partial\eta} + \frac{h_1^2}{r^2}\frac{\partial^2\Phi}{\partial\phi^2}\right]. \tag{G.5}$$

The components of ∇v and associated quantities are

$$[\nabla v] = \begin{bmatrix} \frac{1}{h_1}\frac{\partial v_\xi}{\partial\xi} + \frac{v_\eta}{h_1^2}\frac{\partial h_1}{\partial\eta} & \frac{1}{h_1}\frac{\partial v_\xi}{\partial\eta} - \frac{v_\eta}{h_1^2}\frac{\partial h_1}{\partial\xi} & \frac{1}{r}\frac{\partial v_\xi}{\partial\phi} - \frac{cv_\phi}{rh_1}\sinh\xi\sin\eta \\ \frac{1}{h_1}\frac{\partial v_\eta}{\partial\xi} - \frac{v_\xi}{h_1^2}\frac{\partial h_1}{\partial\eta} & \frac{1}{h_1}\frac{\partial v_\eta}{\partial\eta} + \frac{v_\xi}{h_1^2}\frac{\partial h_1}{\partial\xi} & \frac{1}{r}\frac{\partial v_\eta}{\partial\phi} - \frac{cv_\phi}{rh_1}\cosh\xi\cos\eta \\ \frac{1}{h_1}\frac{\partial v_\phi}{\partial\xi} & \frac{1}{h_1}\frac{\partial v_\phi}{\partial\eta} & \frac{1}{r}\frac{\partial v_\phi}{\partial\phi} + \frac{c}{rh_1}\left(v_\eta\cosh\xi\cos\eta + v_\xi\sinh\xi\sin\eta\right) \end{bmatrix}, \tag{G.6}$$

$$\nabla\times v = \left[\frac{1}{h_1 r}\frac{\partial(rv_\phi)}{\partial\eta} - \frac{1}{r}\frac{\partial v_\eta}{\partial\phi}\right]e_\xi + \left[\frac{1}{r}\frac{\partial v_\xi}{\partial\phi} - \frac{1}{h_1 r}\frac{\partial(rv_\phi)}{\partial\xi}\right]e_\eta + \frac{1}{h_1^2}\left[\frac{\partial}{\partial\xi}(h_1 v_\eta) - \frac{\partial}{\partial\eta}(h_1 v_\xi)\right]e_z, \tag{G.7}$$

$$\nabla\cdot v = \frac{1}{h_1^2 r}\left[\frac{\partial}{\partial\xi}(h_1 r v_\xi) + \frac{\partial}{\partial\eta}(h_1 r v_\eta)\right] + \frac{1}{r}\frac{\partial v_\phi}{\partial\phi}, \tag{G.8}$$

where $\partial h_1/\partial\xi = c^2\sinh\xi\cosh\xi/h_1$ and $\partial h_1/\partial\eta = -c^2\sin\eta\cos\eta/h_1$. The components of $\nabla\cdot T$ are given by

$$\begin{aligned}(\nabla\cdot T)_\xi &= \frac{1}{h_1}\left[\frac{\partial T_{\xi\xi}}{\partial\xi} + \frac{\partial T_{\xi\eta}}{\partial\eta}\right] + \frac{1}{r}\frac{\partial T_{\xi\phi}}{\partial\phi} + \frac{(T_{\xi\eta}+T_{\eta\xi})}{h_1^2}\frac{\partial h_1}{\partial\eta} + \frac{(T_{\xi\xi}-T_{\eta\eta})}{h_1^2}\frac{\partial h_1}{\partial\xi} \\ &\quad + \frac{c}{rh_1}\left[(T_{\xi\xi}-T_{\phi\phi})\sinh\xi\sin\eta + T_{\xi\eta}\cosh\xi\cos\eta\right], \\ (\nabla\cdot T)_\eta &= \frac{1}{h_1}\left[\frac{\partial T_{\eta\xi}}{\partial\xi} + \frac{\partial T_{\eta\eta}}{\partial\eta}\right] + \frac{1}{r}\frac{\partial T_{\eta\phi}}{\partial\phi} + \frac{(T_{\xi\eta}+T_{\eta\xi})}{h_1^2}\frac{\partial h_1}{\partial\xi} + \frac{(T_{\eta\eta}-T_{\xi\xi})}{h_1^2}\frac{\partial h_1}{\partial\eta} \\ &\quad + \frac{c}{rh_1}\left[(T_{\eta\eta}-T_{\phi\phi})\cosh\xi\cos\eta + T_{\eta\xi}\sinh\xi\sin\eta\right], \\ (\nabla\cdot T)_\phi &= \frac{1}{h_1}\left[\frac{\partial T_{\phi\xi}}{\partial\xi} + \frac{\partial T_{\phi\eta}}{\partial\eta}\right] + \frac{1}{r}\frac{\partial T_{\phi\phi}}{\partial\phi} + \frac{T_{\phi\eta}}{h_1^2}\frac{\partial h_1}{\partial\eta} + \frac{T_{\phi\xi}}{h_1^2}\frac{\partial h_1}{\partial\xi} \\ &\quad + \frac{c}{rh_1}\left[(T_{\eta\phi}+T_{\phi\eta})\cosh\xi\cos\eta + (T_{\xi\phi}+T_{\phi\xi})\sinh\xi\sin\eta\right].\end{aligned} \tag{G.9}$$

Letting $T_{ij} \equiv (\nabla v)_{ij}$ in the above equations, we get expressions for $(\nabla^2 v)_i$ (see the see the bracketed terms multiplying μ in Eqns. (G.13) below).

The small strain tensor components are given by

$$\begin{aligned}\epsilon_{\xi\xi} &= \frac{1}{h_1}\frac{\partial u_\xi}{\partial\xi} + \frac{u_\eta}{h_1^2}\frac{\partial h_1}{\partial\eta}, & \epsilon_{\xi\eta} &= \frac{1}{2}\left[\frac{\partial}{\partial\eta}\left(\frac{u_\xi}{h_1}\right) + \frac{\partial}{\partial\xi}\left(\frac{u_\eta}{h_1}\right)\right], \\ \epsilon_{\eta\eta} &= \frac{1}{h_1}\frac{\partial u_\eta}{\partial\eta} + \frac{u_\xi}{h_1^2}\frac{\partial h_1}{\partial\xi}, & \epsilon_{\eta\phi} &= \frac{1}{2}\left[\frac{1}{r}\frac{\partial u_\eta}{\partial\phi} + \frac{\sin\eta}{h_1}\frac{\partial(\csc\eta\, u_\phi)}{\partial\eta}\right], \\ \epsilon_{\phi\phi} &= \frac{1}{r}\frac{\partial u_\phi}{\partial\phi} + \frac{c}{rh_1}\left(u_\eta\cosh\xi\cos\eta + u_\xi\sinh\xi\sin\eta\right), & \epsilon_{\xi\phi} &= \frac{1}{2}\left[\frac{1}{r}\frac{\partial u_\xi}{\partial\phi} + \frac{\cosh\xi}{h_1}\frac{\partial(\operatorname{sech}\xi\, u_\phi)}{\partial\xi}\right].\end{aligned} \tag{G.10}$$

From the above relations, it follows that

$$\operatorname{tr}\boldsymbol{\epsilon} = \frac{1}{h_1^2 r}\frac{\partial(h_1 r u_\xi)}{\partial \xi} + \frac{1}{h_1^2 r}\frac{\partial(h_1 r u_\eta)}{\partial \eta} + \frac{1}{r}\frac{\partial u_\phi}{\partial \phi}.$$

The 'linearized rigid displacements' obtained by using $\boldsymbol{Q}_{\text{Cart}}(\boldsymbol{W}\boldsymbol{x}+\boldsymbol{c})$, where $\boldsymbol{Q}_{\text{Cart}}$ is the orthogonal transformation matrix from Cartesian to oblate spheroidal coordinates, $\boldsymbol{W} \in \text{Skw}$, and $\boldsymbol{x}$ are the Cartesian coordinates, are

$$\begin{aligned}
u_\xi &= \frac{c}{h_1}\left[\sinh\xi\sin\eta(c_4\cos\phi + c_5\sin\phi) + c_6\cosh\xi\cos\eta + \sin\eta\cos\eta(c_1\sin\phi - c_2\cos\phi)\right], \\
u_\eta &= \frac{c}{h_1}\left[\cosh\xi\cos\eta(c_4\cos\phi + c_5\sin\phi) - c_6\sinh\xi\sin\eta + \sinh\xi\cosh\xi(c_2\cos\phi - c_1\sin\phi)\right], \\
u_\phi &= c_5\cos\phi - c_4\sin\phi - \sinh\xi\cos\eta(c_1\cos\phi + c_2\sin\phi) + c_3\cosh\xi\sin\eta,
\end{aligned} \tag{G.11}$$

where c_1–c_6 are constants.

The constitutive relations given by Eqns. (6.36) become

$$\begin{aligned}
\tau_{\xi\xi} &= \lambda(\operatorname{tr}\boldsymbol{\epsilon}) + 2\mu\epsilon_{\xi\xi}, & \tau_{\xi\eta} &= 2\mu\epsilon_{\xi\eta}, \\
\tau_{\eta\eta} &= \lambda(\operatorname{tr}\boldsymbol{\epsilon}) + 2\mu\epsilon_{\eta\eta}, & \tau_{\eta\phi} &= 2\mu\epsilon_{\eta\phi}, \\
\tau_{\phi\phi} &= \lambda(\operatorname{tr}\boldsymbol{\epsilon}) + 2\mu\epsilon_{\phi\phi}, & \tau_{\xi\phi} &= 2\mu\epsilon_{\xi\phi},
\end{aligned} \tag{G.12}$$

while the Navier equations of elasticity (Eqns. (6.37)) are given by

$$\begin{aligned}
\rho_0\frac{\partial^2 u_\xi}{\partial t^2} &= \frac{(\lambda+\mu)}{h_1}\frac{\partial(\operatorname{tr}\boldsymbol{\epsilon})}{\partial\xi} \\
&+ \mu\left\{\frac{1}{h_1^2}\left[\frac{\partial^2 u_\xi}{\partial\xi^2} + \tanh\xi\frac{\partial u_\xi}{\partial\xi} - \tanh^2\xi u_\xi + \frac{\partial^2 u_\xi}{\partial\eta^2} + \cot\eta\frac{\partial u_\xi}{\partial\eta} - \tanh\xi\cot\eta u_\eta\right]\right. \\
&+ \frac{1}{h_1^3}\left[u_\eta\left[\tanh\xi\frac{\partial h_1}{\partial\eta} - \cot\eta\frac{\partial h_1}{\partial\xi}\right] - 2\frac{\partial h_1}{\partial\xi}\frac{\partial u_\eta}{\partial\eta} + 2\frac{\partial h_1}{\partial\eta}\frac{\partial u_\eta}{\partial\xi}\right] \\
&\left. - \frac{u_\xi}{h_1^4}\left[\left(\frac{\partial h_1}{\partial\xi}\right)^2 + \left(\frac{\partial h_1}{\partial\eta}\right)^2\right] - \frac{2}{ch_1}\tanh\xi\operatorname{sech}\xi\csc\eta\frac{\partial u_\phi}{\partial\phi} + \frac{1}{c^2}\operatorname{sech}^2\xi\csc^2\eta\frac{\partial^2 u_\xi}{\partial\phi^2}\right\} \\
&+ \rho_0(\boldsymbol{b}_0)_\xi, \\
\rho_0\frac{\partial^2 u_\eta}{\partial t^2} &= \frac{(\lambda+\mu)}{h_1}\frac{\partial(\operatorname{tr}\boldsymbol{\epsilon})}{\partial\eta} \\
&+ \mu\left\{\frac{1}{h_1^2}\left[\frac{\partial^2 u_\eta}{\partial\xi^2} + \tanh\xi\frac{\partial u_\eta}{\partial\xi} - \cot^2\eta u_\eta + \frac{\partial^2 u_\eta}{\partial\eta^2} + \cot\eta\frac{\partial u_\eta}{\partial\eta} - \tanh\xi\cot\eta u_\xi\right]\right.
\end{aligned}$$

$$+\frac{1}{h_1^3}\left[u_\xi\left[\cot\eta\frac{\partial h_1}{\partial\xi}-\tanh\xi\frac{\partial h_1}{\partial\eta}\right]+2\frac{\partial h_1}{\partial\xi}\frac{\partial u_\xi}{\partial\eta}-2\frac{\partial h_1}{\partial\eta}\frac{\partial u_\xi}{\partial\xi}\right]$$

$$-\frac{u_\eta}{h_1^4}\left[\left(\frac{\partial h_1}{\partial\xi}\right)^2+\left(\frac{\partial h_1}{\partial\eta}\right)^2\right]-\frac{2}{ch_1}\operatorname{sech}\xi\cot\eta\csc\eta\frac{\partial u_\phi}{\partial\phi}+\frac{1}{c^2}\operatorname{sech}^2\xi\csc^2\eta\frac{\partial^2 u_\eta}{\partial\phi^2}\Bigg\}$$

$$+\rho_0(\boldsymbol{b}_0)_\eta,$$

$$\rho_0\frac{\partial^2 u_\phi}{\partial t^2}=\frac{(\lambda+\mu)}{r}\frac{\partial(\operatorname{tr}\boldsymbol{\epsilon})}{\partial\phi}+\mu\Bigg\{\frac{1}{h_1^2}\left[\frac{\partial^2 u_\phi}{\partial\xi^2}+\frac{\partial^2 u_\phi}{\partial\eta^2}+\tanh\xi\frac{\partial u_\phi}{\partial\xi}+\cot\eta\frac{\partial u_\phi}{\partial\eta}\right]$$

$$+\frac{1}{c^2}\operatorname{sech}^2\xi\csc^2\eta\frac{\partial^2 u_\phi}{\partial\phi^2}+\frac{2}{ch_1}\operatorname{sech}\xi\csc\eta\left[\tanh\xi\frac{\partial u_\xi}{\partial\phi}+\cot\eta\frac{\partial u_\eta}{\partial\phi}\right]$$

$$-\frac{1}{h_1^2}(\tanh^2\xi+\cot^2\eta)u_\phi\Bigg\}+\rho_0(\boldsymbol{b}_0)_\phi, \tag{G.13}$$

where the Lamé constants (λ,μ) are related to (E,ν) by means of Eqns. (5.20).

If $\xi=\tilde{\xi}(\boldsymbol{X},t)$, $\eta=\tilde{\eta}(\boldsymbol{X},t)$ and $\phi=\tilde{\phi}(\boldsymbol{X},t)$, using Eqns. (A.13) we have

$$\tilde{\boldsymbol{v}}(\boldsymbol{X},t)=h_1\dot{\tilde{\xi}}\boldsymbol{e}_\xi+h_1\dot{\tilde{\eta}}\boldsymbol{e}_\eta+r\dot{\tilde{\phi}}\boldsymbol{e}_\phi,$$

$$\tilde{\boldsymbol{a}}(\boldsymbol{X},t)=\left[\ddot{\tilde{\xi}}h_1+\left(\dot{\tilde{\xi}}^2-\dot{\tilde{\eta}}^2\right)\frac{\partial h_1}{\partial\xi}+2\dot{\tilde{\xi}}\dot{\tilde{\eta}}\frac{\partial h_1}{\partial\eta}-\frac{rc}{h_1}\dot{\tilde{\phi}}^2\sinh\xi\sin\eta\right]\boldsymbol{e}_\xi$$

$$+\left[\ddot{\tilde{\eta}}h_1+\left(\dot{\tilde{\eta}}^2-\dot{\tilde{\xi}}^2\right)\frac{\partial h_1}{\partial\eta}+2\dot{\tilde{\xi}}\dot{\tilde{\eta}}\frac{\partial h_1}{\partial\xi}-\frac{rc}{h_1}\dot{\tilde{\phi}}^2\cosh\xi\cos\eta\right]\boldsymbol{e}_\eta$$

$$+\left[r\ddot{\tilde{\phi}}+2c\dot{\tilde{\phi}}\left(\dot{\tilde{\xi}}\sinh\tilde{\xi}\sin\tilde{\eta}+\dot{\tilde{\eta}}\cosh\tilde{\xi}\cos\tilde{\eta}\right)\right]\boldsymbol{e}_\phi.$$

H Arbitrary Curvilinear Coordinate Systems

In this appendix, we generalize the procedure presented in Appendix A to arbitrary curvilinear coordinate systems. More detailed treatments may be found in [59] and [209]. In what follows, δ_{ij} is as defined by Eqn. (1.5). As in Appendix A, let $\boldsymbol{x}$ represent the position vector of a point with Cartesian coordinates (x_1, x_2, x_3), and curvilinear coordinates (ξ_1, ξ_2, ξ_3). The position vector is expressed in terms of the curvilinear coordinates as $\boldsymbol{x} = \hat{\boldsymbol{x}}(\xi_1, \xi_2, \xi_3)$, and the inverse transformation is given by $\boldsymbol{\xi} = \hat{\boldsymbol{\xi}}(x_1, x_2, x_3)$. The base vectors of the Cartesian coordinate system are denoted by $(\boldsymbol{i}_1, \boldsymbol{i}_2, \boldsymbol{i}_3)$. From the differential $d\boldsymbol{x} = (\partial\hat{\boldsymbol{x}}/\partial\xi_i)d\xi_i$, we see that a natural way of defining the *covariant basis* $\{\boldsymbol{e}_i\}$ is

$$\boldsymbol{e}_i := \frac{\partial \hat{\boldsymbol{x}}}{\partial \xi_i} = \frac{\partial \hat{x}_m}{\partial \xi_i}\boldsymbol{i}_m. \tag{H.1}$$

The *contravariant basis* (also called the dual or reciprocal basis), $\{\boldsymbol{e}^i\}$, is defined such that

$$\boldsymbol{e}^i \cdot \boldsymbol{e}_j = \delta_{ij}, \tag{H.2}$$

and is given by

$$\boldsymbol{e}^i = \boldsymbol{\nabla}_x \hat{\xi}_i = \frac{\partial \hat{\xi}_i}{\partial x_n}\boldsymbol{i}_n, \tag{H.3}$$

since, as can be seen by applying the chain rule,

$$\boldsymbol{e}^i \cdot \boldsymbol{e}_j = \frac{\partial \hat{x}_m}{\partial \xi_i}\frac{\partial \hat{\xi}_j}{\partial x_n}\boldsymbol{i}_m \cdot \boldsymbol{i}_n = \frac{\partial \hat{x}_m}{\partial \xi_i}\frac{\partial \hat{\xi}_j}{\partial x_n}\delta_{mn} = \frac{\partial \hat{\xi}_j}{\partial x_m}\frac{\partial \hat{x}_m}{\partial \xi_i} = \frac{\partial \hat{\xi}_j}{\partial \xi_i} = \delta_{ij}.$$

Note that the basis vectors $\boldsymbol{e}_i$ or $\boldsymbol{e}^i$ need not be of unit length. By virtue of Eqn. (H.2), the vectors $\{\boldsymbol{e}^i\}$ can be thought of to be the row vectors of $[\boldsymbol{e}_1 \,|\, \boldsymbol{e}_2 \,|\, \boldsymbol{e}_3]^{-1}$ or column vectors of $\begin{bmatrix}\boldsymbol{e}_1\\ \boldsymbol{e}_2\\ \boldsymbol{e}_3\end{bmatrix}^{-1}$.

The covariant and contravariant components of a vector $\boldsymbol{v}$ are defined as $v_i = \boldsymbol{v} \cdot \boldsymbol{e}_i$ and $v^i = \boldsymbol{v} \cdot \boldsymbol{e}^i$, respectively. From these definitions and Eqn. (H.2), it follows that

$$\begin{aligned}\boldsymbol{v} &= v_i\boldsymbol{e}^i = (\boldsymbol{v} \cdot \boldsymbol{e}_i)\boldsymbol{e}^i\\ &= v^i\boldsymbol{e}_i = (\boldsymbol{v} \cdot \boldsymbol{e}^i)\boldsymbol{e}_i.\end{aligned} \tag{H.4}$$

Similarly, the covariant, contravariant and mixed components of a tensor $\boldsymbol{T}$ are defined as $T_{ij} = \boldsymbol{e}_i \cdot \boldsymbol{T}\boldsymbol{e}_j$, $T^{ij} = \boldsymbol{e}^i \cdot (\boldsymbol{T}\boldsymbol{e}^j)$, $T^i_{\cdot j} = \boldsymbol{e}^i \cdot \boldsymbol{T}\boldsymbol{e}_j$ and $T_i^{\cdot j} = \boldsymbol{e}_i \cdot \boldsymbol{T}\boldsymbol{e}^j$. Note that, in general, $T_i^{\cdot j} \neq T^j_{\cdot i}$.

As an example, we have $(\boldsymbol{a} \otimes \boldsymbol{b})_{ij} = a_i b_j$, $(\boldsymbol{a} \otimes \boldsymbol{b})^{i}_{.j} = a^i b_j$ etc. In terms of its components, the tensor $\boldsymbol{T}$ can be expressed as

$$\boldsymbol{T} = T_{ij}\boldsymbol{e}^i \otimes \boldsymbol{e}^j = T^{ij}\boldsymbol{e}_i \otimes \boldsymbol{e}_j = T^{i}_{.j}\boldsymbol{e}_i \otimes \boldsymbol{e}^j = T_{i}^{.j}\boldsymbol{e}^i \otimes \boldsymbol{e}_j.$$

From the above equations, it follows that

$$\boldsymbol{T}\boldsymbol{e}_j = T_{ij}\boldsymbol{e}^i = T^{i}_{.j}\boldsymbol{e}_i,$$
$$\boldsymbol{T}\boldsymbol{e}^j = T^{ij}\boldsymbol{e}_i = T_{i}^{.j}\boldsymbol{e}^i.$$

Let $\boldsymbol{J} := \boldsymbol{i}_k \otimes \boldsymbol{e}_k$, so that $\boldsymbol{J}^{-1} = \boldsymbol{e}^k \otimes \boldsymbol{i}_k$. The components of $\boldsymbol{J}$ and $\boldsymbol{J}^{-1}$ with respect to the Cartesian system are

$$\begin{aligned} J_{ij} &= \boldsymbol{i}_i \cdot (\boldsymbol{i}_k \otimes \boldsymbol{e}_k)\boldsymbol{i}_j = \boldsymbol{e}_i \cdot \boldsymbol{i}_j, \\ (J^{-1})_{ij} &= \boldsymbol{e}^j \cdot \boldsymbol{i}_i. \end{aligned} \tag{H.5}$$

$\boldsymbol{J}$ can be written in matrix form as

$$[J_{ij}] = \begin{bmatrix} \frac{\partial \hat{x}_1}{\partial \xi_1} & \frac{\partial \hat{x}_2}{\partial \xi_1} & \frac{\partial \hat{x}_3}{\partial \xi_1} \\ \frac{\partial \hat{x}_1}{\partial \xi_2} & \frac{\partial \hat{x}_2}{\partial \xi_2} & \frac{\partial \hat{x}_3}{\partial \xi_2} \\ \frac{\partial \hat{x}_1}{\partial \xi_3} & \frac{\partial \hat{x}_2}{\partial \xi_3} & \frac{\partial \hat{x}_3}{\partial \xi_3} \end{bmatrix} = \begin{bmatrix} \boldsymbol{e}_1 \\ \boldsymbol{e}_2 \\ \boldsymbol{e}_3 \end{bmatrix}. \tag{H.6}$$

Using the definition of $\boldsymbol{J}$, we have

$$\boldsymbol{e}_i = \boldsymbol{J}^T\boldsymbol{i}_i, \tag{H.7a}$$
$$\boldsymbol{i}_i = \boldsymbol{J}\boldsymbol{e}^i, \tag{H.7b}$$

from which it follows that

$$\boldsymbol{e}_i = \boldsymbol{J}^T\boldsymbol{J}\boldsymbol{e}^i. \tag{H.8}$$

From Eqns. (H.5), it also follows that

$$\begin{aligned} \boldsymbol{e}_i &= (\boldsymbol{e}_i \cdot \boldsymbol{i}_j)\boldsymbol{i}_j = J_{ij}\boldsymbol{i}_j, \\ \boldsymbol{e}^i &= (\boldsymbol{e}^i \cdot \boldsymbol{i}_j)\boldsymbol{i}_j = (J^{-1})_{ji}\boldsymbol{i}_j, \\ \boldsymbol{i}_i &= (\boldsymbol{i}_i \cdot \boldsymbol{e}_j)\boldsymbol{e}^j = J_{ji}\boldsymbol{e}^j \\ &= (\boldsymbol{i}_i \cdot \boldsymbol{e}^j)\boldsymbol{e}_j = (J^{-1})_{ij}\boldsymbol{e}_j. \end{aligned}$$

The *metric tensors* and their components with respect to the Cartesian basis are

$$\boldsymbol{g} := \boldsymbol{J}\boldsymbol{J}^T = (\boldsymbol{i}_i \otimes \boldsymbol{e}_i)(\boldsymbol{e}_j \otimes \boldsymbol{i}_j) = g_{ij}\boldsymbol{i}_i \otimes \boldsymbol{i}_j; \quad g_{ij} = \boldsymbol{e}_i \cdot \boldsymbol{e}_j, \tag{H.9a}$$
$$\boldsymbol{g}^{-1} = \boldsymbol{J}^{-T}\boldsymbol{J}^{-1} = (\boldsymbol{i}_i \otimes \boldsymbol{e}^i)(\boldsymbol{e}^j \otimes \boldsymbol{i}_j) = g^{ij}\boldsymbol{i}_i \otimes \boldsymbol{i}_j; \quad g^{ij} = \boldsymbol{e}^i \cdot \boldsymbol{e}^j. \tag{H.9b}$$

From the above expressions, it is obvious that $\boldsymbol{g} \in \text{Sym}$, $g_{ik}g^{kj} = [(\boldsymbol{e}_i \cdot \boldsymbol{e}_k)\boldsymbol{e}^k \cdot \boldsymbol{e}^j] = \boldsymbol{e}_i \cdot \boldsymbol{e}^j = \delta_{ij}$ and $g := \det \boldsymbol{g} = (\det \boldsymbol{J})^2$. Note also that, by virtue of Eqns. (H.7) and (H.9),

$$\boldsymbol{J}^{-1}\boldsymbol{g}\boldsymbol{J}^{-T} = \boldsymbol{J}^T\boldsymbol{g}^{-1}\boldsymbol{J} = g_{ij}\boldsymbol{e}^i \otimes \boldsymbol{e}^j = g^{ij}\boldsymbol{e}_i \otimes \boldsymbol{e}_j = \delta_{ij}\boldsymbol{e}_i \otimes \boldsymbol{e}^j = \boldsymbol{e}_i \otimes \boldsymbol{e}^i = \boldsymbol{e}^i \otimes \boldsymbol{e}_i = \boldsymbol{I},$$

i.e., g_{ij}, g^{ij} and δ_{ij} are the covariant, contravariant and mixed components of the identity tensor. A basis $\{e_i\}$ is orthogonal if $g_{ij} = 0,\ i \neq j$, and is orthonormal, if, in addition, each basis vector is of unit length, i.e., $|e_i| = \sqrt{e_i \cdot e_i} = 1$; this in turn implies that for an orthonormal system, $g = I$, so that $J = Q \in$ Orth. It then follows from Eqn. (H.8) that, for an orthonormal basis, $e_i = e^i$. Thus, for an orthonormal curvilinear coordinate system, we need not distinguish between covariant and contravariant components, and one can simply denote the (unit) basis vectors as e_i, as we have already done in Appendix A.

The components of the metric tensor can be used to convert covariant vector and tensor components to contravariant ones and vice versa as we now show. Using Eqns. (H.4), we have

$$e_i = (e_i \cdot e_j)e^j = g_{ij}e^j,$$
$$e^i = (e^i \cdot e^j)e_j = g^{ij}e_j.$$

Thus, if v_i and v^i denote the covariant and contravariant components of v, then

$$v^i = v \cdot e^i = v \cdot (g^{ij}e_j) = g^{ij}v \cdot e_j = g^{ij}v_j, \tag{H.10a}$$
$$v_i = v \cdot e_i = v \cdot (g_{ij}e^j) = g_{ij}v \cdot e^j = g_{ij}v^j. \tag{H.10b}$$

Similarly,

$$\begin{aligned} T_{ij} &= e_i \cdot Te_j = g_{im}g_{jn}T^{mn}, \\ T^{ij} &= e^i \cdot Te^j = g^{im}g^{jn}T_{mn}, \\ T^i_{\cdot j} &= e^i \cdot Te_j = g^{ik}T_{kj} = g_{jk}T^{ik}, \\ T_i^{\cdot j} &= e_i \cdot Te^j = g_{ik}T^{kj} = g^{jk}T_{ik}. \end{aligned} \tag{H.11}$$

The above relations can be written in matrix form in the usual manner by letting the first and second indices denote the row and column of the matrix, respectively, e.g., the last of the above relations can be written as $[v_i] = [g_{ij}][v^j]$ and $[T_i^{\cdot j}] = [g_{ik}][T^{kj}] = [T_{ik}][g_{kj}]^{-1}$.

The components of T^T are related to the components of T as follows:

$$\begin{aligned} (T^T)_{ij} &= e_i \cdot T^T e_j = e_j \cdot Te_i = T_{ji}, \\ (T^T)^{ij} &= e^i \cdot T^T e^j = e^j \cdot Te^i = T^{ji}, \\ (T^T)^i_{\cdot j} &= e^i \cdot T^T e_j = e_j \cdot Te^i = T_j^{\cdot i}, \\ (T^T)_i^{\cdot j} &= e_i \cdot T^T e^j = e^j \cdot Te_i = T^j_{\cdot i}. \end{aligned}$$

Thus, the matrix representation of the covariant and contravariant components of T^T are the transpose of the matrix representation of the respective components of T–for the mixed components, however, this is not true, since, in general, $[T^j_{\cdot i}] \neq [T_j^{\cdot i}]$. For a symmetric tensor S, it follows from the above relations that $S_{ij} = S_{ji}$, $S^{ij} = S^{ji}$, and $S^i_{\cdot j} = S_j^{\cdot i}$. Although, the matrix representation of the covariant and contravariant components of S is symmetric, this is not true of the matrix representation of its mixed components.

The dot product of two vectors u and v is given by

$$u \cdot v = u^i v^j e_i \cdot e_j = g_{ij}u^i v^j = u_i v^i = u^i v_i = g^{ij}u_i v_j.$$

Denoting the scalar triple products $[e_i, e_j, e_k]$ and $\left[e^i, e^j, e^k\right]$ by e_{ijk} and e^{ijk}, respectively, we have

$$e_{ijk} = J_{ip}J_{jq}J_{kr}\left[i_p, i_q, i_r\right] = \epsilon_{pqr}J_{ip}J_{jq}J_{kr} = \epsilon_{ijk}\det J = \epsilon_{ijk}\sqrt{g},$$
$$e^{ijk} = (J^{-1})_{pi}(J^{-1})_{qj}(J^{-1})_{rk}\left[i_p, i_q, i_r\right] = \frac{1}{\sqrt{g}}\epsilon_{ijk} = \frac{1}{g}e_{ijk}.$$

From the above relations, it follows that e_{ijk} and e^{ijk} have the same permutation properties as ϵ_{ijk}, e.g., $e_{ijk} = e_{jki} = -e_{jik}$ etc. Using Eqn. (1.19a) and the above relations, we also have

$$e^{ijk}e_{iqr} = \delta_{jq}\delta_{kr} - \delta_{jr}\delta_{kq}, \tag{H.12}$$

from which it also follows that $e^{ijk}e_{ijm} = 2\delta_{km}$ and $e^{ijk}e_{ijk} = \epsilon_{ijk}\epsilon_{ijk} = 6$.

The cross product of two covariant and contravariant basis vectors is obtained using Eqn. (H.4) as

$$e_i \times e_j = (e_k \cdot e_i \times e_j)e^k = e_{ijk}e^k,$$
$$e^i \times e^j = (e^k \cdot e^i \times e^j)e_k = e^{ijk}e_k.$$

It immediately follows that

$$\begin{aligned} u \times v &= u^j v^k e_j \times e_k = e_{ijk}u^j v^k e^i \\ &= u_j v_k e^j \times e^k = e^{ijk}u_j v_k e_i. \end{aligned}$$

The element of length, element of surface area on a surface parameterized by ξ and η, and with unit normal n, and element of volume are given by

$$ds^2 = dx \cdot dx = e_i \cdot e_j d\xi_i d\xi_j = g_{ij}d\xi_i d\xi_j,$$
$$n dS = \left(\frac{\partial x}{\partial \xi} \times \frac{\partial x}{\partial \eta}\right) d\xi d\eta,$$
$$dV = \left[\frac{\partial x}{\partial \xi_1}, \frac{\partial x}{\partial \xi_2}, \frac{\partial x}{\partial \xi_3}\right] d\xi_1 d\xi_2 d\xi_3 = (\det J)d\xi_1 d\xi_2 d\xi_3 = \sqrt{g}d\xi_1 d\xi_2 d\xi_3.$$

The covariant components of Tu in terms of the components of T and u are given by

$$\begin{aligned} (Tu)_i = e_i \cdot (Tu) &= e_i \cdot \left[(T_k^{\cdot j} e^k \otimes e_j)u\right] = T_k^{\cdot j} u_j e_i \cdot e^k = T_i^{\cdot j} u_j \\ &= e_i \cdot \left[(T_{kj} e^k \otimes e^j)u\right] = T_{kj}u^j e_i \cdot e^k = T_{ij}u^j. \end{aligned}$$

Similarly, one has

$$(RS)_{ij} = R_i^{\cdot k} S_{kj} = R_{ik}S^k_{\cdot j},$$
$$(RS)_i^{\cdot j} = R_i^{\cdot k} S_k^{\cdot j} = R_{ik}S^{kj},$$

and so on.

The trace of a tensor T in component form is given by

$$\operatorname{tr} T = \operatorname{tr}(T_{ij}e^i \otimes e^j) = T_{ij}g^{ij}$$

$$
\begin{aligned}
&= \mathrm{tr}\,(T^{ij} e_i \otimes e_j) = T^{ij} g_{ij} \\
&= \mathrm{tr}\,(T_i^{\cdot j} e^i \otimes e_j) = T_i^{\cdot i} = e_i \cdot (T e^i) \\
&= \mathrm{tr}\,(T^i_{\cdot j} e_i \otimes e^j) = T^i_{\cdot i} = e^i \cdot (T e_i).
\end{aligned}
$$

The tensor inner product of two tensors is given by

$$
R : S = \mathrm{tr}\,(R^T S) = R_{ij} S^{ij} = R^{ij} S_{ij},
$$

with expressions in terms of other components obtained using Eqns. (H.11). The above expression shows that the strain-energy density in linearized elasticity $(\tau : \epsilon)/2$ can be obtained by using the contravariant components of the stress, and the covariant components of the strain.

Choosing $u = e_p$, $v = e_q$ and $w = e_r$ in Eqn. (1.43c), and using the facts that $Te_m = T^i_{\cdot m} e_i$ and $\det T^T = \det T$, we get, analogous to Eqn. (1.52), the relation

$$
e_{pqr} \det T = e_{ijk} T^i_{\cdot p} T^j_{\cdot q} T^k_{\cdot r} = e_{ijk} T_p^{\cdot i} T_q^{\cdot j} T_r^{\cdot k},
$$

Multiplying the above equation by e^{pqr}, we get, analogous to Eqn. (1.50),

$$
\det T = \frac{1}{6} e_{ijk} e^{pqr} T^i_{\cdot p} T^j_{\cdot q} T^k_{\cdot r} = \frac{1}{6} e_{ijk} e^{pqr} T_p^{\cdot i} T_q^{\cdot j} T_r^{\cdot k}.
$$

By choosing $u = e_q \times e_j$ and $v = e^q$ in Eqn. (1.54), we get, analogous to Eqn. (1.60),

$$
\begin{aligned}
(\mathbf{cof}\, T)_{ij} &= \frac{1}{2} e_{imn} e_{jpq} T^{mp} T^{nq}, \\
(\mathbf{cof}\, T)^i_{\cdot j} &= \frac{1}{2} e^{imn} e_{jpq} T_m^{\cdot p} T_n^{\cdot q}.
\end{aligned}
$$

It follows from the above equation and Eqn. (H.12) (or directly by means of Eqn. (1.63)) that

$$
I_2(T) = \mathrm{tr}\,\mathbf{cof}\, T = (\mathbf{cof}\, T)^i_{\cdot i} = \frac{1}{2}\left[T_p^{\cdot p} T_q^{\cdot q} - T_q^{\cdot p} T_p^{\cdot q} \right].
$$

Analogous to Eqn. (1.90), we have $w^i = -e^{ijk} W_{jk}/2$ and $W_{ij} = -e_{ijk} w^k$.

We now discuss the relations between the components of tensors in the curvilinear and Cartesian coordinate systems. Denoting the components in the Cartesian system with an overbar, we have

$$
v_i = v \cdot e_i = J_{ij} v \cdot i_j = J_{ij} \bar{v}_j.
$$

The contravariant components in the curvilinear system can be obtained using Eqn. (H.10a), i.e., $v^i = g^{ij} v_j = g^{ij} J_{jk} \bar{v}_k$, or directly as

$$
v^i = v \cdot e^i = (J^{-1})_{ji} v \cdot i_j = (J^{-1})_{ji} \bar{v}_j.
$$

In matrix form, the above equations can be written as

$$
[v_\alpha] = [J][\bar{v}_a],
$$

$$[v^\alpha] = [J]^{-T}[\bar{v}_a],$$

where Greek and Latin alphabets are used to denote the appropriate components in curvilinear and Cartesian coordinates, respectively.

The transformation relations for the components of a second-order tensor $\boldsymbol{T}$ are

$$T_{ij} = \boldsymbol{e}_i \cdot \boldsymbol{T}\boldsymbol{e}_j = J_{im}J_{jn}\bar{T}_{mn},$$
$$\bar{T}^{ij} = \bar{T}_{ij} = \boldsymbol{i}_i \cdot \boldsymbol{T}\boldsymbol{i}_j = J_{mi}J_{nj}T^{mn},$$
$$T_i^{\cdot j} = \boldsymbol{e}_i \cdot \boldsymbol{T}\boldsymbol{e}^j = J_{im}(J^{-1})_{nj}\bar{T}_{mn},$$
$$T^i_{\cdot j} = \boldsymbol{e}^i \cdot \boldsymbol{T}\boldsymbol{e}_j = (J^{-1})_{mi}J_{jn}\bar{T}_{mn},$$

which in matrix form read

$$[T_{\alpha\beta}] = [J][\bar{T}_{ab}][J]^T, \tag{H.13a}$$
$$[\bar{T}^{ab}] = [\bar{T}_{ab}] = [J]^T[T^{\alpha\beta}][J], \tag{H.13b}$$
$$[T_\alpha^{\cdot\beta}] = [J][\bar{T}_{ab}][J]^{-1}, \tag{H.13c}$$
$$[T^\alpha_{\cdot\beta}] = [J]^{-T}[\bar{T}_{ab}][J]^T. \tag{H.13d}$$

By Eqn. (1.52), we know that $\det \boldsymbol{T}$ is just the determinant of the matrix of components of $\boldsymbol{T}$ in the Cartesian system. Hence, from Eqns. (H.13),

$$\det \boldsymbol{T} = \det[\bar{T}_{ab}] = (\det \boldsymbol{J})^{-2}\det[T_{\alpha\beta}] = (\det \boldsymbol{J})^2 \det[T^{\alpha\beta}] = \det[T_\alpha^{\cdot\beta}] = \det[T^\alpha_{\cdot\beta}].$$

Since

$$(T^{-1})_{ik}T^{kj} = (T^{-1})^{ik}T_{kj} = (T^{-1})^i_{\cdot k}T^k_{\cdot j} = (T^{-1})_i^{\cdot k}T_k^{\cdot j} = \delta_{ij},$$

the matrices of components of the inverse are related to the inverses of the matrices of components as follows:

$$[(T^{-1})_{ij}] = [T^{ij}]^{-1},$$
$$[(T^{-1})^{ij}] = [T_{ij}]^{-1},$$
$$[(T^{-1})_i^{\cdot j}] = [T_i^{\cdot j}]^{-1},$$
$$[(T^{-1})^i_{\cdot j}] = [T^i_{\cdot j}]^{-1}.$$

We now develop the calculus of tensors. Differentiating Eqns. (H.7) with respect to the curvilinear coordinate ξ_j, we get

$$\frac{\partial \boldsymbol{e}_i}{\partial \xi_j} = \frac{\partial \boldsymbol{J}^T}{\partial \xi_j}\boldsymbol{i}_i = \boldsymbol{H}^{(j)}\boldsymbol{e}_i = \boldsymbol{H}_{(j)}\boldsymbol{e}^i, \tag{H.14a}$$
$$\frac{\partial \boldsymbol{e}^i}{\partial \xi_j} = \frac{\partial \boldsymbol{J}^{-1}}{\partial \xi_j}\boldsymbol{i}_i = -\boldsymbol{J}^{-1}\frac{\partial \boldsymbol{J}}{\partial \xi_j}\boldsymbol{J}^{-1}\boldsymbol{i}_i = -[\boldsymbol{H}^{(j)}]^T\boldsymbol{e}^i, \tag{H.14b}$$

where,

$$H^{(j)} := \frac{\partial J^T}{\partial \xi_j} J^{-T} = \frac{\partial e_i}{\partial \xi_j} \otimes e^i, \quad H_{(j)} := \frac{\partial J^T}{\partial \xi_j} J = \frac{\partial e_i}{\partial \xi_j} \otimes e_i. \tag{H.15}$$

$H^{(i)}$ has the following properties:

$$H^{(j)} e_i = \frac{\partial^2 \hat{x}}{\partial \xi_i \partial \xi_j} = H^{(i)} e_j, \quad \left(\text{also } H_{(j)} e^i = H_{(i)} e^j\right), \tag{H.16a}$$

$$W^{(kl)} \equiv \frac{\partial H^{(l)}}{\partial \xi_k} - \frac{\partial H^{(k)}}{\partial \xi_l} + H^{(l)} H^{(k)} - H^{(k)} H^{(l)} = -W^{(lk)} \in \text{Skw}, \tag{H.16b}$$

where the second relation follows by using $\partial^2 g_{ij}/\partial \xi_k \partial \xi_l = \partial^2 g_{ij}/\partial \xi_l \partial \xi_k$ and Eqn. (H.9a). For an orthonormal coordinate system, since $J \in \text{Orth}$, we have $H^{(j)} = H_{(j)} \in \text{Skw}$, and we recover Eqns. (A.4) from Eqns. (H.14).

The gradient of a scalar function Φ can be obtained by noting that

$$(\nabla \Phi, e_i) = \frac{\partial \Phi}{\partial x_j} \frac{\partial \hat{x}_j}{\partial \xi_i} = \frac{\partial \Phi}{\partial \xi_i},$$

from which it follows that

$$\nabla \Phi = \frac{\partial \Phi}{\partial \xi_i} e^i = \frac{\partial \Phi}{\partial \xi_i} (e^i \cdot e^j) e_j = g^{ij} \frac{\partial \Phi}{\partial \xi_i} e_j.$$

The components of ∇v can be obtained from the relation

$$(\nabla v) e_j = \frac{\partial v}{\partial x_k} \frac{\partial \hat{x}_k}{\partial \xi_j} = \frac{\partial v}{\partial \xi_j}, \tag{H.17}$$

from which it follows that the covariant components of ∇v are

$$(\nabla v)_{ij} = e_i \cdot (\nabla v) e_j = \frac{\partial \hat{x}}{\partial \xi_i} \cdot \frac{\partial v}{\partial \xi_j}. \tag{H.18}$$

To get an expression for ∇v in terms of the components of v, we use Eqns. (H.14) to write the right-hand side of Eqn. (H.17) as

$$\begin{aligned} \frac{\partial v}{\partial \xi_j} &= \frac{\partial (v^k e_k)}{\partial \xi_j} = \frac{\partial v^k}{\partial \xi_j} e_k + v^k H^{(j)} e_k = \frac{\partial v^k}{\partial \xi_j} e_k + v^k H^{(k)} e_j \\ &= \frac{\partial (v_k e^k)}{\partial \xi_j} = \frac{\partial v_k}{\partial \xi_j} e^k - v_k \left[H^{(j)}\right]^T e^k. \end{aligned}$$

Thus, ∇v and its covariant components, and $\nabla \cdot v$ are given by

$$(\nabla v)_{ij} = g_{ik} \frac{\partial v^k}{\partial \xi_j} + v^k \left[e_i \cdot H^{(j)} e_k\right] = g_{ik} \frac{\partial v^k}{\partial \xi_j} + v^k \left[e_i \cdot H^{(k)} e_j\right]$$

$$= \frac{\partial v_i}{\partial \xi_j} - v_k \left[e_i \cdot \left[H^{(j)} \right]^T e^k \right] = \frac{\partial v_i}{\partial \xi_j} - v_k \left[e^k \cdot H^{(j)} e^i \right],$$

$$\nabla v = \frac{\partial v}{\partial \xi_j} \otimes e^j = \frac{\partial v_i}{\partial \xi_j} e^i \otimes e^j - v_i \left\{ \left[H^{(j)} \right]^T e^i \right\} \otimes e^j$$

$$= \frac{\partial v^i}{\partial \xi_j} e_i \otimes e^j + v^i \left[H^{(j)} e_i \right] \otimes e^j = \frac{\partial v^i}{\partial \xi_j} e_i \otimes e^j + v^i \left[H^{(i)} e_j \right] \otimes e^j,$$

$$\nabla \cdot v = \operatorname{tr} \nabla v = \frac{\partial v}{\partial \xi_j} \cdot e^j = g^{ij} \frac{\partial v_i}{\partial \xi_j} - v_i e^i \cdot H^{(j)} e^j = \frac{\partial v^i}{\partial \xi_i} + v^i \operatorname{tr} H^{(i)}. \tag{H.19}$$

It follows that $(\nabla \times v)^i = -e^{ijk} (\nabla v)_{jk}$. Also, substituting $v_i = (\partial \Phi / \partial \xi_i)$ in the expression for $\nabla \cdot v$, we get

$$\nabla^2 \Phi = g^{ij} \frac{\partial^2 \Phi}{\partial \xi_i \partial \xi_j} - \frac{\partial \Phi}{\partial \xi_i} e^i \cdot H^{(j)} e^j. \tag{H.20}$$

Similar to Eqn. (H.17), we have

$$(\nabla T) e_k = \frac{\partial T}{\partial \xi_k} = \frac{\partial}{\partial \xi_k} (T^{ij} e_i \otimes e_j) = \frac{\partial T^{ij}}{\partial \xi_k} e_i \otimes e_j + T^{ij} \left[H^{(k)} e_i \right] \otimes e_j + T^{ij} e_i \otimes \left[H^{(k)} e_j \right]$$

$$= \frac{\partial}{\partial \xi_k} (T_{ij} e^i \otimes e^j) = \frac{\partial T_{ij}}{\partial \xi_k} e^i \otimes e^j - T_{ij} \left[H^{(k)} \right]^T e^i \otimes e^j - T_{ij} e^i \otimes \left[H^{(k)} \right]^T e^j,$$

so that

$$\nabla T = \frac{\partial T^{ij}}{\partial \xi_k} e_i \otimes e_j \otimes e^k + T^{ij} \left[H^{(k)} e_i \right] \otimes e_j \otimes e^k + T^{ij} e_i \otimes \left[H^{(k)} e_j \right] \otimes e^k$$

$$= \frac{\partial T_{ij}}{\partial \xi_k} e^i \otimes e^j \otimes e^k - T_{ij} \left[[H^{(k)}]^T e^i \right] \otimes e^j \otimes e^k - T_{ij} e^i \otimes \left[[H^{(k)}]^T e^j \right] \otimes e^k$$

$$\nabla \cdot T = \frac{\partial T^{ij}}{\partial \xi_j} e_i + T^{ij} H^{(j)} e_i + T^{ij} (\operatorname{tr} H^{(j)}) e_i$$

$$= g^{jk} \frac{\partial T_{ij}}{\partial \xi_k} e^i - g^{jk} T_{ij} \left[H^{(k)} \right]^T e^i - T_{ij} \left[e^j \cdot H^{(k)} e^k \right] e^i.$$

Taking $T_{ij} \equiv (\nabla v)_{ij} = \partial v_i / \partial \xi_j - v_k \left[e^k \cdot H^{(j)} e^i \right]$ in the above expression, we obtain $\nabla \cdot (\nabla v) = \nabla^2 v$.

The Christoffel symbol of the first kind is defined as

$$\Gamma_{ijk} := e_k \cdot \frac{\partial e_i}{\partial \xi_j} = e_k \cdot \frac{\partial e_j}{\partial \xi_i} = e_k \cdot H^{(j)} e_i = e_k \cdot H^{(i)} e_j = \Gamma_{jik}.$$

From the above equation, we see that Γ_{ijk} are the (k,i)th and (k,j)th components of the *second-order* tensors $H^{(j)}$ and $H^{(i)}$, respectively. It follows that

$$\frac{\partial e_j}{\partial \xi_i} = H^{(i)} e_j = H^{(j)} e_i = \left[\frac{\partial e_j}{\partial \xi_i} \cdot e_k \right] e^k = \Gamma_{ijk} e^k, \tag{H.21a}$$

$$e_i \cdot \frac{\partial e^j}{\partial \xi_k} = -e^j \cdot \frac{\partial e_i}{\partial \xi_k} = -g^{jl} e_l \cdot \frac{\partial e_i}{\partial \xi_k} = -g^{jl} \Gamma_{ikl}. \tag{H.21b}$$

Using the relations $g_{ij} = e_i \cdot e_j$ and $\partial e_i / \partial \xi_j = \partial e_j / \partial \xi_i$, we obtain

$$\Gamma_{ijk} = \frac{1}{2}\left(\frac{\partial g_{ik}}{\partial \xi_j} + \frac{\partial g_{jk}}{\partial \xi_i} - \frac{\partial g_{ij}}{\partial \xi_k} \right). \tag{H.22}$$

As an application of the above development, let $\boldsymbol{X}$ and $\boldsymbol{x} = \boldsymbol{\chi}(\boldsymbol{X}, t)$ be the position vectors of a point in the reference and spatial configurations with Cartesian coordinates (X_1, X_2, X_3) and (x_1, x_2, x_3), and curvilinear coordinates $\boldsymbol{\beta} \equiv (\eta_1, \eta_2, \eta_3)$ and $\boldsymbol{\xi} \equiv (\xi_1, \xi_2, \xi_3)$, respectively. Also let $\boldsymbol{x} = \hat{\boldsymbol{x}}(\xi_1, \xi_2, \xi_3)$, $\boldsymbol{X} = \hat{\boldsymbol{X}}(\eta_1, \eta_2, \eta_3)$ and $\boldsymbol{\xi} = \hat{\boldsymbol{\xi}}(x_1, x_2, x_3) \equiv \tilde{\boldsymbol{\xi}}(X_1, X_2, X_3, t)$, so that $\boldsymbol{\xi} = \hat{\boldsymbol{\xi}} \circ \boldsymbol{\chi} \circ \hat{\boldsymbol{X}}(\boldsymbol{\beta}) \equiv \hat{\boldsymbol{\chi}}(\eta_1, \eta_2, \eta_3, t)$ represents the mapping from curvilinear material to curvilinear spatial coordinates. Then,

$$\begin{aligned}
\tilde{\boldsymbol{v}} &= \left(\frac{\partial \boldsymbol{\chi}}{\partial t}\right)_{\boldsymbol{X}} = \frac{\partial \hat{\boldsymbol{x}}}{\partial \xi_i}\left(\frac{\partial \tilde{\xi}_i}{\partial t}\right)_{\boldsymbol{X}} = \left(\frac{\partial \tilde{\xi}_i}{\partial t}\right)_{\boldsymbol{X}} \boldsymbol{e}_i, \\
\tilde{\boldsymbol{a}} &= \left(\frac{\partial^2 \tilde{\xi}_i}{\partial t^2}\right)_{\boldsymbol{X}} \boldsymbol{e}_i + \left(\frac{\partial \tilde{\xi}_i}{\partial t}\right)_{\boldsymbol{X}} \left(\frac{\partial \tilde{\xi}_j}{\partial t}\right)_{\boldsymbol{X}} \boldsymbol{H}^{(j)} \boldsymbol{e}_i \\
&= \left(\frac{\partial^2 \tilde{\xi}_i}{\partial t^2}\right)_{\boldsymbol{X}} \boldsymbol{e}_i + \left(\frac{\partial \tilde{\xi}_i}{\partial t}\right)_{\boldsymbol{X}} \left(\frac{\partial \tilde{\xi}_j}{\partial t}\right)_{\boldsymbol{X}} \boldsymbol{H}_{(j)} \boldsymbol{e}^i, \quad \text{(compare with Eqns. (A.13))} \\
\boldsymbol{a} &= \left(\frac{\partial v^i}{\partial t}\right)_{\boldsymbol{x}} \boldsymbol{e}_i + v^j \frac{\partial v^i}{\partial \xi_j} \boldsymbol{e}_i + v^i v^j \boldsymbol{H}^{(j)} \boldsymbol{e}_i \\
&= \left(\frac{\partial v^i}{\partial t}\right)_{\boldsymbol{x}} \boldsymbol{e}_i + v^j \frac{\partial v^i}{\partial \xi_j} \boldsymbol{e}_i + v^i v^j \boldsymbol{H}_{(j)} \boldsymbol{e}^i, \quad \text{(since } (\partial \boldsymbol{e}_i / \partial t)_{\boldsymbol{x}} = \boldsymbol{0}) \\
F^i_{.J} &= \boldsymbol{e}^i \cdot \boldsymbol{F} \boldsymbol{e}_J = \frac{\partial \hat{\xi}_i}{\partial x_m} \frac{\partial \chi_m}{\partial X_N} \frac{\partial \hat{X}_N}{\partial \eta_J} = \frac{\partial \hat{\xi}_i}{\partial \eta_J}, \quad \text{(compare with Eqn. (A.27))} \\
(F^T)^J_{.i} &= F_i^{.J} = g_{ij} g^{JN} F^j_{.N}, \quad \text{where } g^{JN} = \boldsymbol{e}^J \cdot \boldsymbol{e}^N, \\
C_{IJ} &= (F^T)_{Ii} F^i_{.J} = F_{iI} F^i_{.J} = g_{ij} F^j_{.I} F^i_{.J} = g_{ij} F^i_{.I} F^j_{.J}, \quad \text{(since } g_{ij} = g_{ji}) \\
B^{ij} &= F^i_{.I} (F^T)^{Ij} = F^i_{.I} F^{jI} = g^{IJ} F^i_{.I} F^j_{.J}, \\
\bar{F}_{iJ} &= \boldsymbol{i}_i \cdot \boldsymbol{F} \boldsymbol{i}_J = J_{ji} (J^{-1})_{JK} \boldsymbol{e}^j \cdot \boldsymbol{F} \boldsymbol{e}_K = J_{ji} (J^{-1})_{JK} F^j_{.K} \\
\Longrightarrow \det \boldsymbol{F} &= \sqrt{\frac{g}{G}} \det[F^a_{.A}], \text{ where } g = \det[g_{mn}], \; G = \det[g_{MN}], \\
|(\mathbf{cof}\, \boldsymbol{F}) \boldsymbol{n}_0| &= \sqrt{\boldsymbol{n}_0 \cdot (\mathbf{cof}\, \boldsymbol{F})^T (\mathbf{cof}\, \boldsymbol{F}) \boldsymbol{n}_0} = \sqrt{\boldsymbol{n}_0 \cdot (\mathbf{cof}\, \boldsymbol{C}) \boldsymbol{n}_0} = \det \boldsymbol{F} \sqrt{(n_0)_I (C^{-1})^{IJ} (n_0)_J}, \\
\epsilon_{IJ} &= \frac{1}{2}\left(\boldsymbol{e}_I \cdot \frac{\partial \boldsymbol{u}}{\partial \eta_J} + \boldsymbol{e}_J \cdot \frac{\partial \boldsymbol{u}}{\partial \eta_I} \right), \quad \text{(by Eqn. (H.18))} \\
E_{IJ} &= \frac{1}{2}\left[\boldsymbol{e}_I \cdot \frac{\partial \boldsymbol{u}}{\partial \eta_J} + \boldsymbol{e}_J \cdot \frac{\partial \boldsymbol{u}}{\partial \eta_I} + \left(\boldsymbol{e}_K \cdot \frac{\partial \boldsymbol{u}}{\partial \eta_I} \right) \left(\boldsymbol{e}^K \cdot \frac{\partial \boldsymbol{u}}{\partial \eta_J} \right) \right] \\
&= \frac{1}{2}\left[\boldsymbol{e}_I \cdot \frac{\partial \boldsymbol{u}}{\partial \eta_J} + \boldsymbol{e}_J \cdot \frac{\partial \boldsymbol{u}}{\partial \eta_I} + \frac{\partial \boldsymbol{u}}{\partial \eta_I} \cdot \frac{\partial \boldsymbol{u}}{\partial \eta_J} \right],
\end{aligned}$$

where $[(C^{-1})^{IJ}] = [C_{IJ}]^{-1}$ and $e_I = \partial\hat{X}/\partial\eta_I$. In obtaining the expression for E_{IJ}, we have used the fact that $[(\nabla u)^T(\nabla u)]_{IJ} = (\nabla u)^T_{IK}(\nabla u)^K_{.J} = (\nabla u)_{KI}(\nabla u)^K_{.J}$, and the fact that $v = (v \cdot e^K)e_K$.

The curvature tensor R is defined as

$$R := \left[e_i \cdot \{\nabla[(\nabla e_j)e_l]e_k - \nabla[(\nabla e_j)e_k]e_l\}\right] e^i \otimes e^j \otimes e^k \otimes e^l.$$

In component form, we have

$$\begin{aligned}
R_{ijkl} &= e_i \cdot \left[\frac{\partial}{\partial\xi_k}\left(\frac{\partial e_j}{\partial\xi_l}\right) - \frac{\partial}{\partial\xi_l}\left(\frac{\partial e_j}{\partial\xi_k}\right)\right] && \text{(H.23)}\\
&= e_i \cdot \left[\frac{\partial}{\partial\xi_k}\left(\Gamma_{ljp}e^p\right) - \frac{\partial}{\partial\xi_l}\left(\Gamma_{kjp}e^p\right)\right] \qquad \text{(by Eqn. (H.21a))} &&\\
&= \frac{\partial\Gamma_{jli}}{\partial\xi_k} - \frac{\partial\Gamma_{jki}}{\partial\xi_l} + g^{pq}\left[\Gamma_{jkp}\Gamma_{ilq} - \Gamma_{jlp}\Gamma_{ikq}\right] \qquad \text{(by Eqn. (H.21b))} && \text{(H.24)}\\
&= \frac{1}{2}\left(\frac{\partial^2 g_{li}}{\partial\xi_j\partial\xi_k} + \frac{\partial^2 g_{jk}}{\partial\xi_i\partial\xi_l} - \frac{\partial^2 g_{jl}}{\partial\xi_i\partial\xi_k} - \frac{\partial^2 g_{ki}}{\partial\xi_j\partial\xi_l}\right) + g^{pq}\left[\Gamma_{jkp}\Gamma_{ilq} - \Gamma_{jlp}\Gamma_{ikq}\right], && \text{(H.25)}
\end{aligned}$$

where the last step follows from Eqn. (H.22). An alternative expression for the curvature tensor components is obtained from Eqn. (H.23) as follows:

$$\begin{aligned}
R_{ijkl} &= e_i \cdot \left[\frac{\partial}{\partial\xi_k}\left(H^{(l)}e_j\right) - \frac{\partial}{\partial\xi_l}\left(H^{(k)}e_j\right)\right]\\
&= e_i \cdot \left[\frac{\partial H^{(l)}}{\partial\xi_k} - \frac{\partial H^{(k)}}{\partial\xi_l} + H^{(l)}H^{(k)} - H^{(k)}H^{(l)}\right] e_j\\
&= e_i \cdot W^{(kl)} e_j. \qquad \text{(by Eqn. (H.16b))} && \text{(H.26)}
\end{aligned}$$

Since $W^{(kl)} \in \text{Skw}$ and $W^{(kl)} = -W^{(lk)}$, the symmetries $R_{ijkl} = -R_{jikl} = -R_{ijlk}$ are immediately evident from the above equation. If the space is three-dimensional, Eqn. (H.26) can also be written as

$$R_{ijkl} = \left[e_i, \omega^{(kl)}, e_j\right] = -\left[\omega^{(kl)}, e_i, e_j\right],$$

where $\omega^{(kl)}$ is the axial vector of $W^{(kl)}$.

The curvature tensor is zero in a Cartesian coordinate system (since $\Gamma_{ijk} = 0$), and hence zero in any coordinate system in the (Euclidean) space $\Re^3$, i.e.,

$$R_{ijkl} = 0, \quad i,j,k,l = 1,2,3.$$

However, if one considers a curved surface described by ξ_α, $\alpha = 1,2$, then the two-dimensional curvature tensor, $R_{\alpha\beta\gamma\delta}$, $\alpha,\beta,\gamma,\delta = 1,2$, obtained by omitting all terms in which the index 3 occurs at least once, is not zero in general.

As an illustration, let the position vector of a point be given by

$$\hat{x} = \phi(\xi_1,\xi_2) + \xi_3 e_3(\xi_1,\xi_2),$$

where $\boldsymbol{\phi}(\xi_1, \xi_2)$ is a curved surface described by the curvilinear coordinates (ξ_1, ξ_2), and $e_3 = \boldsymbol{\phi}_{,1} \times \boldsymbol{\phi}_{,2} / \left\| \boldsymbol{\phi}_{,1} \times \boldsymbol{\phi}_{,2} \right\|$ (with the comma denoting a partial derivative) is a unit vector normal to the curved surface. The other two covariant basis vectors on the curved surface are given by

$$\boldsymbol{e}_\alpha = \left. \frac{\partial \hat{\boldsymbol{x}}}{\partial \xi_\alpha} \right|_{\xi_3 = 0} = \frac{\partial \boldsymbol{\phi}}{\partial \xi_\alpha}, \quad \alpha = 1, 2.$$

By Eqn. (H.21a), we have

$$\frac{\partial \boldsymbol{e}_j}{\partial \xi_i} = \Gamma_{ijk} \boldsymbol{e}^k = (g^{kl} \Gamma_{ijk}) \boldsymbol{e}_l = \Gamma^l_{ij} \boldsymbol{e}_l = \Gamma^\alpha_{ij} \boldsymbol{e}_\alpha + \Gamma^3_{ij} \boldsymbol{e}_3, \quad \alpha = 1, 2, \tag{H.27}$$

where $\Gamma^k_{ij} := g^{km} \Gamma_{ijm} = \Gamma^k_{ji}$ is the Christoffel symbol of the second kind. The two dimensional counterpart of Eqn. (H.27) is obtained simply by dropping the "out-of-plane" contribution given by the last term, and we denote this new "planar" covariant derivative by a semi-colon, i.e.,

$$\boldsymbol{e}_{\beta;\alpha} := \Gamma^\gamma_{\alpha\beta} \boldsymbol{e}_\gamma, \quad \alpha, \beta, \gamma = 1, 2.$$

The curvature tensor $R_{\alpha\beta\gamma\delta} = \boldsymbol{e}_\alpha \cdot [\boldsymbol{e}_{\beta;\delta\gamma} - \boldsymbol{e}_{\beta;\gamma\delta}]$ formed using the above definition of covariant derivative is not zero in general, since the terms with index 3 have been dropped. However, it does obey the symmetry relations given by Eqn. (2.25), and also the constraint equations known as the Bianchi identities, and given by

$$R_{\alpha\beta\gamma\delta;\eta} + R_{\alpha\beta\eta\gamma;\delta} + R_{\alpha\beta\delta\eta;\gamma} = 0.$$

For orthogonal curvilinear coordinate systems, it is advantageous to normalize the basis vectors so that they are of unit length, and then use the components of tensors referred to this orthonormal basis (as mentioned in Appendix A, such components are referred to as physical components)–this was the procedure used in Appendices A–G. However, just for the sake of illustration, we shall now outline how the component forms can be derived with respect to the covariant and contravariant basis vectors of the cylindrical and spherical coordinate systems. Note that these basis vectors are no longer of unit length. We shall now denote the covariant basis vectors by $(\boldsymbol{e}_r, \boldsymbol{e}_\theta, \boldsymbol{e}_z)$ and $(\boldsymbol{e}_r, \boldsymbol{e}_\theta, \boldsymbol{e}_\phi)$, respectively–these *should not be confused* with the *orthonormal* basis vectors denoted by the same symbols in Appendices B and C.

For the cylindrical coordinate system $(\xi_1, \xi_2, \xi_3) \equiv (r, \theta, z)$ and $\boldsymbol{x} = (r\cos\theta, r\sin\theta, z)$. Using Eqns. (H.1) and (H.2), we get

$$\boldsymbol{e}_r = (\cos\theta, \sin\theta, 0), \qquad \boldsymbol{e}^r = \boldsymbol{e}_r,$$

$$\boldsymbol{e}_\theta = (-r\sin\theta, r\cos\theta, 0), \qquad \boldsymbol{e}^\theta = \frac{1}{r^2} \boldsymbol{e}_\theta,$$

$$\boldsymbol{e}_z = (0, 0, 1), \qquad \boldsymbol{e}^z = \boldsymbol{e}_z.$$

Via Eqns. (H.6), (H.9) and (H.15), this leads to

$$\boldsymbol{J} = \begin{bmatrix} \cos\theta & \sin\theta & 0 \\ -r\sin\theta & r\cos\theta & 0 \\ 0 & 0 & 1 \end{bmatrix}, \qquad \boldsymbol{g} = \begin{bmatrix} 1 & 0 & 0 \\ 0 & r^2 & 0 \\ 0 & 0 & 1 \end{bmatrix}, \qquad \boldsymbol{g}^{-1} = \begin{bmatrix} 1 & 0 & 0 \\ 0 & \frac{1}{r^2} & 0 \\ 0 & 0 & 1 \end{bmatrix},$$

$$H^{(r)} = \begin{bmatrix} \frac{\sin^2\theta}{r} & -\frac{\sin\theta\cos\theta}{r} & 0 \\ -\frac{\sin\theta\cos\theta}{r} & \frac{\cos^2\theta}{r} & 0 \\ 0 & 0 & 0 \end{bmatrix}, \quad H^{(\theta)} = \begin{bmatrix} 0 & -1 & 0 \\ 1 & 0 & 0 \\ 0 & 0 & 0 \end{bmatrix}, \quad H^{(z)} = \mathbf{0}.$$

Thus, from Eqns. (H.19) and (H.20), we get

$$\begin{aligned} \nabla \cdot v &= \frac{\partial v^r}{\partial r} + \frac{\partial v^\theta}{\partial \theta} + \frac{\partial v^z}{\partial z} + v^r \operatorname{tr} H^{(r)} + v^\theta \operatorname{tr} H^{(\theta)} + v^z \operatorname{tr} H^{(z)} \\ &= \frac{\partial v^r}{\partial r} + \frac{\partial v^\theta}{\partial \theta} + \frac{\partial v^z}{\partial z} + \frac{v^r}{r}, \\ \nabla^2 \Phi &= g^{rr}\frac{\partial^2\Phi}{\partial r^2} + g^{\theta\theta}\frac{\partial^2\Phi}{\partial \theta^2} + g^{zz}\frac{\partial^2\Phi}{\partial z^2} - \left[\frac{\partial\Phi}{\partial r}e^r + \frac{\partial\Phi}{\partial\theta}e^\theta + \frac{\partial\Phi}{\partial z}e^z\right] \cdot h \\ &= \frac{\partial^2\Phi}{\partial r^2} + \frac{1}{r^2}\frac{\partial^2\Phi}{\partial\theta^2} + \frac{\partial^2\Phi}{\partial z^2} + \frac{1}{r}\frac{\partial\Phi}{\partial r}, \end{aligned}$$

where

$$h = H^{(r)}e^r + H^{(\theta)}e^\theta + H^{(z)}e^z = \begin{bmatrix} -\frac{\cos\theta}{r} \\ -\frac{\sin\theta}{r} \\ 0 \end{bmatrix}.$$

The equivalence of the above relations with Eqns. (B.4) and (B.6) follows by noting that the physical components (v_r, v_θ, v_z) in Appendix B are given by (v^r, rv^θ, v^z).

For the spherical coordinate system $(\xi_1, \xi_2, \xi_3) \equiv (r, \theta, \phi)$ and $x = (r\sin\theta\cos\phi, r\sin\theta\sin\phi, r\cos\theta)$. Using Eqns. (H.1) and (H.2), we get

$$\begin{aligned} e_r &= (\sin\theta\cos\phi, \sin\theta\sin\phi, \cos\theta), & e^r &= e_r, \\ e_\theta &= (r\cos\theta\cos\phi, r\cos\theta\sin\phi, -r\sin\theta), & e^\theta &= \frac{1}{r^2}e_\theta, \\ e_\phi &= (-r\sin\theta\sin\phi, r\sin\theta\cos\phi, 0), & e^\phi &= \frac{1}{r^2\sin^2\theta}e_\phi, \end{aligned}$$

which on substituting into Eqns. (H.6), (H.9) and (H.15), leads to

$$J = \begin{bmatrix} \sin\theta\cos\phi & \sin\theta\sin\phi & \cos\theta \\ r\cos\theta\cos\phi & r\cos\theta\sin\phi & -r\sin\theta \\ -r\sin\theta\sin\phi & r\sin\theta\cos\phi & 0 \end{bmatrix},$$

$$g = \begin{bmatrix} 1 & 0 & 0 \\ 0 & r^2 & 0 \\ 0 & 0 & r^2\sin^2\theta \end{bmatrix}, \quad g^{-1} = \begin{bmatrix} 1 & 0 & 0 \\ 0 & \frac{1}{r^2} & 0 \\ 0 & 0 & \frac{1}{r^2\sin^2\theta} \end{bmatrix},$$

$$H^{(r)} = \begin{bmatrix} \frac{\cos^2\theta\cos^2\phi + \sin^2\phi}{r} & -\frac{\sin^2\theta\sin\phi\cos\phi}{r} & -\frac{\sin\theta\cos\theta\cos\phi}{r} \\ -\frac{\sin^2\theta\sin\phi\cos\phi}{r} & \frac{\cos^2\theta\sin^2\phi + \cos^2\phi}{r} & -\frac{\sin\theta\cos\theta\sin\phi}{r} \\ -\frac{\sin\theta\cos\theta\cos\phi}{r} & -\frac{\sin\theta\cos\theta\sin\phi}{r} & \frac{\sin^2\theta}{r} \end{bmatrix},$$

$$H^{(\theta)} = \begin{bmatrix} \frac{\cos\theta\sin^2\phi}{\sin\theta} & -\frac{\cos\theta\sin\phi\cos\phi}{\sin\theta} & \cos\phi \\ -\frac{\cos\theta\sin\phi\cos\phi}{\sin\theta} & \frac{\cos\theta\cos^2\phi}{\sin\theta} & \sin\phi \\ -\cos\phi & -\sin\phi & 0 \end{bmatrix},$$

$$H^{(\phi)} = \begin{bmatrix} 0 & -1 & 0 \\ 1 & 0 & 0 \\ 0 & 0 & 0 \end{bmatrix}.$$

Thus, from Eqns. (H.19) and (H.20), we get

$$\begin{aligned} \nabla \cdot v &= \frac{\partial v^r}{\partial r} + \frac{\partial v^\theta}{\partial \theta} + \frac{\partial v^\phi}{\partial \phi} + v^r \operatorname{tr} H^{(r)} + v^\theta \operatorname{tr} H^{(\theta)} + v^\phi \operatorname{tr} H^{(\phi)} \\ &= \frac{\partial v^r}{\partial r} + \frac{\partial v^\theta}{\partial \theta} + \frac{\partial v^\phi}{\partial \phi} + \frac{2v^r}{r} + \cot\theta v^\theta, \\ \nabla^2\Phi &= g^{rr}\frac{\partial^2\Phi}{\partial r^2} + g^{\theta\theta}\frac{\partial^2\Phi}{\partial \theta^2} + g^{\phi\phi}\frac{\partial^2\Phi}{\partial \phi^2} - \left[\frac{\partial\Phi}{\partial r}e^r + \frac{\partial\Phi}{\partial\theta}e^\theta + \frac{\partial\Phi}{\partial\phi}e^\phi\right]\cdot h \\ &= \frac{\partial^2\Phi}{\partial r^2} + \frac{1}{r^2}\frac{\partial^2\Phi}{\partial\theta^2} + \frac{1}{r^2\sin^2\theta}\frac{\partial^2\Phi}{\partial\phi^2} + \frac{2}{r}\frac{\partial\Phi}{\partial r} + \frac{\cot\theta}{r^2}\frac{\partial\Phi}{\partial\theta}, \end{aligned}$$

where

$$h = H^{(r)}e^r + H^{(\theta)}e^\theta + H^{(\phi)}e^\phi = \begin{bmatrix} \frac{(\cos 2\theta - 3)\cos\phi}{2r\sin\theta} \\ \frac{(\cos 2\theta - 3)\sin\phi}{2r\sin\theta} \\ -\frac{\cos\theta}{r} \end{bmatrix}.$$

The equivalence of the above relations with Eqns. (C.4) and (C.7) follows by noting that the physical components (v_r, v_θ, v_ϕ) in Appendix C are given by $(v^r, rv^\theta, r\sin\theta v^\phi)$.

I Matrix Representations of Tensors for Engineering Applications

In engineering applications, it is often convenient, by exploiting existent symmetries, to express tensorial expressions such as $\tau_{ij} = \mathsf{C}_{ijkl}\epsilon_{kl}$ in compact matrix form as $\boldsymbol{\tau}_{\text{engrg}} = \boldsymbol{C}_{\text{engrg}}\boldsymbol{\epsilon}_{\text{engrg}}$, where $\boldsymbol{\tau}_{\text{engrg}}$ and $\boldsymbol{\epsilon}_{\text{engrg}}$ are 6×1 column vectors, and $\boldsymbol{C}_{\text{engrg}}$ is a 6×6 matrix. In this appendix, we outline a systematic procedure due to Nadeau and Ferrari [233] for carrying out such tensor to matrix conversions. In what follows, $\mathbb{I} = \delta_{ik}\delta_{jl}\boldsymbol{e}_i \otimes \boldsymbol{e}_j \otimes \boldsymbol{e}_k \otimes \boldsymbol{e}_l$ and $\mathbb{S} = \frac{1}{2}(\delta_{ik}\delta_{jl} + \delta_{il}\delta_{jk})\boldsymbol{e}_i \otimes \boldsymbol{e}_j \otimes \boldsymbol{e}_k \otimes \boldsymbol{e}_l$ represent the fourth-order identity and symmetrizer tensors, respectively.

Let $\boldsymbol{U}$ and $\boldsymbol{V}$ be arbitrary second-order tensors, and $\mathbb{U}$ and $\mathbb{V}$ be arbitrary fourth-order tensors. The mapping Ψ is defined as

$$\Psi(\boldsymbol{U}) := \begin{bmatrix} U_{11} \\ U_{22} \\ U_{33} \\ U_{12} \\ U_{23} \\ U_{31} \\ U_{21} \\ U_{32} \\ U_{13} \end{bmatrix}, \quad \Psi(\mathbb{U}) := \begin{bmatrix} \mathbb{U}_{1111} & \mathbb{U}_{1122} & \mathbb{U}_{1133} & \mathbb{U}_{1112} & \mathbb{U}_{1123} & \mathbb{U}_{1131} & \mathbb{U}_{1121} & \mathbb{U}_{1132} & \mathbb{U}_{1113} \\ \mathbb{U}_{2211} & \mathbb{U}_{2222} & \mathbb{U}_{2233} & \mathbb{U}_{2212} & \mathbb{U}_{2223} & \mathbb{U}_{2231} & \mathbb{U}_{2221} & \mathbb{U}_{2232} & \mathbb{U}_{2213} \\ \mathbb{U}_{3311} & \mathbb{U}_{3322} & \mathbb{U}_{3333} & \mathbb{U}_{3312} & \mathbb{U}_{3323} & \mathbb{U}_{3331} & \mathbb{U}_{3321} & \mathbb{U}_{3332} & \mathbb{U}_{3313} \\ \mathbb{U}_{1211} & \mathbb{U}_{1222} & \mathbb{U}_{1233} & \mathbb{U}_{1212} & \mathbb{U}_{1223} & \mathbb{U}_{1231} & \mathbb{U}_{1221} & \mathbb{U}_{1232} & \mathbb{U}_{1213} \\ \mathbb{U}_{2311} & \mathbb{U}_{2322} & \mathbb{U}_{2333} & \mathbb{U}_{2312} & \mathbb{U}_{2323} & \mathbb{U}_{2331} & \mathbb{U}_{2321} & \mathbb{U}_{2332} & \mathbb{U}_{2313} \\ \mathbb{U}_{3111} & \mathbb{U}_{3122} & \mathbb{U}_{3133} & \mathbb{U}_{3112} & \mathbb{U}_{3123} & \mathbb{U}_{3131} & \mathbb{U}_{3121} & \mathbb{U}_{3132} & \mathbb{U}_{3113} \\ \mathbb{U}_{2111} & \mathbb{U}_{2122} & \mathbb{U}_{2133} & \mathbb{U}_{2112} & \mathbb{U}_{2123} & \mathbb{U}_{2131} & \mathbb{U}_{2121} & \mathbb{U}_{2132} & \mathbb{U}_{2113} \\ \mathbb{U}_{3211} & \mathbb{U}_{3222} & \mathbb{U}_{3233} & \mathbb{U}_{3212} & \mathbb{U}_{3223} & \mathbb{U}_{3231} & \mathbb{U}_{3221} & \mathbb{U}_{3232} & \mathbb{U}_{3213} \\ \mathbb{U}_{1311} & \mathbb{U}_{1322} & \mathbb{U}_{1333} & \mathbb{U}_{1312} & \mathbb{U}_{1323} & \mathbb{U}_{1331} & \mathbb{U}_{1321} & \mathbb{U}_{1332} & \mathbb{U}_{1313} \end{bmatrix}.$$

Using Eqns. (1.177), one can easily find $\Psi(\boldsymbol{A} \otimes \boldsymbol{B})$, $\Psi(\boldsymbol{A} \boxtimes \boldsymbol{B})$ and so on.

The rules for applying the Ψ-mapping are [233] (for an alternative way of defining the Ψ mapping, see below)

$$\Psi(\boldsymbol{U}+\boldsymbol{V}) = \Psi(\boldsymbol{U}) + \Psi(\boldsymbol{V}), \tag{I.1a}$$

$$\Psi(\mathbb{U}+\mathbb{V}) = \Psi(\mathbb{U}) + \Psi(\mathbb{V}), \tag{I.1b}$$

$$\Psi(\mathbb{U}^T) = [\Psi(\mathbb{U})]^T, \tag{I.1c}$$

$$\Psi(\boldsymbol{U} : \boldsymbol{V}) = [\Psi(\boldsymbol{U})]^T\Psi(\boldsymbol{V}), \tag{I.1d}$$

$$\Psi(\mathbb{U}\boldsymbol{V}) = \Psi(\mathbb{U})\Psi(\boldsymbol{V}), \tag{I.1e}$$

$$\Psi(\mathbb{U}\mathbb{V}) = \Psi(\mathbb{U})\Psi(\mathbb{V}), \tag{I.1f}$$

$$\Psi(\mathbb{U}^{-1}) = \Psi(\mathbb{U})^{-1}, \tag{I.1g}$$

where $\boldsymbol{U} : \boldsymbol{V}$, $\mathbb{U}\boldsymbol{V}$ and $\mathbb{U}\mathbb{V}$ have the component forms $U_{ij}V_{ij}$, $\mathbb{U}_{ijkl}V_{kl}$ and $\mathbb{U}_{ijmn}\mathbb{V}_{mnkl}$, respectively. As an example,

$$\Psi(\mathbb{I}) = \boldsymbol{I}_{9\times 9}, \tag{I.2a}$$

$$\Psi(\mathbb{T}) = \begin{bmatrix} \boldsymbol{I}_{3\times3} & \boldsymbol{0}_{3\times3} & \boldsymbol{0}_{3\times3} \\ \boldsymbol{0}_{3\times3} & \boldsymbol{0}_{3\times3} & \boldsymbol{I}_{3\times3} \\ \boldsymbol{0}_{3\times3} & \boldsymbol{I}_{3\times3} & \boldsymbol{0}_{3\times3} \end{bmatrix}, \tag{I.2b}$$

$$\Psi(\mathbb{S}) = \frac{1}{2}[\Psi(\mathbb{I}) + \Psi(\mathbb{T})] = \begin{bmatrix} \boldsymbol{I}_{3\times3} & \boldsymbol{0}_{3\times3} & \boldsymbol{0}_{3\times3} \\ \boldsymbol{0}_{3\times3} & \frac{1}{2}\boldsymbol{I}_{3\times3} & \frac{1}{2}\boldsymbol{I}_{3\times3} \\ \boldsymbol{0}_{3\times3} & \frac{1}{2}\boldsymbol{I}_{3\times3} & \frac{1}{2}\boldsymbol{I}_{3\times3} \end{bmatrix}, \tag{I.2c}$$

$$\Psi(\mathbb{W}) = \frac{1}{2}[\Psi(\mathbb{I}) - \Psi(\mathbb{T})] = \begin{bmatrix} \boldsymbol{0}_{3\times3} & \boldsymbol{0}_{3\times3} & \boldsymbol{0}_{3\times3} \\ \boldsymbol{0}_{3\times3} & \frac{1}{2}\boldsymbol{I}_{3\times3} & -\frac{1}{2}\boldsymbol{I}_{3\times3} \\ \boldsymbol{0}_{3\times3} & -\frac{1}{2}\boldsymbol{I}_{3\times3} & \frac{1}{2}\boldsymbol{I}_{3\times3} \end{bmatrix}. \tag{I.2d}$$

In accordance with Eqn. (1.186), we have $\Psi(\mathbb{T})\Psi(\mathbb{T}) = \boldsymbol{I}_{9\times9}$, so that $\Psi(\mathbb{T}^{-1}) = [\Psi(\mathbb{T})]^{-1} = \Psi(\mathbb{T})$. Using Eqn. (1.348), we have

$$\Psi\left(\frac{\partial}{\partial \boldsymbol{T}} \operatorname{\mathbf{cof}} \boldsymbol{T}\right) = \begin{bmatrix} 0 & T_{33} & T_{22} & 0 & -T_{32} & 0 & 0 & -T_{23} & 0 \\ T_{33} & 0 & T_{11} & 0 & 0 & -T_{13} & 0 & 0 & -T_{31} \\ T_{22} & T_{11} & 0 & -T_{21} & 0 & 0 & -T_{12} & 0 & 0 \\ 0 & 0 & -T_{21} & 0 & T_{31} & T_{23} & -T_{33} & 0 & 0 \\ -T_{32} & 0 & 0 & T_{31} & 0 & T_{12} & 0 & -T_{11} & 0 \\ 0 & -T_{13} & 0 & T_{23} & T_{12} & 0 & 0 & 0 & -T_{22} \\ 0 & 0 & -T_{12} & -T_{33} & 0 & 0 & 0 & T_{13} & T_{32} \\ -T_{23} & 0 & 0 & 0 & -T_{11} & 0 & T_{13} & 0 & T_{21} \\ 0 & -T_{31} & 0 & 0 & 0 & -T_{22} & T_{32} & T_{21} & 0 \end{bmatrix}.$$

Now, let $\boldsymbol{U}$ and $\boldsymbol{V}$ be symmetric second-order tensors, $\mathbb{U}$ and $\mathbb{V}$ be fourth-order tensors which possess the minor symmetries (i.e., $\mathbb{U}_{ijkl} = \mathbb{U}_{ijlk} = \mathbb{U}_{jikl}$). The mapping χ, which exploits the symmetries in $\boldsymbol{U}$ and $\mathbb{U}$, is defined as

$$\chi(\boldsymbol{U}) := \begin{bmatrix} U_{11} \\ U_{22} \\ U_{33} \\ U_{12} \\ U_{23} \\ U_{31} \end{bmatrix}, \quad \chi(\mathbb{U}) := \begin{bmatrix} \mathbb{U}_{1111} & \mathbb{U}_{1122} & \mathbb{U}_{1133} & \mathbb{U}_{1112} & \mathbb{U}_{1123} & \mathbb{U}_{1131} \\ \mathbb{U}_{2211} & \mathbb{U}_{2222} & \mathbb{U}_{2233} & \mathbb{U}_{2212} & \mathbb{U}_{2223} & \mathbb{U}_{2231} \\ \mathbb{U}_{3311} & \mathbb{U}_{3322} & \mathbb{U}_{3333} & \mathbb{U}_{3312} & \mathbb{U}_{3323} & \mathbb{U}_{3331} \\ \mathbb{U}_{1211} & \mathbb{U}_{1222} & \mathbb{U}_{1233} & \mathbb{U}_{1212} & \mathbb{U}_{1223} & \mathbb{U}_{1231} \\ \mathbb{U}_{2311} & \mathbb{U}_{2322} & \mathbb{U}_{2333} & \mathbb{U}_{2312} & \mathbb{U}_{2323} & \mathbb{U}_{2331} \\ \mathbb{U}_{3111} & \mathbb{U}_{3122} & \mathbb{U}_{3133} & \mathbb{U}_{3112} & \mathbb{U}_{3123} & \mathbb{U}_{3131} \end{bmatrix}. \tag{I.3}$$

As an example, $\mathbb{S}(\boldsymbol{A} \otimes \boldsymbol{B})\mathbb{S} = (\mathbb{S}\boldsymbol{A}) \otimes (\mathbb{S}\boldsymbol{B}) = \frac{\boldsymbol{A}+\boldsymbol{A}^T}{2} \otimes \frac{\boldsymbol{B}+\boldsymbol{B}^T}{2}$ and $\mathbb{S}(\boldsymbol{A} \boxtimes \boldsymbol{B})\mathbb{S}$ are fourth-order tensors with minor symmetries (but not necessarily major symmetry; major symmetry is achieved if $\boldsymbol{A} = \boldsymbol{B}$ in the first case, and $\boldsymbol{A}, \boldsymbol{B} \in$ Sym in the second.). The components of $\mathbb{S}(\boldsymbol{A} \otimes \boldsymbol{B})\mathbb{S}$ are

$$[\mathbb{S}(\boldsymbol{A} \otimes \boldsymbol{B})\mathbb{S}]_{ijkl} = \frac{1}{4}(A_{ij} + A_{ji})(B_{kl} + B_{lk}),$$

and thus its χ-mapping is given by

$$\begin{bmatrix} A_{11}B_{11} & A_{11}B_{22} & A_{11}B_{33} & \frac{A_{11}(B_{12}+B_{21})}{2} & \frac{A_{11}(B_{23}+B_{32})}{2} & \frac{A_{11}(B_{13}+B_{31})}{2} \\ A_{22}B_{11} & A_{22}B_{22} & A_{22}B_{33} & \frac{A_{22}(B_{12}+B_{21})}{2} & \frac{A_{22}(B_{23}+B_{32})}{2} & \frac{A_{22}(B_{13}+B_{31})}{2} \\ A_{33}B_{11} & A_{33}B_{22} & A_{33}B_{33} & \frac{A_{33}(B_{12}+B_{21})}{2} & \frac{A_{33}(B_{23}+B_{32})}{2} & \frac{A_{33}(B_{13}+B_{31})}{2} \\ \frac{(A_{12}+A_{21})B_{11}}{2} & \frac{(A_{12}+A_{21})B_{22}}{2} & \frac{(A_{12}+A_{21})B_{33}}{2} & \frac{(A_{12}+A_{21})(B_{12}+B_{21})}{4} & \frac{(A_{12}+A_{21})(B_{23}+B_{32})}{4} & \frac{(A_{12}+A_{21})(B_{13}+B_{31})}{4} \\ \frac{(A_{23}+A_{32})B_{11}}{2} & \frac{(A_{23}+A_{32})B_{22}}{2} & \frac{(A_{23}+A_{32})B_{33}}{2} & \frac{(A_{23}+A_{32})(B_{12}+B_{21})}{4} & \frac{(A_{23}+A_{32})(B_{23}+B_{32})}{4} & \frac{(A_{23}+A_{32})(B_{13}+B_{31})}{4} \\ \frac{(A_{13}+A_{31})B_{11}}{2} & \frac{(A_{13}+A_{31})B_{22}}{2} & \frac{(A_{13}+A_{31})B_{33}}{2} & \frac{(A_{13}+A_{31})(B_{12}+B_{21})}{4} & \frac{(A_{13}+A_{31})(B_{23}+B_{32})}{4} & \frac{(A_{13}+A_{31})(B_{13}+B_{31})}{4} \end{bmatrix}. \quad \text{(I.4)}$$

The components of $\mathbb{S}(\boldsymbol{A} \boxtimes \boldsymbol{B})\mathbb{S}$ (which are equal to the components of $\mathbb{S}(\boldsymbol{B} \boxtimes \boldsymbol{A})\mathbb{S}$ by virtue of Eqn. (1.174)) are obtained using Eqn. (1.162) as

$$[\mathbb{S}(\boldsymbol{A} \boxtimes \boldsymbol{B})\mathbb{S}]_{ijkl} = \frac{1}{4}\left[A_{ik}B_{jl} + A_{il}B_{jk} + A_{jk}B_{il} + A_{jl}B_{ik}\right]. \quad \text{(I.5)}$$

The χ-mapping of $\mathbb{S}(\boldsymbol{A} \boxtimes \boldsymbol{B})\mathbb{S}$ can be written using the above component form.

As another example, using Eqn. (1.351) (or, alternatively, Eqn. (I.8c)), we get for $\boldsymbol{C} \in$ Sym,

$$\chi\left(\frac{\partial}{\partial \boldsymbol{C}}\mathbf{cof}\,\boldsymbol{C}\right) = \begin{bmatrix} 0 & C_{33} & C_{22} & 0 & -C_{23} & 0 \\ C_{33} & 0 & C_{11} & 0 & 0 & -C_{31} \\ C_{22} & C_{11} & 0 & -C_{12} & 0 & 0 \\ 0 & 0 & -C_{12} & -\frac{C_{33}}{2} & \frac{C_{31}}{2} & \frac{C_{23}}{2} \\ -C_{23} & 0 & 0 & \frac{C_{31}}{2} & -\frac{C_{11}}{2} & \frac{C_{12}}{2} \\ 0 & -C_{31} & 0 & \frac{C_{23}}{2} & \frac{C_{12}}{2} & -\frac{C_{22}}{2} \end{bmatrix}. \quad \text{(I.6)}$$

Let

$$\boldsymbol{H} = \begin{bmatrix} \boldsymbol{I}_{3\times 3} & \boldsymbol{0}_{3\times 3} \\ \boldsymbol{0}_{3\times 3} & \frac{1}{2}\boldsymbol{I}_{3\times 3} \end{bmatrix},$$

$$\boldsymbol{M} = \begin{bmatrix} \boldsymbol{I}_{3\times 3} & \boldsymbol{0}_{3\times 3} \\ \boldsymbol{0}_{3\times 3} & \boldsymbol{I}_{3\times 3} \\ \boldsymbol{0}_{3\times 3} & \boldsymbol{I}_{3\times 3} \end{bmatrix},$$

$$\bar{\boldsymbol{M}} = \begin{bmatrix} \boldsymbol{I}_{3\times 3} & \boldsymbol{0}_{3\times 3} & \boldsymbol{0}_{3\times 3} \\ \boldsymbol{0}_{3\times 3} & \frac{1}{2}\boldsymbol{I}_{3\times 3} & \frac{1}{2}\boldsymbol{I}_{3\times 3} \end{bmatrix}.$$

Note that

$$\bar{\boldsymbol{M}}\boldsymbol{M} = \boldsymbol{I}_{6\times 6}, \quad \text{(I.7a)}$$

$$\boldsymbol{M}^T\boldsymbol{M} = \boldsymbol{H}^{-1}, \quad \text{(I.7b)}$$

$$\bar{\boldsymbol{M}}\bar{\boldsymbol{M}}^T = \boldsymbol{H}, \quad \text{(I.7c)}$$

$$\boldsymbol{M}\bar{\boldsymbol{M}} = \Psi(\mathbb{S}), \quad \text{(I.7d)}$$

$$\bar{\boldsymbol{M}}^T\boldsymbol{H}^{-1} = \boldsymbol{M}. \quad \text{(I.7e)}$$

From Eqns. (I.7a) and (I.7d), it follows that $\bar{M}\Psi(\mathbb{S}) = \bar{M}$.

Let ϕ be a scalar, $U \in$ Sym, and $\mathbb{U}$ be a fourth-order tensor with minor symmetries. Then the rules for transforming from the Ψ to the χ mapping are

$$\chi(\phi) = \Psi(\phi), \tag{I.8a}$$

$$\chi(U) = \bar{M}\Psi(U), \quad \Psi(U) = M\chi(U), \tag{I.8b}$$

$$\chi(\mathbb{U}) = \bar{M}\Psi(\mathbb{U})\bar{M}^T, \quad \Psi(\mathbb{U}) = M\chi(\mathbb{U})M^T. \tag{I.8c}$$

The matrix representation of $\mathbb{S}$ is given by H, i.e.,

$$\chi(\mathbb{S}) = H, \tag{I.9}$$

since $\bar{M}\Psi(\mathbb{S})\bar{M}^T = \bar{M}\bar{M}^T = H$. Similarly, we have

$$\chi(I \otimes I) = \begin{bmatrix} 1 & 1 & 1 & 0 & 0 & 0 \\ 1 & 1 & 1 & 0 & 0 & 0 \\ 1 & 1 & 1 & 0 & 0 & 0 \\ 0 & 0 & 0 & 0 & 0 & 0 \\ 0 & 0 & 0 & 0 & 0 & 0 \\ 0 & 0 & 0 & 0 & 0 & 0 \end{bmatrix}. \tag{I.10}$$

The rules for applying the χ-mapping when $U, V \in$ Sym, and $\mathbb{U}$ and $\mathbb{V}$ have minor symmetries are [233]

$$\chi(U + V) = \chi(U) + \chi(V), \tag{I.11a}$$

$$\chi(\mathbb{U} + \mathbb{V}) = \chi(\mathbb{U}) + \chi(\mathbb{V}), \tag{I.11b}$$

$$\chi(\mathbb{U}^T) = [\chi(\mathbb{U})]^T, \tag{I.11c}$$

$$\chi(U : V) = [\chi(U)]^T H^{-1}\chi(V), \tag{I.11d}$$

$$\chi(\mathbb{U}V) = \chi(\mathbb{U})H^{-1}\chi(V), \tag{I.11e}$$

$$\chi(\mathbb{U}\mathbb{V}) = \chi(\mathbb{U})H^{-1}\chi(\mathbb{V}), \tag{I.11f}$$

$$\chi(\mathbb{U}^{-1}) = H[\chi(\mathbb{U})]^{-1}H, \tag{I.11g}$$

where $U : V$, $\mathbb{U}V$, etc. have the same interpretation as before, and where $\mathbb{U}\mathbb{U}^{-1} = \mathbb{U}^{-1}\mathbb{U} = \mathbb{S}$ (as we shall point out later, $\mathbb{U}$ has three zero eigenvalues, and hence is not invertible in the conventional sense; it is invertible if its action is restricted to symmetric tensors. In this sense $\mathbb{S}^{-1} = \mathbb{S}$ since $\mathbb{S}\mathbb{S} = \mathbb{S}$). The proofs for Eqns. (I.11d)–(I.11f) are as follows:

$$\begin{aligned}
\chi(U : V) &= \Psi(U : V) && \text{(by Eqn. (I.8a))} \\
&= \Psi(U)^T\Psi(V) && \text{(by Eqn. (I.1d))} \\
&= \chi(U)^T M^T M\chi(V) && \text{(by Eqn. (I.8b))} \\
&= \chi(U)^T H^{-1}\chi(V) && \text{(by Eqn. (I.7b))} \\
\chi(\mathbb{U}V) &= \bar{M}\Psi(\mathbb{U}V) && \text{(by Eqn. (I.8b))}
\end{aligned}$$

$$
\begin{aligned}
&= \bar{M}\Psi(\mathbb{U})\Psi(V) && \text{(by Eqn. (I.1e))}\\
&= \bar{M}M\chi(\mathbb{U})M^T M\chi(V) && \text{(by Eqns. (I.8b), (I.8c))}\\
&= \chi(\mathbb{U})H^{-1}\chi(V) && \text{(by Eqns. (I.7a), (I.7b))}\\
\chi(\mathbb{U}\mathbb{V}) &= \bar{M}\Psi(\mathbb{U}\mathbb{V})\bar{M}^T && \text{(by Eqn. (I.8c))}\\
&= \bar{M}\Psi(\mathbb{U})\Psi(\mathbb{V})\bar{M}^T && \text{(by Eqn. (I.1f))}\\
&= \bar{M}M\chi(\mathbb{U})M^T M\chi(\mathbb{V})M^T\bar{M}^T && \text{(by Eqn. (I.8c))}\\
&= \chi(\mathbb{U})H^{-1}\chi(\mathbb{V}). && \text{(by Eqn. (I.7a), (I.7b))}
\end{aligned}
$$

To prove Eqn. (I.11g), we note using Eqns. (I.9) and (I.11f) that

$$H = \chi(\mathbb{S}) = \chi(\mathbb{U}\mathbb{U}^{-1}) = \chi(\mathbb{U})H^{-1}\chi(\mathbb{U}^{-1}),$$

from which the desired result follows.

For a change of basis from $\{e^i\}$ to $\{i_i\}$ via Eqn. (H.7b), with the components of J given by Eqn. (H.6), Eqn. (H.13b) (with T assumed to be symmetric) can be written as

$$\chi(\bar{T}) = \chi\left(\mathbb{S}(J^T \boxtimes J^T)\mathbb{S}\right) H^{-1}\chi(T) = A\chi(T),$$

where

$$
\begin{aligned}
A &= \chi\left(\mathbb{S}(J^T \boxtimes J^T)\mathbb{S}\right) H^{-1}\\
&= \begin{bmatrix}
J_{11}^2 & J_{21}^2 & J_{31}^2 & 2J_{11}J_{21} & 2J_{21}J_{31} & 2J_{11}J_{31}\\
J_{12}^2 & J_{22}^2 & J_{32}^2 & 2J_{12}J_{22} & 2J_{22}J_{32} & 2J_{12}J_{32}\\
J_{13}^2 & J_{23}^2 & J_{33}^2 & 2J_{13}J_{23} & 2J_{23}J_{33} & 2J_{13}J_{33}\\
J_{11}J_{12} & J_{21}J_{22} & J_{31}J_{32} & (J_{11}J_{22}+J_{21}J_{12}) & (J_{21}J_{32}+J_{31}J_{22}) & (J_{31}J_{12}+J_{11}J_{32})\\
J_{12}J_{13} & J_{22}J_{23} & J_{32}J_{33} & (J_{12}J_{23}+J_{22}J_{13}) & (J_{22}J_{33}+J_{32}J_{23}) & (J_{32}J_{13}+J_{12}J_{33})\\
J_{11}J_{13} & J_{21}J_{23} & J_{31}J_{33} & (J_{13}J_{21}+J_{23}J_{11}) & (J_{23}J_{31}+J_{33}J_{21}) & (J_{33}J_{11}+J_{13}J_{31})
\end{bmatrix}.
\end{aligned}
\tag{I.12}
$$

Similarly, for the fourth-order tensor transformation $\bar{\mathbb{T}}^{ijrs} = \bar{\mathbb{T}}_{ijrs} = J_{mi}J_{nj}J_{pr}J_{qs}\mathbb{T}^{mnpq}$, with $\mathbb{T}$ assumed to have both the minor symmetries, i.e., $[\bar{\mathbb{T}}] = \mathbb{S}(J^T \boxtimes J^T)\mathbb{S}[\mathbb{T}]\mathbb{S}(J \boxtimes J)\mathbb{S}$, we have

$$\chi(\bar{\mathbb{T}}) = A\chi(\mathbb{T})A^T. \tag{I.13}$$

For an orthogonal change of coordinate system such as in Eqn. (1.100) or (1.150), we have $J \equiv Q^T$.

As an application of the above relations, consider the constitutive relation $\tau = \mathsf{C}\epsilon$. Applying the mapping χ to both sides of this equation, and using Eqn. (I.11e), we get

$$\chi(\tau) = \chi(\mathsf{C})H^{-1}\chi(\epsilon).$$

Noting from Eqn. (I.3) that $\chi(\tau) \equiv \tau_{\text{engrg}}$, $\chi(\mathsf{C}) \equiv C_{\text{engrg}}$ and $H^{-1}\chi(\epsilon) \equiv \epsilon_{\text{engrg}}$, we see that the above equation is the same as the familiar relation $\tau_{\text{engrg}} = C_{\text{engrg}}\epsilon_{\text{engrg}}$. Another more complicated example considered in [233] is to find the engineering form of the tensor $\mathbb{T} := [\mathbb{S} + \mathbb{E}(\mathbb{C}^m)^{-1}(\mathbb{C}^f - \mathbb{C}^m)]^{-1}$, where $\mathbb{E}$, $\mathbb{C}^m$ and $\mathbb{C}^f$ are fourth-order tensors. We get

$$\begin{aligned}
\chi(\mathbb{T}) &= \chi\left([\mathbb{S} + \mathbb{E}(\mathbb{C}^m)^{-1}(\mathbb{C}^f - \mathbb{C}^m)]^{-1}\right) \\
&= \boldsymbol{H}\left[\chi\left(\mathbb{S} + \mathbb{E}(\mathbb{C}^m)^{-1}(\mathbb{C}^f - \mathbb{C}^m)\right)\right]^{-1}\boldsymbol{H} && \text{(by Eqn. (I.11g))} \\
&= \boldsymbol{H}\left[\chi(\mathbb{S}) + \chi\left(\mathbb{E}(\mathbb{C}^m)^{-1}(\mathbb{C}^f - \mathbb{C}^m)\right)\right]^{-1}\boldsymbol{H} && \text{(by Eqn. (I.11b))} \\
&= \boldsymbol{H}\left[\chi(\mathbb{S}) + \chi\left(\mathbb{E}(\mathbb{C}^m)^{-1}\right)\boldsymbol{H}^{-1}\chi(\mathbb{C}^f - \mathbb{C}^m)\right]^{-1}\boldsymbol{H} && \text{(by Eqn. (I.11f))} \\
&= \boldsymbol{H}\left[\chi(\mathbb{S}) + \chi(\mathbb{E})\boldsymbol{H}^{-1}\chi[(\mathbb{C}^m)^{-1}]\boldsymbol{H}^{-1}\chi(\mathbb{C}^f - \mathbb{C}^m)\right]^{-1}\boldsymbol{H} && \text{(by Eqn. (I.11f))} \\
&= \boldsymbol{H}\left[\chi(\mathbb{S}) + \chi(\mathbb{E})[\chi(\mathbb{C}^m)]^{-1}\left(\chi(\mathbb{C}^f) - \chi(\mathbb{C}^m)\right)\right]^{-1}\boldsymbol{H}. && \text{(by Eqn. (I.11g))}
\end{aligned}$$

As a final example, consider the following eigenvalue problem for a fourth-order tensor $\mathbb{U}$ with both the minor symmetries:

$$(\mathbb{U} - \gamma\mathbb{S})\boldsymbol{V} = \boldsymbol{0}.$$

Applying the χ mapping to both sides and using Eqns. (I.9) and (I.11e), we get

$$[\chi(\mathbb{U}) - \gamma\boldsymbol{H}]\,\boldsymbol{H}^{-1}\chi(\boldsymbol{V}) = \boldsymbol{0}. \tag{I.14}$$

Thus, the characteristic equation is obtained as

$$\det\left[\chi(\mathbb{U})\boldsymbol{H}^{-1} - \gamma\boldsymbol{I}_{6\times 6}\right] = 0. \tag{I.15}$$

Solving the above equation yields six eigenvalues, which on substituting into Eqn. (I.14) can be used to find the eigenvectors. As usual, the eigenvalues can be repeated, and eigenvectors need not be unique. The remaining three eigenvalues are all zero, and the corresponding eigentensors are any three, linearly independent, skew-symmetric second-order tensors. Note that $\det(\mathbb{U}) = \det[\chi(\mathbb{U})\boldsymbol{H}^{-1}]$, so that, in particular, $\det(\mathbb{S}) = \det[\boldsymbol{H}\boldsymbol{H}^{-1}] = 1$.

As an example, let us find the eigenvalues of the elasticity tensor for an isotropic material given by $\mathbb{C} = \lambda\boldsymbol{I}\otimes\boldsymbol{I} + 2\mu\mathbb{S}$. Using Eqns. (I.9) and (I.10), we have

$$\chi(\mathbb{C}) = \begin{bmatrix} \lambda+2\mu & \lambda & \lambda & 0 & 0 & 0 \\ \lambda & \lambda+2\mu & \lambda & 0 & 0 & 0 \\ \lambda & \lambda & \lambda+2\mu & 0 & 0 & 0 \\ 0 & 0 & 0 & \mu & 0 & 0 \\ 0 & 0 & 0 & 0 & \mu & 0 \\ 0 & 0 & 0 & 0 & 0 & \mu \end{bmatrix}.$$

By using Eqn. (I.15), we get the eigenvalues of $\mathbb{C}$ as $(3\lambda + 2\mu, 2\mu, 2\mu, 2\mu, 2\mu, 2\mu)$. Thus, by Theorem 1.6.4, $\mathbb{C}$ is positive definite if and only if $(3\lambda + 2\mu) > 0$ and $\mu > 0$, a result that we have proved in Theorem 5.8.2 by a different method. The eigentensors corresponding to the eigenvalues $3\lambda + 2\mu$ and 2μ are $\boldsymbol{I}/\sqrt{3}$ and $\boldsymbol{S}_0/\|\boldsymbol{S}_0\|$, respectively, where $\boldsymbol{S}_0$ is a traceless symmetric tensor. The eigenvalues/eigentensors for anisotropic elastic materials are discussed in [213].

The Ψ mapping defined at the beginning of this section can be defined in alternative ways. For example,

$$\hat{\Psi}(\boldsymbol{U}) := \begin{bmatrix} U_{11} \\ U_{12} \\ U_{13} \\ U_{21} \\ U_{22} \\ U_{23} \\ U_{31} \\ U_{32} \\ U_{33} \end{bmatrix}, \quad \hat{\Psi}(\mathbb{U}) := \begin{bmatrix} \mathbb{U}_{1111} & \mathbb{U}_{1112} & \mathbb{U}_{1113} & \mathbb{U}_{1121} & \mathbb{U}_{1122} & \mathbb{U}_{1123} & \mathbb{U}_{1131} & \mathbb{U}_{1132} & \mathbb{U}_{1133} \\ \mathbb{U}_{1211} & \mathbb{U}_{1212} & \mathbb{U}_{1213} & \mathbb{U}_{1221} & \mathbb{U}_{1222} & \mathbb{U}_{1223} & \mathbb{U}_{1231} & \mathbb{U}_{1232} & \mathbb{U}_{1233} \\ \mathbb{U}_{1311} & \mathbb{U}_{1312} & \mathbb{U}_{1313} & \mathbb{U}_{1321} & \mathbb{U}_{1322} & \mathbb{U}_{1323} & \mathbb{U}_{1331} & \mathbb{U}_{1332} & \mathbb{U}_{1333} \\ \mathbb{U}_{2111} & \mathbb{U}_{2112} & \mathbb{U}_{2113} & \mathbb{U}_{2121} & \mathbb{U}_{2122} & \mathbb{U}_{2123} & \mathbb{U}_{2131} & \mathbb{U}_{2132} & \mathbb{U}_{2133} \\ \mathbb{U}_{2211} & \mathbb{U}_{2212} & \mathbb{U}_{2213} & \mathbb{U}_{2221} & \mathbb{U}_{2222} & \mathbb{U}_{2223} & \mathbb{U}_{2231} & \mathbb{U}_{2232} & \mathbb{U}_{2233} \\ \mathbb{U}_{2311} & \mathbb{U}_{2312} & \mathbb{U}_{2313} & \mathbb{U}_{2321} & \mathbb{U}_{2322} & \mathbb{U}_{2323} & \mathbb{U}_{2331} & \mathbb{U}_{2332} & \mathbb{U}_{2333} \\ \mathbb{U}_{3111} & \mathbb{U}_{3112} & \mathbb{U}_{3113} & \mathbb{U}_{3121} & \mathbb{U}_{3122} & \mathbb{U}_{3123} & \mathbb{U}_{3131} & \mathbb{U}_{3132} & \mathbb{U}_{3133} \\ \mathbb{U}_{3211} & \mathbb{U}_{3212} & \mathbb{U}_{3213} & \mathbb{U}_{3221} & \mathbb{U}_{3222} & \mathbb{U}_{3223} & \mathbb{U}_{3231} & \mathbb{U}_{3232} & \mathbb{U}_{3233} \\ \mathbb{U}_{3311} & \mathbb{U}_{3312} & \mathbb{U}_{3313} & \mathbb{U}_{3321} & \mathbb{U}_{3322} & \mathbb{U}_{3323} & \mathbb{U}_{3331} & \mathbb{U}_{3332} & \mathbb{U}_{3333} \end{bmatrix}.$$

We have

$$\hat{\Psi}(\mathbb{I}) = \boldsymbol{I}_{9\times 9},$$

$$\hat{\Psi}(\mathbb{T}) = \begin{bmatrix} 1 & 0 & 0 & 0 & 0 & 0 & 0 & 0 & 0 \\ 0 & 0 & 0 & 1 & 0 & 0 & 0 & 0 & 0 \\ 0 & 0 & 0 & 0 & 0 & 0 & 1 & 0 & 0 \\ 0 & 1 & 0 & 0 & 0 & 0 & 0 & 0 & 0 \\ 0 & 0 & 0 & 0 & 1 & 0 & 0 & 0 & 0 \\ 0 & 0 & 0 & 0 & 0 & 0 & 0 & 1 & 0 \\ 0 & 0 & 1 & 0 & 0 & 0 & 0 & 0 & 0 \\ 0 & 0 & 0 & 0 & 0 & 1 & 0 & 0 & 0 \\ 0 & 0 & 0 & 0 & 0 & 0 & 0 & 0 & 1 \end{bmatrix},$$

$$\hat{\Psi}(\mathbb{S}) = \frac{1}{2}\left[\hat{\Psi}(\mathbb{I}) + \hat{\Psi}(\mathbb{T})\right],$$

$$\hat{\Psi}(\mathbb{W}) = \frac{1}{2}\left[\hat{\Psi}(\mathbb{I}) - \hat{\Psi}(\mathbb{T})\right].$$

The rules for $\hat{\Psi}$ are the same as those for Ψ, since they are related by

$$\Psi(\boldsymbol{U}) = \boldsymbol{Q}\hat{\Psi}(\boldsymbol{U}), \qquad \hat{\Psi}(\boldsymbol{U}) = \boldsymbol{Q}^T\Psi(\boldsymbol{U}),$$

$$\Psi(\mathbb{U}) = \boldsymbol{Q}\hat{\Psi}(\mathbb{U})\boldsymbol{Q}^T, \qquad \hat{\Psi}(\mathbb{U}) = \boldsymbol{Q}^T\Psi(\mathbb{U})\boldsymbol{Q},$$

where

$$
\boldsymbol{Q} = \begin{bmatrix}
1 & 0 & 0 & 0 & 0 & 0 & 0 & 0 & 0 \\
0 & 0 & 0 & 0 & 1 & 0 & 0 & 0 & 0 \\
0 & 0 & 0 & 0 & 0 & 0 & 0 & 0 & 1 \\
0 & 1 & 0 & 0 & 0 & 0 & 0 & 0 & 0 \\
0 & 0 & 0 & 0 & 0 & 1 & 0 & 0 & 0 \\
0 & 0 & 0 & 0 & 0 & 0 & 1 & 0 & 0 \\
0 & 0 & 0 & 1 & 0 & 0 & 0 & 0 & 0 \\
0 & 0 & 0 & 0 & 0 & 0 & 0 & 1 & 0 \\
0 & 0 & 1 & 0 & 0 & 0 & 0 & 0 & 0
\end{bmatrix} \in \text{Orth}.
$$

In the presence of symmetries, we obtain from Eqns. (I.8) and the above equations, the relations $\chi(\boldsymbol{U}) = \bar{\boldsymbol{M}}\Psi(\boldsymbol{U}) = \hat{\boldsymbol{M}}\hat{\Psi}(\boldsymbol{U})$, $\chi(\mathbb{U}) = \hat{\boldsymbol{M}}\hat{\Psi}(\mathbb{U})\hat{\boldsymbol{M}}^T$, $\hat{\Psi}(\boldsymbol{U}) = \tilde{\boldsymbol{M}}\chi(\boldsymbol{U})$ and $\hat{\Psi}(\mathbb{U}) = \tilde{\boldsymbol{M}}\chi(\mathbb{U})\tilde{\boldsymbol{M}}^T$, where

$$
\hat{\boldsymbol{M}} := \bar{\boldsymbol{M}}\boldsymbol{Q} = \begin{bmatrix}
1 & 0 & 0 & 0 & 0 & 0 & 0 & 0 & 0 \\
0 & 0 & 0 & 0 & 1 & 0 & 0 & 0 & 0 \\
0 & 0 & 0 & 0 & 0 & 0 & 0 & 0 & 1 \\
0 & \frac{1}{2} & 0 & \frac{1}{2} & 0 & 0 & 0 & 0 & 0 \\
0 & 0 & 0 & 0 & 0 & \frac{1}{2} & 0 & \frac{1}{2} & 0 \\
0 & 0 & \frac{1}{2} & 0 & 0 & 0 & \frac{1}{2} & 0 & 0
\end{bmatrix}, \quad
\tilde{\boldsymbol{M}} := \boldsymbol{Q}^T\boldsymbol{M} = \begin{bmatrix}
1 & 0 & 0 & 0 & 0 & 0 \\
0 & 0 & 0 & 1 & 0 & 0 \\
0 & 0 & 0 & 0 & 0 & 1 \\
0 & 0 & 0 & 1 & 0 & 0 \\
0 & 1 & 0 & 0 & 0 & 0 \\
0 & 0 & 0 & 0 & 1 & 0 \\
0 & 0 & 0 & 0 & 0 & 1 \\
0 & 0 & 0 & 0 & 1 & 0 \\
0 & 0 & 1 & 0 & 0 & 0
\end{bmatrix}.
$$

Note that

$$
\begin{aligned}
\hat{\boldsymbol{M}}\hat{\Psi}(\mathbb{S}) &= \bar{\boldsymbol{M}}\boldsymbol{Q}\boldsymbol{Q}^T\Psi(\mathbb{S})\boldsymbol{Q} = \bar{\boldsymbol{M}}\Psi(\mathbb{S})\boldsymbol{Q} = \bar{\boldsymbol{M}}\boldsymbol{Q} = \hat{\boldsymbol{M}}, \\
\hat{\Psi}(\mathbb{S})\tilde{\boldsymbol{M}} &= \boldsymbol{Q}^T\Psi(\mathbb{S})\boldsymbol{Q}\boldsymbol{Q}^T\boldsymbol{M} = \boldsymbol{Q}^T\Psi(\mathbb{S})\boldsymbol{M} = \boldsymbol{Q}^T\boldsymbol{M} = \tilde{\boldsymbol{M}}.
\end{aligned}
$$

The $\hat{\Psi}$ mapping of $\boldsymbol{A} \boxtimes \boldsymbol{B}$ is given by

$$
\hat{\Psi}(\boldsymbol{A} \boxtimes \boldsymbol{B}) = \begin{bmatrix}
A_{11}B_{11} & A_{11}B_{12} & A_{11}B_{13} & A_{12}B_{11} & A_{12}B_{12} & A_{12}B_{13} & A_{13}B_{11} & A_{13}B_{12} & A_{13}B_{13} \\
A_{11}B_{21} & A_{11}B_{22} & A_{11}B_{23} & A_{12}B_{21} & A_{12}B_{22} & A_{12}B_{23} & A_{13}B_{21} & A_{13}B_{22} & A_{13}B_{23} \\
A_{11}B_{31} & A_{11}B_{32} & A_{11}B_{33} & A_{12}B_{31} & A_{12}B_{32} & A_{12}B_{33} & A_{13}B_{31} & A_{13}B_{32} & A_{13}B_{33} \\
A_{21}B_{11} & A_{21}B_{12} & A_{21}B_{13} & A_{22}B_{11} & A_{22}B_{12} & A_{22}B_{13} & A_{23}B_{11} & A_{23}B_{12} & A_{23}B_{13} \\
A_{21}B_{21} & A_{21}B_{22} & A_{21}B_{23} & A_{22}B_{21} & A_{22}B_{22} & A_{22}B_{23} & A_{23}B_{21} & A_{23}B_{22} & A_{23}B_{23} \\
A_{21}B_{31} & A_{21}B_{32} & A_{21}B_{33} & A_{22}B_{31} & A_{22}B_{32} & A_{22}B_{33} & A_{23}B_{31} & A_{23}B_{32} & A_{23}B_{33} \\
A_{31}B_{11} & A_{31}B_{12} & A_{31}B_{13} & A_{32}B_{11} & A_{32}B_{12} & A_{32}B_{13} & A_{33}B_{11} & A_{33}B_{12} & A_{33}B_{13} \\
A_{31}B_{21} & A_{31}B_{22} & A_{31}B_{23} & A_{32}B_{21} & A_{32}B_{22} & A_{32}B_{23} & A_{33}B_{21} & A_{33}B_{22} & A_{33}B_{23} \\
A_{31}B_{31} & A_{31}B_{32} & A_{31}B_{33} & A_{32}B_{31} & A_{32}B_{32} & A_{32}B_{33} & A_{33}B_{31} & A_{33}B_{32} & A_{33}B_{33}
\end{bmatrix}.
$$

As an application of the above equation, we have

$$
\begin{aligned}
\hat{\Psi}(\boldsymbol{A} : \boldsymbol{B}\boldsymbol{T}^T) &= \hat{\Psi}(\boldsymbol{A})^T\hat{\Psi}(\boldsymbol{B}\boldsymbol{T}^T) \\
&= \hat{\Psi}(\boldsymbol{A})^T\hat{\Psi}((\boldsymbol{I} \boxtimes \boldsymbol{T})\boldsymbol{B})
\end{aligned}
$$

$$
\begin{aligned}
&= \hat{\Psi}(\boldsymbol{A})^T \hat{\Psi}(\boldsymbol{I} \boxtimes \boldsymbol{T}) \hat{\Psi}(\boldsymbol{B}) \\
&= \hat{\Psi}(\boldsymbol{A})^T \boldsymbol{T}_M \hat{\Psi}(\boldsymbol{B}),
\end{aligned}
$$

where

$$
\boldsymbol{T}_M = \hat{\Psi}(\boldsymbol{I} \boxtimes \boldsymbol{T}) = \begin{bmatrix}
T_{11} & T_{12} & T_{13} & 0 & 0 & 0 & 0 & 0 & 0 \\
T_{21} & T_{22} & T_{23} & 0 & 0 & 0 & 0 & 0 & 0 \\
T_{31} & T_{32} & T_{33} & 0 & 0 & 0 & 0 & 0 & 0 \\
0 & 0 & 0 & T_{11} & T_{12} & T_{13} & 0 & 0 & 0 \\
0 & 0 & 0 & T_{21} & T_{22} & T_{23} & 0 & 0 & 0 \\
0 & 0 & 0 & T_{31} & T_{32} & T_{33} & 0 & 0 & 0 \\
0 & 0 & 0 & 0 & 0 & 0 & T_{11} & T_{12} & T_{13} \\
0 & 0 & 0 & 0 & 0 & 0 & T_{21} & T_{22} & T_{23} \\
0 & 0 & 0 & 0 & 0 & 0 & T_{31} & T_{32} & T_{33}
\end{bmatrix}. \tag{I.16}
$$

The derivatives of $\mathbf{cof}\,\boldsymbol{T}$ and $(\mathbf{cof}\,\boldsymbol{T})^T$ in matrix form are

$$
\hat{\Psi}\left(\frac{\partial}{\partial \boldsymbol{T}} \mathbf{cof}\,\boldsymbol{T}\right) = \begin{bmatrix}
0 & 0 & 0 & 0 & T_{33} & -T_{32} & 0 & -T_{23} & T_{22} \\
0 & 0 & 0 & -T_{33} & 0 & T_{31} & T_{23} & 0 & -T_{21} \\
0 & 0 & 0 & T_{32} & -T_{31} & 0 & -T_{22} & T_{21} & 0 \\
0 & -T_{33} & T_{32} & 0 & 0 & 0 & 0 & T_{13} & -T_{12} \\
T_{33} & 0 & -T_{31} & 0 & 0 & 0 & -T_{13} & 0 & T_{11} \\
-T_{32} & T_{31} & 0 & 0 & 0 & 0 & T_{12} & -T_{11} & 0 \\
0 & T_{23} & -T_{22} & 0 & -T_{13} & T_{12} & 0 & 0 & 0 \\
-T_{23} & 0 & T_{21} & T_{13} & 0 & -T_{11} & 0 & 0 & 0 \\
T_{22} & -T_{21} & 0 & -T_{12} & T_{11} & 0 & 0 & 0 & 0
\end{bmatrix},
$$

$$
\hat{\Psi}\left(\frac{\partial}{\partial \boldsymbol{T}} (\mathbf{cof}\,\boldsymbol{T})^T\right) = \hat{\Psi}(\mathbb{T}) \hat{\Psi}\left(\frac{\partial}{\partial \boldsymbol{T}} \mathbf{cof}\,\boldsymbol{T}\right).
$$

For axisymmetric problems, with (r, z, θ) corresponding to the indices 1, 2 and 3, we define the Ψ and $\hat{\Psi}$ maps for $\boldsymbol{U} \in \text{Lin}$, and a fourth-order tensor $\mathbb{U}$ as

$$
\Psi_a(\boldsymbol{U}) := \begin{bmatrix} U_{11} \\ U_{22} \\ U_{12} \\ U_{21} \\ U_{33} \end{bmatrix}, \qquad
\Psi_a(\mathbb{U}) := \begin{bmatrix}
\mathbb{U}_{1111} & \mathbb{U}_{1122} & \mathbb{U}_{1112} & \mathbb{U}_{1121} & \mathbb{U}_{1133} \\
\mathbb{U}_{2211} & \mathbb{U}_{2222} & \mathbb{U}_{2212} & \mathbb{U}_{2221} & \mathbb{U}_{2233} \\
\mathbb{U}_{1211} & \mathbb{U}_{1222} & \mathbb{U}_{1212} & \mathbb{U}_{1221} & \mathbb{U}_{1233} \\
\mathbb{U}_{2111} & \mathbb{U}_{2122} & \mathbb{U}_{2112} & \mathbb{U}_{2121} & \mathbb{U}_{2133} \\
\mathbb{U}_{3311} & \mathbb{U}_{3322} & \mathbb{U}_{3312} & \mathbb{U}_{3321} & \mathbb{U}_{3333}
\end{bmatrix},
$$

$$
\begin{aligned}
\hat{\Psi}_a(\boldsymbol{U}) &:= \begin{bmatrix} U_{11} \\ U_{12} \\ U_{21} \\ U_{22} \\ U_{33} \end{bmatrix} & \hat{\Psi}_a(\mathbb{U}) &:= \begin{bmatrix}
\mathbb{U}_{1111} & \mathbb{U}_{1112} & \mathbb{U}_{1121} & \mathbb{U}_{1122} & \mathbb{U}_{1133} \\
\mathbb{U}_{1211} & \mathbb{U}_{1212} & \mathbb{U}_{1221} & \mathbb{U}_{1222} & \mathbb{U}_{1233} \\
\mathbb{U}_{2111} & \mathbb{U}_{2112} & \mathbb{U}_{2121} & \mathbb{U}_{2122} & \mathbb{U}_{2133} \\
\mathbb{U}_{2211} & \mathbb{U}_{2212} & \mathbb{U}_{2221} & \mathbb{U}_{2222} & \mathbb{U}_{2233} \\
\mathbb{U}_{3311} & \mathbb{U}_{3312} & \mathbb{U}_{3321} & \mathbb{U}_{3322} & \mathbb{U}_{3333}
\end{bmatrix} \\
&= \boldsymbol{Q}_a^T \Psi_a(\boldsymbol{U}), & &= \boldsymbol{Q}_a^T \Psi_a(\mathbb{U}) \boldsymbol{Q}_a,
\end{aligned}
$$

where

$$Q_a = \begin{bmatrix} 1 & 0 & 0 & 0 & 0 \\ 0 & 0 & 0 & 1 & 0 \\ 0 & 1 & 0 & 0 & 0 \\ 0 & 0 & 1 & 0 & 0 \\ 0 & 0 & 0 & 0 & 1 \end{bmatrix} \in \text{Orth}.$$

We have

$$\Psi_a(\mathbb{I}) = \boldsymbol{I}_{5\times5}, \qquad \hat{\Psi}_a(\mathbb{I}) = \boldsymbol{I}_{5\times5},$$

$$\Psi_a(\mathbb{T}) = \begin{bmatrix} 1 & 0 & 0 & 0 & 0 \\ 0 & 1 & 0 & 0 & 0 \\ 0 & 0 & 0 & 1 & 0 \\ 0 & 0 & 1 & 0 & 0 \\ 0 & 0 & 0 & 0 & 1 \end{bmatrix}, \qquad \hat{\Psi}_a(\mathbb{T}) = \begin{bmatrix} 1 & 0 & 0 & 0 & 0 \\ 0 & 0 & 1 & 0 & 0 \\ 0 & 1 & 0 & 0 & 0 \\ 0 & 0 & 0 & 1 & 0 \\ 0 & 0 & 0 & 0 & 1 \end{bmatrix},$$

$$\Psi_a(\mathbb{S}) = \frac{1}{2}\left[\Psi_a(\mathbb{I}) + \Psi_a(\mathbb{T})\right], \qquad \hat{\Psi}_a(\mathbb{S}) = \frac{1}{2}\left[\hat{\Psi}_a(\mathbb{I}) + \hat{\Psi}_a(\mathbb{T})\right],$$

$$\Psi_a(\mathbb{W}) = \frac{1}{2}\left[\Psi_a(\mathbb{I}) - \Psi_a(\mathbb{T})\right], \qquad \hat{\Psi}_a(\mathbb{W}) = \frac{1}{2}\left[\hat{\Psi}_a(\mathbb{I}) - \hat{\Psi}_a(\mathbb{T})\right].$$

For $\boldsymbol{U} \in \text{Sym}$, and $\mathbb{U}$ having minor symmetries, we define

$$\chi_a(\boldsymbol{U}) := \begin{bmatrix} U_{11} \\ U_{22} \\ U_{12} \\ U_{33} \end{bmatrix}, \qquad \chi_a(\mathbb{U}) := \begin{bmatrix} \mathbb{U}_{1111} & \mathbb{U}_{1122} & \mathbb{U}_{1112} & \mathbb{U}_{1133} \\ \mathbb{U}_{2211} & \mathbb{U}_{2222} & \mathbb{U}_{2212} & \mathbb{U}_{2233} \\ \mathbb{U}_{1211} & \mathbb{U}_{1222} & \mathbb{U}_{1212} & \mathbb{U}_{1233} \\ \mathbb{U}_{3311} & \mathbb{U}_{3322} & \mathbb{U}_{3312} & \mathbb{U}_{3333} \end{bmatrix}.$$

It can be shown that [159]

$$\Psi_a(\boldsymbol{A} \otimes \boldsymbol{B}) = \begin{bmatrix} A_{11}B_{11} & A_{11}B_{22} & A_{11}B_{12} & A_{11}B_{21} & A_{11}B_{33} \\ A_{22}B_{11} & A_{22}B_{22} & A_{22}B_{12} & A_{22}B_{21} & A_{22}B_{33} \\ A_{12}B_{11} & A_{12}B_{22} & A_{12}B_{12} & A_{12}B_{21} & A_{12}B_{33} \\ A_{21}B_{11} & A_{21}B_{22} & A_{21}B_{12} & A_{21}B_{21} & A_{21}B_{33} \\ A_{33}B_{11} & A_{33}B_{22} & A_{33}B_{12} & A_{33}B_{21} & A_{33}B_{33} \end{bmatrix},$$

$$\Psi_a(\boldsymbol{A} \boxtimes \boldsymbol{B}) = \begin{bmatrix} A_{11}B_{11} & A_{12}B_{12} & A_{11}B_{12} & A_{12}B_{11} & 0 \\ A_{21}B_{21} & A_{22}B_{22} & A_{21}B_{22} & A_{22}B_{21} & 0 \\ A_{11}B_{21} & A_{12}B_{22} & A_{11}B_{22} & A_{12}B_{21} & 0 \\ A_{21}B_{11} & A_{22}B_{12} & A_{21}B_{12} & A_{22}B_{11} & 0 \\ 0 & 0 & 0 & 0 & A_{33}B_{33} \end{bmatrix},$$

$$\hat{\Psi}_a(\boldsymbol{A} \otimes \boldsymbol{B}) = \begin{bmatrix} A_{11}B_{11} & A_{11}B_{12} & A_{11}B_{21} & A_{11}B_{22} & A_{11}B_{33} \\ A_{12}B_{11} & A_{12}B_{12} & A_{12}B_{21} & A_{12}B_{22} & A_{12}B_{33} \\ A_{21}B_{11} & A_{21}B_{12} & A_{21}B_{21} & A_{21}B_{22} & A_{21}B_{33} \\ A_{22}B_{11} & A_{22}B_{12} & A_{22}B_{21} & A_{22}B_{22} & A_{22}B_{33} \\ A_{33}B_{11} & A_{33}B_{12} & A_{33}B_{21} & A_{33}B_{22} & A_{33}B_{33} \end{bmatrix},$$

$$\hat{\Psi}_a(\boldsymbol{A} \boxtimes \boldsymbol{B}) = \begin{bmatrix} A_{11}B_{11} & A_{11}B_{12} & A_{12}B_{11} & A_{12}B_{12} & 0 \\ A_{11}B_{21} & A_{11}B_{22} & A_{12}B_{21} & A_{12}B_{22} & 0 \\ A_{21}B_{11} & A_{21}B_{12} & A_{22}B_{11} & A_{22}B_{12} & 0 \\ A_{21}B_{21} & A_{21}B_{22} & A_{22}B_{21} & A_{22}B_{22} & 0 \\ 0 & 0 & 0 & 0 & A_{33}B_{33} \end{bmatrix},$$

$$\chi_a(\boldsymbol{A} \otimes \boldsymbol{B}) = \begin{bmatrix} A_{11}B_{11} & A_{11}B_{22} & A_{11}B_{12} & A_{11}B_{33} \\ A_{22}B_{11} & A_{22}B_{22} & A_{22}B_{12} & A_{22}B_{33} \\ A_{12}B_{11} & A_{12}B_{22} & A_{12}B_{12} & A_{12}B_{33} \\ A_{33}B_{11} & A_{33}B_{22} & A_{33}B_{12} & A_{33}B_{33} \end{bmatrix}, \quad \boldsymbol{A}, \boldsymbol{B} \in \text{Sym},$$

$$\chi_a(\mathbb{S}[\boldsymbol{A} \boxtimes \boldsymbol{B}]\mathbb{S}) = \begin{bmatrix} A_{11}B_{11} & A_{12}B_{12} & \frac{A_{11}B_{12}+A_{12}B_{11}}{2} & 0 \\ A_{21}B_{21} & A_{22}B_{22} & \frac{A_{21}B_{22}+A_{22}B_{21}}{2} & 0 \\ \frac{A_{11}B_{21}+A_{21}B_{11}}{2} & \frac{A_{12}B_{22}+A_{22}B_{12}}{2} & \frac{A_{11}B_{22}+A_{12}B_{21}+A_{21}B_{12}+A_{22}B_{11}}{4} & 0 \\ 0 & 0 & 0 & A_{33}B_{33} \end{bmatrix}.$$

It can also be shown that

$$\hat{\Psi}_a(\boldsymbol{A} : \boldsymbol{B}\boldsymbol{T}^T) = \hat{\Psi}_a(\boldsymbol{A})^T(\boldsymbol{T}_M)_a\hat{\Psi}_a(\boldsymbol{B}),$$

where

$$(\boldsymbol{T}_M)_a = \hat{\Psi}_a(\boldsymbol{I} \boxtimes \boldsymbol{T}) = \begin{bmatrix} T_{11} & T_{12} & 0 & 0 & 0 \\ T_{21} & T_{22} & 0 & 0 & 0 \\ 0 & 0 & T_{11} & T_{12} & 0 \\ 0 & 0 & T_{21} & T_{22} & 0 \\ 0 & 0 & 0 & 0 & T_{33} \end{bmatrix}.$$

Let

$$\boldsymbol{H}_a = \begin{bmatrix} 1 & 0 & 0 & 0 \\ 0 & 1 & 0 & 0 \\ 0 & 0 & \frac{1}{2} & 0 \\ 0 & 0 & 0 & 1 \end{bmatrix},$$

$$\boldsymbol{M}_a = \begin{bmatrix} 1 & 0 & 0 & 0 \\ 0 & 1 & 0 & 0 \\ 0 & 0 & 1 & 0 \\ 0 & 0 & 1 & 0 \\ 0 & 0 & 0 & 1 \end{bmatrix},$$

$$\bar{M}_a = \begin{bmatrix} 1 & 0 & 0 & 0 & 0 \\ 0 & 1 & 0 & 0 & 0 \\ 0 & 0 & \frac{1}{2} & \frac{1}{2} & 0 \\ 0 & 0 & 0 & 0 & 1 \end{bmatrix}.$$

Then Eqns. (I.7) and (I.8) hold with $\boldsymbol{M}_a$, $\bar{\boldsymbol{M}}_a$, $\boldsymbol{H}_a$, $\Psi_a(\mathbb{S})$ and $\boldsymbol{I}_{4\times 4}$ in place of $\boldsymbol{M}$, $\bar{\boldsymbol{M}}$, $\boldsymbol{H}$, $\Psi(\mathbb{S})$ and $\boldsymbol{I}_{6\times 6}$. Analogous to Eqn. (I.9), we also have $\chi_a(\mathbb{S}) = \boldsymbol{H}_a$.

The χ_a and $\hat{\Psi}_a$ maps are related by $\chi_a(\boldsymbol{U}) = \hat{\boldsymbol{M}}_a\hat{\Psi}_a(\boldsymbol{U})$, $\chi_a(\mathbb{U}) = \hat{\boldsymbol{M}}_a\hat{\Psi}_a(\mathbb{U})\hat{\boldsymbol{M}}_a^T$, $\hat{\Psi}_a(\boldsymbol{U}) = \tilde{\boldsymbol{M}}_a\chi_a(\boldsymbol{U})$, and $\hat{\Psi}_a(\mathbb{U}) = \tilde{\boldsymbol{M}}_a\chi_a(\mathbb{U})\tilde{\boldsymbol{M}}_a^T$, where

$$\hat{\boldsymbol{M}}_a := \bar{\boldsymbol{M}}_a\boldsymbol{Q}_a = \begin{bmatrix} 1 & 0 & 0 & 0 & 0 \\ 0 & 0 & 0 & 1 & 0 \\ 0 & \frac{1}{2} & \frac{1}{2} & 0 & 0 \\ 0 & 0 & 0 & 0 & 1 \end{bmatrix}, \quad \tilde{\boldsymbol{M}}_a := \boldsymbol{Q}_a^T\boldsymbol{M}_a = \begin{bmatrix} 1 & 0 & 0 & 0 \\ 0 & 0 & 1 & 0 \\ 0 & 0 & 1 & 0 \\ 0 & 1 & 0 & 0 \\ 0 & 0 & 0 & 1 \end{bmatrix}.$$

We also have $\hat{\boldsymbol{M}}_a\hat{\Psi}_a(\mathbb{S}) = \hat{\boldsymbol{M}}_a$ and $\hat{\Psi}_a(\mathbb{S})\tilde{\boldsymbol{M}}_a = \tilde{\boldsymbol{M}}_a$.

Finally, we have

$$\Psi_a\left(\frac{\partial}{\partial \boldsymbol{T}}\operatorname{\mathbf{cof}} \boldsymbol{T}\right) = \begin{bmatrix} 0 & T_{33} & 0 & 0 & T_{22} \\ T_{33} & 0 & 0 & 0 & T_{11} \\ 0 & 0 & 0 & -T_{33} & -T_{21} \\ 0 & 0 & -T_{33} & 0 & -T_{12} \\ T_{22} & T_{11} & -T_{21} & -T_{12} & 0 \end{bmatrix},$$

$$\hat{\Psi}_a\left(\frac{\partial}{\partial \boldsymbol{T}}\operatorname{\mathbf{cof}} \boldsymbol{T}\right) = \begin{bmatrix} 0 & 0 & 0 & T_{33} & T_{22} \\ 0 & 0 & -T_{33} & 0 & -T_{21} \\ 0 & -T_{33} & 0 & 0 & -T_{12} \\ T_{33} & 0 & 0 & 0 & T_{11} \\ T_{22} & -T_{21} & -T_{12} & T_{11} & 0 \end{bmatrix},$$

$$\chi_a\left(\frac{\partial}{\partial \boldsymbol{C}}\operatorname{\mathbf{cof}} \boldsymbol{C}\right) = \begin{bmatrix} 0 & C_{33} & 0 & C_{22} \\ C_{33} & 0 & 0 & C_{11} \\ 0 & 0 & -\frac{C_{33}}{2} & -C_{12} \\ C_{22} & C_{11} & -C_{12} & 0 \end{bmatrix}, \quad \boldsymbol{C} \in \mathrm{Sym}.$$

As an application of the $\hat{\Psi}$ mapping, suppose one wants to solve the eigenvalue problem given by Eqn. (J.36) with an invertible but otherwise arbitrary $\boldsymbol{A}$, then writing Eqn. (J.36) as $\left[\boldsymbol{A}\boxtimes\boldsymbol{I} - \gamma\boldsymbol{I}\boxtimes\boldsymbol{A}^T\right]\boldsymbol{B} = \boldsymbol{0}$, and applying the $\hat{\Psi}$ map, we get

$$\hat{\Psi}\left[\boldsymbol{A}\boxtimes\boldsymbol{I} - \gamma\boldsymbol{I}\boxtimes\boldsymbol{A}^T\right]\hat{\Psi}(\boldsymbol{B}) = \boldsymbol{0}. \tag{I.17}$$

Using Eqn. (I.16), we get

$$\begin{bmatrix} A_{11}\boldsymbol{I}_{3\times 3} - \gamma\boldsymbol{A}^T & A_{12}\boldsymbol{I}_{3\times 3} & A_{13}\boldsymbol{I}_{3\times 3} \\ A_{21}\boldsymbol{I}_{3\times 3} & A_{22}\boldsymbol{I}_{3\times 3} - \gamma\boldsymbol{A}^T & A_{23}\boldsymbol{I}_{3\times 3} \\ A_{31}\boldsymbol{I}_{3\times 3} & A_{32}\boldsymbol{I}_{3\times 3} & A_{33}\boldsymbol{I}_{3\times 3} - \gamma\boldsymbol{A}^T \end{bmatrix}\hat{\Psi}(\boldsymbol{B}) = \boldsymbol{0}. \tag{I.18}$$

The eigenvalues are found using the characteristic equation given by

$$\det \begin{bmatrix} A_{11}\boldsymbol{I}_{3\times3} - \gamma \boldsymbol{A}^T & A_{12}\boldsymbol{I}_{3\times3} & A_{13}\boldsymbol{I}_{3\times3} \\ A_{21}\boldsymbol{I}_{3\times3} & A_{22}\boldsymbol{I}_{3\times3} - \gamma \boldsymbol{A}^T & A_{23}\boldsymbol{I}_{3\times3} \\ A_{31}\boldsymbol{I}_{3\times3} & A_{32}\boldsymbol{I}_{3\times3} & A_{33}\boldsymbol{I}_{3\times3} - \gamma \boldsymbol{A}^T \end{bmatrix} = 0, \tag{I.19}$$

while the eigenvectors are given by $\hat{\Psi}(\boldsymbol{B})$. Solutions $\boldsymbol{B}$ to Eqn. (I.18) for $\gamma = 1$ correspond to matrices that commute with $\boldsymbol{A}$ (which is no longer assumed to be invertible).

J Some Results in n-Dimensional Euclidean Spaces

The results in the main text have mostly been derived for three-dimensional Euclidean spaces. We now discuss the extension of some of these results to n-dimensional Euclidean spaces.

Let $\{e_i\}$ represent the canonical basis for this space. The components of a second-order tensor T with respect to this basis are still given by Eqn. (1.28), but with i and j both ranging from 1 to n. The representation of T in terms of its components is given by Eqn. (1.37) with the dyadic product defined via Eqn. (1.36). The trace is defined as a linear map that also satisfies

$$\text{tr}\,(\boldsymbol{a} \otimes \boldsymbol{b}) = \boldsymbol{a} \cdot \boldsymbol{b} \quad \forall \boldsymbol{a}, \boldsymbol{b} \in V.$$

It then follows using Eqn. (1.37) that Eqn. (1.44) is still valid, with, again, i ranging from 1 to n. Using indicial notation, we get

$$\text{tr}\,(\boldsymbol{RS}) = \text{tr}\,(\boldsymbol{SR}), \tag{J.1}$$

as in the three-dimensional case. The determinant of T is defined as the determinant of the matrix of its components with respect to the canonical basis, while the matrix inner product of two second-order tensors is defined in the same manner as in the three-dimensional case, i.e., via Eqn. (1.47). Equation (1.83) is also valid with 3 replaced by n.

In even dimensions, all the eigenvalues of a skew-symmetric tensor can be complex, as the example $\boldsymbol{W} = \begin{bmatrix} 0 & a \\ -a & 0 \end{bmatrix}$; $a \in \Re$ in dimension $n = 2$ shows. The complex eigenvalues of $\boldsymbol{W} \in \text{Skw}$ are always of the form $\pm i\alpha_i$, where $\alpha_i \in \Re$, with $\{\boldsymbol{n}_i, \hat{\boldsymbol{n}}_i\}$ as the corresponding eigenvectors, which satisfy $\boldsymbol{n}_i \cdot \boldsymbol{n}_j = \hat{\boldsymbol{n}}_i \cdot \hat{\boldsymbol{n}}_j = 0$ for all i and j. To see this, the relation $\boldsymbol{Wn} = \lambda \boldsymbol{n}$ implies $(\hat{\boldsymbol{n}} \cdot \boldsymbol{Wn}) = \lambda(\boldsymbol{n} \cdot \hat{\boldsymbol{n}})$. By taking the dot product of the relation $\boldsymbol{W}\hat{\boldsymbol{n}} = \hat{\lambda}\hat{\boldsymbol{n}}$, and using the fact $\boldsymbol{W}^T = -\boldsymbol{W}$, we get $-(\hat{\boldsymbol{n}} \cdot \boldsymbol{Wn}) = \hat{\lambda}(\boldsymbol{n} \cdot \hat{\boldsymbol{n}})$. Since $\boldsymbol{n} \cdot \hat{\boldsymbol{n}} \neq 0$, we get $\lambda + \hat{\lambda} = 0$, which shows that the real part of the eigenvalues has to be zero, and since the complex eigenvalues of a real tensor have to occur in conjugate pairs, they have to be of the form $\pm i\alpha_i$. In addition, since $i(\alpha_i + \alpha_j)(\boldsymbol{n}_i, \boldsymbol{n}_j) = (\boldsymbol{W}\boldsymbol{n}_i, \boldsymbol{n}_j) - (\boldsymbol{W}\boldsymbol{n}_i, \boldsymbol{n}_j) = 0$, we have $(\boldsymbol{n}_i, \boldsymbol{n}_j) = 0$ (and hence $\hat{\boldsymbol{n}}_i \cdot \hat{\boldsymbol{n}}_j = 0$), for all i, j (this result also holds for orthogonal tensors as shown in Section 1.5). For example, the eigenvector $\boldsymbol{n} = [-i\ 1]$ of $\begin{bmatrix} 1 & i \\ -i & 1 \end{bmatrix}$ satisfies $(\boldsymbol{n}, \boldsymbol{n}) = 0$. When the dimension n is odd, every $\boldsymbol{W} \in \text{Skw}$ has at least one real eigenvalue, and the real eigenvalues can only have a value of zero, as we have just shown above.

We saw in Section 1.5 that when $n = 3$, the axis of a rotation matrix $\boldsymbol{R}$ is a one-dimensional subspace of V. We now discuss if such a result holds in other dimensions. When n is even, 1 need not even be an eigenvalue of $\boldsymbol{R}$, since all the eigenvalues can occur in complex-conjugate pairs; in fact, the proof that $\det(\boldsymbol{R} - \boldsymbol{I}) = 0$ presented in Section 1.5 hinged on the property $\det(-\boldsymbol{A}) = -\det \boldsymbol{A}$ which holds only in odd dimensions (since $\det(\lambda \boldsymbol{A}) = \lambda^n \det \boldsymbol{A}$). Hence, consider the next higher odd dimension, namely $n = 5$.

The matrix $\mathbf{diag}[1,1,1,-1,-1]$, which has 1 as a repeated eigenvalue is an example of a rotation matrix which is not equal to $\boldsymbol{I}$, whose axis is not unique. Similar examples can be constructed for all other odd dimensions higher than 5. Thus, the uniqueness of the axis of a rotation matrix holds only for $n = 3$. Even the results of Theorem 1.8.1 and Problem 18 do not hold in all dimensions. For example, when $n = 2$, $\boldsymbol{T} = \left[\begin{smallmatrix} \alpha & \beta \\ -\beta & \alpha \end{smallmatrix}\right]$; $\alpha, \beta \in \Re$, commutes with all skew-symmetric and proper orthogonal tensors. However, according to Truesdell [334], the results of Theorem 1.8.1 do hold for all $n > 2$. For an improper orthogonal tensor $\boldsymbol{Q}$, -1 is always an eigenvalue irrespective of the dimension n since, using $\det \boldsymbol{Q} = -1$, one can show that $\det(\boldsymbol{Q} + \boldsymbol{I}) = 0$ as in Section 1.5.

The spectral resolution of a symmetric tensor tensor in the n-dimensional case is given by

$$\boldsymbol{S} = \sum_{i=1}^{n} \lambda_i \boldsymbol{e}_i^* \otimes \boldsymbol{e}_i^*. \tag{J.2}$$

Just as in the case when $n = 3$, the $\{\boldsymbol{e}_i^*\}$ are orthonormal in case the eigenvalues are distinct, or can be chosen to be orthonormal when they are repeated. Thus, we have $\boldsymbol{e}_i^* \cdot \boldsymbol{e}_j^* = \delta_{ij}$. The matrix of components of $\boldsymbol{S}$ in the principal frame formed by the $\{\boldsymbol{e}_i^*\}$ is the diagonal matrix $\mathbf{diag}[\lambda_1, \lambda_2, \ldots, \lambda_n]$. Since the $\{\boldsymbol{e}_i^*\}$ are orthonormal, the above equation can be written as

$$\boldsymbol{S} = \sum_{i=1}^{k} \lambda_i \boldsymbol{P}_i, \tag{J.3}$$

where k is the number of distinct eigenvalues of $\boldsymbol{S}$, and the projections $\boldsymbol{P}_i$ obey $\boldsymbol{P}_i \boldsymbol{P}_j = \delta_{ij} \boldsymbol{P}_j$ (no sum on j) and $\sum_{i=1}^{k} \boldsymbol{P}_i = \boldsymbol{I}$. Due to these properties of $\boldsymbol{P}_i$, we have $\boldsymbol{S}\boldsymbol{P}_i = \boldsymbol{P}_i \boldsymbol{S} = \lambda_i \boldsymbol{P}_i$ (no sum on i). Again, as in the $n = 3$ case, the spectral resolution of $\boldsymbol{S}$ is unique, with $\boldsymbol{P}_i$ given by

$$\boldsymbol{P}_i(\boldsymbol{S}) = \begin{cases} \prod_{\substack{j=1 \\ j \neq i}}^{k} \frac{\boldsymbol{S} - \lambda_j \boldsymbol{I}}{\lambda_i - \lambda_j}, & k > 1 \\ \boldsymbol{I}, & k = 1. \end{cases} \tag{J.4}$$

To prove this, note that for $k = 1$, $\sum_{i=1}^{k} \boldsymbol{P}_i = \boldsymbol{I}$ implies that $\boldsymbol{P}_1 = \boldsymbol{I}$, while for $k > 1$,

$$\prod_{\substack{j=1 \\ j \neq i}}^{k} \frac{\boldsymbol{S} - \lambda_j \boldsymbol{I}}{\lambda_i - \lambda_j} = \prod_{\substack{j=1 \\ j \neq i}}^{k} \frac{\sum_{\substack{m=1 \\ m \neq j}}^{k} (\lambda_m - \lambda_j) \boldsymbol{P}_m}{\lambda_i - \lambda_j} = \prod_{\substack{j=1 \\ j \neq i}}^{k} \frac{(\lambda_i - \lambda_j) \boldsymbol{P}_i}{\lambda_i - \lambda_j} = \boldsymbol{P}_i,$$

where the second step is obtained from the first using the fact that $\boldsymbol{P}_i$ is the only common projection among the groups of sums in the numerator. As an example, the projection matrix

$$\hat{\boldsymbol{P}} = \boldsymbol{I} - \boldsymbol{n} \otimes \boldsymbol{n}$$

projects any vector $\boldsymbol{x}$ onto the plane that is perpendicular to $\boldsymbol{n}$, since $\hat{\boldsymbol{P}}\boldsymbol{x} = \boldsymbol{x} - (\boldsymbol{x} \cdot \boldsymbol{n})\boldsymbol{n}$. The above projection matrix can be written in spectral form as $0(\boldsymbol{n} \otimes \boldsymbol{n}) + 1(\boldsymbol{I} - \boldsymbol{n} \otimes \boldsymbol{n})$ showing that 0 (multiplicity 1) and 1 (multiplicity $n-1$) are the eigenvalues of $\hat{P}$ with $\boldsymbol{n}$ and $n-1$ perpendicular vectors in the plane perpendicular to $\boldsymbol{n}$ as the corresponding orthonormal eigenvectors.

The determination of the polar decomposition $\boldsymbol{F} = \boldsymbol{R}\boldsymbol{U} = \boldsymbol{V}\boldsymbol{R}$ (where $\boldsymbol{F}$ is real-valued) in n-dimensional vector spaces is discussed in [153], and has been presented in Section 1.9.4. From the explicit expressions presented, existence and uniqueness of $\boldsymbol{U}$, $\boldsymbol{V}$, and $\boldsymbol{R}$ for the case when $\boldsymbol{F}$ is invertible, are immediately obvious. When $\boldsymbol{F}$ is singular, $\boldsymbol{U}$ and $\boldsymbol{V}$ are still unique (although, of course, singular); however, $\boldsymbol{R}$ is not unique. As an example, if $\boldsymbol{F}$ is such that $F_{11} = 1$ with the other elements zero, then $\boldsymbol{U} = \boldsymbol{V} = \boldsymbol{F}$, but $\boldsymbol{R}$ can have either of the following forms:

$$\begin{bmatrix} 1 & 0 & 0 \\ 0 & \cos\theta & \sin\theta \\ 0 & -\sin\theta & \cos\theta \end{bmatrix}, \quad \begin{bmatrix} 1 & 0 & 0 \\ 0 & \cos\theta & \sin\theta \\ 0 & \sin\theta & -\cos\theta \end{bmatrix}, \quad \theta \in [0, 2\pi].$$

Note that the first class of matrices above are proper orthogonal, while the second one are improper orthogonal. Thus, not only are infinite choices for $\boldsymbol{R}$ possible, but it can be proper or improper as well. This is consistent with the fact that when $\boldsymbol{F}$ is singular, $\det \boldsymbol{F} = \det \boldsymbol{U} = \det \boldsymbol{V} = 0$, so that a value of $\det \boldsymbol{R}$ equal to either of the two possibilities $+1$ or -1 satisfies $\det \boldsymbol{F} = (\det \boldsymbol{R})(\det \boldsymbol{U}) = (\det \boldsymbol{R})(\det \boldsymbol{V})$. The extension of the strategy discussed in [153] to complex-valued tensors using Eqn. (J.22) is straightforward. From this point on until the end of this appendix, we allow tensors to be complex-valued.

The characteristic equation is given by $\det(\boldsymbol{T} - \lambda \boldsymbol{I}) = 0$, which in expanded form reads

$$\lambda^n - I_1\lambda^{n-1} + I_2\lambda^{n-2} + \cdots + (-1)^n I_n = 0. \tag{J.5}$$

The Cayley–Hamilton theorem is given by

$$\boldsymbol{T}^n - I_1\boldsymbol{T}^{n-1} + I_2\boldsymbol{T}^{n-2} + \cdots + (-1)^n I_n \boldsymbol{I} = \boldsymbol{0}. \tag{J.6}$$

To prove this result, let $\boldsymbol{H}$ be the adjoint of $(\alpha\boldsymbol{I} - \boldsymbol{T})$, i.e., $\boldsymbol{H}$ satisfies

$$\begin{aligned} \boldsymbol{H}(\alpha\boldsymbol{I} - \boldsymbol{T}) &= \det(\alpha\boldsymbol{I} - \boldsymbol{T})\boldsymbol{I} \\ &= \left[\alpha^n - \alpha^{n-1}I_1 + \alpha^{n-2}I_2 + \cdots + (-1)^n I_n\right]\boldsymbol{I}. \end{aligned} \tag{J.7}$$

From the above equation, we conclude that $\boldsymbol{H}$ must be of the form

$$\alpha^{n-1}\boldsymbol{I} + \alpha^{n-2}\boldsymbol{H}_1 + \alpha^{n-3}\boldsymbol{H}_2 + \cdots + \boldsymbol{H}_{n-1}. \tag{J.8}$$

Substituting this expression into Eqn. (J.7), and matching the coefficients of α^n, α^{n-1} and so on[1], we get

$$\begin{aligned} I_1\boldsymbol{I} &= (-1)^1\left[\boldsymbol{H}_1 - \boldsymbol{T}\right], \\ I_2\boldsymbol{I} &= (-1)^2\left[\boldsymbol{H}_2 - \boldsymbol{H}_1\boldsymbol{T}\right], \end{aligned}$$

[1] This matching of coefficients can be justified by setting first α to zero, getting equality for the coefficient of 1, then differentiating both sides with respect to α, evaluating them at $\alpha = 0$, thus getting equality for the coefficients of α, and so on.

and so on, or, in general (with $\boldsymbol{H}_0 \equiv \boldsymbol{I}$ and $\boldsymbol{H}_n \equiv \boldsymbol{0}$),

$$I_k \boldsymbol{I} = (-1)^k \left[\boldsymbol{H}_k - \boldsymbol{H}_{k-1}\boldsymbol{T}\right]; \quad k = 1, \ldots, n. \tag{J.9}$$

Thus,

$$\begin{aligned} \boldsymbol{T}^n - I_1\boldsymbol{T}^{n-1} + \cdots + (-1)^n I_n \boldsymbol{I} &= \boldsymbol{T}^n + \left[-\boldsymbol{T}^n + \boldsymbol{H}_1\boldsymbol{T}^{n-1}\right] + \left[\boldsymbol{H}_2\boldsymbol{T}^{n-2} - \boldsymbol{H}_1\boldsymbol{T}^{n-1}\right] + \cdots + \\ &\quad + \left[\boldsymbol{H}_{n-1}\boldsymbol{T} - \boldsymbol{H}_{n-2}\boldsymbol{T}^2\right] - \boldsymbol{H}_{n-1}\boldsymbol{T} \\ &= \boldsymbol{0}, \end{aligned}$$

which completes the proof.

Taking the trace of Eqn. (J.6), we get the expression for the determinant as

$$I_n = \frac{(-1)^{n+1}}{n} \operatorname{tr} \left[\boldsymbol{T}^n - I_1\boldsymbol{T}^{n-1} + \cdots + (-1)^{n-1} I_{n-1}\boldsymbol{I}\right]. \tag{J.10}$$

The following explicit formula for the inverse of $\boldsymbol{T}$ when it exists (i.e., when $I_n \neq 0$), also follows from the Cayley–Hamilton theorem:

$$\boldsymbol{T}^{-1} = \frac{(-1)^{n+1}}{I_n} \left[\boldsymbol{T}^{n-1} + \sum_{i=1}^{n-1} (-1)^i I_i \boldsymbol{T}^{n-i-1}\right],$$

with $\boldsymbol{T}^0 = \boldsymbol{I}$. The Woodbury and Sherman–Morrison formulae given by Eqns. (1.70) and (1.75) remain valid since their proof did not depend on the dimension n.

If one defines $\mathbf{cof}\,\boldsymbol{T}$ as the transpose of $\boldsymbol{H}$, then

$$(\mathbf{cof}\,\boldsymbol{T})^T \boldsymbol{T} = (\det \boldsymbol{T})\boldsymbol{I}. \tag{J.11}$$

It immediately follows from Eqn. (J.11) that $\det(\mathbf{cof}\,\boldsymbol{T}) = (\det \boldsymbol{T})^{n-1}$. From Eqn. (J.6), we get the following explicit expression for $\mathbf{cof}\,\boldsymbol{T}$:

$$\mathbf{cof}\,\boldsymbol{T} = (-1)^{n-1} \left[\boldsymbol{T}^{n-1} - I_1\boldsymbol{T}^{n-2} + \cdots + (-1)^{n-1} I_{n-1}\boldsymbol{I}\right]^T. \tag{J.12}$$

Of course, if $\boldsymbol{T}$ is invertible, then from the definition itself, we have $\mathbf{cof}\,\boldsymbol{T} = (\det \boldsymbol{T})\boldsymbol{T}^{-1}$. Using the linearity of the trace function, and Eqns. (J.1), (J.10) and (J.12), it is easily seen that

$$\det(\boldsymbol{T} + \boldsymbol{U}) = \det \boldsymbol{T} + \mathbf{cof}\,\boldsymbol{T} : \boldsymbol{U} + o(\boldsymbol{U}),$$

so that we obtain the first relation in Eqn. (1.226c) even in the n-dimensional case:

$$\frac{\partial I_n}{\partial \boldsymbol{T}} = \mathbf{cof}\,\boldsymbol{T}. \tag{J.13}$$

From Eqn. (1.220), we get

$$\frac{d}{dt}(\det \boldsymbol{T}) = \mathbf{cof}\,\boldsymbol{T} : \frac{d\boldsymbol{T}}{dt}, \tag{J.14}$$

which is the same as Eqn. (1.222).

To find the explicit formula for the principal invariants of a tensor known as the Newton identity, we present a method that is similar in spirit (in the sense that it is a matrix-based method) to the one in [170]. Using Eqn. (J.14), and the fact that $\boldsymbol{H}^T$ (defined via Eqn. (J.7)) is $\mathbf{cof}\,(\alpha\boldsymbol{I} - \boldsymbol{T})$, we get

$$\frac{d}{d\alpha}\left[\det(\alpha\boldsymbol{I} - \boldsymbol{T})\right] = \boldsymbol{H}^T : \boldsymbol{I} = \operatorname{tr}\boldsymbol{H}.$$

Substituting the expressions for $\det(\alpha\boldsymbol{I} - \boldsymbol{T})$ and $\boldsymbol{H}$ given in Eqns. (J.7) and (J.8), respectively, into the above equation, we get

$$n\alpha^{n-1} - (n-1)\alpha^{n-2}I_1 + \cdots + (-1)^{n-1}I_{n-1} = n\alpha^{n-1} + \operatorname{tr}(\boldsymbol{H}_1)\alpha^{n-2} + \cdots + \operatorname{tr}\boldsymbol{H}_{n-1}.$$

Again matching the coefficients of α^{n-1}, α^{n-2} and so on, we get

$$(n-k)I_k = (-1)^k \operatorname{tr}\boldsymbol{H}_k; \quad k = 1, \ldots, n-1. \tag{J.15}$$

Taking the trace of Eqn. (J.9), we have (with $\boldsymbol{H}_0 \equiv \boldsymbol{I}$ and $\boldsymbol{H}_n \equiv \boldsymbol{0}$)

$$nI_k = (-1)^k \operatorname{tr}\boldsymbol{H}_k - (-1)^k \operatorname{tr}(\boldsymbol{H}_{k-1}\boldsymbol{T}); \quad k = 1, \ldots, n. \tag{J.16}$$

Subtracting Eqn. (J.15) from Eqn. (J.16), we get

$$kI_k = (-1)^{k+1} \operatorname{tr}(\boldsymbol{H}_{k-1}\boldsymbol{T}); \quad k = 1, \ldots, n.$$

Using the recursive formula given by Eqn. (J.9), we obtain

$$\boldsymbol{H}_{k-1} = \boldsymbol{T}^{k-1} - I_1\boldsymbol{T}^{k-2} + \cdots + (-1)^{k-1}I_{k-1}\boldsymbol{I}.$$

Thus, the principal invariants I_k, $k = 1, 2, \ldots, n$, are given by the following recursive formula known as the Newton identity:

$$\begin{aligned} I_k &= \frac{(-1)^{k+1}}{k}\left[\operatorname{tr}\boldsymbol{T}^k - I_1 \operatorname{tr}\boldsymbol{T}^{k-1} + \cdots + (-1)^{k-1}I_{k-1}\operatorname{tr}\boldsymbol{T}\right] \\ &= \frac{(-1)^{k+1}}{k}\operatorname{tr}\left[\boldsymbol{T}^k + \sum_{i=1}^{k-1}(-1)^i I_i \boldsymbol{T}^{k-i}\right]. \end{aligned} \tag{J.17}$$

Eqn. (J.17) is useful for finding higher-order invariants when the lower-order invariants are known. Provided $\boldsymbol{T}$ is invertible, Eqns. (J.6) and (J.17) can be combined to yield the lower-order invariants from the higher-order ones as follows:

$$I_{n-j} = \frac{(-1)^{j+1}}{j}\operatorname{tr}\left[I_n\boldsymbol{T}^{-j} - I_{n-1}\boldsymbol{T}^{-j+1} + \cdots + (-1)^{j-1}I_{n-j+1}\boldsymbol{T}^{-1}\right].$$

The gradient of the k'th invariant is obtained using Eqn. (1.227), and is given by (see [38], [67], [106] and [147] for various proofs)

$$\frac{\partial I_k}{\partial \boldsymbol{T}} = (-1)^{k+1}\left[\boldsymbol{T}^{k-1} + \sum_{i=1}^{k-1}(-1)^i I_i \boldsymbol{T}^{k-1-i}\right]^T, \tag{J.18}$$

or, alternatively, by the recursive formula

$$\frac{\partial I_{k+1}}{\partial \boldsymbol{T}} = I_k \boldsymbol{I} - \boldsymbol{T}^T \frac{\partial I_k}{\partial \boldsymbol{T}}. \tag{J.19}$$

We now present Dui's proof [67] for Eqn. (J.19); Eqn. (J.18) is obtained from Eqn. (J.19) by induction. Let $\gamma \neq -\lambda_k$, where λ_k, $k = 1, 2, \ldots, n$ are the eigenvalues of $\boldsymbol{T}$. Then, expanding $\det(\boldsymbol{T} + \gamma \boldsymbol{I})$ as in Eqn. (J.5), we have

$$\gamma^n + I_1 \gamma^{n-1} + I_2 \gamma^{n-2} + \cdots + I_n = \det(\boldsymbol{T} + \gamma \boldsymbol{I}). \tag{J.20}$$

Taking the gradient of the above equation with respect to $\boldsymbol{T}$ (keeping in mind that γ is *not* an eigenvalue of $\boldsymbol{T}$), and using Eqn. (J.13), we get

$$\frac{\partial I_1}{\partial \boldsymbol{T}} \gamma^{n-1} + \frac{\partial I_2}{\partial \boldsymbol{T}} \gamma^{n-2} + \cdots + \frac{\partial I_n}{\partial \boldsymbol{T}} = \mathbf{cof}\,(\boldsymbol{T} + \gamma \boldsymbol{I}).$$

Multiplying both sides by $(\boldsymbol{T} + \gamma \boldsymbol{I})^T$, and using Eqn. (J.11), we get

$$(\boldsymbol{T}^T + \gamma \boldsymbol{I}) \left[\frac{\partial I_1}{\partial \boldsymbol{T}} \gamma^{n-1} + \frac{\partial I_2}{\partial \boldsymbol{T}} \gamma^{n-2} + \cdots + \frac{\partial I_n}{\partial \boldsymbol{T}} \right] = (\gamma^n + I_1 \gamma^{n-1} + \cdots + I_n) \boldsymbol{I}.$$

Matching the coefficients of the various powers of γ, we get

$$\frac{\partial I_1}{\partial \boldsymbol{T}} = \boldsymbol{I},$$
$$\frac{\partial I_2}{\partial \boldsymbol{T}} = I_1 \boldsymbol{I} - \boldsymbol{T}^T,$$

and, in general, Eqn. (J.19).

Although two tensors $\boldsymbol{A}$ and $\boldsymbol{B}$ do not, in general, commute, we now show that the principal invariants (and hence, the eigenvalues) of $\boldsymbol{AB}$ and $\boldsymbol{BA}$ are the same, i.e.,

$$\mathcal{I}_{\boldsymbol{AB}} = \mathcal{I}_{\boldsymbol{BA}} \tag{J.21}$$

From Eqn. (J.17), we see that the principal invariants of a tensor $\boldsymbol{T}$ are a function of the lower-order invariants and powers of $\boldsymbol{T}$. Using Eqn. (J.1), we have

$$\operatorname{tr}(\boldsymbol{AB})^2 = \operatorname{tr}(\boldsymbol{ABAB}) = \operatorname{tr}(\boldsymbol{BABA}) = \operatorname{tr}(\boldsymbol{BA})^2,$$
$$\operatorname{tr}(\boldsymbol{AB})^3 = \operatorname{tr}(\boldsymbol{ABABAB}) = \operatorname{tr}(\boldsymbol{BABABA}) = \operatorname{tr}(\boldsymbol{BA})^3,$$
$$\cdots$$
$$\operatorname{tr}(\boldsymbol{AB})^n = \operatorname{tr}(\boldsymbol{BA})^n.$$

Thus, from Eqn. (J.17), it follows using induction that Eqn. (J.21) holds. Since the principal invariants are the same, it follows that the characteristic equation, and hence the eigenvalues, of $\boldsymbol{AB}$ and $\boldsymbol{BA}$ are the same.

Given $\dot{\boldsymbol{C}}$, the general expression for $\dot{\boldsymbol{U}}$ has already been presented in Eqn. (1.246).

A tensor $\boldsymbol{N}$ is said to be *nilpotent* of order ν, where $2 \leq \nu \leq n$, if $\boldsymbol{N}^\nu = \boldsymbol{0}$ but $\boldsymbol{N}^{\nu-1} \neq \boldsymbol{0}$. A tensor $\boldsymbol{P}$ is said to be *idempotent* or a *projection* if $\boldsymbol{P}^2 = \boldsymbol{P}$. Generalizing this idea, we say that a tensor $\boldsymbol{P}$ is said to be idempotent of order ν, $3 \leq \nu \leq n$, if $\boldsymbol{P}^\nu = \boldsymbol{P}$, but $\boldsymbol{P}^{\nu-1} \neq \boldsymbol{P}$. If $\boldsymbol{P}$ is

Order	Nilpotent	Idempotent
2	$e_1 \otimes e_2$, $e_1 \otimes e_2 + e_3 \otimes e_4$	$e_1 \otimes e_1$, $e_1 \otimes (e_1 + e_2)$
3	$e_1 \otimes e_2 + e_2 \otimes e_3$, $e_1 \otimes e_2 + e_3 \otimes e_1$	$e_1 \otimes e_2 + e_2 \otimes e_1$
	$e_1 \otimes e_2 + e_2 \otimes e_3 + e_4 \otimes e_5 + e_5 \otimes e_6$	$e_1 \otimes e_2 + e_2 \otimes e_1 + e_3 \otimes e_3$
4	$e_1 \otimes e_2 + e_2 \otimes e_3 + e_3 \otimes e_4$	$e_1 \otimes e_2 + e_2 \otimes e_3 + e_3 \otimes e_1 + e_4 \otimes e_4$

Table J.1 *Some examples of real-valued nilpotent and idempotent tensors of various orders*

idempotent, then $\boldsymbol{I} - \boldsymbol{P}$ is also idempotent. Although, a projection $\boldsymbol{P}$ need not be invertible, $(\boldsymbol{I} + \alpha \boldsymbol{P})$, $\alpha \in \Re$, is always invertible, and the inverse is given by $\boldsymbol{I} - \frac{\alpha}{1+\alpha}\boldsymbol{P}$. Some simple examples of nilpotent and idempotent tensors are presented in Table J.1. If $\boldsymbol{T}$ is a nilpotent or idempotent tensor of order ν, then $\boldsymbol{H}^{-1}\boldsymbol{T}\boldsymbol{H}$, where $\boldsymbol{H}$ is any invertible tensor, is also a nilpotent or idempotent tensor of the same order since

$$\begin{aligned}(\boldsymbol{H}^{-1}\boldsymbol{T}\boldsymbol{H})^\nu &= (\boldsymbol{H}^{-1}\boldsymbol{T}\boldsymbol{H})(\boldsymbol{H}^{-1}\boldsymbol{T}\boldsymbol{H})\cdots(\boldsymbol{H}^{-1}\boldsymbol{T}\boldsymbol{H}) \quad (\nu \text{ times})\\ &= \boldsymbol{H}^{-1}\boldsymbol{T}^\nu \boldsymbol{H}.\end{aligned}$$

This fact can be used to generate more complicated nilpotent or idempotent tensors starting from relatively simple ones as in Table J.1.

All the eigenvalues of a nilpotent tensor are zero since $\boldsymbol{T}^n\boldsymbol{n} = \lambda^n \boldsymbol{n} = \boldsymbol{0}$. Conversely, if all the eigenvalues of a tensor are zero, then the tensor is nilpotent since the characteristic equation is $\lambda^n = 0$, so that by the Cayley–Hamilton theorem $\boldsymbol{T}^n = \boldsymbol{0}$. Thus, a tensor is nilpotent if and only if all its eigenvalues are zero. However, note that this characterization does not distinguish between nilpotent tensors of different orders. By the spectral decomposition (see Eqn. (J.25)), a diagonalizable (in particular, symmetric) tensor is nilpotent if and only if it is the zero tensor. The eigenvalues of an idempotent tensor can only be zero or one, since $\boldsymbol{P}^2\boldsymbol{n} = \lambda^2\boldsymbol{n} = \boldsymbol{P}\boldsymbol{n} = \lambda\boldsymbol{n}$, so that $\lambda(\lambda - 1) = 0$, which in turn implies that λ is either 0 or 1. Similarly, the admissible eigenvalues of an idempotent tensor of order ν can be obtained by finding the roots of the equation $\lambda^\nu = \lambda$.

A natural question that arises at this stage is whether a spectral resolution using eigenvalues and idempotents can be given for complex-valued symmetric and skew-symmetric tensors. The answer, quite surprisingly, is in the negative. For example, the symmetric and skew-symmetric tensors $\begin{bmatrix} 1 & i \\ i & -1 \end{bmatrix}$ and $\begin{bmatrix} 0 & 1 & i \\ -1 & 0 & 0 \\ -i & 0 & 0 \end{bmatrix}$ have all eigenvalues zero! Thus, while all real-valued symmetric and skew-symmetric tensors are diagonalizable, there are *some* complex-valued symmetric and skew-symmetric tensors that are nilpotent. In the complex case, the analogues of symmetric, skew-symmetric and orthogonal tensors are Hermitian $(\boldsymbol{S}^* = \boldsymbol{S})$, skew-Hermitian $(\boldsymbol{W}^* = -\boldsymbol{W})$ and unitary tensors $(\boldsymbol{Q}^* = \boldsymbol{Q}^{-1})$, respectively, where $\boldsymbol{T}^* := \hat{\boldsymbol{T}}^T$. Note that $\hat{\boldsymbol{x}} \cdot \boldsymbol{T}\boldsymbol{y} = \widehat{\boldsymbol{T}^*\boldsymbol{x}} \cdot \boldsymbol{y}$. Similar to (real-valued) symmetric tensors, Hermitian tensors have real eigenvalues, similar to (real-valued) skew-symmetric tensors, the real part of the eigenvalues of skew-Hermitian tensors is zero (although the eigenvalues need not occur in complex conjugate pairs), and similar to (real-valued) orthogonal tensors, unitary tensors have eigenvalues whose magnitude is one. A tensor $\boldsymbol{T}$ is Hermitian if and only if $i\boldsymbol{T}$ is skew-Hermitian. Every complex-valued tensor can be decomposed uniquely as $\boldsymbol{T} = \boldsymbol{A} + i\boldsymbol{B}$, where $\boldsymbol{A} = (\boldsymbol{T} + \boldsymbol{T}^*)/2$ and $\boldsymbol{B} = (\boldsymbol{T} - \boldsymbol{T}^*)/(2i)$ are Hermitian.

Every tensor T can also be decomposed as $T = RU = VR$ where R is unitary ($R^*R = RR^* = I$), and $U = \sqrt{T^*T}$ and $V = \sqrt{TT^*}$ are (Hermitian) positive semi-definite ($\hat{x} \cdot Ux \geq 0\ \forall x$). Similar to the spectral resolution based on the polar decomposition given by Eqn. (1.135), we have

$$T = \sum_{i=1}^{n} \lambda_i n_i \otimes \hat{N}_i, \tag{J.22}$$

where $n_i \cdot \hat{n}_j = N_i \cdot \hat{N}_j = \delta_{ij}$, and the $\{\lambda_i\}$'s which are the square roots of the eigenvalues of T^*T (or TT^*) are nonnegative numbers (strictly positive if T is nonsingular). The representations for R, U and V are

$$R = \sum_{i=1}^{n} n_i \otimes \hat{N}_i,$$
$$U = \sum_{i=1}^{n} \lambda_i N_i \otimes \hat{N}_i,$$
$$V = \sum_{i=1}^{n} \lambda_i n_i \otimes \hat{n}_i.$$

If S is a (complex-valued) symmetric tensor, then Eqn. (J.22) reduces to

$$S = \sum_{i=1}^{n} \lambda_i n_i \otimes n_i = \left(\sum_{i=1}^{n} n_i \otimes e_i \right) \left(\sum_{i=1}^{n} \lambda_i e_i \otimes e_i \right) \left(\sum_{i=1}^{n} e_i \otimes n_i \right) = R\Lambda R^T,$$

where $R = \sum_{i=1}^{n} n_i \otimes e_i$ is a unitary tensor, and $\Lambda = \sum_{i=1}^{n} \lambda_i e_i \otimes e_i$ is a diagonal matrix with the singular values λ_i (which are the square roots of eigenvalues of $\hat{S}S$) along the diagonal.

We have seen via the Cayley–Hamilton theorem that a tensor satisfies an equation that has the same form as its own characteristic equation. The left-hand side of Eqn. (J.6) can be written as

$$p(T) = \prod_{i=1}^{k} (T - \lambda_i I)^{n_i}, \tag{J.23}$$

where k is the number of distinct eigenvalues, $1 \leq n_i \leq n$, and $\sum_{i=1}^{k} n_i = n$. Thus, $p(T)$ is a polynomial that annihilates T. Among all polynomials that annihilate T, the one that has the minimum degree, and whose coefficient of the highest term is unity is called the *minimum polynomial* of T. It can be shown that the minimal polynomial of T must have the form

$$q(T) = \prod_{i=1}^{k} (T - \lambda_i I)^{m_i}, \tag{J.24}$$

where $1 \leq m_i \leq n_i + 1 - r_i$, where r_i is the number of linearly independent eigenvectors[2] corresponding to the eigenvalue λ_i. Thus, if $r := \sum_{i=1}^{k} r_i$ and $m := \sum_{i=1}^{k} m_i$, then $k \leq m \leq$

[2]The number of times n_i an eigenvalue λ_i occurs as a root of the characteristic equation is called the algebraic multiplicity of λ_i. The dimension of the eigenspace corresponding to λ_i, namely r_i, is called the geometric multiplicity of λ_i.

$n+k-r$. For example, if the matrix representation of $\boldsymbol{T}$ with respect to a particular basis is the diagonal matrix $\mathbf{diag}[1,2,2]$, then $p(\boldsymbol{T}) = (\boldsymbol{T}-\boldsymbol{I})(\boldsymbol{T}-2\boldsymbol{I})^2$ and $q(\boldsymbol{T}) = (\boldsymbol{T}-\boldsymbol{I})(\boldsymbol{T}-2\boldsymbol{I})$. For this example, we have $n=3$, $k=2$, $r=3$ and $m=2$ so that $m=n+k-r$. On the other hand, for the tensor $\boldsymbol{T} = \boldsymbol{e}_1 \otimes \boldsymbol{e}_2 + \boldsymbol{e}_3 \otimes \boldsymbol{e}_4$ in four-dimensional space, we have $n=4$, $k=1$, $r=2$ and $m=2$, so that $m < n+k-r$. For diagonalizable tensors discussed below, we have $r=n$ and $k=m$. On the other hand, there are tensors for which the minimal polynomial is equal to the characteristic polynomial. The necessary and sufficient condition for the minimal and characteristic polynomials to be equal, i.e., $m=n$, is that $r_i=1$ for each i, i.e., $r=k$. Summarizing, sufficient (but not necessary) conditions for $m=n+k-r$ to hold are

1. The tensor is diagonalizable, so that $r=n$ and $k=m$.
2. The minimal polynomial and the characteristic polynomial are the same, i.e., $m=n$ and $r=k$.
3. $n \leq 3$.

If a tensor is diagonalizable *and* has minimal and characteristic polynomials the same, then $k=r=n=m$, so that all its eigenvalues are distinct.

A tensor is called *diagonalizable* if there exists an invertible tensor $\boldsymbol{H}$ such that the matrix representation of $\boldsymbol{H}^{-1}\boldsymbol{T}\boldsymbol{H}$ with respect to some basis is diagonal. We have already seen that symmetric tensors are diagonalizable by an orthogonal tensor (whose columns are the orthonormal vectors $\{\boldsymbol{e}_i^*\}$ occurring in the spectral resolution).

The following theorem characterizes diagonalizability for an arbitrary (real or complex-valued) tensor $\boldsymbol{T}$ [125]:

Theorem J.0.3 (Characterization of Diagonalizable Tensors). *The following are equivalent:*

1. *The tensor $\boldsymbol{T}$ is diagonalizable.*
2. *The tensor $\boldsymbol{T}$ has a set of n eigenvectors that are linearly independent, or, equivalently, the geometric multiplicity is the same as the algebraic multiplicity for each eigenvalue of $\boldsymbol{T}$, i.e., $n_i = r_i$ or $n=r$.*
3. *The minimal polynomial of the tensor $\boldsymbol{T}$ is*

$$q(\boldsymbol{T}) = \prod_{i=1}^{k} (\boldsymbol{T} - \lambda_i \boldsymbol{I}),$$

i.e., each $m_i = 1$ (so that $m=k$) in Eqn. (J.24), and $\mathrm{rank}(\boldsymbol{T} - \lambda_i \boldsymbol{I}) = n - n_i$.

Since symmetric tensors are diagonalizable, the results of the above theorem can be directly applied to them. Every diagonalizable tensor can be written as

$$\boldsymbol{T} = \sum_{i=1}^{k} \lambda_i \boldsymbol{P}_i, \tag{J.25}$$

where λ_i, $i = 1, 2, \ldots, k$, are the distinct eigenvalues of $\boldsymbol{T}$, and $\boldsymbol{P}_i$ is given by Eqn. (J.4) with $\boldsymbol{S}$ replaced by $\boldsymbol{T}$, i.e.,

$$\boldsymbol{P}_i(\boldsymbol{T}) = \begin{cases} \prod_{\substack{j=1 \\ j \neq i}}^{k} \frac{\boldsymbol{T} - \lambda_j \boldsymbol{I}}{\lambda_i - \lambda_j}, & k > 1 \\ \boldsymbol{I}, & k = 1. \end{cases} \tag{J.26}$$

By substituting the results of Problem 13 into Eqn. (1.251) (which actually holds for any underlying space dimension n), we get for the case of *distinct* eigenvalues (i.e., $k = n$)

$$\begin{aligned} \frac{\partial \lambda_i}{\partial \boldsymbol{T}} &= \boldsymbol{P}_i^T, \\ \frac{\partial \boldsymbol{P}_i}{\partial \boldsymbol{T}} &= \sum_{\substack{j=1 \\ j \neq i}}^{k} \frac{\boldsymbol{P}_i \boxtimes \boldsymbol{P}_j^T + \boldsymbol{P}_j \boxtimes \boldsymbol{P}_i^T}{\lambda_i - \lambda_j}. \end{aligned} \tag{J.27}$$

An invertible tensor is not necessarily diagonalizable, e.g. $\boldsymbol{T} = \left[\begin{smallmatrix} 1 & 1 \\ 0 & 1 \end{smallmatrix}\right]$, whose minimal polynomial is $(\boldsymbol{T} - \boldsymbol{I})^2$.

Let $\boldsymbol{u}_i$ and $\boldsymbol{v}_i$ denote the (possibly complex) eigenvectors of $\boldsymbol{T}$ and $\boldsymbol{T}^T$, respectively, i.e.,

$$\begin{aligned} \boldsymbol{T}\boldsymbol{u}_i &= \lambda_i \boldsymbol{u}_i, \\ \boldsymbol{T}^T \boldsymbol{v}_i &= \lambda_i \boldsymbol{v}_i, \end{aligned}$$

and let $\boldsymbol{u}_i$ and $\boldsymbol{v}_i$ be normalized such that $\boldsymbol{u}_i \cdot \boldsymbol{v}_j = \delta_{ij}$. Any vector $\boldsymbol{a}$ can be expressed as $\sum_{i=1}^n a_i \boldsymbol{u}_i = \sum_{i=1}^n b_i \boldsymbol{v}_i$. By taking the dot product with $\boldsymbol{v}_j$ and $\boldsymbol{u}_j$, respectively, we get $a_i = \boldsymbol{a} \cdot \boldsymbol{v}_i$ and $b_i = \boldsymbol{a} \cdot \boldsymbol{u}_i$. Thus,

$$\begin{aligned} \boldsymbol{a} &= (\boldsymbol{a} \cdot \boldsymbol{v}_i)\boldsymbol{u}_i = (\boldsymbol{u}_i \otimes \boldsymbol{v}_i)\boldsymbol{a} \\ &= (\boldsymbol{a} \cdot \boldsymbol{u}_i)\boldsymbol{v}_i = (\boldsymbol{v}_i \otimes \boldsymbol{u}_i)\boldsymbol{a}, \end{aligned}$$

and since $\boldsymbol{a}$ is arbitrary, we get $\sum_{i=1}^n \boldsymbol{u}_i \otimes \boldsymbol{v}_i = \sum_{i=1}^n \boldsymbol{v}_i \otimes \boldsymbol{u}_i = \boldsymbol{I}$. Any diagonalizable tensor can be written as

$$\boldsymbol{T} = \sum_{i=1}^{n} \lambda_i \boldsymbol{u}_i \otimes \boldsymbol{v}_i. \tag{J.28}$$

By comparing against Eqn. (J.25), we see that if all eigenvalues are *distinct*, then we have $\boldsymbol{P}_i = \boldsymbol{u}_i \otimes \boldsymbol{v}_i$ (no sum on i). If two or more eigenvalues are repeated, then $\boldsymbol{P}_i$ is the sum of dyadic products of $\boldsymbol{u}_i$ and $\boldsymbol{v}_i$ corresponding to the repeated eigenvalue; for example, if $n = 3$ and $\lambda_2 = \lambda_3$, then $\boldsymbol{P}_1 = \boldsymbol{u}_1 \otimes \boldsymbol{v}_1$ and $\boldsymbol{P}_2 = \boldsymbol{u}_2 \otimes \boldsymbol{v}_2 + \boldsymbol{u}_3 \otimes \boldsymbol{v}_3 = \boldsymbol{I} - \boldsymbol{u}_1 \otimes \boldsymbol{v}_1$. Using Eqn. (1.172), it follows that the projections $\boldsymbol{P}_i = \boldsymbol{u}_i \otimes \boldsymbol{v}_i$ that occur in the spectral decomposition of a diagonalizable tensor with distinct eigenvalues obey

$$\boldsymbol{P}_i \boxtimes \boldsymbol{P}_i^T = \boldsymbol{P}_i \otimes \boldsymbol{P}_i^T \quad \text{(no sum on } i\text{)}. \tag{J.29}$$

For a symmetric tensor, of course, $\boldsymbol{P}_i^T = \boldsymbol{P}_i$.

If $\{e_i\}$ denotes the canonical Cartesian basis, then the diagonalizing tensor $\boldsymbol{H}$ is given by $\sum_{i=1}^n \boldsymbol{u}_i \otimes \boldsymbol{e}_i$ (so that $\boldsymbol{H}^{-1} = \sum_{i=1}^n \boldsymbol{e}_i \otimes \boldsymbol{v}_i$), since, as can be easily verified $\boldsymbol{H}^{-1}\boldsymbol{T}\boldsymbol{H} = \sum_{i=1}^n \lambda_i \boldsymbol{e}_i \otimes \boldsymbol{e}_i$. Equation (J.25) can be derived by noting that

$$\boldsymbol{T} = \boldsymbol{T}\boldsymbol{I} = \boldsymbol{T}\sum_{i=1}^n \boldsymbol{u}_i \otimes \boldsymbol{v}_i = \sum_{i=1}^n (\boldsymbol{T}\boldsymbol{u}_i) \otimes \boldsymbol{v}_i = \sum_{i=1}^n \lambda_i \boldsymbol{u}_i \otimes \boldsymbol{v}_i = \sum_{i=1}^k \lambda_i \boldsymbol{P}_i.$$

Conversely, if $\boldsymbol{H}$ diagonalizes $\boldsymbol{T}$, i.e., $\boldsymbol{H}^{-1}\boldsymbol{T}\boldsymbol{H} = \sum_{i=1}^n \lambda_i \boldsymbol{e}_i \otimes \boldsymbol{e}_i$, then $\boldsymbol{T} = \sum_{i=1}^n \lambda_i (\boldsymbol{H}\boldsymbol{e}_i) \otimes (\boldsymbol{H}^{-T}\boldsymbol{e}_i)$. Let $\boldsymbol{u}_i := \boldsymbol{H}\boldsymbol{e}_i$, and $\boldsymbol{v}_i := \boldsymbol{H}^{-T}\boldsymbol{e}_i$. The vectors $\boldsymbol{u}_i$ are the eigenvectors of $\boldsymbol{T}$ since

$$\boldsymbol{T}\boldsymbol{u}_j = \left[\sum_{i=1}^n \lambda_i (\boldsymbol{H}\boldsymbol{e}_i) \otimes (\boldsymbol{H}^{-T}\boldsymbol{e}_i)\right] \boldsymbol{H}\boldsymbol{e}_j = \lambda_j (\boldsymbol{H}\boldsymbol{e}_j) = \lambda_j \boldsymbol{u}_j.$$

Thus,

$$\boldsymbol{H} = \boldsymbol{H}\boldsymbol{I} = \boldsymbol{H}(\boldsymbol{e}_i \otimes \boldsymbol{e}_i) = (\boldsymbol{H}\boldsymbol{e}_i) \otimes \boldsymbol{e}_i = \boldsymbol{u}_i \otimes \boldsymbol{e}_i,$$

so that the columns of the diagonalizing tensor $\boldsymbol{H}$ are the eigenvectors $\boldsymbol{u}_i$ of $\boldsymbol{T}$.

A tensor is said to be normal if $\boldsymbol{A}\boldsymbol{A}^* = \boldsymbol{A}^*\boldsymbol{A}$; obviously, $\boldsymbol{U} = \sqrt{\boldsymbol{A}^*\boldsymbol{A}} = \sqrt{\boldsymbol{A}\boldsymbol{A}^*} = \boldsymbol{V}$ in this case, so that $\boldsymbol{A} = \boldsymbol{R}\boldsymbol{U} = \boldsymbol{U}\boldsymbol{R}$. The Hermitian and skew-Hermitian parts in the additive decomposition of $\boldsymbol{A}$ also commute. Examples of normal tensors are unitary, Hermitian and skew-Hermitian tensors (and hence, real-valued orthogonal, symmetric and skew-symmetric tensors), as also sums of these tensors with the identity tensor. One of the remarkable facts relating diagonalizability to normality is the following: A tensor is unitarily diagonalizable (i.e., the diagonalizing tensor $\boldsymbol{H}$ is unitary) if and only if it is normal. As an example, we have already seen that a real-valued symmetric tensor is diagonalizable by an orthogonal tensor. In contrast, in the case of a real-valued skew-symmetric tensor, the diagonalizing tensor is unitary but not orthogonal. The spectral resolution of a normal tensor is $\boldsymbol{A} = \sum_{i=1}^n \lambda_i \boldsymbol{n}_i \otimes \hat{\boldsymbol{n}}_i$, where $\boldsymbol{n}_i \cdot \hat{\boldsymbol{n}}_j = \delta_{ij}$, and the eigenvalues λ_i can be complex-valued; the diagonalizing unitary tensor is given by $\boldsymbol{H} = \sum_{i=1}^n \boldsymbol{n}_i \otimes \hat{\boldsymbol{n}}_i$. Using this result, the spectral resolutions of Hermitian, (real-valued) skew-symmetric, skew-Hermitian, (real-valued) orthogonal and unitary tensors are

$$\begin{aligned}
\boldsymbol{S}_H &= \sum_{i=1}^n \lambda_i \boldsymbol{n}_i \otimes \hat{\boldsymbol{n}}_i, \quad \lambda_i \in \Re,\\
\boldsymbol{W} &= \sum_{i=1}^m i\lambda_i (\boldsymbol{n}_i \otimes \hat{\boldsymbol{n}}_i - \hat{\boldsymbol{n}}_i \otimes \boldsymbol{n}_i), \quad \lambda_i \in \Re,\ m \le n/2,\ (\boldsymbol{n}_i \cdot \boldsymbol{n}_j) = 0,\ \forall i,j\\
\boldsymbol{W}_H &= \sum_{i=1}^n i\lambda_i \boldsymbol{n}_i \otimes \hat{\boldsymbol{n}}_i, \quad \lambda_i \in \Re,\\
\boldsymbol{Q} &= \sum_{i=1}^m \lambda_i \boldsymbol{n}_i \otimes \hat{\boldsymbol{n}}_i + \hat{\lambda}_i \hat{\boldsymbol{n}}_i \otimes \boldsymbol{n}_i + \sum_{i=2m+1}^n \pm \boldsymbol{N}_i \otimes \boldsymbol{N}_i, \quad |\lambda_i| = 1,\ m \le n/2,\ (\boldsymbol{n}_i \cdot \boldsymbol{n}_j) = 0,\ \forall i,j\\
\boldsymbol{Q}_H &= \sum_{i=1}^n \lambda_i \boldsymbol{n}_i \otimes \hat{\boldsymbol{n}}_i, \quad |\lambda_i| = 1,
\end{aligned}$$

where some of the λ_i's in the expressions for $\boldsymbol{S}$, $\boldsymbol{W}$ and $\boldsymbol{W}^H$ can be zero, and the $\{\boldsymbol{N}_i\}$'s are real-valued orthonormal unit vectors corresponding to the eigenvalues ± 1. In the above spectral resolutions, if $\boldsymbol{P}_i \equiv \boldsymbol{n}_i \otimes \hat{\boldsymbol{n}}_i$, then $\boldsymbol{P}_i$ is a projection (i.e., $\boldsymbol{P}_i^2 = \boldsymbol{P}_i$) that satisfies

$P_iP_j = 0, i \neq j$, since $(n_i \otimes \hat{n}_i)(n_j \otimes \hat{n}_j) = (\hat{n}_i \cdot n_j)n_i \otimes \hat{n}_j = \delta_{ij} n_i \otimes \hat{n}_j = n_i \otimes \hat{n}_i$ (with no sum over i and j). Thus, similar to Eqn. (J.3), all the above tensors can be represented as $\sum_{i=1}^{k} \lambda_i P_i$, where the λ_i's are their distinct eigenvalues.

We now discuss commuting tensors. Given an arbitrary tensor A (not necessarily nonsingular), we want to find the most general solution for B such that $AB = BA$. For a given A, although linear combinations of $\{I, A, \ldots, A^{n-1}\}$ commute with A, every tensor that commutes with A cannot be expressed as a linear combination of $\{I, A, \ldots, A^{n-1}\}$. For example, $A = e_1 \otimes e_1$ commutes with $B = e_2 \otimes e_3$, which cannot be expressed as a combination of $\{I, A, A^2\}$. Note also that e_3 is an eigenvector of A, but not an eigenvector of B. Thus, although commuting tensors A and B have a common set of eigenvectors, not all eigenvectors are common. Although one can solve this problem of finding the most general form of B numerically as shown towards the end of Appendix I, we show below that an analytical characterization can be found under certain restrictions.

Let the underlying space dimension be n, and let (λ_i, u_i) and (λ_i, v_i), $i = 1, 2, \ldots, k$, be the eigenvalues/eigenvectors of A and A^T. Let the minimal polynomial of A be as given by Eqn. (J.24), and let r_i be the number of linearly independent eigenvectors corresponding to each distinct eigenvalue λ_i, $i = 1, 2, \ldots, k$. Thus, the total number of linearly independent eigenvectors is $r := \sum_{i=1}^{k} r_i$. Writing $AB - BA = 0$ as $(A \boxtimes I - I \boxtimes A^T)B = 0$, we see that B are the eigentensors corresponding to the zero eigenvalues of $A \boxtimes I - I \boxtimes A^T$. Similar to the proof on page 74, it can be shown that $\lambda_i - \lambda_j$, $i, j = 1, 2, \ldots, n$, are the eigenvalues of $A \boxtimes I - I \boxtimes A^T$. The eigenvalues $\lambda_i - \lambda_j$ are zero either when $i = j$, or when $\lambda_i = \lambda_j$, $i \neq j$. Thus, there are at least n zero eigenvalues of $A \boxtimes I - I \boxtimes A^T$, and as we show below, there are at least n linearly independent eigentensors B corresponding to these zero eigenvalues. This shows that there are *at least* n and *at most* n^2 linearly independent elements in the most general form of the commuting tensor.

Let $H^{-1}JH$ be the Jordan canonical form of A where $J = \text{diag}[J_1, J_2, \ldots, J_r]$ is the Jordan matrix with the Jordan blocks J_i, $i = 1, 2, \ldots, r$, along the 'diagonal'. Let $\tilde{n}_i$ be the dimension of each Jordan block, so that $\sum_{i=1}^{r} \tilde{n}_i = n$. Further, let $\tilde{J}_i$ be the Jordan matrix with all the Jordan blocks except the ith Jordan block replaced by zeros, i.e., $\tilde{J}_i = \text{diag}[0, 0, J_i, 0]$. Then a matrix that commutes with A (but which is *not* the most general form of the matrix that commutes with A) is

$$B = \sum_{i=1}^{r} \alpha_i u_i \otimes v_i + \sum_{i=1}^{r} \sum_{j=i+1}^{r} \beta_{ij}(\alpha_{ij} u_i \otimes v_j + \alpha_{ji} u_j \otimes v_i) + H^{-1} \left[\sum_{i=1}^{r} \sum_{j=1}^{\tilde{n}_i - 1} \gamma_{ij} \tilde{J}_i^{j-1} \right] H, \tag{J.30}$$

where $\alpha_i, \alpha_{ij}, \alpha_{ji}$ are arbitrary (complex-valued) constants, $\tilde{J}_i^0 \equiv \text{diag}[0, 0, I_{\tilde{n}_i \times \tilde{n}_i}, 0]$, and

$$\beta_{ij} = \begin{cases} 1 & \text{if } \lambda_i = \lambda_j, \\ 0 & \text{if } \lambda_i \neq \lambda_j. \end{cases} \tag{J.31}$$

In the above expression, λ_i, $i = 1, 2, \ldots, r$, are the eigenvalues corresponding to the linearly independent eigenvectors u_i of A. From Eqn. (J.30), we see that the dimension of the space of commuting tensors corresponding to the first and last terms on the right-hand side is at least n. To see that Eqn. (J.30) is not the most general form of the commuting tensor, consider $A = e_1 \otimes e_2 + e_3 \otimes e_4 + e_4 \otimes e_5$, for which we have $n = 5$, $k = 1$, $r = 2$, $m = 2$

(note that $m < n+k-r$), $\lambda_1 = 0$, $\lambda_2 = 0$, $\boldsymbol{u}_1 = \boldsymbol{e}_1$, $\boldsymbol{v}_1 = \boldsymbol{e}_2$, $\boldsymbol{u}_2 = \boldsymbol{e}_3$, $\boldsymbol{v}_2 = \boldsymbol{e}_5$, $\boldsymbol{H} = \boldsymbol{I}$, so that using Eqn. (J.30), we get

$$\begin{aligned}\boldsymbol{B} = \alpha_1 \boldsymbol{e}_1 \otimes \boldsymbol{e}_2 + \alpha_2 \boldsymbol{e}_3 \otimes \boldsymbol{e}_5 + \alpha_{12} \boldsymbol{e}_1 \otimes \boldsymbol{e}_5 + \alpha_{21} \boldsymbol{e}_3 \otimes \boldsymbol{e}_2 + \gamma_{11}(\boldsymbol{e}_1 \otimes \boldsymbol{e}_1 + \boldsymbol{e}_2 \otimes \boldsymbol{e}_2) \\ + \gamma_{21}(\boldsymbol{e}_3 \otimes \boldsymbol{e}_3 + \boldsymbol{e}_4 \otimes \boldsymbol{e}_4 + \boldsymbol{e}_5 \otimes \boldsymbol{e}_5) + \gamma_{22}(\boldsymbol{e}_3 \otimes \boldsymbol{e}_4 + \boldsymbol{e}_4 \otimes \boldsymbol{e}_5).\end{aligned}$$

However, the most general form of the commuting matrix (which can be obtained using a 5×5 version of Eqn. (I.18) with $\gamma = 1$) has the two additional terms $\boldsymbol{e}_2 \otimes \boldsymbol{e}_1 + \boldsymbol{e}_4 \otimes \boldsymbol{e}_2$ and $\boldsymbol{e}_1 \otimes \boldsymbol{e}_4 + \boldsymbol{e}_2 \otimes \boldsymbol{e}_5$.

Now suppose that the given tensor $\boldsymbol{A}$ is such that $m = n + k - r$ (sufficient conditions for this to hold are stated on page 810). Then, the most general form of $\boldsymbol{B}$ that commutes with $\boldsymbol{A}$ is

$$\boldsymbol{B} = \sum_{i=1}^{r} \alpha_i \boldsymbol{u}_i \otimes \boldsymbol{v}_i + \sum_{i=1}^{r} \sum_{j=i+1}^{r} \beta_{ij}(\alpha_{ij} \boldsymbol{u}_i \otimes \boldsymbol{v}_j + \alpha_{ji} \boldsymbol{u}_j \otimes \boldsymbol{v}_i) + \sum_{i=1}^{n-r} \alpha_{r+i} \boldsymbol{A}^{i-1}, \tag{J.32}$$

with β_{ij} given by Eqn. (J.31). If one wants a real-valued $\boldsymbol{B}$, then some constants in Eqn. (J.32) are appropriately chosen to be complex-conjugates of some others. Since all the elements in the solution given by Eqn. (J.32), including the n-element set

$$\{\boldsymbol{u}_1 \otimes \boldsymbol{v}_1, \boldsymbol{u}_2 \otimes \boldsymbol{v}_2, \ldots, \boldsymbol{u}_r \otimes \boldsymbol{v}_r, \boldsymbol{I}, \boldsymbol{A}, \ldots, \boldsymbol{A}^{n-r-1}\}, \tag{J.33}$$

are linearly independent, the dimension of the space of commuting tensors is

$$d = \sum_{i=1}^{k} [r_i + r_i(r_i - 1)] + n - r = n - r + \sum_{i=1}^{k} r_i^2 \geq n - r + \sum_{i=1}^{k} r_i = n. \tag{J.34}$$

Recall that tensors whose characteristic and minimal polynomials are the same have $k = r$ and $m = n$, and, hence, satisfy the restriction $m = n + k - r$. Thus, if the characteristic and minimal polynomial are the same, then the second term drops out from Eqn. (J.32). Since the highest index of $\boldsymbol{A}$ in the minimal polynomial is $n - 1$, the set $\{\boldsymbol{I}, \boldsymbol{A}, \ldots, \boldsymbol{A}^{n-1}\}$ is linearly independent, so that the linear span of this set is the same as the set in Eqn. (J.33). It follows that the most general form of the commuting tensor $\boldsymbol{B}$ can be written as a polynomial in $\boldsymbol{A}$. Conversely, if the most general form of the commuting tensor is a polynomial is $\boldsymbol{A}$, then, since there have to be at least n elements in this most general set, the set $\{\boldsymbol{I}, \boldsymbol{A}, \ldots, \boldsymbol{A}^{n-1}\}$ is linearly independent, which implies that the minimal polynomial is the same as the characteristic polynomial. Thus, the necessary and sufficient condition for the most general form of the commuting tensor $\boldsymbol{B}$ to be a polynomial in $\boldsymbol{A}$, i.e., $\boldsymbol{B} = \sum_{i=1}^{n} \tilde{\alpha}_i \boldsymbol{A}^{i-1}$, is that $r_i = 1$ for each i (i.e., $k = r$), or, alternatively, if the characteristic and minimal polynomials of $\boldsymbol{A}$ are the same[3].

Now consider the case where $\boldsymbol{A}$ is diagonalizable, so that it can be represented using Eqn. (J.28). Since the eigenvectors $\boldsymbol{u}_i, \boldsymbol{v}_i$, $i = 1, 2, \ldots, n$, of $\boldsymbol{A}$ are now linearly independent, the set

$$\{\boldsymbol{u}_1 \otimes \boldsymbol{v}_1, \boldsymbol{u}_2 \otimes \boldsymbol{v}_2, \ldots, \boldsymbol{u}_n \otimes \boldsymbol{v}_n\}$$

[3]This result is proved for arbitrary $\boldsymbol{A}$, i.e., ones for which $m \leq n + k - r$, in Chapter 4 of [136].

is linearly independent. Hence, the most general form of the commuting tensor is given by

$$B = \sum_{i=1}^{n} \alpha_i u_i \otimes v_i + \sum_{i=1}^{n} \sum_{j=i+1}^{n} \beta_{ij} (\alpha_{ij} u_i \otimes v_j + \alpha_{ji} u_j \otimes v_i). \tag{J.35}$$

Since $r_i = n_i$, where n_i is the algebraic multiplicity of λ_i, $i = 1, 2, \ldots, k$, from Eqn. (J.34), the dimension of the space of commuting tensors is $d = \sum_{i=1}^{k} n_i^2$. For the special case of a diagonalizable A whose eigenvalues are distinct, the last term in Eqn. (J.35) gets excluded, and, since each $u_i \otimes v_i$ is a projection P_i, we get $B = \sum_{i=1}^{n} \alpha_i P_i = \sum_{i=1}^{n} \tilde{\alpha} A^{i-1}$, i.e., B is a polynomial in A. Conversely, if B is a polynomial in A for a diagonalizable A, since the minimum dimension must be n, we have $B = \sum_{i=1}^{n} \tilde{\alpha} A^{i-1} = \sum_{i=1}^{n} \alpha_i P_i$, and hence, the eigenvalues of A must be distinct.

Now consider the slightly more general problem where, given an arbitrary nonsingular A which satisfies the restriction $m = n + k - r$, we want to find (γ, B) such that

$$AB = \gamma BA. \tag{J.36}$$

The reason that we now restrict A to be nonsingular is that, if it is singular, then on multiplying both sides by $(\mathbf{cof}\, A)^T$, the right-hand side goes to zero, so that the value of γ remains undefined. Once again, one can solve this problem for arbitrary invertible A numerically, as shown towards the end of Appendix I. However, if A is such that $m = n + k - r$, then the most general solution for B that satisfies Eqn. (J.36) is given by

$$B = \sum_{i=1}^{r} \alpha_i u_i \otimes v_i + \sum_{i=1}^{r} \sum_{j=i+1}^{r} (\alpha_{ij} u_i \otimes v_j + \alpha_{ji} u_j \otimes v_i) + \sum_{i=1}^{n-r} \alpha_{r+i} A^{i-1},$$

where α_i, α_{ij}, α_{ji} are again arbitrary (complex-valued) constants. The γ corresponding to $B = \alpha_i u_i \otimes v_i$ and $B = \alpha_i A^{i-r-1}$ is 1, while for $B = \alpha_{ij} u_i \otimes v_j$ and $B = \alpha_{ji} u_j \otimes v_i$, we have $\gamma = \lambda_i / \lambda_j$ and λ_j / λ_i, respectively. Thus, the eigenvalues/eigentensors (γ, B) in Eqn. (J.36) are given by $(\lambda_i / \lambda_j, u_i \otimes v_j)$, $i, j = 1, 2, \ldots, r$. Note that the product of the eigenvalues γ is 1.

Two symmetric matrices commute if and only if they are simultaneously diagonalizable, i.e., if $A, B \in \text{Sym}$, and if $AB = BA$, then there exists $Q \in \text{Orth}^+$, such that QAQ^T and QBQ^T are both diagonal, and conversely, if there exists $Q \in \text{Orth}^+$, such that QAQ^T and QBQ^T are both diagonal then $AB = BA$. To see this (we consider the case $n = 3$ for simplicity; the proof can be extended to any n), if $A = \sum_{i=1}^{3} \lambda_i e_i^* \otimes e_i^*$, then if all the eigenvalues of A are distinct, by Eqn. (J.35), $B = \sum_{i=1}^{3} \alpha_i e_i^* \otimes e_i^*$, so that A and B are both diagonal in the $\{e_i^*\}$ basis. If $\lambda_2 = \lambda_3$, say, then by Eqn. (J.35), $B = \sum_{i=1}^{3} \alpha_i e_i^* \otimes e_i^* + \alpha_{12}(e_2^* \otimes e_3^* + e_3^* \otimes e_2^*)$. The tensor $Q = e_1^* \otimes e_1^* + \bar{e}_2 \otimes e_2^* + \bar{e}_3 \otimes e_3^* \in \text{Orth}^+$, where $\{e_1^*, \bar{e}_2, \bar{e}_3\}$ and $\{e_1^*, e_2^*, e_3^*\}$ are both orthonormal bases, diagonalizes both A and B. Finally, if all the eigenvalues of A are repeated, i.e., if $A = \lambda I$, then B is any arbitrary symmetric tensor, and the Q that diagonalizes B, also keeps A diagonal.

To prove the converse, let $QAQ^T = D_1$ and $QBQ^T = D_2$, where D_1 and D_2 are both diagonal matrices. Then $A = Q^T D_1 Q$ and $B = Q^T D_2 Q$, and we have

$$AB = (Q^T D_1 Q)(Q^T D_2 Q) = Q^T D_1 D_2 Q = Q^T D_2 D_1 Q = (Q^T D_2 Q)(Q^T D_1 Q) = BA.$$

The above result can be generalized as follows. Two diagonalizable tensors commute if and only if they are simultaneously diagonalizable by the same tensor $\boldsymbol{H}$. Since the columns of $\boldsymbol{H}$ are eigenvectors, the commuting diagonalizable tensors have a set of n linearly independent common eigenvectors.

Let $\{\boldsymbol{e}_1, \boldsymbol{e}_2, \boldsymbol{e}_3, \ldots, \boldsymbol{e}_n\}$ denote the canonical basis of $\Re^n$. For a given $\boldsymbol{A}$ for which $m = n + k - r$, the following examples find the most general form of a commuting tensor $\boldsymbol{B}$ using Eqn. (J.32) (and checked by using Eqn. (I.18) with $\gamma = 1$):

1. For $\boldsymbol{A} = \boldsymbol{e}_1 \otimes \boldsymbol{e}_1$, $n = 3$, $k = 2$, $r = 3$ ($r_1 = 1$, $r_2 = 2$), $m = 2$, $\lambda_1 = 1$, $\lambda_2 = \lambda_3 = 0$, $\boldsymbol{u}_1 = \boldsymbol{v}_1 = \boldsymbol{e}_1$, $\boldsymbol{u}_2 = \boldsymbol{v}_2 = \boldsymbol{e}_2$, $\boldsymbol{u}_3 = \boldsymbol{v}_3 = \boldsymbol{e}_3$, we get
$$\boldsymbol{B} = \alpha_1 \boldsymbol{e}_1 \otimes \boldsymbol{e}_1 + \alpha_2 \boldsymbol{e}_2 \otimes \boldsymbol{e}_2 + \alpha_3 \boldsymbol{e}_3 \otimes \boldsymbol{e}_3 + \alpha_{23} \boldsymbol{e}_2 \otimes \boldsymbol{e}_3 + \alpha_{32} \boldsymbol{e}_3 \otimes \boldsymbol{e}_2.$$
This example shows that, given a symmetric tensor $\boldsymbol{A}$, the most general form of the commuting tensor $\boldsymbol{B}$ need not be symmetric. From Eqn. (J.34), we get the dimension $d = r_1^2 + r_2^2 = 5$, which matches with the above solution.

2. For $\boldsymbol{A} = \boldsymbol{e}_1 \otimes \boldsymbol{e}_2$, $n = 3$, $k = 1$, $r = 2$, $m = 2$, $\lambda_1 = \lambda_2 = 0$, $\boldsymbol{u}_1 = \boldsymbol{e}_1$, $\boldsymbol{u}_2 = \boldsymbol{e}_3$, $\boldsymbol{v}_1 = \boldsymbol{e}_2$, $\boldsymbol{v}_2 = \boldsymbol{e}_3$, we get
$$\boldsymbol{B} = \alpha_1 \boldsymbol{e}_1 \otimes \boldsymbol{e}_2 + \alpha_2 \boldsymbol{e}_3 \otimes \boldsymbol{e}_3 + \alpha_{12} \boldsymbol{e}_1 \otimes \boldsymbol{e}_3 + \alpha_{21} \boldsymbol{e}_3 \otimes \boldsymbol{e}_2 + \alpha_3 \boldsymbol{I}.$$
Using Eqn. (J.34), we get $d = r^2 + n - r = 5$, which again matches with the above solution.

3. For $\boldsymbol{A} = \boldsymbol{e}_1 \otimes \boldsymbol{e}_2 + \boldsymbol{e}_3 \otimes \boldsymbol{e}_3$, $n = 3$, $k = 2$, $r = 2$ ($r_1 = 1$, $r_2 = 1$), $m = 3$, $\lambda_1 = 0$, $\lambda_2 = 1$, $\boldsymbol{u}_1 = \boldsymbol{e}_1$, $\boldsymbol{u}_2 = \boldsymbol{e}_3$, $\boldsymbol{v}_1 = \boldsymbol{e}_2$, $\boldsymbol{v}_2 = \boldsymbol{e}_3$, we get
$$\boldsymbol{B} = \alpha_1 \boldsymbol{e}_1 \otimes \boldsymbol{e}_2 + \alpha_2 \boldsymbol{e}_3 \otimes \boldsymbol{e}_3 + \alpha_3 \boldsymbol{I}.$$
Since $k = r$, we can also express $\boldsymbol{B}$ as a polynomial in $\boldsymbol{A}$, i.e., as $\tilde{\alpha}_0 \boldsymbol{I} + \tilde{\alpha}_1 \boldsymbol{A} + \tilde{\alpha}_2 \boldsymbol{A}^2$. From Eqn. (J.34), $d = 3$.

4. For $\boldsymbol{A} = \boldsymbol{e}_1 \otimes \boldsymbol{e}_2 + \boldsymbol{e}_2 \otimes \boldsymbol{e}_3$, $n = 3$, $k = 1$, $r = 1$, $m = 3$, $\lambda_1 = 0$, $\boldsymbol{u}_1 = \boldsymbol{e}_1$, $\boldsymbol{v}_1 = \boldsymbol{e}_3$, we get
$$\boldsymbol{B} = \alpha_1 \boldsymbol{e}_1 \otimes \boldsymbol{e}_3 + \alpha_2 \boldsymbol{I} + \alpha_3 \boldsymbol{A}.$$
Since $k = r$, we can also express $\boldsymbol{B}$ as $\tilde{\alpha}_0 \boldsymbol{I} + \tilde{\alpha}_1 \boldsymbol{A} + \tilde{\alpha}_2 \boldsymbol{A}^2$.

5. For $\boldsymbol{A} = \boldsymbol{e}_1 \otimes \boldsymbol{e}_2 + \boldsymbol{e}_3 \otimes \boldsymbol{e}_3 + \boldsymbol{e}_4 \otimes \boldsymbol{e}_4 + \boldsymbol{e}_3 \otimes \boldsymbol{e}_4$, $n = 4$, $k = 2$, $r = 2$ ($r_1 = 1$, $r_2 = 1$), $m = 4$, $\lambda_1 = 0$, $\lambda_2 = 1$, $\boldsymbol{u}_1 = \boldsymbol{e}_1$, $\boldsymbol{v}_1 = \boldsymbol{e}_2$, $\boldsymbol{u}_2 = \boldsymbol{e}_3$, $\boldsymbol{v}_2 = \boldsymbol{e}_4$, we get
$$\boldsymbol{B} = \alpha_1 \boldsymbol{e}_1 \otimes \boldsymbol{e}_2 + \alpha_2 \boldsymbol{e}_3 \otimes \boldsymbol{e}_4 + \alpha_3 \boldsymbol{I} + \alpha_4 \boldsymbol{A}.$$
Since $k = r$, we can also express $\boldsymbol{B}$ as $\tilde{\alpha}_0 \boldsymbol{I} + \tilde{\alpha}_1 \boldsymbol{A} + \tilde{\alpha}_2 \boldsymbol{A}^2 + \tilde{\alpha}_3 \boldsymbol{A}^3$. From Eqn. (J.34), $d = r_1^2 + r_2^2 + n - r = 4$.

6. For $\boldsymbol{A} = \boldsymbol{e}_1 \otimes \boldsymbol{e}_2 + \boldsymbol{e}_3 \otimes (\boldsymbol{e}_3 + \boldsymbol{e}_4) + \boldsymbol{e}_4 \otimes (\boldsymbol{e}_4 + \boldsymbol{e}_5)$, $n = 5$, $k = 2$, $r = 3$ ($r_1 = 2$, $r_2 = 1$), $m = 4$, $\lambda_1 = 0$, $\lambda_2 = 0$, $\lambda_3 = 1$, $\boldsymbol{u}_1 = \boldsymbol{e}_1$, $\boldsymbol{v}_1 = \boldsymbol{e}_2$, $\boldsymbol{u}_2 = \boldsymbol{e}_5 - \boldsymbol{e}_4$, $\boldsymbol{v}_2 = \boldsymbol{e}_5$, $\boldsymbol{u}_3 = \boldsymbol{e}_3$, $\boldsymbol{v}_3 = \boldsymbol{e}_4 + \boldsymbol{e}_5$, we get
$$\begin{aligned}\boldsymbol{B} = {} & \alpha_1 \boldsymbol{e}_1 \otimes \boldsymbol{e}_2 + \alpha_2 (\boldsymbol{e}_5 - \boldsymbol{e}_4) \otimes \boldsymbol{e}_5 + \alpha_3 \boldsymbol{e}_3 \otimes (\boldsymbol{e}_4 + \boldsymbol{e}_5) + \alpha_{12} \boldsymbol{e}_1 \otimes \boldsymbol{e}_5 \\ & + \alpha_{21} (\boldsymbol{e}_5 - \boldsymbol{e}_4) \otimes \boldsymbol{e}_2 + \alpha_4 \boldsymbol{I} + \alpha_5 \boldsymbol{A}.\end{aligned}$$
From Eqn. (J.34), $d = r_1^2 + r_2^2 + n - r = 7$, which matches with the above expression.

7. For

$$A = \begin{bmatrix} 3 & 1 & -1 \\ 2 & 2 & -1 \\ 2 & 2 & 0 \end{bmatrix},$$

we have $n = 3, k = 2, r = 2, \lambda_1 = 2, \lambda_2 = 1, u_1 = (1\,1\,2), u_2 = (1\,0\,2), v_1 = (-2\,0\,1)$, $v_2 = (-1\,1\,0)$, and we get

$$B = \alpha_1 \begin{bmatrix} -2 & 0 & 1 \\ -2 & 0 & 1 \\ -4 & 0 & 2 \end{bmatrix} + \alpha_2 \begin{bmatrix} -1 & 1 & 0 \\ 0 & 0 & 0 \\ -2 & 2 & 0 \end{bmatrix} + \alpha_3 I.$$

Since $k = r$, we can also express B as $\tilde{\alpha}_0 I + \tilde{\alpha}_1 A + \tilde{\alpha}_2 A^2$.

8. For $A = W \in$ Skw (with underlying space dimension 3), since all the eigenvalues are distinct, the most general form of the commuting tensor B is $\alpha_1 I + \alpha_2 W + \alpha_3 W^2$.

The following theorem, which is a generalization of Theorem J.0.3, states that every tensor can be split into a diagonalizable part and a nilpotent part, and that these two parts commute.

Theorem J.0.4 (Jordan Decomposition Theorem). *Let T be a tensor whose minimal polynomial is given by Eqn. (J.24). Then it can be decomposed uniquely as*

$$T = \sum_{i=1}^{k} (\lambda_i P_i + N_i), \tag{J.37}$$

where the λ_i, $i = 1, 2, \ldots, k$, are the distinct eigenvalues of T, and $\{P_i\}$ and $\{N_i\}$ are projections and nilpotent tensors that have the following properties:

1. $\operatorname{rank}(P_i) = n_i$; $\quad (m_i - 1) \le \operatorname{rank}(N_i) \le (n_i - 1)$
2. $\sum_{i=1}^{k} P_i = I$
3. $P_i P_j = \delta_{ij} P_i$ (no sum on i)
4. $P_i N_j = N_i N_j = 0$ *if* $i \neq j$
5. $P_i N_i = N_i P_i = N_i$ (no sum on i)

Obviously, in the expression for T, the part $D = \sum_{i=1}^{k} \lambda_i P_i$ is diagonalizable, and the part $N = \sum_{i=1}^{k} N_i$ is nilpotent. D and N commute as a result of the properties of $\{P_i\}$ and $\{N_i\}$ stated in the theorem. In fact, it can be shown that both D and N are polynomial expressions of T, and the fact that they commute follows directly from this. We also have $N_i = (T - \lambda_i I)P_i = P_i(T - \lambda_i I)$ (so that $N_i N_j = 0$ for $i \neq j$), and $N_i^{m_i} = 0$, i.e., N_i is a nilpotent tensor of order m_i. Note that even if T is real, the individual components $\{P_i\}$ can be complex, since the eigenvalues of T can be complex; however, D and $\{N_i\}$ (and hence

N) are real. Equation (J.37) is known as the Jordan canonical form or the spectral representation of a tensor. Equation (J.4) holds with $S \equiv D$; of course, if T is diagonalizable, then $T = D$.

If $\det T \neq 0$, then, in terms of the spectral resolution, we have

$$T^{-1} = \sum_{i=1}^{k} \frac{1}{\lambda_i} \left[P_i + \sum_{j=1}^{m_i - 1} \left(\frac{-N_i}{\lambda_i} \right)^j \right]. \tag{J.38}$$

The above result can be proved by using the results $TN_i^j = \lambda_i N_i^j + N_i^{j+1}$, $N_i^{m_i} = \mathbf{0}$ and $TP_i = \lambda_i P_i + N_i$ to show that $TT^{-1} = I$.

The multiplicity m_i of λ_i in the minimal polynomial is equal to the smallest l such that $\operatorname{rank}(T - \lambda_i I)^l = \operatorname{rank}(T - \lambda_i I)^{l+1}$. Thus, the minimal polynomial can be determined by examining the ranks of $(T - \lambda_i I)^l$ and $(T - \lambda_i I)^{l+1}$, with l varying from 1 to the upper bound n_i.

We now show how the parts D and N can be obtained as polynomial expressions of T. Let $m = \sum_{i=1}^{k} m_i$. Then $\{I, T, T^2, \ldots, T^{m-1}\}$ is equivalent to the basis

$$\{P_1, P_2, \ldots, P_k, N_1, N_1^2, \ldots, N_1^{m_1 - 1}, \ldots, N_k, N_k^2, \ldots, N_k^{m_k - 1}\}.$$

Thus, we can simply solve for the individual components $\{P_i\}$ and $\{N_i\}$, in terms of T and its powers, and then obtain $D = \sum_{i=1}^{k} \lambda_i P_i$ and $N = \sum_{i=1}^{k} N_i$.

As an example, consider the problem of finding the Jordan decomposition of the matrix

$$T = \begin{bmatrix} \alpha & 1 & 0 \\ 0 & \beta & 1 \\ 0 & 0 & \beta \end{bmatrix}. \tag{J.39}$$

The minimal polynomial for this matrix is given by $q(T) = (T - \alpha I)(T - \beta I)^2$. Since $m_1 = 1$, $N_1 = \mathbf{0}$. Also, since the number of distinct eigenvalues is two, we have

$$\begin{aligned} I &= P_1 + P_2, \\ T &= \alpha P_1 + \beta P_2 + N_2, \\ T^2 &= \alpha^2 P_1 + \beta^2 P_2 + 2\beta N_2, \end{aligned}$$

Solving for P_1, P_2 and N_2, we get

$$P_1 = \frac{\beta^2 I}{(\alpha - \beta)^2} - \frac{2\beta T}{(\alpha - \beta)^2} + \frac{T^2}{(\alpha - \beta)^2} = \begin{bmatrix} 1 & \frac{1}{\alpha - \beta} & \frac{1}{(\alpha - \beta)^2} \\ 0 & 0 & 0 \\ 0 & 0 & 0 \end{bmatrix},$$

$$P_2 = \frac{\alpha(\alpha - 2\beta) I}{(\alpha - \beta)^2} + \frac{2\beta T}{(\alpha - \beta)^2} - \frac{T^2}{(\alpha - \beta)^2} = \begin{bmatrix} 0 & -\frac{1}{\alpha - \beta} & -\frac{1}{(\alpha - \beta)^2} \\ 0 & 1 & 0 \\ 0 & 0 & 1 \end{bmatrix},$$

$$N_2 = -\frac{\alpha\beta I}{(\alpha - \beta)} + \frac{(\alpha + \beta) T}{(\alpha - \beta)} - \frac{T^2}{(\alpha - \beta)} = \begin{bmatrix} 0 & 0 & -\frac{1}{\alpha - \beta} \\ 0 & 0 & 1 \\ 0 & 0 & 0 \end{bmatrix}.$$

Substituting these expressions into the equations $D = \alpha P_1 + \beta P_2$, and $N = N_2$, we get

$$D = \begin{bmatrix} \alpha & 1 & \frac{1}{\alpha-\beta} \\ 0 & \beta & 0 \\ 0 & 0 & \beta \end{bmatrix}, \quad N = \begin{bmatrix} 0 & 0 & -\frac{1}{\alpha-\beta} \\ 0 & 0 & 1 \\ 0 & 0 & 0 \end{bmatrix}.$$

One of the important applications of the Jordan decomposition is that it can be used to yield an explicit formula for the exponential function e^{Tt} (see Section 1.10).

K A Note on Boundary Conditions

This note is motivated by our belief that the boundary conditions that are routinely used, such as the "no-slip" condition in fluid mechanics or the temperature continuity condition at an interface in heat-transfer problems, are too "strong", and should ideally speaking be replaced by weaker conditions that may approximate reality better. To see this, consider two solid objects at different temperatures brought into contact with each other–the temperature continuity condition at the interface requires that the two bodies acquire a common temperature at the contact surface *instantaneously*, which is not possible in practice. An analogous example in the case of fluids is a solid plate moving with some velocity and impinging almost tangentially on the surface of a stationary fluid, in which case, the fluid in contact with the plate has to instantaneously move at the same velocity as that of the plate. Although conditions such as the "no-slip" condition may be good approximations at 'large' times, weaker conditions such as the ones that we propose in the following discussion may be more appropriate to model the initial transients in a given problem.

The boundary condition in the case of fluids should satisfy at least the following two necessary conditions:

1. In the inviscid limit, i.e., as $\mu \to 0$, there should be no relation between the tangential velocities at the solid–fluid or fluid–fluid interfaces.

2. In the same manner as we have required for the other laws, this law also should be frame-indifferent, i.e., it should have the same form in all frames of reference.

We now discuss the Navier slip condition, which is perhaps the most popular among the various proposed slip conditions, and which satisfies both these conditions. Usually, only a "one-dimensional" version of this slip conditions is presented, which we generalize to a three-dimensional version here, for an incompressible Newtonian fluid. We shall assume that the normal velocity at the interface is continuous since this just follows from the principle of mass conservation (see Chapter I of Vol. II). The Navier slip condition can be written as

$$\boldsymbol{t}_t = \frac{\mu}{\beta}\boldsymbol{v}_\Delta, \tag{K.1}$$

where $\boldsymbol{t}_t$ is the tangential component of the traction, μ is the dynamic viscosity, $\boldsymbol{v}_\Delta = \boldsymbol{v}_2 - \boldsymbol{v}_1$ is the jump in tangential velocity at the interface as shown in Fig. K.1, and β is an experimentally determined parameter known as the "slip-length". Since for an incompressible Newtonian fluid, $\boldsymbol{t} = -p\boldsymbol{n} + 2\mu\boldsymbol{D}$, we have

$$\begin{aligned}
\boldsymbol{t}_t &= \boldsymbol{t} - (\boldsymbol{t}\cdot\boldsymbol{n})\boldsymbol{n} \\
&= -p\boldsymbol{n} + 2\mu\boldsymbol{D}\boldsymbol{n} - [-p\boldsymbol{n} + 2\mu(\boldsymbol{n}\cdot\boldsymbol{D}\boldsymbol{n})\boldsymbol{n}] \\
&= 2\mu\,[\boldsymbol{D} - (\boldsymbol{n}\cdot\boldsymbol{D}\boldsymbol{n})\boldsymbol{I}]\,\boldsymbol{n}.
\end{aligned}$$

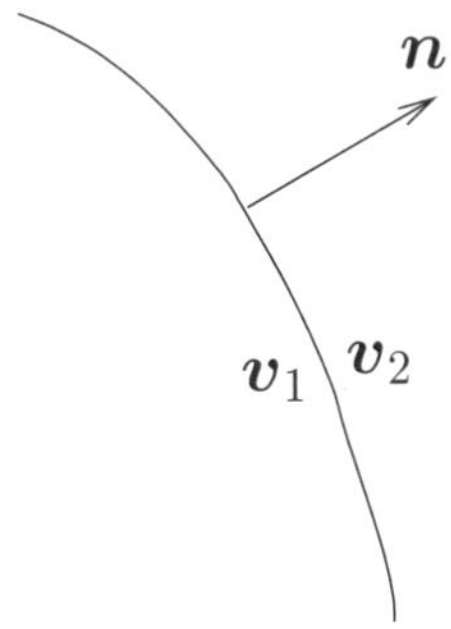

Fig. K.1 *A solid–fluid or fluid–fluid interface with a discontinuity in velocity at the interface.*

Substituting this relation into Eqn. (K.1), we get

$$2\mu\left[D - (n \cdot Dn)I\right] n = \frac{\mu}{\beta} v_\Delta.$$

In the inviscid limit $\mu \to 0$, the above relation reduces to the degenerate form $0 = 0$, so that, as required by inviscid flow theory, there is *no restriction* on the tangential component of the velocity. If $\mu \neq 0$, then the above relation reduces to

$$\left[D - (n \cdot Dn)I\right] n = \frac{1}{2\beta} v_\Delta, \tag{K.2}$$

which is the required slip condition on the tangential component of the velocity. Since $D^* = QDQ^T$, $(v_\Delta)^* = Qv_\Delta$, $n^* = Qn$, and assuming β to be a frame-indifferent parameter, we see that the above slip condition is frame-indifferent.

To compare the solutions obtained with the above slip condition with those obtained using the "no-slip" condition, we now consider a few examples. For fully developed plane Poiseuille flow (see Chapter 5 of Vol. II) through a conduit of height $2h$, we get the solution for the axial velocity as

$$v_z = \frac{Gh^2}{2\mu}\left[1 - \left(\frac{y}{h}\right)^2 + \frac{2\beta}{h}\right], \quad G = -\frac{dP}{dz}.$$

The corresponding solution with the no-slip boundary condition is the same as above, but without the β term. Similarly for Poiseuille flow through a circular pipe of radius R, the solution with the Navier slip condition is

$$v_z = \frac{GR^2}{4\mu}\left[1 - \left(\frac{r}{R}\right)^2 + \frac{2\beta}{R}\right], \quad G = -\frac{dP}{dz}.$$

For plane Couette flow, where the upper surface of the fluid at $y = d$ is in contact with a plate moving with velocity U while the lower surface at $y = 0$ is in contact with a stationary surface, the steady-state solution obtained is $v_z = U(y + \beta)/(d + 2\beta)$, which should be compared with the solution $v_z = Uy/d$ obtained using the no-slip condition.

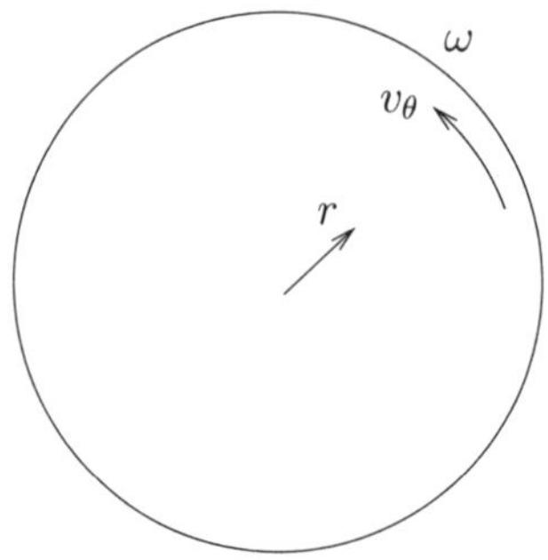

Fig. K.2 *Flow in a rotating cylinder.*

Now consider the case of a stationary fluid in a circular cylinder of radius R in which flow is induced by rotating the cylinder starting from an angular speed of zero, and finally reaching some steady-state value, say ω_∞ (see Fig. K.2). Assuming v_θ to be the only nonzero component, and the flow to be two-dimensional, the governing equation is given by

$$\frac{\partial v_\theta}{\partial t} = \nu \left[\frac{1}{r}\frac{\partial}{\partial r}\left(r\frac{\partial v_\theta}{\partial r} \right) - \frac{v_\theta}{r^2} \right]. \tag{K.3}$$

If ω denotes the angular speed, then the no-slip condition requires that $v_\theta = \omega R$ at $r = R$, so that $\partial v_\theta/\partial t|_{r=R} = \dot{\omega}R$. If ω is such that $\dot{\omega}(0) \neq 0$, then keeping in mind that the fluid is initially stationary, we see that at $t = 0$ and at $r = R$, the left-hand side of Eqn. (K.3) is nonzero, while the right-hand side is zero. This paradox can be resolved by using the Navier slip condition instead of the no-slip condition, so that the fluid need not have a velocity of ωR at the cylinder surface. Note that in this particular problem, the steady-state solutions obtained with or without the no-slip condition are the same, namely, $v_\theta = \omega_\infty r$. If one considers the flow to be outside a rotating solid cylinder, then the steady-state solution is

$$v_\theta = \frac{\omega_\infty R^2}{\left[1 + \frac{2\beta}{R}\right] r},$$

which should be compared with the steady-state solution $v_\theta = \omega_\infty R^2/r$.

Finally, the solution to Stokes' second problem (see Chapter 5 of Vol. II) obtained using the slip condition

$$\left.\frac{\partial u}{\partial y}\right|_{y=0} = \frac{1}{\beta}\left[u|_{y=0} - U\cos\Omega t \right],$$

is given by

$$u = \frac{(\sigma + 1/\beta)U}{4\sigma + 1/\beta} e^{-\sigma y}\cos(\Omega t - \sigma y) + \frac{(\sigma + 1/\beta)U}{4\sigma + 1/\beta} e^{-\sigma y}\sin(\Omega t - \sigma y).$$

The "no-slip" solution is retrieved in the limit as $\beta \to 0$.

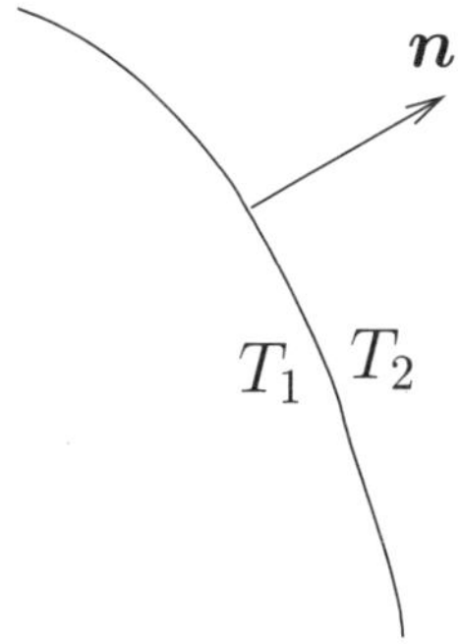

Fig. K.3 *Newton law of cooling.*

For heat-transfer problems, the analogue of the Navier slip condition is the Newton law of cooling given by

$$\boldsymbol{q} \cdot \boldsymbol{n} = \beta(T_1 - T_2), \tag{K.4}$$

where β is again an experimentally determined constant, and $\boldsymbol{n}$ is as shown in Fig. K.3. Note that the boundary condition given by Eqn. (K.4) should be applied to both regions on either side of the interface. Note also that since $\boldsymbol{q}^* \cdot \boldsymbol{n}^* = (\boldsymbol{Q}\boldsymbol{q}) \cdot (\boldsymbol{Q}\boldsymbol{n}) = \boldsymbol{q} \cdot \boldsymbol{n}$, this law satisfies the requirement of frame-indifference. Solutions to special problems when the boundary condition is given by Eqn. (K.4) may be found in [122].

Bibliography

[1] Abbassi, M. M., Torsion of circular shafts of variable diameter, *ASME J. Appl. Mech.*, **22**, 530–532, 1955.

[2] Abeyaratne, R. and C. O. Horgan, The pressurized hollow sphere problem in finite elastostatics for a class of compressible materials, *Int. J. Solids Struct.*, **20(8)**, 715–723, 1984.

[3] Acharya, A., On compatibility conditions for the left Cauchy–Green deformation field in three dimensions, *J. Elasticity*, **56(2)**, 95–105, 1999.

[4] Ahmad, F. and M. A. Rashid, Linear invariants of a Cartesian tensor, *Quart. J. Mech. Appl. Math.*, **62(1)**, 31–38, 2009.

[5] Anand, L., On H. Hencky's approximate strain energy function for moderate deformations, *ASME J. Appl. Mech.*, **46(1)**, 78–82, 1979.

[6] Anand, L., Moderate deformations in extension-torsion of incompressible isotropic elastic materials, *J. Mech. Phys. Solids*, **34(3)**, 293–304, 1986.

[7] Anderson G. L. and C. R. Thomas, A forced vibration problem involving time derivatives in the boundary conditions, *J. Sound Vibrat.*, **14(2)**, 193–214, 1971.

[8] Andrews, D. L. and W. A. Ghoul, Irreducible fourth-rank Cartesian tensors, *Phys. Rev. A*, **25(5)**, 2647–2657, 1982.

[9] Andrianov, I. V. and J. Awrejcewicz, Compatibility equations in the theory of elasticity, *J. Vibrat. Acoustics* **125(2)**, 244–245, 2003.

[10] Angel Y. C., On the static and dynamic equilibrium of concentrated loads in linear acoustics and linear elasticity, *J. Elast.*, **24(1–3)**, 21–42, 1990.

[11] Antman, S. S. and J. E. Osborn, The principle of virtual work and the integral laws of motion, *Arch. Rational Mech. Anal.*, **69(3)**, 231–262, 1979.

[12] Armero, F., Elastoplastic and viscoplastic deformations in solids and structures, *Encyclopedia of Computational Mechanics, Vol. II: Solids and Structures*, Eds. E. Stein, Rene de Borst and T. J. R. Hughes, 227–266, 2004.

[13] Ball, J. M. and D. G. Schaeffer, Bifurcation and stability of homogeneous equilibrium configurations of an elastic body under dead-load tractions, *Math. Proc. Camb. Phil. Soc.*, **94(02)**, 315–339, 1983.

[14] Bakker, M. C. M., M. D. Verweij, B. J. Kooij, H. A. Dieterman, The traveling point load revisited, *Wave Motion*, **29(2)**, 119–135, 1999.

[15] Batista, M., Stresses in a confocal elliptic ring subject to uniform pressure, *J. Strain Anal.*, **34(3)**, 217–221, 1999.

[16] Bauer, H. F., Table of the roots of the associated Legendre function with respect to the degree, *Math. Comput.*, **46(174)**, 601–602/S29-S41, 1986.

[17] Beatty, M. F., Topics in finite elasticity: hyperelasticity of rubber, elastomers, and biological tissues–with examples, *Appl. Mech. Rev.*, **40(12)**, 1699–1734, 1987.

[18] Belik, P. and R. Fosdick, The state of pure shear, *J. Elast.*, **52**, 91–98, 1998.

[19] Bertram, A. and B. Svendsen, On material objectivity and reduced constitutive equations, *Arch. Mech.*, **53(6)**, 653–675, 2001.

[20] Betten, J., Mathematical modelling of materials behavior under creep conditions, *Appl. Mech. Reviews.*, **54(2)**, 107–132, 2001.

[21] Bischoff, J. E., E. M. Arruda and K. Grosh, A new constitutive model for the compressibility of elastomers at finite deformations, *Rubber Chem. Tech.*, **74(4)**, 541–559, 2001.

[22] Blackmore, D. and L. Ting, Surface integral of its mean curvature vector, *SIAM Rev.*, **27(4)**, 569–572, 1985.

[23] Blume, J., Compatibility conditions for a left Cauchy–Green strain field, *J. Elast.*, **21(3)**, 271–308, 1989.

[24] de Boor, C., A naive proof of the representation theorem for isotropic, linear asymmetric stress–strain relations, *J. Elast.*, **15(2)**, 225–227, 1985.

[25] Borodachev, N. M., Three-dimensional elasticity-theory problem in terms of the stress, *Int. Appl. Mech.*, **31(12)**, 991–996, 1995.

[26] Bosch, A. J., The factorization of a square matrix into two symmetric matrices. *Am. Math. Monthly*, **93(6)**, 462–464, 1986.

[27] Bounlanger, P. and M. Hayes, On pure shear, *J. Elast.*, **77(1)**, 83–89, 2004.

[28] Bradley, F. E., Development of an Airy stress function of general applicability in one, two, or three dimensions, *J. Appl. Phys.*, **67(1)**, 225–226, 1990.

[29] Bramble, J. H., The thick elastic spherical shell under concentrated torques, *Proc. London Math. Soc.*, **s3–9(4)**, 492–502, 1959.

[30] Brown, D. K., A computer program to calculate the elastic stress and displacement fields around an elliptical hole under any applied plane state of stress, *Comput. Struct.*, **7(4)**, 571–580, 1977.

[31] Bruhns, O. T., H, Xiao and A. Meyers, Constitutive inequalities for an isotropic elastic strain-energy function based on Hencky's logarithmic strain tensor, *Proc. R. Soc. London Ser. A*, **457(2013)**, 2207–2226, 2001.

[32] Bruhns, O. T., H, Xiao and A. Meyers, Finite bending of a rectangular block of an elastic Hencky material, *J. Elast.*, **66(3)**, 237–256, 2002.

[33] Bustamante, R., Some topics on a new class of elastic bodies, *Proc. R. Soc. Ser. A*, **465(2105)**, 1377–1392, 2009.

[34] Cardoso, J. R., An explicit formula for the matrix logarithm, *S. Afr. Optom.*, **64(3)**, 80–83, 2005.

[35] Carlson, D. E., Unpublished notes on continuum mechanics. Available at http://imechanica.org/node/15845 (courtesy of Professor Amit Acharya).

[36] Carlson, D. E., Linear thermoelasticity, *Handbuch der Physik*, Vol. VIa/2., Berlin Heidelberg: Springer-Verlag, 297–345, 1972.

[37] Carlson, D. E. and A. Hoger, The derivative of a tensor-valued function of a tensor, *Quart. Appl. Math.*, **44(3)**, 409–423, 1986.

[38] Carlson, D. E. and A. Hoger, On the derivatives of the principal invariants of a second-order tensor, *J. Elast.*, **16(2)**, 221–224, 1986.

[39] Carlson, D. E., E. Fried and D. A. Tortorelli, Geometrically-based consequences of internal constraints, *J. Elast.*, **70(1–3)**, 101–109, 2003.

[40] Carroll, M. M., Finite strain solutions in compressible isotropic elasticity, *J. Elast.*, **20(1)**, 65–92, 1988.

[41] Carroll, M. M. and C. O. Horgan, Finite strain solutions for a compressible elastic solid, *Q. Appl. Math.*, **48(4)**, 767–780, 1990.

[42] Chadwick, P., Thermo-mechanics of rubberlike materials, *Phil. Trans. R. Soc. London Ser. A.*, **276(1260)**, 371–403, 1974.

[43] Chadwick, P., *Continuum Mechanics*, New York: Dover Publications, 1976.

[44] Chattarji, P. P., Torsion of epitrochoidal sections, *Z. Angew Math. Mech.*, **89(3/4)**, 135–138, 1959.

[45] Chattarji, P. P., A note on the torsion of circular shafts of variable diameter, *ASME J. Appl. Mech.*, **29**, 477–478, 1957.

[46] Chen, Y. C., Stability of homogeneous deformations of an incompressible elastic body under dead-load surface tractions, *J. Elast.*, **17(3)**, 223–248, 1987.

[47] Chen, Y. and L. Wheeler, Derivatives of the stretch and rotation tensors, *J. Elast.*, **32(3)**, 175–182, 1993.

[48] Chen, S. J. and D. G. Howitt, On the Galerkin vector and the Eshelby solution in linear elasticity, *J. Elast.*, **44(1)**, 1–8, 1996.

[49] Chen, T., A homogeneous elliptical shaft may not warp under torsion, *Acta Mech.*, **169(1–4)**, 221–224, 2004.

[50] Chen, T. and C. J. Wei, Saint-Venant torsion of anisotropic shafts: Theoretical frameworks, extremal bounds and affine transformations, *Quart. J. Mech. Appl. Math.*, **58(2)**, 269–287, 2005.

[51] Cheng, S. and T. Angsirikul, Three-dimensional elasticity solution and edge effects in a spherical dome, *ASME J. Appl. Mech.*, **44(4)**, 599–603, 1977.

[52] Cheng, H-W. and S. Yau, More explicit formulas for the matrix exponential, *Lin. Alg. Appl.*, **262**, 131–163, 1997.

[53] Cheng, S. H., N. J. Higham, C. S. Kenney and A. L. Laub, Approximating the logarithm of a matrix to specified accuracy, *SIAM J. Matrix Anal. Appl.*, **22(4)**, 1112–1125, 2001.

[54] Chiang, C-R., Torsion of a round shaft of variable diameter, *J. Eng. Math.*, **77(1)**, 119–130, 2012.

[55] Chree, C., On long rotating circular cylinders, *Proc. Cambridge Philos. Soc.*, **7**, 283–305, 1892.

[56] Christov, C. I., On frame indifferent formulation of the Maxwell-Cattaneo model of finite-speed heat conduction, *Mech. Res. Commun.*, **36(4)**, 481–486, 2009.

[57] Chung, D. T., C. O. Horgan and R. Abeyaratne, The finite deformation of internally pressurized hollow cylinders and spheres for a class of compressible materials, *Int. J. Solids Struct.*, **22(12)**, 1557–1570, 1986.

[58] Ciarlet, P. G., *Three-Dimensional Elasticity*, Elsevier Science Publishers, North Holland, 1988.

[59] Ciarlet, P. G., An introduction to differential geometry with applications to elasticity, *J. Elast.*, **78–79(1–3)**, 1–215, 2005.

[60] Ciarlet, P. G., On rigid and infinitesimal rigid displacements in three-dimensional elasticity, *Math. Models Meth. Appl. Sci.*, **13(11)**, 1589–1598, 2003.

[61] Coleman, B. D. and W. Noll, The thermodynamics of elastic materials with heat conduction and viscosity, *Arch. Rational Mech. Anal.*, **13(1)**, 245–261, 1963.

[62] Coleman, B. D. and V. J. Mizel, Existence of caloric equations of state in thermodynamics, *J. Chem. Phys.*, **40(4)**, 1116–1125, 1964.

[63] Conway, J. B., *A Course in Functional Analysis*, Springer, New York, 1990.

[64] Dai, H. H., Model equations for nonlinear dispersive waves in a compressible Mooney–Rivlin rod, *Acta Mech.*, **127(1–4)**, 193–207, 1998.

[65] Dempsey, J. P., The wedge subjected to tractions: a paradox resolved, *J. Elast.*, **11(1)**, 1–10, 1981.

[66] Donnell, L. H.,, Stress concentrations due to elliptical discontinuities in plates under edge force, *Theodore von Karman Anniversary Volume, California Institute of Technology, Pasadena*, 293–309, 1941.

[67] Dui, G., M. Jin and M. Huang, On the derivation for the gradients of the principal invariants, *J. Elast.*, **75(2)**, 193–196, 2004.

[68] Dui, G. and Y. Chen, A note on Rivlin's identities and their extension, *J. Elast.*, **76(2)**, 107–112, 2004.

[69] Edstrom, C. R., The vibrating beam with nonhomogeneous boundary conditions, *ASME J. Appl. Mech.*, **48(3)**, 669–670, 1981.

[70] Edwards, R. H., Stress concentrations around spheroidal inclusions and cavities, *ASME J. Appl. Mech.*, **18(1)**, 19–30, 1951.

[71] Ehlers, W. and G. Eipper, The simple tension problem at large volumetric strains computed from finite hyperelastic material laws, *Acta Mech.*, **130(1–2)**, 17–27, 1998.

[72] Ericksen, J. L., Deformations possible in every compressible, isotropic, perfectly elastic material, *J. Math. Phys.*, **34(2)**, 126–128, 1955.

[73] Ericksen, J. L., Equilibrium theory of liquid crystals, *Advances in Liquid crystals, Vol. 2*, Ed. G. H. Brown, pp. 233–298, Academic Press, New York, 1976.

[74] Eshelby, J. B., The determination of the elastic field of an ellipsoidal inclusion, and related problems, *Proc. R. Soc. London Ser. A*, **241(1226)**, 376–396, 1957.

[75] Eskin, G. and J. Ralston, On the inverse boundary value problem for linear isotropic elasticity, *Inverse Problems*, **18(3)**, 907–921, 2002.

[76] Filon, L. N. G., On the elastic equilibrium of circular cylinders under certain practical systems of load, *Phil. Trans. R. Soc. of London Ser. A.*, **198**, 147–233, 1902.

[77] Filon, L. N. G., On an approximate solution for the bending of a beam of rectangular cross-section under any system of load, with special reference to points of concentrated or discontinuous loading, *Phil. Trans. R. Soc. London Ser. A.*, **201**, 63–155, 1903.

[78] Fish, M. J., A particular boundary value problem, *J. Math. Phys.*, **14**, 262–273, 1935.

[79] Fisher, H. D., Discussion on [?], *ASME J. Appl. Mech.*, **49(2)**, 459–460, 1982.

[80] Folias, E. S. and J. J. Wang, On the three-dimensional stress field around a circular hole in a plate of arbitrary thickness, *Comput. Mech.*, **6(5–6)**, 379–391, 1990.

[81] Fosdick, R. and G. Royer-Carfagni, The constraint of local injectivity in linear elasticity theory, *Proc. R. Soc. London Ser. A*, **457(2013)**, 2167–2187, 2001.

[82] Fosdick, R., F. Freddi and G. Royer-Carfagni, Bifurcation instability in linear elasticity with the constraint of local injectivity, *J. Elast.*, **90(1)**, 99–126, 2008.

[83] Freddi, F. and G. Royer-Carfagni, From non-linear elasticity to linearized theory: Examples defying intuition, *J. Elast.*, **96(1)**, 1–26, 2009.

[84] Freiberger, W., The uniform torsion of an incomplete tore, *Austral. J. Sci. Res. Ser. A*, **2(3)**, 354–375, 1949.

[85] Fulmer, E. P., Computation of the matrix exponential, *Am. Math. Monthly*, **82(2)**, 156–159, 1975.

[86] Gallier, J. and D. Xu, Computing of exponentials of skew-symmetric matrices and logarithms of orthogonal matrices, *Int. J. Robotics Auto.*, **17(4)**, 1–11, 2002.

[87] Gao X. L., A general solution of an infinite elastic plate with an elliptic hole under biaxial loading, *Int. J. Pres. Ves. Piping*, **67(1)**, 95–104, 1996.

[88] Galmudi, D. and J. Dvorkin, Stresses in anisotropic cylinders, *Mech. Res. Comm.*, **22(2)**, 109–113, 1995.

[89] Georgiadis H. G., D. Vamvatsikos, I. Vardoulakis, Numerical implementation of the integral-transform solution to Lamb's point-load problem, *Comput. Mech.*, **24(2)**, 90–99, 1999.

[90] Gerhardt T. O., and S. Cheng, Truncated hollow spheres, *ASCE J. Eng. Mech.*, **109(3)**, 885–895, 1983.

[91] Geymonat, G. and F. Krasucki, Beltrami's solutions of general equilibrium equations in continuum mechanics, *C. R. Acad. Sci. Paris, Ser I*, **342(5)**, 359–363, 2006.

[92] Ghosh A. K., Axisymmetric vibration of a long cylinder, *J. Sound Vibrat.*, **186(5)**, 711–721, 1995.

[93] Goldberg, M. A. and M. Sadowsky, Stresses in an ellipsoidal rotor in a centrifugal force field, *J. Appl. Mech.*, **26(4)**, 549–552, 1959.

[94] Goldberg, M. A., V. L. Salerno and M. A. Sadowsky, Stress distribution in a rotating spherical shell of arbitrary thickness, *J. Appl. Mech.*, **28(1)**, 127–131, 1961.

[95] Golovchan, V. T., Torsion of a cylinder of finite length having a cylindrical cavity, *Prikladnaya Mekhanika*, **8(3)**, 37–41, 1972.

[96] Gong S. X. and S. A. Meguid, On the elastic fields of an elliptical inhomogeneity under plane deformation, *Proc. R. Soc. London Ser. A*, **443(1919)**, 457–471, 1993.

[97] Goodier, J. N., Concentration of stress around spherical and cylindrical inclusions and flaws, *Trans. ASME, Applied Mech.*, **55(7)**, 39–44, 1933.

[98] Goss R. N., Center of flexure of a triangular beam, *Proc. Am. Math. Soc.*, **1(6)**, 744–750, 1950.

[99] Grant D. A., Beam vibrations with time-dependent boundary conditions, *J. Sound Vibrat.*, **89(4)**, 519–522, 1983.

[100] Green, A. E., Three-dimensional stress systems in isotropic plates. I, *Phil. Trans. R. Soc. London Ser. A*, **240(825)**, 561–597, 1948.

[101] Gregory, R., D. and I. Gladwell, The cantilever beam under tension, bending or flexure at infinity, *J. Elast.*, **12(4)**, 317–343, 1982.

[102] Guo, Z. H., The representation theorem for isotropic, linear asymmetric stress–strain relations, *J. Elast.*, **13(2)**, 121–124, 1983.

[103] Guo, Z. H., An alternative proof of the representation theorem for isotropic, linear asymmetric stress–strain relations, *Quart. Appl. Math.*, **41(1)**, 119–123, 1983.

[104] Guo Z. H., Rates of stretch tensors, *J. Elast.*, **14(3)**, 263–267, 1984.

[105] Guo, Z. H. and P. Podio-Guidugli, A concise proof of the representation theorem for linear isotropic, tensor-valued mappings of a skew argument, *J. Elast.*, **21(3)**, 317–320, 1989.

[106] Guo, Z. H., Derivatives of the principal invariants of a 2-nd order tensor, *J. Elast.*, **22(2)**, 185–191, 1989.

[107] Gupta, A., D. J. Steigmann and J. S. Stolken, On the evolution of plasticity and incompatibility, *Math. Mech. Solids*, **12(6)**, 583–610, 2007.

[108] Gurtin, M. E., *An Introduction to Continuum Mechanics*, San Diego:Academic Press, 1981.

[109] Gurtin, M. E., *The Linear Theory of Elasticity*, Handbuch der Physik VIa/2, Berlin Heidelberg:Springer-Verlag, 1972.

[110] Gurtin, M. E., A short proof of the representation theorem for isotropic, linear stress–strain relations, *J. Elast.*, **4(3)**, 243–245, 1974.

[111] Gurtin, M. E., A generalization of Beltrami stress functions in continuum mechanics, *Arch. Rational Mech. Anal.*, **13(1)**, 321–329, 1963.

[112] Gurtin, M. E. and L. Anand, The decomposition $\boldsymbol{F} = \boldsymbol{F}^e\boldsymbol{F}^p$, material symmetry, and plastic irrotationality for solids that are isotropic-viscoplastic or amorphous, *Int. J. Plast.*, **21(9)**, 1686–1719, 2005.

[113] Hadjesfandiari, A. R. and G. F. Dargush, Analysis of bi-material interface cracks with complex weighting functions and non-standar quadrature, *Int. J. Solids Struct.*, **48(10)**, 1499–1512, 2011.

[114] Hardiman, N. J., Elliptic elastic inclusion in an infinite elastic plate, *Quart. J. Mech. Appl. Math.*, **7(2)**, 226–230, 1954.

[115] Hartmann, S. and P. Neff, Polyconvexity of generalized polynomial-type hyperelastic strain energy function for near-incompressibility, *Int. J. Solids Struct.*, **40(11)**, 2767–2791, 2003.

[116] Hay G. E., The method of images applied to the problem of torsion, *Proc. London Math. Soc.*, **s2-45(1)**, 382–397, 1939.

[117] Hayes M. and T. J. Laffey, Pure shear–A footnote, *J. Elast.*, **92(1)**, 109–113, 2008.

[118] Higgins, T. J., A comprehensive review of Saint-Venant's torsion problem, *Am J. Phys.*, **10(5)**, 248–259, 1942.

[119] Higgins, T. J., Stress analysis of shafting exemplified by Saint Venant's torsion problem, *Experimental Stress Analysis*, **3(1)**, 94–101, 1945.

[120] Hill, R., On uniqueness and stability in the theory of finite elastic strain, *J. Mech. Phys. Solids*, **5(4)**, 229–241, 1957.

[121] Hill, R., Constitutive inequalities for isotropic elastic solids under finite strain, *Proc. R. Soc. London Ser. A*, **314(1519)**, 457–472, 1970.

[122] Hill, J. M. and J. N. Dewynne, *Heat Conduction*, Boston: Blackwell Scientific Publications, 1987.

[123] Hillman, A. P. and H. E. Salzer, Roots of $\sin z = z$. *Phil. Mag.*, **34 (235)**, 575–575, 1943.

[124] Hiramatsu, Y. and Y. Oka, Determination of the tensile strength of rock by a compression test of an irregular test piece, *Int. J. Rock Mech. Min. Sci.*, **3(2)**, 89–99, 1966.

[125] Hoffman, K. M. and R. Kunze, *Linear Algebra*, New Jersey: Prentice Hall, 1971.

[126] Hoger, A. and D. E. Carlson, Determination of the stretch and rotation in the polar decomposition of the deformation gradient, *Quart. Appl. Math.*, **42(1)**, 113–117, 1984.

[127] Hoger, A. and D. E. Carlson, On the derivative of the square root of a tensor and Guo's rate theorems, *J. Elast.*, **14(3)**, 329–336, 1984.

[128] Hoger, A., The stress conjugate to logarithmic strain, *Int. J. Solids Struct.*, **23(12)**, 1645–1656, 1987.

[129] Holl D. L. and D. H. Rock, The flexure and torsion of a beam whose cross-section is a limacon, *Z. Angew Math. Mech.*, **19(3)**, 141–145, 1939.

[130] Holzapfel, G. A. and J. C. Simo, Entropy elasticity of isotropic rubber-like solids at finite strains, *Comput. Methods Appl. Mech. Eng.*, **132(1)**, 17–44, 1996.

[131] Horgan, C. O. and S. C. Baxter, Effects of curvilinear anisotropy on radially symmetric stresses in anisotropic linearly elastic solids, *J. Elast.*, **42(1)**, 31–48, 1996.

[132] Horgan, C. O., Equilibrium solutions for compressible nonlinearly elastic materials. In: Y. B. Fu and R. W. Ogden, Eds., *Nonlinear elasticity: Theory and applications*, Cambridge: Cambridge University Press, 2001.

[133] Horgan, C. O., On the torsion of functionally graded anisotropic linearly elastic bars, *IMA J. Appl. Math.*, **72(5)**, 556–562, 2007.

[134] Horgan, C. O. and J. G. Murphy, A generalization of Hencky's strain-energy density to model the large deformations of slightly compressible solid rubbers, *Mech. Mater.*, **41(8)**, 943–950, 2009.

[135] Horibe, T., E. Tsuchida, Y. Arai and N. Kusano, Stresses in an elastic strip having a circular inclusion under tension, *J. Solid Mech. Mater. Eng.*, **2(7)**, 900–911, 2008.

[136] Horn, R. A. and C. R. Johnson, *Topics in Matrix Analysis*, Cambridge: Cambridge University Press, 1991.

[137] Huber, A., The elastic sphere under concentrated torques, *Quart. Appl. Math.*, **13(1)**, 98–102, 1955.

[138] Hunter, S. C., *Mechanics of Continuous Media*, Chichester: Ellis Horwood Limited, 1983.

[139] Inglis, C. E., Stresses in a plate due to the presence of cracks and sharp corners, *Trans. Institution of Naval Architects*, **60**, 219–230, 1913.

[140] Iyengar, K. T. and M. K. Prabhakara, A three-dimensional elasticity solution for rectangular prism under end loads, *Z. Angew Math. Mech.*, **49(6)**, 321–332, 1969.

[141] Itskov, M., On the theory of fourth-order tensors and their applications in computational mechanics, *Comp. Meth. Appl. Mech. Eng.*, **189(2)**, 419–438, 2000.

[142] Itskov, M., The derivative with respect to a tensor: Some theoretical aspects and applications, *Z. Angew Math. Mech.*, **82(8)**, 535–544, 2002.

[143] Itskov, M. and N. Aksel, A closed-form representation for the derivative of non-symmetric tensor power series, *Int. J. Solids Struct.*, **39**, 5963–5978, 2002.

[144] Itskov, M., Application of the Dunford-Taylor integral to isotropic tensor functions and their derivatives, *Proc. R. Soc. London Ser. A*, **459(2034)**, 1449–1457, 2003.

[145] Jain, R., K. Ramachandra and K. R. Y. Simha, Rotating anisotropic disc of uniform strength, *Int. J. Mech. Sci.*, **41(6)**, 639–648, 1999.

[146] Jain, R., K. Ramachandra and K. R. Y. Simha, Singularity in rotating orthotropic discs and shells, *Int. J. Solids Struct.*, **37(14)**, 2035–2058, 2000.

[147] Jaric, J. P., On the gradients of the principal invariants of a second-order tensor, *J. Elast.*, **44(3)**, 285–287, 1996.

[148] Jaric, J. P., On the representation of symmetric isotropic 4-tensors, *J. Elast.*, **51(1)**, 73–79, 1998.

[149] Jeffrey, G. B., Plane stress and plane strain in bipolar co-ordinates, *Phil. Trans. R. Soc. London Ser. A.*, **221**, 265–293, 1921.

[150] Jerphagnon, J., Invariants of the third-rank Cartesian tensor: Optical nonlinear susceptibilities, *Phys. Rev. B*, **2(4)**, 1091–1098, 1970.

[151] Jerphagnon, J., D. Chemla and R. Bonneville, The description of the physical properties of condensed matter using irreducible tensors, *Adv. Phys.*, **27(4)**, 609–650, 1978.

[152] Jiang, X. and R. W. Ogden, On azimuthal shear of a circular cylindrical tube of compressible elastic material, *Q. J. Mech. Appl. Math.*, **51(1)**, 143–158, 1998.

[153] Jog, C. S., On the explicit determination of the polar decomposition in n-dimensional vector spaces, *J. Elast.*, **66(2)**, 159–169, 2002.

[154] Jog, C. S., The accurate inversion of Vandermonde matrices, *Comput. Math. Appl.*, **47(6–7)**, 921–929, 2004.

[155] Jog, C. S., Derivatives of the stretch, rotation and exponential tensors in n-dimensional vector spaces, *J. Elast.*, **82(2)**, 175–192, 2006.

[156] Jog, C. S. and P. P. Kelkar, Nonlinear analysis of structures using high performance hybrid elements, *Int. J. Num. Meth. Eng.*, **68(4)**, 473–501, 2006.

[157] Jog, C. S., A concise proof of the representation theorem for fourth-order isotropic tensors, *J. Elast.*, **85(2)**, 119–124, 2006.

[158] Jog, C. S., The explicit determination of the logarithm of a tensor and its derivatives, *J. Elast.*, **93(2)**, 141–148, 2008.

[159] Jog, C. S. and R. Bayadi, Stress and strain-driven algorithmic formulations for finite strain viscoplasticity for hybrid and standard finite elements, *Int. J. Num. Meth. Eng.*, **79(7)**, 773–816, 2009.

[160] Jog, C. S. and Phani Motammari, An energy-momentum conserving algorithm for nonlinear transient analysis within the framework of hybrid elements, *J. Mech. Materials Struct.*, **4(1)**, 157–186, 2009.

[161] Jog, C. S., Improved hybrid elements for structural analysis, *J. Mech. Materials Struct.*, **5(3)**, 507–528, 2010.

[162] Jog, C. S. and R. K. Pal, A monolithic strategy for fluid-structure interaction problems, *Int. J. Num. Meth. Eng.*, **85(4)**, 429–460, 2011.

[163] Jog, C. S., The equations of equilibrium in orthogonal curvilinear reference coordinates, *J. Elast.*, **104(1)**, 385–395, 2011.

[164] Jog, C. S. and K. D. Patil, Conditions for the onset of elastic and material instabilities in hyperelastic materials, *Arch. Mech.*, **83(5)**, 661–684, 2013.

[165] Jog, C. S. and H. P. Cherukuri, A reexamination of some puzzling results in linearized elasticity, *Sadhana*, **39(1)**, 137–147, 2014.

[166] Jog, C. S. and I. S. Mokashi, A finite element method for the Saint-Venant torsion and bending problems for prismatic beams, *Comput. Struct.*, **135**, 62–72, 2014.

[167] Jog, C. S. and A. Nandy, Conservation properties of the trapezoidal rule in linear time domain analysis of acoustics and structures, *ASME J. Vibration Acoustics*, **137(2)**, 021010, 2015.

[168] Johnson, C., *Numerical Solution of Partial Differential Equations by the Finite Element Method*, Cambridge: Cambridge University Press, 1990.

[169] Jones, D. L. and C. P. Burke, Zonal harmonic series expansions of Legendre functions and associated Legendre functions, *J. Phys. A*, **23(14)**, 3159–3168, 1990.

[170] Kalman, D., A Matrix Proof of Newton's Identities, *Math. Mag.*, **73(4)**, 313–315, 2000.

[171] Kantorovich, L. V. and V. I. Krylov, *Approximate Methods of Higher Analysis*, Groningen:P. Noordhoff Limited, 1958.

[172] Kawashima, K., E. Tsuchida and I. Nakahara, Stresses in an elastic circular cylinder having a spherical inclusion under tension, *Theo. Appl. Mech.*, **27**, 79–89, 1979.

[173] Kearsly, E. A., Asymmetric stretching of a symmetrically loaded elastic sheet. *J. Elast.*, **22(2)**, 111–119, 1986.

[174] Knowles, J. K. and E. Sternberg, On the ellipticity of the equations of nonlinear elastostatics for a special material, *J. Elast.*, **5(3–4)**, 341–361, 1975.

[175] Knowles, J. K., On the representation of the elasticity tensor for isotropic materials, *J. Elast.*, **39(2)**, 175–180, 1995.

[176] Knowles, J. K., *Linear Vector Spaces and Cartesian Tensors*, New York: Oxford University Press, 1998.

[177] Kolossoff, M. C., Sur la torsion des primes ayant pour base un triangle rectangle, *Comptes Rendus*, **178**, 2057–2060, 1924.

[178] Kondo, M., Eine methode zur losung der drehungsspannungen der walztrager von den rechtwinkligen und gleichschenkligen dreieck-formigen querschnitten, *J. Jap. Soc. Mech. Eng.*, **36(194)**, 408–416, 1933.

[179] Korsgaard J., On the representation of symmetric tensor-valued isotropic functions, *Int. J. Eng. Sci.*, **28(12)**, 1331–1346, 1990.

[180] Krawietz, A., A comprehensive constitutive inequality in finite elastic strain, *Arch. Rational Mech. Anal.*, **58(2)**, 127–149, 1975.

[181] Kutsenko, G. V. and A. F. Ulitko, Elastic equilibrium of an ellipsoid under the influence of concentrated forces, *Int. Appl. Mech.*, **9(4)**, 359–364, 1973.

[182] Kutsenko, G. V. and A. F. Ulitko, An exact solution of the axisymmetric problem of the theory of elasticity for a hollow ellipsoid of revolution, *Int. Appl. Mech.*, **11(10)**, 1029–1032, 1975.

[183] Langhaar, H. L., Torsion of curved beams of rectangular cross-section, *ASME J. Appl. Mech.*, **19(1)**, 49–53, 1952.

[184] Lekhnitskii, S. G., *Anisotropic plates*, New York: Gordon Breach, 1968.

[185] Leslie, F. M., Some constitutive equations for liquid crystals, *Arch. Rational Mech. Anal.*, **28(4)**, 265–283, 1968.

[186] Leslie, F. M., Some thermal effects in cholesteric liquid crystals, *Proc. R. Soc. London Ser. A.*, **307(1490)**, 359–372, 1968.

[187] Leslie, F. M., Theory of flow phenomena in liquid crystals, *Advances in Liquid crystals, Vol. 4*, Ed. G. H. Brown, pp. 1–81, New York: Academic Press, 1979.

[188] Levine, H. S. and J. M. Klosner, Axisymmetric elasticity solutions of spherical shell segments, *ASME J. Appl. Mech.*, **38(1)**, 197–208, 1971.

[189] Levinson M., The simply supported rectangular plate: An exact, three-dimensional, linear elasticity solution, *J. Elast.*, **15(3)**, 283–291, 1985.

[190] Ling, C. B., On the stresses in a notched plate under tension, *J. Math. Phys.*, **26**, 284–289, 1947.

[191] Ling, C. B., K. L. Yang, On symmetrical strain in solids of revolution in spherical coordinates, *Trans. J. Appl. Mech.*, **18(4)**, 367–370, 1951.

[192] Ling, C. B., Torsion of a circular cylinder having a spherical cavity, *Quart. Appl. Math.*, **10**, 149–156, 1952.

[193] Ling, C. B., Stresses in a circular cylinder having a spherical cavity under tension, *Quart. Appl. Math.*, **13(4)**, 381–391, 1956.

[194] Lion, A., On the large deformation behaviour of reinforced rubber at different temperatures. *J. Mech. Phys. Solids*, **45(11/12)**, 1805–1834, 1997.

[195] Little, R. W., *Elasticity.*, New Jersey: Prentice-Hall Inc., 1973.

[196] Little, R. W. and S. B. Childs, Elastostatic boundary region in solid cylinders. *Quart. Appl. Math.*, **25(3)**, 261–274, 1967.

[197] Liu, I-Shih, *Continuum Mechanics*, Berlin: Springer-Verlag, 2002.

[198] Liu, I-Shih, On the transformation property of the deformation gradient under a change of frame, *J. Elast.*, **71(1–3)**, 73–80, 2003.

[199] Liu, I-Shih, On Euclidean objectivity and the principle of material frame-indifference, *Continuum Mech. Thermodyn.*, **16(1–2)**, 177–183, 2004.

[200] Liu, I-Shih, Further remarks on Euclidean objectivity and the principle of material frame-indifference, *Continuum Mech. Thermodyn.*, **17(2)**, 125–133, 2005.

[201] Lamb, H., On the propagation of tremors over the surface of an elastic solid, *Phil. Trans. R. Soc. London Ser. A.*, **203**, 1–42, 1904.

[202] Love, A. E. H., The propagation of wave-motion in an isotropic elastic medium, *Proc. London Math. Soc.*, **2(1)**, 291–344, 1904.

[203] Love, A. E. H., *A Treatise on the Mathematical Theory of Elasticity*, New York: Dover, 4'th ed., 1927.

[204] Lubarda, V. A., Constitutive theories based on the multiplicative decomposition of deformation gradient: Thermoelasticity, elastoplasticity, and biomechanics *Appl. Mech. Rev.*, **57(2)**, 95–108, 2004.

[205] Macdonald, H. M., On the torsional strength of a hollow shaft, *Proc. Cambridge Phil. Soc.*, **8**, 62–68, 1893.

[206] MacSithigh, G. P., Energy-minimal finite deformations of a symmetrically loaded elastic sheet, *Q. J. Mech. Appl. Math*, **39(1)**, 111–124, 1986.

[207] Malvern, L. E., *Introduction to the Mechanics of a Continuous Medium*, New Jersey: Prentice Hall Inc., 1969.

[208] Malyi, V. I., One representation of the conditions of the compatibility of deformations, *PMM U.S.S.R.*, **50(5)**, 679–681, 1986.

[209] Marsden, J. E. and T. J. R. Hughes, *Mathematical Foundations of Elast.*, New York: Dover, 1994.

[210] Martins, L. C. and P. Podio-Guidugli, A new proof of the representation theorem for isotropic, linear constitutive relations, *J. Elast.*, **8(3)**, 319–322, 1978.

[211] Mathias, R., Evaluating the Frechet derivative of the matrix exponential, *Num. Math.*, **63(1)**, 213–226, 1992.

[212] Maunsell, F. G., Stresses in a notched plate under tension, *Phil. Mag.*, **21(142)**, 765–773, 1936.

[213] Mehrabadi, M. M. and S. C. Cowin, Eigentensors of linear anisotropic elastic materials, *Quart. J. Mech. Appl. Math.*, **43(1)**, 15–41, 1990.

[214] Meleshko, V. V. and A. M. Gomilko, Infinite systems for a biharmonic problem in a rectangle, *Proc. R. Soc. London Ser. A*, **453(1965)**, 2139–2160, 1997.

[215] Meleshko, V. V., Equilibrium of an elastic finite cylinder: Filon's problem revisited, *J. Engg. Math.*, **46(3–4)**, 355–376, 2003.

[216] Meleshko, V. V., Selected topics in the history of the two-dimensional biharmonic problem, *Appl. Mech. Rev.*, **56(1)**, 33–85, 2003.

[217] Meleshko, V. V., Thermal stresses in an elastic rectangle, *J. Elast.*, **105(1–2)**, 61–92, 2011.

[218] Merodio, J. and R. W. Ogden, A note on strong ellipticity for transversely isotropic linearly elastic solids, *Quart. J. Mech. Appl. Math.*, **56(4)**, 589–591, 2003.

[219] Meschke, G. and W. N. Liu, A re-formulation of the exponential algorithm for finite strain plasticity in terms of Cauchy stresses, *Comput. Methods Appl. Mech. Eng.*, **173(1)**, 167–187, 1999.

[220] Meyers, A., P. Schiebe and O. T. Bruhns, Some comments on objective rates of symmetric Eulerian tensors with application to Eulerian strain rates, *Acta Mech.*, **139(1–4)**, 91–103, 2000.

[221] Michell, J. H., On the direct determination of stress in an elastic solid, with application to the theory of plates, *Proc. London Math. Soc.*, **1(1)**, 100–124, 1899.

[222] Michell, J. H., Elementary distributions of plane stress, *Proc. London Math. Soc.*, **32**, 35–61, 1900.

[223] Min-zhong, W. and W. Lu-nan, Derivation of some special stress function from Beltrami-Schaefer stress function, *Appl. Math. Mech.*, **10(7)**, 665–673, 1989.

[224] Mindlin, R. D. and L. E. Goodman, Beam vibrations with time-dependent boundary conditions, *ASME J. Appl. Mech.*, **17(4)**, 377–380, 1950.

[225] Moler, C. and C. V. Loan, Nineteen dubious ways to compute the exponential of a matrix, twenty-five years later, *SIAM Rev.*, **45(1)**, 3–49, 2003.

[226] Morinaga, K. and T. Nono, On the non-commutative solutions of the exponential equation $e^x e^y = e^{x+y}$, *J. Sci. Hiroshima Univ. (A)*, **17**, 345–358, 1954.

[227] Morinaga, K. and T. Nono, On the non-commutative solutions of the exponential equation $e^x e^y = e^{x+y}$, II, *J. Sci. Hiroshima Univ. (A)*, **18**, 137–178, 1954.

[228] Muller, I., On the frame dependence of stress and heat flux, *Arch. Rational Mech. Anal.*, **45(4)**, 241–250, 1972.

[229] Muller, I., Two instructive instabilities in non-linear elasticity: Biaxially loaded membrane, and rubber balloons, *Meccanica*, **31(4)**, 387–395, 1996.

[230] Murdoch, A. I., On objectivity and material symmetry for simple elastic solids, *J. Elast.*, **60(3)**, 233–242, 2000.

[231] Murdoch, A. I., Objectivity in classical continuum physics: a rationale for discarding the 'principle of invariance under superposed rigid body motions' in favor of purely objective considerations, *Continuum Mech. Thermodyn.*, **15(3)**, 309–320, 2003.

[232] Murdoch, A. I., On criticism of the nature of objectivity in classical continuum physics, *Continuum Mech. Thermodyn.*, **17(2)**, 135–148, 2005.

[233] Nadeau, J. C. and M. Ferrari, Invariant tensor-to-matrix mappings for evaluation of tensor expressions, *J. Elast.*, **52(1)**, 43–61, 1998.

[234] Nakamura, G. and G. Uhlmann, Global uniqueness for an inverse boundary value problem arising in elasticity, *Inven. Math.*, **118(1)**, 457–474, 1994.

[235] Nakamura, G. and G. Uhlmann, Correction to–Global uniqueness for an inverse boundary value problem arising in elasticity, *Inven. Math.*, **152(1)**, 205–207, 2003.

[236] Noda, N. and Y. Moriyama, Stress concentration of an ellipsoidal inclusion of revolution in a semi-infinite body under biaxial tension, *Arch. Appl. Mech.*, **74(1–2)**, 29–44, 2004.

[237] Noll, W., *The Foundations of Mechanics and Thermodynamics*, Berlin: Springer-Verlag, 1974.

[238] Norris, A. N., Quadratic invariants of elastic moduli, *Quart. J. Mech. Appl. Math.*, **60(3)**, 367–389, 2007.

[239] Ogden, R. W., Large deformation isotropic elasticity: on the correlation of theory and experiment for compressible rubberlike solids, *Proc. R. Soc. London Ser. A*, **326(1567)**, 565–584, 1972.

[240] Ogden, R. W., *Nonlinear elastic deformations*, New York: Dover, 1997.

[241] Ortiz, M., R. A. Radovitzky and E. A. Repetto, The computation of the exponential and logarithmic mappings and their first and second linearizations, *Int. J. Num. Meth. Eng.*, **52(12)**, 1431–1441, 2001.

[242] Ostrosablin, N. I., Compatibility conditions of small deformations and stress functions, *J. Appl. Mech. Tech. Phys.*, **38(5)**, 774–783, 1997.

[243] Ostrosablin, N. I., Comments of the publication "Compatibility conditions of small deformations and stress functions", *J. Appl. Mech. Tech. Phys.*, **40(3)**, page 549, 1999.

[244] Padovani, C., On the derivative of some tensor-valued functions, *J. Elast.*, **58(3)**, 257–268, 2000.

[245] Payne, L. E., Torsion of composite sections, *Iowa State College J. Sci.*, **23**, 381–395, 1949.

[246] Pearcy C., A complete set of unitary invariants for 3×3 complex matrices, *Trans. Am. Math. Soc.*, **104(3)**, 425–429, 1962.

[247] Peng, S. H. and W. V. Chang, A compressible approach in finite element analysis of rubber-elastic material, *Comput. Struct.*, **62(3)**, 573–593, 1997.

[248] Penn R. W., Volume changes accompanying the extension of rubber, *Trans. Soc. Rheol.*, **14(4)**, 509–517, 1970.

[249] Pennisi S. and M. Trovato, On the irreducibility of Professor G. F. Smith's representations for isotropic functions, *Int. J. Eng. Sci.*, **25(8)**, 1059–1065, 1987.

[250] Pericak-Spector, K. A. and S. J. Spector, On the representation theorem for linear, isotropic tensor functions, *J. Elast.*, **39(2)**, 181–185, 1995.

[251] Pericak-Spector, K. A., J. Sivaloganathan and S. J. Spector, The representation theorem for linear, isotropic tensor functions in even dimensions, *J. Elast.*, **57(2)**, 157–164, 1999.

[252] Peyraut, F., Z. Q. Feng, Q. C. He, N. Labed, Robust numerical analysis of homogeneous and non-homogeneous deformations, *Appl. Num. Math.*, **59(7)**, 1499–1514, 2009.

[253] Piero, G. D., Some properties of the set of fourth-order tensors, with application to elasticity, *J. Elast.*, **9(3)**, 245–261, 1979.

[254] Podio-Guidugli, P., The Piola–Kirchhoff stress may depend linearly on the deformation gradient, *J. Elast.*, **17(2)**, 183–187, 1987.

[255] Polya, G., Torsional rigidity, principal frequency, electrostatic capacity and symmetrization, *Quart. Appl. Math.*, **6(3)**, 267–277, 1948.

[256] Polya, G. and A. Weinstein, On the torsional rigidity of multiply connected cross-sections, *Ann. Math.*, **52(1)**, 154–163, 1950.

[257] Poschl T., Bisherige losungen des torsionsproblems fur drehkorper, *Z. Angew Math. Mech.*, **2(2)**, 137–147, 1922.

[258] Power, L. D. and S. B. Childs, Axisymmetric stresses and displacements in a finite circular bar, *Int. J. Eng. Sci.*, **9(2)**, 241–255, 1971.

[259] Putzer, E. J., Avoiding the Jordan canonical form in the discussion of linear systems with constant coefficients, *Am. Math. Monthly*, **73(1)**, 2–7, 1966.

[260] Qi, L., Eigenvalues and invariants of tensors, *J. Math. Anal. Appl.*, **325(2)**, 1363–1377, 2007.

[261] Ramachandra Rao, B. S., C. S. Kale and R. P. Shimpi, The sector problem in plane elastostatics, *Int. J. Eng. Sci.*, **11(5)**, 531–542, 1973.

[262] Ramachandra Rao, B. S., A. K. Kandya and S. Gopalacharyulu, Axial eigenfunctions for the axisymmetric problem of an elastic circular cylinder, *Int. J. Eng. Sci.*, **14(1)**, 99–112, 1976.

[263] Rayleigh, L., On waves propagated along the plane surface of an elastic solid, *Proc. London Math. Soc.*, **17**, 4–11, 1885.

[264] Reese, S. and P. Wriggers, Material instabilities of an incompressible elastic cube under triaxial tension, *Int. J. Solids Struct.*, **34(26)**, 3433–3454, 1997.

[265] Reese, S., On material and geometrical instabilities in finite elasticity and elastoplasticity, *Arch. Mech.* **52(6)**, 969–999, 2000.

[266] Reese, S. and S. Govindjee, A theory of finite viscoelasticity and numerical aspects, *Int. J. Solids Struct.*, **35(26–27)**, 3455–3482, 1998.

[267] Reissner, E., Note on the problem of St. Venant flexure, *Z. Angew Math. Phys.*, **15(2)**, 198–200, 1964.

[268] Rivlin, R. S., Universal relations for elastic materials, Rendiconti di Matematica, Serie VII, Roma, **20**, 35–55, 2000.

[269] Rivlin R. S. and M. F. Beatty, Dead loading of a unit cube of compressible isotropic elastic material, *Z. Angew Math. Phys.*, **54(6)**, 954–963, 2003.

[270] Robbins, C. I. and R. C. T. Smith, A table of roots of $\sin z = -z$. *Phil. Mag.*, **39 (299)**, 1004–1005, 1948.

[271] Robert, M. and L. M. Keer, An elastic circular cylinder with displacement prescribed at the ends–axially symmetric case, *Quart. J. Mech. Appl. Math.*, **40(3)**, 339–363, 1987.

[272] Rosati, L., Derivatives and rates of the stretch and rotation tensors, *J. Elast.*, **56(3)**, 213–230, 1999.

[273] Rosati, L., A novel approach to the solution of the tensor equation $\boldsymbol{A}\boldsymbol{X} + \boldsymbol{X}\boldsymbol{A} = \boldsymbol{H}$, *Int. J. Solids Struct.*, **37(25)**, 3457–3477, 2000.

[274] Rostamian, R., The completeness of Maxwell's stress function representation, *J. Elast.*, **9(4)**, 349–356, 1979.

[275] Saccomandi, G. and R. C. Batra, Additional universal relations for transversely isotropic elastic materials, *Math. Mech. Solids*, **9(2)**, 167–174, 2004.

[276] Sadowsky, M. A. and E. Sternberg, Stress concentrations around a triaxial ellipsoidal cavity, *ASME J. Appl. Mech.*, **16(2)**, 149–157, 1949.

[277] Sadowsky, M. A. and E. Sternberg, Pure bending of an incomplete torus, *ASME J. Appl. Mech.*, **20(2)**, 215–226, 1953.

[278] Saito, H., The axially symmetrical deformation of a short circular cylinder, *Trans. Jap. Soc. Mech. Engr.*, **18(68)**, 21–28, 1952.

[279] Saito, H., On the stress distribution in a rotating circular disk of constant thickness, *Trans. Jap. Soc. Mech. Engr.*, **18(75)**, 40–43, 1952.

[280] Saravanan, U. and K. R. Rajagopal, Inflation, extension, torsion and shearing of an inhomogeneous compressible elastic right circular annular cylinder, *Math. Mech. Solids*, **10(6)**, 603–650, 2005.

[281] Saravanan, U. and K. R. Rajagopal, On some finite deformations of inhomogeneous elastic solids, *Math. Proc. R. Irish Academy*, **107A(1)**, 43–72, 2007.

[282] Scheidler, M., The tensor equation $\boldsymbol{AX} + \boldsymbol{XA} = \boldsymbol{\Psi}(\boldsymbol{A}, \boldsymbol{H})$, with applications to kinematics of continua, *J. Elast.*, **36(2)**, 117–153, 1994.

[283] Seegar M. and K. Pearson, De Saint-Venant solution for the flexure of cantilevers of cross-Section in the form of complete and curtate circular sectors, and on the influence of the manner of fixing the built-in end of the cantilever on its deflection, *Proc. R. Soc. London Ser. A*, **96(676)**, 211–232, 1919.

[284] Seth, B. R., On flexure in prisms with cross-sections of uniaxial symmetry, *Proc. London Math. Soc.*, **s2-37(1)**, 502–511, 1934.

[285] Seth, B. R., On flexure of beams of triangular cross-section, *Proc. London Math. Soc.*, **s2-41(5)**, 323–331, 1936.

[286] Seth, B. R., On the flexure of a hollow shaft-I, *Proc. Math. Sci.*, **4(5)**, 531–541, 1936.

[287] Shames, I. H. and C. Dym, *Energy and Finite Element Methods in Structural Mechanics*, New York: Hemisphere Publishing Co., 1985.

[288] Sheehan J. P. and L. Debnath, Transient vibrations of an isotropic elastic sphere, *Pure and Applied Geophysics*, **99(1)**, 37–48, 1972.

[289] Shepherd, W. M., Torsion of a cracked shaft, *Engineering*, **128(3313)**, 39–39, 1929.

[290] Shepherd, W. M., The torsion and flexure of shafting with keyways or cracks, *Proc. R. Soc. London Ser. A*, **138(836)**, 607–634, 1932.

[291] Shepherd, W. M., The flexure of a prism with cross-section bounded by a cardioid, *Proc. R. Soc. London Ser. A*, **154(882)**, 500–509, 1936.

[292] Shield, R. T., Deformations possible in every compressible, isotropic, perfectly elastic material, *J. Elast.*, **1(1)**, 91–92, 1971.

[293] Shnaid I., Thermodynamically consistent description of heat conduction with finite speed of heat propagation, *Int. J. Heat Mass Transfer*, **46(20)**, 3583–3863, 2003.

[294] Sibirskii K. S., A minimal polynomial basis of unitary invariants of a square matrix of the third order, *Matematicheskie Zametki*, **3(3)**, 291–296, 1968.

[295] Silhavy, M., *The Mechanics and Thermodynamics of Continuous Media*, Berlin: Springer-Verlag, 1997.

[296] Simo, J. C., Algorithms for static and dynamic multiplicative plasticity that preserve the classical return mapping schemes of the infinitesimal theory, *Comput. Methods Appl. Mech. Eng.*, **99(1)**, 61–112, 1992.

[297] Singh, R. P. and C. S. Jog, A hybrid finite element formulation for flexible multibody dynamics, *J. Multi-body Dynamics, Proc. Institution Mech. Eng. Part K*, DOI: 10.1177/1464419315569622.

[298] Smith, O. K., Eigenvalues of a symmetric 3×3 matrix, *Comm. ACM*, **4(4)**, 168–168, 1961.

[299] Snell, C., The twisted sphere, *Mathematika*, **4(02)**, 162–165, 1957.

[300] Sokolnikoff, I. S., *Mathematical Theory of Elasticity*, Florida: Robert E. Krieger Publishing Company, 1987.

[301] Sokolnikoff, I. S. and E. S. Sokolnikoff, Torsion of regions bounded by circular arcs, *Bull. Am. Math. Soc.*, **44(3)**, 384–387, 1938.

[302] Southwell, R. V., Castigliano's principle of minimum strain-energy and the conditions of compatibility for strain, *S. Timoshenko, 60th Anniversary Volume*, 1938, 211–217.

[303] Sparrow, E. M., Laminar flow in isosceles triangular ducts, *A.I.Ch.E. J.*, **8(5)**, 599–604, 1962.

[304] Srinivasan, T. P., Decomposition of tensors representing physical properties of crystals, *J. Phys.: Condens. Matter*, **10(16)**, 3849–3496, 1998.

[305] Stephenson, R. A., On the uniqueness of the square-root of a symmetric, positive definite tensor, *J. Elast.*, **10(2)**, 213–214, 1980.

[306] Sternberg, E. and F. Rosenthal, The elastic sphere under concentrated loads, *ASME J. Appl. Mech.*, **19**, 413–421, 1952.

[307] Sternberg, E., Three-dimensional stress concentrations in the theory of elasticity, *Appl. Mech. Rev.*, **11(1)**, 1–4, 1958.

[308] Sternberg, E., On the integration of the equations of motion in the classical theory of elasticity, *Arch. Rational Mech. Anal.*, **6(1)**, 34–50, 1960.

[309] Stevenson, A. C., Flexure with shear and associated torsion in prisms of uni-axial and asymmetric cross-sections, *Phil. Trans. R. Soc. London Ser. A.*, **237(776)**, 161–229, 1938.

[310] Stevenson, A. C., The torsion and flexure solutions for the elliptic limacon cross-section, *Proc. London Math. Soc.*, **s2–45(1)**, 126–143, 1939.

[311] Stevenson, A. C., Complex potentials in two-dimensional elasticity, *Proc. R. Soc. London Ser. A.*, **184(997)**, 129–179, 1945.

[312] Stevenson, A. C., The centre of flexure of a hollow shaft, *Proc. London Math. Soc.*, **s2–50(1)**, 536–549, 1949.

[313] Stewart, I. W., *The Static and Dynamic Continuum Theory of Liquid Crystals*, London: Taylor and Franscis, 2004.

[314] Straughan, B., *Heat waves*, New York: Springer, 2011.

[315] Sussman, T. and K. J. Bathe, A finite element formulation for nonlinear incompressible elastic and inelastic analysis, *Comput. Struct.*, **26 (1/2)**, 357–409, 1987.

[316] Svendsen, B. and A. Bertram, On frame-indifference and form-invariance in constitutive theory, *Acta Mech.*, **132(1–4)**, 195–207, 1999.

[317] Swan, G. W., The semi-infinite cylinder with prescribed end-displacements, *SIAM J. Appl. Math.*, **16(4)**, 860–881, 1968.

[318] Sylvester, J. and G. Uhlmann, A global uniqueness theorem for an inverse boundary value problem, *Ann. Math.*, **125(1)**, 153–169, 1987.

[319] Tang, S., Elastic stresses in rotating anisotropic disks, *Int. J. Mech. Sci.*, **11(6)**, 509–517, 1969.

[320] Tarantino, A. M., Homogeneous equilibrium configurations of a hyperelastic compressible cube under equitriaxial dead-load tractions, *J. Elast.*, **92(3)**, 227–254, 2008.

[321] Tarn J-Q., Stress singularity in an elastic cylinder of cylindrically anisotropic materials, *J. Elast.*, **69(1–3)**, 1–13, 2002.

[322] Tarn, J-Q., Tseng W-D. and Chang H-H., A circular elastic cylinder under its own weight, *Int. J. Solids Struct.*, **46(14)**, 2886–2896, 2009.

[323] Thompson, T. R. and R. W. Little, End effects in a truncated semi-infinite cone, *Quart. J. Mech. Appl. Math.*, **23(2)**, 185–195, 1970.

[324] Timoshenko, S. P. and J. N. Goodier, *Theory of Elasticity*, New York: McGraw-Hill Book Company, 1970.

[325] Ting, T. C. T., The wedge subjected to tractions: a paradox re-examined, *J. Elast.*, **14(3)**, 235–247, 1984.

[326] Ting, T. C. T., Determination of $\boldsymbol{C}^{1/2}$, $\boldsymbol{C}^{-1/2}$ and more general isotropic tensor functions of $\boldsymbol{C}$, *J. Elast.*, **15(3)**, 319–323, 1985.

[327] Ting, T. C. T., New expressions for the solution of the matrix equation $\boldsymbol{A}^T\boldsymbol{X} + \boldsymbol{X}\boldsymbol{A} = \boldsymbol{H}$, *J. Elast.*, **45(1)**, 61–72, 1996.

[328] Ting, T. C. T., The remarkable nature of radially symmetric deformation of spherically uniform linear anisotropic elastic solids, *J. Elast.*, **53(1)**, 47–64, 1999.

[329] Ting, T. C. T., New solutions to pressuring, shearing, torsion and extension of a cylindrically anisotropic elastic circular tube or bar, *Proc. R. Soc. London Ser. A*, **455(1989)**, 3527–3542, 1999.

[330] Tokovyy, Y. V., K. M. Hung and C. C. Ma, Determination of stresses and displacements in a thin annular disk subjected to diametral compression, *J. Math. Sci.*, **165(3)**, 342–354, 2010.

[331] Tolf, G., Saint-Venant Bending of an Orthotropic Beam, *Composite Struct.*, **4(1)**, 1–14, 1985.

[332] Tortorelli, D. A., A generalized formulation of elastodynamics: Small on rigid, *J. Elast.*, **105(1–2)**, 349–363, 2011.

[333] Tranter, C. J., The application of the Laplace transformation to a problem on elastic vibrations, *Phil. Mag.*, **33(223)**, 614–622, 1942.

[334] Truesdell, C. *A First Course in Rational Continuum Mechanics*, London: Academic Press, 1977.

[335] Truesdell, C. *Rational Thermodynamics*, New York: Springer-Verlag, 1984.

[336] Truesdell, C. *An Idiot's Fugitive Essays on Science*, New York: Springer-Verlag, 1984.

[337] Truesdell, C. and W. Noll *The Non-linear Field Theories of Mechanics*, Handbuch der Physik 3, Berlin: Springer-Verlag, 1965.

[338] Tsamasphyros, G. and P. S. Theocaris, On the solution of the sector problem, *J. Elast.*, **9(3)**, 271–281, 1979.

[339] Tsuchida, E. and I. Nakahara, Three-dimensional stress concentration around a spherical cavity in a semi-infinite elastic body, *Bull. Jap. Soc. Mech. Eng.*, **13(58)**, 499–508, 1970.

[340] Tsuchida, E. and T. Uchiyama, Stresses in an elastic circular cylinder with a prolate spheroidal cavity under tension, *Bull. Jap. Soc. Mech. Eng.*, **22(166)**, 476–482, 1979.

[341] Tsuchida, E. and T. Uchiyama, Stresses in an elastic circular cylinder with an oblate spheroidal cavity or an internal penny-shaped crack under tension, *Bull. Jap. Soc. Mech. Eng.*, **23(175)**, 1–8, 1980.

[342] Uchiyama, T. and E. Tsuchida, Stresses in an elastic circular cylinder with a prolate spheroidal cavity under torsion, *J. Elast.*, **20(1)**, 41–52, 1988.

[343] Ulitko, A. F., Stress state of a hollow sphere loaded by concentrated forces, *Prikladnaya Mekhanika*, **4(5)**, 38–45, 1968.

[344] Vallee, C., Q. He and C. Lerintiu, Convex analysis of the eigenvalues of a 3D second-order symmetric tensor, *J. Elast.*, **83(2)**, 191–204, 2006.

[345] Vujosevic, L. and V. A. Lubarda, Finite-strain thermoelasticity based on multiplicative decomposition of deformation gradient, *Theo. Appl. Mech.*, **28-29**, 379–399, 2002.

[346] Wang, W. and M. Z. Wang, Constructivity and completeness of the general solutions in elastodynamics, *Acta Mech.*, **91(3–4)**, 209–214, 1992.

[347] Wang X. and Y. Gong, A theoretical solution for axially symmetric problems in elastodynamics, *Acta Mech. Sinica*, **7(3)**, 275–282, 1991.

[348] Washizu, K., A note on the conditions of compatibility, *J. Math. Phys.*, **36(4)**, 306–312, 1958.

[349] Watson, G. N., *A Treatise on the Theory of Bessel Functions*, London: Cambridge University Press, Second Edition, 1966.

[350] Wermuth, E. M. E., Two remarks on matrix exponentials, *Lin. Alg. Appl.*, **117**, 127–132, 1989.

[351] Wheeler, L. T. and E. Sternberg, Some theorems in classical elastodynamics, *Arch. Rational Mech. Anal.*, **31(1)**, 51–90, 1968.

[352] Wheeler, L. T., On the derivatives of the stretch and rotation with respect to the deformation gradient, *J. Elast.*, **24(1–3)**, 129–133, 1990.

[353] Wolfram, S., *Mathematica*, Champaign, Illinois: Wolfram Research Inc., 2014.

[354] Wood, J. A., The chain rule for matrix exponential functions, *College Math. J.*, **35(3)**, 220–222, 2004.

[355] Xiao, H., Unified explicit basis-free expressions for time rate and conjugate stress of an arbitrary Hill's strain, *Int. J. Solids Struct.*, **32(22)**, 3327–3340, 1995.

[356] Xiao H., O. T. Bruhns, and A. Meyers, On objective corotational rates and their defining spin tensors, *Int. J. Solids Struct.*, **35(30)**, 4001–4014, 1998.

[357] Xiao H., O. T. Bruhns, and A. Meyers, Existence and uniqueness of the integrable-exactly hypoelastic equation $\overset{o}{\boldsymbol{\tau}}{}^{*} = \lambda(\mathrm{tr}\,\boldsymbol{D})\boldsymbol{I} + 2\mu\boldsymbol{D}$ and its significance to finite inelasticity, *Acta Mech.*, **138(1)**, 31–50, 1999.

[358] Yavari, A., Compatibility equations of nonlinear elasticity for non-simply connected bodies, *Arch. Rational Mech. Anal.*, **209(1)**, 237–253, 2013.

[359] Young, A. W., E. M. Elderton, K. Pearson, On the torsion resulting from flexure in prisms with cross-sections of uniaxial symmetry only, *Draper' Company Memoirs*, **Technical Ser. VII**, 1918.

[360] Zheng, X. and P. Palffy-Muhoray, Eigenvalue decomposition for tensors of arbitrary rank, *Electronic-Liquid Crystal Communication*, 2007.

[361] Zidi, M., Combined torsion, circular and axial shearing of a compressible hyperelastic and prestressed tube, *ASME J. Appl. Mech.*, **67(1)**, 33–40, 2000.

Index